International Conference on Innovative Technology, Engineering and Sciences (iCITES 2018)

IOP Conference Series: Materials Science and Engineering
Volume 342

Pekan, Malaysia
1 – 2 March 2018

ISBN: 978-1-5108-6118-3
ISSN: 1757-8981

Printed from e-media with permission by:

Curran Associates, Inc.
57 Morehouse Lane
Red Hook, NY 12571

CURRAN ASSOCIATES INC. proceedings.com

Some format issues inherent in the e-media version may also appear in this print version.

Printed by Curran Associates, Inc. (2018)

Institute of Physics
Dirac House, Temple Back
Bristol BS1 6BE UK

Phone: 44 1 17 929 7481
Fax: 44 1 17 920 0979

techtracking@iop.org

Additional copies of this publication are available from:

Curran Associates, Inc.
57 Morehouse Lane
Red Hook, NY 12571 USA
Phone: 845-758-0400
Fax: 845-758-2633
Email: curran@proceedings.com
Web: www.proceedings.com

TABLE OF CONTENTS

IOP Conf. Series: Materials Science and Engineering **342** (2018) 011001 doi:10.1088/1757-899X/342/1/011001

Preface

The International Conference on Innovative Technology, Engineering and Sciences 2018 (iCITES 2018) held on the 1st and 2nd March 2018 in Universiti Malaysia Pahang (UMP), Pekan is an inaugural conference jointly organised by the UMP Alumni Society and UMP Holdings Sdn. Bhd, in an effort to promote key innovation in Green & Frontier Materials, Innovative Robotics & Automation, Renewable & Green Energy, Sustainable Manufacturing as well as Data Analytics. This conference is aimed at building a platform that allows interested experts to share and discuss their latest researches, ideas, and survey reports in the theoretical and practical areas of the aforementioned fields.

This volume contains the accepted papers presented at iCITES 2018. We would like to thank all the authors who have contributed to this volume and also not forgetting the distinguished keynote speakers (Prof. Dr. Zahari Taha, Prof. Dr. Ing. Maurice Kettner, Dr. Syed Hussein Syed Alwi, Dr. Ali Awad Abbas, and Assoc. Prof. Shahrul Na'im Sidek), the reviewers, session chairs, sponsors (SCANIA Malaysia, Intel Malaysia, Tourism Pahang), our collaborators (UMP Students Affairs & Alumni Department, Innovative Manufacturing, Mechatronics & Sports Laboratory and Hochschule Karlsruhe-Technik & Wirtschaft) as well as all the conference committee for their hard work in making iCITES 2018 a success.

iCITES 2018 Organising Committee
7th March 2018

Peer review statement

All papers published in this volume of *IOP Conference Series: Materials Science and Engineering* have been peer reviewed through processes administered by the proceedings Editors. Reviews were conducted by expert referees to the professional and scientific standards expected of a proceedings journal published by IOP Publishing.

IOP Conf. Series: Materials Science and Engineering **342** (2018) 012001 doi:10.1088/1757-899X/342/1/012001

Sustainable materials used as stone column filler: A short review

Azhani Zukri[1,2] and Ramli Nazir[2]

[1]Fakulti Kejuruteraan Awam & Sumber Alam, Universiti Malaysia Pahang, Lebuhraya Tun Razak, Gambang, Pahang, Malaysia
[2]Fakulti Kejuruteraan Awam, Universiti Teknologi Malaysia, Skudai, Johor, Malaysia

E-mail: engr.azhani@gmail.com

Abstract. Stone columns (also known as granular piles) are one of the methods for soft soil stabilization and typically used to increase bearing capacity and stability of slope.; Apart from decreasing the compressibility of loose and fine graded soils, it also accelerates the consolidation effect by improving the drainage path for pore water pressure dissipation and reduces the liquefaction potential of soils during earthquake event. Stone columns are probably the most "natural" ground treatment method or foundation system in existence to date. The benefit of stone columns is owing to the partial replacement of compressible soil by more competent materials such as stone aggregate, sand and other granular materials. These substitutes also act as reinforcement material, hence increasing overall strength and stiffness of the soft soil. Nowadays, a number of research has been conducted on the behaviour and performance of stone columns with various materials utilized as column filler replacing the normal aggregate. This paper will review extensively on previously conducted research on some of the materials used as stone column backfill materials, its suitability and the effectiveness as a substitute for regular aggregates in soft soil improvement works.

1. Introduction

A great deal of previous research in soft soil improvement has been found to employ the stone column method with their primary focus engaging towards settlement and deformation behaviours. The aim of stone column construction are to improve the soil characteristics, support the structure and infrastructure overlying both in very soft to firm cohesive soils and also loose silty sands having greater than 15% of fine aggregates [1]. Most of the stone columns analysed in this study are found to utilize normal aggregate as column filler. Column filler materials normally consist of stone aggregates between 20 to 75 mm, gravel and sand are compacted into a vertical hole generally of 0.6 to 1.0 meter in diameter and 15 to 20 meters deep. Crushed stones or gravel for column backfills should be clean, hard, and un-weathered—free from organics, trash or other deleterious materials. Three main criteria are taken into consideration when selecting an efficient backfill material, i.e. availability, suitability, and economy. A mixture of crushed stone and sand might also be used in the proportion of 1:0.2 to 0.5 by volume [2]. When columns are formed with granular fill, their load capacities become highly dependent on the strength of the fill material and the confining stress of the surrounding soil [3]. To create required strength capacity with stone columns, the granular materials consisting of stone or stone sand should be compacted [4]. Generally stone columns constructed using ramming or vibro-replacement method either by wet process or dry process, depends on the availability of equipment and its application.

2. Stone Column

According to Han and Ye, stone columns and sand compaction columns have relatively lower load capacities and stiffness compared to rigid columns (e.g., vibro-concrete columns) and as such they are categorized as flexible columns. Granular columns have excessive bulging problem especially at low confining stress near ground surface, which makes it unsuitable for very soft soil with undrained shear strength lower than 15 kPa. The bulging will occur within 2.5 to 3 times the length of column, measured from the top of the stone column [5]. The problem can be solved by using geosynthetic-encased granular columns where the geosynthetic increases the stiffness of the columns thus improving the soft soil behaviour in term of deformation and strength. However, the geosynthetic is more expensive and takes time to be installed as compared with granular columns without encasement. Several other researches have already been published on encased granular columns, either using geosynthetic, geotextile, etc. However , the stone column design method is still in empirical phase, where based on past experiences and in need of few field trials before any further execution [6]. Hence, no well-defined or well-structured guidelines and/or codes are available to date.

Nowadays, various potential materials have been identified and studied to be adopted as column fillers. It requires further evaluation to validate the performance of the stone columns in soil treatment. For example, quarry dust was selected as a filler material in stone columns to improve clayey soils, however, its performance was not as good as that of gravel. The materials used in stone column construction are required to be free-draining, hard, and inert and comply with acceptable criteria in terms of material type, grading, hardness and chemical stability. In pursuing environmental sustainability in ground treatment, there is an increasing desire to use recycled materials for vibro-stone column techniques [7]. For example, recycled materials can be classified as Recycled Aggregate, from the processing of construction waste materials or Secondary Aggregates, from the bi-products of industrial processes [8]. These aggregate sources have been used historically for lower grade applications such as bulk fills and hard standing, which require low levels of specification. However, after processing, recycled aggregate materials could be utilized for higher grade applications such as stone column construction. Readily available recycled materials in construction industries such as crushed concrete, recycled aggregates, tyre chips, railway ballast and crushed glass, have big potential to be used in stone column construction. More recent studies have also incorporated materials such as PFA, fly ash, bottom ash, and quarry dust as column filler materials in the performance evaluation of stone columns that will be used for soft ground treatment.

3. Previous researches on stone column filler materials

Many researches have been performed to evaluate the effectiveness of other materials either manufactured or recycle materials to substitute the normal aggregate in stone column construction. Apart from the type of material used, it is important to adopt proper and effective compaction technique and equipment that match with the selected granular backfill material to achieve the highest relative density and thus optimum improvements effects.

The research conducted by Juran *et al.* provided a preliminary basis for a fundamental understanding of the cementation effect on the performance of the granular column and a qualitative evaluation of the group effect on the settlement response of the reinforced soil [9]. Triaxial compression tests were conducted on composite-reinforced soil samples made of annular, normally consolidated kaolin representing the soft soil, reinforced by both cemented and untreated columns of river sand. In addition to that, Daya D. and Niranjana K. evaluated the influence of column material in the performance of stone column made up of conventional aggregate and gravel through numerical analysis using PLAXIS 3D software [10]. The results are then have been validated through load tests and show clear improvement in the load carrying capacity of the column when gravel is used. Some studies on the effectiveness of gravel as stone column filler were carried out by many researchers such as Tander Y.

K. *et al.,* A. Marto *et al.*, A. Zahmatkesh & A.J. Choobbasti, Mahmoud Ghazavi & Samira Ebrahimi and many more [11][12][13][14].

In the meantime, Andreou *et al.* conducted a study on the influence of several controlling parameters, which are the drainage conditions, grain size of the stone column materials, the confining pressure of the soil, and the rate of deformation in the design of stone columns through a series of laboratory experiments on two types of material; sand and gravel. The experimental results showed the advantages of this ground improvement technique, particularly on the strength characteristics of the foundation soil, even with a relatively low replacement factor compared with that used in practice. The study found that as the confining pressure increases, the improvement in the specimen with the sand column is more pronounced [15].

The consolidation effect of sand and aggregate as stone column filler material under distributed load has been evaluated using a unit cell [16]. An axisymmetric consolidation model using Plaxis software simulation was used to compare the increasing rate of consolidation for both materials. The study found that sand can be considered a suitable filler material to increase the consolidation rate, thus shortening the time of the consolidation process. Therefore, the desired settlement and bearing capacity can be achieved in a short time. Meanwhile, the study conducted by Shadi S. Najjar found that the drained shear strength parameters unaffected by the sand column reinforcement, except for fully penetrating columns with high area replacement ratios. The study was performed using consolidated undrained triaxial tests on normally consolidated kaolin specimens based on various parameters such as diameter and height of the sand columns, the type of columns geotextile encased versus non-encased and the effective confining pressure [17]. In addition, the research conducted by A. Namur K.S. *et al.* revealed that for both floating and end bearing types, the sand columns at low relative density exhibited higher bearing improvement ratios and lower settlement reduction ratios compared to stone columns. On the other hand, a reverse behavior was noticed, when the backfill material was placed at "dense state" soil condition [18].

Other than sand and gravel, Silica-Manganese slag which is a by-product from ferro-alloy industries also can be used as the stone column material. In the study conducted by Prasad *et al.* on marine clay stabilized by stone column filler, load carrying capacities of the floating stone columns reinforced with circular discs placed at D/2 spacing showed better performance than the columns with discs placed at D spacing, where D is column diameter. The circular discs were placed at two different spacing (D and D/2) over varied reinforcement depths (0.25L, 0.5L, 0.75L and L) [19].

Pulverised fuel ash (PFA), is a by-product of pulverised fuel (typically coal) fired power stations and can be used as engineering fill and as a component for concrete. Several researchers established the study on PFA as stone column material to stabilized soft clay and peat soil. A. Muntohar *et al.* conducted a study on peat soil treated by PFA column using two types of models; Model A and Model B, where Model A represents geometrical block while Model B, column group. Soil and soil-column parameters have been back-calculated using numerical analysis based on the field static loading test. It was found that both models used in the study are reliable in simulating the field static-loading test for column-treated peat. Essentially, Model B shows higher stability to failure if compared to Model A. The effects of PFA on strength behaviour of stone column also investigated by M. Shakri *et al.* and found that PFA really helps in process of increasing the unconfined compression strength of columns [20]. A laboratory study on bearing capacity of treated stone column with PFA as filler material has been performed by Nazaruddin A.T. *et al.* through UCT test and Plate Loading Test and concluded that the used of PFA mixture on stone column also improve the bearing capacity and reduce settlement on soft soil [21].

A series of experimental study were conducted to understand the behaviour of a group of three columns and seven columns built using six different filler materials including clay, quarry dust, sea sand, river sand, gravel, and stones, besides observing the influence of these column materials to stone column performance [22]. Finite element analysis using 15-noded triangular elements via the Plaxis software package was carried out to compare the load settlement behaviour with the developed test model. From the study, the researcher concluded that among the different stone column materials, stones are found to be more effective from both the single column test and group column test. Moreover, the study found that quarry dust is also effective in improving the load deformation characteristics of the soil. Therefore, quarry dust can also be used as an alternative material to construct stone column fillers, which is in fact efficient and economical as it is cheap and easily available.

Apart from that, the performance of stone column constructed with two different filler materials (gravel and quarry dust) have been evaluated through Priebe's method by using the unit cell concept [23]. Researchers found that load bearing capacity increases by 14% when gravel is used as a reinforcing material as compared to that of quarry dust for a given stone column dimension. The improvement factor obtained from the study was also higher when gravel is used as reinforcing material, than when quarry dust is used. The improvement factor determination was also established by Pivarc using the values calculated from numerical and laboratory models, which were then compared with the improvement factors as acquired from Priebe´s theory [24]. The stone column filler used in the study was made from gravel or crushed stones arranged by a vibrator. The study concluded that the agreement of the results between the improvement factor calculated through Priebe`s method, the FE method, and the laboratory experiments, is evidently satisfactory.

Spent Railway Ballast is high quality aggregate (typically granite or limestone) taken from beneath railway tracks, when excessive fines occur after continued attrition under repeated railway carriage loads. The study on the mechanisms and interactions of granular material "ballast" in isolated or grouped stone columns with the surrounding ground has been investigated by Baba *et al.* [25]. The settlement performance of various configurations of columns beneath rigid square footing was examined using finite element modelling and found that the column material properties (stiffness, friction angle), which is influenced by column arrangement (column position), and the behaviour law obeyed by both columns and soils, increase the settlement reduction factor. The study also proved that the column spacing is more relevant, such that when floating columns reach critical length, their behaviour become similar to end-bearing columns.

Stone columns were constructed to improve the deformation characteristics of treated soil especially to reduce primary settlement, whereby limestone quarries ranging from 2.50 to 10 mm in size were used as column filler materials. In this study, Jadid managed to use the calculated modulus value within a depth of approximately six meters from the existing grade and the information obtained from the post-improved CPT test results for more than six meters deep, to estimate the allowable bearing pressure for the planned shallow foundations. This approach ultimately reduced foundation settlement and improved bearing capacity, hence reducing footing size requirements, which enabled shallow footing construction and stabilization of the slope [26]. Meanwhile, Xueyi Liu *et al.* focused towards investigating the stress concentration ratio on stone columns of different backfill materials modified by sand, lime and cement under confined conditions using model tests of stone column performed inside cylindrical container of 300 mm in diameter and 350 mm in height. The results revealed that the values of stress concentration ratio, ranges from 4 for columns with crushed stone as backfill material, to 7.6 for columns with crushed stone stabilized with 5% cement. However, the presence of dry lime did not show any significant increase in the stress concentration [27]. In addition, the study on the behaviour of geosynthetic encased quarry waste column has been conducted by K.Balan *et al.*, P. Ancy Genu C George, K. Beena, D. Isaac and S. Siva Gowri Prasad [28][29][30][31][32].

Stone dust is a multipurpose material and well suited to a yard or passageway surface. It is also a great choice for the sub-base in laying paving blocks and slabs, and for jointing natural stone, such as slate. The study on stone dust utilized in stone column construction has been performed by S. Rani and P. Kumar [33]. The properties of stone dust in this study is similar to normal aggregate, therefore it is suitable to use as stone column filler and found to be acceptable in soil improvement work, where the soil bearing capacity tend to be increased.

Many studies focused on the better way to expand the usage of industrial by-products to alleviate disposal problems such as steel slag, bottom ash, PFA and so on. Large quantities of steel slag were deposited in yards, causing environmental pollution. Like other metallurgical slags, steel slags exhibit a great potential to be used as aggregate in highway construction. The physical properties of steel slag were found to be acceptable and meet the aggregate standard specifications for Malaysian road works [34]. A Vibro-flotation and stone column application is the latest application of steel slag [35]. Steel slag has been used in stone column construction in Saudi Arabia since 1995. The stone column constructed using steel slag need to extend down to considerable depth (end bearing) or hard stratum, which means that this kind of materials are not suitable for floating columns [36]. Steel slag has an extensive record of use as a material in the sand compaction pile (SCP) by vibration compaction as a marine construction method, mainly for ports and harbours. The study conducted by K. Onda *et al.* summarized that steel slags are satisfactory for use in the sand compaction pile method by static compaction for liquefaction countermeasures [37]. There is a weakness in the use of steel slag because of this material is quite abrasive and could lead to increase the maintenance cost of vibro equipment. In addition the expansive potential issue also should be noted.

Coal bottom ash (CBA) is the waste from coal burning and possesses similar properties to sand, making it a suitable stone columns filler material. The excess bottom ash production in Malaysia lead to several environmental problems due to the high production of electricity [38]. A research on CBA used to improve the shear strength parameters of clay soil was conducted through Consolidated Undrained Triaxial Tests with pore pressure measurements by A. Marto *et al.*. Results indicated that the installation of CBA columns in clay specimens increased the apparent cohesion but did not show any significant difference in effective friction angle. The apparent cohesion of soft clay reinforced with partially penetrating columns is higher than that of fully penetrating columns. It was also found that during consolidation, the dissipation of pore water pressure accelerated due to the presence of bottom ash columns, which draws the conclusion that consolidation and shear strength of soft clay could be improved by installing bottom ash columns [39].

On the other hand, shear strength improvement of soft clay reinforced with encapsulated single bottom ash column was also proven from the Unconfined Compression Test in a laboratory. A Polyester Non-woven Geotextile Needle-punched Fabric (MTS 130) was used to encase the bottom ash reinforced kaolin clay column. This encasement significantly improved the overall shear strength of the specimens [40]. The investigation on bearing capacity of soft clay installed with singular and group of encased bottom ash columns also has been done by A. Marto *et al.*, V. R. Raju and M. Hassan *et al.* [41][42][43].

Other materials tested for its effectiveness as stone column material is crushed polypropylene (PP). The use of PP as stone column material was investigated by Muzamir *et al.* [44]. Since PP is a definite waste material, currently disposed in large quantities into landfills, the cost of soil improvement can be reduced significantly. A total of 28 unconfined compression tests were conducted on kaolin samples reinforced with crushed polypropylene. In brief, shear strength parameters are affected by the diameter and height of the columns, and in the case of PP being introduced into the column, it brings better improvement to the shear strength.

Shredded waste tyres were also proposed as an alternative to normal aggregates in constructing stone columns as described in the research conducted by Ayothiraman and Soumya . Different mix proportions of stone aggregates and tyre chips were considered in this study. The study involved several series of triaxial tests on samples of 50 mm diameter and experiments onto physical stone column model of 1/12 scale factor installed in a kaolin clay bed of uniform consistency. This clearly shows that waste tyre chips can be used as partial replacement of stone aggregates up to approximately 60% in stone columns [45][46]. In the meantime, the investigation on strength improvement of stone columns using tyre chips and aggregates through plate load test was carried out by A. Akhitha and found that the lesser the amount of stone aggregate used the more it will reduce the overall cost of a stone column and incite better waste utilisation and preservation of natural resources [47]. Hence, tyre chips are indirectly proven to cost less aside from being environmental friendly. However, these findings need to be further verified on a full-scale environment, to develop design guidelines before implementing the use of tyre chips in the field.

The use of fly ash aggregate as replacement for conventional aggregate material was investigated through consolidation tests in order to evaluate the suitability of fly ash aggregate as column filler material in problematic clays. From the experiment, it was found that the time required for 90% consolidation of stone aggregate and fly ash aggregate column in soft clay are comparable regardless of pressure increment and number of columns. Hence it can be concluded that fly ash aggregate may be effectively utilized as column material in the place of conventional coarse aggregate for improving soft clay. Research on evaluation of fly ash aggregate as stone column materials for soft clay improvement also has been done by S. Vidhyalakshmi *et al.* and Saravanan [48][49][50][51].

Vibro-stone columns are a common form of ground improvement technique that typically utilizes primary aggregates but in recent years, due to sustainable practices, recycled aggregates have been used as an alternative. Some of the recycled aggregates and secondary aggregates available to date have been the focus of interest, especially for its suitability in the construction of vibro-stone columns. These include steel slag, spent railway ballast, crushed concrete, demolition materials and 'waste' rock. However, Serridge has the opinion that crushed concrete and recycled (spent) railway ballast offer the greatest potential for use in vibro-stone columns [52]. In the research conducted by Amini, a selection of primary (granite) and three recycled aggregates (crushed concrete and brick, incinerated bottom ash aggregate type 1 and 2), which are commonly used in the construction of vibro-stone columns, were compared in the actual context of the installation and loading of a single stone column in soft clay [53]. The findings showed that despite the various results given by the aggregate index tests, the recycled aggregates can still be used as stone column fillers. The crushing of the aggregates during the installation can affect the behaviour of the column even more than during loading. Also, the contamination of the column material with fines can significantly reduce the performance of the stone columns under static loading. The installation time for each layer of aggregates should be reasonable to allow the aggregates to be sufficiently compacted, while retaining the aggregates from damage by crushing or affected by overtreatment. The use of recycled concrete aggregate can be considered in the construction of granular columns and play an important role in sustainable geotechnical design as proved by S. Demir *et al.*, C. Serridge and T. Robert *et al.*. [46][47][48].

Other waste materials such as crushed concrete and crushed glass, or manufactured materials such as lightweight expanded clay aggregate can indeed become secondary aggregates, which are prospective materials for replacing regular aggregate in stone column construction. Many researchers have explored the potential of crushed waste glass as a replacement for sand in concrete. However, this material has never been found to be used as a stone filler. Crushed glass of 20/40 mm grade in vibro-stone columns would still have sharp edges that may cause problems in handling and instigate potential health and safety issues [54]. Recycled glass characteristics indicate that it is suitable to be used as a fill (backfill) material in structural and geotechnical applications. Recycled glass on its own or in a mixture with natural or recycled aggregates (such as crushed rock and crushed concrete) can be used in a range of road work applications including sub base, embankments material and drainage media in roads [55]. Nevertheless, this material still needs to be further examined in order to verify its suitability as stone column filler materials. The manufactured material that is similar to normal aggregate in terms of their physical properties such as lightweight expanded clay aggregate (LECA) tend to be more appropriate as stone aggregate replacement in stone column construction. Furthermore, many studies have been performed to evaluate the potential used of LECA in Geotechnical applications, especially in embankment and bridge construction, road and highway pavement, slope filling and so on [56][57][58][59][60]. However, it is worth noting that no studies have been found using this material as stone column fillers replacing the conventional aggregate. A summary of previous studies conducted to evaluate the performance of granular materials as a filler of stone column replacing the normal aggregates is shown in Table 1.

IOP Conf. Series: Materials Science and Engineering **342** (2018) 012001 doi:10.1088/1757-899X/342/1/012001

Table 1. Previous studies conducted to evaluate the performance of granular materials as a filler of stone column replacing the normal aggregates.

Researcher, Year	Filler material	Finding
Andreou *et al*., 2008	Comparison between sand and gravel	As the confining pressure increases, the strength characteristics improvement in the specimen with the sand column is more pronounced.
Dipty Sarin Isaac, 2009	Comparison between clay, quarry dust, sea sand, river sand, gravel, and stones	Stones are found to be more effective from both the single column test and group column test
Shadi S. Najjar, 2010	Sand	The Undrained strength significantly improved even for area replacement ratios less than 18%, while drained shear strength parameters were found to be relatively unaffected by the sand column reinforcement, except for end bearing columns with high area replacement ratios.
Namur K.S. Al Saudi *et al*., 2016	Sand	The sand columns at low relative density exhibited higher bearing improvement ratios and lower settlement reduction ratios, while a reverse behaviour was noticed, when the backfill material was placed at "dense state"
Jayalekshmi *et al*., 2010 K.Balan *et al*.,2013 A Genu C George, 2016 S. Siva Gowri Prasad, 2016	Comparison between gravel and quarry dust	Load bearing capacity increases and soil improvement factor was higher when gravel is used as a reinforcing material as compared to that of quarry dust.
Ismail *et al*., 2011	Sand	The consolidation rate increased.
Pivarc, 2011	Crushed stones (gravel)	Improvement factors was higher for s/d=2, followed by s/d=3 and 4, where s and d is spacing and diameter of stone column, respectively.
Ayothiraman & Soumya, 2011 A. Akhitha, 2017	Tyre chips	Strength improved, however, need to be verified on a full-scale environment.
Nazaruddin A.T. *et al*., 2013 Muntohar *et al*., 2013 Shakri M.S. *et al*. (2014)	PFA	Improve the bearing capacity and reduction of soft soil settlement.
Prabhu, 2013 S. Vidhyalakshmi *et al*., 2009 Saravanan, 2013	Fly ash aggregate	Compression index and coefficient of volume compressibility found to be lower than that of clay without column, thus improved the settlement of soft soil.
Daya D. & Niranjana K., 2017	Comparison between normal aggregate and gravel	The column made up of gravel giving more load carrying capacity compare to conventional aggregate column

IOP Conf. Series: Materials Science and Engineering **342** (2018) 012001 doi:10.1088/1757-899X/342/1/012001

Tandel Y. K. *et al.*, 2012 A. Marto *et al.*, 2013 A. Zahmatkesh & A.J. Choobbasti, 2010	Gravel	Improved bearing capacity and reduced settlement of soft clay.
Prasad *et al.*, 2015	Silica-Manganese slag	Reinforced with circular configuration placed at D/2 spacing showed better performance than the columns with discs placed at D spacing, where D is stone column diameter.
Amini, 2015	Recycle aggregate	Can be used as column filler but not very good due to construction and handling problem
A. Marto, 2014	Coal bottom ash	Consolidation and shear strength of soft clay could be improved by installing bottom ash columns
A. Marto *et al.*, 2016 M. Hassan *et al.*, 2016	Coal bottom ash	Bearing capacity improved
M. Hassan *et al.*, 2015	crushed polypropylene (PP)	Shear strength parameters are affected by the diameter and height of the columns.
K. Onda *et al.*, 2014	Comparison between Steel slag and natural sand	In comparison with natural sand, which is used in conventional SCP, the improvement effect and workability (diameter of pile, construction time) of these steel slags were similar and comparable to those of natural sand.

4. Conclusion

The partial replacement of compressible soil with more competent and promising materials such as stone aggregate, sand, and other granular materials, presents several benefits in the construction of stone columns. It also acts as reinforcing materials that increase the overall strength and stiffness of soft soils. Until now, a number of researches have been accomplished pertaining to stone column behaviour and performance with various materials used as column fillers. Most of the alternative materials used as stone column fillers or backfill materials can be produced from waste and recycled materials. These materials have been established based on experiments, numerical analysis and/or physical modelling as conducted in a laboratory. Therefore, the findings need to be further verified in a full-scale environment, to develop design guidelines, specifications and QA testing, prior to implementing any of these replacement materials to be conventionally utilized as stone column fillers. If this is fulfilled, then these materials could offer excellent and sustainable solutions by utilizing potential wastes effectively while reducing demands for primary aggregates. These alternative materials for stone column fillers offer high efficiency apart from being economical, as the materials are cheap, easily available and environmental friendly.

From this review paper, materials such as fly ash, coal bottom ash, crushed polypropylene (PP), recycled aggregate, tyre chips, Pulverized Fuel Ash (PFA), Silica-Manganese slag, quarry dust, stone dust, limestone, river sand, etc. have all been recognized to be utilized as stone column materials replacing its traditional aggregate materials. Nonetheless, no study or research has been found with respect to the usage of crushed waste glass and lightweight expanded clay aggregates (LECA), which are categorized as alternative granular materials to be used as stone column backfills. Therefore, it is suggested that studies can be made to assess the effectiveness of these two materials in the stone column construction as one of soft soil improvement method.

References

[1] Barksdale, R.D.and Bachus R C 1983 Design and construction of stone columns *FHWA/RD-83/026, Federal Highway Administration, Washington, D.C.*

[2] Vahedian a., Mahini S and Aghdaei S a. 2014 A Short State-of-the-Art Review on Construction and Settlement of Soft Clay Soil Reinforced with Stone ColumnA Short State-of-the-Art Review on Construction and Settlement of Soft Clay Soil Reinforced with Stone Column *International Journal of Engineering and Technology* **6** 420–6

[3] Han J 2015 *Principles and Practice of Ground Improvement* (John Wiley & Sons, Inc. Hoboken, New Jersey)

[4] Dheerendra Babu M R, Nayak S and Shivashankar R 2013 A Critical Review of Construction, Analysis and Behaviour of Stone Columns *Geotechnical and Geological Engineering* **31** 1–22

[5] Das B M Principles *of Foundation Engineering, SI* ed 7th Edition (USA: Cengage Learning)

[6] Ambily A P and Gandhi S R 2004 Experimental and Theoretical Evaluation of Stone Column in Soft Clay *Icgge* 201–6

[7] Serridge C J 2005 Achieving sustainability in vibro stone column techniques *Proceedings of the Institution of Civil Engineers - Engineering Sustainability* **158** 211–222

[8] Tranter R, Jefferson I and Ghataora G 2008 The use of recycled aggregate in vibro-stone columns - A UK perspective *GeoCongress 2008: Characterization, Monitoring, and Modeling of GeoSystems, March 9, 2008 - March 12, 2008* 630–7

[9] Juran B I and Riccobono O 1991 Reinforcing Soft Soils With Artificially Cemented Compacted-Sand Columns *Journal of Geotechnical Engineering* **117** 1042–60

[10] Daya Dath N K 2017 A Study On Conventional Aggregate And Gravel As Columnar Inclusions In Clay *International Conference on GEOTECHNIQUES FOR INFRASTRUCTURE PROJECTS* (Thiruvananthapuram) pp 1–5

[11] Tandel Y k., Solanki C H and Desai A K 2012 Numerical modelling of encapsulated stone column-reinforced ground *International Journal of Civil, Structural, Environmental and Infrastructure Engineering* **2** 82–96

[12] Aminaton Marto Farshad Helmi, Nima Latifi, Mohsen Oghabi R M 2013 Performance analysis of reinforced stone columns using finite element method *Electronic Journal of Geotechnical Engineering (EJGE)* **18 B** 315–23

[13] A. Zahmatkesh A J C 2010 Settlement Evaluation Of Soft Clay Reinforced By Stone Columns, Considering The Effect Of Soil Compaction *International Journal of Research & Reviews in Applied Sciences* **3** 159–66

[14] Ghazavi M and Nazari Afshar J 2013 Bearing capacity of geosynthetic encased stone columns Geotextiles and Geomembranes 38 26–36

[15] Andreou P, Frikha W, Frank R, Canou J, Papadopoulos V and Dupla J-C 2008 Experimental study on sand and gravel columns in clay *Proceedings of the Institution of Civil Engineers - Ground Improvement* **161** 189–98

[16] M. A. Mohamad Ismail L. K. Ng and N.A. Ramli H M Y 2011 Consolidation of Sand and Aggregate as Stone Column Material *Electronic Journal of Geotechnical Engineering (EJGE)* **15** 2705–11

[17] Shadi S. Najjar Tarek Maakaroun S S 2010 Effect of Sand Columns on the Undrained Load Response of Soft Clays *JOURNAL OF GEOTECHNICAL AND GEOENVIRONMENTAL ENGINEERING* **136** 1263–77

[18] Namur K.S. Al Saudi Nawres A.A. Rajab, G Tanyrbergenova A G 2015 Sand and stone columns in soft soil at different relative densities *Japanese Geotechnical Society Special Publication* **2** 2121–6

[19] Siva S, Prasad G, Harish Y and Satyanarayana P V V 2015 Stabilization of Marine Clays with Geotextile Reinforced Stone Columns Using Silica-Manganese Slag as a Stone Column Material *ISSN || International Journal of Computational Engineering Research* **9** 2250–3005

[20] Shakri M S, Hafez M A, Adnan M A and Nazaruddin A T 2014 Effects of use of PFA on strength of stone column and sand column *Electronic Journal of Geotechnical Engineering* **19 L** 3745–55

[21] Nazaruddin A.T. Hafez M.A. S M S 2013 Laboratory Study on Bearing Capacity of Treated Stone Column *Electronic Journal of Geotechnical Engineering* **18** 5871–80

[22] Dipty Sarin Isaac G M S 2009 Suitability of Different Materials for Stone Column Construction *Electronic Journal of Geotechnical Engineering (EJGE)* **14** 1–12

[23] Jayalekshmi S. Harish Paravastu S V 2010 Ground Improvement using Stone Columns for Infrastructural Development *CISTUP @ CiSTUP 2010* (Bangalore: Indian Institute of Science)

[24] Pivarc J 2011 Stone columns – determination of the soil improvement factor *SOLVAK Journal of Civil Engineering* **19** 17–21

[25] Khadija.Baba Latifa.Ouadif N N 2015 Stone Columns In Silty Clay Soil: A Numerical Model *3rd Turkish Conference on Earthquake Engineering and Seismology* 1–10

[26] Jadid M N 2013 A Practical Approach For Computing Soil Bearing Capacity Under Shallow Foundations Using Vibro-Replacement Method *International Refereed Journal of Engineering and Science (IRJES)* **2** 54–62

[27] Xueyi Liu ; A. M. Shubber Ammar ; Hongsong Lin J R 2009 The Stress Concentration Ratio of Stone Columns under Confined Condition *Second International Conference on Transportation Engineering* 4247–55

[28] K. Balan P K J and S T 2015 "Studies on natural geotextile reinforced quarry waste columns for soft soil stabilizatio *Indian Journal of Geosynthetics and Ground Improvement* **4** 3–11

[29] Ancy Genu C George P K C 2016 Studies On Natural Geotextile Reinforced Quarry Waste Column For Improving Soft Soil *International Journal of Science, Engineering and Technology Research* **5**

[30] Siva Gowri Prasad S and Satyanarayana P V V 2016 Improvement of soft soil performance using stone columns improved with circular geogrid discs *Indian Journal of Science and Technology* **9**
[31] Beena K S 201AD Ground Improvement using Stone Columns *International Conferences on Recent Advances in Geotechnical Earthquake Engineering and Soil Dynamics.* (California: Scholars' Mine) pp 1–7
[32] Isaac D S and Girish M S 2009 Suitability of different materials for stone column construction *Electronic Journal of Geotechnical Engineering* **14 M** 1–12
[33] Rani S S and Kumar P S P 2015 A Study of Behaviour on Stone Column in Homogenous Soil *International Journal of Science and Research (IJSR)* **5** 1215–9
[34] Oluwasola E A, Hainin M R and Aziz M M A 2014 Characteristics and utilization of steel slag in road construction *Jurnal Teknologi* **70** 117–23
[35] Holiday, K. A 1997 Steel Slag-The High Performance Industrial Aggregate. *Proceedings of the XIIth World Meeting of the International Road Federation* (Toronto)
[36] Wang G C 2016 *The Utilization of Slag in Civil Infrastructure Construction* (Woodhead Publishing)
[37] Onda K, Honda H and Yoshitake H 2014 Examination of application of steel slag to sand compaction pile method *JFE Technical Report* **19** 41–8
[38] Ramzi N I R, Shahidan S, Maarof M Z and Ali N 2016 Physical and Chemical Properties of Coal Bottom Ash (CBA) from Tanjung Bin Power Plant *IOP Conference Series: Materials Science and Engineering* **160** 12056
[39] Aminaton Marto Masayuki Hyodo, Ahmad Mahir Makhtar M H 2014 Shear Strength Parameters and Consolidation of Clay Reinforced with Single and Group Bottom Ash Columns *Arabian Journal for Science and Engineering* **39** 2641–2654
[40] Muzamir Hasan Mohd Nor, Suzillco Suki N P 2016 Shear Strength Of Soft Clay Reinforced With Single Encased Bottom Ash Columns *ARPN Journal of Engineering and Applied Sciences* **11** 1562–9
[41] Muzamir Hasan Norhayani Pangee N A H 2016 Strength Of Soft Clay Reinforced With Square And Triangular Pattern Encapsulated Bottom Ash Columns *Second International Conference on Science, Engineering & Environment*
[42] Aminaton Marto Choy Soon Tanb, Fauziah Kasima, Nor Zurairahetty Mohd Yunusa, Razieh Moradia N A R 2016 Bearing capacity of soft clay installed with singular and group of encased bottom ash columns *Jurnal Teknologi (Sciences & Engineering)* **78** 105–110
[43] V R Raju 2011 Ground Improvement Using Vibro Techniques in FlyAsh Deposits *National Conference on Recent Advances in Ground Improvement Techniques* pp 0–12
[44] Muzamir Hasan Noor Amirah Noor Shahrudeen, Anas Mohammed Husian Kassim N Y 2015 Strength of Soft Clay Reinforced with Group Crushed Polypropylene (PP) Columns *Electronic Journal of Geotechnical Engineering (EJGE)* **20** 12308–991
[45] Ayothiraman R and Soumya S 2015 Model tests on the use of tyre chips as aggregate in stone columns *Proceedings of the Institution of Civil Engineers - Ground Improvement* **168** 187–93
[46] R. Ayothiraman S S 2011 Use Of Shredded Tyre Chips As Aggregates In Stone Column: An Experimental Study *Proceedings of Indian Geotechnical Conference* 711–4
[47] Akhitha A. A S 2017 Strength Improvement Of Stone Columns Using Tyre Chips And Aggregates *International Conference on GEOTECHNIQUES FOR INFRASTRUCTURE PROJECTS*
[48] Saravanan S priya 2013 Study On Sintered Flyash Aggregate As Columnar Inclusions On Soft Clay *International Journal of Combined Research & Development* **1**
[49] S. Vidhyalakshmi KE.A. Palaniappan V K S 2009 Studies on the Behaviour of Flyash Aggregate as Column Material in Soft Clay *India Geotechnical Society* 280–3
[50] S.Ayyappan M.Sundaram K H 2010 Investigation of Engineering Behavior of Soil, Polypropylene Fibers and Fly Ash -Mixtures for Road Construction *International Journal of Environmental Science and Development* **1** 171–5

IOP Conf. Series: Materials Science and Engineering **342** (2018) 012001 doi:10.1088/1757-899X/342/1/012001

[51] K. E. A. Palaniappan S P 2013 Improving Soft Clay Soil Using Fly Ash As Material Of Column *International Journal of Engineering Research & Technology (IJERT)* **2** 1828–32
[52] Robert Tranter Gurmel Ghataora I J 2008 The use of recycled aggregate in vibro-stone columns- A UK Perspective *GeoCongress 2008* 630–7
[53] Amini R 2015 *Physical Modelling Of Vibro Stone Column Using Recycled Aggregates* (University of Birmingham)
[54] Robert T, Ian J and Gurmel G 2017 The Use of Recycled Aggregate in Vibro-Stone Columns, A UK Perspective *GeoCongress 2008*
[55] Vikas Kumar K. Mani, Deepak Gupta A K C 2014 *Environmental Sustainability: Concepts, Principles, Evidences and Innovations* ed P (Dr. . G C Mishra (New Delhi: EXCELLENT PUBLISHING HOUSE)
[56] Anand J. Puppala Raja Yenigalla, Ekarut Archeewa B C 2013 Settlement Mitigation Using Light Weight Fill Embankment Systems *Geo-Congress 2013* 582–92
[57] A.J. Valsankar T A H 1987 Model Tets on Peat-Geotextile-Lighweight Aggregate System. *Geotextile and Membranes* **5** 251–60
[58] Siamak Boudaghpour S H 2008 A Study on Light Expended Clay Aggregate (LECA) in a Geotechnical View and its Application on Greenhouse and Greenroof Cultivation *INTERNATIONAL JOURNAL OF GEOLOGY* **2** 59–63
[59] Hoff I, Watn A and Oiseth E Light Weight Aggregate (LWA) Used in Road Pavements 1–10
[60] Davies L, Fry C and Bull J 104 - Lightweight backfill materials in integral bridge construction 47–54

IOP Conf. Series: Materials Science and Engineering **342** (2018) 012002 doi:10.1088/1757-899X/342/1/012002

Numerical modelling techniques of soft soil improvement via stone columns: A brief review

Azhani Zukri[1,2] and Ramli Nazir[2]

[1]Fakulti Kejuruteraan Awam & Sumber Alam, Universiti Malaysia Pahang, Lebuhraya Tun Razak, Gambang, Pahang, Malaysia
[2]Fakulti Kejuruteraan Awam, Universiti Teknologi Malaysia, Skudai, Johor, Malaysia

E-mail: engr.azhani@gmail.com

Abstract. There are a number of numerical studies on stone column systems in the literature. Most of the studies found were involved with two-dimensional analysis of the stone column behaviour, while only a few studies used three-dimensional analysis. The most popular software utilised in those studies was Plaxis 2D and 3D. Other types of software that used for numerical analysis are DIANA, EXAMINE, ZSoil, ABAQUS, ANSYS, NISA, GEOSTUDIO, CRISP, TOCHNOG, CESAR, GEOFEM (2D & 3D), FLAC, and FLAC 3. This paper will review the methodological approaches to model stone column numerically, both in two-dimensional and three-dimensional analyses. The numerical techniques and suitable constitutive model used in the studies will also be discussed. In addition, the validation methods conducted were to verify the numerical analysis conducted will be presented. This review paper also serves as a guide for junior engineers through the applicable procedures and considerations when constructing and running a two or three-dimensional numerical analysis while also citing numerous relevant references.

1. Introduction

There are a number of methods to analyze the geotechnical problems starting from closed-form analytical methods to the finite element (FE) based numerical methods. Each method has its own advantages and disadvantages. The solution will be accurate only if all the following conditions were satisfied [1].

i. Equilibrium of forces: external forces that should be exactly equal to the internal forces,
ii. Compatibility: the relation between displacements at different locations within a continuum also called as continuity equations,
iii. Constitutive behaviour: stress-strain equations of the material, and
iv. Boundary conditions: force or displacement conditions.

Finite Element (FE) analysis is the most widely used method to analyse the stone column-improved soil behaviour. The best way to represent the real conditions of soft soil behaviour is through a three-dimensional modelling; however, for practical purposes, either the axisymmetric or plane strain model is generally used. Many studies have utilised the axisymmetric model for the unit cell model of uniformly loaded stone column groups or plane strain models (Barksdale & Bachus, 1983). The numerical analyses of stone columns generally need a complex modelling of the soil reinforcement system. As stated by Canizal *et. al.* there are five main approaches for modelling stone columns numerically [2];

i. Unit cell approach: In the axisymmetric model, only one column and its surrounding soil consisting of a "unit cell" are modelled. (Balaam & Booker, 1981),
ii. Plane strain method: The numbers of columns are modelled as stone trenches that are usually used under continuous loads such as embankments. (Van Impe & De Beer, 1983),
iii. Axial symmetry technique: Stone rings are modelled instead of cylindrical columns to represent the columns under circular loads such as tanks and circular foundation. (Elshazly, 2008),
iv. Homogenization technique: The composite soil parameters are used to model the improved homogeneous soil with stone columns. (Schweiger, 1989), and
v. Full 3D Model: The most complex system to simulate the stone column-improved soil behaviour is three dimensional numerical models. (Weber et al., 2008).

The first four approaches are considered as the two-dimensional (2D) analyses.

Table 1 below summarize the suitability of some of these models to study on improvement of soft soil stabilized with non-encased stone column.

Table 1. Suitability of simplifies geometrical models to study different features of stone column treatment [3].

Model	Settlement	Consolidation	Stability
Unit cell	Completely suitable	Completely suitable	Not suitable
Stone trenches	Moderate suitable	Moderate suitable	Moderate suitable
Homogenization	Moderate suitable	Slightly suitable	Slightly suitable
3D model	Completely suitable	Completely suitable	Completely suitable

2. Finite element analysis (FE)

FE analysis can be applied in order to determine the complicated parameters that would be difficult to measure through experimental work. In addition, FE analysis is an effective way of alternative process for laboratory investigation studies, especially in the case of large physical models, which leads to saving the time and costs associated with the construction of physical models. However, the selection of suitable constitutive model to represent the actual soft soil behaviour is not simple since soils are complex material consisting of soil particles and void filled with water and/or air. It is very important to choose appropriate model when running the analysis so that the result produced will match the behaviour of soil since its behaviour is not always linear. Nowadays, the numerical modelling software was equipped with numerous constitutive models for modelling the stress-strain behaviour of soil starting from simple to more complex based on required analysis needs. In addition, several sets of laboratory tests should be done to collect all the important parameters regarding compression and shearing behaviour of soils, such as physical tests, triaxial and oedometer test. This parameters are essential for numerical analysis to be conducted. Despite the fact that, no soil constitutive model available that can describe the complex behaviour of real soils under all conditions completely. Other than that, the soil model should be calibrated through sensitivity analysis to define the most appropriate meshes, boundary condition, type of analysis; drained or undrained and soil interface interaction that will be utilised in the numerical analysis.

For example, the latest version of Plaxis 3D software is come up with 8 constitutive models for different conditions and type of soil analysis [4]. So that, the engineers should choose 'right' constitutive model which provides a reasonable fit to data obtained from laboratory tests. It is highly recommended to conduct various measurement comparisons along with additional experiment models and/or full-scale tests to ascertain the degree of reliability in the models in order to adjust and refine them to each type of different modelling application [5]. Poul V. Lade proposed three main categories for the evaluation and selection of suitable constitutive model to be used in numerical analysis [6];

i. Category 1: Model fulfil all the conditions 1 to 5 as tabulated in the Table 2 below.
ii. Category 2: Model lacks one or more conditions.
iii. Category 3: Model is deficient of several conditions as shows in the table below.

Table 2. Model conditions for model selection evaluation.

Condition No	Description
1	Theoretically sound framework that is sufficiently transparent and accessible to anticipate and evaluate the model performance.
2	Model includes effects of confining pressure.
3	Model can handle 3D conditions.
4	Straight forward to find the parameters, which may be determined from conventional experiment.
5	Models exhibit the overall high quality o fit with the observed behavior.

Even correct models were chosen, it should be emphasized that there are approximations within level of accuracy such as, approximation in the finite element method, assumptions made about the constitutive soil response and the detailed explanation of the numerical model and its boundary conditions.

2.1. Previous study on two-dimensional analysis

A comparison between two-dimensional numerical modelling (Plaxis V8 Software) and analytical and laboratory model has been done by Pivarc to estimate the settlement improvement factor of soft soil improved by stone columns [7]. All the models in this study were prepared by the vibro-replacement method, which means that the soil is removed from a hole and not compacted to the sides such as in the vibro-displacement method. For the numerical modelling, the Mohr-Coulomb (MC) elastic-perfectly plastic criteria was chosen to model a soft soil with fifteen node elements applied. The boundary conditions along the vertical boundaries of the axially symmetrical model were fixed for the lateral deformations to allow vertical deformation. The study found that the results of the improvement factor calculated by the Priebe`s method, numerical method, and results obtained from the laboratory experiments were in good agreement, show that MC model selected in this analysis was appropriated.

Joel Gniel and Abdelmalek Bouazza predicted the site behavior of geogrid encased stone through small scale testing and two-dimensional numerical modelling using the PLAXIS software [8]. A numerical study was used to replicate the results of a small-scale testing in order to establish a method of modelling column installation and column behavior based on basic laboratory tests. Following this, the modelling process was scaled up to predict the behavior of columns installed on site including the impact of varying different parameters. For clay, the Soft Soil (SS) constitutive model in PLAXIS most closely matched the compressibility observed in the oedometer tests while for sand, the Hardening Soil (HS) model most closely matched the test behavior. The fiberglass mesh was modelled using an elastic tensile element with the tensile strength and stiffness derived from the uniaxial tensile testing of the mesh. Results from the small-scale testing and a detailed numerical study have indicated that fully encased stone columns with geogrid could extend the ground improvement technique to extremely soft soils.

A. Zahmatkesh and A. J. Choobbasti evaluated the settlement of soft clay reinforced by stone columns considering the effect of soil compaction [9]. The study was conducted using the two-dimensional numerical analysis then the results were verified with those available in the literature. A drained analysis was carried out using the Mohr-Coulomb's criterion for soft clay, stones, and sand. The interface elements were used between the stone column and soft clay to evaluate the installation effect due to soil compaction activity. The column installation was simulated to calculate the stresses due to the compaction of soil. From the analysis, the variation of stress in the soft soil due to installation of the column with distance from the column is significantly reduced. Likewise, the research conducted by J. Castro et al. on the installation effects of stone columns in natural soft clay using numerical modelling found that the stresses near the column were related to the initial undrained shear strength of the soil. For normal column spacing, the soil is in a plastic state, which may be negative for a subsequent loading process due to over consolidation or positive if the yield surface has been notably expanded (strain hardening) [10]. Other studies on numerical analysis on the behavior of soft clay improved by stone columns with considering the installation effects also have been performed by Ellouze S *et. al.*[11], Sexton and McCabe [12], Guetif Z *et. al.* [13] and Castro J *et. al.* [14].

Meanwhile, the analysis of settlement and bearing capacity of stone columns and geogrid encased stone columns with various diameters was conducted using data obtained from an experimental investigation using FE software Plaxis 2D. An axisymmetric model of the stone column was configured while the load-settlement response of the stone column was defined using SS and MC model. Detailed parametric analyses were performed by varying the diameter of the stone column and stiffness of the geogrid used for encasement. The results were found to be in reasonable agreement between the experimental investigation and the FE method [15].

Muntohar et al. presented a numerical analysis of the pulverised fuel ash (PFA) column-treated peat and validated with the field static-loading test results [16]. Back analysis was performed to determine the material parameters and soil stiffness surrounding soil and soil-column. The peat and column-treated peat were modelled as Mohr-Coulomb model while the elastic behaviour was chosen for the loading plate. According to the basic design, the peat and column treated peat were assigned as undrained behaviour. The loading plate wall was modelled as the beam element with 10 mm thickness. The simulation results in this study showed that the elasto-plastic Mohr-Coulomb model for the peat and PFA column-treated peat can be applied to model the load-settlement behaviour. The study found that both models were reliable to simulate the field static-loading test for column-treated peat.

The stress concentration factor in stone columns was determined through a numerical modelling where two-dimensional plane strain models were utilized to specify the concentration of stress in the stone columns and reduction in settlement by improving the soil with stone columns [17]. In the scope of this study, drained case was defined for all materials including the soft silty clay, clayey sand, stones, and the sand drainage-stress for the generation of the FE model. Mohr-Coulomb failure criterion considering the elasto-plastic behaviour was utilised for all materials. The study concluded that the behaviour of change of the ratio with a modulus of elasticity of soil in floating and end bearing columns was similar to the rigid foundations while the stress concentration factors were almost constant at different flexible foundation pressures. However, no validation process conducted in the study. In order to verify the findings accuracy, experimental model or analytical calculation can be performed.

The behaviour of piled raft foundation considering pile-soil, raft-soil, pile-soil-pile, and raft-soil-pile interactions have been analyzed using 2D numerical method [18]. The settlements of the system were estimated to satisfy the compatibility of two separate models which were the flexible raft model supported by springs representing soils and piles and the group-pile model embedded in the layers of soil having different properties. The adequacy of the analysis was compared with the analysis result of

Plaxis 3D. The constitutive models used in Plaxis 3D analysis were the hardening-soil model for soil and linear-elastic model for the raft and piles. The interface between piles and soil in Plaxis 3D was simulated with elements of fictitious thickness showing elasto-plastic behaviour based on the Coulomb failure criteria. The elastic modulus of soil was assumed to increase with depth and estimated using the similar method proposed by Schmertmann and Hartmann (1978). Meanwhile, the effect of the stiffness of the outer and inner piles was verified through a series of experimental load tests. The numerical analysis was well adopted to identify the piled raft interaction using the FE package Plaxis 2D [19].

Other researchers, analysed a piled raft foundation on soft soil using Plaxis 2D by considering the various numbers of piles. The raft and piles were modelled as plate elements, while the entire soil material models were Mohr-Coulomb. The design process for piled rafts in this study involved three stages namely, to assess the required raft thickness, to assess the required pile length in the piled raft system and to obtain the optimum number of piles. From the analysis, it was found that the addition of piles could reduce the settlement but after reaching a certain number of piles, the increasing number of piles showed that the settlement tended to be constant. The researchers concluded that for an economic design, it is necessary to consider the optimum number of piles in piled raft foundation system based on the allowable settlements [20]. The effects of pile length and number of piles on the settlement of the piled raft foundation were investigated using two-dimensional analysis and found that the settlement was reduced with the increase in length and number of piles. In this study, the plane strain analysis was taken for analyzing the load-settlement behaviour where the raft was modelled as a plate element, while the piles were modelled as a series of beam elements. The Mohr-coulomb model was adopted for the soil medium while the linear elastic model was adopted for the raft and pile. This numerical analysis was validated by experimental results [21].

An analysis was carried out using the Mohr-Coulomb criterion model for stone columns and embankment fills (drained condition) and soft soil clay (undrained condition) to evaluate the bearing capacity of the square foundation on the ground improved with small groups of stone columns. All the analyses were performed using the axisymmetric idealization of a cylindrical unit cell consisting of the stone column and soft soil under the embankment fills. Meanwhile, the boundaries for unit cell model was used where the vertical and horizontal displacements in the bottom boundaries were restricted while only the horizontal displacement in the lateral boundaries was restricted [22].

The study on floating stone columns were conducted using the 2D FE analyses with the influence of different key parameters such as column length, area replacement ratio, loading intensity, and post installation lateral earth pressure on the settlement ratio, then the design chart developed from this study was validated by a case study where a reasonable agreement was obtained. The undrained analysis was implemented in this study and was verified using analytical method [23][24]. Other than that, the optimum length of stone column (floating column) also can be predicted using numerical method. The finite element study by Tan et al. [25] and Wood [26] showed the existence of the optimum length, Lo_p for stone columns which is the column length where lengthening it will not significantly contribute to the reduction of settlement. Two dimensional numerical method was utilized to analysis the load-settlement behavior. The researcher established the method to predict the stone column optimum length which can be used for both homogenous and non-homogenous ground condition. The settlement predictions made by the proposed method are validated with the 3 dimensional finite elements results and field measurement [27].

A series of numerical analysis was performed to investigate the reliability of the axisymmetrical concentric ring model to be used in problems where the stone column was adopted as the ground improvement method to support a small foundation. This study adopted the model by Elshazly et al. (2008) where the thickness and radius of the ring are adjusted to give the correct equivalent area of the stone columns. The 2D analysis was executed using the FE programme PLAXIS where the HS model was used for both the soft clay and column material. The 3D analysis was done as a validation process to the 2D analysis results. Generally, the 2D and 3D analyses display the same deformation mode of failure despite some discrepancies in detail [23]. Similar modelling simulation was used by others (Ambily and Gandhi [28]; Elshazly, Hafez, & Mossaad [29]; Nazari Afshar [30]; Mitchell and Huber [31]. The accuracy of the developed models were verified by experimental models conducted in a laboratory.

Other studies on the utilization of two-dimensional numerical software were also conducted by Lan *et. al.* [32]; P. Mohanty and M. Samanta [33]; and Ornek *et. al.* [34]. Table 3 represents summary of the numerous previous two dimensional numerical studies conducted. It is shows that most of the study used MC model with drained analysis, which are more suitable in term of short time consumed for running the analysis and limited soil parameters.

Table 3. Summary of 2D Numerical Study, Constitutive Model, Analysis and Validation Method Used.

Bil	Author Name, Year	Software Name	Constitutive model*		Type of Analysis for Soil	Validation method	Focused study
			Soft Soil	Stone filler			
1	J. Pivarc, 2011	Plaxis 2D	MC	MC	Drained	Experimental model & Analytical (Priebe`s Method)	Settlement
2	Gniel & Bouazza, 2009	Plaxis 2D	SS	HS	Drained	Small Scale model	Settlement
3	A. Zahmatkesh & A. J. Choobbasti, 2010	Plaxis 2D	MC	MC	Drained	Previous study/ literature	Settlement
4	J. Castro *et. al.*, 2013	Plaxis 2D	S-CLAY1 & S-CLAY1S		Drained	Modified Cam Clay model	Settlement, consolidation properties
5	A. Marto *et. al.*, 2013	Plaxis 2D	SS	MC	Drained	Experimental model	Load-settlement
6	Muntohar *et. al.*, 2013	Plaxis 2D	MC	MC	Undrained	Field static-loading test	Load-settlement
7	Yildiz M., 2013	Plaxis 2D	MC	MC	Drained	Not mentioned	Stress concentration
8	Joon-Shik Moon & W. P., 2015	Plaxis 2D	HS	LE	Drained	3D numerical	Settlement
9	Elwakil A. Z. & Azzam W. R., 2016	Plaxis 2D	MC	MC	Drained	Experimental load tests	Settlement, bearing capacity
10	R.Radhika, 2015	Plaxis 2D	MC	LE	Drained	Experimental result	Settlement
11	Mohammadreza Jaberi Nasab A A, 2015	Plaxis 2D	MC	MC	Undrained	Experimental result	Bearing capacity
12	K.S. Ng & S.A. Tan, 2014	Plaxis 2D	MC	MC	Undrained	Analytical (Priebe`s Method & α-β Method)	Consolidation settlement
13	K.S. Ng, 2013	Plaxis 2D	HS	HS	Undrained	3D numerical	Settlement
14	Nima Rostami Alkhorshid, 2010	Plaxis 2D	MC	MC	Undrained	3D numerical	Settlement

15	M. Y. Fattah & Q. G. Majeed, 2011 [35]	CRISP2D	MC	MC	Undrained	Not mentioned	Bearing capacity
16	Sexton B G *et.al.*, 2016 [36]	Plaxis 2D	SSC	HS	Drained	Field study	Settlement
* LE: Linear-elastic Model, MC: Mohr-Coulomb Model, SS: Soft Soil Model, HS: Hardening Soil Model, MCC: Modified Cam Clay model, SSC: Soft Soil Creep							

2.2. Previous study on three-dimensional analysis

A computer programme incorporating the proposed analytical method was developed and the adequacy of analysis was compared with the analysis result of Plaxis 3D (Brinkgreve & Broere, 2006). Plaxis 3D programme is based on three-dimensional FE method and its versatility and adequacy for both foundation design and research were verified by many engineers. This numerical modelling is a simple and effective alternative to approach the real behaviour of soils reinforced by stone columns as it allows settlement analysis, lateral deformation, and vertical and horizontal stresses in order to understand the behaviour of columns and soil. It also has the advantage of integrating the settlements of the underlying layers, especially those of least resistance.

The study conducted by Nima Rostami Alkhorshid investigated the performance of stone columns and encased sand columns in soft compressible clay using FE analyses. A drained analysis for stones and sand columns was carried out using the Mohr-Coulomb criterion whereas, for the soft clay, the undrained analysis was performed. In a full-scale three-dimensional FE analysis, the bulging and punching failure and vertical settlement of stone columns were analysed [37]. Three dimensional analysis was used to verify the two dimensional analysis in this study.

A numerical analysis was carried out by using the geotechnical FE software, Plaxis 3D Foundation to investigate the influence of raft thickness, pile length and spacing, and a number of piles on piled raft foundation behaviour. In this analysis, only one-quarter of the raft was modelled due to symmetry. The numerical analysis was conducted to verify the in-situ measured values [38]. Similar study on the piled raft foundation behaviour using the three-dimensional analysis was established by several researchers; Hussein H. Karim *et. al.* [39]; E.Y. N Oh *et. al.* [40], Sreechithra P. & Niranjana K. [41]; K. Watcharasawe [42]; Comodromos [43].

Noura Nehab *et. al.* established the relationship between the column spacing, properties, and length and the settlement improvement of column group using FE modelling (Plaxis 3D Foundation) and took a "Bouregreg Valley" project as a case study. The study concluded that the comparison of Plaxis 3DF and analytical methods in the case of isolated column gave early confidence in the ability of Plaxis 3DF to capture stone column behaviour [44].

Meanwhile, a three-dimensional FE analysis was conducted by Micheal M. Killeena and Bryan A. Mc Cabeb to identify the effect of variables of column arrangement, spacing, length, and Young's modulus of the column material in the design process and the interactions between them [45]. A simplified method was proposed to relate the settlement of small groups to a reference unit cell settlement predicted by the current analytical approaches. 15-node wedge isoperimetric elements and lateral boundaries of the computational domain allowed vertical movement that was a restricted radial movement. Both the stone columns and host soft clay were modelled using the elasto-plastic HS model. Three analytical design methods by Balaam and Booker [46], Priebe [47], and Pulko *et. al.* [48] were used to validate the numerical analysis.

A series of three-dimensional numerical analysis was carried out to evaluate the settlement of mat foundation rested on soft peat reinforced by a group of cement columns [49]. The analysis was carried out using soft soil creep (SSC) constitutive model considering the elasto-plastic behaviour of peat, while the cement columns were simulated as the embedded pile. This model is appropriate for near-normally consolidated clay, clayey silt, and peat. A drained analysis was assumed for soft soil. It was found that the results from the analytical methods were comparable to those by using the FE analysis.

The research by Yogendra K. *et. al.* addressed a parametric study on the behaviour of embankment resting on the Geosynthetic Reinforced Stone Column (GRSC). The effects of stone column spacing to diameter ratio, deformation of stone column material and embankment fill, geosynthetic stiffness, the height of embankment, and soft clay thickness on the settlement of embankment were investigated by employing the three-dimensional FE modelling [50]. For the hydraulic boundary condition, the phreatic level was set at the top surface of the soft clay layer to generate a hydrostatic pore water pressure profile in the domain while the zero pore pressure boundary condition was applied at the top. The left and right boundary were assumed impervious to consider the fact that no flow entered or left the symmetry plane, while right boundary was too far from the embankment to have a significant influence on the results. The FE mesh was developed using 10-node tetrahedral elements to represent soft clay, stone column, and embankment fill, while the geosynthetic reinforcement was modelled as the geogrid element.

Recent research conducted by Akhila Shaji *et. al.* focused on comparative study of how an embankment of soft clay behaves in the absence of any reinforced technique and with the installation of the stone column using commercial software PLAXIS 3D. Mohr-Coloumb criterion was used for the analysis while drainage conditions suitable for developing effective pore pressure and stiffness parameters were used. The variation of deformation, stress, and strain with ultimate load acting on soil was obtained. From this study, the stone column was proven to be a better reinforced technique in slope stabilisation [51]. However, the researcher did not performed any validation method to proof the outcomes of this study. Validation can be made either using selected analytical method or others numerical software, which is easier compared to experimental model or full scale test.

The study focused on the numerical modelling of a small group of laboratory-modelled reinforced stone columns considering parameters like area replacement ratio, the stiffness of reinforcement material, and reinforcement length. The bottom face of the clay bed was considered as fixed. Mohr–Coulomb failure criterion was adopted for the stone column and clay, which has the linearly elastic-perfectly plastic behaviour. The geosynthetics were modelled as geogrid element available in PLAXIS 3D having axial stiffness only. The numerical analysis was conducted to verify the physical modelling established in the laboratory [52].

Other studies were conducted to utilise the three-dimensional numerical software and later validated through experimental works, analytical analysis or compared to actual case studies or contrariwise by Hasan et al.[53], Hasan M and Samadhiya N.K [54], Weber, T M Springman, S M [55], Sreechithra P. and Niranjana K [56], S Muzammil *et.al.* [57], Y Tandel *et. al.* [58] and L Keykhosropur *et. al.* [59]. Table 4 tabulates summary of the numerous previous three dimensional numerical studies conducted.

Table 4. Summary of 3D Numerical Study, Constitutive Model, Analysis and Validation Method Used.

Bil	Author Name, Year	Software Name	Constitutive model*		Type of analysis	Validation method	Focused study
			Soft Soil	Stone filler			

1	Morteza Esmaeili & Seyed Mehrab Hakimpour, 2015 [60]	FLAC3D	MC	MC	Undrained	Centrifuge test results	Liquefaction
2	Shaymaa Kadhim, 2007 [61]	FLAC3D	MC	MC	Drained	Experimental load tests	Slope stability
3	K. Watcharasawe & P. Jongpradist, 2014	Plaxis 3D	HS	MC	Undrained	Literature	Settlement, bearing capacity
4	Alireza Roshan & Issa Shooshpasha, 2014	Plaxis 3D	MC	-	Drained	Not mentioned	Settlement
5	Noura Nehab et. al., 2016	Plaxis 3D	MC	LE	Drained	Analytical method	Settlement, bearing capacity
6	Kongpop Watcharasawe et. al., 2017	Plaxis 3D	HS	LE	Undrained	Field static-loading test	Bearing capacity, settlement
7	Micheál M.Killeena & Bryan A.Mc Cabeb, 2014	Plaxis 3D	HS	HS	Drained	Three different analytical method	Settlement
8	Ali Dehghan Banadaki, 2012	Plaxis 3D	SSC	-	Drained	Analytical method	Settlement
9	Akhila Shaji et. al., 2016	Plaxis 3D	MC	MC	Drained	Not mentioned	Deformation, stress, and strain
10	Hasan *et. al.* (2016)	Plaxis 3D	MC	MC	Drained	Experimental result	Load-Settlement
11	Basim Jabbar Abbas, 2016 [62]	Plaxis 3D	MC	MC	Drained	Analytical Brom's Method	Settlement
12	Sexton & McCabe, 2016 [63]	Plaxis 3D	SSC	HS	Drained	Field Study	Settlement
13	Martin Gab & Helmut F. Schweiger, 2009 [64]	Plaxis 3DF	HS, MCC	HS	Drained	Field Test	Settlement

* LE: Linear-elastic Model, MC: Mohr-Coulomb Model, SS: Soft Soil Model, HS: Hardening Soil Model, MCC: Modified Cam Clay model, SSC: Soft Soil Creep

3. Conclusion

From the review paper, it can be concluded that the numerical modelling software is the most popular and a good method to be utilised for the stimulation and difficult analysis of geotechnical behaviours. The software was equipped with different types of constitutive model to be selected during analysis based on the different conditions of the structure, soil type, and construction method and liquefaction phenomena. In term of analysis, 3D provide more precise results at all desired angle, compared to 2D which only presents the result based on plain strain model. However, the 3D analyses require longer execution times than the 2D analyses and relatively expensive. A major problem is to know how accurate the results from the numerical modelling correspond to what is happening in the field and actual geotechnical behaviour. Therefore, the analysis needs a further step for the verification and validation purposes. Verification is to certify that the computer model has been developed correctly, while validation is to ensure that the model fulfils its use. From the review, the numerical modelling can be validated using other numerical methods, analytical solution using established methods, experimental work or physical modelling, an actual case study and full-scale or field tests. The first two validation methods are simpler, faster and more economical than physical modelling and full-scale tests. However, physical modelling and full-scale testing can provide even more accurate verification though not economical and require more time to perform testing.

References

[1] K. Rajagopal B M B 2016 *A Primer on Numerical and Physical Modelling in Geotechnical Engineering* vol Technical (New Delhi: Indian Geotechnical Society,)

[2] J. Canizal A. Cimentada, A. Da Costa, M. Miranda, C. Sagaseta J C 2012 Theoretical analyses of laboratory tests of kaolin clay improved with stone columns *ISSMGE - TC 211 International Symposium on Ground Improvement IS-GI Brussels* vol III, ed N H Nicolas Denies (Brussels, Belgium) pp 373–81

[3] Castro J 2017 Modeling Stone Columns *Materials* **10** 782

[4] Plaxis 2010 Material Models Manual *Plaxis* 202

[5] Ti K S, Gue See S, Huat B B ., Noorzaei J and Saleh M 2009 A Review of Basic Soil Constitutive Models for Geotechnical Application *Electronic Journal of Geotechnical Engineering* **14** 18

[6] Lade P V 2005 Overview of Constitutive Models for Soils *Soil Constitutive Models* 1–34

[7] Pivarc J 2011 Stone columns – determination of the soil improvement factor *SOLVAK Journal of Civil Engineering* **19** 17–21

[8] Joel Gniel A B 2009 Predicted Site Behaviour Of Geogrid Encased Stone Columns *Australian Geomechanics* **44** 11–6

[9] A. Zahmatkesh A J C 2010 Settlement Evaluation Of Soft Clay Reinforced By Stone Columns, Considering The Effect Of Soil Compaction *International Journal of Research & Reviews in Applied Sciences* **3** 159–66

[10] Castro J, Karstunen M, Sivasithamparam N and Sagaseta C 2013 Numerical analyses of stone column installation in Bothkennar clay *Installation Effects in Geotechnical Engineering* (CRC Press) pp 212–8

[11] Ellouze S, Bouassida M, Bensalem Z and Znaidi M N 2017 Numerical analysis of the installation effects on the behaviour of soft clay improved by stone columns *Geomechanics and Geoengineering* **12** 73–85

[12] Sexton B G and McCabe B A 2015 Modeling stone column installation in an elasto-viscoplastic soil *International Journal of Geotechnical Engineering* **9** 500–12

[13] Guetif Z, Bouassida M and Debats J M 2007 Improved soft clay characteristics due to stone column installation *Computers and Geotechnics* **34** 104–11

[14] Castro J, Karstunen M and Sivasithamparam N 2014 Influence of stone column installation on settlement reduction *Computers and Geotechnics* **59** 87–97

[15] Aminaton Marto, Farshad Helmi, Nima Latifi, Mohsen Oghabi R M 2013 Performance analysis of reinforced stone columns using finite element method *Electronic Journal of Geotechnical Engineering (EJGE)* **18 B** 315–23

[16] Muntohar Rahman, M. E., Hashim, R., Islam, M. S. A S 2013 A Numerical Study of Ground Improvement Technique Using Group of Soil-Column on Peat *Journal of Science and Technology* 625–34

[17] Yildiz M 2013 *Determination Of Stress Concentration Factor In Stone Columns By Numerical Modelling* (MIDDLE EAST TECHNICAL UNIVERSITY)

[18] Joon-Shik Moon W P 2015 Analysis of Piled Raft Foundation Behavior Considering Raft-Pile-Soil Interaction *Conf. on Advances in Civil, Structural and Mechanical Engineering* 64–7

IOP Conf. Series: Materials Science and Engineering **342** (2018) 012002 doi:10.1088/1757-899X/342/1/012002

[19] Elwakil A Z and Azzam W R 2016 Experimental and numerical study of piled raft system *Alexandria Engineering Journal* **55** 547–60

[20] Paravita Sri Wulandari D T 2015 Analysis of piled raft foundation on soft soil using PLAXIS 2D *The 5th International Conference of Euro Asia Civil Engineering Forum (EACEF-5)* 363 – 367

[21] R.Radhika P.Soundrapandiyan S P J 2015 Parametric Study and Numerical Analysis of Piled Raft Foundation on Soft Clay *International Journal For Research In Emerging Science And Technology* **2** 90–7

[22] Mohammadreza Jaberi Nasab A A 2015 Numerical Analysis of the Bearing Capacity of Stone Columns Impoved Ground *International Journal of integrative Sciences, Innovation and Technology* **4** 1–5

[23] Shien N K 2013 *Numerical Study Of Floating Stone Columns* (National University Of Singapore)

[24] Ng K S and Tan S A 2014 Design and analyses of floating stone columns *Soils and Foundations* **54** 478–87

[25] Sun S A T ; K S N ; J 2014 Column Group Analyses for Stone Column Reinforced Foundation *From Soil Behavior Fundamentals to Innovations in Geotechnical Engineering* vol 233, ed H R E Olson ((ASCE), Reston) pp 597–608

[26] Wood D M, Hu W and Nash D F T 2000 Group effects in stone column foundations: model tests *Géotechnique* **50** 689–98

[27] Ng K S and Tan S A 2015 Settlement Prediction of Stone Column Group *International Journal of Geosynthetics and Ground Engineering* **1** 33

[28] Ambily A P and Gandhi S R 2007 Behavior of Stone Columns Based on Experimental and FEM Analysis *Journal of Geotechnical and Geoenvironmental Engineering* **133** 405–15

[29] Elshazly H A, Hafez D H and Mossaad M E 2008 Reliability of Conventional Settlement Evaluation for Circular Foundations on Stone Columns *Geotechnical and Geological Engineering* **26** 323

[30] Ghazavi M and Nazari Afshar J 2013 Bearing capacity of geosynthetic encased stone columns *Geotextiles and Geomembranes* **38**

[31] K. M J and R. H T 1985 Performance of a Stone Column Foundation *Journal of Geotechnical Engineering* **111** 205–23

[32] Lan L V T · Y L Z · S X D · P V A · H T 2015 Numerical analysis of the influence of replacement area ratio in foundation reinforced with soil cement columns *Electronic Journal of Geotechnical Engineering (EJGE)* **20** 3821–8

[33] Mohanty P and Samanta M 2015 Experimental and Numerical Studies on Response of the Stone Column in Layered Soil *International Journal of Geosynthetics and Ground Engineering* **1** 27

[34] Murat Ornek Mustafa Laman, Abdulazim Yildiz A D 2012 Numerical Analysis Of Circular Footings On Natural Clay Stabilized With A Granular Fill *Acta Geotechnica Slovenica* 61–75

[35] Fattah M Y and Majeed Q G 2012 Finite Element Analysis of Geogrid Encased Stone Columns *Geotechnical and Geological Engineering* **30** 713–26

[36] Sexton B G, McCabe B A, Karstunen M and Sivasithamparam N 2016 Stone column settlement performance in structured anisotropic clays: the influence of creep *Journal of Rock Mechanics and Geotechnical Engineering* **8** 672–88

[37] Alkhorshid N R 2012 *Numerical Analysis of Soft Clay Reinforced with Stone Columns* (Eastern Mediterranean University)

[38] Alireza Roshan I S 2014 Numerical Analysis of Piled Raft Foundations in Soft Clay *Electronic Journal of Geotechnical Engineering (EJGE)* **19** 4541–54

[39] Hussein H. Karim Mudhafar K. Hameedi M R A-Q 2013 Numerical Analysis of Piled Raft Foundation on Clayey Soil *Eng. & Tech. Journal* **31** 1297–312

[40] E.Y. N Oh Q. M. Bui, M. Huang, C. Surarak, A. S. Balasurbamaniam D G L 2009 Numerical Analysis of Piled Raft Foundation in Sandy and Clayey Soils. *Proceedings of the 17th International Conference on Soil Mechanics and Geotechnical Engineering*

[41] P. S and K. N 2017 Numerical Analysis of Pile Raft System *International Conference on Geotechniques For Infrastructure Projects* (Thiruvananthapuram, India: Indian Geotechnical Society)
[42] Watcharasawe K, Jongpradist P and Kitiyodom P 2014 Numerical Analyses of Piled Raft Foundation in Soft Soil using 3D-FEM *International Conference on Advances in Civil Engineering for Sustainable Development* (Thailand: Suranaree University of Technology) pp 219–24
[43] Comodromos E M, Papadopoulou M C and Laloui L 2015 Contribution to the design methodologies of piled raft foundations under combined loadings *Canadian Geotechnical Journal* **53** 559–77
[44] Noura Nehab Latifa Ouadif, Lahcen Bahi K B 2016 Three-Dimensional Modeling Of A Group Of Stone Columns In "Bouregreg Valley" Soft Ground *ARPN Journal of Engineering and Applied Sciences* **11** 14537–44
[45] Killeen M M and McCabe B A 2014 Settlement performance of pad footings on soft clay supported by stone columns: A numerical study *Soils and Foundations* **54** 760–76
[46] Balaam N P and Booker J R 1985 Effect of stone column yield on settlement of rigid foundations in stabilized clay *International Journal for Numerical and Analytical Methods in Geomechanics* **9** 331–51
[47] Priebe H J 1995 The Design of Vibro Replacement
[48] Pulko B, Majes B and Logar J 2011 Geosynthetic-encased stone columns: Analytical calculation model *Geotextiles and Geomembranes* **29** 29–39
[49] Ali Dehghan Banadaki Nazri Ali K A 2012 Initial Settlement of Mat Foundation on Group of Cement Columns in Peat – Numerical Analysis *Electronic Journal of Geotechnical Engineering (EJGE)* **17** 2243–53
[50] Yogendra K. Tandel C H S & A K D 2013 3D FE Analysis of an Embankment Construction on GRSC and Proposal of a Design Method *ISRN Civil Engineering* 1–11
[51] Akhila Shaji Haripriya M., Rajeswari K. R., Geethu Thomas A A 2016 Stability Analysis Of Soft Clay Stabilized Using Stone Column *International Conference on "Structural Engineering & Construction Management" SECM-2016* **3**
[52] Yogendra Tandel Chandresh Solanki, Atul Desai M J 2015 Numerical analyses of laboratory-modelled reinforced stone column *CURRENT SCIENCE (00113891)* **108** 1354–62
[53] Hasan M and Samadhiya N K 2016 Experimental and Numerical Analysis of Geosynthetic-Reinforced Floating Granular Piles in Soft Clays *International Journal of Geosynthetics and Ground Engineering* **2**
[54] Hasan M S N K 2016 3d Numerical Analysis Of Granular Piles With Internal Horizontal Geogrid Strips In Layers *Indian Geotechnical Conference* (Chennai, India: IIT Madras) pp 1–4
[55] Weber T M and Springman S M 2009 Numerical modelling of stone columns in soft clay under an embankment 305–11
[56] Sreechithra P 2017 Numerical Analysis of Pile Raft System 3–6
[57] Muzammil S P, Varghese R M and Joseph J 2017 Performance of Stone Column under Circular **6** 34–8
[58] Tandel Y k., Solanki C H and Desai A K 2012 Numerical modelling of encapsulated stone column-reinforced ground *International Journal of Civil, Structural, Environmental and Infrastructure Engineering* **2** 82–96
[59] Keykhosropur L, Soroush A and Imam R 2012 3D numerical analyses of geosynthetic encased stone columns *Geotextiles and Geomembranes* **35**
[60] Esmaeili M and Hakimpour S M 2015 Three Dimensional Numerical Modelling of Stone Column to Mitigate Liquefaction Potential of Sands *Journal of Seismology and Earthquake Engineering* **17** 127–40
[61] Kadhim S 2007 Stability Analysis of Geotextile Encased Sand Columns The Dissertation Committee for Shaymaa Kadhim certifies that this is the approved version of the following dissertation : Stability Analysis of Geotextile Encased Sand Columns

[62] Abbas B J 2016 The Settlement Evaluation of Improved Soft Clay using Sand Columns and Partial Replacement Technique *International Journal of Engineering Research & Technology (IJERT)* **5** 348–55

[63] Sexton B G and McCabe B A 2016 Stone column effectiveness in soils with creep: a numerical study *Geomechanics and Geoengineering* **11** 252–69

[64] Martin Gab H F S 2008 Geotechnics of Soft Soil focus on Ground Improvement *CRC Press* ed M L Minna Karstunen (London, UK: CRC Press/Balkema) p 444p.

IOP Conf. Series: Materials Science and Engineering **342** (2018) 012003 doi:10.1088/1757-899X/342/1/012003

Automated egg grading system using computer vision: Investigation on weight measure versus shape parameters

Ahmad Fakhri Ab Nasir, Siti Suhaila Sabarudin, Anwar P P Abdul Majeed and Ahmad Shahrizan Abdul Ghani

Innovative Manufacturing, Mechatronics and Sport Laboratory, Faculty of Manufacturing Engineering, Universiti Malaysia Pahang, Pekan, Malaysia

Email: afakhri@ump.edu.my

Abstract. Chicken egg is a source of food of high demand by humans. Human operators cannot work perfectly and continuously when conducting egg grading. Instead of an egg grading system using weight measure, an automatic system for egg grading using computer vision (using egg shape parameter) can be used to improve the productivity of egg grading. However, early hypothesis has indicated that more number of egg classes will change when using egg shape parameter compared with using weight measure. This paper presents the comparison of egg classification by the two above-mentioned methods. Firstly, 120 images of chicken eggs of various grades (A–D) produced in Malaysia are captured. Then, the egg images are processed using image pre-processing techniques, such as image cropping, smoothing and segmentation. Thereafter, eight egg shape features, including area, major axis length, minor axis length, volume, diameter and perimeter, are extracted. Lastly, feature selection (information gain ratio) and feature extraction (principal component analysis) are performed using *k*-nearest neighbour classifier in the classification process. Two methods, namely, supervised learning (using weight measure as graded by egg supplier) and unsupervised learning (using egg shape parameters as graded by ourselves), are conducted to execute the experiment. Clustering results reveal many changes in egg classes after performing shape-based grading. On average, the best recognition results using shape-based grading label is 94.16% while using weight-based label is 44.17%. As conclusion, automated egg grading system using computer vision is better by implementing shape-based features since it uses image meanwhile the weight parameter is more suitable by using weight grading system.

1. Introduction

The egg production industry has become a large industry owing to the demand and requirement by consumers, particularly in Malaysia. Egg is widely used worldwide and is healthy and safe for consumers. According to the report from the Ministry of Agriculture and Agra 2011 (Report 2011), the consumption of eggs per capita in Malaysia is 295 pieces per year [1]. The average production of eggs in the country is approximately 25 million per year [1]. The production rate increases to 37 million during festival seasons, such as Hari Raya, Chinese New Year and school holidays [1]. The high demand of eggs by consumers comes with high expectation for eggs with good quality. Many production companies are competing to produce eggs with extra nutrients and good shapes. Chicken eggs are rated and graded into AA, A, B, C, D, E and F on the basis of their weight. The grade classification based on weight is shown in Table 1 [1].

Table 1. Grade classification based on weight.

Grade	Weight (g)
AA	More than 70
A	65–70
B	60–65
C	55–60
D	50–55
E	45–50
F	Below 40

The incremental percentage of egg demand in Malaysia has required improvement in egg grading systems. Classifying eggs on the basis of their grade becomes easier than conventional method, fast and efficient by implementing an automated system. Egg grading cannot be perfectly and continuously conducted by human operators because of eye fatigue and increasing error as a result of over wearing. An automated system for egg grading can be a solution.

Many methods using computer vision are introduced to accelerate the speed of egg grading systems. These methods can be categorised as grading using egg size or shape parameter. They differ from using egg weight in egg grading by egg suppliers. Researchers have continued performing egg grading using egg size parameter despite the said difference. As mentioned earlier, many countries, including Malaysia, use weight in egg grading. Therefore, the classification process based on weight measure versus shape parameter is conducted in detail in this study.

2. Related Works

The literature on egg grading shows that computer vision systems are used for egg dirt inspection, egg blood spot detection and egg crack detection [2-8]; however, these topics are beyond the scope of the current study. Two case studies in [9] and [10] are considered. All algorithm and experimental setup details are illustrated in the next section.

An egg grading system generally consists of two sections, namely, hardware and software designs. The hardware design includes the equipment and components needed to carry eggs and capture their images, such as a conveyor to carry the eggs, a camera to capture their images and a lamp for lighting in capturing the egg images. The software design involves the algorithms that process the egg images. The software design starts when the egg image is already captured by the camera. Then, the image will be sent into the pattern recognition system for processing, and the result will determine the egg grade.

2.1. Case Study 1

In this study, a grade classification and dirt inspection system for eggs is developed using a combination of image processing techniques. A few tools are used for the hardware system. An illumination box is built to enable light adjustment and light noise removal. Two bulb lamps are attached on top of the box. The dimension of the box is 10 cm × 10 cm × 10 cm. A webcam camera is installed to capture egg images, and the lens camera is attached 200 mm from the egg. Figure 1 shows the overview of the image capturing system.

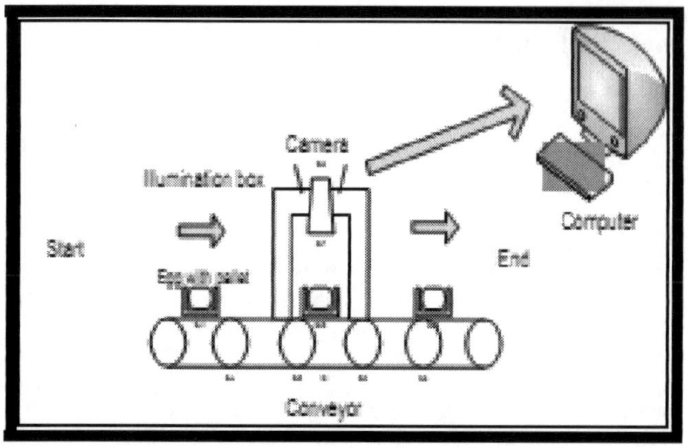

Figure 1. Overview of the system [9].

A database is developed to classify eggs. Pre-processing of images involves four steps: (i) converting colour images to greyscale; (ii) image thresholding processing (greyscale to binary); (iii) removing noises in binary images; and (iv) sharpening, smoothening and filling the region and hole of binary images. Two major features, area of white pixels and diameter of egg, are used in this system (Figure 2). A total of 60 eggs from 6 different grades (AA-E) are used in the experiment. The classifier for classification is not discussed in detail herein. The experimental results are illustrated in Figure 3. The system tests the cleanliness and dirtiness of eggs and executes classification and dirt inspection of eggs simultaneously.

Grade	Diameter (cm)	White pixel area
AA	>= 15.0	>= 79000
A	14.5 - 14.9	>= 72000 - < 79000
B	14.0 - 14.4	>= 66000 - < 72000
C	13.5 - 13.9	>= 61000 - < 66000
D	13.0 - 13.4	>= 56000 - < 61000
E	12.5 - 12.9	>= 48000 - < 56000

Figure 2. Egg diameter vs. pixel area [9].

Grade	Classification (%)
AA	90%
A	90%
B	80%
C	90%
D	80%
E	80%

Figure 3. Classification results [9].

2.2. Case Study 2

An algorithm for indirect measurement of egg weight is designed using machine vision system and image analysis. The hardware setup uses a digital camera VIDO AU CC540HDN, light source system and work surface. The steps of the proposed algorithm are shown in Figure 4.

Seven features are used as geometric features of egg. Amongst them, four features, namely, Perimeter (*P*), Area (*S*), Major Axis Length (*B*) and Minor Axis Length (*A*), are calculated using the 'regionprops' function in MATLAB. Other features are executed on the basis of the previous basic features, as illustrated in the next paragraphs.

Egg volume (V_C) is calculated using the following equation:

$$V_C = \frac{2\pi}{3} AB \tag{1}$$

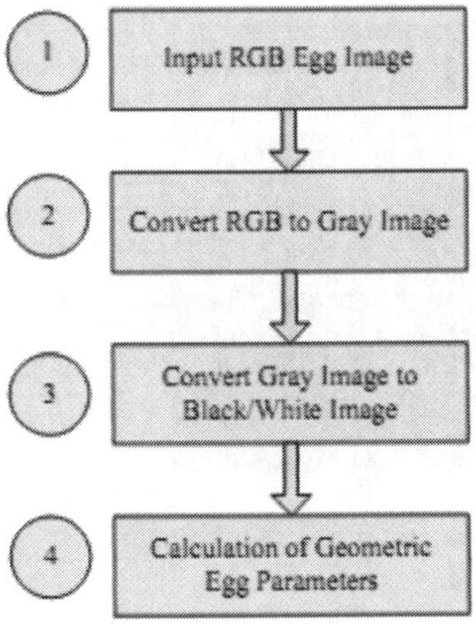

Figure 4. Block diagram of the method [10].

Two coefficients of egg shape (K_1 and K_2) are calculated using the following equation:

$$K_1 = \frac{P^2}{S} \text{ and } K_2 = \frac{A}{B} \quad (2)$$

In this study, coefficients (regression analysis) are used to measure the distance and difference of features. The results indicate that Perimeter, Major and Minor Axis Length and the two shape coefficients are significant features for indirect measurement of egg weight. Unfortunately, the features are not labelled on the basis of true weight classes. Classification, that is, assigning class labels, is essential to develop an effective automated egg grading system. After a feature extraction or selection procedure finds a proper representation, a classifier can be constructed from the labelled training data by using many possible approaches [11]. Then, this classifier is used to predict the class label of the test data. Classification is a type of supervised learning algorithm.

Supervised learning is a process of assigning a function to some desired categories as determined from the supervised training data. Here, the training data consist of a set of training examples, in which each set consists of a pair composed of an input object and a desired output value. A supervised learning algorithm learns from this training pair relationship and produces an inferred function. Specifically, in supervised learning, a teacher provides a category label or cost for each pattern in the training set that is used as a classifier. In this case, a 'true' category label is not inserted in examining the relationship between features. Therefore, significant features are only concluded on the basis of their weight without identifying the desired egg class. Hence, a complete egg grading system using statistical pattern recognition approach is developed to investigate features.

3. System Development

Firstly, digital images are taken as preliminary data. Then, image processing and egg feature measurement are conducted. Lastly, classification process is carried out.

3.1. Image Acquisition

A digital camera (a FUJIFILM camera of 8.1 MP) is used to capture images of graded eggs. An illumination box with a dimension of 100 cm x 100 cm is developed for egg placement to enable light adjustment and light noise removal. Two bulb lamps with yellow light are attached on top of the box. The distance between the lens and egg is approximately 200 mm. Figure 5 shows the overview of the hardware setup.

Figure 5. Image acquisition environment and setup.

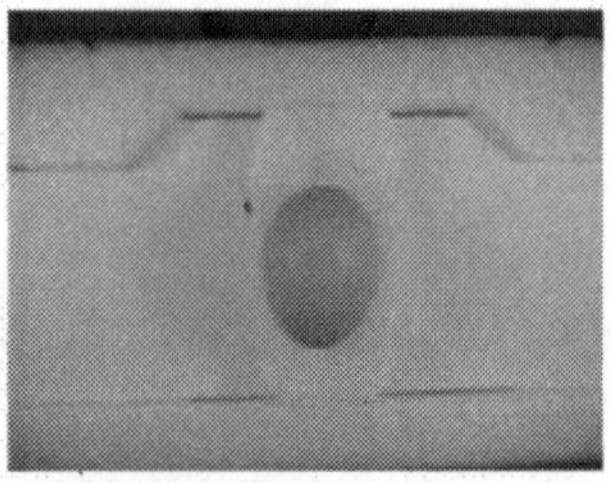

Figure 6. Original image.

A total of 120 egg images that consisted of 30 eggs from each of four grades (A-D) are captured. The sample of image is shown in Figure 6.

3.2. Image Processing Tasks

Before geometric features can be measured, the images should be processed first to obtain the final image containing only [1, 0] binary values. Firstly, original *RGB* images are cropped to remove unwanted background. This procedure is essential because the image should show only the centre and a small amount of the total area (Figure 6). Accordingly, the amount of time executions for the next tasks can be reduced. This process finds the centre of image pixel firstly and then creates a rectangle box that covers the area of egg. Lastly, the box rectangle size is cropped, and the targeted image is saved in the database.

Secondly, the cropped image is converted to greyscale image. Thresholding process is executed on greyscale image by using manually setup level. In this case, the ideal level of 150 is used to convert to binary image. The smoothing of binary images involves two stages. The hole in the image is filled firstly and then the image is smoothened using the 'bwareaopen' function. The function checks the connection of pixels to its neighbourhood with low number of pixels. By implementing this function, some noises outside the egg can be removed given that the connected pixels present extremely lower number of pixels compared with those of the targeted image. The result of all image processing tasks is shown in Figure 7.

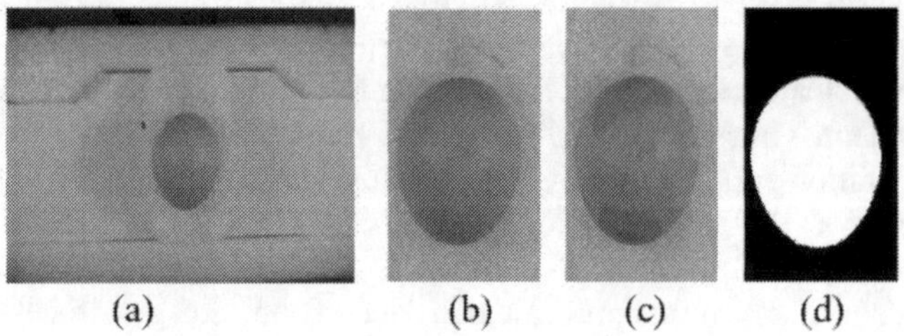

(a) (b) (c) (d)

Figure 7. Processed egg image, (a) original image, (b) cropped image, (c) greyscale image, (d) final binary image.

Seven features are used as in Case Study 2. These features are Area (*S*) denoted as {f1}, Major Axis Length (*B*) denoted as {f2}, Minor Axis Length (*A*) denoted as {f3}, Volume (V_C) denoted as {f4}, Perimeter (*P*) denoted as {f6} and coefficients of egg shape (K_1 denoted as {f7} and K_2 denoted as {f8}). One feature, which is Diameter (*D*) denoted as {f5}, is added. Diameter is a scalar that specifies the diameter of a circle with the same area as a region [11]. Diameter is computed as $4A/\pi$, where A is the egg area.

3.3. *Classification Tasks*

Two feature representation models, namely, principal component analysis (PCA) and information gain ratio (IGR), are considered.

PCA is a statistical method for reducing the dimension of variables. PCA is useful when implemented on a large number of variables with some redundancy. In this case, redundancy indicates that some of the variables are correlated with one another. The general form for the formula to compute scores on the first component extracted in PCA is expressed in the following equation:

$$C_1 = b_{11}(X_1) + b_{12}(X_2) + \ldots b_{1p}(X_p) \tag{3}$$

where C_1 is the subject score on principal component (PC) (the first component extracted); b_{1p} is the regression coefficient (or weight) for observed variable p, as used in creating first PC; and X_p is the subject score on observed variable p.

IGR is usually used to determine the capability of each single feature to separate the given dataset. Entropy is commonly used in the information theory measure and characterises the purity of an arbitrary collection of examples. If A is an attribute (feature) and C is the class, then the following equations provide the entropy of the class before and after observing the attribute:

$$H(C) = -\Sigma_{c \in C} p(c) \log_2 p(c) \tag{4}$$

$$H(C|A) = -\Sigma_{a \in A} p(a) \Sigma_{c \in C} p(c|a) \log_2 p(c|a) \tag{5}$$

The amount by which the entropy of the class decreases reflects the additional information about the class provided by the attribute and is called information gain. Each attribute A_i is assigned a score on the basis of the information gain between itself and the class, as follows:

$$IG_i = H(A_i) + H(C) - H(A_i, C) \tag{6}$$

k-nearest neighbour (k-NN) is used for classification purposes. k-NN is a method for classifying objects on the basis of closest training examples in the feature space. k-NN is a type of example-based learning or lazy learning where the function is only approximated locally and all the computation is deferred until classification. This algorithm is a simple machine learning algorithm in which an object is classified using a majority vote of its neighbours. The object is then assigned to the class, which is the most common amongst its k-NN. The neighbours are taken from a set of objects for which the correct label is known. The learning in this model is based on storing all the training instances that correspond to points in an n-dimensional space along with their class labels, and classification is delayed until a new instance arrives.

4. Experimental Results

A total of 120 egg images from four classes are used in the experiments with 10-fold cross validation for classifier evaluation. The performance evaluation is examined by the recognition accuracy through the presented formula:

$$RC = \frac{M_C}{M_T} \tag{7}$$

where M_C presents the number of correct classified eggs in each class, and M_T is the number of all eggs in all classes. The experiments are conducted on four different ways:

(i) The egg features {f1:f8} are used as starting features to find the best single feature using IGR theory.

(ii) The sorted features are {f5 = 0.655, f4 = 0.643, f3 = 0.571, f2 = 0.531, f7 = 0, f6 = 0, f8 = 0}, from the top to the bottom ranking features. Two starting features of all features (original) and two of the best features of > 0.5 information gain value {f5, f4, f3, f2} are used for classification process.

(iii) The feature extraction using PCA is implemented on seven features {f1:f8} for comparison with other feature representation techniques. Three PCs are retained and selected as starting features. The three PCs cover 95% of total variance for the seven features.

(iv) The classification process uses five different distance functions in linear *k*-NN: (a) Euclidean distance, (b) Manhattan distance, (c) Minkowski distance, (d) Filtered distance and (e) Chebychev distance. One nearest neighbour 1-NN is set for the entire experiments.

Table 2 shows the average recognition accuracy using weight-based grading as a referred class (egg grading). As shown in Table 2, the recognition accuracy for PCA is better than using the combination of all original features {f1:f8} and selected several best single features {f5, f4, f3, f2} by at least 2%–4%. Comparing the distance function methods in *k*-NN shows some result differences among them, thereby increasing the recognition accuracy. However, the accuracy result for this dataset only achieves a maximum of 51.67%. As predicted at the beginning of this study, the recognition accuracy should be low because the classification process is conducted by referring to supplier's egg grading labelling, that is, using weight measure. Therefore, a new egg class labelling based on shape measure is obtained through simple *k*-means clustering for solving the above-mentioned problem.

Table 2. Average recognition accuracy (weight-based grading).

No.	Method	Starting Features	Recognition Accuracy (%)					Average (%)
			(a)	(b)	(c)	(d)	(e)	
(ii)	Original features	{f1:f8)	41.67	39.17	41.67	40.83	45.83	41.83
	IGR	{f5, f4, f3, f2}	41.67	37.50	41.67	38.33	43.33	40.50
(iii)	PCA	{f1:f8}	42.50	40.00	42.50	51.67	44.17	44.17

Table 3 shows that the first 30 images in the database, which are supposed to be in grade A in accordance with weight measure, are changed to multiple classes using shape measure after performing simple *k*-means clustering. Five eggs are clustered to class 0, two eggs to class 1, three eggs to class 3 and others (majority) are classified as class 2. Many class changes also occur for other egg grades in the database. The classification experiments are repeated using a new cluster. Table 4 shows the recognition accuracy using shape-based grading. The classification results are increased dramatically compared with the previous results of weight-based grading. The best recognition results are achieved consistently using original features compared with using PCA and single feature set. PCA can fully improve the low recognition results for this type of dataset (Table 2). When PCA is implemented on good recognition results (Table 4), the results are insignificantly improved.

For both datasets, using the best single features (based on IGR theory) also does not help improve the results. The reason is that the single feature alone is filtered without considering the combination of feature subsets. In the future, wrapper feature selection approach can to be used provide improved combination of feature subsets.

Table 3. New cluster assigned for egg grade A.

Image No.	New Cluster Assigned	Image No.	New Cluster Assigned	Image No.	New Cluster Assigned
1	0	11	2	21	2
2	0	12	2	22	2
3	2	13	3	23	2
4	2	14	2	24	0
5	2	15	2	25	2
6	2	16	2	26	3
7	2	17	2	27	3
8	2	18	2	28	0
9	2	19	2	29	0
10	2	20	1	30	1

Table 4. Average recognition accuracy (shape-based grading).

No.	Method	Starting Features	Recognition Accuracy (%)					Average (%)
			(a)	(b)	(c)	(d)	(e)	
(ii)	Original features	{f1:f8)	94.17	95.00	94.17	93.33	94.17	94.16
	IGR	{f4, f3, f2}	82.50	83.33	82.50	83.33	77.50	81.83
(iii)	PCA	{f1:f8}	91.67	92.50	91.67	96.00	90.83	92.53

5. Conclusions

In this study, an automated egg grading system using computer vision is developed. Two methods using weight and shape measure are comprehensively investigated. The experimental results show that the recognition results are better when the classification process is conducted on the basis of a new cluster (shape-cluster). By contrast, the recognition results of weight-cluster are extremely low since the shape measure is not related to weight measure (class). Previous researchers have supported the strong relation between the two. However, in this study, the relationship is poor as the classification results are extremely low by using weight-based grading. Whether the egg class will change using shape-based grading is unclear. If the answer is yes, then this system can be implemented to help accelerate the speed of grading. Otherwise, further studies are needed to find the best features (based on egg image) that are nearly perfectly correlated with weight measure.

Acknowledgments

The authors are grateful to the Universiti Malaysia Pahang for providing the funding under internal research grant [RDU1703159] to support this research.

References

[1] Anim M 2013 Telur ayam gred dan harga http://animhosnan.blogspot.my/2013/07/telur-ayam-gred-dan-harga.html
[2] Patel V C, Mcclendon R W and Goodrum J W 1998 Color Computer Vision and Artificial

Neural Networks for the Detection of Defects in Poultry Eggs *Artificial Intelligence Review* **12** pp 163–176

[3] Mertens K, De Ketelaere B, Kamers B, Bamelis F R, Kemps B J, Verhoelst E M, De Baerdemaeker J G and Decuypere E M 2005 Dirt Detection on Brown Eggs by Means of Color Computer Vision *Poultry Science* **84** pp 1653–1659

[4] Dehrouyeh M H, Omid M, Ahmadi H, Mohtasebi S S and Jamzad M 2010 Grading and Quality Inspection of Defected Eggs Using Machine Vision *International Journal of Advance Science and Technology* **16** pp 43–50

[5] Loredana Lunadei, Luis Ruiz-Garcia, Luigi Bodria and Riccardo Guidetti 2011 Automatic Identification of Defects on Eggshell Through a Multispectral Vision System *Food Bioprocess Technology* **5** pp 3042–3050

[6] Yoon S C, Lawrence K C, Jones D R and Heitschmidt G W 2012 Improved modified pressure imaging and software for egg micro-crack detection and egg quality grading *Applied Engineering in Agriculture* **28** pp 283–293

[7] Arivazhagan S, Newlin Shebiah R, Hariharan Sudharsan, Rajesh Kannan R and Ramesh R 2013 External and Internal Defect Detection of Egg using Machine Vision *Journal of Emerging Trends in Computing and Information Sciences* **4** pp 257–262

[8] Mahmoud Omid, Mahmoud Soltani, Mohammad Hadi Dehrouyeh, Seyed Saeid Mohtasebi and Hojat Ahmadi 2013 An expert egg grading system based on machine vision and artificial intelligence techniques *Journal of Food Engineering* **118** pp 70–77

[9] Ibrahim R, Mohd Zin Z, Nadzri N, Shamsudin M Z and Zainudin M Z 2012 Egg's Grade Classification and Dirt Inspection Using Image Processing Techniques *Proceedings of the World Congress on Engineering* vol 11 (London: Newswood Limited) pp 1–3

[10] Alikhanov D, Penchev S, Georgieva Ts, Moldajanov A, Shynybaj Z and Daskalov P 2015 Indirect Method for Egg Weight Measurement Using Image Processing *International Journal of Emerging Technology and Advanced Engineering* **5** pp 30–34

[11] A Fakhri A Nasir, M Nordin A Rahman, Nashriyah Mat, A Rasid Mamat and Ahmad Shahrizan Abdul Ghani 2017 Ficusdeltoidea (Jack) Moraceae Varietal Identification Using Statistical Recognition Approach *World Applied Sciences Journal* **35** pp 82–88

Assessing the implementation of bias correction in the climate prediction

Nurul Nadrah Aqilah Tukimat[1,2]

[1]Faculty of Civil Engineering and Earth Resources, Universiti Malaysia Pahang, 26300 Gambang, Pahang
[2]Centre for Earth Resources Research Management (CERRM), Universiti Malaysia Pahang, 26300 Gambang, Pahang

E-mail: nadrah@ump.edu.my

Abstract. An issue of the climate changes nowadays becomes trigger and irregular. The increment of the greenhouse gases (GHGs) emission into the atmospheric system day by day gives huge impact to the fluctuated weather and global warming. It becomes significant to analyse the changes of climate parameters in the long term. However, the accuracy in the climate simulation is always be questioned to control the reliability of the projection results. Thus, the Linear Scaling (LS) as a bias correction method (BC) had been applied to treat the gaps between observed and simulated results. About two rainfall stations were selected in Pahang state there are Station Lubuk Paku and Station Temerloh. Statistical Downscaling Model (SDSM) used to perform the relationship between local weather and atmospheric parameters in projecting the long term rainfall trend. The result revealed the LS was successfully to reduce the error up to 3% and produced better climate simulated results.

1. Introduction

Nowadays, the global warming and climate changes issues become most significant threat to the whole countries. The rises in the global temperature creates other disasters such as drought, flood, thunderstorm, and flash flood in everywhere and anytime. These changes also to have continuous impacts on different aspects of society, including health, agriculture, water resources, and energy demand [1].Thus, the forecasting of the future changes of the climatic events become significant to provide beneficial data input especially when involve with the planning and managing of water resources.

Statistical Downscaling Model (SDSM) has been introduced by Wilby et al. [2] to downscale the spatial and temporal resolution which provided by General Circulation Model (GCMs) into finer resolution scale suit to the local climatic information. The main purpose is to look forward the changes pattern of climate trend affected by the greenhouse gases (GHGs) emissions and the implication to the hydrological system. The SDSM generally is considered by many researchers because the model applied simpler mathematical equation that is easier to understand the relationship pattern between predictor and predictand. Besides, the model does not require high computational demand to view the simulation result because the output is presented in the finer resolutions. Thus, it could reduce the total cost of the project but at the same time has higher ability to produce better climate simulation. This model has been applied widely in Malaysia as a statistical climate model in various sectors and purposes [3–5].

However, the accuracy of the climate simulation in the SDSM depends on the predictors selections. The predictors should have better association with the local surface climate to generate good agreement during calibration and validation process. Some criteria that are essential for the predictors to be reliably simulated by GCM are readily available to achieve from GCM output, and strongly correlated with the surface variable of interest. Disagreement between model and observations can be related to an inadequate selection of the value of some parameters that are not precisely known. Besides, inaccuracy in the estimation and forecasting will create errors in the sampling measurement and processing and produce unreliable simulation results.

Previous study applied various methods in effort to achieve the closest calibration among predictor and predictand such as Multiple Correlation Matrix (M-CM), Classical Least Square (CLS), Multiple Linear Regression (MLR), Nonlinear Programming (NLP), singular value decomposition analysis, and Canonical Correlation Analysis (CCA) [6,7]. Another way to overcome the simulation gaps is with the bias correction (BC) treatment. Mahmood et al. [8] applied Linear Scaling (LS) to forecast future temperature and detected the hotter climate in the future relative to the present climate. Chen et al. [9] revealed the Quantile Mapping (QM) and Local Intensity Scaling (LOCI) produced better performances particularly in the downscaling of precipitation extremes than the traditional multiple linear regression moethods. Bennett et al. [10] proved the quantile mapping BC able to improve the spatial correlation between modelled and observed of seasonal and annual rainfall. Meanwhile Lafon et al. [11] compared the performances of 4 different BC methods there are linear, nonlinear, gamma, and empirical distribution-based quantile mapping to reduce the RCM-simulated precipitation gaps for seven catchments spread across Great Britain. The results showed that the mean and standard deviation of daily precipitation can be corrected robustly while the correction of skewness and kurtosis of daily precipitation are much more sensitive to the choice of a bias correction method and the selection of a particular calibration period. Thus, LS method was applied in this study to treat the gaps in the simulation and future projection results. It is very significant to determine the importance of the bias treatment in the climate simulation.

2. Methodology of the Study

The study have been analysed as stated in the Figure 1 below. LS method was selected to treat the gaps between observed and simulated results produced by SDSM modelling. The selection of this method because of the simplicity, accuracy, parameter considerations and previous literatures in maintaining the reliability of the results after treatment. The equation of the LS and LOCI methods as shown below:

$$\text{LS method} = P_{raw}\frac{\mu(P_{obs,m})}{\mu(P_{raw,m})} \tag{1}$$

where $P_{raw}$and P_{Obs} are the raw and observe rainfall on the mth month and μ is the average rainfall for i month.

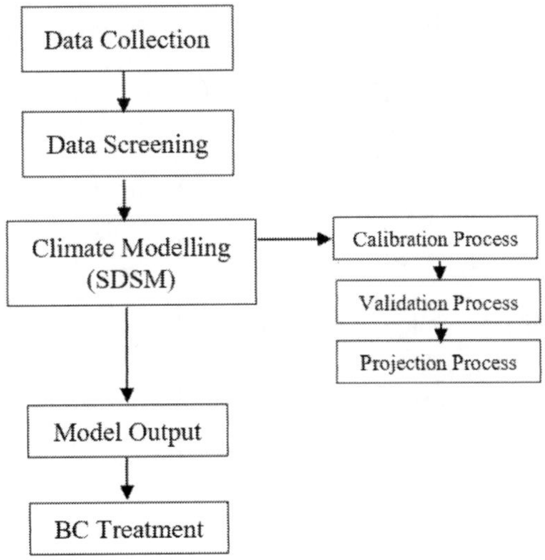

Figure 1. Methodology of the Study.

The SDSM model was chosen to simulate the future rainfall at the grid box of 28X x 34Y. This model is widely used in the context of hydrological issue due to climate scenarios because it provides station scale climate information from grid resolution GCM-scale as input based on multiple regression techniques. These regressions emerged from the empirical statistical relationship between atmospheric circulation pattern (predictors) and local scale parameters (predictand). Besides, the simplicity and computationally undemanding of SDSM tools made it acceptable and recognizable in terms of climate model projection. Daily rainfall data and daily atmospheric predictors from NCEP reanalysis data calibrates and validates in the period year 1982 - 2011 to develop the downscaling relationship. Then, the GCMs data are utilized to generate future trend for interval year periods of Δ2050. The A2 scenario chosen for this study provides an upper bound on future emissions, and it is selected from an impacts-and-adaptation point of view; if it is adaptable to large climate change, it will have no problem with smaller climate change and lower end scenario, although low emissions scenario gives less information from this point of view. As stated by Tukimat and Alias [12], the A2 scenario has more potential to develop bigger impacts of global warming compared to other emission groups.

2.1. Study Area

The study have been tested at Lubuk Paku rainfall station (3527092) and Temerloh rainfall station (3424081). These stations have been located in the Pahang River Basin which are influencing the water level in the Pahang River. The rainfall distributions at this region is influenced by the wind direction and 2 monsoons pattern known as Northeast (Oct – Mar) and Southwest (Jun – Aug) monsoons. The annual rainfall is 2400mm/year with average temperature of 28°C. The highest and lowest rainfall intensity is focused on Dec and Jan, respectively. The average wind speed is 5mph with the relative humidity of 85%.

3. Results and Discussions

3.1. Calibration and Validation Processes

About 30-years length of historical data have been considered at Lubuk Paku station (year 1982 to 2011) and Temerloh station (year 1975 to 2004) to forecast the future rainfall changes in year 2040 to 2069. The SDSM model was applied to downscale the GCMs resolution suit to the local climate condition. Based on the screening process, each locations requires different atmospheric predictors to provide better predictand-predictor's relationships. The selection of the predictors were based on the correlation value between predictand-predictor relationships for every month. Table 1 shows the predictor selection for each stations.

Table 1. Predictors selection for each rainfall stations.

Predictor	Station	
	Lubuk Paku	Temerloh
Surface airflow strength (p_f)	/	/
850hPa airflow strength (p8_f)	/	
Specific Humidity (shum)	/	/
Zonal velocity (p_u)		/
Meridional velocity (p_v)		/
850hpa Relative humidity (r850)		/

Figure 2 and Figure 3 express the performances of simulated values in the calibration and validation processes at Lubuk Paku and Temerloh stations. Both stations were produced good simulated results in the calibrations but not in the validation. Calibration and validation have difference processes where the calibration is the stage of the equation development from 15years historical record using required variables meanwhile the validation is the stage to validate the reliability of that equation using another 15 years length of historical records. Referring to the Lubuk Paku station, the error was focused on Nov and Dec with 5%. Even very good calibration but the error become bigger in the validated results. Meanwhile in Temerloh station, both processes produced significant errors where BC needs to improve the quality of the simulated result.

Table 2 show the percentage error of every month for both stations. The table shows the SDSM was successfully to produce good simulated results with less than 15% error as proved that the selected predictors had good association with the local stations. At Lubuk Paku station, the bigger gaps were identified start from July to Oct during validation process. Meanwhile for Temerloh station was started from Nov to Feb. The average errors were 7.6% for Lubuk Paku station and 9.9% for Temerloh station. Even the bias was not in critical however the treatment was still significant to ensure the reliability of the projected results.

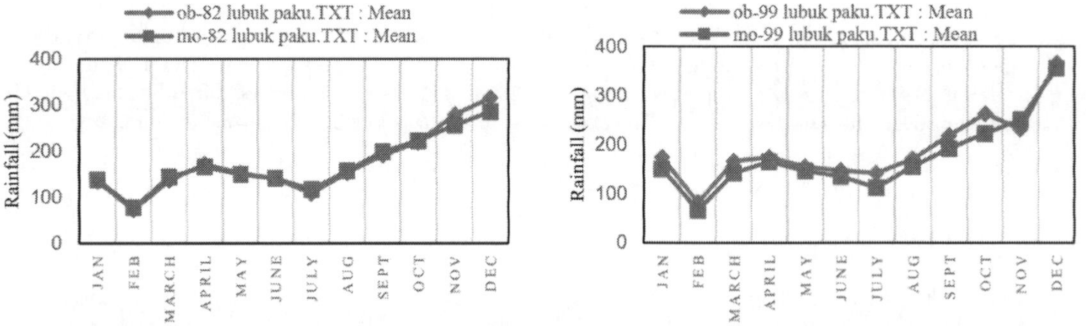

Figure 2. Calibrated (1982-1996) and validated (1997-2011) results without BC treatment at Lubuk Paku station.

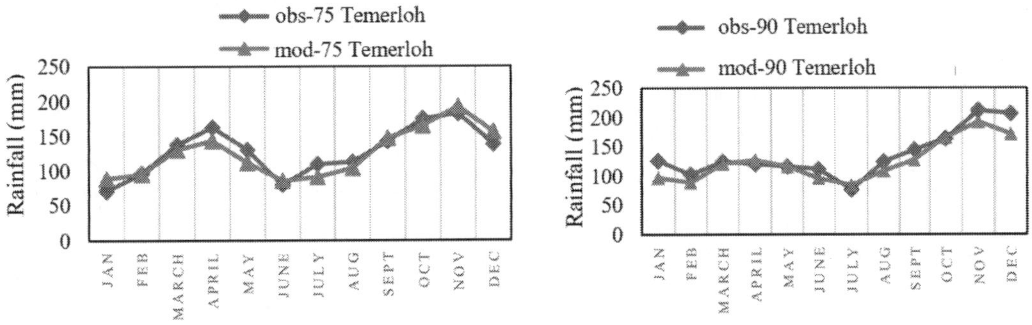

Figure 3. Calibrated (1975-1989) and validated (1990-2004) results without BC treatment at Temerloh station.

Table 2. Error (%) between simulated and observed values.

	ERROR (%)											
	Jan	Feb	Mar	Apr	May	Jun	Jul	Aug	Sep	Oct	Nov	Dec
Lubuk Paku	8.0	11.2	9.8	4.1	3.5	4.2	14.1	5.7	7.9	8.5	8.2	6.3
Temerloh	20.1	7.0	4.0	8.3	7.3	11.0	13.2	10.2	7.2	3.5	7.3	14.8

3.2. BC Treatment Implementation and Climate Projection Results

In this study, LS method was implemented to treat the gaps between daily observed and simulated values. Graph in the Figure 4 shows the performances of the corrected simulation values after considered the BC at both stations. In general, the LS method was successfully to treat the simulated results more closely to the observed for every month. For Lubuk Paku station, the error was successfully to reduce to 4.4%, about 3.2% was treated by BC. Meanwhile, the gaps of the simulated value at Temerloh station was also successfully to reduce to 5.3%, about 4.6% difference from untreated results. It shows that the BC treatment is significant to treat the gaps in the simulation process.

Figure 5 shows the projected climate results after considered the BC treatment at both stations. In the projection results, the correction factors values were considered to consistent for every station. Based on the results, the rainfall intensity at both locations were expected to change in the c2050s. The rainfall patterns are also expected differ from the historical trend. At Lubuk Paku, the rainfall intensity was expected to reduce in average 12%. The heavy rainfall was estimated to occur started from Nov and then declined in the rest months. Differ in Temerloh station whereby the future rainfall trend was estimated to be fluctuated through a year. The highest rainfall trend was expected occur in June during Southwest monsoon and then pick up again in Nov and Jan during Northeast monsoon. The annual rainfall in the Δ2050s is expected to rise up to 40% becomes 2212mm/year compared to the historical.

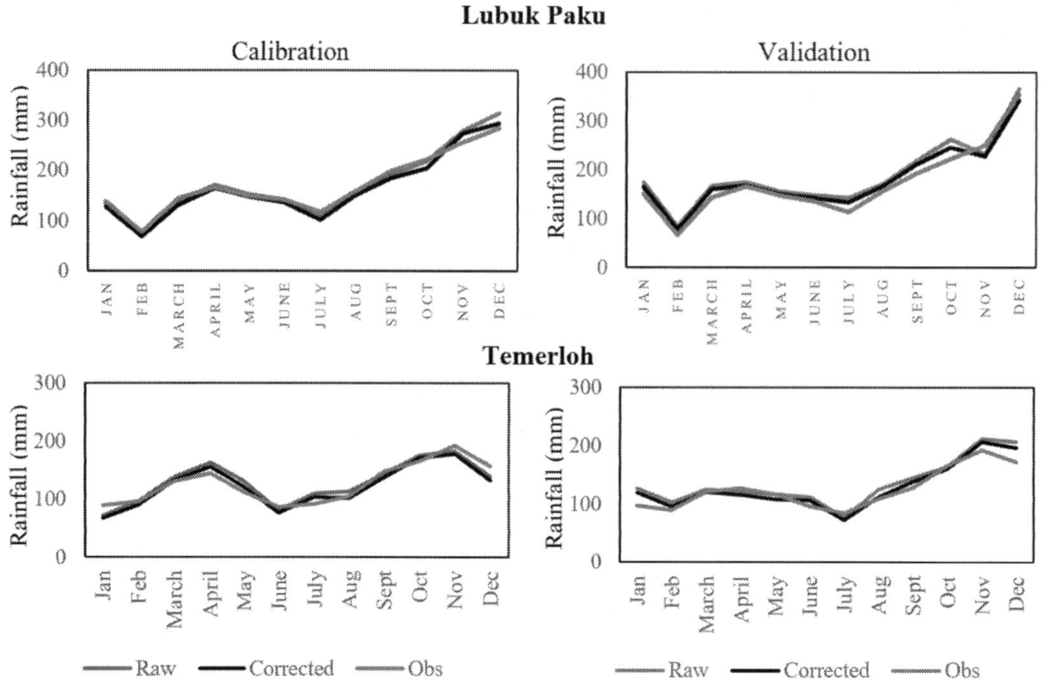

Figure 4. Calibrated (1982-1996) and validated (1997-2011) results with LS method.

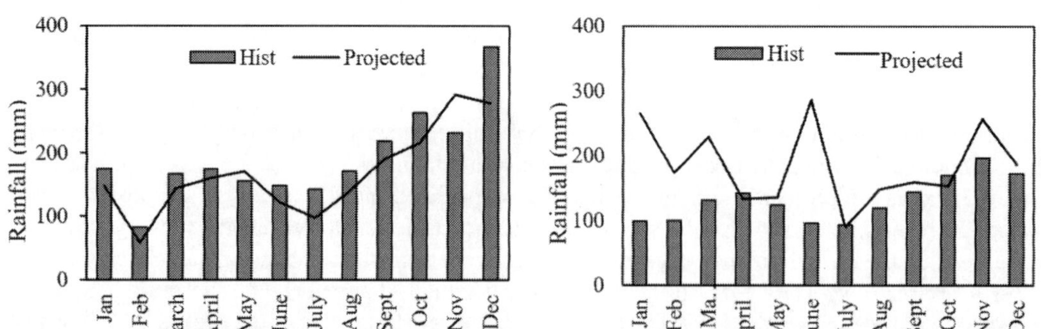

Figure 5. Projected rainfall (2040 to 2069) at Lubuk Paku and Temerloh stations with BC treatment.

4. Conclusion

An approaching of the statistical climate modeling nowadays becomes significant with the technologies growth. The simulation results produced from this technology can be as an importance data input especially for the long term planning and management in the water resources and land development. Thus, the model accuracy have been controlled to enhance the reliability of the production results. In this case study, LS method was applied and successfully treated the large biases in the simulation. Based on the result, the error was successfully to reduce to 4.4% and 5.3% in Lubuk Paku and Temerloh stations, respectively. The consistent LS values were used to treat the projected climate result in year

Δ2050 at both locations. It proved that the BC treatment is significant to treat the gaps in the simulation process.

Acknowledgement
This research was supported by Drainage and Irrigation Department (DID) and Ministry of Higher Education (MHE) for grant vot RDU150113.

References
[1] Raneesh K Y and Thampi S G 2013 Bias correction for RCM predictions of precipitation and temperature in the Chaliyar River Basin *J. Climatol. Weather Forecast.* 1–7
[2] Wilby R L, Hay L E and Leavesley G H 1999 A comparison of downscaled and raw GCM output: implications for climate change scenarios in the San Juan River basin, Colorado *J. Hydrol.* **225** 67–91
[3] Tukimat N N A, Harun S and Shahid S 2017 Modeling Irrigation Water Demand in a Tropical Paddy Cultivated Area in the Context of Climate Change *J. Water Resour. Plan. Manag.* **143** 5017003
[4] Tukimat N N A and Harun S 2015 Climate change impact on rainfall and temperature in Muda irrigation area using multicorrelation matrix and downscaling method *J. Water Clim. Chang.* **6** 647–60
[5] Hamidon N, Harun S, Malek M A, Ismail T and Alias N 2015 Prediction of paddy irrigation requirements by using statistical downscaling and cropwat models: A case study from the Kerian Irrigation Scheme in Malaysia *J. Teknol.* **76** 281–8
[6] Tukimat N N A and Harun S 2013 Multi-correlation matrix (M-CM) for the screening complexity in the statistical downscaling model (SDSM) *Int. J. Eng. Sci. Innov. Technol* **2** 331–42
[7] Busuioc A, Giorgi F, Bi X and Ionita M 2006 Comparison of regional climate model and statistical downscaling simulations of different winter precipitation change scenarios over Romania *Theor. Appl. Climatol.* **86** 101–23
[8] Mahmood R, Jia S and Babel M S 2016 Potential impacts of climate change on water resources in the kunhar river basin, Pakistan *Water* **8** 23
[9] Chen J, Brissette F P, Chaumont D and Braun M 2013 Finding appropriate bias correction methods in downscaling precipitation for hydrologic impact studies over North America *Water Resour. Res.* **49** 4187–205
[10] Bennett J C, Grose M R, Post D A, Ling F L N, Corney S P and Bindoff N L 2011 Performance of quantile-quantile bias-correction for use in hydroclimatological projections *MODSIM 2011-19th International Congress on Modelling and Simulation-Sustaining Our Future: Understanding and Living with Uncertainty* pp 2668–75
[11] Lafon T, Dadson S, Buys G and Prudhomme C 2013 Bias correction of daily precipitation simulated by a regional climate model: a comparison of methods *Int. J. Climatol.* **33** 1367–81
[12] Tukimat N N A and Alias N A 2016 Assessment the potential of SRES scenario for Kuala Sala, Malaysia *J. Mech. Civ. Eng* **13** 6–12

IOP Conf. Series: Materials Science and Engineering **342** (2018) 012005 doi:10.1088/1757-899X/342/1/012005

Integration of Mahalanobis-Taguchi system and traditional cost accounting for remanufacturing crankshaft

M Y Abu[1], N S Norizan[1] and M S Abd Rahman[2]

[1]Faculty of Manufacturing Engineering, Universiti Malaysia Pahang
[2]Faculty of Engineering Technology, Universiti Teknikal Malaysia Melaka

E-mail: myazid@ump.edu.my

Abstract. Remanufacturing is a sustainability strategic planning which transforming the end of life product to as new performance with their warranty is same or better than the original product. In order to quantify the advantages of this strategy, all the processes must implement the optimization to reach the ultimate goal and reduce the waste generated. The aim of this work is to evaluate the criticality of parameters on the end of life crankshaft based on Taguchi's orthogonal array. Then, estimate the cost using traditional cost accounting by considering the critical parameters. By implementing the optimization, the remanufacturer obviously produced lower cost and waste during production with higher potential to gain the profit. Mahalanobis-Taguchi System was proven as a powerful method of optimization that revealed the criticality of parameters. When subjected the method to the MAN engine model, there was 5 out of 6 crankpins were critical which need for grinding process while no changes happened to the Caterpillar engine model. Meanwhile, the cost per unit for MAN engine model was changed from MYR1401.29 to RM1251.29 while for Caterpillar engine model have no changes due to the no changes on criticality of parameters consideration. Therefore, by integrating the optimization and costing through remanufacturing process, a better decision can be achieved after observing the potential profit will be gained. The significant of output demonstrated through promoting sustainability by reducing re-melting process of damaged parts to ensure consistent benefit of return cores.

1. Introduction

In 19th century, remanufacturing was applied during World War II when the manufacturers converted their usual product to the products of military. This shown that the remanufactured products keep increasing in used as the countries encounter with materials shortage. Remanufacturing is "the used parts are replaced" and "testing is carried out to specification of manufacturer and standards of original production" [1]. The end of life (EOL) products are disassembled into parts then they will be cleaned and inspected followed by process of repair and replacement for the worn-out parts. Finally, the product is reassembled in purpose to function like a new product. Several benefits of remanufacturing such as: (1) More profitable than traditional manufacturing because used less labor skills, energy, materials and cost of disposal, (2) The work environment in assembly lines is dull in traditional manufacturing while in remanufacturing, the work environment is more diverse. Workers have a set of broader skill and higher fulfillment of work, (3) The critical information on design of product and project developments are provided by activities of remanufacturing and failure mode information can be acquire through it,

(4) Product that has high-quality and cheaper can be obtained by customers. The remanufactured product prices are 30% to 40% less than the new products and (5) The lead times is shorter for the remanufactured products [2]. Mahalanobis-Taguchi System (MTS) is a method for diagnosis of behavior patterns and forecast corresponding to sets of multivariate patterns that proposed by Genichi Taguchi. The application of MTS has been reported in hospital [3], green manufacturing [4], green technologies [5], decision making [6] and computational [7]. [8] examined relationship between the MTS with the health condition of cooling fan and induction motor. [9] presented an effect of MTS on inspection time in electronics industry for flip-chip bumping height inspection. [10] analyzed the capability of MTS as a response detector for the brain on a single-trial (ST) basis. [11] discussed the application of MTS for identifying the optimum hardness profile to avoid failures of the shafts to reduce in wasting of money. In order to make a right decision, [12] presented applied MTS for analysis of software defects, in particular fault-prone software modules and can help analyze which of the variables describing the software modules that contribute the most to the abnormality. In work by [13], the authors described the case of quality awards on the effect of performance aggregation based on MTS. [14] applied MTS approach to hospital is an effective method that determines the decision for the obstructive sleep apnea (OSA) patients. Due to competitive pressure forces from manufacturers to produce more products with shorter life span and better quality, yet at a lower cost, [15] presented a significant factor to determine the best configuration in the cutting process by applied MTS. In order to improve the agriculture, [16] assessed how the implementation of MTS can assist in making decision to identify the classification within each category. [17] proved that, the application of MTS is a practical method in making decision. [18] discussed the implementation of MTS in machining in order to improve the energy saving and efficiency of the production. [19] presented a combination of MTS and self-organization mapping (SOM) system to monitoring the bearing condition, indicate the current degradation state and tracking the degradation trend dynamically. The traditional cost accounting (TCA) are used to allocate the overhead of manufacturing to produce units of product. Industries are mostly using TCA because of the cost of product that act as a key function and also assist in accomplish the goals and to apply principles like estimate competitors, growth and improvement targets, synchronize between quantities and processes and measurement and control.

Therefore, it is very important for this work to identify critical parameters using MTS to improve the quality inspection. Factorial effect graph has been developed to observe which parameters were critical to the remanufacturing process. Eventually the cost was estimated using TCA by considering those critical parameters for better profit.

2. Methodology

This work selected crankpin of crankshafts as a subject matter and the number of sample is shown in table 1. The data were collected from a remanufacturing industry which located at Rawang, Malaysia.

Table 1. Collection data of crankpin.

Engine model	Number of sample	Remanufacturable tolerance (mm)	
		Lower limit	Upper limit
Caterpillar	70	117.000	117.043
MAN	84	112.020	112.040

Subsequently, this work need to develop a clustering between both engine model to indicate their Mahalanobis Distance (MD) value. Higher the MD value indicated that the sample have a distinctive pattern from the reference sample while, smaller value of MD indicated that the sample have closer pattern to the reference sample. In this work, MAN engine model have been selected as a reference sample due to the pattern was closer to the original crankshaft. The calculation of MD as shown in equation (1) whereby Y represented the normalized Y_1 and Y_2, A represented the inverse matrix, Y^T represented transposition of Y and k represented number of variables.

$$\text{Mahalanobis distance} = \frac{YAY^T}{k} \quad (1)$$

The MD for each sample of both engine model was then used as response to evaluate the criticality of parameters. The diameters of crankpin will be used as input and the MD will be used as output. The data should be normal or densely populated in the medium range to apply this method by creating a histogram. The unit data is the highest sample while signal data is the remaining samples number. Normalization process is required in order to develop a dimensionless condition of the signal data. The normalization is used to make the data more flexible by diminishing their redundancy. Equation (2) and equation (3) are used to calculate the normalization.

$$X_{ij} = \acute{x}_{ij} - \bar{x}_j \quad (2)$$

$$M_i = \acute{y}_i - m_0 \quad (3)$$

Proportional coefficient β and SN ratio η are determined using equation (4) and equation (5).

$$\text{Proportional coefficient, } \beta_1 = \frac{M_1X_{11} + M_2X_{21} + \cdots + M_lX_{l1}}{r} \quad (4)$$

$$\text{SN ratio, } \eta_1 = \begin{cases} \frac{\frac{1}{r}(S_{\beta 1} - V_{el})}{V_{el}} & (when\ S_{\beta 1} > V_{el}) \\ 0 & (when\ S_{\beta 1} \leq V_{el}) \end{cases} \quad (5)$$

Positive value of SN ratio η indicated that the parameter was significant for the first filtration and it will be considered for the second filtration. Proportional coefficient, β represented the steepness of the line in a scatter diagram. The integrated estimate value is then calculated as shown in equation (6) by using the significant parameter which have been evaluated during first evaluation. In addition, the calculation also based on the orthogonal array which have been assigned from number of original parameters consideration.

$$\text{Integrated estimate value, } \widehat{M}_i = \frac{\eta_1 \times \frac{X_{i1}}{\beta 1} + \eta_2 \times \frac{X_{i2}}{\beta 2} + \cdots + \eta_k \times \frac{X_{ik}}{\beta k}}{\eta_1 + \eta_2 + \cdots + \eta_k} \quad (6)$$

As a result, equation (7) contributed to the criticality of parameter for the second filtration. However, it can be observed from factorial effect graph.

$$\text{Integrated estimate SN ratio, } \eta = 10 \log \left[\frac{\frac{1}{r}(S_\beta - V_e)}{V_e} \right] \quad (7)$$

When the parameters used by the SN ratio is larger and when the parameter is not used with the SN ratio is smaller, so the level of contribution to be positive and is classified as a critical parameter. In other words, on factorial effect graph, descending the line indicated that the parameter is significant. The critical parameters were considered into the costing using TCA. The cost-driver value for the appropriate duration is determined and the predetermined overhead rate is calculated. Predetermined overhead rate is used to implement the overhead cost to product costs for a period. Finally, implement the overhead cost to the product by using predetermined overhead rate. For each activity, overhead costs assigned to the products obtained by multiplying the predetermined overhead rate. Then, the actual cost of producing a single unit of crankshaft was estimated based on critical parameters.

3. Result and discussion

At this stage, this work was developed MD value for each sample of both engine model. According to the figure 1 (a) which specifically for Caterpillar engine model, since the sample numbers of remanufacturable (35) and rejected (35) of crankpin diameters are equal, the distribution of sample inside the scatter diagram is stable for both regions. The MD value for this engine model with 70 samples was between 774002.21 to 798279.92 with the average of 787586, which is far from the unit data. According to figure 1 (b), the distribution of samples inside the scatter diagram is quite fair because the sample numbers of remanufacturable (40) and repairable (44) of crankpin diameter are approximately equal. Obviously, there was no sample distribution belonging to the rejected group. Consequently, the threshold of MD for MAN engine model with 84 samples is between 0.092534 to 7.446333 with an average of 1.0, which is the unit data in this crankshaft classification. The overall classification can be seen as shown in figure 1 (c) which combined for both engine model. This explained that the pattern of Caterpillar engine model was obviously distinctive.

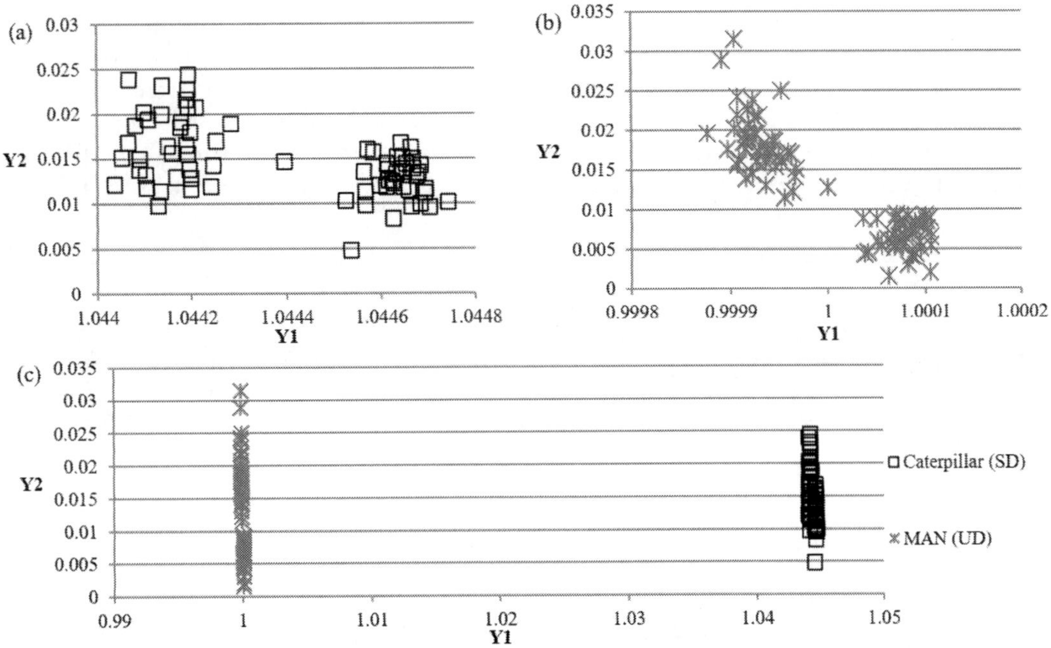

Figure 1. Scatter diagram of (a) Caterpillar, (b) MAN and (c) combination of both engine model.

The parameter evaluation was started with Caterpillar engine model by considering 6 crankpin diameters ranging from 117.0002 mm to 117.0428 mm as inputs and MD ranging from 788974.3 to 798279.9 as outputs. From that, a histogram as shown in figure 2 was constructed and the highest peak is between 796000 to 797000 of MD with the frequency of 9 and a bit skewing to the right.

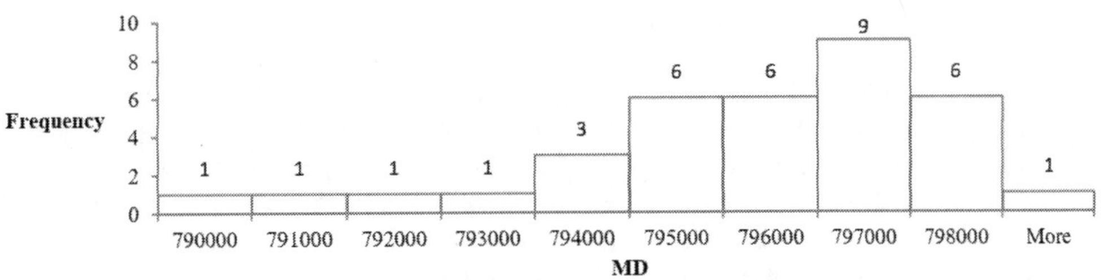

Figure 2. Histogram of Caterpillar.

MAN engine model was considering 6 crankpin diameters ranging from 112.020 mm to 112.0399 mm as inputs and MDs ranging from 0.092534 to 2.495802 as outputs with 84 samples belonging to the remanufacturable and hybrid crankshafts. From that, a histogram as shown in figure 3 been constructed and the highest peak with the frequency of 18 and a bit skewing to the left.

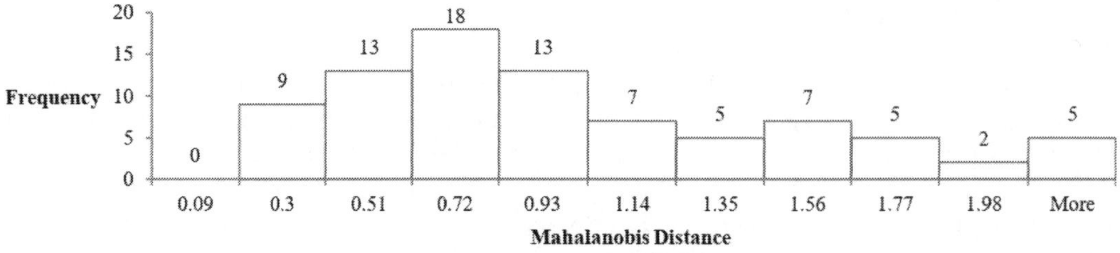

Figure 3. Histogram of MAN.

The result of first filtration for Caterpillar engine model based on the value of proportional coefficient β and SN ratio η as shown in figure 4. Ascending the line to the right indicates that the parameter has positive value of proportional coefficient β. However, their steepness is increasing from diameter 6, 2, 3, 4, 5 and 1. Then, SN ratio η indicates the relationship between the parameter and output: the larger the SN ratio η is, the stronger the relationship. The relationship started stronger from diameter 2, 6, 4, 1, 3 and 5. Therefore, those crankpins diameter are well suited to the purpose of integrated estimate value.

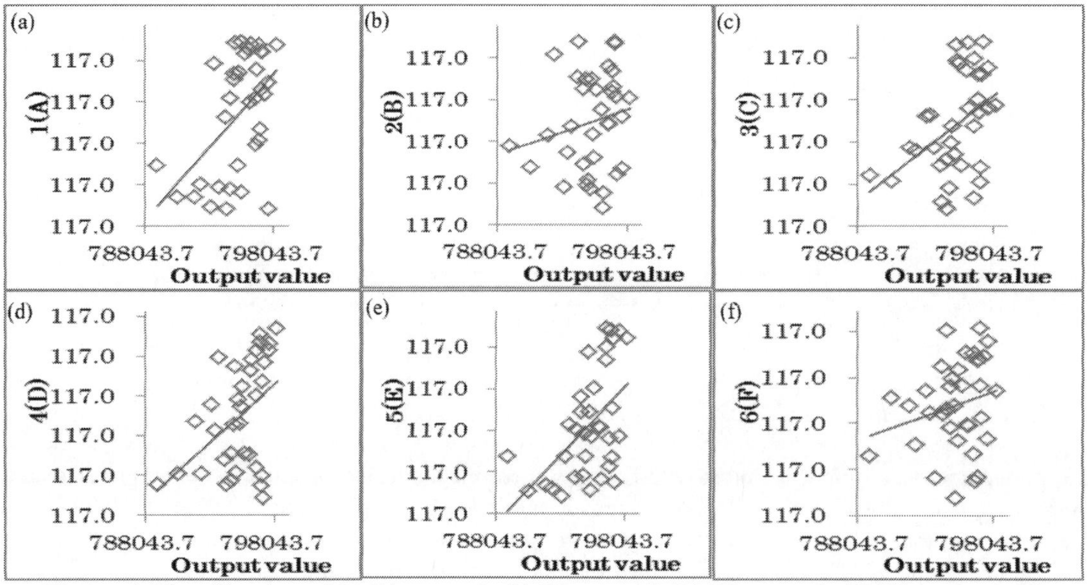

Figure 4. First filtration of (a) Diameter 1, (b) Diameter 2, (c) Diameter 3, (d) Diameter 4, (e) Diameter 5 and (f) Diameter 6.

With respect to figure 5, the steepness for MAN engine model was increasing from diameters 6, 4 and 5. Descending the line to the right indicates that the parameter has negative value of proportional coefficient β. However, their steepness is increasing from diameters 2, 1 and 3. Then, SN ratio η indicates the relationship between the parameter and output: the larger the SN ratio η is, the stronger the relationship. The relationships get stronger starting from diameters 2, 6, 4, 5, 1 and 3. Therefore, these crankpin diameters are well suited for the purpose of integrated estimate value.

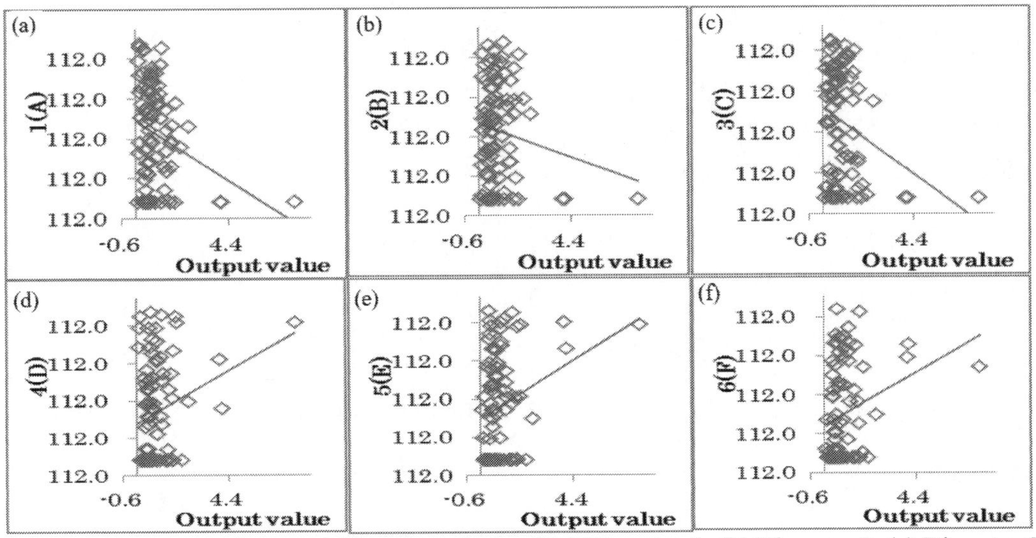

Figure 5. Scatter of output MD and crankpin of (a) Diameter 1, (b) Diameter 2, (c) Diameter 3, (d) Diameter 4, (e) Diameter 5 and (f) Diameter 6.

By using those values of proportional coefficient β and SN ratio η from figure 4, this work obtained the value of integrated estimate value. Thus, the larger the SN ratio η is, the greater the degree of its contribution to the integrated estimates of MD value. The parameter evaluation was performed using L_8 type orthogonal array and the result was translated into a factorial effect graph as shown in figure 6. Descending the line from left (level 1) to right (level 2) indicated that the parameter has an effect (degree of contribution) of elevating the output. When the crankpin diameter 5 is used (level 1) with a larger relationship (SN ratio = 67.2 db) to the output and when the parameter is not used (level 2) with a smaller relationship (SN ratio = 70.6) to the output, the larger degree of contribution (3.38 db) of the parameter will be gained (critical). It means that this crankpin diameter 5 has the highest contribution to the output.

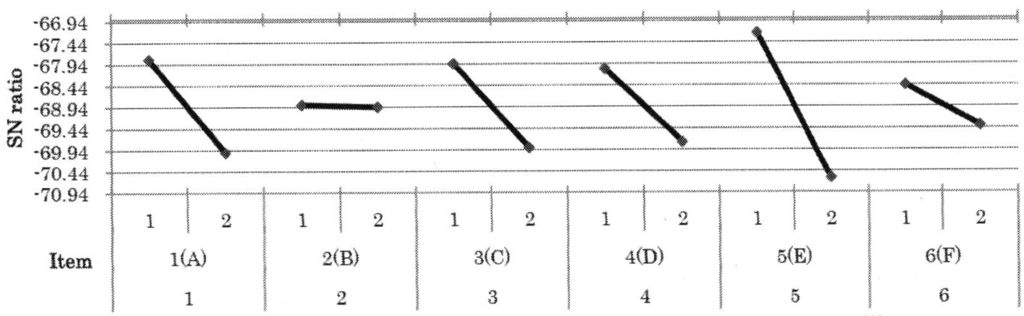

Figure 6. Evaluation importance through factorial effect graph of Caterpillar.

Therefore, among those 6 crankpins diameter of 35 remanufacturable crankshafts, these were 100% critical. Furthermore, no improvement can be made on inspection quality and no saving with respect to the cost, time and energy for the time being. In addition, this work concluded their criticality according to their contribution from diameter 2, 6, 4, 3, 1 and finally 5. By considering the figure 7 for MAN engine model, among the 6 crankpin diameters of 84 remanufacturable and hybrid crankshafts, only 83% are critical and 17% are non-critical. Furthermore, some improvements can be made on the inspection quality and savings with respect to the cost, time and energy expended can be made. In addition, this research can increasingly rank their criticality according to their contribution from diameters 4, 5, 6, 1 and finally 3.

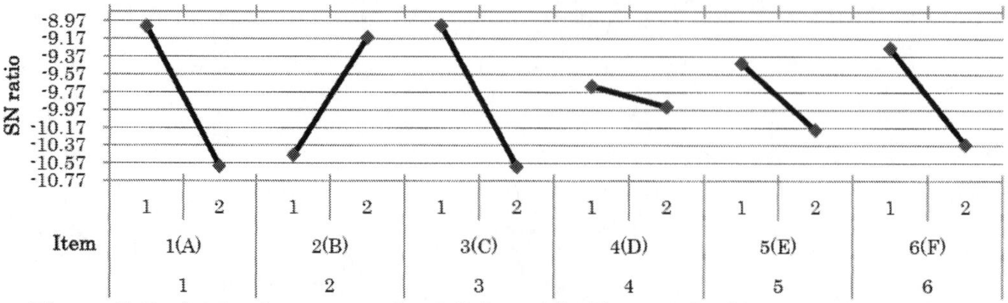

Figure 7. Evaluation importance through factorial effect graph of MAN engine model.

Table 2 showed the Caterpillar engine cost per unit of crankshaft equal to MYR1323.09 which is identical to the yearly overhead to be MYR 10608.33 but when divided by 35 units of crankshaft, the overhead per unit crankshaft equal to MYR 303.09. Time taken to produce each unit of crankshaft is 13.6 hours.

Table 2. Cost by TCA system from inspection of traditional (Caterpillar).

Process	Duration per crankpin (hour)	Number of crankpin (unit)	Duration per crankshaft (hour)	Man power	Working rate/hour (MYR)	Cost per unit (MYR)	Number of crankshaft (unit)	Annual cost per unit (MYR)
Inspection	0.2		1.2			90.00		
Grinding	0.7	6	4.2			315.00		
Polishing	0.6		3.6	1	75	270.00		
Cleaning			1.0			75.00		
Leak testing			2.5			187.50		
Packaging			1.1			82.50		
Total direct labor						1020.00	35	35700.00
Overhead						303.09		10608.33
Total						1323.09		46308.33

Table 3 shows the MAN engine cost per unit of crankshaft equal to MYR 1401.29 which is identical to the yearly overhead to be MYR 10608.33 but when divided by 84 units of crankshaft, the overhead per unit crankshaft equal to MYR 126.29. Time taken to produce each unit of crankshaft is 17.0 hours.

Table 3. Cost by TCA system from inspection of traditional (MAN).

Process	Duration per crankpin (hour)	Number of crankpin (unit)	Duration per crankshaft (hour)	Man power	Working rate/hour (MYR)	Cost per unit (MYR)	Number of crankshaft (unit)	Annual cost per unit (MYR)
Inspection	0.3		1.8			135.00		
Grinding	0.9	6	5.4			405.00		
Polishing	0.8		4.8	1	75	360.00		
Cleaning			1.2			90.00		
Leak testing			2.6			195.00		
Packaging			1.2			90.00		
Total direct labor						1275.00	84	107100.00
Overhead						126.29		10608.33
Total						1401.29		117708.33

Table 2 and table 3 was then been compared with the costing with optimization using MTS. Since the critical parameters of Caterpillar engine model have not changed, subsequently this work proceed with the comparison for MAN engine model. Table 4 showed the cost per unit of crankshaft equal to

MYR1251.29 which is identical to the yearly overhead to be MYR10608.33 but when divided by 84 units of crankshaft, the overhead per unit crankshaft equal to MYR126.29.

Table 4. Cost by TCA system from MTS (MAN).

Process	Duration per crankpin (hour)	Number of crankpin (unit)	Duration per crankshaft (hour)	Man power	Working rate/hour (MYR)	Cost per unit (MYR)	Number of crankshaft (unit)	Annual cost per unit (MYR)
Inspection	0.3		1.5			112.50		
Grinding	0.9	5	4.5			337.50		
Polishing	0.8		4.0	1	75	300.00		
Cleaning			1.2			90.00		
Leak testing			2.6			195.00		
Packaging			1.2			90.00		
Total direct labor						1125.00	84	94500.00
Overhead						126.29		10608.33
Total						1251.29		105108.33

This work proved that, the contribution of MTS was obviously optimized the process of remanufacturing on identification on the critical parameters.

4. Conclusion

By utilizing the MTS, the critical and non-critical parameter can be identified in the remanufacturing process. Reduction of parameters number can improved the inspection quality. As a result, the integration of MTS and TCA has proven that the cost per unit of remanufactured crankshaft of MAN engine model has been reduced from MYR1401.29 to MYR1251.29. The company also can reduce a lot of waste after identified the non-critical parameters especially the costs to other processes of remanufacturing.

Acknowledgement

The authors would like to be obliged to Universiti Malaysia Pahang for providing financial assistance under RDU170387.

References

[1] Zhang TZ, Chu JW, Wang XP, Liu XH and Cui PF 2011 Development pattern and enhancing system of automotive components remanufacturing industry in China *Resources, Conservation and Recycling* **55** 613-22

[2] Ijomah WL 2010 The application of remanufacturing in sustainable manufacture *Waste and Resource Management* **163** 157-63

[3] Haldar NAH, Khan FA, Ali A and Abbas H 2017 Arrhythmia classification using Mahalanobis distance based improved Fuzzy C-Means clustering for mobile health monitoring systems *Neurocomputing* **220** 221-35

[4] Reséndiz E and Rull-Flores CA 2013 Mahalanobis–Taguchi system applied to variable selection in automotive pedals components using Gompertz binary particle swarm optimization *Expert Systems with Applications* **40** 2361-5

[5] Li C, Xiao Q, Tang Y and Li L 2016 A method integrating Taguchi, RSM and MOPSO to CNC machining parameters optimization for energy saving *Journal of Cleaner Production* **135** 263-75

[6] Ketkar M and Vaidya OS 2014 Evaluating and Ranking Candidates for MBA program: Mahalanobis Taguchi System Approach *Procedia Economics and Finance* **11** 654-64

[7] Reséndiz E, Moncayo-Martínez LA and Solís G 2013 Binary ant colony optimization applied to variable screening in the Mahalanobis–Taguchi System *Expert Systems with Applications* **40** 634-7

[8] Jin X and Chow TWS 2013 Anomaly detection of cooling fan and fault classification of induction motor using Mahalanobis–Taguchi system *Expert Systems with Applications* **40** 5787-95
[9] Yang T and Cheng YT 2010 The use of Mahalanobis–Taguchi System to improve flip-chip bumping height inspection efficiency *Microelectronics Reliability* **50** 407-14
[10] Liparas D, Laskaris N and Angelis L 2013 Incorporating resting state dynamics in the analysis of encephalographic responses by means of the Mahalanobis–Taguchi strategy *Expert Systems with Applications* **40** 2621-30
[11] John B 2014 Application of Mahalanobis-Taguchi system and design of experiments to reduce the field failures of splined shafts *International Journal of Quality & Reliability Management* **31** 681-97
[12] Liparas D, Angelis L and Feldt R 2012 Applying the Mahalanobis-Taguchi strategy for software defect diagnosis *Automated Software Engineering* **19** 141-65
[13] Beshah B 2014 The effect of aggregating multivariate performance parameters in a quality award process *International Journal of Quality & Reliability Management* **31** 618-28
[14] Wang PC, Su CT, Chen KH and Chen NH 2011 The application of rough set and Mahalanobis distance to enhance the quality of OSA diagnosis *Expert Systems with Applications* **38** 7828-36
[15] Limon-Romero J, Tlapa D, Baez-Lopez Y, Maldonado-Macias A and Rivera-Cadavid L 2016 Application of the Taguchi method to improve a medical device cutting process *International Journal Advance Manufacturing Technology* **87** 3569-77
[16] Iquebal AS, Pal A, Ceglarek D and Tiwari MK 2014 Enhancement of Mahalanobis–Taguchi System via Rough Sets based Feature Selection *Expert Systems with Applications* **41** 8003-15
[17] Deepa N and Ganesan K 2016 Mahalanobis Taguchi system based criteria selection tool for agriculture crops *Sadhana* **41** 1407-14
[18] Li C, Xiao Q, Tang Y and Li L 2016 A method integrating Taguchi, RSM and MOPSO to CNC machining parameters optimization for energy saving *Journal of Cleaner Production* **135** 263-75
[19] Hu J, Zhang L and Liang W 2013 Dynamic degradation observer for bearing fault by MTS–SOM system *Mechanical Systems and Signal Processing* **36** 385–400

IOP Conf. Series: Materials Science and Engineering **342** (2018) 012006 doi:10.1088/1757-899X/342/1/012006

Costing improvement of remanufacturing crankshaft by integrating Mahalanobis-Taguchi System and Activity based Costing

M Y Abu[1], E E Mohd Nor[1] and M S Abd Rahman[2]

[1]Faculty of Manufacturing Engineering, Universiti Malaysia Pahang
[2]Faculty of Engineering Technology, Universiti Teknikal Malaysia Melaka

E-mail:myazid@ump.edu.my

Abstract. Integration between quality and costing system is very crucial in order to achieve an accurate product cost and profit. Current practice by most of remanufacturers, there are still lacking on optimization during the remanufacturing process which contributed to incorrect variables consideration to the costing system. Meanwhile, traditional costing accounting being practice has distortion in the cost unit which lead to inaccurate cost of product. The aim of this work is to identify the critical and non-critical variables during remanufacturing process using Mahalanobis-Taguchi System and simultaneously estimate the cost using Activity Based Costing method. The orthogonal array was applied to indicate the contribution of variables in the factorial effect graph and the critical variables were considered with overhead costs that are actually demanding the activities. This work improved the quality inspection together with costing system to produce an accurate profitability information. As a result, the cost per unit of remanufactured crankshaft of MAN engine model with 5 critical crankpins is MYR609.50 while Detroit engine model with 4 critical crankpins is MYR1254.80. The significant of output demonstrated through promoting green by reducing re-melting process of damaged parts to ensure consistent benefit of return cores.

1. Introduction

In the early 20th century, remanufacturing has been proven by industry and it expended by the need to reuse military machinery and vehicles during and after the Second World War [1]. In China, Volkswagen established a remanufacturing plant for modern engine at Shanghai [2] which offered an inferior market price and same quality of new products, in addition remanufactures existing product into a new product with same warranty [3]. Remanufacturing is a chain of manufacturing phases which performed on the transformation industrial process that refurbished and recycled of a product in order to return it to an exceptional performance and it was not worse compare to the original product [4]. Economic and environmental potential, both are importance in remanufacturing industry which provides cost savings up to 90% of raw materials, ratio range of six to one for original production and remanufacturing in energy consumption, high efficiency and able to process high quality cores [5]. In fact, the manufacturing process of large and expensive engine parts are energy intensive and generate large amounts of environmental emissions especially during the critical steps in remanufacturing [6], because there is no available method to decide which critical crankshaft are eligible to that process forward it will generate a lot of waste with respect to the energy, material, time and cost [7].

Mahalanobis distance (MD) and principle of Taguchi method, both is used in Mahalanobis-Taguchi System (MTS) approach by constructing a measurement of abnormality of patterns and make accurate prediction [8], where the interdependence between the variables is influential in pattern analysis. Hence, the class imbalance problem will be resolved and the data distribution is not affected. A vector analysis such as orthogonal arrays (OA) and signal-to-noise ratios (SN) is used to consolidate data from many parameters of a system into a single group or dimension [9]. By integrating the MTS, it has provided profit worth considering in many various applications [10]. Work by [11] applied MTS to present a multi-criteria evaluation framework and ranking candidates for MBA program by considering several criteria. [12] presented the MTS to designate the health condition of cooling fan and induction motor based on vibration signal by anomaly detection wherein the different unbalanced electrical faults of induction motor can be classified with a higher accuracy. [13] implemented the Gompertz binary particle swarm optimization algorithm to solve the problem of combinatorial optimization from variable selection which detected by MTS in the manufacturing process of automotive pedals components. [14] solved the combinatorial optimization problem of variable selection with applied the binary ant colony optimization algorithm in the application of the MTS.

In recent year, common approaches to service and product costing have been focus to continue and diverse condemnation. Nevertheless, the level of change in management accounting system has created to stimulate and several improvements have created as much attentiveness as activity based costing (ABC) which has developed during the last span as one means of addressing the flaws of the traditional cost accounting (TCA) method [15] which only assigns costs to product based on an average overhead rate. Therefore, ABC are introduced as allocates the costs of manufacturing a product according to the activities needed to produce the item. The application of ABC has been reported in organizational, hospital, supply chain, cellular manufacturing, farm management, transportation, decision making, radiotherapy and in numerous energy and environmental issues. Work by [16] performed the 0-1 mixed integer programming using ABC system and life cycle assessment (LCA) which assign resources and funding for energy saving activities to each green building through appropriate cost drivers. [17] proposed a cost estimation of safety by using activity of risk assessment for the initial stages of construction bidding phase since the fatal construction accidents are major problem in construction industry and construction project scheduling by focusing on construction activities. Furthermore, [18] resolved the strategic decision making under resource constraints and the carbon footprint factor which is the perspective of sustainable transport and the criterion of decreasing traffic energy consumption are the most significant evaluation factors by the connection with ABC evaluation and LCA. [19] proved that the ABC models for inter-firm cost accounting are appropriate in the context of supply chain environment by developing the conceptual framework. Therefore, it is hard to find any previous works that integrating the MTS and ABC to improve simultaneously quality and costing system.

This work identifies critical variables in the remanufacturing process in order to improve the quality inspection. Factorial effect graph has been developed to observe which variables were critical to the remanufacturing process. Meanwhile, the structure of costing system has been improved to obtain accurate cost of product and to gain accurate profitability.

2. Methodology

This work selected crankpin of crankshafts as a subject matter and the number of sample is shown in table 1. The data were collected from a remanufacturing industry which located at Rawang, Malaysia.

Table 1. Data collection for crankpin of crankshaft.

Engine Model	No. of sample	Remanufacturable tolerance (*mm*)	
		Lower limit	Upper limit
Detroit	59	114.262	114.305
MAN	84	112.020	112.040

Subsequently, this work need to develop a clustering between both engine model to indicate their Mahalanobis Distance (MD) value. Higher the MD value indicated that the off-target sample have a distinctive pattern from the target sample while, smaller value of MD indicated that the off-target sample have closer pattern to the target sample. MAN engine model have been selected as the target sample due to the pattern was closer to the original crankshaft while Detroit engine model as the off-target sample. The calculation of MD as shown in equation (1) whereby Y represented the normalized Y_1 and Y_2, A represented the inverse matrix, Y^T represented transposition of Y and k represented number of variables.

$$\text{Mahalanobis distance} = \frac{YAY^T}{k} \quad (1)$$

The MD for each sample of both engine model was then used as response to evaluate the criticality of variables. The diameters of crankpin were used as input and the MDs were used as output. The data should be normal or densely populated in the medium range to apply this method by creating a histogram. The target data is the highest sample while off-target data is the remaining samples number. Normalization process is required in order to develop a dimensionless condition of the off target data. The normalization is used to make the data more flexible by diminishing their redundancy. Equation (2) and equation (3) are used to calculate the normalization.

$$X_{ij} = \acute{x}_{ij} - \bar{x}_j \quad (2)$$

$$M_i = \acute{y}_i - m_0 \quad (3)$$

Proportional coefficient β and SN ratio η are determined using equation (4) and equation (5).

$$\text{Proportional coefficient, } \beta_1 = \frac{M_1X_{11} + M_2X_{21} + \cdots + M_lX_{l1}}{r} \quad (4)$$

$$\text{SN ratio, } \eta_1 = \begin{cases} \dfrac{\frac{1}{r}(S_{\beta 1} - V_{el})}{V_{el}} & (when\ S_{\beta 1} > V_{el}) \\ 0 & (when\ S_{\beta 1} \le V_{el}) \end{cases} \quad (5)$$

Positive value of SN ratio η indicated that the variable was significant for the first filtration and it will be considered for the second filtration. Proportional coefficient, β represented the steepness of the line in a scatter diagram. The integrated estimate value is then calculated as shown in equation (6) by using the critical variable which have been evaluated during first evaluation. In addition, the calculation also based on the orthogonal array which have been assigned from number of original parameters consideration.

$$\text{Integrated estimate value, } \widehat{M}_i = \frac{\eta_1 \times \frac{X_{i1}}{\beta 1} + \eta_2 \times \frac{X_{i2}}{\beta 2} + \cdots + \eta_k \times \frac{X_{ik}}{\beta k}}{\eta_1 + \eta_2 + \cdots + \eta_k} \quad (6)$$

As a result, equation (7) contributed to the criticality of variables for the second filtration. However, it can be observed from factorial effect graph.

$$\text{Integrated estimate SN ratio, } \eta = 10 \log \left[\frac{\frac{1}{r}(S_\beta - V_e)}{V_e} \right] \quad (7)$$

When the parameters were used by the SN ratio is larger and when the parameter is not used with the SN ratio is smaller, so the level of contribution to be positive and is classified as a critical variable. In other words, on factorial effect graph, descending the line indicated that the variable is significant. The critical variables were considered into the costing using ABC. To perform the method, this work need to identify centres with their activities that give a major contribution to the production process. Then, resources cost is assigned to activity centres. This is actually done by estimating an amount of overhead for that year to be apportioned in percentage to those activity centres. The costs of the resources are already recorded in the existing accounting system. Then, activity centre cost is assigned to activities. The cost just apportioned in the second stage should be apportioned to the different types of product in terms of percentages based on their activity load. Similarly, overhead also would be apportioned by using percentage for such categories as wages, depreciation, consumables, energy, building cost and other costs. Fourthly, the cost per unit of activity driver is estimated. An activity driver is a cost driver used to estimate the cost of an activity consumed by the cost object. The activity driver should be proportional to the activity cost or activities that consume costs. Finally, is preparing a bill of activities for a unit of crankshaft. A cost object is any product for which management wants a separate measure of the cost. Cost per unit of activity driver would multiply with annual quantity of cost driver consumed to obtain the total of annual cost. The outcome is a total cost for each product that consists of a list of the activities used by the product and their costs.

3. Result and discussion

At this stage, this work was developed MD value for each sample of both engine model. According to the figure 1 (a) which specifically for MAN engine model, the distribution of samples inside the scatter diagram is quite fair because the sample numbers of remanufacturable (40) and repairable (44) of crankpin diameter are approximately equal. Obviously, there was no sample distribution belonging to the rejected group. Consequently, the MD value for Man engine model with 84 samples is between 0.092534 to 7.446333 with an average of 1.0, which is the unit data in this crankshaft classification. According to the figure 1 (b), since the sample numbers of remanufacturable (29) of crankpin diameter are greater than the rejected (16) and the repairable (14), the distribution of samples inside the scatter diagram is dense on the right side. Then, the threshold of MD for Detroit engine model with 59 samples is between 153277.39 to 164993.55 with an average of 160844.42, which is closer to the target sample. The overall classification can be seen as shown in figure 1 (c) which combined for both engine model. This explained that the pattern of Detroit engine model was obviously distinctive.

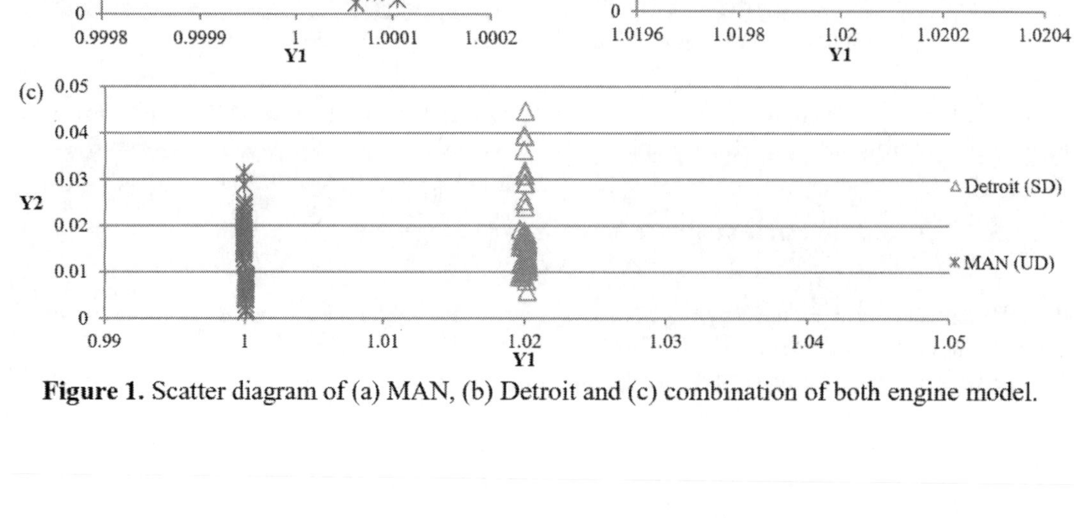

Figure 1. Scatter diagram of (a) MAN, (b) Detroit and (c) combination of both engine model.

The parameter evaluation was started with the result of first filtration for MAN engine model based on the value of proportional coefficient β and SN ratio η as shown in figure 2. Ascending the line to the right indicates that the parameter has positive value of proportional coefficient β. However, their steepness is increasing from diameters 6, 4 and 5. Then, SN ratio η indicates the relationship between the parameter and output: the larger the SN ratio η is, the stronger the relationship. The relationships get stronger starting from diameters 2, 6, 4, 5, 1 and 3. Therefore, these crankpin diameters are well suited for the purpose of integrated estimate value.

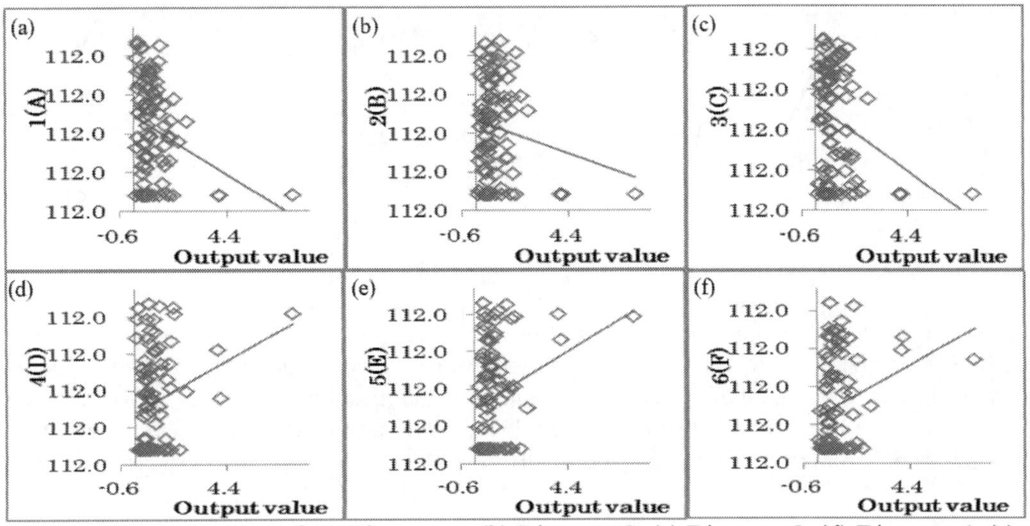

Figure 2. First filtration of (a) Diameter 1, (b) Diameter 2, (c) Diameter 3, (d) Diameter 4, (e) Diameter 5 and (f) Diameter 6.

With respect to the figure 3 of Detroit engine model, the steepness is increasing from diameter 3, 6, 4, 2, 5 and 1 while their relationship gets stronger starting from diameter 5, 4, 2 and 1. Therefore, those crankpin diameters are well suited to the purpose of integrated estimate value except for diameter 3 and 6 because both values are -6.20 db and -3.02 db respectively. When SN ratio η turned negative due to the $S_\beta \leq V_e$, the value will be treated as zero and not considered into integrated estimate value.

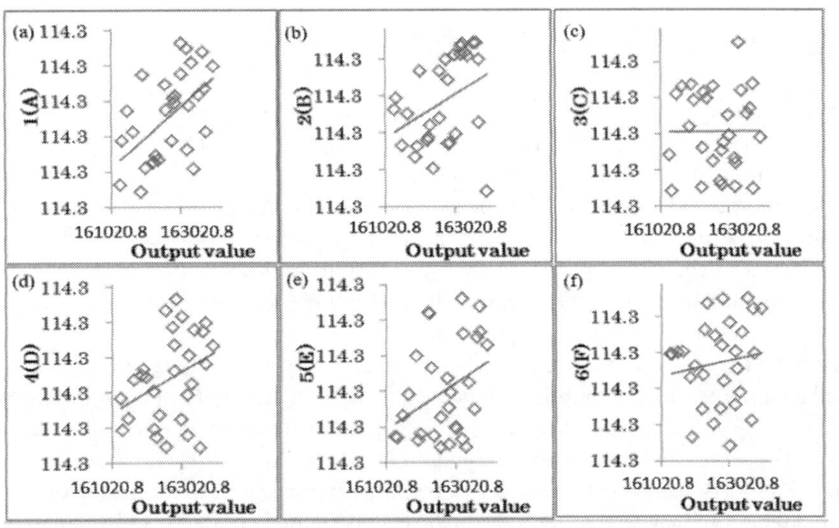

Figure 3. Scatter of output MD and crankpin of (a) Diameter 1, (b) Diameter 2, (c) Diameter 3, (d) Diameter 4, (e) Diameter 5 and (f) Diameter 6.

By considering the figure 4 for MAN engine model, among the 6 crankpin diameters of 84 remanufacturable and hybrid crankshafts, only 83% are critical and 17% are non-critical. Furthermore, some improvements can be made on the inspection quality and savings with respect to the cost, time and energy expended can be made. In addition, this work can increasingly rank their criticality according to their contribution from diameters 4, 5, 6, 1 and finally 3.

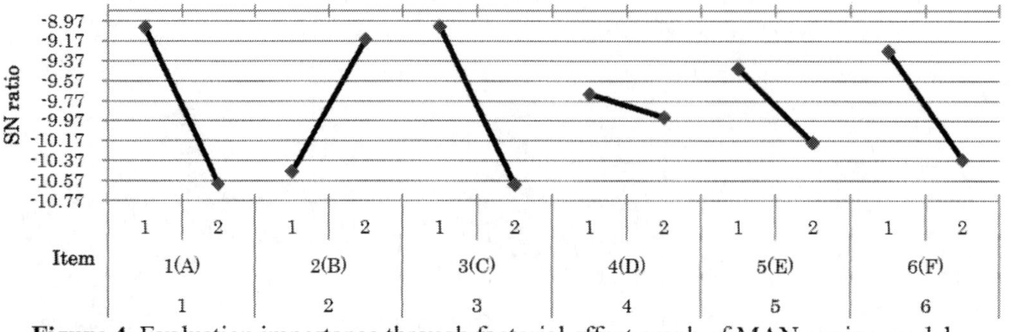

Figure 4. Evaluation importance through factorial effect graph of MAN engine model.

With respect to Detroit engine model, among those 6 crankpins diameter of 29 remanufacturable crankshafts, only 67% are critical while 33% are non-critical as shown in figure 5. Furthermore, some improvements can be made on the quality of inspection including savings with respect to the cost, time and energy expended. In addition, this research is able to increasingly rank their criticality according to their contributions from diameters 5, 4, 2 and finally 1.

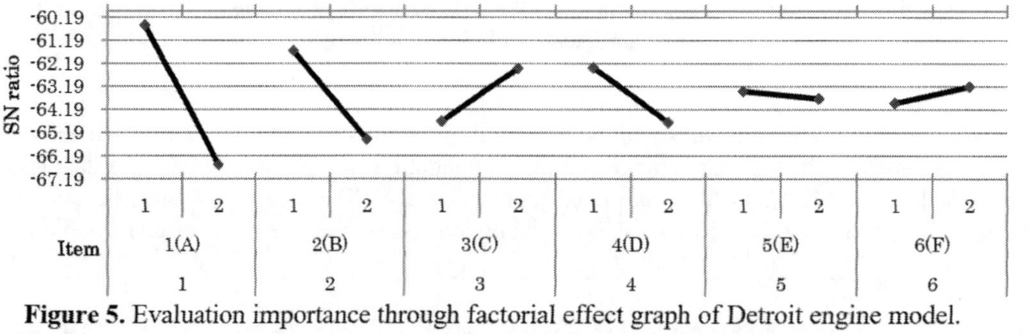

Figure 5. Evaluation importance through factorial effect graph of Detroit engine model.

Subsequently, those significant variables were considered into the costing using ABC. There were six departments which really affected to the industry's business such as administration, corporate management, factory management, quality control, remanufacturing, and sales and dispatch. Activity driver is a cost driver used to estimate the costs of an activity consumed by the cost objects. In selecting an activity driver, there must be a linear relationship between the activity costs and the activity drivers, otherwise the costs of the object will be distorted. Activity drivers can be more than one per activity. Estimation of the quantity of activity driver is obtained for each activity for the year so that the cost per unit of activity driver may be calculated as shown in table 2. The longer the machine and setup hours are, the larger the costs of the grinding activity. Continuous usage of the grinding machines to remove excess metal on the surface of crankpins indicates that maintenance or depreciation costs would increase. Similarly, the protective personal equipment being used such as coveralls, safety shoes and goggles would also increase the consumable costs. The longer the machine and setup hours, the larger the costs of polishing. Similarly, continuous usage of the polishing machines to remove excess metal on the surface of crankpins indicates that maintenance or depreciation costs would increase. The coveralls, safety shoes and goggles used tend to increase consumable costs.

Table 2. Activities and activity cost per unit of activity drivers.

Activity centre with activities	Activity cost	Activity driver	Predict quantity of activity driver	Cost per unit of activity driver (RM)
Administration				
Process receivables	15000	No. of invoices	800 units	18.75
Process payables	24000	No. of purchase orders	750 units	32.00
Operating IT system	26250	No. of CPU time-hours	9000 hours	2.92
Preparing payroll	9750	No. of pay slips	300 persons	32.50
	75000			
Corporate management				
Providing continuity for organization	19500	None available	Facility activity	-
Appointing a chief executive	17250	None available	Facility activity	-
Governing the organization	15000	None available	Facility activity	-
Acquiring resources	23250	None available	Facility activity	-
	75000			
Factory management				
Program production	16500	No. of man hours	26000 hours	0.64
Expediting order	14000	No. of grinding machines	12 units	291.67
		No. of polishing machines	12 units	291.67
		No. of cleaning machines	11 units	318.18
		No. of man power	20 persons	175.00
Managing plant	13000	No. of crankshafts	450 units	28.89
Training production personnel	6500	No. of training-hours	550 hours	11.82
	50000			
Quality control				
Tools and equipment maintenance	18750	No. of tools-hours	2700 hours	6.95
Hardness measurement	11250	No. of crankshafts	700 units	16.07
Inspection process	15000	No. of crankpins	3100 units	4.84
Leak testing	17250	No. of testing-hours	5000 hours	3.45
Training on quality	12750	No. of training-hours	950 hours	13.42
	75000			
Remanufacturing				
Grinding	33000	No. of machine hours	11000 hours	1.50
		No. of setup hours	8320 hours	1.98
Polishing	32000	No. of machine hours	11000 hours	1.46
		No. of setup hours	7780 hours	2.06
Cleaning	35000	No. of machine hours	4000 hours	4.38
		No. of setup hours	8240 hours	2.13
	100000			
Sales and dispatch				
Process sales order	8740	No. of sales orders	250 units	34.96
Packaging plan	15200	No. of man hours	5160 hours	2.95
Packaging training	14060	No. of training-hours	880 hours	15.98
	38000			
Total cost	413000			

A bills of activity is prepared for the MAN engine model as shown in Table 3. The crankshafts have an estimated cost of MYR609.05 per unit after considering 84 units of crankshaft with 5 critical crankpins. This research maintains the similar number of annual operating IT hour (2112 hour) based on the assumption of work of 8 hours per day and 22 days per month. This explained that, ABC is able to estimate the cost of product accurately based on the number of activities used.

Table 3. Activity-based product cost (MAN).

Activity Level	Activity	Cost per unit of activity driver (MYR)	Annual quantity of cost driver consumed	Annual Cost (MYR)	Cost per unit of product (MYR)
Unit	Measure hardness	16.07	84 units	1350.09	
Unit	Inspection process	4.84	420 units	2013.75	

Unit	Grinding-machine hours	1.50	756 units	1134.00	
Unit	Grinding-setup hours	1.98	1512 hours	2997.54	
Unit	Polishing-machine hours	1.46	210 hours	305.55	
Unit	Polishing-setup hour	2.06	1428 hours	2938.11	
Unit	Cleaning-machine hours	4.38	672 hours	2940.00	
Unit	Cleaning-setup hour	2.13	1334 hours	2856.00	
Unit	Perform leak testing	3.45	336 hours	1159.20	
Unit	Packing plan	2.95	176.4 hours	519.50	
Batch	Program production	0.64	6437.4 hours	4087.75	
Batch	Expedite order-grinding machines	291.67	2 units	583.34	
Batch	Expedite order-polishing machines	291.67	2 units	583.34	
Batch	Expedite order-cleaning machines	318.18	2 units	636.37	
Batch	Expedite order-man powers	175.00	4 persons	700.00	
Facility	Manage Plan	28.89	29 units	2426.76	
Total remanufacturing activity cost				**27249.28**	**324.40**
Unit	Tools and equipment maintenance	6.95	151.2 hours	105.09	
Batch	Process receivables	18.75	179 units	3356.25	
Batch	Process Payable	32.00	179 units	5728.00	
Batch	Operate IT system	2.92	2112 hours	6156.48	
Batch	Prepare payroll	32.50	26 persons	845.00	
Batch	Process sales order	34.96	110 units	3845.60	
Facility	Training production personnel	11.82	72 hours	850.86	
Facility	Training on quality	13.42	72 hours	966.24	
Facility	Packaging training	15.98	72 hours	1150.38	
Total non-remanufacturing activity cost				**23948.90**	**285.10**
Total crankpin cost per unit					**609.50**

A bill of activities is prepared for the Detroit engine model as shown in table 4. The crankshaft has an estimated cost of MYR1254.80 per unit after taking into consideration 29 units of crankshaft with 4 crankpins. For all those bills of engine model, in order to make a fair judgement, this research keep similar unit of grinding, polishing and cleaning machines of expediting order which is 2. This proved that all the activities required to produce the remanufactured were transparently listed. In order to increase the accuracy of the cost, the company might perform other analysis to find the critical activity driver as well.

Table 4. Activity-based product cost (Detroit).

Activity Level	Activity	Cost per unit of activity driver (MYR)	Annual quantity of cost driver consumed	Annual Cost (MYR)	Cost per unit of product (MYR)
Unit	Measure hardness	16.07	29 units	466.10	
Unit	Inspection process	4.84	116 units	561.15	
Unit	Grinding-machine hours	1.50	266.8 units	400.20	
Unit	Grinding-setup hours	1.98	487.2 hours	965.88	
Unit	Polishing-machine hours	1.46	92.8 hours	135.03	
Unit	Polishing-setup hour	2.06	464 hours	954.68	
Unit	Cleaning-machine hours	4.38	255.2 hours	1116.50	
Unit	Cleaning-setup hour	2.13	440.8 hours	936.70	
Unit	Perform leak testing	3.45	127.6 hours	440.22	
Unit	Packing plan	2.95	78.3 hours	230.59	
Batch	Program production	0.64	2212.7 hours	1405.07	
Batch	Expedite order-grinding machines	291.67	2 units	583.34	
Batch	Expedite order-polishing machines	291.67	2 units	583.34	
Batch	Expedite order-cleaning machines	318.18	2 units	636.37	
Batch	Expedite order-man powers	175.00	4 persons	700.00	
Facility	Manage Plan	28.89	29 units	873.81	

Total remanufacturing activity cost				**10952.96**	**377.69**
Unit	Tools and equipment maintenance	6.95	60.9 hours	422.95	
Batch	Process receivables	18.75	200 units	3750.00	
Batch	Process Payable	32.00	200 units	6400.00	
Batch	Operate IT system	2.92	2112 hours	6156.48	
Batch	Prepare payroll	32.50	26 persons	845.00	
Batch	Process sales order	34.96	140 units	4894.40	
Facility	Training production personnel	11.82	72 hours	850.86	
Facility	Training on quality	13.42	72 hours	966.24	
Facility	Packaging training	15.98	72 hours	1150.38	
Total non-remanufacturing activity cost				**25436.30**	**877.21**
Total crankpin cost per unit					**1254.80**

4. Conclusion

The critical and non-critical variables are identified for inspection whereby MAN engine model had 83% critical and 17% non-critical while Detroit engine model had 67% critical and 33% non-critical. Subsequently, the cost for MAN and Detroit engine model are MYR609.50 and MYR1254.80 respectively by considering the critical variable. This work has proposed a good idea in order to improve the quality inspection and costing system but at the same time reducing a lot of waste. The application of ABC has proved that the costing was totally relying on the number of activities. Meanwhile, the method also very transparent to observe any cost drivers which directly related to the production of remanufactured crankshaft. Therefore, integration of MTS and ABC has a great impact to the development of remanufacturing industries in Malaysia.

Acknowledgement

The authors would like to be obliged to Universiti Malaysia Pahang for providing financial assistance under RDU170387.

References

[1] Ridley SJ and Ijomah WL 2015 A novel pre-processing inspection methodology to enhance productivity in automotive product remanufacture: an industry-based research of 2196 Engines *Journal of Remanufacturing* **5** 1-12.

[2] Chen M 2005 End-of-life vehicle recycling in China: now and the future *The Journal of The Minerals, Metals & Materials Society* **57** 20-6.

[3] Piñeyro P and Viera O 2015 The economic lot-sizing problem with remanufacturing: analysis and an improved algorithm *Journal of Remanufacturing* **5** 1-13.

[4] Zhichao L, Afrinaldi F, Zhang HC and Jiang Q 2016 Exploring optimal timing for remanufacturing based on replacement theory *CIRP Annals - Manufacturing Technology* **65** 447–50.

[5] Matsumoto M, Yang S, Martinsen K and Kainuma Y 2016 Trends and research challenges in remanufacturing *International Journal of Precision Engineering and Manufacturing-Green Technology* **3** 129-42.

[6] Dong JH, Xue BS, Xue N, Wang HP and Li HY 2013 Development of remaining life prediction of crankshaft remanufacturing core **1** 91-6.

[7] Kareem B 2015 Evaluation of failures in mechanical crankshafts of automobile based on expert opinion *Case Studies in Engineering Failure Analysis* **3**:25–33.

[8] Haldar NAH, Khan FA, Ali A and Abbas H 2016 Arrhythmia classification using mahalanobis distance based improved fuzzy c-means clustering for mobile health monitoring systems *Neurocomputing* **220** 221-35.

[9] Pal A and Maiti J 2010 Development of a hybrid methodology for dimensionality reduction in mahalanobis taguchi system using mahalanobis distance and binary particle swarm

optimization *Expert Systems With Applications* **37** 1286–93.
[10] Holcomb S 2016 Mahalanobis Taguchi System (MTS) for pattern recognition, prediction and optimization *International Conference of Modeling and Simuation* **34** 1–8.
[11] Ketkar M and Vaidya OS 2014 Evaluating and ranking candidates for MBA program : mahalanobis Taguchi system approach *Procedia Economics and Finance* **11** 654–64.
[12] Jin X and Chow TWS 2013 Anomaly detection of cooling fan and fault classification of induction motor using mahalanobis-taguchi system *Expert System Applications* **40** 5787-95.
[13] Reséndiz E and Rull-flores CA 2013 mahalanobis-taguchi system applied to variable selection in automotive pedals components using gompertz binary particle swarm optimization *Expert Systems with applications* **40** 2361–65.
[14] Reséndiz E, Moncayo-martínez LA and Solís G 2013 binary ant colony optimization applied to variable screening in the mahalanobis-taguchi system *Expert systems with applications* **40** 634-37.
[15] Tsai WH, Chen HC, Leu JD, Chang YC and Lin TW 2013 A product-mix decision model using green manufacturing technologies under activity-based costing *Journal of Cleaner Production* **57** 178–87.
[16] Tsai WH, Yang CH, Chang JC and Lee HL 2014 An activity-based costing decision model for life cycle assessment in green building projects *European Journal of Operational Research* **238** 607-19.
[17] Gurcanli GE, Bilir S and Sevim M 2015 Activity based risk assessment and safety cost estimation for residential building construction projects *Safety Science* **80** 1–12.
[18] Yang CH, Lee KC and Chen HC. 2016 Incorporating carbon footprint with activity-based costing constraints into sustainable public transport infrastructure project decisions *Journal of Cleaner Production* **133** 1154–66.
[19] Schulze M, Seuring S and Ewering C 2012 Applying activity-based costing in a supply chain environment *Intern. Journal of Production Economics* **135** 716–25.

Formulation of lubricating grease using Beeswax thickener

N Suhaila A Japar, M Aizudin A Aziz and Mohd Najib Razali

Faculty of Chemical and Natural Resources Engineering,
Universiti Malaysia Pahang, Kuantan, Pahang, Malaysia

E-mail: maizudin@ump.edu.my

Abstract. The issues on environmental pollution has brought the industries to seek the alternative green solutions for lubricating grease formulation. The significant challenges in producing modified grease are in which considering the chosen thickener as one of the environmental friendly material. The main purposes of the current research were to formulate lubricant grease using different types of base oils and to study the effect of thickener on the formulated lubricant grease. Used oil and motor oil were used as the base oils for the grease preparation. Beeswax and Damar were used as thickener and additive. The grease is tested based on its consistency, stability and oil bleeding. The prepared greases achieved grease consistency of grade 2 and 3 except for grease with unfiltered used oil. Grease formulated with used oil and synthetic oil tend to harden and loss its lubricating ability under high temperature compared to motor oil' grease. Grease modification using environmental friendly thickener were successfully formulated but it is considered as a low temperature grease as the beeswax have low melting point of 62°C-65°C.

1. Introduction

Most of the lubricants used nowadays lead to environmental pollution [1]. The environmental legislation by government bodies as well as higher consumer awareness are influencing the design process of the lubricating grease to employ more sustainable and environmentally friendly components [2]. Due to this scenario, a new market revolving around eco-friendly products is developing very rapidly, where consumers exhibits higher tendency to use new products and is also willing to pay for green products to conserve the environment from negative effects [3]. Several studies has been done related to the environmental friendly lubricating grease through the utilization of eco-friendly thickener such as cellulose derivatives and beeswax [4,5].

Grease is a semi-fluid product which comprises of a thickening agent along with other elements that provide the grease with special properties dispersed in a lubricating base oil. The common percentage compositions for base oils, thickeners and additives are 70-95%, 3-30% and 0-10% respectively [6]. Many studies has been done related to eco-friendly grease, where most researchers utilized vegetable oil as the base oil [7–9]. In contrast, the amount of studies done on thickeners are scarce. Thickeners are known to play a vital role in grease composition for which it determines the consistency of grease.

Used oils are generally generated during the engines service and maintenance however, it is often contaminated with impurities in the course of usage and handling. In addition, used oils often points out as a serious pollution problems due to irresponsible act of illegal dumping and unmanageable disposal

IOP Conf. Series: Materials Science and Engineering **342** (2018) 012007 doi:10.1088/1757-899X/342/1/012007

resulting in groundwater, surface water and soil contamination[10]. Thus, in this study, used oils were utilized as one of the base oils to prevent the aforementioned problems.

The thickener's main function is to remain in contact with or hold lubrication from moving or leaking out under the force of gravity and often referred to as a sponge that holds the lubricants [11]. Beeswax is a natural substance obtained from the honeycombs of bees after the honey has been removed by draining or centrifuging. It possesses characteristic of insoluble in water, sparingly soluble in alcohol and very soluble in ether. In this study, beeswax was used as the thickeners in grease composition. Damar (*Shorea dipterocarpaceae*) are solid resins, generally hard and durable, were used as grease additive. It provides a unique depth of gloss, superior color and gloss retention, adhesion and flexibility [12], and it has melting point ranging from 85°C-115°C.

Lubricating grease functions in reducing the wear and tear of the surfaces by avoiding direct metal to metal contact between the rubbing surfaces and lowering the expansion of metal due to frictional heat and destruction of material. Motor oils and used oils, beeswax as well as damar were used to produce grease. This research study was carried out to formulate grease using different types of base oils and to study the effect of thickener on the formulated grease.

2. Methodology

2.1. Materials

Different types of base oils were used which are motor oils and used oils. The grade of the motor oils chosen were Toyota Premium Mineral (PM), Toyota Synthetic Formulation (SF) and Perodua Semi Synthetic (SS). Used oil collected from workshop was selected to be another alternative of base oils. Beeswax was chosen to act as a thickener, is a natural substance obtained from the honeycombs of bees after the honey has been removed by the process of draining or centrifuging meanwhile the additive, Damar (*Shorea dipterocarpaceae*) a solid resin is the sap from naturally wounds to the surface of the tree.

2.2. Preparation of grease samples

The compositions to produce 200g of grease samples were fixed with ratio of 14:5:1 of base oil, thickener and additive, respectively. 50g of beeswax was slurred together with half of the amount of base oil (70g) in a 500-mL beaker at continuous stirring rate. The speed of the mixer was adjusted according to the viscosity of the paste and for each of the sample, the mixing was continued until a smooth paste was formed at a constant temperature of 50°C-60°C. The other half amount of base oil and Damar (10g) were added slowly into the paste and the process continuous until all the materials blend together. The samples for each formulated grease were labelled as follows in accordance to its base oils types:

Table 1. Grease samples identification.

Type of base oil	Sample identification
Toyota Premium Mineral	PM
Toyota Synthetic Formulation	SF
Perodua Semi-Synthetic	SS
Used oil (unfiltered)	WO
Used oil (filtered)	FWO

2.3. Consistency test

The tests on each sample were performed using the SKF Grease Test Kit TGKT 1. A fixed grease volume was spread between the two glass plates for 15 seconds by means of the weight. The consistency of the grease strain was observed and evaluated using calibrated measuring scale NLGI grade. This test method was in accordance of ISO 2137 which specifies methods for determining the consistency of lubricating greases when only small samples were available. The results of NLGI for each grease indicates their consistency level [13].

Table 2. NLGI classification system based on consistency.

NLGI Number	Worked Penetration, tenth of millimeter	Consistency
000	445-475	Very soft
00	400-430	
0	355-385	Soft
1	310-340	Soft
2	265-295	Creamy texture (buttery)
3	220-250	Semi-solid
4	175-205	Stiff
5	130-160	Stiff
6	85-115	Hard solid

2.4. *Stability test*

The test was done as complied with the Standard Test Method for Oil Separation from Lubricating Grease during Storage (ASTM-D1742). It is known as oil separation test in which it determines the stability of the grease formulated. All grease samples were stored in enclosed container and the initial level of the grease is observed and recorded. The samples were left for 2 months at room temperature. The amount of oil separated was measured using the measuring cylinder and weighed. Theoretically, the lesser the oil separated, the better the stability or quality of the grease. The sample is considered stable if and only if the amount of the oil separated less than 4% [14].

2.5. *Oil bleeding test*

The tests on each sample were performed using the SKF Grease Test Kit TGKT 1. A fixed amount of grease was put on a piece of blotter paper provided. The samples were heated for two hours at constant temperature of 60°C according to the recommended operating condition of the grease test kit [15]. The base oil released from the grease created a stained on the paper. The bleed area and the percentage difference between bleed area of fresh and used samples were calculated by using equation (1) and (2).

$$S_{...} = 0.785 \times (D^2_{AV...} - 100) \quad (1)$$

$$\%Diff = 100 \times \frac{(S_{Used} - S_{Fresh})}{S_{Fresh}} \quad (2)$$

3. Results and discussion

3.1. Modified grease appearances

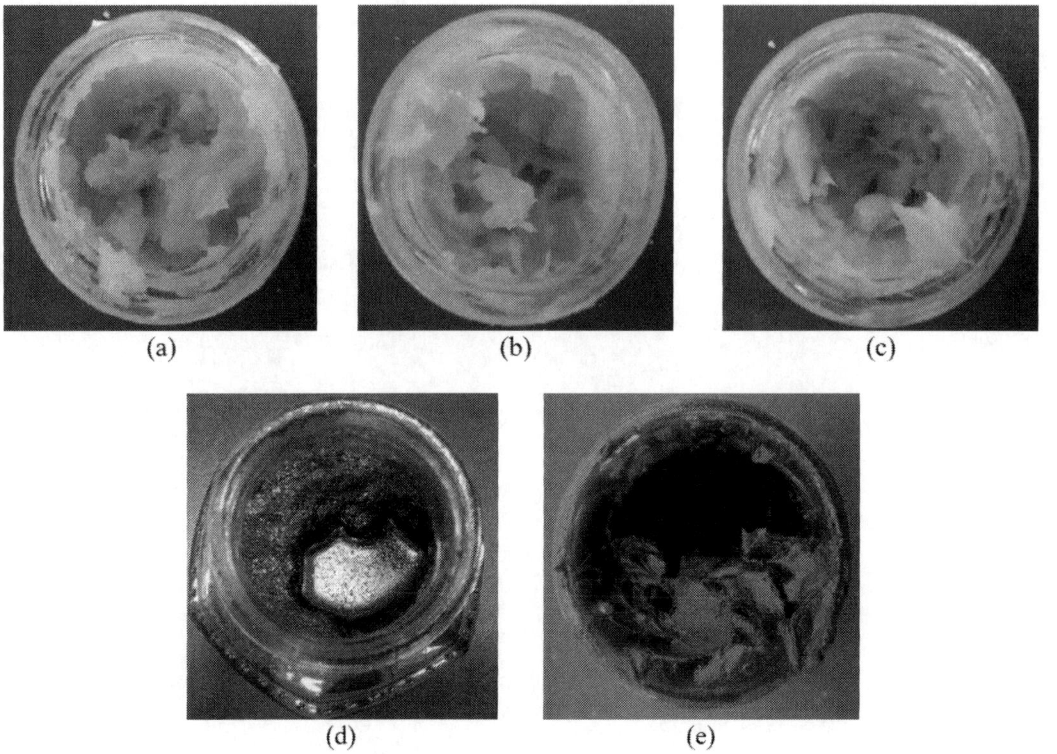

Figure 1. The appearances of modified grease samples. (a) Toyota premium mineral-oil-based grease, (b) Toyota synthetic-oil-based grease, (c) Perodua semi-synthetic-oil-based grease, (d) unfiltered used-oil-based grease, (e) filtered used-oil-based grease.

The modified grease samples prepared were semi-solid and their color depend on the color of base oil which vary from light yellow to black as in Figure 1. The properties of modified greases are as in the following subtopics.

3.2. Formulated lubricant grease consistency

The outcomes from the testing done upon each of the grease samples in accordance to the NLGI consistency grade (Table 2). The grades defined the consistency of the greases and their appearance at 25°C. SF-based grease, SS-based grease, and FWO-based grease were recorded NLGI grade 2 and PM-based grease of grade 3, where grease with these consistencies were most commonly used in general industrial, automotive and others [16]. Grease with NLGI Grade 2 were suitable for rolling bearings moderately loaded with medium speed applications. Grease of this class of consistency are formulated to give a good balance of properties required for easy pumping through dispensing systems [17]. In addition, most of the multipurpose grease also of this grade of NLGI 2-3 consistency.

Table 3. Consistency of prepared modified grease.

Sample	NLGI	Consistency
PM	3	Firm
SF	2	Buttery
SS	2	Buttery
WO	00	Semifluid
FWO	2	Buttery

Comparing the WO-based grease and the other samples, very large difference in consistency observed for WO-based grease of which it is of NLGI grade 00. WO-based grease is softer than the other prepared greases. By comparing WO-based grease and FWO-based grease, FWO-based grease is formulated with filtered used oil. The presence of contaminants in the unfiltered used oil (WO) might be affecting the grease consistency for which most of these contaminants is removed before being used in FWO-based grease formulation [18]. Therefore, it was inferred that the WO-based grease consistency can be increase by decreasing the oil-to-thickener ratio [19].

3.3. *Oil separation during storage*

Greases tends to release their base oils during storage (static bleeding). Static oil bleeding can be affected by storage conditions and is more pronounced if the grease is soft in consistency (NLGI grades 00, 0 and 1) [20]. Oil separation often relates to the base oil and thickener combination with kinematic viscosity of the base oil [21]. All grease samples were stored at room temperature for 2 months and were observed for any changes of the behavior and properties. Table 4 shows oil separation only occurred to WO-based grease.

Table 4. Grease oil separation upon storage at 25°C.

Sample	Oil separated (g)	%loss	Description
PM	-	-	Stable
SF	-	-	Stable
SS	-	-	Stable
WO	16.20	8.10	Unstable
FWO	-	-	Stable

WO-based grease was found to be unstable and recorded excessive oil separation which is more than 4%. Contaminants present in the unfiltered used oil can draw out the base oils from the thickener system over time [20]. However, the oil puddle on top of the grease does not result in the grease being unsuitable for use. It can be either removed by decanting the free oil from the surface or by manually stirring it back into the grease.

Prepared grease other than WO-based grease shown to have very good oil separation resistance. From the appearance of the other grease samples, the base oil and the thickening agent chosen seems to have a strong physically-acting bond forces that build a stable three-dimensional network structure [22]. Interestingly, filtered used oil-base grease (FWO) has good oil separation comparable to the fresh motor oils-base grease. This is might be due to the pre-treatment of used oil that was done such as settling, filtration and dehydration for which most of the contaminants in the used oil were being removed.

3.4. *Oil bleeding of the formulated grease*

This test method is not suitable for greases softer than NLGI grade 1 consistency due to the tendency for softer greases to sweep away while the test is performed. Therefore, the testing was not applicable on WO-based grease of grade 00. Comparison of grease bled between fresh samples and used samples grease were made. bled area difference (%Diff) between the used and the fresh sample were calculated to indicated either the lubricating ability of the grease. Positive values mean the grease lubricating ability is increased or vice versa.

Based on the results in table 5, PM-based grease and SS-based grease shows the positive value in the percentage of area different which indicates that the grease lubricating ability can be maintained at high temperature. Other greases indicated reduction on their lubricating ability. SF-based grease and FWO-based grease losses its lubricating ability when subjected to high temperature for which it causes these greases harden as time goes. The rate of oil released from the grease will increase with time and vary based on the temperature at which it is stored [23]. It was observed that the formulated greases have low working temperature range which was found to have similarity with Kreivaitis' [5] study, as he stipulated that the dropping-point temperature of the beeswax-thickened greases were within the range of 52°C to 66°C. However, the formulated grease high temperature tolerance can be improved by adding proper additive to increase its dropping point temperature properties.

Table 5. Oil bleeding test at 60°C.

Sample	$D_{AV\ fresh}$ (mm)	$D_{AV\ used}$ (mm)	S_{Fresh} (mm^2/s)	S_{Used} (mm^2/s)	%Diff
PM	34.0	36.5	828.96	967.31	16.68
SF	45.0	43.0	1511.13	1372.97	-9.14
SS	40.0	42.0	1177.50	1306.24	10.93
WO	N/A	N/A	N/A	N/A	N/A
FWO	35.5	34.0	910.79	828.96	-8.98

Table 6 shows the data analyzed from the tests carried out on each of the grease samples. The tests carried out includes the consistency test, stability test and oil bleeding test. Consistency test were done to determine the NLGI consistency grade for grease samples. In stability test, any changes in grease samples were analyzed after 2 months of storage at room temperature. Oil bleeding tests were carried out for two hours at constant temperature of 60°C where the base oil released from the grease of fresh and used samples created oil stained on the paper. Both the diameter of the stained and the bleeding properties were measured.

Table 6. Overall results of formulated lubricant grease analysis.

Sample	Color	Appearance	NLGI	Stability @ 25°C	%Diff oil bleed @ 25°C
PM	Light yellow	Firm	3	Stable	16.68
SF	Light yellow	Buttery	2	Stable	-9.14
SS	Light yellow	Buttery	2	Stable	10.93
WO	Black	Semifluid	00	Unstable with 8% oil separated	Not applicable
FWO	Dark brown	Buttery	2	Stable	-8.98

4. Conclusion

Beeswax-thickened greases are potentially an environmental friendly type of grease. Based on the result obtained, most of the formulated grease were successfully formulated by utilizing beeswax as thickener with NLGI consistency range of grade 2 – 3. However, these greases are not meant to be subjected to high temperature as it tends to lose it lubricating ability and harden as it exposed to high temperature for a period of time.

IOP Conf. Series: Materials Science and Engineering **342** (2018) 012007 doi:10.1088/1757-899X/342/1/012007

References

[1] Diphare M J, Pilusa J, Muzenda E and Mollagee M 2013 A Review of Waste Lubricating Grease Management *2nd Int. Conf. Environ. Agric. Food Sci. August 25-26, 2013 Kuala Lumpur*
[2] Kajdas C 2000 Major Pathways for Used Oil Disposal and Recycling *Tribotest J.* **7**
[3] Ong J W, Goh G G G, Goh C Y, Too M S M and Goh L P 2015 Green Product Pricing and Malaysian Consumers' Willingness To Pay *Kaji. Malaysia* **33** 191–210
[4] Núñez N, Martín-Alfonso J E, Valencia C, Sánchez M C and Franco J M 2012 Rheology of new green lubricating grease formulations containing cellulose pulp and its methylated derivative as thickener agents *Ind. Crops Prod.* **37** 500–7
[5] Kreivaitis R, Padgurskas J and Gumbyt M 2014 An assessment of beeswax as a thickener for environmentally friendly lubricating grease production *Lubr. Sci.*
[6] Wright J 2008 Grease Basics *Noria*
[7] Abdulbari H A, Rosli M Y, Abdurrahman H N and Nizam M K 2011 Lubricating grease from spent bleaching earth and waste cooking oil : Tribology properties *Int. J. Phys. Sci.* **6** 4695–9
[8] El-Adly R A, Bedier A H, Hussein M F, Ismail E A and El-emary M M 2014 Jojoba and Castor Oils as Fluids for the preparation of bio greases: A Comparative Study *Int. J. Sci. Eng. Res.* **5** 755–62
[9] Nagendramma P and Kumar P 2015 Eco-Friendly Multipurpose Lubricating Greases from Vegetable Residual Oils *Lubricants* **3** 628–36
[10] Srivastava S P 2014 Rerefining and recycling of used lubricating oil *Development in Lubricant Technology* (Hoboken: John Wiley & Sons, Inc.) pp 299–307
[11] Reid-Peters S 2011 Synthetic Basestocks - Synergy with Greases *Lube Eur. Lubr. Ind. Mag.* 21–4
[12] Coppen J J W 1995 *Gums, resins and latexes of plant origin* (Rome: Food and Agriculture Organization of the United Nations)
[13] ASTM D217-17 2017 *Standard Test Methods for Cone Penetration of Lubricating Grease* vol 5(West Conshohocken, PA: ASTM International)
[14] Lugt P M 2013 *Grease Lubrication in Rolling Bearings* (Chichester: John Wiley & Sons, Inc.)
[15] SKF 2009 SKF Grease Test Kit TKGT 1: Instruction Manual 1–40
[16] Rizvi S Q A 2008 *A Comprehensive Review of Lubricant Chemistry, Technology, Selection, and Design*
[17] Neale M J 1995 *The Tribology Handbook* (Oxford: Butterworth-Heinemann)
[18] Herguth B 2002 Grease Analysis - Monitoring Grease Serviceability and Bearing Condition *Mach. Lubr.*
[19] Doyle D 2015 General Grease Overview and Bearing Lubrication *ALS Tribol.*
[20] Ludwig L G 2012 Storing Grease to Avoid Bleed and Separation *Mach. Lubr.*
[21] Rezasoltani A and Khonsari M M 2016 On Monitoring Physical and Chemical Degradation and Life Estimation Models for Lubricating Greases *Lubricants* **4** 24
[22] Franco J M 2015 Influence of Base Oil Polarity on the Transient Shear Flow of Biodegradable Lubricating Greases *Lubricants* **3** 611–27
[23] Sander J and Mcdaniel E 2007 Grease Characterization : Are All Greases Lithium Greases ? *Lubrication and Lubricants* pp 431–7

IOP Conf. Series: Materials Science and Engineering **342** (2018) 012008 doi:10.1088/1757-899X/342/1/012008

Application of natural seaweed modified mortar for sustainable concrete production

M N I Siddique and A W Zularisam

Faculty of Engineering Technology, University Malaysia Pahang (UMP), Lebuhraya Tun Razak, 26300 Gambang, Kuantan, Pahang, Malaysia.

E-mail: mdnurul@ump.edu.my

Abstract. The effect of seaweed such as Eucheuma Cottonii (gel) and Gracilaria Sp. modified mortar on the properties of sustainable concrete was investigated. Pre-experiment and main-experiment was conducted to carry out this study. Pre-experiment was conducted to study the compressive strength of the sustainable concrete. The main-experiment studied the compressive and splitting strength. Results showed that seaweed modified mortar yielded satisfactory compressive and splitting strength of 30 MPa and 5 MPa at 28 days.

1. Introduction

Sustainable concrete plays significant role in ecological development [1]. Global sustainability is driving a number of studies for achieving green construction [2]. As concrete is becoming the standard construction material currently, green concrete is the requirement for the sustainable development [3]. Therefore, a range of studies has been performed on manufacturing green concrete [4]. The outcomes of these studies made sustainable concrete as 'green 'concrete and less carbon emission material [5]. Currently, the world is facing two serious problems such as lack of sustainability and emission control [6].

Green concrete requires durability and longer life time [7]. Polymer adhesives can improve the mechanical properties of concrete [8]. However, Polymers can be made more durable through polymerization of monomers [9]. Concrete composed of polymers is known as polymer concrete [10]. If polymer is mixed together with cement and aggregates, then the mixture is called polymer modified concrete. Polymer modified sustainable concrete has great binding strength [11]. Hence, it can be used for maintenance work [12].

Currently, rubber is being added to the polymer modification process to produce high quality synthetic resin which in turn can produce synthetic resin modified concrete. [13]. Although the use of organic polymer has already been studied in the development of polymer modification process the use of carbohydrate polymer hasn't been reported till to date [14]. The present study aims to explore the use of seaweed in the polymer modification process. Seaweed such as Eucheuma Cottonii (gel) and Gracilaria Sp. (powder) were introduced in the present work as these species comprise agarans and carrageenans. Eucheuma Cottonii is classified as polysaccharides which comprises kappa carrageenan. This species can be used in emulsification, condensation, and stabilization purpose [1]. Gracilaria Sp. is classified as polysaccharide that comprises agarose and can be used to prepare strong gel [15]. In addition, these two species contain rheological properties as thickening agents that may be used as epoxy resin in polymer modification process [16]. After considering all these benefits, seaweed modified mortars are likely to attain outstanding bonding and durability to attain sustainability. Therefore, the present work examined the compressive and splitting tensile strength of seaweed modified concrete.

2. Materials and methods

Materials utilized in the present study were mortar and seaweed. Two specific species of seaweeds were employed in this study such as Eucheuma Cottonii (gel) and Gracilaria Sp. (powder). The Eucheuma Cottonii was collected from raw heated seaweed and Gracilaria Sp. was collected from the powder.

In the present work, mortar mix composition of cement: sand: water was 1: 1: 0.5. However, the proportion of seaweed in the mix design was listed in Table 1.

Table 1. Proportion of gel and powder seaweed in mortar.

Experiments	% seaweed gel of the cement weight	% seaweed powder of the cement weight
Pre-experiment	0.1; 0.6;1.1;6	0.1; 0.6;1.1,6
Main-experiment	-	0.1; 0.3; 0.6;1.1; 2.1, 5.1

2.1. *Pre-experiment*

This study was carried out in two parts, pre-experiment and main experiment. Pre-experiment was performed to study the compressive strength of seaweed modified mortar at the end of 7 and 14 days curing. The proportions of seaweed used in the mix design are shown in Table 1. The outcomes of pre-experiment shall define which seaweed is going to be employed in the subsequent experiment.

2.2. *Main-experiment*

After the completion of pre-experiment, the main-experiment was conducted. The main-experiment examined compressive and splitting tensile strength of seaweed modified mortar. The proportions of seaweed used in the main-experiment are shown in Table 1. Each of the compressive strength of the samples was measured after 7, 14, and 28 days while splitting tensile strength was determined after 28 days. The results of seaweed modified mortars shall be weighted to control specimens.

2.3. *Specimen test*

Samples were made according to ASTM C-305 [17]. Cube-samples measuring 50 mm x 50 mm x 50 mm were used in the pre-experiment. Cube-samples measuring 50 mm x 50 mm x 50 mm and cylinder-samples measuring 150 mm x 300 mm were used in the main-experiment. Five samples were prepared for each mix design. Compressive strength of the samples was measured by ASTM C-39. However, splitting tensile strength was determined by ASTM C-496.

3. Results and discussions

In the pre-experiment, seaweed gel modified concrete attained high compressive strength after 7 days of curing. The highest compressive strength of 30 MPa was achieved by CS-07-0.6 gel (Figure 1). It was observed that after 14 days of curing compressive strength of CS-14-1.1 gel (22 MPa) was reduced compared to CS-14-1.1 powder (25 MPa) (Figure 2). However, compressive strength of CS - 07 and 14 powder showed stable performance.

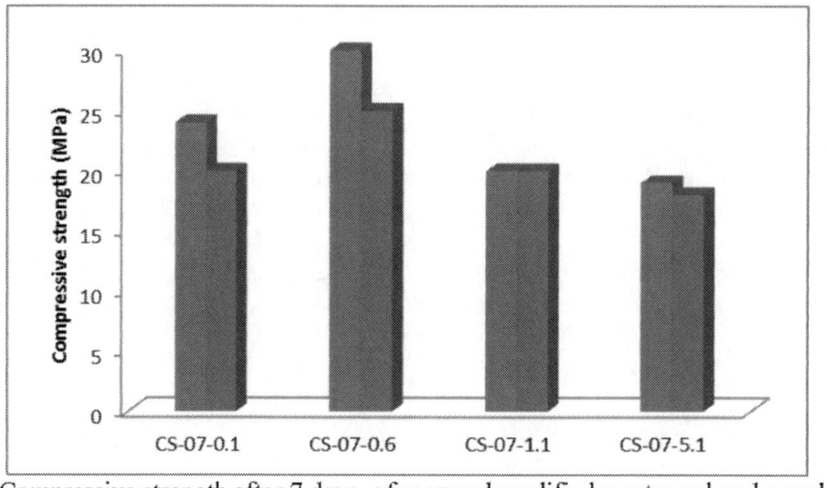

Figure 1. Compressive strength after 7 days, of seaweed modified mortar gel and powder, where (blue) gel and (red) powder.

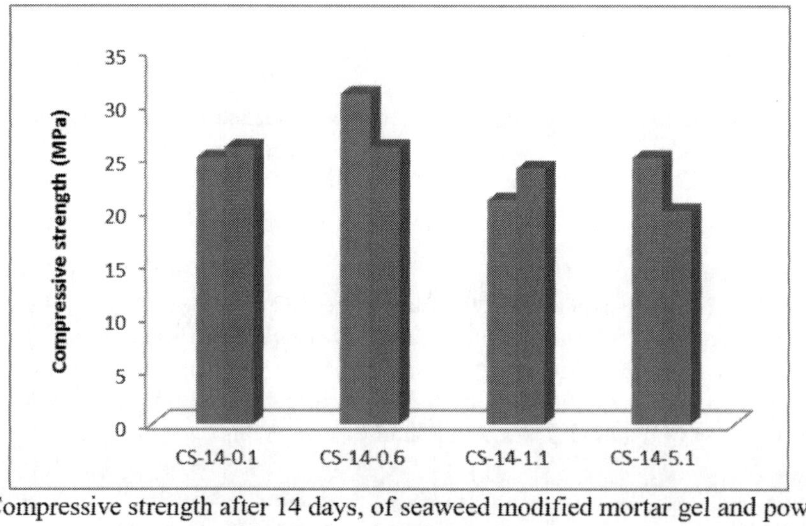

Figure 2. Compressive strength after 14 days, of seaweed modified mortar gel and powder, where (blue) gel and (red) powder.

Seaweed gel and seaweed powder (Gracilaria Sp.) modified mortar have properties of gelling and thickening. Therefore, Gracilaria Sp. formed stronger gel compared to that of Seaweed gel. For this reason Gracilaria Sp. Gel modified mortar increased the bonding strength of the specimen. The outcomes of the pre-experiment showed that the seaweed powder of Gracilaria Sp. becomes the suitable natural polymer of the main-experiment.

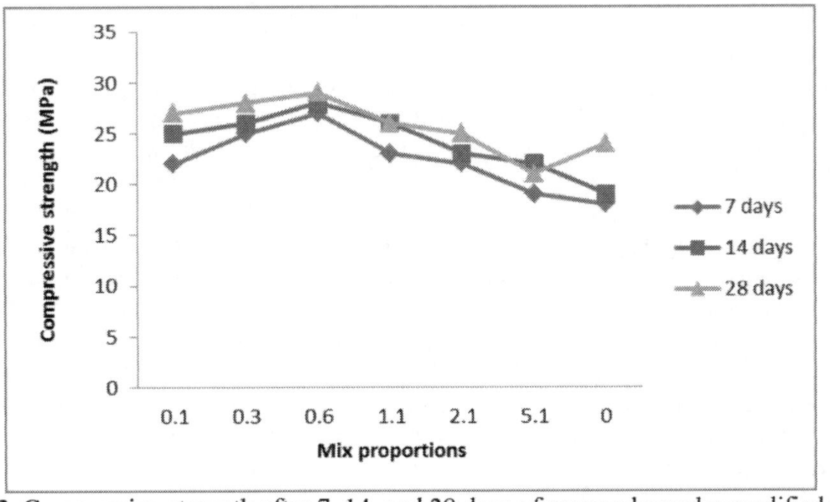

Figure 3. Compressive strength after 7, 14, and 28 days of seaweed powder modified mortar.

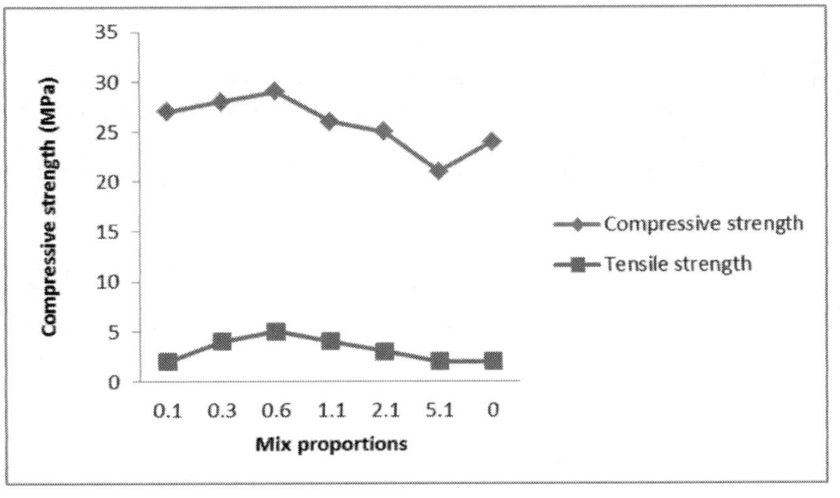

Figure 4. Compressive strength and splitting tensile strength of seaweed powder modified mortar.

Main-experiment examined both compressive and splitting strength of the different mix proportions of seaweed modified mortar and control samples. It has been observed that mix design of CS-0.6 has the maximum compressive strength of 26 MPa after 7 days; 27 MPa after 14 days; and 28 MPa after 28 days (Fig 3). It can be seen that compressive strength of CS-0.1, CS-0.3, and CS-0.6 was steadily increased after 7 days, 14 days and 28 days. On the other hand, compressive strength of CS-1.1, CS-2.1, and CS-5.1 was slowly declined after 7 days, 14 days and 28 days. Control samples have attained lower compressive strength (17 MPa after 7 days; 18 MPa after 14 days; and 24 MPa after 28 days) than other mix designs, except of CS-6 after 28 days. From the outcomes of the main-experiment it was observed that higher proportion of mortar sample (CS-1, CS-2.1, and CS-5.1) did not increased the bonding strength. Therefore, the optimum mix proportion was selected CS-0.6.

As the mortar produced outstanding compressive strength, the addition of seaweed powder was likely to improve its tensile strength. It was observed from Fig. 4 that the desired splitting tensile strength was attained by mix proportion of CS-0.6. It was observed that CS-0.6 achieved the maximum splitting tensile strength of 5 MPa. Moreover, control samples attained 2 MPa of splitting tensile strength. It could be due to the formation of strong gel and improved the bonding strength which made the performance seaweed powder modified mortar outstanding [1].

Seaweed modified concrete possess outstanding binding and adhesion strength with the aggregates. For this reason, it can be used for maintenance works. As added to the cement, seaweed fills the porous portions and improves its performance [2]. Gracilaria Sp. and Seaweed gel contain identical features for binding and condensing. Due to the agarans and carrageenans of these two species they can be considered handy for cement hydration phase. Gracilaria Sp. contains lesser shrinkage but more ductility than Seaweed gel [1]. It may be due to the outcomes of pre-experiment achieved improved compressive strength with seaweed powder than seaweed gel. It is noticeable that raw Seaweed gel was prior to be used with the mortar and achieved cement hydration. Gracilaria Sp powder was active in binding and condensing. As Gracilaria Sp. contains minor shrinkage and abundant ductility, it can enhance bonding capacity. Therefore, it is clear that main-experiment outcomes showed improved compressive and tensile strength for CS-0.6. Results revealed that high proportion of seaweed (CS-1.1, CS-2.1, and CS-5.1) did not increase compressive strength of the mortar. It can be observed that lower proportion of seaweed (CS-0.1, CS-0.3, and CS-0.6) increased compressive strength of the mortar. As Gracilaria Sp. contains lower shrinkage and abundant ductility, the splitting tensile strength of seaweed modified mortar was improved. In addition, the mix proportion of CS-0.6 was the optimum.

4. Conclusions

Seaweed gel and Gracilaria Sp. (powder) may be applied as natural polymer. These two species showed great performance in binding and condensing and can improve bonding strength. Results showed that Gracilaria Sp. powder modified mortar achieved improved compressive and splitting tensile strength than control samples. CS-0.6 was the optimal mix proportion for better mortar performance. The commercial application of this technology is expected to be encouraging for the production of sustainable concrete.

Acknowledgements
The authors are grateful to the Faculty of Engineering Technology at the University of Malaysia, Pahang for providing uninterrupted entrée to their lab services. This study was made possible by the RDU-160315 grant.

References

[1] Liu B and Liang D 2017 Effect of mass ratio of asphalt to cement on the properties of cement modified asphalt emalsion mortar *Constr. Build. Mater.* **134** 39-43
[2] Siddique M N I, Munaim M S A M and Zularisam A W 2014 Mesophilic and thermophilic biomethane production by co-digesting pretreated petrochemical wastewater with beef and dairy cattle manure *J. Ind. Eng. Chem.* **20** 331-337
[3] Siddique M N I, Munaim M S A M and Zularisam A W 2015 Influence of flow rate variation on bio-energy generation during anaerobic co-digestion *J. Ind. Eng. Chem.* **27** 44-49
[4] Siddique M N I, Munaim M S A M and Zularisam A W 2015 Feasibility analysis of anaerobic co-digestion of activated manure and petrochemical wastewater in Kuantan (Malaysia) *J. Clean. Prod.* **106** 380-388
[5] Siddique M N I, Munaim M S A M and Zularisam A W 2017 The combined effect of ultrasonic and microwave pre-treatment on bio-methane generation from co-digestion of petrochemical wastewater *J. Clean. Prod.* **145** 303-309
[6] Siddique M N I, Munaim M S A M and Zularisam A W 2015 Role of biogas recirculation in enhancing petrochemical wastewater treatment efficiency of continuóus stirred tank reactor *J. Clean. Prod.* **91** 229-234
[7] Siddique M N I, Munaim M S A M and Zularisam A W 2013 Cycle time minimization during renewable methane production from H2O2 enhanced anaerobic Co-digestion of

Petrochemical wastewater in CSTR *Asian. J. Chem.* **25** 1140-1148

[8] Siddique M N I, Munaim M S A M and Zularisam A W 2013 Role of H_2O_2 enhanced anaerobic co-digestion of Petrochemical wastewater on cycle time minimization during biomethanation *Sci. Res.Essay.* **8**(22) 996-1009

[9] Siddique M N I and Zularisam A W 2012 Continuous stirred tank reactor application for renewable methane from anaerobic biodegradation of Petrochemical wastewater *Eng. E-Transac.* 7 23-27

[10] Singh L, Siddique M N I and Zularisam A W 2012 Optimization of sustained hydrogen production from palm oil mill effluent using immobilized mixed culture *Int. J. Civ. Eng. Geo-Environ.* **3** 71-76

[11] Siddique M N I and Zularisam A W 2012 Sustainable bio-methane generation from petrochemical wastewater using CSTR *Int. J. Eng. Tech. Sci* **1** 1-5.

[12] Syukor A R, Zularisam A W, Said M I M and Siddique M N I. 2015 Bio-sorption of Chromium using Aquatic Plant, Lepironia Articulata *Adv. Environ. Biol.* **1** 13

[13] Siddique M N I and Zularisam A W 2014 Enhanced production of renewable methane from petrochemical wastewater using CSTR *Proc. Int. Conf. on advances in applied science and environmental engineering ASEE-2014* 02-03August 2014 Kuala Lumpur Malaysia

[14] Siddique M N I and Zularisam A W 2014 Intensified CSTR for bio-methane generation from petrochemical wastewater *Proc. Int. Conf. on chemical, environmental and biological sciences CEBS-2014* 17-18, Sep 2014 Kuala Lumpur Malaysia.

[15] Siddique M N I and Zularisam A W 2016 High rate bio-methane generation from Petrochemical wastewater using improved CSTR *Proc. Int. Conf. on Water Resources and Renewable Energy Development* Jan 25-26 2016 Istanbul Turkey.

[16] Siddique M N I, Munaim M S A M and Zularisam A W 2013 Bioenergy production from co-digestion of chemically-treated petrochemical wastewater and activated manure sludge *Ener. Edu. Sci. Technol.* **31**(1) 80-90

[17] American Society for testing and Materials, Standard practice for Mechanical Mixing of hydraulic cement pastes and mortars of plastic consistency.ASTM international west Conshohocken, Pa.,(ASTM C 305-306).

Development of Sustainability Assessment Tool for Malaysian hydropower industry: A case study

Faiz Mohd Turan, Kartina Johan and Muhammad Irfan Abu Sofian

Faculty of Manufacturing Engineering, Universiti Malaysia Pahang, 26600 Pekan, Pahang, Malaysia

E-mail: faizmt@ump.edu.my

Abstract. This research deals with the development of sustainability assessment tools as a medium to assess the performance of a hydropower project compliances towards sustainability practice. Since the increasing needs of implementing sustainability practice, developed countries are utilizing sustainability tools to achieve sustainable development goals. Its inception within ASEAN countries including Malaysia is still low. The problem with most tools developed from other countries is that it is not very comprehensive as well as its implementation factors are not suitable for the local environment that is not quantified. Hence, there is a need to develop a suitable sustainable assessment tool for the Malaysian hydropower industry to comply with the sustainable development goals as a bridging gap between the governor and the practitioner. The steps of achieving this goal is separated into several parts. The first part is to identify sustainable parameters from established tools as a model for comparison to enhance new parameters. The second stage is to convert equivalent quantification value from the model to the new developed tools. The last stage is to develop software program as a mean of gaining energy company feedback with systematic sustainable reporting from the surveyor so as to be able to integrate sustainability assessment, monitoring and reporting for self-improved reporting.

1. Introduction

Sustainability can be stated as new implementation that can cater the needs of current situation without sacrificing future needs of next generation [1]. Sustainable development has become the main concern among energy industrial player [2]. During the past, quality of services has become the main interest in expending business venture. The last decade seen major changes in a company policy when producing business services [3]. The introduction of Sustainable Development Goals (SDGs) by the United Nations (UN) act as the main factor of improving the sustainability development [4-5]. The Bursa Malaysia as the authority that governs Malaysian-based public listed companies (PLCs) wants to emphasize the corporate transparency on environmental, society and corporate governance related issues. They introduced requirements listing for a new sustainability framework to the amended Sustainable Reporting Guide and Toolkit. Since its deception in 2015, PLCs are required to enhance corporate social responsibility (CSR) on sustainability reporting. The new implementation plays a major role towards more sustainable business practices and increasing competitiveness in the market [6-7].

The sustainability assessment (SA) tools can be recognized as a significant tool to assist towards a sustainability reporting production in addition to aid transformation towards sustainability [8-9]. It quantifies parameters that previously only qualifying the effects towards sustainability. SA tools assist decision-makers to decide the very best option they have to create a more sustainable society. The SA have a definite target towards sustainable development through plan, system or activity. The SA thinking has been projected from several initially assessment method focusing on environmental. The Environmental Impact Assessment (EIA) as the first generation address the increasing pressure on human environment, following the economic and social transformation and resulted in public environmental concerns [1]. Strategic Environmental Assessment (SEA) as the later generation for assessment of Policies, Plans and Programs (PPPs). These assessments lead to extension of environmental assessment through SA. Therefore, SA is generally considered with same definition of EIA-driven that specify a model that is integrating social, economic and environmental impacts. Thus, SA is a crucial aspect to secure a long-term value creation for company and society. There are several green environment practices assessment tools which have matured remarkably since the introduction of the Green Project Management (GPM) P5 method [10].

The above examples indicate that there are many assessment tools focusing on triple bottom line: people, planet, profit, product sustainability perspective and process sustainability perspective. The focus of current practices is turned to developing assessment frameworks and tools for sustainability practices. This can be regarded as another step towards fulfilment Bursa Malaysia requirement of pursuing sustainable development at the local level.

There is a scant amount of research evaluating their sustainability performance and effectivity. The current sustainable assessment approach includes 3P (People, Planet, Profit) elements but they are individually assessed. It caused many individual reports to be produced. In terms of sustainability impact, it causes difficulty of criteria measurement, hard to control and thus become impossible to self-improve as the weak area is hard to be determined [11]. Most assessment focuses on Life-Cycle Assessment (LCA) method. The assessment are only concerns with environment, social and financial [12-13]. There is a need to unify the sustainability criterions for preparation of sustainability reporting. The most significant review regarding sustainability practices is that researches had been conducted thoroughly in western countries while it is more or less new among developing countries including Malaysia.

There are problems associated with the selection of criteria and transferability of SA tools to other contexts [14]. These studies are mainly focused on one of the existing tools, and are primarily aiming at providing a general introduction of the tools. They could also be mentioned with associated problems concerning weighting, criteria selection, and lack of a systems approach. However, there is still a lack of in-depth critical evaluation of the SA tools. Thus, the flaw in the reporting tool should be reassessed by the introduction Systematic Sustainability Assessment (SSA) providing guidelines to the industry to measure their level of sustainability compliance [15]. The resulting assessment tool will help the Malaysian hydropower industry to comply on sustainability reporting based reassessed weak criterions area.

This general objective of the project is to encourage sustainability perception in the Malaysian hydropower industry. It measures the current practices of sustainability compliances based on their feedback result.

i. To investigate the level of sustainability compliance using GPM P5 standard.
ii. To quantify sustainability parameters related to hydropower industry in Malaysia.
iii. To develop software program that are able to compute the level of sustainability compliances.

This project begins with an introduction about the background of assessment practices, the significance of conducting assessment in the Malaysian hydropower industry, previous researches on SA tools, and aims of the study. An overview of proposed SSA tools, and selected tools for further analyses are briefly introduced. The criteria used for analyses and framework designed for this research will be presented. In addition, tools are analysed against the framework designed. Each sub-section of this part of the project deals with one of the five criteria identified in methodology. In each sub-section, first the justification for choosing the criterion, its importance, and its optimal state are described. At the end of the project, the findings of this study are discussed and makes some suggestions for consideration in the future refinements.

2. Methodology

Generally, Table 1 shows the relationship of the research methodology that need to be carried out to fulfil all the objectives of this study. The selection of the method used for each objective as shown in Table 1 should meet the requirement of the study in order to gain as much as information to complete the project and also to avoid error occurs in the future. Each method is arranged systematically and in detail according to what should be done first and then followed by the next steps which require the information from the previous method. This means, each method is interrelated and interdependent with each other.

Table 1. Relation between objectives and methodology.

Objectives	Methodology
To investigate the level of sustainability compliance using GPM P5 standard.	1. Define phase - Observation - Problem statement - Survey 2. Measure phase - Sample inspection - Experimental study
To quantify sustainability parameters related to hydropower industry in Malaysia.	3. Analyse phase - Statistical analysis 4. Verification phase - Sustainability performance evaluation
To develop software program that are able to compute the level of sustainability compliances	5. Develop desktop program - Visual studio community2015 6. Validation - Evaluation of program model - Sustainable measurement index

2.1. Data Collection

Data collection is of utmost important to test the developed program in the project to analyse its capability before making improvements from the problems that occurs in the Malaysian industry. The next step of data analysis is the quantification using "Weighing scale" and "Criteria based" [16]. Results from the calculated technique is to be interpreted from data tabulation on the line graph.

The data gathering process was conducted from several departments of the tested energy company. Data input using the developed program was distributed to the listed departments. Data was then being inserted in the scoring input window. The chosen correspondence came from several departments that is related to the Malaysian hydropower industry.

The "Weighting scale" is used during the distribution of the research questionnaire where the 6 value alluding to the highly positive impact meanwhile the 0 rating scale will show the highly negative impact of the sustainability compliance level in those companies. The departments involved in this field of study are:

i. Department of Compliance and Enforcement
ii. Department of Project Construction
iii. Department of Technical
iv. Department of Environmental Quality
v. Department of Quality

The P5 concept integration matrix describes in the following paragraph:

i. Product impacts – objectives and efforts, lifespan and servicing

ii. Process impacts – maturity and efficiency
iii. Society (People) – labour practices and decent work, society and customers, human rights, ethical behaviour
iv. Environment (Planet) – transport, energy, water, waste
v. Financial (Profit) – return on investment, business agility, economic simulation

3. Case Study Results

3.1. Overall Sustainability Score

The main point of results at this part is where the research feedback at the energy company is being mapped on the scoring board which has been injected by the GPM P5 Standard as the guideline. It can be seen on the scoring board in Figure 1 where it comprises of the five sustainability elements which are correlated to each other. Apart from that, the relation of the environmental, social and financial elements with the product and process part in the waste management system are being illustrated in the pie chart at each of the departments.

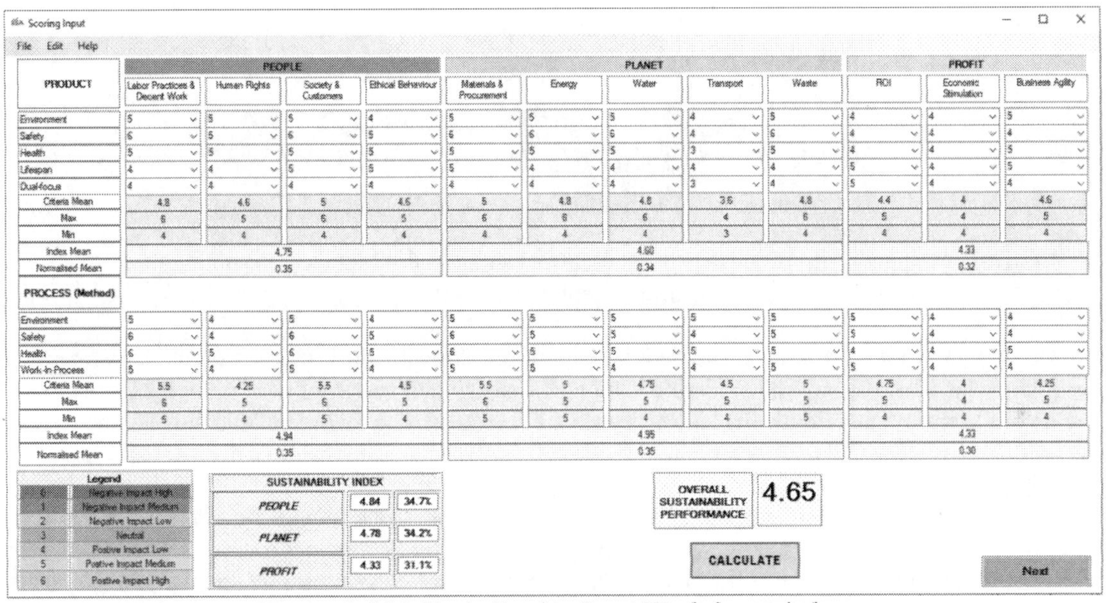

Figure 1. SSA Tools Scoring Input Worksheet window.

Compliance and Enforcement department and the department of quality shows the result of the sustainability assessment where it is still at the partially complied level. Here, the overall percentage result of the research survey of the both departments marks on the 53.67% only which are not quite a good compliance level of sustainability appraisal. In addition, the example of division percentage of the three elements that parallel to the other two parts of Product and Process for Compliance and Enforcement department are also being illustrated in the pie chart as Figure 2 and Figure 3.

Eventually, for the department of Quality, the percentage outcomes from the pie chart (refer Appendix) of the 3P's factor (People, Planet and Profit) when interconnected to Process aspect are 32%, 35%, and 33% subsequently. Aside from that, the relation of the 3P's factor (People, Planet and Profit) with the Product itself comprises of 34%, 33% and 33% respectively.

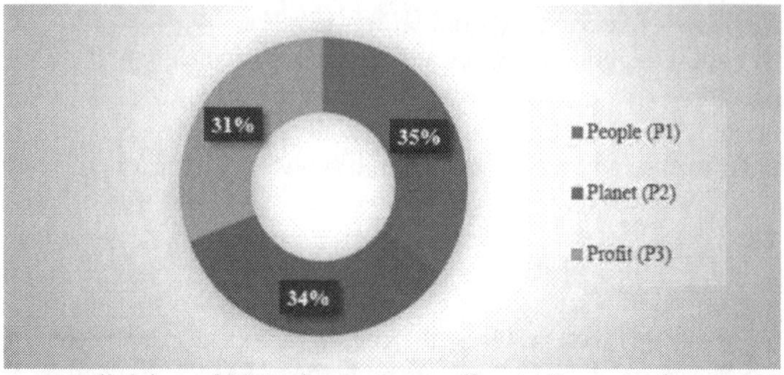

Figure 2. Percentage division of 3P's element according to Process element (Compliance and Enforcement department).

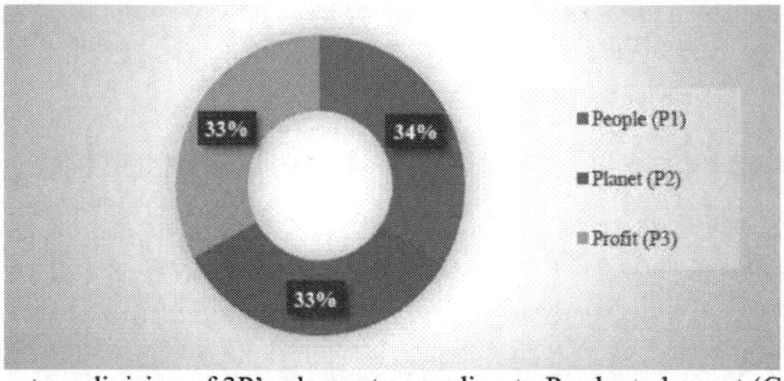

Figure 3. Percentage division of 3P's element according to Product element (Compliance and Enforcement department).

Moving on to the next point is the departments of the Project Construction and the department of Technical which have shown the moderate level of the sustainability compliance in their assembly line of the production. Their implementation degree of the "green" assessment that covers those five elements is only at 53.33%. This situation displayed us that the assessment performance is does not thoroughly practiced.

For the Technical department, the results exhibit the society, nature and economical aspect to be 34%, 35% and 31% correspondingly when connected to the Process component. In addition, the yielding results of those elements when being mapped into the pie chart diagram shows the percentage of the society to be 37%, nature to be 30% and economy to be 30% when interconnected to the Product factor. The outcome from the pie chart that embraces the relation of all sustainability constituents shows that the relation of Process part to the 3P's components which are People, Planet and Profit lies on the 35%, 34% and 31% respectively. Meanwhile, the calculated percentage that associated to the Product part along with the People, Planet and Profit are 35%, 34% and 31% respectively.

Consequently, the highest level of the overall project sustainability score is only 53.8% which also resembles the moderate level of the sustainability assessment only. The outcome from the figure of the percentage division for human, earth and the financial earning related with the Process constituent's marks on 35%, 35% and 30% correspondingly while the three elements when being interconnected with the Product will show the reading of 35% for human, 33% for our earth nature and 32% for the financial side.

3.2. Criteria Based Analysis. Here, at this section the data discussion will continue to be emphasized on the criterion based towards the sustainability compliance in the environmental management system for Malaysian context industry. By referring back to the formulated data that has been normalised earlier, the gap difference of the human's perception can be integrated to the sustainability assessment level of compliance by finding the value of the standard deviation from the raw data collected.

The data result has been sorted according to the each of the criteria that will interconnected to the Product and Process elements. It will demonstrate on how the data being normalised to get the spreading tendency of the data. This outcome is the utmost important peak point that will help to understand more about the relation of the sustainability assessment in the environmental management system.

3.2.1. Process. The value of the data that covers from the five departments that have been mentioned earlier are being sorted according to the twelve criterions. The maximum and minimum value of the data can be determined directly from the categorised element while the average number at each of the criteria are being computed. These calculated values indicate that the lower the value of the difference between the maximum and minimum value, the better the result is and vice versa.

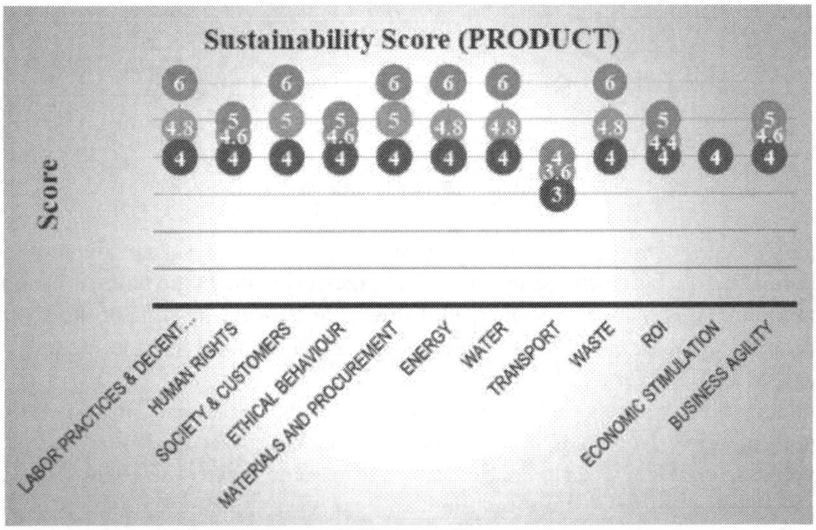

Figure 4. Process line graph.

Referring to the graph in Figure 4, it shows that the gap difference of the maximum rating value of the Society and Customers that is from the People group is on the uppermost rank amongst all of the criterion which marks on 1.0 value. It shows that, this part of element is essentially needed the enhancement of improving the understanding of the sustainability concept in the environmental management system. Meanwhile, the zero value of the disparity between the maximum and minimum number of the rating values are from the Planet element which are (Water) and (Waste). It demonstrates that these parts are having no gap difference of understanding among the twelve criterion listed.

3.2.2. Product. Graphically, the result in Figure 5 shows that there are several criteria having the same value of the gap difference of the understanding about the sustainability assessment in their company scope of the environmental management system. The value of the Ethical & Behaviour from the People portion, Energy from the Planet division and Return of Investment (ROI) along with the Economic Stimulation from the Profit part is 0.8 which is similar to each other.

In contrast from the Product line graph in Figure 5, the gap distinction of understanding in the terms of Product is the superlative one which is zero value by means there is no variance of ideas in the sustainability concept in the Malaysian hydropower industry. Thus, this situation will make the process of reaching the main goals of this research study to be easier where the lower the gap difference of understanding variation, the better the results of the project study will be achieved.

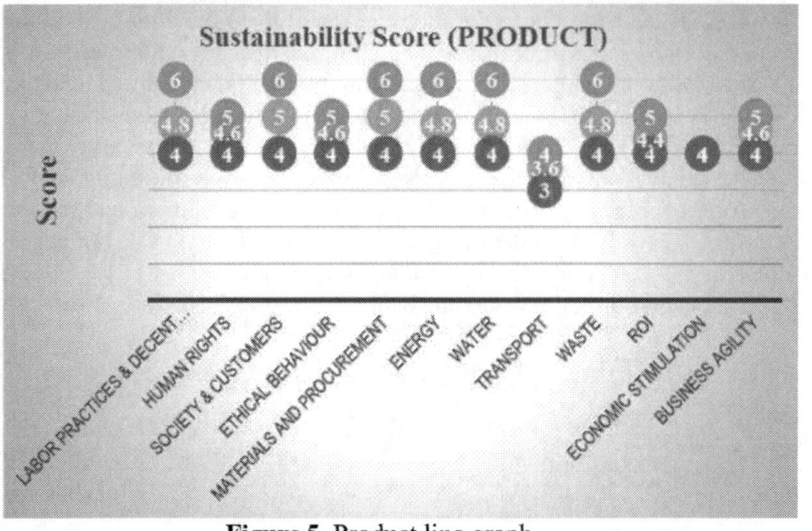

Figure 5. Product line graph.

3.2.3. Statistical Analysis. Moving on to the next part of this topic is the significant value of the standard deviation. The standard deviation can be defined as a numerical value in the units of the observed values that measures the spreading tendency of the data. Hence, the greater the value of the standard deviation at each of the criterion, it will indicate the challenging task in aligning them to the targeted objectives of this study will become harder and vice versa.

Concerning to Process bar graph in the Figure 6, it represents the standard deviation value according to each criteria from those elements in the sustainability assessment in the declining order. The highest value of the spreading tendency goes to the Society and Customer from the People group by 0.41 which is the highest rate among others. In addition, there are two types of the criteria from the Planet element which are Water and Waste that are having similar value of the dispersion of the data which indicates zero value. This situation shows that the outcomes from this part are showing such a good quality of sustainability measurement.

In contrast, the overall dispersion of the cumulative data is showing quite an enormous divergence in their understanding level of the sustainability compliance. It can be clearly seen when the disparity between the first highest value of the criteria which are Society and Customer along with the Transport part in the Planet group with the last two criterion which are previously stated is quite a huge difference value. 0.41 value marks that this restraint is compulsory to be reduced or even removed in the future.

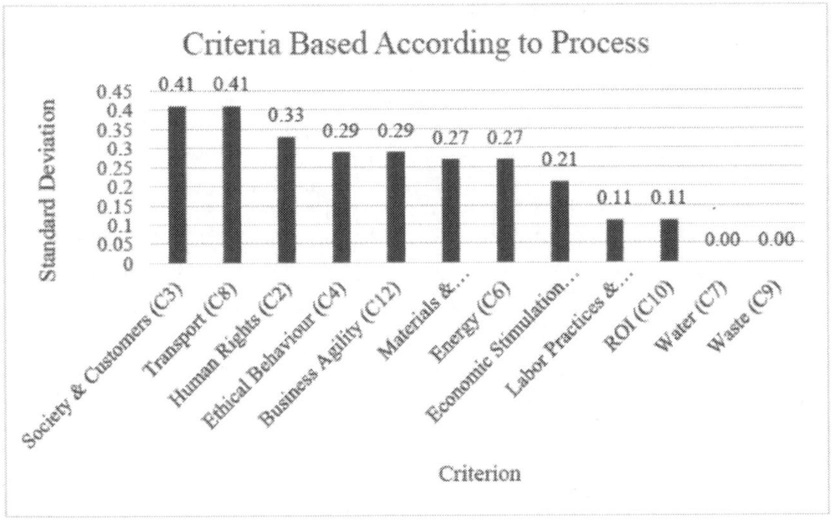

Figure 6. Process bar graph illustrating the rank value of the standard deviation.

Continuing to the Product bar graph in Figure 7, the Energy part which comes from the Planet criteria is laying on the topmost of the ranking. The value of the spreading tendency from this criteria is 0.33 only. Consequently, the overall overview of the criteria based graph in terms of the Product is portraying a slightly minimal value of the standard deviation obtained from the formulated data.

In contrast, there is only one criteria in this data that is having zero value of the dispersion data which is Society and Customers that comes from the group of People. It indicates that most of the interpretation from the respondents seems unbalance even though its having lower value of the spreading. Hence, the overall observation portraying that there is quite a challenging mission of this research study in order to succeed the goals.

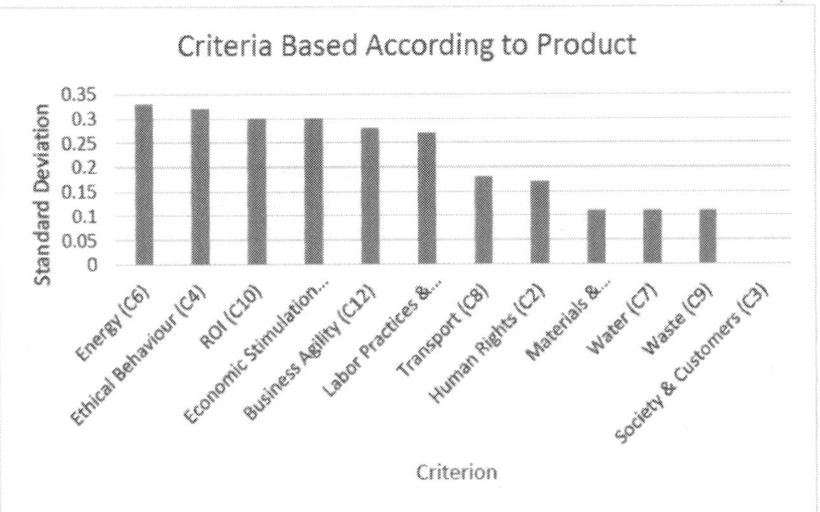

Figure 7. Product bar graph illustrating the rank value of the standard deviation.

The example of SSA tool dashboard from the program is shown in Figure 8.

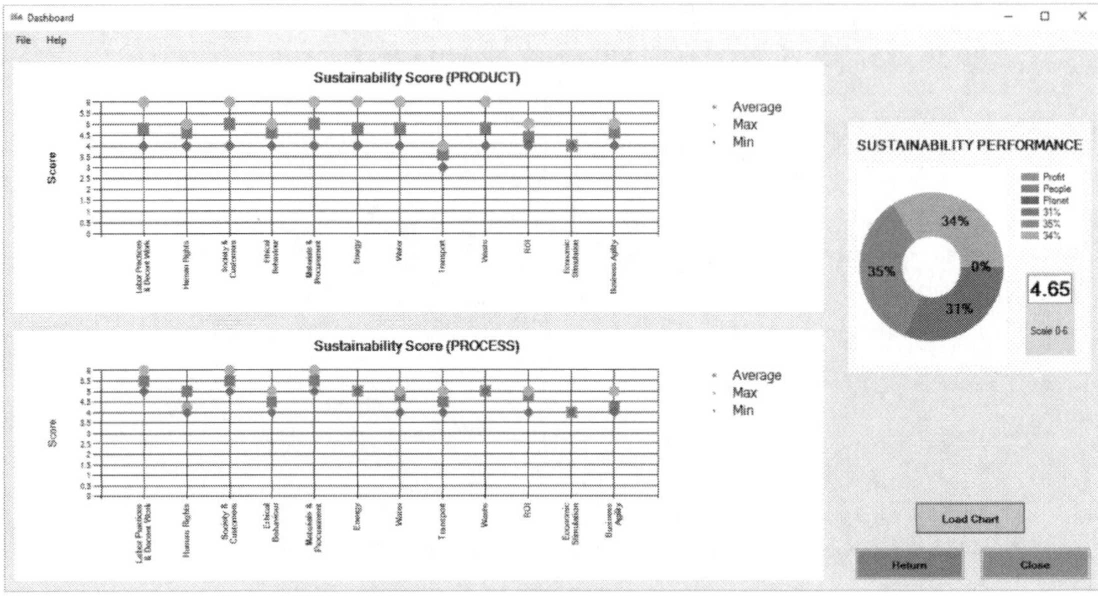

Figure 8. SSA Tools Dashboard Results window.

4. Conclusions

Conclusively, this research work is focusing on the sustainability assessment in the environmental management system for Malaysian hydropower of industry. By means, it represents on how to integrate the current assessment for the sustainability concept for the environmental management system especially in hydropower industry. The existed approaching systems are not still in the deficiency level; hence it acquires a new solution to overcome this recent problem.

Most of the respondents that involve in this project research are having diverge of understanding about the sustainability compliance. The ideas towards this assessment are only restricted on the existed sustainability tool of measure which does not comprises of the product and the process elements. Since the new integrated sustainability system approach is introducing the five components into the assessment, the outcomes will surely across-the-board of the current assessment in the sustainability practices.

In addition, it will help to fill the knowledge gap towards the sustainability compliance concept and provides new opportunities for further studies. Hopefully, the findings of this study will become a benchmarking purposes and as a point of references for the identifications in Malaysian hydropower industry.

Acknowledgments

The authors would like to give special thanks to Research & Innovation Department, Universiti Malaysia Pahang, Malaysia for funding this research project (RDU160308).

References

[1] Sharifi A and Murayama A 2013 A Critical Review of Seven Selected Neighborhood Sustainability Assessment Tools *Environmental Impact Assessment Review* **15**

[2] Hydropower Sustainability Assessment Protocol (November 2010), International Hydropower Association.

[3] Ngai E W T, Chau D C K, Lo C W H and Lei C F 2014 Design and Development of A Corporate Sustainability Index Platform for Corporate Sustainability Performance Analysis *Journal of Engineering and Technology Management* **34** 15.

[4] UN (2013). Indicators of sustainable development; guidelines and methodologies. United

Nations Commission on Sustainable Development.
[5] Global Reporting Initiative (2015). Linking the SDGs and GRI, 6.
[6] B Malaysia 2015 Sustainability Reporting Guide. Available: http://www.bursamalaysia.com/
[7] Abdulrazak S R and Ahmad F S 2014 Sustainable Development: A Malaysian Perspective *Social and Behavioral Sciences* **164** 5.
[8] Lin C, Madu C N, Kuei C H, Tsai H L and Wang K N 2015 Developing An Assessment Framework for Managing Sustainability Programs: A Analytic Network Process Approach *Expert Systems with Applications*, **14**.
[9] Fargonali, M., Minicis, D. M., & Tronci, M. (2014). Design Management for Sustainability: An integrated approach for the Development of Sustainable Products *Journal of Engineering and Technology Management* **34** 17
[10] The GPM Global Standard P5 of sustainability in Project Management, first edition, GPM Global, 2014. Available www.greenprojectmanagement.org
[11] Koc S, & Durmaz V 2015 Airport Corporate Sustainability: An Analysis of Indicators Reported in the Sustainability Practices *Social and Behavioral Sciences* **13**
[12] Buxel H, Esenduran G and Griffin S 2015 Strategic Sustainability: Creating Business Value with Life Cycle Analysis *Business Horizons* **14**
[13] Cinelli M, Coles S R and Kirwan K 2014 Analysis of the Potentials of Multi Criteria Decision Analysis Methods to Conduct Sustainability Assessment *Ecological Indicators* **46**
[14] Turan F M and Johan K 2016 Assessing Sustainability Framework of Automotive-related Industry in the Malaysian Context based on GPM P5 Standard *ARPN Journal of Engineering and Applied Sciences* **11** 7606 - 7611
[15] Turan F M, Johan K, Lanang W N S W and Nur N H M 2016 Development of Systematic Sustainability Assessment (SSA) for the Malaysian Industry *IOP Conf. Series: Materials Science and Engineering* **160** 012047
[16] Turan F M, Johan K and Nur N H M 2016 Criteria Assessment Model for Sustainable Product Development *IOP Conf. Series: Materials Science and Engineering* **160** 012004

Open circuit V-I characteristics of a coreless ironless electric generator for low density wind power generation

Akhtar Razali, Fadhlur Rahman, Syaiful Azlan, Mohd Razali Hanipah and Mohd Azri Hizami

Faculty of Mechanical Engineering, Universiti Malaysia Pahang, 26600 Pekan. Pahang.

E-mail: akhtar@ump.edu.my

Abstract. Cogging is an attraction of magnetism between permanent magnets and soft ironcore lamination in a conventional electric ironcore generator. The presence of cog in the generator is seen somehow restricted the application of the generator in an application where low rotational torque is required. Cog torque requires an additional input power to overcome, hence became one of the power loss sources. With the increasing of power output, the cogging is also proportionally increased. This leads to the increasing of the supplied power of the driver motor to overcome the cog. Therefore, this research is embarked to study fundamentally about the possibility of removing ironcore lamination in an electric generator. This research deals with removal of ironcore lamination in electric generator to eliminate cog torque. A confinement technique is proposed to confine and focus magnetic flux by introducing opposing permanent magnets arrangement. The concept is then fabricated and experimentally validated to qualify its no-load characteristics. The rotational torque and power output are measured and efficiency is then analyzed. Results indicated that the generator produced RMS voltage of 416VAC at rotational speed of 1762 RPM. Torque required to rotate the generator was at 2Nm for various rotational speed. The generator has shown 30% lesser rotational torque compared to the conventional ironcore type generator due to the absent of cogging torque in the system. Lesser rotational torque required to rotate has made this type of generator has a potential to be used for low wind density wind turbine application.

1. Introduction

Ironcore lamination in a generator is used to confine and guide magnetic flux as to boost its efficiency to the maximum. Ironcore as what the name implies, consists of lamination of highly permeability core material which can increase the density of the magnetic flux in over several thousand. The advantage of confining and guiding the flux however giving a major drawback to the generator. When the ironcore lamination is exposed to the magnetic field, magnetic particle in the core tends to line up with the magnetic field of the permanent magnets. When the magnets rotated, its magnetic field changed in direction. This induces continuous movement of the ironcore's magnetic particle for them to align with the change of magnetic field direction and coincidently produces molecular friction. This in turn produces heat and is then transmitted/distributed to the ironcore lamination and windings. Heat causes increases in winding resistance and at the same time retains electromagnetism in the windings and ironcore laminations.

Electromagnetism in the ironcore lamination acted as a magnet and react (attract and repel) with the permanent magnets' field. The more power is generated from the generator, the more power is required to maintain its rotation as the input torque increased proportional to the power output. Therefore, the more heat is generated, the stronger the electromagnetism exists in the core causes unwanted attraction force with the permanent magnets field. Attraction force between permanent magnets and ironcore lamination in a generator or also known as 'cog' is seen as one of the major inefficiency in the generator [1-2].

Cog as the name implies, increases spinning torque where this indirectly increases the amount of work/energy used to spin the generator. Energy to overcome cog is basically proportional to the output power produced. The larger the output power is produced, the larger the amount of torque is required to spin (or maintaining the rotation speed). The presence of cog in the system however has made this conventional generator is far from being usable in low torque application [3-4]. Continuous and consistent power is required to overcome cog and the power varies when rotational speed varies.

There were many attempts made elsewhere to reduce cog to increase the application boundary of a generator [5]. Most of the efforts were focused on the ironcore optimization in design to minimize the effect of cog [6]. However, the presence of cog is still inevitable [7]. Effort made by removing iron core material from an electric motor had demonstrated great success more than a decade ago. The success story of this ironless motor has been widely shared by both among the researchers and industries. The idea behind this achievement is by removing the iron core lamination and replace it with non-ferrous material. The ironless motor works flawlessly and one of the major advantage came from it is outstanding positional accuracy and repeatability due to no cogging affecting the positioning.

Similar idea may be applied to demonstrate coreless generator design. Effort made on this subject however, is still lacking [8-10]. The idea of removing ironcore lamination material may cause non-concentrated flux and leads to deterioration in output power efficiency [3]. However, there is a method which may be used to concentrate and focus magnetic flux to create denser magnetic field. An additional permanent magnets arrangement added to the generator may be a solution to address this issue. The absence of ironcore lamination in the system also represent cog free rotation. This provides advantage in terms of very low starting torque and less counter electromotive force is produced.

2. Material and equipment

In the open circuit test for the ironless coreless electricity generator, a 2hp three-phase induction motor was used to function as the prime mover to rotate the rotor of the generator. This induction motor had a rating no-load rotational speed of 1400RPM, input voltage of 220V to 240V for delta connection, input voltage of 380V to 415V for star connection, rating current of 5.64A to 6.15A for delta connection and rating current of 3.26A to 3.56A for star connection with efficiency of 75.5% if the load usage was below half of the motor but 78.5% if the load usage was above three quarters of the motor. Table 1 shows the specification of the induction motor.

Table 1. Specification of the three-phase induction motor used in the experiment.

	Type of connections	
	Star connection	Delta connection
Horsepower	2hp	
No-load rotational speed	1400RPM	
Efficiency (<1/2)	75.5%	
Efficiency (>3/4)	78.5%	
Input voltage	380V-415V	220V-240V
Rating current	3.26A-3.56A	5.64A-6.15A

Since the motor was using alternating current supply to power up, the inverter was used to convert the direct current supply into alternating current supply so that the motor could operate. Based on the data obtained from the manufacturer datasheet of the inverter, the inverter had an efficiency of 96% when working within the rated load. Wires were used to connect the power source to the inverter. They were also used to connect the inverter to the motor. To measure the wave frequency and the voltage output of the three-phase circuit of the ironless coreless electricity generator, the oscilloscope was used. The oscilloscope was connected to the end of the three-phase circuit by using wires. The tachometer was used to measure the rotational speed of the rotors of the generator. Figure 1 shows the block diagram for the connection of the ironless coreless electricity generator in open circuit test.

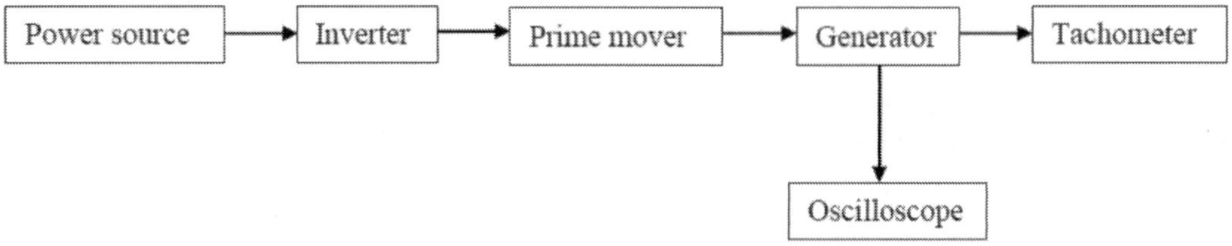

Figure 1. Block diagram for the connection of open circuit test.

3. Experimental setup and procedures

Before connecting the circuit to the induction motor, it was first aligned with the ironless coreless electricity generator. The center of motor shaft should be around 189mm from the surface of test bed to align them with the shaft of the ironless coreless electricity generator. The plate which supported the motor was then tightened using M8 bolts and nuts after the height of motor was set. The distance between the ironless coreless electricity generator and the motor was aligned horizontally by adjusting the generator. The bottom surface of the ironless coreless electricity generator was then locked on the test bed using M8 bolts and nuts. After the induction motor was well aligned with the generator, the connection between the induction motor and the generator was tightened with specifically made socket using socket cup point grub screws.

After the test bed was completely set up (Figure 2), the three-phase induction motor was connected using star connection as it could support higher voltage. The inverter was connected to power supply and the motor after the ironless coreless electricity generator and the motor were in place. The three-phase circuit of the ironless coreless electricity generator was then connected to the oscilloscope. When all the mechanical assembly and electrical assembly were done, the experiment began. After the power supply was switched on, a specified amount of power transferred to the motor was set by the inverter. The input power of the motor was controlled by the inverter using frequency input in the inverter. Once the setting was done and the rotational speed of the rotor was stabilized, the rotational speed of the rotor was observed using tachometer and recorded. To clearly obtain the graphical output wavelength from the oscilloscope, the x-axis was set to 10 milliseconds per column and y-axis was set to 50V per division, depending on the output voltage of the experiment. When the rotors were rotating, wave frequency and the output voltage were produced (see Figure 3). When everything was observed to have been completed, screenshots of the oscilloscope were taken and the data were transferred into the computer for data extraction process.

IOP Conf. Series: Materials Science and Engineering **342** (2018) 012010 doi:10.1088/1757-899X/342/1/012010

Figure 2. Experimental test bed and setup.

4. Open circuit tests results and discussions

In the open circuit test, there were fifteen experiments carried out with various rotational speed settings Table 2. Figure 3 show the results of experiment screenshots from the oscilloscope with three colors to indicate different waveforms in the graph. Each color represented the output for U-phase voltage output, V-phase voltage output and W-phase voltage output, respectively. The x-axis represented times while the y-axis represented output voltage. The results showed the three-phase voltage output in the open circuit test was pure sinusoidal wave. Normally, the experimental results for other generators would not show such outcome without having done the augmentation on the output of the electricity generator. This was the selling point of this ironless coreless electricity generator because without adding extra equipment to produce or modify the output became pure sinusoidal wave, the losses in the system would be minimal. The statement regarding to the losses was supported by the findings of Tamura's research on calculation method of losses and efficiency of wind generators [11].

Table 2. Rotational speed and output voltage relation at various speed point.

Load Test	Rotational Speed (RPM)	Wave Frequency (Hz)	Output Voltage (V_{rms})	Rotational Speed Characteristic Group
1	166	22.81	39.4	Low
2	195	27.11	47.2	Low
3	262	35.16	61.3	Low
4	294	39.62	69.1	Low
5	398	53.19	92.7	Low
6	427	58.21	101.0	Low
7	465	62.81	109.0	Low
8	493	66.62	116.0	Low
9	528	69.64	121.0	Medium
10	560	75.64	131.0	Medium
11	690	92.42	159.0	Medium
12	740	100.90	163.0	Medium
13	1500	210.10	356.0	High
14	1671	228.80	403.0	High
15	1762	238.70	416.0	High

IOP Conf. Series: Materials Science and Engineering **342** (2018) 012010 doi:10.1088/1757-899X/342/1/012010

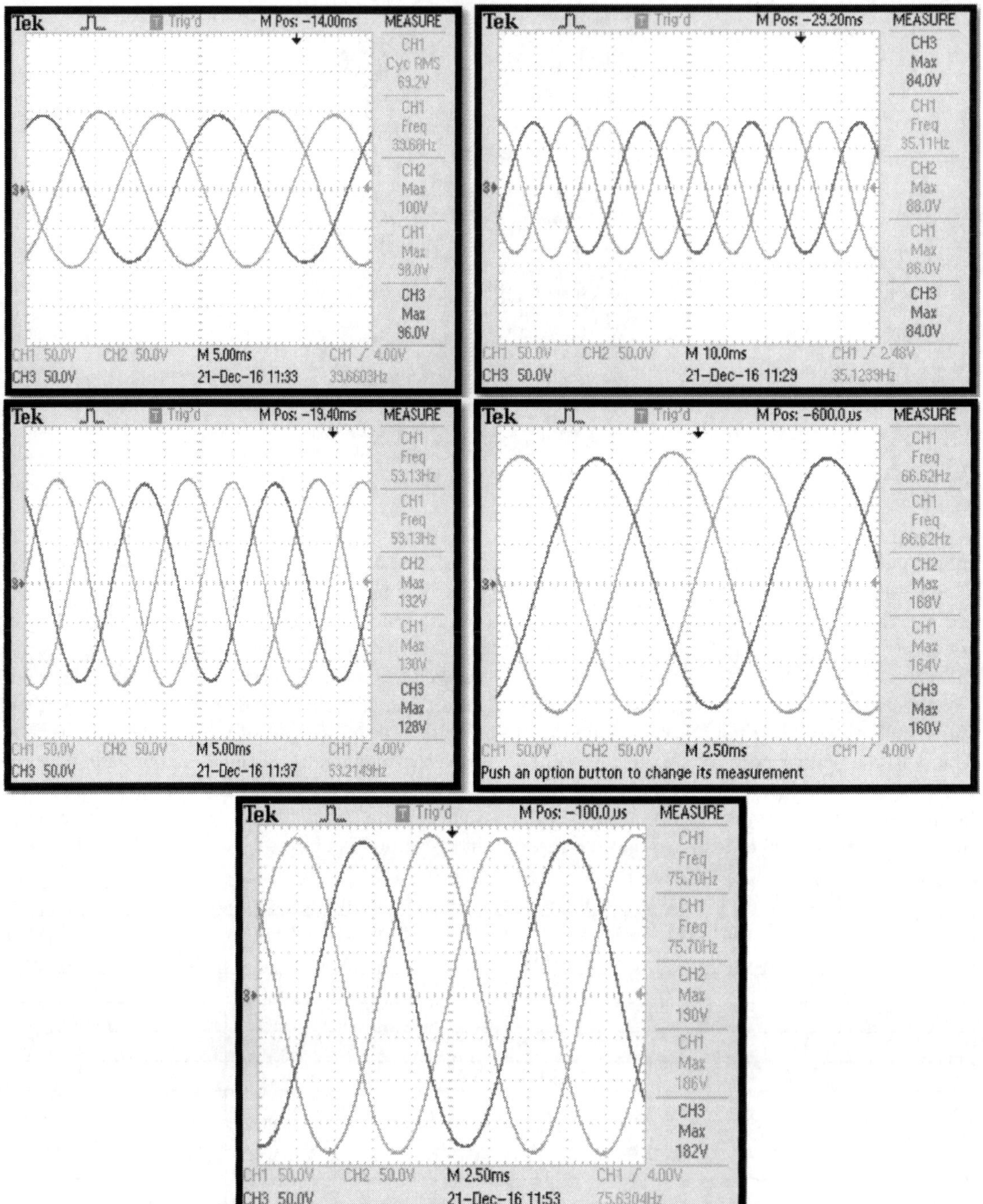

Figure 3. Phase voltages and characteristics for various rotational speed.

Table 2 summarizes the results of open circuit test from experiment 1 to 15. The experiments were categorized into three main groups denoted as low, medium and high speed. Low speed denotes speed between 0-499rpm while medium and high speeds are ranged between 500-999rpm and above 1000rpm respectively. At low speed 493rpm, the output voltage produced is at 116Vrms while medium speed at 740rpm produces 163Vrms. The maximum speed rotated by the induction motor is at 1700rpm where at this point, the output voltage was recorded at 416Vrms.

As shown in the Table 2, it was noted that the frequency of wave increased and the output voltage produced by the three-phase circuit of the ironless coreless electricity generator increased within the time frame when the rotational speed of the rotors of the ironless coreless electricity generator increased. This is because when the rotors of the generator were spinning faster, the frequency of magnetic flux cutting in the three-phase circuit on the stator of the ironless coreless electricity generator increased proportionally. When there was more magnetic flux cut by the three-phase circuit in the stator, the voltage per phase and wave frequency of each phase produced by the ironless coreless electricity generator also increased. This explanation for the open circuit test is similar and agrees with that of Chalmers and Spooner's research findings on an axial flux permanent magnet generator for a gearless wind energy system [12].

Figure 4 shows mechanical torque plot at various speed parameter. The RMS torque is rated at nearly 2Nm as recorded. This amount of torque is due to the force required to get the generator spun at constant speed without the presence of cogging torque. It is about 30% reduction of torque compared to the ironcore type AFPM generator [13].

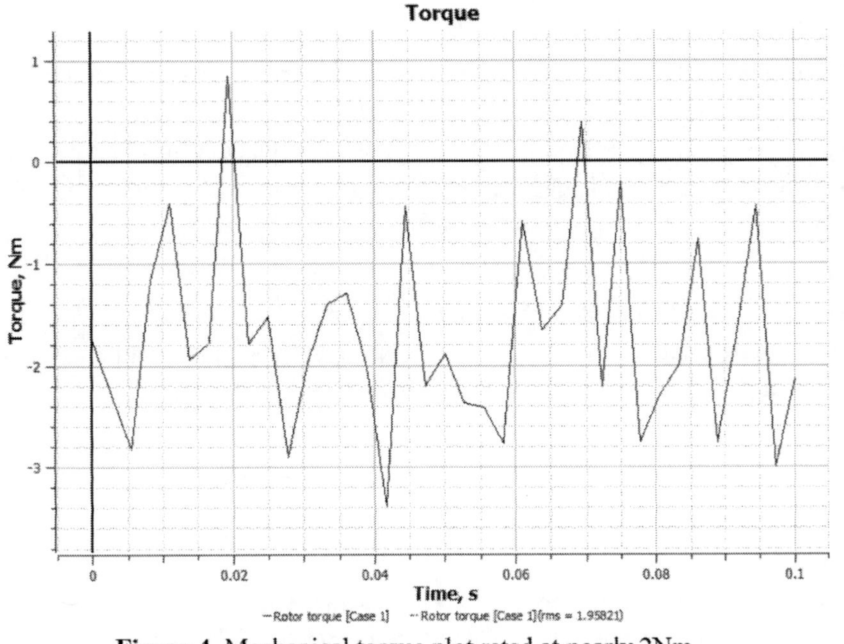

Figure 4. Mechanical torque plot rated at nearly $2Nm_{rms}$.

5. Conclusions

It can be concluded that the maximum voltage generated at maximum motor driver speed of 1762rpm was found to be at $416V_{rms}$. Pure sinusoidal voltage characteristics were observed throughout the tested low, medium and high rotational speed. The rated torque at various rotational speed was found at nearly 2Nm which is around 30% lesser than what conventional AFPM generator has to offer.

Acknowledgement

The authors would like to express their warmest thank you to the Kementerian Pendidikan Tinggi Malaysia (KPT Malaysia) and the Universiti Malaysia Pahang (UMP) for providing the financial support through the Fundamental Research Grant Scheme (FRGS by the KPT) to realize this project through the research grants RDU160134 (FRGS) and RDU160387 (UMP Grant).

References

[1] Zhang Z, Chen A, Matveev A, Nilssen R and Nysveen A 2013 High-power generators for offshore

wind turbines *Energy Procedia* **35** 52–61

[2] Ting C-C and Yeh L-Y 2014 Developing the full-field wind electric generator *Int. J. Electr. Power Energy Syst.* **55** 420–8

[3] Ahmed D and Ahmad A 2013 An optimal design of coreless direct-drive axial flux permanent magnet generator for wind turbine *Journal of Physics: Conference Series* vol 439 (IOP Publishing) 12039

[4] Kobayashi H, Doi Y, Miyata K and Minowa T 2009 Design of axial-flux permanent magnet coreless generator for the multi-megawatts wind turbines *EWEC2009*

[5] Kurt E, Gör H and Demirtaş M 2014 Theoretical and experimental analyses of a single phase permanent magnet generator (PMG) with multiple cores having axial and radial directed fluxes *Energy Convers. Manag.* **77** 163–72

[6] Gor H and Kurt E 2016 Waveform characteristics and losses of a new double sided axial and radial flux generator *Int. J. Hydrogen Energy* **41** 12512–24

[7] Gör H and Kurt E 2016 Preliminary studies of a new permanent magnet generator (PMG) with the axial and radial flux morphology *Int. J. Hydrogen Energy* **41** 7005–18

[8] Mahmoudi A, Kahourzade S, Rahim N A, Ping H W and Uddin M N 2013 Design and prototyping of an optimised axial-flux permanent-magnet synchronous machine *IET Electr. Power Appl.* **7** 338–49

[9] Virtic P and Avsec J 2011 Analysis of coreless stator axial flux permanent magnet synchronous generator characteristics by using equivalent circuit *Prz. Elektrotechniczny* **87** 208–11

[10] Zikowski Ł and Koczara W 2011 Design And Analysis Of Axial-Flux Coreless Permanent Magnet Disk Generator *Journal of Electrical Engineering*

[11] Tamura J 2012 Calculation method of losses and efficiency of wind generators *Wind energy conversion systems* (Springer) pp 25–51

[12] Chalmers B J and Spooner E 1999 An axial-flux permanent-magnet generator for a gearless wind energy system *IEEE Trans. Energy Convers.* **14** 251–7

[13] Kim C-E, Jang J-K and Joo S-J 2014 The influence of stator pole shape and its arrangements on cogging torque for double-sided AFPM generator *Electrical Machines and Systems (ICEMS), 2014 17th International Conference on* (IEEE) pp 463–6

IOP Conf. Series: Materials Science and Engineering **342** (2018) 012011 doi:10.1088/1757-899X/342/1/012011

V-I characteristics of a coreless ironless electric generator in a closed-circuit mode for low wind density power generation

Akhtar Razali, Fadhlur Rahman, Yap Wee Leong, Mohd Razali Hanipah and Mohd Azri Hizami

Faculty of Mechanical Engineering, Universiti Malaysia Pahang, 26600 Pekan. Pahang.

E-mail: akhtar@ump.edu.my

Abstract. This research deals with removal of ironcore lamination in electric generator to eliminate cog torque. A confinement technique is proposed to confine and focus magnetic flux by introducing opposing permanent magnets arrangement. The generator was fabricated and experimentally validated to qualify its loaded characteristics. The rotational torque and power output are measured and efficiency is then analyzed. At 100Ω load, the generator power output increased with the increased of rotational speed. Nearly 78% of efficiency was achieved when the generator was rotated at 250rpm. At this speed, the generator produced RMS voltage of 81VAC. Torque required to rotate the generator was found to be 3.2Nm. The slight increment of mechanical torque to spin the generator was due to the counter electromotive force (CEMF) existed in the copper windings. However, the torque required is still lower by nearly 30% than conventional AFPM generator. It is there concluded that this generator is suitable to be used for low wind density power generation application.

1. Introduction

Ironcore lamination in a generator is used to confine and guide magnetic flux as to boost its efficiency to the maximum. Ironcore as what the name implies, consists of lamination of highly permeability core material which can increase the density of the magnetic flux in over several thousand. The advantage of confining and guiding the flux however giving a major drawback to the generator. When the ironcore lamination is exposed to the magnetic field, magnetic particle in the core tends to line up with the magnetic field of the permanent magnets. When the magnets rotated, its magnetic field changed in direction. This induces continuous movement of the ironcore's magnetic particle for them to align with the change of magnetic field direction and coincidently produces molecular friction. This in turn produces heat and is then transmitted/distributed to the ironcore lamination and windings. Heat causes increases in winding resistance and at the same time retains electromagnetism in the windings and ironcore laminations.

Electromagnetism in the ironcore lamination acted as a magnet and react (attract and repel) with the permanent magnets' field. The more power is generated from the generator, the more power is required to maintain its rotation as the input torque increased proportional to the power output. Therefore, the more heat is generated, the stronger the electromagnetism exists in the core causes unwanted attraction force with the permanent magnets field. Attraction force between permanent magnets and ironcore lamination in a generator or also known as 'cog' is seen as one of the major inefficiency in the generator [1-2].

Cog as the name implies, increases spinning torque where this indirectly increases the amount of work/energy used to spin the generator. Energy to overcome cog is basically proportional to the output power produced. The larger the output power is produced, the larger the amount of torque is required to spin (or maintaining the rotation speed). The presence of cog in the system however has made this conventional generator is far from being usable in low torque application [3-4]. Continuous and consistent power is required to overcome cog and the power varies when rotational speed varies.

There were many attempts made elsewhere to reduce cog to increase the application boundary of a generator [5]. Most of the efforts were focused on the ironcore optimization in design to minimize the effect of cog [6]. However, the presence of cog is still inevitable [7]. Effort made by removing iron core material from an electric motor had demonstrated great success more than a decade ago. The success story of this ironless motor has been widely shared by both among the researchers and industries. The idea behind this achievement is by removing the iron core lamination and replace it with non-ferrous material. The ironless motor works flawlessly and one of the major advantage came from it is outstanding positional accuracy and repeatability due to no cogging affecting the positioning.

Similar idea may be applied to demonstrate coreless generator design. Effort made on this subject however, is still lacking [8-10]. The idea of removing ironcore lamination material may cause non-concentrated flux and leads to deterioration in output power efficiency [3]. However, there is a method which may be used to concentrate and focus magnetic flux to create denser magnetic field. An additional permanent magnets arrangement added to the generator may be a solution to address this issue. The absence of ironcore lamination in the system also represent cog free rotation. This provides advantage in terms of very low starting torque and less counter electromotive force is produced.

2. Material and equipment

In the closed-circuit test for the ironless coreless electricity generator, a 2hp three-phase induction motor was used to function as the prime mover to rotate the rotor of the generator. This induction motor had a rating no-load rotational speed of 1400RPM, input voltage of 220V to 240V for delta connection, input voltage of 380V to 415V for star connection, rating current of 5.64A to 6.15A for delta connection and rating current of 3.26A to 3.56A for star connection with efficiency of 75.5% if the load usage was below half of the motor but 78.5% if the load usage was above three quarters of the motor. Table 1 shows the specification of the induction motor.

Table 1. Specification of the three-phase induction motor used in the experiment.

	Type of connections	
	Star connection	Delta connection
Horsepower	2hp	
No-load rotational speed	1400RPM	
Efficiency (<1/2)	75.5%	
Efficiency (>3/4)	78.5%	
Input voltage	380V-415V	220V-240V
Rating current	3.26A-3.56A	5.64A-6.15A

Since the motor was using alternating current supply to power up, the inverter was used to convert the direct current supply into alternating current supply so that the motor could operate. Based on the data obtained from the manufacturer datasheet of the inverter, the inverter had an efficiency of 96% when working within the rated load. Wires were used to connect the power source to the inverter. They were also used to connect the inverter to the motor. To measure the wave frequency and the voltage output of the three-phase circuit of the ironless coreless electricity generator, the oscilloscope was used. The oscilloscope was connected to the end of the three-phase circuit by using wires. The tachometer was used to measure the rotational speed of the rotors of the generator. DC electronic load was used to analyse the currents and power output when the generator was subjected to load. Figure 1 shows the block diagram for the connection of the ironless coreless electricity generator in closed circuit test.

3. Experimental setup and procedures

Before connecting the circuit to the induction motor, it was first aligned with the ironless coreless electricity generator. The center of motor shaft should be around 189mm from the surface of test bed to align them with the shaft of the ironless coreless electricity generator. The plate which supported the motor was then tightened using M8 bolts and nuts after the height of motor was set. The distance between the ironless coreless electricity generator and the motor was aligned horizontally by adjusting the generator. The bottom surface of the ironless coreless electricity generator was then locked on the test bed using M8 bolts and nuts. After the induction motor was well aligned with the generator, the connection between the induction motor and the generator was tightened with specifically made socket using socket cup point grub screws.

After the test bed was completely set up (Figure 2), the three-phase induction motor was connected using star connection as it could support higher voltage. The inverter was connected to power supply and the motor after the ironless coreless electricity generator and the motor were in place. The three-phase circuit of the ironless coreless electricity generator was then connected to the oscilloscope. When all the mechanical assembly and electrical assembly were done, the experiment began. After the power supply was switched on, a specified amount of power transferred to the motor was set by the inverter. The input power of the motor was controlled by the inverter using frequency input in the inverter. Once the setting was done and the rotational speed of the rotor was stabilized, the rotational speed of the rotor was observed using tachometer and recorded. To clearly obtain the graphical output wavelength from the oscilloscope, the x-axis was set to 10 milliseconds per column and y-axis was set to 50V per division, depending on the output voltage of the experiment. When the rotors were rotating, wave frequency and the output voltage were produced. When everything was observed to have been completed, screenshots of the oscilloscope were taken and the data were transferred into the computer for data extraction process. DC electronic load was used to give the generator load and measure the output current and power generated.

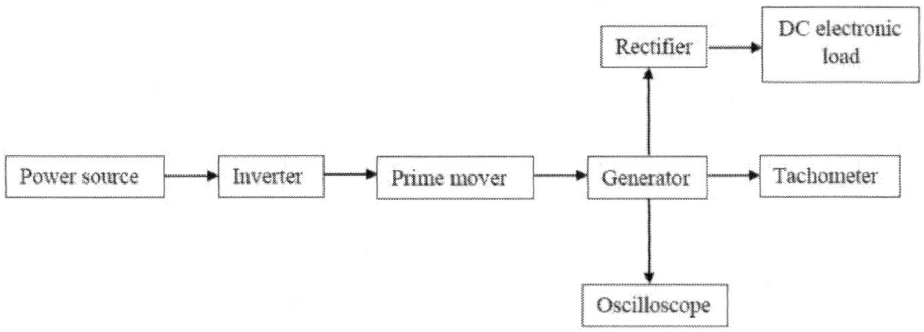

Figure 1. Block diagram for the connection of closed circuit test.

Figure 2. Experimental test bed and setup.

4. Closed circuit tests results and discussions

There were 12 sets of experiments carried out in the various rotational speed constant load tests starting from 104RPM al the way up to the 250RPM. It can be observed that the mechanical torque required to rotate the generator was nearly 3Nm throughout the speeds. However, there is slight fluctuation of mechanical torque due to the force required to overcome rotor inertia at low speed is slightly larger than at higher speed. At rotational speed of 250RPM, output power produced was 66.1W and the output RMS voltage was 81V. At this voltage, the output current of 0.814A was produced, at mechanical torque of 3.24Nm. The generator efficiency was found at 77.95%. Table 2 summarizes the results of power input and output, mechanical torque, rotational speed and efficiency of the generator. As shown in the Table 2, it was noted that the power, current and output voltage produced by the three-phase circuit of the ironless coreless electricity generator increased when the rotational speed of the rotors of the ironless coreless electricity generator increased. This is because when the rotors of the generator were spinning faster, the frequency of magnetic flux cutting in the three-phase circuit on the stator of the ironless coreless electricity generator increased proportionally. When there was more magnetic flux cut by the three-phase circuit in the stator, the voltage per phase produced by the ironless coreless electricity generator also increased [12]. The significant amount of torque required to spin the generator is due to the force required to get the generator spun at constant speed and overcoming the counter electromotive force generated by the copper windings. However, the torque is still 30% lesser if compared to the ironcore type AFPM generator [13]. Figure 3 shows efficiency plot for the generator. The efficiency increased with the increasing of the rotational speed and the increment starts to slow down when it hits 200rpm. Beyond 238rpm, the increment of efficiency is no longer significant. At this point, the it can be concluded that with the increasing of rotational speed, the generator operates at nearly 78% of efficiency. The inefficiency is due to the electromagnetic losses and as well as mechanical losses due to friction between rotational bodies and support bearings used in the system.

Table 2. Results for torque and efficiency in various rotational speed constant load tests.

Speed (RPM)	Voltage (V)	Current (A)	Power input (W)	Power output (W)	Torque (Nm)	Efficiency (%)
104	33	0.300	32.83	10.7	3.01	32.59
118	38	0.360	37.47	13.8	3.03	36.83
131	43	0.410	39.28	17.3	2.86	44.04
145	46	0.460	41.53	21.4	2.73	51.53
158	51	0.500	44.72	25.2	2.70	56.35
171	55	0.546	49.64	30.3	2.77	61.03
184	60	0.592	52.47	35.9	2.72	68.41
198	65	0.636	58.63	41.3	2.83	70.43
211	69	0.682	63.42	46.9	2.87	73.95
224	72	0.727	68.20	52.1	2.91	76.39
238	77	0.771	76.32	59.5	3.06	77.96
250	81	0.814	84.80	66.1	3.24	77.95

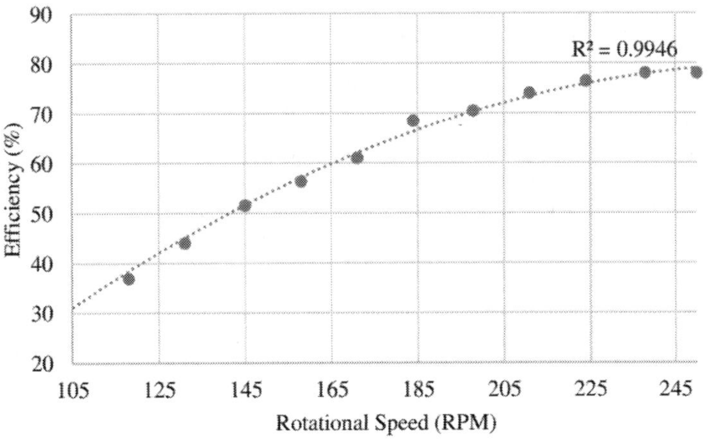

Figure 3. Efficiency of the system at various rotational speed.

5. Conclusions

It can be concluded that the maximum efficiency achieved by the generator is nearly 78%. At this efficiency, the output voltage tested was 81VAC. The rated torque at this rotational speed was found at 3.2Nm which is around 30% lesser than what conventional AFPM generator has to offer.

Acknowledgement

The authors would like to express their warmest thank you to the Kementerian Pendidikan Tinggi Malaysia (KPT Malaysia) and the Universiti Malaysia Pahang (UMP) for providing the financial support through the Fundamental Research Grant Scheme (FRGS by the KPT) to realize this project through the research grants RDU160134 (FRGS) and RDU160307 (UMP Grant).

References

[1] Zhang Z, Chen A, Matveev A, Nilssen R and Nysveen A 2013 High-power generators for offshore wind turbines *Energy Procedia* **35** 52–61

[2] Ting C-C and Yeh L-Y 2014 Developing the full-field wind electric generator *Int. J. Electr. Power Energy Syst.* **55** 420–8

[3] Ahmed D and Ahmad A 2013 An optimal design of coreless direct-drive axial flux permanent magnet generator for wind turbine *Journal of Physics: Conference Series* vol 439(IOP Publishing)p 12039

[4] Kobayashi H, Doi Y, Miyata K and Minowa T 2009 Design of axial-flux permanent magnet coreless generator for the multi-megawatts wind turbines *EWEC2009*

[5] Kurt E, Gör H and Demirtaş M 2014 Theoretical and experimental analyses of a single phase permanent magnet generator (PMG) with multiple cores having axial and radial directed fluxes *Energy Convers. Manag.* **77** 163–72

[6] Gor H and Kurt E 2016 Waveform characteristics and losses of a new double sided axial and radial flux generator *Int. J. Hydrogen Energy* **41** 12512–24

[7] Gör H and Kurt E 2016 Preliminary studies of a new permanent magnet generator (PMG) with the axial and radial flux morphology *Int. J. Hydrogen Energy* **41** 7005–18

[8] Mahmoudi A, Kahourzade S, Rahim N A, Ping H W and Uddin M N 2013 Design and prototyping of an optimised axial-flux permanent-magnet synchronous machine *IET Electr. Power Appl.* **7** 338–49

[9] Virtic P and Avsec J 2011 Analysis of coreless stator axial flux permanent magnet synchronous generator characteristics by using equivalent circuit *Prz. Elektrotechniczny* **87** 208–11

IOP Conf. Series: Materials Science and Engineering **342** (2018) 012011 doi:10.1088/1757-899X/342/1/012011

[10] Zikowski Ł and Koczara W 2011 Design And Analysis Of Axial-Flux Coreless Permanent Magnet Disk Generator *Journal of Electrical Engineering*

[11] Tamura J 2012 Calculation method of losses and efficiency of wind generators *Wind energy conversion systems* (Springer) pp 25–51

[12] Chalmers B J and Spooner E 1999 An axial-flux permanent-magnet generator for a gearless wind energy system *IEEE Trans. Energy Convers.* **14** 251–7

[13] Kim C-E, Jang J-K and Joo S-J 2014 The influence of stator pole shape and its arrangements on cogging torque for double-sided AFPM generator *Electrical Machines and Systems (ICEMS), 2014 17th International Conference on* (IEEE) pp 463–6

Maximum power point tracking analysis of a coreless ironless electric generator for renewable energy application

Akhtar Razali, Fadhlur Rahman, Yap Wee Leong, Mohd Razali Hanipah and Mohd Azri Hizami

Faculty of Mechanical Engineering, Universiti Malaysia Pahang, 26600 Pekan. Pahang.

E-mail: akhtar@ump.edu.my

Abstract. The magnetism attraction between permanent magnets and soft ironcore lamination in a conventional electric ironcore generator is often known as cogging. Cogging requires an additional input power to overcome, hence became one of the power loss sources. With the increasing of power output, the cogging is also proportionally increased. This leads to the increasing of the supplied power of the driver motor to overcome the cog. Therefore, this research is embarked to study fundamentally about the possibility of removing ironcore lamination in an electric generator to see its performance characteristic. In the maximum power point tracking test, the fabricated ironless coreless electricity generator was tested by applying the load on the ironless coreless electricity generator optimization to maximize the power generated, voltage and the current produced by the ironless coreless electricity generator when the rotational speed of the rotor increased throughout the test. The rotational torque and power output are measured, and efficiency is then analyzed. Results indicated that the generator produced RMS voltage of 200VAC at rotational speed of 318 RPM. Torque required to rotate the generator was at 10.8Nm. The generator had working efficiency of 77.73% and the power generated was at 280W.

1. Introduction

Attraction force between permanent magnets and ironcore lamination in a generator or also known as 'cog' is seen as one of the major inefficiency in the generator [1-2]. Cog as the name implies, increases spinning torque where this indirectly increases the amount of work/energy used to spin the generator. Energy to overcome cog is basically proportional to the output power produced. The larger the output power is produced, the larger the amount of torque is required to spin (or maintaining the rotation speed). The presence of cog in the system however has made this conventional generator is far from being usable in low torque application [3-4]. Continuous and consistent power is required to overcome cog and the power varies when rotational speed varies.

There were many attempts made elsewhere to reduce cog to increase the application boundary of a generator [5]. Most of the efforts were focused on the ironcore optimization in design to minimize the effect of cog [6]. However, the presence of cog is still inevitable [7]. Effort made by removing iron core material from an electric motor had demonstrated great success more than a decade ago. The ironless motor works flawlessly and one of the major advantage came from it is outstanding positional accuracy and repeatability due to no cogging affecting the positioning. Due to being rated as high efficiency, it has received widest coverage of application due to its advantages compared to the conventional rotary motor. Similar idea may be applied to demonstrate coreless generator design. Effort made on this subject however, is still lacking [8-10]. The idea of removing ironcore lamination material may cause non-concentrated flux and leads to deterioration in output power efficiency [3]. However, there is a method which may be used to concentrate and focus magnetic flux to create denser magnetic field. An additional permanent magnets arrangement added to the generator may be a solution to address this issue. The absence of ironcore lamination in the system also represent cog free rotation. This provides advantage in terms of very low starting torque and less counter electromotive force is produced.

2. Material and equipment

In the open circuit test for the ironless coreless electricity generator, a 2hp three-phase induction motor was used to function as the prime mover to rotate the rotor of the generator. This induction motor had a rating no-load rotational speed of 1400RPM, input voltage of 220V to 240V for delta connection, input voltage of 380V to 415V for star connection, rating current of 5.64A to 6.15A for delta connection and rating current of 3.26A to 3.56A for star connection with efficiency of 75.5% if the load usage was below half of the motor but 78.5% if the load usage was above three quarters of the motor. Table 1 shows the specification of the induction motor.

Table 1. Specification of the three-phase induction motor used in the experiment.

	Type of connections	
	Star connection	Delta connection
Horsepower	2hp	
No-load rotational speed	1400RPM	
Efficiency (<1/2)	75.5%	
Efficiency (>3/4)	78.5%	
Input voltage	380V-415V	220V-240V
Rating current	3.26A-3.56A	5.64A-6.15A

Since the motor was using alternating current supply to power up, the inverter was used to convert the direct current supply into alternating current supply so that the motor could operate. Based on the data obtained from the manufacturer datasheet of the inverter, the inverter had an efficiency of 96% when working within the rated load. Wires were used to connect the power source to the inverter. They were also used to connect the inverter to the motor. To measure the wave frequency and the voltage output of the three-phase circuit of the ironless coreless electricity generator, the oscilloscope was used. The oscilloscope was connected to the end of the three-phase circuit by using wires. The tachometer was used to measure the rotational speed of the rotors of the generator.

3. Experimental setup and procedures

Before connecting the circuit to the induction motor, it was first aligned with the ironless coreless electricity generator. The center of motor shaft should be around 189mm from the surface of test bed to align them with the shaft of the ironless coreless electricity generator. The plate which supported the motor was then tightened using M8 bolts and nuts after the height of motor was set. The distance between the ironless coreless electricity generator and the motor was aligned horizontally by adjusting the generator. The bottom surface of the ironless coreless electricity generator was then locked on the test bed using M8 bolts and nuts. After the induction motor was well aligned with the generator, the connection between the induction motor and the generator was tightened with specifically made socket using socket cup point grub screws.

After the test bed was completely set up (Figure 1), the three-phase induction motor was connected using star connection as it could support higher voltage. The inverter was connected to power supply and the motor after the ironless coreless electricity generator and the motor were in place. The three-phase circuit of the ironless coreless electricity generator was then connected to the oscilloscope. When all the mechanical assembly and electrical assembly were done, the experiment began. After the power supply was switched on, a specified amount of power transferred to the motor was set by the inverter. The input power of the motor was controlled by the inverter using frequency input in the inverter. Once the setting was done and the rotational speed of the rotor was stabilized, the rotational speed of the rotor was observed using tachometer and recorded. 4 channels DC electronic load with maximum voltage power of 80V/channel and 300W respectively was used to give the generator load and measure the output current and power generated.

Figure 1. Experimental test bed and setup.

In the maximum power point tracking test, the fabricated ironless coreless electricity generator was tested by applying the load on the ironless coreless electricity generator optimization to maximize the power generated, voltage produced and the current produced by the ironless coreless electricity generator when the rotational speed of the rotor increased throughout the test. Nine experiments were conducted for the maximum power point tracking test on the ironless coreless electricity generator.

4. Maximum point tracking tests results and discussions

Table 2 shows tabulated results of power input to inverter, power input to motor, power input to generator and power output by the generator in Watts (W). Throughout the rotational speeds, it can be observed that there are losses occurred when the power is transmitted from the source to the generator. As the rotational speeds increased, the efficiency of the generator in producing electricity is increased. Since the DC electronic load used has a maximum power measuring capability at 80V and 300W, therefore the experiments have to be stopped when the power output generated was at 280W. Any increment in rotational speed and power produced beyond this threshold could damage the instrument used. Between the speed range of 167 to 318RPM, the optimum load was found to be at around 35Ω. Voltage produced was found doubled at 100VAC when the rotational speed was at 318RPM if compared to the 167RPM. The increasing of rotational speed while at the same time increases the voltage output, it also increases the output currents generated by the generator. As voltage was doubled when the rotational speed got doubled from 167RPM to 318RPM, the currents also similar trend increased from 1.44A to 2.80A. As the trends of voltage and currents are increased when the rotational speed is increased, the mechanical torque was also found increased. The increased in counter electromotive force (CEMF) existed in the copper windings has made more power is required to overcome the attraction force between the coils and magnets. The increased torque also reflects the increment of power input to the generator throughout all experiments.

Table 2. Results for power input/output, torque, voltage, currents and efficiency in maximum power point tracking test.

Speed (RPM)	Power input to inverter (W)	Power input to motor (W)	Power input to generator (W)	Power output (Generator) (W)	Torque (Nm)	Efficiency (%)	Volt (V)	Currents (A)	Load (ohm)
167	168	161.28	121.77	72	6.96	59.13	50	1.44	34.40
189	190	182.40	137.71	92	6.96	66.81	56	1.64	34.96
207	222	213.12	160.91	118	7.40	73.33	64	1.84	35.20
223	263	252.48	190.62	140	8.16	73.44	70	2.00	34.40
240	305	292.80	221.06	164	8.77	74.19	76	2.16	34.60
258	344	330.24	249.33	192	9.22	77.01	82	2.34	34.96
277	397	381.12	287.75	220	9.92	76.46	88	2.50	34.96
305	445	427.20	322.54	250	10.10	77.51	94	2.66	34.72
318	497	477.12	360.23	280	10.82	77.73	100	2.80	35.08

Figure 2 illustrates the power generated by the ironless coreless electricity generator versus the rotational speed of the rotor graph. As shown in the Figure 2, the graph is increase linearly as the output power generated by the ironless coreless electricity generator increased from 75W to 380W when the rotational speed of the rotors of the generator increased from 170RPM to 318RPM. Figure 3 and Figure 4 show the current produced by the ironless coreless electricity generator versus rotational speed of the rotor graph and the voltage produced by the ironless coreless electricity generator versus rotational speed of rotor graph, respectively. As illustrated in the Figure 3, the output current produced by the generator increased from 1.5A to 2.8A when the rotational speed of the rotors of the generator increased from 170RPM to 318RPM. Figure 4 provides an explanation to justify when the rotational speed of rotors of the generator rose from 170RPM to 318RPM, the output voltage generated by the ironless coreless electricity generator increased from 52V to 100V.

Figure 5 shows the graph for the optimum resistive load used by the ironless coreless electricity generator versus rotational speed graph. From the Figure 5, it is clear that when the resistive load was around 35 ohms, the ironless coreless electricity generator produced highest output power, output voltage and output current when the rotational speed of the rotors of the generator was 170RPM to 318RPM. Figure 6 illustrates the torque of the ironless coreless electricity generator versus rotational speed graph. The torque of the ironless coreless electricity generator was increase linearly from around 6.8Nm to around 10.5Nm when the rotational speed of the ironless coreless electricity generator increased from 170RPM to 318RPM. Figure 7 shows the efficiency of the ironless coreless electricity generator versus the rotational speed graph. The efficiency of the ironless coreless electricity generator increased gradually from around 59% until around 78% and after achieving 78% efficiency, the generator became less efficient as the rotational speed of the rotors increased from 170RPM to 318RPM.

IOP Conf. Series: Materials Science and Engineering **342** (2018) 012012 doi:10.1088/1757-899X/342/1/012012

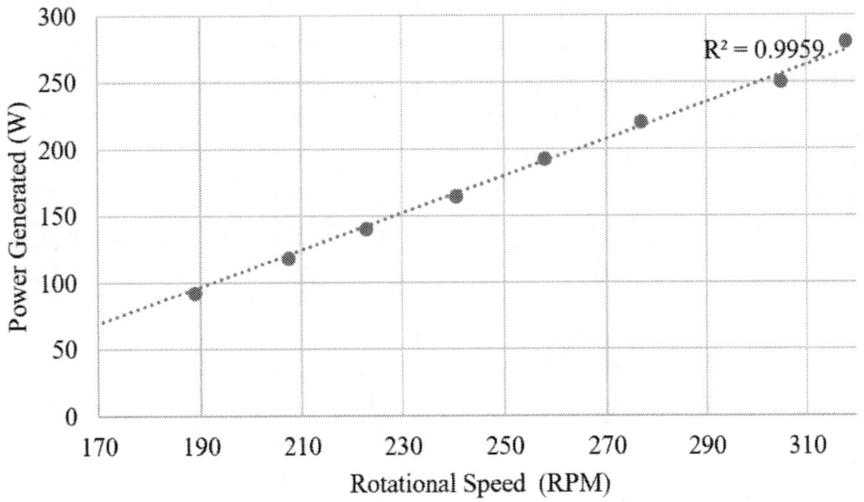

Figure 2. Power generated by the ironless coreless electricity generator versus rotational speed graph for maximum power point tracking test.

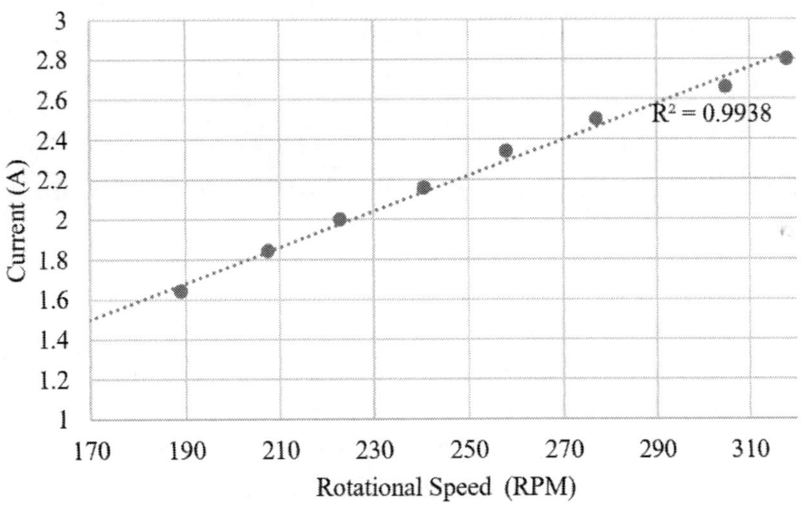

Figure 3. Current versus rotational speed graph for maximum power point tracking test.

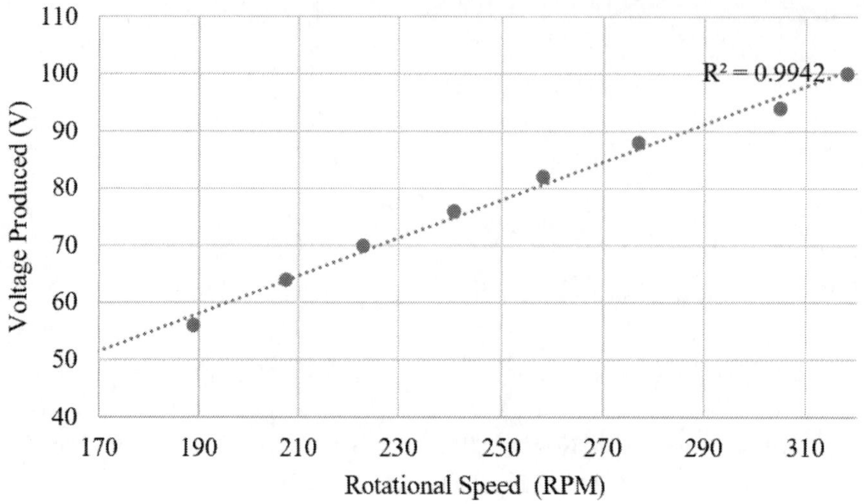

Figure 4. Voltage produced by the ironless coreless electricity generator versus rotational speed graph for maximum power point tracking test.

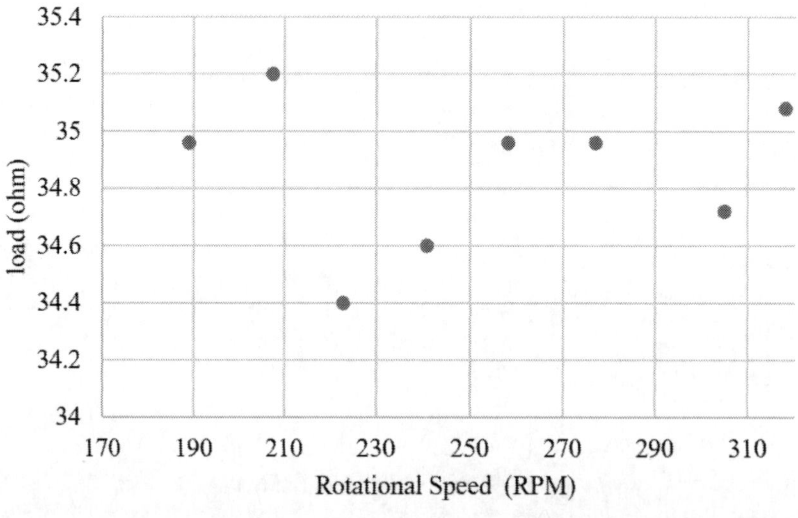

Figure 5. Optimum resistive load versus rotational speed graph for maximum power point tracking test.

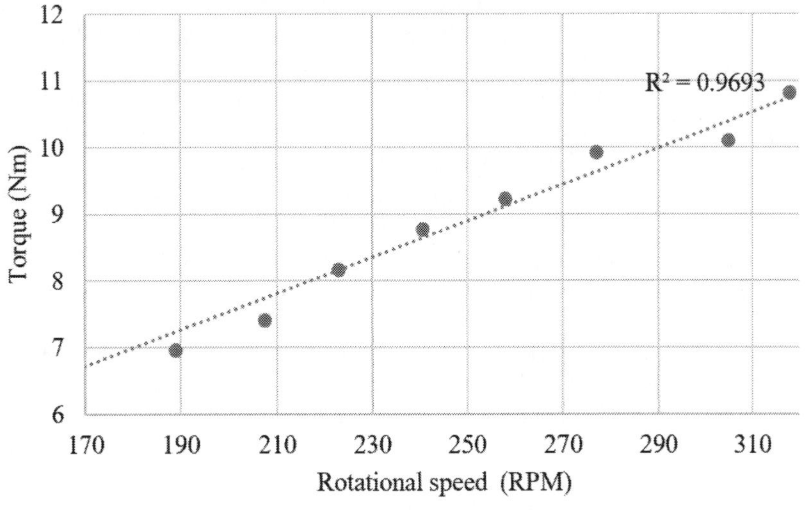

Figure 6. Torque versus rotational speed graph for maximum power point tracking test.

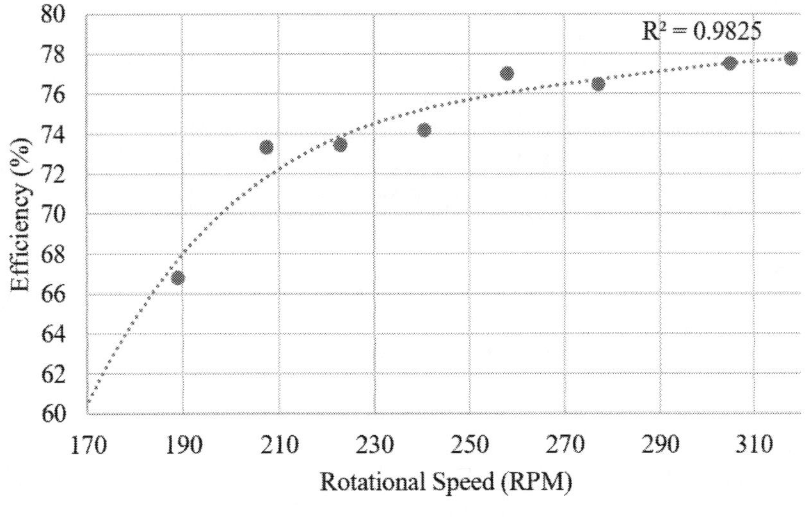

Figure 7. Efficiency versus rotational speed graph for maximum power point tracking test.

The reason to conduct the maximum power point tracking test was because the maximum power point could be achieved by the permanent magnet synchronous generator by varying the generator's load [11]. The results of the maximum power point tracking test were similar with the results of the various rotational speed constant load tests. This is because the after several times and some efforts to obtain the optimum resistive load which was most suitable for the fabricated ironless coreless electricity generator, the difference of the resistive load used by the ironless coreless electricity generator was small throughout the maximum power point tracking test. Since the resistive load used by the ironless coreless electricity generator was almost constant, the only thing that would affect the output of the ironless coreless electricity generator would be the rotational speed of the rotor. For the fabricated ironless coreless electricity generator in this research, the optimum range of the resistive load was between 34Ω and 35Ω.

In the maximum power point tracking test, the torque of the ironless electricity generator increased when the rotational speed increased. When the rotational speed of the ironless coreless electricity generator increased, the power output of the generator also increased. As the power output of the generator became higher, the counter electromotive force between the rotors and stator of the generator would likely be higher. With this, the resistance of rotors and stator would increase, thus, more force would be required to operate the generator and thus, producing a higher torque. According to [12], in their research, when the power was higher, the torque density would increase and the power density would decrease.

It was noted that in the constant load various rotational speed tests and the maximum power point tracking test, the efficiency of the ironless coreless electricity generator only increased to around 78% as the rotational speed of the rotors of the generator kept increasing. This is because the ironless coreless electricity generator itself had reached its maximum efficiency when its efficiency was around 78%. It had been anticipated that the efficiency of the ironless coreless electricity generator would still be around 78% although there would be increment on the rotational speed of the rotor of the ironless coreless electricity generator itself. The trend of the efficiency of fabricated ironless coreless electricity generator was in good agreement with that in [13] research as the efficiency of DFIG3G, DDSG and DDPMG in their research would stay constant when these generators reached their maximum capabilities. Due to the limitation on instrumentation, the fabricated generator cannot reach rotational speed to as high as the [14]. But it is believed that there is still more potential and performance that can be harnessed from the generator.

5. Conclusions

The ironless coreless electricity generator showed evidence that it could achieve up to 78% efficiency. In the maximum power point tracking test, after several attempts and efforts, the results obtained from this test were that the optimum resistive load was around 34 ohms to 35 ohms as the ironless coreless electricity generator would be able to produce maximum amount of output power, output voltage, output current, torque and efficiency throughout the rotational speed.

Acknowledgement

The authors would like to express their warmest thank you to the Kementerian Pendidikan Tinggi Malaysia (KPT Malaysia) and the Universiti Malaysia Pahang (UMP) for providing the financial support through the Fundamental Research Grant Scheme (FRGS by the KPT) to realize this project through the research grants RDU160134 (FRGS) and RDU160307 (UMP Grant).

References

[1] Zhang Z, Chen A, Matveev A, Nilssen R and Nysveen A 2013 High-power generators for offshore wind turbines *Energy Procedia* **35** 52–61

[2] Ting C-C and Yeh L-Y 2014 Developing the full-field wind electric generator *Int. J. Electr. Power Energy Syst.* **55** 420–8

[3] Ahmed D and Ahmad A 2013 An optimal design of coreless direct-drive axial flux permanent magnet generator for wind turbine *Journal of Physics: Conference Series* vol 439(IOP Publishing)p 12039

[4] Kobayashi H, Doi Y, Miyata K and Minowa T 2009 Design of axial-flux permanent magnet coreless generator for the multi-megawatts wind turbines *EWEC2009*

[5] Kurt E, Gör H and Demirtaş M 2014 Theoretical and experimental analyses of a single phase permanent magnet generator (PMG) with multiple cores having axial and radial directed fluxes *Energy Convers. Manag.* **77** 163–72

[6] Gör H and Kurt E 2016 Preliminary studies of a new permanent magnet generator (PMG) with the

axial and radial flux morphology *Int. J. Hydrogen Energy* **41** 7005–18

[7] Gor H and Kurt E 2016 Waveform characteristics and losses of a new double sided axial and radial flux generator *Int. J. Hydrogen Energy* **41** 12512–24

[8] Mahmoudi A, Kahourzade S, Rahim N A, Ping H W and Uddin M N 2013 Design and prototyping of an optimised axial-flux permanent-magnet synchronous machine *IET Electr. Power Appl.* **7** 338–49

[9] Virtic P and Avsec J 2011 Analysis of coreless stator axial flux permanent magnet synchronous generator characteristics by using equivalent circuit *Prz. Elektrotechniczny* **87** 208–11

[10] Zikowski Ł and Koczara W 2011 Design And Analysis Of Axial-Flux Coreless Permanent Magnet Disk Generator *Journal of Electrical Engineering*

[11] Abdullah M A, Yatim A H M, Tan C W and Saidur R 2012 A review of maximum power point tracking algorithms for wind energy systems *Renew. Sustain. energy Rev.* **16** 3220–7

[12] Zhang S, Tseng K-J, Vilathgamuwa D M, Nguyen T D and Wang X-Y 2011 Design of a robust grid interface system for PMSG-based wind turbine generators *IEEE Trans. Ind. Electron.* **58** 316–28

[13] Polinder H, Van der Pijl F F A, De Vilder G-J and Tavner P J 2006 Comparison of direct-drive and geared generator concepts for wind turbines *IEEE Trans. energy Convers.* **21** 725–33

[14] Hosseini S M, Agha-Mirsalim M and Mirzaei M 2008 Design, prototyping and analysis of a low-cost disk permanent magnet generator with rectangular flat-shaped magnets *Iran. J. Sci. Technol.* **32** 191

IOP Conf. Series: Materials Science and Engineering **342** (2018) 012013 doi:10.1088/1757-899X/342/1/012013

Performance and optimum characteristics by finite element analysis of a coreless ironless electric generator for low wind density power generation

Akhtar Razali, Fadhlur Rahman, Yap Wee Leong, Mohd Razali Hanipah, and Mohd Azri Hizami

Faculty of Mechanical Engineering, Universiti Malaysia Pahang, 26600 Pekan. Pahang.

E-mail: akhtar@ump.edu.my

Abstract. Cogging is an attraction of magnetism between permanent magnets and soft ironcore lamination in a conventional electric ironcore generator. The presence of cog in the generator is seen somehow restricted the application of the generator in an application where low rotational torque is required. Cog torque requires an additional input power to overcome, hence became one of the power loss sources. With the increasing of power output, the cogging is also proportionally increased. This leads to the increasing of the supplied power of the driver motor to overcome the cog. Therefore, this research is embarked to study fundamentally about the possibility of removing ironcore lamination in an electric generator. This research deals with removal of ironcore lamination in electric generator to eliminate cog torque. A confinement technique is proposed to confine and focus magnetic flux by introducing opposing permanent magnets arrangement. There were several parameters analysed using the JMAG Designer. Transient response analysis was used in the JMAG Designer. The parameters analysed were the number of coil turns per phase, gap distance between the magnet pairs as well as the magnet grade used. These few parameters were analysed under the open circuit condition. Results showed with the increasing of gap distance, output voltage produced decreased. The increment of number of turns in the coils and higher magnet grades used, these increased the output voltage of the generator. With the help of these results, a reference point is established to get optimum design parameter for fabrication of working prototype.

1. Introduction

Ironcore lamination in a generator is used to confine and guide magnetic flux as to boost its efficiency to the maximum. Ironcore as what the name implies, consists of lamination of highly permeability core material which can increase the density of the magnetic flux in over several thousand. The advantage of confining and guiding the flux however giving a major drawback to the generator. When the ironcore lamination is exposed to the magnetic field, magnetic particle in the core tends to line up with the magnetic field of the permanent magnets. When the magnets rotated, its magnetic field changed in direction. This induces continuous movement of the ironcore's magnetic particle for them to align with the change of magnetic field direction and coincidently produces molecular friction. This in turn produces heat and is then transmitted/distributed to the ironcore lamination and windings. Heat causes increases in winding resistance and at the same time retains electromagnetism in the windings and ironcore laminations.

Electromagnetism in the ironcore lamination acted as a magnet and react (attract and repel) with the permanent magnets' field. The more power is generated from the generator, the more power is required to maintain its rotation as the input torque increased proportional to the power output. Therefore, the more heat is generated, the stronger the electromagnetism exists in the core causes unwanted attraction force with the permanent magnets field. Attraction force between permanent magnets and ironcore lamination in a generator or also known as 'cog' is seen as one of the major inefficiency in the generator [1-2].

Cog as the name implies, increases spinning torque where this indirectly increases the amount of work/energy used to spin the generator. Energy to overcome cog is basically proportional to the output power produced. The larger the output power is produced, the larger the amount of torque is required to spin (or maintaining the rotation speed). The presence of cog in the system however has made this conventional generator is far from being usable in low torque application [3-4]. Continuous and consistent power is required to overcome cog and the power varies when rotational speed varies.

There were many attempts made elsewhere to reduce cog to increase the application boundary of a generator [5]. Most of the efforts were focused on the ironcore optimization in design to minimize the effect of cog [6]. However, the presence of cog is still inevitable [7]. Effort made by removing iron core material from an electric motor had demonstrated great success more than a decade ago. The success story of this ironless motor has been widely shared by both among the researchers and industries. The idea behind this achievement is by removing the iron core lamination and replace it with non-ferrous material. The ironless motor works flawlessly and one of the major advantage came from it is outstanding positional accuracy and repeatability due to no cogging affecting the positioning.

Similar idea may be applied to demonstrate coreless generator design. Effort made on this subject however, is still lacking [8-10]. The idea of removing ironcore lamination material may cause non-concentrated flux and leads to deterioration in output power efficiency [3]. However, there is a method which may be used to concentrate and focus magnetic flux to create denser magnetic field. An additional permanent magnets arrangement added to the generator may be a solution to address this issue. The absence of ironcore lamination in the system also represent cog free rotation. This provides advantage in terms of very low starting torque and less counter electromotive force is produced.

2. Procedures

Electromechanical analysis could help gain the result of magnetic flux exerted by the magnet from the magnetic flux density distribution of the electricity generator design and the magnetic flux on the coil. The ironless electricity generator was designed using CAD software. The proposed design for the ironless coreless electricity generator in this research was double-rotor single-stator configuration. Within the stator, there were 12 coils which were joined using three-phase star connection. As for the rotors, 16 magnets were embedded on each rotor. The ironless coreless electricity generator model was then exported to the JMAG designer software when design was completed. The ironless coreless electricity generator model was analysed by using JMAG designer analysis software. Transient analysis was used to examine the model because it was able to simulate real life situation in which the ironless coreless electricity generator would be operating.

After choosing the type of magnets to be used in the analysis, the direction of the magnetic force was defined. Since the connection for the ironless coreless electricity generator model has to be in star three-phase connection, the circuit connection for the analysis on the model was set accordingly. The voltage probe was attached on the end of star three-phase connection so that the voltage for each phase could be shown. As for the open-loaded test, there was no load applied on the circuit. Meanwhile, for the closed-loaded test, the load was applied on the end of the star connection. After the connection for the analysis was defined, the rotation motion which needed to be applied on the rotors of the model was defined. It was set to 500 rotational per minute as the rotational speed would be the constant variable for this experiment. The U-phase, V-phase and W-phase FEM coils were defined accordingly and set to clockwise direction to avoid directional error on the circuit. These coils were then synchronised with the circuit. The method of how to set the phase for the FEM coils are readily available in the JMAG designer official website, application note section.

As for the meshing type in this analysis, the standard meshing with auto meshing method was chosen as the software is set to the appropriate meshing size by default for the ironless coreless electricity generator model. The step control for the analysis on the model for this research was set based on Equation 1 where t_1 is desired end time of the analysis and t_0 is start time of analysis. For the case on this research, the regular intervals step control was chosen. It is because by using this type of step control, the regularity of the simulated situation for the model could be shown. The number of steps used was defined as 37 steps and the number of division was set to 36 divisions. The reason for the setting to be made in such a way was because the first steps of the simulation were not so significant since the generator just started to operate. Another reason is the simulation had to run in full cycle, as 36 divisions of analysis were able to finish it with 37 steps. The reason that the simulation just ran in one full cycle is because to save the time on running the analysis on the model since the outcome of the simulation would be the same. After all the setting was done, the analysis on the ironless coreless electricity generator model was started.

$$Time\ Interval\ (s) = \frac{t_1 - t_0}{Division} \tag{1}$$

Results such as the magnetic flux density contour plot, circuit voltage and magnetic flux of FEM coil were obtained in the open-loaded circuit test as there was no load applied on the analysis on the model. As for the closed-loaded circuit, the joule loss, current and electric power could be obtained since there was load applied on the analysis on the model and the current generated was flowing through the applied load. Several parameters were tested on the generator. These parameters were the magnet grade used on the model, number of coil turns per phase and gap distance between the magnets. The reason to test the parameter was to gain knowledge on how these parameters would affect the output of the model. It would also help to get the optimized parameter which could be used on the fabrication works of the ironless coreless electricity generator.

3. Results and discussions

3.1 Number of Coil Turns per Phase

There were four sets of number of coil turns per phase done in this section: 500 coil turns per phase; 1000 coil turns per phase; 1500 coil turns per phase; and, 2000 coil turns per phase. Other parameters such as magnet grade used, material used on rotor and stator and gap distance between the magnet pairs remained constant so that the best results for the number of coil turns per phase could be obtained. The magnet grade used in this analysis was N48 Neodymium magnet, while the gap distance between the magnet pairs was set to 12mm and the material used on the rotor and stator was aluminium. The only concern was how number of coil turn per phase effects the outcome, other parameters just act as the constant variables. Figure 1 shows the magnetic flux density contour plot on the ironless electricity generator in JMAG designer software. The purple colors signify that there was no magnetic flux distribution within the area as the value of the magnetic flux density was 0T, while the lighter the blue color signifies the more magnetic flux distribution happened within the area as the magnetic flux density value fluctuated between 4.4737×10^{-2}T and 3.5789×10^{-1}T. It is good result as it shows very less magnetic flux flow outwards the rotor and stator.

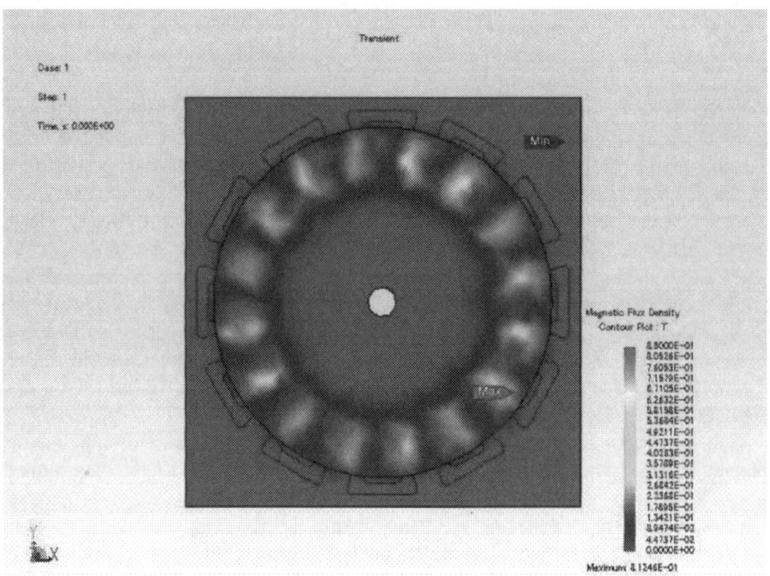

Figure 1. Magnetic Flux Density Contour Plot for 500 Coil Turns per Phase Analysis in JMAG designer software.

Table 1 shows the summary of the finite element analysis on the number of coil turns per phase. It is apparent that the torque of the rotors of the ironless coreless electricity generator remained constant when the coil turns per phase increased. The value of the rotor torque in RMS was around 2.391Nm. The magnetic flux within U-phase coil, V-phase coil and W-phase coil as well as circuit voltage for U-phase coil, V-phase coil and W-phase coil showed an upward trend as the number of coil turns per phase increased. By comparing the results for 500 coil turns per phase and 1000 coil turns per phase, the output precisely increased one-fold as observed in the 500 coil turns per phase analysis. As can be seen in the table, the RMS value for the magnetic flux per phase in the 500 coil turns per phase was approximately 0.334Wb/phase and this increased to 0.668Wb/phase for the magnetic flux per phase in the 1000 coil turns per phase. The magnetic flux value per phase further increased to approximately 1.003Wb/phase and 1.337Wb/phase in the 1500 coil turns per phase and in the 2000 coil turns per phase, respectively. As for the value of circuit voltage, an upward trend was also observed when the coil turns per phase increased. In the 500 coil turns per phase, the RMS value was approximately 132V/phase and later increased to 264V/phase in the 1000 coil turns per phase. The value of the circuit voltage further increased to approximately 396V/phase and 528V/phase in the 1500 coil turns per phase and in the 2000 coil turns per phase, respectively.

Table 1. Summary of the finite element analysis on the number of coil turns per phase.

	Number of turns per phase (coils)			
	500	1000	1500	2000
Gap distance between the magnet	12mm			
Magnet grade used	N48 Neodymium Magnet			
Rotational speed of rotor	500RPM			
Rotor Torque (Nm_{rms})	2.391	2.391	2.391	2.391
Magnetic flux for U-phase coil (Wb_{rms})	0.335	0.669	1.004	1.339
Magnetic flux for V-phase coil (Wb_{rms})	0.334	0.668	1.003	1.337
Magnetic flux for W-phase coil (Wb_{rms})	0.338	0.676	1.013	1.351
Circuit voltage for U-phase coil (V_{rms})	131.531	263.061	394.592	526.123
Circuit voltage for V-phase coil (V_{rms})	132.844	265.687	398.531	531.374
Circuit voltage for W-phase coil (V_{rms})	131.233	262.466	393.700	524.933

3.2 Gap Distance between the Magnet Pairs

Table 2 shows the summary of the finite element analysis on the gap distance between the magnet pairs. What is interesting in these data is that the torque of the rotors of the ironless coreless electricity generator showed a downward trend as the gap distance between the magnet pair increased. During 14mm gap distance test, the torque RMS value of the rotors was 2.951Nm. When the gap distance increased to 16mm, the torque RMS value of the rotors decreased to 2.893Nm. The torque RMS value of the rotors was then further decreased to 2.838Nm when the gap distance increased to 18mm. In 20mm gap distance test, the torque RMS value of the rotors dropped significantly to 2.367Nm. The RMS value of torque of the rotors further dropped to 2.291Nm during the 22mm gap distance test. The torque RMS of the rotors decreased to 2.098Nm when the gap distance increased to 24mm. During the 26mm gap distance analysis, the RMS value of the torque of the rotors of the ironless coreless electricity generator decreased to 1.819Nm.

In the gap distance between magnet pair analysis, the magnetic flux and the circuit voltage of the three-phase coil of the ironless coreless electricity generator was in decreasing trend while the gap distance between the magnet pairs is increase. As shown in 14mm gap distance test, the magnetic flux RMS value per phase was 0.415Wb/phase. The magnetic flux RMS value decreased to approximately 0.386Wb/phase for 16mm gap distance test. The magnetic flux RMS value further decreased to around 0.361Wb/phase for 18mm gap distance test. When the experiment proceeded with the 20mm gap distance test, the magnetic flux RMS value per phase decreased to 0.337Wb/phase. For the 22mm gap distance test, the magnetic flux RMS value was reduced to 0.335Wb/phase. The magnetic flux RMS value per phase then further dropped to 0.294Wb/phase when the 24mm gap distance test was carried out. During the 26mm gap distance test, the lowest value of magnetic flux RMS value per phase amongst all the test results was obtained, which was approximately 0.274Wb/phase.

In the 14mm gap distance test, the circuit voltage RMS value was approximately 163V/phase. The circuit voltage RMS value decreased to 152V/phase when 16mm gap distance test was carried out. The circuit voltage RMS then further dropped to 142V/phase during the 18mm gap distance test. In 20mm gap distance test, the circuit voltage RMS value was approximately 132V/phase, indicating a lower value compared to that of the 18mm gap distance test. For 22mm gap distance test, the circuit voltage RMS value was 131V/phase. The RMS value for the circuit voltage then continued to decline to 116V/phase in 24mm gap distance test. In the 26mm gap distance test, the circuit voltage RMS was 109V/phase, which was the lowest amongst all values in the gap distance tests.

3.3 Magnet Grade Used on Magnet Pairs

There were five sets of analysis on the magnet grade test. The magnet grades used were N42, N45, N48, N50 and N52 Neodymium. Other parameters such as the number of coil turns per phase, gap distance between the magnet pairs and material used on the rotor and stator remained constant so that the results for magnet grade tests could be obtained accurately. The number of coil turns per phase was set to 500 coil turns per phase and the gap distance between the magnet pairs was set to 20mm.

Table 3 shows the summary of the finite element analysis on magnet grade used on magnet pairs. From Table 3, the torque of the rotors of the ironless coreless electricity generator shows an upward trend when the magnet grade used was increased. When N42 Neodymium Magnets was used in the analysis, the RMS value of torque of the rotors was 1.958Nm. The RMS value of the torque of the rotors then increased to 2.103Nm when N45 Neodymium Magnet was used. The RMS value of the torque of the rotors further increased to 2.165Nm when the N48 Neodymium Magnet was used in the analysis. When the N50 Neodymium Magnet and the N52 Neodymium Magnet were used in the magnet grade analysis, the RMS values of torque of the rotors rose to 2.293Nm and 2.391Nm, respectively.

In the magnet grade analysis, the magnetic flux and the circuit voltage in the three-phase coil showed an increasing trend from N42 Neodymium magnet to N52 Neodymium magnet. As indicated in the magnetic flux RMS value in the N42 Neodymium magnet test, a value of around 0.303Wb/phase was obtained. The magnetic flux RMS value increased to 0.314Wb/phase in the N45 Neodymium magnet test and further increased to 0.318Wb/phase in the N48 Neodymium magnet test. The N50 Neodymium magnet test produced a value of approximately 0.328Wb/phase for the magnetic flux RMS value while the N52 Neodymium magnet test produced a value of 0.335Wb/phase, which was the highest value amongst all values in the five tests.

Meanwhile, the circuit voltage RMS value for in the N42 Neodymium magnet test was 119V/phase. This value increased to 123V/phase in the N45 Neodymium magnet test and slightly increased to 125V/phase in the N48 Neodymium magnet test. The circuit voltage RMS values in the N50 Neodymium and N52 Neodymium magnet tests continued to increase to 129V/phase and 131V/phase, respectively.

The circuit voltage RMS value in N52 Neodymium magnet test was the highest amongst all values in the five magnets grade test.

Table 2. Summary of finite element analysis on the gap distance between magnet pairs.

	Gap Distance (mm)						
	14mm	16mm	18mm	20mm	22mm	24mm	26mm
Number of coil turns per phase	500						
Magnet grade used	N48 Neodymium Magnet						
Rotational speed	500RPM						
Rotor Torque (Nm_{rms})	2.951	2.893	2.838	2.367	2.291	2.098	1.819
Magnetic flux for U-phase coil (Wb_{rms})	0.416	0.386	0.361	0.337	0.335	0.294	0.274
Magnetic flux for V-phase coil (Wb_{rms})	0.415	0.385	0.358	0.336	0.334	0.294	0.275
Magnetic flux for W-phase coil (Wb_{rms})	0.418	0.388	0.362	0.340	0.338	0.296	0.277
Circuit voltage for U-phase coil (V_{rms})	162.989	151.330	141.315	131.901	131.531	115.031	107.439
Circuit voltage for V-phase coil (V_{rms})	164.736	152.881	142.304	133.684	132.844	116.692	109.364
Circuit voltage for W-phase coil (V_{rms})	162.932	151.248	141.110	132.625	131.233	115.526	108.128

Table 3. Summary of the finite element analysis on magnet grade used on magnet pairs.

	N42	N45	N48	N50	N52
Number of coil turns per phase			500		
Gap distance between magnet pairs			20mm		
Rotational speed of rotor			500RPM		
Rotor torque (Nm_{rms})	1.958	2.103	2.165	2.293	2.391
Magnetic flux for U-phase coil (Wb_{rms})	0.303	0.314	0.318	0.328	0.335
Magnetic flux for V-phase coil (Wb_{rms})	0.302	0.313	0.318	0.327	0.334
Magnetic flux for W-phase coil (Wb_{rms})	0.306	0.316	0.321	0.331	0.338
Circuit voltage for U-phase coil (V_{rms})	119.031	123.338	125.159	128.800	131.531
Circuit voltage for V-phase coil (V_{rms})	120.217	124.570	126.408	130.086	132.844
Circuit voltage for W-phase coil (V_{rms})	118.760	123.060	124.876	128.509	131.233

3.4 Discussion on the Simulation Results

During the test on coil turns per phase, the torque on the rotors of the ironless coreless electricity generator was remained constant when the coil turns per phase increase. It was because there was no adjustment made on the rotors, for example, to make the rotors heavier, and thus, the torque of the rotors remained constant. Both magnetic flux within the circuit and circuit voltage showed an upward trend with the increase in the coil turns per phase. This is because as the number of coil turns per phase increased, there would be increasing number of magnetic flux cutting process by the coil and thus, magnetic flux and circuit voltage within the three-phase coil would also increase. An increase in coil turns would cause the output of the generator, the cost as well as the size to increase. Since in this research was concerned with developing a table size ironless coreless electricity generator, 500 coil turns per phase was chosen because it required less space and minimal cost.

The circuit voltage RMS value declined while the gap distance between the magnets pairs increased in the gap distance analysis. The reason behind this is because when the distance between magnet pairs increased, the magnetic strength between the magnet pairs decreased, which caused the quality of the magnetic flux cut to drop and thus, magnetic flux and circuit voltage generated in three phase coils would decrease. If the gap distance was too close, there could be a chance that rubbing might occur between the rotor and stator, which would reduce the efficiency of the ironless coreless electricity generator. By considering this error, the gap distance between the magnet pairs of 26mm was chosen to prevent such problem from happening.

The magnetic flux RMS value and circuit voltage RMS value showed increasing trends with the increase in the grade of the magnet used in the magnet grade analysis. The reason behind it is because amongst these five grades of magnet, N52 had the strongest magnetic field and it did produce better quality magnetic flux during the three-phase coil cutting through the magnetic field. In the market, N50

Neodymium magnet and N52 Neodymium magnets can be easily obtained. N52 Neodymium magnet was the choice for making the electricity generator because it could produce better quality of the magnetic flux compared to that of the other grades. Although using N52 Neodymium magnet on ironless coreless electricity generator did increase the torque of the generator, the output of the generator did compensate on such weakness.

During the gap distance analysis, the torque of the rotors of the ironless coreless electricity generator showed downward trend as the gap distance increased. The reason behind it is because when the magnet pairs moved further away, the magnetic force between it became weakened and made the pulling force between the rotors weaker. In the magnet grade analysis, the torque of the rotors of the ironless coreless electricity generator increased when the magnet grade used on the analysis was stronger. Such phenomena occurred because when the stronger magnet was used on the rotors, it made the pulling force between the rotors became stronger. When the pulling force between the rotors was stronger, more energy would be required to spin. Thus, this would increase the torque of the rotors.

4. Conclusions

The results in this finite element analysis showed that the optimum parameters to be used were: 1000 coil turns per phase; 14mm gap distance between the magnet pairs; N52 Neodymium magnet grade; and, plastic as the material used on rotor and stator. However, due to the consideration of the manufacturing process tolerance such as deformation after machining, 500 coil turns per phase, 26mm gap distance between the magnet pairs, N52 Neodymium magnet grade used on rotor and stator was chosen. Another finite element analysis would also be conducted on the optimum parameter design so that the results could be used as the datum to fabricate the ironless coreless electricity generator.

Acknowledgement

The authors would like to express their warmest thank you to the Kementerian Pendidikan Tinggi Malaysia (KPT Malaysia) and the Universiti Malaysia Pahang (UMP) for providing the financial support through the Fundamental Research Grant Scheme (FRGS by the KPT) to realize this project through the research grants RDU160134 (FRGS) and RDU160307 (UMP Grant).

References

[1] Zhang Z, Chen A, Matveev A, Nilssen R and Nysveen A 2013 High-power generators for offshore wind turbines *Energy Procedia* **35** 52–61

[2] Ting C-C and Yeh L-Y 2014 Developing the full-field wind electric generator *Int. J. Electr. Power Energy Syst.* **55** 420–8

[3] Ahmed D and Ahmad A 2013 An optimal design of coreless direct-drive axial flux permanent magnet generator for wind turbine *Journal of Physics: Conference Series* vol 439 (IOP Publishing) 12039

[4] Kobayashi H, Doi Y, Miyata K and Minowa T 2009 Design of axial-flux permanent magnet coreless generator for the multi-megawatts wind turbines *EWEC2009*

[5] Kurt E, Gör H and Demirtaş M 2014 Theoretical and experimental analyses of a single phase permanent magnet generator (PMG) with multiple cores having axial and radial directed fluxes *Energy Convers. Manag.* **77** 163–72

[6] Gor H and Kurt E 2016 Waveform characteristics and losses of a new double sided axial and radial flux generator *Int. J. Hydrogen Energy* **41** 12512–24

[7] Gör H and Kurt E 2016 Preliminary studies of a new permanent magnet generator (PMG) with the axial and radial flux morphology *Int. J. Hydrogen Energy* **41** 7005–18

[8] Mahmoudi A, Kahourzade S, Rahim N A, Ping H W and Uddin M N 2013 Design and prototyping of an optimised axial-flux permanent-magnet synchronous machine *IET Electr. Power Appl.* **7** 338–49

[9] Virtic P and Avsec J 2011 Analysis of coreless stator axial flux permanent magnet synchronous generator characteristics by using equivalent circuit *Prz. Elektrotechniczny* **87** 208–11

[10] Zikowski Ł and Koczara W 2011 Design And Analysis Of Axial-Flux Coreless Permanent Magnet Disk Generator *Journal of Electrical Engineering*

Colour and spreadability of Neem (Azadirachta Indica A. juss) ointment and cream formulations

Azierah Zawiyyah and Mohd Shamsul Anuar

Department of Process and Food Engineering, Faculty of Engineering, Universiti Putra Malaysia, 43400 UPM Serdang, Selangor

E-mail: mshamsul@upm.edu.my

Abstract. Herbal plants are a major source of raw material for traditional medicines. Recently there has been an increase of interest to study the therapeutic potential of herbal plants as herbal care products. In this study, a preliminary study on the formulation of neem (Azadirachta Indica) ointment and cream have been conducted. The neem leaves were extracted and formulated into ointment and cream. The raw neem extract is added into the ointment and cream bases at four different concentrations (0% w/w, 0.5% w/w, 1% w/w and 2% w/w) and stored at three different storage temperatures (4°C, 25°C and 45°C). The semambu ointment and cream formulated were evaluated in terms of their colour and spreadability. It has been found that the extract content and storage temperature influence the colour and spreadability of the formulated neem ointment and cream.

1. Introduction

Herbal plants are popularly used for their therapeutic, medicinal, aromatic or savory qualities. One or more parts of herbal plants contain and produce a variety of chemical substances and unique bioactive ingredients that can are advantageous to the human body. It has been reported that not less than 80% of the world's populations use herbal medicine for some part of their primary health care [1]. The demand of herbal medicine that is derived mainly from herbal plants is growing exponentially as the productions of these health care products require a great amount of natural ingredients from herbal plants. This also brings up the interest in the design of formulations which contain active natural herbal ingredients that acts as vehicle to deliver the drugs to specific targets in the body.

Pharmaceutical semisolid such as ointment and cream are the examples formulation which can be use as the vehicle to deliver the herbal medicament. The herbal plants extract which acts as the active ingredient is added to the semisolid to give the therapeutic effect. Natural extract of aloe vera, hena, tomato and cucumber are some of example for the most commercial and well known pharmaceutical semisolid products. Neem or semambu (Azadirachta indica A. Juss) as it is known in Malaysia, is one of the most useful traditional medicinal plant which is well known in India (Neem) and neighbouring countries [2]. In Malaysia, semambu leaves are traditionally used to cure mild skin diseases such as measles and scabies. Besides medicinal potential, this plant has also other potential such as for use as a pesticide [3] and pollution control properties [4, 5].

There were a number of recent works that studied the formulation of azadirachta indica based substances into ointments and creams but some important properties such as colour and spreadibility were not investigated and quantified [6, 7] especially at different storage temperatures. Hence, the objectives of this present work are to formulate semambu topical ointments and cream, and to evaluate their physical sensorial properties in terms of colour and spreadability when stored at different storage temperatures.

2. Material and methods

2.1 Preparation of semambu leaves

Fresh healthy semambu leaves (Azadirachta Indica A. juss) were collected from the Faculty of Design and Architecture, Universiti Putra Malaysia. The semambu leaves were destalked. Then the leaves were spread evenly and allowed to dry in an oven at 45^{0}C for 15 hours, where this method had been reported that the semambu extraction leave can still preserved relatively high Total Phenolic Content (TPC) values [8]. The dried semambu leaves were then ground to fine powder and stored in air tight containers.

2.2 Preparation of semambu leaf extract

The dried semambu powders were weight for approximately 5 g. Then, the powder was extracted with 200 ml methanol solution for 5 hours by using Soxhlet apparatus (Electrothemal, U.K). Then the extract solutions were filtered using a Whatman No.1 filter paper. The resulting semambu leaves extracts were evaporated at 45^{0}C temperature using a rotary evaporator (Laborota 4001 Efficient, Heidolph, Germany) to obtain the respective extracts. The concentrated extracts were stored in a refrigerator at -4^{0}C until used for further analysis [8].

2.3 Formulation of semambu ointment and cream

For semambu ointment formulation, the required quantities of petroleum jelly were placed in the water bath and allowed to melt at 70^{0}C. After melting, the liquid petroleum jelly was stirred gently at the temperature of 70^{0}C and the semambu leave was added. The mixture were then cooled and solidified forming the semambu ointment. Meanwhile, for semambu cream, the semambu leave extract was directly mixed with the aqueous cream without heating. The semambu leave extract were added at four different concentrations (0% w/w, 0.5% w/w, 1% w/w and 2% w/w). The different concentrations of semambu ointment and semambu cream prepared were stored at 4°C, 25°C, and 45°C and the physical tests were conducted after 1 week of storage period.

2.4 Physical analyses

2.4.1 Colour

The colour of the samples was measured by using a Colour Reader (Konica Minolta Sensing, Japan). L*, a*, and b* values were measured for each sample. The L* value represent the relative brightness, ranging from total black (L*=0) to total white (L*=100) and a* and b* are the chromaticity coordinates. Where, the chromaticity coordinate a* represents the different between red (positive value, +a*) and green (negative value, -a*) and the chromaticity coordinate b* represents the difference between yellow (positive value, +b*) and blue (negative value, -b*). The meter was calibrated against a standard white plate before each measurement. This method for evaluating the colour of materials has been previously used for herbs and food products [8, 9].

2.4.2 Spreadability

The objective the spreadability characterization for the different concentrations of the semambu ointments and semambu creams is to mimick the process of applying them onto the human skin. The spreadability of the samples was measured using a texture analyzer (Stable Micro Systems TA XT Plus, UK). TTC Spreadability Rig was used to evaluate the spreadability in terms of the firmness properties of the samples. The TTC Spreadability Rig is a set of precisely matched male and female perspex 90° cones. Tests were carried out by fillong the sample in the female cones sample holders immediately after removal from storage. Then the tested samples in female cones were locked into the base holder precisely centred under the matching upper male cone probe. During the experiment, the male cone probe is programmed to penetrate the samples in the female cones holder and the sample is squeezed out from the female cones. A force versus time data of penetration of the male probe through the ointment or cream in the female probe was obtained. When the male TTC Spreadability Rig probe penetrated into the female probe which contains the sample, the force applied increases up to the point of maximum penetration depth. The maximum force value can be taken as the 'firmness' at the specified depth.

3. Results and discussions

3.1 Colour

Figure 1 illustrates the lightness (L*value) for the semambu ointments. Pure petroleum jelly exhibit relatively higher lightness in comparison to ointments containing semambu extracts. There are no apparent differences in terms of lightness between the semambu ointments with 0.5%, 1% and 2% concentrations. There is a slight increase in the L* value when the storage temperature increases from 4°C to 45°C for ointments containing 0.5% and 2% semambu extracts.

Figure 2 illustrates the chromaticity coordinate a* value for semambu ointment where this value represented the differences between red (positive value) and green colour (negative value). It can be observed that as the storage temperature increases, the a* value becomes more positive, indicating that the ointment is more towards the red colour.

As can be observed in Figure 3, all four different concentrations of semambu ointments give a positive b* value. This indicates that the semambu ointment is toward the yellow colour. Increasing semambu extract concentration decreases the b* value. As the temperature increases from 4°C to 25°C, the b* values decreases. But as the temperature is further increased to 45°C, the changes were not apparent. This shows that the b* value are affected by low storage temperature changes between 4°C and 25°C.

In general, the lightness of the semambu cream is decreasing as the added semambu extract concentration increases as shown in Figure 4. There are generally no apparent differences in terms of lightness between semambu creams with 0.5% and 1% semambu extract concentrations except at the highest storage temperature of 45°C. The lowest lightness is observed for the cream with 2% semambu extract concentration at 45°C storage temperature.

Figure 5 represents the chromaticity a* for semambu cream. All the four different concentrations of the semambu creams give negative a* values indicating towards green colour when stored at 4°C and 25°C. A higher storage temperature of 45°C produces semambu creams that exhibit higher redness in their colours; positive a* values. Hence, storage temperature also affects the a* value where the a* value generally increases (becomes more positive) with increasing storage temperature.

Figure 6 illustrates that all four different concentrations of semambu cream give a positive b* value. This indicates that the semambu cream is toward the yellow colour. In general, as the concentration of the semambu extract increases, the yellow colour is decreasing approaching blue colour denoted by the decrease in the b* values except at the lowest storage temperature of 4°C.

In terms of the colour lightness or L* value, the obvious differences between the semambu ointment and semambu cream are that the effect of semambu extract concentrations is more pronounced for the semambu cream formulations depicted by the higher L* value fluctuations (Figure 6) compared to semambu ointments (Figure 1). For the a* value, semambu ointment gives positive values at all the storage temperatures but for semambu cream only the semambu cream at storage temperature of 45°C gives a positive (red colour) result. Meanwhile, for the b* value which represent the differences between yellow (positive value) and blue colour (negative value), both semambu ointment and semambu cream yield positive b* value results but the semambu creams exhibit relatively higher positive b* value (yellow colour) then the semambu ointments.

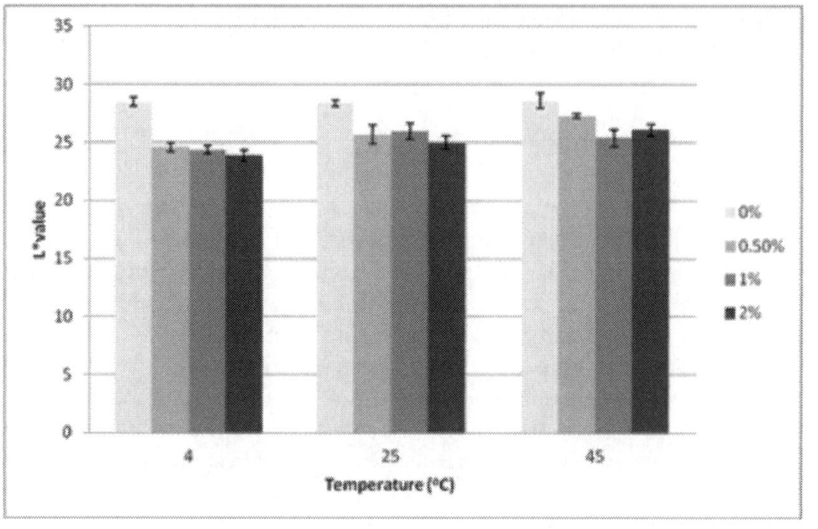

Figure 1. L* value versus storage temperature for semambu ointments containing 0% w/w, 0.5% w/w, 1% w/w and 2% w/w semambu extracts.

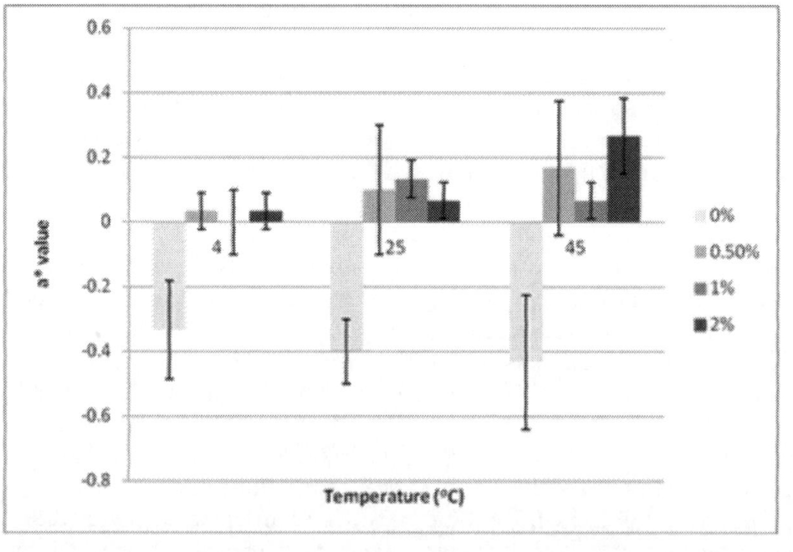

Figure 2. a* value versus storage temperature for semambu ointments containing 0% w/w, 0.5% w/w, 1% w/w and 2% w/w semambu extracts.

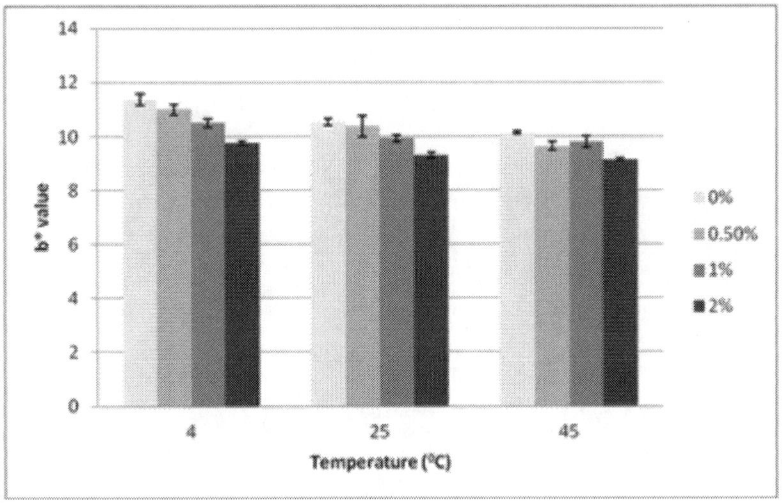

Figure 3. b* value versus storage temperature for semambu ointments containing 0% w/w, 0.5% w/w, 1% w/w and 2% w/w semambu extracts.

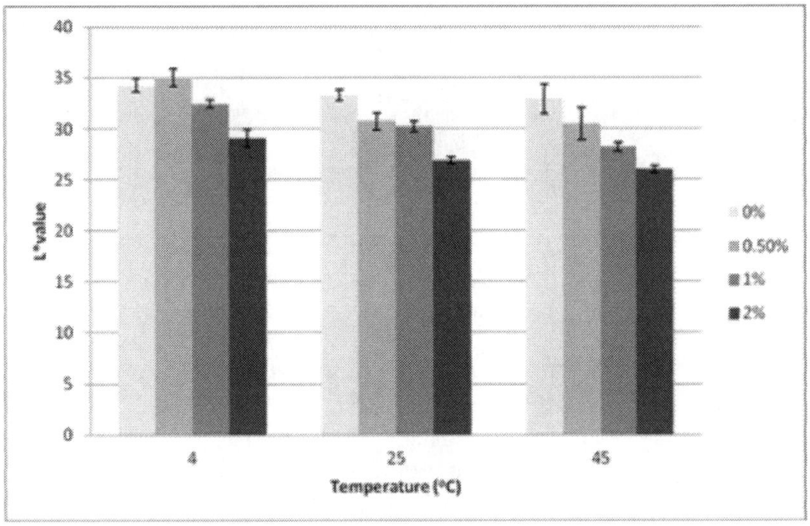

Figure 4. L* value versus storage temperature for semambu creams containing 0% w/w, 0.5% w/w, 1% w/w and 2% w/w semambu extracts.

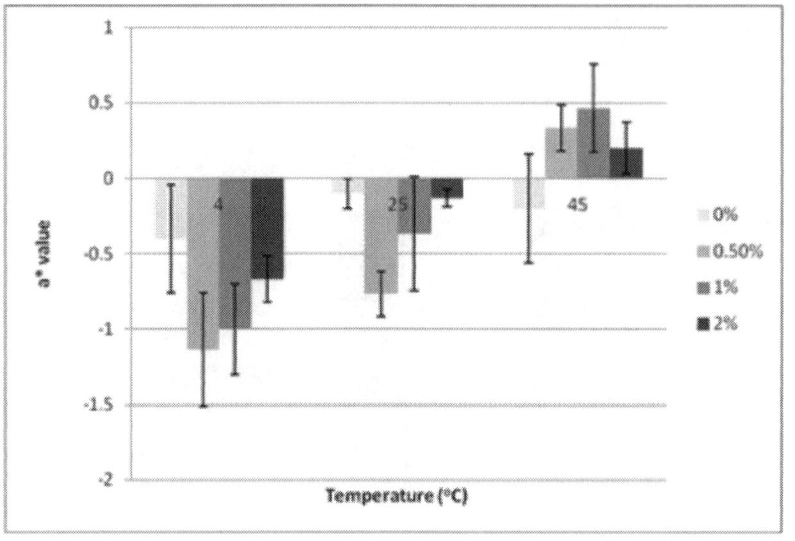

Figure 5. a* value versus storage temperature for semambu creams containing 0% w/w, 0.5% w/w, 1% w/w and 2% w/w semambu extracts.

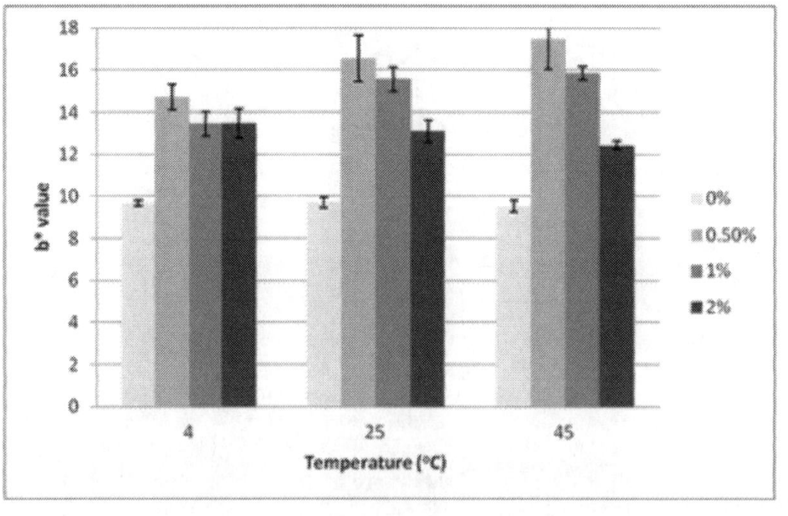

Figure 6. b* value versus storage temperature for semambu creams containing 0% w/w, 0.5% w/w, 1% w/w and 2% w/w semambu extracts.

3.2 Spreadability
The spreadability properties can be determined by measuring the firmness of the sample. The firmness of the semambu ointments and semambu cream at three different storage temperatures for four different semambu extract concentrations are presented in Figure 7 and Figure 8.

As the storage temperature of the semambu ointment increases, the firmness of the semambu ointments decrease exponentially as illustrated in Figure 7. At storage temperatures of 25°C and 45°C, the firmness of the semambu ointments are lower compared to the semambu ointments stored at 4°C. The ointment network structure degrades as the temperature increases which resulting the semambu ointment firmness decreases as the storage temperature increases [10]. Figure 7 also illustrates that the firmness of the semambu ointments is inversely proportional to the added semambu extract concentrations, where the higher the concentration of the semambu extract, the lower the firmness of the ointment formulation. This shows that the spreadability of the formulated semambu ointments increases with increasing semambu extract concentration. The amount of aqueous semambu extracted added will influence the properties of the ointment as the network structure of the ointment base has been disrupted [11]. Hence, the semambu ointment that exhibit the highest spreadability is the ointment which has the highest amount of semambu extract added, 2% w/w and being stored at highest storage temperature of 45°C. There is also a relatively higher difference in the ointment firmness when the storage temperature increases from 4°C to 25°C in comparison to those observed between the storage temperature increases from 25°C to 45°C.

Figure 8 illustrates that the firmness for semambu creams containing 0.5% w/w semambu extract exhibit a relatively linear decrease as the storage temperature increases. The effect of increasing storage temperature on the spreadability of the semambu creams is more pronounced as the concentration of the semambu extract increases. Higher semambu extract concentrations and storage temperatures lowers the semambu creams firmness hence enhances their spreadabilities. Thus, the semambu cream exhibiting the highest spreadability contains 2% w/w of semambu extract stored at 45°C.

The firmness of the semambu ointments are relatively higher compared to semambu creams at lower storage temperatures of 4°C and 25°C . This shows that the semambu ointment has a firmer structure and thus lower spreadability relative to the semambu cream at these storage temperatures. Both formulations yield similar trends where the spreadability increases with the increase in the added semambu extract concentration and storage temperature.

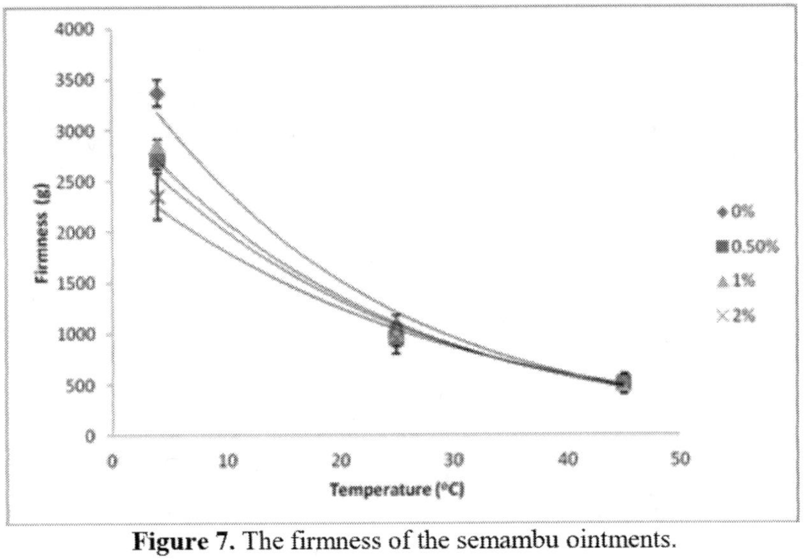

Figure 7. The firmness of the semambu ointments.

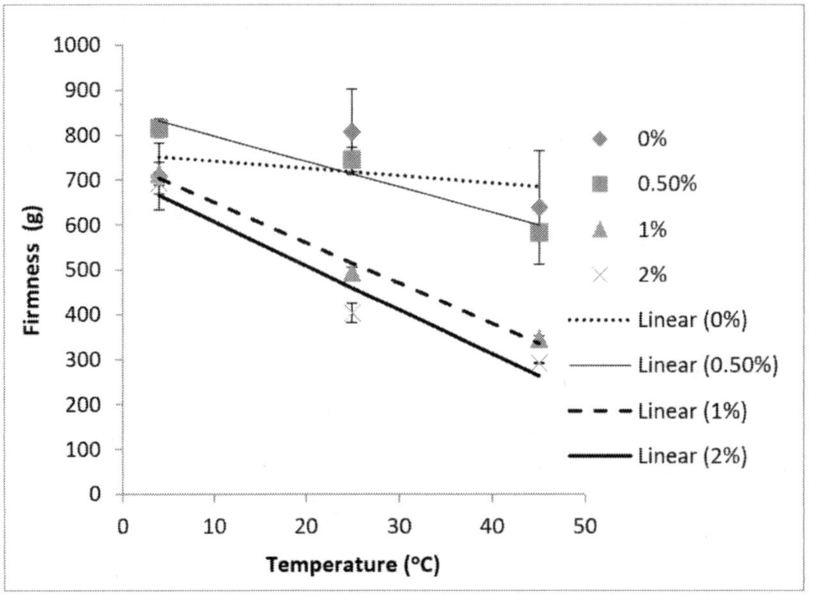

Figure 8. The firmness of the semambu creams.

4. Conclusions

The physical properties of the semambu ointment and semambu cream formulations were evaluated in terms of their colour and spreadability. In terms of colour, semambu ointment shows a more stable formulation compared to semambu cream when subjected to a variation in the storage temperature and semambu extract concentration. From the observation, it can be concluded that the firmness for the semambu ointment and semambu cream decreases as the storage temperature increases. In addition, the semambu ointment has a firmer structure and low spreadability properties compared to the semambu cream.

Acknowledgements

The authors acknowledges the financial sponsorship from Universiti Putra Malaysia in funding this research project.

References

[1] Ekor M 2013 The growing use of herbal medicines: issues relating to adverse reactions and challenges in monitoring safety *Frontiers in Pharmac.* **4** 177

[2] Maan P, Yadav K S and Yadav N P 2017 Wound Healing Activity of *Azadirachta indica* A. Juss Stem Bark in Mice *Pharmacognosy Mag.* **13**(Suppl 2) S316–S320

[3] Chaudhary S, Kanwar R K, Sehgal A, Cahill D M, Barrow C J, Sehgal R and Kanwar J R 2017 Progress on *Azadirachta indica* Based Biopesticides in Replacing Synthetic Toxic Pesticides *Frontiers in Plant Sci.* **8** 610

[4] Bhattacharyya K G and Sharma A 2004 Azadiachta indica leaf powder as an effective biosorbent for dyes: a case study with aqueous congo red solutions *J. Environ. Management* **71** 217-229

[5] Ibrahim M B and Sani S 2015 Neem (azadirachta indica) leave for removal of organic pollutants *J. Geoscience Environ. Protection* **3** 1-9

[6] Banerjee K, Thiagarajan N and Thiagarajan P 2016 Azadirachta indica A. Juss based emollient cream for potential dermatological applications *Indian J. Pharm. Sci.* **78** 320-325

[7] Sawant, S E and Tajane M D 2016 Formulation and evaluation of herbal ointment containing Neem and Tumeric extract *J. Sci. Innov. Res.* **5** 149-151

[8] Sejali S N F and Anuar M S 2011 Effects of drying methods on phenolic contents of neem (azadirachta indica) leaf powder *J. Herbs, Spices & Med. Plants*, **17** 119-131
[9] Tarlak F, Ozdemir M and Melikoglu M 2016 Computer vision system approach in colour measurements of foods: Part II. validation of methodology with real foods *Food Sci. Technol.* **36** 499-504
[10] Pena L E, Lee B L and Stearns J F 1994 Structural rheology of a model ointment *Pharmac. Res.* **11** 875-881
[11] Park K and Song K W 2010. Rheological evaluation of petroleum jelly as a base material in ointment and cream formulations with respect to rubbing onto the human body *Korea-Australia Rheol. J.* **22** 279-289

IOP Conf. Series: Materials Science and Engineering **342** (2018) 012015 doi:10.1088/1757-899X/342/1/012015

The trend of road traffic crashes at urban signalised intersection

Nurul Farhana Nasarrudin and Intan Suhana Mohd Razelan

Faculty of Civil Engineering and Earth Resources, Universiti Malaysia Pahang, Kuantan

E-mail: ana_nasarrudin@yahoo.com

Abstract. Road traffic crash is one of the main contributing factors for deaths in the world. Intersection is listed as the second road type which road crashes occurred frequently. Hence, the traffic light was installed to minimise the road crashes at intersection. However, the crashes are still occurring and arising. The objective of this study was to exhibit the trend of road crashes at the signalised intersections. The data of road crashes for the past 6 years were analysed using descriptive analysis. The results showed that the road traffic crashes at three- and four-legged signalised intersection recorded the increasing trend. In conclusion, this finding shows that the road traffic crashes for these types of signalised intersection in Malaysia is rising. It is also one the contributors to the increasing number of crashes in Malaysia. This finding will encourage the local authority to conduct awareness programs on the safety at the signalised intersection.

1. Introduction

Road traffic crash is one of the main contributing factors for deaths in the world [1]. It is estimated that almost 1.20 million lives were lost due to road traffic crashes. These days, road accidents are the common issues to be reported in the newspaper. In Malaysia, road traffic crashes are becoming a severe problem. According to the statistics from the Royal Malaysia Police, the total number of road traffic crashes in Malaysia was increasing from the year of 2006 to 2015, where 341,252 crashes were reported in 2006 and it increased to 489,606 crashes in 2015. The increasing number of road traffic crashes in Malaysia emerged as one of the big and unsolved problems, since according to Mustafa [2] this problem has been considered as the Malaysian Government's responsibility.

There are five types of road in Malaysia which are the straight road, bend, roundabout, interchanges, and intersections. As mentioned by the previous studies, the straight road was listed as the most common type of road where the crashes frequently occurred. As indicated in Figure 1, about 67 % of road traffic crashes occurred at the straight road. It is the highest when compared with other road types.

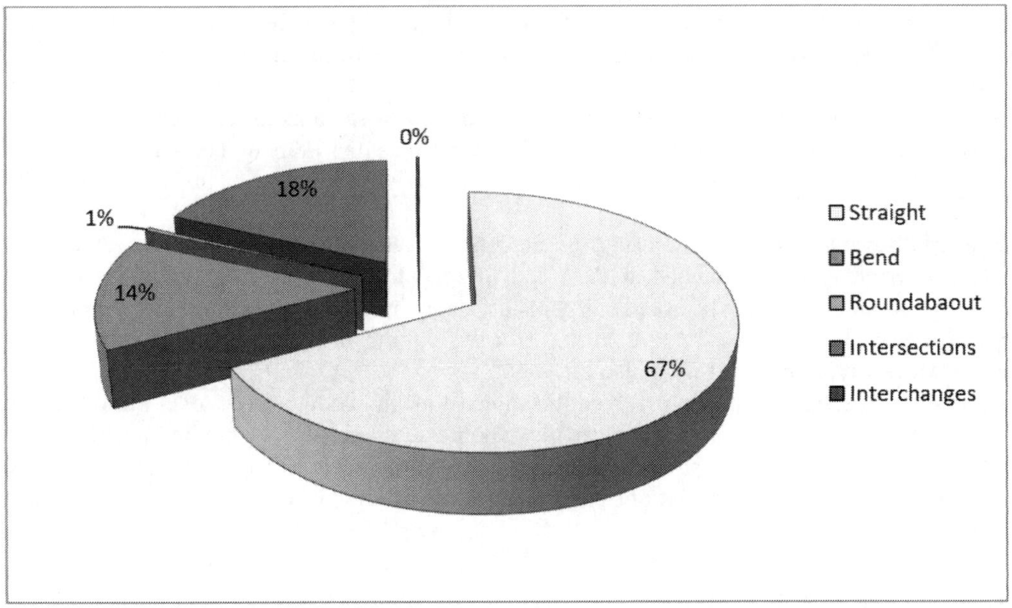

Figure 1. Road traffic crashes according to the different road types.

The intersections were categorised as the second road type where the road traffic crashes frequently occurred, followed by the bend road and roundabout with 14 and 1 %, respectively. However, the road traffic crashes at intersections cannot be neglected. Malaysia is one of the developing countries, where the road traffic crashes are commonly involved the intersections, especially at the urban areas [3].

Therefore, the traffic signal was implemented to reduce the road traffic crashes and improve the traffic safety at urban intersection. Nevertheless, severe road traffic crashes are still occurring and arising at the urban signalised intersection [4]. This resulted in listing the urban signalised intersection as the most dangerous and risky location of road network [5]. According to [6] there were four primary factors that caused the road traffic crashes to occur, which were the human, vehicles, weather condition, and roadway factors.

There are two types of signalised intersection that mostly located in the urban areas, which are the three- and four-legged signalised intersections. The road traffic crashes at both signalised intersections contribute to the increasing number of road traffic crashes in Malaysia. However, both of the signalised intersections have different rate or number of road crashes. Many previous research stated the difference in term of the rate of road traffic crashes between the three- and four-legged signalised intersection. A research done by Abdel-Aty and Wang [7] stated that the three-legged signalised intersection had lower road traffic crash rate as compare to four-legged signalised intersection.

Therefore, there is a need to study about the trend of road traffic crashes for different types of signalised intersections. It is important to researchers find out about the trend road traffic crashes at the three- and four-legged signalised intersections in Malaysia. Other than that, it is also important to find out about the different rate of road traffic crashes at both signalised intersections. The main purpose of this study was to exhibit the trend of road traffic crashes for different types of signalised intersections.

2. Methodology

Pahang was chosen to represent Malaysia in this study since Pahang was ranked the fifth out of thirteen states in Malaysia, with the highest number of death from the year of 2007 until 2016. Pahang consists of 11 districts which are Kuantan, Pekan, Temerloh, Jerantut, Raub, Bentong, Lipis, Cameron

Highlands, Rompin, Maran, and Bera. From these 11 districts of Pahang, Kuantan was selected to represent Pahang. This is because Kuantan recorded the highest number of road traffic crashes in 2015 and 2016.

The 6 years data on road traffic crashes at the signalised intersections in Kuantan (from 2010 until 2015) were collected from the Royal Malaysian Police of Kuantan branch. The data of road traffic crashes were sorted based on the type of signalised intersections which were three- and four-legged signalised intersections.

The descriptive analysis was done using the Statistical Package for Social Science (SPSS) in this study. The descriptive analysis analysed the trend of road traffic crashes for both types of urban signalised intersections. The analysis consisted of the summarisation, collection, and presentation of data obtained from the sample. Other than that, the histogram, measures of central tendency, and measures of variation (dispersion) were also included.

The measures of central tendency involved the value of mean, median, and mode. From these three values, the data distribution can be decided as whether symmetry distribution, positive skewed distribution, or negative skewed distribution. Meanwhile, the measures of variation involved the value of range, variance, and standard deviation. The measures of variation provided a complete overview of the dataset distribution.

3. Results and Discussion

Descriptive analysis provides the value of mean, median, mode, standard deviation, sample variance, maximum, minimum, range, kurtosis, and skewness as depicted in Table 1.

Table 1. Descriptive analysis.

	Three-legged Signalised Intersection		Four-legged Signalised Intersection	
Descriptive	**Statistics**	**SE**	**Statistics**	**SE**
Mean	105.67	30.631	241.00	28.190
Median	95.00		219.00	
Mode	44		209	
Standard Deviation	75.030		69.051	
Sample Variance	5629.467		4768.000	
Minimum	44		185	
Maximum	248		377	
Range	204		192	
Kurtosis	1.712	0.845	2.071	0.845
Skewness	3.456	1.741	4.652	1.741

The measures of central tendency are also known as the measures of average. It includes the median, mode, and mean values. The values can be used to describe the data distribution. From Table 3.1, the values of mean, median, and mode for three-legged signalised intersection are 105.67, 95.00, and 44, respectively. Meanwhile, the values of mean, median, and mode for four-legged signalised intersection are 241.00, 219.00, and 209, respectively. It can be concluded that the sample is positively-skewed distribution since the value of mean > median > mode.

Other than that, the skewness and kurtosis values indicate the distribution symmetriness. The skewness value for three- and four-legged signalised intersections sample indicate the positive skewness values which are 1.712 and 2.071, respectively. The positive value shows positively-skewed distribution which the scores clustered at the low values. The kurtosis values for three- and four-legged signalised intersections samples indicate the positive kurtosis values which are 3.456 and 4.562. Other than that, the data distribution is also illustrated in histogram.

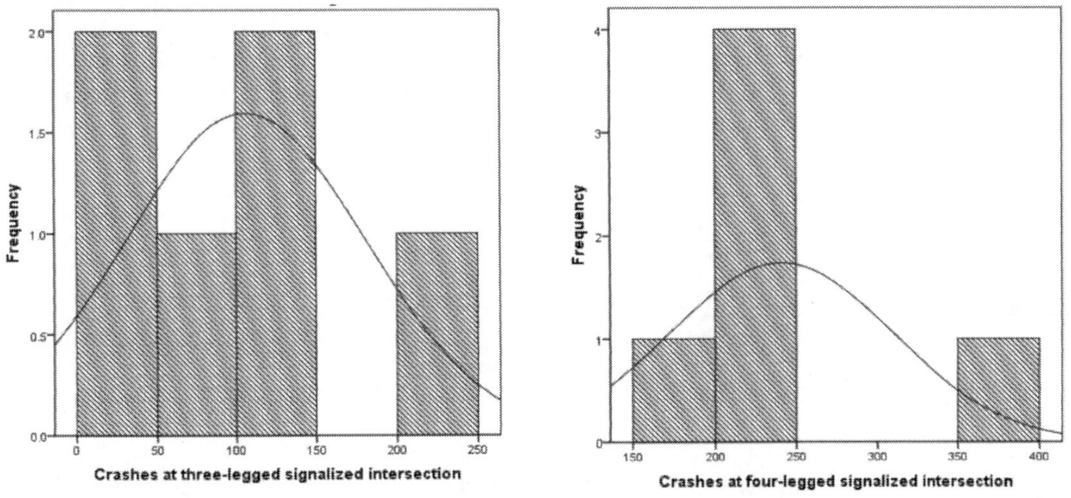

Figure 2(a). The data distribution histogram for crashes at three-legged signalised intersection

Figure 2(b). The data distribution histogram for crashes at four-legged signalised intersection

Figure 2(a) and Figure 2(b) indicate that the curve lines are skewed to the right and the peaks distribution are quite high. Hence, it can be explained that the data distribution for three- and four-legged signalised intersections are positively-skewed or right-skewed distribution and positive kurtosis. This indicates that there are a lot of signalised intersections with crashes of less than 150 crashes for three-legged signalised intersection and 250 crashes for four-legged signalised intersections. Other than that, it is also shows that the four-legged signalised intersection is more skewed than the three-legged signalised intersection. The four-legged signalised intersection is having higher number of crashes than three-legged signalised intersection. This is because the number vehicle conflict points at four-legged signalised intersection are higher than the three-legged signalised intersection [7]. This also can be explained by the need of drivers to pay more attention since the movements and attractions between vehicles at four-legged signalised intersection are higher.

Meanwhile, the measures of variation (dispersion) explains about the spread of data values. The range is different between maximum and minimum values in one data set. From Table 3.1, the range for three-legged signalised intersection is 204. It can be concluded that the highest number of road traffic crashes at three-legged signalised intersection is 248 crashes and the lowest number of road traffic crashes at three-legged signalised intersection is 44 crashes. Meanwhile, the range for four-legged signalised intersection is 192. It can be said that the highest number of road traffic crashes at four-legged signalised intersection is 377 crashes and the lowest number of road traffic crashes at four-legged signalised intersection is 185 crashes. It can be seen that the range number of crashes at three-legged signalised intersection is higher than the four-legged signalised intersection.

Besides, from Table 1, the values of variance and standard deviation for three-legged signalised intersection is larger than those of four-legged signalised intersection. Hence, the data of the three-legged signalised intersection is more dispersed compared to the four-legged signalised intersection, since the S2 and S (three-legged signalised intersection) > S2 and S (four-legged signalised intersection). The larger values of variance and range for the three-legged signalised intersection explain that the data of road traffic crashes is more dispersed from the mean value where the number of road traffic crashes is different each year. Meanwhile, the number of road traffic crashes at four-legged signalised intersection is almost the same each year since the value of variance is smaller than three-legged signalised intersection. It is also can be confirmed in Figure 2, where the number of crashes at four-legged signalised intersection is almost the same, unlike the three-legged signalised intersection. This is presumably because of the situation at four-legged signalised intersection which is more controlled since the number of crashes is higher than three-legged signalised intersection. Meanwhile, the movements at three-legged signalised intersection is probably less controlled since the conflict point and number of crashes are fewer than four-legged signalised intersection.

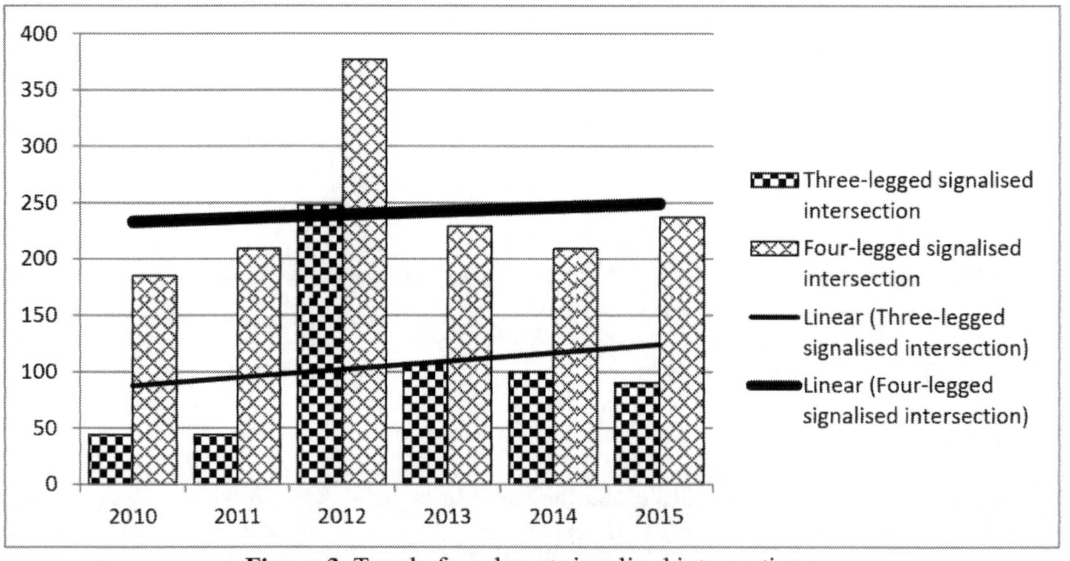

Figure 3. Trend of crashes at signalised intersections.

Figure 2 shows the trend of crashes at the three- and four-legged signalised intersection in Kuantan urban area from 2010 to 2015. The three-legged signalised intersection has recorded 44 crashes in 2010 and 2011. In 2012, the number of crashes increased to 248 crashes. Then, it decreased to 108 crashes in the year of 2013, followed by 100 crashes in 2014, and 90 crashes in 2015. Meanwhile, the number of road traffic crashes at four-legged signalised intersection has recorded 185 crashes in 2010 and it increased to 209 crashes in 2011. In the year of 2012, the number of crashes increased to 377 crashes. Then, it decreased to 229 crashes in 2013, followed by 209 crashes in 2014, and 237 crashes in 2015.

Other than that, Figure 3 depicts that the number of road traffic crashes at four-legged signalised intersection is higher than the three-legged signalised intersection. This is because the four-legged signalised intersection are having more vehicle conflict point as compared to the three-legged signalised intersection, which contributed to the higher number of movements [7]. Based on the trend observed, it also can be concluded that the trend of road traffic crashes at both signalised intersection is increasing from the year of 2010 to 2015, despite the irregular ups and downs. The increasing trend of road traffic crashes at both signalised intersections is probably due to the contribution of primary factors of road crashes such as the human, vehicle, and environment factors. This finding is also supported by Hawa and Akmalia [8], where the road traffic crashes at Malaysia signalised intersection recorded the increasing trend because of road environment situation since the drivers have to deal with different task within the intersection area [9].

4. Conclusion

To sum up, according to the past 6 years recorded data (2010 – 2015), the trend of road traffic crashes at three- and four-legged urban signalised intersection shows increment. This kind of trend observed might be contributed by the previously mentioned four primary factors. Both types of signalised intersections show the positively-skewed data distribution. The three- and four-legged signalised intersections recorded the range number of 204 and 192 crashes, respectively. Besides, according to the obtained values of variance, it can be concluded that the number of crashes quite differ for each year at the three-legged signalised intersection, unlike the four-legged signalised intersection which shows similar pattern each year. For the past 6 years (2010 – 2015), it can be observed that there is an increased trend for the crashes at signalised intersection. Regardless of that, further investigations with longer period needed to be done since this study only consider the 6 years data so that the data of trend for road traffic crashes at signalised intersection obtained can be generalised to other location in Malaysia. The information obtained can be used by the local authority to organise campaigns and awareness programs on the prevention of crashes at signalised intersection.

Acknowledgements

The authors would like to express their gratitude to UMP for providing grant entitled 'Smart Traffic System for Universiti Malaysia Pahang Pekan Campus (RDU160363) that has enabled the authors to carry out research on the above issues

References

[1] Peden M, Scurfield R, Sleet D, Mohan D, Hyder A A, Jarawan E and Mathers C D 2004 World report on road traffic injury prevention

[2] Mustafa M N 2005 Overview of current road safety situation in Malaysia *Highw. Plan. Unit, Road Saf. Sect. Minist. Work.* 5–9

[3] Archer J and Vogel K 2000 The traffic safety problem in urban areas *Res. Rep.*

[4] Polders E, Daniels S, Hermans E, Brijs T and Wets G 2015 Crash patterns at signalized intersections *Transp. Res. Rec. J. Transp. Res. Board* 105–16

[5] Wang X 2006 No Title

[6] Munteanu P L, Roşu M, Panaitescu V and Pungă A 2014 Human and environmental factors contributing to fatal road accidents in a Romanian population *Rom J Leg Med* **22** 97–100

[7] Abdel-Aty M and Wang X 2006 Crash estimation at signalized intersections along corridors: analyzing spatial effect and identifying significant factors *Transp. Res. Rec. J. Transp. Res. Board* 98–111

[8] JAMIL H M and SHABADIN A 2012 A case study of the prevalence and characteristics of red light runners in Malaysia *Inj. Prev.* **18** A201–A201

[9] Wong S C, Sze N-N and Li Y-C 2007 Contributory factors to traffic crashes at signalized intersections in Hong Kong *Accid. Anal. Prev.* **39** 1107–13

Real-time slicing algorithm for Stereolithography (STL) CAD model applied in additive manufacturing industry

F A Adnan, F R M Romlay and M Shafiq
Faculty of Mechanical Engineering, Universiti Malaysia Pahang, 26600, Pekan, Pahang, Malaysia

E-mail: fb11002.ump@gmail.com

Abstract. Owing to the advent of the industrial revolution 4.0, the need for further evaluating processes applied in the additive manufacturing application particularly the computational process for slicing is non-trivial. This paper evaluates a real-time slicing algorithm for slicing an STL formatted computer-aided design (CAD). A line-plane intersection equation was applied to perform the slicing procedure at any given height. The application of this algorithm has found to provide a better computational time regardless the number of facet in the STL model. The performance of this algorithm is evaluated by comparing the results of the computational time for different geometry.

1. Introduction

Recently introduced Digital-Light-Processing (DLP) 3D printer utilizes mask projection by DLP projector onto photopolymer resin to cure the liquid form resin into solid model [1-2]. The mask projection is proven to be much faster and it provides uniform curing process due to each layer is being projected at once compared to SLA 3D Printer [3-4] . CAD file such as STereoLithography (STL) is considered as de facto in 3D printing. However, the STL file must be sliced into layers of 2D contours (as in Figure 1) before the contour projection process takes place [5-7]. In the past, researchers proposed several methods of this pre-processing stage of stereolithography such as uniform slicing and adaptive slicing methods [8]. All the mentioned methods are performed before the process of contour projections, and often affect the output of the printing process which is called cusp height or staircase effect due to differential slicing height.

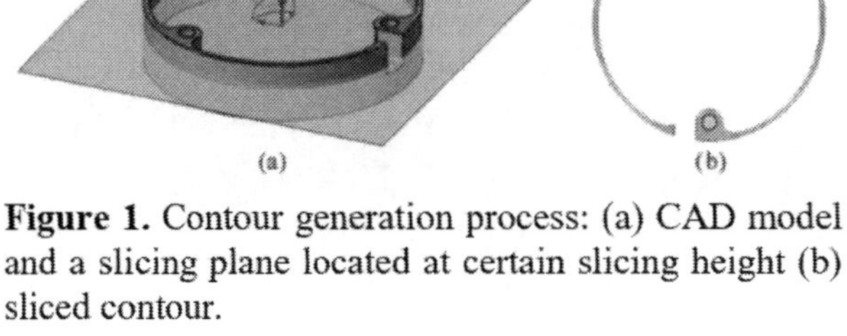

Figure 1. Contour generation process: (a) CAD model and a slicing plane located at certain slicing height (b) sliced contour.

This paper proposed a real-time slicing algorithm for contour mask generation with low computational time. The terms real-time in this context refers to the ability of the algorithm to compute a single layer of contour with respect to the changing slicing height.

2. STL Format

STL format is considered as a de facto in rapid prototyping. It was developed back in 1987 by 3D Systems for 3D CAD model conversion. The conversion involves tessellation process which forms multiple triangular facet of XYZ to represent the surface feature of the geometry. There are two types of STL format: binary STL and ASCII STL format. The study only focused on ASCII STL format. An ASCII STL format always begins with the *solid name* syntax where name is an optional and often omitted with spaces, then followed by *facet* syntax with its normal vector. The *outer loop* marks the beginning of vertices data which is used as P_1, P_2, and P_3 respectively in the algorithms. The *n* and *v* is a floating number with numbering format of *sign-mantissa-"e"-sign-exponent,* e.g. "2.999381e-002".

```
solid name
facet normal ni nj nk
   outer loop
      vertex v1x v1y v1z
      vertex v2x v2y v2z
      vertex v3x v3y v3z
   endloop
endfacet
endsolid name
```

The *endfacet* syntax marks the end of the facet. An STL file may consists of more than one facet, usually thousands, depending on the complexity of the geometry. If there is another facet, new *facet* syntax can be located after the previous *end facet* syntax. The STL file will have *endsolid name* syntax as the terminator of the file [9].

3. Algorithms

3.1 Facet Class

The proposed algorithm stored each facet in a list of facet class to be read by the slicing algorithm. The class stored information such as the facet vertices (P_1, P_2, P_3), maximum and minimum Z-height of the facet. The maximum/minimum Z value is later used to filter out other facets except the ones which intersect with the slicing plane by comparing $z_{min} \leq z_{slice} \leq z_{max}$ for each facet in the list.

3.2 Slicing Algorithm

The slicing algorithm is a method of converting each triangular facet into each respective line segment. These line segments can later be connected into contour lines by using contour generation algorithm [10]. Facets in STL format are arbitrary and in various orientations. Figure 2 and Table 1 describe every possible interaction between the slicing plane and the facets [5,11]. Each interaction must be handled properly since a line segment only requires 2 distinct points.

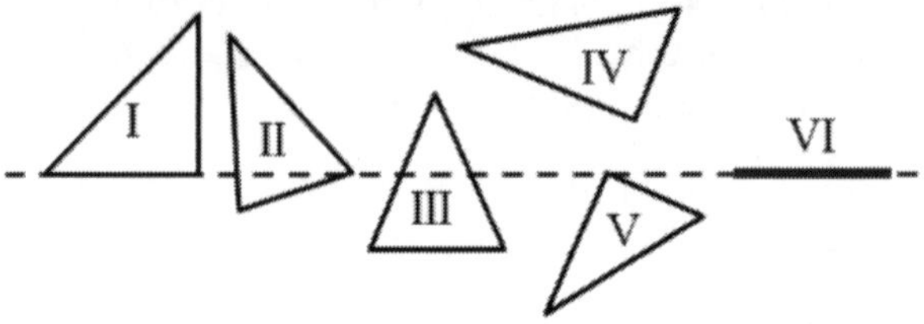

Figure 2. Possible orientation and position of facets.

Table 1. Definition of facet-plane intersection.

Case	Interaction of facet and plane	Possible Point
I	Line through one side of the facet	4
II	Line bisecting the facet through one vertex	3
III	Line bisecting the facet through two sides	2
IV	No intersection	0
V	Vertex intersection	2
VI	Parallel intersection	6

Table 1 classifies possible number of point that will be generated for each case of interaction mentioned above when sliced. Only Case I, II, and III will form the correct line segment. The rest of the cases are ignored. In Case V, the generated point will be the same and will not produce a line segment. For Case VI, this is considered as redundancy because in actual model, there will be another Case I facet at the exact position but in perpendicular (Case I) to the one in Case VI. Hence, Case VI must be ignored to eliminate overlapping line segment. Lastly, Case IV, there is no intersection and the algorithm will ignore it and proceed to search the next intersecting facet.

Intersections at vertex will produce two intersection points due to the facet edges that share the same vertex. This occurs in Case I, II, V, and VI. Hence, these cases must be handled properly. First, the proposed algorithm checks whether the edge is in parallel or it intersects with the plane to isolate Case VI and eliminates the coinciding edge in Case I. Case II is handled by eliminating one recurring intersection point. Case V can be identified when there exist only two intersection points which are similar.

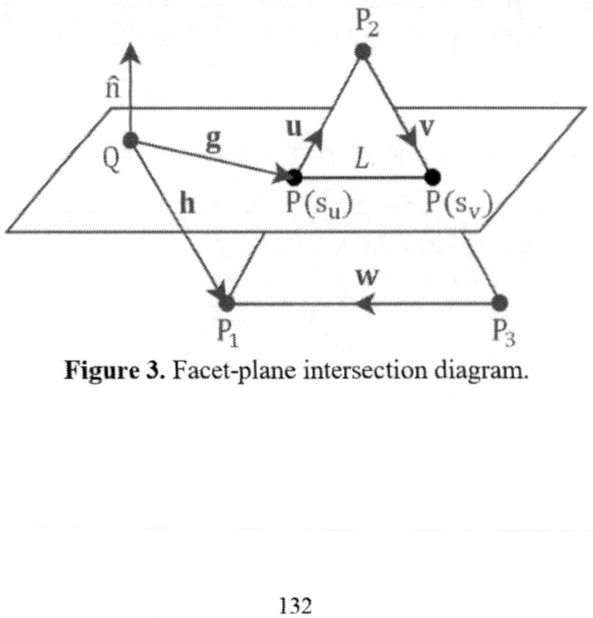

Figure 3. Facet-plane intersection diagram.

The proposed slicing algorithm is fundamentally based on line-plane intersection equation. Consider one side of the facet as line **L** that connects two vertices. In 3D, a line is either parallel to a plane or intersects it in a single point (see Figure 3). Parametric equation of a line can be written as:

$$P(s) = P_o + s(P_f - P_o) \tag{1}$$

and $Q = (x, y, z_{slice})$ is a point that exists on the plane where x and y can be any point on the plane (usually set at the origin), z_{slice} is the slicing height and the unit vector $\hat{n} = (0, 0, 1)$ is the plane normal (for the case of slicing with respect to Z axis). Changing the value of z_{slice} will affect the slicing height and resulted in different line segment. The algorithm first checks whether intersection exist between the lines and the plane by using the dot product criterion, $\hat{n} \cdot \mathbf{u} = 0$, $\hat{n} \cdot \mathbf{v} = 0$, $\hat{n} \cdot \mathbf{w} = 0$. These criterions will also handle both Case I and Case VI by eliminating the parallel line in the facet. As seen in the Fig.3, the $\mathbf{u}$, $\mathbf{v}$, and $\mathbf{w}$ is the direction vector for each side of the facet in clockwise order (P_1, P_2, P_3, P_1). The dot product will become zero if there is no intersection which also means that the line is parallel to the plane and will not intersect with the plane. Should the line intersect with the plane such as the line between point P_1 and P_2 where $P_o = P_1$, $P_f = P_2$ and $s = s_u$, the $\hat{n} \cdot \mathbf{u} \neq 0$. At the intersection point, the direction vector from $P(s)$ to Q is $\mathbf{g} = \mathbf{h} + s\mathbf{u}$ where vector $\mathbf{h} = P_o - Q$. Since the vector $\mathbf{g} = P(s) - Q$ lies on the plane, previous criterion can be used, hence it is known that $\hat{n} \cdot \mathbf{g} = \hat{n} \cdot (\mathbf{h} + s\mathbf{u}) = 0$. Solving this will get:

$$s = \frac{-\hat{n} \cdot \mathbf{h}}{\hat{n} \cdot \mathbf{u}} = \frac{\hat{n} \cdot (Q - P_o)}{\hat{n} \cdot (P_f - P_o)} \tag{2}$$

By using Equation 3 and the s_u value obtained from Equation 4, the equation now becomes:

$$P(s_u) = P_1 + s_u(P_2 - P_1)$$

However, the value s_u must be verified to be within the range $0 \leq s \leq 1$ to ensure that the intersection point exists only within line. The same method is applied to the rest of the lines for which the criterion is fulfilled. In Figure 3 the $P(s_u)$ and $P(s_v)$ will generate a line segment L that will be used in the contour generation algorithm.

4. Experimental Results

The mentioned algorithm is developed in VB.NET programming language using Microsoft Visual Studio 2012 and executed on LENOVO IdeaPad i3-2350M CPU at 2.30GHz 6GB RAM and running on Windows 7 SP1 Home Premium (64-bit) operating system. The algorithm is tested with multiple STL model with different complexity and its results are shown in Table 2.

IOP Conf. Series: Materials Science and Engineering **342** (2018) 012016 doi:10.1088/1757-899X/342/1/012016

Table 2. Result of slicing algorithm.

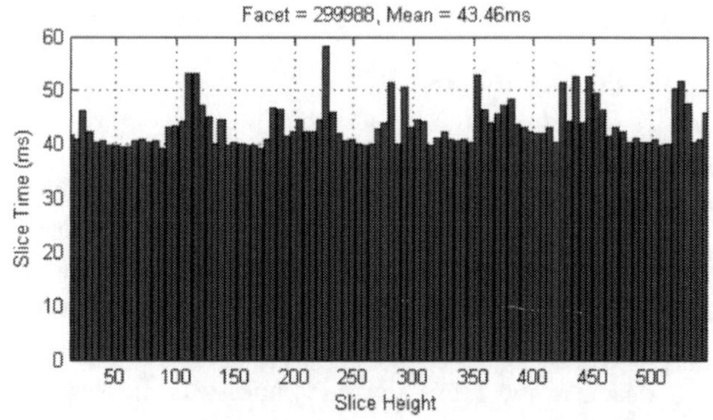

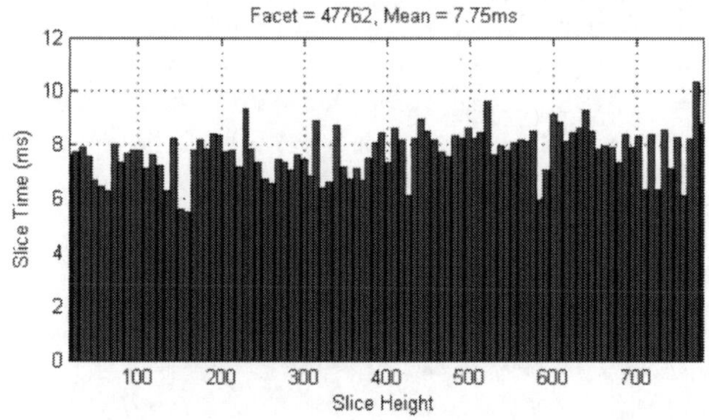

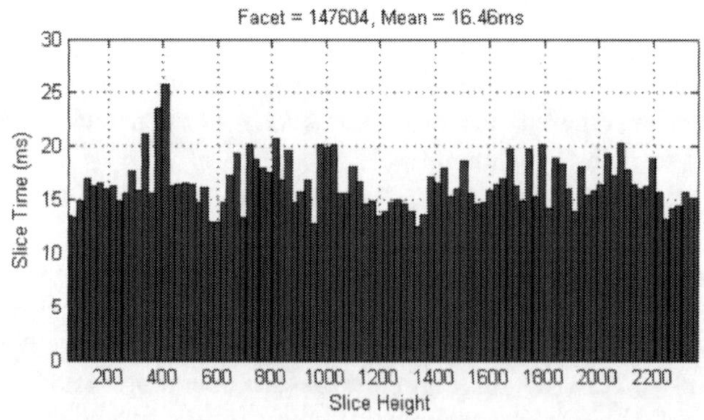

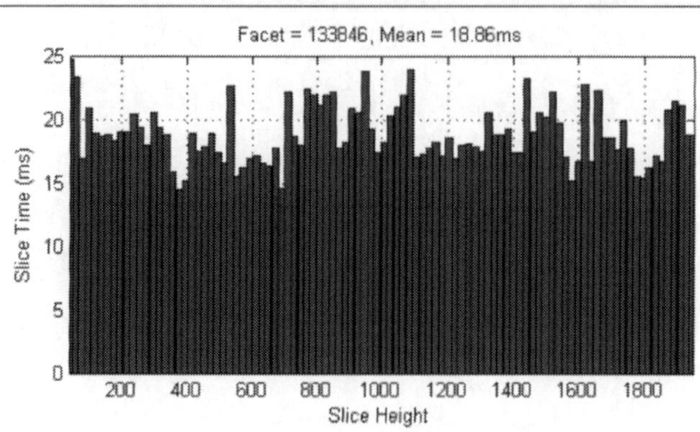

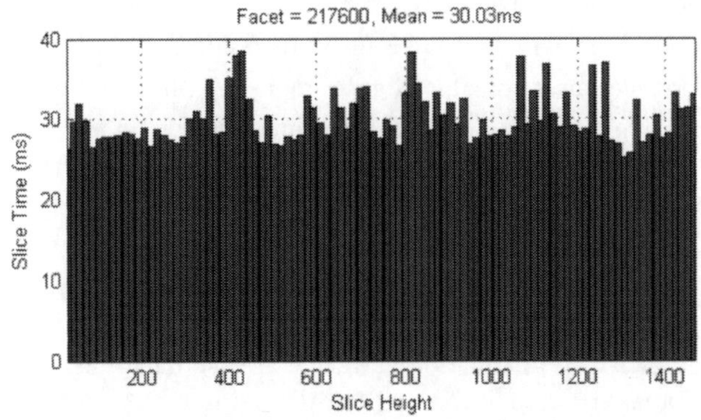

As shown in Table 2, the results consist of slicing time at different slicing height starting from the bottom of the geometry until the top of each respective model. The consistency in the performances of each slicing height are noticeable. Hence, it can be represented as mean average slicing time as shown in the graph next to the 3D model. Each model contains different facet number indicating the complexity of the model. The graphs shown above are limited to 100 slices of contour. However, this can be adjusted regardless any value depending on the DLP 3D printing machine resolution. Based on each individual facet number and mean average, a graph of facet number versus mean can be plotted to analyse its correlation as shown in Figure 4 below.

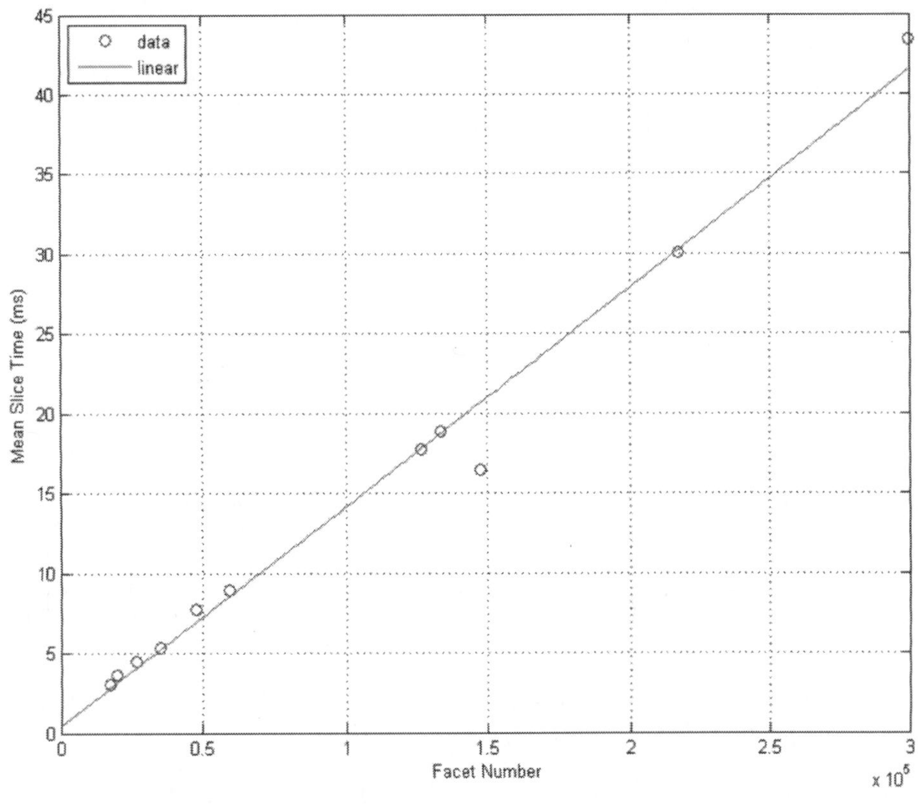

Figure 4. Graph of Facet Number vs. Mean.

As seen in Figure 4, mean slice time is linearly increase with respect to the facet number. More complex geometry will result in longer slicing time. According to Big-O notation which is considered as standard measures for algorithm complexity, the mentioned slicing algorithm runs on $O(n)$ that tells increasing number of elements will increase the algorithm execution time.

5. Conclusion

This paper presented real-time slicing algorithm which runs on $O(n)$ that able to slice an STL model at any given height directly with low computational time. The algorithm is fundamentally based on line-plane intersection equation and several case handlings as previously mentioned. The paper also focused on real-time evaluation of the slicing process so that the responses at each slicing heights are measured. Based on the results, it is found that, the slicing time of a geometry is consistent regardless the slicing height. However, for complex model, the slicing time will linearly increase.

Usually, a slicing algorithm is paired with a contour generation algorithm to connect each sliced line segments into single/multiple closed contour loop. This can be the extended work for the proposed slicing algorithm.

Acknowledgement

This research was funded by Universiti Malaysia Pahang (UMP) research grant number RDU160387. Credits to everyone whom directly and indirectly involved in helping us to achieve all the specified objectives of this research.

References

[1] Limaye A S and Rosen D W 2007 Process planning method for mask projection micro-stereolithography *Rapid Prototyp. J.* **13** 76–84

[2] Katal G, Tyagi N and Joshi A 2013 Digital Light Processing and its Future Applications *Int. J. Sci. Res.* **3** 1–16

[3] Chia H N and Wu B M 2015 Recent advances in 3D printing of biomaterials *J. Biol. Eng.* **9** 4

[4] Zhao X 2009 Process planning for thick-film mask projection micro stereolithography

[5] Topçu O, Taşcıoğlu Y and Ünver H Ö 2011 A Method for Slicing CAD Models in Binary STL Format *6th Int. Adv. Technol. Symp.* 141–8

[6] Minetto R, Volpato N, Stolfi J, Gregori R M M H and da Silva M V G 2017 An optimal algorithm for 3D triangle mesh slicing *Comput. Des.* **92** 1–10

[7] Pan X, Chen K and Chen D 2014 Development of rapid prototyping slicing software based on STL model *Proc. 2014 IEEE 18th Int. Conf. Comput. Support. Coop. Work Des. CSCWD 2014* 191–5

[8] Zhou M Y, Xi J T and Yan J Q 2004 Adaptive direct slicing with non-uniform cusp heights for rapid prototyping *Int. J. Adv. Manuf. Technol.* **23** 20–7

[9] Huang S H, Zhang L C and Han M 2002 An effective error-tolerance slicing algorithm for STL files *Int. J. Adv. Manuf. Technol.* **20** 363–7

[10] Zhang Z and Joshi S 2015 An improved slicing algorithm with efficient contour construction

using STL files *Int. J. Adv. Manuf. Technol.* **80** 1347–62

[11] Brown A C and De Beer D 2013 Development of a stereolithography (STL) slicing and G-code generation algorithm for an entry level 3-D printer *IEEE AFRICON Conf.*

The classification of hunger behaviour of *Lates Calcarifer* through the integration of image processing technique and *k*-Nearest Neighbour learning algorithm

Z Taha[1], M A M Razman[1], A S Abdul Ghani[1], A P P Abdul Majeed[1], R M Musa[1], F A Adnan[1], M F Sallehudin[2] and Y Mukai[2]

[1]Innovative Manufacturing, Mechatronics, and Sports Lab (iMAMS), Faculty of Manufacturing Engineering, Universiti Malaysia Pahang, Pekan Campus, 26600 Pekan, Pahang, Malaysia.
[2]Kulliyyah of Science, International Islamic University Malaysia, Jalan Sultan Ahmad Shah, Bandar Indera Mahkota, 25200, Kuantan, Pahang, Malaysia.

E-mail: azraai@ump.edu.my

Abstract. Fish Hunger behaviour is essential in determining the fish feeding routine, particularly for fish farmers. The inability to provide accurate feeding routines (under-feeding or over-feeding) may lead the death of the fish and consequently inhibits the quantity of the fish produced. Moreover, the excessive food that is not consumed by the fish will be dissolved in the water and accordingly reduce the water quality through the reduction of oxygen quantity. This problem also leads the death of the fish or even spur fish diseases. In the present study, a correlation of Barramundi fish-school behaviour with hunger condition through the hybrid data integration of image processing technique is established. The behaviour is clustered with respect to the position of the school size as well as the school density of the fish before feeding, during feeding and after feeding. The clustered fish behaviour is then classified through *k*-Nearest Neighbour (*k*-NN) learning algorithm. Three different variations of the algorithm namely cosine, cubic and weighted are assessed on its ability to classify the aforementioned fish hunger behaviour. It was found from the study that the weighted *k*-NN variation provides the best classification with an accuracy of 86.5%. Therefore, it could be concluded that the proposed integration technique may assist fish farmers in ascertaining fish feeding routine.

1. Introduction

In understanding fish characteristics, one should consider the various factors that lead to its behaviour such as the movement activity, sensory cues, and the circadian rhythm. The locomotor activity of the fish either individually or in groups has been reported to increase in the event when it is anticipating food [1]. The hunger demand that affects the fish movement has caught the attention of researchers throughout the years. It was inferred that fish scatters in a larger area when they are hungry, and they move away from the walls of confinement [2]. A behavioural pattern or in general term the circadian rhythm is an endogenous type which is a biological conduct that the fish encounter. It emulates similar patterns with sleeping representation and mostly associated with feeding rhythms [3]. Researchers have labeled this pattern as free running when the routine preserve for a period of 24 hours.

Modelling the hunger behaviour requires an essential form of evidence on the necessary amount of food and the timing gaps between each meal for the feeding time to be assessed. The locomotor activity of

the fish usually decreases after it has repleted [4]. It has been reported that data gathered from a school of fish replicates the individual fish movement as the movement does not vary significantly [5]. Moreover, it has also been demonstrated in a study that fish also are sensitive towards ultradian rhythm when they are hungry [6]. An automated feeder was used in the aforementioned study, to serve food based on the demand of the fish. In the present study, a similar concept is applied will be discussed in great length in the subsequent section. To gather the adequate amount of food during the experiment, the usage of a self-demand feeder is necessary, as suggested in [7]. Due to the myriad number of variables such as social interactions, species and environmental shifts that have been discovered to influence the fish hunger behaviour, there is a need for a standared experimental setup. Different type of fish species was investigated on the circadian rhythm throughout the years with respect to the Asian sea bass or commonly known as Barramundi. For instance, the growth rate of the fish in correlation to light sensitivity was investigated [8]. The resting time, as well as the active motion between genders of zebrafish or Danio rerio and razor fish or Xyrithchys novacula with relation to feeding state, has also been investigated [9-10].
Several classification techniques have been employed due to its superiority against conventional statistical concept [11-12]. For instance, SVM classification was performed on raiboow trout or *Oncorhynchus mykiss* in observing the muscle activity either in the satiated or hungry state [13]. *k*-Nearest Neighbour (*k*-NN) algorithm has also been demonstrated to be able to detect abnormal fish trajectories [11]. This study aims at evaluating the classification ability of *k*-NN in identifying the hunger behaviour based on two distinct parameters namely the location of the centre of gravity and of the school size of *Lates Calcarifer*.

2. Materials and Methods

This section describes the experimental setup that includes the fish species, the quantity of the fish, laboratory condition and the classification method.

2.1 Fish

The Asian sea bass species was selected in the present study as it depicts the typical circadian rhythm on food demand of day and night cycle. The fish was acquired from Fisheries Research Institute, Gelang Patah Johor, Malaysia, with a quantity of 20 juveniles sea bass that is stationed at the International Islamic University Malaysia laboratory.

2.2 Experiment Setup

The amount of water filled into the tank is 130 L where the fish is being placed, and the dimension of the tank is 92cm x 46cm x 46cm. The fish then triggered the suspended infrared sensor that is installed on the self-demand feeder when the school reaches slightly below the water surface. Each time the sensor is prompted, approximately 0.5g of pellets will drop with a size of 3 mm in diameter, and this continues until the sensor was not triggered from any of the fish. Data for generated feeding moments and timestamp will be sent via microcontroller to the computer and recorded for clustering purposes with the tags of Before Feeding (BF), During Feeding (DF) and After feeding (AF). The location of parameters above if the centre of gravity (COG) and the box size is captured using a webcam modeled Logitech, C270h. The video is then converted into .avi file and evaluated to Roborealm software for evaluation as in figure 1. Hence, each parameter had (x, y) coordinates in pixels which then indicates the position of the fish at the given period. The clustering was made from the evaluated parameters of BF, DF, and AF and is then classified via the *k*-NN models.

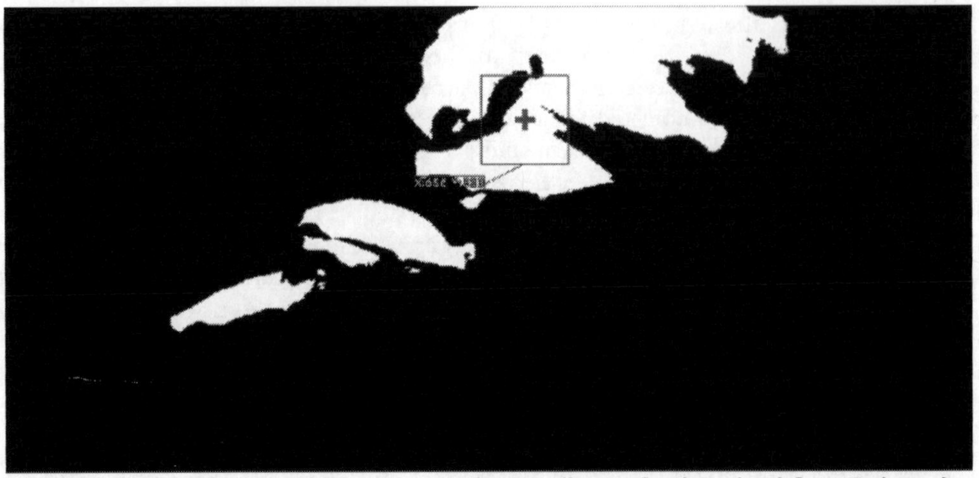

Figure 1. COG of the box which gives x and y coordinates for the school from Roborealm.

2.3 Classification method

In this study, three different *k*-NN variations viz. cosine, cubic and weighted are investigated. The number of neighbours, *k* employed for all variations is 10, whilst the distance metrics used are cubic, cosine, and Euclidean, respectively. The equal distance (no weight assigned) are used for all the evaluated variations except for the weighted variation, in which an inverse squared distance weight is imposed. In addition, a fivefold cross-validation procedure was used for model training and testing [14]. The performance of the models were assessed and evaluated through its classification accuracy (CA) via MATLAB 2016a (Mathworks Inc., Natick, USA).

3. Results and Discussion

Table 1. *k*-NN classification results.

k-NN variation	CA (%)
Cubic	68.8
Cosine	78.5
Weighted	86.5

It is apparent from table 1 that the weighted *k*-NN variation is able to provide reasonably accurate classification with a classification accuracy of 86.5 %. Furthermore, it could also be seen that the cubic variation performs the worst amongst all variation tested. Nevertheless, it is worth to mention that the ability of the weighted *k*-NN is dependent of the parameters selected i.e. the centre of gravity as well as the school density, suggesting that the selected parameters are nontrivial or vital in describing the feeding behaviour of *Lates Calcarifer*. The importance of the aforementioned parameters i.e. COG and box size density has been reported in a previous study for understanding fish feeding behaviour [10]. Figure 2 depicts the confusion matrix of the tested models.

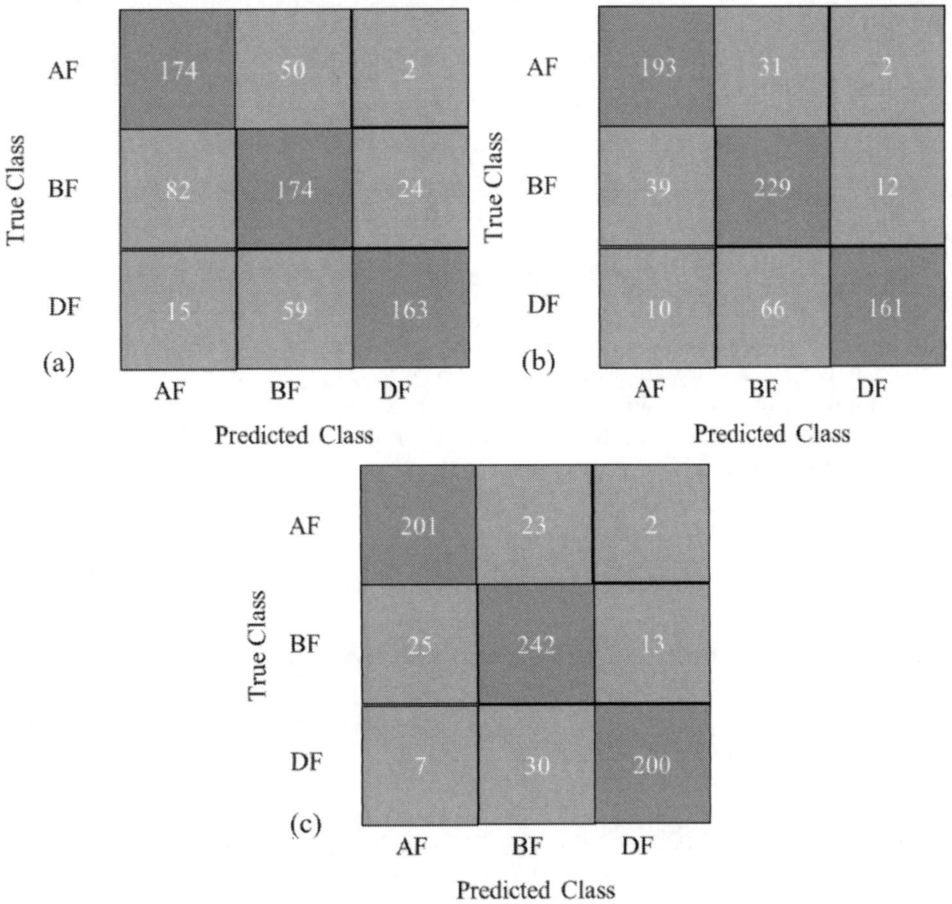

Figure 2. Confusion matrix for the k-NN variations (a) Cubic; (b) Cosine; (c) Weighted.

4. Conclusion

It could be concluded from the present investigation that weighted *k*-NN variation is able to provide a reasonably accurate classification of fish feeding behaviour with regards to the abovementioned parameters examined viz. centre of gravity and school fish density. Further study could be carried out by considering other relevant parameters that may explain the hunger behaviour of *Lates Calcarifer* as well as the effectiveness of other forms of classification techniques.

Acknowledgement

This research work is also supported by the University Research Grant titled "*Dual Image Fusion Technique for Enhancement of Underwater Image Contrast*" (RDU170392), Faculty of Manufacturing Engineering, Universiti Malaysia Pahang.

References

[1] Cavallari N, Frigato E, Vallone D, Fröhlich N, Lopez-Olmeda J F, Foà A, Berti R, Sánchez-Vázquez F J, Bertolucci C and Foulkes N S 2011 A Blind Circadian Clock in Cavefish Reveals that Opsins Mediate Peripheral Clock Photoreception *PLOS Biology* **9** e1001142

[2] Hansen M J, Schaerf T M and Ward A J W 2015 The effect of hunger on the exploratory behaviour of shoals of mosquito fish *Behaviour* **152** 1659-77

[3] Beale A, Guibal C, Tamai T K, Klotz L, Cowen S, Peyric E, Reynoso V H, Yamamoto Y and Whitmore D 2013 Circadian rhythms in Mexican blind cavefish Astyanax mexicanus in the lab and in the field **4** 2769
[4] Priyadarshana T, Asaeda T and Manatunge J 2006 Hunger-induced foraging behavior of two cyprinid fish: Pseudorasbora parva and Rasbora daniconius *Hydrobiologia* **568** 341-52
[5] Chapman B B, Morrell L J and Krause J 2010 Unpredictability in food supply during early life influences boldness in fish *Behavioral Ecology* **21** 501-6
[6] F. Javier Sanchez-Vázquez J A M a S Z 1995 Circadian Rhythms of Feeding Activity in Sea Bass, Dicentrarchus labrax L.: Dual Phasing Capacity of Diel Demand-Feeding Pattern *Journal of Biological Rhythms* **10** 11
[7] Biswas G, Thirunavukkarasu A R, Sundaray J K and Kailasam M 2010 Optimization of feeding frequency of Asian seabass (Lates calcarifer) fry reared in net cages under brackishwater environment *Aquaculture* **305** 26-31
[8] Harpaz S, Hakim Y, Barki A, Karplus I, Slosman T and Tufan Eroldogan O 2005 Effects of different feeding levels during day and/or night on growth and brush-border enzyme activity in juvenile Lates calcarifer reared in freshwater re-circulating tanks *Aquaculture* **248** 325-35
[9] Ariyomo T O and Watt P J 2015 Effect of hunger level and time of day on boldness and aggression in the zebrafish Danio rerio *Journal of Fish Biology* **86** 1852-9
[10] Josep Alós M M-B a A C-C 2017 Repeatability of circadian behavioural variation revealed in free-ranging marine fish *Royal Society Publishing* **4** 11
[11] Cortes C and Vapnik V 1995 Support-vector networks *Machine Learning* **20** 273-97
[12] Dutta M K, Sengar N, Kamble N, Banerjee K, Minhas N and Sarkar B 2016 Image processing based technique for classification of fish quality after cypermethrine exposure *LWT - Food Science and Technology* **68** 408-17
[13] Cubitt K F, Williams H T, Rowsell D, McFarlane W J, Gosine R G, Butterworth K G and McKinley R S 2008 Development of an intelligent reasoning system to distinguish hunger states in Rainbow trout (Oncorhynchus mykiss) *Computers and Electronics in Agriculture* **62** 29-34
[14] Taha Z, Musa R M, P.P. Abdul Majeed A, Alim M M and Abdullah M R 2018 The identification of high potential archers based on fitness and motor ability variables: A Support Vector Machine approach *Hum. Mov. Sci.* **57** 184–93

IOP Conf. Series: Materials Science and Engineering **342** (2018) 012018 doi:10.1088/1757-899X/342/1/012018

Multi-actuators vehicle collision avoidance system - Experimental validation

Umar Zakir Abdul Hamid[1, 2], Fakhrul Razi Ahmad Zakuan[1, 2], Khairul Akmal Zulkepli[1], Muhammad Zulfaqar Azmi[1, 2], Hairi Zamzuri[1*], Mohd Azizi Abdul Rahman[1] and Muhammad Aizzat Zakaria[3]

[1] Vehicle System Engineering iKohza, Malaysia-Japan International Institute of Technology, Universiti Teknologi Malaysia, Kuala Lumpur, Malaysia
[2] Moovita Pte Ltd, 317 Outram Road, #01-60, Holiday Inn Atrium, 169075 Singapore
[3] Faculty of Manufacturing Engineering, Universiti Malaysia Pahang, Malaysia

E-mail: umar@moovita.com

Abstract. The Insurance Institute for Highway Safety (IIHS) of the United States of America in their reports has mentioned that a significant amount of the road mishaps would be preventable if more automated active safety applications are adopted into the vehicle. This includes the incorporation of collision avoidance system. The autonomous intervention by the active steering and braking systems in the hazardous scenario can aid the driver in mitigating the collisions. In this work, a real-time platform of a multi-actuators vehicle collision avoidance system is developed. It is a continuous research scheme to develop a fully autonomous vehicle in Malaysia. The vehicle is a modular platform which can be utilized for different research purposes and is denominated as Intelligent Drive Project (iDrive). The vehicle collision avoidance proposed design is validated in a controlled environment, where the coupled longitudinal and lateral motion control system is expected to provide desired braking and steering actuation in the occurrence of a frontal static obstacle. Results indicate the ability of the platform to yield multi-actuators collision avoidance navigation in the hazardous scenario, thus avoiding the obstacle. The findings of this work are beneficial for the development of a more complex and nonlinear real-time collision avoidance work in the future.

1. Introduction

Road fatalities remain one of the major death factors in South East Asian region [1]. It is estimated that the hazardous occurrences killed more than 300,000 people annually, i.e. 25% of the road fatalities total globally [2]. In a report by the Institute of Highway Safety of the United States (IIHS), the automation of the vehicle safety feature is reported to have the ability to reduce the casualties figure [3]. This has catalyzed the development of autonomous vehicle globally. The automated driving field is expected to transform the automotive sectors due to its disruptive innovation nature [4]. The advent of autonomous vehicles is mentioned in many future automotive roadmaps, and the technology is expected to be marketed to the general audience as early as in the year 2025 [5 - 6]. In addition to the aforementioned progress, many major scientific bodies such as the Fédération Internationale des Sociétés d'Ingénieurs des Techniques de l'Automobile (FISITA), Institute of Electrical and Electronics Engineers (IEEE) and

Society of Automotive Engineers (SAE) are organizing academic and industrial conferences which are focusing on the autonomous vehicle development [7- 9]. However, to the best of the authors' knowledge, the autonomous vehicle field is yet to be widely discussed in Malaysia, despite the rapid progress done globally [10-11]. Thus, this work is a part of a continuous research in Malaysia by Universiti Teknologi Malaysia, with the support from PROTON Holdings Berhad to reduce road fatalities and developing a fully autonomous vehicle technology. Most of the preliminary works by the authors involve simulation works as well as hardware implementations of other Advanced driver-assistance systems (ADAS), which include the Blind Spot Monitoring System, Side Collision Avoidance, Lane Change and Lane Departure Assistance Systems [12-16]. Despite this, prior to this work, the development of a multi-actuators platform on a real vehicle for frontal collision avoidance system has not yet been done by the authors. Since frontal collision avoidance is an important feature of an autonomous vehicle, consequently, this work focuses on the development and experimental validation of a vehicle collision avoidance system. A vehicle of modular nature which can be upgraded for various research purposes is used as the research platform. It is important to be noted that since the focus of this work is to develop a collision avoidance (CA) system for the autonomous vehicle, the paper's discussion will be limited to the said topic. The paper is organized as follows. Section 2 briefly describes the platform architecture and design. Section 3 discusses the development of the collision avoidance system, where it consists of modular longitudinal and lateral tracking strategies. In section 4, the experimental designs are discussed, followed by its results. In the final section, the conclusions of this work are jotted down as well as the future research suggestions.

2. Experimental Platform Design and Architecture

Figure 1. Development of iDrive Research Platform.

An experimental platform, which is denominated as Intelligent Drive project (iDrive) is utilized for the development of a comprehensive multi-actuators collision avoidance system in this work. It is originated from Proton Exora. Since the work is the continuation of the previous studies done by the authors, for brevity, the detailed design of iDrive will not be written and can be found in [12], [13] and [14]. iDrive is designed with several modules, which include the IMU Module, Perception Module, Monitoring Module as well as the Collision Avoidance System Module. IMU Module aids the vehicle to obtain its current motion states using xSense IMU unit. For the environmental and obstacle detection, iDrive perceives its surrounding using Smartmicro UMRR-0A Radar for frontal obstacle detection, and a couple of Continental SRR-208 for rear blind spot monitoring. The authors discuss the Collision Avoidance system module in the next section. The modules are connected to each other by several means of protocols, which include User Datagram Protocol (UDP), Universal Serial Bus (USB) and Controller

Area Network Bus (CAN). MATLAB and dSPACE MicroAutobox are integrated into the platform to allow rapid online computation of the systems. More details on iDrive's parameters are available in [12-14].

3. Vehicle Collision Avoidance System Architecture

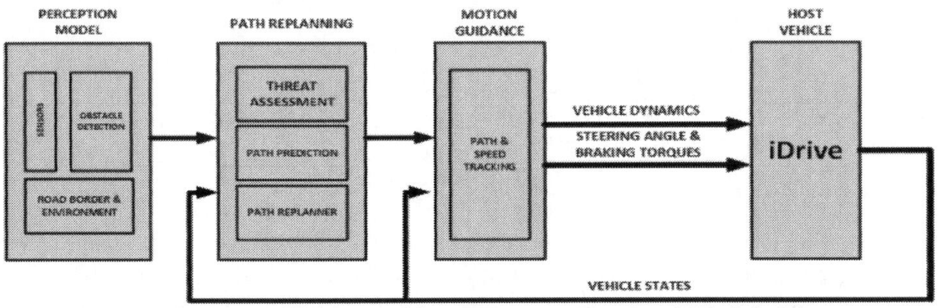

Figure 2. iDrive Collision Avoidance Systems Top-level Module View.

In this section, the vehicle collision avoidance system architecture is discussed, where it is incorporated with a Guidance and Navigation Control System, which is designed from the scenario-based behaviour of the human driver in unwanted events [17]. The architecture is modelled using hierarchically layered architectures and consists of a high-level planning to the low-level control workflow. The high level is depicted in Figure 2, where it consists of the perception and path replanning modules. To allow feasible CA navigation during risky scenarios, the vehicle is assigned with multi-actuators interventions, i.e. active front steering (AFS) and active braking torques. A brushless DC motor with a rotary encoder is adapted for the AFS installation. For the braking actuator, Firgelli Auto linear sensor's installation allows for the active braking intervention. As shown in Figure 2, the perception module is responsible for the environmental sensing of the host vehicle. It includes the obstacle detection. The findings of the module will be sent to the path replanning block. If the threat assessment threshold is violated by the host vehicle, the replanned path will be tracked by the motion guidance. The required steering angle and braking torques for the CA navigation are then fed to the iDrive. The details on the path replanning and motion guidance blocks are written in the next subsections.

3.1. Lateral Motion Controller (Active Steering Intervention)

Both AFS and Active Braking intervention is activated in relation to the threshold of Potential Field Strategy Threat Assessment. The formulation of the Potential Field is formulated based on the work of [16], [18] and [19]. The decision-making formulation for the CA maneuver is written below, where the vehicle distance to the frontal obstacle is measured by the front radar. Once the vehicle violated the threat assessment and safe-distance threshold, its current x coordinate, x_{cur} will be equivalent to x_{start}, which indicates the point where the steering and braking interventions begin. x_{start} is formulated in accordance with the safe distance of the host vehicle to the expected collision point, which subsequently allows the research platform to maintain the safe time metrics during the avoidance [15].

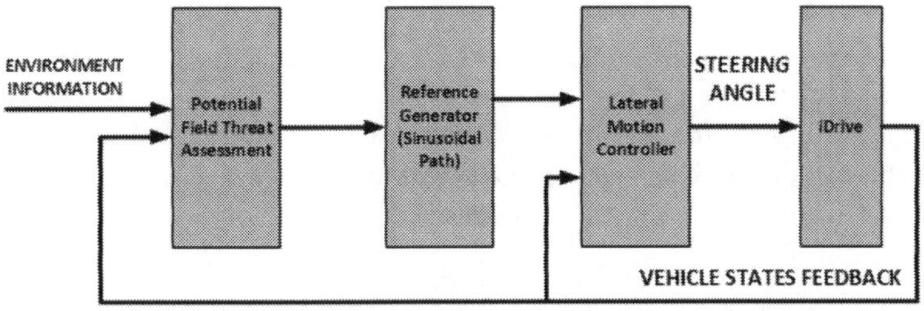

Figure 3. Active Steering Intervention Formulation.

$$\delta_F = \begin{cases} \delta_F & if\ x_{cur} \geq x_{start} \\ \delta_{CA} & if\ x_{cur} \leq x_{start} \end{cases} \tag{1}$$

With the violation, the current steering angle of iDrive, δ_F is replaced with an emergency collision avoidance actuation, δ_{CA}. The formulation of the reference steering angle is formulated with a sinusoidal path lane change strategy [20]. The strategy formulated the δ_{CA} by initially calculating the desired avoidance lateral acceleration as follows:

$$a_{lat} = \begin{cases} \frac{2\pi d}{t_{LC}^2} \sin\left(\frac{2\pi}{t_{LC}}(t - t_{lat})\right), & if\ t \subset |t_{lat}, t_{LC} + t_{lat}| \\ 0, & otherwise \end{cases} \tag{2}$$

where a_{lat} refers to the desired lateral acceleration during the maneuver, t_{LC} depicts the total time taken to complete the lane change and t is the time for the whole formulation to be computed and d is the lateral distance to the road centerline. Deriving from Equation 2, the δ_{CA} is derived as written below:

$$\delta_{CA} = \frac{2\pi d_y}{t_{CA}^2} \sin\left(\frac{2\pi}{t_{ca}}(t - t_{init})\right), if\ t \subset |t_{init}, t_{CA} + t_{init}| \tag{3}$$

t_{CA} in Equation 3 refers to the total time for the complete avoidance navigation taken by the steering, t is the elapsed time for the maneuver while d is the lateral distance. t_{init} refers to the time for the whole formulation to be computed. This replanning module can be considered as an offline path planning, where the path is already predefined. In the event of obstacle's existence, the replanned steering angle will be tracked by the actuators.

3.2. Longitudinal Motion Controller (Active Braking Intervention)

As for the active braking intervention, which tracks the vehicle desired speed in the occurrence of obstacles' risk, the algorithms formulation is:

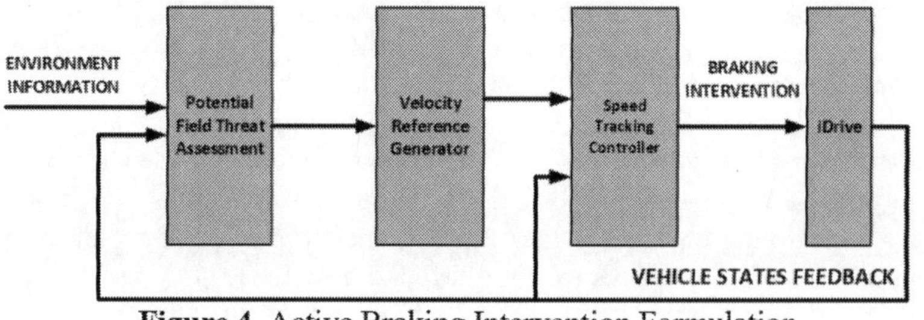

Figure 4. Active Braking Intervention Formulation.

$$F_B = \begin{cases} k \cdot F_B\ if\ x_{cur} \leq x_{start} \\ 0\ if\ x_{cur} \geq x_{start} \end{cases} \tag{4}$$

where F_B stands for the full braking actuation, while k is the braking ratio. The value of k is obtained heuristically. In the case when fully braking intervention is desirable, k is set to 1. Similar to the AFS, the Active Braking intervention is formulated in relation to the Potential Field Threat Assessment strategy.

IOP Conf. Series: Materials Science and Engineering **342** (2018) 012018 doi:10.1088/1757-899X/342/1/012018

3.3. Motion Guidance Modules Validation

Prior to the collision avoidance performance experimental validation, simple scenario preliminary evaluations are done to verify the performance of the sensors and low-level controllers (actuators). The preliminary tests are done in Universiti Teknologi Malaysia, Kuala Lumpur campus area. Two scenarios are done, where each of the tests requires the vehicle to mitigate the collision with single AFS and single Active Braking intervention, respectively. Due to the safety issues and limited size of the testing area, the vehicle speed is set at low speed (below $10\ km/h$). This is to allow sufficient time for the host vehicle to mitigate the collisions [15]. The safe-distance threshold is set to a value of $15\ m$. Once the threshold is violated, the CA intervention begin. Both scenarios serve to verify iDrive's ability to act as formulated in Equations 1-4. The experiment is considered successful once the vehicle managed to outputs needed steering or braking actions to steer and avoid the frontal obstacle.

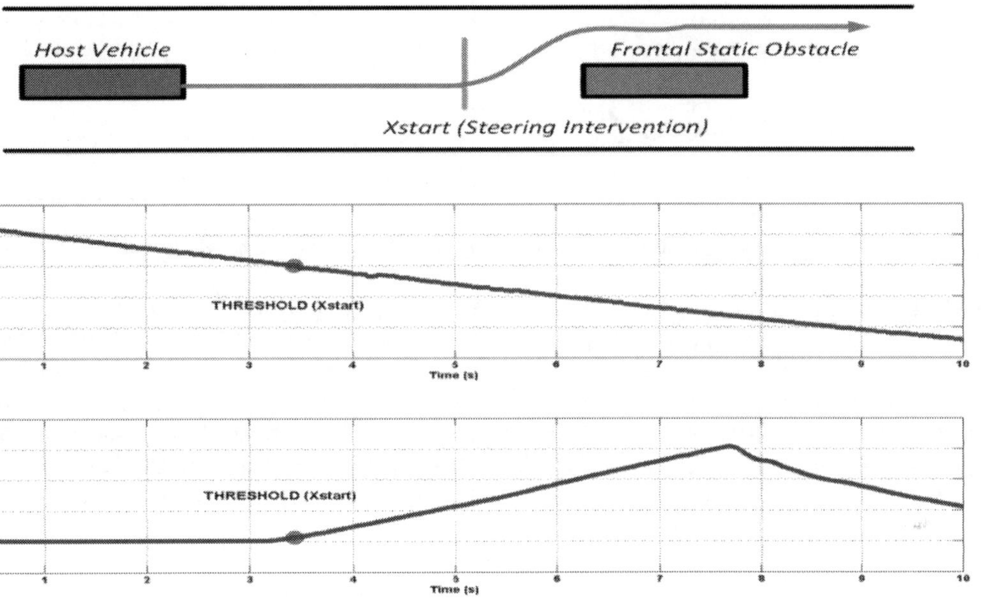

Figure 5. Active Front Steering Collision Avoidance Evaluation.

For AFS, the steering controller is shown to successfully tracked the desired steering angle in Figure 5 by yielding δ_{CA} when the vehicle reaches x_{start} (Equations 1-3). The safe distance for the active braking intervention evaluation is set to be 5 m. In Figure 6, it is shown that iDrive is able to actuate the active braking in the occurrence of the obstacle. As can be seen, the host vehicle fully stops after the braking, thus mitigating the collisions. iDrive successfully yield 100% braking actuation to allow the vehicle for a full deceleration once the 5 m threshold distance is violated. Both of the pseudo-obstacle avoidance scenarios validation in Figures 5-6 show that the vehicle hardware features are able to steer and brake in a timely manner in relation to the obstacle detection by the sensors. Thus, iDrive is expected to perform well as a multi-actuators CA platform for the CA experiment, which will be discussed in the next section.

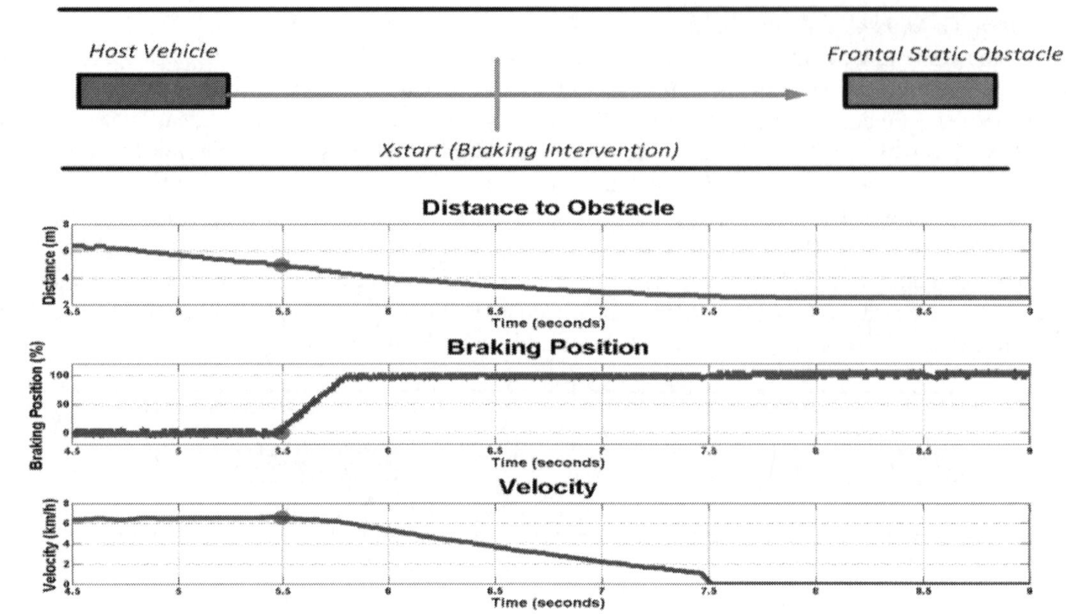

Figure 6 Active Braking Intervention for Collision Avoidance Evaluation.

4. Controlled Environment Collision Avoidance Setup (Experimental Design)

The collision avoidance experiment was done to validate the iDrive's ability as a multi-actuators CA platform. The test is done in a standard football field-sized open space in Universiti Teknologi Malaysia, Johor Bahru campus. The host vehicle initially navigates with the maximum speed of $50\,km/h$. A dummy obstacle is created using Polyvinyl chloride-based (PVC) pipes. The obstacle dimension is the same as a Honda Jazz dimension. Its size in relation to the iDrive is depicted in Figure 7. Its initial position is located $80\,m$ in front of the host vehicle initial whereabouts. The avoidance's idea is similar to Figure 5's scenario, but with a higher speed and required multi-actuators intervention during the CA navigation. The host vehicle needs to provide multi-actuation interventions (braking and steering), relative to the Risk Measurement by the Potential Field strategy. Once the threshold is violated, the host vehicle will initially decelerate, before tracking the reference steering angle. The reference trajectories are as formulated in Equations 1-4. Due to the limited length of the work, the results only discuss the ability of the iDrive as the multi-actuators CA platform to avoid the collision with the obstacle.

Figure 7. Obstacle's dimension relative to the iDrive.

5. Results and Discussions

Figure 8 shows the still frame images of the experiment's video recording. The host vehicle was navigating in a straight line when a frontal static obstacle was detected. The CA navigations are yielded

IOP Conf. Series: Materials Science and Engineering **342** (2018) 012018 doi:10.1088/1757-899X/342/1/012018

in the second and third sequence of Figure 8, and as can be seen, the human driver is not holding the steering wheel during the maneuver.

Figure 8. Still frame images from the experiment's video recording.

The results in Figure 9 shows that the Potential Field successfully provide risk measurement in relation to the relative distance between the host vehicle and obstacle, provided by the frontal sensor. The highest collision risk with the frontal obstacle is at $16\,s$ of the experiment elapsed time. The host vehicle subsequently yielded the desired braking torques to allow for its deceleration at $12.5 - 13.8$ seconds of elapsed experimental time in Figure 10. The steering also provided an emergency actuation based on the sinusoidal reference trajectory, allowing the host vehicle to avoid the frontal obstacle to the left side when the risk fields threshold is violated (Figure 11).

IOP Conf. Series: Materials Science and Engineering **342** (2018) 012018 doi:10.1088/1757-899X/342/1/012018

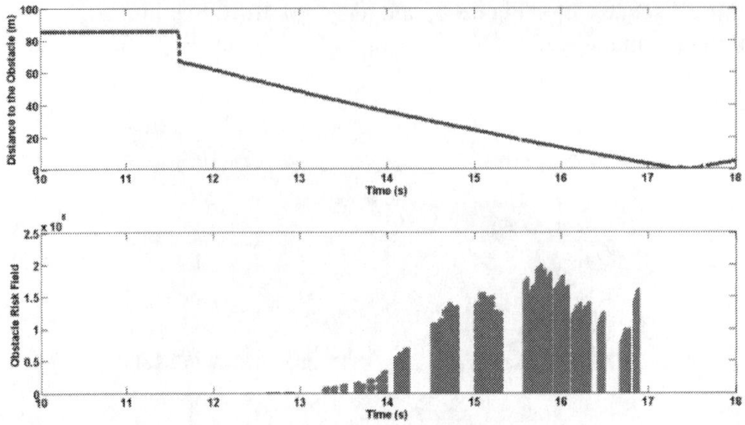

Figure 9. Potential Field Risk Assessment for Collision Avoidance Experiment.

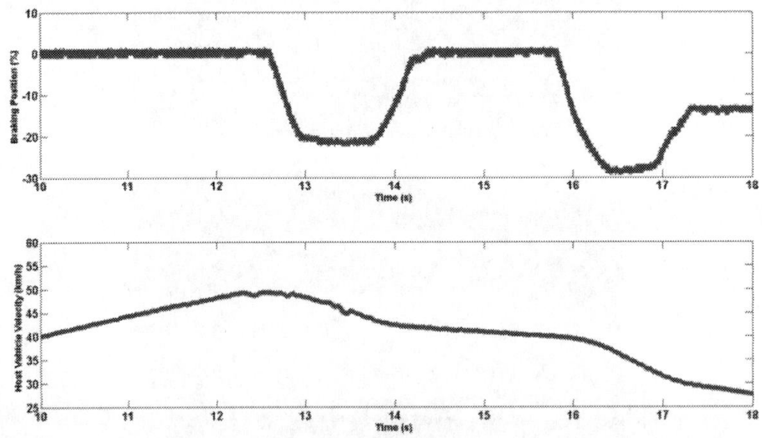

Figure 10. Active Braking Intervention for Collision Avoidance Experiment.

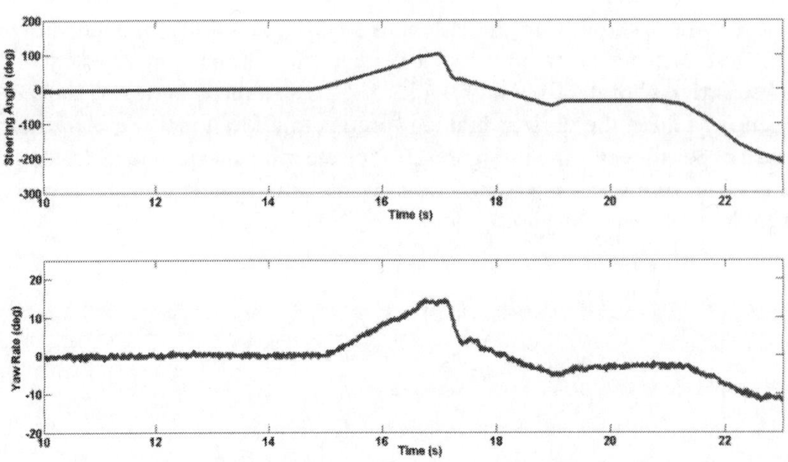

Figure 11. Active Front Steering Actuation for Collision Avoidance Experiment.

However, several issues are identified in this experiment (Figures 9-11), which include:

1. In Figure 9, around 11.8 seconds, the signal of the radar is interrupted for a while due to the detection of another obstacle. Hence, it portrays a noise in the distance to the obstacle measurement reading. Thus, for safety reason, for a more complex scenario, more isolated areas are required for validation.
2. The braking torques are also shown to be providing unstable intervention after the maneuver (Figure 10, after $16\ s$). This is due to the absence of a reliable reference generator for the motion replanning.
3. Potential Field only considers the risk of obstacle, but it does not include the consideration of environmental risk (which include lane departure risk). Thus, after the maneuver, in Figure 11 (at $19 - 22\ s$), iDrive's steering is constantly showing evasive maneuver actuations though there are no obstacle, allowing lane departure.
4. The lane departure mentioned in number (4) above is also caused by the high velocity of the host vehicle ($50\ km/h$), which demands a nonlinear tracking controller to accommodate the vehicle dynamics.
5. There are needs to design a more reliable system as the scenario only considers static obstacle.

Relying on these findings, the development of a more reliable collision avoidance will be considered in the future work of this research.

6. Conclusions

As detailed discussions on the results have been jotted in the previous section, a brief conclusion of this work is written in this section. In this work, a multi-actuators collision avoidance system research platform for the autonomous vehicle is developed (iDrive). The platform acts by providing multi-actuation collision avoidance maneuver (steering and braking interventions) during the emergence of an obstacle. To validate the performance, a real-time experiment is done where iDrive produced the desired emergency navigation, thus allowing for the collision to be avoided. This work is important since the results can be further implemented in the Malaysian-marketed cars in the future, as well as allowing more research works on the topics of the autonomous vehicle. Future works include the study on the optimal path during the avoidance as well as mimicking the human behavior. This is to ensure that the platform will abide the New Car Assessment Program for Southeast Asian Countries (ASEAN NCAP) before being marketed to the general audiences.

Acknowledgement

The work presented in this study is funded by Ministry of Higher Education, Malaysia and Research University Grant, Universiti Teknologi Malaysia. VOTE NO: 13H73. This work is also supported by PROTON Holdings Berhad (VOTE NO: 4C099) and Universiti Malaysia Pahang, Research Grant RDU 170371. The authors would like to thank Pongsathorn Raksincharoensak and Yuichi Saito from the Smart Mobility Research Center, Tokyo.

References

[1] Naci H, Chisholm, D and Baker, T D 2009 Distribution of road traffic deaths by road user group: a global comparison. *Injury prevention* **15** (1) 55-59

[2] World Health Organization, Regional Office for South-East Asia., 2016 Road safety in the South-East Asia Region 2015. http://searo.who.int/, accessed: 2017.09.29

[3] Insurance Institute for Highway Safety Highway Loss Data Institute - Automation and crash avoidance.http://www.iihs.org/iihs/topics/t/automation-and-crash-avoidance/topicoverview accessed: 2017.09.29

[4] Rahman A A, Hamid U Z A and Chin T A 2017 Emerging Technologies with Disruptive Effects: A Review. *PERINTIS eJournal* **7**(2) 111-128

[5] Hamid U Z A, Pushkin K, Zamzuri H, Gueraiche D and Rahman M A A 2016 Current Collision Mitigation Technologies for Advanced Driver Assistance Systems–A Survey. *PERINTIS eJournal* **6**(2)

[6] Kelkel R 2015 Predicting consumers' intention to purchase fully autonomous driving systems: which factors drive acceptance? (Doctoral dissertation), Universidade Católica Portuguesa.

[7] FISITA World Congress. https://www.fisita-congress.com/ accessed: 2017.09.29

[8] IEEE 20th International Conference on Intelligent Transportation Systems http://www.itsc2017.org/, accessed: 2017.09.29

[9] From ADAS to Automated Driving, SAE http://www.sae.org/events/adas/ accessed: 2017.09.29

[10] Eichberger A, Markovic G, Magosi Z, Rogic B, Lex C and Samiee S 2017 A Car2X sensor model for virtual development of automated driving. *International Journal of Advanced Robotic Systems* **14**(5) https://doi.org/10.1177/1729881417725625

[11] Jorlöv S, Bohman K, and Larsson A 2017 Seating Positions and Activities in Highly Automated Cars–A Qualitative Study of Future Automated Driving Scenarios. In *International Research Conference on the Biomechanics of Impact.*

[12] Zakuan F R A, Zamzuri H, Rahman M A A, Yahya W J, Hazima N, Ismail F, Zakariya M S, Zulkepli K A and Zulfaqar Azmi M 2017 Threat Assessment Algorithm for Active Blind Spot Assist System Using Short Range Radar Sensor. *ARPN Journal of Engineering and Applied Sciences* **12** 4270-4275

[13] Azmi M Z, Zamzuri H, Rahman M A A, Yahya W J, Razi F, Zakuan A, Zulkepli K A, Ismail N H F and Zakariya M S 2017 Steering Intervention Strategy for Side Lane Collision Avoidance. *ARPN Journal of Engineering and Applied Sciences* **12** 4265-4269

[14] Zulkepli K A, Zamzuri H, Rahman M A A, Yahya W J, Aizzat M, Zakaria M Z A, Zakuan F R A and Faezaa N H, 2017. i-Drive: Modular System Architecture and Hardware Configuration for an Intelligent Vehicle Research Platform. *ARPN Journal of Engineering and Applied Sciences* **12** 4259-4264

[15] Hamid U Z A, Ariff M H M, Zamzuri H, Saito Y, Zakaria M A, Rahman M A A and Raksincharoensak P 2017 Piecewise Trajectory Replanner for Highway Collision Avoidance Systems with Safe-Distance Based Threat Assessment Strategy and Nonlinear Model Predictive Control. *Journal of Intelligent & Robotic Systems* 1-23.

[16] Hamid U Z A, Zamzuri H, Yamada T, Abdul Rahman M A, Saito Y and Raksincharoensak P 2017 Modular design of artificial potential field and nonlinear model predictive control for a vehicle collision avoidance system with move blocking strategy. *Proceedings of the Institution of Mechanical Engineers, Part D: Journal of Automobile Engineering*, https://doi.org/10.1177/0954407017729057

[17] Falcone P, Borrelli F, Asgari J, Tseng H E and Hrovat D 2007 Predictive active steering control for autonomous vehicle systems. *IEEE Transactions on control systems technology* **15**(3) 566-580

[18] Matsumi R, Raksincharoensak P and Nagai M 2013 Autonomous braking control system for pedestrian collision avoidance by using potential field. *IFAC Proceedings Volumes* **46**(21) 328-334

[19] Raksincharoensak P, Ehira T, Shimono K and Tagawa Y 2016 December. Autonomous Vehicle Trajectory Planning and Control Based on Virtual Disturbance Compensation via Simulation of Feedback Control Systems. In *Advanced Vehicle Control: Proceedings of the 13th International*

Symposium on Advanced Vehicle Control (AVEC'16), September 13-16, 2016, Munich, Germany (p. 189). CRC Press.

[20] Butakov V A and Ioannou P 2015 Personalized driver/vehicle lane change models for ADAS. *IEEE Transactions on Vehicular Technology* **64**(10) 4422-4431

IOP Conf. Series: Materials Science and Engineering **342** (2018) 012019 doi:10.1088/1757-899X/342/1/012019

The application of *k*-Nearest Neighbour in the identification of high potential archers based on relative psychological coping skills variables

Zahari Taha[1], Rabiu Muazu Musa[1,2], Anwar P P Abdul Majeed[1], Mohamad Razali Abdullah[2], Muhammad Muaz Alim[1] and Ahmad Fakhri Ab Nasir [1]

[1]Innovative Manufacturing, Mechatronics and Sports Laboratory, Faculty of Manufacturing Engineering, Universiti Malaysia Pahang, 26600 Pekan, Pahang, Malaysia
[2]Faculty of Applied Social Sciences, Universiti Sultan Zainal Abidin, 21300 Kuala Terengganu, Terengganu, Malaysia

E-mail: rabiumuazu@imamslab.com

Abstract. The present study aims at classifying and predicting high and low potential archers from a collection of psychological coping skills variables trained on different k-Nearest Neighbour (k-NN) kernels. 50 youth archers with the average age and standard deviation of (17.0 ±.056) gathered from various archery programmes completed a one end shooting score test. Psychological coping skills inventory which evaluates the archers level of related coping skills were filled out by the archers prior to their shooting tests. k-means cluster analysis was applied to cluster the archers based on their scores on variables assessed k-NN models, i.e. fine, medium, coarse, cosine, cubic and weighted kernel functions, were trained on the psychological variables. The k-means clustered the archers into high psychologically prepared archers (HPPA) and low psychologically prepared archers (LPPA), respectively. It was demonstrated that the cosine k-NN model exhibited good accuracy and precision throughout the exercise with an accuracy of 94% and considerably fewer error rate for the prediction of the HPPA and the LPPA as compared to the rest of the models. The findings of this investigation can be valuable to coaches and sports managers to recognise high potential athletes from the selected psychological coping skills variables examined which would consequently save time and energy during talent identification and development programme.

1. Introduction

Various researchers have endeavoured to provide insights on the impacts of psychological skills to the advancement of athlete's performance in different sports. Abdullah et al. studied the roles of psychological factors on the performance of elite soccer players and established that although, psychological factors alone could not forecast the performance of elite soccer players, but play an essential part in improving their performance [1]. Likewise, Lazarus [2] and Salim et al. [3] inferred that psychological elements such as stress and worry could be a significant constituent in hindering athletes performance. They further emphasised that athletes who are unable to cope with stress and adversity effectively might likely suffer from performance declining and overall psychological well-being issues. Furthermore, Pasaei et al. [4] stated that in the contemporary archery competitions there are presents of various noises, tensions, and anxiety, and as such selective attention through the application of the psychological coping skills during the competition become paramount for better performance. This implies that the athletes with higher psychological coping skills are likely to outperform their opponents with lesser coping skills.

The employment of machine learning or artificial intelligence has gained popularity in prediction and classification in sports and exercise due to its superiority over conventional means. Bayesian network (BN) analysis was utilised to investigate a semi-professional team's performance and collective efficacy based on a number of psychological features [5]. In a different study, Fuster-Parra et al. [6] also employed BN in correlating negative psychological features with sportive behaviours of young team players in competitive sports.

It is evident from the literature that the employment of machine learning could be a useful tool for prediction as well as classification. The k-Nearest Neighbour (k-NN) is a non-parametric regression and classification method that was initially developed by Fix and Hodges [13]. It is regarded as one of the simplest forms of supervised machine learning algorithms that has been successfully utlised for a number of classification problems in different fields [15–19].

2. Material and methods

2.1. Participants

A complete number of 50 archers were selected to take part in in this study. The participants consisted of 37 male and 13 female youth archers between the age's range of 13-20 with a mean and standard deviation of (17.0 ± 0.56) drawn from different archery programmes in Malaysia. The archers were under a development program for preparing both at university and the state level and consequently targeted to be preferred to state and national archers respectively. Written consent was obtained, and all the archers signed consent forms. All the procedures, protocol, and equipment for this study were authorised by the Research Ethics Board of the Terengganu Sports Institute (ISNT) with a memo number 04-04/T-01/Jid 2.

2.2. Psychological coping skills inventory assessment

The Athletic Coping Skills Inventory (ACSI), an instrument for assessing athlete's psychological skills, produced by Smith et al. [7] was selected. This instrument is considered apt for evaluating the psychological skills of the archers in this study because of its connection to the nature of archery game as an individual sport rather than a team sport. The instrument was distributed to the archers prior to the shooting test, and their responses were compiled and analysed. A simulated shooting competition area was set up, and all the archers' shoot six arrows (one end) over a distance of 50 meters. All the archers were given trials of four arrows shot before recording the final six arrows scores.

2.3. Data analysis

2.3.1. Clustering: k-means cluster analysis. In the current study, the k-means clustering algorithm, which is essentially a type of unsupervised learning, is used to separate the classes of the related performance variables assessed via Orange 2.7. It has been reported in the literature that the k-means clustering method is more reliable than the hierarchical agglomerative clustering in catering large datasets [8]. This is primarily due to the fact that it operates on actual observations rather than the dissimilarity measures employed in hierarchical clustering. The number of clusters, k is selected to be two as the data is consists of HPPA and LPPA.

2.3.2. Classification: k-Nearest Neighbour (k-NN). In this study, six variations of k-NN are examined i.e. fine, medium, coarse, cosine, cubic and weighted. The number of neighbours, k for fine and coarse are 1, and 100, respectively whilst for the rest of the variations is selected to be 10. The Euclidean distance metrics is employed in the fine, medium, coarse and weighted k-NN variations, whilst the cubic and cosine utilises a special case of the Minkowski distance and the cosine distance, respectively. The no weights are attributed to the distance for all variations of k-NN with an exception for the weighted k-NN, in which, the weight is the squared inverse of the distance.

2.3.3. Model training and testing. A fivefold cross-validation method was utilised in this study. This form of validation technique is desirable as it mitigates the notion of overfitting through partitioning the dataset into a number of folds and estimating the accuracy of each fold. The data (50 observations) is randomly split into five subsets, and for each iteration, one of the five subsets are used as the testing data, whilst the remaining four will be used as the training data [9]. Then, the average performance over all the folds is then computed. The SVM analysis and assessment were conducted by means of MATLAB 2016a (Mathworks Inc., Natick, USA).

2.3.4. Model evaluation. The variations of the SVM employed in this study are evaluated by means of classification accuracy (ACC), sensitivity (SENS), specificity (SPEC), precision (PREC), error rate (ERR) as well as Matthew's correlation coefficient (MCC). The confusion matrix allows the observation of correctly classified and misclassified observations that transpires between the defined classes.

3. Results

3.1. Clustering

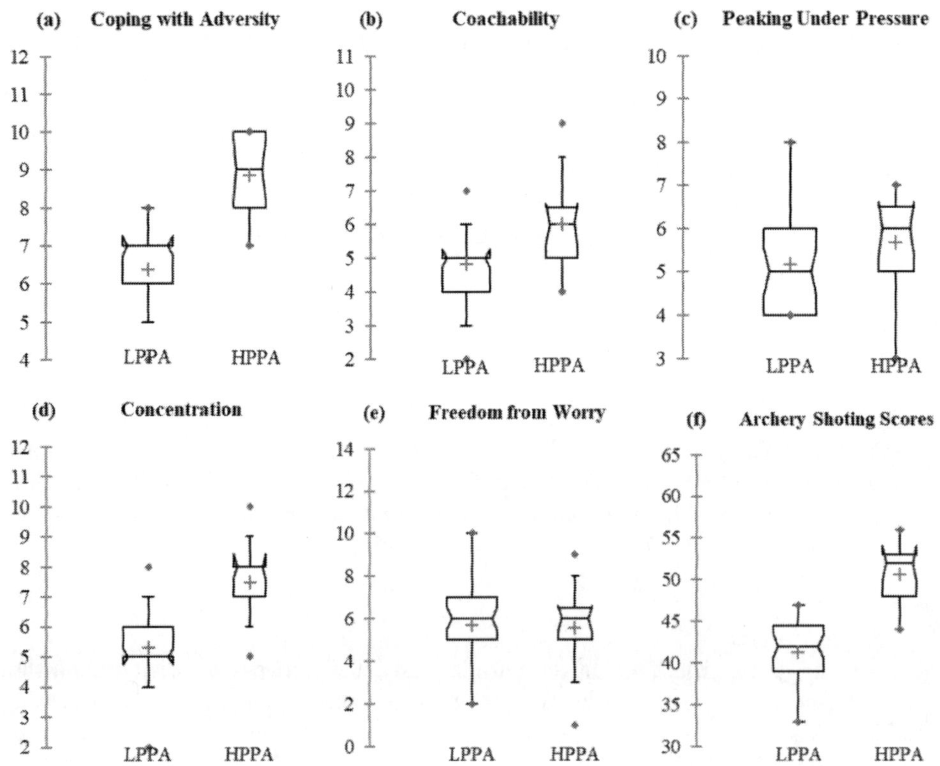

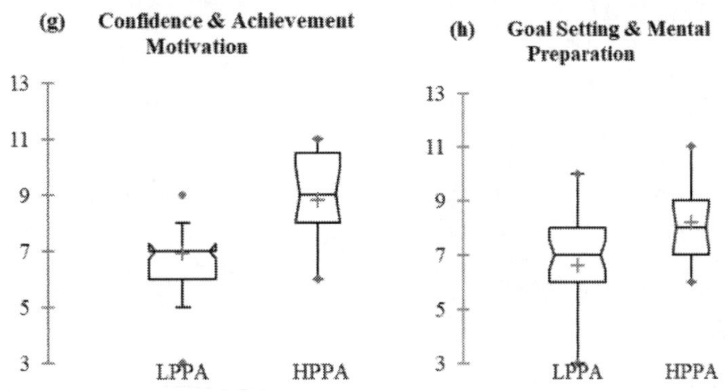

Figure 1. Comparisons of performance differences of the archers based on ACSI (a) Coping with Adversity; (b) Coachability; (c) Peaking Under Pressure; (d) Concentration; (e) Freedom from Worry; (f) Archery shooting score; (g) Confidence and Achievement Motivation; (h) Goal Setting and Mental Preparation.

The performance differences of the archers based on the eight performance variables assessed that were clustered via *k*-means are shown in figure 1. It could be seen from the box plots that the mean performances of HPPA are greater than LPPA across all the seven psychological coping skills measured in the study except for 'freedom from worry' (figure 1(e)). This observation is non-trivial as worry is often observed to be inversely proportional towards performance. Therefore, the aforementioned variables are essential attributes that allow for an accurate discrimination between the HPPA and LPPA.

3.2. Classification

Table 1. Model evaluation.

k-NN variation	ACC (%)	SENS (%)	SPEC (%)	PREC (%)	ERR (%)	MCC
Fine	92	97.37	75	92.5	8	0.7729
Medium	86	100	41.67	84.44	14	0.5932
Coarse	76	100	0	76	24	-
Cosine	94	97.87	83.33	94.87	6	0.8320
Cubic	84	100	33.33	82.61	16	0.5248
Weighted	90	100	58.33	88.37	10	0.7180

It could be observed from table 1 that the cosine *k*-NN variation is able to produce exceptional classification through the evaluation of all assessment parameters as compared to the rest of the *k*-NN variations. It is also apparent that the fine-based model could offer a reasonably accurate classification with an error rate of 8 %. Conversely, the coarse *k*-NN model is rather unsuitable for predicting the correct classification of HPPA and LPPA, as it produces the highest error rate at 24% on top of a poor negative class accuracy shown via the SPEC evaluation that in turn produces an uncorrelated model as demonstrated through the MCC. It is evident through the present study that the cosine-based model may be used for the purpose of talent identification for archers based on the predefined psychological coping skills. The confusion matrix of the evaluated *k*-NN models are illustrated in figure 2.

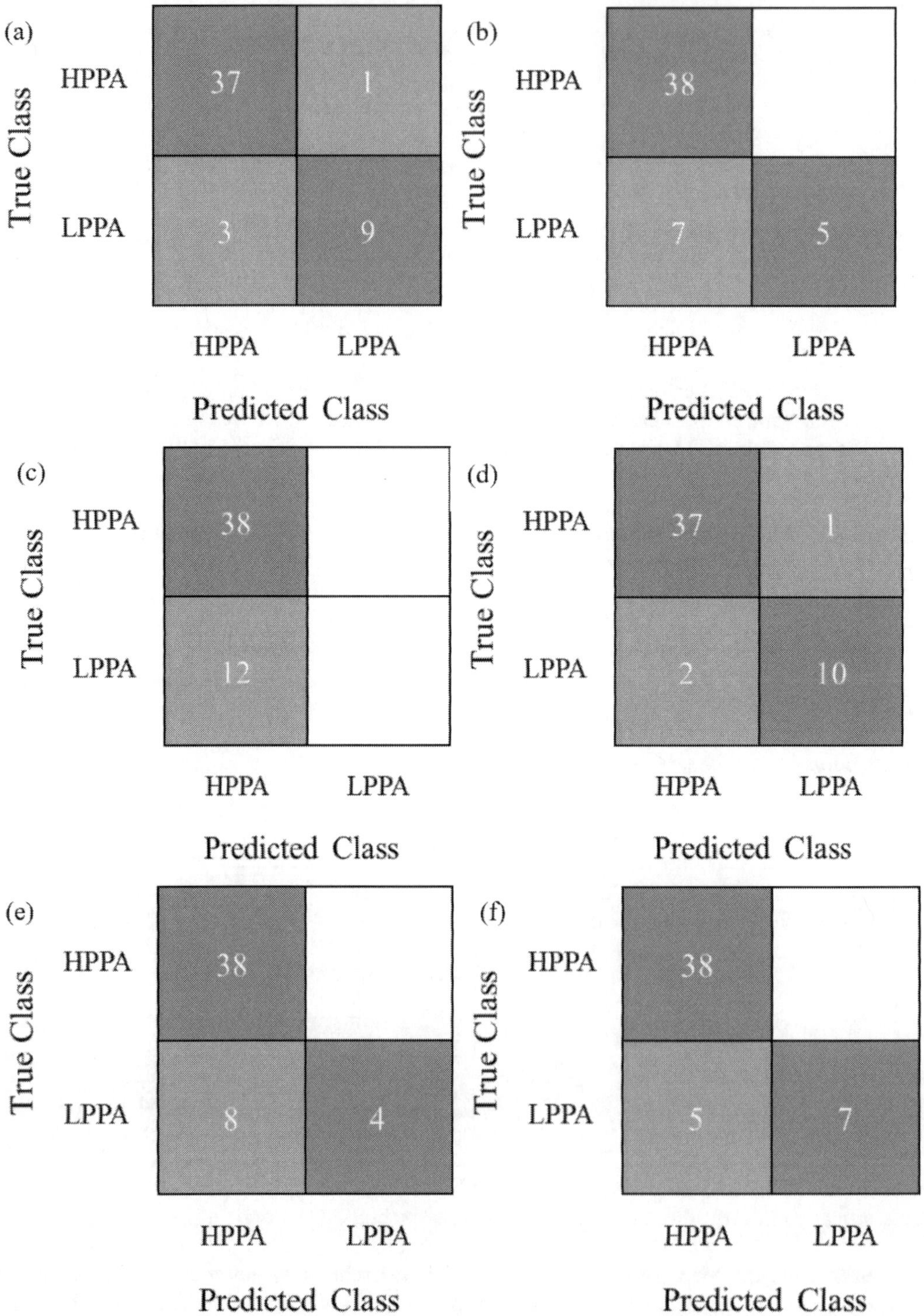

Figure 2. Confusion matrix. (a) Fine; (b) Medium; (c) Coarse; (d) Cosine; (e) Cubic; (f) Weighted

4. Discussion

The present study has demonstrated that the selected psychological coping skills variables established can predict or classify well the performance of the archers either as LPPA or HPPA. The shooting scores allow for the clustering of the archers with respect to measured psychological coping skills. This was carried out by means of *k*-means as illustrated in the box plot (Figure 1). It was shown that the cosine-based model is able to classify the LPPA and HPPA well.

The result accentuated the need for mental aptitudes, for example, dependence and self-motivational strategies in the sport. Archery is closed skill-based sport, in which the archer is required to compete against another archer. For an archer to get a higher score, he or she should have the ability to maintain confidence as well as self-motivation. Evidence in sports and exercise science demonstrates that one's attention to capability or self-assurance is the significant factor prompting the achievement of any type of sport [11].

Additionally, it has been depicted that confidence and self-motivation are the amongst the qualities of a winner. They are the secret elements that all superior athletes seem to achieve, regardless of what level they contest since positive mental attitude keeps an athlete working hard irrespective of how often he/she may fail or how many hindrances might come in his/her direction [12]. Confidence and self-motivation could give a team or an ordinary athlete the determination and focus on defeating a stronger opponent. These mental abilities could lead athletes to the greater level of performance in their domain. Likewise, inadequate or lack of confidence and self-motivation could render athletes perform below their natural potential. Poor confidence and self-motivation could diminish an athlete's indulgence of the sport and turn him/her into a dropout and a frequent loser [1].

5. Conclusion

The present study has successfully shown that the selected psychological coping skills variables, i.e. coping with adversity, coachability, peaking under pressure, concentration, freedom from worry, confidence and achievement motivation, goal setting and mental preparation does to a certain degree influence the performance of the archers. The study also has demonstrated that the employment of such machine learning methods allow coaches to appropriately identify high potential athletes in the sport of archery by taking into consideration the aforementioned elements. The current findings are non-trivial to coaches and sports managers in identifying talents with significantly little effort and cost.

Acknowledgement

This work is funded by the National Sports Institute of Malaysia (ISNRG: 8/2014-12/2014).

References

[1] Abdullah M R, Musa R M, Maliki A B H M B, Kosni N A and Suppiah P K 2016 Role of psychological factors on the performance of elite soccer players *J. Phys. Educ. Sport* **16** 170
[2] Lazarus R S 2000 How emotions influence performance in competitive sports *Sport Psychol.* **14** 229–52
[3] Smith R E and Christensen D S 1995 Psychological skills as predictors of performance and survival in professional baseball *J. Sport Exerc. Psychol.* **17** 399–415
[4] Pasaei N, Abdoli B, Kazem S, Mousavi V and Aslankhani M 2013 Does Predict Personality Introversion-Extraversion Dimension the Archer's Performance? *Middle-East J. Sci. Res.* **13** 760–3
[5] Fuster-Parra P, García-Mas A, Ponseti F J and Leo F M 2015 Team performance and collective efficacy in the dynamic psychology of competitive team: A Bayesian network analysis *Hum. Mov. Sci.* **40** 98–118
[6] Fuster-Parra P, García-Mas A, Ponseti F J, Palou P and Cruz J 2014 A Bayesian network to discover relationships between negative features in sport: a case study of teen players *Qual. Quant.* **48** 1473–91

IOP Conf. Series: Materials Science and Engineering **342** (2018) 012019 doi:10.1088/1757-899X/342/1/012019

[7] Smith R E, Schutz R W, Smoll F L and Ptacek J T 1995 Development and validation of a multidimensional measure of sport-specific psychological skills: The Athletic Coping Skills Inventory-28 *J. Sport Exerc. Psychol.* **17** 379–98
[8] Collins B M, Lydersen J M, Fry D L, Wilkin K, Moody T and Stephens S L 2016 Variability in vegetation and surface fuels across mixed-conifer-dominated landscapes with over 40 years of natural fire *For. Ecol. Manage.* **381** 74–83
[9] Taha Z, Musa R M, P.P. Abdul Majeed A, Alim M M and Abdullah M R 2018 The identification of high potential archers based on fitness and motor ability variables: A Support Vector Machine approach *Hum. Mov. Sci.* **57** 184–93

The employment of Support Vector Machine to classify high and low performance archers based on bio-physiological variables

Zahari Taha[1], Rabiu Muazu Musa[1,2], Anwar P P Abdul Majeed[1], Mohamad Razali Abdullah[2], Muhammad Amirul Abdullah[1], Mohd Hasnun Arif Hassan[1] and Zubair Khalil[1]

[1]Innovative Manufacturing, Mechatronics and Sports Laboratory, Faculty of Manufacturing Engineering, Universiti Malaysia Pahang, 26600 Pekan, Pahang, Malaysia
[2]Faculty of Applied Social Sciences, Universiti Sultan Zainal Abidin, 21300 Kuala Terengganu, Terengganu, Malaysia

E-mail: anwarmajeed@imamslab.com

Abstract. The present study employs a machine learning algorithm namely support vector machine (SVM) to classify high and low potential archers from a collection of bio-physiological variables trained on different SVMs. 50 youth archers with the average age and standard deviation of (17.0 ±.056) gathered from various archery programmes completed a one end shooting score test. The bio-physiological variables namely resting heart rate, resting respiratory rate, resting diastolic blood pressure, resting systolic blood pressure, as well as calories intake, were measured prior to their shooting tests. k-means cluster analysis was applied to cluster the archers based on their scores on variables assessed. SVM models i.e. linear, quadratic and cubic kernel functions, were trained on the aforementioned variables. The *k*-means clustered the archers into high (HPA) and low potential archers (LPA), respectively. It was demonstrated that the linear SVM exhibited good accuracy with a classification accuracy of 94% in comparison the other tested models. The findings of this investigation can be valuable to coaches and sports managers to recognise high potential athletes from the selected bio-physiological variables examined.

1. Introduction

Research has indicated that archery sport involves conscious breathing adjustment and synchronization with arm actions [1]. Shooting process in the sport of archery has been described as the concurrent associations between breathing; gross motor control of body positioning; fine-motor control of the archer's fingers, joints, limbs, feet, and cheek; and the handling of perceptual cues related with the target, the sights and the peep or string [2].

Despite the evidence that archery is not perceived as a physically intensive compared to other sports such as basketball, rugby, cycling or soccer, yet, there is still a demand for the archers to take their nutritional status into account. For instance, it has been found that breakfast is an essential meal of the day. When an archer is planning for a practice or competition that could extend for a long time, he/she could deliver a better performance after taken a healthy breakfast [3]. Moreover, evidence has established that in the course of archery performance, an archer needs to have a high degree of attention and relaxation, and as such there is need to consider a number of cardiac parameters such as resting heart rate, resting systolic blood pressure, resting diastolic blood pressure as well as the pulmonary parameters which comprise of resting respiratory rate [4]. The aforementioned physiological variables are important for influencing success in the sport of archery.

Recently, different types of machine learning techniques have been utilized for forecasting and classifying a diversity of human performance variables owing to its innate advantages against conventional means [5]. For example, artificial neural networks has been applied to predict energy expenditure [6] as well as to classify activity types [7]. Random forest classifier, in another perspective, has been employed to also classify activities from wrist-worn accelerometers [8] as well as classifying the playing positions of elite junior football players [9] amongst others. Nonetheless, the usage of Support Vector Machine (SVM) in classifying potential archers based on a number of bio-physiological performance variables has yet been revealed. Thus, this study aims at assessing the effectiveness of the preceding classifier in determining high and low potential archers.

2. Material and methods

2.1. Participants

50 archers were enlisted in this investigation. The archers consisted of 37 male and 13female youth between the age's range of 13-20 with a mean, and standard deviation of (17.0 ± 0.56) gathered from various archery programme in Malaysia. The archery shooting test was completed prior to the previous fitness testing utilizing a simulated competition shooting area of 50 meters range. The archers were granted four trials shot before taking the final six arrows scores. Informed consent was obtained in accordance with the procedures of the Ethical Review Board of the University Sultan Zainal Abidin, Malaysia.

2.2. Nutritional assessment

The nutritional status of the archers was evaluated using their three days dietary re-call. To achieve this aim, the researcher developed a form containing columns for the dietary of three days and the column for a time the meal is consumed. The form contained the serving size of each meal taken. The archers were enlightened on how to measure the quantity of each meal taken to ensure the accuracy of the actual dietary intake. The researcher instructed the archers to fill the form whenever they take their meals and drinks. The data collected for all their dietary intake of three days (two-week days and a weekend) was assessed using a Nutritionist Pro software to determine their dietary status.

2.3. Resting Heart Rate, Diastolic and Systolic Blood Pressure Measurement

Resting heart rate (RHR), resting systolic blood pressure (RSBP) and resting diastolic blood pressure (RDBP) of the archers were evaluated using Omron automatic Blood Pressure Monitor Device (HEM-7120) which has been reported as a reliable tool for measuring the related parameters [10]. The all specified cardiac parameters were measured while the archers were in resting condition and sitting comfortably on a chair, the device is placed at the level of chest, and the arm strap was attached securely on the left arm of the archers after which the strap was inflated automatically by the device, and monitor displayed result of Blood pressure, pulse rate measurement was noted. All measurements were completed early in the morning prior to any exercises or warm up.

2.4. Resting Respiratory Rate Measurement

The resting respiratory rate (RRR) was measured while the archers were comfortably sitting on a chair in an erect position. The palpating the respiratory movements of archer's chest was assessed manually by a specialist therapist for a period of one minute. A stop watch was used for one-minute time setting to ensure that the reading was collected within one minute as suggested by the preceding researchers [11]. The assessments were conducted three times, and the maximum value was taken.

2.5. Data analysis

2.5.1. Clustering: k-means cluster analysis. The k-means clustering technique is employed to cluster the archers based on their performances in the selected variables and their shooting scores in which two classes, i.e. high potential archers (HPA), and low potential archers (LPA) are clustered.

2.5.2. Classification: Support Vector Machine. SVM is utilised to acquire the optimal hyperplane that could correctly classify the data into two distinct categories that are represented by HPA and LPA in the present study. Three different kernel functions are investigated namely fine, medium and coarse RBF. The scale of the fine, medium and coarse RBF is defined by $0.25*P^{0.5}$, $P^{0.5}$, $4*P^{0.5}$, respectively, where P is the number of predictors viz five. The fivefold cross-validation technique was employed in the present investigation [12]. The performance of the SVM models are evaluated through the classification accuracy (CA), sensitivity (SENS), specificity (SPEC), precision (PREC), error rate (ERR) as well as Matthew's correlation coefficient (MCC) of the aforesaid models.

3. Results and discussion

3.1. Clustering

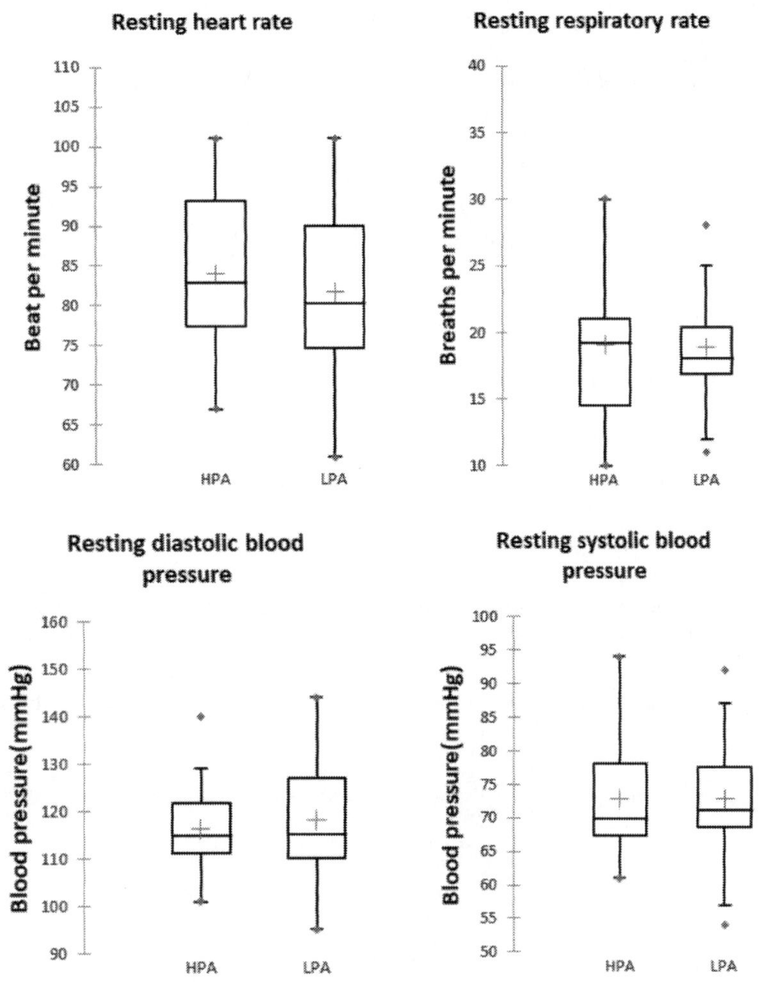

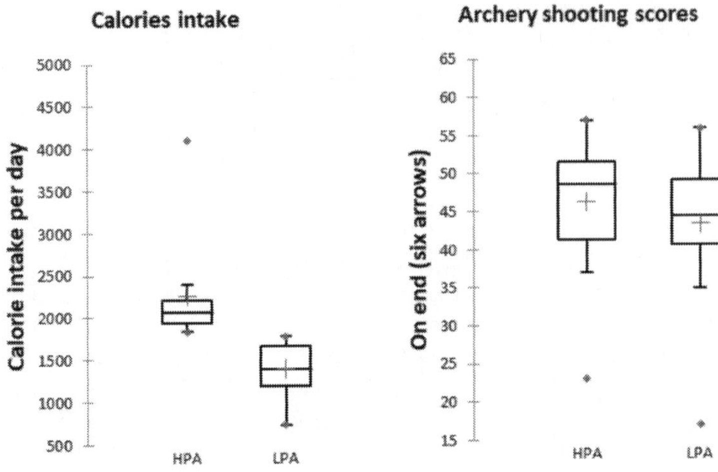

Figure 1. Comparisons of performance differences of the archers based on the evaluated bio-physiological variables.

The performance differences of the archers based on the six performance variables assessed that were clustered via *k*-means are shown in figure 1. It could be seen from the box plots that the mean performances of HPA are greater than LPA across all the five bio-physiological parameters measured in the study except for 'resting diastolic blood pressure (RDBP)'. This observation is non-trivial as RDBP is should be inversely proportional towards performance. Therefore, the aforementioned variables are essential attributes that allow for an accurate discrimination between the HPA and LPA.

3.2. Classification

Table 1. Model evaluation.

SVM Models	ACC (%)	SENS (%)	SPEC (%)	PREC (%)	ERR (%)	MCC
Linear	94	88.89	96.88	94.12	6	0.8690
Quadratic	88	83.33	90.63	83.33	12	0.7396
Cubic	84	77.78	87.50	77.78	16	0.6528

It could be observed from table 1 that the linear SVM variation is able to produce exceptional classification accuracy of 94% through the evaluation of all assessment bio-physiological parameters. Furthermore, the model has a high MCC of 0.8690 which indicates the high correlation of the parameters assessed with respect to the classification of the HPA and LPA. The quadratic model on the other hand does provide a reasonably accurate classification of the archers with a CA of 88 %, whilst the cubic provides a CA of 84 %. It is evident through the present study although all models does provide good accuracy, nonetheless that the linear-based model provides the best classification for the purpose of talent identification for archers based on the predefined bio-physiological variables. The confusion matrix of the evaluated SVM models are illustrated in figure 2.

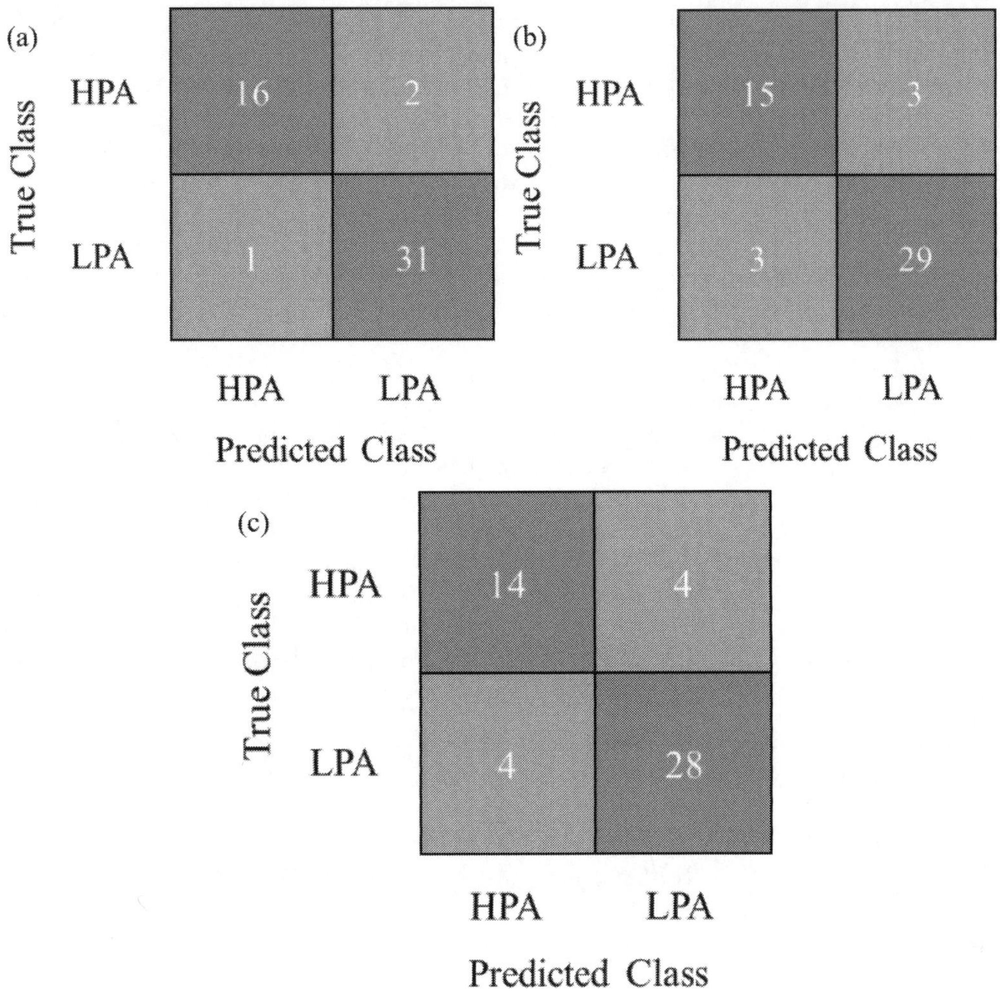

Figure 2. Confusion matrix (a) Linear; (b) Quadratic; (c) Cubic.

The present study has demonstrated that the selected bio-physiological variables examined can predict or classify reasonably the performance of the archers either as HPA or LPA. The shooting scores allow for the clustering of the archers with respect to measured bio-physiological variables. This was implemented via *k*-means as illustrated in the box plot (Figure 1).

It has been established from the investigation that nutrition is a vital component of archery performance. Since, sporting activity entails training and competition. The development of certain necessary variables such as skill, strength, power, speed, endurance, as well physical training could be accomplished when there is a good nutrition [13]. Furthermore, it was described that an archer has to consider optimum nutrition during both training and competition. If an archer plans on expending more than an hour shooting, it is recommended for him/her to have water, Gatorade, and refreshments with them [14]. It is equally essential to keep themselves "fueled" which can provide energy to the functioning muscle during shooting. An archer may become feeble and lose their vigour during competition due to the lack of eating. Moreover, an archer finishes shooting at the end of a long day; he/she may require eating an appropriate meal to refuel the body. Similarly, nutrition is likewise crucial in the pre, during, and post of the shooting process. Similarly, the present study to exemplify that resting heart rate, resting diastolic and systolic blood pressure could differentiate the performance level of the archers (see Figure 1).

The findings have also indicated that the capacity of lung function is integral to the archery performance. This discovery is consistent with the of the previous researchers who investigated the specific pulmonary parameter in all India inter-university archery competition and found that the capacity of lung function was high on all archers compare to normal men [4]. Moreover, it was inferred that the elite archers' respiratory frequency is higher as compared to that of novices and as such elite archers have a better and consistent performance during archery competition [15]. The present study shows the high, and medium performance archers have recorded higher levels of resting respiratory rate. The results explained further that the HPA have a good exhale control and autonomic balance which is essential for achieving optimum lung frequency during rest and in turn contribute to a better breathing control and postural stability during archery performance.

4. Conclusion

The present research has examined the impact of different bio-physiological variables in the effective performance of archery sport. It has been found from the study that a combination of bio-physiological variables specifically resting respiratory rate, calories intake and archery shooting scores could determine the grouping of the archer's performance. The study has also revealed that the utilisation of machine learning algorithms, in particular, the variation of SVM models is able to reasonably predict the class of the archers based on the chosen performance variables. In addition, the finding of this study could be useful to coaches and sports managers to identify the bio-physiological variables that influence the performance of the sport. Future study should consider other related performance variables correlated with the sport as well as other non-conventional classification techniques.

Acknowledgement

This work is funded by the National Sports Institute of Malaysia (ISNRG: 8/2014-12/2014) and the publication of the manuscript is funded by Universiti Malaysia Pahang (RDU 1703251)

References

[1] Bartlett D and Leiter J C 2012 Coordination of breathing with nonrespiratory activities *Compr. Physiol.*

[2] Chung G K W K, Delacruz G C, de Vries L F, Bewley W L and Baker E L 2006 New Directions in Rifle Marksmanship Research. *Mil. Psychol.* **18** 161

[3] Mayén A-L, Marques-Vidal P, Paccaud F, Bovet P and Stringhini S 2014 Socioeconomic determinants of dietary patterns in low-and middle-income countries: a systematic review *Am. J. Clin. Nutr.* ajcn-089029

[4] Thakare V 2015 Comparative study of peak expiratory flow rate of archery players participated in all India inter university archery competition

[5] Robertson S 2017 Improving load/injury predictive modelling in sport: The role of data analytics *J. Sci. Med. Sport* **18** e25–6

[6] Montoye A H K, Begum M, Henning Z and Pfeiffer K A 2017 Comparison of linear and non-linear models for predicting energy expenditure from raw accelerometer data *Physiol. Meas.* **38** 343–57

[7] Hagenbuchner M, Cliff D P, Trost S G, Van Tuc N and Peoples G E 2015 Prediction of activity type in preschool children using machine learning techniques *J. Sci. Med. Sport* **18** 426–31

[8] Pavey T G, Gilson N D, Gomersall S R, Clark B and Trost S G 2017 Field evaluation of a random forest activity classifier for wrist-worn accelerometer data *J. Sci. Med. Sport* **20** 75–80

[9] Woods C T, Veale J, Fransen J, Robertson S and Collier N F 2017 Classification of playing position in elite junior Australian football using technical skill indicators *J. Sports Sci.* 1–7

[10] Abdullah M R, Eswaramoorthi V, Musa R M, Maliki M, Husin A B, Kosni N A and Haque M 2016 The Effectiveness of Aerobic Exercises at difference Intensities of Managing Blood Pressure in Essential Hypertensive Information Technology Officers. *J. Young Pharm.* **8**

[11] Han D and Ha M 2015 Effect of pelvic floor muscle exercises on pulmonary function *J. Phys. Ther. Sci.* **27** 3233–5

[12] Taha Z, Musa R M, P.P. Abdul Majeed A, Alim M M and Abdullah M R 2018 The

identification of high potential archers based on fitness and motor ability variables: A Support Vector Machine approach *Hum. Mov. Sci.* **57** 184–93

[13] Martorell M, Capó X, Sureda A, Tur J A and Pons A 2013 Effects of docosahexaenoic acid diet supplementation, training, and acute exercise on oxidative balance in neutrophils *Appl. Physiol. Nutr. Metab.* **39** 446–57

[14] Vargas A, Parizzi S V, Fiamoncini R L and Navarro F 2010 Utilização da creatina no treinamento de força-revisão sistemática *Rev. Bras. Nutr. Esportiva* **4** 5

[15] Neumann D L and Thomas P R 2009 The relationship between skill level and patterns in cardiac and respiratory activity during golf putting *Int. J. Psychophysiol.* **72** 276–82

IOP Conf. Series: Materials Science and Engineering **342** (2018) 012021 doi:10.1088/1757-899X/342/1/012021

Design, simulation and analysis a microstrip antenna using PU-EFB substrate

S N S Mahmud[1], M A Jusoh[1*], S E Jasim[1, 2], A H Zamani[1] and M H Abdullah[1]

[1]Faculty of Industrial Sciences & Technology, Universiti Malaysia Pahang, 26300 Gambang Kuantan, Pahang, Malaysia

[2]Technical Institute/Al Haweejah, Northern Technical University, Al Haweejah-36007, Kirkuk, Republic of Iraq

Email: ashry@ump.edu.my

Abstract. A low cost, light weight and easy to fabricate are the most important factor for future antennas. Microstrip patch antennas offer these advantages and suitable for communication and sensor application. This paper presents a design of simple microstrip patch antenna working on operating frequency of 2.4 GHz. The designed process has been carried out using MATLAB and HFSS software by entering 2.3 for the dielectric constant of PU-EFB. The results showed that high return loss, low bandwidth and good antenna radiation efficiency of which are -21.98 dB, 0.28 dB and 97.33%, respectively.

1. Introduction

Wireless transmission technology, a microstrip patch antenna was the most popular, especially in microwave systems because of their attractive features. The microstrip patch antenna was small in size, light weight and easy to fabricate. It is also give high performance and low cost in production[1,2]. The basic configuration of the microstrip patch antenna shown in Figure 1.

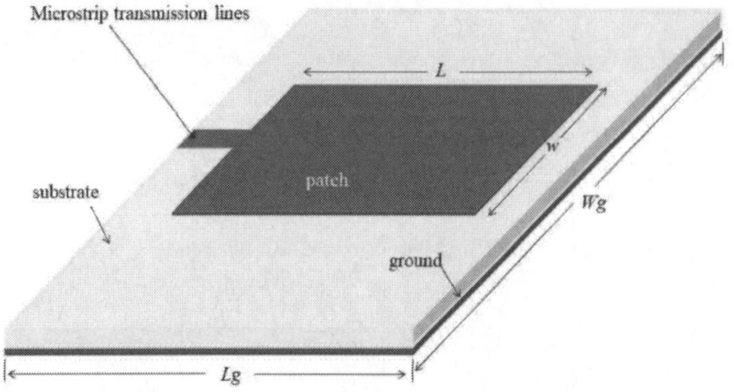

Figure 1. Rectangular Microstrip patch antenna configuration.

The geometry of a microstrip patch antenna consists of a patch on one side of a dielectric substrate and has a ground plane on the other side. The patch and ground are made of conducting materials such as copper and will connect. Feeding mechanism plays an important role in the design of microstrip patch antennas. A microstrip patch antenna can be fed either by coaxial probe or by microstrip feedline.[3]. In this study, microstrip patch antenna was using line feed for feeding technic. Microstrip feedline is easy to fabricate as it is a conducting strip connecting to the patch and therefore can be considered as extension of the patch[4]. The designing of patch antennas via using different software's has become most popular in recent years. This may be due to their useful dealings with much parameter, which extremely plays an important role in designing high performance antenna[5]. In this study, an antenna design was used polyurethane filled empty fruit bunch composite (PU-EFB) composite as a substrate[6]. The production of PU-EFB used low cost material. The main demand for technology in communication is the use of low cost and produce high-performance product. This study will designs and simulate an antenna to assess the effectiveness of dielectric material used as a substrate.

2. Patch Antenna design

The antenna is designed by using PU-EFB composites as substrate material with dielectric constant 2.3[6]. In designing the antenna, there are three steps follow as shown in Figure 2. First step is to calculate the dimension of the antenna using MATLAB2009a software. Follow by geometrical design by HFSS-13 software. The last step is to simulate the obtained antenna design using HFSS-13.

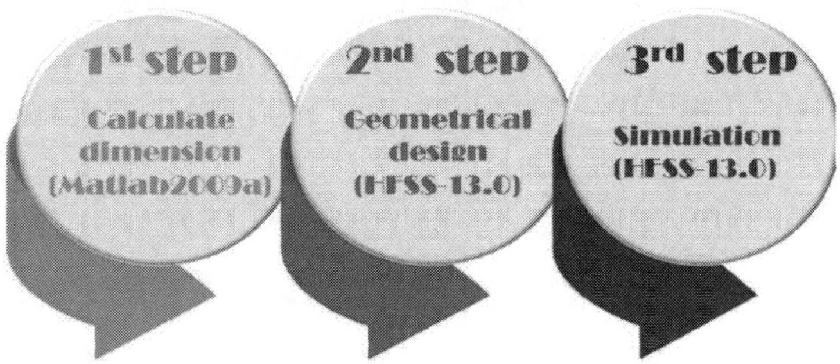

Figure 2. Steps in antenna's design and simulation.

2.1 Calculate dimension using MATLAB2009a

In calculate the dimension of antenna, other variables to be determined are the thickness of the substrate, dielectric constant of the substrate and the frequency of operation[2,7]. Here, the thickness value used was 3.0 mm, the dielectric constant was 2.3 and the frequency of operation was 2.4 GHz. The values that have to calculate in antenna design are

- Width of patch
- Length of patch
- Width of ground
- Length of ground

The width,w of the Microstrip patch antenna is given as:

$$w = \frac{c}{2f_o}\sqrt{\frac{2}{\varepsilon_r + 1}} \tag{1}$$

Where,
f_0 = operation frequency
c = speed of light
ε_r = dielectric constant of the substrate

In calculation of the length, L of the patch, there are several other computations included. The first unknown to be determine is the effective dielectric constant.[3] The effective value of dielectric constant εeff will be closer to the value of the actual dielectric constant εr of the substrate. The value of the dielectric constant of the substrate is given by:

$$\varepsilon_{eff} = \frac{\varepsilon_r + 1}{2} + \frac{\varepsilon_r - 1}{2}\left[1 + 12\left(\frac{h}{w}\right)\right]^{-\frac{1}{2}} \tag{2}$$

h = thickness of the substrate
ε_{eff} = effective dielectric constant of the substrate

The value for effective length can be calculated by using equation:

$$L_{eff} = \frac{c}{2 f_o \sqrt{\varepsilon_{eff}}} \tag{3}$$

Next to be calculating is length extension, ΔL. Because of fringing effects, electrically the micro strip antenna looks larger than its actual physical dimensions[7]] .The length extension is given by equation:

$$\Delta L = 0.412h \frac{(\varepsilon_{eff} + 0.3)\left(\frac{w}{h} + 0.264\right)}{(\varepsilon_{eff} - 0.258)\left(\frac{w}{h} + 0.8\right)} \tag{4}$$

The actual length of patch (L) is obtained by

$$L = L_{eff} - 2\Delta L \tag{5}$$

In designing patch antenna, it must have a finite ground plane, with the type of material being carried out. The ground has the same dimensions on the substrate but they are larger than the dimensions of the patch by six times the substrate thickness around the periphery[2,5]. The dimension ground plane could be calculated by equation

$$L_g = 6h + L \tag{7}$$

$$W_g = 6h + w \tag{8}$$

Where
w = width of the patch antenna
L = length of the patch antenna
W_g = width of ground plane
L_g = length of ground plane

By using the parameter of dielectric constant of the substrate which is 2.3, thickness of the substrate 3.0 mm and operation frequency is 2.4 GHz, the dimension for patch and ground plane has been calculated via MATLAB 2009b software. The result for patch and ground shows in Figure 3.

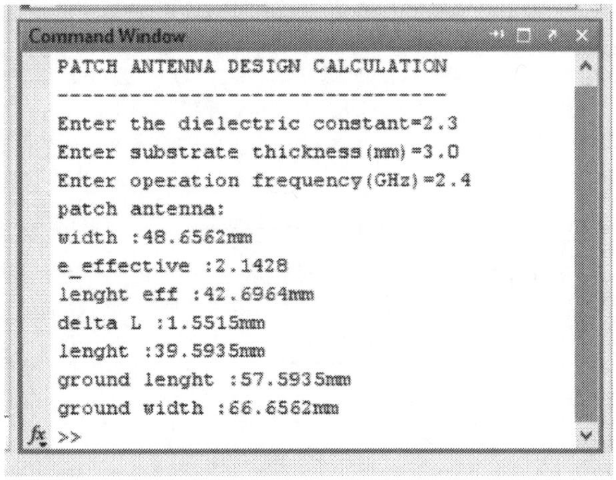

Figure 3. Dimensions result for patch antenna obtained from MATLAB2009a.

Through MATLAB2009a programming, the dimensions for antenna obtained are list in the Table 1.

Table 1. Dimensions value for patch antenna design from MATLAB2009a software.

Parameter	Dimension (mm)
w = width of the patch antenna	48.656
L = length of the patch antenna	39.593
W_g= width of ground plane	57.593
L_g= length of ground plane	66.656

2.2 Geometrical design and simulation using HSFF-13.0

By using the dimensions in the table, the HFSS-13.0 software will be used to design the geometry for the antenna. HFSS stands for High Frequency Structural Simulation, commercial finite element method for electromagnetic structure from Ansys Corp. One of several commercial tools used for antenna design. The patch antenna is designed to be located at origin coordinates x-y plane while the height of the substrate lies in z-direction. Feeding technique used in antenna was microstrip line feed. The geometry design illustrates in Figure 4.

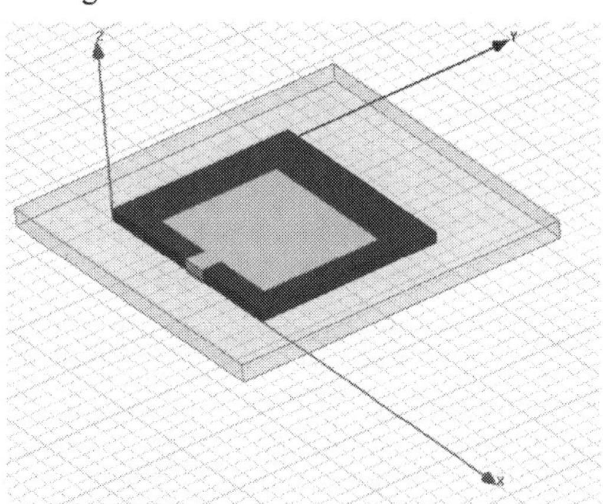

Figure 4. The image of patch antenna from HFSS-13 software.

In this type of feed technique, a conducting strip is connected directly to the edge of the microstrip patch. Through this method, the fabrication process is easier [8]. A polyurethane filled empty fruit bunch (PU-EFB) composite with a thickness of 3.0 cm and an approximate dielectric constant of 2.3

was used. By using data from MATLAB calculations and predefined values, this antenna needs to be adjusted so that the antenna can operate at a 2.4 GHz resonant value. All data from HSFF-13 simulated provided in table 2.

Table 2. Data obtained from HFSS-13.

Parameter	Material	Dimension (mm)		
		Width	Length	Height
Ground	Copper	65	55	0.1
Substrate	PU-EFB	65	55	3.0
Patch	Copper	45	36	0.1
Feed line	Copper	7	20	0.1

3. Result and Discussion

The performance of designed antenna determine by a few parameters such as return loss, bandwidth, VSWR, radiation pattern, directivity and antenna gain.

3.1 Return loss

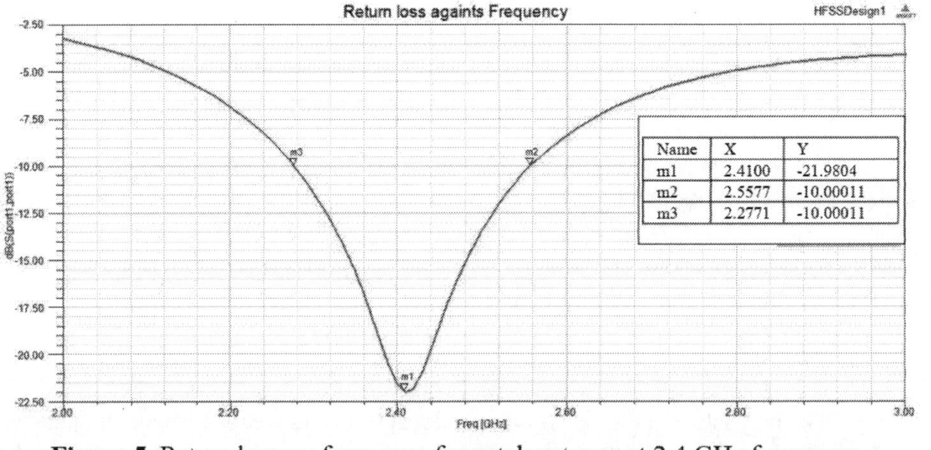

Figure 5. Return loss vs. frequency for patch antenna at 2.4 GHz frequency.

The return loss represents how much feeding power that was reflected back at the port of patch antenna as a result of the mismatches between the transmission line and the feeding points[5]. When the antenna and transmission line are not perfectly matched, reflections at the antenna port travel back towards the source and cause a standing wave to form[1] Good antenna must have high return loss[9]. From figure 5 implies that the antenna radiates best at the chosen resonance frequency, 2.4 GHz, where the return loss is equal to –21.98 dB.

3.2 Bandwidth

The bandwidth of an antenna refers to the range of frequencies over which the antenna can operate correctly. The antenna's bandwidth is the number of Hz for which the antenna will exhibit an SWR less than 2:1[10]. The value of bandwidth can be by referring to the graph return loss against frequency at point 10 dB[11]. As shown in graph return loss against frequency in figure 5, the value of bandwidth is 0.28 dB. Different types of antennas have different bandwidth limitations[7].

3.3 Voltage Standing Wave Ratio (VSWR)

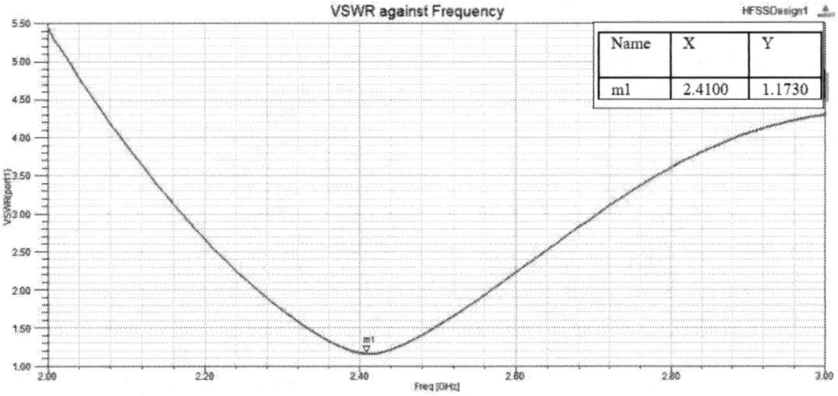

Figure 6. VSWR against frequency plot of the patch antenna.

VSWR shows how much power is reflected back from the antenna towards the source. The good value that accepted for antenna is 1≤ VSWR ≥2 [1,12]. If the VSWR value equals 1, means all of the given power to the antenna is transmitted. From figure 6, the value for designed antenna is 1.17.

3.4 Radiation Pattern

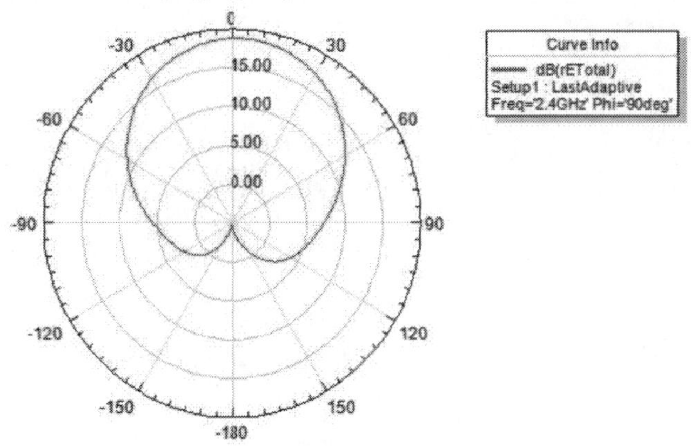

Figure 7(a). 2D radiation pattern at 2.4 GHz frequency.

The energy radiated by antenna is represented by radiation pattern of the antenna. There are a few types of radiation pattern such as 3D radiation pattern, 2D radiation pattern and Lobe formation radiation pattern. Figure 7(a) present the radiation pattern of 2D radiation pattern at 2.4 GHz at phi equals 90deg. The radiation pattern of an antenna has a normal radiation distribution to its surface and it gives an image nature of the value and direction of radiation, by which the antenna emits or receives the electromagnetic waves[5]. Figure 7(b) shows the elevation pattern for designed antenna in 3D radiation pattern

IOP Conf. Series: Materials Science and Engineering **342** (2018) 012021 doi:10.1088/1757-899X/342/1/012021

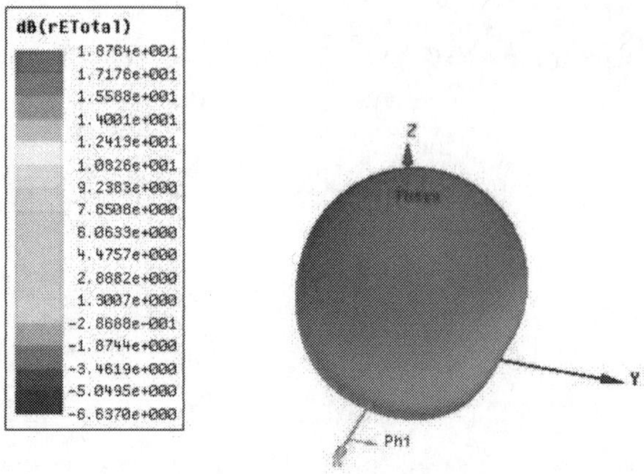

Figure 7(b). 3D radiation pattern at 2.4 GHz frequency.

3.5 The directivity
Directivity is the ability of an antenna to focus energy in a particular direction when transmitting, or to receive energy better from a particular direction when receiving [7]. An antenna that radiates equally in all directions would have effectively zero directionality, and the directivity of this type of antenna would be 1 (or 0 dB)[1]. Figure 8 shows the directivity of designed antenna at 2.4GHz

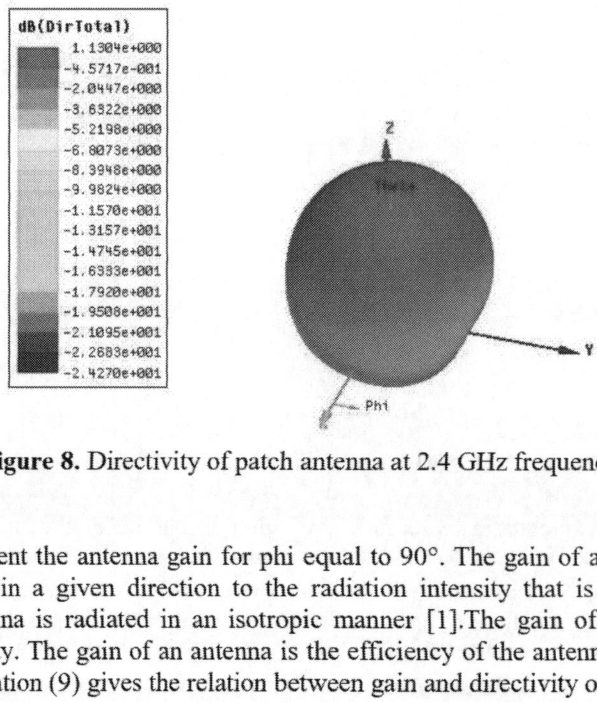

Figure 8. Directivity of patch antenna at 2.4 GHz frequency.

3.6 The Antenna gain
Figure 9 shows represent the antenna gain for phi equal to 90°. The gain of an antenna is defined as the ratio of intensity in a given direction to the radiation intensity that is obtained if the power accepted by the antenna is radiated in an isotropic manner [1].The gain of the antenna is closely related to the directivity. The gain of an antenna is the efficiency of the antenna as well as directional capabilities [1,7]. Equation (9) gives the relation between gain and directivity of an antenna.

$$G = \eta * D \tag{9}$$

Where, η = the efficiency, D = the directivity

An antenna is not a lossless system, and some energy entering the antenna terminals is lost due to heat losses etc. The gain of an antenna describes the rates of power transmitted to peak radiation through isotropic source.

IOP Conf. Series: Materials Science and Engineering **342** (2018) 012021 doi:10.1088/1757-899X/342/1/012021

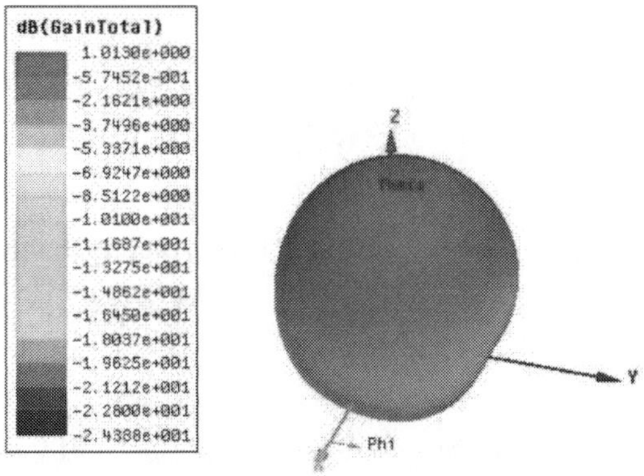

Figure 9. 3D gain total of patch antenna at 2.4 GHz.

4. Conclusion

A microstrip antenna for PU-EFB substrate bands has been successfully designed in a single patch with 50 Ω probe feed. The antenna simulation give the value for return loss below -10dB[12] which is the value for the design antenna is -21.98dB. Antenna radiation efficiency is 97.33% in the designated for frequency 2.4GHz. The gain of the antenna is in conformity to work perfectly and directivity of the proposed antenna shows that it has a good directional performance.

Acknowledgments

Authors acknowledge financial support from Ministry of Higher Education Malaysia (MOHE) under grant PRGS 151307 and RDU1214407.

References

[1] Bhattacharya A 2013 Design , simulation and analysis of a Penta Band Microstrip Patch Antenna with a Circular Slot **2** 1868–72
[2] Irianto A, Mutiara A B and Shappe A A 2011 Designing and Manufacturing Microstrip Antenna for Wireless Communication at 2 . 4 GHz **3** 1–6
[3] Kb Y I P 2003 Design Formula for Inset Fed Microstrip Patch Antenna **3** 5–10
[4] Mandal A, Ghosal A and Majumdar A 2012 Analysis of Feeding Techniques of Rectangular 26–31
[5] Jasim S E, Jusoh M A, Mazwir M H and Mahmud S N S 2015 Finding The Best Feeding Point Location of Patch Antenna Using HFSS **10** 17444–9
[6] Mahmud S N S, Jusoh M A, You K Y, Salim N, Shaheen S and Sutjipto A G E 2017 Structural and Dielectric Properties of Polyurethane Palm Oil Based Filled Empty Fruit Bunch **6495** 259–64
[7] Constantine A. Balanis 2005 *Antenna Theory: Analysis Design* (John Wiley & Sons Inc)
[8] Bisht S, Saini S, Prakash V and Nautiyal B 2014 Study The Various Feeding Techniques of Microstrip Antenna Using Design and Simulation Using CST Microwave Studio . **4** 318–24
[9] Anon Kraus-Antennas-2nd.Edition-1988.pdf
[10] Kaushal V, Singh T, Kumar V and Kumar A 2014 A Comprehensive Study of Antenna Terminology Using HFSS **7109** 24–9
[11] Wu B, Wang W, Pacheco J, Chen X, Grzegorczyk T, Kong J A and Art P 2005 Substrate to Enhance Gain 295–328
[12] Hasan A Al 2017 Design and Simulation of a Novel Dual Band Microstrip Antenna for LTE-3 and LTE-7 Bands **8** 223–8

Design of 2.5 GHz broad bandwidth microwave bandpass filter at operating frequency of 10 GHz using HFSS

S E Jasim[1, 2], M A Jusoh[1], S N S Mahmud[1] and A H Zamani[1]

[1]Faculty of Industrial Sciences & Technology, Universiti Malaysia Pahang, 26300 Gambang Kuantan, Pahang, Malaysia
[2]Technical Institute/Al Haweejah, Northern Technical University, Al Haweejah-36007, Kirkuk, Republic of Iraq

E-mail: ashry@ump.edu.my

Abstract. Development of low losses, small size and broad bandwidth microwave bandpass filter operating at higher frequencies is an active area of research. This paper presents a new route used to design and simulate microwave bandpass filter using finite element modelling and realized broad bandwidth, low losses, small dimension microwave bandpass filter operating at 10 GHz frequency using return loss method. The filter circuit has been carried out using Computer Aid Design (CAD), Ansoft HFSS software and designed with four parallel couple line model and small dimension (10 × 10 mm2) using LaAlO3 substrate. The response of the microwave filter circuit showed high return loss -50 dB at operating frequency at 10.4 GHz and broad bandwidth of 2.5 GHz from 9.5 to 12 GHz. The results indicate the filter design and simulation using HFSS is reliable and have the opportunity to transfer from lab potential experiments to the industry.

1. Introduction

Development of low losses and broad bandwidth microwave bandpass filter is an active area of research for selection the high-fidelity signals within the broadcast radio in radio frequency (RF) and microwave communication systems [1-3]. Microwave bandpass filter is the essential part in the microwave circuits and wireless communication systems, which is implemented to achieve a unique and accurate radiation frequency [4]. A number of reports have been published on the microwave circuits such as the antennas, filters, phase shifters, multiplexers, couplers and delay lines with high performance and low cost [5, 6]. Bandpass filters with broad bandwidth, miniaturized size, low-power consumption, and low cost are desired for microwave broadcast application [7].

Through the distribution of the microwave filter components over the entire circuits, microwave filter still occupies a lot of space. Their large dimension is due to the filter circuit distributed, which is commonly made from several resonators, which their dimensions are proportional to the guided wavelength. Hence, numerous researcher made quite efforts to develop the topology of miniaturized resonators [8]. Miniaturization of filters could be realized through change the filter geometry by replacing the 3-dimensional structure waveguide filter with 2-dimensional planar waveguide structure (i.e. microstrip line structure). Several microwave microstrip bandpass filters have been designed with surface area larger than 2 cm2 [9]. However, broad bandwidth properly maintained through the coupling of two lumped elements to form the couple transmission line [10]. Many bandpass filters have been designed using numerus types of the lumped elements such as, compact eight-pole [11], four-pole cross-coupled [12], and parallel-connected network [13]. However, all broad bandpass filters were designed with large circuit area from 225 to 5900 mm2, which have been designed and simulated using different software such as, IE, EM and E6 [10-13]. Bandpass filters with small size (100 mm2), broad bandwidth and low losses, designed and simulated using HFSS software has not investigated until now.

In this work, a cascade of parallel-coupled four lines has been adopted to realize the miniaturized microwave bandpass filter of broad bandwidth, low losses and high performance, bandpass filter is designed to allow for the central frequency 10.0 GHz to pass through and attenuate other frequencies outside. The microwave filter is designed with parallel coupling line model (short strip lines) using high-frequency structural simulator (HFSS) Ansoft software [14]. This paper describes an efficient method used to design the microwave bandpass filter characteristic with broad bandwidth, high operation frequency, the detailed process has explained in this article.

2. Filter Design and Simulation

Microwave filter were designed with four parallel couple lines, which assigned as a perfect conductor and the LaAlO3 substrate is assigned dielectric material with dielectric constant 23.5 [2]. The HFSS software has been developed to design most of kind of the microwave circuit components, it is offering an engineering task, which is able to obtain the physical dimensions of any microwave component with the specifications of high performance [15]. The data of the dimensions such as, width, height, and radius inside the command dialog windows of the software could be edited to realize the desired high-performance results. The band pass filter is consisted of dual number of the geometrical components (i.e. tape), for that it could be duplicated to create the symmetrical microwave filter. The microwave filter circuit is designed with an enclosed air body with the circuit components listed in table 3.1. The microwave bandpass filter was designed by click on the menu bar and choose the desired shape and followed by editing the characteristics of the positions and shapes in the command tab windows [16, 17]. Then by editing the data of the specific dimensions such as, width, height, and radius inside the command dialog windows of the software. The bandpass filter consists of individual number of the geometrical components that can be duplicated to create a symmetric device. The filter is designed using the coupling resonator model, which considered to be in symmetrical geometry. There are several ways to create coupling resonator between the microstrip lines, the basic principle is to design the coupled lines using two microstrip parallel end lines at least within subsection from the length of the microstrip line, which equals to the 1/4 or 1/8 of the wavelength. The coupled line of the filter has designed using four parallel tapes, which assigned as a perfect conductor.

Table 1. The component of the designed microwave filter system.

NO.	The designed part	Assigned material
1	Air body	Vacuum
2	Feed1 (Coax outer diameter)	Vacuum
3	Feedpin1 (Coax inner diameter)	Perfect conductor
4	Substrate ($LaAlO_3$)	Dielectric material
5	Ground plane	Perfect conductor
6	Resonators (2 in number)	Perfect conductor

The first part is the enclosure air body, which includes the whole filter circuits, while the rest parts designed with duplicate mode as illustrated before. The x-y coordinates center of the filter was assigned at the top center of the substrate as illustrated in figure 1. Table 2, shows the adaptive solutions in the HFSS software solution step are selected with the parameters. Table 3, shows the frequency setup details in the HFSS software and the start and stop frequency. Table 4, presents the designed part dimensions of the microwave filter circuit.

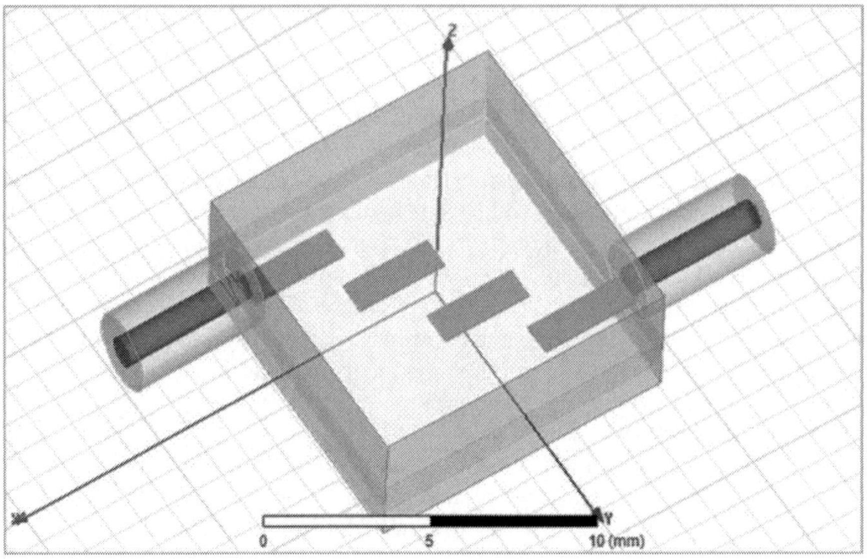

Figure 1. The image of microwave filter system from the HFSS Software.

Table 2. Adaptive solution in the HFSS software.

Sweep type	Maximum number of passes	Maximum Delta
Fast	15	0.01

Table 3. The frequency setup details in the HFSS software.

Start frequency	Stop frequency	Step Size
8.0 GHz	12.0 GHz	0.001

Table 4. The designed part dimensions of the bandpass filter.

The part	Dimension (mm^2)	Height (mm)
Feed1 (Coax Outer Diameter)	1.2 (in Radius)	4.5
Feedpin1 (Coax Inner Diameter)	0.45 (in Radius)	4.5
Substrate ($LaAlO_3$)	10 × 10 (in Area)	1.0
Ground Plane	10 × 10 (in Area)	0.05
L1 Resonator	3.3 × 1.0	0.05
L2 Resonator	3.0 × 1.0	0.05

3. Results and Discussion

Figure 2, summarizes the frequency response characteristics of the simulated microwave filter by HFSS, the performance of the filter can be presented in term reflection coefficient (S11) and transmission coefficient (S12) as a function of frequency. The return loss is calculated from S11 and insertion loss from S12. The typical filter should be having high amount of return loss, lower insertion loss, broad bandwidth, good matching electrical connecting impedance and high selective frequency. Obviously, the filter has around 10.5 GHz operation frequency, the filter has broad band width of 2.5 GHz and it applicable to pass the frequencies from 9.5 GHz to 12 GHz. Furthermore, the performance

IOP Conf. Series: Materials Science and Engineering **342** (2018) 012022 doi:10.1088/1757-899X/342/1/012022

is well showed return loss more than -50 dB at the 10.5 GHz frequency and the lowest return loss -12 dB for 11.4 GHz with approximately zero insertion losses.

In general, the results of simulation of a filter depend on line width, line height, line coupling length, and the distance between the lines [16, 17]. As can be seen from Figure 2, the optimized filter has an operation frequency of ~10 GHz with a bandwidth of ~2.5 GHz, transmission frequencies in the 9.5 – 12 GHz range, a return loss more than -12 dB, and approximately no insertion losses. On the other hand, Kumar et al [18] recently designed microwave filter using finite modeling with 4-parallel coupled line pairs with a spacing < 1 mm apart using the HFSS software. These filters on the FR4 substrate (which is a composite material composed of flame resistant woven fiberglass cloth with an epoxy resin binder) of dielectric constant ~4.2 and thickness ~1.58 mm, gave much inferior performance (operating frequency ~2.48 GHz, bandwidth ~0.6 GHz, higher insertion loss ~2.2 dB, and low return loss ~-12 dB). On the other hand, we have undertaken an extensive optimization procedure in the present work that yielded high performance with wider bandwidth, high return loss for couple line distance >1 mm.

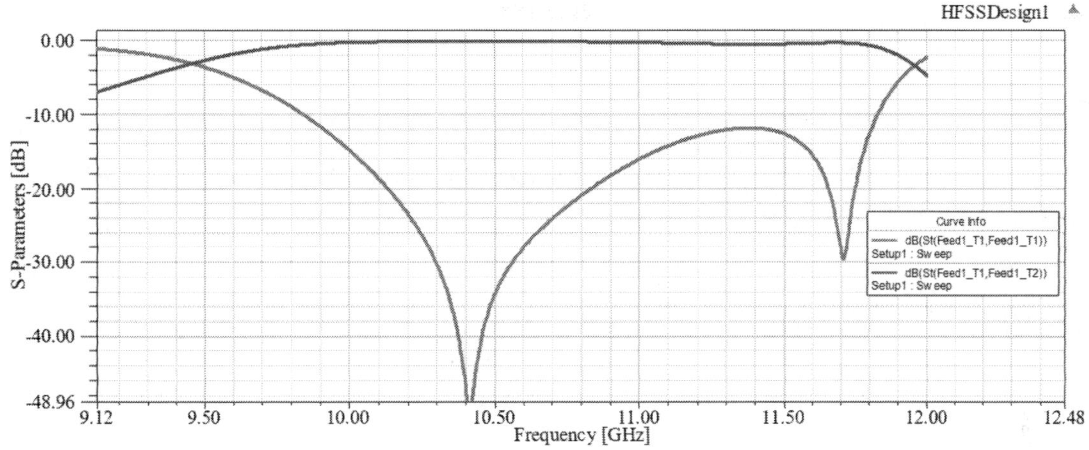

Figure 2. HFSS plot of the S-parameters versus frequency with frequency center 10.2 GHz

4. Conclusion

This work presents an innovative method used to design high return loss, wider bandwidth, and high performance miniaturized microwave filter. The filter has designed with short parallel couple line distance more than 1 mm. The microwave filter has been designed and simulated using the return loss method to find the physical dimensions of the circuit components. The microwave bandpass filter has operation frequency center at 10.5 GHz, broad bandwidth of 2.5 GHz, low return loss of -50 dB. The microwave bandpass filter is designed on the low RF losses and small area LaAlO3 substrate. The simulation result of the designed filter has been investigated using HFSS and found it depends on the circuit physical dimensions such as, line width, line height, line coupling long, and the distance between the lines [19].

Acknowledgments

This research is funded by the Ministry of Education Malaysia from Ministry of Higher Education Malaysia (MOHE) under grant PRGS 151307 and RDU1214407.

IOP Conf. Series: Materials Science and Engineering **342** (2018) 012022 doi:10.1088/1757-899X/342/1/012022

5. References

[1] Nisenoff M 2011 Microwave superconductivity part 1: History, properties and early applications in Microwave Symposium Digest (MTT) *IEEE MTT-S International* 1-4

[2] Mansour R R 2002 Microwave superconductivity *Microwave Theory and Techniques, IEEE Transactions on,* **50** 750-759

[3] Ribadeneira-Ramírez J, Martínez G, Gomez-Barquero D, and Cardona N 2016 Interference analysis between digital terrestrial television (DTT) and 4G LTE mobile networks in the digital dividend bands, *IEEE Transactions on Broadcasting* **62** 24-34

[4] Marimuthu J 2004 Design of microwave bandpass filter with novel parallel coupled grooved microstrip structure *Thesis, unpublished dissertation, Universiti Technologi Malaysia* 32-45

[5] Newman N and Lyons W G 1993 High-temperature superconducting microwave devices: fundamental issues in materials, physics, and engineering *Journal of Superconductivity* **6** 119-160

[6] Weigel R Valenzuela A and Russer P 1993 YBCO superconducting microwave components *Applied Superconductivity* **1** 1595-1604

[7] Liu A Yu A and Zhang Q 2006 Broad-band band-pass and band-stop filters with sharp cut-off frequencies based on series CPW stubs," in *Microwave Symposium Digest IEEE MTT-S International*, 353-356

[8] Lee J S 2009 Microwave resonator filters for advanced wireless systems Thesis University of Michigan

[9] Marimuthu J 2004 Design of Microwave Bandpass Filter with Novel Parallel Coupled Grooved Microstrip Structure," *Master's Thesis, unpublished dissertation, Universiti Technologi Malaysia*

[10] Bonache J Gil I Garcıa-Garcıa J and Martın F 2007 Compact microstrip band-pass filters based on semi-lumped resonators," *IET Microwaves, Antennas & Propagation* **1** 932-936

[11] Bai D Du J Zhang T and He Y 2013 A compact high temperature superconducting bandpass filter for integration with a Josephson mixer," *Journal of Applied Physics* **114** 133906

[12] Wang L Hsieh C H and Chang C C 2005 Cross-coupled narrow-band filter for the frequency range of 2.1 GHz using YBCO resonators with artificial magnetic pinning lattices *IEEE transactions on applied superconductivity* **15** 1040-1043

[13] Zhang T Yang K Zhu H Zhou L Jiang M Dang W *et al.*2015 Miniaturized HTS linear phase filter based on neighboring CQ units sharing resonators *Superconductor Science and Technology* **28** 105012

[14] Toossi A Moghadas H Daneshmand M and Sameoto D 2015 Bonding PMMA microfluidics using commercial microwave ovens *Journal of Micromechanics and Microengineering* **25** 085008

[15] Zhang Y 2006 Modeling and design of microwave-millimeterwave filters and multiplexers *Thesis University of Maryland, College Park* 22-45

[16] Flanner M S 2011 Microwave Filter Design: Coupled Line Filter *Master's Thesis of Science, California State University, Chico* 19-45

[17] Sun L P 2011 Dual-Mode Dual-Band Microstrip Bandpass Filter *Thesis Faculty of Engineering and Science, Universiti Tunku Abdul Rahman* 23-40

[18] Mukesh Kumar S K 2014 Designing of Half Wavelength Parallel- Edge Coupled Line Bandpass Filter Using HFSS *International Journal of Advanced Research in Computer Science and Software Engineering* **4** 876-882

[19] Flanner M 2011 Microwave Filter Design: Coupled Line Filter *Thesis California State University, Chico* 20-35

Behavior of Double-Web Angles Beam to column connections

K Al Fakih, S C Chin and S I Doh

Faculty of Civil Engineering & Earth Resources, Universiti Malaysia Pahang, 26300 Gambang, Pahang, Malaysia.

E-mail: PAC16001@stdmail.ump.edu.my

Abstract. This paper contains the study performed on the behavior of double-web angles by using finite element analysis computer package known as "Abaqus". The aim of this present study was simulating the behavior of double-web angles (DWA) steel connections. The purpose of this article is to provide the basis for the fastest and most economical design and analysis and to ensure the required steel connection strength. This study, started used review method of behavior of steel beam-to-column bolted connections. Two models of different cross-section were examined under the effect of concentrated load and different boundary conditions. In all the studied case, material nonlinearity was accounted. A sample study on DWA connections was carried out using both material and geometric nonlinearities. This object will be of great value to anyone who wants to better understand the behavior of the steel beam to column connection. The results of the study have a field of reference for future research for members of the development of the steel connection approach with simulation model design.

1. Introduction

Many fabricators, manufacturers, and erectors prefer to make a simple connection that can be easily assembled and installed, like bolted joints instead of welded joints. Welding requires more time, combined with a highly skilled workforce. Therefore, many steel manufacturers prefer bolted joints in this area. Steel joints are classified into two types: the first type is partially restrained (PR) and the second type is completely restrained (FR). In the construction of steel joints, it is assumed that the FR joints have sufficient rigidity to maintain the angles between the members of the connection. It is also assumed that the PR connections do not have sufficient stiffness to maintain the angles between the members of the connection (AISC 2011). These connections are leads to the overload resulting from other factors such as welds, loads and gravity. The available experimental results are used to find useful limit states and to improve a rational design procedure for bolted links similar to the links that are considered in this study. The proposed study focuses on the understanding of the behavior of steel joints exposed to shear or moment loads using the finite element method (FEM). The objective of this study is to theoretically provide the moment-rotation characteristics and the corresponding parameters of the semirigid connections between beam and column using the ABAQUS computer software. It affects the flexibility of the connections both of the distribution of forces and of the deformation in columns and beams of the frame. The frame must be calculated in the analysis. It is believed that the flexible analysis of linear distortions is relatively small and that equilibrium equations can be formulated with respect to the initial technique. Several types of research have been conducted on the behavior and design of steel joints. The objective was to provide better understanding and better techniques for evaluating between beams and column joints under the influence of shear and moment loads.

Figure 1 shows the two kinds of connection; the first type is partially restrained (PR) such as top-set angle connection and the second type is completely restrained (FR) such as extended end plate steel connection and the top-set angle with double web angle connection. In addition, can be seen that EP connection and TSA - DWA connections are amongst the stiffest connection [10].

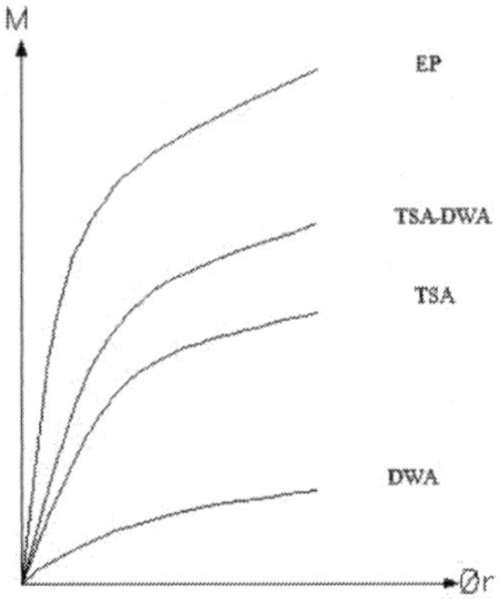

Figure 1. Typical moment rotation curve of beam-to-column connections [10].

Prabha et al. focused on the influence of many factors in the beam/column connection especially in the double web angle connection (DWA), such as air gap effect, angular length, angular thickness and gauge length on the behavior of the links. Three-dimensional (3D) connection moments, plastic-elastic nonlinear finite element analysis, to evaluate the behavior of the DWA connection. Analysis by changing the number of elements in a set of trace models. The proposed equation is compared with the experimental results and the free Morris model was found in full agreement [1]. The design process and the results obtained were presented with the numerical simulation of the upper seat angle with double angle web angles [2]. This was followed by finite element modeling using the ABAQUS program, which is based on previous experimental results. The reference researcher prepared the numerical models of four different configurations according to the program. numerical results from available experimental tests. The distortions and behavior of the virtual sample are the nearly same as those of the existing tests. Ahmed and Hasan proposed a research study for nonlinear finite element analysis (FE) of known set-top angle connections [3]. Performing the simulation 16 samples tests from available references. Compared to the results of the analysis of finite elements with the available experimental results to examine the applicability of the model, the results were close to each other. Sulong et al., nonlinear models were developed for various types of connections using component methods to simulate temperature changes and steel connections at different light load conditions. The model can represent the combination of the typical connection types, the top-set angle with double web angle connection, the double web angle connection, top-seat, extended end plate steel connection. Verification of the proposed connection model is carried out by comparing the analytical simulations with the current results of common tests and by testing joints and lower seats insulated at high-temperature conditions. The results demonstrate the efficiency of identifying the convergence of the proposed model with the power of the communication features and the results of the laboratory test [4]. Abolmaali et al., presented two types of double web angle semi-rigid connections to know the cyclic behavior of such connections: welded with bolted connections and bolted-bolted connections and prestressed bolts for load testing. Cyclic loading was applied to twenty test samples prepared for each test sample using the load control in the initial cycle. Fault modes and moment-rotation hysteresis circuits are provided for all test cases. The results of the experimental tests are verified for all test samples and the cyclic behavior of each test sample is discussed. Two failure modes were observed for the DWA semi-rigid bolted-bolted connections: excessive rotation due to bearing damage and angled bending of the beam bar. Furthermore, this study presented the final moment values, initial rigidity, failure modes and final rotation capacity of the links [5]. Xinwu, examined the behavior of semi-rigid connections under cyclic

inversion load [6]. Liew, et al., studied in part1 and 2, the connection parameters were developed and generalized in the context of angular links, TSA links, DWA with TSA links [7,8]. Hong et al., examined the effect of the angular thickness and the gauge space of bolt, and find the stress distribution and moment-rotation of double-web angle connections of each sample. Samples exposed to shear loads are considered to have a structural strain formed of a perfectly elastic-plastic, that was assumed. ABAQUS software is used to analyze the nonlinear behavior of double-web angled links. Experimental tests were performed with the same material and geometric conditions to verify the analysis of the finite element model. According to the test results, the element models were verified by comparing the moment-rotation curves with the deformed shapes derived from the analysis of model and the available results from literature [9]. The behavior of a connection is explained by the moment-rotation curve. Moment-rotation curves for some connections usually use PRC as shown in Figure 1.
There are many researchers who have studied the behavior of the column connection to beam and simulate it using computer software [11-19].

2. Modeling and Simulation

2.1. Experimental Specimen Description
Based on that experimental data obtained from studies reported by Lewitt et al, which studied the DWA connections [11]. In order to know the behavior of the DWA connection and to test the predictions of the connection model, it was studied and compared with the with the available full-scale test results of connections that have material and component compatibility. Connection components A36 are made of steel and 19.1 mm are made of A325 screws. The connection details of the DWA are shown in Figure 2. In this analysis, the flexible frame study carried out by Lewitt et al., Al, was considered [11]. A summary of the DWA connection geometries is as shown in Tables 1 to 3.

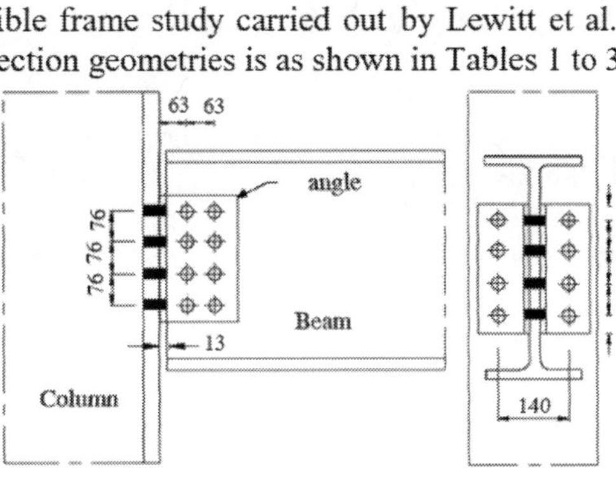

Figure 2. Typical TSA connections [11].

Table 1. Connection geometry of DWE [11]

Test Ref	Beam size mm*kg/m	Column size mm*kg/m	Gage mm	Web-Angle size mm	Fastener mm
FK-3	W310x40	W250x73	66.67	152.4x102x9.5x216	A325-19.1
FK-5	W530x92	W130x97	63.5	152.4x102x11.11x368	A325-19.1

Table 2. Columns geometric dimensions and yield stresses [11]

Test Ref	d mm	b_f mm	t_w mm	t_f mm	A mm	L mm	F_y MPa
FK-3	254	254	8.6	14	5740	1000	248
FK-5	307.8	304.8	9.9	15.4	12320	1000	248

Table 3. Beams geometric dimensions and yield stresses [11]

Test Ref	d mm	b_f mm	t_w mm	t_f mm	A mm	L mm	F_y MPa
FK-3	310	165	5.84	9.65	4995	1200	248
FK-5	533	209	10.2	15.6	9480	1200	248

IOP Conf. Series: Materials Science and Engineering **342** (2018) 012023 doi:10.1088/1757-899X/342/1/012023

2.2. Finite Element Modeling

The ABAQUS program can organize and analyze complex problems of connections, including structural problems and three-dimensional structures. In such software, technology lies in finding the right form: the development of the mesh - arrangement. It should balance the size of the mesh - an arrangement between the need for a fine mesh to give an accurate distribution of stress and time reasonable analysis. The perfect solution is to use a fine mesh in high-pressure areas and a coarser mesh in the remaining areas. In all types of connection, the connection is symmetrically arranged about the vertical medial line as shown in Figure 3.

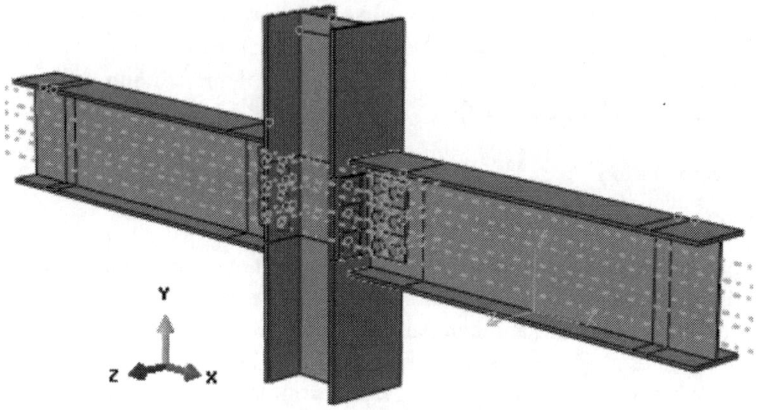

Figure 3. DWA Connections.

2.2.1. Types of elements. In the finite element model, the three-dimensional solid brick element (3D) in ABAQUS was used as shown in Figure 4; consists of eight nodes and each node has three degrees of freedom. This element has the ability to model plasticity, has a large deformation and stress-strain behavior and uses a lower integration method. Bolts, beams, columns and corner elements are modeled using C3D8 elements for the parts. The C3D8 elements are three-dimensional solid hexahedral elements with eight nodes with three degrees of freedom x, y, and z, or 1,2 and 3 respectively, as shown in Figure 4. Referring Figure 5 shows the solid elements C3D8 of the 3D iso-Parametric. 3D iso-parametric solid continuum elements with higher order models are capable of modeling curved boundaries. The elements were numerically integrated, although the C3D8 elements were linear with respect to geometry.

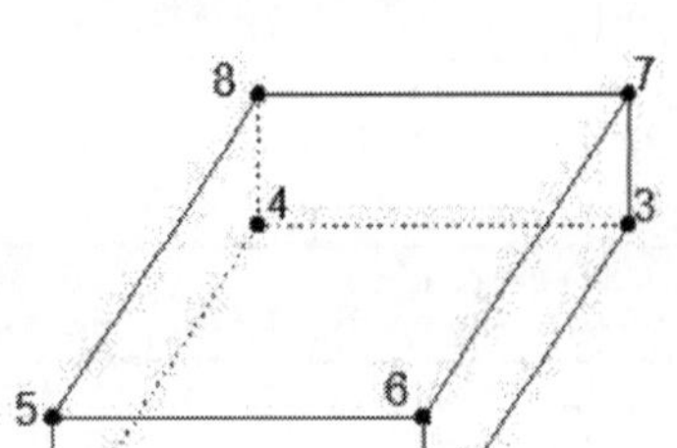

Figure 4. Bolt Solid elements.

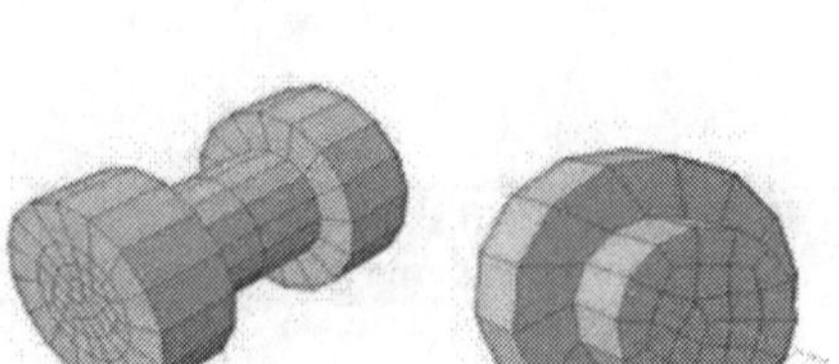

Figure 5. Bolt mesh arrangement.

The complete beam model with final arrangement of mesh discretization is shown in Figure 6.

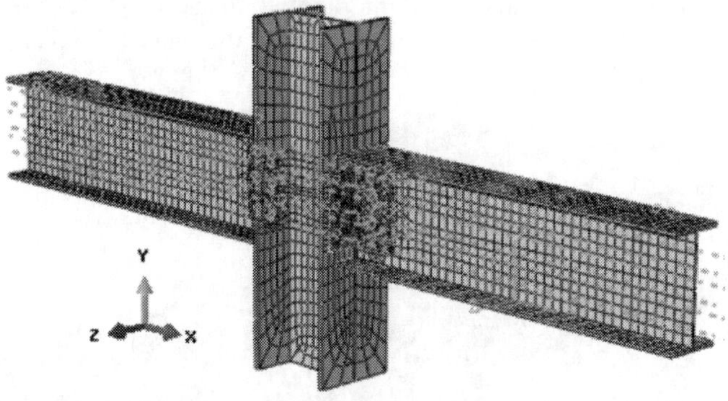

Figure 6. Complete beam models with final arrangement.

2.2.2. *Nonlinear Materials Behaviour*. Nonlinear material definitions play an important role in finite element analysis because each definition of the different parts of the connection must be considered carefully. This means that you need to simulate all the parameters using the basic elements of connection, physical models, and material parameters such as on the beams, columns, and web-angles, flanges, bolts, and nuts. Non-linear material occurs when the stress-strain relationship is not linear so that the steel yield becomes plastic. In fact, carbon steel ASTM A36 mild steel used in the analysis of Azizinamini, et al., may be of different strain-stress diagrams by applying different loading rates and temperature [19]. Otherwise, the loading rate is a specific property. Through the experimental research that has been investigated for analysis in this property where the rate of a load on the connection that has been investigated in the case of a load of monotonous. It was found that the steel takes the lowest yield point and follows almost the exact path of the stress-strain chart as those found in any steel design manual. The figure that explains areas such as stress-strain diagram is shown in Figure 7 for convenience. This graph explains the three phases of the mild carbon steel behaviour from the initial loading whit failure.

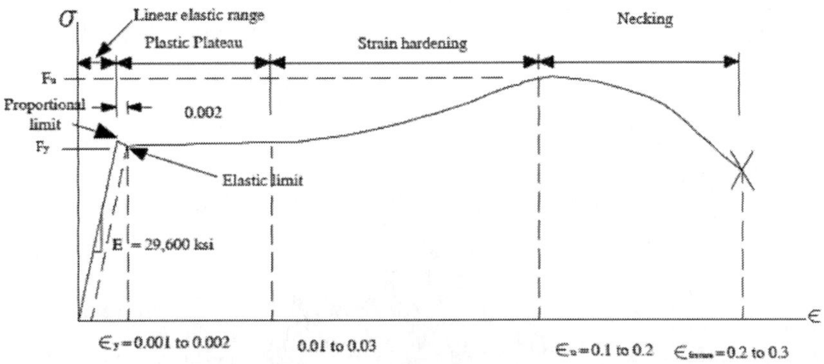

Figure 7. A typical Stress – Strain Diagram for Mild-Carbon Steel [19].

All elements were identified as elastic isotropic properties. The data of test materials were obtained from available all steel tests and the curves allowed stress-strain based on actual values instead of theoretical values. For plastic, elastic data bolts and other elements are defined as shown in Table 4. Von Misses yield performance criteria are used for all kinds of steel connection.

Table 4. Material non-linearity properties.

Tyap of element	Yield stress MPa	Ultimate stress MPa	Modulus of Elasticity Gap	Poisson ritio
Beam, Column and plate	248	483	207	0.3
Bolts	634	828	210	0.3

IOP Conf. Series: Materials Science and Engineering **342** (2018) 012023 doi:10.1088/1757-899X/342/1/012023

2.2.3. Boundary Condition. The boundary conditions applied to such connections prevent displacement in X, Y, Z directions. Fig. 8 shows the support boundaries at the end of the package (curve U1 = U2 = U3 = 0).

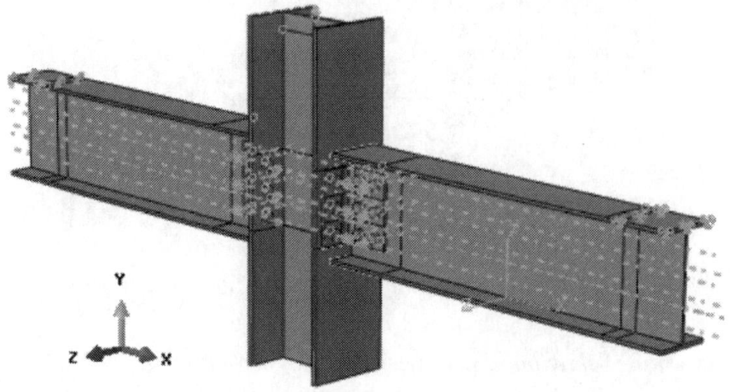

Figure 8. Boundary conditions and loads

2.2.4. Loads. The load was loaded via a concentrated load, at a point in the center of the model. The load is added to the load control file, in the add loading window to achieve the desired size of the bending moments of the connections. The step of the load factor was set to 1 and the maximum load factor was set at 100 at each iteration with the maximum variation of the load factor change automatically. The load applied to sample FK-3 was 36 KN and 144 KN for the sample FK-5 [11].

3. Results

In all cases, the inputs and outputs are based on the description in this section for load factor 1 to 100 set. The maximum load factor of change in each iteration automatically.

3.1. Specimen FK-3.

Using the 1.0 scale factor to make the displacement visible, the deformities shape von-misses stress are plot as shown in Figure 9 and Figure 10.

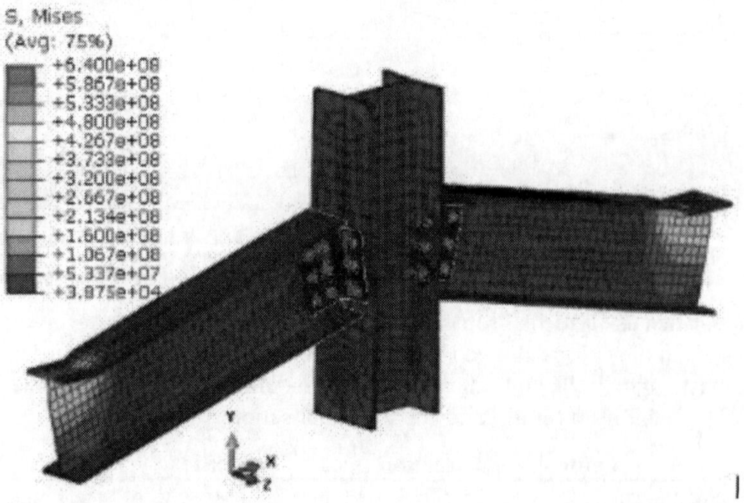

Figure 9. Deformation and Von-Misses stress plot of specimen FK-3.

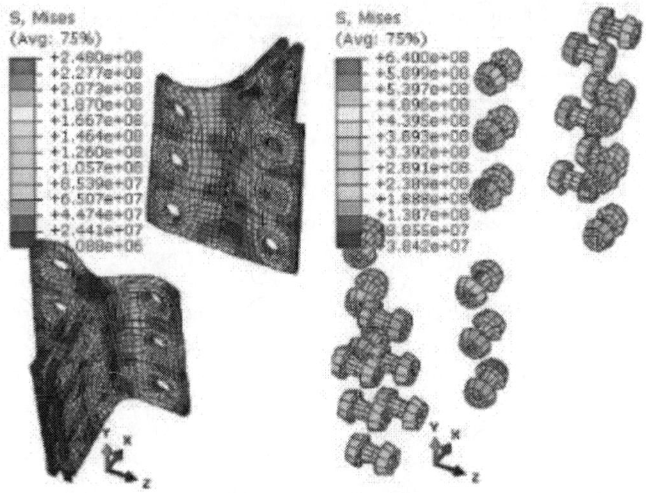

Figure 10. Connection deformation and Von-Misses stress plot of specimen FK-3.

The thick shell element stresses and solid element cannot be in in a single frame because they have different degrees of freedom. Node 6 from the beam was selected for finding the displacement analysis at this node. In addition, the node 6 selected was located at the top of the beam end at 1.2 m from the end support as shown in Figure 11.

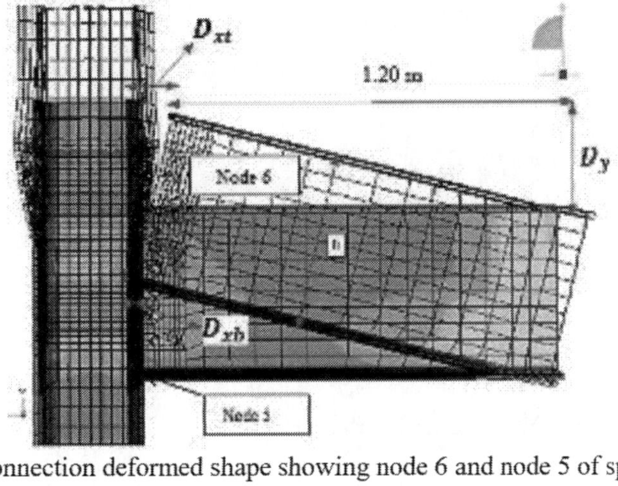

Figure 11. Connection deformed shape showing node 6 and node 5 of specimen DWE.

The connection rotates at Ø as shown in Figure 11.
The connection rotation is defined as

$$\text{Ø} = \tan^{-1}\left[\frac{\Delta x}{h}\right] \quad (1)$$

$$\Delta x = [D_{xt} - D_{xb}] \quad (2)$$

Where:
Ø = the relative rotation of connection.
D_{xt}= the top horizontal displacement of node 6.
D_{xb}= the bottom horizontal displacement of node 5.
h = the beam depth (0.310 m).
The rotation Ø in radian unit, while linear measurements are in meters' unit.
The applied moment was calculated using the following equation:
M = Load factor x 0.5 * L
Where (M) is applied moment in kN.m acting 1.2 m from the support, (L) is the length of the beam.
Numerical results for the calculated loads and displacements are shown in Table 5.

IOP Conf. Series: Materials Science and Engineering **342** (2018) 012023 doi:10.1088/1757-899X/342/1/012023

Table 5. Load versus and displacement (D_y) for specimen FK-3.

Load-Step	Displacement -D_y (m)	Load (kN)
1	0.0000000	0.000
2	0.0000180	0.141
3	0.0000370	0.281
4	0.0000655	0.492
5	0.0001083	0.809
6	0.0001718	1.283
7	0.0002670	1.995
8	0.0004100	3.063
9	0.0006251	4.666
10	0.0009479	7.067
11	0.0014313	10.670
12	0.0021568	16.078
13	0.0060526	24.188
14	0.0098908	27.140
15	0.0155946	30.092
16	0.0226904	33.048
17	0.0307499	36.000

The total vertical displacement against the total load factor graph for node 6 as shown in Figure 12.

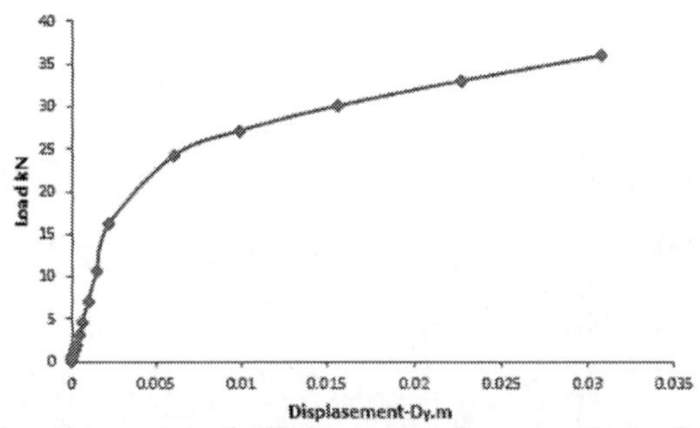

Figure 12. Load -versus vertical displacement D_y at node 6 of specimen FK-3.

Full computer results and the calculated Moment and rotations Ø are shown in Table 6.

Table 6 Calculated moments and corresponding rotations for specimen FK-3.

Load-Step	D_{xt} (m)	D_{xb} (m)	$[\Delta_x/h]$	M (kN.m)	Ø (rad)
1	0.0000000	0.0000000	0.0000000	0.000	0.0000000
2	0.0000004	-0.0000048	0.0000167	0.074	0.0000167
3	0.0000009	-0.0000098	0.0000343	0.148	0.0000343
4	0.0000015	-0.0000173	0.0000607	0.258	0.0000606
5	0.0000025	-0.0000286	0.0001003	0.425	0.0001003
6	0.0000040	-0.0000453	0.0001591	0.674	0.0001591
7	0.0000062	-0.0000705	0.0002474	1.047	0.0002473
8	0.0000096	-0.0001082	0.0003798	1.608	0.0003796
9	0.0000146	-0.0001650	0.0005791	2.449	0.0005788
10	0.0000220	-0.0002501	0.0008779	3.710	0.0008775
11	0.0000333	-0.0003776	0.0013252	5.602	0.0013246
12	0.0000501	-0.0005688	0.0019964	8.441	0.0019954
13	0.0000826	-0.0016284	0.0055196	12.699	0.0055168
14	0.0001000	-0.0026792	0.0089651	14.249	0.0089605
15	0.0001235	-0.0042432	0.0140861	15.799	0.0140790
16	0.0001518	-0.0061905	0.0204590	17.350	0.0204486
17	0.0001838	-0.0084037	0.0277016	18.900	0.0276875

IOP Conf. Series: Materials Science and Engineering **342** (2018) 012023 doi:10.1088/1757-899X/342/1/012023

The moment-rotation (M-Ø) curve for specimen FK-3 is shown in Figure 13. These diagrams are used to calculate the rotation of the beam and the column.

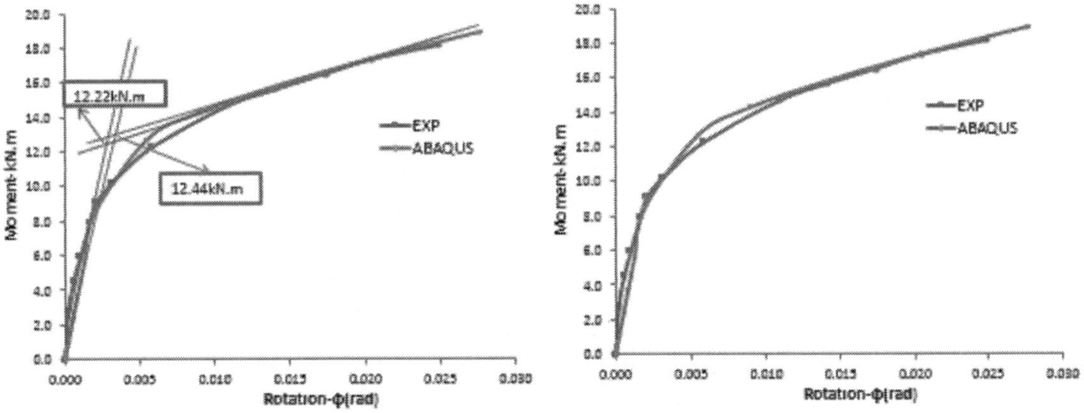

Figure 13. Moment-Rotation (M-Ø) curve for specimen FK-3.

3.2. Specimen FK-5.

Using the 1.0 scale factor to make the displacement visible, the deformities shape von-misses stress are plot as shown in Figure 14 and Figure 15.

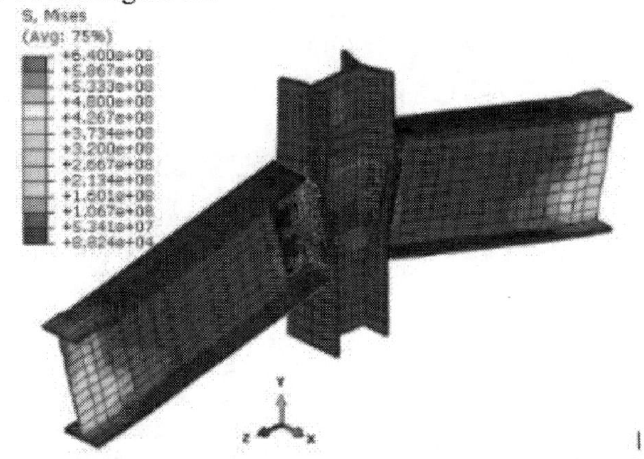

Figure 14. Deformation and Von-Misses Stress of connection plot of specimen FK-5.

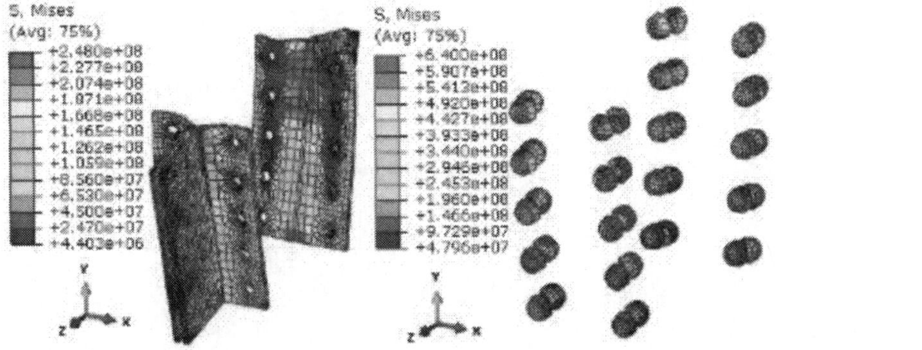

Figure 15. Connection deformation and Von-Misses Stress plot of specimen FK-5.

Numerical results for the calculated loads and displacements are shown in Table 7.

Table 7. Load versus and displacement (D_y) for specimen FK-5.

Load-Step	Displacement -D_y (m)	Load (kN)
1	0.0000000	0.000
2	0.0003211	9.000
3	0.0006580	18.000
4	0.0011517	31.507
5	0.0018886	51.754
6	0.0050622	82.123
7	0.0071695	93.514
8	0.0114548	110.606
9	0.0133329	117.014
10	0.0167126	126.619
11	0.0242357	141.034
12	0.0260660	144.000

The total vertical displacement against the total load factor graph for specimen FK-5 as shown in Figure. 16. These diagrams are used to calculate the rotation of the beam and the column.

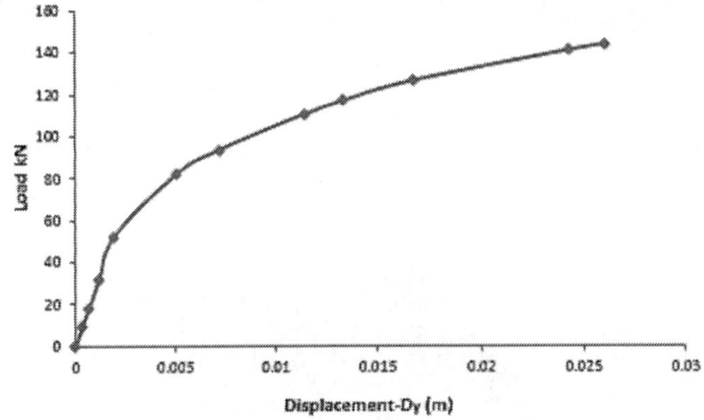

Figure 16. Load -versus vertical displacement D_y for specimen FK-5.

The result, calculated rotations Ø and moments from the computer analysis are shown in Table 8.

Table 8 Calculated moments and corresponding rotations for specimen FK-5.

Load-Step	D_{xt} (m)	D_{xb} (m)	$[\Delta_x/h]$	M (kN.m)	Ø (rad)
1	0.00000000	0.00000000	0.000000	0.000	0.000000
2	0.00011042	-0.00002933	0.000262	5.400	0.000262
5	0.00022926	-0.00005749	0.000538	10.800	0.000538
4	0.00040318	-0.00009866	0.000942	18.904	0.000941
5	0.00066314	-0.00015955	0.001544	31.052	0.001543
6	0.00177142	-0.00043705	0.004143	49.274	0.004141
7	0.00250457	-0.00062543	0.005872	56.108	0.005869
8	0.00398712	-0.00102393	0.009402	66.364	0.009397
9	0.00463671	-0.00120039	0.010951	70.209	0.010946
10	0.00580824	-0.00151561	0.013741	75.972	0.013734
11	0.00843281	-0.00221190	0.019971	84.620	0.019961
12	0.00907059	-0.00238372	0.021490	86.400	0.021479

Experimental data were obtained from Lewitt et, al., [11]. This data was modeled and analyzed by the Abaqus program. The M-Ø diagram is drawn for both the large-scale experimental test results and the Abaqus analysis results, and the yield moments of the links are determined by the intersection of the two tangents as shown in Figure 17.

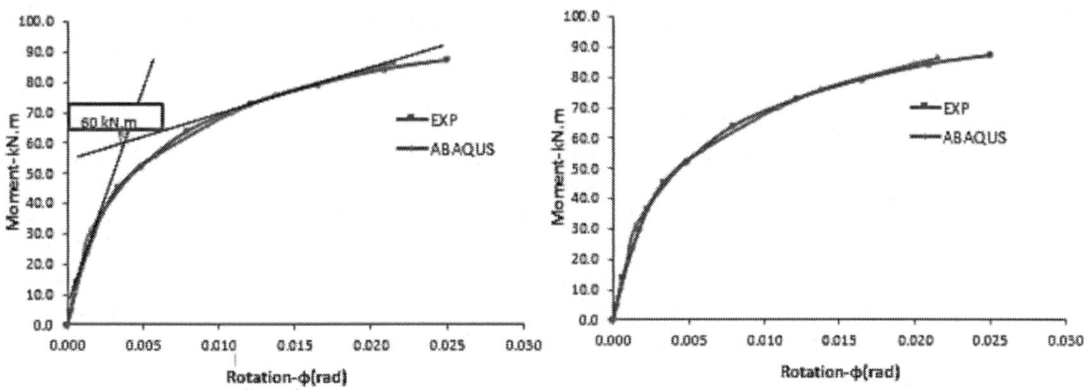

Figure 17. Moment-Rotation (M-Ø) curve for specimen FK-5.

4. Discussion of Results

Based on the charts shown in Figure 13, Figure 17 and analysis results; summary of the results for both specimens of DWA connections are shown in Table 9.

Table 9. Summary of the results for specimens FK-3 and FK-5.

Test Ref	Yield moment Kn.m		Ultimate moment Kn.m		Displacement m	
	Aba	Exp	Aba	Exp	Aba	Exp
FK-3	12.44	12.22	18.9	18.11	0.03	0.025
FK-5	60	60	86	87	0.026	0.025

Referring to Figure 13, Figure 17, and Table 8, it was found that the determination of yield moment of the connection according to computer analysis software; ABAQUS, is against the experiment, with a difference of 1.80 % for the specimen FK-3 while, in the specimen, FK-5 was 0.00 %. In addition, from the ABAQUS software analysis, the yield moment was higher than the experimental test and the ultimate moment is higher than the experimental test for the specimen FK-3. Whereas, in the specimen FK-5, the yield moment was almost equal than the experimental test and the ultimate moment is higher than the experimental. Furthermore, the ultimate moment values from computer analysis were very close from the experiment for both specimens. moreover, the displacement of computer analysis is more than the experience by a very simple difference. Finally, it can also be noted that the computer model seems to be stiffer than the experimental model at lower loads up to the yield moments, after which the experimental model showed stiffer than the ABAQUS numerical results in the specimen FK-5. The results of this study identify weaknesses in connections that can be overcome sand strengthened to improve the behaviour of such connections. For example, the diameter of the bolts of the specimen FK-3 and FK-5 connection, which fails in the fault mode at the all angles connection, can be increased from 19 mm to 20 mm. This solution increases the strength of communication, especially since the rest of the elements of communication did not collapse and has the ability to withstand longer. Moreover, the angular thickness of the specimen FK-3 connections, which fails in the fault mode at the top of web angle, can be increased from 9.5 to 11 mm. In addition, for the specimen FK-5 with an angular thickness of 11.1 mm and a failure mode at the top of web angle and bolts, the thickness can be increased from 11.1 mm to 12 mm.

5. Conclusion

This paper reports on the DWA connections, a finite element model using ABAQUS in order to investigate the performance of DWA connections steel at room temperature. Validation of the numerical modelling against all representative experimental results conducted on the moment-rotation relationship, final deformation and yield line pattern of the DWA connections shows good agreements exist. For this reason, this numerical analysis method can be used safely to predict the behavior of DWA steel beam-column connections at room temperature. The following conclusions can be drawn. The difficulty of numerical modeling contact interactions taking into account the material and non-linear geometric effects has been successfully solved. The proposed DWA model is verified to simulate DWA steel connections up to the maximum moment at room temperature. Accordingly, modeling DWA can be used to predict stress distribution of high strength steel DWA connections both at ambient temperature, which is not easy to achieve in full-scale tests, via inputting proper mechanical properties of high strength steels at ambient temperature. By the current modeling approach, the first critical component of DWA links can be identified, only the occurrence of component breakdown and subsequent failures of other components cannot be predicted. Finally, the load bearing capacity, as well as rotation capacity of DWA connections, is dependent on the thickness of angle and numbers of bolts. By the way, all connection studies, that use finite element analysis using computer software provides a lot of effort and money, as compared to a large-scale test, and can produce a complete picture of the behavior of joints and deformation, tension and distribution of power and stress.

Acknowledgments
The authors would like to thank the two anonymous reviewers for their constructive comments that have enhanced the quality of the paper. This work was supported by the Ministry of Higher Education and the University of Malaysia Pahang (UMP) and the authors would like to thank them for their support.

References
[1] Prabha P, Rekha S, Marimuthu V, Saravanan, M, Palani G S and Surendran M 2015 Modified Frye–Morris polynomial model for double web-angle connections *International Journal of Advanced Structural Engineering (IJASE)* **7**(3) 295-306
[2] Ghindea M, Catarig A and Ballok R I 2015 Semi-Rigid Behaviour Of Bolted Connections Using Angle Cleats Part 1. Development Of 3-D Finite Element Model *Buletinul Institutului Politehnic din lasi. Sectia Constructii, Arhitectura* **61**(4) 53
[3] Ahmed A and Hasan R 2015 Effect and evaluation of prying action for top-and seat-angle connections *International Journal of Advanced Structural Engineering* (IJASE) **7**(2) 159-169
[4] Sulong N R, Elghazouli A Y, Izzuddin B A and Ajit N 2010 Modelling of beam-to-column connections at elevated temperature using the component method *Steel and Composite Structures*, **10**(1) 23-43
[5] Abolmaali A, Kukreti A R and Razavi H 2003 Hysteresis behavior of semi-rigid double web angle steel connections *Journal of Constructional Steel Research* **59**(8) 1057-1082
[6] Xinwu W 2007 Experimental research on hysteretic behavior of top-seat and web angles connections. In 5th WSEAS International Conference on Environment, Ecosystems and Development, Tenerife, Spain (pp. 14-16)
[7] Liew J R, White D W and Chen W F 1993 Limit states design of semi-rigid frames using advanced analysis: Part 1: Connection modeling and classification *Journal of Constructional Steel Research* **26**(1) 1-27
[8] Liew J R, White D W and Chen W F 1993 Limit states design of semi-rigid frames using advanced analysis: Part 1: Connection modeling and classification *Journal of Constructional Steel Research* **26**(1) 1-27
[9] Hong K, Yang J G and Lee S K 2002 Moment–rotation behavior of double angle connections subjected to shear load *Engineering Structures* **24**(1) 125-132
[10] Chen W F 2011 Semi-rigid connections handbook. J. Ross Publishing, pp 23-32.
[11] Lewitt C W, Chesson Jr E, & Munse W H 1966 Restraint characteristics of flexible riveted and bolted beam-to-column connections. University of Illinois Engineering Experiment Station. College of Engineering. University of Illinois at Urbana-Champaign.
[12] Pirmoz A, Khoei A S, Mohammadrezapour E and Daryan A S 2009 Moment–rotation behavior of bolted top–seat angle connections *Journal of Constructional Steel Research* **65**(4) 973-984
[13] Danesh F, Pirmoz A and Daryan A S 2007 Effect of shear force on the initial stiffness of top and seat angle connections with double web angles *Journal of Constructional Steel Research* **63**(9), 1208-1218
[14] Hean L S, Sulong N R and Jameel M 2016 Effect of axial restraints on top-seat angle connections at elevated temperatures *KSCE Journal of Civil Engineering* **20**(6) 2375-2383

IOP Conf. Series: Materials Science and Engineering **342** (2018) 012023 doi:10.1088/1757-899X/342/1/012023

[15] Daryan A S and Yahyai M 2009 Behavior of bolted top-seat angle connections in fire *Journal of Constructional Steel Research* **65**(3) 531-541

[16] Stelmack T W, Marley M J and Gerstle, K. H 1986 Analysis and tests of flexibly connected steel frames *Journal of Structural Engineering* **112**(7) 1573-1588

[17] Kukreti A R and Abolmaali A S 1999 Moment-rotation hysteresis behavior of top and seat angle steel frame connections *Journal of Structural Engineering* **125**(8) 810-820

[18] Dai X H, Wang Y C and Bailey C G 2010 Numerical modelling of structural fire behaviour of restrained steel beam–column assemblies using typical joint types *Engineering Structures* **32**(8) 2337-2351

[19] Azizinamini A and Radziminski J B 1989 Static and cyclic performance of semirigid steel beam-to-column connections. *Journal of Structural Engineering* **115**(12) 2979-2999.

IOP Conf. Series: Materials Science and Engineering **342** (2018) 012024 doi:10.1088/1757-899X/342/1/012024

Resin bleed improvement on surface mount semiconductor device

Indra Kumar Rajoo[1], Suraya Mohd Tahir[1], Faieza Abdul Aziz[1] and Mohd Shamsul Anuar[2]

[1]Department of Mechanical and Manufacturing Engineering, Faculty of Engineering, Universiti Putra Malaysia, 43400 UPM Serdang, Selangor.
[2]Department of Process and Food Engineering, Faculty of Engineering, Universiti Putra Malaysia, 43400 UPM Serdang, Selangor.

Email: su_mtahir@upm.edu.my

Abstract. Resin bleed is a transparent layer of epoxy compound which occurs during molding process but is difficult to be detected after the molding process. Resin bleed on the lead on the unit from the focused package, SOD123, can cause solderability failure at end customer. This failed unit from the customer will be considered as a customer complaint. Generally, the semiconductor company has to perform visual inspection after the plating process to detect resin bleed. Mold chase with excess hole, split cavity & stepped design ejector pin hole have been found to be the major root cause of resin bleed in this company. The modifications of the mold chase, changing of split cavity to solid cavity and re-design of the ejector pin proposed were derived after a detailed study & analysis conducted to arrive at these solutions. The solutions proposed have yield good results during the pilot run with zero (0) occurrence of resin bleed for 3 consecutive months.

1. Introduction

A semiconductor company in Malaysia continues to receive customer return for SOD 123 package from various customers and having internal detection with complaints of their undesirable lead colour. Resin bleed is a transparent, brownish layer of mold-epoxy resins which occurs on the surface of the leads as shown in Figure 1, which happened after the encapsulation (molding) process. It is not visible immediately after molding process because resin is a colorless component of the molding compound. It can be only observed after the tin-plating process. The area experiencing the resin bleed will not be plated with tin-plating. The criteria set for resin bleed is that if the bleed is more than 0.1 micron in any direction at any location on the lead and bottom of the lead, the unit is a reject unit. Resin bleed can cause solderability issue on the customer side.

Resin bleed observed in the SOD123 package could cause solderability issue on the PCB board during mounting by the customer. This failed unit if returned to this semiconductor company would be considered as a customer complaint. A significant amount of money and resources needed for the submission of an 8D (Eight Disciplines of Problem Solving) to the customer for the complaint. 8D (Eight Disciplines of Problem Solving) is a meticulous process used to solve complex problems.

IOP Conf. Series: Materials Science and Engineering **342** (2018) 012024 doi:10.1088/1757-899X/342/1/012024

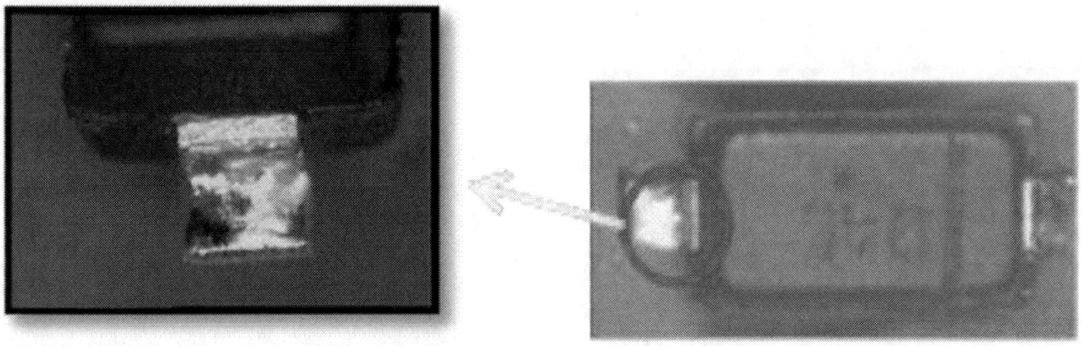

Figure 1. Example of resin bleed.

Resin bleed occurs because the resin which is a colorless adhesive material mixed with the silica filler to form the mold compound bleeds out of the cavity during molding to the lead thus coating the lead with a colourless layer.

There are various reasons that can cause resin bleed and does not need to be confined to only the molding process. Turng and Wang [1] clearly stated that the variance formed between the mold plate and the lead frame caused resin bleed on the leads. Kerk, Tey, and Hu [2] also stressed that a lead frame with no full flatness on the surface of the copper lead frame could cause resin bleed. Since the compound flows through the lead, the resin tend to stick on the lead. Tong, Wong and Yu [3] mentioned that in current practice, the setting of molding parameters was done by engineers in a trial-and-error manner could be one of the reasons for resin bleed.

Numerous methods were proposed by researcher to encounter resin bleed. Kerk *et al.* [2] stated that resin bleed could be solved by using chemically etched copper lead frames. Resin bleed could also be solved by electrical deflash and high pressure water jet. This is a process that softens the adhesion resulting the removal of organic compound on the surface of the frame as proposed by Marvin and Gavino [4]. Paquet [5] stated that the control of the injection molding parameters, such as transfer speeds, mold temperature, clamping force and transfer pressure, could generally help to eliminate the emergence of resin bleed and void. To eliminate resin bleed, Chiang and Su [6] applied Fuzzy quality loss function (FQLF) to the qualitative responses, since the molding defects cannot be simply represented by the relationship between molding conditions and mathematical models. Another comprehensive and proven method to stop resin bleed is to modify the lead frame to have a control step down in the perimeter of the cavity as explained by Colin, Boon, and Shugumar [7]. This step-down would act as a drain to trap all the resin which tends to bleed out. Carrie and Lim [8] proposed that placing an adhesive tape on the exposed pads side to prevent resin bleed during encapsulation which would be able to prevent mold compound resin bleeding to the exposed pads.

The current control for resin bleed in this company is to perform weekly imprint check on the SOD123 molds to check for mold balancing. If there was a weak imprint trace as in Figure 2, it showed the mold was imbalanced. The mold would then be repaired to improve the mold balance which would be verified through visual inspection on an imprint paper as in Figure 3.

IOP Conf. Series: Materials Science and Engineering **342** (2018) 012024 doi:10.1088/1757-899X/342/1/012024

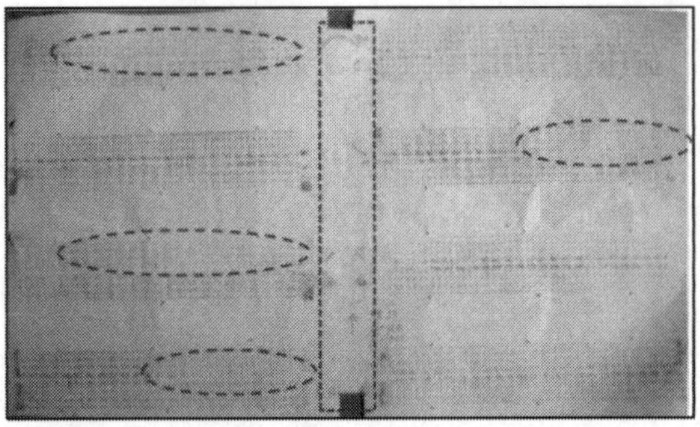

Figure 2. All the highlighted areas are having low/weak imprint mark.

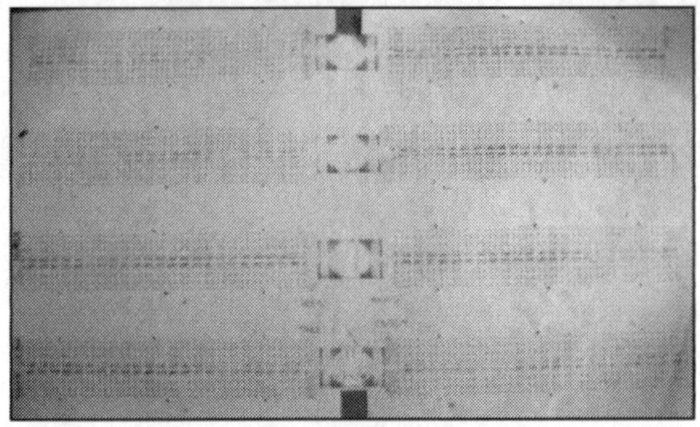

Figure 3. The mold is having a good imprint overall.

Since the resin bleed was due to weak clamping at the lead area, it was possible to increase mold clamping pressure which increased the lead impression, eventually reducing the bleed. But this is not a recommended solution as it could lead to other issues such as lead flash and smash lead. Approach of increasing the setting of the high pressure water jet (HPWJ) to remove the resin bleed was also considered but was not a viable choice as it would affect the package reliability.

2. Methodology

2.1 Measurement system analysis (MSA) study- Attribute gage repeatability & reproducibility (GR & R)

An attribute Gage Repeatability & Reproducibility (GR&R) study was done to measure the efficiency of the pre-gate operators to detect resin bleed reject and also to avoid over-rejection /escapee during pre-gate process. 15 serialized samples consisted of reject and accept units were given to the all three shift pre-gate operators. There were two possibilities; either the operator are over-rejecting or under rejecting. To solve this problem, all the 6 operators were given training and the GR&R study is done again. When the study was conducted after this training, the result of the study showed a positive improvement.

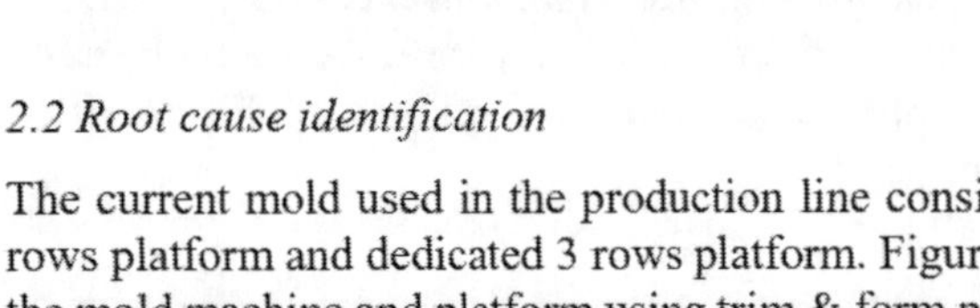

Figure 4: Attribute GR&R study result (before training)

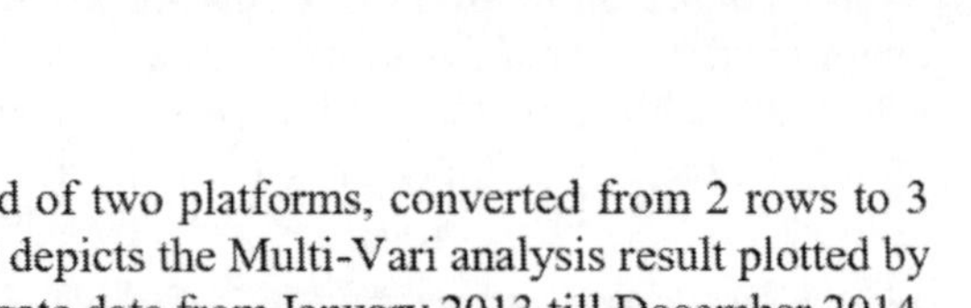

Figure 5: Attribute GR&R study result (after training)

2.2 Root cause identification

The current mold used in the production line consisted of two platforms, converted from 2 rows to 3 rows platform and dedicated 3 rows platform. Figure 6 depicts the Multi-Vari analysis result plotted by the mold machine and platform using trim & form pre-gate data from January 2013 till December 2014. It could be seen from figure 6 that the converted 3 row molds were having a higher contribution of resin bleed compared to the original 3 row platform. The first phase of further analysis would be on the high occurrence of resin bleed in converted 3 row platform and the second phase would be on occurrence of resin bleed in dedicated 3 row platform.

IOP Conf. Series: Materials Science and Engineering **342** (2018) 012024 doi:10.1088/1757-899X/342/1/012024

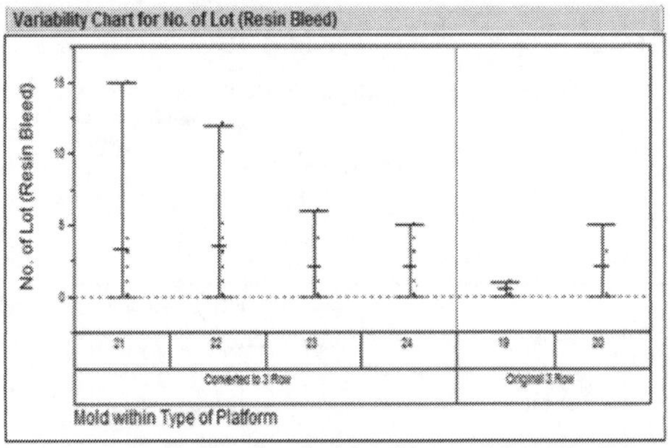

Figure 6. Analysis of resin bleed by mold & platform.

2.2.1 Comparison between converted and dedicated 3 rows platform. Initially SOD123 lead frame has 2-row. In year 2002, there was an increase in lead frame density for SOD123 from 96 units to 144 units whereby the number of rows in a lead frame was increased from 2 rows to 3 rows. To cater to this change, the mold parts were modified. The mold cavities were new but the chases as shown in Figure 7 were re-worked to cater to the new mold cavities. The reworked chase has many unwanted holes which caused excess compound to sip under the chase when there was a compound leak causing the mold surface to be imbalanced.

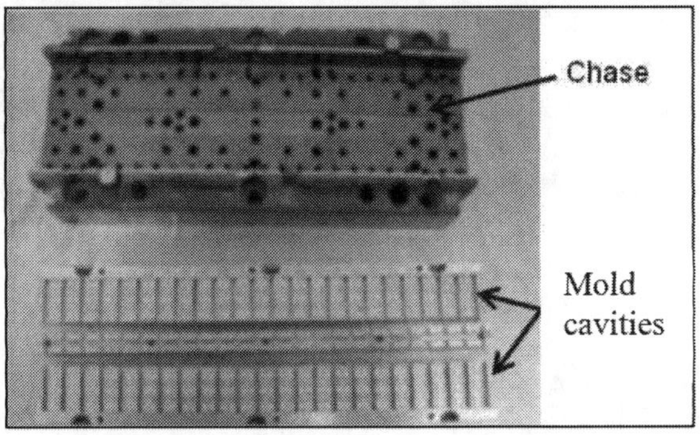

Figure 7. Dismantled chase & mold cavity.

2.2.2 Analysis converted 3 row platforms. Based on the cause & effect diagram in Figure 8, there were four main possible categories to cause resin bleed which were material, machine, human handling and method.

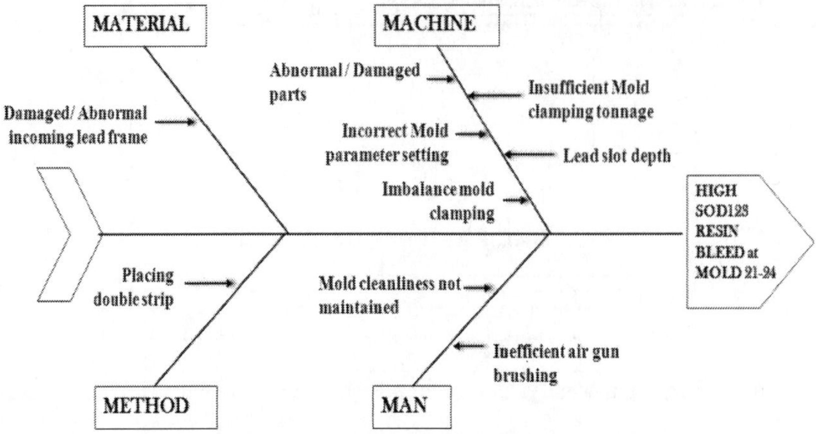

Figure 8. Cause & Effect diagram.

From the cause & effect diagram, the cause & effect matrix was done as in Table 1 and it was found out that the most probable factors were as below;

a) Imbalance mold clamping
b) Abnormal/Damage parts

Table 1. Cause & Effect Matrix.

		Y1	Y2		
	Output Indicators >>>>>	Resin Bleed Occurred to Particular Mold (21 - 24)	Resin Bleed Occurred to Random Mold		
	Importance >>>>>	10	3		
No.	Input / Process Indicators	-- Correlation of Input to Output ----		Total	Rank
1	Abnormal/damaged incoming lead	0	1	3	
2	Placing double strip	0	1	3	
3	Mold cleanliness not maintained	0	1	3	
4	Inefficient air gun brushing	0	1	3	
5	Imbalance mold clamping	9	3	99	1
6	Incorrect Mold parameter setting	0	1	3	
7	Abnormal /Damaged parts	3	9	57	2
8	Insufficient Mold clamping tonnage	0	1	3	
9	Lead slot depth	0	1	3	
	SCALE: 0 = NONE, 1 = LOW, 3 = MODERATE, 9 = STRONG				

2.2.3 Analysis on dedicated 3 row platform. Mold #19 & #20 were basically identical to each other except that mold #19 was newer than mold #20. It was observed that mold #20 was using split cavity as shown in Table 2.

Table 2. Comparison between Mold 19 & Mold 20.

	Mold #19	**Mold #20**
Chase	Originally 3 row	Originally 3 row
Cavity	Solid Cavity	Rework split cavity & solid cavity

2.2.4 5-Why Analysis & Solution Selection. 5-Why Analysis as in Figure 9 and Figure 10 was used on the most probable factors to further investigate the root cause derived from the cause and effect matrix which were imbalanced mold clamping and abnormal/damage parts. The possible counter measures were evaluated and the best solution was selected.

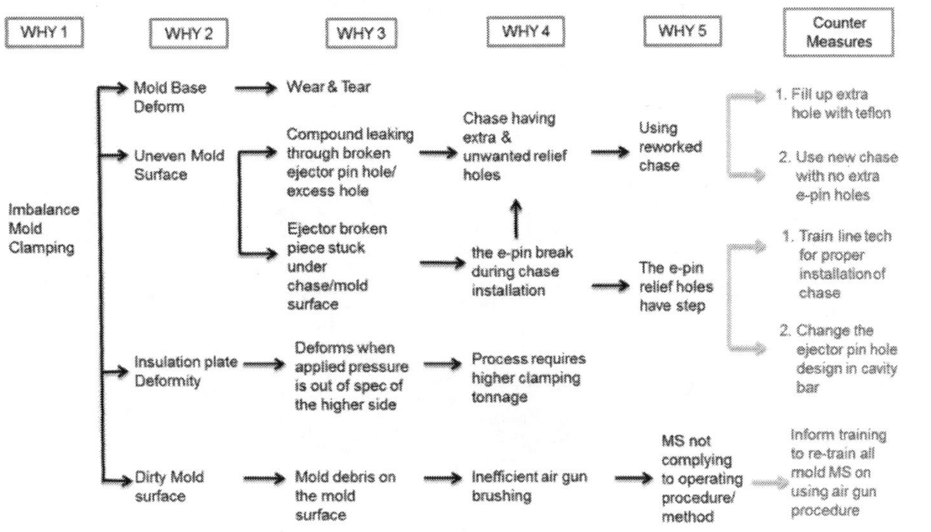

Figure 9. 5-Why Analysis on imbalance mold clamping.

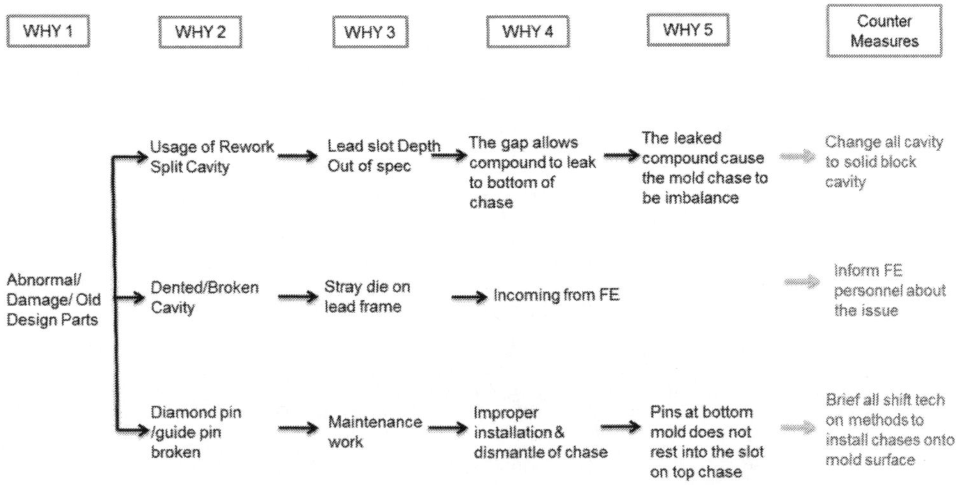

Figure 10. 5-Why Analysis on abnormal/damage/old design parts.

2.2.5 Rework split cavity. The lead slot depth of the cavity by design must be within the range of 0.094 mm to 0.097 mm. Lead slot was the position where the lead rest in the mold cavity. The lead needed to be clamped to ensure the compound did not not flow out to the lead during the molding process. Analysis was done after measuring the lead slot depth at 30 positions in one set of split rework cavity. The Chi Square value and t-value as in Figure 11 shows there were significant differences between the actual reading in the split rework cavity & the specification.

The lead slot depth was critical as an excessive clamp force on the lead could cause smash lead and low clamp force on the lead could cause resin bleed. Based on the specification, the maximum allowable lead slot depth was 0.097 mm which gave a minimum height difference between lead thickness and the lead slot depth of 0.012 mm. Based on the measurements taken on the split rework cavity, the lead slot depth was above 0.10 mm. This gave the height difference between lead thickness and the lead slot depth less than 0.012 mm. This would cause the resin from the compound to bleed out onto the lead.

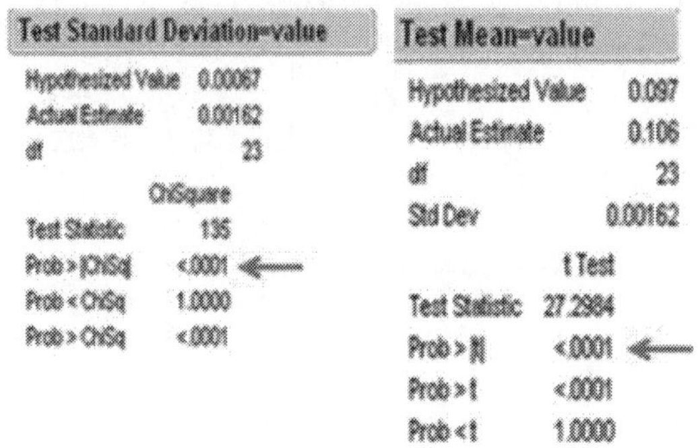

Figure 11. Analysis result on the split rework cavity lead slot depth.

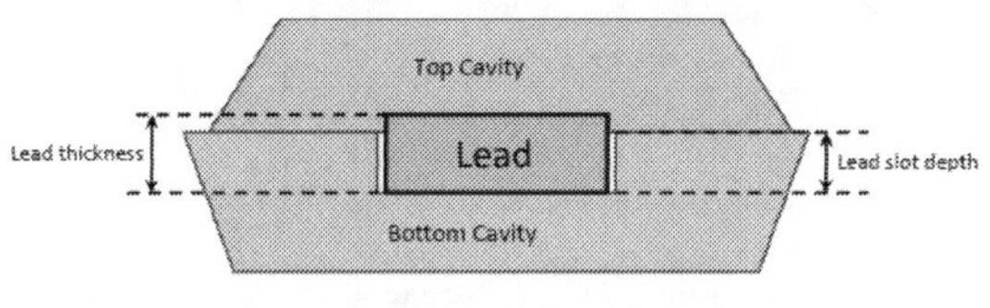

Figure 12. Illustration of Lead slot and lead.

3. Results and discussion

3.1 Generating & evaluating solution

Table 3. Generating and evaluating the solutions.

Possible solution	Evaluation Result	Conclusion
1. Fill up extra hole in the chase with teflon	• Filling up the extra hole using teflon cause the chase to be imbalance when placed onto the mold base	-Not Applicable
2. Use new chase with no extra e-pin holes	• There is no gap for the compound to leak under the chase	-Applicable
3. Train line tech for proper installation of chase	• Training done but still the installation is human dependent. Still has OFE	-Not Applicable
4. Change the ejector pin hole design in cavity bar	• Installation of the chase is more easier and less hassle since the hole is tapered	-Applicable
5. Change all cavity to solid block cavity	• Better imprint result & easier to control mold impression	-Applicable

Some solutions in Table 3 were deemed as "Not Applicable" as they were not systemic solutions but only temporary solutions which can lead to the occurrence of resin bleed. The new chase eliminated the teflon used as in rework chase to close the excess holes which contributed to mold imbalance. The new chase as in Figure 13 with no extra holes as in the new design ensured no mold compound would leaked onto the mold base. This would ensure the mold base would always be flat.

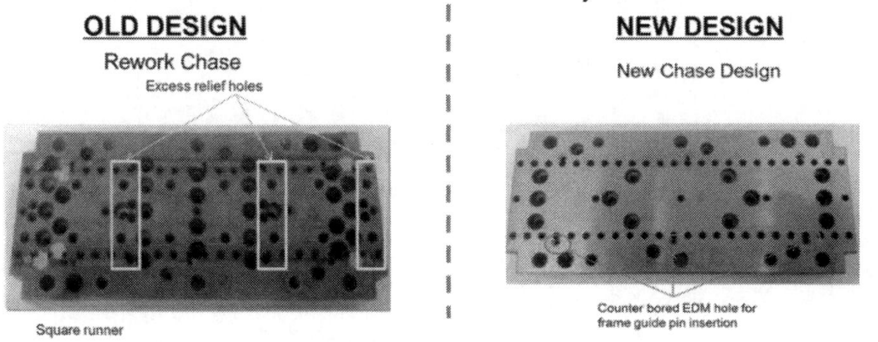

Figure 13. Comparison of old and new mold chase design.

The split cavity as in Figure 14 allowed the compound to leak and also possessed potential misalignment which could cause the lead frame to be misaligned. Once the lead frame was misaligned, resin bleed could occur.

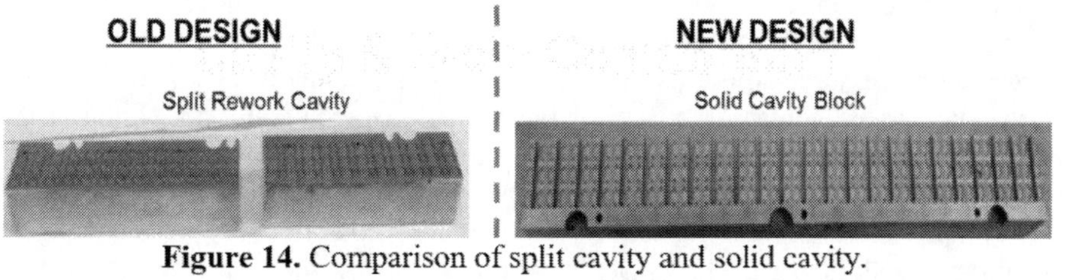

Figure 14. Comparison of split cavity and solid cavity.

The function of ejector pin was to eject the lead frame from the mold surface once the molding process. The current design of the ejector pin holes as in Figure 15 which was a stepped design caused many ejector pin breakages. This pin then would fall back on the mold base causing the mold to be imbalanced.

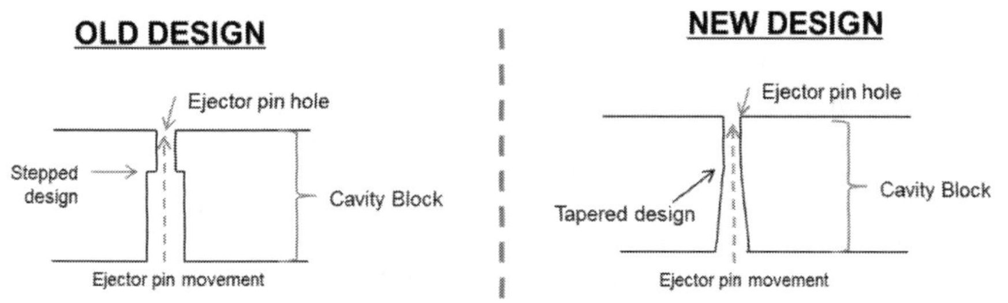

Figure 15. Comparison of stepped design ejector pin hole and tapered ejector pin hole.

3.2 Pilot run & solution implementation

All the three selected solutions were put for trial run at Mold #21. Mold #21 was selected because it was showing high contribution to the occurrence of resin bleed cases cumulatively from January 2013 to December 2014. The pilot run was conducted for 3 months and the effectiveness of these solutions would then be measured with the detection at the in-process inspection.

Pilot run result – After 3 months of pilot run, it was found that by implementing the selected solutions, the reduction of the resin bleed cases detected during inspection could be observed as illustrated in Figure 16. Hence, it could be concluded that the three solutions were able to reduce the resin bleed occurrence in all the molds.

IOP Conf. Series: Materials Science and Engineering **342** (2018) 012024 doi:10.1088/1757-899X/342/1/012024

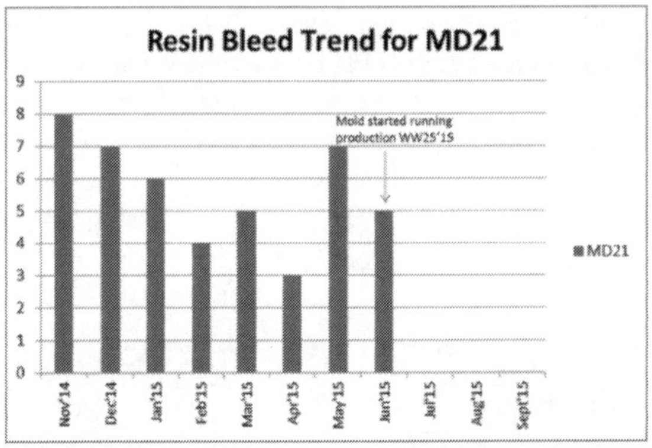

Figure 16. Pilot run result at Mold #21 with selected solutions from Feb'11 to Apr'11.

4. Conclusion

Three root causes have been identified and improvement actions were implemented to address these root causes of resin bleed in SOD123. The major root causes of resin bleed in SOD123 package during molding were usage of rework chase, usage of abnormal/damage parts (split cavity) and ejector pin hole design in the cavity block which lead to breakage of ejector pins. All of these root causes pointed to the mold compound leaking to the mold base causing the mold to be imbalanced. When the mold became imbalanced, mold compound tended to leak onto the lead of the package causing resin bleed. To tackle the identified root cause, three solutions were identified and evaluated. The solutions were new mold chase with no extra holes, replacing all split cavities with 1-piece solid cavity and re-designing the ejector pin hole from step design to tapered design. After the implementation of these refurbishments with the selected solutions, the occurrence of resin bleed was reduced significantly. The pilot run done on mold #21 which was having the highest occurrences showed zero (0) occurrence of resin bleed for three consecutive months (July 2015 to September 2015) from the implementation of the selected solutions hence proving the effectiveness of these methods. With the pilot run results showing that internal occurrence was good for three consecutive months, it showed that the issue had been resolved and customer would be receiving no reject units.

References

[1] Turng L S and Wang V W 1993 On the simulation of microelectronic encapsulation with epoxy molding compound *J. of Reinfor. Plast. and Comp.* **12** 506-519

[2] Kerk S L, Tay S C and Hu S J 1989 Effect of inherent deformations of leadframes on bleedability of plastic dip. *J. of Elect. Mater.* **18** 117-121

[3] Tong K W, Kwong C K and Yu K M 2004 Intelligent process design system for the transfer moulding of electronic packages *Int. J. of Prod. Res.* **42** 1911-1931

[4] Marvin P and Gavin C 2008 Characterization of electro-chemical deflash and high pressure water jet through MPCpS. *33rd IEEE/CPMT Int. Electronics Manufacturing Technology Conference (IEMT)* Penang 1-5

[5] Paquet M C 1998 Molding challenges of LOC packages with large devices *IEEE Transactions on Components, Packaging, and Manufacturing Technology: Part C* **21** 142-147

[6] Chiang T L and Su C T 2003 Optimization of TQFP molding process using neuro-fuzzy-GA approach *Eur. J. of Oper. Res.* **147** 156-164

[7] Colin B, Boon O C and Shugumar B 1997 Resin bleed minimization through process optimization and leadframe modification *Company Internal Publication*

[8] Carrie C and Lim L Y 2006 Study of QFN pre-tape material characteristics & impacts to assembly Performance *Thirty-First IEEE/CPMT Int. Electronics Manufacturing Technology Symposium* Petaling Jaya 220-22

Degradability studies of PLA nanocomposites under controlled water sorption and soil burial conditions

H Norazlina[1], A A Hadi[1], A U Qurni[1], M Amri[1], S Mashelmie[1] and Y Kamal[2]

[1]Faculty of Chemical Engineering Technology, TATI University College, Jalan Panchor, Teluk Kalong, 24000 Kemaman, Terengganu Darul Iman, Malaysia
[2]Faculty of Chemical and Natural Resources Engineering, University of Malaysia Pahang, 26300 Gambang, Pahang Darul Makmur, Malaysia

Abstract. Polymer blended nanocomposites based on polylactic acid (PLA) were prepared via a simple melting process and investigated for its biodegradation behaviour. The treated CNTs were surface modified by using acid treatment and characterisations of composites were done by using Fourier Transform Infra-Red (FTIR) and UV-Vis. FTIR spectra and UV-Vis peak confirmed the surface modification of CNTs. The water uptake and weight loss behaviour based on CNTs and m-CNTs loading at different temperatures (25°C and 45°C) were studied. It was found that the water absorption and weight loss of nanocomposites increased by the incorporation of CNTs and m-CNTs. Moisture induced degradation of composite samples was significant at elevated temperature. The addition of treated CNTs successfully reduced the water uptake and weight loss of nanocomposites due to less hydrolytic effect of water on nanocomposites. In soil burial test, the weight loss increases with addition of nanofiller. The loading of m-CNT reduced the ability of nanocomposites degradation.

1. Introduction

Biodegradable polymer is originated from renewable sources; as a result, it has a negative effect on environment as compared to petroleum based materials. Polylactic acid (PLA) offers several smart features that make it highly competitive with conventional non-biodegradable materials, such as good mechanical properties and the processability by standard techniques [1]. In the presence of water, PLA experiences important physical modifications: water plasticises and swells the material, its glass transition temperature is lowered and mobile and thus crystallisation kinetics is augmented. On a longer timescale, hydrolysis reduces - molecular weight, thus the physical and mechanical properties are also affected [2]. The hydrolytic degradation, above all, has been variously studied and the main factors affecting this phenomenon are quite well defined [3–6]. Surprisingly, on checking the literature, it can be noticed that when temperatures close to the glass, transition of PLA were studied, the effects of water were mostly assessed at times when the hydrolysis induced significant changes in the material, after a few days or weeks [4,7]. Obviously, it was difficult to understand to which extent other phenomena taking place in the material, such as opacification, crystallinity evolution or vice versa can develop independently [7,8]. Microbial degradation of PLA should be studied for packaging of foods containing microorganisms including lactic acid bacteria, and fungi for its possible abilities of PLA degradation. Several studies have been reported on soil degradation of PLA composites such as blending with natural filler [9-12] and carbon-based materials [13-15]. This work aims to study the water sorption and weight loss within a given time and at different temperatures, in order to analyse the phenomenon with water in PLA. Besides, the trend in weight loss after soil burial in 180 days was focused. Also, the effect of surface modification of CNTs is one of the viewpoints needed to cover in this study.

2. Materials and methods

2.1 Chemicals and materials

Industrial grade multi-wall carbon nanotubes (CNTs) with >90wt% purity and ash <1.5wt% were supplied by Global Science Resources, Malaysia. Their main characteristic is outer diameter close to 20–40 nm and length between 10–30 μm. PLA was provided from NatureWorks, USA with grade 3251D, while 95-97% of H_2SO_4 and 65% of HNO_3 were supplied by Merck KGaA, Darmstadt, Germany.

2.2 CNTs modification

The CNTs samples were modified by acid treatment. First, CNT 10wt% was immersed in 20% nitric acid (HNO_3) to remove impurities. This carboxylic group was used as reaction precursor in the functionalisation. The mixtures of CNT and HNO_3 were stirred for a few minutes and sonicated for 2 h, then diluted in water, filtered till to neutral pH, washed several times with deionised water and then dried overnight in oven at 60°C. Next, the dried mixtures of CNT were mixed together with pure 50 mL HNO_3 and 150mL sulfuric acid (H_2SO_4) to reflux for 100 minutes. Finally, the CNTs were again filtered to neutral pH, washed several times with deionised water and then dried overnight in vacuum oven at 60°C, before it was used to prepare melt-blending with polymer matrix.

2.3 Nanocomposites preparation

PLA nanocomposites were prepared by melt-blending method. The PLA/CNTs and PLA/m-CNTs with weight ratios of 99.5/0.5%, 99.0/1.0%, 98.5/1.5% and 98.0/2.0% were incorporated. A Benchtop Two-Roll Mill (LRM-M-100) was used to disperse conductive filler in the polymer matrix. Blending process was carried out at 4 rpm for 20 min and at 170°C, which was slightly above the melting point of PLA.

2.4 Nanocomposites characterisations

Fourier Transmission Infrared Spectroscopy (FTIR) analysis was carried out by using a Perkin-Elmer Model FTIR spectrometer and KBr method. The transmission of infrared spectra was obtained in the range between $400cm^{-1}$ to $4000cm^{-1}$ at room temperature. The spectra of the blends were used to determine the types of bonding in the blends.

In measuring the different molecule absorbs at difference specific wavelengths of electromagnetic spectrum region, a UV-VIS test was run by using UV-VIS machine, i.e. Shimadzu UV-1800. This method also shows absorption bands from functional group which exits in sample. Visible wavelengths cover a range from approximately 400 to 800 nm. The samples were put in the quartz cuvettes and the differences in the absorption band were observed.

2.5 Water uptake and weight loss tests

The samples were cut into 10 square-shaped. The samples were dried at 60°C for to reach a constant weight. Then, the samples were immersed into 50 ml beaker, each containing 30 ml of deionised water (DI) and placed in a laboratory in room temperature for up 28 days. The experiments took place in two temperature conditions, 25°C and 45°C. In every 7 days, the samples were taken from DI immersion and wiped with tissue paper to remove the water on surface. Water mass uptake of the samples was measured periodically. The percentage change of water uptake was calculated by Equation 1:

$$\Delta Wt = \frac{W_f - W_i}{W_i} x\ 100 \qquad (1)$$

where, ΔWt is moisture uptake, W_f and W_i are the mass of the specimen during and before aging, respectively.

After weighing, the same samples were dried in oven at 60°C to eliminate moisture. Then, the samples were weighed and the weight loss was recorded as in Equation 2. The samples were returned into water bath and the same procedure was repeated until 28 days immersion was completed.

$$\Delta W_{loss} = \frac{W_i - W_f}{W_i} x\ 100 \quad (2)$$

2.6 Soil burial test

The blended samples were buried under loamy sand-type of soil. The 10 square-shaped of samples were buried in the loamy sand at a depth of 15 cm, for 6 months. After 30, 60, 90, 120, 150 and 180 days of incubation, samples were cleaned by washing in distilled water, drained on tissue paper and then dried in oven until constant weight was reached. The progress of nanocomposite degradation was estimated by weight loss in the samples during the experiments performed. The calculation of weight loss is as shown in Equation 2.

3. Results and discussions

3.1 FTIR analysis

The method used to functionalise the pristine CNTs in this study was the acid treatment method which is described in the methodology section. Through this process, CNTs were oxidised and purified by eliminating impurities such as carbon and metal catalysts. The characteristic bands because of generated functional groups were observed in the spectrum of pure CNTs and surface treated CNTs is shown in Figure 1. The acid treated CNTs showed new peaks in comparison with FTIR spectrum of the untreated CNTs, which lacked hydroxyl and carbonyl groups. The peak at 3767 cm^{-1} was attributed to free hydroxyl groups. The peaks around 3467 cm^{-1}, 2899 cm^{-1} and 1382 cm^{-1} were assigned to the O-H band of carboxylic acid group, and the peak at 1648 cm^{-1} was assigned to the C=O band for asymmetric carboxyl groups, demonstrating that there have been introduction of hydroxyl and carboxyl groups on the nanotube surface. This peak may also link to aromatic C=C stretches, which explained the appearance in both pure and modified CNTs. In addition, the peak appeared at around 1106 cm^{-1}, apparently corresponded to the stretching mode C–O functional group of carboxylic acid. All new peaks appeared in m-CNTs confirmed the modification of CNTs.

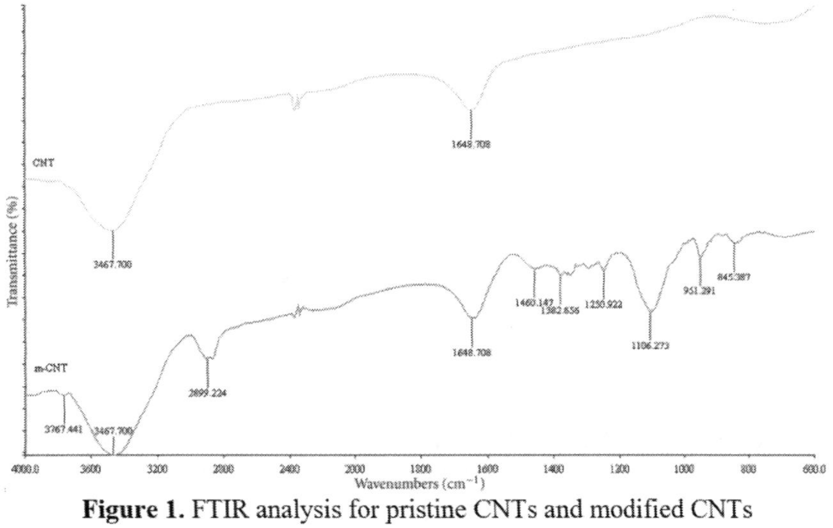

Figure 1. FTIR analysis for pristine CNTs and modified CNTs

3.2 UV-Vis Analysis

Figure below shows the different absorption peaks between pure CNT and modified CNT (m-CNT) when dissolved in deionised water. The spectrum absorption peak for pure CNT is at 325 nm and 385 nm, attributed to π-π* conjugation of aromatic C–C bonds and n-π* transition of carboxylic group C=O, respectively. The spectrum absorption peak for m-CNT appears at 692 nm, due to π-π* exciton-like transitions according to Beers Lambert Law. The change in absorbance at the given wavelength is due to the change in concentration. The fact that, the untreated CNTs do not show excitation state suggests that substantial surface oxidation did not occur.

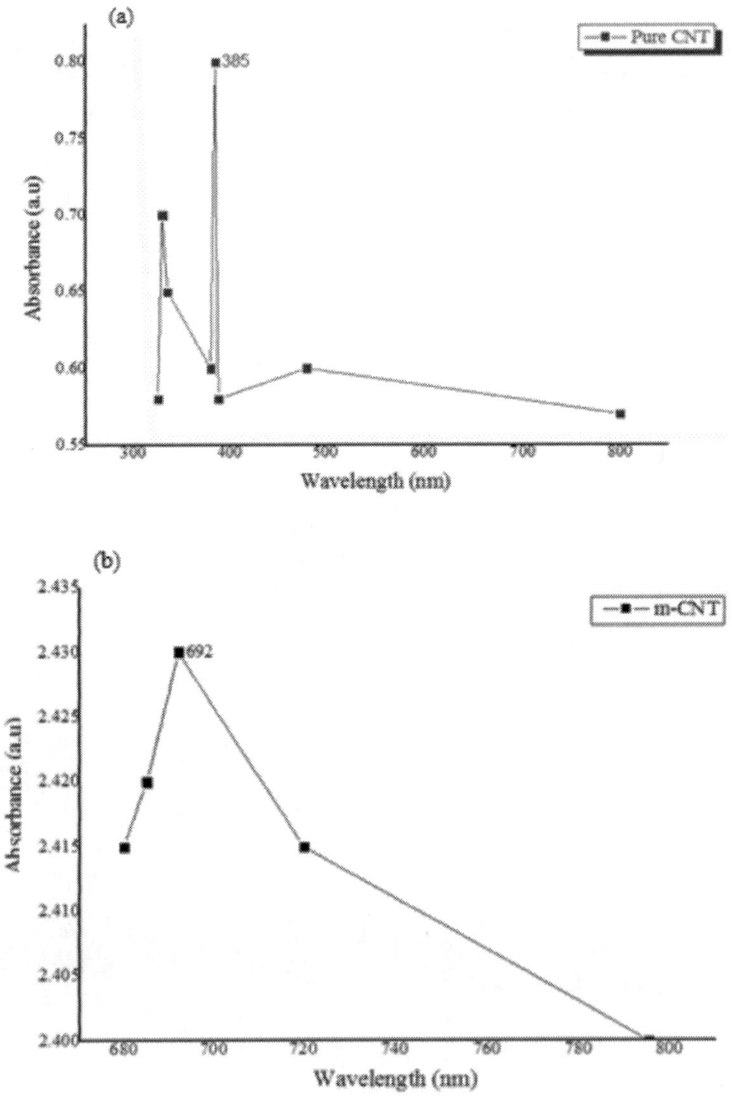

Figure 2. The UV-VIS result for a) pristine CNTs, and b) modified CNTs

3.3 Water uptake of nanocomposites

Figures 3 and Figure 4 show the water absorption behaviour for nanocomposites with unmodified CNTs and modified CNTs at different temperatures. The neat PLA indicated hydrophilicity behaviour by absorbing minimum water for both at 25°C and 45°C. Obviously, by adding filler loading in all nanocomposites, arising rise in water sorption trend can be observed, parallel to the Fickian diffusion behaviour. In Fick's Law order, the concentration of gradient is the driving force for diffusion and amount of the component diffused in function of time [16]. In general, the water uptake of polymer is determined by two theories: (i) the free volume theory (controlled by the moisture diffusion through substrates) and (ii) the interaction theory (controlled by chemical bonding, e.g. H-bond at the polar site) [17,18]. The nanofiller absorbs water due to the presence of hydroxyl groups which absorb water through the formation of hydrogen bonding [19,20]. All composites showed sharp a sharp increment in water sorption in the beginning and then remained constant after 21 days. The water uptakes at 45°C are higher than at room temperature which reached to 1.951% for 2.0% CNTs loading as compared to 25°C, the water uptake was 1.208%. Higher temperature seems to accelerate the moisture uptake behaviour. This can be explained, at higher temperature, the diffusion process was successfully achieved when water molecules gaining higher energy. Other studies have also reported a similar trend for the ageing of polymer composites at elevated temperature [21,22].

Chemical surface modification can alter polarity and surface tension of the fibres. Hence, this would improve interfacial bonding and decrease the moisture absorption [23]. This was proven by values of water uptake as shown by m-CNTs nanocomposites in Figure 4(a) and Figure 4(b) as compared to untreated CNTs. For 2% m-CNTs loading, the weight loss in room temperature condition reduces to 1.194% while at 45°C, the value decreases to 1.525%. Right chemical treatment improves the filler quality, increases the filler yields and filler's hydrophobicity and reduces swelling. This finding was similar to Chow et. al (2011) result which their study concluded the water sorption of PMMA/5HA treated zirconate coupling agent (ZCA) was lower than untreated PMMA/5HA composite [18]. This was due to the enhancement in the interfacial bonding between PMMA and ZCA treated HA which leads to better aqueous solution barrier and thus lowering the simulated body fluid (SBF) diffusion. Another finding was reported by Eng and co-workers (2014), where the silane treated OPMF observed lower water sorption rate because modified fibre is less hydrophilic as numbers of hydrophilic hydroxyl groups reduce by react with silane which leads to exclusion of water from substrate [16].

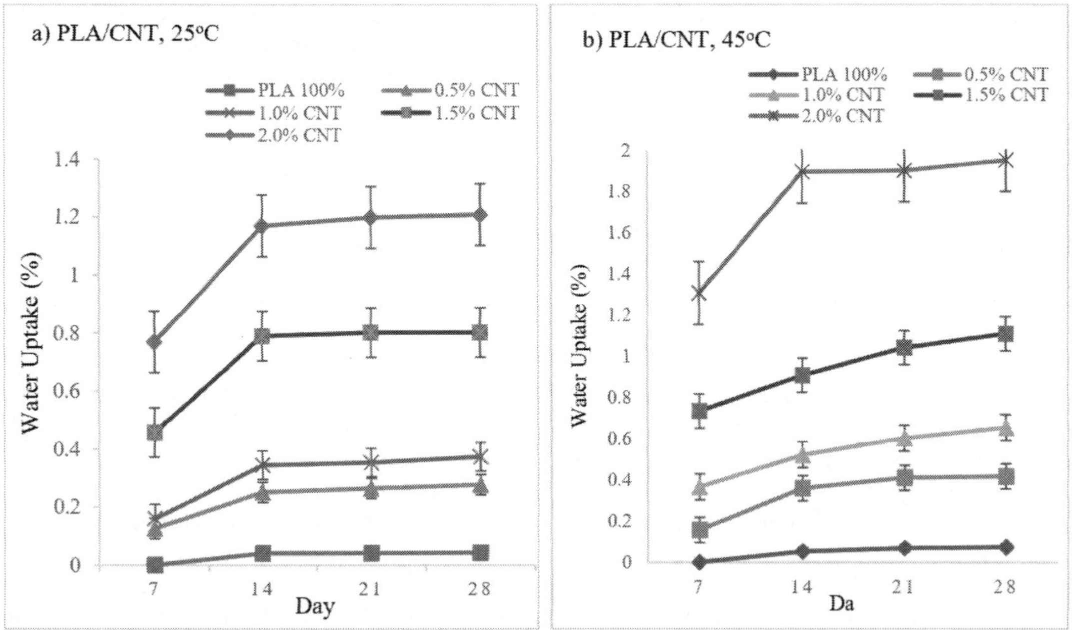

Figure 3. Water sorption trend in 28 days observation for PLA/CNTs nanocomposites at a) 25°C and b) 45°C

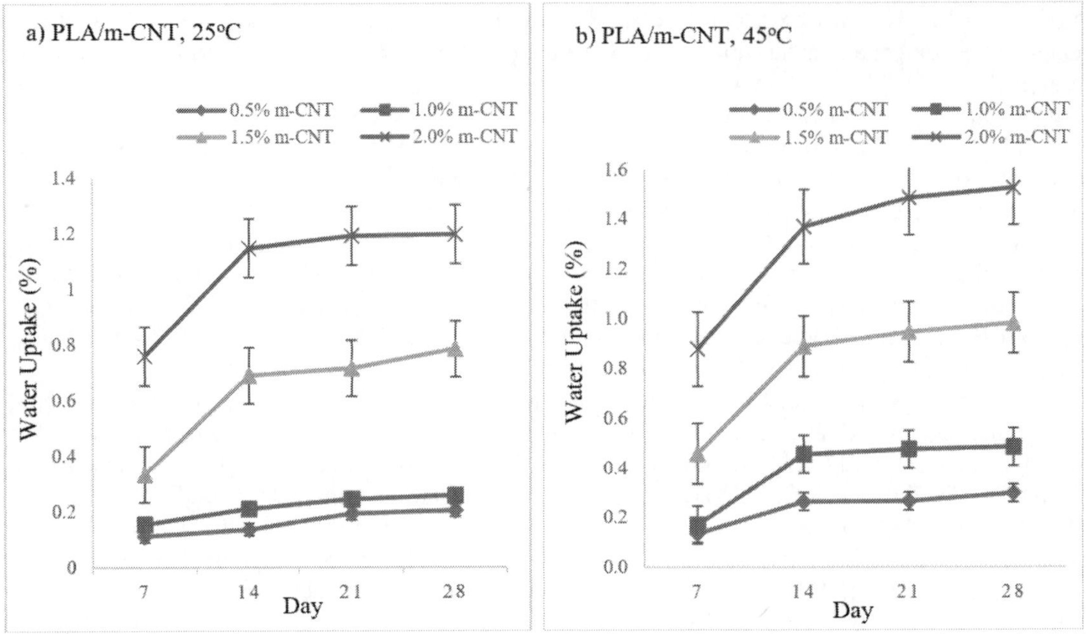

Figure 4. Water sorption trend in 28 days observation for PLA/m-CNTs nanocomposites at a) 25°C and b) 45°C

3.4 Weight loss behaviour after water sorption

PLA, one of the most well-known biodegradable polyesters, has been investigated in the pharmaceutical and biomedical applications, such as orthopaedic fixture materials, degradable sutures, absorbable fibres, and con- trolled-release devices, etc. Biodegradable polyesters are known to be degraded by the uptake of water followed by the hydrolysis of ester bonds [24]. The phenomenon may be caused by two reasons; (i) water molecules permeate into amorphous area in degradation, (ii) most of molecular chain of amorphous area is ruptured by random hydrolysis. Degradation studies were carried out in deionised water at 25°C and 45°C. Figure 5 describes the weight loss of PLA and PLA/CNTs nanocomposites as a function of time. At room temperature, PLA showed no degradation effect after 28 days compared with PLA/CNTs nanocomposites. The weight loss of PLA/CNTs with 0.5%, 1.0%, 1.5% and 2.0% CNTs were observed to be 0.191%, 0.247%, 0.721% and 1.013%. Along this degradation, the weight loss rate of PLA/CNTs at 45°C was higher than at 25°C. In that 4 weeks, the weight loss of PLA/CNTs increased twice for 1.5% and 2.0% CNTs. These changes can be related to the cleavage of PLA chains to shorter fragments which are caused by the hydrolysis of the ester linkage, followed by interchain transesterification and depolymerisation by back-biting (intra-molecular transesterification .

The weight loss of treated CNTs nanocomposites are lower than untreated CNTs as seen in Figure 6, which could be attributed to the superior combinations between the m-CNTs and PLA matrix. The enhanced interfacial interactions gave rise to improved dispersion and the wetting of CNTs in polymer matrix, enhancing the 'bridging' effect between matrix and nanofiller. Chemical modification on the filler surface is applied to promote adhesion between nanofiller and polymer matrix, hence to enhance the dispersion. This can provide compatibilisation with the polymer matrix by bridging, increasing specific interfacial areas and wrapping. This indicates that the m-CNTs is able to hinder the water diffusion into PLA/m-CNTs nanocomposites, avoid deterioration the bonding between PLA and m-CNTs which in turn, can prevent formation of microcavities [25]. The microcavities act as stress concentrators and initiate matrix cracking, leading to the reduction of molecular weight and mechanical properties.

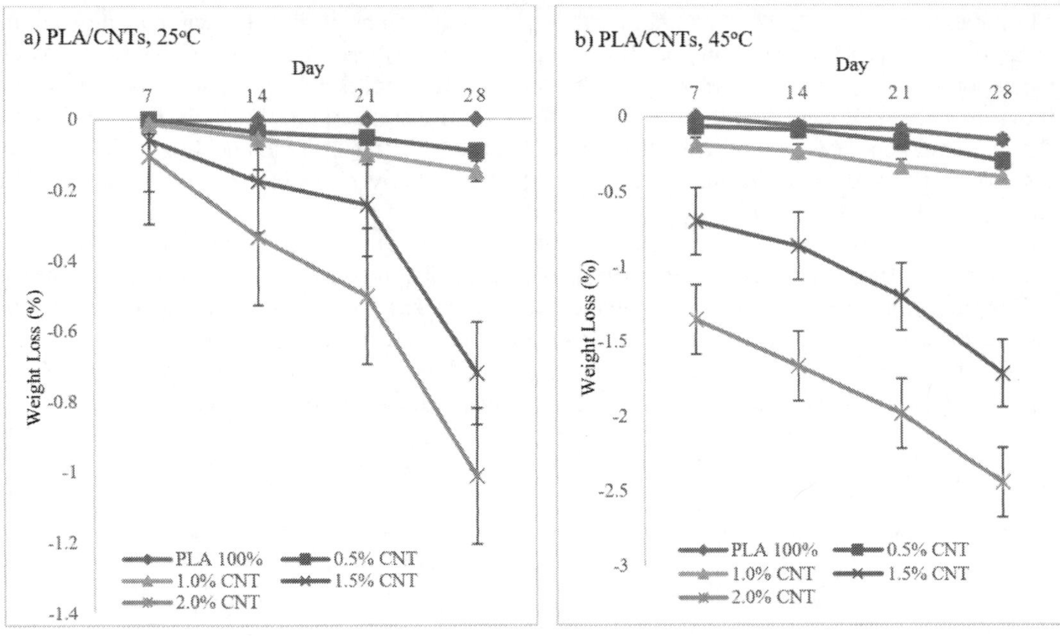

Figure 5. Weight loss of PLA/CNTs nanocomposites at different temperature; a) 25°C and b) 45°C

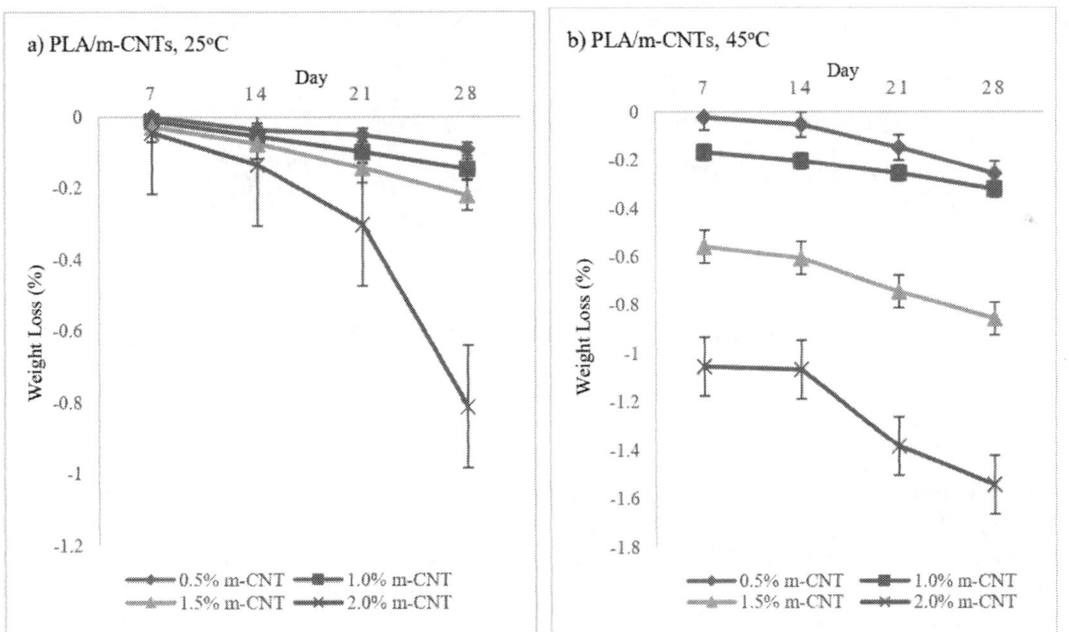

Figure 6. Weight loss of PLA/m-CNTs nanocomposites at different temperature; a) 25°C and b) 45°C

3.5 Soil burial effect on weight loss

PLA is one of the hydro-biodegradation polymers that is well-known for being bioassimilable. The speed of biodegradations depends on temperature, humidity, number and type of microbes [26]. Figure 7 presents the biodegradability of the buried specimens in terms of weight loss. The neat PLA shows least degradation which was only 0.043% weight loss after 6 months as compared to PLA nanocomposites. However, a contrary phenomenon was observed by Wu et al. (2010), where they found the degradation rate of neat PLA was higher than PLA nanocomposites [13]. This indicates that the presence of CNTs could retard the degradation of PLA. Obviously, for both untreated and treated CNTs nanocomposites, the weight loss increased with loading of nanofiller. The PLA/CNTs recorded by addition of 0.5%, 1.0%, 1.5% and 2.0% CNTs, the weight losses are 0.601%, 1.120%, 2.367% and 4.876%. This finding was similar with other researches [14,27]. The CNTs assemblies could be readily released from the nanocomposites surface due to the lack of adhesiveness of CNTs to PLA phase, which accelerated the degradation of PLA by the formation of porous structure.

In treated nanocomposites, the rate of degradation significantly reduced and was proven by mass in loss as 0.563%, 0.952%, 1.949% and 4.502% by nanofiller loading. This indicated that surface modification can influence the bio-disintegrability of CNTs. The compatibility in composites resulted by enhancement of interfacial bonding between PLA and nanofillers [28]. With addition of m-CNTs, it overcomes the problem of phase separation or aging in PLA/m-CNTs blends. The addition of m-CNTs will prevent the diffusion of short molecular chains, which will lead to loss of molecular weight.

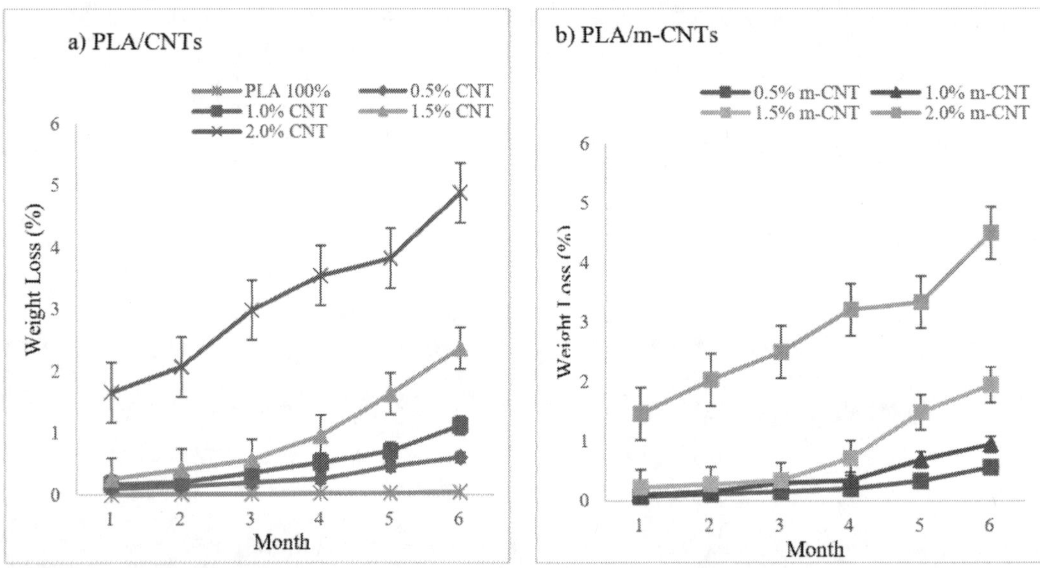

Figure 7. Weight loss trend of nanocomposites with a) untreated CNTs and b) treated CNTs after 6 months' soil burial

4. Conclusion

The water absorption of PLA nanocomposites increased with the incorporation of CNTs and m-CNTs. This attributed to the fragmentation and chain scission of polymer initiated by water diffusion under the water absorption environment (especially at immersion temperature, T = 45°C) owing to the molecular weight reduction. The water uptake of PLA/m-CNTs nanocomposites was lower than that of PLA/CNTs due to surface modification of m-CNTs which could restrict the diffusivity of water. The decreasing rate in weight loss after soil burial analysis for PLA/m-CNTs shows the improvement of surface modification of nanofiller.

Acknowledgement

The authors gratefully acknowledge the Faculty of Chemical Engineering Technology, TATI University College (TATIUC) and University Malaysia of Pahang (UMP) for providing laboratory facilities and giving supports in this study.

6. References

[1] Lim L-T, Auras R and Rubino M 2008 *Prog Polym Sci.* **8** 820
[2] Gorrasi G and Pantani R 2013 *Polym Degrad Stab.* **98** 1006
[3] Iñiguez-Franco F, Auras R, Burgess G, Holmes D, Fang X, Rubino M and Soto-Valdez H 2016 *Polym.* **99** 315
[4] Benali S, Aouadi S, Dechief A-L, Murariu M and Dubois P 2015 *Nanocompos* **1** 51
[5] Stloukal P, Kalendova A, Mattausch H, Laske S, Holzer C and Koutny M 2015 *Polym. Test.* **41** 124
[6] Huang Y, Zhang C, Pan Y, Zhou Y, Jiang L and Dan Y 2013 *Polym. Degrad. Stab.* **98** 943
[7] Xu H, Yang X, Xie L and Hakkarainen M 2016 *Biomacromolecules* **17** 985
[8] Fortunati E, Armentano I, Iannoni A, Barbale M, Zaccheo S, Scavone M, Visai L and Kenny JM 2012 *J. Appl. Polym. Sci.* **124** 87
[9] Shanshan L, Yanhua Z, Jiyaou G and Haiyan T 2017 *Coll. Surf. B: Biointer.* **7765** 30572
[10] Rodrigues RA, Tofanello A, Nantes IL and Rosa DS 2015 *ACS Sustain. Chem. Eng.* **3** 2756
[11] Zuo YF, Gu J, Qiao Z, Tan H, Cao J and Zhang Y 2015 *Int. J. Biol. Macromol.* **72** 391
[12] Rajesh G, Prasad AR and Gupta AJ 2015 *Reinf. Plast. Comp.* **34** 951
[13] Wu D, Wu L, Zhou W, Zhang M and Yang T 2010 *Polym. Eng. Sci.* **50** 1721
[14] Tsuji H, Kawashima Y, Takikawa H, and Tanaka S 2007 Polymer **48** 4213
[15] Li SM, Garreau H, and Vert M 1990 *J. Mater. Sci. Mater. Med.* **1** 131
[16] Eng CC, Ibrahim NA, Zainuddin N, Ariffin H and Yunus WMZW 2014 *J. Polym. Sci.* **2014** 1
[17] Tham W, Poh B, Ishak ZM and Chow W 2017 *J Thermoplast Compos Mater.* **30** 416
[18] Chow WS, Tham WL and Ishak ZAM 2012 *J. Thermo. Compos. Mater.* **25** 165
[19] Abdul Khalil HPS, Ismail H, Ahmad MN, Ariffin A and Hassan K 2001 *Polym Int.* **50** 395
[20] Dhakal HN, Zhang ZY and Richardson MOW 2007 *Compos. Sci. Technol.* **67** 1674
[21] Stazi F, Giampaoli M, Rossi M and Munafò P 2015 *Compos Struct.* **133** 404
[22] Doroudgarian N, Pupure L, and Joffe R 2015 *Polym. Compos.* **8** 1510
[23] Choi K, Choi M-C, Han D-H, Park T-S and Ha C-S 2013 *Eur. Polym. J.* **49** 2356
[24] Najafi N, Heuzey MC, Carreau PJ and Wood-Adams PM 2012 *Polym. Degrad. Stab.* **97** 554
[25] Tham WL, Poh T, Arifin Z, Ishak M and Chow WS 2016 *J. Eng. Sci.* **12** 13
[26] Valentina S, Pietro R,Santina R and Marco DR 2008 *Trends Food Sci. Technol.* **19** 634
[27] Pamula E, Blazewicz M, Paluszkiewicz C and Dobrzynski P 2001 *J. Mol. Struct.* **596** 69
[28] Liu L, Yu J, Cheng L and Yang X 2009 *Polym. Degrad. Stab.* **94** 90

IOP Conf. Series: Materials Science and Engineering **342** (2018) 012026 doi:10.1088/1757-899X/342/1/012026

Compaction behaviour and mechanical strength of lactose-sodium starch glycolate and lactose-croscarmellose sodium binary tablets

Nur Ashikin Yaakub[1], Mohd Shamsul Anuar[1] and Suraya Mohd Tahir[2]

[1]Department of Process and Food Engineering, Faculty of Engineering, Universiti Putra Malaysia, 43400 UPM Serdang, Selangor.
[2]Department of Mechanical and Manufacturing Engineering, Faculty of Engineering, Universiti Putra Malaysia, 43400 UPM Serdang, Selangor.

E-mail: mshamsul@upm.edu.my

Abstract. The focus of this study is to elucidate the effects of adding super disintegrants (SSG and Acdisol) to a filler (lactose) in terms of the compaction behaviour and mechanical strength of the formed binary tablets. The tablets were formed in a uniaxial die compaction process with compaction pressures ranging from 37.7MPa to 150.7 MPa. Consequently, the findings indicated that the increasing of the compaction pressure and the percentage mass composition of the super disintegrants would led to the increased in the strength of the tablets as well as their plastic energies, where this was more apparent for the case of the binary lactose/Acdisol tablets. In addition, as the compaction pressure increased, the maximum ejection pressure required to eject the tablet from the die cavity also increased. In contrast, a decreased in the maximum ejection pressure was observed as the composition of both super disintegrants increased in the lactose-super disintegrant binary tablets. In conclusion, the addition of super disintegrant; SSG with lactose and Acdisol with lactose; would enhanced the mechanical strength of lactose based tablets especially for the case of acdisol-lactose binary tablets in the experimental conditions adopted in this current work.

1. Introduction

Compaction of binary powder mixtures is an important area of research in the pharmaceutical tablet industries [1]. Tablets should be formulated in the correct proportion of active ingredients and excipients so that the tablet can disintegrate and dissolve at the right time and right place to release the drug in the human body [2]. Super disintegrants are added to decrease the dissolution time of tablets [3]. Two types of commonly used super disintegrants are sodium starch glycolate (SSG) and croscarmellose sodium (Acdisol) [3]. Apart from the intended purpose of increasing the spontaneity of the tablet dissolution, it will also inevitably affect the compaction behaviour during processing as well as the final tablet mechanical strength [4, 5]. The compaction behaviour such as the plastic deformation reflected in the calculated plastic energy under the force-displacement profile in the loading-unloading stages of the compaction process influences the final tablet mechanical integrity [6]. The final tablet mechanical integrity is important to ensure that the formed tablet can withstand the subsequent processing and handling throughout its lifetime before consumption by the end user. The ease of pushing or extruding the tablet from the die cavity is also important as a measure of the die wall friction between the tablet and the die wall, where low die wall friction hence low maximum ejection forces are favoured in the compaction process to minimize wear and work [7].

Hence, in this current study, the effectiveness of the binary tablet was analyzed in terms of its compaction behaviour and mechanical strength characteristic. Lactose was chosen as the model filler excipient, where SSG and Acdisol super disintegrant powders were added individually to lactose powder to form a model binary tablet containing a filler and a super disintegrant.

2. Material and methods

2.1 Materials

A common pharmaceutical filler excipient; α-lactose monohydrate powder, and common super disintegrant agents; sodium starch glycolate and croscarmellose sodium were used in the making of the tablets. Lactose powders were received from Meggle Group Wasserburg, Germany, sodium starch glycolate (SSG) from Yung Zip Chemical Ind.CO., Taichung, Taiwan and croscarmellose sodium (Acdisol) from FMC Biopolymer, Philadephia, USA.

2.2. Uniaxial die compaction process

The powders that have been received were sieved and only powders containing particles with sizes below 125 mm sieve size were used for tabletting. Lactose/SSG and and Lactose/Acdisol were mixed at different percentage of weight/weight of Lactose/SSG and Lactose/Acdisol which are 100/0 % 80/20 %, 60/40 %, 40/60 %, 20/80 %, and 0/100%. 1.0g of the sieved powder was then compacted inside a 13mm stainless steel cylindrical uniaxial die (Specac, UK) at 37.7MPa, 75.4MPa, 113.0MPa and 150.7 MPa compaction pressures using a universal testing machine (model 3382, Instron, Canton MA, U.S.A)., therefore producing 1 g, 13mm diameter tablets. The cross head loading speed used was 0.1 mm/s and during the unloading stage was 0.0167 mm/s. The cross head speed during ejection was 0.1 mm/s.

2.3 Compaction behaviour analysis

2.3.1 Loading-unloading force displacement profile: plastic energy. The force-displacement profile is the force and displacement data recorded during compaction involving the loading and unloading of the force onto the powder bed inside the die. This profile can be used to investigate the behaviour of a powder during compression. The area under the force-displacement curve represented the work or energy involve during the compaction process. The plastic work represents the area constituted between the loading and unloading curves and yield a quantitative description regarding the plastic deformation of the powders during compression [4, 6, 8, 9, 10].

2.3.2 Ejection force displacement profile: maximum ejection pressure. The application of force in order to extrude the tablet out from the die cavity is due to the existence of the die wall friction. The ejection force-displacement profile is the force and displacement data recorded when the tablet is extruded or ejected from the die cavity. The maximum force recorded during this ejection stage is normally associated with the 'static' friction [11]. It is considered detrimental to the mechanical integrity of the tablet and increases the possibility of tablet damages occurring upon ejection from the die cavity [12].

2.4 Mechanical strength analysis

In order to measure the mechanical strength of the tablet, the indirect tensile strength method or Brazilian method was used. The test was conducted 24 hours after the tablet was formed. The tablet was place on its side and compressed using a universal testing machine (model 5566, Instron, Canton MA, USA) at a speed of 0.0116mm/s [10]. The load cell was then programmed to move down compressing the tablet until the tablet fracture diametrically. The indirect tensile strength of the tablet, σ_t, was calculated by using the equation [13]:

$$\sigma_t = \frac{2P}{\pi DT} \tag{1}$$

where P=breaking force
D= diameter of tablet
t= height of the tablet

The diameter and the thickness of the tablet were measured by using a digital vernier calliper (Mitutoyo, Japan).

3. Result and discussion

3.1 Plastic energy

Figure 1 illustrates that as the compaction pressure increased, the plastic energy of Lactose-SSG binary tablet increased. The highest plastic energy recorded was for pure SSG tablets formed at the highest compaction pressure of 150.7MPa, while the lowest was also for the pure SSG tablets. Similar to Lactose/SSG binary tablets, Lactose/Acdisol binary tablets also recorded a similar trend of increasing plastic energy with higher compaction pressure being used for tablet formation (Figure 2). The highest plastic energy recorded was for pure Acdisol tablet formed at 150.7MPa compaction pressure, while the lowest plastic energy is at 0% of Acdisol which is pure Lactose tablet formed at 37.7MPa compaction pressure.

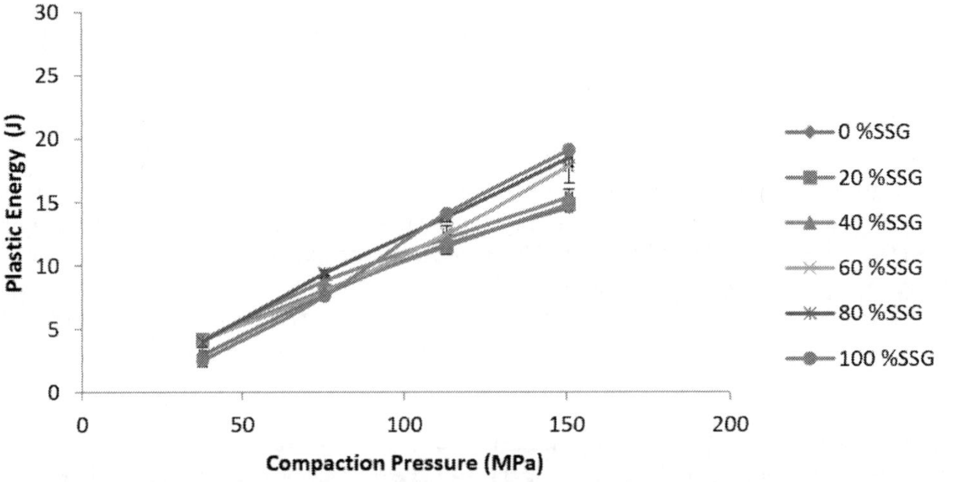

Figure 1. Plastic energy of lactose-SSG binary tablets.

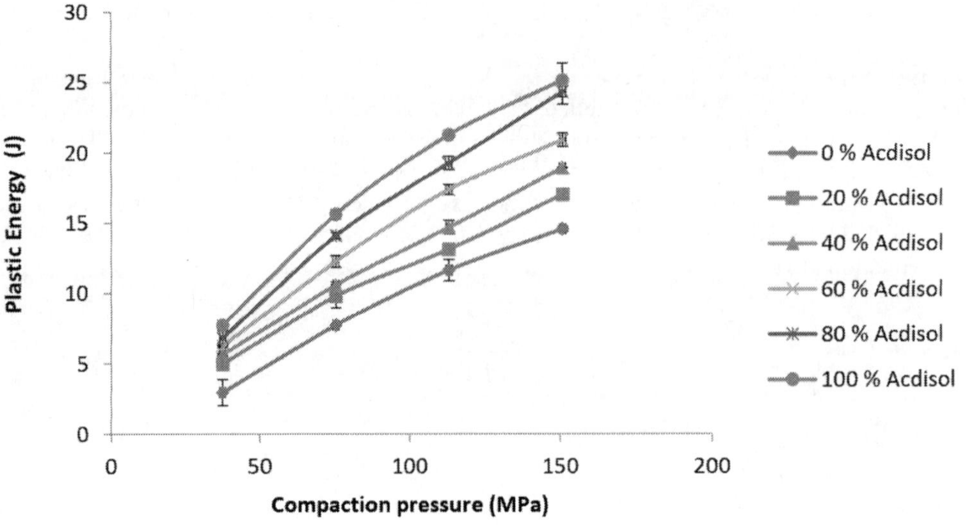

Figure 2. Plastic energy of lactose-Acdisol binary tablets.

In general, from 0% to 40% composition of SSG, the trends shows a rather constant plastic energy values (Figure 3). At further addition of SSG up to 80% composition resulted in a slightly higher plastic energy values at the higher compaction pressures of 113 MPa and 150.7 MPa. In contrast, slight reductions in the plastic energy values were observed at the lower compaction pressures of 37.7 MPa and 75.4 MPa. The addition of Acdisol increased the plastic energy as shown in Figure 4. Hence, it can be deduced that the addition of Acdisol increases the plastic energy during tablet formation more than the addition of SSG, based on the data observed in Figures 3-4.

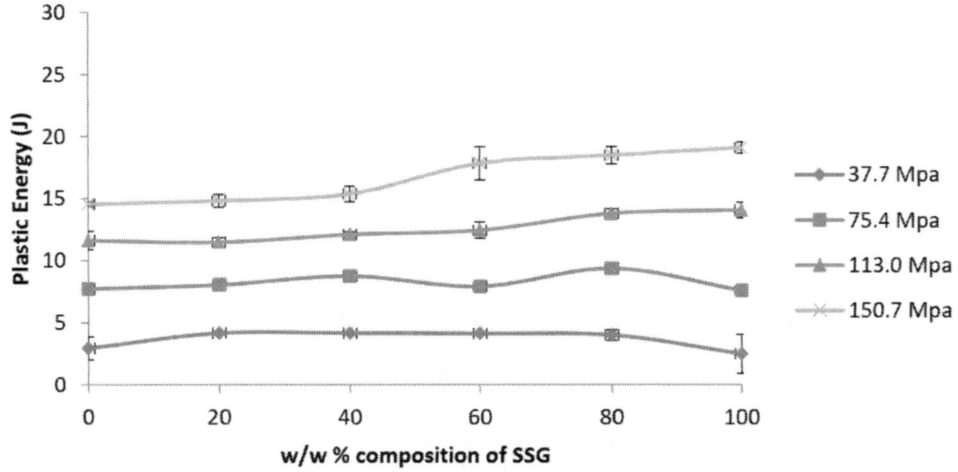

Figure 3. Plastic energy of SSG tablets by composition of SSG.

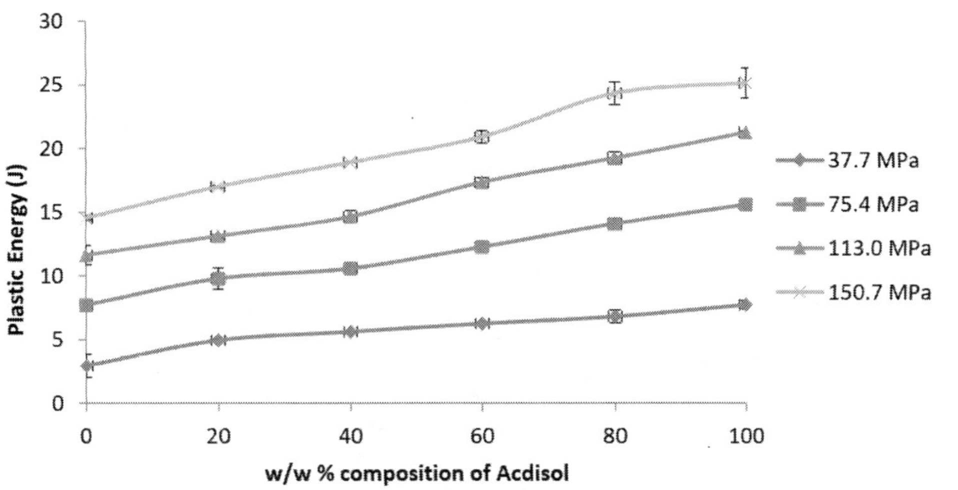

Figure 4. Plastic Energy of Acdisol tablets by composition of Acdisol.

3.2 Maximum ejection pressure

After the loading-unloading stages, the application of force was needed in order to eject the tablet from the die. This stage is called the ejection stage. Based on Figure 5, pure Lactose tablets recorded the highest maximum ejection pressure needed to remove the tablets from the die, especially at compaction pressures above the lowest compaction pressure of 37.7MPa. Low maximum ejection pressures were recorded for both pure Acdisol and SSG tablets, with no apparent dependency on the compaction pressures used during tablet formation.

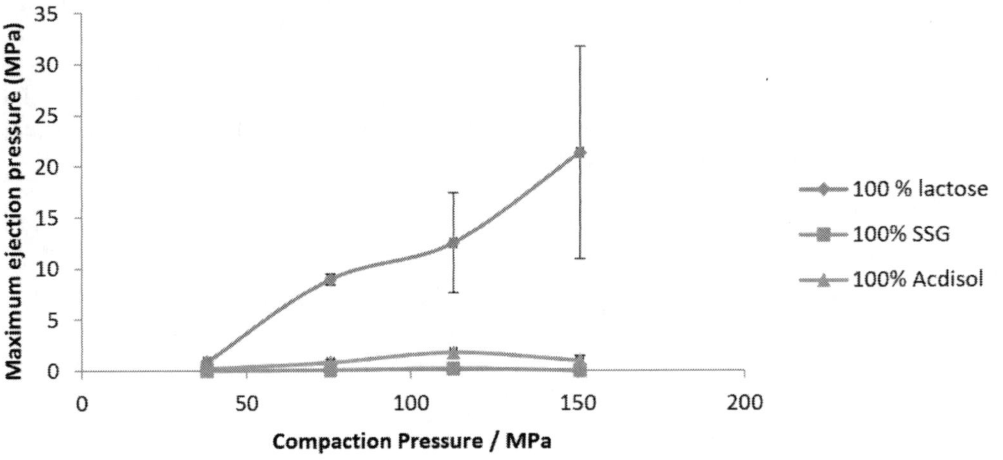

Figure 5. Maximum ejection pressures for pure Lactose, SSG and Acdisol tablets at different compaction pressures.

The presence of resistance to motion of the tablet when it is pushed during ejection is due to its internal stored elastic energy when it undergoes compression in the loading stage of the compaction process [14]. A recompression of the already formed tablet during the ejection stage corresponding to this maximum ejection force has also been shown to occur [14] due to the internal stored elastic energy of the tablet. Hence, it can be deduced that pure Lactose tablets exhibited higher internal stored elastic energies at the start of the ejection stage, which increased when the compaction pressure increased. Meanwhile, the low maximum ejection pressures of Acdisol and SSG pure tablets reflected the low internal stored elastic energies possessed by them at the start of the ejection stage.

Figure 6 illustrates the decreased in the maximum ejection pressure with the increase in the SSG composition in the Lactose-SSG binary tablets. A similar trend could also be observed for the Lactose-Acdisol binary tablets maximum ejection pressure values as the composition of Acdisol added was increased (Figure 7).

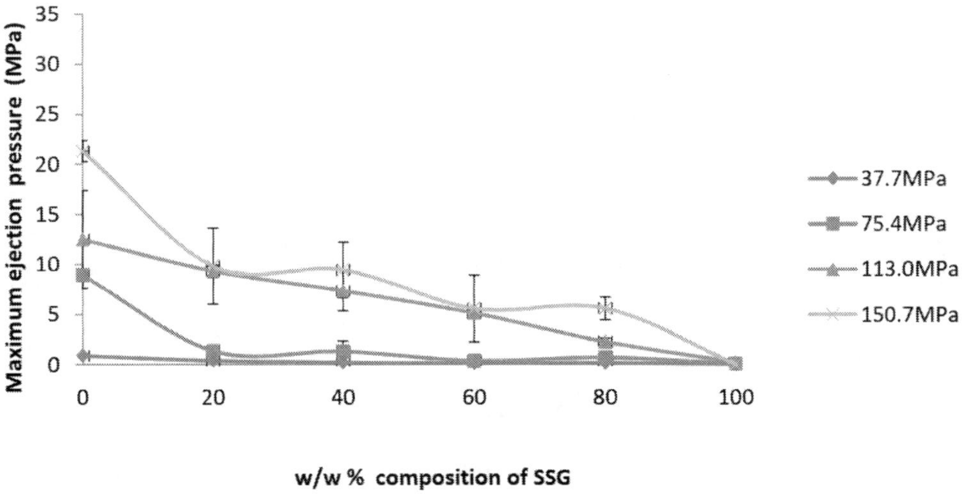

Figure 6. Maximum ejection pressure for Lactose-SSG binary tablets at different compositions of SSG.

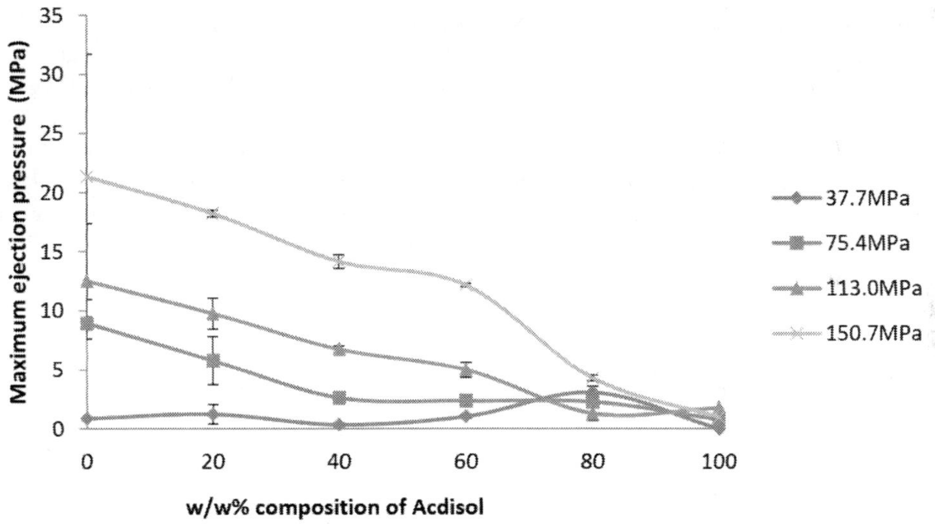

Figure 7. Maximum ejection pressure for lactose-SSG binary tablets at different compositions of Acdisol.

3.3 Tablet mechanical strength

Based on Figure 8, the pure tablets having the highest mechanical integrity in terms of its tensile strength were pure Acdisol tablets, at all compaction pressures used in this work. Acdisol tablets displayed increasing tensile strengths in tandem with increasing compaction pressures used during the tablet formation. Meanwhile, pure lactose and SSG tablets exhibited apparently similar magnitude of low tensile strength values at lower compaction pressures of 37.7 MPa and 75.4 MPa. However, when the compaction pressure increased to 113 MPa and 150.7 MPa, pure Lactose tablets showed relatively higher tensile strength in comparison to pure SSG tablets. Hence, the tensile strength of pure SSG tablets can be regarded to be relatively uninfluenced by the range of compaction pressures used in this study.

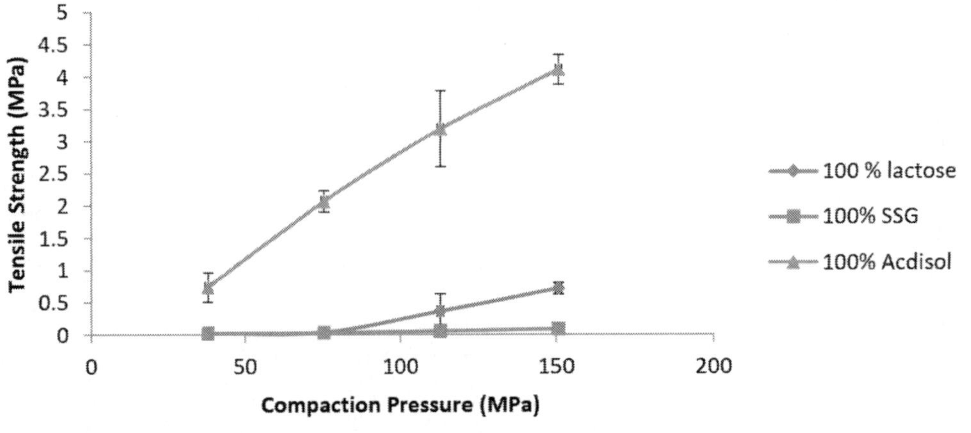

Figure 8. Tensile strength of pure tablets at different compaction pressures.

The relationship between tensile strength and compaction pressure for Lactose-SSG and Lactose-Acdisol binary tablets could be observed in Figures 9-10. The influence of the compaction pressure on the tablet tensile strength was not apparent for Lactose-SSG binary tablets in comparison to Lactose-Acdisol binary tablets. In addition, the latter exhibited a clear dependency of its tensile strength on the compaction pressure, where higher compaction pressures produced mechanically stronger Lactose-Acdisol binary tablets. Furthermore, Lactose-Acdisol binary tablets possessed higher tensile strengths in comparison with Lactose-SSG binary tablets. This was due to the relatively higher plastic energies of Lactose-Acdisol binary tablets than Lactose-Acdisol binary tablets (Figures 1-4). Higher plastic energies provided a qualitative indication of stronger inter-particulate bondings due to higher plastic deformations enhancing bonding area (Shamsuddin et al. 2012; Zahraa et al. 2013; Mohammed et al. 2005). This in turn produced mechanically stronger tablets as for the case of Lactose-Acdisol binary tablets.

The addition of SSG to lactose did not affect much the tensile strength of the lactose-SSG tablets in comparison to the addition of Acdisol to lactose, as the latter increased the tensile strength of the lactose-Acdisol binary tablets especially at compaction pressures larger than 37.7 MPa (Figures 11-12).

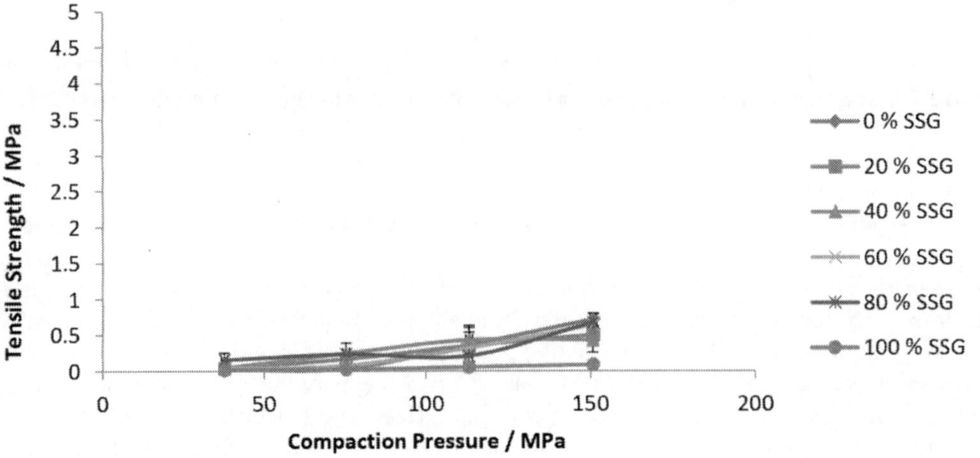

Figure 9. Tensile strengths of lactose-SSG binary tablets at various compaction pressures.

IOP Conf. Series: Materials Science and Engineering **342** (2018) 012026 doi:10.1088/1757-899X/342/1/012026

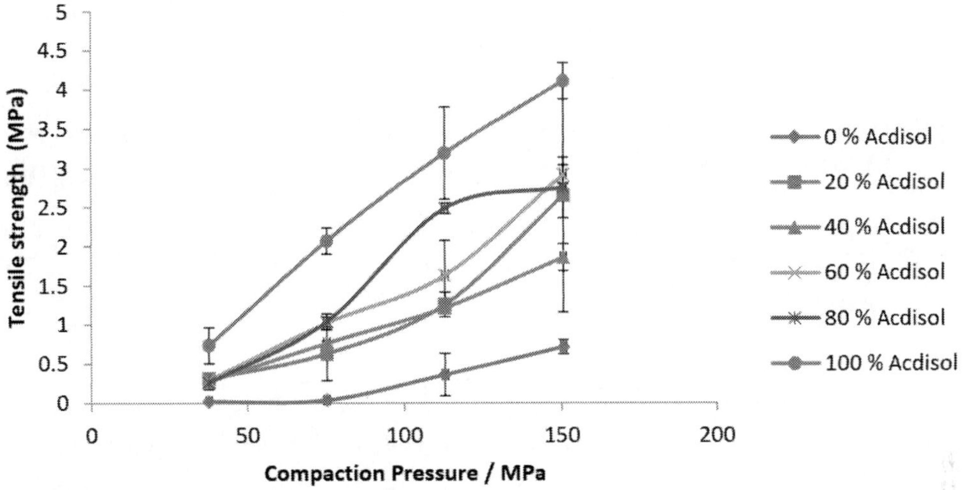

Figure 10. Tensile strengths of lactose-Acdisol binary tablets at various compaction pressures.

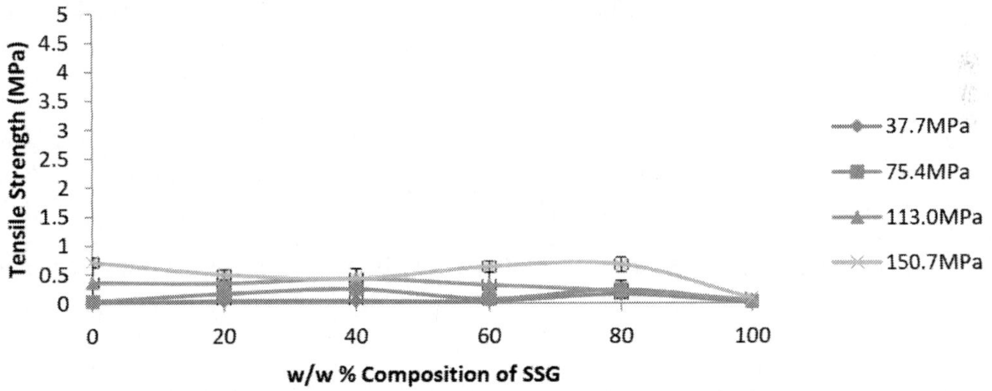

Figure 11. Tensile strengths of lactose-SSG binary tablets at different compositions of SSG.

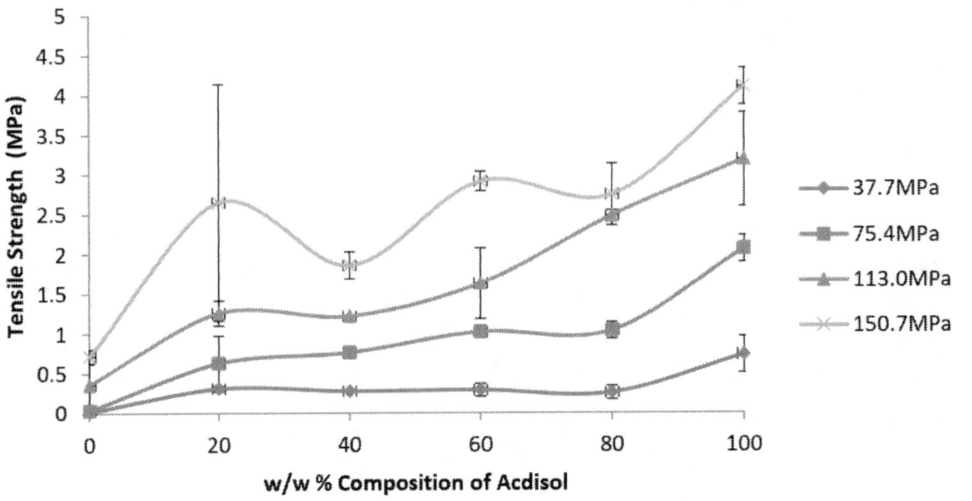

Figure 12. Tensile strength of lactose-Acdisol binary tablets at different composition of Acdisol.

4. Conclusion

The addition of super disintegrant to lactose affected the compaction behaviour as well as the mechanical strength of the formed binary tablets. The observed effects were dependent upon the compaction pressure used to form the tablets as well as the amount of super disintegrant used in making the binary tablets. The increased in the compaction pressure and the percentage mass composition of the super disintegrants would led to the increased in the strength of the tablets as well as their plastic energies, where this was more apparent for the case of the binary lactose/Acdisol tablets. Also, the maximum ejection pressure required to eject the tablet from the die cavity also increased in tandem with the increasing compaction pressure used in the study. In contrast, a decreased in the maximum ejection pressure was observed as the composition of both super disintegrants increased in the lactose-super disintegrant binary tablets. In conclusion, the addition of super disintegrant; SSG with lactose and Acdisol with lactose; would enhanced the mechanical strength of lactose based tablets especially for the case of acdisol-lactose binary tablets in the experimental conditions adopted in this current work.

Acknowledgements

The authors acknowledges the financial sponsorship from Universiti Putra Malaysia in funding this research project.

References

[1] Sun C C 2016 A classification system for tableting behaviors of binary powder mixtures.*Asian J. of Pharmaceut. Sci.* **11** 486-491

[2] Aulton M E and Taylor K M G 2017 *Aulton's pharmaceutics: design and manufacture of medicines* (Edinburgh: Elsevier)

[3] Desai P M, Liew C V and Heng P W S 2016 Review of disintegrants and the disintegration Phenomena *J. Pharm. Sci.* **105** 2545-2555

[4] Zahraa A A M, Anuar M S, Taip F S, Amin M C I, Tahir S M and Ali B M 2013 Deformation and mechanical characteristics of compacted binary mixtures of plastic (microcrystalline cellulose), elastic (sodium starch glycolate) and brittle (lactose monohydrate) pharmaceutical excipients *Particul. Sci. Technol.* **31** 561-567

[5] Shamjuddin A, Anuar M S and Tahir S M 2014 Characteristics of tabletted roselle (hibiscus sabdariffa linn.) with addition of sodium starch glycolate *Particul. Sci. Technol.* **32** 384-391

[6] Shamsudin I S, Anuar M S and Tahir S M 2012 Compaction of sweet potato (Ipomoea Batatas L.) and stevia rebaudiana food powders *Particul. Sci. Technol.* **30** 136-144
[7] Takeuchi H, Nagira S, Yamamoto H and Kawashima Y 2004 Die wall pressure measurement for evaluation of compaction property of pharmaceutical materials *Int. J. Pharm.* **274** 131-138
[8] Alderborn G and Nyström C 1996 *Pharmaceutical compaction technology* (New York: Marcel Dekker)
[9] Mohammed H 2004 *Contact mechanical aspects of pharmaceutical compaction* Ph.D. Thesis (London: Imperial College London)
[10] Mohammed H, Briscoe B J and Pitt K G 2005 The interrelationship between the compaction behaviour and the mechanical strength of pure pharmaceutical tablets *Chem. Eng. Sci.* **60** 3941-3947
[11] Macleod H M and Marshall K 1977 The determination of density distributions in ceramic compacts using autoradiography *Powder Technol.* **16** 107-122
[12] Anuar M S and Briscoe B J 2010 Detrimental consequences of paracetamol tablet elastic relaxation during ejection *Drug Dev. Indust. Pharm..* **36** 972-979
[13] Fell J T and Newton J M 1970 Determination of tablet strength by diametral compression test *J. Pharm. Sci.* **59** 688-691
[14] Anuar M S and Briscoe B J 2009 The elastic relaxation of starch tablets during ejection *Powder Technol.* **195** 96-104

Effect of stabilization temperature during pyrolysis process of P84 co-polyimide-based tubular carbon membrane for H_2/N_2 and He/N_2 separations

N Sazali[1,2], W N W Salleh[1,2], A F Ismail[1,2], N H Ismail[1,2], F Aziz[1,2], N Yusof[1,2] and H Hasbullah[1,2]

[1]Advanced Membrane Technology Research Centre (AMTEC), Universiti Teknologi Malaysia, 81310 Skudai, Johor Darul Takzim, Malaysia.
[2]Faculty of Chemical and Energy Engineering (FCEE), Universiti Teknologi Malaysia, 81310 Skudai, Johor Darul Takzim, Malaysia.

E-mail: hayati@petroleum.utm.my; afauzi@utm.my

Abstract. In this study, the effect of stabilization temperature on the performance of tubular carbon membrane was being investigated. P84 co-polyimide-based tubular carbon membrane will be fabricated through the dip-coating technique. The tubular carbon membrane performance can be controlled by manipulating the pyrolysis conditions which was conducted at different stabilization temperatures of 250, 300, 350, 400, and 450°C under N_2 environment (200 ml/min). The prepared membranes were characterized by using scanning electron microscopy (SEM), x-ray diffraction (XRD), and pure gas permeation system. The pure gas of H_2, He, and N_2 were used to determine the permeation properties of the carbon membrane. The P84 co-polyimide-based tubular carbon membrane stabilized at 300°C demonstrated an excellent permeation property with H_2, He, and N_2 gas permeance of 1134.51±2.87, 1287.22±2.86 and 2.98±1.28GPU, respectively. The highest H_2/N_2 and He/N_2 selectivity of 380.71±2.34 and 431.95±2.61 was obtained when the stabilization temperature of 450°C was applied. It is concluded that the stabilization temperatures have protrusive effect on the carbon membrane properties specifically their pore structure, and eventually their gas separation properties.

1.Introduction

Nowadays, hydrogen invention from biomass is the greatest favourable development because of the waste decreasing matter and the concerns for the sustainable growth. A common technology used in producing hydrogen (H_2) from natural gas is by steam reforming [1]. However, H_2 production (as an alternative of consuming fossil fuel-like steam methane reforming) offers a boundless chance for H_2 generation, but the process typically produces a humidified gas consist of H_2 and CO_2 where hydrogen content usually ranges between 35 and 65 (vol.) %, else may also contain volatile organic compounds (VOC), H_2S and NH_3 [2]. The purification process of H_2 from the main product stream is compulsory prior usage. The current technologies for H_2 purification are pressure swing adsorption (PSA), cryogenics, palladium (Pd) membranes [3]. Conventional PSA or cryogenics will not be appropriate for minor scale H_2 productions from fermentation processes due to the predictable high costs. Although the palladium membranes have even now been commercialized for H_2 purification such as H_2 Pure™ and HYSEP, the membrane invention cost is still high, and the period of usage is quite short due to the H_2 embrittlement, which leads to a high investment spending (CAPEX). Some literature has already reviewed the status of different polymeric membranes for hydrogen purification [4]. The problems and tasks with gas separation membranes to be practical for the composition of fermentatively produced H_2 gas has been reviewed recently by Bakonyi et. al., [4]. They stated that the highest H_2/CO_2 gas selectivity (1.62) could be achieved at feed pressure, separation temperature and recovery value of ~2.2 bar, 55 °C and 0.6, respectively, using a 65 vol% hydrogen/35 vol% carbon-dioxide binary mixture. Li et al. equated the H_2-selective and CO_2-selective membrane materials and processes, and find out the forthcoming path on the advance of energy-efficient membranes [5]. With respect to the non-porous polymeric and mixed matrix membranes, it is relatively challenging to accomplish a high H_2 purity (>99%) obligation with low selectivity (<10) [6,7]. Additionally, the composition of the biohydrogen gas mixture, its contamination (e.g., H_2S) content was initiate to significantly affect the polymeric PDMS membrane performance reported by Bakonyi et al. [8]. They also mention that polymeric membranes were frequently tried under ideal laboratory conditions in the literature; yet, the hydrogen generated by fermentation process is a part of a complex biogas stream, leading to the possible transformations in the elements behaviour. A hybrid polymeric membrane/absorber system might be a solution to achieve the separation requirement for biohydrogen purification reported by Bakonyi et al. [9], the process complexity and economic feasibility should be further investigated. The above defined tasks to the development of an alternative membrane material with high H_2/N_2 selectivity and good chemical stability showing to the impurities (e.g., H_2S). This contaminant which are frequently occurred in the raw biohydrogen gas mixture, and can carry down the H_2 purification cost, and possibly increase the H_2 production. Furthermore, carbon membranes are also predictable to tolerate the impurities (e.g., H2S) when visible to a fermentation gas stream. This statement is based on experience accomplished with a carbon membrane pilot tested at a biogas plant in Southern Norway [10]. In this study, our goal is to fabricate polyimide based tubular carbon membrane for both H_2 purification and flue gas treatment. To the best of our expertise, there are only a few number of publications that approach tubular carbon membranes for H_2/N_2 separation. Helium is a valuable gas with limited properties, applications and still the prominent industrial cause of helium is natural gas. At the present time, the demand of Helium (He) for aerospace and cryogenic technology applications are obviously increased. Regrettably He is quite occasional, and this shortage hinders the enhancement in both research and industry [11]. It is of massive consequence to accumulate a competent method to separate He as this would generate a chance for recovering He from characteristic gas with lower grouping of He in an economically practical process. Alternations in the membrane configurations as well as hybridization of membranes with cryogenic and pressure swing adsorption processes are among the valuable options for optimizing techno-economic viability of helium separation and purification. However, there is some experimental information on He purification using pressure swing adsorption utilizing adsorbents, for instance activated carbon, are reachable [12,13]. According to previous study, there are difficulties to examined He/N_2 separation these membranes are bounded by the exchange between the membrane permeability and selectivity whereby high He selectivity was achieved by sacrificing permeability or vice-versa [13,14]. In overall, inorganic membranes such as carbon membrane have shown good performance in helium separation, but production of such membranes has not yet become viable on the commercial scale. Consecutively, inorganic and mixed matrix membranes (MMMs) are recommended as feasible candidates to overwhelm these difficulties [15]. Hence, this study can provide convenient understandings on fabrication of tubular carbon membrane for H_2/N_2 and He/N_2 separations.

Carbon membrane is one of the capable materials in enhancing gas separation performance [16]. Carbon membrane is one of the potential methods that provide better selectivity, and their higher thermal and chemical stability; comparing to their polymeric counterpart [17]. Different precursors would require different stabilization conditions due to the chemical structure and thermal behaviour of those precursor materials. Based from previous study, carbon membranes are more suitable to be applied at high temperature (500-900°C) [18]. Cipriani et al. have completed pyrolysis on a thin film polyacrylonitrile (PAN) at a stabilization temperature below 400°C where the final structure of the samples resembled graphitised structures [19]. Chemical reaction is started during the isothermal step when stabilization process take place, indicating that most of nitriles take part in the cyclization and trimerization reactions below 400°C [19]. Barbosa-Coutinho et al. revealed that a severe degradation of the polyetherimide-based carbon membrane was detected when stabilization temperature above 500°C was applied. They have proved that membrane exposure under maximum temperature of stabilization were decisive in the final membrane morphologic characteristics and properties. In addition, when the stabilization temperature was above 500°C, an intensive degradation of the fiber was observed. This is because the level of membrane acquaintance below maximum temperature of stabilization was significant with the final membrane morphologic characteristics and properties [20].

Besides, Hameed et al. had studied the effect of oxidative stabilization and pyrolysis processes on the structure, mass and mechanical properties of polyacrylonitrile (PAN) precursor fibres. The results indicated that after stabilization at a maximum temperature of 255°C and pyrolysis at a maximum temperature of 800°C, the tensile strength and modulus of the fibres had increased [21].

Table 1 summarized the pyrolysis conditions and types of gas separation used by previous researchers in carbon membranes preparation. Despite their impact on carbon membrane properties, researches specifically on the stabilization conditions are very limited in the current literature since most researches focusing on pyrolysis conditions. Based on the author's knowledge, the effects of stabilization temperatures on P84 co-polyimide-based tubular carbon membranes have not been reported in the current literature. Therefore, in this study, the effects of stabilization temperature on the gas permeation properties of the P84 co-polyimide-based tubular carbon membranes were explored. It is expected that the findings from this study would help bridge the literature gap in the understanding of the influence of stabilization temperature for the improvement of gas separation properties of the carbon membrane.

Table 1. Carbon membrane reported by previous researchers.

Carbon Membrane		Pyrolysis Conditions		Applications	Ref.
Material	Configuration	Stabilization Temp (°C)	Pyrolysis Temp (°C)		
Sulfonated poly (aryl ether ketone), SPAEK	Flat sheet	250	800	H_2 and CO_2 separation	[22]
Matrimid 5218	Tubular support	300	850	CO_2 separation	[23]
Polyacrylonitrile/ Polyvinylpyrrolidone	Flat sheet	200	800	Super capacitor performance	[24]
Phenolic novolac resin	Tubular support	150-300	600-900	N_2 separation	[25]
Poly (amic acid)/ zeolite ZSM-5	Flat sheet	100-350	600	O_2/N_2 separation	[26]
polyacrylonitrile (PAN)	Flat sheet	Below 400	800	Chemical reactions	[19]

2. Experimental procedures

2.1 Materials

BTDA-TDI/MDI (P-84) co-polyimide are marketable bought from Sigma Aldrich (CAS#: 58698-66-1) was select as a precursor to produce carbon membranes. N-Methyl-2-Pyrrolidone (NMP) was used as a solvent and purchased from Merck (Germany). Tubular support membrane contains of TiO2 (4.5-5.5mm), with a coating of ZrO2 (2-3nm) on the inner surface was purchased from Shanghai Gongtao Ceramics Co., Ltd. The dimensions of the tubular support were 8cm in length and 13mm in diameter. The porosity of the support was 40-50% with an average pore size of 0.2μm.

2.2 Carbon Membrane Preparation

The dope solution was prepared by stirred of 15wt% of P84 co-polyimide and 85wt% of NMP for 24 hours. Tubular carbon membrane was prepared through dip-coating technique; where tubular support were dip-coated with the dope solution for 15 min to form a uniform layer of polymeric precursor over the outside surface of the tubular support. The membrane was then dried at 80°C for 24 hours. Next, the membranes were immersed in methanol for 2 hours, followed by drying for 24 hours. The dip-coating cycles were repeated three times to remove any pinhole occurrences on the surface of the prepared tubular membranes. The prepared tubular precursor membranes were then placed inside a Carbolite horizontal tubular furnace for stabilization and pyrolysis processes. In this study, two steps of the heat treatment were employed. In the first step, the membranes were heated at different stabilization temperatures (250, 300, 350, 400 and 450°C) with a heating rate of 3°C/min under N_2 atmosphere (200ml/min). In the second step, the membranes were pyrolyzed up to 800°C at the same heating rate and gas flow as mentioned in step 1. The nomenclature of the resultant carbon membranes was given in the form of (CM-stabilization temperature); CM-250/CM-300/CM-350/CM-400/CM-450. Carbon membranes without tubular support were also prepared through similar procedures for characterization purposes.

2.3 Membrane Characterization
The morphological structures of the P84 co-polyimide-based membranes were viewed via scanning electron microscopy (SEM), model JEOL JSM-5610LV. A small piece of the tested membranes was cut and conditioned in liquid nitrogen to leave an unreformed structure to be mounted on sample stubs. The microstructure properties of the prepared carbon membranes were examined on X'Pert PRO X-ray diffractometer (XRD) from PANalytical with the diffraction angle of 2θ, between 10° and 50°. Ni-filtered CuKα radiation with a wavelength of λ= 1.54 Å was applied in the experiments. The interplanar distance (d-spacing) of the carbon membranes were calculated by the Bragg equation [27,28,29].

2.4 Gas Permeation Test
The pure gas permeation system used for gas permeation test can be found elsewhere [23,29]. The prepared membrane was placed in a stainless steel module prior the gas test. O-rings rubber was used so that leaks can be prohibited when the membranes were housed within the module. Pure gas H_2 (0.289 nm), N_2 (0.364 nm) and He (0.260nm), were fed individually at a trans membrane pressure of 8 bars into the module. The permeance, P/l (GPU) and selectivity, α of the membranes were calculated using equations below:

Permeance, P:

$$(P/l)_i = \frac{Q_i}{\Delta p.A} = \frac{Q}{\eta \pi D l \Delta P} \quad (1)$$

Selectivity, α :

$$\alpha_{A/B} = \frac{P_A}{P_B} = \frac{(P/l)_A}{(P/l)_B} \quad (2)$$

where P/l is the membrane permeance, Qi the gas i's volumetric flow rate of at standard temperature and pressure (cm3 (STP/s), p the difference in pressure between the feed the permeation side of the membrane (cmHg), A the area of membrane surface (cm2), n the fiber quantity in module, D the membrane's outer diameter (cm), and l the membrane's effective length (cm). The selectivity is defined as the permeation ratio of fast gas permeation to slow gas permeation.

3. Results and discussion
3.1 Morphological structure analysis
Figure 1 (a) presented the cross-section microphotographs of the precursor membrane, while Figure 1 (b) to (f) showed the cross-section microphotographs of carbon membranes prepared at various stabilization temperatures. Due to the difficulty in preparing SEM samples for the supported carbon

membrane, self-supported carbon membrane prepared without tubular support was used for SEM characterization. Referring to Figure 1 (a), precursor membranes have sponge-like structures. From Figure 1(b) to (f), all prepared carbon membranes with different stabilization temperatures are observed to have a pore structure with more open pores [30]. This phenomenon is likely due to the stabilization of thermoplastic polymers in air permits oxygen scaffolds to be made between aromatic molecules that repress the reworking and development of aromatic crystallites amidst the pyrolysis step. Principally, during the stabilization step, the precursor would experience various physical and substance changes because of a mixture of exothermic chemical reactions, including decay, cyclisation, dehydrogenation, oxidation, cross-linking and fracture [31]. Therefore, after pyrolysis at elevated temperature, the resulted carbon membrane became more porous as stabilization temperature increase. Work by Hosseini and Chung (2009) suggested that the porous structure allows high permeability while the molecular sieving network provides the efficient size and shape selection of molecules [32]. Furthermore, as the stabilization temperature increased, the level of decomposition and cross-linking in the membrane expanded, which brought about distinctive micropore structures of the carbon membranes that can be utilised to tune gas division [31].

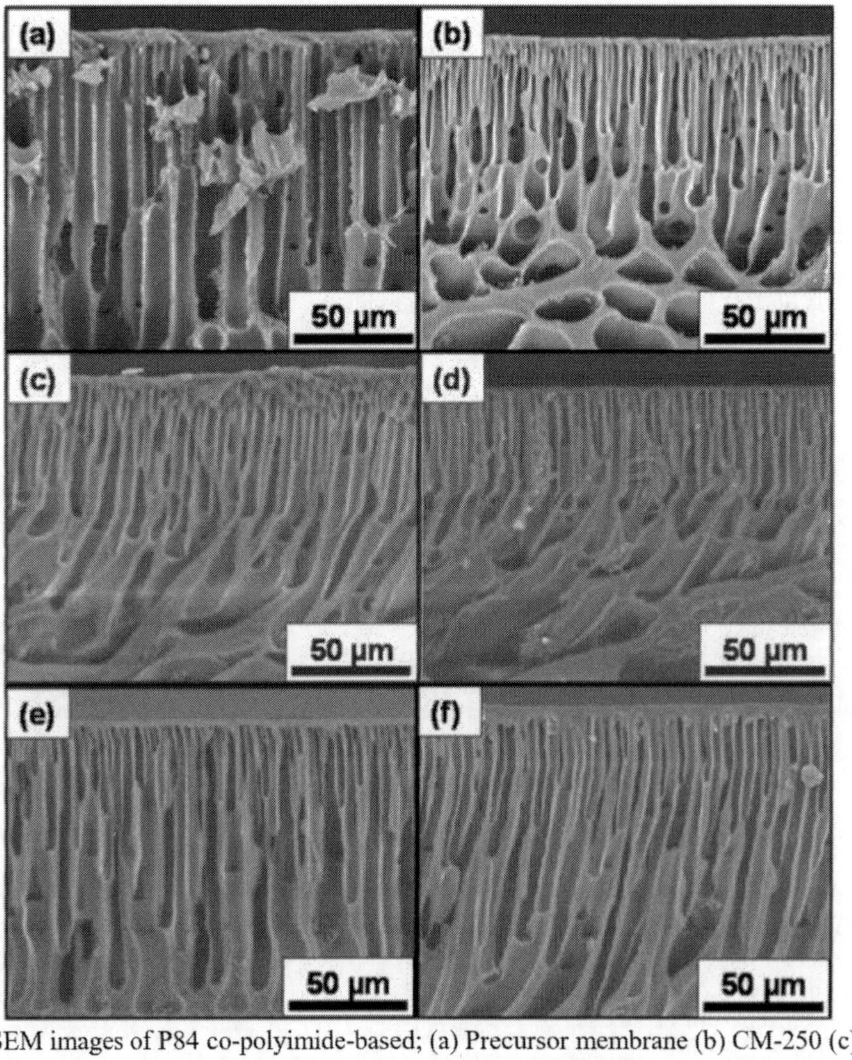

Figure 1. SEM images of P84 co-polyimide-based; (a) Precursor membrane (b) CM-250 (c) CM-300 (d) CM-350 (e) CM-400 (f) CM-450.

3.2 Wide-angle X-ray diffraction pattern

The XRD pattern of the prepared membrane was shown in Figure 2. The XRD spectra were shifted accordingly in order to show a clear x-ray diffraction result without affecting the diffraction angle. The average d-spacing between the individual layers of carbon was calculated by adopting Bragg's Law. These d-spacing values were determined at the maxima of broad peaks and only reflect the average space between the centres of the chain segments in the polymer matrix [32]. The reflection at 2θ=23.02 (d-spacing: 0.498nm) can be observed for polymeric precursor membrane (P84 co-polyimide), while the carbon membrane reflection for all stabilized temperature from CM-250 until CM-450 were at 2θ=24.75 (d-spacing: 0.402nm), 2θ=24.81 (d-spacing: 0.384nm), 2θ=26.23 (d-spacing: 0.369nm), 2θ=26.75 (d-spacing: 0.351nm), and 2θ=27.03 (d-spacing: 0.351nm), respectively. The pattern for all carbon membranes indicated that there was not much difference. Based on the data, the d-spacing value decreased as the stabilization temperature increased from 250 until 450°C. This was due to the formation of more compact structures with higher packing density, and the d-spacing increment may coincide to the evolution of a more ordered microstructure with efficiently packed graphite layers [31].

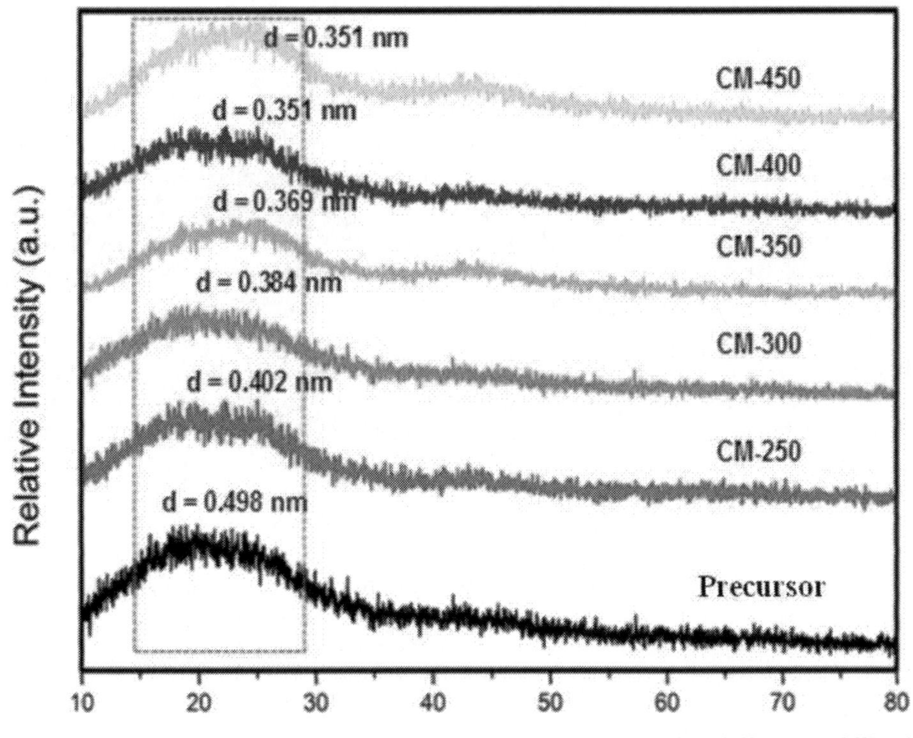

Figure 2. XRD spectra of precursor and carbon membrane prepared at different stabilization temperatures.

3.3 Gas permeation results

The gas permeation properties of the prepared tubular carbon membranes were measured using the gas permeation test apparatus at room temperature. The permeance of three pure gases with different molecular sizes, H_2 (0.289 nm), N_2 (0.364 nm) and He (0.260nm) over the resulted tubular carbon membranes prepared at different stabilization temperatures were shown in Table 2. As demonstrated in table below, the stabilization temperature during pyrolysis process is principally affect the performance of the consequent tubular carbon membrane. The results indicated that carbon membrane with stabilization temperature of 300°C showed superior selectivity compared to the others stabilized temperature. This was due to its pore system contains of wide-ranging openings with narrow constriction [28]. The partition is created by the passage of smaller gas atoms (He and H_2) over the pores while the bigger gas particles are stuck, because of the transport mechanism. Additionally, tubular carbon membranes offer an outstanding performance due to the prearrangement of pore structure by cross-linking and volatilization of minor vaporous particles delivered by thermal degradation during stabilization process. Based on previous study, the maximum ideal selectivity values obtained for He/N_2 and H_2/N_2 gas couples were up to 350 and 600, respectively [12].

Table 2. Gas permeation properties of P84 co-polyimide-based tubular polymeric and carbon membranes.

Membrane	Permeability			Selectivity	
	N_2	H_2	He	H_2/N_2	He/N_2
Polymeric	0.67±2.81	32.67±2.94	38.30±4.22	48.76±2.76	57.17±1.19
CM-250	2.65±2.43	975.81±1.28	1027.33±2.41	368.23±2.28	387.67±3.11
CM-300	2.98±1.28	1134.51±2.87	1287.22±2.86	380.71±2.34	431.95±2.61
CM-350	2.31±1.54	772.81±4.32	802.44±5.11	334.55±2.39	347.38±3.12
CM-400	2.26±3.22	742.55±2.83	787.21±2.19	328.56±4.51	348.32±2.19
CM-450	2.11±3.29	682.87±1.98	726.77±4.31	323.64±1.65	344.13±7.66

Furthermore, the degree of deterioration and cross-linking in the membrane was increased as the stabilization temperature was increased. Therefore, it will result in changed micropore structures of resultant carbon membranes; as noticeable via the SEM images. It is suggested that the gas permeation properties of the carbon membrane could be controlled by the stabilization conditions. Previous researchers have agreed that the microstructure of carbon membranes (pore size, pore volume and pore distribution) could be tailored by monitoring the heat treatment process.

4. Conclusion

The gas permeation properties of the P84 co-polyimide-based tubular carbon membranes were obviously improved in magnitude compared to the polymeric precursor membranes. This was due to the formation of pore structures by cross-linking and volatilization of small gaseous molecules produced by minor thermal degradation during stabilization process. Stabilization temperatures have protrusive effect on the membrane properties specifically their pore structure, and eventually their gas separation properties. Tubular carbon membrane that stabilized at 300°C possessed superior gas separation for H_2/N_2 and He/N_2 gas pairs of 380.71±2.34 and 431.95±2.61, respectively. The overall results demonstrated that the P84 co-polyimide-based carbon membrane is a promising material for H_2/N_2 and He/N_2 separation application.

Acknowledgments

The authors gratefully acknowledge the financial support from the Ministry of Higher Education and Universiti Teknologi Malaysia under Higher Institution Centre of Excellence Scheme (Project Number: R.J090301.7846.4J188), Fundamental Research Grant Scheme (Project Number: R.J130000.7846.4F846) and Research University Grant Scheme (Project Number: Q.J130000.2546.03G69). The authors would also like to acknowledge technical and management support from Research Management Centre (RMC), Universiti Teknologi Malaysia.

References

[1] Haider S 2018 *Sep. Purif. Technol.* **190** 177-189
[2] Levin D B, Pitt L and Love M 2004 *Int. J. of Hyd. Energy* **29** 173-18
[3] Guo Y, Jin Y, Wu H, Zhou L, Chen Q, Zhang X and Li X 2014 *Int. J. of Hyd. Energy* **39** 21044-21052
[4] Bakonyi P, Nemestóthy N and Bélafi-Bakó K 2013 *Int. J. of Hyd. Energy* **38** 9673-9687
[5] Li P, Wang Z, Qiao Z, Liu Y, Cao X, Li W, Wang J and Wang S 2015 *J. Membr. Sci.* **495** 130-168

[6] Mohamad I N, Rohani R Mastar@Masdar, M S Mohd Nor and M T Md Jahim 2016 *Int. J. of Hyd. Energy* **41** 4474-4488
[7] Zhuang G L, Tseng H-H and Wey M Y 2014 *Int. J. of Hyd. Energy* **39** 17178-17190
[8] Bakonyi P 2013 *Int. J. of Hyd. Energy* **38** 15092-15099
[9] Bakonyi P, Bogdán F, Kocsi V, Nemestóthy N, Bélafi-Bakó K and Buitrón G 2016 *Sep. Purif. Technol* **157** 222-228
[10] J.A. Lie, X. He, I. Kumakiri, H. Kita and M.-B. Hägg 2017- Books 405–431
[11] Scholes C A and U Ghosh, 2016 *J. Membr. Sci.* **520** 221-230
[12] Scholes C A, G W Stevens and S E Kentish 2012 *Fuel* **96** 15-28
[13] Favvas E P 2015 *Sep. Purif. Technol.* **142**176-181
[14] Favvas E P 2014 *Sep. Purif. Technol.* 2014 **122** 262-269
[15] Ye P *2016 AIChE Journal* **62** 2833-2842
[16] He X and M-B Hägg 2012 *J. Membr. Sci.* **390-391** 23-31
[17] Lee H-J 2007 *J. Membr. Sci.*, **296** 139-146
[18] Fu Y-J 2011 *Micropor Mesopor Mat,* **143** 78-86
[19] Cipriani E 2016 *Polym. Degrad. Stab.* **123** 178-188
[20] Barbosa-Coutinho E, V M M Salim and C Piacsek Borges 2003 *Carbon* **41** 1707-1714
[21] Hameed N 2016 *Polym. Degrad. Stab.* **128** 39-45
[22] Xiao Y 2010 *Carbon,***48** 408-416
[23] Sazali N 2015 *Int. J. of Hyd. Energy* **3** 167-171
[23] Fan H 2014 *J Energy Chem* **23** 684-693
[25] Lee P-S 2016 *Micropor Mesopor Mat* **224** 332-338
[26] Liu Q 2006 *Chem. Commun.* **11** 1230-1232
[27] Ismail N H, Salleh W N W, Sazali N and Ismail A F 2017 *Sep. Sci.Technol.* 1-13
[28] Ismail N H, Salleh W N W, Sazali N and Ismail A F 2017 J. *Ind. Eng. Chem.* **45** 145-149
[29] Sazali N, Salleh, W N W Ismail and A. F. 2017 *Int. J. of Hyd. Energy* **42** 9952-9957
[30] Zhang K, J D Way, 2011, *J. Membr. Sci.*, **369** 243-249
[31] Salleh W N W and A F Ismail 2013 *Sep. Sci. Technol.* **48** 1030-1039
[32] Hosseini S S and T S Chung 2009 *J. Membr. Sci.* **328** 174-185

Isolation and identification of soil fungi isolates from forest soil for flooded soil recovery

Nor Hazwani Aziz and Norazwina Zainol

Faculty of Chemical and Natural Resources Engineering, Universiti Malaysia Pahang, Lebuhraya Tun Razak, 26300 Gambang, Kuantan, Pahang, Malaysia

E-mail: norhazwani.aziz@gmail.com

Abstract. Soil fungi have been evaluated for their ability in increasing and recovering nitrogen, phosphorus and potassium content in flooded soil and in promoting the growth of the host plant. Host plant was cultivated in a mixture of fertile forest soil (nutrient-rich soil) and simulated flooded soil (nutrient-poor soil) in an optimized soil condition for two weeks. The soil sample was harvested every day until two weeks of planting and was tested for nitrogen, phosphorus and potassium concentration. Soil fungi were isolated by using dilution plating technique and was identified by Biolog's Microbial Systems. The concentration of nitrogen, phosphorus, and potassium was found to be increasing after two weeks by two to three times approximately from the initial concentration recorded. Two fungi species were identified with probability more than 90% namely *Aspergillus aculeatus* and *Paecilomyces lilacinus*. Both identified fungi were found to be beneficial in enhancing plant growth and increasing the availability of nutrient content in the soil and thus recovering the nutrient content in the flooded soil.

1. Introduction

Flooding is a natural disturbance which occurs almost everywhere worldwide. Flooding disturbs crop and forages production globally due to the detrimental effects that it provokes on most terrestrial plants [1]. The frequency of flooding events was predicted to be increasing globally by climate change models, making flooding a major environmental threat for plants [2]. Crop damages due to unseasonal and severe flooding events amount to a huge extent in yield losses. Despite the variety of plant tolerant to flooding, there is significant variation in the vulnerability of most crops to wet conditions. Flooding can damage plants by several means. Excessive moisture in soil decreases oxygen levels as plant growth under flooding condition is trying to sustain aerobic respiration of submerged [3]. The inhibition of respiration leads to the build-up of toxic compounds which ultimately suffocate and damage plant [4].

Flooding may also affect nutrient content in the soil. Plants will suffer from nutrient insufficiencies as a result of reduced uptake of nutrients, denitrification, and leaching of mobile nutrients and also ions dilutions in flooded soil [5]. Prolonged waterlogging decreases the concentration of nitrogen, phosphorus and potassium in the soil. Nitrogen appears to have been denitrified and lost from the system [6]. Since phosphorus uptake depends on microbes in most plant, prolonged waterlogging has reduced microbial activity which in turn affects the absorption of phosphorus into the plant roots [7]. Furthermore, soil potassium is less available in soils that remain wet since wet soils are more prone to compaction, which restricts plant root growth and uptake of soil potassium [8].

Application of fungi in increasing soil nutrient content has been widely discussed by many authors. Fungi strongly influence ecosystem structure and functioning and therefore plays a significant role in many ecological services [9]. Soil fungi perform crucial roles within the soil with respect to the cycling of nutrients, disease suppression and water dynamics where all of which help plants become healthier and more vigorous [10]. The use of fungi in recovering nutrient content is currently a favorable method for many agricultural practices over the chemical fertilizers. Since the use of chemical fertilizers and pesticides has created the problem of environmental pollution [11], the use of fungi in agricultural practices seems to be a promising technique to farmers as this technique are an environmentally friendly and are cost-effective. In addition, over application of chemical fertilizer resulting in reduced colonization of plant roots with symbiotic fungi thus reducing the beneficial effect of soil fungi.

In recent years, application of microorganism as biofertilizer has arisen as a significant component of the integrated nutrient supply system and hold a pronounced potential to recover crop productions through environmentally improved nutrient supplies [12]. However, the application in practice, in some way, has not accomplished persistent outcome. The mechanisms and interactions among these microbes still are not well understood, especially in real applications. Hence this study aims to isolate and identify the soil fungi and to evaluate the effect of inoculating fertile soil containing naturally occurring soil fungi and infertile flooded soil with the aim to naturally recover the nutrient in that particular soil.

2. Materials and Method

2.1. Soil collection and preparation

A pot experiment was carried out to determine the effect of naturally occurring soil fungi which were commonly found in the fertile soil when inoculated with infertile flooded soil in recovering the nutrient in the soil. The fertile soil was collected at a secondary forest zone located in Universiti Malaysia Pahang (UMP), Kuantan while the simulated flooded soil was prepared under greenhouse by submerging the collected fertile soil in the water for two weeks [13]. The water was then removed from the soil. The ability of soil fungi in recovering the nutrient in flooded soil was evaluated by the help of onion as a host plant and by analyzing the nutrient content in the soil every day until two weeks of planting.

Five factors were previously selected in a preliminary study namely pH of the soil, soil water content, ratio of fertile to flooded soil, light supply and depth of soil in order to determine the most contributing factor for flooded soil recovery by using the concept of two levels factorial via Design Expert software (Ver 71.6). The analyzed data showed that water content, light supply and depth of soil were the important contributing factors for nutrient recovery in flooded soil. The best condition for flooded soil recovery was determined at pH4, with 28mL water content, 1:5 ratio of fertile forest soil to simulated flooded soil, depth of soil at 5 cm and with the presence of light.

The onion was planted in an optimized soil condition in five pot of 650 g mixed soil with 1:5 ratio of fertile forest soil to the simulated flooded soil each with 28 mL water content, at pH4 and under the presence of light. Onion bulb was planted in 5 cm deep. The soil was harvested for fungi isolation and nutrient testing every day until two weeks of planting and the initial concentration of each nutrient was recorded prior to experimentation.

2.2 Soil sample analysis

The soil was tested for nitrogen, phosphorus and potassium content by using Hach Spectrophotometer. The soil sample was crushed by mortar and pestle first to get the fine structure of the soil. The soil then was diluted with distilled water by two dilution factor for nitrogen and phosphorus testing and three dilution factor for potassium testing. The diluted soil then was filtered by Whatman filter paper to get the clear soil solution before been tested for the presence of nitrogen, phosphorus and potassium. Both nitrogen and phosphorus samples were digested first with DRB 200 reactor before been tested with Hach. Nitrogen was analysed by using persulfate digestion method whereas phosphorus was tested by molybdovanadate method. Meanwhile, potassium was analysed by using tetraphenylborate method. The procedures for each method were referred from DR2800 Hach Spectrophotometer manual.

2.3 Isolation of fungi

The isolation of fungi from soil was carried out in order to identify each fungus up to species level. The serial dilution plating method was used to dilute the soil sample as described by [14] with the purpose of minimizing the fungi in the soil in each dilution. The dilution of soil sample was conducted in two replicate where each replicate was diluted six times and labelled as 10-1 until 10-6. 50 g of soil was added in 100 mL 85% NaCl solution and was thoroughly shake to mix the solution. The solution then was diluted to a series of prepared vials containing 9 mL 85% NaCl solution. 9 mL of the soil-NaCl solution was transferred to the first vial by using a pipette. Subsequently, another 9 mL of the solution from the first vials was transferred to the second vial and the steps continued until the last vial. 0.1 mL of the solution in each vial was pipetted into the prepared PDA plate. The solution was then spread on the plate by using a hockey stick and incubated at room temperature for seven days.

The colony of fungi that appeared in the plate after incubation was then isolated in a new plate. A pure culture of each colony type on each plate was obtained and maintained. The maintenance was done by sub-culturing each of the different colonies onto another new PDA plate. Each colony were cut into 3 mm pieces with a sterilized razor blade, surface sterilized in 70 % ethanol. The colonies were incubated again at room temperature for five days. Each plate of single fungi colony was then sent to UMP Central Laboratory for identification purposes.

2.4 Identification of fungi

Ten pure sample culture was sent to UMP Central Laboratory for identification purpose. The identification was done by Biolog Microbial Identification System as described by Dobranic and Zak [15]. Biolog Microplate tests the ability of a microorganism to assimilate or oxidize compounds from a preselected panel of different carbon sources. The test yields a characteristics pattern of reddish-orange wells and turbidity changes, which constitutes a Metabolic Fingerprint. The Microplate test panel provides a standardized micro method using 95 biochemical tests to identify a broad range of fungi including both filamentous and yeast form.

The identification result by Biolog's Microbial was performed by reading and comparing between plate's phenotypic fingerprint and Biolog's Microbial Identification Systems Database. The result will show no ID if the plate's phenotypic fingerprint produced not matched to the database more than 50%. The no ID result may be due to the isolated fungi was possibly a new species as the profile cannot be found in the database, thus a larger database is needed for further identification by using the molecular technique.

3. Results and discussion

3.1 Nutrient concentration

The average concentration of nitrogen, phosphorus and potassium from the harvested soil sample in five pots was 10.7, 6.5 and 15.2 mg/L respectively. The concentration of all nutrients increased approximately by two to three times from the recorded initial nutrient concentration. The concentration of nutrients shows an increasing trend from day zero until day fourteen. Figure 1 demonstrates the trends of all nutrient from day zero until day fourteen. This finding suggests that application of soil fungi to recover the infertile flooded soil can be implemented for soil nutrient recovery.

A similar observation has also been recorded by Wu et al. [16], who studied the effects of microorganisms containing nitrogen-fixer, phosphorus and potassium solubilizers and arbuscular mycorrhizal fungi on the growth of maize. Inoculation of soil with microorganisms resulted in an increase in phosphorus and potassium uptake to different degrees when compared with the control. Higher nutrient concentration was recorded in pots inoculated with the combination of nitrogen-fixer, phosphorus and potassium solubilizers and arbuscular mycorrhizal fungi.

Meanwhile, according to Rashid et al. [17],co-inoculation of bacteria and fungi with or without organic fertilizer are the most beneficial for restoring the soil fertility and organic matter content rather than single inoculum. Bacterial and fungal inocula have a potential to restore the fertility of degraded land through various processes. These microorganisms increase the nutrient bioavailability through nitrogen fixation and mobilization of key nutrients (phosphorus, potassium and iron) to the crop plants while remediating soil structure by improving its aggregation and stability.

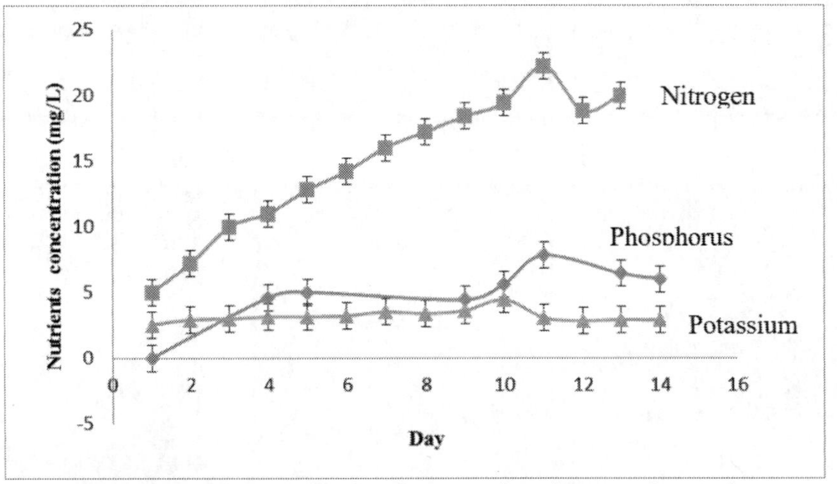

Figure 1. Nutrient profile for nitrogen, phosphorus and potassium

3.2 Identification of fungi

Every single colony of different shapes and colours appeared on the PDA plate after the first incubation was subcultured and maintained on another new PDA plate. Nine single colonies were successfully isolated on a new plate and were presented in Figure 2. The plate was labelled as Colony 1 until Colony 9. Of nine sample plate sent for identification, only five plates have been identified with probability more than 90%. Meanwhile, another five plate shows no ID result. The isolates were identified by Biolog's Microbial identification up until genus level. Only two species were identified, namely *Paecilomyces lilacinus* and *Aspergillus aculeatus*. The result shows that *Aspergillus aculeatus* was the dominant species found in the soil sample.

IOP Conf. Series: Materials Science and Engineering **342** (2018) 012028 doi:10.1088/1757-899X/342/1/012028

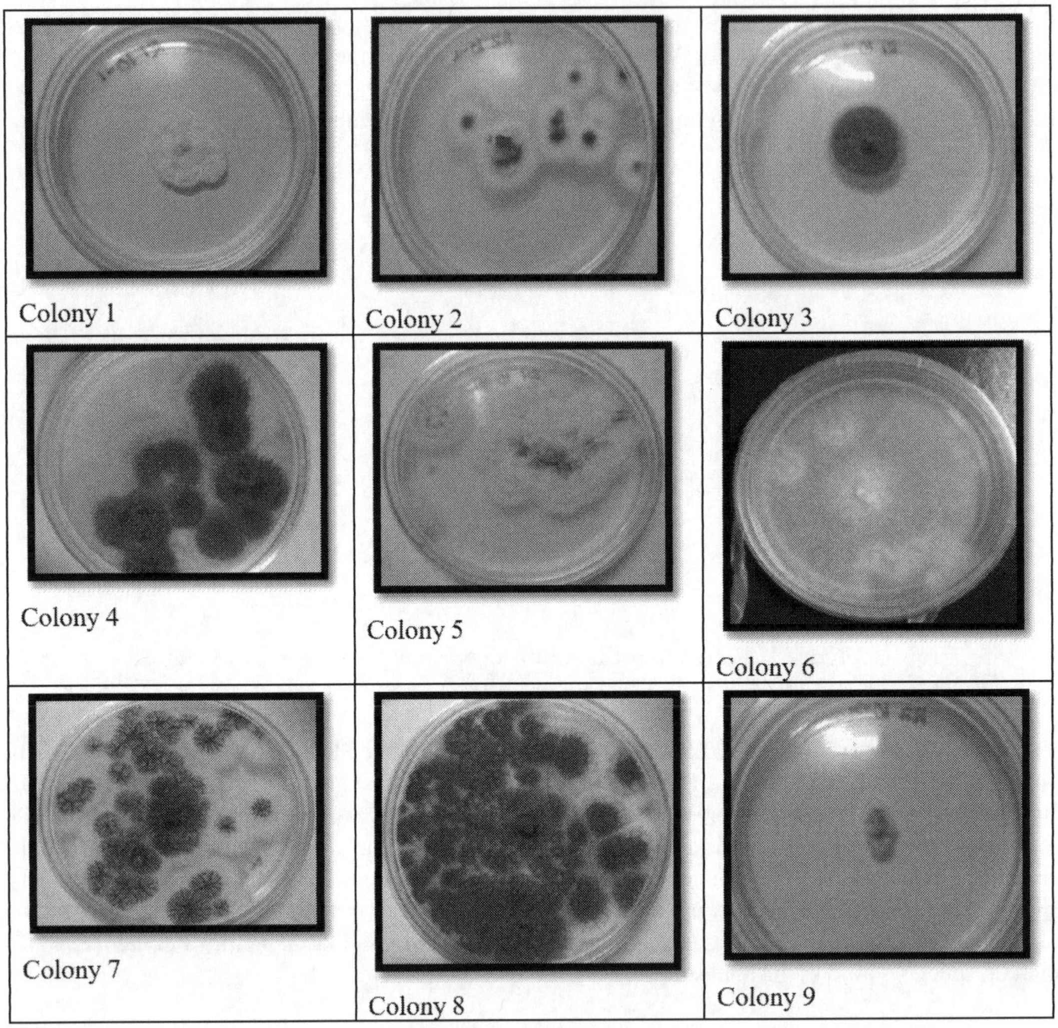

Figure 2. Single isolated colony of fungi in PDA plate.

A. aculeatus was classified under the phylum Ascomycota, Trichomaceae family and under the genus of *Aspergillus*. *A. aculeatus* is a ubiquitous species that is commonly isolated from soil and rotting fruit. Due to its ability to degrade plant cell walls rapidly, *A. aculeatus* isolates have been used to produce a number of important industrial enzymes such as cellulases, hemicellulases, and proteases that are used commercially in the food and feed industries. The biochemical and catalytic properties of several extracellular glycosyl hydrolases from *A. aculeatus* have been studied in detail by numerous researchers. In addition, several *A. aculeatus* polysaccharide degrading enzymes have been studied at the structural level by X-ray crystallography.

Paecilomyces lilacinus was categorized under the phylum Ascomycota, Trichomaceae family and under the genus of *Paecilomyces*. *P. lilacinus* was originally described by the taxonomist in the early 19th under the name *Penicillium lilacinum*. Another taxonomist then transferred the species to *Paecilomyces* in 1974. Until now *P. lilacinus* was declared as synonyms to *P. lilacinum*. *P. lilacinus* is a saprophytic soil fungus and can be found in a wide range of habitats. It has a high frequency of occurrence in the tropics and sub tropic and can be found in most of the agricultural soils.

P. lilacinus has drawn many research attentions due to its promising effect in parasitizing and controlling the population of nematodes. *P. lilacinus* is one the most widely tested biological control agent for management of plant-parasitic nematodes. The use of *P. lilacinus* as a biological control agent of plant-parasitic nematodes has been discussed and described by many authors. Those include as early as reported by Dube and Smart Jr [18], Kiewnick and Sikora [19] to the latest as described by Wagh and Pramanik [20]. All researchers showed successful nematode control by *P. lilacinus* and thus can act as potential bio control agent of a nematode.

Phosphorus (P) and potassium (K) were both found abundant in soil, however, they exist in the form that is unavailable for the plant to absorb. Plants can only absorb inorganic phosphorus [21], thus the mineralization of organic phosphorus is an important soil process since it results in the release of inorganic phosphorus to the soil solution. Mineralization is facilitated by soil microorganisms, and almost half of the microorganisms present in soil and on plant roots possess the ability to mineralize organic phosphate [22]. Many soil fungi, predominantly of the genera *Aspergillus* and *Penicillium*, have been shown to possess the ability to bring insoluble phosphates in soil into a soluble form by secreting organic acids [23]. These acids lower the pH and bring about the dissolution of bound forms of phosphates. Growth promotion and increased uptake of phosphorus by plants inoculated with P-solubilizing fungi have been reported by many investigators.

Chhonkar and Subba Rao [24] evaluated the phosphate-solubilizing ability of different isolates of fungi associated with legume root nodules under in vitro conditions. Among the fungi tested, isolates of *Penicillium lilacinum*, *Aspergillus* sp., *A flavus*, *A. niger*, *A. terreus* and *A. nidulans* solubilized insoluble tricalcium phosphate. Several other studies have described the ability of *A. aculeatus* in solubilizing phosphate and potassium in the soil. Narshian and Patel [25] used *A. aculeatus* isolates to test the availability of named fungi in solubilizing rock phosphate and found out that *A. aculeatus* was capable to solubilize all natural form of phosphorus tested. The study of mineralization of organic phosphates via *A. aculeatus* by Narshian and Patel [26] resulting in high acid phosphatase activity in Pikovskaya's broth which indicates that high percentage of phosphorus being released. In addition, *A. niger* has been formerly studied by Cerezine et al. [27] followed by Caravaca et al. [28] as rock phosphate solubilizer. Kang et al. [29] reported that a significant amount of insoluble rock phosphate was solubilized by *Aspergillus* sp as evident by increased soluble phosphorus concentration. This followed by Singh et al. [30] who also tested and prove the ability of two *A. niger* strains as good phosphate solubilizer. Microbial rock potassium solubilization was first discovered by Muentz [31]. Prajapati [32-33] characterized both *A.niger* and *A. terreus* as potassium solubilizing fungi. Inoculation of the isolates with soil treated with insoluble potassium showed a significant increase in the concentration of potassium.

Penicillium pinophilum have been used by Fan et al. [34] in evaluating the ability of isolates to improve the growth of the strawberry plant. *P. pinophilum* formed arbuscular mycorrhizae with the roots of strawberry and the interaction not only improve the plant growth but also enhance the nutrient uptake as well as the rate of photosynthesis of the plant. The inoculation has increased the nitrogen and phosphorus content and also shorten the blossom and ripening period of strawberry.

Nitrogen availability in almost all ecosystems, including agricultural conditions, is a limiting factor that limits primary productivity and nitrogen cycling [35]. A plant needs a very large amount of nitrogen for photosynthesis hence the rate of photosynthesis can be directly linked to nitrogen availability [36]. Plants can absorb nitrogen in the form of inorganic nitrogenous compounds. Fixed nitrogen from the atmosphere in terrestrial systems may be bound in organic matter and must be subsequently mineralized to releases inorganic nitrogen compound that can be used by plants [37]. Nitrogen conversion is largely mediated by fungal decomposer in the soil [38].

Several soil fungi are able to exploit the energy stored in the redox potential of organic nitrogen [39]. The exo enzymes for example proteases and peptidase released by some fungi can break down organic matter and subsequently capture nitrogen-containing compounds, thus providing a direct path from organically bound nitrogen in the soil to the plant [40]. Protease has been reported to be found in several *Aspergillus* species. Olutiola and Nwaogwugwu [41] reported that *A. aculeatus* is able to produce protease with optimum activity expressed at pH 7. Several other *Aspergillus* species that can produce protease are as described by Ruann Janser and Harumi Sato [42] in *A. oryzae* and by Farnell et al. [43] in *A. fumigatus*.

4. Conclusions

Depending on the type of interaction between fungi and plant, fungi have been proven to have a beneficial effect to plant. Inoculating fertile soil containing microorganisms with flooded soil has significantly increased the nutrient content in the soil. The current study shows an increase of nutrient concentration by two to three times from recorded initial nutrient concentration which suggests that applying fungi in restoring infertile flooded soil can be employed successfully. The application of fungi in agricultural practices should be implemented and replaced the conventional chemical fertilization method. *A. aculeatus* was found to be beneficial in increasing the bioavailability of phosphorus content in the soil by its potential in the solubilizing insoluble form of phosphorus to a form that plant can uptake. Meanwhile, *P.lilacinus* was found to be a very good biological agent for controlling nematodes.

Acknowledgments

The author wishes to acknowledge the Universiti Malaysia Pahang for funding the project under grant RDU170346.

References

[1] Colmer T D and Vosenek L A C J 2009 Flooding tolerance: suites of plant traits in variable environments *Functional Plant Biology* **36** 665-681.

[2] Arnell N and Liu C 2001 *Climatic Change 2001*: hydrology and water resources. Report from the Intergovernmental Panel on Climate Change. Available at http://www.ipcc.ch/

[3] Vartapetian B and Jackson M 1997 Plant adaptation to anaerobic stress *Annals of Botany* **3** 3-20

[4] Gustavo G S 2012 Flooding stress on plants: anatomical, morphological and physiological responses. In *Botany*, Mworia, J. (eds), InTech

[5] Drew M C 1991 Oxygen deficiency in the root environment and plant mineral nutrition *Plant Life under Oxygen Deprivation* ed MB Jackson, DD Davies and H Lambers (The Hague: Academic Publishing) pp 301-316.

[6] Kreuzwieser J Fu¨rniss S and Rennenberg H 2002 The effects of flooding on the N metabolism of flood tolerant and sensi-tive tree species.*Plant, Cell and Environment* **25** 1039-1050.

[7] Ellis J R 1998 Post flood syndrome and vesicular-arbuscular mycorrhizal fungi *Journal of Production Agriculture* **11** 200-204.

[8] Lipiec J and Stepniewski W 1995 Effects of soil compaction and tillage systems on uptake and losses of nutrients *Soil Tillage Research* **35** 37–52.

[9] Orgiazzi A, Lumini E, Nilsson R H, Girlanda M and Vizzini A 2012 Unravelling soil fungal communities from different Mediterranean land-use backgrounds *PLoS One* **7** (4): e34847.

[10] Jenkins A 2005 Soil fungi. In: *Soil biology basics, information series, NSW Department of Primary Industries*. Url: http://www.dpi.nsw.gov.au/__data/assets/pdf_file/0020/41645/Soil_fungi.pdf.

[11] Suhag M 2016 Potential of biofertilizers to replace chemical fertilizers *International Advanced Research Journal in Science, Engineering and Technology* **3**: 163-167.

[12] Berruti A, Lumini E, Balestrini R and Bianciotto V 2016 Arbuscular mycorrhizal fungi as natural biofertilizers: Let's benefit from past successes *Frontiers in Microbiology* **6** 1559.

[13] Unger I M, Kennedy A C and Muzika R M 2009 Flooding effects on soil microbial communities. *Applied Soil Ecology* **42** 1-8.

[14] Waksman S A 1922 A method for counting the number of fungi in the soil *Journal of Bacteriology* **7** 339-341.

[15] Dobranic J K and Zak J C 1999 A microtiter plate procedure for evaluating fungal functional diversity *Mycologia* **91** 756-765.

[16] Wu S C, Ca Z H, Li Z G, Cheung K C and Wong M H 2005 Effects of biofertilizer containing N-fixer, P and K solubilizers and AM fungi on maize growth: a greenhouse trial *Geoderma* **125** 155-166.
[17] Rashid M I, Mujawar L H, Shahzad T, Almeelbi T and Ismail I M 2016 Bacteria and fungi can contribute to nutrients bioavailability and aggregate formation in degraded soils *Microbiological Research* **183** 26-41.
[18] Dube B and Smart Jr G C 1987 Biological Control of *Meloidogyne incognita* by *Paecilomyces lilacinus* and *Pasteuria penetrans Journal of Nematology* **19** (2) 222-227.
[19] Kiewnick S and Sikora R A 2006 Biological control of the root-knot nematode *Meloidogyne incognita* by *Paecilomyces lilacinus* strain 251 *Biological Control* **38** (2) 179-187.
[20] Wagh V H and Pramanik A 2014 Evaluation of *Paecilomyces lilacinus* for management of root knot nematode *Meloidogyne Incognita*, in tomato *Supplement on Plant Pathology* **9** (1) 407-411.
[21] Rendig V V and Taylor H M 1989 *Principles of soil-plant interrelationships* (New York: McGraw-Hill Publishing Company).
[22] Cosgrove D J 1967 Metabolism of organic phosphates in soil *Soil biochemistry* (vol 1) ed A D McLaren G H Peterson (New York: Marcel Dekker) pp 216-228.
[23] Whitelaw M A 1999 Growth promotions of plants inoculated with phosphate-solubilizing fungi *Advances in Agronomy* **69** 99-151.
[24] Chhonkar P K and Subba Rao NS 1967 Phosphate solubilization by fungi associated with legume root nodules *Canadian Journal of Microbiology* **13** 749-753.
[25] Narsian V and Patel H H 2000 *Aspergillus aculeatus* as a rock phosphate solubilizer *Soil Biology and Biochemistry* **32** (4) 559-565.
[26] Narsian V and Patel H H 2002 *Aspergillus aculeatus* as an organic phosphate mineralizer. *Indian Journal of Agricultural Sciences* **72** (3) 177-179.
[27] Cerezine P C, Nahas E and Banzatto D A 1988 Soluble phosphate accumulation by *Aspergillus niger* from fluorapatite *Applied Microbiology and Biotechnology* **29** 501-505.
[28] Caravaca F, Alguacil M M, Azcon R, Parlade J, Torres P and Roldan A 2005 Establishment of two ectomycorrhizal shrub species in a semiarid site after in situ amendment with sugar beet, rock phosphate and *Aspergillus niger Microbial Ecology* **49** 73-82.
[29] Kang S C, Piyush Pandey, Rajat Khillon and Maheshwari D K 2008 Process of rock phosphate solubilization by *Aspergillus* sp PS 104 in soil amended medium *Journal of Environmental Biology* **29** (5) 743-746.
[30] Singh S M, Yadav L, Singh S K, Singh P, Singh P N and Ravindra R 2011 Phosphate solubilizing ability of Arctic *Aspergillus niger* strains. *Polar Research* **30** 7283.
[31] Muentz 1890 Surla décomposition desroches etla formation de la terrearable *CR Academic Science* **110** 1370-1372.
[32] Prajapati K B and Modi H A 2012 Isolation and characterization of potassium solubilizing bacteria from ceramic industry soil *CIBTech Journal of Microbiology* **1** 8-14.
[33] Prajapati K B, Sharma M C and Modi H A 2013 Growth promoting effect of potassium solubilizing microorganisms on *Abelmoscus esculantus International Journal of Agricultural Science* **3** (1) 181-188.
[34] Fan Y, Luan Y and Yu K 2008 Arbuscular mycorrhizae formed by *Penicillium pinophilum* improve the growth, nutrient uptake and photosynthesis of strawberry with two inoculum-types *Biotechnology Letters* **30** (8) 1489-1494.
[35] Boring L R, Swank W T, Waide J B and Henderson G S 1988 Sources, fates, and impacts of nitrogen inputs to terrestrial ecosystems: review and synthesis *Biogeochemistry* **6** 119–159.
[36] Yoneyama K, Xie X, Kusumoto D, Sekimoto H, Sugimoto Y and Takeuchi Y 2007 Nitrogen deficiency as well as phosphorus deficiency in sorghum promotes the production and exudation of 5-deoxystrigol, the host recognition signal for arbuscular mycorrhizal fungi and root parasites *Planta* **227** 125-132.

[37] Van Der Heijden M G A, Bardgett R D and Van Straalen N M 2008 The unseen majority: soil microbes as drivers of plant diversity and productivity in terrestrial ecosystems *Ecology Letters* **11** 296-310.

[38] Hodge A and Fitter A H 2010 Substantial nitrogen acquisition by arbuscular mycorrhizal fungi from organic material has implications for N cycling *Proceedings of the National Academy of Sciences U.S.A.* **107** 13754–13759.

[39] Sinsabaugh R L 2010 Phenol oxidase, peroxidase and organic matter dynamics of soil *Soil Biology and Biochemistry* **42** 391-404.

[40] Schimel J P and Bennett J 2004 Nitrogen mineralization: challenges of a changing paradigm *Ecology* **85** 591-602.

[41] Olutiola P O and Nwaogwugwu R I 1982 Growth, sporulation and production of maltase and proteolytic enzymes in *Aspergillus aculeatus Transactions of the British Mycological Society* **78** (1) 105-113.

[42] Ruann Janser Soares de Castro and Helia Harumi Sato 2014 Protease from *Aspergillus oryzae*: biochemical characterization and application as a potential biocatalyst for production of protein hydrolysates with antioxidant activities *Journal of Food Processing*.

[43] Farnell E, Rousseau K, Thornton D J, Bowyer P and Herrick S E 2012 Expression and secretion of *Aspergillus fumigatus* proteases are regulated in response to different protein substrates *Fungal Biology* **116** (9) 1003-1012.

IOP Conf. Series: Materials Science and Engineering **342** (2018) 012029 doi:10.1088/1757-899X/342/1/012029

Filament winding technique, experiment and simulation analysis on tubular structure

Ma Quanjin[1,2*], M R M Rejab[1,2], Jiang Kaige[3], M S Idris[1] and M N Harith[1]

[1]Structural Materials & Degradation Focus Group, Faculty of Mechanical Engineering, Universiti Malaysia Pahang, 26600 Pekan, Pahang, Malaysia
[2]School of Mechanical Engineering, Ningxia University, Yinchuan,750021,China
[3]Material R&D Supervisor Engineering, Zoomlion Co., Ltd, Hunan Province, China.

E-mail: neromaquanjin@gmail.com/maquanjin123@sina.com

Abstract. Filament winding process has emerged as one of the potential composite fabrication processes with lower costs. Filament wound products involve classic axisymmetric parts (pipes, rings, driveshafts, high-pressure vessels and storage tanks), non-axisymmetric parts (prismatic nonround sections and pipe fittings). Based on the 3-axis filament winding machine has been designed with the inexpensive control system, it is completely necessary to make a relative comparison between experiment and simulation on tubular structure. In this technical paper, the aim of this paper is to perform a dry winding experiment using the 3-axis filament winding machine and simulate winding process on the tubular structure using CADWIND software with 30°,45°,60° winding angle. The main result indicates that the 3-axis filament winding machine can produce tubular structure with high winding pattern performance with different winding angle. This developed 3-axis winding machine still has weakness compared to CAWIND software simulation results with high axes winding machine about winding pattern, turnaround impact, process error, thickness, friction impact etc. In conclusion, the 3-axis filament winding machine improvements and recommendations come up with its comparison results, which can intuitively understand its limitations and characteristics.

1. Introduction

The composite material has been gradually recognized by the many advanced material industries and researchers as a potential engineering material, which can replace the conventional and general materials in producing light-weight and low-cost products. Filament winding is a mature manufacturing process in the composite material, which can produce a wide range of moderate and high mechanical performance parts. For more than 30 years the primary aim of filament winding process and development, this technique has been evolved into automated advanced composite materials with different structures. The industries and companies are looking for the light-weight, high mechanical performances and cost savings. There are many types and sophisticated development in modern filament winding for the initial process, which collects different type fibres are continuously wound on a rotation supporting form or mandrel with the different winding angle. The first filament winding machine was developed in 1950 in accordance with the revolutions of the filament winding application [1]. Filament wound applications are applied to make tube-shaped or pipe-shaped production such as high-pressure storage tanks, rocket motor cases, and launch tubes. For commercial applications, filament wound products can be used for commercial equipment such as golf club shafts and fishing rods [2].

Filament winding machine also has been developed based on its high mechanical performance products. Filament winding machine structure has been evolved from 2 axes classical lathe-type winder towards high axes winders with an accelerated degree of freedom, which can promote filament winding machine processing capability. With advances in technology, standard industrial robots have been put it into use in the composite fabrication process. Figure 1 shows COMEC company product filament winding robot. It overcomes the filament winding process weakness in a proper way, which lightens the filament winding development orientation.

For the sake of filament winding process development, many researchers have focused on improving filament winding machine axes, mechanical performances, and material properties experiments on filament wound composite products. It can be divided into two parts, which are filament winding machine development and filament wound composite product experiments. F. H. Abdalla et. al have demonstrated that a 2-axis filament winding machine was designed and fabricated as pipes and round shape specimens [3]. M. R. M. Rejab et. al have proposed four axes filament winding machine modification and machine structure [4]. Saad Mutasher et.al performed experimental investigations on small-scale automated filament winding machine controlled by a single PIC 18F452 microcontroller. Winding test of the prototype shows that it is capable of producing winding angles in the range of 40 to 80 degree[5]. The filament winding process has faced limiting fabrication obstacles, which can try to design complex geometries of the composite structure. E. Vargas Rojas et. al have discussed the challenges and validates a generic mathematical model that contributes either to wind complex shapes or solve common filament winding disadvantages on basis of an integrated strategy [6]. However, there are no related studies examined using 3-axis filament winding machine experiment products compared with CADWIND software simulation results, which can have a good understanding and comparison results to improve the tested machine process capability.

Another winding research branch is to perform and analyze the filament wound composite products, which can get high mechanical performances. A. Zuraida et. al have investigated the experimental and finite element analysis for hybrid and non-hybrid glass and carbon/epoxy composite tubes subjected to quasi-static lateral indentation loading. A useful comparison was done for the tested composite tubes between the finite element and the experimental results, which shows that carbon/epoxy tubes resist higher load with less deformation the glass and hybrid glass-carbon tubes [7]. Xiaolong Jia et.al have noted that the cylinder with end reinforcing layer had higher compressive properties and sequentially showed brittle fracturing model, local buckling model and transverse shearing mode with winding angle increasing [8]. Previous research has shown that composite products can have quite potential properties to replace common material in the next time.

Figure 1. Filament winding robot prototype [9].

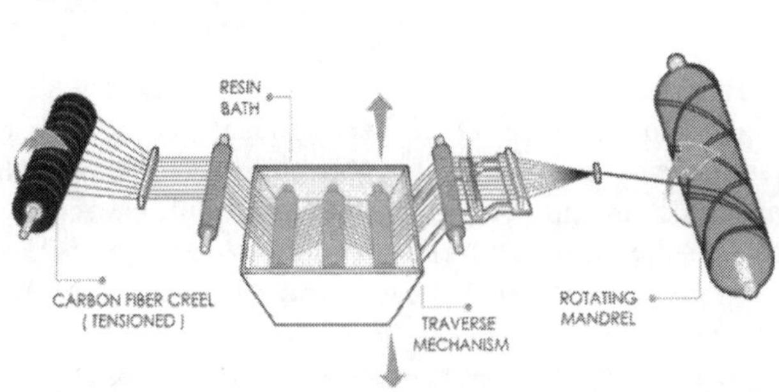

Figure 2. Filament winding process schematic [10].

1.1 Filament winding

Filament winding is a process in which the resin and fibres through the resin bath, which is wound on a different shape of the mandrel with controllable speed. In filament winding method, fibre strands or tows are wound and steadily passed through the resin bath. In the resin bath, fibre strands are impregnated with the polyesters or epoxy. Then related resin impregnated strands are wound on a rotating mandrel in a controlled system with specific fibre orientation, which can implement different winding angle based on customers' requirements.

The schematic of filament winding process is shown in Figure 2. Fibres are collected and put in the order through a comb, and then fibre strands are wetted with the resin in resin bath. Furthermore, fibre with the resin passes through another comb and a pay-out eye device. During the process, traverse mechanism can move forward and backward in the delivery system. In addition, fibres with impregnated resin are wound around a rotating mandrel at various winding angles to satisfy mechanical requirements, such as strength, elasticity, ductility, stiffness and fatigue strength. The tubular structure is then cured at ambient temperature or special temperature and the mandrel is removed. To remove the mandrel from the composite products, hydraulic rams may be used on different filament winding machine. Hollow pipe and other shape parts are fabricated via filament winding technique. Fibre tension force is critical specification in filament winding because compaction is achieved by the fibre tension, which is a necessary parameter should be under control. The fibre tension will affect the properties of the composite products. There are many tension device types such as magnetic or friction brakes, electronic rewind, rotating scissor bars and high-performance solenoids. Filament winding has another main advantage is that a high fibre volume fraction can be achieved in the composite material with filament winding process technique.

1.2 Filament winding pattern generation

Filament winding technique has three winding patterns, which are the hoop, helical and polar winding. Most filament wound products are produced by means of the helical winding pattern. In helical winding method, the fibre tow is wound on the mandrel with a moderate angle to the longitudinal axis. The winding process focuses on a tubular structure specimen, which can summarize as follow. The main aim is to cover a cylindrical mandrel completely with a constant thickness of the fibre tow. At the start of the dry winding process, the fibre tow is deposited at the left side of the mandrel and wound with a constant winding angle to the right side of the mandrel. The first movement path is named a passage, which can be illustrated in Figure 3 (a).

At the end of this passage, the transverse motion of the pay-out eye unit is stopped and prepared to reverse with the fibre tow to change the direction. During the reverse motion, the mandrel can still keep the rotation, which can wind between the end of the first passage and the beginning of the second passage. The fibre tow can be wound toward the left side from the right side of the mandrel. The two passages are called a circuit. It can be demonstrated in Figure 3 (b). At the end of each circuit, it has a constant value to add between next successive circuits, which are plotted in Figure 3 (c). The final winding pattern depends on increasing value, which can directly control the rotation angle between each passage. The rotation angle at each side of the mandrel is a means of controlling filament winding structure based on the filament winding angle. Filament winding pattern can be presented by a regular diamond-shaped cell, which is shown in Figure 3 (d). In the diamond-shaped cell, it is simple to come up with laminated +φ/-φ fibre orientation.

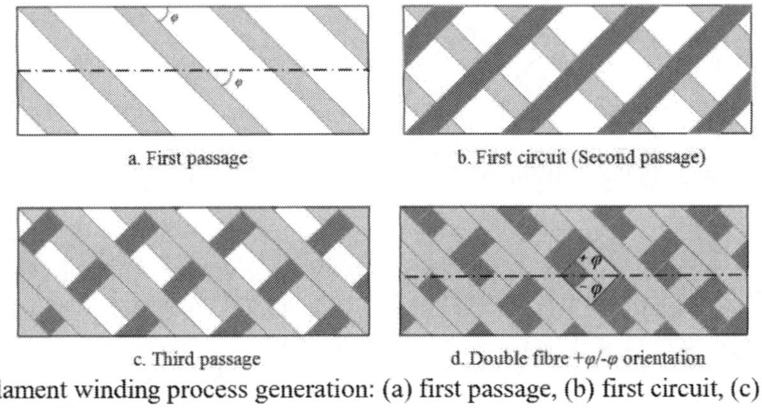

Figure 3. Filament winding process generation: (a) first passage, (b) first circuit, (c) third passage, (d) double fibre +φ/-φ fibre orientation.

IOP Conf. Series: Materials Science and Engineering **342** (2018) 012029 doi:10.1088/1757-899X/342/1/012029

1.3 Filament winding angle and turnaround zone

The fibre is wound and completed one full cycle through entire mandrel length, which one circuit is formed. One layer is the number of patterns required to cover the entire mandrel surface with fibre. The mathematical calculation should formulate and demonstrate winding angle, which can refer to the winding pattern, winding circuit[5]. It has two methods to calculate the winding angle, which is displacement calculation and velocity calculation. Because of 3-axis filament winding machine control system, displacement calculation method has been chosen to implement expected winding angle. The length of the filament wound cylinder is 210mm, $X_1=X_2=X_3=210$mm. Based on the triangle formula, 60° winding angle, 45°winding angle, and 30° winding angle calculation schematics are shown in Figure 4.

$$Y_1 = \tan 30^0 \times X_1 = 121.245mm \tag{1}$$

$$Y_2 = \tan 45^0 \times X_2 = 210mm \tag{2}$$

$$Y_3 = \tan 60^0 \times X_3 = 363.730mm \tag{3}$$

Mandrel shape has chosen a tube, almost winding parameters are fixed by the fibre tow and winding pattern. Their parameters are the tubes radius R=19mm, the winding angle and the bandwidth of the tow b=2mm. With the following equation, it can have access the number of circuits P necessary to obtain complete mandrel coverage with a double layer [11]:

$$P = \frac{2\pi R \cos \varphi}{b} \tag{4}$$

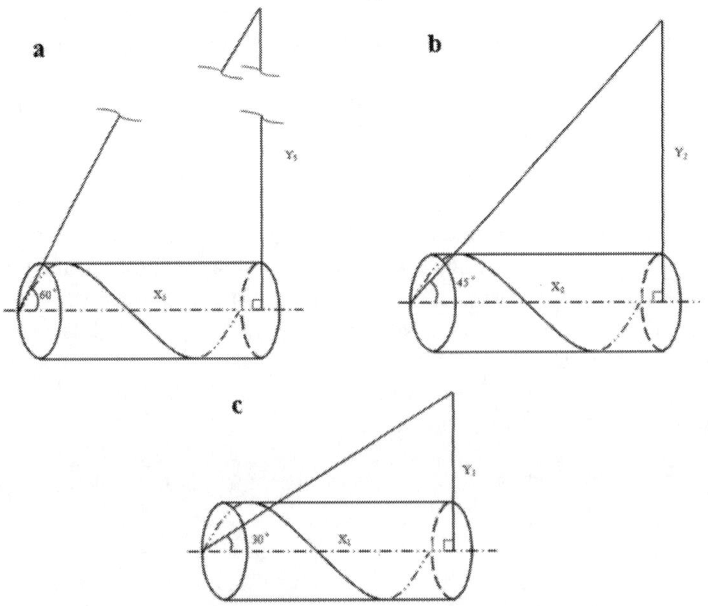

Figure 4. Filament winding angle displacement calculation schematic: (a) 60° winding angle, (b) 45° winding angle, (c) 30° winding angle.

In the above basic equation, it is obvious that the number of a circuit should be an integer. Therefore, the bandwidth value can be adjusted slightly. As an example, the winding angle is calculated using φ=45°. Other filament winding angles can be calculated with the same method, which can only change the winding angle value. Figure 5 is presented filament winding parameter calculation diagram, which can easily handle calculation formulation.

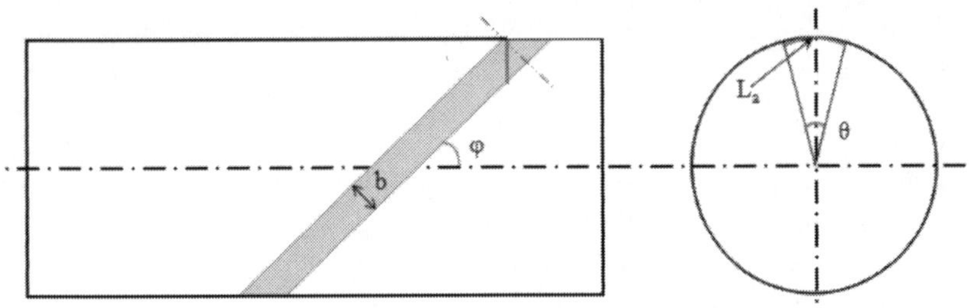

Figure 5. Filament winding parameter calculation schematic.

Mandrel circumference C_m: $$C_m = \pi \times d = 3.14 \times 38 = 119.320mm \quad (5)$$

Curve length L_a: $$L_a = \frac{towwidth}{\cos\varphi} = \frac{2}{\cos 45^0} = 2.828mm \quad (6)$$

Rotation angle θ: $$\theta = \frac{360^0 \times L_a}{C_m} = \frac{360^0 \times 2.828}{119.320} = 8^\circ 31' \quad (7)$$

One circuit the rotation angle can increase: $$2\theta = 16^\circ 62' \quad (8)$$

Circuit number N_c: $$N_c = \frac{360^\circ}{2\theta} = \frac{360^\circ}{16^\circ} = 22.5 \approx 23 \quad (9)$$

The turnaround area in the filament winding process is the area where the fibre orientation starts to change the winding angle in order to come back to wind another side[12]. The winding angle can gradually increase until hoop winding and then reduce to expected winding angle. Most wasted filament wound products happen in the turnaround area, which is an inevitable process. It can lower the wasted material to reduce the turnaround zone area and process time. This can be achieved by adding the hoop winding on both sides, which can decrease the fibre slippage condition. The turnaround zone method schematic is shown in Figure 6.

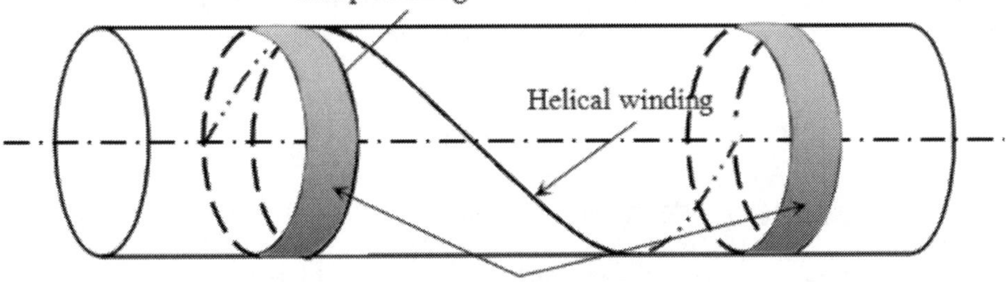

Figure 6. Turnaround zones diagram.

2. Methodology

Research methodology in this technical paper can be separated into three steps, which will describe details as follows:

Step 1: Calculation and write filament winding codes

- Based on the 3-axis filament winding machine parameter settings, filament winding codes should be properly written with different winding angles.
- Before taking the next step, the operator needs to check and run without wool yarn or carbon fibre. It is quite necessary to run the machine that can decrease mistakes.

Step 2: Perform filament winding experiments with 30°, 45° and 60° winding angle

• Prepare the experiment components such as the 3-filament winding machine, a wool yarn spool, power plug, USB wire, the PC, protractor, Universal G-code Sender software (UGS) and tested filament winding codes
• Connect USB wire and power plug, then fix the wool yarn one side on the mandrel with packaging tape. Next upload the different winding angle UGS codes in UGS software *File mode* sub-interface
• Finish filament winding process completely and measure the winding angle using the protractor
• Check the mandrel coverage rate, repeated pattern, product shape.
Step 3: Filament winding simulation using CADWIND software
• Have a brief introduction to CADWIND software, and learn the method to get proper simulation results
• Based on the previous filament winding experiment parameters and condition, the operator sets suitable parameter simulation values and click *"Start winding"* button. Eventually, the build-up of the different winding angle tubular products can be shown with the isometric view and front view.
Step 4: Comparison 3-axis filament winding machine tubular structure results with CADWIND simulation results
• Compare the 3-axis filament winding machine manufacturing capacity with CADWIND simulation process
• Compare experiment results with CADWIND simulation results about mandrel coverage rate, repeated characteristic pattern, winding structure and turnaround area pattern.
• Make a conclusion about 3-axis filament winding machine working capacity, which can demonstrate its advantages and disadvantages, products quality. The most important point is that it can give many pertinent suggestions on the 3-axis filament winding machine improvement, CADWIND software evaluation and related UGS codes optimization.

45° winding codes have been explained as follows, which set an example in this paper to write the winding codes. Based on the 3-axis filament winding machine system, Y-axis is the mandrel rotation. The 1 revolution means that the mandrel can rotate 1 full rotation, which moves mandrel circumference length. The 45° winding codes can be written based on the above calculation. The one circuit of the 45° winding code is shown in Figure 6. Based on the equation (1-9) calculation result, the total winding code can repeat 23 times to build the winding code.

$$\frac{1}{C_m} = \frac{N_2}{Y_2} \tag{10}$$

Y-axis rotation step of 45 winding angle N_2: $N_2 = 1.76$ (11)

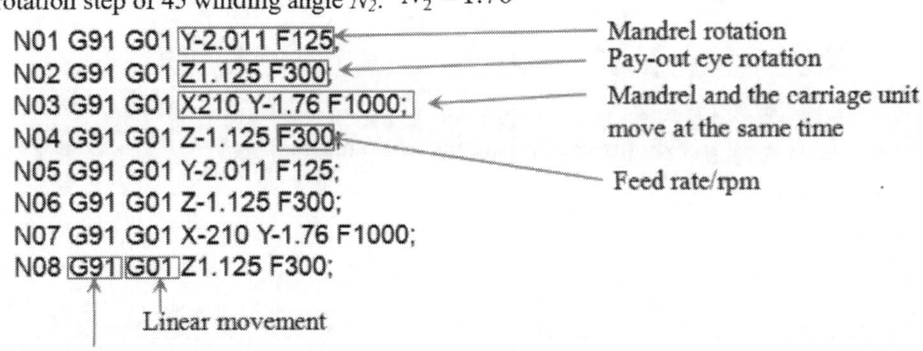

Figure 6. 45° winding code interpretation

3. Experiment using the 3-axis filament winding machine

The 3-axis filament winding machine has been designed with the inexpensive control system, which is still in process testing state. The main objective is to use the portable, high accurate 3-axis filament winding with the inexpensive control system, which can fabricate the filament wound

IOP Conf. Series: Materials Science and Engineering **342** (2018) 012029 doi:10.1088/1757-899X/342/1/012029

composite tubes with different winding angles. The 3-axis filament winding machine is used in experiment process, which can perform dry winding. The 3-axis filament winding machine is developed by the author, which is strongly supported by Human Engineering Group Lab, Faculty of Mechanical Engineering, Universiti Malaysia Pahang.

3.1 Experiment setup

The 3 axes experimental filament winding machine should be put on a flat table or workbench. Figure 8 presents the tested machine on the operating platform. This machine is used to apply carbon fibre (Pyrofil TR30S 3K), but it will increase the experiment costs and waste too much carbon fibre for further tests. This kind of carbon fibre type width is 2mm and the approximate thickness is 0.13mm of the 3K tow. It can decrease the machine movement accuracy rating to a certain extent. However, a pink wool yarn is chosen as the dry winding testing filament, which has approximate 1mm width. Figure 9 displays the wool yarn ball details, which is fully convenient to perform dry winding process especially in fibre path and repeated winding pattern.

Figure 8. 3-axis filament winding machine.

Figure 9. 100% wool yarn [16].

The experiment sets upon a proper position, which can simply perform filament winding process. The setting structure can be shown in Figure 10. The wool yarn string should be pasted on the mandrel, which can use packaging tape. Figure 11 illustrates preparation work on the mandrel. An A4 paper covers on the aluminium mandrel, which can provide a constant surface friction. The wool yarn path plotted in Figure 12 clearly shows the wool yarn trajectory through the resin bath (without resin), nip roller device, comb guide, guide roller, pay-out eye and pay-out eye comb.

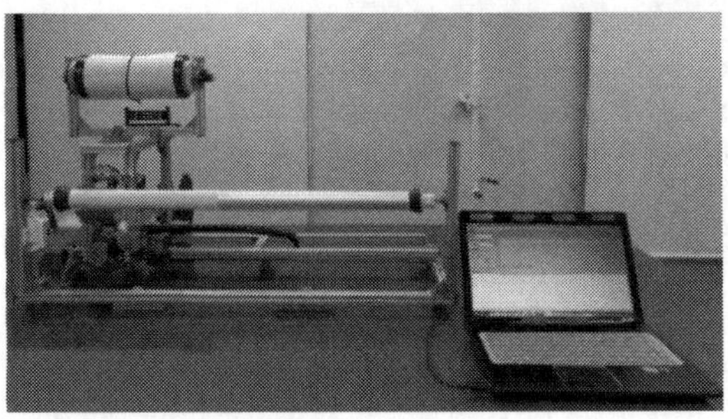

Figure 10. Experiment position setup.

Figure 11. Paste wool yarn on the mandrel.

Figure 12. The wool yarn trajectory on the carriage unit.

3.2 Perform experiments with 30°, 45°, and 60° winding angle

3.2.1 Experiment with 30° winding angle. The 30° winding angle UGS codes save as text. Then upload the 30° winding angle codes on *File Mode* and click "Send" button, the 3-axis filament winding machine can automatically perform filament winding process with 30° winding angle. Figure 13 shows the upload interface when it accomplishes. Figure 14 supplies detection result using the protractor at 30° winding angle.30° winding angle experiment product is illustrated in Figure 15, which appears some voids and fibre slippage.

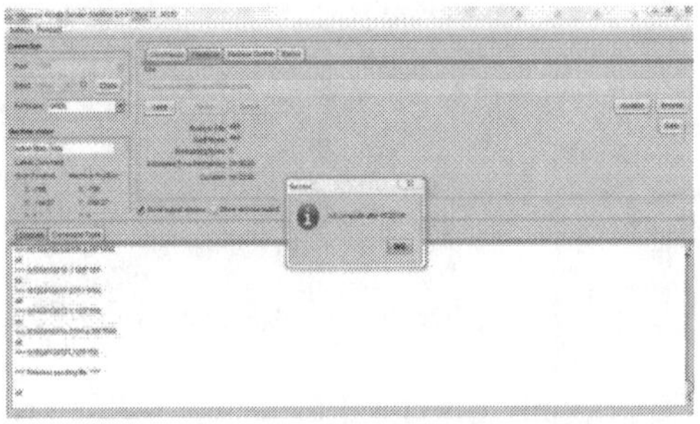

Figure 13. Complete interface with 30° with winding angle.

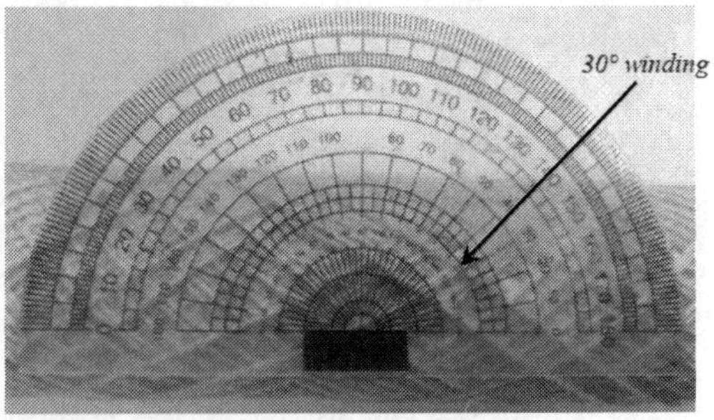

Figure 14. Measurement using the protractor 30° winding angle.

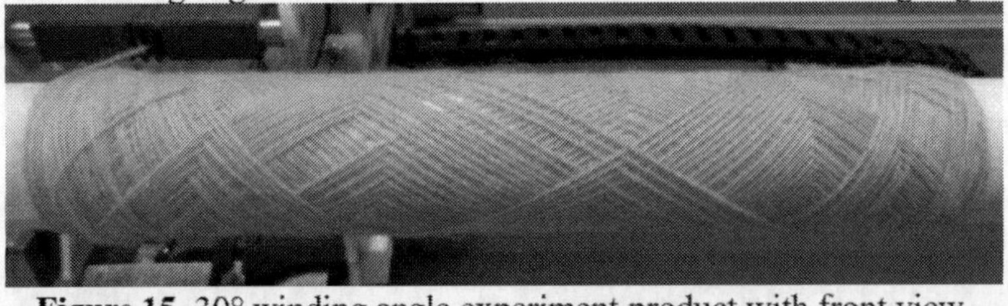

Figure 15. 30° winding angle experiment product with front view.

3.2.2 Experiment with 45° winding angle. The 45° winding angle UGS codes should be saved as text. Then upload the 30° winding angle codes on *File Mode* and click "Send" button, the tested filament winding machine can automatically implement filament winding process with 45° winding angle. Figure 16 shows the upload interface when it accomplishes, which has a longer time compared to 30° winding angle. Figure 17 supplies measurement result using the protractor at 30° winding angle. 45° winding angle experiment product is illustrated in Figure 18, which can simply recognize diamond-shaped unit cell.

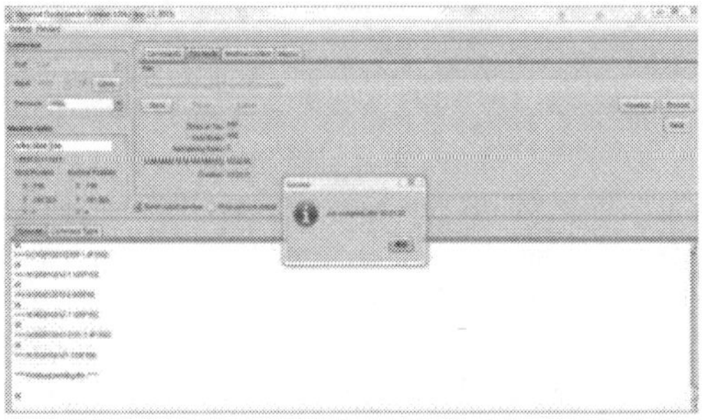

Figure 16. Complete interface with 45° winding angle

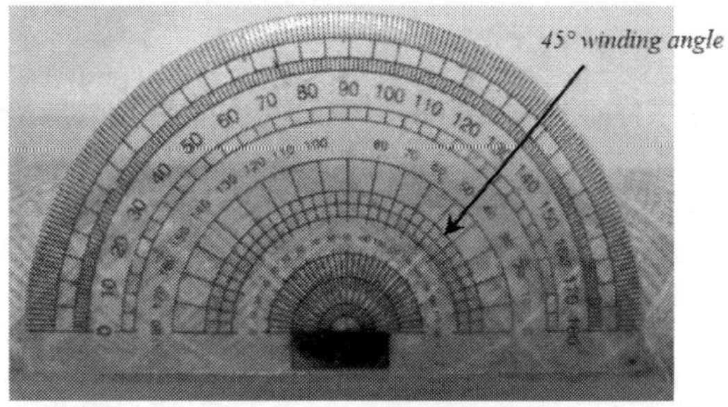

Figure 17. Measurement using the protractor with 45° winding angle.

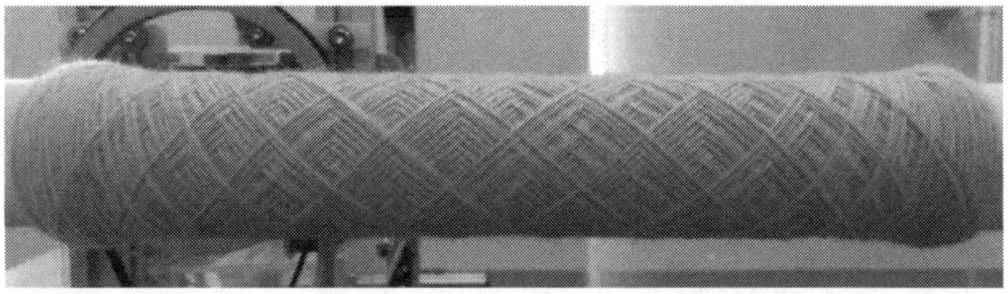

Figure 18. 45° winding angle experiment product with front view.

3.2.3 Experiment with 60° winding angle. The 60° winding angle UGS codes save as text. Next upload the 60° winding angle codes on *File Mode* and click “Send” button, the 3-axis filament winding machine can willingly carry out filament winding process with 60° winding angle. Figure 19 shows the implementation interface. Figure 20 provides measurement result using the protractor at 60° winding angle. 60° winding angle experiment product is exhibited in Figure 21, which have a good winding pattern generation with diamond-shaped cell unit.

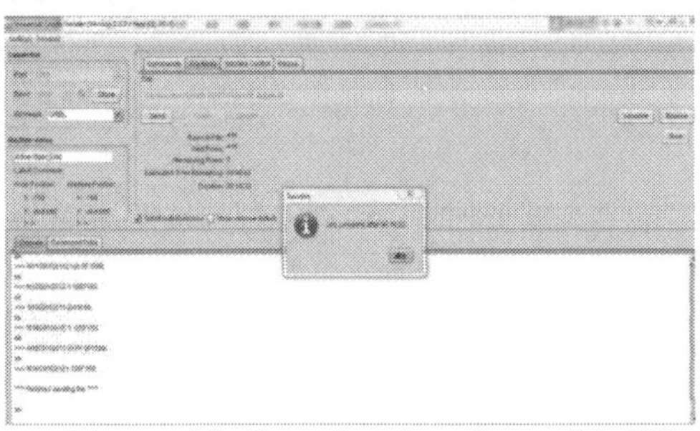

Figure 19. Complete interface with 60° winding angle.

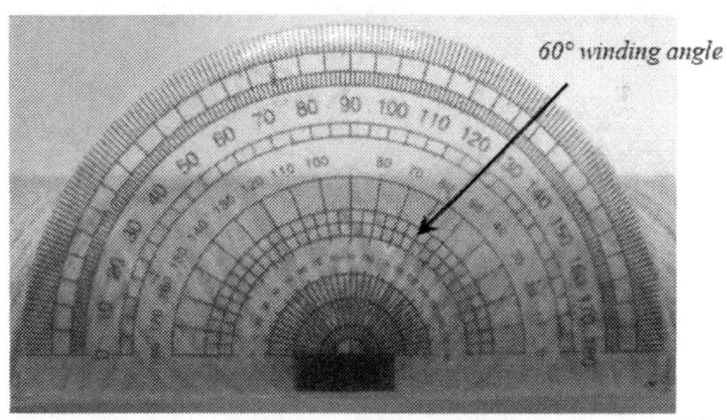

Figure 20. Measurement using the protractor with 60° winding angle.

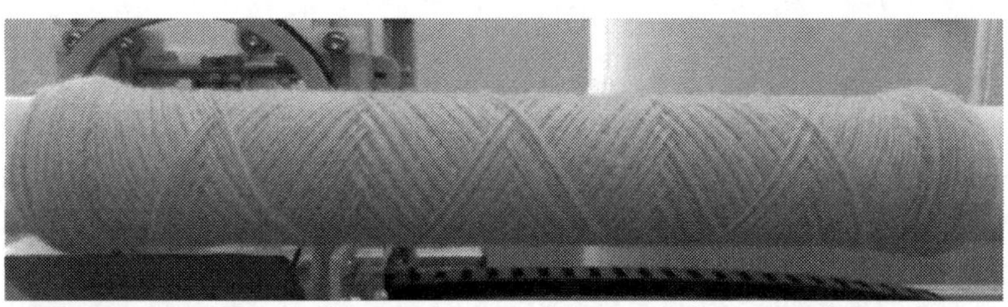

Figure 21. 60° winding angle experiment product with front view.

4. Filament winding simulation

The CNC technique and computer evolution have provided a critical leap in filament winding. It supplied the competence to control the fibre placement, mandrel rotation, and control system accuracy, which makes the possibility of interaction between the CNC machine and the innovated simulation software. This interaction can allow the CNC machine to make more complex shapes and have a high precision for products. During the years many CAD/CAM software and environments were developed which would enable a user of a CNC-controlled filament winding machine the possibility of designing parts for different mandrel shapes and produce their own parts[13].

In the filament winding process research field, some related CAD/CAM software for the filament winding applications have been applied. Such as ComposicaD^{TM}, CadPath^{TM}, fibreGraphix^{TM}, CadFil^{TM} and EasyWinder^{TM}. This kind of software can provide degrees of control systems codes before the process and different symmetric or unsymmetric filament wound products, product material data collection and analysis output, filament winding simulation process and FEA (Finite Element Analysis) mesh output and graphical user interface. Besides above available software, Matlab^{TM} software can also simulate filament winding process via Pathwind, which is the software that aims to simulate filament winding along axis-symmetric mandrels. It can generate paths and patterns for different mandrel geometries and tow properties as well as generate CNC code for filament winding machines. This software was first developed as part of a Master thesis while working in INEGI[12].

4.1 CADWIND Software introduction

Filament winding is one of the more sophisticated methods of producing composite products, which can calculate the complex movements of up to six axes machine on CADWIND software. The development of CADWIND, which simulates on the basis of a physical model the winding process on the computer, calculates the fibre lay-up on the mandrel and automatically generates the related program codes for the six-axis filament winding machine[1]. CADWIND has progressed development from academic research projects to Material S.A company standard for different kinds of filament wound products. Over 300 research institutes, universities and companies are using CADWIND software to simulate filament winding process.

4.2 CADWIND simulation

The procedure working principle of CADWIND consists of three steps. Details are as follows[14]:
Step 1: Creating the mandrel model;
Step 2: Calculating the winding process;
Step 3: Generating the part program code;

Step 1: Creating the mandrel model
1. Choose the menu *FILE/CREATE MANDREL/CIRCULAR CROSS-SECTION*
2. Enter the parameters are shown in Figure 22.

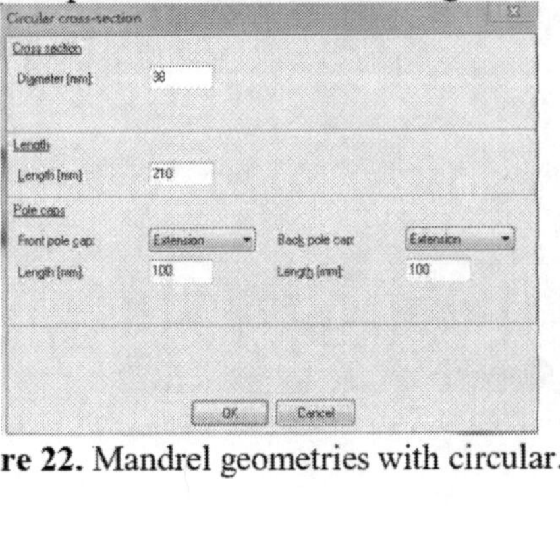

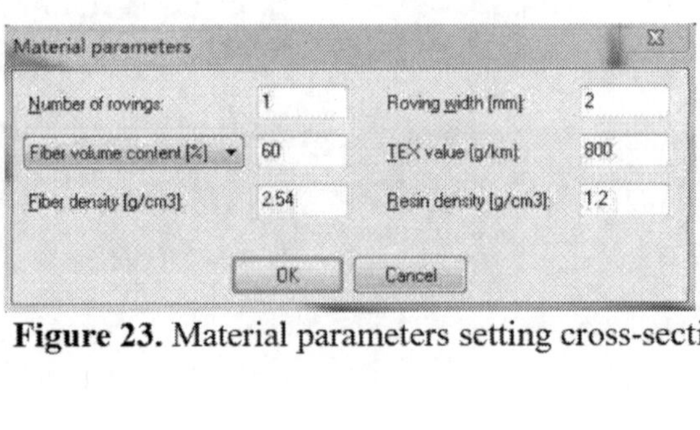

Figure 22. Mandrel geometries with circular. **Figure 23.** Material parameters setting cross-section.

Step 2: Winding process
Choose *WINDING/SELECTION* from the menu, selection non-geodesic parameters are present in Figure 23
Step 3: Choose *WINDING/NON-GEODESIC PATTERN* from the menu, enter the following parameters are provided in Figure 24

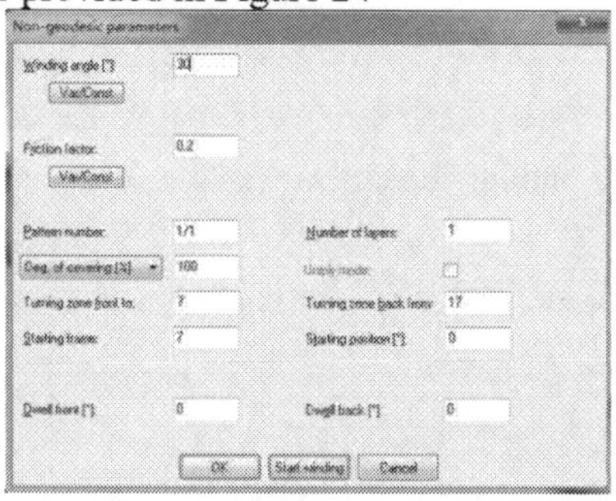

Figure 24. Non-geodesic parameters setting.

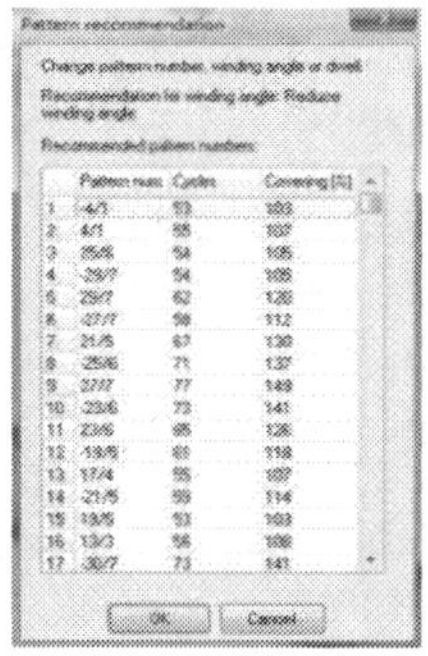

Figure 25. Pattern recommendation.

Then click on "Start winding", the program calculates the winding pattern according to the entered winding and material parameters. About pattern number, system default parameter is 1. Based on different setting values, pattern number can be calculated by the software itself, which can select a proper pattern number is noted in Figure 25. The first fibre trajectory is visible, the complete laminate will be wound on the mandrel.

4.3 30°, 45°, and 60° winding angle filament wound cylinders simulation

Based on the above the working principle of CADWIND software, it is sufficient to perform different winding angle simulation to check and compare with 3 axes filament winding machine experiment products. This simulation will cover 30°, 45°and 60° three different type winding angles, which will fabricate the tubular products with 210mm length and 38 mm inner diameter.

4.3.1 Simulation with 30° winding angle. Simulation parameters have been shown in Table 1, which is the related setting for 30° winding angle. Figure 26 illustrates the 30° winding angle tubular specimen with the isometric view. Figure 27 is the front view of 30° winding angle tubular specimen, which is clear to observe the fibre path, repeated pattern, and coverage.

Table 1. 30° winding angle tubular specimen details.

Core model size/mm			Winding angle/°	Fiber bandwidth/mm	Friction coefficient	Remark
Diameter	Winding length	Extension length				
38	210	100	30	2	0.2	
Note: Winding length refers to the length of the fiber covered area with a 30 degree winding angle Extension length refers to the fiber on mandrel with 30 degree winding angle, which coverage beyond the ends of the length. It can provide adequate space for changing fiber direction.						

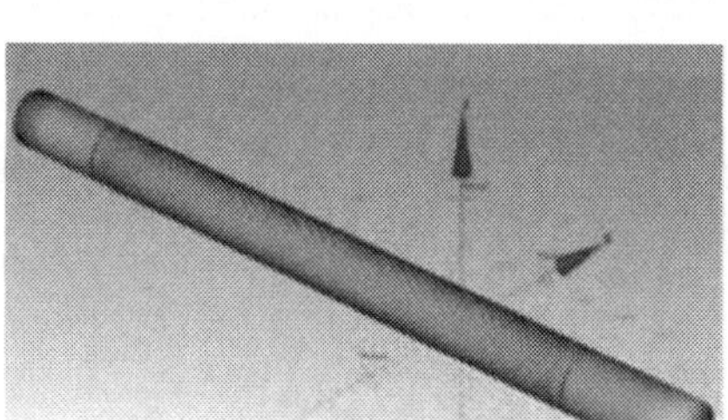

Figure 26. 30° winding angle tubular specimen

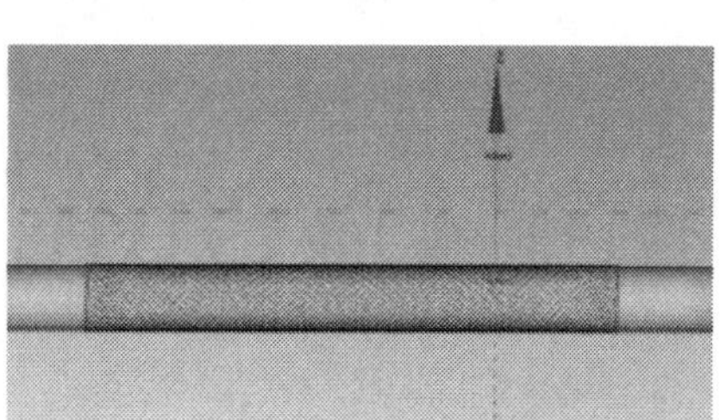

Figure 27. 30° winding angle tubular specimen

with the isometric view. with the front view.

4.3.2 Simulation with 45° winding angle. 45° winding angle simulation parameters have been presented in Table 2, which is the vital setting for 45° winding angle. Figure 28 illustrates the 45° winding angle tubular specimen with an isometric view. Figure 29 is the front view of 45° winding angle tubular specimen, which can have a brief observation on fibre path, repeated characteristic pattern, and coverage.

Table 2. 45° winding angle tubular specimen details.

Core model size/mm			Winding angle/°	Fiber bandwidth/mm	Friction coefficient	Remark
Diameter	Winding length	Extension length				
38	210	100	45	2	0.2	
Note: Winding length refers to the length of the fiber covered area with a 30 degree winding angle Extension length refers to the fiber on mandrel with 30 degree winding angle, which coverage beyond the ends of the length. It can provide adequate space for changing fiber direction.						

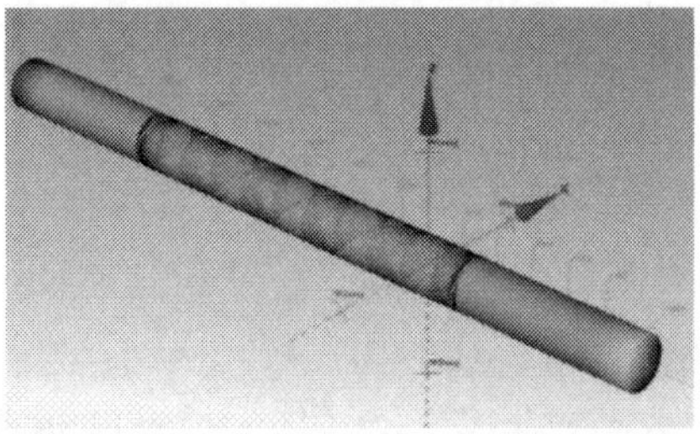

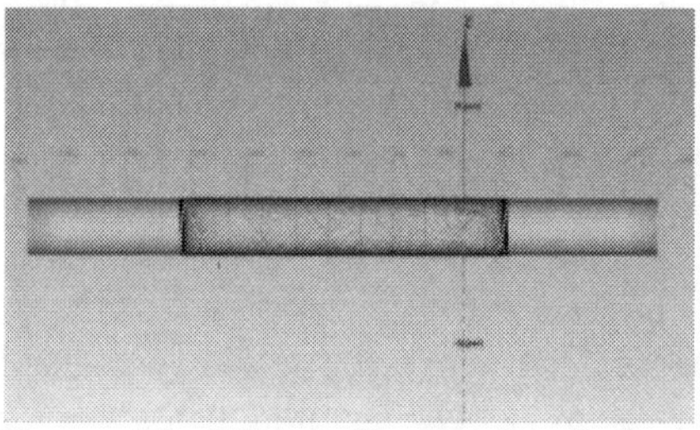

Figure 28. 45° winding angle tubular specimen with the isometric view. **Figure 29.** 45° winding angle tubular specimen with front view.

4.3.3 Simulation with 60 winding angle. 60° winding angle simulation parameters have been provided in Table 3, which supplies setting values for 60° winding angle. Figure 30 illustrates the 60° winding angle tubular specimen with an isometric view. Figure 31 is the front view of 60° winding angle tubular specimen, which can have an accurate view of fibre path, repeated characteristic pattern, and coverage.

Table 3. 60° winding angle tubular specimen details.

Core model size/mm			Winding angle/°	Fiber bandwidth/mm	Friction coefficient	Remark
Diameter	Winding length	Extension length				
38	210	100	60	2	0.2	
Note: Winding length refers to the length of the fiber covered area with a 30 degree winding angle Extension length refers to the fiber on mandrel with 30 degree winding angle, which coverage beyond the ends of the length. It can provide adequate space for changing fiber direction.						

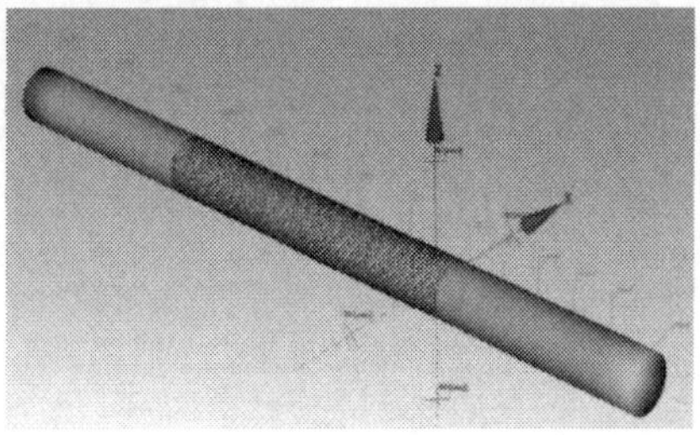

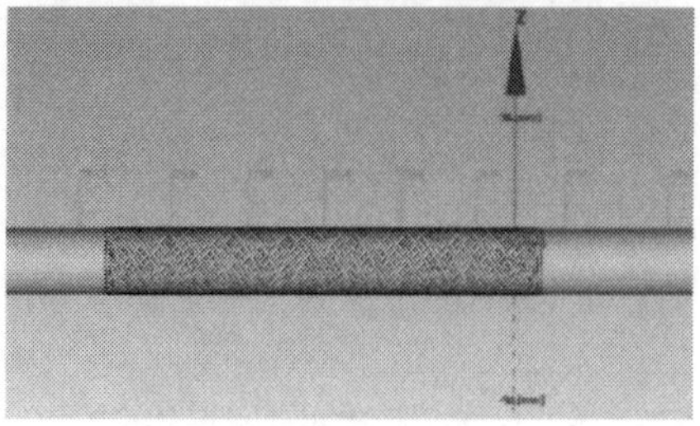

IOP Conf. Series: Materials Science and Engineering **342** (2018) 012029 doi:10.1088/1757-899X/342/1/012029

Figure 30. 60° winding angle tubular specimen with with isometric view.

Figure 31. 60° winding angle tubular specimen with front view.

5. Experiment and CADWIND software simulation comparison

In comparison with the 3-axis filament winding machine experiment and CADWIND software simulation, which refer to filament winding process level, mandrel coverage rate, repeated characteristic pattern, the change degree of specimen shape, specimen surface smoothness, turnaround impact and other parameters. Experiment products and simulation specimen photos are shown in Table 4. The comparison results are presented in Table 5.

Table 4. Experiment and simulation winding pattern view.

Filament winding process / Winding angle	*Experiment Results*	*Simulation Results*
30 winding angle	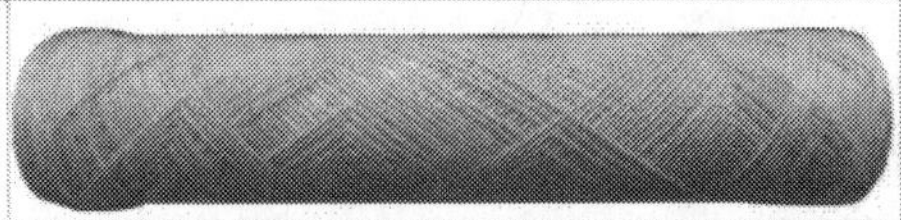	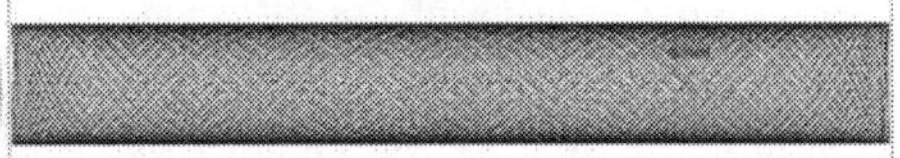
45 winding angle		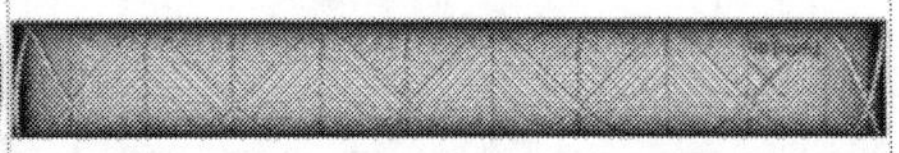
60 winding angle		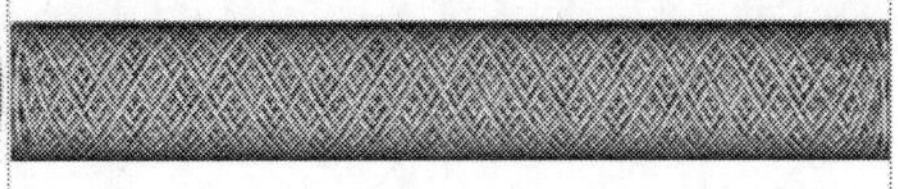

Table 5. Comparison specification results.

Comparison specification	Experiment process	Simulation process
Filament winding process level	High	High
Mandrel coverage rate	75%—85%	100%(controlled)
Repeated characteristic pattern	One pattern	More than two patterns
Degree of specimen shape change	15%—24%	Less 5%
Specimen surface smoothness	Moderate-High	High
Turnaround impact	High	Low
Machine axes	3 axes	4 axes
Friction impact	Unstable value	Constant value
Process error	$<$15%	1—2%
Control system code	Calculation and write	Generate automatically
Process time	Long time (based on filament winding angle)	Short time
Fibre slippage impact	Low-Moderate	No effects
Winding angle accuracy	Moderate-High	High
Control system	Arduino UNO and CNC shield	CNC control system

Experiment and simulation results can present more ideas on the 3-axis filament winding machine and CADWIND software. As for the 3 axes filament winding machine, the machine can provide a good winding pattern with high winding angle. The low winding angle can appear more voids, which can be modified by filament winding codes. The turnaround zones on both sides can cause friction, which can affect the winding angle. This problem obviously happened on 30 winding angle. In the real winding process, the turnaround zone needs to take into consideration on effect on winding angle. The machine also can reduce the fibre slippage through lower the process speed. In short, the 3-axis filament machine can wind as the winding codes, but it needs to improve winding condition. As for the CADWIND software, it is the sufficient and reliable software to simulate winding condition. Software function is trying to simulate the real winding condition as close as possible. The fibre tow thickness and turnaround zones issues need to exploit and add in this software, which can complete the software function and simulation results. There is no doubt that CADWIND software is efficient in filament winding technique field, which can meet general requirements for winding parameters.

6. Conclusion and recommendations

6.1 Conclusion

The 3-axis filament winding machine has been successfully designed and fabricated, which can have a sufficient processing capacity to produce certain composite material products. Based on the experiment and simulation results, the machine provides a good filament winding process and a moderate filament wound composite cylinders. This 3-axis filament winding machine can evaluate its capability from machine structure and control system perspectives. In terms of machine design structure, it can meet the requirements in mandrel rotation and carriage unit movement. In control system perspective, it can simply control the machine using UGS software with an inexpensive control system. But this 3-axis filament winding machine still needs further improvements before selecting it as final machine development to fabricate filament wound composite cylinders for the compression test. It also provides the development and optimization orientation about the machine from comparison results.

6.2 Recommendations

It is suggested that the following improvements of the 3-axis filament winding machine should be carried out afterward:

a) Improvement of UGS codes

The current machine carriage unit changes the direction in turnaround zone, which needs to have one more round to rotate. It can effectively reduce the fibre slippage movement. But on the other hand, it will waste much extra fibre and resin. To remove one more round in the turnaround zone, it needs another available way to deal with fibre slippage especially in 30° filament winding angle process.

b) Control the fibre tension and friction

During the filament winding machine process, a certain fibre tension should be provided with tension device or manual in experiments. The friction force mainly affects the filament winding products with the certain winding angle on the mandrel surface. It should be calculated and tested in the future, which can improve the filament winding angle accuracy.

c) Control system replacement

This machine is controlled by Arduino UNO and CNC shield with low costs and inexpensive control system characteristic. However, it has limitations about the control system. The upload related G-code is limited to support the format, which cannot write more general type of C-codes to control the machine. It is better to use PLC control system to replace it, which can separate control each axis rotation and movement. It is quite convenient to test and perform different mandrel shape.

Acknowledgement

The authors are grateful to the Ministry of Higher Education and Universiti Malaysia Pahang for funding this research with FRGS/1/2014/TK01/UMP/02/3 and RDU1603108. The main author also thanks Material R&D Supervisor Engineer Jiang Kaige for his CADWIND 2011 version software simulation part and interrelated software guidance, who works in *Zoomlion Co., Ltd, China*.

References

[1] Laval C 2006 *Reinforced Plastics* **50** 34-37
[2] Shen F C 1995 *Materials Chemistry and Physics* **42** 96-100
[3] Abdalla F H, Mutasher S A, Khalid Y A, Sapuan S M, Hamouda A M S, Sahari B B and Hamdan, M M 2007 *Materials & Design* **28** 234-239
[4] Rejab M R M, Kadirgama K, Noor M M, Sani M S M and Daud R 2008 in *International Conference on Science & Technology: Application in Industry & Education*
[5] Mutasher S, Mir-Nasiri N, and Lin L C 2012 *Journal of Engineering Science and Technology,* **7** 156-168
[6] Rojas E V, Chapelle D, Perreux D, Delobelle B and Thiebaud F 2014. *Composite Structures,* **116** 805-813
[7] Zuraida A, Khalid A A, and Ismail A 2007 *Materials & Design* **28** 71-77
[8] Jia X, Chen G, Yu Y, Li G, Zhu J, Luo X, Duan C, Yang X and Hui D 2013 *Composites Part B: Engineering* **45** 1336-1343
[9] C Innovation 2016 *Filament winding robot*. Available: http://www.comecinnovative.it/products/filament-winding-robot
[10] Benedict T 2016 *Filament spin evloution winds up ligher, stiffer and stronger carbon rims*. Available:www.bikerumor.com/2016/11/23/filament-spin-evolution-winds-lighter-stiffer-stronger-carbon-rims/
[11] Rousseau J, Perreux D and N. Verdiere N 1999 *Composites Science and Technology* **59** 1439-1449
[12] Power B 2016 Master thesis, University of Porto
[13] Koussios S 2004 Filament Winding. A Unified Approach
[14] M. S. A. CADWIND Process simulation system for filament winding [Online].

IOP Conf. Series: Materials Science and Engineering **342** (2018) 012030 doi:10.1088/1757-899X/342/1/012030

Multivariable PID controller design tuning using bat algorithm for activated sludge process

Nur Atikah Nor'Azlan[1], Nur Asmiza Selamat,[2] and Nafrizuan Mat Yahya[1]

[1] Faculty of Manufacturing Engineering, Universiti Malaysia Pahang, Pahang, Malaysia
[2] Faculty of Electrical Engineering, Universiti Teknikal Malaysia Melaka, Melaka, Malaysia

E-mail: nuratikahnorazlan@gmail.com

Abstract. The designing of a multivariable PID control for multi input multi output is being concerned with this project by applying four multivariable PID control tuning which is Davison, Penttinen-Koivo, Maciejowski and Proposed Combined method. The determination of this study is to investigate the performance of selected optimization technique to tune the parameter of MPID controller. The selected optimization technique is Bat Algorithm (BA). All the MPID-BA tuning result will be compared and analyzed. Later, the best MPID-BA will be chosen in order to determine which techniques are better based on the system performances in terms of transient response.

1. Introduction

Nowadays, the health of natural ecosystems has been giving more attention cause of the effect of the human development in very different ways [1] by increased environmental awareness in terms of water pollution anticipation. The constricted laws and requirements toward quality of water are fortunately acting like a driving force for the development of wastewater treatment plants (WWTPs) [2]. The complex, interrelated and highly nonlinear of its biological, physical and chemical phenomena make the WWTP become more difficult to be controlled while optimizing operating and management costs [3].

Meanwhile, the activated sludge process (ASP), a biological processes are usually popular methods used to remove carbon as well as components of nitrogenous from wastewater beforehand it being released [4]. ASP has been a widely cased study in the automatic control perspective, for example by Yong et al. [5] where the concentration of ammonia in the fluent of the wastewater plant is reducing by the implementation of cascading PI-like controller with feed-forward actions. The model predictive control (MPC) by Holenda et al. [6] proposed that MPC method also can be used for ASP system. There are significant benefits that can be obtained of using the MPC system which is decreasing of more than 25% in power usage and an increase in plant efficiency [7] but it is only determining and controlling the dissolved oxygen concentration.

A previous study was done by Shen et al. where a multiple input approach is implemented, by the recycle flow rates, the oxygen transfer coefficient of three aerated tanks and a complementary carbon source. Fuzzy controller has proved its efficiency to be implemented at WWTPs for improving the denitrification or nitrification process but these method gives a high cost and behaves relatively rough toward its control actions [8].

WWTP also is categorized as a complex system. Therefore, effective control methods need to be implemented for economic and environmental reasons, especially its ASP because it involves a biological process. Unfortunately, an increasing claim for a more stable effluent water quality [1] makes a scalar PID based control systems are become insufficient anymore due to the complexity of the system such as interrelated and highly nonlinear of its biological, physical and chemical phenomena. Besides, when the system becomes more complex, the process of tuning controllers also becomes more difficult [3].

In an attempt to overcome that problem the suitable controller tuning methods and optimization techniques are investigated in this paper in order to obtain an optimal solution to get the best values of the parameter. A type of MPID system where a combination more than one variable at the input or output is being investigated in this paper. Then, MPID methods considered are Davison, Penttinen-Koivo, Maciejowski and Proposed Combined. The optimization approach also being used in this paper for tuning the MPID of ASP where optimization is the process of finding an ideal solution to get the best values of parameters for the problem under stated conditions by executing the procedure in comparing several solutions till an optimum solution is found using an optimization algorithm [9]. The main objectives of this paper are to gain parameter tuning of ASP using optimization technique of BA and comparing the system performances in term of transient response.

This paper is divided into several sections. Section 2, an activated sludge system used is briefly described. Then, a multivariable PID control tuning method is presented in section 3. Section 4 contains a brief explanation on BA. The objective function used is shown in section 5. In section6, simulation results of MPID-BA are discussed. This paper ends with a conclusion and future works.

2. Activated Sludge Process

Activated sludge process (ASP) is the most widely used for the biological wastewater treatment plant and a type of secondary treatment where a high level of elimination of biodegradable organic pollutants are given to keep receiving water quality that clarification alone cannot provide. The ASP also can speeds up decomposition by adding an activated sludge into the wastewater where the activated sludge particles hold many living organisms that can feed on the incoming wastewater [10].

The non-linear Activated Sludge Process that has been selected and used for this research is taken from [11]. Activated sludge processes are biological that removes pollutant from the wastewater. The system comprises of aerator and settler as shown in figure 1. The bioreactor (aerator) includes secondary clarifier to maintain the biomass in the system while producing high-quality effluent. Part of settler output is recycled to allow the right concentration of microorganism the aerated tank.

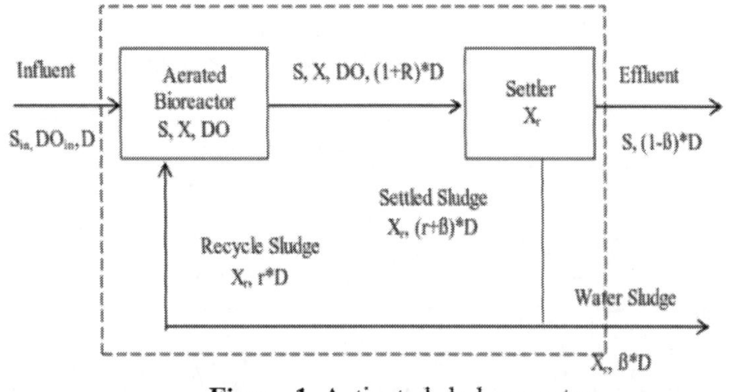

Figure 1. Activated sludge reactor.

The model was derived based on component mass balance equation which yields a set of the non-linear differential equation given by equation (1) until (4).

$$Mass\ balance\ equation = Input - Output \pm Reaction \tag{1}$$
$$\dot{X}(t) = \mu(t)X(t) - D(t)(1+r)X(t) + rD(t)X_r(t) \tag{2}$$
$$\dot{S}(t) = -\frac{\mu(t)}{Y}X(t) - D(t)(1+r)S(t) + D(t)S_{in} \tag{3}$$
$$\dot{C}(t) = -\frac{K_0\mu(t)}{Y}X(t) - D(t)(1+r)C(t) + K_{La}(C_s - C(t) + D(t)C_{in} \tag{4}$$
$$\dot{X}_r(t) = D(t)(1+r)X(t) + D(t)(\beta + r)X_r(t) \tag{5}$$
$$\mu = \mu_{max}\frac{S}{K_s+S} \times \frac{C}{K_c+C} \tag{6}$$

where (t), $S(t)$, $X_r(t)$ and $C(t)$ are the state variables which represents the concentration, biomass, substrate, dissolved oxygen and recycled biomass respectively. S_{in} and C_{in} represent concentrations substrate and dissolved oxygen in the fluent stream. The ratio of recycled and waste flow to the influent flow rate are given by r andβ. Specific growth rate μ produced cell mass Y. K_0 is constant, while C_s and K_{La} is maximum dissolved oxygen concentration and oxygen mass transfer coefficient. The relationship between maximum growth rate to substrate and to dissolve oxygen coefficient is given by Monod equation as shown in equation (5). μ_{max} is the maximum specific growth rate, K_s is the affinity constant and K_c is the saturation constant as shown in equation (6).

The following step is to linearize the nonlinear model. For simplification, the nonlinear is then linearized about its operating point yielding A, B, C and D matrices in the form of state space equation as in equation (7).

$$\begin{bmatrix}\dot{X}\\ \dot{S}\\ \dot{C}\\ \dot{X}_r\end{bmatrix} = \begin{bmatrix}-0.0990 & 0.1234 & 0.2897 & 0.0495\\ -0.05077 & -0.3129 & -0.4457 & 0\\ -0.02538 & -0.0943 & -0.1975 & 0\\ -0.02538 & 0 & 0 & -0.066\end{bmatrix}\begin{bmatrix}X\\ S\\ C\\ X_r\end{bmatrix} + \begin{bmatrix}-82.17 & 0\\ 134 & 0\\ -9.083 & 0.06994\\ 8\times10^{-5} & 0\end{bmatrix}\begin{bmatrix}D\\ W\end{bmatrix} \tag{7}$$
$$Y = \begin{bmatrix}1 & 0 & 0 & 0\\ 0 & 1 & 0 & 0\\ 0 & 0 & 1 & 0\\ 0 & 0 & 0 & 1\end{bmatrix}\begin{bmatrix}X\\ S\\ C\\ X_r\end{bmatrix} + \begin{bmatrix}0 & 0\\ 0 & 0\\ 0 & 0\\ 0 & 0\end{bmatrix}\begin{bmatrix}D\\ W\end{bmatrix}$$

The state space then manipulated into transfer functions matrix form $G(s)$ which shown in (8).

$$G(s) = \begin{bmatrix}g_{11} & g_{12}\\ g_{21} & g_{22}\end{bmatrix} \tag{8}$$

where,

$$g_{11} = \frac{134(s+2.0049)(s+0.1864)(s+0.0117)}{(s+1.9956)(s+0.2573)(s+0.2014)(s+0.0076)}$$
$$g_{12} = \frac{-0.0312(s+0.1863)(s+0.0117)}{(s+1.9956)(s+0.2573)(s+0.2014)(s+0.0076)}$$
$$g_{21} = \frac{-9.0830(s+1.4447)(s+0.1942)(s+0.0048)}{(s+1.9956)(s+0.2573)(s+0.2014)(s+0.0076)}$$
$$g_{22} = \frac{0.0699(s+2.802)(s+0.1993)(s+0.0074)}{(s+1.9956)(s+0.2573)(s+0.2014)(s+0.00076)}$$

In the multivariable system, interaction effect is taken into consideration, hence RGA calculation is compulsory to determine which input to use for which controller. RGA is calculated and the result is shown in equation (9). Since the result near to diagonal matrix, hence it can be concluded that manipulated variable 1 (MV1) controlled by the control variable 1 (CV1) and manipulated variable 2 (MV2) controlled by the control variable 2 (CV2).

IOP Conf. Series: Materials Science and Engineering **342** (2018) 012030 doi:10.1088/1757-899X/342/1/012030

$$\Lambda = \begin{bmatrix} 1.05 & -0.05 \\ -0.05 & 1.05 \end{bmatrix} \quad (9)$$

3. Multivariable PID turning method

This section will discuss the method used to tune the parameter of MPID controller. MPID controller is used due to consideration of multivariable system. Then, four types of MPID tuning have been selected which are Davison, Penttinen-Koivo, Maciejowski and Proposed Combined. These methods had been chosen due to its ability in dealing with interaction in simpler ways and it also required only step or frequency test [12]. The aims are to obtain the substrate and dissolved oxygen concentrations at the desired level.

3.1. Davison method

Davison is one of the MPID tuning methods and it is known to diagonalize plant at low frequency. In Davison, only integral is taken into consideration and the expression is given by equation (10).

$$u(s) = K_i \frac{1}{s} e(s), \quad K_i = \varepsilon G^{-1}(0) \quad (10)$$

where K_i is the integral feedback gain, $G(s)$ is the open loop transfer function matrix and ε as the scalar tuning parameter. Parameter tuning, ε Davison is adjusted in order to obtain a better closed loop system performance.

3.2. Penttinen-Koivo method

For the Penttinen-Koivo method, the plant is diagonalized at high frequency. Based on the controller given in equation (11), proportional and integral terms are taken into accounts. It also can be observed that the equation is an extension of Davison method

$$u(s) = (K_p + K_i) \frac{1}{s} e(s) \quad (11)$$

$$K_i = \varepsilon G^{-1}(0) \quad (12)$$

$$K_p = (CB)^{-1} \rho \quad (13)$$

K_i and K_p are given by equation (12) and (13). C and B value can be obtained from output and input matrices in state space equations. In this method, two scalar parameter can be tune which is ε and ρ.

3.3. Maciejowski method

Maciejowski method also uses the proportional and integral term of PID, it differs with Penttinen-Koivo method in the way of Maciejowski try to diagonalize the system near to bandwidth. The expression of the controller can be given by equation (14).

$$K = \left(K_p + K_i \frac{1}{s}\right) \quad (14)$$

$$K_p = \rho G^{-1}(j\omega_b) \quad (15)$$

$$K_i = \varepsilon G^{-1}(j\omega_b) \quad (16)$$

$$J(K, \theta) = \left[G(j\omega_b)K - e^{j\theta}\right]^T \left[G(j\omega_b)K - e^{j\theta}\right], \quad \theta = diag(\theta_i, \ldots, \theta_n) \quad (17)$$

where K_p is the proportional gain, K_i is an integral gain, ω_b is the bandwidth frequency. The $G^{-1}(j\omega_b)$ results in a complex number, with that a real approximation given in equation (17) are necessary. In this method, three parameters can be tune which are ε, ρ, and $j\omega_b$. But for the purposed of the study, only scalar tuning will be tuned.

3.4. Proposed Combined method

The Proposed Combined method is introduced by [11]. In this method, it uses the criteria of Maciejowski in diagonalizing system near bandwidth frequency. The proposed combined method is being developed to reduce the difficulties in finding the suitable bandwidth frequency. Nevertheless, it shows similar behavior with Maciejowski. The expression of the proposed combined controller is shown in (18).

$$u(s) = e(s)(\rho K + \varepsilon K)\frac{1}{s} \tag{18}$$
$$K = [\alpha G(0) + (1-\alpha)CB]^{-1} \tag{19}$$

where K is given in equation (19) and α is a constant value between [0, 1]. By that, three scalar tuning will be tuned in this method which is α, ε, and ρ.

4. Bat algorithm

Bat algorithm is a type of meta-heuristic swarm algorithm intelligent optimization algorithms. It was introduced by Yang in 2010 [13] and was inspired by the behavior of microbats which use echolocation pulses with different emission and sound. Besides, it is based on the echolocation capability of microbats guiding them on their foraging actions [14].

The approximate rules of BA can be categorized into three [13,14]. First, all the bats use echolocation to sense distance, and they also know the difference between food or prey and background barrier in some unknown way. Second, bats fly randomly with velocity at the position with a fixed frequency, varying wavelength, and the loudness to search for prey by automatically adjust the wavelength of their emitted pulses and also adjust the rate of pulse emission depending on the proximity of their target. Lastly, although the loudness can vary in many ways, assumed that the loudness varies from a large to a minimum constant value. Figure 2 shows the BA flowchart.

The step of BA is given as follows [14]:

1) Initialize the initial population of bats.
2) Setting the parameters: pulse frequency, pulse rate, loudness.
3) Generate the position and velocity for initial bats.
4) If the result of rand>pulse rate is yes, select the solution among best solution.
5) If the result of rand>loudness and Fnew<=fmin is yes, accept the new solution.
6) Rank the bats and determine the best one.
7) Go back to step 3 and repeat all the step until stopping criteria is met

The velocity and position equation is given by equation (20) and (21) respectively. All the parameter value are given in table 1.

$$Q_{new} = Q_{min} + (Q_{min} - Q_{max}) \times random\ no. \tag{20}$$
$$V_{i,k+1} = V_{i,k} + \left(P_{i,k} - X_{i,k}\right) \times Q_{new} \tag{21}$$
$$X_{i,k+1} = P_{i,k} + V_{i,k+1} \tag{22}$$

where Q_{new}, V_i^k , X_i^k and $P_{i,k}$ are the new frequency of bats, the velocity of the ith individual at iteration k, best position of the ith individual at iteration k and position of the ith individual at iteration k respectively.

Table l. Parameter initialization in BA.

Initialization			
Population size	50		
No. of iteration	100		
Search range	0-10		
BA initialization			
Loudness	0.95	Max. frequency	2
Pulse rate	0.9	Min. frequency	0

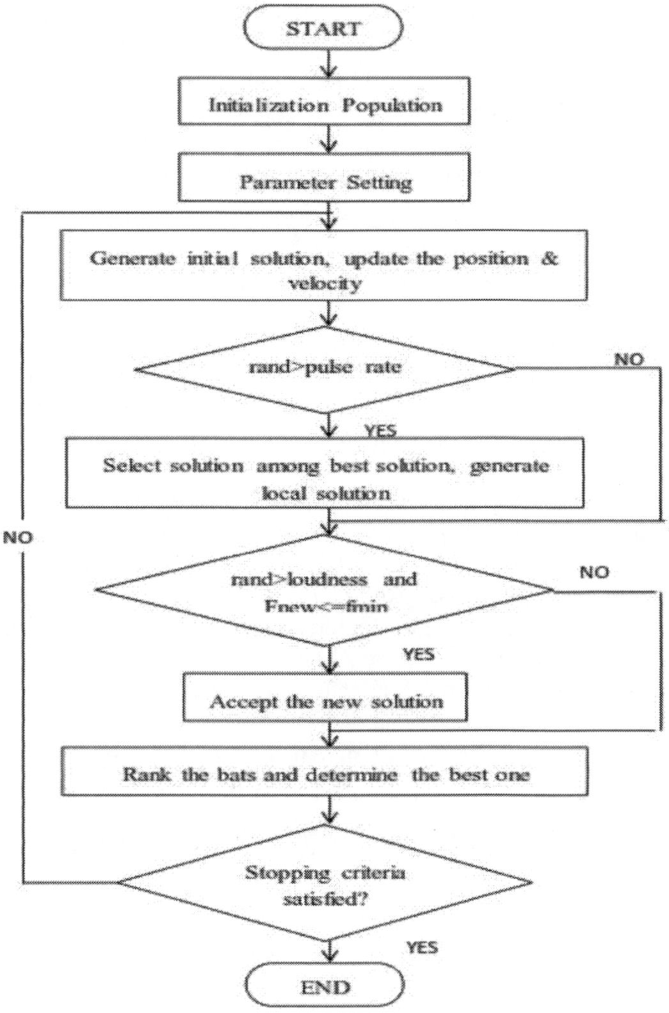

Figure 2. Flowchart of Bat Algorithm.

BA is a stochastic algorithm, where different result obtained for every time the algorithm is executed, even though the same initial point is used. Hence, each optimization technique was executed for 20 times to make a fair observation for further analysis and the result was selected based on the execution that gives the best performances index. The repeated process that done in BA to obtain optimum result caused the stopping criteria to be introduced. For this study, the stopping criteria used is when the maximum iteration is reached. The repetition of the algorithm will stop when the maximum number of iteration is achieved.

5. Objective function

The objective function is to evaluate the performance index of the system where it can be found in various forms such as time-domain specifications, frequency domain specifications and time-integral performance [15]. It is also representing system criteria that desired by the user. In this study, no specific criteria of the system as long as it gives the optimum value of the objective function. Thus, the smaller the error or the objective function, the better the system or performance index.

IOP Conf. Series: Materials Science and Engineering **342** (2018) 012030 doi:10.1088/1757-899X/342/1/012030

Time- integral performance or more specifically Integral Time Square Error (ITSE) has been chosen as the objective function. ITSE expression is given by equation (23) and from the equation, ITSE only required system error and the answer represents the area of the output and desired output. This selection is due to ITSE behavior that provides a better dynamic performance with good settling time.

$$ITSE = \int_0^t (e_i(t))^2 t\,dt \tag{23}$$

$$e(t) = y(t) - r(t) \tag{24}$$

where $e(t)$ is an error in equation (24), $y(t)$ is output of system and $r(t)$ is desired output.

6. Simulation result

The simulation of BA to tune the MPID controller for ASP was carried out by using MATLAB/SIMULINK software and the MPID control methods as previously mentioned was implemented to ASP transfer function. This method is called offline tuning method because the scalar tuning parameter is obtained first using the simulation of MPID-BA tuning using MATLAB. Then, the simulation results obtained are used to run closed loop nonlinear ASP model in SIMULINK.

The best parameter tuning is selected based data execution that gives the smallest fitness function (ITSE value). The smaller the ITSE value, the better the system performance. The simulation result obtained is shown in figure 3 and the data of closed loop performance are given in table 2. Each MPID control methods are being compared by the transient response in terms of rise time, settling time and percentage of overshoot to make an analysis on which methods will the best performances for the ASP system.

Table 2. MPID tuning parameter data using BA,

Controller Tuning	BA							
	Davison		Penttinen-Koivo		Maciejowski		Proposed	
Fitness Function	132.6092		0.0075		1.5119		9.8716	
Epsilon, ε	10.2245		10.0027		7.9816		9.9900	
Rho, ρ	-		4.1857		9.9988		6.0562	
Alpha, α	-		-		-		0	
Output	S	DO	S	DO	S	DO	S	DO
Rise Time, T_R	0.814	0.274	0.476	0.285	0.596	0.144	0.249	0.415
Settling Time, T_S	28.1	3.75	2.10	1.15	3.48	1.40	1.65	1.01
Overshoot, OS%	73.0	48.4	2.4	10.6	8.66	1.36e-6	11.1	6.2e-6

*S: Substrate; DO: Dissolved Oxygen

Based on the time response data, it can be seen clearly that Davison on BA algorithm also give the worst response with the longer settling time and rise time. The reason for it is due to the missing proportional term in its equation where it only used the integral term. The advantages of having the proportional term in a system are where it can help to reduce the rise time and settling time. The Davison also give the worst result in terms of overshoot and fitness function compared to the others.

While, the other three methods control tuning of BA algorithm, Penttinen-Koivo, Maciejowski and Proposed Combined method shows the similar results. The similarity of the result is because of the same property of equation that they are used, which is the proportional and integral term. However, the Proposed Combined method gives the better performance compared to Maciejowski even though they are having similarities in the expression. This is because the complexity of finding suitable bandwidth in Maciejowski method has been reducing by the Proposed Combined method.

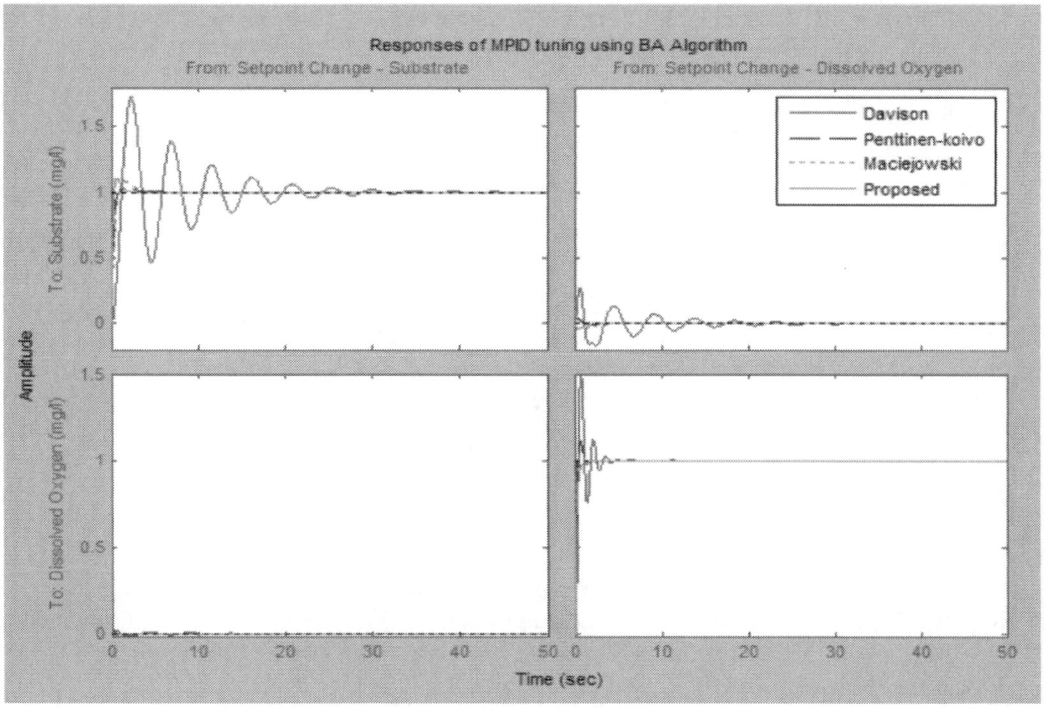

Figure 3. System performances of MPID-BA.

7. Conclusion and future works

This project shows that Davison method gives the worst performance. However, Penttinen-Koivo, Maciejowski and Proposed Combined method produce quite similar results. This is because Davison's method involves integral term only while the other methods consider the part of the proportional and integral terms. The importance of having proportional term in MPID is where the rise time and settle time can be reduced.

This study also proves that by using the optimization technique, a system can be tuned to obtain better performance responses. The result also concludes that the BA optimization with the Proposed Combined method is the best way to tune multivariable PID of activated sludge process. It is because the BA has the advantage with its ability to further find a solution with a lower fitness function. Thus, the ASP performance can be improved in order to have a great wastewater treatment plant.

For this study, the BA optimization has only covered the velocity at position solution of the system with a fixed frequency. It can be improved by continuing the simulation using the others criteria of BA which are by varies the loudness, pulse rate and wavelength for the bat to search a better solution.

Besides, this study can be better if the ASP system with MPID-BA is further testing by including a disturbance in the system. This testing process can determine whether the MPID controller can resist the disturbances or not. It also can prove that whether the system can fix the corner caused by the disturbances.

References

[1] M Mulas, S Tronci, F Corona, H Haimi, P Lindell and M Heinonen 2015 Predictive control of an activated sludge process: An appilcation to wastewater treatment plant *J. Process Control* **35** pp 85-100

[2] G Olsson, ICA and M Mulas 2012 A subjective review: Activated sludge process *Water Res.* **46** pp 1585-1624

[3] M A Shannon, P W Bohn, M Elimelech, J G Georgiadis, B J Marinas 2008 Science and technology for water purification in the coming decades *Nature* **452** pp 301-10
[4] G Olsson and B Newell 1999 *Wastewater treatment systems;Modeling, diagnosis and control* IWA Publishing **1** pp 15-19
[5] M Yong, P Yongzhen and W Shuying 2005 Feedforward-feedback control of dissolved oxygen concentration in a predenitrification system *Bioprocess and Biosystems Engineering* **27** pp 223-8
[6] B Holenda, E Domokos, A Redey and J Fazakas 2008 Dissolved Oxygen Control of The Activated Sludge Wastewater Treatment Process using Model Predictive Control *Computer and Chemical Engineering* **32** pp 1270-8
[7] H Hong-Gui, Q Hu-Hai and Q Jun-Fei 2014 Nonlinear Multiobjective Model-predictive Control Scheme for Wastewater Treatment Process *J. of Process Control* **24** pp 47-59
[8] I. Santin, C pedret, R Vilanova and M Menses 2016 Advanced Decision Control System for Effluent Violations Removal in Wastewater Treatment Plants *Control Engineering Practice* **49** pp 60-75
[9] S Kirkpatrick C D Gelatt and M P Vechchi 1983 Optimization by simulated annealing *Science* **220** pp 671-80
[10] M C Razali, N A WAhab, P Balaguer, M F Rahmat and S I Samsudin 2013 Multivariable PID controllers for dynamic process *IEEE 9th Int. Coll. on Sig. Proc. and Its Appl.* pp 1-5
[11] N A Wahab, M R Katebi and J Balderud 2007 Multivariable PID control design for wastewater systems *Proc. on 15th Mediterranean Conf. on Control and Automation* pp1-6
[12] P Balaguer, N A Wahab, M R Katebi and R Vilanova 2008 Multivariable PID control tuning: A controller validation approach *IEEE Sig. Proc. and Its Appl.* pp289-294
[13] X S Yang 2010 A new metahueristic bat inspired algorithm *Nature Inspired Cooperative Strategies for Optimization (NICSO 2010)* Springer pp 65-74
[14] G Wang and L Guo 2013 A novel hybrid bat algorithm with harmony search for global numerical optmization *Journal of Appl. Math.* **2013** pp 1-21
[15] N A Selamat, N A Wahab and S Sahlan 2013 Particle swarm optimization for multivariable PID controller tuning *IEEE 9th Int. Coll. on Sig. Proc. and Its Appl.* pp 170-5

IOP Conf. Series: Materials Science and Engineering **342** (2018) 012031 doi:10.1088/1757-899X/342/1/012031

An ergonomics prototype of adjustable chin stands aid for visual mechanical inspection at electronic manufacturing-based company in Kuantan, Malaysia

Nurainaa Elias and Nafrizuan Mat Yahya

Faculty of Manufacturing Engineering, University Malaysia Pahang, 26600 Pekan, Malaysia

Email: nurainaa_elias@yahoo.com

Abstract. Chin stands aid is a device designed to reduce fatigue on the chin during the Visual Mechanical Inspection (VMI) task for operators in TT Electronic Sdn Bhd, Kuantan, Malaysia. It is also used to reduce cycle time and also improve employee well-being in terms of comfort. In this project, a 3D model of chin stands aid with an ergonomics approach is created using SOLIDWORKS software. Two different concepts were designed and the best one is chosen based on the Pugh concept selection method, concept screening and also concept scoring. After the selection of concepts is done, a prototype of chin stands aid will be developed and a simulation of the prototype is performed. The simulation has been executed by using Workbench ANSYS software as a tool. Stress analysis, deformation analysis, and fatigue analysis have been done to know the strength and lifespan of the product. The prototype also has been tested to know the functionality and also comfortability for the user to use the chin stands aid.

1. Introduction

Ergonomics is an interdisciplinary field of study that seeks to design tools and equipment for optimizing the interface between humans and systems [1]. Ergonomics is also known as the process of understanding and improving human communications with types of products, equipment, environments, and systems. Ergonomics principles are used to improve the "fit" between the worker and the workplace. Targets for ergonomics improvement are divided into two broad categories which are physical issues, such as reaches and exertion, and cognitive issues, such as mental overload and confusing displays [2].

Many theories, principles, methods, and data relevant to the workstation design have been generated through ergonomics research over the years. Poor workstation design can lead to fatigued, frustrated, and hurting workers [3]. As a result, the productivity of the product and quality will decrease [4].

Often in an industry, the workstation is designed in an arbitrary manner [1], giving little consideration to the anthropometric measurements of the anticipated user. Small changes in workstation dimension can have a considerable impact on worker productivity, and occupational health and safety. Inadequate posture from an improperly designed workstation causes static muscle efforts and then eventually resulting in acute localized muscle fatigue, and consequently in decreased performance and productivity, and in an enhanced possibility of operator-related health hazards [5]. Therefore, workstation design plays an important role to improve productivity and safety.

In the process that mostly required the ergonomics workstation design are cropping process, laser marking process and also Visual Mechanical Inspection (VMI). VMI refers to the simple examination of a component for detects using the human eyes and is aided by artificial tools. VMI is usually used because it is simple, easy to apply and quickly carried out, but good eyesight, concentration and also illumination. However, TT Electronic Sdn Bhd, which located in Kuantan, Malaysia has some problem with VMI. At VMI, the major problem that occurred is an inappropriate workspace that can increase fatigue to the workers, especially on the chin part.

So, for this project, a prototype of adjustable chin stands aid will be proposed to reduce fatigue on the chin during VMI task for operators in TT Electronic Sdn Bhd, Kuantan, Malaysia. At the same time, employees can be more productive to reduce cycle time and also improve employee well-being in terms of comfort. As a result, human well-being can be improved, increased efficiency, produced a defect-free product and also improved profits for the company.

2. Method

The drawing is divided into three steps. The first step is sketching. All the ideas for the chin stand aid sketch on the paper to generated design for idea selection after the design is being chosen. Then, the sketch will be evaluated to set the new concept of the product. The last step is to draw the selected sketch by using SOLIDWORKS software.

2.1. Design

Generally, the chin stands aid can be divided into two main mechanisms which are supported mechanism and also lifting mechanism. For the first design, scissor lift is chosen as the lifting mechanism and rubber chin support for the support mechanism. For the second design, screw lift is chosen as the lifting mechanism and silicone pad for the support mechanism. Figure 1(a) and Figure 1(b) shows the sketching design and the mechanism of the concepts.

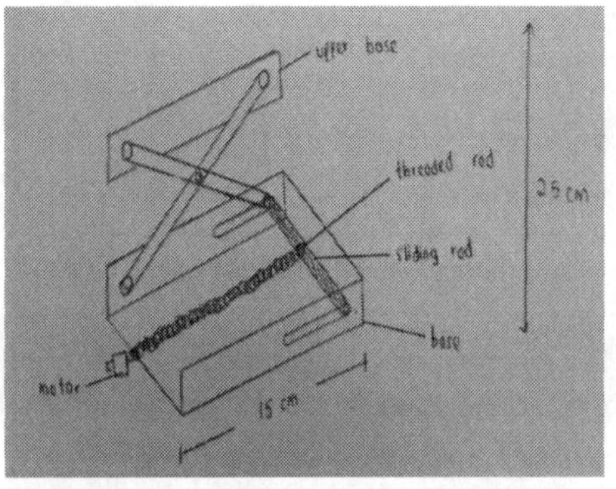

Figure 1(a). Concept A. Scissor lifts as a lifting mechanism.

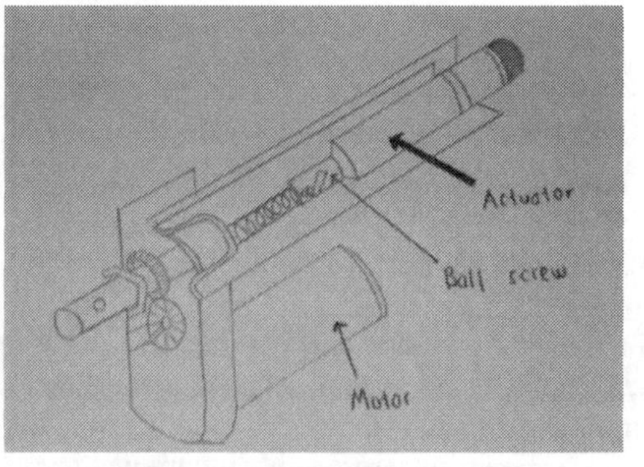

Figure 1(b). Concept B. Linear actuator as a lifting mechanism.

2.2. Concept generation and evaluation

From the two concepts, the best one should be decided as it will be the design of this project. Three sets of evaluation will be used to compare the design. The one design with the best specification will be built. The three sets of evaluation include a Pugh concept selection method [6], concept screening and concept scoring [7].

From the Table 1, the advantages and disadvantages of the design can be outlined. The criteria or the characteristics of the product to be fabricated are the important things to be considered before fabrication process. According to the table, a study of the concept selection shows the concept B scores highest positive signs. So the concept B is the best selection to fabricate.

Table 1. Pugh Concept Selection Method.

Selection criteria	Concept A	Concept B
Safety	-	+
Load capacity	-	+
Adjustable height	+	+
Durability	-	+
Stability	-	+
Manufacturing cost	+	-
Ease of handling	+	+
Ease to manufacture	+	+
Lightweight	+	-
High strength	-	+
$\sum +$	5	8
$\sum -$	5	2
Net score	0	6
Ranking	2	1

Notes: '+' is for better, '-' is for worse

Table 2 shows the result concept screening. The result is evaluated from the two concept design. All the character that has for the concept is inserted in the table to see the concepts that have the characteristic needed for the chin stands aid.

Table 2. Concept Screening.

Concept	Height adjustable	Longevity	Automatic	Cost	Size
A	√		√	√	
B	√	√	√		√

Table 3 shows the result of the concept scoring. The marks are given based on the concept idea. The products that have the best characteristics are given 9 marks, the second best is given 6 and the worse is given 3 marks. The concept products that have the highest mark will proceed to produce and it will add to the character of other product that will improve the quality of the product.

Table 3. Concept Scoring.

Concept	Flexible Use	Ergonomics	Total weight	Appearance	Safety	Total marks	Rankings
A	6	6	9	6	3	30	2
B	6	9	6	6	9	36	1

Notes: '3' is for worse, '6' is for medium, '9' is for good

From the comparison between concepts, Concept B is chosen as the finalize design. It is because the load capacity is more than Concept A. Although the stroke length of Concept A is more than Concept B, screw lift easier to have a stable linear path when moving compared to the scissor lift. Besides that, screw lift has been using the available product while scissor lift has not been used.

IOP Conf. Series: Materials Science and Engineering **342** (2018) 012031 doi:10.1088/1757-899X/342/1/012031

2.3. Finalize product
Figure 2(a) shows the image of the chin stands aid that is generated from the SOLIDWORKS. It shows the complete parts of the design that assembled in the SOLIDWORKS software. The prototype consists of lifting and support mechanism. Screws are used to connect the support and lifting mechanism. The material and the size chosen are referred to the average anthropometric from workers at VMI station and the characteristics of the material itself.

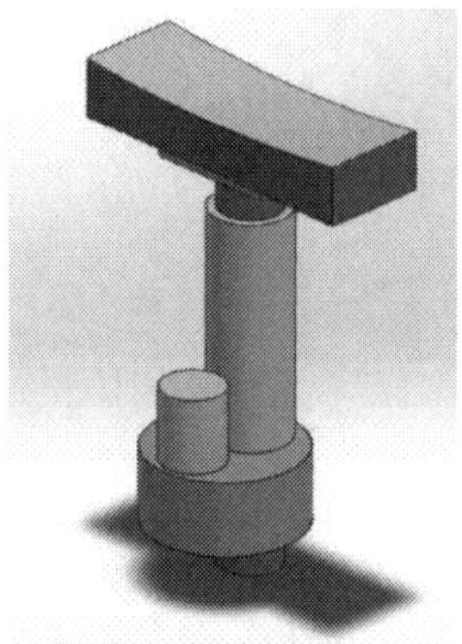

Figure 2(a). Isometric view of chin stands aid using SOLIDWORKS software.

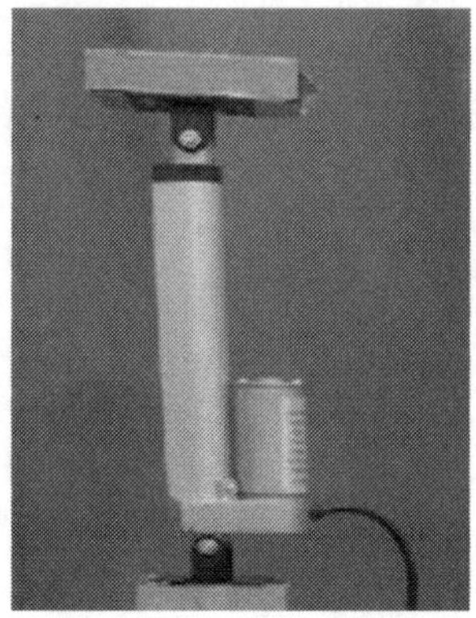

Figure 2(b). A prototype of chin stands aid.

After completing the fabrication process, the prototype as shown in Figure 2(b) must be going through the analysis process. At this stage, all data about the product is gathered. It is important to classify the product before it enters a market.

3. Performance analysis of proposed prototype
After completing the fabrication process, the prototype must be going through the analysis process. At this stage, all data about the product is gathered. It is important to classify the product before it enters the market.

3.1. Experimentation and testing of prototype
The chin stands aid is safe for the user. Figure 3 shows the chin stands aid that been tested by one of the operators at TT Electronic Sdn Bhd. Figure 3 (a) is the front view of the chin stands aid. It shows how the operator used the prototype during working while Figure 3 (b) shows the side view of the chin stands aid.

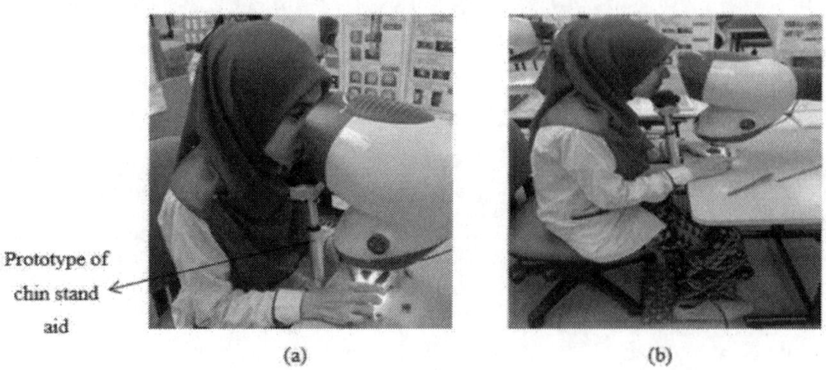

Figure 3. Testing the prototype.

The chin stands aid also have been used by the operators in VMI station to test the comfortability of the product and also measure the parameters such as time taken to adjust the product, strength and also the functionality of the chin stands aid. Table 4 shows the result of the experiment.

Table 4. Result from testing of prototype.

	Age	Height (cm)	Mass of chin (kg)	Time is taken to adjust the chin stands aid (second)	Comfortability
Operator 1	30	160	4.9	6.92	YES
Operator 2	26	156	2.3	5.10	YES
Operator 3	49	164	4.5	7.21	YES
Operator 4	35	157	3.9	6.94	YES
Operator 5	47	155	5.1	7.05	YES

From the result, it is proved that the chin stands aid is user-friendly because of it suitable to use for every age ranges. The height of the chin stands aid is adjustable. Therefore, it can fix with user height so that the users feel more comfortable when using the product. The maximum load (mass of chin) that has been tested is 5.1kg. So, for the analysis, this value will be used to know the lifespan of the chin stands aid. Besides that, the average time taken to adjust the prototype is only less than 8 seconds. So the operators will not waste time to wait for the chin support to be lifted. The maximum time taken is only 7.21seconds. For the comfortability, all the operator agreed that the product is comfortable to use and can reduce fatigue, especially on the chin during VMI process.

3.2. Simulation analysis

Import the drawing of chin stand aid into the ANSYS software to make failure and structural analysis. Stress life type of failure analysis is used for this analysis. Stress life is based on S-N curves (Stress – Cycle curves). Stress life is concerned with total life and does not distinguish between initiation and propagation [8]. In terms of cycles, stress life typically deals with a relatively high number of cycles. A high number of cycles refer to more than 105 (100,000) cycles [9]. Stress life traditionally deals with relatively high numbers of cycles [10] and therefore addresses High Cycle Fatigue (HCF), greater than 105 cycles inclusive of infinite life. For the loading type, constant amplitude (zero-based), proportional loading is used because this loading is of constant amplitude because only one set of FE stress results along with a loading ratio is required to calculate the alternating and mean values. And for the mean stress correction (stress life), Goodman theory is used [11].

In the fatigue analysis, a lifespan of the chin stand aid can be known. When subjected to millions of small repeated loads, a structure may have a slow growth of surface cracks that can cause material strength degradation and sudden failure [12]. Therefore, it is important to know the lifespan of the chin stand aid in maintaining the safety. The result of the fatigue life analysis can be seen in Figure 4. The minimum life for the chin stand aid is 100 months.

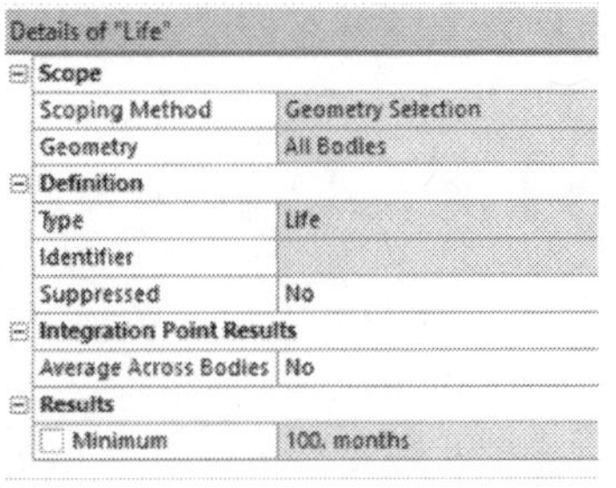

Figure 4. Detail of fatigue life from ANSYS software.

IOP Conf. Series: Materials Science and Engineering **342** (2018) 012031 doi:10.1088/1757-899X/342/1/012031

4. Conclusion

In this work, the prototype which has been proposed meets the objective of the project. This is because after done some analysis like stress analysis and so on, it is proved that the prototype has high strength and also can support certain weight capacity. So this prototype can help operators in VMI station to reduce the fatigue on the chin, especially when working. Other than that, the concept of the product can help them effectively in order to fulfill their need to have chin stands aid that can help to do VMI on product smoothly. In the nutshell, using Concept B for the chin stands aid is better and safe than Concept A.

Acknowledgements

This work was supported by TT Electronics Sdn Bhd. The authors would like to thank TT Electronics Sdn Bhd for encouraging this research.

References

[1] MacLeod D 2006 *The Ergonomics Kit for General Industry* (Boca Raton: Taylor and Francis)
[2] Pheasant S 2002 *Bodyspace: Anthropometry, Ergonomics and the Design of Work* (London: Taylor and Francis)
[3] Kroemer K 2008 *Fitting the Human Introduction to Ergonomics* (Boca Raton: CRC Press)
[4] McCormick E and Sanders M 2001 *Human Factors in Engineering and Design* (New York: McGraw Hill)
[5] Kodak 2003 Human Factors Section Eastman Kodak Company *Ergonomic Design for People at Work* (New York: John Wiley and Sons)
[6] Pugh S and Clausing D 1996 *Creating innovative products using total design: the living legacy of Stuart Pugh*. (New York: Addison-Wesley Longman Publishing)
[7] Yin Y, Vanides J, Ruiz-Primo M, Ayala C and Shavelson R 2005 Comparison of two concept-mapping techniques: Implications for scoring, interpretation, and use. *Journal of Research in Science Teaching* ***42***(2), 166-184
[8] Rust W and Schweizerhof K 2003 Finite element limit load analysis of thin-walled structures by ANSYS (implicit), LS-DYNA (explicit) and in combination. *Thin-walled structures* ***41***(2), 227-244
[9] Stolarski T, Nakasone Y and Yoshimoto S 2011 *Engineering analysis with ANSYS software* (New York: Butterworth-Heinemann)
[10] Youdong Y and Zhebo Z 2006 Model Analysis of Involute Gear by Finite Element Method Based on ANSYS [J]. *Journal of Mechanical Transmission* ***5***, 21
[11] Liu Y, Tian C, Liu T, and Kong F 2015 January Modal Analysis of Worm and Worm Gear Based on ANSYS Workbench. In *MATEC Web of Conferences*, 34
[12] Yu D, Al-Yafawi A, Park S and Chung S 2010 Finite element based fatigue life prediction for electronic components under random vibration loading. In *Electronic Components and Technology Conference (ECTC), Proceedings 60th* (pp. 188-193)

Review of coal bottom ash and coconut shell in the production of concrete

S K Faisal, P N Mazenan, S Shahidan and J M Irwan

Jamilus Research Center, Faculty of Civil and Environment Engineering, Universiti Tun Hussein Onn Malaysia, 86400 Parit Raja, Johor

E-mail: faisalsh@uthm.edu.my

Abstract. Concrete is the main construction material in the worldwide construction industry. High demand of sand in the concrete production have been increased which become the problems in industry. Natural sand is the most common material used in the construction industry as natural fine aggregate and it caused the availability of good quality of natural sand keep decreasing. The need for a sustainable and green construction building material is required in the construction industry. Hence, this paper presents utilization of coal bottom ash and coconut shell as partial sand replacement in production of concrete. It is able to save cost and energy other than protecting the environment. In summary, 30% usage of coal bottom ash and 25% replacement of coconut shell as aggregate replacement show the acceptable and satisfactory strength of concrete.

1. Introduction

An utilization of industrial and agricultural waste produced by industrial processes has been the focus on waste reduction. Therefore, researchers and scientist are in search of developing alternate binders that are ecofriendly and contribute towards waste management [1,2]. One of the alternative waste materials is coal bottom ash. Non-combustible material from the burning of coal in furnace of coal fired thermal power plants had produced the coal bottom ash [3]. From the burning process of coal, 80 to 90% of the product will become fly ash and other 10 to 20% of product will become bottom ash [4]. Coal bottom ash (CBA) is a by-product produced from municipal solid waste incinerators and coal fueled power stations [5]. Physical characteristics of coal bottom ash are porous surface texture, coarse, glassy, granular and grayish in color that collected from bottom of furnaces that burn coal [4], [5]. Coal bottom ash is industrial waste that disposed after electricity process.

Coconut shell is categorized as solid disposal waste that produce from agricultural activities. Domestic waste volume that represents by coconut shell are more than 60% which it leads to the serious disposal problems for local environment. However, these wastes can be used as potential material or replacement material in the construction industry in developing country where abundant agricultural and industrial waste are discharged and it will give the double advantages of reduction on the cost of construction material other than disposal of wastes [6].

2. Literature survey on coal bottom ash

Coal bottom ash can be used as sand replacement since it has similar particle size distribution as fine aggregate. Numerous researchers have explored for utilizing coal bottom ash in concrete production.

IOP Conf. Series: Materials Science and Engineering **342** (2018) 012032 doi:10.1088/1757-899X/342/1/012032

Singh and Siddique [7] carried out the research to study the effect of coal bottom ash on workability and strength of concrete. This study used coal bottom ash 0%, 20%, 30%, 40%, 50%, 75% and 100% as sand replacement with 0.45 and 0.50 of water-cement ratio. Results were taken at age 28, 90, 180 and 365 days. Conclusion made by author stated that 30% and below volume of coal bottom ash without superplasticizer were considered for workability and strength properties. In the other hand, concrete with superplasticizer was recommended to use up less than 50% of coal bottom ash. Researchers also mentioned that coal bottom ash is suitable to be used as partial sand replacement in production of concrete.

Rafieizonooz et al [8] have substituted the fine aggregate with coal bottom ash and cement with fly ash. Proportions used in this study were 0%, 20%, 50%, 75% and 100% of coal bottom ash and 20% for fly ash in the concrete production with 0.55 of water-cement ratio. The specimens were curing and tested at age 7, 28, 91 and 180 days. Conclusions made with this study is the workability of concrete reduced due to the replacement of sand with coal bottom ash. In addition, flexural and splitting tensile strength of concrete with 75% of coal bottom ash and 20% of fly ash higher than control.

For the experiment carried out by Ibrahim et al [9], the researchers study the effect of sand replacement with in self-compacting concrete and investigate the effect of coal bottom ash on split tensile strength. This study replaced sand by 0%, 10%, 20% and 30% with various of water-cement ratio of 0.35, 0.40 and 0.45. The specimens then cured in tap water and tested on the age of 7, 14 and 28 days. Results found that the workability reduced with the increases of coal bottom ash but the slump flow time increases due to coal bottom ash absorbed the mixture water contents. Researchers also found that 10% of coal bottom ash in self-compacting concrete resulted in highest tensile strength.

Raju [10] had done the study towards the effect of use of coal bottom ash as partial replacement of fine aggregates in various percentages. Percentages used for the replacement level in this study were 0%, 5%, 10%, 15%, 20%, 25% and 30% with 0.4 of water cement ratio. Researchers found that 5% volume of coal bottom ash in shows the higher strength than control mix. However, the workability of concrete decreases with the utilization of coal bottom ash in concrete.

Siddique [11] has explored effect on the properties of self-compacting concrete containing coal bottom ash as sand replacement. The study was replaced fine aggregate by 0%, 10%, 20% and 30%. The strength of the concrete tested on days 7, 28, 90 and 365. The increases volume of coal bottom ash decreased the compressive strength of concrete. In addition, water absorption and sorptivity increased with the increases of coal bottom ash volume.

Kadam and Patil [12] used the volume from 0% to 100% of coal bottom ash with 10% interval to study the effect on the properties of concrete with utilization of coal bottom ash as sand replacement with 0.45 water cement ratio. The strength of the specimen was taken at 7, 28, 56 and 112 days. In summary, the compressive strength of the concrete increases with 20% replacement and further replacement decreases the strength of specimen. Moreover, split tensile and flexural strength increase with the replacement level of coal bottom ash from 10% to 30%.

Table 1. Summary of previous work on coal bottom ash.

Authors & year	Percentage of replacement	Application of material	Findings
Singh & Siddique, 2016	0%, 20%, 30%, 40%, 50%, 75% and 100%	Concrete	30% and below volume of coal bottom ash without superplasticizer were considered for workability and strength properties. Up to 50% was recommended with superplasticizer.
Rafieizonooz, Mirza, Salim, Hussin, &	0%, 20%, 50%, 75% and 100%	Concrete	Workability of concrete reduced due to the replacement of sand with coal bottom ash.

Khankhaje, 2016			Strength of concrete with 75% coal bottom ash and 20% fly ash higher than control mix.
Ibrahim, Hamzah, Jamaluddin, Ramadhansyah, & Fadzil, 2015	0%, 10%, 20% and 30%	Self-compacting concrete	Workability reduced with the increases of coal bottom ash but the slump flow time increases due to coal bottom ash absorbed the mixture water contents. 10% replacement level resulted in highest tensile strength of SCC.
Raju, Paul, & Aboobacker, 2014	0%, 5%, 10%, 15%, 20%, 25% and 30%	Concrete	Workability of concrete decreases with the utilization of coal bottom ash in concrete. 5% of coal bottom ash in concrete shows the higher strength than control mix.
Siddique, 2013	0%, 10%, 20% and 30%	Self-compacting concrete	Slump flow in the range 627mm to 673mm with flow time less than 6s. Compressive strength decreased with the increases of coal bottom ash. Water absorption and sorptivity increased with the volume increases of coal bottom ash.
Kadam & Patil, 2013	0%, 10%, 20%, 30%, 40%, 50%, 60%, 70%, 80%, 90% and 100%	Concrete	Compressive strength increases with 20% replacement. Further replacement decreases the strength of specimen. Split tensile and flexural strength increase with 10% to 30% sand replacement.

3. Literature survey on coconut shell

Most of the researcher used coconut shell as the coarse aggregate replacement in the production of concrete. The aim of Kanojia and Jain's [15] study was to identify the effects of coconut shell in concrete as partial replacement the conventional coarse aggregate. This study replaced the aggregate with coconut shell by 0%, 10%, 20%, 30% and 40% with 0.55 water to cement ratio. The strength of the specimen was taken at age 7 and 28 days. In summary, 40% replacement of coarse aggregate in concrete decrease the strength about 22% in the 28 days strength. Coconut shell concrete shows low strength development at early age but it shows rapid strength development at later age. Moreover, concrete became lighter with replacement of coarse aggregate with coconut shell.

Kakade and Dhawale [16] had done the study towards the properties of concrete containing coconut shells in terms of characteristic compressive and tensile strength. Percentages used for the replacement level in this study were 0%, 25% and 50% with 0.46 of water cement ratio. The test was run after 7 and 28 days of specimens in curing tank. In conclusion, researchers found that 25% usage of coconut shell as replacement of aggregate satisfies the requirement for structural lightweight concrete. However, 50% replacement of coarse aggregate with coconut shell can be used for less important work

Rao et al [17] have substituted coarse aggregate with coconut shell with proportions of 0%, 10% and 20%. The aim of this study was to determine the strength of coconut shell concretes at different replacement level and study the transport properties of concrete with coconut shells as coarse aggregate replacement. The specimens were tested for compression strength and split tensile strength at age 3, 7 and 28 days. The conclusion draws by researchers stated that the workability of concrete decrease with the addition of coconut shells. However, the workability increases with the addition of fly ash as cement replacement. Moreover, the replacement of coconut shells decreased the strength properties of concrete.

Kambli and Mathapati [18] has compared the concrete containing coconut shell as coarse aggregate with conventional concrete. Proportions used in this study were 0%, 10%, 20%, 30% and 40%. The specimens were tested with compressive strength for 7 and 28 days. It is found that the coconut shell can be used with proportions varying from 0 to 30% whereas 30% replacement of coarse aggregate in M20 grade concrete shows 23 Mpa strength in 28 days. Other than that, researchers found that the coconut shell can reduce the material cost in construction because of low cost and abundant agricultural waste other than it has potential as lightweight aggregate in.

The main objectives of study carried by Dahiya and Dharni [19] is to encourage the use of waste products as construction materials in low-cost housing. This study compared the properties of normal concrete with the full replacement of coarse aggregate concrete with 0.47 water cement ratio. The concrete tested after 7 days for compressive strength. The strength of concrete containing coconut shells is 12.44 Mpa which is lower than normal concrete where the strength is 19.36 Mpa. However, the concrete can be used at places where low strength economical concrete is needed.

Yerramala and C [20] have conducted experimental investigation by replacing coarse aggregate with 0%, 10%, 15% and 20% of coconut shell as aggregate replacement to study the properties of concrete with coconut shell. Tests were carried out at age 1, 7 and 28 days. Workability of the concrete decreases as the coconut shell is added. Moreover, both compressive and split tensile strength of concrete containing coconut shell decreased compared to control concrete. Even though, there is many advantages of using construction and agriculture wastes as replacement material in concrete, however the concrete produced must comply with the standard. Hence, the determination of properties of the substitution materials is very important because it can gives effect to the concrete strength either it can increases or decreases the concrete strength [21,22].

Table 2. Summary of previous work on ceramic waste powder.

Authors & year	Percentage of replacement	Application of material	Findings
Kanojia & Jain, 2017	0%, 10%, 20%, 30% and 40%	Concrete	40% replacement of coarse aggregate in concrete decrease the strength about 22% in the 28 days strength.
Kakade & Dhawale, 2015	0%, 25% and 50%	Lightweight concrete	25% replacement of aggregate with coconut shell satisfies the requirement for structural lightweight concrete. 50% replacement can be used for less important work.
Rao, Swaroop, Rao, & Bharath, 2015	0%, 10% and 20%	Concrete	The replacement of coarse aggregate with coconut shell decrease the workability of concrete. Strength of concrete deceased with the replacement of coconut shell as coarse aggregate.

Kambli & Mathapati, 2014	0%, 10%, 20%, 30% and 40%	Concrete	30% replacement of coarse aggregate in M20 grade concrete shows 23 Mpa strength in 28 days. Coconut shell can be used with proportions varying from 0-30%
Dahiya & Dharni, 2015	0%, and 100%	Concrete	Strength of concrete containing coconut shell are recorded at 12.44Mpa lower than normal concrete which 19.36Mpa.
Yerramala & C, 2012	0%, 10%, 15% and 20%	Concrete	Workability decrease with the replacement of coconut shell. Both compressive and split tensile strength of concrete containing coconut shell decreased compared to control concrete.

4. Conclusion

From the researchers discussed above, it is clearly shown that coal bottom ash and coconut shell are suitable to be used in the construction industry and more economic on the making of concrete. Coal bottom ash and coconut shell are found to be suitable as partial replacement of aggregate in concrete production. The researchers suggested up to 30% replacement of coal bottom ash as sand replacement in the concrete production. For coconut shell, the replacement suggested among the researchers is up to 25% and further replacement only suitable for less important work.

References

[1] Khalid F S, Azmi N B, Mohd Sumandi K A S and Mazenan P N 2017 *Mechanical properties of concrete containing recycled concrete aggregate (RCA) and ceramic waste as coarse aggregate replacement* AIP Conference Proceedings 1891, 020079

[2] Sheikh Khalid F, Herman H S and Azmi N B 2017 Properties of Sugar Cane Fiber on the Strength of the Normal and Lightweight Concrete *MATEC Web of Conferences* **103** 01021

[3] Singh M and Siddique R 2013 Effect of coal bottom ash as partial replacement of sand on properties of concrete *Resour. Conserv. Recycl.* **72** pp 20–32

[4] Abubakar A U and Baharudin K S 2013 Tanjung Bin Coal Bottom Ash: From Waste to Concrete Material *Adv. Mater. Res.* **705** pp 163–168

[5] Safiudin M, Jumaat M Z, Salam M A, Islam M S and Hashim R 2010 Utilization of solid wastes in construction materials *Int. J. Phys. Sci.* **5** pp 1952–1963

[6] Gunasekaran K, Annadurai R, and Kumar P S 2012 Long term study on compressive and bond strength of coconut shell aggregate concrete *Constr. Build. Mater.* **28** pp 208–215

[7] Singh M and Siddique R 2016 Effect of coal bottom ash as partial replacement of sand on workability and strength properties of concrete *J. Clean. Prod.* **112** pp 620–630

[8] Rafieizonooz M, Mirza J, Salim M R, Hussin M W, and Khankhaje E 2016 Investigation of coal bottom ash and fly ash in concrete as replacement for sand and cement *Constr. Build. Mater.* **116** pp 15–24

[9] Ibrahim M H W, Hamzah A F, Jamaluddin N, Ramadhansyah P J and Fadzil A M 2015 Split Tensile Strength on Self-compacting Concrete Containing Coal Bottom Ash *Procedia - Soc. Behav. Sci.* **195** pp 2280–2289

[10] Raju R, Paul M M and Aboobacker K A 2014 Strength Performance of COncrete Using Bottom Ash As Fine Aggregate *Int. J. Research Eng. Technol.* **2** pp 111–122

[11] Siddique R 2013 Compressive strength, water absorption, sorptivity, abrasion resistance and

permeability of self-compacting concrete containing coal bottom *Constr. Build. Mater.* **47** pp 1444–1450

[12] Kadam M P and Patil Y D 2013 Effect of coal bottom ash as sand replacement on the properties of concrete with different w/c ratio *Int. J. Adv. Technol. Civ. Eng. ISSN* **1** pp 2231–5721

[13] Singh M and Siddique R 2016 Effect of coal bottom ash as partial replacement of sand on workability and strength properties of concrete *J. Clean. Prod.* **112** pp 620–630

[14] Siddique R 2013 Compressive strength, water absorption, sorptivity, abrasion resistance and permeability of self-compacting concrete containing coal bottom ash *Constr. Build. Mater.* **47** pp 1444–1450

[15] Kanojia A and Jain S K 2017 Performance of coconut shell as coarse aggregate in concrete *Constr. Build. Mater.* **140** pp 150–156

[16] Kakade S A and Dhawale A W 2015 Light Weight Aggregate Concrete by Using Coconut Shell *Int. J. Tech. Res. Appl.* **3** pp 127–129

[17] Rao K V, Swaroop A H L, Rao P K R and Bharath C N 2015 Study on Strength Properties of Coconut Shell Aggregate *Int. J. Civ. Eng. Technol.* **6** pp 42–61

[18] Kambli P S and Mathapati S R 2014 Compressive Strength of Concrete by Using Coconut Shell *IOSR J. Eng.* **4** pp 1–7

[19] Dahiya H and Dharni N 2015 Concrete with Crushed Coconut Shell as Coarse Aggregate *J. Mech. Civ. Eng.* **1** pp 15–19

[20] Yerramala A and R C 2012 Properties of Concrete with Coconut Shells as Aggregate Replacement *Int. J. Eng. Invent.* **1** pp 21–31

[21] Sheikh Khalid F, Shah Herman H, Azmi N B and Juki M I 2017 Sand Cement Brick Containing Recycled Concrete Aggregate as Fine-Aggregate Replacement *MATEC Web of Conferences* **103(5)** 01016

[22] Khalid F S, Juki M I, Othman N and Ibrahim M H W 2017 Pull-Out Strength Of Polyethylene Terephthalate Bottle Fibre In Concrete *Matrix Malaysian Construction Research Journal (MCRJ)* **21(1)** 75-85

IOP Conf. Series: Materials Science and Engineering **342** (2018) 012033 doi:10.1088/1757-899X/342/1/012033

A preliminary survey on mode choice and its effect in users' satisfaction on their journey to the railway station

Nur Fahriza Mohd Ali[1,2*], Ahmad Farhan Mohd Sadullah[3] and Adnan Zulkiple[4]

[1]Pusat Pengajian Kejuruteraan Awam, Kampus Kejuruteraan, Universiti Sains Malaysia, 14300 Nibong Tebal, Pulau Pinang, Malaysia
[2]Innovative Manufacturing, Mechatronics & Sports Laboratory (iMAMS), Fakulti Kejuruteraan Pembuatan, Universiti Malaysia Pahang, 26600 Pekan, Pahang, Malaysia
[3]Pejabat Naib Canselor, Aras 6, Bangunan Canselori, Universiti Sains Malaysia, 11800 USM, Pulau Pinang, Malaysia
[4]Fakulti Kejuruteraan Teknologi, Block A2, Universiti Malaysia Pahang, Lebuhraya Tun Razak, 26600 Gambang, Kuantan, Pahang, Malaysia

E-mail: fahriza90@gmail.com

Abstract. This paper focuses on two lines of investigation with regard to mode choice to Klang Komuter Station. Firstly, the profile of the access modes on journeys to the railway station is analysed. Secondly, the relationship of users' mode choice towards overall perception on traveling from home to the railway station is estimated. The data collection was conducted via Revealed Preferences / Stated Preferences (RP/SP) Survey. Meanwhile, the analysis that was implemented in this study was correspondence analysis. This paper discussed more on journey purposes and the effects of distances from home to the railway, users' trip purposes and travel time between car and bus that was found to have an important effect on the users' mode choice and their satisfaction on their journey to the railway station. The results show that users were more satisfied to reach the station by car instead of the bus.

1. Introduction

Vehicles and transportation play a significant role owing to the increase in the development rate of our country. An efficient transport network is essential for a developing nation as economic activities apart from others are dependent on it. Ample accessibility and efficient transport systems can increase the productivity, social opportunities, as well as provide economic benefits through better accessibility to markets, additional investments and employment.

A railway journey has been always a part of a journey chain that includes different modes of transport to access and later from the railway station. It is necessary to ensure that door-to-door journey is implemented when using rail in order to achieve a continuous travel by integrating the modes involved in railway journey. Therefore, railway journey will be seen as a feasible and attractive alternative which more users urged to move out from cars. Integrated and efficient modes play an important role towards users' satisfaction as they are commuting from their respective point of origin to reach their desired destination.

The railway is an alternative mode to avoid road congestion during morning peak hour. Traffic congestion is not supportive towards transport mobility. Congestion can be defined as the slow movement of transports on the road when there are hindrances along the road that force the vehicles to move below their speed limits. In other words, traffic congestion is a condition of road networks due to the increasing demand among users and is characterized by longer trip times, slower speeds, and additional in vehicles queuing along the road. The traffic condition was affected by the number of cars crowding on the roads which these cars slowing down the traffic movement [1] and [2].
As identified by [3] the congestion leads to the increase of travel time in a journey from an origin to destination. Travel time may become more variable and unpredictable as congestion increases. This situation causes the travel time is difficult to be predicted while commuting during morning peak hour. In addition, the new Park and Ride under government's project near to the Klang Komuter Station have been forecasted to increase the number of cars coming to the town and coincidentally making the traffic congestion worse. Thus, this study explained what causes users to choose cars instead of the bus during morning peak hour and the information revealed can help the government to develop a strategy that can help users in making decisions either to choose bus or car in order to reach their destination.
Travel time consists of walking time, waiting time, and in-vehicle time. The combinations of these components affect the travel time of a transport either it will increase or decrease. The integration of these components is significantly important to achieve door-to-door travel. According to [3], a continuous travel from home is essential to ensure that public transport is viable and attractive mode instead of a car.

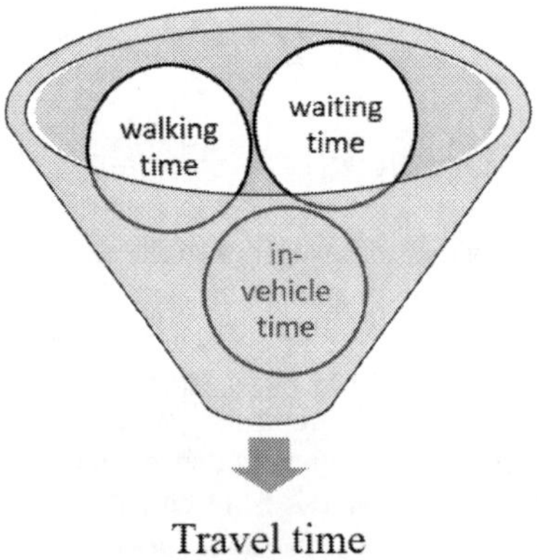

Figure 1. Travel time consists of waiting time, walking time, and in-vehicle time.

The importance in attracting users to switch their mode of transport from car to bus during morning peak hour is to ensure that they will have a more quality of life, including healthy lifestyle (reduce stress), a proper time management (avoid road congestion) as well as to promote sustainable transportation.

2. Literature Review

2.1. Revealed Preference / Stated Preference Survey (RP/SP) Survey
The result of Stated Preference (SP) survey is used to assess how respondents' choices vary to different hypothetical situations. SP is a survey technique deal with measuring and understanding the preferences underlying people's stated choices; based on how they react to hypothetical situations. In this research, respondents are introduced with choices between hypothetical but realistic alternatives, with each alternative being described in terms of their characteristics or "attributes". As noted above, the most common type of questions is, where the respondent is asked to choose, rank or rate different alternatives based on their attributes (e.g., travel time, travel cost, and wait time), thus giving information on the basis of the choice is made. For a choice-based SP survey is aiming to get a choice on the preferred alternative. In a ranking experiment, the respondent has to rank the alternatives in order of preference.

In a rating experiment each alternative must be classified using a scale which measures its attractiveness to the respondent.
In general transportation planning, the purpose of surveys is meant to capture users' current travel behaviour. For instance, a survey is conducted to get data regarding the actual mode a user is using, actual travel times, destinations, and so forth. This is known as Revealed Preference (RP) data, as the user is currently experiencing that behaviour and making a choice based on his or her knowledge of the available travel options. Revealed preference (RP) survey is a method of analysing choices made by users, such as for comparing preferred mode choice to reach Klang Komuter Station. This survey focused on assumption that the preferences of users on mode choice can be revealed by their journey purposes. Revealed preference survey tries to understand the preferences of a user among mode choices, while users were trying to reach the railway station during morning peak hour. For instance, if the users preferred to drive cars to reach the railway station instead of taking a bus, where both modes are available to commute them, it is revealed that users directly prefer cars instead of a bus. However, there might be other factors that involved users' preferences towards their mode choice.

2.2. Correspondence analysis
Correspondence analysis is a descriptive/exploratory technique designed to analyse simple two-way and multi-way tables containing some measure of correspondence between the rows and columns. Correspondence analysis is applied to contingency tables (i.e. cross-tabulations). Its primary goal is to transform a table of numerical information into a graphical display, in which each row and each column is depicted as a point. In a typical correspondence analysis, a cross-tabulation table of frequencies is first standardised, so that the relative frequencies across all cells sum to 1.0. One way to state the goal of a typical analysis is to represent the entries in the table of relative frequencies in terms of the distances between individual rows and/or columns in a low-dimensional space.

2.3. Mode choice
Mode choice analysis is the third step in the conventional four-step transportation forecasting model. The four steps, in order, are trip generation, trip distribution, mode choice analysis, and route assignment. Mode choice analysis allows the respondents or users to determine their preferred mode of transport that will be used, and the results of modal share.

2.4. Waiting Time
Waiting time is one of the travel time's components that is well known as the most influential factor in opting for bus as users have to reach their workplace at a stipulated time. As long as the performance of services for public transportation increases by reducing the travel time of the bus, users will be triggered to shift from cars to bus for their daily journey. Besides, the provision of bus should promise door-to-door approach from home to the station.
According to [4], it can be concluded that modal shift to public transports can be promoted by increasing the number of public transport that consequently reduce the waiting time at the bus stop; therefore, enabling the users to reach their destination in time. Another research carried out by [5] founds that waiting time is the most arduous component of the public transport travel time. The increasing in waiting time at the bus stop is the factor that urges users away from public transport. [5] Walker (2009) added, according to a study by [6] on the implementation of two modes which are trams and buses, the trams are practically evermore frequent than buses. Therefore, [5] Walker (2009) creates a concept of "frequent bus service" which is by increasing the bus frequency can reduce the waiting time for public transport.

2.5. In-vehicle time
In-vehicle time is also known as sitting time in a transport from initial point to the final point. Typically, although travellers "dislike" all time spent travelling, they dislike walking and waiting parts of the journey more than in-vehicle journey time, and thus would be willing to pay more to avoid them. The

first study in Britain to estimate walk and wait time values was that of [7]. He found that, "walking and waiting times are worth between two and three times in-vehicle time".

2.6. Walking distance
According to [8], a station will be more accessible if there are other transit interchanges such as buses, light rail transits or monorails because the distance that passengers need to walk will be less. For example, a bus stop that is located far away from home can be inconvenient for those who do not have other transportation that brings them to the bus stop. Furthermore, with the assistance of shuttle car services or feeder bus will eventually reduce users' walking distance as well as users' travel time from home to the station.
As proven by [8], public transport accessibility is associated with a certain number that is highly related to walking distance or walking time. Walking accessibility is defined as how easy it is for users to access bus stops by the mode of walking [8]. Walking is well known as a green and main mode for a user to access public transport. According to [8], walking and public transportation interrelated; which a proper and attractive walking conditions are necessary for pedestrians to use public transport, since most public transport requires users to walk at one or both ends of their trips. The research carried out by [8], agreed that 500 meters is the accessible walking distance which users are willing to access to the bus stop or to the Klang Komuter Station since Malaysia is a hot-climate country. Whereas, [9] also supported that walking distance that can be tolerated by users is a maximum of 500 meters.

2.7. Summary
From the literature, it can be summarized that this research paper focus on users' mode choice in order to reach Klang Komuter Station from home and the main characteristics of the users. The analysis also reveals the factors that influenced users' mode choice to access railway station and how this will affect their satisfaction on the journey from home to the railway station.

3. Methodology
3.1 Origin-Destination (OD) Survey
The data for OD Survey was collected at Klang Komuter Station. The Klang Komuter Station is a unique Traffic Analysis Zone because of the Park and Ride Station is built in the town center, which is against the norm where Park and Ride is located outskirt of the town. Therefore, it will encourage users to bring along their cars when traveling to the station.
The reliability of an O-D Survey results depends on its sampling scheme and sample size. The O-D Survey was conducted by randomly select 500 respondents who are willing to fill the O-D Survey form. The respondents who are selected to fill the forms should be from different gender, race, and age. The collection of the sampling data was done throughout the normal working days, from 7 AM and 7 PM. The OD Survey is also important to investigate which route heading to the Klang Komuter Station is highly congested due to cars crowding on the road during morning peak hour.

3.2 Revealed Preference / Stated Preference Survey (RP/SP) Survey
The data was taken among users of Klang Komuter Station during weekdays morning peak hour from 7 AM until 9 AM. The users were categorized into five (5) journey purposes which are working, studying, visiting, shopping, and banking etc.

The relationships of journey purposes with parameters including preferred waiting time (PreferredWT), preferred walking distance (PreferredWD), preferred in-vehicle time (PreferredIVT) between bus and cars, rating towards travel time between bus and cars, and distance from users' home to the Klang Komuter Station were investigated. These five (5) aforementioned parameters were selected for this preliminary survey in order to give a basic overview on the factors that affects users' mode choice and their satisfaction on the journey from home to the railway station.

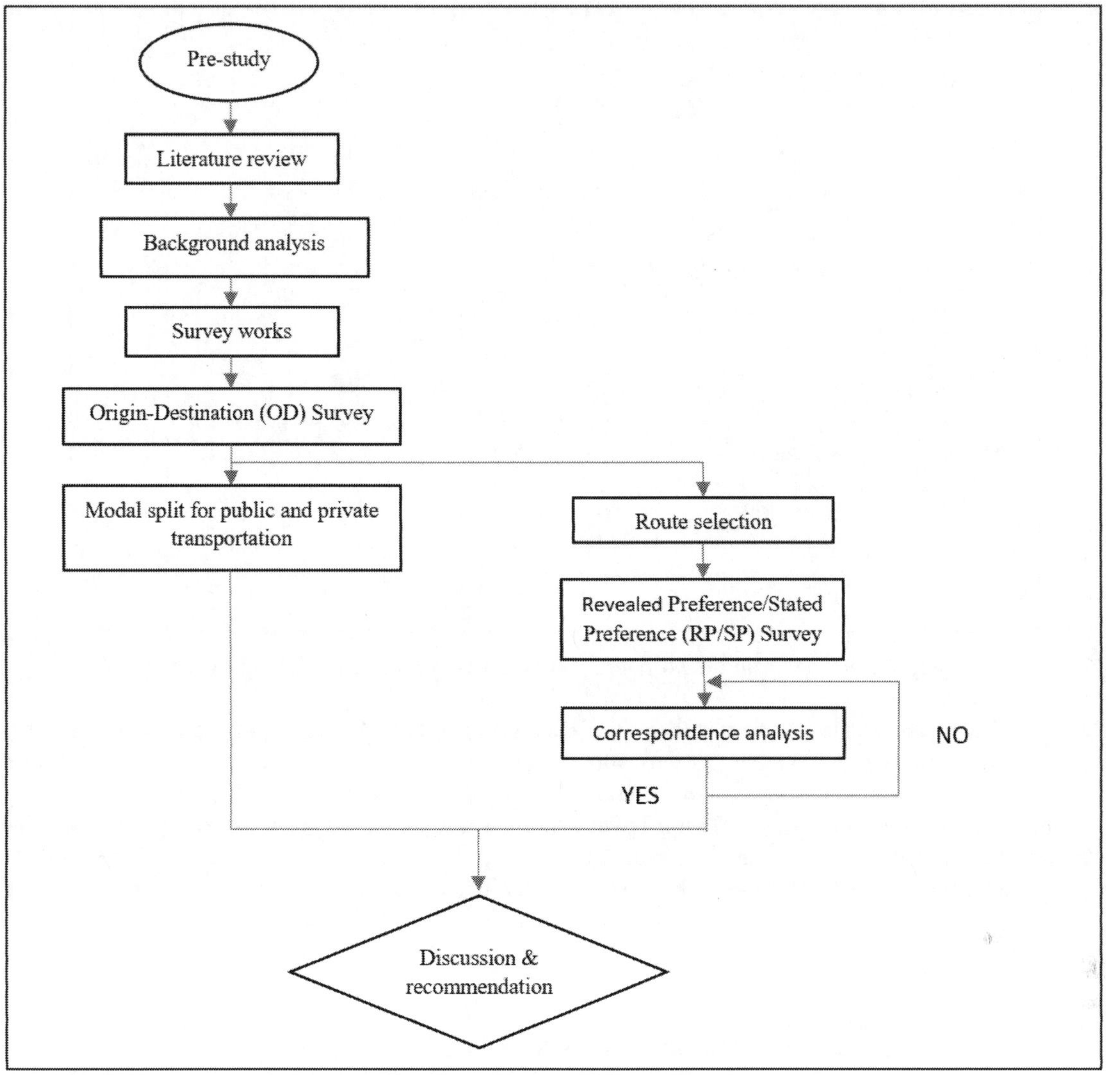

Figure 2. Flow chart of methodology used in the research.

4. Results

4.1 O-D Survey results

The survey site has been divided into 8 zones to facilitate the process of categorization of users from each origin.

IOP Conf. Series: Materials Science and Engineering **342** (2018) 012033 doi:10.1088/1757-899X/342/1/012033

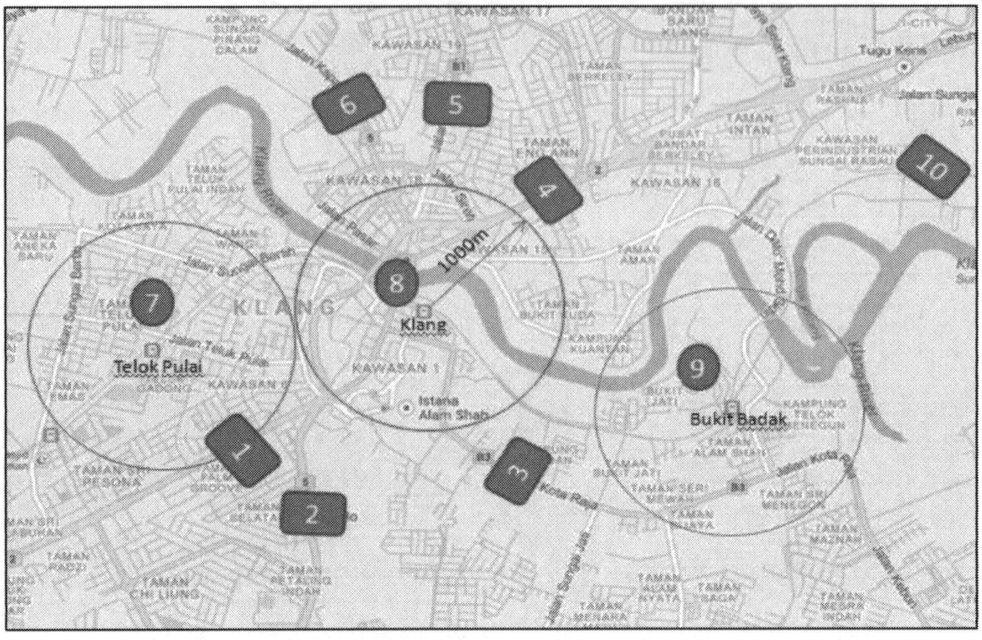

Figure 3. The Origin-Destination Survey was conducted at Klang Town.

The study area was divided into ten zones which are zone 1 (route from Port Klang to Klang Town), zone 2 (route from Banting to Klang Town), zone 3 (route from Taman Sri Muda to Klang Town), zone 4 (route from Lebuhraya Persekutuan to Klang Town), zone 5 (route from Klang Sentral to Klang Town), zone 6 (route from Kapar to Klang Town), zone 7 (1000 meters radius with center of Teluk Pulai Komuter Station), zone 8 (1000 meters radius with center of Klang Komuter Station), Zone 9 (1000 meters radius with center of Bukit Badak Komuter Station), and the last was zone 10 (the KTM Komuter route heading to Kuala Lumpur Central from Klang Komuter Station). Whereas, the routes that heading to the station can be listed as zone 1, 2, 3, 4, 5, and 6, which some of them have to cross the Klang Town in order to reach the station.

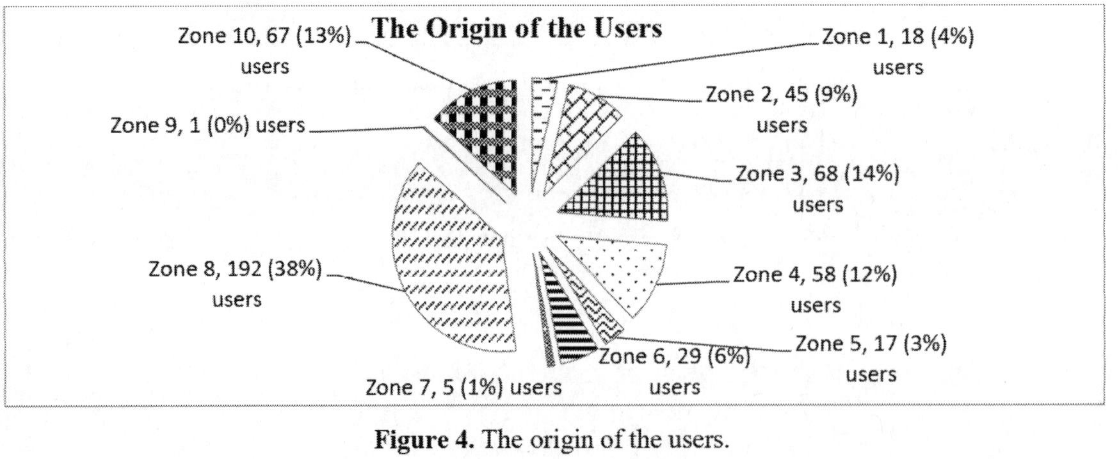

Figure 4. The origin of the users.

Figure 4 shows ten origins of Klang Komuter users. The highest number of users (38%) comes from Zone 8 which locates the Klang Komuter Station. The second highest (14%) comes from Zone 3 whilst the third highest (14%) come from Zone 10. Then, the result is followed by users come from Zone 4 (12%) and Zone 2 (9%). Meanwhile, the rest of the users come from Zone 6 (6%), Zone 1 (4%), Zone

5 (3%) and Zone 7 (1%). Finally, users come from Zone 9 only consists of one user. Hence, from this investigation, Zone 3 is selected for this study as most of Klang Komuter users come from this route to Zone 8.

4.2. Revealed Preference/Stated Preference (RP/SP) Survey by using Correspondence Analysis
The respondents for Revealed Preference/Stated Preference (RP/SP) Survey is consists of users from Zone 3 (OD Survey). The total respondents of RP/SP Survey are 252 users. The number of female users (142 users) at the Klang Komuter Station is higher than male users (110 users). The survey is conducted during morning peak hour from 7 AM until 9 AM and the users are chosen from the origin of Zone 3. Most of the female users are ferried by their husband to the station. This is possibly because the husband and wife are working in different workplaces; therefore, they would like to reduce the travel cost by implementing carpooling.

4.2.1. Modal split of male users. The split of male users and their mode to reach the Klang Komuter Station were found that car users (87%) are higher than bus users (13%). This research believes that the majority of male users prefers to drive to reach the Klang Komuter Station as they are concerned regarding the travel time from home to the station. In addition, if they happen to opt the bus service, they must wait for bus arrival at the bus stop which will consume plenty of their time. Although car users have to undergo the traffic congestion to reach the railway station, most of them are satisfied with the driving since they are more flexible and can travel at any time they preferred.

4.2.2. Modal split of female users. Meanwhile, the split of female users and their mode to reach the Klang Komuter Station can be found that only 12% of female users commute by bus to reach the station, conversely, 88% of female users commute by cars. The result also suggests an increase of car users and a decrease in bus users among females. This research believes that the majority of female users prefers to drive to reach the Klang Komuter Station since they are concerned about the travel time from home to the station as well. Also, if they commute by bus, they must wait for bus arrival at the bus stop which will consume a lot of their time too. It would be detrimental to them if they arrive at their workplace late as there are rules of the company that they must adhere to. Although car users consequently experience traffic congestion to reach their workplace, many of them are still satisfied with the driving as they are more flexible and can leave for work at any time they prefer. In addition, some of them happen to feel even better for having their own privacy whilst driving. As compared to female, male users seem more likely to commute by bus for working, perhaps they can relax and avoid traffic congestion, especially during morning and evening peak hours.

4.2.3. Journey purpose during morning peak hour. The highest journey purpose of Klang Komuter users during morning peak hour is for working which consists of 47% of users. The second highest journey purpose of Klang Komuter users is for studying (35%); meanwhile, the third highest is for visiting (10%). After that, it is followed by banking (6%). Lastly, the lowest journey purpose of Klang Komuter users is shopping, which consists of 2% of users. As a conclusion, the Klang Komuter is opted by the majority of users as a transportation hub for working and studying during normal weekdays from 7 AM until 9 AM. Hence, the aim in opting the commuter to reach the workplaces, schools, and universities is to avoid congestion on the road during morning peak hour.

4.2.4. Distances from home to Klang Komuter Station. The distances from home (Zone 3) to the Klang Komuter Station were grouped into five categories viz. less than or equal to 5 km, 6 km to 10 km, 11 km to 15 km, 16 km to 20 km, and 20 km above. The distance of the highest number of Klang Komuter users comes from less than or equal to 5 km (150 users), followed by 85 users that travel about 6 km to 10 km. Then, there are 9 users, which the distance from home is between 16 km to 20 km. The result is followed by 6 of users, which the distance from home is more than 20 km. Meanwhile, the lowest number of Klang Komuter users is only two users with the distance is between 11 km to 15 km from home to Klang Komuter Station.

4.3. Correspondence Analysis

This study used the combination of Revealed Preference/Stated Preference (RP/SP) Survey to investigate users' preference on transport mode choice. The survey has been analysed by the Correspondence Analysis to reveal the relationship between journey purposes and their preferences. The results are important for the government and service provider to plan for better facilities in order to improve time consumption for each stage of a trip from home (origin) to the station (destination).

Table 1. Correspondence table of the journey purposes and preferred waiting time.

	Preferred Waiting Time (PreferredWT)								
	1	2	3	4	5	6	7	8	
Journey Purpose	< 5 min	6-10 min	11-15 min	16-20 min	21-25 min	26-30 min	31-40 min	> 40 min	Active Margin
1 Work	69	49	0	0	0	0	0	0	118
2 Study	0	18	46	23	2	0	0	0	89
3 Visit	0	0	0	1	6	18	1	0	26
4 Shopping	0	0	0	0	0	0	8	4	12
5 Banking, etc.	0	0	0	0	0	0	0	7	7
Active Margin	69	67	46	24	8	18	9	11	252

From Table 1, users with working purpose show the highest tendency towards preferred waiting time for bus arrival at the bus stop which is less than 5 minutes. Meanwhile, for the study purposes, the tendency towards preferred waiting time for bus arrival at the bus stop is within 11 to 15 minutes. The users with visiting purpose tend to wait for bus arrival within 26 to 30 minutes while for shopping the tendency towards preferred waiting time for bus arrival at the bus stop is within 31 to 40 minutes. Lastly, the purposes of the users' journey of banking and other activities have a tendency towards preferred waiting time for bus arrival at the bus stop more than 40 minutes. It is apparent that time management is greatly important for those users with working purposes since they have to make their arrival at working place punctually. As a consequence of not having the provision of a bus that meets their needs, the users choose to make their journey by cars.

Table 2. Correspondence table of journey purposes and preferred walking distance.

	Preferred Walking Distance (PreferredWD)										
	1	2	3	4	5	6	7	8	9	10	
Journey Purpose	≤100 m	101-150 m	151-200 m	201-250 m	251-300m	301-350 m	351-400 m	401-450 m	451-500 m	>500 m	Active Margin
1 Work	96	22	0	0	0	0	0	0	0	0	118
2 Study	0	33	21	19	16	0	0	0	0	0	89
3 Visit	0	0	0	0	3	5	8	5	5	0	26
4 Shopping	0	0	0	0	0	0	0	0	3	9	12
5 Banking, etc.	0	0	0	0	0	0	0	0	0	7	7
Active Margin	96	55	21	19	19	5	8	5	8	16	252

From Table 2, the highest number of users preferred walking distance to the bus stop at less than or equal to 100 meters for working purpose. Meanwhile, for the study purposes, the tendency towards preferred walking distance to the bus stop is within 101 to 150 meters. The purpose of the users' journey which is for visiting has a tendency towards preferred walking distance to the bus stop within 351 to 400 meters while for shopping purposes, the tendency is more than 500 meters. From the correspondence analysis, it can be understood that users with working purposes were more likely to punctually arrive at their working place. Therefore, users would not likely to spend their time on walking which they end up choose cars as their preferred mode.

Table 3. Correspondence table of journey purpose and preferred delay of travel time between bus and cars.

	Delay of Travel Time							
	1	2	3	4	5	6	7	
Journey Purpose	< 5 min	6-10 min	11-15 min	16-20 min	21-25 min	26-30 min	31-40 min	Active Margin
1 Work	32	68	18	0	0	0	0	118
2 Study	0	0	46	25	18	0	0	89
3 Visit	0	0	0	0	21	1	4	26
4 Shopping	0	0	0	0	0	4	8	12
5 Banking, etc.	0	0	0	0	0	2	5	7
Active Margin	32	68	64	25	39	7	17	252

From Table 3, the purpose of the users' journey which is working has the highest tendency towards the preferred delay of travel time between cars and bus within 6 to 10 minutes. Meanwhile, for the study purposes the tendency towards the preferred delay of travel time between cars and bus within 11 to 15 minutes. The purpose of the users' journey which is visiting tends towards preferred delay of travel time between cars and bus within 21 to 25 minutes while for shopping purposes the tendency towards preferred delay of travel time between cars and bus is within 31 to 40 minutes. From the correspondence analysis, it can be understood that users with working purposes will accept the delay of travel time between bus and car no later than 15 minutes.

Table 4. Correspondence table of journey purpose and users' view towards car's travel time.

	Rating towards Travel Time							
	1	2	3	4	5	6	7	
Journey Purpose	Strongly agree	Agree	Somewhat agree	Neither agree / disagree	Somewhat disagree	Disagree	Strongly disagree	Active Margin
1 Work	89	29	0	0	0	0	0	118
2 Study	0	53	27	9	0	0	0	89
3 Visit	0	0	2	13	11	0	0	26
4 Shopping	1	0	0	0	7	2	2	12
5 Banking, etc.	0	0	0	0	0	6	1	7
Active Margin	90	82	29	22	18	8	3	252

From Table 4, the purpose of the users' journey which is working has strongly agreed that the travel time of a car is lower than a bus. Meanwhile, for the study purposes, the users agree that travel time of cars is lower than bus. The purpose of the users' journey which is visiting were neither agreed nor disagree that travel time of a car is lower than bus while for banking and other activities purpose, the users disagree that travel time of cars lower than a bus. From the corresponding analysis, it can be understood that users with working purposes agree that travel time of a car is lower than a bus; therefore users most attracted towards cars than the bus as their main mode to reach the destination. The results indicated that the improvement towards bus provision is important to ensure that users were attracted to choose bus instead of a car as their mode choice.

Table 5. Correspondence table of journey purposes and distance from users' home to the Klang Komuter Station.

	Distance					
	1	2	3	4	5	
Journey Purpose	≤5 km	6 -10 km	11 -15 km	16 – 20 km	>20km	Active Margin
1 Work	118	0	0	0	0	118
2 Study	32	57	0	0	0	89
3 Visit	0	26	0	0	0	26
4 Shopping	0	2	2	5	3	12
5 Banking, etc.	0	0	0	4	3	7
Active Margin	150	85	2	9	6	252

From Table 5, the distance for working purposes from home to Klang Komuter is less than or equal to 5 km. Meanwhile, for study purposes, the distance from home to the station is within 6 to 10 km while for shopping purposes, the distance from home to Klang Komuter within 16 to 20 km. From the correspondence analysis, it can be understood that those users (especially workers and students) who come to the station during morning peak hour live near to the station (less than or equal to 5 km and within 6 to 10 km). It can be concluded that most users who were getting to the railway station are those who live near to the station between 5 to 10 km or less than such distance. Therefore, the improvement in accessibility and to reach the station is necessary to ensure that users are satisfied with their journey.

4.4. Findings from Revealed Preference/Stated Preference (RP/SP) Survey

Revealed Preference/Stated Preference (RP/SP) Survey can be applied to investigate users' preference on transport mode choice, especially in travel time aspect, as focused on this study. The results also can help the government to establish a future strategy for better facilities in order to improve the time consumption for each stage of a trip from home (origin) to the station (destination). In addition, the results from RP/SP Survey also can help the service provider to gain information on bus travel time compared to a car; therefore they can make some improvement in the provision of bus services. Hence, users will find that bus is way more attractive than the car in term of travel time.

5. Discussion

The result from this study shows that commuting to work and journeys to schools or studies are the main reasons for traveling during weekdays, meanwhile non-work and non-studies journey purposes account for about one-fifth of the journeys. The access from home to the station profile is different for those who live up to 5 km from the station and those who live further away. Distance from home to the station affects the journey purposes of users during morning peak hour. The number of workers is the highest (refer Table 5) compare to other journey purposes for distance up to 5 km from the station.

According to [10], the research found that the demand for inter-city rail travel is relatively high elasticity of demand with respect to the distance from home to the station. Therefore, they suggest that the access to the station can be improved by lowering the distance which can attract more users. The fact that shows from this study is in line with the previous study which most of the users (150 users) were origin from below to 5 kilometers to the Klang Komuter Station. Railway stations are usually located relatively far from housing area which getting to the station has become a great issue to be discussed by the service provider and government as well in order to increase the number of rail users. Therefore, a systematic and convenient public transport especially bus should be well provided in order to attract users to choose bus instead of car while travelling to the railway station.

As expected, journey purposes affect the mode choice, it appears, mainly when the highest number of users (refer Table 4) with working purpose strongly agree that cars is more accessible and faster to reach the train station as compared to a bus. The park and ride is provided near to station which motivates the workers to bring along their cars to the station. Some of the workers and students also agree that traveling by cars is faster and accessible than a bus since they can manage their travel time and arrival time at the final destination as well. However, a few of the respondents that preferred car as their travel mode revealed that they were fetched by family members or by carpooling in order to access the rail station. According to [13], the accessibility of a station might perform as an important element in determining either railway is chosen as a travel alternative. Meanwhile, [11] added that access and egress to the railway station should exceed an absolute maximum threshold or users will not use the public transport system. An improvement towards current travel time for a bus is necessary to ensure users' satisfaction while traveling by bus to reach the railway station. Thus, users will be triggered to choose bus as their main mode instead of car which is absolutely will help in overcome user's problem especially traffic congestion while making their way to reach working place during peak hour.

Besides, the users surveyed with working purposes are most likely to consider to commute by bus if the difference of travel time between bus and car is between 6 to 10 minutes or at least less than 15 minutes (refer Table 3). This situation might happen since workers have to reach their working place punctually or they will be in trouble if they arriving late at their office. As compared to students, their timetable might be flexible which they can compromise with the larger gap of travel time between bus and car. Furthermore, economical factor or that a car is not available to them so that it might affect towards their decision in considering bus as their mode choice to commute from home to the railway station. Therefore, a campaign or scheme to attract more students to choose bus should be considered since it is noticeable that when free public transport was offered for students, it will lure those who currently use cars to move out from it. Thus, this method is positively improving the public transport that connected to the railway station.

Meanwhile, the research shows that journey purposes affect the willingness of the users to reach bus stop or railway station since the users surveyed are less likely to walk to the station. Apparently, most users with working purposes will only consider walking for about 100 meters or less in order to reach the bus stop or to the railway station (refer Table 2). Compare to other studies in which was conducted in Netherlands by [12], the result shows that the access-environment is materializing into an access to station profile that is dominated by walking, bicycling, and public transports. An assumption that can be made regarding this situation is due to factors such geography and hot weather which are unconducive to attract users to reach the bus stop or station by walking. According to the results presented, it is suggested to the service provider or government to provide more convenient facilities such as walking pedestrian with rooftops so that users will be more interested to walk while commuting to the bus stop or railway station.

Lastly, journey purposes seem to give a great impact towards preferred waiting time at the bus stop. Most workers (refer Table 1) are willing to wait for bus arrival at the bus stop less than 5 minutes. This analysis clarifies that the reasons of slow traffic movement during morning peak hour is due to the majority of users bring along their cars to reach their destination since waiting time at the bus stop is usually unpredictable. Therefore, it is crucial to provide bus services that meet users need to lure those users move out from their cars. Further studies should be conducted to investigate the factors that influence users preferred waiting at the bus stop, either due to inconvenient facilities, safety, the bus schedule or else. Waiting time is one of travel time components that is important to be well managed in order to attract users to switch mode from cars to a bus.

6. Conclusion

The research found that there is a concrete reason for the users to choose cars instead of bus during morning peak hour. The travel time plays an important reason while users were making their decision on mode choice. Besides, those workers need to arrive punctually at their working place which becomes one of the main reasons that influences their decision making to choose cars as they can manage their travel time by driving their own vehicles.

References

[1] Stradling S ., Meadows M . and Beatty S 2000 Helping drivers out of their cars Integrating transport policy and social psychology for sustainable change *Transp. Policy* **7** 207–15
[2] Stopher P R 2004 Reducing road congestion: a reality check *Transp. Policy* **11** 117–31

[3] Fosgerau M and Engelson L 2011 The value of travel time variance *Transp. Res. Part B Methodol.* **45** 1–8
[4] KII M and HANAOKA S 2003 Comparison of Sustainability Between Private and Public Transport Considering Urban Structure *IATSS Res.* **27** 6–15
[5] Walker 2009 11 th conference on competition and ownership in land passenger transport Delft University of Technology (The Netherlands) 20-25 September 2009 How flexible should bus service be ? frequent networks as a tool for permanent change 20–5
[6] Infrastucturist 2009 36 Reasons Streetcars Are Better Than Buses « CincyStreetcar Blog
[7] Quarmby D A 1967 Choice of travel mode for the journey to work: some findings *J. Transp. Econ. Policy* 273–314
[8] Din M A M, Karim M R and Saritha P 2006 The Aspect Of Walking Accessibility In The Development Of GIS-Based Transit System Modelling In Kuala Lumpur *Int. Cartogr. Assoc.*
[9] Matulin M, Bošnjak I and Šimunovič L 2009 Different approaches to the modal split calculation in urban areas *Conf. Proc. ICTS 2009 Transp. Marit. Logist. Sci.* 1–7
[10] Wardman M and Tyler J 2000 Rail network accessibility and the demand for inter-urban rail travel *Transp. Rev.* **20** 3–24
[11] Krygsman S, Dijst M and Arentze T 2004 Multimodal public transport: An analysis of travel time elements and the interconnectivity ratio *Transp. Policy* **11** 265–75
[12] Givoni M and Rietveld P 2007 The access journey to the railway station and its role in passengers' satisfaction with rail travel *Transp. Policy* **14** 357–65
[13] Rietveld P 2000 The accessibility of railway stations: The role of the bicycle in The Netherlands *Transp. Res. Part D Transp. Environ.* **5** 71–5

IOP Conf. Series: Materials Science and Engineering **342** (2018) 012034 doi:10.1088/1757-899X/342/1/012034

Performance of magnetorheological elastomer based green epoxidized natural rubber/sucrose acetate isobutyrate hybrid matrix

Muntaz Hana Ahmad Khairi[1], Saiful Amri Mazlan[2], Siti Aishah Abdul Aziz[3], Ubaidillah[4] and Salihah Tan Shilan[5]

[1,2,3,5] VSE Research Laboratory, Malaysian-Japan International Institute of Technology Universiti Teknologi Malaysia, Jalan Sultan Yahya Petra, 54100 Kuala Lumpur, MALAYSIA
[4] Mechanical Engineering Department, Faculty of Engineering, Universitas Sebelas Maret, Jl. Ir. Sutami 36A Kentingan Jebres Surakarta 57126, INDONESIA

E-mail: hana5700@gmail.com

Abstract. This study introduces a sucrose acetate isobutyrate (SAIB) as a novel additive of magnetorheological elastomers (MREs). The MREs utilized an epoxidized natural rubber (ENR) as the matrix and carbonyl iron particles (CIPs) as their filler. The CIPs were fixed at 60 wt%. The viscosity of the compound was observed using a viscometer. Meanwhile, the microstructures were observed by using field emission scanning electron microscope (FESEM). Rheological properties regarding shear storage modulus were measured by using a rheometer (MCR 302, Anton Paar). The experimental results demonstrated that the MREs-based ENR/SAIB had a decrement in their viscosity by 40% reduction. Moreover, the magnetorheological (MR) effect increased by 23% as the increment of magnetic fields. The morphological photograph showed that the CIPs embedded well within the matrix. The fabricated MREs samples were strain dependent, where all MREs samples exhibit the deteriorating trend when increasing the strain amplitude.

1. Introduction

Magnetorheological elastomers (MREs) belongs to smart materials whose rheological properties can be altered under the influence of magnetic field. MREs are fabricated by embedding soft magnetic particles such as carbonyl iron particles (CIPs) in a matrix such as silicone rubber and rubber. MREs can be manufactured in the presence or absence of an external magnetic field [1]. Aligned structures namely anisotropic MREs, are being established during the cross-linking process in the presence of an applied magnetic field due to the induced interactions among particles [1]. On the other hand, the absence of external magnetic during the curing; makes the structure of MREs becoming isotropic [3]. Having field-dependent modulus and damping properties, MREs have a broad range of possible applications such as vibration absorbers [2–6] sensing devices [7] and morphing structure [8]. The MR effect, which calculated from the storage modulus, is usually adopted to evaluate the MRE performance [9].

IOP Conf. Series: Materials Science and Engineering **342** (2018) 012034 doi:10.1088/1757-899X/342/1/012034

The hardness of the matrix has a significant influence on the MR effect [10]. In general, the degree of freedom of movement for the magnetic particles is governed by the matrix elastomer, and the MR effect is usually smaller in a hard matrix than in a soft matrix. Lokander and Stenberg [8] have found that utilization of plasticizer in MREs enhanced the shear storage modulus. The introduction of plasticizer to the elastomeric materials be able to act as a lubricant and hence can raise the mobility of magnetic particles in the matrix and at the same time modify magnetic particles properties [9].
Therefore, in this study, sucrose acetate isobutyrate (SAIB) is utilized in the MREs preparation process to alter the viscosity of the ENR. This ester also can improve the affinity between the CIPs and the matrix. Field emission scanning electron microscopy (FESEM) was used to observe the distribution and the morphology of the CIPs in the matrix. The shear storage modulus and MR effect were further investigated using a rheometer.

2. Experimental Procedure

2.1 Fabrication of MREs and characterization

The isotropic MREs samples used in this work are prepared by compounding CIPs having an average diameter of 6μm (BASF, type CN) and epoxidized natural rubber (ENR). The formula for the rubber compound is 100 phr ENR, 5 phr zinc oxide, 2 phr stearic acid, 19 phr carbon black FEF N550, 5phr tudalen 65, 2.3 phr sulfur and 0.8 phr CBS. (Rubber and rubber additive supplied by Malaysian Rubber Board). The rubber matrix and rubber additive were mixed homogeneously using two roll mills to produce rubber compound. The compound then added with CIPs content of 60 wt% and SAIB content of 8 wt%. MREs sample without SAIB also prepared as a benchmark to the MREs-based ENR/SAIB. The MREs sample then was put into the mold under the high pressure, and the temperature was fixed at 150° C for 10 minutes. The viscosity of the non-vulcanized MREs was measured according to ASTM D 1646 at 100°C on the Mooney viscometer (MV2000 VS Alpha Technologies). The microstructure observation was performed at an accelerating voltage of 1kV with 1000 times magnification. The influence of the ramped magnetic flux on the storage modulus and loss modulus of MREs was measured by the rheometer (Anton Paar, model Physica MCR 302). The samples were evaluated under constant strain amplitudes of 1%, excitation frequency of 1 Hz. On the other hand, for the strain sweep test, the strain amplitude was set from 0.01 to 10% ramped linearly. Its frequency and magnetic field were fixed at 1 Hz and 0.2T, respectively.

3. Results and discussion

3.1 Viscosity Observation

Figure 1 shows the viscosity of the non-vulcanized MREs-based ENR and ENR/SAIB at a constant temperature of 100°C. The viscosity of the non-vulcanized MREs-based ENR was 43.2 MU and 25.9 MU with the addition of the SAIB. For rubber without the SAIB, the CIPs are hard to disperse due to the high viscosity of the rubber matrix. With the addition of the SAIB, the CIPs were found to disperse well in the rubber matrix. SAIB is capable of changing the viscosity of the non-vulcanized rubber matrix by reducing crosslink density between the rubber bonding, and thus improve the fluidity of the rubber matrix on the CIPs which affects the properties of the fabricated MREs. This effect was useful to the MREs properties to enhance the MR effect due to MR effect was more substantial in soft matrix compare to hard matrix since it makes impossible the CIPs to displace when subjected to the magnetic field [11].

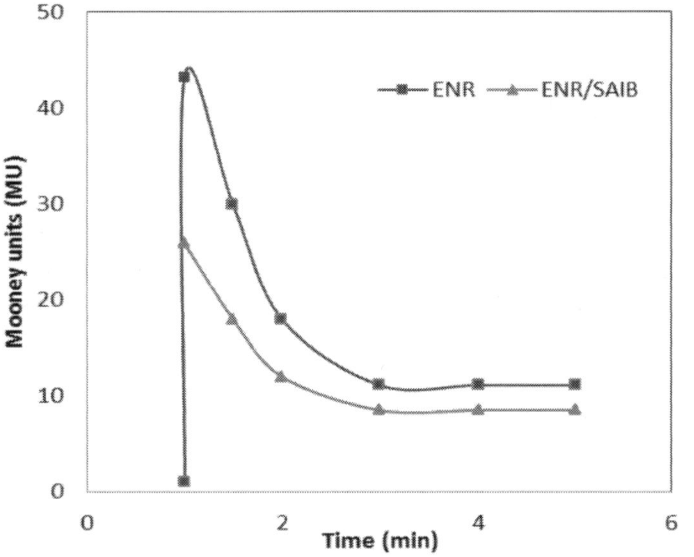

Figure 1. Viscosity of MREs based ENR and ENR/SAIB.

3.2 Microstructure Observation

Figure 2 shows the FESEM images of the MREs-based ENR and ENR/SAIB. The images verify that the CIPs are dispersed randomly in the elastomer matrix. Thus, the fabricated MREs exhibits an isotropic condition for both MREs samples. Apparently, there have significant differences in the texture of the matrix when SAIB was added to the rubber matrix. It increased the porosity of the rubber matrix by reducing the crosslink density between the rubber chains. It is noted that addition of SAIB can enhance the dispersion of the CIPs during the mixing process even though a few agglomerations still occur. Overall, in these FESEM images, no such gaps and cavities are seen obviously, indicating that CIPs have been well embedded.

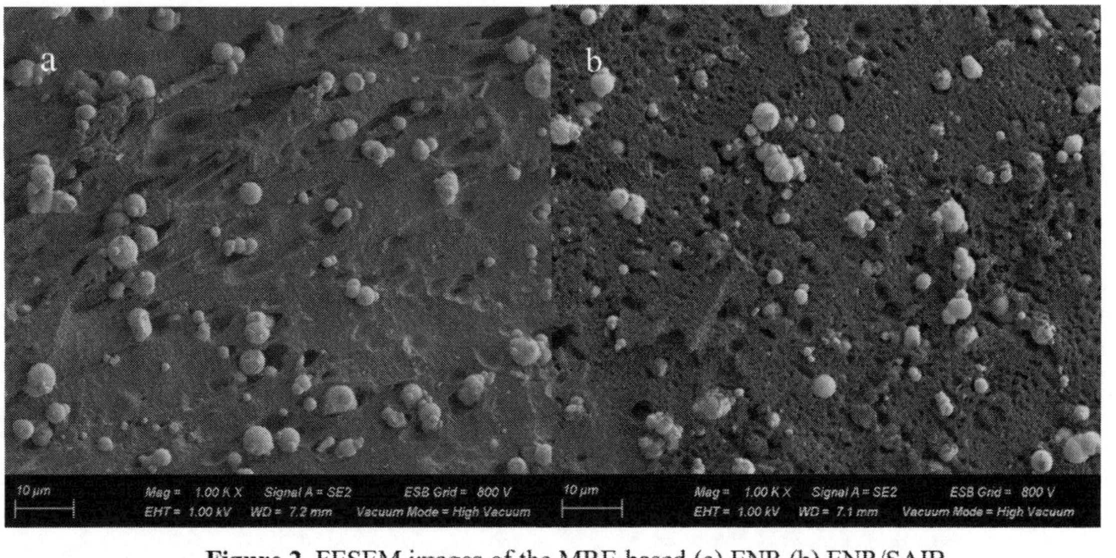

Figure 2. FESEM images of the MRE-based (a) ENR (b) ENR/SAIB.

3.3 Rheological Properties

3.3.1 Effect of Magnetic Field. The influence of the ramped magnetic flux on the storage and loss modulus of MREs was evaluated under constant strain amplitudes of 1%, excitation frequency of 1 Hz, and the temperature of 25°C. The ramp magnetic flux was realized by driving an electric current from 0 to 5A in a linear fashion. The flux densities that were generated to exceed more than 0.7 T for all samples. Figure 3 shows the relationship between the shear storage, loss modulus and SAIB content under the increasing strengths of the magnetic field. As the magnetic field increased, the shear storage modulus of both samples MREs-based ENR and ENR/SAIB were inclined while the CIPs weighed the same of 60 wt%. This proves that both MRE samples exhibited noticeable increment of MR effects. On the other hand, for the loss modulus, the trend was similar to the both samples. Even though both MREs samples showed an inclined trend as the magnetic field increased, however MRE-based ENR/SAIB showed improvement in the MR effect. The MR effects are clearly shown in Table 1 where all the data were collected from the experimental results as shown in Figure 3. The MR effect of MREs- based ENR/SAIB increased from 13 to 23% under the magnetic field of about 0.8T compared to MREs-based ENR. Meanwhile, the zero-field storage modulus of MREs-based ENR and ENR/SAIB was 0.9 and 0.70 MPa %, respectively.

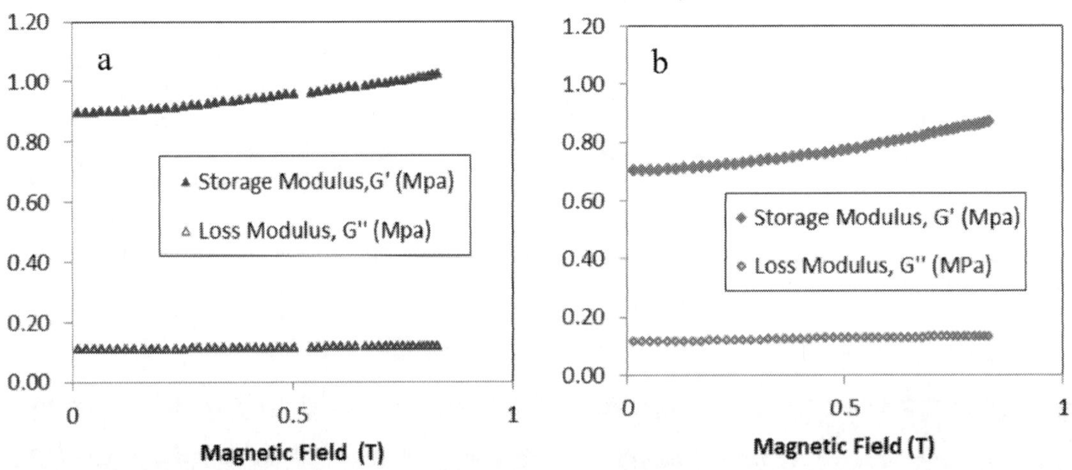

Figure 3. Magnetic field effect on the storage modulus and loss modulus for the MREs samples (a) ENR (b) ENR/SAIB.

Table 1. The zero-field storage modulus (G_0), the magnetically induced modulus (ΔG) and the MR effect of the MREs-based ENR and ENR/SAIB.

Sample	G_0(Mpa)	ΔG (Mpa)	MR effect (%)
ENR	0.90	0.12	13
ENR/SAIB	0.70	0.16	23

3.3.2 Effect of Strain Amplitude. Figure 4 shows the influence of the shear strain amplitude for both samples MRE-based ENR and ENR/SAIB on the shear storage modulus and loss modulus. The initial storage modulus decreased for the MRE-based ENR/SAIB due to the lessening of MREs viscosity that consequently enhance mobility between molecular rubber chains. Meanwhile, the storage modulus diminished with the strain amplitude for both MREs-based ENR and ENR/SAIB samples showed a decreasing trend in the transition region, This non-linear behavior of filled rubber is recognized as the Payne effect [12]. This effect can be defined as the decrease of the storage or in-phase shear modulus with the increasing amplitude of oscillation [12,13]. Furthermore, the Payne effect increased with increasing concentration of filler material in the composite [14]. Sample without SAIB showed a dramatic decrease compared to the sample that contains SAIB ester. Incorporation of SAIB lowering the Payne effect by contributing to the effective interaction between SAIB and the MRE, thus, enhancing the molecules bonding. Therefore, MRE with low viscosity with the addition of SAIB ester gives significant influence improving the Payne effect caused by the filler such as carbon black and CIPs in the MRE. On the other hand, for the effect of the increasing strain on the loss modulus, MRE-based ENR/SAIB showed the slight enhancement in the loss modulus compared to the MRE-based ENR which showed dramatically developed the loss modulus as the strain percentage increased. This fact was resulted from the decent interaction and affinity between ENR and SAIB that reduced the viscous flow of the MRE material.

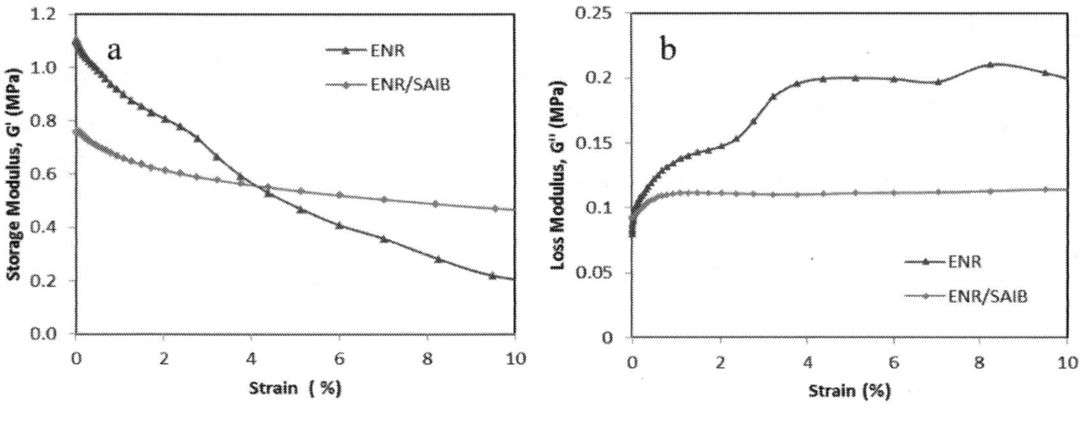

Figure 4. Strain effect of the MREs-based ENR and ENR/SAIB samples (a) storage modulus (b) loss modulus.

4. Conclusion

The effect of SAIB ester as an additive on the performance of MREs based on ENR has been experimentally investigated. The addition of the SAIB ester reduced the viscosity of the rubber matrix, and CIPs have dispersed well with only a few agglomerations structures. The MRE-based ENR/SAIB have had the lower zero-field modulus, and the MR effect of MRE based ENR/SAIB had increased from 13% to 23% when SAIB ester was added to the rubber matrix while the loss modulus is about the same for both samples. On the other hand, shear storage modulus is strain dependent, where both MRE-based ENR and ENR/SAIB samples exhibit the declined trend when increasing the strain amplitude.

Acknowledgement

The research described in this paper was supported by PDRU (Vote no: 04E02), Universiti Teknologi Malaysia under GUP Grant (Vote no:13H55) and Ministry of Higher Education, Malaysia. The authors are also grateful for the support from SHERA project 2017 under Universitas Sebelas Maret and Institut Teknologi Bandung, Indonesia.

IOP Conf. Series: Materials Science and Engineering **342** (2018) 012034 doi:10.1088/1757-899X/342/1/012034

References

[1] An J S, Kwon S H, Choi H J, Jung J H and Kim Y G 2017 *Compos. Struct.* **160** 1020–6
[2] Ubaidillah, Imaduddin F, Li Y, Mazlan S A, Sutrisno J, Koga T, Yahya I and Choi SB 2016 *Smart Mater. Struct.* **25** 115002
[3] Liao G J, Gong X L, Xuan S H, Kang C J and Zong L H 2012 *J. Intell. Mater. Syst. Struct.* **23** 25–33
[4] Hairuddin K, Mazlan S A, Ubaidillah, Zamzuri H and Nor N M 2014 A *Appl. Mech. Mater.* **660** 763–7
[5] Li Y, Li J, Li W and Samali B 2013 *Smart Mater. Struct.* **22** 35005
[6] Li Y en Li J 2015 *J. Intell. Mater. Syst. Struct.* **26** 1861–70
[7] Ubaidillah, Joko S, Agus P and Mazlan S A 2015 *Adv. Eng. Mater.* **17** 563–97
[8] Nguyen V Q, Ahmed A S and Ramanujan R V 2012 Morphing Soft Magnetic Composites *Adv. Mater.* **24** 4041–54
[9] Ge L, Gong X L, Fan Y C and Xuan S H 2013 P *Smart Mater. Struct.* **22** 115029
[10] Wu J K, Gong X L, Fan Y C and Xia H S 2012 *J. Appl. Polym. Sci.* **123** 2476–84
[11] Boczkowska A and Awietjan S F 2011 *Transactions on Engineering Sciences* **72** 263–74
[12] Payne A 1962 *J. Appl. Polym. Sci.* **VI** 57–63
[13] Payne A 1962 *J. Appl. Polym. Sci.* **VI** 368–72
[14] Sorokin V V, Ecker E, Stepanov G V, Shamonin M, Monkman G J, Kramarenko E Y and Khokhlov A R 2014 *Soft Matter* **10** 8765–76

Properties of fresh and hardened sustainable concrete due to the use of palm oil fuel ash as cement replacement

Hussein M Hamada[1], Gul Ahmed Jokhio[1], Fadzil Mat Yahaya[1] and Ali M Humada[2]

[1] Faculty of Civil Engineering & Earth Resources, Universiti Malaysia Pahang, 26300, Gambang, Malaysia
[2] Electricity Production Directorate of Salahaldeen, Ministry of Electricity, 34007, Baiji, Iraq

E-mail: Enghu76@gmail.com

Abstract: Palm oil fuel ash (POFA) is a by-product resulting from the combustion of palm oil waste such as palm oil shell and empty fruit bunches to generate electricity in the palm oil mills. Considerable quantities of POFA thus generated, accumulate in the open fields and landfills, which causes atmospheric pollution in the form of generating toxic gases. Firstly, to protect the environment; and secondly, having excellent properties for this purpose; POFA can be and has been used as partial cement replacement in concrete preparation. Therefore, this paper compiles the results obtained from previous studies that address the properties of concrete containing POFA as cement replacement in fresh and hardened states. The results indicate that there is a great potential to using POFA as cement replacement because of its ability to improve compressive strength, reduce hydration heat of cement mortar and positively affect other fresh and hardened concrete properties. The paper recommends that conducting further studies to exploit high volume of POFA along with other additives as cement replacement while maintaining high quality of concrete can help minimize CO_2 emissions due to concrete.

1. Introduction and background

In recent years, the world has witnessed a high demand on the energy sources. The industrial development worldwide has led to the consumption of large quantities of resources such as coal, natural gas, and petrochemicals. In addition to their limited availability and considerably high production costs, these fuels are the main contributors to the global warming and environmental pollution. Furthermore, these resources are non-renewable. This situation has led scientists and academics to focus their efforts towards researching the potential of using alternative and renewable materials. The ideal attributes sought in these alternatives are that these should be easily available, easy to use without any barriers, and friendly towards the environment. Palm oil waste materials are such renewable sources, which can be used in various industries because of the chemical composition being rich in silica and alumina.

On the other hand, there is a growing demand to produce cement to meet construction requirements. The production of cement worldwide is currently 2.9 billion tons a year; this quantity is expected to rise to

4 billion tons in the next ten years [1]. In order to produce one ton of Ordinary Portland Cement (OPC), about one ton of CO_2 is released into the atmosphere causing pollution. A study on the causes of air pollution reported that the CO_2 emission generated from cement manufacture constituted 7% of total CO_2 generated worldwide. This contributes to the climate change that has implications on the quality of future human life [1-3]. Therefore, there is an urgent need to use alternative materials that have high pozzolanic activity, which makes these materials suitable as replacement of cement in the preparation of concrete. The use of these materials can partially help solve the problems associated with cement production such as high production cost, energy consumption, and pollution. This will also mitigate the dumping of these waste materials into landfills.

Malaysia is considered the main contributor in the palm oil industry and this production continues to increase with time due to expansion in the plantation of palm oil trees. It is the first country in terms of export of palm oil products [4]. Approximately 90% of the palm oil is used in food, while 10% is used to produce soap [5]. The increasing demand of the palm oil products has led to the plantation of palm oil increase considerably, from 400 ha in 1920 to more than 4.5 million ha in 2008 [6]. During the production of palm oil, large quantities of waste materials are generated such as palm oil shell, palm oil clinker, and empty fruit, as shown in Figure 1. These wastes can be used as biomass fuel to generate electricity in the palm oil mills. Each palm oil tree consists of 14-15% fibre, 23% empty fruits, 6–7% palm oil shells, 6–7% palm kernel, and only 21% palm oil [7]. It means that for producing one kg of palm oil, there is 5kg dry biomass generated [6]. The palm oil waste has become one of the challenges facing palm oil industry due to accumulation of these wastes in huge amounts beside palm oil mills and in the landfills. The palm oil waste causes generation of undesirable toxic gases such as CO_2, thus contributing to environmental pollution.

A B C

Figure 1. Biomass wastes from palm oil mills. (A) Empty fruit bunches (EFB), (B) Mesocarp fibre and (C) Palm kernel shells.

However, few of the studies mentioned above have taken into consideration the utilization of POFA in various concrete forms such as foamed concrete, self-compaction concrete, and other types to achieve the desired benefits from it in concrete production. Therefore, large quantities of POFA are still disposed into open fields and landfills, which causes environmental pollution. Many methods have been employed by the palm oil mills to dispose palm oil waste. These trial measures have resulted in two more challenges, which include the contamination of ground water levels [8] and spreading of these wastes away easily by wind, which causes smog [9]. In connection with the efforts to effectively dispose palm oil waste and reduce environmental pollution due to cement production, this study is aimed at compiling the research on the properties of fresh and hardened concrete containing POFA as reported by previous studies to demonstrate the potential of POFA to be used as cement replacement during concrete production.

2. POFA Properties

This section presents the chemical composition and physical properties of POFA, both the unground and ground types.

2.1 Chemical Composition

Many researchers have examined the chemical composition of POFA [10, 11]. It has been reported that the differences in the chemical composition of POFA as in Table 1 is due to many factors such as variety of waste material sources and temperature of burning waste. Silica makes the most proportion in the POFA composition; it is responsible for pozzolanic activity. The highest content of silica from among palm tree parts results from burning kernel [12, 13].

Table 1. Chemical composition of POFA as reported in literature.

Ref.	SiO_2	Al_2O_3	Fe_2O_3	CaO	MgO	Na_2O	K_2O	SO_3	$SiO_2 + Al_2O_3 + Fe_2O_3$	LOI
[11]	57.71	4.56	3.30	6.55	4.23	0.50	8.27	0.25	65.57	10.52
[14]	67.09	6.12	5.92	5.58	3.06	0.11	5.45		79.13	2.20
[15]	55.5	9.2	5.6	12.4	4.6	0.0	0.0	2.3	70.3	7.9
[16]	60.42	4.26	3.34	11.00	5.31	0.18	5.03	0.45	68.02	2.55
[17]	67.72	3.71	4.71	5.57	4.04	0.16	7.67	1.07	76.14	6.20
[18]	54.0	0.9	2.0	12.9	4.9	1.0	13.5	4.0	56.9	3.7
[19]	53.5	1.9	1.1	8.3	4.1	---	6.5	2.36	56.5	20.9
[1]	47.22	2.24	2.65	6.48	5.86	1.22	11.86	9.19	52.11	5.42
[20]	53.5	1.9	1.1	8.3	4.1	1.3	6.5	--	56.5	18.0
[12]	51.18	4.61	3.42	6.93	4.02	0.06	5.52	0.36	59.21	21.6
[21]	63.41	5.55	4.19	4.34	3.74	0.16	6.33	0.91	73.15	6.20
[22]	54.80	7.40	4.47	14.00	4.14	---	---	0.71	66.67	9.3
[23]	43.60	11.40	4.70	8.40	4.80	0.39	3.50	2.80	59.7	18.00
[24]	65.3	2.5	1.9	6.4	3.0	0.3	5.7	0.4	69.7	10.0
[25]	59.62	2.54	5.02	4.92	4.52	0.76	7.52	1.28	67.18	8.25
[26]	57.8	4.6	3.3	6.6	4.2	0.5	8.3	0.3	65.7	10.1
[27]	60.42	4.26	3.34	11	5.31	0.18	5.03	0.45	68.02	2.55

The silica content ranges between 43.6% and 67.72% of the total components in POFA. Silica content contributes mainly to the pozzolanic reaction and is considered a suitable partial replacement of cement. Whereas, the ($SiO_2 + Al_2O_3 + Fe_2O_3$) ranges between 52.11 % and 79.13 % of the total components of POFA, which is responsible for formation of additional calcium hydro silicate C-H-S gels in concrete mixtures [28-30].

2.2 Physical Properties

Physical properties of POFA as presented in Table 2 also differ greatly as reported by several researchers due to the variety of conditions. High burning temperature changes the colour of particles from dark black to grey. Due to the carbon burning in the POFA particles, pozzolanic activity of POFA increases because of reduction in carbon and increased fineness of POFA. In terms of shape, POFA particles have spherical shape and the specific gravity is less than that of cement [31]. The surface area of POFA particles increases when fineness of the particles is increased by grinding mill. In contrast to Unground POFA (UGPOFA), the ground POFA (GPOFA) has the ability to resist high thermal deterioration due to the fineness of the particles, which act as micro-fillers successfully to avoid concrete degradation [32]. The particle size of GPOFA is smaller than that of OPC, while particle size of OPC is smaller than that of UGPOFA. The un-burnt carbon causes absorption of high quantity of water and superplasticizer and thus minimizes the concrete workability. In order to remove the carbon particles, GPOFA should be heated at high temperature of up to 500 °C for one hour [12, 14]. Average particle size can be achieved when POFA particles are collected in a 45 μm sieve. In order to get high fineness POFA, it should be ground to increase the surface area and reduce the particle size to obtain GPOFA. Many researchers have classified the POFA particle size into three, large, medium, and small depending on the particle size resulting from grinding process. The grinding process increases pozzolanic activity due the increased specific gravity and fineness of particles [33].

Table 2. Physical properties of POFA as reported in literature.

Ref.	Specific gravity	Blain fineness (m^2/kg)	Retained on 45 μm sieve (%)	Median particle size d50 μm
[34]	2.42	493	4.98	14.58
[35]	2.15	750	15.00	20.00
[10]	2.33	1244	1.50	10.10
[15]	2.53	---	1.70	10.70
[36]	2.25	1180	1.00–3.00	7.20
[37]	2.39	1228	4.30	12.30
[12]	2.42	435	---	15.76
[14]	2.56	450	---	22.53
[16]	2.6	---	---	1.069
[17]	2.2	1720	---	18.46
[18]	2.42	4930 cm^3/gm	33	---
[19]	2.42	4935 cm^3/gm	4.98	14.58
[1]	2.31	458.2	---	16.08

The specific gravity of POFA is less than that of cement, and ranges between 2.15 and 2.6 as presented in Table 2. Whereas, the median particle size of POFA varies due to the difference in the grinding process and machine used for this purpose; it ranges between 1.069 μm and 22.53 μm.

3. Properties of Fresh Concrete Containing POFA

The properties of fresh concrete, containing GPOFA and UGPOFA, such as workability and heat of hydration are discussed in detail in the following sub-sections.

3.1 Workability

Workability is one of the main properties of fresh concrete, which can be determined by conducting a slump test. Many studies have been conducted to find out the impact of POFA on the workability of fresh concrete. Most of these studies have reported that increasing POFA content leads to decrease in concrete workability. Tay et al. [38] showed that reduction of concrete workability is due to increased quantity of POFA in concrete production. Also, reduction of segregation in concrete occurs and the compacting factor improves, ranging between 0.93 and 0.97. Awal et al. [25] used high volume GPOFA to replace cement by 50%, 60% and 70%; and the slump value was reported as 160, 115, 90, and 80 mm, respectively. Increasing GPOFA resulted in decreasing slump in concrete. Some researchers have used Nano silica in concrete containing UGPOFA in order to reduce utilization of superplasticizer to achieve a required workability [39]. Aldahdooh et al. [40] reported that using GPOFA in concrete as cement replacement increases the concrete workability. Therefore, increase of workability is associated with minimizing carbon content and reducing loss of ignition (LOI) in the GPOFA. On the other hand, Rangbar et al. [41] reported that GPOFA cannot be used as a replacement of cement in self-compacted concrete by more than 20%, and if used it may have a negative impact on other properties such as workability and segregation. A reduction in concrete workability value occurs when more quantity of POFA is added to the cement paste and concrete mortar [42]. In order to get acceptable values of workability, chemical admixtures should be added to achieve the required slump, similar to slump in OPC [43-45].

Due to the coarse pozzolan of the POFA particles, demand of water also reduces, which causes decrease of workability. Crushed and agglomerated POFA particle shapes require higher quantity of water than the natural mix in order to decrease inter particle friction and thus increase the concrete workability

[45]. There was a reduction in the concrete workability due to addition of Nano silica which results in high water demand [46]. Farzadnia et al. [47] used workability of concrete as indicator to water consumed by using Nano silica and/or POFA as cement replacement; the water demand increased when more than 30% of POFA was added to the mix; workability decreased from 115.5mm to 112mm. Many studies have been conducted on the replacement of cement by POFA and it has been reported that adding further quantity of POFA contributes to decrease in the workability value of fresh concrete [14, 48]. Addition of high amounts of POFA causes reduction in the compressive strength of the concrete samples in addition to reducing the workability [47]. Yusuf et al. [27] reported that the workability of concrete can be enhanced by adding H_2O/Na_2O to the concrete mix but it reduces the compressive strength.

3.2 Heat of Hydration

The hydration temperature of a concrete sample describes its chemical characteristics [19]. Concrete hydration heat is basically one of the concrete hardening properties and is measured through a test as shown in Figure 2. Rising heat of hydration usually causes undesirable problems in the concrete mass and adversely affects durability performance [23]. Adding POFA to concrete causes decrease in hydration heat of the mixture [25, 43]. Reduction in the hydration heat of concrete containing POFA is an important advantage to the areas with hot climates such as Malaysia, Indonesia, and Thailand; where the temperatures may exceed 40°C. Lim et al. [19] studied the effect of replacement of cement by high volume of Nano POFA on the hydration heat and microstructure. The authors noted that there is increase in the hydration heat during the initial reading and for the both of GPOFA and UPOFA concretes, while significantly different hydration heat was recorded in the control sample. Generally, in comparison to the control sample, adding POFA, regardless the size, to the concrete mix has potential to reduce the hydration heat successfully; even though the fineness of the binder has a notable impact on the hydration heat of cement mortar [25, 49].

GPOFA and UPOFA play a significant role not only in reducing hydration heat, but also delaying the peak temperature when mixed with concrete mortar [25]. Cubical plywood with the side dimension of 500 mm was internally insulated with 76 mm thick expanded polystyrene acting as the insulator as shown in Figure-2. Four concrete mixes were prepared, one with 100% OPC and the others with 50%, 60% and 70% POFA as cement replacement of the total weight. The concrete was cast into the cubical plywood in order to test heat of hydration. Sata et al. [10] reported that concrete containing 80% POFA has the ability to decrease hydration temperature by 30% if compared with control cement mortar sample. The hydration heat of concrete containing POFA depends mainly on the calcium hydroxide $Ca(OH)_2$ generated during cement hydration process. Cement hydration heat is affected chemically and physically by adding some pozzolanic materials, which have the ability to decrease and delay the hydration heat [50-52].

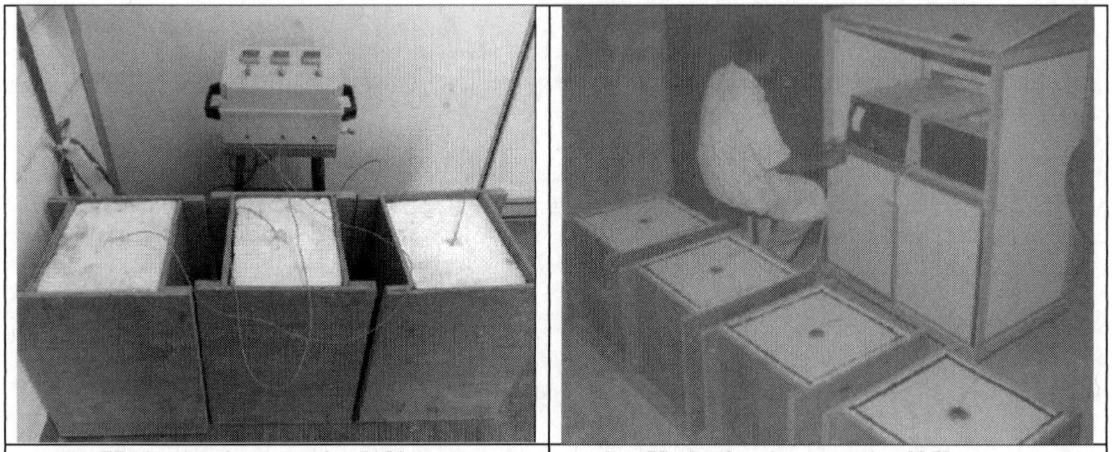

a) Hydration heat test by [19] b) Hydration heat test by [25]

Figure 2. Hydration heat of concrete.

4. Properties of Hardened Concrete Containing POFA

4.1 Compressive Strength

Many studies have been undertaken to identify the compressive strength of concrete incorporating POFA as partial cement replacement. Researchers have utilized POFA as cement replacement in various quantities due to the high silica content in it, which makes a good pozzolanic activity, which is considered favourable in concrete mixtures, as presented in Table 3. Unground POFA has coarse particles, whereas fine POFA called ground POFA has better characteristics. However, compressive strength is usually affected by increased POFA content in the concrete composites. Show [38] and Tay [9] showed that compressive strength decreased for concrete mixtures containing 10%, 20%, 30%, 40% and 50% unground POFA as cement replacement. In terms of the effect of POFA on concrete properties in the long run, the researchers noted that there is no much difference in compressive strength values between the concrete containing POFA and the control mixtures. Although concrete containing POFA has lower hydration rate at earlier age and thus reduced compressive strength [38], however, POFA with fine particles of about 8 μm as median size replacing about 20% of cement replacement has been reported to result in compressive strength higher than the corresponding value for control concrete at 90 days of curing age [26]. Increasing fineness of POFA particles to below 7.4 μm has various advantages in terms of increased pozzolanic reaction and compressive strength up to 99% of compressive strength of the control mix at 90 days curing can be achieved [11].

Researchers have used many experimental tests to investigate the compressive strength development in early and late ages; some of them used 0.4 as constant water/binder ratio, and up to 50% as cement replacement with raw and sieved POFA as presented in Table 3. Awal and Nguong [53] reported that the strength of concrete containing fine POFA is double to that in case of coarse POFA. The high compressive strength in fineness POFA is because of the high surface area of fine particles, which contributes by increasing pozzolanic activity. Sata et al. [54] studied the properties of concrete containing 10%, 20% and 30% of fine POFA particles as cement replacements and different ratios of w/b 0.50, 0.55, 0.60, after 14 days curing. The authors claimed that the concrete mixes with 10% fine POFA as cement replacement has higher compressive strength than other mixes for all the ratios of w/b.

Ground POFA is more useful than unground POFA to improve the compressive strength of recycled aggregate concrete [15]. For the concrete mix containing 20% cement replacement with fine POFA with medium particle size (10.7 μm), the compressive strength was 7% lower than the control mix. Fine POFA was responsible for increasing the compressive strength of concrete due to the high pozzolanic reaction. On the other hand, Sata et al. [10] studied effects of replacement of cement by 20% of POFA to prepare high strength concrete by using cylindrical samples with 100 mm diameter and 200 mm height. The compressive strength for concrete containing 20% POFA was higher than that in both the specimens containing 5% silica fume and the control samples. In terms of the compressive strength for light weight foamed concrete, Lim et al. [29] claimed that the concrete containing 10% and 20% POFA had compressive strengths of 7.17 and 7.06 MPa respectively; these are more than that of the control concrete, which was 6.50 MPa only.

Table 3. Compressive strength of concrete specimens as reported in literature.

Ref.	Compressive strength for concrete containing POFA as cement replacement							W/b
	0%	10%	20%	30%	40%	50%	60%	70%	
[55]	68.9	76.5	58.3	48.3	43.3				0.35
[12]	91.4		98.3		104.2		98.1		0.27
[25]	160					115	90	80	0.48
[9]	35.5	35.5	29.5	25.5	20.2	17.8			0.6
[56]	75	79.3	77.3	72.8	66.5				0.35
[57]	65	67.5	67	65.5					0.3
[58]						35.7	29.5	27.0	0.5
[18]	75	75	72	66.7	61.5				0.35
[38]	34.40	29.76	14.23	15.30	10.33	10.47			0.6

Johari et al. [12] showed that incorporation of ultrafine POFA in concrete as cement replacement reduces the compressive strength of high strength concrete at the early age. After one day of casting, the HSC containing ultrafine POFA with percentages of 20%, 40%, and 60% recorded relative strengths of 71.3%, 57.5% and 41.1%, respectively, of the control concrete sample. Whereas, at 7 days the compressive strengths increased to 100.6%, 97.7%, and 91.4%. The compressive strength of concrete containing various percentages of ultrafine POFA at 28 days increased to become 7.5%, 14% and 7.3% greater than the control concrete sample. Concrete containing ultrafine POFA acquired greater strength with time due to high pozzolanic reaction in POFA that produces additional amounts of C-S-H thus improving the compressive strength. Tangchirapat et al. [24] reported that the concrete containing high fineness POFA at 10%, 20%, and 30% had compressive strengths higher than that in the control concrete mix, which were 58.5, 59.5, and 58.8 respectively. This is due to the fact that high fineness POFA has high silica oxide (SiO_2) content, this helps improve the reaction between calcium hydroxide and silica oxide to produce considerable quantity of calcium hydro silicate C-H-S gels, which leads to increasing the compressive strength in concrete samples [48]. Hussin et al. [59] reported that high pozzolans is due to the fineness POFA particles. A median particle size of ultrafine POFA is about 2 μm, which has better specifications than normal POFA as cement replacement in the pozzolanic mineral admixture.

4.2 Modulus of Rupture

Modulus of rupture is one of the methods used to measure the tensile strength of concrete. It is a measure of an unreinforced concrete slab or beam to resist failure in bending. It is measured by loading (150mm x 150mm) concrete beams with a span length at least three times the depth. Many researchers have investigated the modulus of rupture of concrete with POFA; some have found that increasing POFA proportions in concrete as cement replacement is inversely proportional and leads to decreasing the flexural strength. Lim et al. [29] showed inverse results regarding flexural strength of light weight concrete at 90 days when the cement was replaced by 10% and 20% POFA; whereas the flexural tensile strength was 25% and 23%, respectively, higher than that of the control concrete samples. Increase in the flexural strength is due to formations of additional C-S-H gels by the pozzolanic reaction resulting from high silica content in the POFA particles.

Bashar et al. [17] investigated effects of adding steel fibre in three volume fractions (0.25%, 0.5%, and 0.75%) on the concrete containing high volume of POFA as 90% cement replacement and palm oil shell (POS) as coarse aggregate; significant improvements in flexural tensile strength was noted ranging 7-18% higher than the control mix. The concrete samples with more steel fibres achieved higher value of flexural strength. The flexural strength was 8%, 9% and 12% higher for samples containing 0.25%, 0.50% and 0.75% steel fibres, respectively. Awal and Mohammadhosseini [60] mixed carpet fibres with POFA to enhance the flexural tensile strength; they observed that increase in the flexural strength was due to the control of the cracks by fibres that may have occurred otherwise in the concrete sample. Ranjbar et al. [41] reported that the self-compacting concrete containing POFA was weak in the flexural tensile strength, the weakness was due to the high porous structure in concrete and weakness of bond between paste and aggregate. Altwair et al. [61] reported that there was a significant relationship between flexural strength and the first cracking strength increasing with POFA content and water-binder ratios, whereas the crack width decreased significantly and the number of cracks increased with increasing content of POFA in concrete.

4.3 Tensile Splitting Strength

Tensile Splitting Strength of concrete is one of the significant properties of hardened concrete. The concrete is cast as a cylindrical sample to test the tensile splitting strength of concrete. The value of tensile splitting strength is necessary to identify the suitable elements in concrete structures subjected to transverse shear, effects of shrinkage and temperature, and torsion. The value is required to design the liquid retaining structures, runway slabs, roadways, and pre-stressed concrete structures [62]. Sata et al. [48] investigated different proportions of POFA ranging between 10-30% as cement replacement with constant water cement ratio at 0.28. The results indicated the increase in the tensile splitting strength value for concrete containing 10-20% POFA and higher than control mix at age 90 days, whereas the control concrete and POFA concrete was similar in compressive strength. Jaturapitakkul et al. [37] reported that the nature of pozzolanic reaction resulting from the ground POFA increased together with the fineness of POFA particles, age of mortar and cement replacement. Lim et al., [29] reported that the tensile splitting strength value for concrete containing 10% - 20% POFA is higher than the control concrete by 19% and 9%, respectively, at 90 days. This increase is because of the formation of C-H-S gel resulting from the pozzolanic reaction.

4.4 Drying Shrinkage

Drying shrinkage is the deformation in the hardened concrete mixture due to the loss of capillary water. It causes an increase in tensile stress, which may lead to cracking, internal warping, and external deflection, before the concrete is subjected to any kind of loading [63]. This occurs in the hardened concrete specimens and leads to micro-cracks formation in the concrete. This may affect adversely the durability and mechanical properties and cause hazardous damages to the structure members [32]. Tay [9] investigated the effect of concrete containing 10% - 50% POFA as cement replacement on the drying shrinkage; the drying shrinkage did not change in two cases, in the control specimens and concrete containing 10% POFA at 28 days, while the drying shrinkage in the samples with other percentages of POFA (20% - 50%) were higher than that of control mix. Tangchirapat et al. [24] reported that the drying shrinkage started at the first three months of concrete age. They measured the drying shrinkage in the starting of six months for control specimens, which was 557×10^{-6} micro-strain, whereas, the drying shrinkage was 525, 505 and 494×10^{-6} micro-strain for the concrete containing 10%, 20% and 30% POFA, respectively.

5. Conclusion

The literature review has established that large quantities of palm oil waste is still being transferred into landfills and not being utilized. There is a significant concern regarding the accumulation of these wastes, which causes dangerous environmental issues. Many researches have been conducted to benefit from these wastes especially POFA in concrete production as partial cement replacement. Some researchers used

POFA with cement to benefit from its chemical and physicals properties. POFA is rich in silica and alumina, which makes it suitable for replacing cement partially, the chemical and physical properties of concrete containing POFA at different percentages have been illustrated. The effects of using POFA in concrete can be summarized as:

- Workability of concrete decreases with the use of POFA as cement replacement and may require the use of admixtures
- Heat of hydration of concrete containing POFA is reduced, which is an advantage specially in the regions with higher temperature ranges
- Compressive strength of concrete decreases if unground POFA is used as cement replacement, however, the same is increased when ground POFA is used
- Modulus of rupture and tensile splitting strength of concrete have been reported to increase with partial replacement of cement by POFA due to high pozzolanic activity resulting in more C-H-S gels
- Drying shrinkage of concrete is not affected when smaller proportions of POFA are used as cement replacement, however, for higher proportions of the same, an increase in the drying shrinkage has been reported

This paper concludes that exploiting the POFA in concrete preparation should be focused upon due to its appropriate characteristics. The future recommendation is that emphasis should be laid on using high volume POFA by grinding the particles into finer sizes to prepare concrete. Further studies should be conducted to improve the POFA characteristics as well as adding some additives and further grinding to get smaller particles, which have potential to replace cement with high volume quantity of POFA.

References

[1] M. Karim, M. F. M. Zain, M. Jamil, and F. Lai, "Fabrication of a non-cement binder using slag, palm oil fuel ash and rice husk ash with sodium hydroxide," *Construction and Building Materials,* vol. 49, pp. 894-902, 2013.

[2] A. M. Al Bakri, O. A. Abdulkareem, A. Rafiza, Y. Zarina, M. Norazian, and H. Kamarudin, "Review on Processing of low calcium fly ash geopolymer concrete," *Australian Journal of Basic and Applied Sciences,* vol. 7, pp. 342-349, 2013.

[3] A. Islam, U. J. Alengaram, M. Z. Jumaat, I. I. Bashar, and S. A. Kabir, "Engineering properties and carbon footprint of ground granulated blast-furnace slag-palm oil fuel ash-based structural geopolymer concrete," *Construction and Building Materials,* vol. 101, pp. 503-521, 2015.

[4] S. Yusoff, "Renewable energy from palm oil–innovation on effective utilization of waste," *Journal of cleaner production,* vol. 14, pp. 87-93, 2006.

[5] T. Mahlia, M. Abdulmuin, T. Alamsyah, and D. Mukhlishien, "An alternative energy source from palm wastes industry for Malaysia and Indonesia," *Energy Conversion and Management,* vol. 42, pp. 2109-2118, 2001.

[6] F. Sulaiman, N. Abdullah, H. Gerhauser, and A. Shariff, "An outlook of Malaysian energy, oil palm industry and its utilization of wastes as useful resources," *Biomass and bioenergy,* vol. 35, pp. 3775-3786, 2011.

[7] M. N. Dalimin, "Renewable energy update: Malaysia," *Renewable energy,* vol. 6, pp. 435-439, 1995.

[8] K. Foo and B. Hameed, "An overview of landfill leachate treatment via activated carbon adsorption process," *Journal of hazardous materials,* vol. 171, pp. 54-60, 2009.

[9] J.-H. Tay, "Ash from oil-palm waste as a concrete material," *Journal of Materials in Civil Engineering,* vol. 2, pp. 94-105, 1990.

[10] V. Sata, C. Jaturapitakkul, and K. Kiattikomol, "Utilization of palm oil fuel ash in high-strength concrete," *Journal of Materials in Civil Engineering,* vol. 16, pp. 623-628, 2004.

[11] W. Tangchirapat, T. Saeting, C. Jaturapitakkul, K. Kiattikomol, and A. Siripanichgorn, "Use of waste ash from palm oil industry in concrete," *Waste Management,* vol. 27, pp. 81-88, 2007.
[12] M. M. Johari, A. Zeyad, N. M. Bunnori, and K. Ariffin, "Engineering and transport properties of high-strength green concrete containing high volume of ultrafine palm oil fuel ash," *Construction and Building Materials,* vol. 30, pp. 281-288, 2012.
[13] A. Zeyad, M. Johari, N. M. Bunnori, K. Ariffin, and N. M. Altwair, "Characteristics of treated palm oil fuel ash and its effects on properties of high strength concrete," in *Advanced Materials Research*, 2013, pp. 152-156.
[14] C. Chandara, E. Sakai, K. A. M. Azizli, Z. A. Ahmad, and S. F. S. Hashim, "The effect of unburned carbon in palm oil fuel ash on fluidity of cement pastes containing superplasticizer," *Construction and Building Materials,* vol. 24, pp. 1590-1593, 2010.
[15] W. Tangchirapat, S. Khamklai, and C. Jaturapitakkul, "Use of ground palm oil fuel ash to improve strength, sulfate resistance, and water permeability of concrete containing high amount of recycled concrete aggregates," *Materials & Design,* vol. 41, pp. 150-157, 2012.
[16] M. O. Yusuf, M. A. M. Johari, Z. A. Ahmad, and M. Maslehuddin, "Influence of curing methods and concentration of NaOH on strength of the synthesized alkaline activated ground slag-ultrafine palm oil fuel ash mortar/concrete," *Construction and Building Materials,* vol. 66, pp. 541-548, 2014.
[17] I. I. Bashar, U. J. Alengaram, M. Z. Jumaat, A. Islam, H. Santhi, and A. Sharmin, "Engineering properties and fracture behaviour of high volume palm oil fuel ash based fibre reinforced geopolymer concrete," *Construction and Building Materials,* vol. 111, pp. 286-297, 2016.
[18] W. Kroehong, T. Sinsiri, and C. Jaturapitakkul, "Effect of palm oil fuel ash fineness on packing effect and pozzolanic reaction of blended cement paste," *Procedia Engineering,* vol. 14, pp. 361-369, 2011.
[19] N. H. A. S. Lim, M. A. Ismail, H. S. Lee, M. W. Hussin, A. R. M. Sam, and M. Samadi, "The effects of high volume nano palm oil fuel ash on microstructure properties and hydration temperature of mortar," *Construction and Building Materials,* vol. 93, pp. 29-34, 2015.
[20] M. Ariffin, M. Bhutta, M. Hussin, M. M. Tahir, and N. Aziah, "Sulfuric acid resistance of blended ash geopolymer concrete," *Construction and Building Materials,* vol. 43, pp. 80-86, 2013.
[21] A. Islam, U. J. Alengaram, M. Z. Jumaat, and I. I. Bashar, "The development of compressive strength of ground granulated blast furnace slag-palm oil fuel ash-fly ash based geopolymer mortar," *Materials & Design,* vol. 56, pp. 833-841, 2014.
[22] M. A. A. Rajak, Z. A. Majid, and M. Ismail, "Morphological characteristics of hardened cement pastes incorporating nano-palm oil fuel ash," *Procedia Manufacturing,* vol. 2, pp. 512-518, 2015.
[23] A. A. Awal and M. W. Hussin, "Effect of palm oil fuel ash in controlling heat of hydration of concrete," *Procedia Engineering,* vol. 14, pp. 2650-2657, 2011.
[24] W. Tangchirapat, C. Jaturapitakkul, and P. Chindaprasirt, "Use of palm oil fuel ash as a supplementary cementitious material for producing high-strength concrete," *Construction and Building Materials,* vol. 23, pp. 2641-2646, 2009.
[25] A. A. Awal and I. Shehu, "Evaluation of heat of hydration of concrete containing high volume palm oil fuel ash," *Fuel,* vol. 105, pp. 728-731, 2013.
[26] P. Chindaprasirt, S. Homwuttiwong, and C. Jaturapitakkul, "Strength and water permeability of concrete containing palm oil fuel ash and rice husk–bark ash," *Construction and Building Materials,* vol. 21, pp. 1492-1499, 2007.
[27] M. O. Yusuf, M. A. M. Johari, Z. A. Ahmad, and M. Maslehuddin, "Effects of H 2 O/Na 2 O molar ratio on the strength of alkaline activated ground blast furnace slag-ultrafine palm oil fuel ash based concrete," *Materials & Design,* vol. 56, pp. 158-164, 2014.

[28] M. Ahmad, R. Omar, M. Malek, N. M. Noor, and S. Thiruselvam, "Compressive strength of palm oil fuel ash concrete," in *Proceedings of the international conference on construction and building technology, Kuala Lumpur, Malaysia*, 2008, pp. 297-306.

[29] S. K. Lim, C. S. Tan, O. Y. Lim, and Y. L. Lee, "Fresh and hardened properties of lightweight foamed concrete with palm oil fuel ash as filler," *Construction and Building Materials,* vol. 46, pp. 39-47, 2013.

[30] J. Mirza, M. Riaz, A. Naseer, F. Rehman, A. Khan, and Q. Ali, "Pakistani bentonite in mortars and concrete as low cost construction material," *Applied Clay Science,* vol. 45, pp. 220-226, 2009.

[31] A. A. Awal, I. Shehu, and M. Ismail, "Effect of cooling regime on the residual performance of high-volume palm oil fuel ash concrete exposed to high temperatures," *Construction and Building Materials,* vol. 98, pp. 875-883, 2015.

[32] B. S. Thomas, S. Kumar, and H. S. Arel, "Sustainable concrete containing palm oil fuel ash as a supplementary cementitious material–A review," *Renewable and Sustainable Energy Reviews,* vol. 80, pp. 550-561, 2017.

[33] C. Jaturapitakkul, K. Kiattikomol, W. Tangchirapat, and T. Saeting, "Evaluation of the sulfate resistance of concrete containing palm oil fuel ash," *Construction and Building Materials,* vol. 21, pp. 1399-1405, 2007.

[34] A. Awal and S. I. Abubakar, "Properties of concrete containing high volume palm oil fuel ash: ashort-term investigation," *Malaysian Journal of Civil Engineering,* vol. 23, pp. 164-176, 2011.

[35] S. Rukzon and P. Chindaprasirt, "An experimental investigation of the carbonation of blended portland cement palm oil fuel ash mortar in an indoor environment," *Indoor and Built Environment,* vol. 18, pp. 313-318, 2009.

[36] P. Chindaprasirt and S. Rukzon, "Pore structure changes of blended cement pastes containing fly ash, rice husk ash, and palm oil fuel ash caused by carbonation," *Journal of Materials in Civil Engineering,* vol. 21, pp. 666-671, 2009.

[37] C. Jaturapitakkul, J. Tangpagasit, S. Songmue, and K. Kiattikomol, "Filler effect and pozzolanic reaction of ground palm oil fuel ash," *Construction and Building Materials,* vol. 25, pp. 4287-4293, 2011.

[38] J.-H. Tay and K.-Y. Show, "Use of ash derived from oil-palm waste incineration as a cement replacement material," *Resources, conservation and recycling,* vol. 13, pp. 27-36, 1995.

[39] H. Noorvand, A. A. A. Ali, R. Demirboga, H. Noorvand, and N. Farzadnia, "Physical and chemical characteristics of unground palm oil fuel ash cement mortars with nanosilica," *Construction and Building Materials,* vol. 48, pp. 1104-1113, 2013.

[40] M. Aldahdooh, N. M. Bunnori, and M. M. Johari, "Development of green ultra-high performance fiber reinforced concrete containing ultrafine palm oil fuel ash," *Construction and Building Materials,* vol. 48, pp. 379-389, 2013.

[41] N. Ranjbar, A. Behnia, B. Alsubari, P. M. Birgani, and M. Z. Jumaat, "Durability and mechanical properties of self-compacting concrete incorporating palm oil fuel ash," *Journal of Cleaner Production,* vol. 112, pp. 723-730, 2016.

[42] J. M. Paris, J. G. Roessler, C. C. Ferraro, H. D. DeFord, and T. G. Townsend, "A review of waste products utilized as supplements to Portland cement in concrete," *Journal of Cleaner Production,* vol. 121, pp. 1-18, 2016.

[43] E. Khankhaje, M. W. Hussin, J. Mirza, M. Rafieizonooz, M. R. Salim, H. C. Siong, *et al.*, "On blended cement and geopolymer concretes containing palm oil fuel ash," *Materials & Design,* vol. 89, pp. 385-398, 2016.

[44] B. H. Nagaratnam, M. E. Rahman, A. K. Mirasa, M. A. Mannan, and S. O. Lame, "Workability and heat of hydration of self-compacting concrete incorporating agro-industrial waste," *Journal of Cleaner Production,* vol. 112, pp. 882-894, 2016.

[45] N. Ranjbar, M. Mehrali, A. Behnia, U. J. Alengaram, and M. Z. Jumaat, "Compressive strength and microstructural analysis of fly ash/palm oil fuel ash based geopolymer mortar," *Materials & Design,* vol. 59, pp. 532-539, 2014.
[46] L. Senff, J. A. Labrincha, V. M. Ferreira, D. Hotza, and W. L. Repette, "Effect of nano-silica on rheology and fresh properties of cement pastes and mortars," *Construction and Building Materials,* vol. 23, pp. 2487-2491, 2009.
[47] N. Farzadnia, H. Noorvand, A. M. Yasin, and F. N. A. Aziz, "The effect of nano silica on short term drying shrinkage of POFA cement mortars," *Construction and Building Materials,* vol. 95, pp. 636-646, 2015.
[48] V. Sata, C. Jaturapitakkul, and K. Kiattikomol, "Influence of pozzolan from various by-product materials on mechanical properties of high-strength concrete," *Construction and Building Materials,* vol. 21, pp. 1589-1598, 2007.
[49] T. Mehmannavaz, M. Ismail, S. Radin Sumadi, M. A. Rafique Bhutta, M. Samadi, and S. M. Sajjadi, "Binary effect of fly ash and palm oil fuel ash on heat of hydration aerated concrete," *The Scientific World Journal,* vol. 2014, 2014.
[50] X.-Y. Wang and H.-S. Lee, "Modeling the hydration of concrete incorporating fly ash or slag," *Cement and concrete Research,* vol. 40, pp. 984-996, 2010.
[51] Gutteridge W A and Dalziel J A, 1990 *Cement and Concrete Research* **20** 778-782
[52] Cyr M, Lawrence P, and Ringot E 2005 *Cement and Concrete research* **35** 719-730
[53] Awal A and Nguong S K 2010 A short-term investigation on high volume palm oil fuel ash (POFA) concrete," *Proceedings of the 35th Conferenece on our World in Concrete and Structure,* pp. 185-92
[54] Sata V, Jaturapitakkul C, and Rattanashotinunt C 2010 *Journal of materials in civil engineering* **22** 1033-1038
[55] Sujivorakul C, Jaturapitakkul C, and Taotip A 2011 *Journal of materials in civil engineering* **23** 1281-1288
[56] Sinsiri T, Kroehong W, Jaturapitakkul C, and Chindaprasirt P 2012 *Materials & Design* **42** 424-433,
[57] Chindaprasirt P, Chotetanorm C and Rukzon S 2010 *Journal of Materials in Civil Engineering* **23** 499-503
[58] Awal A A and Shehu I 2015 *Construction and Building Materials* **76** 214-220
[59] Hussin M W, Ismail M A, Budiea A and Muthusamy K 2009 *Malaysian Journal of Civil Engineering* **21** 180-194
[60] Awal A A and Mohammadhossein H 2016 *Journal of Cleaner Production* **137** 157-166
[61] Altwair N M, Johari M M and Hashim, S S 2012 *Construction and Building Materials* **37** 518-525
[62] Sivakumar A and Santhanam M 2007 *Cement and Concrete Composites* **29** 603-608
[63] Ropke J C 1982 *Concrete problems: causes and cures* McGraw-Hill

Effect of various filler types on the properties of porous asphalt mixture

Nurul Athma Mohd Shukry[1], Norhidayah Abdul Hassan[1], Mohd Ezree Abdullah[2], Mohd Rosli Hainin[1], Nur Izzi Md Yusoff[3], Ramadhansyah Putra Jaya[1] and Azman Mohamed[1]

[1] Faculty of Civil Engineering, Universiti Teknologi Malaysia, Johor, Malaysia
[2] Faculty of Civil and Environmental Engineering, Universiti Tun Hussein Onn Malaysia, Johor, Malaysia
[3] Department of Civil & Structural Engineering, Universiti Kebangsaan Malaysia, Selangor, Malaysia

E-mail: hnorhidayah@utm.my

Abstract. The open structure of porous asphalt exposes a large surface area to the effects of air and water, which accelerates the oxidation rate and affects the coating properties of the binder. These factors may influence the adhesive strength of the binder-aggregate and lead to cohesive failure within the binder film, contributing to aggregate stripping and moisture damage. The addition of fillers in asphalt mixtures has been identified to stiffen the asphalt binder and improve mixture strength. This study investigates the effect of various filler types (hydrated lime, cement, and diatomite) on the properties of porous asphalt. Compacted samples of porous asphalt were prepared using Superpave gyratory compactor at the target air void content of 21%. Each sample was incorporated with 2% of filler and polymer-modified binder of PG76. The morphology and chemical composition of fillers were investigated with a field emission scanning electron microscope (FESEM) and energy dispersive X-ray (EDX) analysis. The properties of porous asphalt were evaluated in terms of permeability, abrasion loss, resilient modulus, and indirect tensile strength. All mixtures were found to show high permeability rates. Mixtures with hydrated lime exhibited lower abrasion loss compared to mixtures with cement and diatomite. The use of diatomite increases the resistance of the mixtures to rutting and moisture damage compared to other fillers as shown by the enhanced resilient modulus and indirect tensile strength.

1. Introduction

Porous asphalt has been used widely due to its ability to allow rainwater to drain quickly from the pavement surface through its pore structure. Given this ability, porous asphalt is utilized in wearing courses with approximately 50 mm thickness and placed over the impermeable asphalt surface as a possible solution for road safety improvements in wet conditions and for the reduction of traffic noise. In terms of safety benefits during rain events, porous asphalt was acknowledged to reduce splash and spray because of rapid dry surfaces and minimize the risk of hydroplaning and wet skidding, thereby improving night visibility [1]. Porous asphalt is designed with open-graded aggregate gradations that consist of a large proportion of coarse aggregates with a limited amount of fine aggregates to create larger quantities of interconnected voids of more than 18%, to allow water to penetrate through the voids [2,3].

Despite its safety and environmental benefits, the performance and service life of porous asphalt can be affected by its poor structural durability. The life of a porous surface is expected to be shorter than that of a conventional asphalt surface because of the deterioration by runoff, air infiltration, subsequent stripping and oxidation, and hardening of the binder [4]. On the other hand, the open gradation and high air-void content lead the porous asphalt mixture to poor durability due to less stone-on-stone contact caused by the inappropriate gradation and low density in porous asphalt mixture, which results in a lower performance than normal dense-grade mixture [1]. Moreover, the open structure that facilitates water drainage exposes the pores of porous asphalt to air, water, and clogging materials that erode the binder film and eventually affects the strength of the binder-aggregate bonding [5]. In tropical countries such as Malaysia, which experiences frequent high rainfall intensity, porous structures are subjected to water-induced problems. In addition, issues of high traffic impact stress due to rapid development in infrastructure and road construction exert a profound effect on the durability of the porous asphalt layer, thereby worsening the deterioration of the asphalt pavement. These factors cause loss of bonding in the binder–aggregate system as a result of adhesive and cohesive failures in the porous asphalt, leading to stripping. In turn, stripping contributes to a decline in the performance and service life of the pavement. Stripping failure is defined as the separation or detachment of the aggregate and asphalt binder because of the loss of adhesion between these two materials usually in the presence of moisture; such failure is typically accompanied by gradual loss of strength over the years, which causes distress manifestations including ravelling, rutting, shoving, corrugation, and cracking [6].

The utilization of filler in the asphalt mixture has long been recognized by asphalt-paving technologists. Despite the small proportion it represents in the mixture, the use of filler is proven essential in improving the performance of asphalt mixtures to various distresses. The presence of filler can positively affect the overall mixture performance including strength, stability, workability, resilient modulus, resistance to moisture, resistance to permanent deformation, and aging characteristics [7,8]. The indirect tensile strength test by Huang *et al.* [9] showed that the said strength of the asphalt mixture increases as the filler content rises. This result is in agreement with the findings of Ahmed *et al.* [10], where higher filler content improves the mixture tensile strength. This higher tensile strength implies that mixtures with filler are capable of withstanding larger tensile stress before failure [11]. Another study revealed that the results for resilient modulus also indicated that mixtures with filler show higher resilient modulus values, which leads to stiffer mixtures with superior load-spread capacity and greater resistance to fatigue cracking and permanent deformation [7, 12].

Hydrated lime is the most commonly used additive in asphalt mixtures. It is known as an active filler that provides good adhesive bonding at the binder–aggregate interface, which effectively controls water sensitivity and resists stripping [13,14]. Li *et al.* [15] also investigated the use of diatomite as the filler in porous asphalt mixtures. Their study resulted in improved interface adhesion and low-temperature cracking resistance. On this basis, diatomite is considered a potential filler in porous asphalt, and the extension of the study is necessary. Ordinary Portland cement is also a possible additive, but limited research exists regarding the use of cement as filler in asphalt mixtures [14]. Therefore, this study evaluates the effect of different filler types on the properties of porous asphalt through permeability, abrasion loss, resilient modulus, and indirect tensile strength.

2. Experimental

2.1. Materials

The crushed granite aggregate used in this experiment was supplied by Hanson Quarry Products in Kulai, Johor. The gradation limit of the combined aggregate for porous asphalt was selected according to the Standard Specification for Porous Asphalt from JKR specifications [16] (see Figure 1). The aggregate gradation employed for the porous asphalt mixture in this study was Grading B with nominal maximum aggregate size of 14 mm.

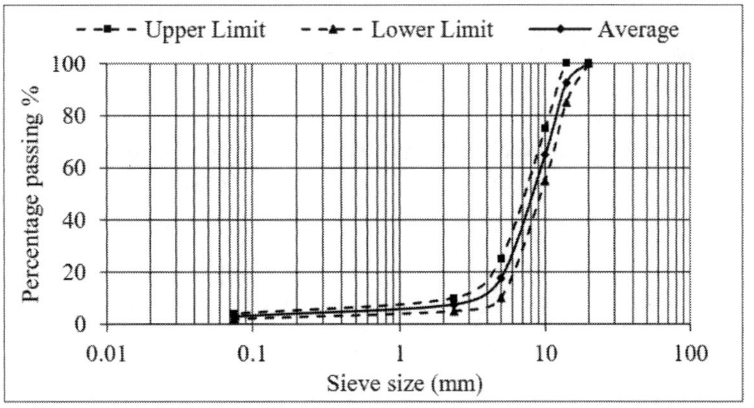

Figure 1. Plotted Malaysian gradation limits for porous asphalt.

The aggregate was tested for specific gravity and water absorption. The polymer modified binder PG76 was used as a binder for the mixture design and sample preparation as per JKR specification [16]. PG76 is highly suitable for the efficient performance of the porous asphalt mixture because it is modified with polymer and exhibits outstanding high viscosity property. The properties for aggregates and the binder utilized in this study are presented in Table 1. Three types of filler were selected for sample preparation: hydrated lime, cement, and diatomite (which passed the 75 μm sieve size). Specific gravities are 2.72 (hydrated lime), 3.2 (cement), and 2.2 (diatomite).

Table 1. Materials properties.

Materials	Properties	Value
Coarse aggregate	Specific Gravity Bulk	2.695
	Water Absorption (%)	0.520
Fine aggregate	Specific Gravity Bulk	2.427
	Water Absorption (%)	2.048
Bitumen PG76	Penetration at 25°C (dmm)	40.6
	Softening point (°C)	70
	Viscosity at 135°C (Pa. s)	2.8
	Specific gravity	1.030

2.2. Mixture design and sample preparation

The design binder content of the porous asphalt mixture was determined by 1) the average of the upper limit from the binder draindown test, 2) the lower limit from the Cantabro test, and 3) the target air void content of 21%. The 21% air void content was selected to prevent over-compaction on the samples (particularly for those with diatomite filler) such that the number of gyrations obtained would not exceed the maximum compaction of 75 gyrations, as recommended in the Superpave system. The binder draindown test is used to quantify the sufficient bitumen film thickness for coating the aggregate particles, whereas the Cantabro test is applied to evaluate the resistivity of the mixture against stripping or aggregate loss. Therefore, the design binder content used for the mixture prepared with hydrated lime and cement was 5%. The mixture with diatomite filler had the highest design binder content (5.25%) because of its high surface area, high porosity, and the high absorptive capacity of the material, which allows it to adsorb higher amounts of asphalt binder compared to other fillers.

The Superpave gyratory compactor was employed to produce compacted cylindrical samples of 100 mm diameter and an approximate thickness of 50 mm, with 725 g of blended aggregate. The samples were mixed at 180 °C. Then, the loose mixtures were conditioned in an oven for 2 h at the compaction temperature of 170 °C to allow the absorption of the asphalt binder into the aggregate. The machine was set at a loading pressure of 600 kPa and an external angle of gyration of 1.25°. A few compaction trials were conducted at various numbers of gyrations (i.e., 20, 40, 60, and 80) to ascertain the desired number of gyrations. As shown in Figure 2, the number of gyrations to achieve the target of 21% air voids content is 40, 45, and 58 for the mixture with hydrated lime, cement, and diatomite, respectively. All materials were added for 2% by mass of total aggregates as a part of the mineral filler.

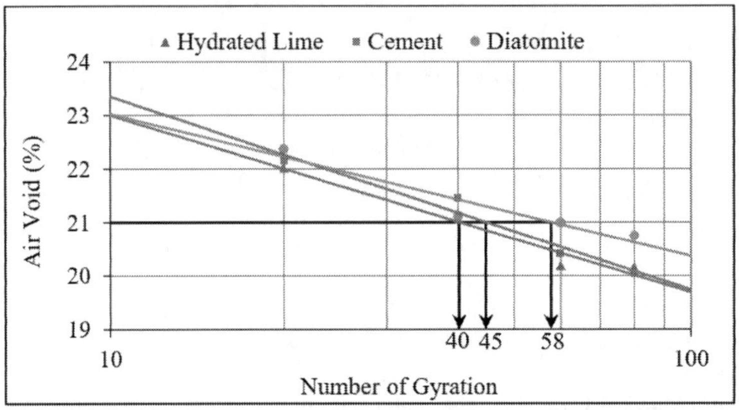

Figure 2. Plot of air void content versus number of gyration.

2.3. *Laboratory tests*

2.3.1. FESEM and EDX. Filler surface textures were examined using a field emission scanning electron microscope (FESEM). This machine captures surface images with a scan using a high energy beam of electrons in a raster scan pattern. It generates images of the sample at the microscale to enable the characterisation of microstructural properties. The samples of hydrated lime, cement, and diatomite powder were prepared in small quantities and sputtered with an extremely thin layer of carbon or metallic coating to render their surfaces conductive. The information in FESEM images is important for verifying some filler properties, such as specific surface area, particle distribution, and porosity. The chemical compositions of all the three filler types were evaluated quantitatively through energy dispersive X-ray (EDX) analysis and provided in the form of element weight percentage. This analysis was used to map the distribution of specific elements (Al, K, and Na) that characterized the various minerals in the material.

2.3.2. Permeability rate. Permeability tests were performed to determine the relative permeability of the compacted samples of porous asphalt mixtures. The falling head permeameter was used and permeability was measured in terms of the discharge time in seconds, which indicates the time required for a specified volume of water to permeate through a compacted sample. The time needed for the water level to fall between two designated points from 60 to 20 on the graduated cylinder was ascertained. The coefficient of water permeability, k, a product of Darcy's Law equation, was calculated using equation (1).

$$k = \frac{al}{At} ln\left(\frac{h_1}{h_2}\right) \quad (1)$$

where k is the coefficient of water permeability (cm/s), a is the inside cross-sectional area of inlet standpipe (cm^2), l is the thickness of the test sample (cm), A is the cross-sectional area of the test sample (cm^2), t is the average elapsed time of water flow between timing marks (s), h_1 is the hydraulic head on the sample at time t_1 (cm), and h_2 is the hydraulic head on the sample at time t_2 (cm).

2.3.3. Abrasion loss. The Cantabro test was conducted to determine the particle loss of the porous asphalt under abrasion. This test measures the resistance of the compacted samples to stone loss at high frequencies using abrasion machines. Abrasion loss, which was calculated through the percentage of weight loss, indicates the durability of the sample. In this study, the results of abrasion loss were also utilized to determine the design binder content of the porous asphalt with an average loss of mass no more than 15%. The percentage of abrasion loss (L) of each sample was computed using the following equation:

$$L = (M_1 - M_2 / M_1) \times 100 \tag{2}$$

where L is the abrasion loss (%), M_1 is the initial weight of the test sample (g), and M_2 is the final weight of the test sample (g).

2.3.4. Resilient modulus. A resilient modulus test was performed to measure the stiffness modulus of the asphalt mixtures using the repeated-load indirect tension test. The test is a non-destructive assay for investigating the recoverable strain of a mixture under repeated stress. Results with high resilient modulus value indicate that the asphalt mixture is stiff and the recoverable strain from repeated vehicle loads is smaller. With this method, the effect of temperature and load on resilient modulus were examined. In this study, the resilient modulus test was performed according to ASTM D4123 [17]. The assessment was conducted at temperatures of 25, 30, and 35 °C, loading frequencies of 1 Hz, and load duration of 0.1 s as recommended in the JKR specifications [16].

2.3.5. Moisture susceptibility test. The paragraph text follows on from the subsubsection heading but should not be in italic. The modified Lottman test is the most commonly used assay for determining the moisture susceptibility of bituminous mixtures. This test was performed according to AASHTO T283 [18]. Samples of dry and moisture conditioned subsets were fabricated, with each subset consisting of three samples. The dry or control subsets (unconditioned) were left at 25 °C in an incubator for 2 h before testing. The wet subsets (conditioned) were preconditioned using vacuum saturation for approximately 15 min to achieve a saturation level of 55% to 80%. After saturation, the samples were wrapped in leak-proof plastic bags containing 10 ml of water. The samples were then placed in a freezer at a temperature of −18 °C for 16 h, followed by sample immersion in a water bath at 60 °C for 24 h. Prior to indirect tensile testing, the conditioned samples were placed in a water bath at 25 °C for another 2 h. The degree of saturation, S, was determined using the following equation:

$$S = W_{SSD} - W_D / V_a \tag{3}$$

where S is the degree of saturation (%), W_{SSD} is the weight of the saturated surface dry sample after vacuum saturation (g), W_D is the weight of the dry sample (g), and V_a is the volume of air voids in the sample (cm^3). The indirect tensile strength (ITS) test was performed according to ASTM D6931 [19]. A loading rate of 50 mm/min was applied using a Marshall loading machine with steel-loading strips. The maximum load was recorded and the tensile strengths of the asphalt mixtures were calculated using equation (4):

$$ITS = 2000F / \pi hD \tag{4}$$

where ITS is the indirect tensile strength (kPa), F is the maximum load (N), h is the height of the specimen (mm), and D is the diameter of the specimen (mm). The ITS values for unconditioned and conditioned samples were utilized to calculate the tensile strength ratio (TSR). The TSR values indicate the resistance of an asphalt mixture to moisture damage. The ratio is calculated as in equation (5):

$$TSR = (ITS_{wet} / ITS_{dry}) \times 100, \quad (5)$$

where TSR is the tensile strength ratio (%), ITS_{wet} is the average ITS of the moisture-conditioned subset (kPa), and ITS_{dry} is the average ITS of the dry subset (kPa).

3. Results and discussion

3.1. FESEM and EDX analyses

The surface texture of hydrated lime, cement, and diatomite are compared in Figure 3. The cement texture appears dense and compact. For hydrated lime, the particles exist in a bonded cloud-of-smoke shape. Unlike hydrated lime and cement, the diatomite particle has a porous structure in the form of a single grain. The porous structure of diatomite shows that it has a high surface area that leads to an increase in design binder content due to its elevated adsorptive capacity. These findings can improve drainability and prevent binder draindown. Table 2 presents the chemical composition of the three fillers. Both hydrated lime and cement have high contents of calcium and oxygen that form calcium oxides (CaO). Cement consists mainly of limestone, clay, and gypsum in a high-temperature process. Hydrated lime, or calcium hydroxide ($Ca(OH)_2$), is produced by treating CaO with water. It is well known as a filler that enhances the chemical bonds between the aggregate and binder in asphalt mixtures [20]. Diatomite primarily consists of silica and oxygen in the form of silicon dioxide (SiO_2), which is known for its hardness. According to Cong *et al.* [21], its high silica content and honey-comb silica structure are the most important features of diatomite because they can generate useful characteristics, such as high absorptive capacity, chemical stability and low bulk density.

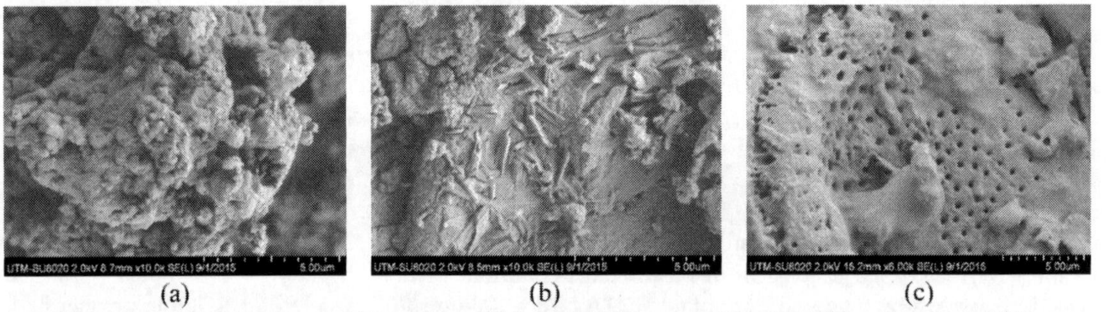

(a) (b) (c)

Figure 3. FESEM images of (a) hydrated lime at 10000× magnification, (b) cement at 10000× magnification, and (c) diatomite at 6000× magnification.

Table 2. EDX results of fillers.

Element	Hydrated lime (wt. %)	Cement (wt. %)	Diatomite (wt. %)
Al	0.2	1.3	1.4
Au	6.2	8.9	8.9
Ca	43.8	37.8	-
C	12.8	9.1	17.4
F	-	3.0	6.7
Fe	0.4	-	-
Mg	0.8	0.5	0.3
Na	-	-	1.7
O	35.7	33.0	31.8

Si	0.2	6.3	31.9

3.2. Permeability rate

A permeability test was conducted to investigate the ability of a porous asphalt mixture to drain water through its porous structure. Results from the mixtures containing hydrated lime, cement, and diatomite were compared as shown in Figure 4. Based on the outcomes, the permeability rate of the mixture with hydrated lime was higher than that of the mixtures with cement and diatomite. However, all mixtures exhibited higher permeability rates than the recommended minimum k value of 0.116 cm/sec for porous asphalt [22]. In addition, the three mixtures exhibited minimal differences from one other. Such small differences might be due to the void distribution and interconnectivity within the mixtures.

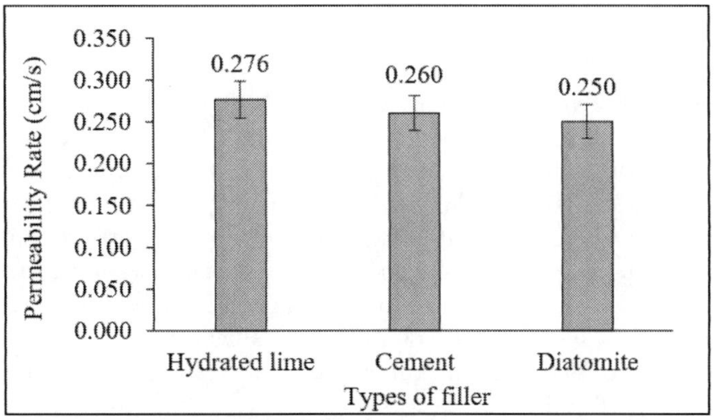

Figure 4. Permeability test result.

3.3. Abrasion loss resistance

Figure 5 presents the results for abrasion loss using the Cantabro test. In this study, the maximum Cantabro particle loss allowed was 25% for unaged porous asphalt mixtures. The abrasion loss of samples prepared with hydrated lime is 21.5%, which is lower than the said loss for samples incorporating cement (33.2%) and diatomite (27.9%). According to Rodriguez-Hernandez *et al.* [23], a lower abrasion loss indicates that the samples possess better cohesion and resistance to ravelling. The current findings demonstrate that the mixture with hydrated lime achieved better results than cement and diatomite, proving that hydrated lime has a good filler property. Given its chemically active nature, which produces a strong physical–chemical interaction with the asphalt binder, hydrated lime is highly effective in enhancing the chemical bonds between the aggregates and the binder in the asphalt mixtures.

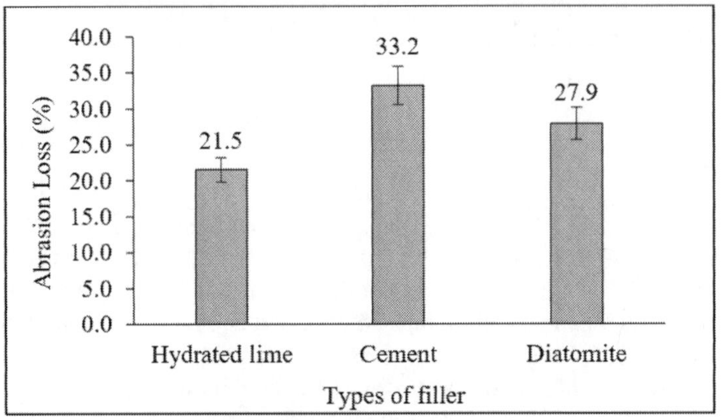

Figure 5. Abrasion loss result.

3.4. Resilient modulus

Results from the resilient modulus tests executed at 25, 30, and 35 °C are shown in Figure 6. The resilient modulus values for all mixture types decreased from 25 °C to 35 °C. According to Faghri *et al.* [24], the material for porous asphalt mixtures is sensitive to temperature. When the temperature increased from 25 °C to 35 °C, the resilient modulus of samples with hydrated lime, cement, and diatomite were reduced by 51%, 64%, and 59%, respectively. The observed significant decrease in resilient modulus at high temperatures is related to the reduced viscosity and softening of asphalt binders, causing the binder to lose its ability to join the aggregates. Therefore, the recoverable strain is increased under applied loads, thereby resulting in lower resilient modulus. According to McDaniel *et al.* [25], the stiffness values of porous asphalt are typically lower than those of dense-graded mixtures given different aggregate gradations. The lack of fine aggregate or mastic in porous asphalt can cause durability problems because the mastic provides strength to the mixture. For porous asphalt, the strength is more dependent on the stone-on-stone contact of the coarse aggregates, which refers to the strong aggregate interlock and good adhesion between aggregates and the polymer-modified binder. Results reveal that, at all test temperatures, the mixture containing diatomite has the highest resilient modulus compared to those with hydrated lime and cement. This finding indicates that mixtures with diatomite are the least susceptible to temperature changes and have the greatest increase in stiffness. Such characteristic is due to the different binder content of mixtures with diatomite, where binder adhesion has a substantial influence on the resilient modulus. As the optimum binder content of diatomite mixtures is slightly higher, the mixture has enough binder content to improve inter-aggregate adhesion and cause a smaller recoverable strain.

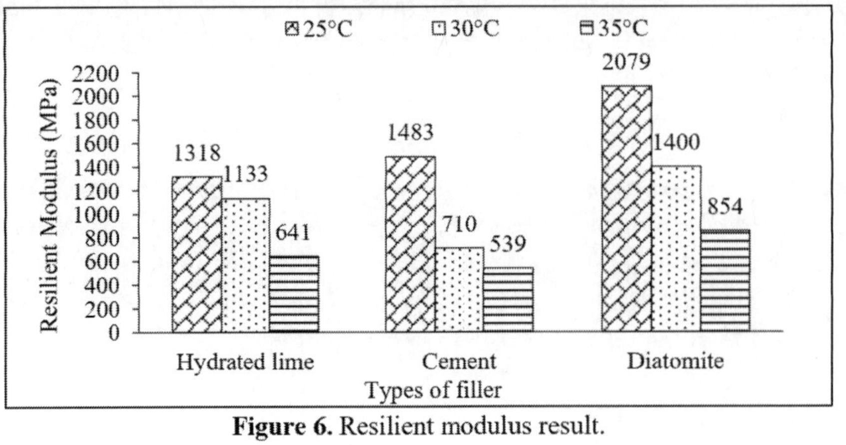

Figure 6. Resilient modulus result.

3.5. Modified Lottman test

The ITS and TSR for both unconditioned (dry) and conditioned (wet) samples are shown in Figure 7. As indicated, unconditioned mixtures containing diatomite exhibit the highest tensile strength values. Similar trends were observed for the ITS of conditioned mixtures. Those outcomes suggest that the ITS for conditioned samples is lower than the ITS for unconditioned samples. This observation indicates that deterioration occurred in the mixtures and that moisture conditioning exerts a significant effect on reducing the tensile strength of the mixtures. Samples prepared with diatomite have greater ITS than those prepared with hydrated lime and cement. Therefore, samples incorporating diatomite exhibit greater resistance to moisture damage compared to samples containing hydrated lime and cement. As disclosed in Figure 7, slight differences in TSR values are observed among the mixtures prepared with hydrated lime (88.8%), cement (90.6%), and diatomite (91.9%). The TSR values for mixtures containing diatomite are slightly higher than those of the other two mixtures. However, a minimum requirement of 80% for TSR value was specified for Superpave mix designs in AASHTO T283. Results confirm that all mixtures demonstrate TSR values greater than 80%, indicating that hydrated lime, cement, and diatomite could improve moisture resistance.

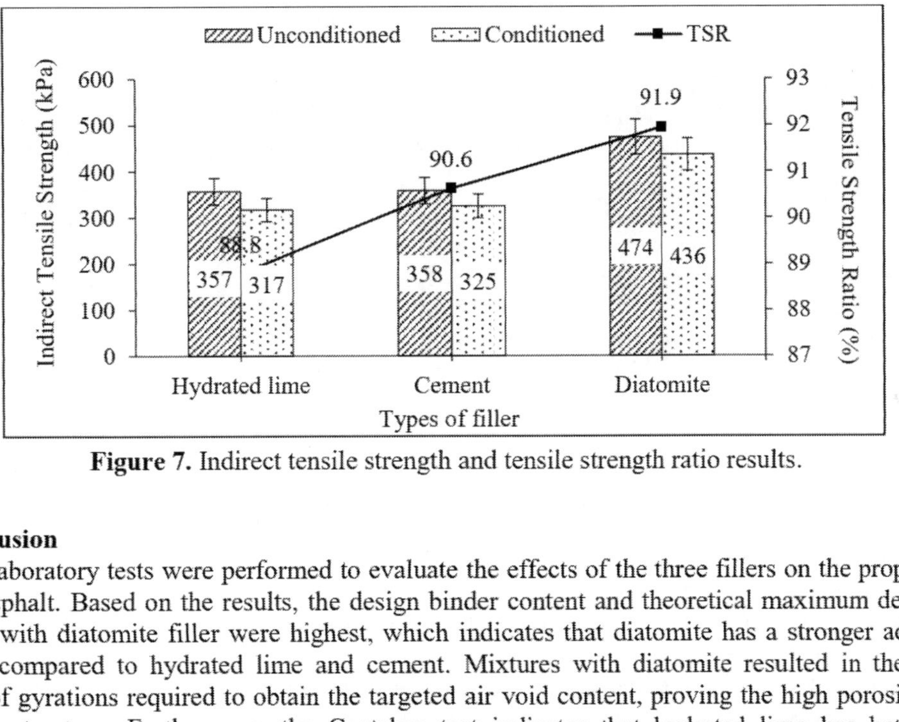

Figure 7. Indirect tensile strength and tensile strength ratio results.

4. Conclusion

Various laboratory tests were performed to evaluate the effects of the three fillers on the properties of porous asphalt. Based on the results, the design binder content and theoretical maximum density for mixtures with diatomite filler were highest, which indicates that diatomite has a stronger adsorption capacity compared to hydrated lime and cement. Mixtures with diatomite resulted in the highest number of gyrations required to obtain the targeted air void content, proving the high porosity of the diatomite structure. Furthermore, the Cantabro test indicates that hydrated lime has better filler properties than cement and diatomite. The lowest abrasion loss from hydrated lime suggests that it provides better cohesion and resistance to ravelling. This feature is due to its strong physical–chemical interaction with asphalt binder. The resilient modulus values were the highest for mixtures containing diatomite at all test temperatures indicating that diatomite improves the stiffness of the porous asphalt mixture more than the mixtures containing hydrated lime and cement. Additionally, the mixture with diatomite shows the highest values for indirect tensile strength and tensile strength ratio. However, the tensile strength ratio for all three mixtures complied with specifications, indicating good resistance to moisture damage. Overall, porous asphalt with diatomite achieved better results in most of the tests, followed closely by the mixtures with hydrated lime, and then by the mixtures with cement.

Acknowledgements
The support provided by Malaysian Ministry of Higher Education (MOHE) and Universiti Teknologi Malaysia (research grant no. Q.J130000.2522.11H76 and R.J130000.7822.4F436) for this study is highly appreciated.

References
[1] Putman B J and Kline L C 2012 Comparison of mix design methods for porous asphalt mixtures *Journal of Materials in Civil Engineering* **24** (11) 1359–67
[2] Alvarez A E, Martin A E. and Estakhri C 2010 Drainability of permeable friction course mixtures *Journal of Materials in Civil Engineering* **22** (6) 556–64
[3] Cetin A 2013 Effects of crumb rubber size and concentration on performance of porous asphalt mixtures *International Journal of Polymer Science* (2013)010
[4] Scholz M and Grabowiecki P 2007 Review of permeable pavement systems *Building and Environment* **42** 3830–36
[5] Aman M Y, Shahadan Z and Noh M Z M 2014 A comparative study of anti-stripping additives in porous asphalt *Jurnal Teknologi (Sciences & Engineering)* **70** (7) 139–45
[6] Haghshenas H F, Khodaii A and Saleh M 2015 Long term effectiveness of anti-stripping agents *Construction and Building Materials* **76** 307–12
[7] Muniandy R, Aburkaba E and Mahdi L M J 2013 Effect of mineral filler type and particle size on asphalt-filler mastic and stone mastic asphalt laboratory measured properties *Australian Journal of Basic and Appled Sciences* **7** (11) 475–87
[8] Chuanfeng Z, Yong Q, Dan L, Ting Z, Xingyang L and Shi Z 2013 Effects of anti-stripping agents on the microscopic strength of mineral aggregate contact surface *Construction and Building Materials* **49** 627–34
[9] Huang B, Shu X and Chen X 2007 Effects of mineral fillers on hot-mix asphalt laboratory-measured properties *International Journal of Pavement Engineering* **8** (1) 1–9
[10] Ahmed H Y, Othman A M and Mahmoud A A 2006 Effect of using waste cement dust as a mineral filler on the mechanical properties of hot mix asphalt *Assiut University Bulletin for Environmental Researches* **9** (1) 51–60
[11] Behiry A E A El-M 2013 Laboratory evaluation of resistance to moisture damage in asphalt mixtures *Ain Shams Engineering Journal* **4** (3) 351–63
[12] Iskender E and Aksoy A 2012 Field and laboratory performance comparison for asphalt mixtures with different moisture conditioning systems *Construction and Building Materials* **27** (1) 45–53
[13] Jahromi S G 2009 Estimation of resistance to moisture destruction in asphalt mixtures *Construction and Building Materials* **23** (6) 2324–31
[14] Huang B, Shu Z, Dong Q and Shen J 2010 Laboratory evaluation of moisture susceptibility of hot-mix asphalt containing cementitious fillers *Journal of Materials in Civil Engineering* **22** (7) 667–73
[15] Li Y, Huang S and Ding Q 2011 Effects of diatomite filler on the performance of porous asphalt mixtures *The 24th ICTPA Annual Conference & NACGEA International Symposium on Geo-Trans* 27-29 May (Los Angeles USA) pp 1–9
[16] Jabatan Kerja Raya (JKR) 2008 *Standard Specification for Road Works – Section 4: Flexible Pavement* Public Works Department, Ministry of Works Malaysia (Kuala Lumpur)
[17] ASTM D4123 1995 *Standard test method for indirect tension test for resilient modulus of bituminous mixtures* (Philadelphia: ASTM International)
[18] AASHTO T283 2007 *Standard method of test for resistance of compacted asphalt mixtures to moisture-induced damage* (Washington DC: AASHTO Provisional Standards)
[19] ASTM D6931 2012 *Standard test method for indirect tensile (IDT) strength of bituminous mixtures* (Philadelphia: ASTM International)

[20] Kavussi A and Bakhtiari J 2014 Application of different testing methods for evaluating effects of hydrated lime in bituminous mixes *Procedia Engineering* **77** 148–54
[21] Cong P, Chen S and Chen H 2012 Effects of diatomite on the properties of asphalt binder *Construction and Building Materials* **30** 495–9
[22] Mallick R B, Kandhal P S, Cooley L A and Watson D E 2000 *Design, construction, and performance of new-generation open-graded friction courses* NCAT Report No. 2000-01 National Center for Asphalt Technology Auburn University
[23] Rodriguez-Hernandez J, Andrés-Valeri V C, Calzada-Pérez M A, Vega-Zamanillo Á and Castro-Fresno D 2015 Study of the raveling resistance of porous asphalt pavements used in sustainable drainage systems affected by hydrocarbon spills *Sustainability* **7** 16226–36
[24] Faghri M, Sadd M H, Cardin J, Daly P, Park K, Chiharu T and Goncalves E 2002 *Performance improvement of open-graded asphalt mixes* Report No. URITC 00-44 University of Rhode Island Transportation Center Rode Island
[25] McDaniel R S, Thornton W D and Dominguez J G 2004 *Field evaluation of porous asphalt pavement* Final Report SQDH 2004–3 North Central Superpave Center, Purdue University West Lafayette Indiana May 2004

IOP Conf. Series: Materials Science and Engineering **342** (2018) 012037 doi:10.1088/1757-899X/342/1/012037

Effect of garlic solution to *Bacillus* sp. removal

N Zainol and S R Rahim

Faculty of Chemical and Natural Resources Engineering, Universiti Malaysia Pahang, Lebuhraya Tun Razak, 26300 Kuantan, Pahang, Malaysia

E-mail: azwina@ump.edu.my

Abstract. Biofilm is a microbial derived sessile community characterized by cells that are irreversibly attached to a substratum or interface to each other, embedded in a matrix of extracellular polymeric substances that they have produced. *Bacillus* sp. was used as biofilm model in this study. The purpose of this study is to determine the effect of Garlic solution in term of ratio of water and Garlic solution (W/G) and ratio of Garlic solution to *Bacillus* sp. (GS/B) on *Bacillus* sp removal. Garlic solution was used to remove *Bacillus* sp. In this study, Garlic solution was prepared by crushing the garlic and mixed it with water. the Garlic solution was added into *Bacillus* sp. mixture and mixed well. The mixture then was spread on nutrient agar. The *Bacillus* sp. weight on agar plate was measured by using dry weight measurement method. In this study, initially Garlic solution volume and Garlic solution concentration were studied using one factor at time (OFAT). Later two-level-factorial analysis was done to determine the most contributing factor in *Bacillus* sp. removal. Design Expert software (Version 7) was used to construct experimental table where all the factors were randomized. Bacilus sp removal was ranging between 42.13% to 99.6%. The analysis of the results showed that at W/G of 1:1, *Bacillus* sp. removal increased when more Garlic solution was added to *Bacillus* sp. Effect of Garlic solution to *Bacillus* sp. will be understood which in turn may be beneficial for the industrial purpose.

1. Introduction

Bacillus sp. have been long used as biofilm model to investigate biofilm [1]. Garlic solution can inhibit much more strongly towards bacterial strains compared to antibiotics [2, 3]. [4] found that the garlic extract inhibited the growth and killed most of the organisms tested. [3] believed that garlic can inhibit *Bacillus subtilis*, *Escherichia coli*, and *Saccharomyces cerevisiae.* Therefore, garlic is used to remove *Bacillus sp.* The concentration of garlic (biocide) also can be defined as the ratio of water and garlic to remove microbes. Concentration of biocide is one of important aspects that should be considered when it is used to remove microbes as it affects the efficacy [4] concentration of biocide usually used as they can remove microbes greater and rapidly compare than low concentation of biocide. The activity of inhibitory natural biocides is directly proportional to the concentration of natural biocide [5]. According to [6], as the concentration of the garlic extraction increase, the diameter of inhibition of microbes increase too. As the garlic solution concentration increases, the inhibitory effect increases. In other words, the inhibitory effect of the extract is proportional to its concentration [5]. As the ratio of biocides extract increases to microbes, it can inhibit the microbes higher [7]. [8] reported that with garlic extract, an initial lag phase of 30 to 60 min occurred at concentrations of 0.34 to 2.75 mg/ml. Lag phases may reflect delay in uptake of garlic molecules and/or of toxic effect upon metabolic processes. However, at the higher concentrations used, a potential for rapid bactericidal action of garlic extract was indicated by complete loss of viability during the first 2 h. The objectives is to study the effect of garlic solution to *Bacillus* sp. removal. One factor at time (OFAT) analysis and response surface method (RSM) were used to evaluate the significance of W/G and GS/B values. Factorial analysis via two level factorial is used to allow chemist to study the effect of each factor on response and study the correlation between the factors [9].

2. Materials and Method

2.1. Garlic solution preparation

Garlic was obtained from a grocery shop in Gambang. The garlic sample was peeled, washed and prepared into two samples with different ratio of water to garlic (W/G).

2.2. Bacillus sp. preparation

Bacillus sp. from the previous study has been used in this research [10]. *Bacillus* sp. was streaked on the surface of the nutrient agar in the petri dish by swabbing it across quadrant number 1 by using sterile inoculation loop. It was repeated until quadrant number 4 before incubating it at 37°C for 24 hours [11].

2.3 Experimental setup for Bacillus sp. removal

Bacillus sp. was mixed with nutrient broth and agitated in incubator shaker for 1 hour at 37°C and 100 rpm. Then the *Bacillus* sp. broth was mixed with garlic solution. Then the mixture was spread on nutrient agar. The mixture was incubated in the incubator shaker at 37°C for 12 hours. Finally the *Bacillus* removal was calculated [12].

2.4 Experimental set-up for one factor at time (OFAT)

2.4.1 Effect of garlic solution volume on Bacillus sp. One colony of *Bacillus* sp. was mixed with 10 mL of broth and was shaken for 1 hour at 37°C. The *Bacillus* broth was mixed with 1 mL of garlic solution and slightly shaken before incubate it. The initial OD of mixture was analysed and recorded. The mixture then was incubated at 37°C and the OD of the mixture after 4 hours. The steps were repeated by using 3 mL of garlic solution.

2.4.2 Effect of garlic solution on the growth of Bacillus sp. Five samples of garlic solution were prepared according to their ratio as shown in Table 1. Half plate of *Bacillus* sp. was mixed with 50 mL of broth and was shaken in incubator shaker at 37°C for 1 hour. 1 mL of 1:1 of garlic solution sample was spread onto the agar plate by using sterile inoculation loop. Then 1 mL of sample that have been agitate before also was spread on the same plate. The agar plate was incubated at 37°C for 24 hours.

Table 1. Volume of water and garlic solution needed for each ratio.

Ratio of water to garlic	Volume of water (mL)	Volume of garlic solution (mL)
1:1	25	25
1:2	25	50
1:3	25	75
1:4	25	100
1:5	25	125

2.5 Experimental setup for Response Surface Method (RSM) analysis

Design Expert software (Version 7.1.3, Stat-Ease, Inc., Minneapolis, MN) program was used in

RSM analysis. The experimental table was constructed in two level factorial of response surface methodology (RSM). Results from OFAT were utilized to determine the range for RSM analysis. Four selected factors were : ratio of water to garlic (W/G) , ratio of garlic solution to *Bacillus* sp. (GS/B), agitation speed and reaction time between *Bacillus* sp. and garlic. These factors were studied to determine their effects on the *Bacillus* sp. removal by using a 2^4 factorial design. Table 2 shows the factors and it levels where low level indicates the lowest range of the factors and high level indicates the highest range of the factors. Experimental design table was constructed by using the Design Expert Software V7 and experimental data was analyzed using the same software. For 1:1 of W/G, the garlic solution was prepared by blending 25 mL of garlic and 25 mL of distilled water. Meanwhile, for 1:5 of W/G, the garlic solution was prepared by blending 125 mL of garlic and 25 mL of water. For 1:1 of GS/B, 10 mL *Bacillus* broth was mixed with 10 mL of garlic solution. On the other hand, for 1:3 of GS/B, 30 mL *Bacillus* broth was mixed with 10 mL of garlic solution.

Table 2. The levels of selected factors.

Factors	Low level (-1)	High level (+1)
Ratio of water and garlic solution	1:1	1:5
Ratio of garlic solution to *Bacillus* sp.	1:1	1:3
Reaction time between garlic solution and *Bacillus* sp. (hours)	12	24
Agitation speed (rpm)	50	100

2.6. Analysis of Bacillus sp. weight

The weight of *Bacillus* sp. on the agar plate was measured by using dry weight measurement method to determine the amount of *Bacillus* sp. for each run. First, by using sterile inoculation loop, 1µL of the sample that has been shaken before was spread onto the agar in petri dish. The petri dish was incubated in the incubator at 37°C for 24 hours. After 24 hours, all the *Bacillus* sp. growth on the agar was scraped out into the 10 mL broth in the centrifuge tube. It should be noted that all the empty centrifuge tube need to be weighed before proceeding to the next step. Then, the mixture was centrifuged at 5000 rpm for 15 minutes to separate *Bacillus* sp. and the broth. The broth was discarded from the centrifuge tube and the centrifuge tube was placed at 100°C. The dry centrifuge tube was weighed to determine the weight of *Bacillus* sp.

3. Results and Discussion

3.1 Effect of garlic solution volume to Bacillus sp.

This study was conducted to determine the effect of different garlic solution (GS) volume towards *Bacillus* sp. For effect of GS volume to *Bacillus* sp., the result was analysed by OD analysis for each sample. Figure 1 shows the result of by adding different volume of garlic had effect on *Bacillus* sp. growth after 4 hours. There was a small reduction of *Bacillus* sp. growth when there is no garlic added into the sample. However, huge reduction of *Bacillus* sp. growth can be seen when 3 mL of garlic added into the sample which is from 3.0000 to 0.2257 followed by 1 mL garlic added with value 0.7620 to 0. This is because when the GS volume was increasing, it reacts with *Bacillus* sp. stronger since it contains more garlic. [13] mentioned that when optical density reduced, this is because the cells had become less motile. OD initial for 3 mL of garlic added is the highest because of the concentration of GS itself is high and cause the OD reading high.

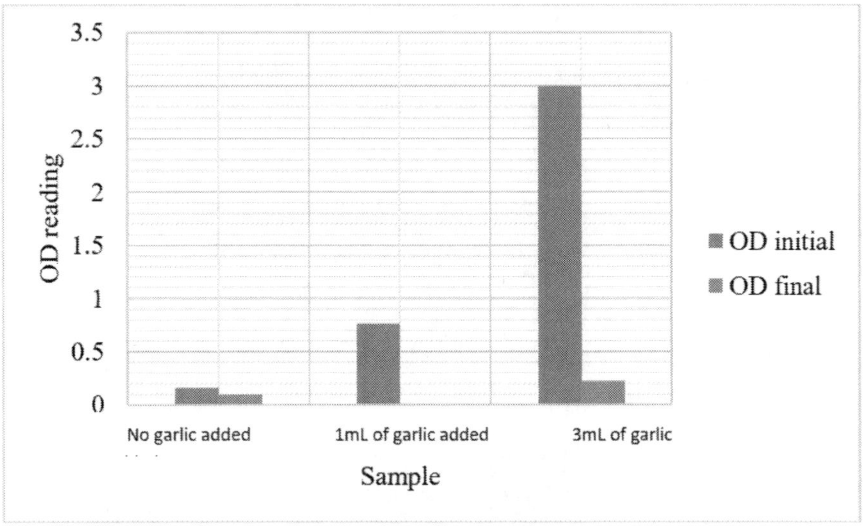

Figure 1. Data analysis of OD on *Bacillus* sp.

3.2 Effect of garlic solution on the growth of Bacillus sp.

This preliminary study was carried out to study the effect of water to garlic ratio (W/G) on *Bacillus* sp. growth. The value to study the W/G were 1:1, 1:2, 1:3, 1:4, and 1:5. 1:1 was set as low level of WG, meanwhile 1:5 was set as high level. Figure 2 shows the effect of W/G on *Bacillus* sp. weight. From the figure, it shows that when W/G increases, the *Bacillus* sp. weight decreases. As the level of WG increasing, the concentration of GS increased. It means that the *Bacillus* sp. weight decreases as the GS concentration increases. The data follow the result from [6] in which *Bacillus* sp. growth decreased as GS concentration increased. Since allicin compound in garlic can inhibit the growth of *Bacillus* sp., as the GS increased, the growth of *Bacillus* sp. decreased. Figure 3 and 4 showed the different of *Bacillus* sp. growth on different W/G.

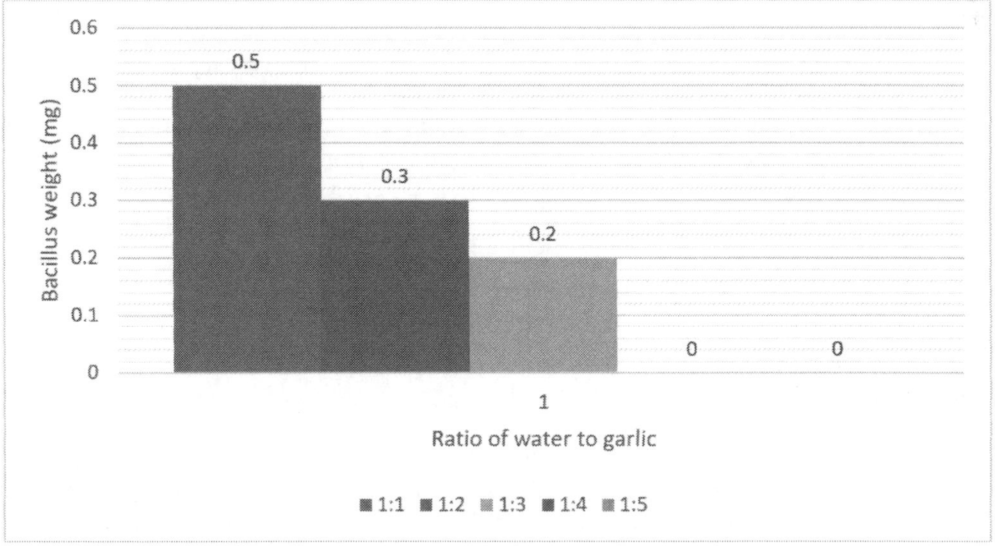

Figure 2. The effect of W/G on weight of *Bacillus* sp.

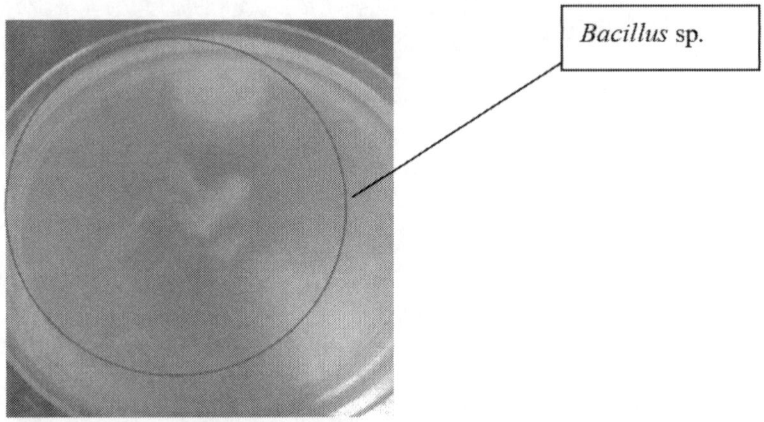

Figure 3. *Bacillus* sp. growth for W/G at 1:1

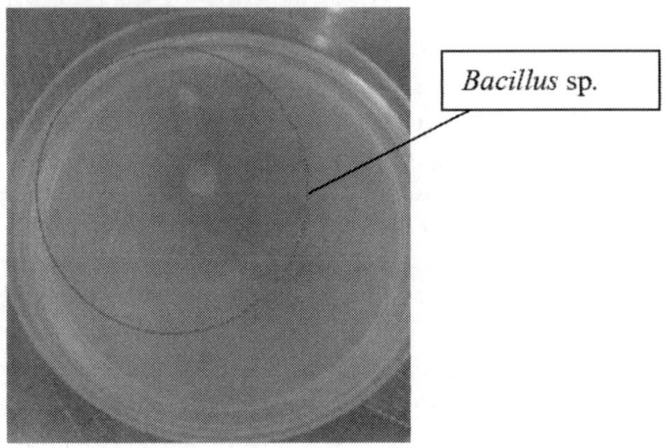

Figure 4. *Bacillus* sp. growth for W/G at 1:2

3.3 RSM analysis

Two level factorial of RSM analysis was used to analyze the process factor by correlating the interaction between input and output variable (equation 1). It also used to provide a clear understanding of the interactions between all selected factors and *Bacillus* sp. removal. In OFAT analysis, *Bacillus* sp. removal was increased when garlic solution and WG increased.

$$\% \textit{ Bacillus} \text{ sp removal} = 86.74 + 4.31*A - 6.32*B + 5.64*C - 0.21*D + 5.90*AB + 1.19*AC + 1.56*AD + 2.66*BC + 3.43*CD - 6.46*ACD + 6.12*BCD \quad (1)$$

where
A-ratio of water to garlic
B-ratio garlic to *Bacillus*
C-agitation speed
D-reaction time between *Bacillus* and garlic

Bacillus sp removal is strongly affected by allicin in garlic due to antimicrobial properties of the allicin. According to the software the suggested best condition for *Bacillus* sp. removal is as mentioned in Table 3 with the predicted *Bacillus* sp removal at 80.65%. However, from the experiment, the *Bacillus* sp. removal was higher than that the predicted (88.46%).

Figure 5 shows the effect of ratio of garlic solution to *Bacillus* sp (GS/B) on *Bacillus* sp. removal.

Based on Table 2, for GS/B, 1:1 was set as low level and 1:3 was set as high level. Negative effect is when the factor is not proportional to the response value. From Figure 5, GS/B gave negative effects towards the respond. Therefore, when the value of GS/B is increasing, the *Bacillus* sp. removal is decreasing. GS/B was increased by increasing the *Bacillus* sp. concentration. To get higher *Bacillus* sp. removal, the concentration of *Bacillus* sp. should be lower in which GS/B should be on low level. *Bacillus* sp. removal was higher when GS/B at 1:1 with value 97.2%, meanwhile, it reduced to 79.7% when GS/B was at 1:3. This is because when the concentration of *Bacillus* sp. was higher than the allicin concentration in garlic, the garlic cannot react or inhibit *Bacillus* sp. since the amount of GS was insufficient to kill *Bacillus* sp. In order to inhibit *Bacillus* sp., the allicin concentration in garlic should be same or higher than the *Bacillus* sp. concentration [5].

Figure 7 shows the interaction between (W/G) and (GS/B). For Factor A and B, 1:1 was considered as low level, and, 1:5 and 1:3 respectively were considered as high level. The *Bacillus* sp. removal was highest when GS/B at 1:1 and W/G at 1:1 with value 97.2%. On the other hand, the *Bacillus* sp. removal was lowest when GS/B and W/G at 1:1 and 1:5 respectively with value 78%. However, the *Bacillus* sp. removal did not really affected by W/G factor when the GS/B was at high value (1:3). The concentration of garlic was increasing when W/G was increasing. The *Bacillus* sp. removal was proportional to the concentration of antimicrobial agent (Figure 6). The activity of inhibitory natural biocides is directly proportional to the concentration of natural biocide [5].

Table 3. Data from validation experiment.

Factor 1 A: ratio water to garlic	Factor 2 B: ratio garlic solution to *bacillus*	Factor 3 C: agitation speed (rpm)	Factor 4 D: reaction time (hours)	Predicted *Bacillus* sp. removal (%)	Experiment *Bacillus* sp. removal (%)
1:1	1:3	50	12	80.65	88.46

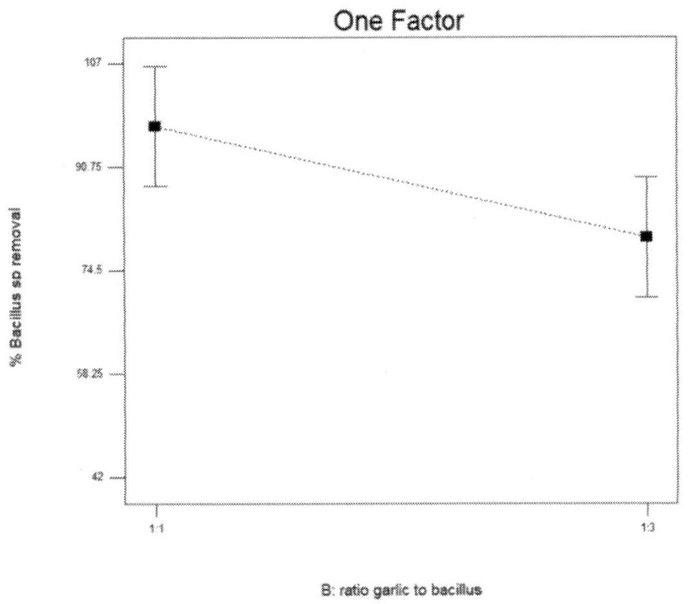

Figure 5. One factor plot of the interaction between GS/B on *Bacillus* sp removal.

IOP Conf. Series: Materials Science and Engineering **342** (2018) 012037 doi:10.1088/1757-899X/342/1/012037

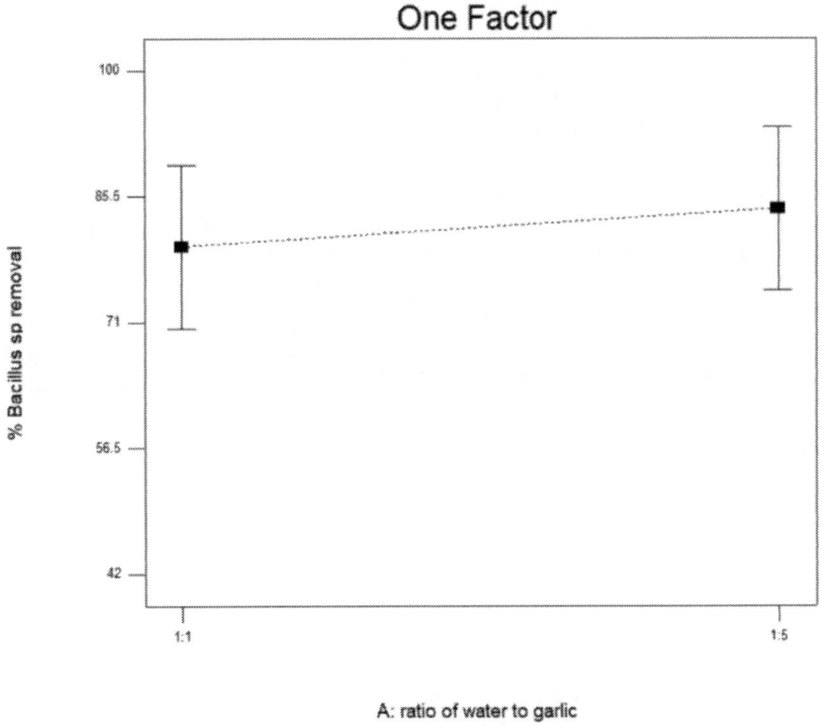

Figure 6. One factor plot of the interaction between W/G on Bacillus sp removal.

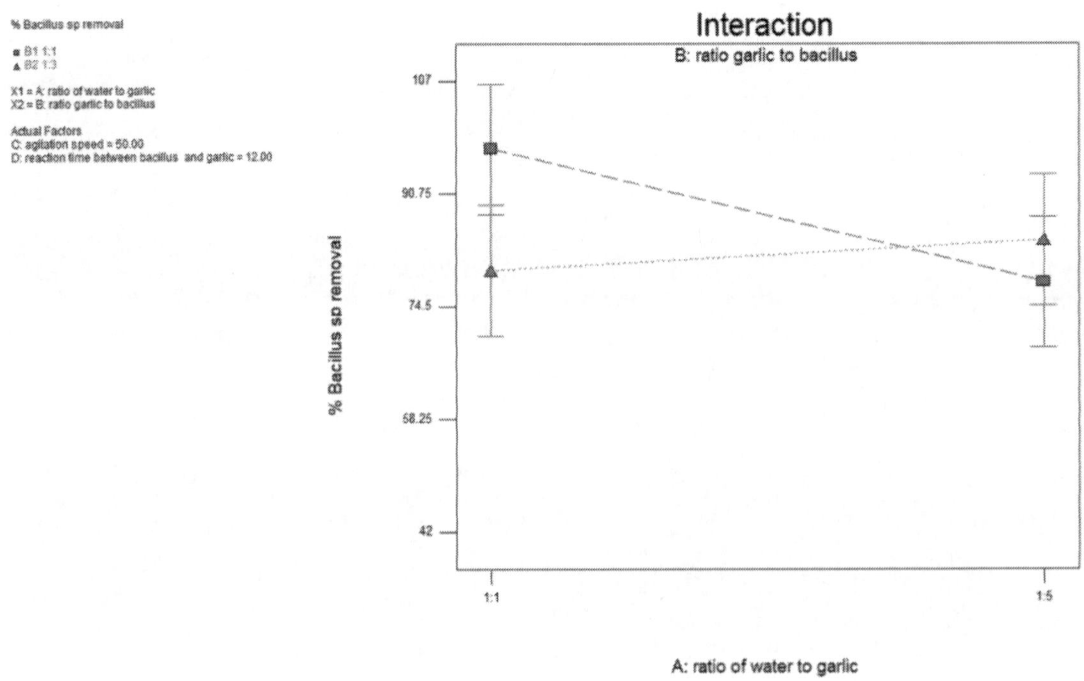

Figure 7. Analysis of interaction effects (Factor AB) on *Bacillus* sp. removal.

4. Conclusion

The purpose of this study is to determine the effect of garlic solution to *Bacillus* sp removal. The effect of garlic solution was studied in term of ratio of water and garlic solution (W/G) and ratio of garlic solution to *Bacillus* sp. (GS/B). Garlic content could increased the *Bacillus* sp. removal. *Bacillus* sp removal is strongly affected by allicin in garlic due to antimicrobial properties of the allicin. The concentration of garlic was increasing when W/G was increasing. The *Bacillus* sp. removal was proportional to the concentration of antimicrobial agent. However, GS/B gave negative effects towards the response where the value of GS/B is increasing when the *Bacillus* sp. removal is decreasing The maximum FA yield predicted by RSM was 80.65% with best condition at W/G at 1:1, GS/B at 1:3, agitation speed at 50 rpm and reaction time at 12 hours . However, the real value from experimental shows higher yields of 88.46% with the best condition. Results from OFAT and RSM do agreed with one another. Results from this study shows the potential of garlic solution application for *Bacillus* sp. removal. Garlic solution can replace other commercial biocides especially in food industry.

Acknowledgement

The author wish to acknowledge the Universiti Malaysia Pahang for funding the project under grant RDU170346.

References

[1] Vlamakis H, Chai Y, Kolter R, Beauregard PB and Losick R 2013 *Nature Review: Microbiology* **11(3)** 157-68
[2] Bakri IM and Douglas CW 2004 *Arch Oral Biol.* **50(7)** 645-51
[3] Lai PK and Roy J 2004 *Current Medical Chemistry.* **11(11)** 1451-60
[4] Russell AD and McDonnell G 2000 *Journal of Hospital Infection* **44(1)** 1–3
[5] Soliman KM and Badeaa RI 2002 *Food and Chemical Toxicology* **40(11)** 1669-1675
[6] Durairaj S, Srinivasan S and Lakshmanaperumalsamy P 2009 *Electronic Journal of Biology* **5(1)** 5-10
[7] Nguefack J, Leth V, Amvam Zollo PH and Mathur SB 2004 *International Journal of Food Microbiology* **94(3)** 329-334
[8] Ross ZM, O'Gara EA, Sleightholme HV, Maslin DJ and Hill DJ 2001 *Applied and Environmental Microbiology* **67(1)** 475-480
[9] Anderson M and Whitcomb P 2000 Chapter 3: Two-Level Factorial Design. *DOE Simplified: Practical Tools for Effective Experimentation*
[10] Abdul Samad K and Zainol N 2017 *Biocatalysis and Agricultural Biotechnology* **10** 9–12
[11] Peterson SB, Yasuhiko I, Borlee BR, Murakami K, Harrison JJ and Colvin KM 2011 *Biofilm Infections* 251-266
[12] Zainol N and Rahim SRR 2017 *Journal of Chemical Engineering and Industrial Biotechnology* **1** 18-28
[13] Houshmand B, Mahjour F and Dianat O 2013 *Indian J Dent Res.* **24(1)** 71-5

IOP Conf. Series: Materials Science and Engineering **342** (2018) 012038 doi:10.1088/1757-899X/342/1/012038

Kinetic study on ferulic acid production from banana stem waste via mechanical extraction

Norazwina Zainol, Nasratun Masngut and Muhamad Khairi Jusup

Faculty of Chemical & Natural Resources Engineering, Universiti Malaysia Pahang, 26300 Gambang, Pahang, MALAYSIA.

E-mail: azwina@ump.edu.my

Abstract. Banana is the tropical plants associated with lots of medicinal properties. It has been reported to be a potential source of phenolic compounds such as ferulic acid (FA). FA has excellent antioxidant properties higher than vitamin C and E. FA also have a wide range of biological activities, such as antioxidant activities and anti-microbial activities. This paper presents an experimental and kinetic study on ferulic acid (FA) production from banana stem waste (BSW) via mechanical extraction. The objective of this research is to determine the kinetic parameters in the ferulic acid production. The banana stem waste was randomly collected from the local banana plantation in Felda Lepar Hilir, Pahang. The banana stem juice was mechanically extracted by using sugarcane press machine (KR3176) and further analyzed in high performance liquid chromatography. The differential and integral method was applied to determine the kinetic parameter of the extraction process and the data obtained were fitted into the 0th, 1st and 2nd order of extraction process. Based on the results, the kinetic parameter and R^2 value from were 0.05 and 0.93, respectively. It was determined that the 0th kinetic order fitted the reaction processes to best represent the mechanical extraction.

1. Introduction

Banana is one of the tropical fruits that well known to be associated with many medicinal properties [1]. This fruit has been reported to be the potent source of phenolic compounds which is the ferulic acid (FA). FA has excellent antioxidant properties which are higher than those of vitamin C and E that makes it as an important ingredient in the industry. Thus, it have a wide range of biological activities, such as antioxidant activities, protection against coronary heart diseases, anti-inflammatory, anti-tumour, anti-mutagenic, anti-carcinogenic and anti-microbial activities that will be helpful in pharmaceutical industry [2]. FA also an important ingredient in the cosmetic industries as skin moisturizer, UV light protection, and has the ability to inhibit melanin formation.

FA can be extracted from different sources using several methods such as solvent extraction, enzymatic reaction, mechanical extraction and chemical reaction [3]. Solvent extraction was commonly used due to its simplicity, efficiency and also it wide applicability. The yield of the chemical extraction depends on the type of solvents used such as methanol and ethanol [4]. However, this extraction method has its drawbacks because the extractant may contain some non-phenolic substances such as sugar, organic acids and fats. For mechanical extraction, the sugar cane press machine was used to extract the FA from the banana stem waste. The sugar cane press machine is a well establish equipment in the plant material processing that were used to extract the sugar from sugar cane and fresh oil palm frond (OPF) particularly [5]. Previous researcher reported that this method was easy to handle since it does not involve any chemicals. The mechanical extraction also takes shorter time than other method of extraction and high yield since the sugarcane press machine doing the mechanical press on the banana stem waste and produce high amount of juice in a shorter time. The mechanical extraction was cheaper than solvent extraction since it does not involve any chemicals and catalysts [6]. The purposes of kinetic study were to determine the factors that could control the fermentation process, increasing the productivity, determine the quality of the product, reducing the operating costs and eliminate disturbances. The factor was the kinetic parameter which was the reaction rate constant (k). However, to determine the reaction rate constant the reaction order of the extraction process must be determined whether it is 0th, 1st or 2nd order reaction by using integral and differential method [7].

The objectives of this research were to extract the banana stem juice containing FA from banana stem waste via mechanical extraction and to determine the value of reaction rate constant (k) in the FA production. There were several scopes of study need to be done in order to achieve the objectives of this research. Based on the data obtained from the quantification of FA in the juice, the reaction rate constant (k) and the order of the reaction will be determined whether the extraction was following 0th, 1st or 2nd order. This research also studied the effect of the storage time on the reaction order and the reaction rate constant (k). R^2 was calculated using linear regression method.

2. Materials and Methods

2.1 Banana stem collection

Banana stem was collected randomly on October during a normal day from local banana plantation located near Felda Lepar Hilir. The outer stem and the pith of banana stem were separated from each other and cut into four parts with dimension of 66.0 cm (length) x 5.5 cm (width).

2.2 Extraction processes

The cut banana stems were pressed using the sugar cane press machine (KR3176) with three rotating cylinder rolling for juice extraction. The juice contained of FA was centrifuged at 15,000 rpm for 15 minutes at 4°C (Thermo Fisher Scientific) and the supernatant was filtered using nylon membrane filter with the pore size 0.45 μm (Cole Parmer) [6].

2.3 Analysis of FA using HPLC

High Performance Liquid Chromatography (HPLC) was applied to separate and quantify the ferulic acid in banana stem juice, following method by [6]. A simple stability-indicating HPLC-DAD method was validated for the determination of ferulic acid (FA) in polymeric microparticles. Chromatographic conditions consisted of a RP C18 column (250mm × 4.60 mm, 5 μm, 110 ° A) using a mixture of methanol and water pH 3.0 (48: 52 v/v) as mobile phase at a flow rate of 1.0 mL/min with UV detection at 320 nm. The developed method was validated as per ICH guidelines with respect to specificity, linearity, limit of quantification, limit of detection, accuracy, precision, and robustness provided suitable results regarding all parameters investigated [8].

2.4 Kinetic study of the mechanical extraction of banana stem

The purpose of this study was to determine the reaction rate constant (k). Here two methods were adopted which were the integral method and the differential method. For integral method, the rate equation was determined from the method of initial rates. The rate was normally verified by comparing the concentrations measured over extended time with the integrated form of the rate equation [7]. Equation 1-3, shows the integral method:

$$C_A = C_{Ao} - kt \quad \text{(0}^{\text{th}}\text{ order extraction process)} \qquad (1)$$

$$ln\, C_A/C_{Ao} = kt \quad \text{(1}^{\text{st}}\text{ order extraction process)} \qquad (2)$$

$$1/C_A - 1/C_{Ao} = kt \quad \text{(2}^{\text{nd}}\text{ order extraction process)} \qquad (3)$$

The next three equations (Equation 4-6), shows the differential method:

$$dC_A/dt = -k \quad (0^{th} \text{ order extraction process}) \tag{4}$$

$$dC_A/dt = -kC_A \quad (1^{st} \text{ order extraction process}) \tag{5}$$

$$dC_A/dt = -kC_A^2 \quad (2^{nd} \text{ order extraction process}) \tag{6}$$

where,
C_A = Concentration of FA
k = reaction rate constant
C_{Ao} = initial concentration of FA
t = storage time

The initial concentration of FA was obtained from the banana stem juice right after it was mechanically extracted. After being extracted, the final concentration of FA was observed based on the optimum condition of the FA. The optimum condition of FA was determined based on the temperature of the pre-treatment and the storage time of FA. Based on the previous study, the optimum condition of the extraction of FA were at 25°C of pre-treatment temperature and 24-hour storage time. The value of reaction rate constant can be calculated after all parameter (the storage time, initial and final concentration of ferulic acid) was determined. After that, the order of the reaction was determined whether the extraction process is following 0th, 1st, or 2nd order.

3. Results and Discussion

3.1 FA profile from mechanical extraction

The graph of the average concentration of FA against storage time (Figure 1) shows that the concentration of FA increases from the beginning to 14 hour of storage time. The concentration decreases afterwards until 24 hour. Storage time is referring to the sampling time after the juice was extracted from the banana stem waste. The graph was observed following the quadratic trend. From Figure 1, the maximum and the minimum of FA concentration was at 14 hour and at initial condition (0 hour) which was 0.0024 g/L and 0.00052 g/L, respectively. Based on the graph trend, the FA concentration increased as the storage time increased and started to decrease after reached a certain value. This is due to the FA was exposed to the surrounding temperature.

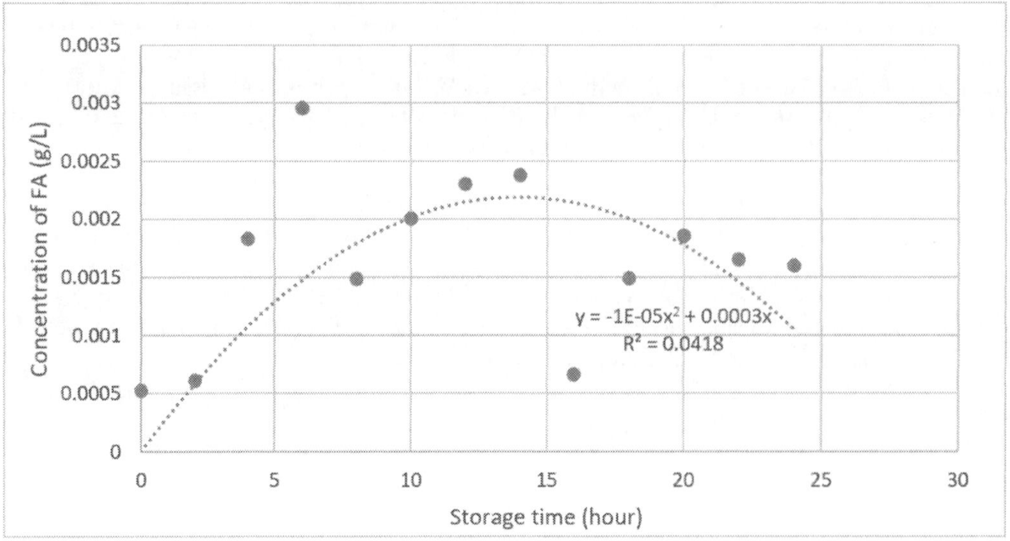

Figure 1. Graph of Average Concentration of FA against Storage time.

3.2 Solution of kinetic constant

3.2.1 Fitting to 0th order equation by using integral and differential method. The graph of FA concentration against storage time was fitted into three different equations of integral and differential methods. For the 0th order, both methods show a linear trend (Figure 2a & b). It was observed that FA concentration was increased linearly from the beginning to 24 hour of storage time. R^2 values for differential and integral method were calculated to be 0.9305 and 0.8305, respectively. The kinetic parameter values determined from here were 0.053 (for differential) and 0.0533 (for integral), respectively. Based on the literature, both differential and integral methods followed the general trend of the 0th order reaction in which the rate was independent of concentration [7]. Therefore, in this research the trend of the graph does not depend on the ferulic acid (FA) concentration since no chemicals were used in addition to raw materials as reactant.

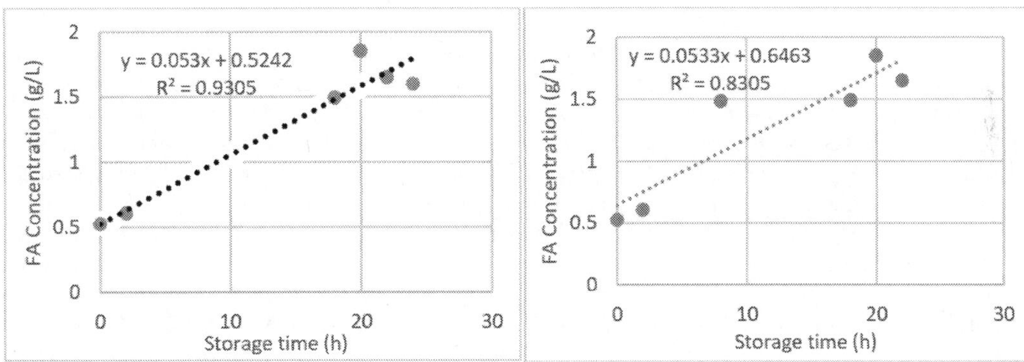

(a) (b)

Figure 2. The trend of FA concentration against storage time for 0th order reaction using (a) integral method, (b) differential method.

3.2.2 Fitting to 1st order equation by using integral and differential method. For 1st order equation, the result for both differential (Figure 3a) and integral (Figure 3b) methods show an increasing trend. FA concentration was increased in both methods from 0 hour to 24 hour. The values of R^2 for both methods were 0.8727 and 0.811, respectively. The kinetic parameter values for 1st order extraction process were 0.05317 (integral method) and 0.0517 (differential method), respectively. In a 1st order reaction, the reaction is directly proportional to one of the reactants concentration [7]. However, this research was an extraction process and does not using any chemicals as reactants. Therefore, the nature of the extraction process in this work cannot be represented by the 1st order reaction.

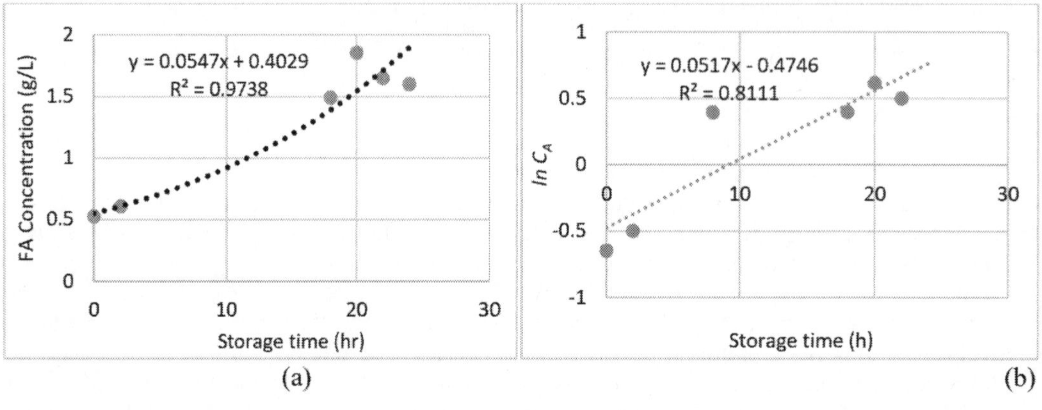

Figure 3. The trend of FA concentration against storage time for 1st order reaction using (a) integral method, (b) differential method.

3.2.3 Fitting to 2nd order equation by using integral and differential method. Using the 2nd order equation, the trend for differential method (Figure 4a) shows a smooth exponential trend, while the integral method (Figure 4b) shows a negative linear trend. For differential method, the FA concentration increased exponentially from 0 hour to 24 hour while for integral method, the $1/C_A$ was decreased linearly from 0 hour to 24 hour. The R^2 for both methods were 0.8375 and 0.8586, respectively. The kinetic parameter values for 2nd order extraction were 0.0627 and 0.0567, respectively. In general, the simplest type of second-order reaction has a reaction rate proportional to the square of one reactant concentration. Another type of second-order reaction has a reaction rate that is proportional to the product of two reactants concentration [7]. For this research, the 2nd order extraction process does not follow any of these general trends for both differential and integral method.

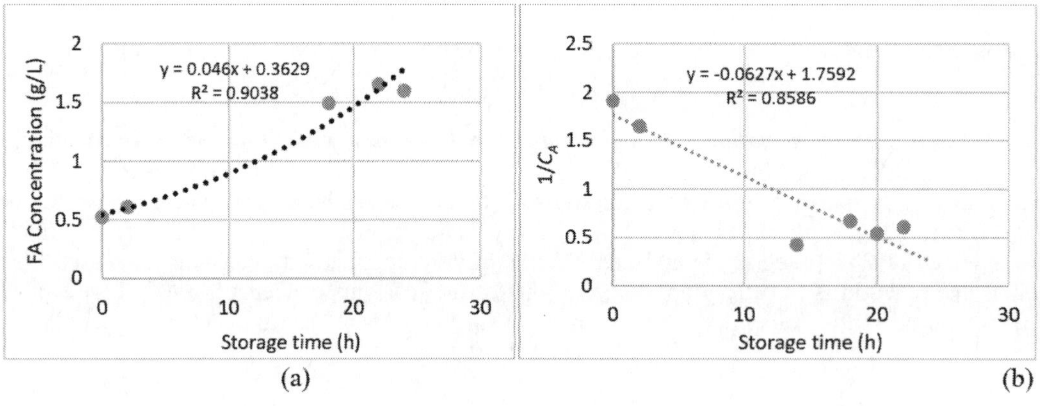

Figure 4. The Graph of FA concentration against time by using differential method (a) and using integral method (b).

3.3 Selection of the reaction order

In order to determine the kinetic value, an analysis must be done to select which reaction order gives the best result by comparing its kinetic value with the value of R^2. Table 1 shows the comparison of reaction order using different solving method with its kinetic value. Based on the table, suitable reaction order for this research was 0th order since the graph follows the general trend for both differential method and integral method. For 1st and 2nd order reaction, both differential and integrals method does not follow the general trend of the graph. This was due to this research was merely an extraction process without involving any chemicals as reactant or catalyst. Thus, the nature of the extraction process for both reactions cannot be represented and the reactions were not suitable to be selected. Comparing the three reaction order (0th, 1st and 2nd), 0th order reaction was the most suitable one. Thus, the 0th order reaction was selected for this research with its kinetic parameter and R^2 values were 0.05 and 0.93, respectively by using differential method.

Table 1. Comparison of reaction order on different solving method.

Reaction order	R^2 value		Kinetic parameter	
	Differential method	**Integral method**	**Differential method**	**Integral method**
0th	**0.9305**	0.8305	**0.0530**	0.0533
1st	0.8727	0.8111	0.05317	0.0517
2nd	0.8375	0.8586	0.0567	0.0627

3.4 Comparison with other research

Table 2 shows the comparison on different types of extraction and its reaction order, respectively. In this research, the type of extraction used was the same with Sabarish et al. (2016) which was the mechanical extraction, but differ on the type of machine used. The raw material used for this research was banana stem waste while for Sabarish et al. (2016) used rubber seed. In this research, the extraction was performed to extract the ferulic acid (FA) without using any chemicals as a reactant or catalyst to facilitate the extraction process. However, Sabarish et al. (2016) used methanol as reactant and sulphuric acid as catalyst in order to extract the oil from rubber seed followed by acid esterification of the extracted rubber seed oil. From the experimental results, the extraction order of this research was 0th while Sabarish et al. (2016) was 1st order for the forward reaction and 2nd order for the reverse reaction.

Table 2. Comparison with different type of extraction and raw material with other researchers.

Research	Type of extraction	Raw material	Reaction order
This research	Sugarcane press machine	Banana stem waste	0th order
Sabarish et al., 2016	Hydraulic press	Rubber seed	1st order
Adnadjevic et al., 2017	Conventional and microwave heating	Guarana seed	1st order
Patil & Akamanchi, 2017	Microwave assisted process intensification	*Nothapodytes nimmoniana*	2nd order
Zhang et al., 2016	Ultrasound-assisted extraction	Chinese chive	1st order

In studied by [10], conventional heating and microwave heating were used to extract tanine from guarana seed. They compared the kinetic analysis of isothermal extraction of caffeine from guarana seed under conventional heating and microwave heating. In their experiment, chemicals were used as solvents which were water extract, hydrochloric acid (HCl) solution, lead acetate ($Pb(CH_3COO)_2$) solution and sulphuric acid. They found out that the reaction order for both conventional heating and microwave heating were following 1st order with kinetic parameters of 4.11 (1/min) and 6.6 (1/min), respectively.

[11] used microwave assisted process intensification to extract camptothecin from *Nothapodytes nimmoniana* plant. They made an attempt to fit the extraction kinetics data with different kinetic models such as 2nd order rate model, power law model and two site kinetic models. From the experimental results, the 2nd order rate model appears to be the best fit to explain the microwave assisted kinetics.

[12] adopted the ultrasound-assisted extraction (UAE) for polysachharide extraction from Chinese chive. They developed a modified kinetic model since the traditional kinetic models cannot reflect the real extraction process of Chinese chive polysaccharide. They compared the performance of the UAE with the conventional hot-water extraction. Phenol-sulphuric acid method was used to determine the content of the polysaccharides for each experiment. From the experimental results, the reaction order for both ultrasound-assisted extraction (UAE) and hot-water extraction was following the 1st order extraction process.

The reaction order (0th, 1st, 2nd, …nth) will depends on the condition of the extraction processes. Some extraction processes used chemicals as reactants or catalyst in order to facilitate and improve the efficiency of the process. The usage of these chemicals affected the extraction process so as the reaction order of the extraction process itself.

4. Conclusion

The kinetic parameter of ferulic acid production from banana stem waste via mechanical extraction has been determined in this study. Based on the results, the kinetic parameter and R^2 value were 0.05 and 0.93, respectively. In this research, it was found out that the 0th order extraction process was the most suitable reaction orders to represent the behavior of this extraction process.

Acknowledgement

The author wish to acknowledge the Universiti Malaysia Pahang for funding the project under grant RDU150381

References

[1] Debjit B, Sampath KP, Umadevi M and Duraivel S 2012 *Journal of Pharmacognosy and Phytochemistry* **1(3)** 51–63
[2] Ongphimai N, Lilitchan S, Aryusuk K, Bumrungpert A and Krisnangkura K 2013 *Chiang Mai Journal of Sciencie* **40(4)** 636–642
[3] Kumar N and Pruthi V 2014 *Biotechnology Reports* **4(1)** 86–93
[4] Dai J and Mumper RJ 2010 *Molecules* **15(10)** 7313–7352
[5] Ismail SN and Zainol N 2014b *Asian Journal of Microbiology, Biotechnology and Environmental Sciences* **16(3)** 479–484
[6] Ismail SN and Zainol N 2014a *International Journal of Applied Engineering Research* **9(20)** 6823–6833
[7] Fogler HS 2006 *Elements of Chemical Reaction Engineering* (4th ed.) Prentice Hall.
[8] Nadal JM, Toledo MG, Pupo YM, Padilha PJ, Farago PV and Zanin SMW 2015 *Journal of Analytical Methods in Chemistry* **2015** 1–10
[9] Sabarish CS, Sebastian J and Muraleedharan C 2016 *Procedia Technology* **2016** 1006–1013
[10] Adnadjevic B, Koturevic B and Jovanovic J 2017 *Chemical Engineering Research and Design* **118** 61–70
[11] Patil DM and Akamanchi KG 2017 *Ultrasonics Sonochemistry* **98** 60–67
[12] Zhang WN, Zhang HL, Lu CQ, Luo JP and Zha, X. Q. 2016 *Food Chemistry* **212** 274–281

Estimation of solar collector area for water heating in buildings of Malaysia

Nallapaneni Manoj Kumar[1], K Sudhakar[1,2] and M Samykano[1]

[1]Faculty of Mechanical Engineering, Universiti Malaysia Pahang, 26600 Pekan, Pahang, Malaysia.
[2]Energy Centre, Maulana Azad National Institute of Technology Bhopal, M.P., India

E-mail: sudhakar@ump.edu.my

Abstract. Solar thermal energy (STE) utilization for water heating at various sectorial levels became popular and still growing especially for buildings in the residential area. This paper aims to study and identify the solar collector area needed based on the user requirements in an efficient manner. A step by step mathematical approach is followed to estimate the area in Sq. m. Four different cases each having different hot water temperatures (45°, 50°C, 55°C, and 60°C) delivered by the solar water heating system (SWHS) for typical residential application at Kuala Lumpur City, Malaysia is analysed for the share of hot and cold water mix. As the hot water temperature levels increased the share of cold water mix is increased to satisfy the user requirement temperature, i.e. 40°C. It is also observed that as the share of hot water mix is reduced, the collector area can also be reduced. Following this methodology at the installation stage would help both the user and installers in the effective use of the solar resource.

1. Introduction

Residential water heating is an important one that everyone has to adopt, this helps in controlling the effects caused by unseen pathogens on human life [1]. There are few conventional and nonconventional methods used for water heating. Conventional water heating is done mainly by burning charcoal, and wood. However, the use conventional fuels cause adverse effects like the release of greenhouse gases (GHG), dust pollutions, etc. To avoid the serious and dangerous effects (climate change, global warming, and environmental pollutions, etc.) of GHG emissions, mitigation methods are to be practiced [2-4]. The best way of avoiding this GHG emission is shifting towards solar energy. Solar energy can be harnessed for electricity generations to cater the ever-increasing energy demands relating to various applications like street lighting, water pump sets, University buildings, industries areas, residential houses, and powering airports. Several authors have also conducted research studies to analyze the performance the PV systems when they are integrated into a building and studies related to floating PV systems are also emerging. The second way of harnessing the solar energy is the use of thermal collectors, where the solar energy is converted into useful thermal energy in a variety of applications by means of solar collectors [5, 6]. Sizing of the SWHS requires a lot of input data, which may not be possible to get all the times. Moreover the efficiently designed system will perform better in delivering the hot water. The problem of identifying the collector areas required as per the share of cold and hot water mix at required temperature levels of the users is still unaddressed. In this paper, an application of solar thermal energy conversion is chosen i.e., solar water heating systems (SWHS) to reduce the energy consumption in the residential zones is considered. This paper is aimed to propose a simple method for estimating the solar collector's area based on the user input requirement and as well the collector delivery water temperatures.

2. Description of solar water heating system

Solar water heating system (SWHS) works by converting the solar energy into heat energy with the help of thermal collectors. These collectors will face in the direction of the sun to capture as much as sun's energy. These were mainly used for heating water in residential areas and industrial sectors. The components of SWHS are a solar collector, storage tank, circulating tubes, pumps, sensors and controls. The schematic view of the solar water heating system (SWHS) is shown in Figure 1. For most of the SWHS, flat plate solar thermal collectors were used. These collector devices were generally mounted on the building roof facing towards the sun. It is equipped with circulating tubes in it attached to the absorber plate of the collector. This helps in carrying the antifreeze solution or water for capturing the heat energy from the sun and deliver the same energy to the storage tank. The storage tank is a kind of modified heat exchanger with insulations providing no loss of heat energy to a maximum extent. This tank helps in storing the hot waters that are heated by the solar collectors. From the storage tanks, hot water is made to flow through the domestic taps or other means where users can access. Other components like pumps will help to pump the cold waters to the collector inlets and sensors and control are used to sense various parameters at the system level [7-13].

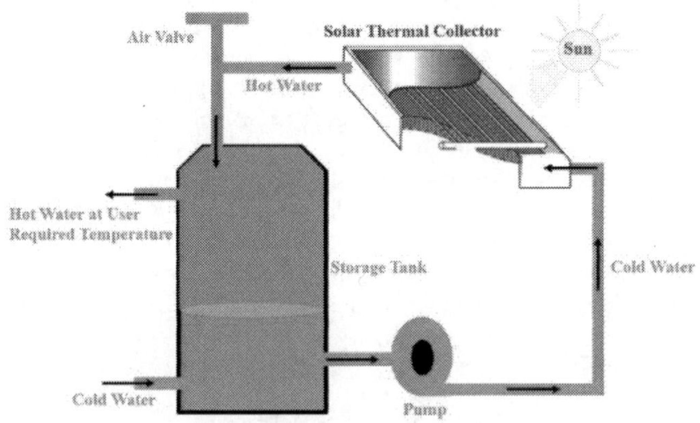

Figure 1. Schematic View of Solar Water Heating System (SWHS).

3. Methodology

3.1. Study location

Study location is chosen to be the Kuala Lumpur City in Malaysia. It is one among the 140 cities that are globally ranked by economic intelligence unit for best living conditions and its rank is 70 [14]. The weather conditions of Kuala Lumpur is tropical experiencing hot and humid climates with heavy rain falls all around the year. The water temperatures in Kuala Lumpur city will not have many variations. The maximum, minimum, and average water temperatures are 30.1°C, 29°C, and 27.1°C respectively shown in Figure 2. [15]. Monthly water temperature details in study location are shown in Table 1. Hot water consumption details for a typical residential family staying at Kuala Lumpur is considered for this study i.e. 840 Liter/day.

Table 1. Study location details.

Description	Value

Name of the Study Location	Kuala Lumpur
Latitude	°C
Longitude	°C
Hot Water Consumption at the Site	840 Litre/day
Minimum Temperature of the Water	27.1 °C
Maximum Temperature of the Water	30.1 °C
Average Temperature of the Water	29 °C

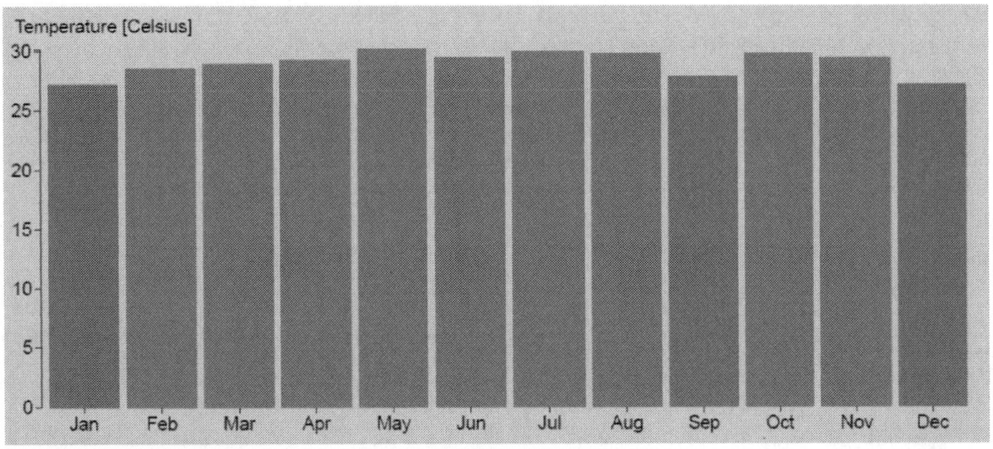

Figure 2. Water Temperature at Kuala Lumpur City [15].

3.2. Calculation procedure

The following methodology is applied to estimate the suitable collector size required for heating water with the help solar thermal energy.

- Step-1: A survey based study is conducted to analyze the hot water consumption patterns per day at the study location.
- Step-2: Estimate the possible water losses during the process that is in the form of hot water losses.
- Step-3: Estimate the final hot water requirements per day considering the heat losses.
- Step-4: Calculate the share of the cold and hot waters to be mixed. This shared value depends on many other parameters such as hot water temperature possible in the solar collector, cold water temperature at the site, usable of hot water temperature requirement by the user.
- Step-5: Estimate the size of solar collector which depends up on the hot water share and the solar water heater maximum delivery capacity.
- Step-6: Size estimation of the storage tank to store the water.

3.3. Total hot water requirement

Total hot water requirement is calculated by conducting extensive surveys related to hot usage at a site [16] where solar water heating systems (SWHS) is about to propose or equip. The surveyed information cannot consider as the final value unless and until water loss parameter is considered. Hence, the total hot water requirement calculated using Eq. 1. [17] Will be the sum of the hot water requirements at the site plus water lost.

$$W_t = W_{r\_site} + W_l \tag{1}$$

Where, W_t is the total hot water required in Litre after considering losses, W_{r_site} is the quantity of water used by number of households for typical tasks in Liter, and W_l is the water lost in Litre.

3.3.1. Hot water consumption details in a residential building. As said earlier, hot water consumptions are calculated based on the surveys conducted among the group of people or a group of apartments or a community. In this study, a typical urban residential building with 'n' number of person's and their water requirements are considered and tabulated in Table 2. This table shows that, residential building consumes hot water in different tasks i.e. 60 Litre per person per day for Bathing, 10 Litre per person per day for Wash Basin, 15 litres per person per day for Kitchen Wash, 10 Litre per person per day for Clothes Wash, 10 Litre per person per day for other usages.

Table 2. Water consumption details.

Description	Water/Person/Day (Litre)	No. of Persons (n)	Water Requirement at the Site (Litre/day)
Bathing	70	6	420
Wash Basin	15	6	90
Kitchen Wash	25	6	150
Clothes Wash	20	6	120
Others	10	6	60

$$W_{r\_site} = \left[(W_{task_1} \times n_1) + (W_{task_2} \times n_2) + (W_{task_3} \times n_3) + \cdots\cdots + (W_{task_n} \times n_n)\right] \quad (2)$$

Hot water requirement at the site is calculated with the help of Eq. 2. Hot water consumption can be defined to be the sum of the water consumed by households in all different tasks.

Where, W_{r_site} is the quantity of water used by a number of households for typical tasks in Litre n_1, n_2, n_3.......n_n are the number of persons in the residential house, W_{task_1}, W_{task_2}, W_{task_3}......W_{task_n} are the various tasks where hot water is used.

3.3.2. Water losses. As discussed earlier in a thermal energy system, loss factor is the most crucial one to be considered during the system analysis. As our application is heating water, there are possibilities for water losses in the form of heat or even sometimes leakage of the water itself. If this is not counted in water requirement section, there is the possibility of miss match between the final hot water requirements. Hence, in this study 20 % is considered as water loss percentage throughout the system [17]. The Effective amount of water that is to be added to W_{r_site} is calculated using Eq. 3.

$$W_l = \frac{\text{Water Loss Percentage (\%)}}{100} \times W_{r\_site} \quad (3)$$

Where, W_l is the water lost in litre, and W_{r_site} is the quantity of water used by number of households for typical tasks in Litre.

3.4. Share of cold and hot water mix.

Once total hot water required is estimated, then the share of cold and hot water mix is calculated. Before calculating this, the ratio of cold water quantity to hot water quantity is to found using Eq. 4. Based on the obtained ratio from Eq. 4, [17] share of the cold and hot water is calculated using Eq. 5. And Eq. 6. Respectively.

$$\frac{Q_{water\_cold}}{Q_{water\_hot}} = \frac{T_{hot\_water\_SWHS} - T_{user\_required}}{T_{user\_required} - T_{cold\_water\_site}} \quad (4)$$

$$S_{water\_cold} = \left(\frac{Q_{cold}}{Q_{cold}+Q_{hot}}\right) \times W_t \tag{5}$$

$$S_{water\_hot} = \left(\frac{Q_{hot}}{Q_{cold}+Q_{hot}}\right) \times W_t \tag{6}$$

Where, Q_{water_cold} is the quantity of cold water in Litre, Q_{water_hot} is the quantity of hot water in Litre, $T_{hot\_water\_SWHS}$ is the hot water temperature in °C that a solar water heating system delivers, $T_{user\_Required}$, is the required water temperature in °C by the users or the households, $T_{cold\_water\_site}$, is the cold water temperature in °C at the site or location, S_{water_cold} is the share of cold water mix in Litre, S_{water_hot} is the share of hot water mix in Litre, W_t is the total hot water required in Litre at a temperature of $T_{user\_Required}$.

3.5. Solar collector area

Solar collector area represents, space where the solar energy is made to concentrate and converted into useful energy. It is measured in Sq. m. The area of the collector is calculated using Eq. 7 [17].

$$Solar\ Collector\ Area = 2 \times \left(\frac{S_{water\_hot}}{W_{delivery\_SWHS}}\right) \tag{7}$$

Where, S_{water_hot} represents the share of hot water mix in Litre at a temperature of $T_{hot\_water\_SWHS}$, $W_{delivery\_SWHS}$ represents the quantity of hot water that can be delivered by solar water heating system in Litre.

3.6. Size of the storage tank.

Size of the storage tank is one of the essential parts of the solar water heating system, which is used for storing the waters. If a continuous usage of the hot water is possible, then the storage tank capacity can be minimised. If it is not the case of continuous usage, storage tank with capacity equals to water requirements can be built. Eq. 8. Is used to calculate the size of the storage tank.

$$Volume\ of\ the\ storage\ tank\ in\ Cubic\ meter = \frac{W_t}{1000} = \frac{W_{r\_site}+W_l}{1000} \tag{8}$$

4. Results and discussion

Results include the required collector areas estimations for heating the required quantity of water as per the user requirement. Here, maintaining the temperature as per user choice is an important aspect to be noticed. If the selected 250 litre system could deliver around 200 litres per day at certain temperatures, see in Figure 3. But most probably this temperature would vary based on the seasons and intensity of captured energy from the sun. Hence, four different temperatures of hot water in °C that a 250 litre system can deliver is considered and they are 45°C, 50°C, 55°C, and 60°C. Hence the results are briefed in four different cases. Also, the correlations between the calculated areas and temperature levels were estimated using curve fit options for result verification. Before proceeding into each case, the total water requirement is calculated using the Eq. 1, Eq. 2, and Eq. 3. From Eq. 2, the quantity of water used by the number households is calculated as follows:

$$W_{r_{site}} = \left[(W_{task_1} \times n_1) + (W_{task_2} \times n_2) + (W_{task_3} \times n_3) + \cdots\cdots\cdots + (W_{task_n} \times n_n)\right]$$

$$W_{r\_site} = [(70 \times 6) + (15 \times 6) + (25 \times 6) + (20 \times 6) + (10 \times 6)]$$

$$W_{r\_site} = (420 + 90 + 150 + 120 + 60) = 840\ Litre/day$$

From Eq. 3, the lost water in terms of heat and other leakage is estimated as follows considering lost percentage as 20 %.

$$W_l = \frac{Water\ Loss\ Percentage\ (\%)}{100} \times W_{r\_site}$$

$$W_l = \frac{20}{100} \times 840 = 168\ Litre/day$$

From Eq. 1, the total hot water required at 40°C is calculated as follows:

$$W_t = W_{r\_site} + W_l$$

$$W_t = 840 + 168 = 1008\ Litre/day$$

Hence, the total quantity of hot water required by the house holds is 1008 Litre/day at 40°C.

4.1. Case-I: 250 litre system delivers 200 litre at 45°C

In Case-I, if the 250-litre system can deliver hot water of 200 litres per day at 45°C temperatures, then the optimal collector area opted for meeting the user requirement is as follows.

From Eq. 4, cold water to hot water mix is calculated as follows. Now $T_{hot\_water\_SWHS}$ becomes 45°C, $T_{user\_required}$ is 40°C, and $T_{cold\_water\_site}$ is 27.1°C.

$$\frac{Q_{water\_cold}}{Q_{water\_hot}} = \frac{T_{hot\_water\_SWHS} - T_{user\_required}}{T_{user\_required} - T_{cold\_water\_site}}$$

$$\frac{Q_{water\_27.1°C}}{Q_{water\_45°C}} = \frac{45 - 40}{40 - 27.1} = \frac{5}{12.9}$$

From the ratio 5/12.9, the numerator becomes the cold water portion, and the denominator becomes the hot water portion, but to get the individual share in Litre, the following mathematical approach is adopted shown in Eq. 5 for cold water share, and Eq. 6 for hot water share.

$$S_{water\_cold} = \left(\frac{Q_{cold}}{Q_{cold} + Q_{hot}}\right) \times W_t$$

$$S_{water\_27.1°C} = \left(\frac{5}{5 + 12.9}\right) \times 1008 = 281.564\ Litre/day$$

$$S_{water\_hot} = \left(\frac{Q_{hot}}{Q_{cold} + Q_{hot}}\right) \times W_t$$

$$S_{water\_45°C} = \left(\frac{12.9}{5 + 12.9}\right) \times 1008 = 726.436\ Litre/day$$

From the above, it is clear that, to get hot water of 1008 Litre/day at 40°C, cold water of 281.564 Litre/day at 27.1°C has to be mixed with hot water of 726.436 Litre/day at a temperature of 45°C delivered by the SWHS. From Eq. 7, the solar collector area is estimated as follows:

$$Solar\ Collector\ Area = 2 \times \left(\frac{S_{water\_hot}}{W_{delivery\_SWHS}}\right)$$

$$Solar\ Collector\ Area = 2 \times \left(\frac{726.436}{200}\right) = 7.26436 \sim 7.3\ Sq.m$$

Solar collector area is found to be around 7.3 Sq. m for capturing the solar energy and convert into useful energy in the form of hot water i.e. 726.436 Litre/day at 45°C.

4.2. Case-II: 250 litre system delivers 200 litre at 50°C

In Case-II, if the 250-litre system can deliver hot water of 200 litres per day at 50°C temperatures, then the optimal collector area opted for meeting the user requirement is as follows.
From Eq. 4, cold water to hot water mix is calculated as follows. Now $T_{hot\_water\_SWHS}$ becomes 50°C, $T_{user\_required}$ is 40°C, and $T_{cold\_water\_site}$ is 27.1°C.

$$\frac{Q_{water\_cold}}{Q_{water\_hot}} = \frac{T_{hot\_water\_SWHS} - T_{user\_required}}{T_{user\_required} - T_{cold\_water\_site}}$$

$$\frac{Q_{water\_27.1°C}}{Q_{water\_50°C}} = \frac{50 - 40}{40 - 27.1} = \frac{10}{12.9}$$

From the ratio 10/12.9, the numerator becomes the cold water portion, and the denominator becomes the hot water portion, but to get the individual share in Litre, the following mathematical approach is adopted shown in Eq. 5 for cold water share, and Eq. 6 for hot water share.

$$S_{water\_cold} = \left(\frac{Q_{cold}}{Q_{cold} + Q_{hot}}\right) \times W_t$$

$$S_{water\_27.1°C} = \left(\frac{10}{10 + 12.9}\right) \times 1008 = 440.175\ Litre/day$$

$$S_{water\_hot} = \left(\frac{Q_{hot}}{Q_{cold} + Q_{hot}}\right) \times W_t$$

$$S_{water\_50°C} = \left(\frac{12.9}{10 + 12.9}\right) \times 1008 = 567.825\ Litre/day$$

From the above, it is clear that, to get hot water of 1008 Litre/day at 40°C, cold water of 440.175 Litre/day at 27.1°C has to be mixed with hot water of 567.825 Litre/day at a temperature of 50°C delivered by the SWHS. From Eq. 7, the solar collector area is estimated as follows:

$$Solar\ Collector\ Area = 2 \times \left(\frac{S_{water\_hot}}{W_{delivery\_SWHS}}\right)$$

$$Solar\ Collector\ Area = 2 \times \left(\frac{567.825}{200}\right) = 5.67825 \sim 5.7\ Sq.m$$

Solar collector area is found to be around 5.7 Sq. m for capturing the solar energy and convert into useful energy in the form of hot water i.e. 577.825 Litre/day at 50°C.

4.3. Case-III: 250 litre system delivers 200 litre at 55°C

In Case-III, if the 250-litre system can deliver hot water of 200 litres per day at 55°C temperatures, then the optimal collector area opted for meeting the user requirement is as follows.

From Eq. 4, cold water to hot water mix is calculated as follows. Now $T_{hot\_water\_SWHS}$ becomes 55°C, $T_{user\_required}$ is 40°C, and $T_{cold\_water\_site}$ is 27.1°C.

$$\frac{Q_{water\_cold}}{Q_{water\_hot}} = \frac{T_{hot\_water\_SWHS} - T_{user\_required}}{T_{user\_required} - T_{cold\_water\_site}}$$

$$\frac{Q_{water\_27.1°C}}{Q_{water\_55°C}} = \frac{55 - 40}{40 - 27.1} = \frac{15}{12.9}$$

From the ratio 15/12.9, the numerator becomes the cold water portion, and the denominator becomes the hot water portion, but to get the individual share in Litre, the following mathematical approach is adopted shown in Eq. 5 for cold water share, and Eq. 6 for hot water share.

$$S_{water\_cold} = \left(\frac{Q_{cold}}{Q_{cold} + Q_{hot}}\right) \times W_t$$

$$S_{water\_27.1°C} = \left(\frac{15}{15 + 12.9}\right) \times 1008 = 541.935\ Litre/day$$

$$S_{water\_hot} = \left(\frac{Q_{hot}}{Q_{cold} + Q_{hot}}\right) \times W_t$$

$$S_{water\_55°C} = \left(\frac{12.9}{15 + 12.9}\right) \times 1008 = 466.065\ Litre/day$$

From the above, it is clear that, to get hot water of 1008 Litre/day at 40°C, cold water of 541.935 Litre/day at 27.1°C has to be mixed with hot water of 466.065 Litre/day at a temperature of 55°C delivered by the SWHS. From Eq. 7, the solar collector area is estimated as follows:

$$Solar\ Collector\ Area = 2 \times \left(\frac{S_{water\_hot}}{W_{delivery\_SWHS}}\right)$$

$$Solar\ Collector\ Area = 2 \times \left(\frac{466.065}{200}\right) = 4.66065 \sim 4.6\ Sq.m$$

Solar collector area is found to be around 4.6 Sq. m for capturing the solar energy and convert into useful energy in the form of hot water i.e. 466.065 Litre/day at 55°C.

4.4. Case-IV: 250 litre system delivers 200 litre at 60°C .

In Case-IV, if the 250-litre system can deliver hot water of 200 litres per day at 60°C temperatures, then the optimal collector area opted for meeting the user requirement is as follows.

From Eq. 4, cold water to hot water mix is calculated as follows. Now $T_{hot\_water\_SWHS}$ becomes 60°C, $T_{user\_required}$ is 40°C, and $T_{cold\_water\_site}$ is 27.1°C.

$$\frac{Q_{water\_cold}}{Q_{water\_hot}} = \frac{T_{hot\_water\_SWHS} - T_{user\_required}}{T_{user\_required} - T_{cold\_water\_site}}$$

$$\frac{Q_{water\_27.1°C}}{Q_{water\_60°C}} = \frac{60 - 40}{40 - 27.1} = \frac{20}{12.9}$$

From the ratio 20/12.9, the numerator becomes the cold water portion, and the denominator becomes the hot water portion, but to get the individual share in Litre, the following mathematical approach is adopted shown in Eq. 5 for cold water share, and Eq. 6 for hot water share.

$$S_{water\_cold} = \left(\frac{Q_{cold}}{Q_{cold} + Q_{hot}}\right) \times W_t$$

$$S_{water\_27.1°C} = \left(\frac{20}{20 + 12.9}\right) \times 1008 = 612.766\ Litre/day$$

$$S_{water\_hot} = \left(\frac{Q_{hot}}{Q_{cold} + Q_{hot}}\right) \times W_t$$

$$S_{water\_60°C} = \left(\frac{12.9}{20 + 12.9}\right) \times 1008 = 395.234\ Litre/day$$

From the above, it is clear that, to get hot water of 1008 Litre/day at 40°C, cold water of 612.766 Litre/day at 27.1°C has to be mixed with hot water of 395.234 Litre/day at a temperature of 60°C delivered by the SWHS.

From Eq. 7, the solar collector area is estimated as follows:

$$Solar\ Collector\ Area = 2 \times \left(\frac{S_{water\_hot}}{W_{delivery\_SWHS}}\right)$$

$$Solar\ Collector\ Area = 2 \times \left(\frac{395.234}{200}\right) = 3.95234 \sim 4.0\ Sq.m$$

Solar collector area is found to be around 4.0 Sq. m for capturing the solar energy and convert into useful energy in the form of hot water i.e. 395.234 Litre/day at 60°C.

4.5. Calculated volume of the storage tank

Volume of the storage tank is expressed in cubic meters, and is almost equal to the total amount of water required including the water lost due to heat and leakage losses.

Volume of the storage is calculated using the Eq. 8.

$$Volume\ of\ the\ storage\ tank\ in\ Cubic\ meter = \frac{W_t}{1000} = \frac{W_{r\_site} + W_l}{1000}$$

$$Volume\ of\ the\ storage\ tank\ in\ Cubic\ meter = \frac{1008}{1000} = 1.008\ Cubic\ meter/day$$

4.6. Discussion

Summary of the results is shown in Table 3. Four different cases considered with different temperatures of the water delivery from SWHS (Four different Systems) is analysed and collector area is estimated. As the temperature of the hot water delivered by the system is increased, the share of cold water is increased. This variation is observed due to the fixed temperature that kept as the user requirement i.e. 40°C. Nowadays solar collectors are capable of delivering the hot waters at higher temperatures also, this is because of the advancement in the material used for design of solar collectors.

Table 3. Estimated required solar collector areas.

Parameters	Case-I	Case-II	Case-III	Case-IV
Cold Water Temperature at Site (°C)	27.1	27.1	27.1	27.1
Quantity of Hot Water (Litre/day) at 40°C	1008	1008	1008	1008
Required Hot Water Temperature (°C)	40	40	40	40
Hot Water Temperature at Delivery point of SWHS (°C)	45	50	55	60
Share of Cold Water at 27.1°C	281.564	440.175	541.935	612.766
Share of Hot Water at Defined Temperatures of Each Case	726.436	567.825	466.065	395.234
Solar Collector Area (Sq. m)	7.3	5.7	4.6	4.0

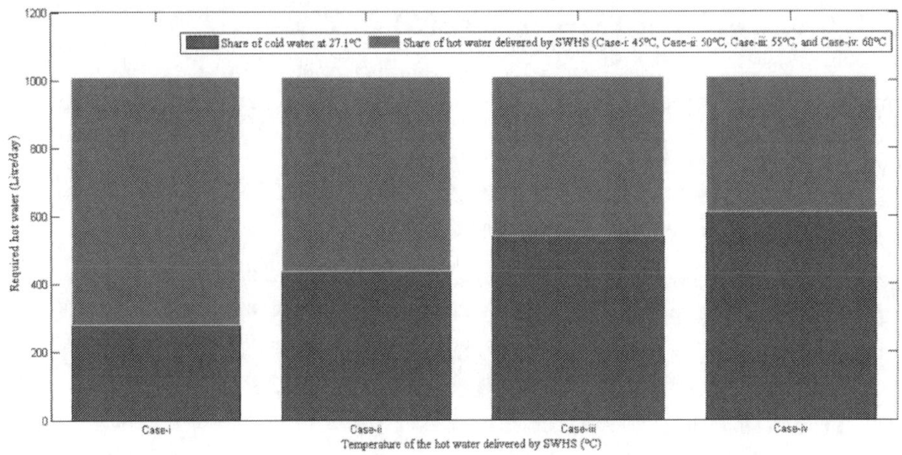

Figure 3. Share of cold water and hot water mix variations.

5. Conclusion

A method for solar collector area estimation is proposed in this work. This method is based on the three different parameters hot water delivery temperatures, cold water temperature at the location, usable temperature. From this study the following conclusions were drawn:

- As the cold water to hot water mix varies, the required area for the solar collector is also varied.
- The temperatures of the hot water delivered by the SWHS will also have the influence on the contributions of the hot and cold water mix
- This method allows the user to estimate the required collector areas effectively and easily. Moreover, it is simple to understand.
- Hence this would help the owners or the installers to choose the collector's areas as per their requirements.

References

[1] Brazeau, R. H., & Edwards, M. A. (2013). Role of Hot Water System Design on Factors Influential to Pathogen Regrowth: Temperature, Chlorine Residual, Hydrogen Evolution, and Sediment. Environmental Engineering Science, 30(10), 617–627. http://doi.org/10.1089/ees.2012.0514

[2] N. M. Kumar, M.S.P Subathra, and O. D. Cota, "Waste Energy recovery in Iron and Steel Industries for CO2 Emission reduction: A Case Study," Journal of Environmental Research and

Development". Vol. 10(1), pp. 149-156, 2015.
[3] N. M. Kumar and G. H. K. Kumar, "Power generation using Green Technology in Iron and Steel industry," 2015 IEEE 9th International Conference on Intelligent Systems and Control (ISCO), Coimbatore, 2015, pp. 1-6. doi: https://doi.org/10.1109/ISCO.2015.7282346
[4] Manoj Kumar Nallapaneni, "Impact of clean development mechanism on eco-friendly energy recovery technology," Procedia Technology, vol. 21, pp. 54-58, 2015. https://doi.org/10.1016/j.protcy.2015.10.009
[5] Deepak Bishoyi, K. Sudhakar: Modeling and performance simulation of 100 MW LFR based solar thermal power plant in Udaipur India. Resource-Efficient Technologies 03/2017;, DOI:10.1016/j.reffit.2017.02.002.
[6] Deepak Bishoyi, K. Sudhakar, Modeling and performance simulation of 100MW PTC based solar thermal power plant in Udaipur India, Case Studies in Thermal Engineering, Volume 10, 2017, Pages 216-226, http://dx.doi.org/10.1016/j.csite.2017.05.005.
[7] A. H. Fanney, B. P. Dougherty, A Photovoltaic Solar Water Heating System, Transactions of the ASME, Journal of Solar Energy Engineering, Vol. 119, pp. 126-133, MAY 1997.
[8] Dejene Assefa Hagos, Alemayehu Gebremedhin, Björn Zethraeus, Solar Water Heating as a Potential Source for Inland Norway Energy Mix, Journal of Renewable Energy, Volume 2014, Article ID 968320, 11 pages. http://dx.doi.org/10.1155/2014/968320
[9] N. V. Ogueke, E. E. Anyanwu, and O. V. Ekechukwu, A review of solar water heating systems, Journal of Renewable and Sustainable Energy, 1 (2009), 043106.
[10] Tomas Matuska, Borivoj Sourek, Performance Analysis of Photovoltaic Water Heating System, International Journal of Photoenergy, Volume 2017, Article ID 7540250, 10 pages https://doi.org/10.1155/2017/7540250
[11] Nosa Andrew Ogie, Ikponmwosa Oghogho, Julius Jesumirewhe, Design and Construction of a Solar Water Heater Based on the Thermosyphon Principle, Journal of Fundamentals of Renewable Energy and Applications, Vol. 3 (2013), Article ID 235592, 8 pages doi:10.4303/jfrea/235592
[12] Yoann Raffenel, Enrico Fabrizio, Joseph Virgone, Eric Blanco, Marco Filippi, Integrated solar heating systems: From initial sizing procedure to dynamic simulation, Solar Energy 83 (2009) 657–663.
[13] Richard O'Hegarty, Oliver Kinnane, Sarah McCormack, A Simplified Procedure for Sizing Solar Thermal Systems; Based on National Assessment Methods in the UK and Ireland, Energy Procedia 62 (2014) 647 – 655.
[14] The Global Liveability Report 2017, Economic Intelligence Unit. https://www.eiu.com/public/topical_report.aspx?campaignid=liveability17
[15] Water Temperature - Kuala Lumpur, Malaysia. https://www.watertemperature.org/Kuala-Lumpur-Malaysia-City.html 2/
[16] John C. Evarts, Lukas G. Swan, Domestic hot water consumption estimates for solar thermal system sizing, Energy and Buildings 58 (2013) 58–65. http://dx.doi.org/10.1016/j.enbuild.2012.11.020
[17] User's handbook on SOLAR WATER HEATERS. www.undp.org/content/dam/india/.../user's_handbook_on_solar_water_heaters.pdf

Sunglass detection method for automation of video surveillance system

Tasriva Sikandar, Wan Nur Azhani W Samsudin, Kamarul Hawari Ghazali, Izzeldin I Mohd and Mohammad Fazle Rabbi

Faculty of Electrical and Electronics Engineering, Universiti Malaysia Pahang, 26600 Pekan, Pahang, Malaysia

E-mail:tasrivasikandar@gmail.com; nurazhani@ump.edu.my; kamarul@ump.edu.my; izzeldin@ump.edu.my; fzrabbi@gmail.com

Abstract. Wearing sunglass to hide face from surveillance camera is a common activity in criminal incidences. Therefore, sunglass detection from surveillance video has become a demanding issue in automation of security systems. In this paper we propose an image processing method to detect sunglass from surveillance images. Specifically, a unique feature using facial height and width has been employed to identify the covered region of the face. The presence of covered area by sunglass is evaluated using facial height-width ratio. Threshold value of covered area percentage is used to classify the glass wearing face. Two different types of glasses have been considered *i.e.* eye glass and sunglass. The results of this study demonstrate that the proposed method is able to detect sunglasses in two different illumination conditions such as, room illumination as well as in the presence of sunlight. In addition, due to the multi-level checking in facial region, this method has 100% accuracy of detecting sunglass. However, in an exceptional case where fabric surrounding the face has similar color as skin, the correct detection rate was found 93.33% for eye glass.

1. Introduction

With the flourishing advances in autonomous systems, video surveillance cameras integrated with image processing techniques are replacing human operators. These automatic video surveillance cameras have been used to detect suspicious persons, their appearance, and their activity in mass gathering places such as shopping malls, airports, rail stations and automated teller machines etc. One of the most important detection in suspicious persons' appearance is detection of sunglasses. Committing crime by wearing sunglass is very common, as it is one of the most available accessories to cover face. Therefore, various existing approaches [1-4] attempted to detect sunglass using video surveillance. The algorithm in [1] made use of facial components (eye, nose, mouth) feature to detect face covered by sunglasses. The facial components were identified using Viola Jones face detector and then SVM classifier was used to verify if the detected regions are covered by sunglass. However, the method in [1] failed to detect eye shaped sunglasses and reflecting surface of sunglasses, since the facial component detector used in the algorithm misguidedly detect these scenario as eye location [5]. The algorithm in [2] detected facial components using Gabor features and verified the presence of occlusion by sunglass using SVM. The use of Gabor features incurs high computational complexity, since Gabor feature is high dimensional and also needs additional dimensionality reduction scheme

[6]. Furthermore, [3] proposed another method using both elliptic face shape feature and skin color feature of face for sunglass detection. The algorithm detected head using a head-shoulder template and then detected face using ellipse fitting method [7]. In addition, [3] applied thresholding classification approach using skin color area ratio (SCAR) value of eye area to verify the presence of sunglasses. However, the head-shoulder template assumption is adopted from the experience of craftsmanship and not from an established theory or experiment [8]. In addition, the use of ellipse fitting algorithm in this approach requires redundant edge points' calculation through iterations and thus, suffers from high computational complexity. Similar to the approach of [3], the algorithm in [4] also utilized elliptic shape and skin color. In addition, the algorithm made use of Omega shape and face templates. The algorithm determined the exact head region by an energy-based algorithm and ellipse fitting. Then, the algorithm optimally approximates the Omega shape formed by the head and shoulder with a Gaussian function. An elliptical head tracker was used to match the head region. At the end, the method integrated two classifiers for verification. The classifiers were based on SCAR value and face template. However, this technique gives in general decision such as "normal" and "abnormal" for faces covered with different accessories. The algorithm does not give specific decision on detection of covered area by sunglass. In addition, this method requires a huge amount of mathematical computations.

From the above discussion it can be summarized that there are a number of limitations in the previous sunglass detection algorithms such as, erroneous detection of eye location, bulk and complex mathematical computations and unspecified or in general decision regarding covered area. To alleviate these problems, this study proposes a sunglass detection algorithm for video surveillance camera. The proposed algorithm utilizes skin color, elliptic face shape and facial height-width ratio(ρ) feature of face. Note that, ρ has been found as a unique feature and can be calculated by equation (1) [9].

$$\rho = 3WH^{-1} \qquad (1)$$

where W and H represent the facial width and height, respectively as shown in figure 1. In addition it has been found that, value of ρ varies within a range of 1.44 and 2.12 for different races of people [10,11]. Furthermore, in our previous study, behavior of ρ value for different nature of covered faces (*e.g.* uncover, middle portion cover, lower half cover, upper half cover face) was analyzed [12]. The analysis showed that, the range, $1.44 \leq \rho \leq 2.12$ holds for uncover and middle portion covered faces

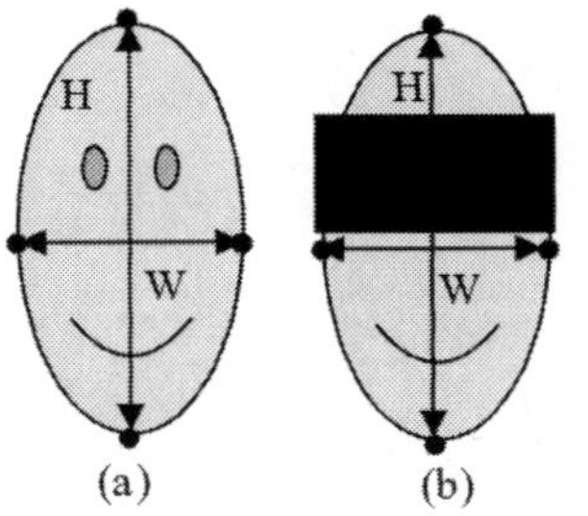

Figure 1. (a) Uncover (b) middle portion cover.

as shown in figure 1. Consequently, other nature of covered faces exceeds the range $1.44 \leq \rho \leq 2.12$. It is important to note that, the nature of sunglass covered face is considered as middle portion covered. By employing three facial features, the proposed algorithm requires only four edge points (top most, bottom most, left most and right most point) around facial region to evaluate presence of a sunglass. Thus the proposed method can avoid complex computation for redundant edge-points extraction of ellipse fitting method. Moreover, the algorithm includes two inspection levels to assure specific location and detection of sunglass. Also, the multilevel inspection involves less complex mathematical derivations and thresholding classification approach using covered area percentage (*CAP*). The threshold value of *CAP* has been validated in terms of Average-Max-Min chart. Performance of the proposed method has been demonstrated in terms of correct detection rate. The correct detection rate is determined using equation (2) similar to [13].

$$Correct\ Detection\ Rate = \frac{N_{correct}}{N_{correct} + N_{incorrect}} \times 100 \tag{2}$$

where $N_{correct}$ is number of correctly detecting a case and $N_{incorrect}$ is number of incorrectly detecting the case.

The rest of the article is organized as follows. Section 2 presents the step by step description of proposed algorithm. Description of experiment including data acquisition process, experiment demonstration and discussion on result are presented in Section 3. Finally some concluding remarks have been given in Section 4.

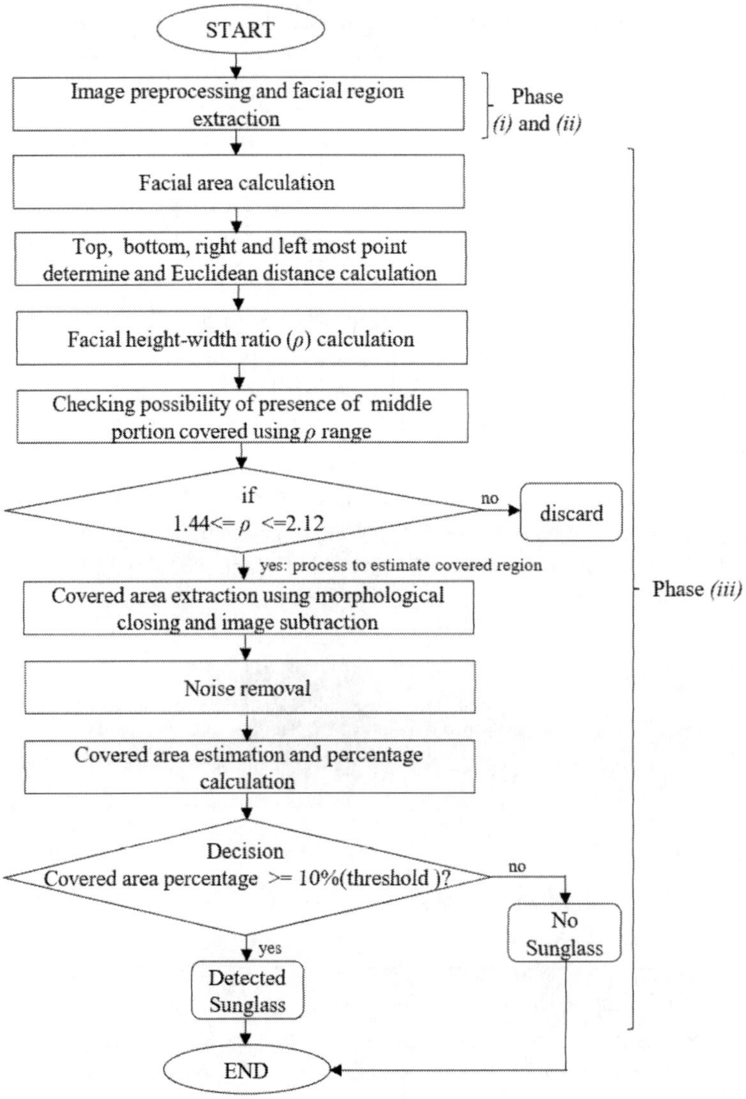

Figure 2. Flow chart of the algorithm.

2. Proposed Algorithm

Figure 2 illustrates the flow chart of the proposed algorithm that includes three phases; (*i*) image preprocessing, (*ii*) facial region extraction and (*iii*) sunglass detection. Size and illumination of input image will be normalized and reduced in the first phase. Next, facial region will be extracted using skin color information in the second phase. Finally, sunglass will be detected using the ρ feature and the area of covered face. Specifically, this phase employs two inspection levels. The first inspection level checks for the presence of middle portion covered area using ρ range. If the face image passes the first inspection with ρ range, the proposed method extracts the covered area using morphological closing and estimates the percentage of covered area. Finally, the face image experience the second inspection level using covered area percentage value to give decision for sunglass detection. Each step of the flow chart is described in details in the following subsections.

2.1. *Image preprocessing*

The input RGB images are size normalized by down sampling the original size of input image to 250x250 pixel using MATLAB function *imresize()*. Facial region is detected using skin color information in YCbCr color space, since using skin color information in YCbCr is an effective approach to detect facial region [14,15]. Note that, influence of illumination causes noise during image conversion process. Furthermore, the performance of the skin pixel classification can be improved by applying color correction process [14]. Therefore, the input RGB image is illumination compensated by using Gray world algorithm [16] prior to converting into YCbCr.

2.2. *Facial region extraction*

Skin pixel classification is performed applying thresholding approach on pixel values. Threshold value range for Cb and Cr is defined as Cb= [77 127] and Cr= [133 173]. This range has been found to be robust against different skin tone and under influence of fabric color in our previous article [17]. The facial region is extracted in the form of a binary image, B_{face}, with skin color pixel as 1(white). Then, the B_{face} is further processed to detect sunglass.

2.3. *Detection of sunglasses*

In the proposed algorithm, the area of facial region is calculated and denoted as $Area_{face}$, with the total number of 1 (white) pixel in B_{face}. Following the previous studies, the method considers the facial region extracted using skin color as elliptical shape. Then the algorithm determines four points, *i.e.* the top most point T_{point} , the bottom most point B_{point}, the left most point L_{point} and the right most point R_{point}, from the facial region in B_{face}. The algorithm calculates Euclidean distance for $d(T_{point}, B_{point})$ and $d(L_{point}, R_{point})$. The method considers $d(T_{point}, B_{point})$ and $d(L_{point}, R_{point})$ as facial region's height H and width W, respectively. The method is also used to calculate the ρ according to Equation 1 using the determined facial region's width W and height H. The algorithm verifies the calculated ρ by checking its range to specify the nature of covered face. If the calculated ρ value is found within the range $1.44 \leq \rho \leq 2.12$ then the face image is either uncover or middle portion cover. The algorithm then further evaluates the facial region to detect sunglass. The algorithm applies morphological closing operation on B_{face} to fill in the gaps in facial region. To preserve the circular nature of face shape and to fill in the largest gap inside the face the algorithm uses a disk shape structuring element of radius of 25 pixels for morphological closing. The covered region is extracted by subtracting B_{face} from the morphologically closed image. The extracted region is presented as 1 (white) pixel in the resultant image of subtraction operation. The noise is removed from the resultant image using *bwareaopen()* function of MATLAB to eliminate the extra pixels from the image after introducing morphological closing. The area of covered region, $Area_{cr}$ is calculated with the number of 1 (white) pixels of extracted covered region. The covered area percentage (*CAP)* is obtained using equation (3).

$$CAP = (Area_{cr} \times (Area_{cr} + Area_{face})^{-1}) \times 100 \tag{3}$$

If the estimated covered area percentage is $CAP > 10\%$ (threshold), it is considered that sunglass is detected. This threshold value is proven, in our experiments, to be optimal for distinguishing among uncover face, eyeglasses and sunglasses wearing faces.

Table 1. Summary of Dataset.

Cases	Lens color	Frame color	No of images
Uncover face	NA	NA	90
Eyeglasses wearing face	Transparent	Transparent, Black, Brown, Blue	90
Sunglasses wearing face	Black, Brown, Blue	Black, Grey, Brown, Blue	90
		Total images =	270

3. Experiment Design

3.1. Data acquisition

To conduct an experiment in a practical scenario we have constructed a dataset as shown in table 1. As shown in the table, three face covering cases are considered; (*i*) uncover face (*ii*) eyeglasses wearing face (*iii*) sunglasses wearing face. The lenses of the eyeglasses are transparent in color with different colored frames such as transparent, black, brown and blue. The lenses and frames of the sunglasses are mostly black, grey, brown and blue. In this experiment 270 images (90 images per case) were from 30 subjects (15 male and 15 female). Note that, indoor environment with two illumination conditions were considered; (*i*) room illumination and (*ii*) room illumination in the presence of sunlight (window side). The images are taken in three sessions; (*i*) session 1: Morning (8.00 am-10.00 am) (*ii*) session 2:Mid-day (12.00 am-2.00pm) and (*iii*) session 3: Afternoon (4.00pm-6.00pm). The subjects are instructed to sit on a chair which was placed in front of a wall white background. In addition, the subjects had to keep their back upright and face straight. We set the camera on a tripod, which is placed at a distance of 2 feet from the subject. Only frontal face of the subject has been captured. Since the proposed method down sample the images to 250x250 pixels, the method is independent to the camera resolution. The images are captured using a 5MP camera. The experiment data was analyzed with MATLABR2014a software in an Intel(R) Core(TM)i5-2400CPU, 3.10GHz computer. Some sample images of three cases are given in figure 3.

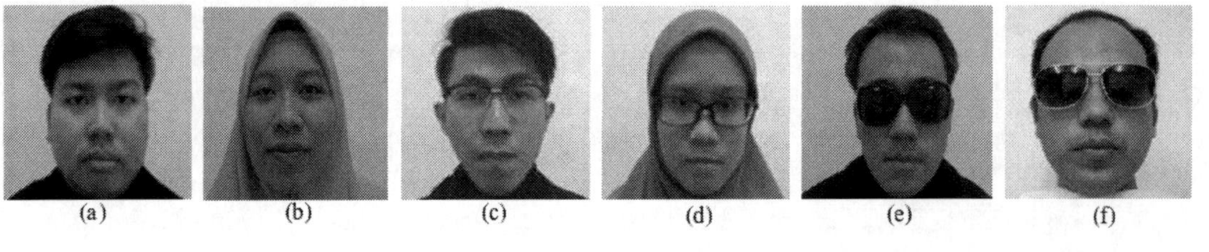

Figure 3. Sample images of dataset captured under different illumination and session (a)-(b) uncover face (c)-(d) two types of eyeglasses wearing face (e)-(f) two types of sunglasses wearing face.

3.2. Experiment demonstration

This sub-section presents demonstration of the proposed algorithm on the sample images of three cases. A message box is used to show the detection result in terms of the ρ value estimation, covered area percentage estimation and the status according to the estimation. Figure 4 and figure 5 show the cases of uncover and eyeglasses wearing face, respectively. Both the cases are detected as "No Sunglass". For both cases ρ value is within the range $1.44 \leq \rho \leq 2.12$, but the covered area percentage (*CAP*) is below threshold 10%. Figure 6 shows the case of sunglasses wearing face and the case is detected as "Sunglass detected". In this case, ρ value is within the range $1.44 \leq \rho \leq 2.12$ as well as the *CAP* is greater than the threshold 10%.

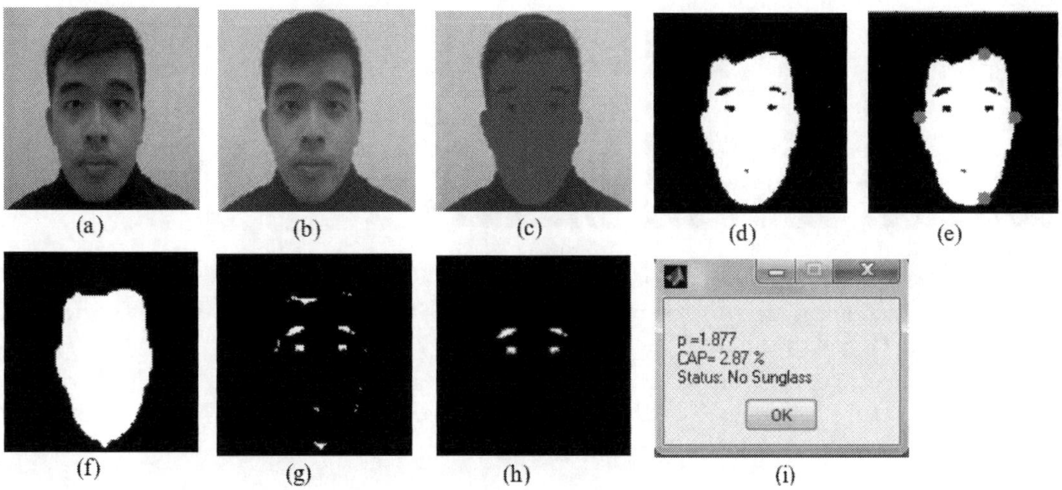

Figure 4. Application of proposed algorithm on Uncover facial case (a) input RGB image (b) illumination compensation (c) skin pixel (blue) in YCbCr color space (d) extracted facial region (e) top, bottom, left and right most point determination and ρ calculation (f) morphologically closed image (g) extracted covered area after image subtraction (h) covered area after noise removal (i) result: "No Sunglass" (right detection) since ρ value is within the range $1.44 \leq \rho \leq 2.12$ and *CAP*< 10%.

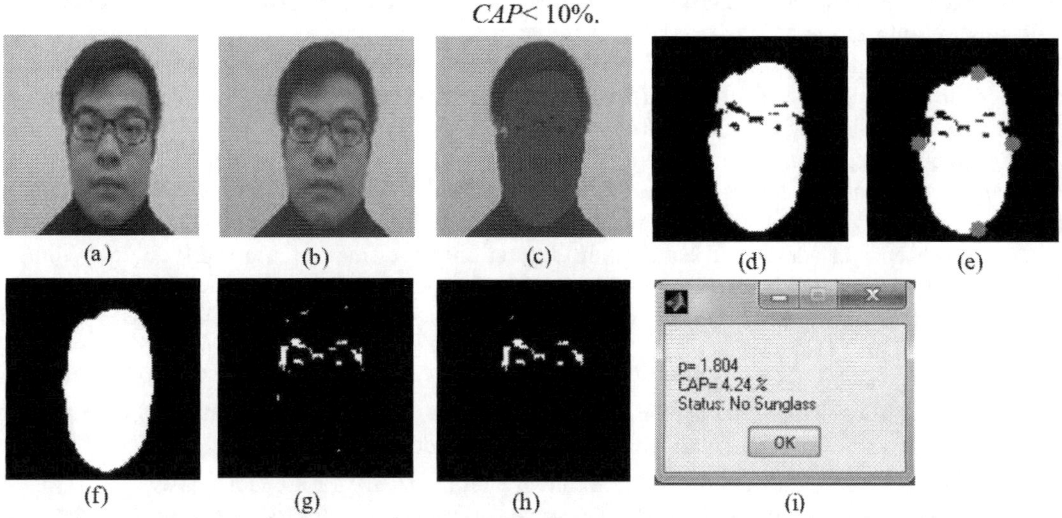

Figure 5. Application of proposed algorithm on Eyeglasses wearing facial case (a) input RGB image (b) illumination compensation (c) skin pixel (blue) in YCbCr color space (d) extracted facial region (e) top, bottom, left and right most point determination and ρ calculation (f) morphologically closed

image (g) extracted covered area after image subtraction (h) covered area after noise removal (i) result: "No Sunglass" (right detection) since ρ value is within the range $1.44 \le \rho \le 2.12$ and *CAP*< 10%.

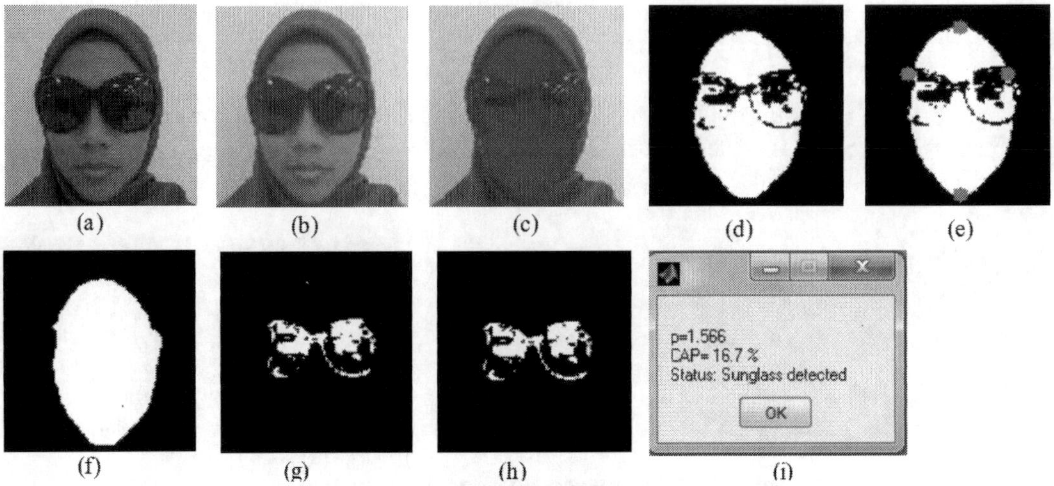

Figure 6. Application of proposed algorithm on Sunglasses wearing facial case (a) input RGB image (b) illumination compensation (c) skin pixel (blue) in YCbCr color space (d) extracted facial region (e) top, bottom, left and right most point determination and ρ calculation (f) morphologically closed image (g) extracted covered area after image subtraction (h) covered area after noise removal (i) result: "Sunglass detected" (right detection) since ρ value is within the range $1.44 \le \rho \le 2.12$ and *CAP*> 10%.

3.3. *Discussion on experiment results*

The experiment evaluates *CAP* value for the total 270 images of three facial cases including 90 uncover face, 90 eyeglasses face and 90 sunglasses face. An Average-Max-Min chart has been presented in figure 7. As demonstrated in the figure, the average *CAP* value for uncover, eyeglasses face and sunglasses face are 2.32%, 4.47% and 16.53%, respectively. From this result it is evident that, average *CAP* value for sunglass wearing faces is much higher than that of uncovered and eyeglasses wearing faces. Moreover, the *CAP* value ranges from around 1% to 3% for uncover face, 2% to 6.5% for eyeglasses faces and 10% to 18.5% for sunglass wearing faces. This implies that, for both uncover and eyeglasses face the *CAP* value remains less than 6.5%. On the other hand, the *CAP* value of sunglasses wearing faces is greater than 10%.

One of the reasons behind the high *CAP* for sunglass and low *CAP* for eyeglasses is the lens and frame color. Although the lenses of eyeglasses are transparent, the frames are in different color *i.e.* transparent, black, brown and blue. As a result the algorithm can easily detect skin color pixel through the transparent lens. In addition, the algorithm detects only the small area of dark colored frame as covered and hence the *CAP* is lower for eyeglasses face. The explanation is justified by the experiment demonstration given in figure 5. On the other hand, the lens and frame color of sunglasses are black, grey, brown and blue. Therefore, the algorithm cannot detect skin color pixel through the dark colored lens and detects a wider area as covered. This results in the higher *CAP* value for sunglass wearing faces. The explanation is justified by the experiment demonstration given in figure 6. Therefore, the algorithm detects a facial case with ρ value ranging between $1.44 \le \rho \le 2.12$ as "Sunglass detected", when the algorithm also find *CAP* value greater than 10% for the facial case.

The proposed algorithm achieves 100% correct detection rate for sunglass detection with the above conditions. Figure 8 demonstrates correct detection performance of the proposed method. As depicted in the figure, the correct detection rate for uncover face and eyeglass wearing face is 93.33%. Due to some rare facial cases, where hijab fabric color surrounding the face is similar to the skin tone, the hijab is misdirected as face area [17]. Therefore, the elliptical shape of extracted facial region becomes distorted. This leads to a misapprehend ρ value and thus, the facial case is discarded during ρ value checking. However, according to our previous study [17], this limitation can be addressed by analyzing influence of different shades of hijab/niqab fabric color on skin and non-skin pixel classification and, by designing a uniform color component thresholding approach which is robust against fabric color influence. In addition, the proposed method is experimented with uniform background in the present study. Future research will be carried out for outdoor dynamic environment.

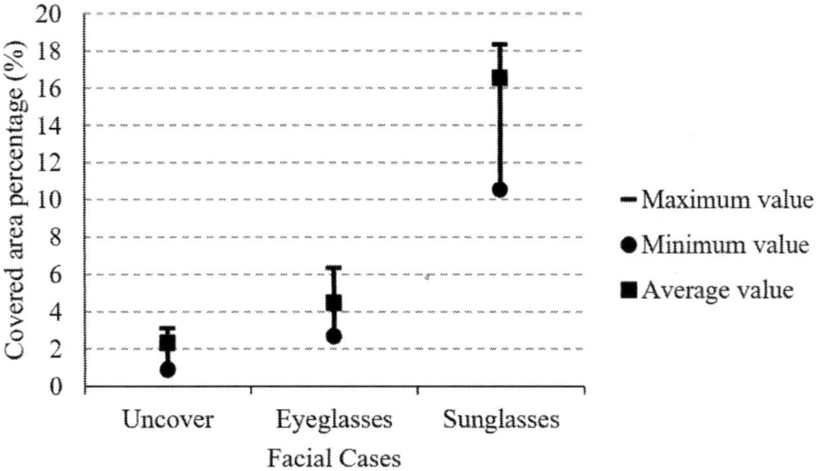

Figure 7. Average-Max-Min chart plot of covered area percentage (*CAP*) value for all the images of three facial cases.

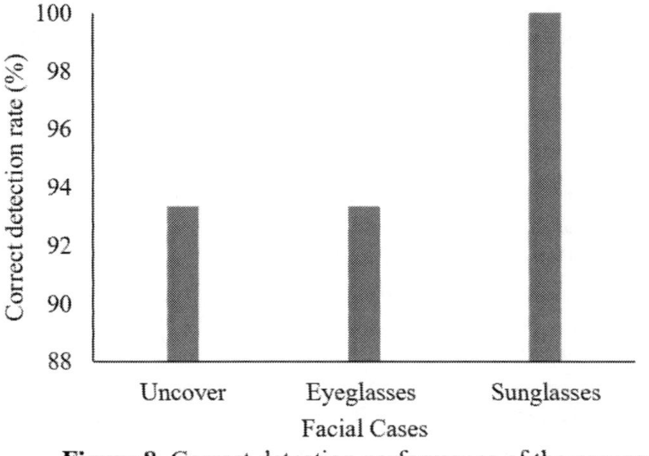

Figure 8. Correct detection performance of the proposed method.

4. Conclusion

In this paper sunglass detection method has been proposed for automation of surveillance camera. Covered area of face was calculated using a novel facial geometry called height-width ratio. The experiment was conducted in practical scenario of both room illumination and in the presence of sunlight. The result of the experiment revealed that, the sunglass can be detected in both scenarios. In addition, due to the multi-level checking on facial area, the proposed method can differentiate normal eye glass and sunglass with high detection accuracy of 93.33% and 100%, respectively. A variety of shapes for sunglass frame will be considered in future study. In addition, future research will embark on fixing the limitation of misapprehended ρ value due to similarity of hijab fabric color and skin tone of the face.

Acknowledgement
This research was supported by Universiti Malaysia Pahang (UMP) research grant scheme (No. RDU 1703256)

References
[1] Jae Kyu S, Eum S, Gi H, Li G, Kim G, Kim J 2012 Recognizability assessment of facial images for automated teller machine applications *Pattern Recognit.* **45.5** 1899–1914.
[2] Min R, Hadid A, Jean-Luc D 2014 Efficient Detection of Occlusion prior to Robust Face Recognition. *Sci World J.*
[3] Kim G, Suhr JK, Jung HG, Jaihie K 2010 Face Occlusion Detection by using B-spline Active Contour and Skin Color Information. *In: In Control Automation Robotics & Vision (ICARCV), 2010 11th International Conference on IEEE* pp 627–632.
[4] Zhang T, Li J, Jia W, Sun J, Yang H 2017 Fast and robust occluded face detection in ATM surveillance *Pattern Recognit Let.*
[5] Eum S, Suhr JK, Kim J 2011 Face recognizability evaluation for ATM applications with exceptional occlusion handling. *In: Computer Vision and Pattern Recognition Workshops (CVPRW), 2011 IEEE Computer Society Conference on IEEE* pp 82–89.
[6] Yan H, Wang P, Chen WD, Liu. J 2015 Face Recognition Based on Gabor Wavelet Transform and Modular 2DPCA *In: In International Conference on Power Electronics and Energy Engineering (PEEE 2015)* pp 245–248.
[7] Halır R, Flusser J 1998 Numerically stable direct least squares fitting of ellipses *In: In Proc 6th International Conference in Central Europe on Computer Graphics and Visualization WSCG* pp 125–132.
[8] Xia Y, Zhang B, Coenen F 2015 Face Occlusion Detection Based on Multi-task Convolution Neural Network. Fuzzy Syst Knowl Discov (FSKD) 12th Int Conf on IEEE pp 375–379.
[9] Prendergast PM 2012 *Facial Proportions Adv Surg Facial Rejuvenation* (Springer Berlin Heidelberg) pp 15–22.
[10] Bird BM, Cid VS, Geniole SN, Welker KM, Zilioli S, Maestripieri D, et al 2016 Evolution and Human Behavior Does the facial width-to-height ratio map onto variability in men's testosterone concentrations ? *Evol Hum Behav.* **37.5** 392–398.
[11] Mccormick CM, Carre JM 2008 In your face : facial metrics predict aggressive behaviour in the laboratory and in varsity and professional hockey players. *Proc R Soc London B Biol Sci.* vol 275.1651 pp 2651–2656.
[12] Sikandar T, Ghazali KH, Rabbi MF 2017 Covered Face Detection with Facial Width-to-Height Ratio for ATM Surveillance Camera *Image Vis Comput.*under review
[13] Yi M 2011 Abnormal Event Detection Method for ATM Video and Its Application. *Adv Res Comput Educ Simul Model* 186–192.
[14] Kakumanu P, Makrogiannis S, Bourbakis N 2007 A survey of skin-color modeling and detection methods. *Pattern Recognit.* **40.3** 1106–1122.
[15] Basha K, Ganesan P, Kalist V, Sathish BS, Mary JM 2015 Comparative Study of Skin Color Detection and Segmentation in HSV and YCbCr Color Space *Procedia - Procedia Comput Sci.* vol 57 pp 41–48.
[16] Kovac J, Peer P, Solina F 2003 Human skin color clustering for face detection *IEEE* pp 144-148.
[17] Sikandar T, Ghazali KH, Mohd II, Rabbi MF. Skin Color Pixel Classification for Face Detection with Hijab and Niqab 2017 In: ICISPC 2017 *Proceedings of the International Conference on Imaging, Signal Processing and Communication* (ACM New York, NY, USA ©2017) pp 1–4.

IOP Conf. Series: Materials Science and Engineering **342** (2018) 012041 doi:10.1088/1757-899X/342/1/012041

Statistical Analysis of Compressive and Flexural Test Results on the Sustainable Adobe Reinforced with Steel Wire Mesh

Gul A. Jokhio[1], Sharifah M. Syed Mohsin[1] and Yasmeen Gul[2]

[1]Faculty of Civil Engineering and Earth Resources, University Malaysia Pahang, Malaysia

[2]Faculty of Built Environment, University Technology Malaysia, Malaysia

E-mail: jokhio@ump.edu.my

Abstract. It has been established that Adobe provides, in addition to being sustainable and economic, a better indoor air quality without spending extensive amounts of energy as opposed to the modern synthetic materials. The material, however, suffers from weak structural behaviour when subjected to adverse loading conditions. A wide range of mechanical properties has been reported in literature owing to lack of research and standardization. The present paper presents the statistical analysis of the results that were obtained through compressive and flexural tests on Adobe samples. Adobe specimens with and without wire mesh reinforcement were tested and the results were reported. The statistical analysis of these results presents an interesting read. It has been found that the compressive strength of adobe increases by about 43% after adding a single layer of wire mesh reinforcement. This increase is statistically significant. The flexural response of Adobe has also shown improvement with the addition of wire mesh reinforcement, however, the statistical significance of the same cannot be established.

1. Introduction

Adobe is commonly associated with low-cost construction as it requires relatively less skills [1]. It has been used all over the world, especially the dry regions, for over 10 thousand years [2-4]. Adobe structures have been found in South Asia, Africa, Europe, and Central Asia [5]. Some structures of a very high religious importance [6], especially domed structures, as well as regular traditional houses [5-11] have been built with this material. Evidence has been found of the use of mud bricks in protective structures for thousands of years [12], and approximately 30% of the present world population lives in earthen structures [13]. A rich cultural heritage of earth building can be found today in Africa, Iran, Afghanistan, Yemen, Iraq, Syria, Spain, Germany, England, France, Portugal, Italy, Denmark, and Sweden [14].

For the past about 30 years, the interest of researchers and academicians has increased in the use of this material [15]. It is evident from the amount of literature being published in this regard which has increased to about ten times in the past decade compared to what was published in the previous 10 years [16]. This increase in the interest in adobe is due to the advantages of this material in terms of economy and sustainability. However, adobe suffers from strength issues and is generally considered weak when subjected to adverse loading conditions such as earthquake [17-18]. Moreover, due to the lack of technical knowledge available, it is difficult to get adobe structures insured at a reasonable price, which hinders the progress of this material as a mainstream construction material and technique [14]. In addition to the above, adobe also suffers from class culture as it is considered only for the poor. Sometimes, very rich people build some adobe structure as a fashion statement, however, the dominant middle class stays away from adobe construction [19]. All these facts notwithstanding, the need of energy-efficient sustainable housing development makes the search of alternative and sustainable material inevitable. Materials and building technologies need to be environment friendly, energy efficient, and affordable [20-22]. This calls for the promotion of adobe as much as possible [23], and it is needed that resources be allocated for further research in this field.

There are no strict guidelines available for the preparation of adobe. Generally it is prepared by unskilled labourers using different methods and a variety of mix proportions. Therefore, it is difficult to ascertain the mechanical properties of any trial mix of adobe while it is being prepared. In the literature, a wide range of mechanical properties of adobe has been reported, which is due to the above-mentioned issues. Some researchers have studied historical buildings for their in-situ characterisation and constituent properties [6, 24, 25]. A few standards for the experimental evaluation of the mechanical properties can be found in the literature [26-30], however, these standards generally don't agree and provide different, often times contrasting, procedures and criteria making it difficult to comply [26, 31-32]. Therefore, towards attaining the goal of bringing adobe to the mainstream, researchers must focus on improving its mechanical properties. One of the several approaches for doing that is to add reinforcement to adobe. Different types of reinforcements have been studied such as bamboo fibres etc. The present study focuses on the utilization of steel wire mesh as reinforcement for adobe. Compressive and flexural tests in this regard were carried out and the present paper reports the statistical analysis results and significant conclusions drawn from the same.

2. Experimental Setup and Methodology

The experimental setup for the present study consisted of two types of tests i.e. compressive strength test and flexural strength test. The details of specimens, materials, preparation procedures and tests conducted have been reported elsewhere [33]. A total of 22 samples were prepared and tested in both categories. There were 12 cubes for testing of compressive strength, 6 out of which were control specimens whereas the remaining 6 were reinforced with wire mesh. The mix design for the samples was not changed as it was not the focus of this study. The ingredients of the mixture consisted of sand, silt, clay, and Kenaf (hibiscus cannabinus) fibre. The proportion of different constituents used in the preparation of the mixture has been reported elsewhere [33]. Once the adobe material mix was ready, the specimens were prepared by pouring the mixture into moulds. The pouring process was carried out in steps with every next material layer added only after the sufficient compaction of the previous layer was ensured. The compressive strength tests were carried out on the Universal Testing Machine by adapting for adobe the British standard for compressive strength tests for concrete cubes (BS EN 12390:3-2002). It was tried to adhere to the standard as much as possible, however, the compaction procedures and mix pouring had to be modified due to the nature of adobe material. The breaking load was noted as the maximum load when the specimen was subjected to the test in the machine. The test setup for flexural strength test consisted of a three point loading system. The two end supports were placed about 50 mm inside the edge of the beam resulting in an effective span of 400 mm. The test setup was mounted on the Universal Testing Machine. Deflection sensors were used to measure the beam deflection in order to plot the load-deflection curve. The load and deflection values were recorded for each specimen from the start of the test until failure at regular intervals. The details of both the compressive and the flexural test set ups have been reported elsewhere [33]. However, for a quick referral, these details are being re-produced here.

A total of 12 cubes were tested for compressive strength and 10 prismatic samples were tested for flexural strength. The specimen naming convention has been adapted in such a way that the first letter of the name represents whether it is a 'compressive sample' (C) or a 'flexural sample' (F). The second letter represents whether it is a 'plain sample' (P) or 'reinforced with wire mesh' (W). The last digit is just the serial number of the sample which is 1 to 6 for compressive and 1 to 5 for flexural specimens. Proportions of the materials used to prepare adobe mix are given in Table 1, whereas, the specimens ready for testing are shown in Figure 1. The wire mesh reinforcement was provided by adding a single wire mesh layer such as shown in Figure 2. The test setups for compressive and flexural tests are shown in Figures 3 and 4.

Table 1. Proportions of constituents in adobe mixture used.

S. No.	Name of Constituent	Average Size of Particles	Proportion
1.	Sand	2 mm	50%
2.	Silt	0.06 mm	20%
3.	Clay	0.002 mm	25%
4.	Kenaf fibre	-	5%

Figure 1. Cube and prismatic samples ready for testing (cubes size: 150 mm × 150 mm × 150 mm; prismatic specimens Size 100 mm × 100 mm × 500 mm)

Figure 2. Wire mesh reinforcement layer prepared for a cubic specimen

Figure 3. A cube sample being tested for compressive strength

Figure 4. A prismatic sample being tested for flexural strength

3. Results and Discussion

3.1 Compressive Strength Test Results

The results of the compressive strength test showed that on average, the total load taken by the samples without any reinforcement was about 4.97 kN resulting in a compressive strength of 0.22 MPa. The inclusion of the wire mesh reinforcement, however, increased the average total load to 7.12 kN resulting in compressive strength of 0.31 MPa. Therefore, the addition of a single layer of wire mesh reinforcement increases the compressive strength of adobe samples by approximately 43%. The summary of the results of the compressive strength test is presented in Table 2.

An interesting observation was made in that the cubes reinforced with wire mesh retained their shape even after failure and did not totally dismantle. This behaviour is different from the unreinforced cubic samples, which after failure are totally destroyed. It is hypothesized that the walls made with adobe bricks reinforced with wire mesh will tend to retain their shape even after failure, resulting in an improved response to earthquake and other similar disasters.

The statistical analysis was carried out on the compressive strength test results. The analysis of variance results suggest that the difference between the results of unreinforced samples and those reinforced with wire mesh are statistically significant. Table 3 presents the results of the analysis of variance. It can be seen that the p-value between groups is 0.002365, which is much smaller than the 0.005 value generally accepted to indicate statistical significance. It can also be seen that the F value is 16.314, which is much higher than the critical value for the same (4.9646). A higher value of F is again an indicator of statistical significance. The variance in the compressive strength test results is also very low, which is at 0.000314 for unreinforced samples and 0.003044 for reinforced samples. This establishes the reliability of the results and the conclusions inferred from the same.

3.2 Flexural Strength Test Results

The unreinforced adobe samples showed little or no flexural resistance at all. The comparison of the normalized stress-strain curve of un-reinforced prismatic specimens was made with the stress strain curve for adobe suggested by Illampas et al. [36]. The experimental results were taken as the average load of all the 5 samples against each deflection value. This comparison has been shown in Fig 5. It can be seen that while the average of the experimental results does display the tendency of the proposed model, the variation throughout the load-deflection history is relatively large, thus reducing the reliability of this association.

It can be observed in Fig 5 that the initial pre-peak response is varying highly, whereas, the post-peak softening is closer to the proposed model. The overall value of coefficient of determination between the two data sets is only 27.35%, whereas the coefficient of determination for the softening branch only is 71%. This can in part be attributed to the fact that the overall flexural strength of adobe is very low resulting in some discrepancies in the initial measurement of the same.

Table 2. Summary of compressive strength test results.

S. No.	Sample Name	Maximum Load (kN)	Compressive Strength (MPa)
1	CP-1	5.3	0.2356
2	CP-2	4.9	0.2178
3	CP-3	4.7	0.2089
4	CP-4	4.6	0.2044
5	CP-5	5.6	0.2489
6	CP-6	4.7	0.2089
Average for Unreinforced Samples		4.9667	0.2207
Standard Deviation for Unreinforced Samples		0.3983	0.0177
7	CW-1	5.8	0.2578
8	CW-2	6.8	0.3022
9	CW-3	6.9	0.3067
10	CW-4	9.5	0.4222
11	CW-5	6.8	0.3022
12	CW-6	6.9	0.3067
Average for Samples Reinforced with Wire Mesh		7.1167	0.3163

Average for samples Reinforced with Wire Mesh	1.2416	0.0552
% Increase for Samples Reinforced with Wire Mesh	43.2883%	

Table 3. Analysis of variance for compressive strength test results.

SUMMARY

Groups	*Count*	*Sum*	*Average*	*Variance*
Unreinforced Samples	6	1.3245	0.22075	0.000314
Samples Reinforced with Wire Mesh	6	1.8978	0.3163	0.003044

ANOVA

Source of Variation	*SS*	*df*	*MS*	*F*	*P-value*	*F crit*
Between Groups	0.027389	1	0.027389	16.31408	0.002365	4.964603
Within Groups	0.016789	10	0.001679			
Total	0.044178	11				

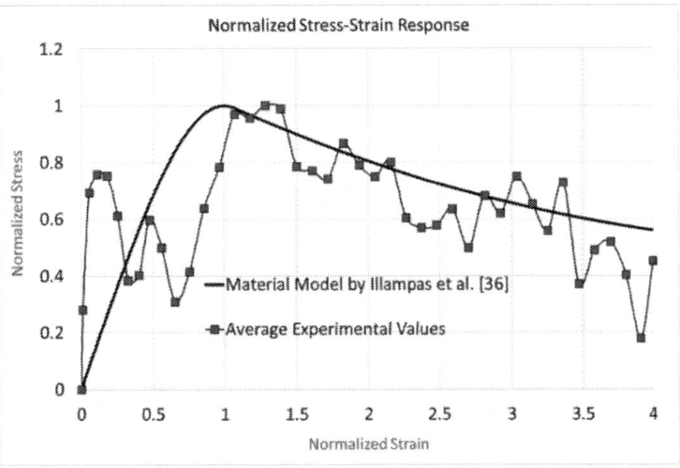

Figure 5. Stress-strain response for unreinforced adobe beams.

The addition of a single layer of wire mesh reinforcement, however, significantly improves the flexural response of the prismatic adobe specimens. The comparison has been presented in Table 4 as well as in Fig 6, where the average of actual load-deflection response of the unreinforced specimens is compared with that of the specimens reinforced with wire mesh. The addition of wire mesh reinforcement has not only increased the flexural strength of adobe samples to about 3 times the original values, the response of the reinforced specimens also appears more reliable and can be represented by an idealized tri-linear curve. The slope and the relevant coefficients of determination for the three parts of the idealized curve are given in Table 5.

IOP Conf. Series: Materials Science and Engineering **342** (2018) 012041 doi:10.1088/1757-899X/342/1/012041

Although the coefficient of determination for the stage 3 or the proposed tri-linear model is numerically low, it is due to the nature of this calculation that in the case of zero or a very small slope, the coefficient of determination is generally low. It does not directly indicate the lack of significance of the relationship.

The analysis of variance results as shown in Table 6 present an interesting read. While it can be seen that there is a large difference, about three times, between the average flexural load of unreinforced samples (approx. 100 N) and the average flexural load of the samples reinforced with wire mesh (approx. 300 N). However, there is also a very large variance in the results, which is 3105.314 for unreinforced samples and 37485.7 for reinforced samples. This large variance indicates that the reliability of these results cannot be established. Therefore, the analysis of variance suggests that the difference between the two categories, although large, is not statistically significant. This is evident through the p-value, which is 0.061. The value of F (4.74) is also slightly less than the critical value for the same (5.32).

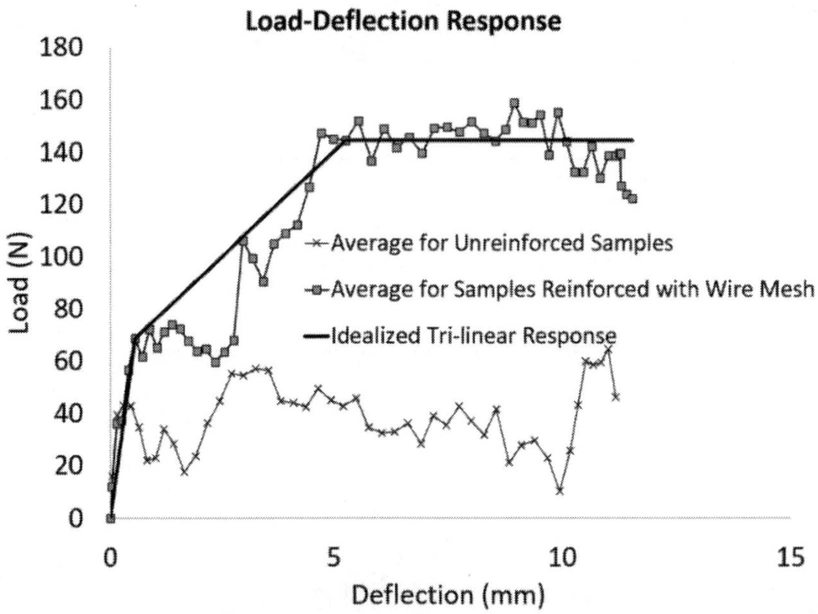

Figure 6. Comparison of load-deflection response of unreinforced samples and samples reinforced with wire mesh along with idealized tri-linear load-deflection curve

Table 4. Summary of flexural strength test results.

S. No.	Sample Name	Maximum Load (N)
1	FP-1	97.947
2	FP-2	194.381
3	FP-3	70.655
4	FP-4	52.157
5	FP-5	80.968
Average for Unreinforced Samples		295.4208
7	FW-1	278.985
8	FW-2	628.02
9	FW-3	241.08
10	FW-4	134.337
11	FW-5	194.682
Average for Samples Reinforced with Wire Mesh		99.2216

Table 5. Summary of compressive strength test results.

Stage of Load-Deflection Curve	Slope (Modulus of Elasticity in MPa)	Coefficient of Determination (%)
Stage 1	121.98	93.4
Stage 2	16.13	81.2
Stage 3	0	26.5
Overall		90

Table 6. Analysis of variance for flexural strength test results.

SUMMARY

Groups	*Count*	*Sum*	*Average*	*Variance*
Unreinforced Samples	5	496.108	99.2216	3105.314
Samples Reinforced with Wire Mesh	5	1477.104	295.4208	37485.4

ANOVA

Source of Variation	*SS*	*df*	*MS*	*F*	*P-value*	*F crit*
Between Groups	96235.32	1	96235.32	4.741741	0.061095	5.317655
Within Groups	162362.8	8	20295.35			
Total	258598.2	9				

4. Conclusions

From the results obtained through experiments in the present research, the following conclusions and recommendations are made:

- The addition of a single layer of wire mesh reinforcement increases the compressive strength of adobe samples by about 43%.
- The difference between the compressive strength of adobe without reinforcement and that of adobe reinforced with wire mesh is statistically significant.
- The load-deflection response of adobe in case of flexural loading is generally highly unreliable, however, with the addition of wire mesh reinforcement, the flexural strength of adobe can be increased about 3 times.
- The statistical significance of the difference of flexural strength of reinforced and unreinforced adobe samples cannot be established.

Acknowledgements

The authors acknowledge the research environment and infrastructure provided by University Malaysia Pahang. The research funding for the present study was also provided by University Malaysia Pahang through research grant number RDU160370, for which the authors are highly grateful.

References

[1] M. Blondet and G. V. Garcia M., Adobe Construction, Tech. Rep., (2003).
[2] Q. Angulo-Ibanez, A. Mas-Tomas, V. Galvan-LLopis and J. L. Santolaria-Montesinos, Const & Bldg Mat, 30, 389-399 (2012).
[3] L. Keefe, Earth Buildings - Methods, materials, repair and conservation, (Taylor & Francis, 2005).
[4] T. Morton, Earth masony - design and construction guidelines, (HIS BRE Press, 2008).
[5] E. Fodde, Int J of Arch Heritage, 3, 2, 145-68, (2009).
[6] E. Adorni, E. Coisson and D. Ferretti, Const & Bldg Mat, 40, 1-9, (2013).

IOP Conf. Series: Materials Science and Engineering **342** (2018) 012041 doi:10.1088/1757-899X/342/1/012041

[7] A. Chasagnoux, L'architecture voute iranienne-modelisation et simulation par elements finis. PhD thesis (in French), (Universite de Nantes, 1996).
[8] R. Ghirshman, S. Gilbert and J. Emmons, Persian Art, The Parthian and Sassanian Dynasties, 249{BC}-{AD}651., (Golden Press, 1962).
[9] M. Hejazi, Historical buildings of Iran: their architecture and structure., (Queen Mary and Westfield College University of London, 1997).
[10] A. U. Pope, Persian Architecture, (George Braziller, 1965).
[11] H. E. Wulff, The traditional crafts of Persia, (MIT Press, 1966).
[12] H. Binici, O. Aksogan, D. Bakbak, H. Kaplan and B. Isik, Const & Bldg Mat, 23, 2, 1035-1041, (2009).
[13] F. Fratini, E. Pecchioni, L. Rovero and U. Tonietti, App Clay Sc, 53, 3, 509-516, (2010).
[14] M. C. J. Delgado and I. C. Guerrero, Const & Bldg Mat, 20, 9, 679-690, (2006).
[15] R. Illampas, I. Ioannou and D. C. Charmpis, Int J of Arch Heritage, 7, 2, 165-188, (2013).
[16] F. Pacheco-Torgal and S. Jalali, Const & Bldg Mat, 29, 512-519, (2012).
[17] M. Blondet and G. V. Garcia M., Earthquake Resistant Earthen Buildings?, (2004).
[18] Y. Gul, G. A. Jokhio and E. Kakar, 6th Int Civ Engg Cong, (2013).
[19] L. Baker, Mud, 1, (2015).
[20] K. M. Roufechaei, A. H. Abu Bakar and A. A. Tabassi, J of Cleaner Prod, 65, 380-388, (2014).
[21] S. H. Sameh, J of Cleaner Prod, 65, 362-373, (2014).
[22] G. Seyfang, En Pol, 38, 12, 7624-7633, 2010.
[23] H. Niroumand, M. F. Zain and M. Jamil, Proc - Soc and Behav Sc, 89, 0, 231-236, (2013).
[24] E. Nodarou, C. Frederick and A. Hein, J of Arch Sc, 35, 11, 2997-3015, (2008).
[25] E. Quagliarini and S. Lenci, J of Cul Heritage, 11, 3, 309-314, (2010).
[26] R. Walker and H. Morris, Proc of the Aust Str Engg Conf, Auckland, vol. 1, pp. 477-84, 1998.
[27] P. Walker, The Australian earth building handbook, (Standards Australia, 2002).
[28] NZS, Materials and workmanship for earth buildings, (Standards New Zealand, 1998).
[29] NMAC, New Mexico earthen building materials code, (New Mex Admin Code, 2009).
[30] P. Walker, Masonry Int, 1-6, (1996).
[31] B. Briccoli, L. Rovero and U. Tonietti, Proc of the 10th Int Conf on the Study of Conserv of Earthen Arch Heritage, Mali, 253-8, (2008).
[32] J. C. Morel, A. Pkla and P. Walker, Const & Bldg Mat, 21, 2, 303-309, (2007).
[33] G. A. Jokhio, Y. M. Y. Al-Tawil, S. M. Syed Mohsin, Y. Gul, and N. I. Ramli, MUCET 2017, Penang, Malaysia

IOP Conf. Series: Materials Science and Engineering **342** (2018) 012042 doi:10.1088/1757-899X/342/1/012042

Inspired by design and driven by innovation. A conceptual model for radical design driven as a sustainable business model for Malaysian furniture design

Wan Zaiyana Mohd Yusof and Puteri Fadzline Muhamad Tamyez

Faculty of Industrial Management, Universiti Malaysia Pahang, 26300, Kuantan, Pahang

E-mail:wan0069@gmail.com; fadzline@ump.edu.my

Abstract. The definition of innovation does not help the entrepreneurs, business person or innovator to truly grasp what it means to innovate, hence we hear that government has spend millions of ringgit on "innovation" by doing R & D. However, the result has no avail in terms of commercial value. Innovation can be defined as the exploitation of commercialization of an idea or invention to create economic or social value. Most Entrepreneurs and business managers, regard innovation as creating economic value, while forgetting that innovation also create value for society or the environment. The ultimate goal as Entrepreneur, inventor or researcher is to exploit innovation to create value. As changes happen in society and economy, organizations and enterprises have to keep up and this requires innovation. This conceptual paper is to study the radical design driven innovation in the Malaysian furniture industry as a business model which the overall aim of the study is to examine the radical design driven innovation in Malaysia and how it compares with findings from Western studies. This paper will familiarize readers with the innovation and describe the radical design driven perspective that is adopted in its conceptual framework and design process.

1. Introduction

Europe is well known with the "Italian style" luxury furniture and Indonesiawith its teak furniture and internationally-acclaimed German designer *Klaus Kummer* reveals that there is a lack of Malaysian identity in Malaysian furniture as the furniture industry prefers to look to the West for furniture inspiration. This notion is consistent where Malaysian furniture are in lesser originality due to the commonality of manufacturers to copy other existing product designs [1].

Most firms design for the West which does not reflect the culture of their own. Statement made by Chief Executive Officer, Malaysian Furniture Promotion Council also confirmed that according to experts'perceptions" Malaysian furniture industry needs a clear vision, strategy, and direction, to upgrade its design capabilities urgently and to be better organized and coordinated [2]. Malaysia has lacked of design expertise, R&D and brand identity as compared to Italy and Thailand. Therefore, such a phenomenon has created a concern of whether the current manufacturers have been equipped with the right innovation. Although there have been a few studies on design innovation in Malaysia, there is less focus on radical design innovation and the radical design innovation literature is mainly based on data from Western samples from Western developed country. Little is known about the radical design innovation in East Asia countries [3,4,5]. This raises the question of whether radical design innovation can be generalized to other cultures.

There are still questions raised as to how well Western findings could be generalized to other societies with a dissimilar culture. This is an important question to be addressed so that radical design innovation policies could be developed to suit the Eastern context. Thus, this study address this gap in the radical design innovation literature.In Italy, numerous studies revealed that the theory of Radical Design Driven Innovation is proven to drive markets along with significant profit margins and brand value. The Italians' approach is more on a knowledge research process aiming to develop new meanings and languages [6]. They are very much dependent on Design Driven Laboratories (DDL) where this phase is carried out within organizations and active cooperation by the designers in order to gain influence on customers to perceive the projected product. Design fairs in Italy function as a market test to gain feedback from visitors on developed prototypes [7]. Excellent examples of radical design driven innovation that prove meanings are *Nintendo, Apple, Artemide, Whole Foods Markets* and *Alessi*.

Therefore, in order for Malaysia to compete with Italy in furniture design, Malaysia needs to have a robust framework as a unique design innovation strategy. This innovation often look against the dominant aesthetic standards and the process involves a tacit, invisible-no-method, no tools, no steps and mainly based on networks on uncodified interaction among agents of innovation [8]. However, questions the usefulness of radical design driven innovation as excellent systems applied by the Italian companies as it will only to be applied on firms that has a stable and diffuse technological competencies [9]. Not only that, he discussed that the adoption of design driven innovation is beneficial to firms that are already reputable in the market and credible along with strong complementary assets. Additionally, Tether (2005)also added that firms that are applying this type of innovation are only suitable for young unknown designer to channel through established and credible firms for a better adaption to this innovation. Therefore, it is important to re-evaluate and complement the theory to cater the emotion, cultural and social needs of users.

2. Purpose of Inquiry and Inquiry Questions

This conceptual paper is intended to identify and understand the radical design driven innovation in Malaysia and how it compares with findings from Western studies. Thus, this research aims to highlight the radical design driven innovation in the Malaysian furniture industry as a business model. Therefore, the inquiry questions that arise for this paper are:

I. Innovation Approach in New Product Development Process that had been widely accepted and practiced in current product development processes. How far do the furniture manufacturers in Malaysia practice this approach?

II. What are the factors that could be identified in relation to propose the Radical Design Innovation Framework?

III. How can Radical Design Innovation Framework be enhanced to align with the Malaysia context?

3. Significance of Study

This conceptual paper is significant because it will provide valuable information in understanding and formulating ways to enrich the radical design innovation and is much in relevance to the Government policy. It benefits the furniture industry further as Malaysia Design Council (MRM) spearheads design programmes to pave a future of innovation and design for Malaysia. Theoretically, the results are intended to fill the gaps. It is an attempt to investigate the influence of radical design innovation on their competitiveness in the Malaysian furniture industry. This paper will also provide new insights about how radical design innovation is understood in an Eastern developing country, and to understand to what extent Western findings can be generalized to an Eastern context and make recommendations for future research.

4. Delimitations and Limitations

This conceptual paper is delimited by several elements. First, this paper only reviews design driven innovation. This is delimiting because it excludes other innovation strategies was in place and other innovation strategies were being implemented as Design Driven innovation was found successful strategies in business model. Second, the theoretical framework uses ; the design approach that embraces design, innovation and strategy and the business modeling approach that embraces strategy, business model and innovation. This is delimiting because when understanding innovation, one must go into other subjects such as strategy, politics, and economics of a company which in this study the companies that will be selected are those who won the Malaysia Pride award.

5. Literature Review

5.1. Defining Design and the Changing Notion of Design

Serious attention towards design being worthy of in-depth, scientific investigation was triggered in the early 1960s by two conferences, one held in London in 1962 and one held in Birmingham in 1965 [10], both of which raised interesting debates on the need to develop a science of design. Design was a much broader field, which encompassed any type of planning "aimed at changing existing situations into preferred ones" (Simon, 1969). [12] provides the following sentence: "Design is when designers design a design to produce a design".While [13], defined design is conceived as part of a problem-solving activity beginning with the perception of a gap in a user experience, leading to a plan for a new artifact, and resulting in the production of that artifact. The artifact can be any result of intentional creation, including physical goods, software, and services. By starting with a reflection on the semantics of the term 'design', [14] suggests that its etymology goes back to the Latin 'de + signare' and refers to "making something, distinguishing it by a sign, giving it significance, designating its relation to other things, owners, users or goods" (2003:157). Moreover, 'design' can be a noun or a verb, that is, "a product which has been made by humans may be referred to as 'a design', while the activity that produced this product is the act of 'designing' [15].

According to systems designer and educator [16], design has "different meanings in different domains, [is] used in different ways by researchers in the same domain, and [is] found in the literature referring to concepts at different levels of abstraction" (p. 347).Many others in a wide range of disciplines have noted there are a multitude of definitions of design as two mentioned above [17,18,19,20]. However, It is widely recognized that design is an important factor contributing to firms' success because of its potential to boost their competitiveness [21,22]. Design is closely associated with success in NPD performance. Regardless of the type of strategy pursued, design has a statistically significant and substantial association with new product development performance [23].

An integration of brand strategy and design innovation will tend to perform superior market dominance [24]. However, national efforts are given focus in promoting design culture in the Malaysian furniture industry through the formulation of Furniture & Wood Design Roadmap (FWDR) 2012-2020. It is carried out to create a portfolio of new exciting designs that have international market appeal that can be manufactured and marketed by Malaysian furniture manufacturers who have strong product branding/identity reinforce the nation's furniture industry image. This will place ourselves as one of the countries recognized for good design (green and sustainable designs) [25].The search for a Malaysian identity is a retrospective effort in innovation and it is possible to turn to the traditional arts and culture as a source of reference [26]. This is carried out by coming up with designs that have elements of Malaysian culture to be sold all over the world. Ultimately, firms should be able to create different designs that will appropriately fit to numerous markets around the world. Asian designs would be respected by the world over through that status [27].

However, [28] stated that functionality gives lack of fulfillment to users and are more concern on products that have the language of emotion (imagination, myths and ritual) which effect their decisions in buying a certain product. Copying other designs for the European market is not sufficient to define 'design' in the context of the Malaysian furniture industry.

5.2. Innovation Concepts and Typology

According to [29] in their analysis of concepts, aspects, definitions and types of innovation, they concluded that; the challenge will be to refine the innovation definition and classification of innovation types and streamline them into a usable and understandable set of definitions, concepts and types which are of use for academics and practitioners. For private sector practitioners this is of utmost importance since it shows that too many different concepts appeared in the last years which are more of marketing and advertising style rather than adding real value to company operations. In the political sphere such understanding should emphasize the potential impacts of innovation for the given political and societal goals but these needs to bea clear communication beyond the respective communities involved. Overall innovation needs to be considered as a long time investment. There is still the assumption that actors (companies or countries) investing in innovation are the most successful ones but in reality the tie dimension of impact from innovation is neglected.

Innovation is important for business as it creates economic value, the return on investment spent on the invention and it is about competing effectively in the market. Value creation is the raison d'être of innovation. Hence innovation impacts on enterprises whom making money while doing something good for the environment and for society. [30] also emphasized that for most Entrepreneurs and business managers, are creating "Enterprise Value" which innovation is about creating something of economic value that increases revenue, reduces cost, improves the customer experience, makes a profit and thereby increases shareholder value. Therefore, if there is no "Enterprise Value", then there is no innovation and it is merely invention which add no value to the organization.

5.2.1. Strategies of Innovation. A new product development is a process of generation and integration of knowledge and it can be identified by two principal sources of knowledge: the knowledge about technology and the availability of new technologies solutions and the knowledge about customers' needs. [8] identifies it as the Market pull strategy referred to design is called user-centered design. User-centered design approach states that product development should start from a deep analysis of user needs. A company can successfully innovate by asking users about their needs or by observing them as they use existing products and by tracking their behavior in consumption processes. In the market-pull approach the market and the consumer play a central role.

The technology-push approach argues that the source of innovation comes from the research and development activities of the company and the identification and development of new technologies based on the previous research work. The technology-push approach is based on the company and the development of new technologies that subsequently drives the innovation process.

The design driven innovation approach, the semantic dimension that guides the innovation process plays a more important role rather than the market or the technology. According to that, a third source of knowledge has to be added to the knowledge about user needs and the one about technological opportunities, such as *"the knowledge about the signs that can be used to deliver a message to the user and about the socio-cultural context in which the user will give meaning to those signs"* [14]. In a design-driven strategy the crucial aspect of innovation concerns the capability to understand, anticipate and influence the emergence of new product meanings. Verganti defines the radical design driven innovation as *"an innovation where novelty of message and design language is significant and prevalent compared to novelty of functionality and technology"*.

5.3 Theory of Radical Design-Driven Innovation. This innovation is led by Robert Verganti from Politecnico De Milano in Italy whom is the apostle of radical innovation of meaning. This is supported by [31] that socially recognized notions of identity as true products and become tokens of the symbolic exchange of meaning. This radical innovation focuses on exploring new market potential and boost brand image. Correspondingly, introduced a new interpretation of "Function versus Form" where product gives functionality and 'form' as giving new meanings for customers in terms of emotional and socio-cultural needs. This strategy aims at radically change the emotional and symbolic content of products (i.e their meanings and languages), through a deep understanding of broader changes in society, culture and technology. Rather than being pulled by user requirement, design driven innovation is pushed by a firm's vision about possible new product meanings and languages could diffuse in society. In this case, semantic is given more paramount importance in products rather than aesthetics.

5.4 Summary of Theoretical Framework

This conceptual paper integrates design innovation approach and business modeling approach which make it necessary to give a theoretical framework on the relevance of Radical Design Innovation. The model is appropriate for this inquiry because it allows the reader to understand how Radical Design Innovation is formed.

5.5 Overview of the Conceptual Framework

To address the main issues of the present work, different and various literature streams (studies on innovation, on design, on strategy and on business models) have been leveraged. What is interesting are the links between these different approaches. Two main "approaches" can be recognized:

(1) The design approach that embraces design, innovation and strategy.

(2) The business modeling approach that embraces strategy, business model and innovation.

The first approach was developed by [32]. The framework comprises of two components; which are Strategic Design Subsystem and Operational Subsystem.Strategic Design is described as identifying the recent trend in both the latest technology and lifestyle which leads to an intensive design discourse. This is followed by interpreting these sensing outcomes by finalizing the most suitable idea which aligns with vision of the company. Finally, the chosen idea will be addressed as a high level concept that consists of its main elements which are Product Vision, Design Statement, Future / Anticipated Socio-Cultural Context.

The second subsystem is Operational Subsystem, which comprises of developing the new product (NPD) that includes Technology Searching/Development, Product Language Design and Product Development. The second element is Communication, where they crucially design the ultimate strategy to communicate the new value embedded in the product to attract future customers.

The second approach was inspired from the product semantic framework of [33]. It shows a circular process of meaningful relations between strategists (entrepreneur and managers), business model and customers-stakeholders. Since "people understand the world through bracketing and chunking experience into meaningful units", the company's business model is seen as the experience of customers and stakeholders and meaning always refers to the business model perceived in its enacted context. In this sense, paraphrasing the Krippendorff "form follows meaning", we may affirm that experience follows meaning. The business model's meaning strategy becomes a mediating construct between technology and economic value [34] through its semantic value. To accomplish this task, a creative process is required based on experiences and the creation of meanings, due to a fluid intelligence able in imagining new languages possibilities [35] and proposed as in Figure 1;

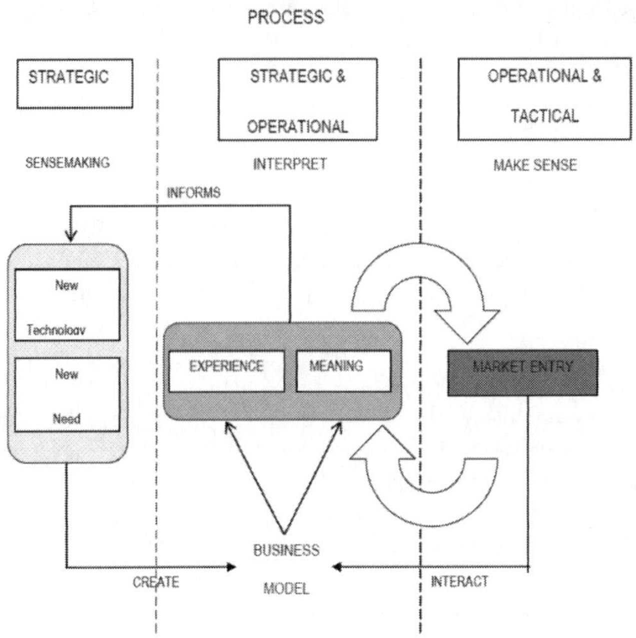

Figure 1. Conceptual framework of Radical Design Driven Innovation.

6. Conclusions

Business models constitute, in such a perspective, frames that managers and firms develop to organize not only "the way they make money" but also the way they convey meanings, beliefs, values and organizational culture. The sense making and making sense processes for meanings creation are traits of union concepts between the design approach and the business modeling approach. It is therefore possible to extend the design-driven innovation approach [33,8] to strategic business modeling and propose the concept of Meaning Strategy as a strategy aiming to make sense of an enterprise business model. The meaning strategy is a strategy that shapes the business model with the aim to convey a precise meaning. It is possible to act conveying new meanings in old building blocks or building new building blocks in order to convey an old or a new meaning. Therefore, in order to answer the research questions in this study the two approaches of; the design approach that embraces design, innovation and strategy and the business modeling approach that embraces strategy, business model and innovation will be abducted.

References

[1] Valipoor, S, & Ujang, B (2011).Challenges of Sustainable Design in Malaysian Furniture Industry. 2011 International Conference on Environment and Industrial Innovation (IPCBEE 2011), vol.12.

[2] Au Leck Chai (2012) New Quality To Mark Malaysian Furniture –Malaysia Pride: Seminar On Sabah Furniture Industry –A Potential Hub for Outdoor Furniture . Kota Kinabalu, Sabah.

[3] Academy of Sciences (2015). *Furniture Industry Sector*.(Academy of Sciences Report).Mega Science 3.0.

[4] Ratnasingam, J., Ramasamy, G., Ioras, F., Kaner, J., and Wenming, L. (2012). Production potential of rubberwood in Malaysia: Its economic challenges, Not Bot Horti Agrobo 40(2), 317-322.

[5] Abdullah, I. H. I. (2012).Malaysia Design Industry. (M. R. Malaysia, Ed.)Design Seminar and Workshop: *Malaysia-Korea Design Sharing: Design As A Strategic Tool for Innovation*. Seri Pacific Hotel, Kuala Lumpur, Malaysia.

[6] Dell 'Era, C., Marchesi, A., &Verganti, R. (2010).Mastering Technologies in Design-Driven Innovation.*Research Technology Institute*, 53(2), 12–23.
[7] Dell'Era, C. and Verganti, R. (2009), Design-driven laboratories: organization and strategy of laboratories specialized in the development of radical design-driven innovations. R&D Management, 39: 1–20. doi:10.1111/j.1467-9310.2008.00541.x.
[8] Verganti, R. (2008). Design, Meanings, And Radical Innovation: A meta model And A Research Agenda. *Journal of Product Innovation Management*, *25*(5), 436-456. doi: 10.1111/j.1540-5885.2008.00313.x
[9] Tether, B. S. (2005) 'Do Services Innovate (Differently)?: Insights from the European Innobarometer Survey', Industry and Innovation, 12.2, pp. 153-184.
[10] Maffei, S., (2010). The project is a clinical practice. The perspective of design. In: Fabbri, T.M. (Ed.), Organize: Concepts and Methods. Carocci, Rome, pp. 265
[11] Simon, H.A., 1969. The Sciences of the Articial. MIT Press, London.
[12] John Heskett (2001) Heskett, J.(2001). Past, Present, and Future in Design for Industry. *Design Issues*, 17(1), 18-26.
[13] Ulrich, K. T. (2011), Design Is Everything?.*Journal of Product Innovation Management*, *28*: 394–398. doi:10.1111/j.1540-5885.2011.00809.x
[14] Verganti, R. (2003), Design as brokering of languages: Innovation strategies in Italian firms. *Design Management Journal14*: 34–42. doi:10.1111/j.1948-7169.2003.tb00050.x
[15] Beltagui, A., Riedel, J., Livesey, F., Demian, P., Moultrie, J.(2008). Briefing note 1: What is design? *A review of definitions. Design Scoreboard - A novel approach to measuring the UK's National Design Performance*. 1-19.
[16] Love, T. (2002). Constructing a coherent body of theory about designing and designs: some philosophical issues. *Design Studies*, 23(3), 345-361.
[17] Bamford, G. (1990). Design, science, and conceptual analysis. In: *Architectural Science and Design in Harmony* (pp. 229-238). Joint ANZAScA/ADTRA Conference, Sydney, July 10–12, 1990.
[18] Galle, P. (2008). Candidate world views for design theory. *Design Studies*, 29(3), 267-303
[19] Poggenpohl, S., Chayutsahakij, P., &Jeamsinkul, C. (2004). Language definition and its role in developing a design discourse. *Design Studies*, 25(6), 579-605.
[20] Ralph & Wand, 2009). P. Ralph,& Y. Wand. (2009). A Proposal for a Formal Definition of the Design Concept. *Design Requirements Engineering: A Ten-Year Perspective, 14*, 103-136.
[21] Hertenstein, J.H., Platt, M.B., Veryzer, R.W. (2013). What is 'good design'? An investigation of the complexity and structure of design. Des. Manag. J. 8, 8
[22] Roy, R., Riedel, J.C., (1997). Design and innovation in successful product competition. Technovation 17, 537.
[23] Roper,S., Love, J.H and Vahter, P. (2012). The value of design strategies for new product development: Some econometric evidence. ISSN 0964-9328 – CSME Working Papers
[24] Tamyez, P.F, Nor, N.M., & Nasir, S. J.A (2014).*The Effects of Brand Orientation, Brand Distinctiveness and Design Innovation on the Brand Performance of the Malaysian Furniture Manufacturing Firms*. 1st AAGBS International Conference on Business Management 2014 (AiCoBM 2014), Penang, Malaysia, 6-8 April 2014.
[25] Hajah Robiyah Hj. Husin (2012) Furniture Industry In Malaysia - Status & Market Prospects. Seminar on Sabah Furniture Industry – A Potential Hub for Outdoor Furniture By: Malaysian Timber Industry Board . Kota Kinabalu, Sabah.
[26] Bajuri, M. T. bin. (1997). Industrial Design Scenario in Malaysia. *International Council Society of Industrial Designers Conference*. Toronto, Canada.
[27] Bajuri, M. T. bin. (2002). Speech of 2002 Malaysia Good Design Mark. Kuala Lumpur: Majlis Rekabentuk Malaysia (MRM).
[28] Jensen, R. (1999). The Dream Society: How the coming shift from information to imagination will transform your business (New York: McGraw-Hill).

IOP Conf. Series: Materials Science and Engineering **342** (2018) 012042 doi:10.1088/1757-899X/342/1/012042

[29] Kotsemir,M., & Abroskin, A, (2013). Innovation Concepts And Typology – An Evolutionary Discussion . Working Papers. Series: Science, Technology And Innovation Wp Brp 05/Sti/2013
[30] Vivekarajah, S (2013). *Blue Sky Innovation: Killer innovations that changed the world of business*..First Enteprise Kuala Lumpur, Malaysia.ISBN 978-967-11767-1-9
[31] Margolin, V., & Buchanan, R. (1996). The Idea of Design. Cambridge, Mass: The MIT Press.
[32] Kembaren, P., Simatupang, TM., Larso,D., & Wiyancoko, D (2014).Design Driven Innovation Practices in Design-preneur led Creative Industry. ***Journal of Technology Management & Innovation****, 9* (3) .
[33] Krippendorff, K., (1989). On the essential contexts of artifacts or on the proposition that "design is making sense (of things)". Des. Issues (5), 9
[34] Chesbrough, H. and Rosenbloom, R.S. (2002). 'The role of the business model in capturing value from innovation: Evidence from Xerox Corporation's technology spin-off companies', *Industrial and Corporate Change*, 11(3), 529- 555.
[35] Rullani, E. (2004) . The Economy of Knowledge. Creativity and Value in Network Capitalism. Rome: Carocci

Fabrication of hydrophobic compressed oil palm trunk surface by sol-gel process

Syafiqah Muzakir[1], Nurjannah Salim[1,*], Nurul Huda Abu Bakar[1], Rasidi Roslan[1], Lim Wan Sin[1] and Rokiah Hashim[2]

[1] Faculty of Industrial Sciences & Technology, Universiti Malaysia Pahang, Lebuhraya Tun Razak, 26300 Gambang, Kuantan, Pahang, Malaysia
[2] Division of Bioresource, Paper and Coatings Technology, School of Industrial Technology, Universiti Sains Malaysia, 11800 Minden, Penang, Malaysia

E-mail: njannah@ump.edu.my

Abstract. Improvement of the robustness of hydrophobic surfaces is crucial to achieving commercial applications of these surfaces in such various areas as self-cleaning, water repellency and corrosion resistance. Compressed oil palm trunk (OPT) panel is one of potential product which can be used as panelling and indoor furniture application. By adding hydrophobic properties to compressed oil palm trunk panel might increase the application of compressed oil palm trunk especially for outdoor application. In this study, fabrication is using the sol-gel technique. Sol-gel was prepared by adding ethanol with Hexadecyl Trimethyl Ammonium Bromide (CTAB) solution with Tetraethyl Orthosilicate (TEOS) with surface modification of chlorotrimethylsilane (CTMS). The surface with hydrophobic coating was undergone surface analysis with contact angle machine with the aid of software SCA 20 and the determined of the morphology of surface with scanning electron microscope (SEM). The produced compressed oil palm trunk surfaces exhibited promising hydrophobic properties with a contact angle of 104° and the relatively better mechanical robustness.

1. Introduction

Nowadays, in Malaysia due to the shortage of raw material in wood industry, the people working struggle to obtain the raw material at competitive price. OPT is one of the wood that available abundantly and less expensive raw material compared to wood [14]. The utilization of OPT could reduce the traditional wood's demand from forests. Malaysia could achieve a potential to convert these trunks into value-added product such as plywood, particleboard and medium density fiberboard [8]. Compressed wood is a type of treatment practiced aim to increase dimensional stability and strength. Wood compression techniques have been applied to solid wood, wood chip, and wood veneer that involved platen pressing [1,2,15]. Compression process was also studied as means to reduce glue consumption in panel products [1,3]. The concept of compressed wood could therefore be applied to OPT which has not been reported yet. As our wood industry having shortage supplies of timber, many kind of research have been done to find the alternative substitute to the timber sources [10]. However as stated by H'ng (2013), the compressed OPT itself has some limitation such as low in strength, less durability, low dimensional stability and high portions of thin-walled parenchyma tissue present in the trunk. One possible way to solve the problem of absorption water is to create highly water-repellent surface. This behavior exists in nature known as 'lotus effect' named after the leaves of lotus leaves which had excellent water repelling capacity [4].

The hydrophobic surface is measured by the contact angle which varies with the surface energy and roughness of the surface. Surface energy is usually depend on the surface chemistry, chemical composition and the atomic arrangement near or at the surface [12]. The surface could be said as superhydrophobic when the contact angle is greater than 150°, while it is considered as hydrophobic if the contact angle is greater than 90° and hydrophilic when it is less than 90°. [16]. Hydrophobic coating is widely been used due to its potential application in self-cleaning, anti-icing and transportation of liquids [16]. Based on some of research that has been reported there are various types of technique that developed for creating hydrophobic surface. These techniques can be chemical etching, chemical vapor deposition, sol-gel process, electrospinning, layer-by-layer (LBL) assembly and dip coating [13].

As described by Li (2016) sol-gel process was the process considered a convenient and economically strategy for the production of hydrophobic surface. The reason was due to sol-gel process is able to engineer surface structure and roughness easily. Hydrophobic surface was obtained by a combination of hydrolysis and condensation [4]. Typically the hydrophobic surface from sol-gel involving the process separately making the topography at single or dual scale from formation of silica nanoparticles, and followed by applying a low surface energy coating on surface [9]. Sol-gel was a process involving the transition a system from a liquid "sol" (mostly colloidal) into a semi-solid "gel" phase. Dip coating method is one of the technique for sol-gel process. Dip coating method is effectively for synthesis of transparent and uniform films. Dip coating method also usually used to prepare silica films [11].

This study aims to create a hydrophobic surface by sol-gel process. The step for preparation of hydrophobic surface are as follows; (i) fabrication of silica nanoparticles on compressed OPT surface by dip-coating technique and (ii) surface modification of silica nanoparticles with CTMS. The microstructure and chemical composition of hydrophobic surface were described by scanning electron microscope (SEM) and energy dispersive spectrometer (EDS) and the hydrophobic property of the modified compressed OPT was measured by contact angle (CA) measurements.

2. Experimental

2.1. Materials

OPT were obtained from local plantation in Northen Malaysia. Ethanol, hexadecyl trimethyl ammonium bromide (CTAB) tetraethyl orthosilicate (TEOS), chlorotrimethylsilane (CTMS) were purchased from BT Science Sdn. Bhd. All Chemical used were not undergo purification process.

2.2. Preparation of compressed OPT

OPT were cut into small pieces with a dimension of 20cm x 20cm x 4cm in tangential direction. The samples were wrapped in a plastic bag and kept in a freezer before further test to avoid contamination and fungal attack before further use.

The raw OPT will firstly be steamed in the close system using autoclave at a temperature of 130°C for 120min. Small amount of water is added into the base of the autoclave machine for steaming purpose. The solid OPT samples will then put in a steel basket and the steaming conditions will be set.

After steaming, samples will be put in the oven at 50°C temperature until dried. It will approximately took about 3 days to dry. The moisture content of the samples at this stage is about 15% to 20%. A slow drying will be needed to avoid the oil palm sample from warping and to retain as much as possible the properties of oil palm trunk. After drying process is completed, process will continued with making compressed OPT. Ten replicates of compressed OPT were produces.

The manufacturing process for making compressed OPT is based on the optimum condition obtained from the previous research. The compressed OPT will be pressed for 60 minutes at 200°C temperature with 11.04MPa hot press machine pressure. Before coating the surface of compressed OPT, they will be cleaned with ethanol to remove possible contamination. The coating procedure will be carried out using dip coating technique. All the compressed OPT lumbers made will be conditioned in conditioning room with a temperature of 21°C and a relative humidity of 65% before cutting into each size of test sample which is 5cm x 5cm x 10cm.

2.3. *Preparation of sol-gel*

Sol-gel was prepared by adding 50 ml of ethanol with 3 ml hexadecyl trimethyl ammonium bromide (CTAB) solution with vigorously stir with temperature 40 °C for 30 min. Next 25 ml tetraethyl orthosilicate (TEOS) was added dropwise continually stirred for 90 min. After that the sol-gel prepared was aging for 45 min at room temperature before coating on OPT.

2.4. *Fabrication of sol-gel and CTMS on compressed oil palm trunk*

Dip coating technique was used for the coating process the compressed OPT. The compressed OPT after dip coating was dried under room temperature for 5 min before dried in oven at temperature 110 °C for 30 minutes. The dip coating process will be repeated 5 times before surface modification with chlorotrimethylsilane (CTMS). After completed dip coating, the previous surface coating with sol-gel was modified with CTMS. CTMS which contain -CH3 group to form covalent bond with substrate surface. 5 ml of CTMS was used for the modifying process with temperature 40 °C.

2.5. *Characterizations*

The surface with hydrophobic coating was undergoes surface analysis with contact angle machine with the aid of software SCA 20 and the determined of the morphology of surface with scanning electron microscope (SEM).

3. Results and discussion

3.1. *Wettability of hydrophobic surface*

The wettability can be measured by getting the contact angle of water on the modified OPT surface. The compressed OPT is a kind of hydrophilic material when the reading of the contact angle of the water droplet is 21.25 degree. But after the compressed OPT surface modified by silica nanoparticles and CTMS, the surface can achieve contact angle of 111.25° and 103.9° at 0 and 60 sec respectively as shown in Table 1.

By this silica nanoparticles and CTMS modification, the surface properties of the compressed OPT has been improved by increasing of the surface roughness and the surface free energy. Compressed OPT is expected to be rough a result packing of monosized silica as shown in Fig. 1. The solution of CTMS contains –CH3 groups that form covalent chemical bonds with the substrate surface, although precursor containing this group is easily adsorbed onto any hydroxylated solid surfaces through covalent bonds, hydrogen bonds and van der Waals forces. Fig. 2 presents the schematic of surface chemistry after self-assembly with CTMS.

Generally, the wettability of smooth surface can be described by Young equation:

$$\cos\theta = \frac{\sigma_{sv} - \sigma_{sl}}{\sigma_{lv}} \qquad (1)$$

Wenzel equation was applied to the contact angle θw of liquid on rough homogenous surface by modified Eq. (1), where σ_{sl}, σ_{lv} and σ_{sv} are the interfacial tensions of the solid-liquid, liquid-vapor phases and solid-vapor, respectively.

$$\cos\theta w = r\left(\frac{\sigma_{sv} - \sigma_{sl}}{\sigma_{lv}}\right) = r\cos\theta \qquad (2)$$

IOP Conf. Series: Materials Science and Engineering **342** (2018) 012043 doi:10.1088/1757-899X/342/1/012043

Where r is the roughness factor. It is defined as the ratio of the actual surface area of the rough surface to the geometric projected area, and it is always larger than 1. The Cassie equation was employed to apply to low energy surfaces and heterogeneous roughness which exhibit superhydrophobic for further understanding of the wettability of the hydrophobic wood surface,

$$\cos \theta c = f\,(\cos\theta + 1) - 1 \qquad (3)$$

The fraction of air pocket that contacting with liquid at the superhydrophobic surface can be calculated by applying the Cassie's equation where f is the fraction of the solid surface contacting with liquid, the fraction of air in contact with contacting with liquid at the surface is $1\text{-}f$, θ, and θ_c represent the water contact angle on rough and smooth surfaces respectively. Here, the value of water contact angle θ_c on the hydrophobic compressed OPT surface is 103.9°, and according to a report by Shang *et al* (2004), the water contact angle θ on the smooth surface modified by CTMS is 85°. Therefore, by referring to Cassie equation, the fraction of air in contact with liquid at the surface is calculated as 0.69.

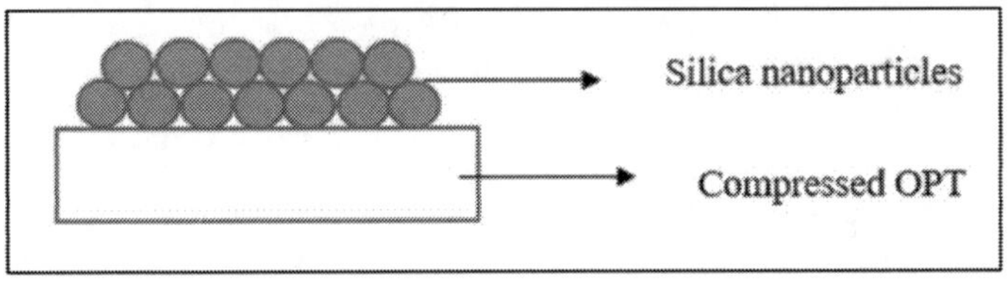

Figure 1. Schematic of silica nanoparticles deposited on compressed OPT surface.

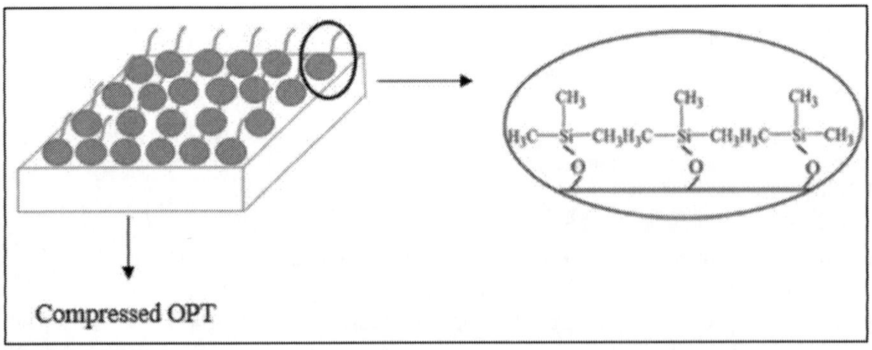

Figure 2. Schematic of compressed OPT surface after modified with CTMS.

IOP Conf. Series: Materials Science and Engineering **342** (2018) 012043 doi:10.1088/1757-899X/342/1/012043

Table 1. A table of images and contact angle of OPT before and after modification.

Sample	CA left (°)	CA right (°)	Average (°)
OPT before modification 	19.8	22.7	21.25
OPT after modification at 0 sec 	111.9	110.6	111.25
OPT after modification at 60 sec 	104.5	103.3	103.9

3.2. The chemical composition and morphology of the hydrophobic surface

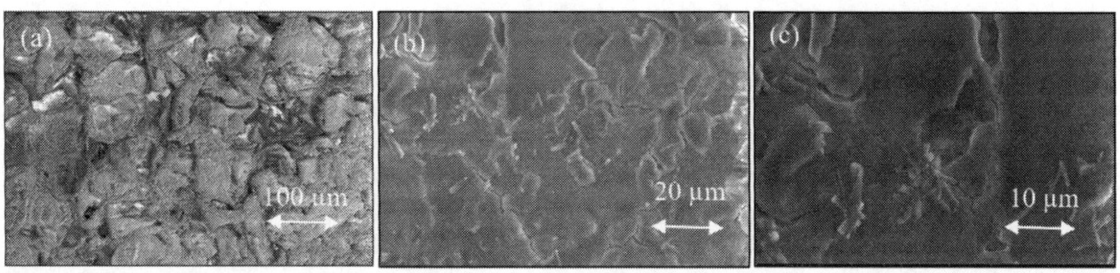

Figure 3. SEM images of OPT surface (a) before modification and the one modified with silica nanoparticles and CTMS at (b) low and (c) high magnification.

The morphology of the hydrophobic surface is characterized by scanning electron microscopy (SEM). Figure 3 shows the scanning electron microscopy images of OPT surface before and after modified with silica nanoparticles and CTMS at low and high magnification. Figure 3a shows that compressed OPT is a heterogeneous and porous material. Figure 3b shows that the silica nanoparticles deposited on the OPT cell walls and filled into the pits of the OPT surface. The silica nanoparticles are adhered onto wood surface through the chemical bond between hydroxyl groups of OPT surface and silica nanoparticles. Figure 3c demonstrate that many of sphere particles randomly distribute and stack over the wood surface with diameters ranging between 300 to 500 nm, and their aggregation between each sphere is not apparent. The wood surface is roughening intensely due to the arrangement of spheres that creates a number of cavity or interspaces between each silica sphere. The treated sample cannot be wet by water droplet and presents the property of hydrophobicity since the combination of the two key factors which is the surface roughness and a layer of low surface energy of CTMS are applied. Thus, a larger amount of air trapped into the cavities or interspaces of the OPT surface, and the water droplet primarily contacts with the trapping air.

The crucial information regarding the chemical composition of hydrophobic compressed OPT surface are observed by Electron dispersive X-ray spectrometer (EDS). Figure 4 represents the EDS spectrum of compressed OPT surface. It is clearly shows that the elements such as C, O and Si are found in the EDS spectrum after silica nanoparticles and CTMS deposited on the compressed OPT surface. The Si element is derived from silica nanoparticles and the C element is derived from CTMS. Thus, it can be confirmed that the silica nanoparticles and CTMS are successfully deposited which contribute to surface free energy and surface roughness.

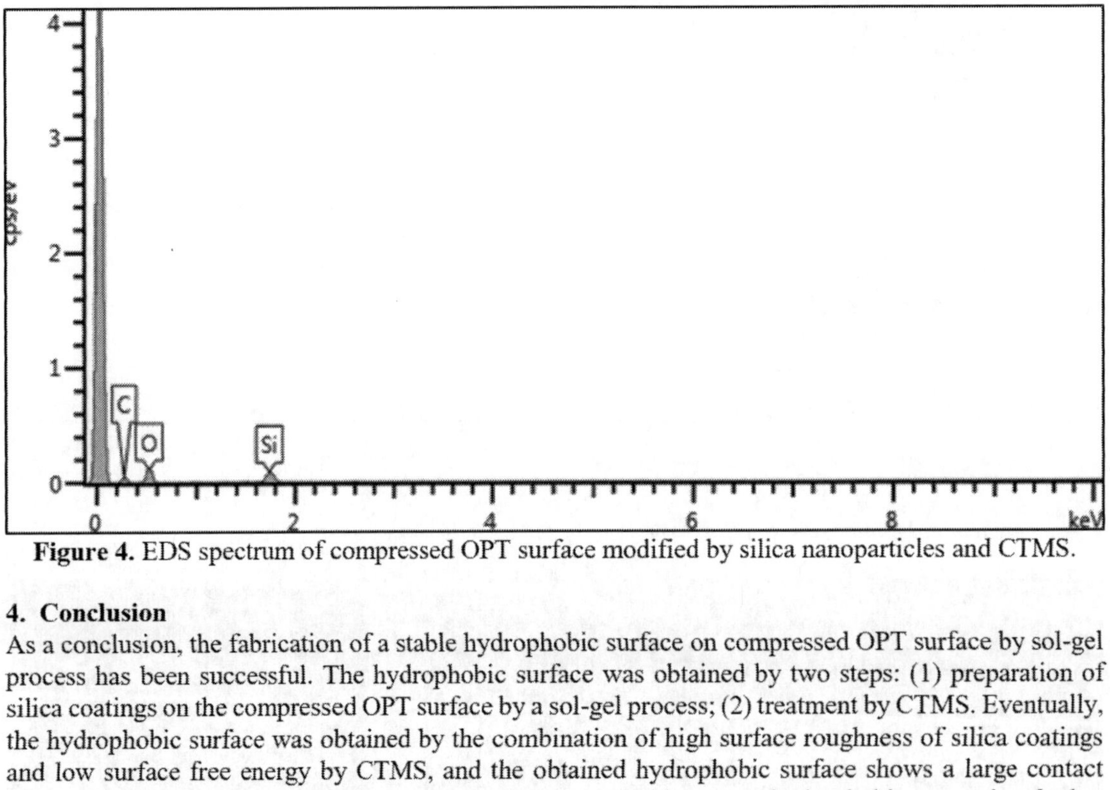

Figure 4. EDS spectrum of compressed OPT surface modified by silica nanoparticles and CTMS.

4. Conclusion

As a conclusion, the fabrication of a stable hydrophobic surface on compressed OPT surface by sol-gel process has been successful. The hydrophobic surface was obtained by two steps: (1) preparation of silica coatings on the compressed OPT surface by a sol-gel process; (2) treatment by CTMS. Eventually, the hydrophobic surface was obtained by the combination of high surface roughness of silica coatings and low surface free energy by CTMS, and the obtained hydrophobic surface shows a large contact angle of 103.9°. However, since the contact angle does not achieve superhydrophobic properties, further study should be implemented for a better results.

Acknowledgement

This research was supported by Ministry of Higher Education (MOHE), Malaysia given through the research grants RDU 150106 and UIC 170710. Special thanks are also due to the staffs and members of the Faculty of Science and Technology Industries (FIST) laboratory, University of Pahang for their support and relentless help provided throughout the period of research.

References

[1] Adachi, K., Inoue, M., Kannyama, K., Rowell, R.M., Kawai, S. 2004. Water removal of wet veneer by roller pressing. J. Wood Science **50** 479-83.

[2] Bekhta, P., Hiziroghlu, S. and Shepelyuk, O. 2009. Properties of plywood manufactured from compressed veneer as building material. Materials and Design **30** 947-953.

[3] Bekhta, P.A.and Marutzky, R. 2007. Reduction of glue consumption in the plywood production by using previously compressed veneer. Holz als Roh-und werkstoff **65**(1):87-8.

[4] Bengtsson, E. (2013). Creating super hydrophobic surfaces for moisture protection of biobased composites: Citeseer

[5] H.M., Shang, Y., Wang, S.J., Limmer, T.P., Chou, K., Takahashi, G.Z., & Cao. 2004. Optically transparent superhydrophobic silica-based films. Thin Solid Films **472**(2005): 37-43

[6] H'ng, P., Chai, L., Chin, K., Tay, P., Eng, H., Wong, S., Wong, W., Chow, M., & Chai, E. (2013). Urea formaldehyde impregnated oil palm trunk as the core layer for three-layered board. Materials & Design, **50** 457-462.

[7] Inoue, M., Norimoto, M., Tanahashi, M., and Rowell, R. M. 1993. Steam or heat fixation of compressed wood. Wood Fiber Science **25**(3): 224-235.

[8] Jumhuri, N., Hashim, R., Sulaiman, O., Nadhari, W. N. A. W., Salleh, K. M., Khalid, I., Saharudin, N. I., & Razali, M. Z. (2014). Effect of treated particles on the properties of particleboard made from oil palm trunk. Materials & Design, **64** 769-774.

[9] Li, Q., Yan, Y., Yu, M., Song, B., Shi, S., & Gong, Y. (2016). Synthesis of polymeric fluorinated sol–gel precursor for fabrication of superhydrophobic coating. Applied Surface Science, **367** 101-108.

[10] Nordin, N. A., Sulaiman, O., Hashim, R., Salim, N., Sato, M., & Hiziroglu, S. 2013. Properties of laminated panels made from compressed oil palm trunk.Composites Part B: Engineering, **52** 100-105.

[11] Rao, A. V., Latthe, S. S., Nadargi, D. Y., Hirashima, H., & Ganesan, V. (2009). Preparation of MTMS based transparent superhydrophobic silica films by sol–gel method. Journal of colloid and interface science, **332**(2), 484-490

[12] Shang, H., Wang, Y., Takahashi, K., Cao, G., Li, D., & Xia, Y. (2005). Nanostructured superhydrophobic surfaces. Journal of materials science, **40**(13), 3587-3591.

[13] Shang, Q., & Zhou, Y. (2016). Fabrication of transparent superhydrophobic porous silica coating for self-cleaning and anti-fogging. Ceramics International, **42**(7), 8706-8712.

[14] Sulaiman, O., Salim, N., Nordin, N. A., Hashim, R., Ibrahim, M., & Sato, M. (2012). The potential of oil palm trunk biomass as an alternative source for compressed wood. BioResources, **7**(2), 2688-2706.

[15] Unsal, O., Kartal, S.N., Cendan, Z., Avango, R.A., Clausen, C.A., III, F.G. 2009. Decay and termite resistance, water absorption and swelling of thermally compressed wood panels. I. Biodeterioration and Biodegradation **63** 548-552.

[16] Wang, N., Xiong, D., Pan, S., Deng, Y., Shi, Y., & Wang, K. (2016). Superhydrophobic paper with superior stability against deformations and humidity. Applied Surface Science, **389** 354-360.

The effect of power intensity properties of microwave modified oil palm trunk lumber

Anis Izzati Ibrahim[1], Nurjannah Salim[1,*], Rasidi Roslan[1], Mohammad Ashry Jusoh[1] and Rokiah Hashim[2]

[1] Faculty of Industrial Sciences & Technology, Universiti Malaysia Pahang, Lebuhraya Tun Razak, 26300 Gambang, Kuantan, Pahang, Malaysia
[2] Division of Bioresource, Paper and Coatings Technology, School of Industrial Technology, Universiti Sains Malaysia, 11800 Minden, Penang, Malaysia

E-mail: njannah@ump.edu.my

Abstract. In the decade, oil palm (*Elaeis guineensis*) in Malaysia is one of the conventional sources that will be rising, and the rate of biomass will considerably increase in yet to come. Presently, oil palm biomass is going through research and development and appears to be the most sustainable alternative. Investigations on oil palm biomass have been conducted to support in draw out waste of oil palm and in the meantime can help economic yield to the country. This study was expected to estimate the effect of power intensity properties of microwave modified oil palm trunk lumber. Microwave treatment of oil palm trunk samples was set of connections by using a microwave operating at 2.45 GHz with the liberated process input power intensity (600-1000W) were studied under the given condition. Impact and compression of the samples were tested. The analysis of properties of the fresh material and dry samples was employed by scanning electron microscopy. Oven drying technique also was involved as a comparison of the conventional drying process in this research. Based on the outcomes of this study, both drying methods improved the characteristics of the specimens.

1. Introduction

In timber-based manufacturing, the unavailability of wood as a raw material has recently become a significant apprehension. Many plants that produce timber-based products, especially plywood and lumber, already shut down. Wood-based industries are now fronting a difficult with the supply of raw materials, not only from natural forests but also rubber plantations. Year after year, it is expected that at least 20 million tonnes of solid wood are used in timber-based manufacturing. Currently, supplies of wood are becoming restricted, and oil palm trunk could be potential substitutes source to replace wood [1]. Oil palm trunk (OPT) is especially promising, as it can be used as a value-added product as well as in future wood-based industries [2]. However, fresh oil palm trunk contains high humidity that could reach as high as 500% based oven-dry weight, and when degraded it could pollute the surrounding environment [3]. Before oil palm trunks could be used as raw materials for manufacture products, the oil palm trunk was usually condensed to smaller sizes in chips form and dried to moisture content around 7-10% [4]. Due to high moisture content, the oil palm trunk is very liable to fungal attack and will worsen quickly. Thus, desiccating the oil palm trunk to adequate moisture content should be done as soon as possible.

IOP Conf. Series: Materials Science and Engineering **342** (2018) 012044 doi:10.1088/1757-899X/342/1/012044

Oil palm is monocotyledonous species that does not have cambium, growth ring and ring cells in contrast to tropical woody plants. However, it consists of mostly parenchymatous ground materials and vascular bundles [5]. Vascular bundle and parenchyma tissues in oil palms are anatomically different from wood species. Oil palm trunks have some unique characteristics, including being accomplished of holding high moisture content (1.5 to 2.5 times the weight of the dry material), low cellulose and lignin content, and high contents of water-soluble [6]. However, oil palm trunk has very low natural durability due to the existence of sugar and starch in oil palm parenchyma tissues. Due to very high sugar and starch contents, oil palm trunk is very susceptible to the attack by genetic agents such as termites and fungi. Ramle et al. [6] determined that parenchyma cells have high capacity in absorbing water of water absorption as compared to the vascular bundle that could be connected to their anatomical structure [7].

The properties and feature of the product produced are also profoundly influenced by the moisture content. The primary objective of drying is to diminish the moisture to a certain level at which on further drying, no remarkable change in volume and weight of the material will take place. Also, the low moisture content will allow the materials to be kept at a longer time without any fungal attack [3]. Drying of biomass like oil palm trunk is a complex process of heat and mass transfer. The mass loss during drying mostly following the falling rate curves. The water loss from the bulk of the fibrous material is controlled by the capillary action and internal diffusion of water molecules. This loss of water caused the shrinkage of the biomass [8]. Kiln drying process, oven drying process, and air drying process are the conventional drying methods used in industry. These dryers cause drying defects such as warping, raise grain, collapse with low recovery and require the considerable amount of energy and long drying times to obtain high-quality woods, even when using laboratory [1, 9]. Microwave technique is believed more efficient, fast, and also can reduce the drying times and at the same time can increase the production. Thus, it is considered that microwave drying process can reduce the moisture content with fewer defects to the trunk itself [10].

Strong prominence has currently been done to invent and to lessen the cost of drying [11]. Microwave processing technologies are developing rapidly in all industries [12]. Microwave drying would be new-fangled in the wood industry, but it has been used in our daily life including food processing and telecommunication. The application of microwave technology has been proved to improve drying time and could increase the drying rate compared with the conventional drying methods [13, 14]. Researchers indicate that microwave drying with proper selection of power input, the weight of drying material and drying time could increase the drying rate. As a result, it could save up to 50% of energy and significantly decrease volatile organic compound emissions when compared with the conventional drying methods. It can be concluded that the microwave energy consumption is relatively higher than oven drying based on watt/h. Considering much lower time will make the total energy consumed to be significantly lower than conventional methods. Thus, this study aimed to optimize the drying conditions using a microwave to enhance a complete drying, avoid burning, shrinkage, and swelling and increase the permeability of oil palm trunk so it could be improved to a value-added material [13].

2. Material and methods

Oil palm trunk samples were harvested from a local plantation in Northern Malaysia. The trunk were sawn into round disks 10 cm in diameter before they was reduced to approximate dimensions of 16 cm x 6 cm x 4 cm with the utility of laboratory-type band saw. Then, the oil palm samples were cut into specific size for each method and testing. The moisture content and weight of oil palm trunk sample were taken during experimental work.

Table 1. Amount of sample and techniques used in experimental work.

Type of method	Amount of sample
Microwave drying	3 : (600 W, 800 W, 1000 W) for 10 min
Oven drying	1 : (105 °C, 24 hours)
Fresh oil palm trunk	1 : Fresh from freezer to avoid fungal attack

2.1. Oven drying process
The sample with preferred weight and initial moisture of 50% was oven dried (Model Memmert Universal Oven UF55) at 105°C to compare the effect of drying with microwave heat source. The drying was varied for 24 hours as shown in table 1.

2.2. Microwave radiation process
There were three different microwave treatment combinations of power intensity with one fresh sample as a control. All of the oil palm trunk blocks were standardize to the size of 165 mm x 65 mm x 40 mm as that is the most excellent size for the block to be placed inside the microwave space. The three combinations of microwave power and exposure time were stated in table 1.

2.3. Evaluation of mechanical properties of the samples
The samples were cut from the trunk to its specific size for impact testing and compressive strength testing. Mechanical features in the form of impact testing of the samples were determined for each condition of power intensity of microwave drying as well as oven and fresh sample. The impact strength of the oil palm samples has been identified by an Izod impact tester instrument. Minimum of 5 samples with dimensions of 60 mm x 13 mm x 5 mm have been used for the mechanical tests, and the value has been presented in this paper. Ten samples with measurement 6 cm x 2cm x 2cm from each panel were used for compression test using 5KN Celtron compression testing. The moisture content and weight of each sample were stately before they were experiencing microwave and oven drying process.

2.4. Scanning Electron Microscopy (SEM) analysis
The cross-section of oil palm trunk samples with dimensions of 1 cm x 1 cm was taken from each sample for testing. The samples were observed for microscopic analysis using a scanning electron microscope Quanta 450.

3. Result and discussion

3.1. Moisture content measurement
The moisture content of oil palm trunk samples before and after drying process was summarized in table 2. This indicates that each of the microwave treatment combinations and oven drying process can significantly reduce the moisture content.

Table 2. The moisture content measurement of drying process treatment.

Type of sample	Moisture content after (%)
Oven drying process	0.61
Microwave drying (600 W)	31.18
Microwave drying (800 W)	11.78
Microwave drying (1000 W)	6.88

From table 2, oven drying process has the highest reduction of the moisture content of oil palm trunk sample. From 50% to 0.6% of moisture content, oven drying process could be the right method for drying industry. Nevertheless, microwave drying reveals more effectiveness and improvement of oil palm trunk sample. In this study, microwave drying in 600-watt shows the highest amount of reduction in moisture content percentage of moisture reduction of approximately 31.18% while microwave treatment of 800-watt power produced the better rate of moisture reduction compare to 600-watt. However, microwave drying at 1000-watt power for 10 minutes with a percentage of moisture reduction approximately 6.88% probably is the better combination of power intensity and exposure time to oil palm trunk sample in drying trade. This is because when microwave treatment of 1000-watt took place, the sample is dried out consistently and shrinkage did not occur. It was observed that there is no shrinkage occurs in the samples. It is also noted that to obtain nearly 30-40% of moisture content in dried oil palm trunk samples, only 10 min microwave drying time is required. Hence, microwave drying was found to be very useful for the removal of moisture in short time.

During microwave treatment, water is drawn from oil palm trunk cells permitting the material to dry. The microwave heating process took about 10 minutes to complete. This is much quicker than other ways and means which may take several weeks, such as air drying, oven drying, and kiln drying. The dissimilarity of moisture content inside the main part of the samples results in the variety of microwave absorbed energy. Microwave drying consists typically of three different periods: warm up, water evaporation and heating up. Water vaporized faster during drying process when the moisture content is very high. During the heating-up period, some of the microwave energy was used to heat up the strands. The temperature of strands would sometimes be so high that the strand surfaces would burn. Accordingly, many commercial wood drying operations impose an extended period of air drying to reduce the occurrence of drying defects before main drying stage [15, 16]. During microwave aeration, the warm-up period was shorter, and the boiling point was reached faster, which might increase the drying speed. In microwave drying, the interactions of water molecules inside wood generate heat. When the drying temperature is close to the boiling point, it will cause burning of the sample due to surface dehydration. However, the microwave-dried sample was utterly acceptable in behavior and appearance of the sample [17].

3.2. Impact strength

Figure 1 shows the impact strength of five oil palm samples in term of the H5 farming hammer. There were some types of hammer in which H1 to H5. The differences of all farming hammer due to their weight and momentum of striker strikes sliding, hitting the test material in the middle, at the bottom of its swing.

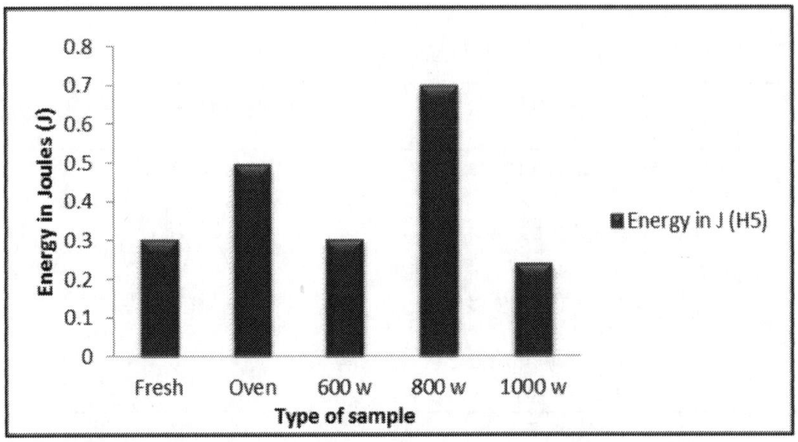

Figure 1. Impact strength of 5 samples of oil palm in Joules (J).

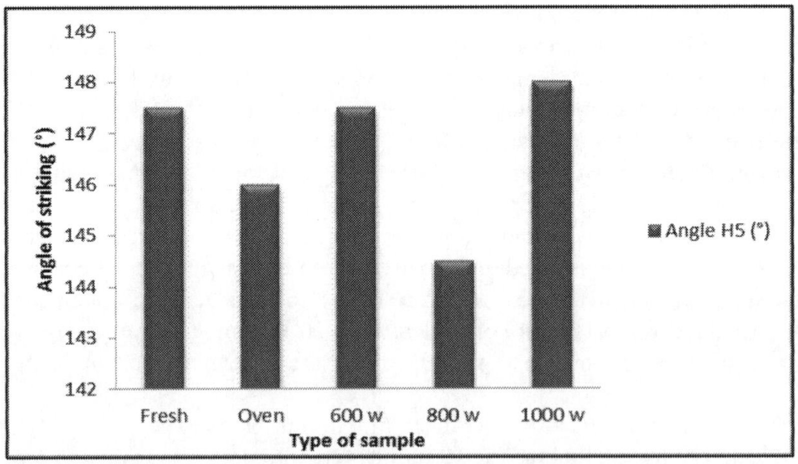

Figure 2. Angle of striking of oil palm samples (°).

From the graph, it demonstrated the amount of energy engaged by microwave 800W material during breakage was the highest among all the samples. The absorbed energy was a measure of a given material's toughness and the more upper resulting numbers, the more robust the material. Among of three samples of microwave radiation process, the sample of 1000 W was the lowest compared to 600 W and 800 W. This is because higher input power intensity probably shrunk the cell structure of oil palm trunk and also reduced the strong point of the sample. From previous research, it was predicted that 1000 W would create the highest moisture content in the samples as more top power intensity input results in the temperature increase at even higher moisture content. However, in this study, it was verified that exposure time also influenced the moisture reduction [1]. For oven drying process and fresh sample, it showed a massive difference. The energy of the fresh sample was the lowest energy because the fresh sample had more moisture content and influenced the toughness of the sample. Meanwhile, for oven drying sample, the moisture content was utterly the lowest because the sample was dried entirely at 105°C for 24 hours. To compare between an oven drying and fresh sample, ultimately the energy of oven sample more higher than the fresh sample. However, the amount of energy of 800 W microwave drying was more significant than oven sample. So, the conclusion showed that the microwave drying significantly more efficient and save power consumption as well as the drying process do not take more time.

Figure 2 display the angle of striking of Izod impact test in which the only one to determine the imperfections in material likely to fracture by "shock." Once the analysis was complete, the hammer was automatically seized and reverted to its starting point. This allows the starting angle to be changed according to the test, which enables optimum test parameters, such as impact speed and striking angle, to measure. From Figure 2, 800 W microwave drying technique and oven drying have less angle striking. The angle of hitting of the hammer for Izod test = 90°. The test specimen absorbed energy when it breaks. So the hammer does not rose high during the test if the material absorbed more energy. From the graph of angle striking of five types of samples, it can eventually say, the higher the energy consumed by an element when fracture, the lower the angle striking of the test specimen. However, there was some imperfection data due to the sample has high angle hitting but less energy absorbed during its break. A material's toughness was a dynamic in its capability to absorb energy during distortion. The impact rate of an element can also change with temperature. Generally, at lower temperatures, the impact energy of the material was decreased. The size of the sample may also affect the value of the Izod impact test because it may consent a diverse number of failures in the material, which can act as stress risers and lower the impact energy.

3.3. Compressive strength

Figure 3 shows the compressive strength of microwave drying, oven drying and the fresh sample of oil palm trunk. In the place that can be seen from the graph, there was a comparison of the compressive strength of parallel to grain and perpendicular to the grain oil palm trunk sample.

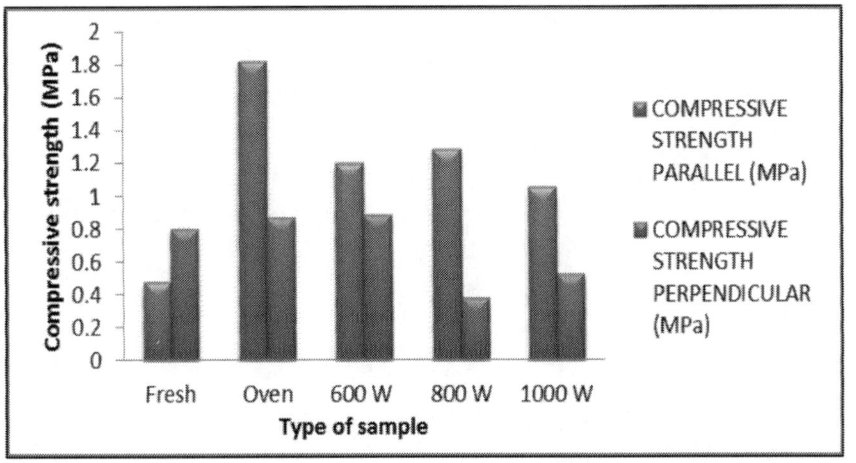

Figure 3. The compressive strength of parallel and perpendicular to the oil palm grain.

The amount of compression strength which undergo treatment variants 600 W 800 W and 1000 W parallel and perpendicular of grain were 1.2 MPa, 1.28 MPa, 1.05 MPa and 0.88 MPa, 0.38 MPa, 0.52 MPa respectively. Oven drying sample has an amount of 1.82 MPa for parallel grain and 0.87 MPa for perpendicular grain. Furthermore, the fresh sample displayed 0.48 MPa for parallel grain and 0.80 MPa for perpendicular grain. The compressive strength of parallel grain at every drying treatment was completely different in comparison to the perpendicular of grain included the fresh sample. Based on the graph, the parallel grain of drying samples has higher strength compared to the perpendicular grain. The perpendicular to grain compressive strong point of timber is known to be much lower than the strength parallel to the grain. Many timber structures, however, rely on this property especially in bearings that frequently occur in building practice. The linear elastic-plastic behavior of structural timber loaded perpendicular to grain has been a problematic issue for decades which is reflected in the differences between the prediction models in structural design codes over the world [18]. Other than that, the compressive strength of parallel grain oven drying shows the most upper power. This is because the oven was drying and completely removed the moisture content sample to 1%. However, the oven drying process takes a lot of time and power consuming throughout the process.

For the compressive strength of parallel grain, there was a less significant difference between each treatment of microwave treatments for 600 W, 800 W and 1000 W. For 800 W of microwave technique clearly has excellent strength compared to 600 W and 1000 W. This could be the concerns of the higher drying rate in microwave treatment compared to the conventional method. According to [14], microwave drying process usually undergoes three distinct stages; a warm-up step where the temperatures are rising, evaporation stage where the temperature is constant or plateau while the moisture evaporates, and heating-up stage which the water is done evaporate but increase in the temperature of the specimen body. The same stages were also occurrence in convective drying. However, in convective drying the warm-up stage and to achieve the plateau temperature (boiling point of water) take a longer time. Microwave treatment enables shortening both of these stages resulting in a better drying rate. With better drying rate, the material becomes less hygroscopic as stated by [19]. Lignocellulosic material that treated with high temperature was less absorption and desorption rate as compared to untreated. Less hygroscopic materials contain less moisture bound in their fibers which improves the mechanical properties of the samples [20].

3.4. Scanning electron microscopy (SEM)

The effects of drying techniques of oil palm trunk samples were perceived by SEM analysis. Figure 4 displays the SEM analysis of (a) fresh oil palm trunk, (b) oven drying and (c) (d) (e) microwave radiation process.

IOP Conf. Series: Materials Science and Engineering **342** (2018) 012044 doi:10.1088/1757-899X/342/1/012044

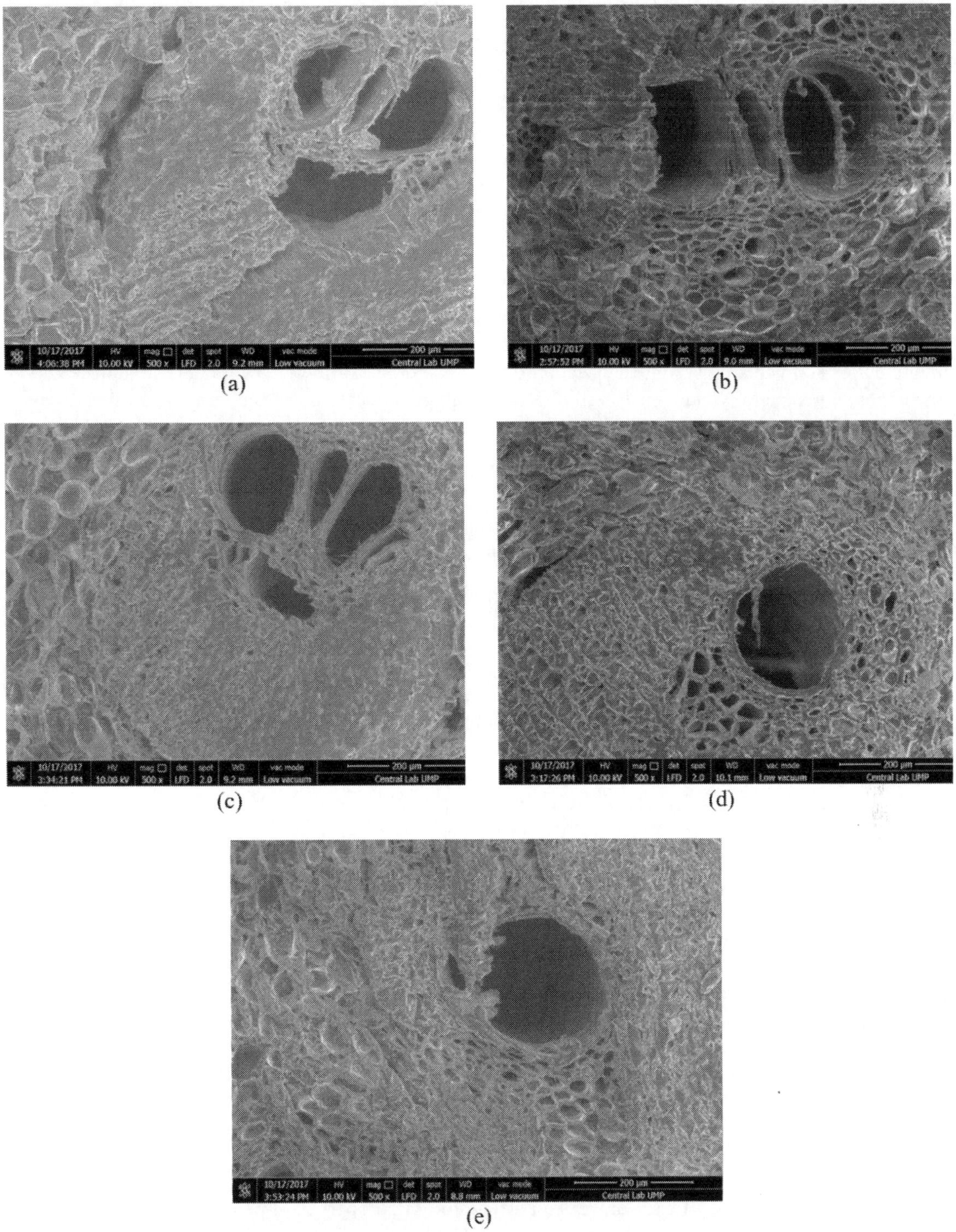

Figure 4. SEM micrographs of oil palm trunk vascular bundle (x500) (a) fresh sample (b) oven drying process (c) 600 W microwave drying process (d) 800 W microwave drying process (e) 1000 W microwave drying process.

The logs are composed of vascular bundle and parenchymatous tissue which act as transporting and food storing organs, respectively [21]. From the figure, the vascular bundle of fresh sample smaller than oven drying sample. Regarding for microwave drying technique, it showed the vascular bunch of 1000 W and 800 W bigger than the vascular bundle of 600 W. The shape was formed probably due to the result of temperature and power intensity of drying process respectively. Hence, the higher the temperature and heat transfer, the bigger the vascular bundle of oil palm trunk sample. The vascular bundle was found more in the outer zone of the trunks compared to the central zones, the fiber becomes shorter and having a smaller diameter and thicker cell walls..

4. Conclusion

The properties of oil palm trunk wood were improved with the treatment of microwave radiation process. However, oven drying sample also showed improvement after treatment. But microwave drying technique was more efficient and can reduce the cost of drying. The sample of microwave treatment of 800 watts power intensity with 10 minutes duration showed the highest strength value in the series of treatment which consists of different wattage power. Hence, the sample could be used as a reference to enhance the strength properties of oil palm trunk wood.

Acknowledgments

The authors would like to express warm appreciation to the Ministry of Higher Education (MOHE), Malaysia for the support given through the research grants RDU 150106 and UIC 170710. Special thanks are also to the staffs and members of the Faculty of Science and Technology Industries (FIST) laboratory, the University of Pahang for their support and relentless help provided throughout the research.

References

[1] Rahman N A A, Nordin K, Saptri M F, Jamaludin M A, 2013 The Effect of Microwave Treatment on Physical Properties of Oil Palm Trunk (*Elaeis guineensis*), in: Business Engineering and Industrial Applications Colloquium (BEIAC) 2013, pp. 339-343

[2] Khalil H A, Fazita M N, Bhat A, Jawaid M, Fuad N N, 2010 Development and material properties of new hybrid plywood from oil palm biomass, *Materials & Design*, **31**(1) 417-424

[3] Nadhari W N A W, Hashim R, Sulaiman O, Jumhuri N, 2014 Drying kinetics of oil palm trunk waste in control atmosphere and open air convection drying, *International Journal of Heat and Mass Transfer*, **68** 14-20

[4] Hashim R, Nadhari W N A W, Sulaiman O, Sato M, Hiziroglu S, Kawamura F, Sugimoto T, Seng T G, Tanaka R, 2012 Properties of binderless particleboard panels manufactured from oil palm biomass, *BioResources*, **7**(1) 1352-1365

[5] Baskaran M, Hashim R, Sudesh K, Sulaiman O, Hiziroglu S, Arai T, Kosugi A, 2013 Influence of steam treatment on the properties of particleboard made from oil palm trunk with addition of polyhydroxyalkanoates, *Industrial crops and products*, **51** 334-341

[6] Ramle S F M, Sulaiman O, Hashim R, Arai T, Kosugi A, Murata Y, Mori Y, 2012 Characterization of parenchyma and vascular bundle of oil palm trunk as function of storage time, *Lignocellulose Journal*, **1**(1)

[7] Jumhuri N, Hashim R, Sulaiman O, Nadhari W N A W, Salleh K M, Khalid I, Saharudin N I, Razali M Z, 2014 Effect of treated particles on the properties of particleboard made from oil palm trunk, *Materials & Design*, **64** 769-774

[8] Leskovšek U, Medved S, 2011 Heat and moisture transfer in fibrous thermal insulation with tight boundaries and a dynamical boundary temperature, *International Journal of Heat and Mass Transfer*, **54**(19) 4333-4340

[9] Ratanawilai T, Nuntadusit C, Promtong N, 2015 Drying Characteristics Of Rubberwood by Impinging Hot-Air and Microwave Heating, *Wood Research*, **60**(1) 59-70

[10] Sulaiman O, Salim N, Nordin N A, Hashim R, Ibrahim M, Sato M, 2012 The potential of oil palm trunk biomass as an alternative source for compressed wood, *BioResources*, **7**(2) 2688-2706

[11] Vermaas H, 1995 Drying eucalypts for quality: Material characteristics, pre-drying treatments, drying methods, schedules and optimisation of drying quality, *South African Forestry Journal*, **174**(1) 41-49

[12] Torgovnikov G, Vinden P, 2012 Microwave wood modification technology and its applications, *Forest Products Journal*, **60**(2) 173-182

[13] Amouzgar P, Khalil H A, Salamatinia B, Abdullah A Z, Issam A, 2010 Optimization of bioresource material from oil palm trunk core drying using microwave radiation; a response surface methodology application, *Bioresource technology*, **101**(21) 8396-8401

[14] Wang S, Du G, Zhang Y, 2005 Microwave wood strand drying: energy consumption, VOC emission and drying quality, in: *Proceeding of the 3rd Inter-American Drying conference*, Montreal, Canada, 10pp

[15] Shulman H S, 2003 Microwaves in high-temperature processes, *Industrial heating*, **70**(3) 59-61

[16] Nair K M, Thomas S, Groeninckx G, 2001 Thermal and dynamic mechanical analysis of polystyrene composites reinforced with short sisal fibres, *Composites Science and Technology*, **61**(16) 2519-2529

[17] Khalil H A, Amouzgar P, Jawaid M, Abdullah C, Issam A, Zainudin E, Paridah M, Hassan A, 2013 Physical and thermal properties of microwave-dried wood lumber impregnated with phenol formaldehyde resin, *Journal of Composite Materials*, **47**(28) 3565-3571

[18] Leijten A, 2016 The bearing strength capacity perpendicular to grain of norway spruce–Evaluation of three structural timber design models, *Construction and Building Materials*, **105** 528-535

[19] Taib M, Azman M N, Jamaludin M A, Abu Kassim M, 2012 Flexural properties of bio-composite made from hydrothermally treated kenaf (Hibiscus cannabinus) fibres, in: Advanced Materials Research, *Trans Tech Publ*, pp. 2409-2412

[20] Saptri M F, Nordin K, Rahman N A A, Jamaludin M A, Ahmad M, 2013 Strength performance of outer layer oil palm trunk wood treated with microwave energy, in: Business Engineering and Industrial Applications Colloquium (BEIAC) 2013, pp. 695-697

[21] Lamaming J, Hashim R, Leh C P, Sulaiman O, Sugimoto T, Nasir M, 2015 Isolation and characterization of cellulose nanocrystals from parenchyma and vascular bundle of oil palm trunk (*Elaeis guineensis*), *Carbohydrate polymers*, **134** 534-540

IOP Conf. Series: Materials Science and Engineering **342** (2018) 012045 doi:10.1088/1757-899X/342/1/012045

Electrical study on Carboxymethyl Cellulose-Polyvinyl alcohol based bio-polymer blend electrolytes

M A Saadiah[1,2] and A S Samsudin[2]

[1] Department of Chemistry, Centre for Foundation Studies, International Islamic University Malaysia, 26300 Gambang, Pahang
[2] Advanced Material Group, Faculty of Industrial Sciences & Technology, Universiti Malaysia Pahang, Lebuhraya Tun Razak, 26300, Kuantan, Pahang

E-mail: ahmadsalihin@ump.edu.my

Abstract. The present work deals with the formulation of bio-materials namely carboxymethyl cellulose (CMC) and polyvinyl alcohol (PVA) for bio-polymer blend electrolytes (BBEs) system which was successfully carried out with different ratio of polymer blend. The biopolymer blend was prepared via economical & classical technique that is solution casting technique and was characterized by using impedance spectroscopy (EIS). The ionic conductivity was achieved to optimum value 9.12 x 10^{-6} S/cm at room temperature for sample containing ratio 80:20 of CMC:PVA. The highest conducting sample was found to obey the Arrhenius behaviour with a function of temperature. The electrical properties were analyzed using complex permittivity ε^* and complex electrical modulus M^* for BBEs system and it shows the non-Debye characteristics where no single relaxation time has observed.

1. Introduction

Bio-polymer blend electrolytes (BBEs) have drawn great attention due to their prospective applications in electrochemical devices for instance rechargeable batteries, fuel cells, super capacitors, and solar cells [1-2]. Various types of polymers from natural and synthetic sources have been studied extensively to pursue the development of BBEs system. Among the natural polymer, chitosan [3], carrageenan [4], polyactides [5], and carboxymethyl cellulose [6] are favorable in the preparation of polymer electrolytes. Whereas from synthetic sources, polyethylene oxide (PEO) [7], polyvinyledene fluoride (PVdF) [8], polyacrylonitrile (PAN) and polyvinyl alcohol (PVA) [9] are also suitable polymers as they offer relatively higher conductivity. Carboxymethyl cellulose is a cellulose derivative from the regenerated cellulose $[C_6H_{10}O_5]n$ with backbone consist of D-glucose residues linked by β-1,4-linkage [10]. CMC polymer binder is certain to provide multipoint interaction in building stable solid biopolymers (SBEs). Owing to its characteristic, CMC is widely used as filming, emulsification, suspension, water maintaining, bind and inspissation [11]. Another potential candidate is the polyvinyl alcohol (PVA) which is claimed to give advantage in the application of composite films due to its hydrophilicity properties. PVA has the structure where its carbon chain backbone has attached hydroxyl group (-OH) to methane carbon, making it to be the source of intermolecular attraction forces which is significant to form bio-polymer blends electrolytes (BBEs) [12].

Blending is a technique that involves mixing of two or more different polymers without the formation of covalent bonding [13]. The ionic conductivity of the polymer blend can be enhanced compared to single polymer [14]. The blending study has been focusing on intermolecular interaction in governing the ionic conductivity [15]. The incorporation of CMC into carrageenan as reported by Rudzhiah [16] has proved that blending can increase the ionic conductivity via the formation of H-bonding. On top of that, CMC has ability to form hydrogen bonds due to the existence of characteristic functional group carboxylate anion (-COO$^-$) and hydroxyl (–OH). Therefore, it is expected for the BBEs of CMC and PVA to be the platform with tailored individual properties in numerous applications. The blending method is believed to improve the ionic conductivity properties by changing the structural behavior [17].

Ionic conductivity is the most crucial property of polymer electrolyte, thus, researchers have shown an increasing interest to discover the interactions between the two polymers and to achieve the desired performance in different applications. A study on the dielectric properties of CMC [18] and PVA [19-20] has been reported previously. In the case of CMC, the electric structure is unipolar thus having similar dielectric properties like other polyelectrolytes. PVA is well known to have chain flexibility attributed to the intra- or inter-molecular attraction which is focusing at the hydroxyl (-OH) group. The advantage offered by PVA can accommodate the segmental motion and rotation to the relaxation process during formation of H-bonding between the two polymers.

Most of the BBEs system offers low ionic conductivity at room temperature. Knowledge about electrical properties of the polymer electrolyte is beneficial in order to study the interaction during blending which is expected to happen in this present work. The present work is an attempt to study the electrical properties carboxymethyl cellulose (CMC)-polyvinyl alcohol (PVA) as the bio-polymer blends electrolytes (BBEs) which were prepared by using solution casting method at various ratio compositions. The BBEs system was analyzed via the electrical impedance spectroscopy (EIS) for the evaluation of electrolyte composition that exhibits the highest electrical conductivity at ambient temperature. The increase in chain flexibility of the CMC/PVA based BBEs is the key factor for the enhancement of the ionic conductivity.

2. Materials and methods

2.1. Preparation of sample

Polymers studied in the present work were procured from Acros Organic Co. for CMC and Merck Co. for the PVA. The molecular weight of CMC and PVA of approximately 90000 and 70000 were used in this work. In the preparation of BBEs solution, total weighed amounts of granules PVA and powder CMC, 2.0 g was dissolved directly in distilled water. The dissolution was done at temperature 55 °C under constant stirring. The CMC/PVA solutions were prepared at different weight percentage 100/0, 90/10/ 80/20, 70/30, 60/40, 50/50 and 40/60 (%/%). The notation of sample compositions is tabulated in Table 1. These solutions were casting into glass petri dishes and put in the oven for about 5 hours at 60 °C. The process was accompanied by further drying in the desiccator to ensure all solvent has been evaporated.

Table 1. List of designation of CMC/PVA with their composition respectively.

Sample Designation	CMC/PVA blends (%)
C1	100/0
C2	90/10
C3	80/20
C4	70/30
C5	60/40
C6	50/50
C7	40/60

2.2. Characterization

2.2.1. Impedance analysis. The dielectric properties of the bio-polymer blend CMC/PVA electrolyte film with different ratios were tested by using HIOKI 3532-50 LCR Hi-Tester over the frequency range from 50 Hz to 1 MHz at room temperature. The BBEs sample was cut into small disks with are of 2 cm diameter and sandwiched between two stainless steel electrodes of the sample holder. The bulk resistance (R_b) value was obtained from the plot of imaginary impedance (Z_i) versus real impedance (Z_r) and the ionic conductivity, σ was determined using equation (1) below:

$$\sigma = \frac{t}{R_b A} \qquad (1)$$

where t is a thickness of the sample and A (cm^2) the cross-sectional area of the BBEs film. The thickness of the bio-polymer electrolytes was found to be in the range of 0.0093 to 0.0914 cm using a digital thickness gauge (DML3032)

2.2.2. Complex permittivity (ε)*. Complex permittivity is important to explain the mechanism and interactions of ions in the BBEs system [15]. The complex permittivity can be defined as:

$$\varepsilon^* = \varepsilon_r + j\varepsilon_i \qquad (2)$$

where ε_r is the dielectric constant (real part of complex permittivity) and ε_i is the dielectric loss (imaginary of complex permittivity), can be calculated using equation as follows:

$$\varepsilon_r = \frac{Z_i}{\omega C_o(Z_r^2+Z_i^2)} \qquad (3)$$

$$\varepsilon_i = \frac{Z_r}{\omega C_o(Z_r^2+Z_i^2)} \qquad (4)$$

2.2.3. Complex modulus(M)*. The complex modulus is needed to highlight the suppression of electrode polarization and bulk dielectric behaviour of the BBEs system. The real modulus, M_r and imaginary modulus, M_i was calculated using the following equation:

$$M_r = \frac{\varepsilon_r}{(\varepsilon_r^2 + \varepsilon_i^2)} \qquad (5)$$

$$M_i = \frac{\varepsilon_i}{(\varepsilon_r^2 + \varepsilon_i^2)} \qquad (6)$$

3. Results and discussion

3.1. Impedance studies

Figure 1 depicts the Cole-Cole plot of CMC/PVA based BBEs films at room temperature. The Cole-Cole plot of BBEs containing 0, 10, 20, 30 and 40 percent composition of PVA showed by Fig 1(a-e) shows an incomplete semicircle curve. Two regions could be observed as semicircle at high frequency and spike at low frequency. These impedance plots are frequently used to separate the electrical polarization that affects the selected impedance frequency and the impedance of bulk material as reported earlier [21].

IOP Conf. Series: Materials Science and Engineering **342** (2018) 012045 doi:10.1088/1757-899X/342/1/012045

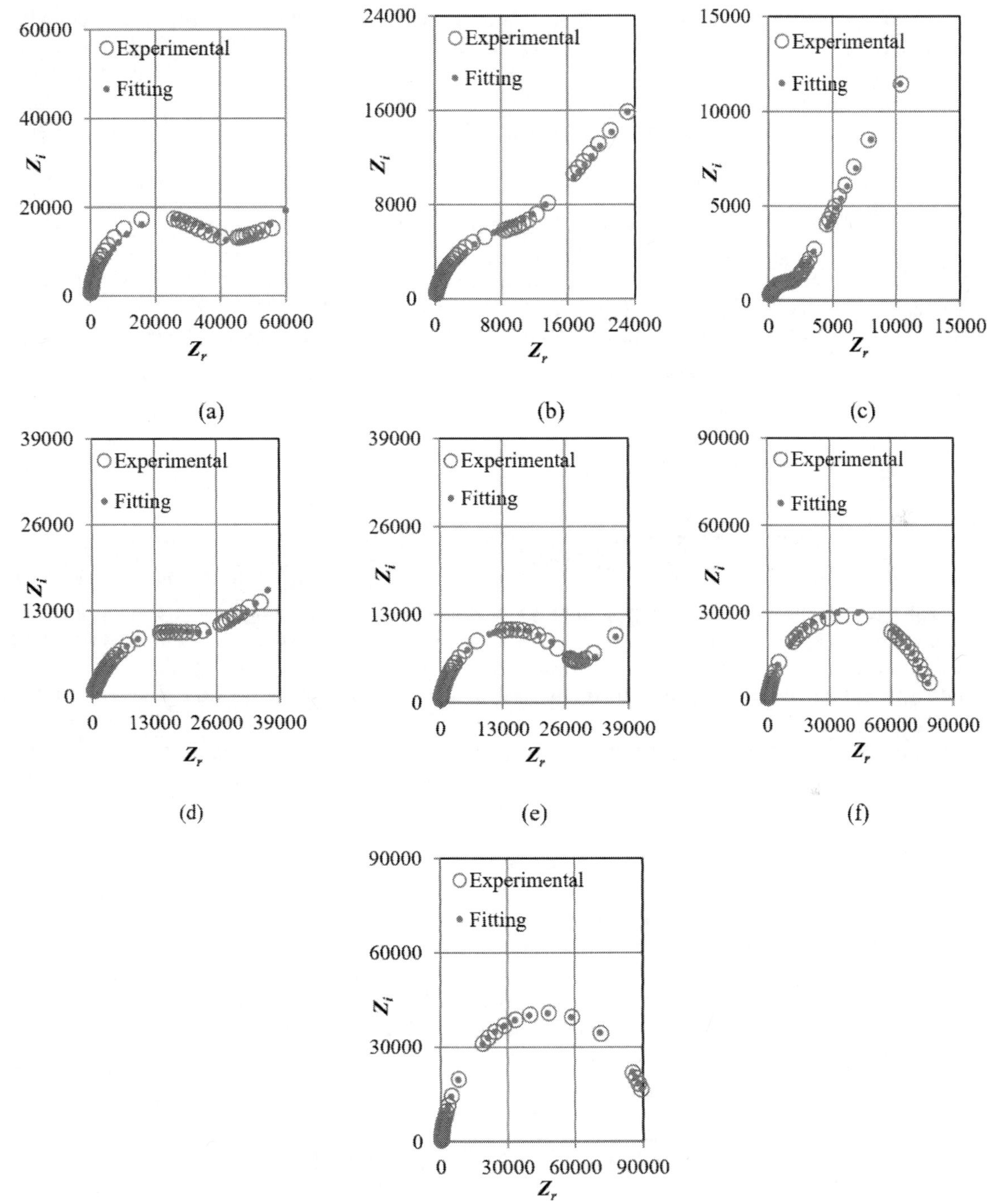

Figure 1. Cole-Cole plot of CMC/PVA based bio-polymer blend electrolytes system for composition (a) 100:0 (b) 90:10 (c) 80:20 (d) 70:30 (e) 60-40 (f) 50:50 (g) 40:60.

In this work, the impedance data for the shows semicircle and spike that can be represented by a parallel combination of bulk resistance (R_b) and constant phase element (CPE). The R_b was determined based on the interception between higher and lower frequency from the Cole-Cole plot of complex impedance [22-23]. CPE was utilized in a model to replace the capacitor to balance the heterogeneity of the BBEs system studied in this work. The impedance of CPE (Z_{CPE}) can be represented by the following equation:

$$Z_{CPE} = 1/k(j\omega)^p \text{ where } 0 \leq p \leq 1 \quad (7)$$

Or

$$Z_{CPE} = k\,[\cos(p\pi/2) - j\sin(p\pi/2)]\,/\omega^p \quad (8)$$

Variable of k^{-1} corresponds to the capacitance value of the CPE element, ω is the angular frequency where, $\omega = 2\pi f$ (f is frequency), and p is referred to the deviation of the vertical axis in the Z_r versus Z_i plot. The values of Z_r and Z_i associated to the equivalent circuit can be termed as follows:

$$Z_r = \frac{R_b + R_b^2 k_1^{-1}\omega^{P_1}\cos(\frac{\pi p_1}{2})}{1+2R_b k_1^{-1}\omega^p\cos\left(\frac{\pi p_1}{2}\right)+R_b^2 k_1^{-2}\omega^{2p_1}} + \frac{\cos\pi p_2/2)}{k_2^{-1}\omega p_2} \quad (9)$$

$$Z_i = \frac{R_b^2 k_1^{-1}\omega^{P_1}\sin(\frac{\pi p_1}{2})}{1+2R_b k_1^{-1}\omega p_1\cos\left(\frac{\pi p_1}{2}\right)+R_b^2 k_1^{-2}\omega^{2p_1}} + \frac{\sin\pi p_2/2)}{k_2^{-1}\omega p_2} \quad (10)$$

Table 2 lists the parameter of the circuit elements for all BBEs composition studied at room temperature. The formation of semicircle could be explained by the occurrence of a capacitor and a resistor, which is in parallel series. Meanwhile, an inclined spike was due to the effect of electrode polarization, which is the characteristic of ions diffusion process for the bio-polymer blends electrolyte system [24]. This diffusion happens among ions when PVA was incorporated into CMC and causes the insulating properties start to change into semi-conducting material. As shown in Figure 1 (c-f), when PVA was incorporated into the CMC the tilted spike has inclined at an angle lesser than 90° along the real axis which might be attributed to heterogeneity between the electrode-biopolymer electrolyte interface and resulting to a decrease in ionic conductivity [25]. These result support previous research conducted by [26], which has explained on the electrical properties of CMC, doped with NH_4Br.

Figure 1 (f) and (g) shows the only semicircle curve after the addition of more than 50 percent composition of PVA into the BBEs system. This is directly related to the bulk resistance and bulk capacitance of the BBEs system [27]. The part of semicircle shown in the Cole-Cole plot can be demonstrated as a parallel resistor attributed to the movable ions inside the polymer matrix and capacitor due to the motionless polymer chain circuit network. The addition of more PVA in C5 until C7 will destruct the dielectric polarization and lead to greater hindrance to the polymer resulting in an increase in R_b and decreasing the ionic conductivity.

Table 2. List of parameters of circuit elements for all BBEs system composition at room temperature.

Sample	p_1 (rad)	C_1(F)	p_2 (rad)	C_2 (F)
C1	0.850	5.5 x10^8	0.430	5.58 x10^5
C2	0.775	4.56 x10^8	0.495	9.90 x10^5
C3	0.745	5.82 x10^7	0.585	1.44 x10^6
C4	0.749	9.60 x10^7	0.499	8.35 x10^5
C5	0.810	1.40 x10^8	0.472	2.05 x10^5
C6	0.839	2.60x10^8	-	-
C7	0.901	8.85 x10^8	-	-

3.2. Conductivity studies

There were numerous efforts adopted for the development of the bio-polymer blend electrolytes in order to enhance the ionic conductivity. CMC is the most potent candidates as host polymer due to low cost, eco-friendly and can offer better chemical and physical properties [22, 25]. Despites the advantages, CMC has limitation as the conductance property is relatively low. Thus, one of the alternatives to increase the conductivity of CMC is the incorporation of high dielectric permittivity dipolar polymer such as PVA. As shown in Figure 2, the ionic conductivity of CMC/PVA based BBEs has increased until the composition 80:20 which suggests the optimum amount of PVA. The electron donating group, carboxylate anion of CMC backbone mainly coordinates with the PVA and produces the ion-dipolar complexes via intra- or inter-chain hopping between the functional group the two polymers during the blending process. This interaction in such blend matrix was able to increase the flexibility of polymer chain via process of forming or breaking the ion-dipolar bonds that exist between the CMC/PVA which served as the host polymer. Consequently, the conductivity of the CMC/PVA can achieve the optimum value, which is 9.12 x 10^{-6} S/cm due to higher flexibility of polymer chain that can fasten the segmental dynamic of the BBEs system studied in this present work.

The result obtained similar to chitosan/PVA based bio-polymer blend electrolyte conducted by Buraidah [12] which managed to obtain conductivity of 1.77 x 10^{-6} S/cm. For sample containing more than 20% of PVA the optimum concentration has exceeded because at this concentration, ion association is more significant than ion dissociation and disables the interaction between CMC and PVA [28]. This will decrease the number of mobile ions within the bulk of the BBEs and therefore decreasing the conductivity. The plot of log conductivity versus reciprocal temperature is shown in figure 3. The correlation between conductivity and temperature for the highest conducting sample, C3 obeys the Arrhenius law with the regression value, R^2 approaching to 1 indicating the BBEs system was thermally assisted [29].

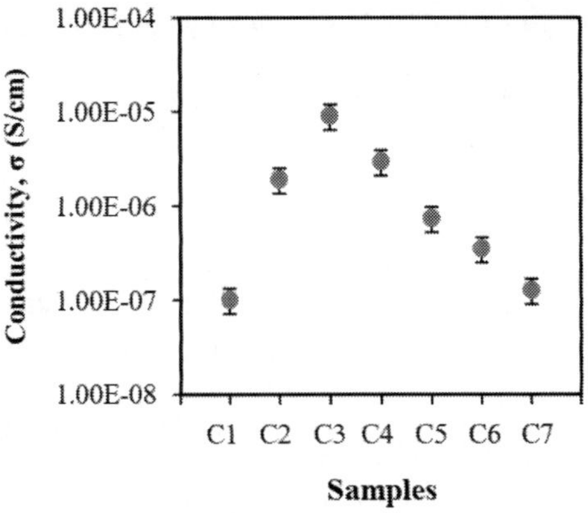

Figure 2. The ionic conductivity of CMC/PVA BBEs system at room temperature.

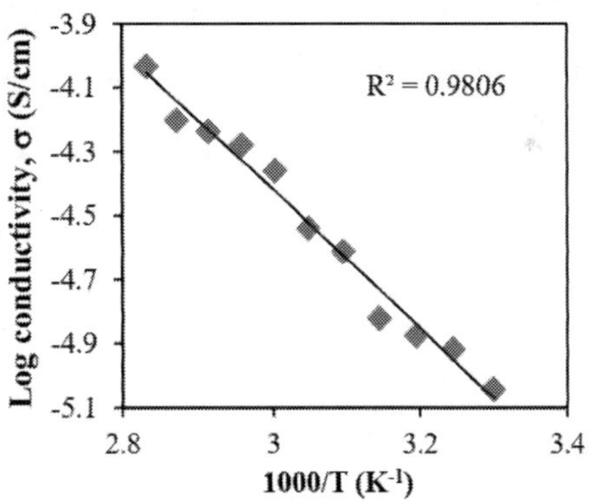

Figure 3. Temperature dependence for the highest conductivity CMC/PVA (80:20 compositions).

3.3. Electrical studies

The dielectric properties may vary in different types of polymer electrolyte system attributed to several factors including frequency of applied electrical field, temperature, structural characteristics or other external factors. Figure 4 (a) and (b) shows the frequency dependence of the real part of the dielectric constant (ε_r) for all composition of CMC/PVA bio-polymer blend electrolytes system which is also known as stored charge in any material. In the studied range of frequency of 50Hz to 1 MHz, it can be observed ε_r and ε_i increases sharply from low frequency contributed by electrode, molecular and ionic polarization. The dielectric constant, ε_r was found to increase for the sample C3 which indicated that charge carrier concentration has increased in the space charge accumulation area and later increasing the conductivity [30]. Sample C3 shows the highest value of ε_r may be due to an increase in the heterogeneity in the polymer blend which could increase the free volume. At high frequency, both dielectric properties were showed decreasing trend explained the decreasing rearrangement of dipolar groups and polar segments of the polymer which is no longer capable to react to the applied electric field hence causes electrical relaxation process to happen [31].

Modulus formalism is important to analyze the ionic conductivity or in further analysis as the loss peak could not be detected in the dielectric loss spectrum. It is a useful application to analyze the electrical response and relaxation of polymer electrolyte system. The real (M_r) and imaginary (M_i) part of complex modulus are derived from their complex permittivity and shown in figure 5. In figure 5 (a), the M_r values were increased proportionally with the increase of frequency. According to Ramesh [32], the non-zero values of M_r at low frequency suggest that the BBEs system was negligible from electrode polarization. Plot of imaginary modulus (M_i) able to evaluate the effect of the smallest capacitance and the largest resistance, hence determine the relaxation process attributed to short or long-range movement of charge carrier as reported by Sinclair and West [33]. The occurrences of relaxation peaks in the M_i plots in figure 5(b) reveals the characteristic of the BBEs system as ionic conductors [29]. The peak position and height of C1 sample at high frequency can be distinguished from other samples in both M_i and M_r spectrum may be due to the bulk effect as CMC is exceptionally stiff which in turn resulting to a decrease in chain mobility. This peaking curve also attributed to the relaxation phenomenon in the polymer electrolyte system.

Compared with CMC, the CMC/PVA based BBEs contribute much more to the M_r of the blend, which becomes broadening and lowering down which means that stiffness of CMC has decreased as PVA was added [34]. At lower frequency, there was a large association between the values of capacitance with electrodes which can be seen by a plateau peak for M_r and this present work result was in agreement with work done by Samsudin [28]. Thus, this confirmed that the behaviour of BBEs system work in this study is non-Debye.

Figure 6 illustrates the variation of real modulus (M_r) and imaginary modulus (M_i) parts of electrical modulus for the highest conductivity sample containing 80:20 compositions of CMC/PVA. Both figure 6(a) and (b) show long tails at lower frequency indicating that no relaxation peaks at that region. Figure 6(a) shows an increasing trend at higher frequency with no definitive peaks. In addition, the highest peak frequency of the highest conducting BBEs sample, C3, presented in Figure 6(b) which was corresponding to the shortest relaxation time [29]. Accordingly, the relaxation time decreases as temperature increases. At higher temperature the ion mobility of the polymer chain has increased and hence increasing the conductivity. This result also supports the work done by Pradhan [35].

Dielectric loss tangent (tan δ) spectrum of CMC/PVA in Figure 7 depicts the dielectric relaxation process which is derived from the ratio $\varepsilon_i/\varepsilon_r$. Sample C1 until C5 exhibits a relaxation peak at high frequency corresponding to the conductivity relaxation process, whereas sample C6 and C7 decreased sharply from lower to higher frequency. The dielectric and conductivity relaxation was evaluated by the reorientation process of dipolar molecules in the polymer which is represented by a peak in the tangent dielectric loss spectrum. The highest conducting sample, S3 exhibits the shortest relaxation time at higher frequency due to the increasing number of segmental motion [36]. Meanwhile slower relaxation process was indicated by higher PVA composition more than 50 percent composition which gives reflection to a lower conductivity. Figure 8 presented the plot of tan δ versus log frequency. The trend was clearly observed where tan δ shifts to higher frequency and the height of the peak also increased when temperature increased. Based on the plot of tan δ, the relaxation time, τ can be determined by using the following equation:

$$\tau = 1/\omega_{peak} \tag{11}$$

where ω_{peak} is the angular frequency of the relaxations peak. The temperature dependence of relaxation time for C3 BBEs is shown by figure 9 with regression value (R^2) of 0.9764. The plot shows good linearity and confirms the conductivity of temperature dependence of the present work BBEs system follows the Arrhenius law. The effect of temperature also suggests that there was no phase transition in the BBEs containing 80:20 composition of CMC/PVA and this result was in line with previous study on bio-polymer blend electrolytes of starch and methylcellulose [37].

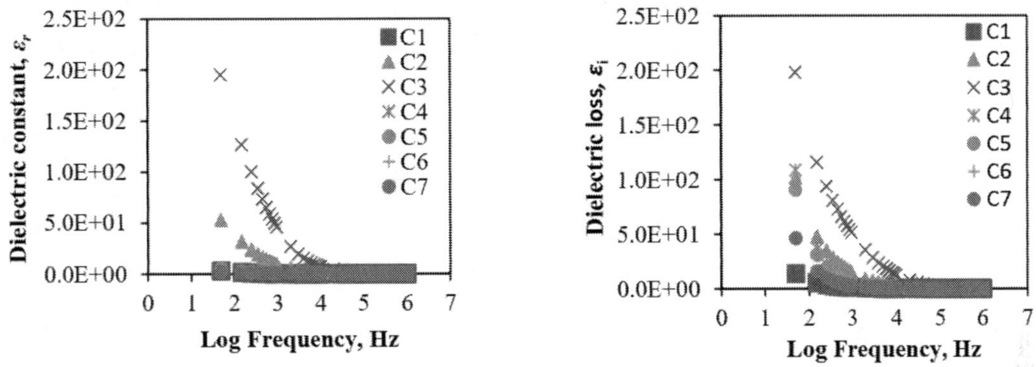

(a) (b)

Figure 4. Frequency dependence of (a) dielectric constant (b) dielectric loss for various composition of CMC/PVA at room temperature, 303 K.

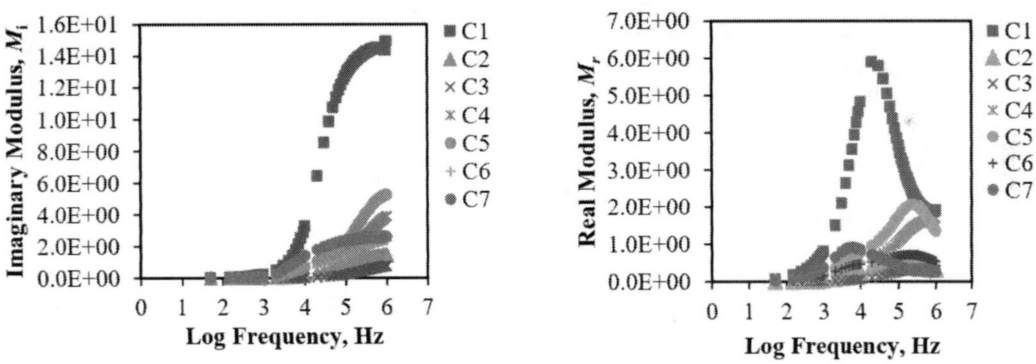

(a) (b)

Figure 5. Frequency dependence of (a) real modulus (b) imaginary modulus for various composition of CMC/PVA at room temperature, 303 K.

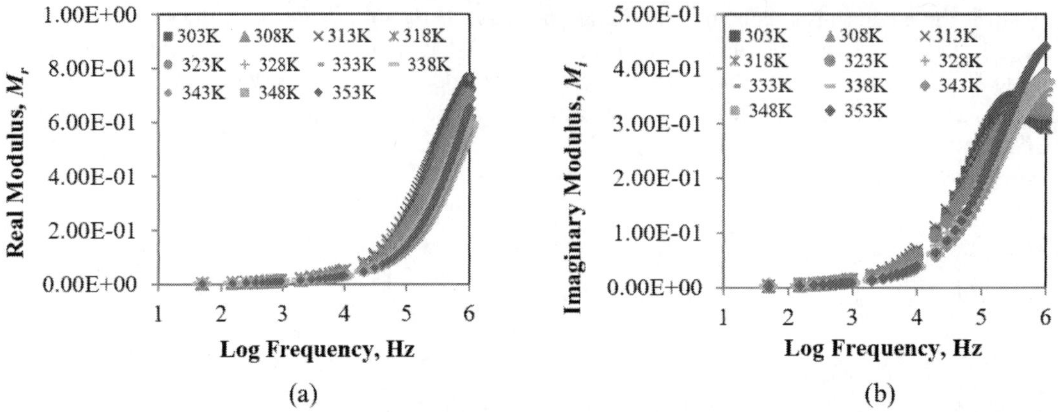

Figure 6. Frequency dependence of (a) real modulus (b) imaginary modulus for the highest conductivity CMC/PVA (80:20 compositions) at different temperature.

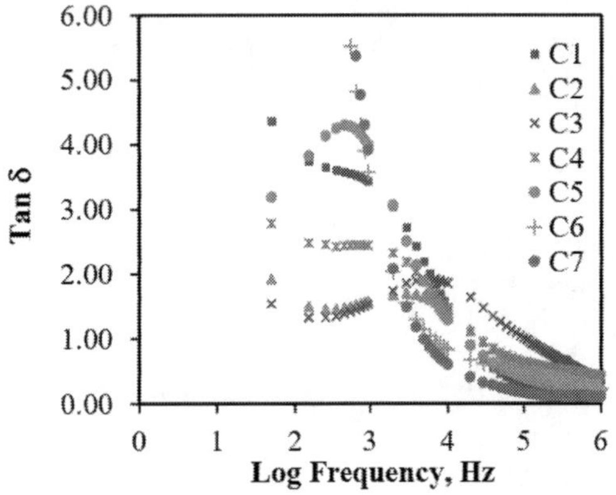

Figure 7. The dependence of tan δ on frequency for all compositions of CMC/PVA.

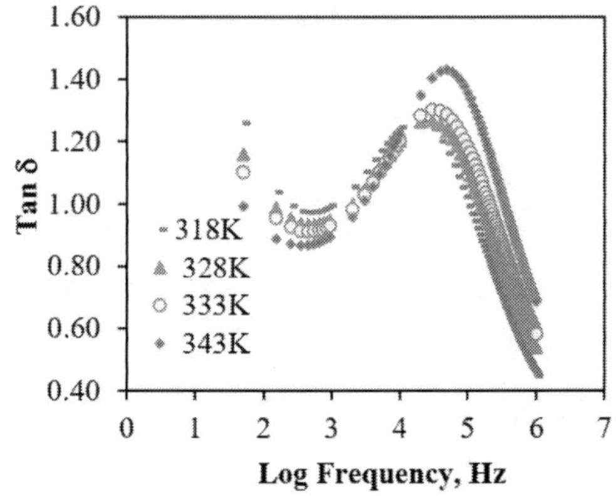

Figure 8. The dependence of tan δ on frequency of the highest conductivity CMC/PVA (80:20 compositions) at different temperature.

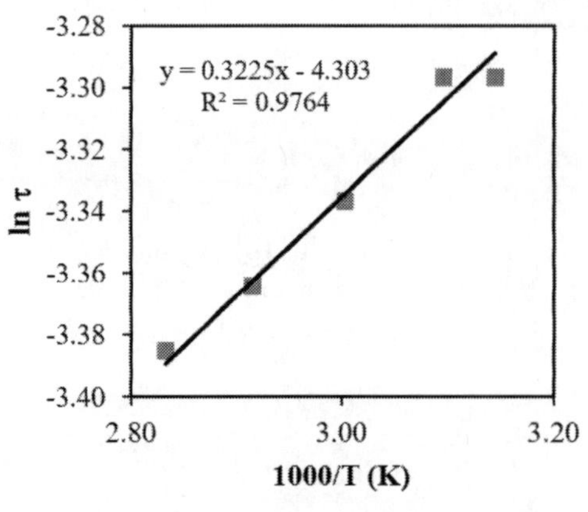

Figure 9. The temperature dependence of relaxation time of highest conductivity CMC/PVA (80:20 compositions).

4. Conclusion

The study of complex permittivity ε^* and complex electrical modulus M^* for BBEs system is a powerful approach to investigate the characteristic of molecular and ionic interaction towards the enhancement of ionic conductivity. The bio-polymer blend electrolytes containing CMC and PVA in a ratio of 80:20 prepared via casting solution method shows the highest ionic conductivity among all the test samples at ambient room temperature which is 9.12 x 10^{-6} S/cm. The dielectric constant and dielectric loss decrease with increasing frequency that is explained by the accumulation of ions contributed by electrode, molecular and ionic polarization. The occurrence of peak in the imaginary modulus (M_i) confirmed the BBEs system is non-Debye characteristic. According to temperature dependence study of the highest conductivity, CMC/PVA system is thermally activated and follows the Arrhenian characteristics where R^2~1. Based on the result obtained, the conductivity of CMC/PVA based BBEs is still relatively lower compared to the reported conductivity of polymer electrolytes as this is the preliminary study. Thus, the conductivity of CMC/PVA can be enhanced with the addition of salt and/or plasticizer.

Acknowledgement
The authors would like to thank MOHE for FRGS (RDU170115), Faculty of Industrial Sciences and Technology, University Malaysia Pahang, for the help and support given for the completion of this work.

References
[1] Park C.H., Park M., Yoo S.I, Joo S.K. 2006 *J. Power Sources* **58** 1442-1446
[2] Chen J., Asano M., Maekawa Y., Yoshida M. 2008 *J. Membrane Science* 319 1-4
[3] Shukur M.F., MAjid N.A., Ithnin R., and Kadir M.F.Z 2013 *Ionics*, **20** 997-999
[4] Mobarak N.N, Ramli N, Ahmad A, and Rahman M.Y.A 2013 *Solid State Ionics* **224** 51-57
[5] Tan C.H., Ahmad A., and Anuar F.H. 2014 *Asian Journal of Chemistry* **26** 230-236
[6] Samsudin A.S., and Isa M.I.N 2012 *J. Applied Sciences* **12** 174-179
[7] Mohamad Sri M.N.S, Buraidah M.H., and Teo L.P. 2017 *Materials Today* **4** 5161-5168
[8] Muthuvinayagam M. and Gopinathan C. 2015 *Polymer* **68**122-130
[9] Abdel-Baset T.A., and HAssen A. 2016 *Physica B: Condensed Matter* **499** 24-28
[10] Kariman M.E.S 2007 *J. of Macromolecule Science* **44** 619-624
[11] Wach R.A, Mitomo H., Yoshii H. and Kumo T. 2011 *J. Appl. Polymer Science* **81** 3000
[12] Buraidah M.H. and Arof A.K. 2011 *J. of Non-Crystalline Solids* **357** 3261-3266
[13] Shukur M.F and Kadir M.F.Z 2015 *Electrochimica Acta* **158** 152-165
[14] Sandoval C., Castro C., Gargallo L., Radic D., and Freire J. 2005 *Polymer* **46** 10437-10442
[15] Utracki L.A. 2002 *The Canadian J. of Chem. Eng.*, **80** 1008-1016
[16] Rudzhiah S., Ahmad A., Ahmad I., and Mohamed N.S 2015 *Electrochimica Acta* **175** 162-168
[17] El-Sayed S., Mahmoud K., Fatah A., and Hassen A. 2011 *Physica B* **406** 4068-4076.
[18] El-Sayed A.M. 2014 *Nucl. Isntrum. Methods Phys. Res. B* **321** 41-48
[19] Sengwa R.J and kaur K. 2000 *Polym. Int.* **49** 1314
[20] Shejar B.C, Veeravazhithi, Sakthivel S., Mangalaraj D. and Narayandass S.K 1999 *Thin Solid Films* **348** 122
[21] Shukur M.F., Ithnin R. and Kadir M.F.Z 2014 *Ionics* **20** 977-999
[22] Abraham, K.M., Alamgir, M. and Hoffman, D.K. 1995 *J. Electrochem Soc.* **142** 638-687.
[23] Appetecchi, G.B., De Paolis, F.A. and Scrosati, B.A. 1999. *J. Electroanal. Chem.* **463** 248–252
[24] Samsudin A.S., Kahirul W.M. and Isa M.I.N 2012 *J. Non-Crystalline Solids* **358** 1104-1112
[25] Hema H., Selvasekarapandian S., Nithya H., Sakunthala H. and Arunkumar D. 2008 *Ionics* **15** 487-491
[26] Rasali, N., Muzakir, S., & Samsudin, A. 2017. *Makara Journal Of Technology* **21** 65-69
[27] Malathi J., Kumaravadivel M., Brahmanandhan G.M, Hema M., Baskaran and R.,Selvasekarapandian 2010 *J. Non-Crystalline Solids* **356** 2277-2281
[28] Kadir M.F.Z, Majid S.R and Arof A.K. 2010 *Electrochim. Acta* **55** 1475-1482
[29] Fadzallah I.A., Majid S.R., Careem M.A and Arof A.K 2014 *Ionics* **20** 969-975

IOP Conf. Series: Materials Science and Engineering **342** (2018) 012045 doi:10.1088/1757-899X/342/1/012045

[30] Sudhakar Y.Nand Selvakumar M. 2013 *J. Applied Electrochem.* **43** 21
[31] Chai M.N. and Isa M.I.N. 2013 *J. Current Engineering Research* **1** 1-5
[32] Ramesh S., Yuen T.F., and Shen C.J 2008 *J. Non-Crystalline Solids* **351** 583-594
[33] Sinclair D.C and West A.R. 1989 *J.Applied Physiscs* **66** 3850
[34] Zakaria N.A., Yahya S.Y.S, Isa M.I.N, Mohamed N.S and Subban R.H.Y 2010 *J. Advanced Material* **93** 429-432
[35] Pradhan D.K., Choudhary R.N.P. and Samataray B.K 2008 *Express Polym Lett.* **2** 630-638
[36] Bergman R., Brodin A., Engberg, and Lu Q. 1995 *Electrochim. Acta* **40** 2049
[37] El Kader M.F.H., and Ragab H.S. 2012 *Ionics* **19** 361-369

IOP Conf. Series: Materials Science and Engineering **342** (2018) 012046 doi:10.1088/1757-899X/342/1/012046

Comparative studies of Titanium Dioxide and Zinc Oxide as a potential filler in Polypropylene reinforced rice husk composite

M Awang and W R Wan Mohd
School of Ocean Engineering, Universiti Malaysia Terengganu, 21300 Kuala Terengganu, Terengganu, Malaysia

E-mail: wroslinamohd@gmail.com

Abstract. Arising global environmental issues have triggered the search of new products and processes that are compatible with the environment while maintaining novel properties of materials. In this work, green composites containing rice husk (RH), polypropylene (PP), and incorporated with two different fillers namely titanium dioxide (TiO_2) and zinc oxide (ZnO) were prepared using an internal mixer and were injected into desired specimen by using an injection molding method. Mechanical properties of the composite were studied using Instron universal testing machine with load cell of 30kN capacity. Morphological of tensile fractured surface of composites was observed using scanning electron microscopy (SEM). The results show that the composites with the addition of TiO_2 gave an excellent mechanical properties than the composites filled with ZnO. Furthermore, morphological image of PP/RH/TiO_2 also shows a good interaction occurred between polymer matrix and RH particles as compared to that of PP/RH/ZnO.

1. Introduction

The novelty of materials is evaluated based on their mechanical characteristics such as tensile properties, compression properties, impact properties and flexural properties. These characteristics are essential to figure out material ability to withstand extreme and critical conditions resulted to their engineering performance [1].

Polymer composites reinforced natural fiber have attracted a broad attention of researchers in conjunction with arising concern towards environmental issues. This is due to a novel attribute of natural fiber which is renewable, biodegradable and sustainability. They also provide some excellent features such as light weight, low cost, eco-friendly and naturally degradable composites without eliminating its rigidity properties [2]. Rice husk is one of the natural fibers which comes from major agro-waste by-product with an abundance of sources and become a staple food in Malaysia. It is decomposed of cellulose 35 %, lignin 20%, hemicellulose 25% and ash 17% (silica 94%) by weight [3]. However, there are some shortcomings of this natural fiber where hydrophilic properties which belongs to lignocellulosic materials made their mechanical properties become lower than their synthetic counterpart. Another reason that led to low mechanical properties of the composites is poor interfacial bonding between the lignocellulosic material and the hydrophobic polymer matrix. Comprehensive studies has been made on polyolefin such as polypropylene and polyethylene and various natural reinforcing fillers, in conjunction with various chemicals that could affect the interface [4,5]. This drawback can be alleviated by the use of inorganic filler such as metal oxide which is TiO_2, ZnO, $CaCO_3$ and others. The main purpose of utilization of inorganic filler in polymer composite not only limited to cost reduction but it also to enhance the performance of mechanical properties such as rigidity, dimensional stability and toughness [6].

TiO_2 is an inexpensive and non-toxic semiconductor and extensively used to enhance the properties of polymers [7]. TiO_2 and ZnO have drawn broad attention in recent years due to their wide application especially in functional devices, pigments, cosmetics and ultraviolet (UV) absorber and antibacterial agent [7–14]. An incorporation of TiO_2 and ZnO into polymers does not only produce functional capabilities such as photostabilization and antibacterial agent but it also can enhance mechanical properties by providing a strong interfacial interaction between polymers matrix and filler [15]. The use of TiO_2 and ZnO with PP composite have been researched by a number of researchers in order to investigate the physical and mechanical properties [7,16-21]. To date, there is no research has been conducted relating to substitution of TiO_2 or ZnO filler into PP reinforced RH composites.

In this study reported herein, the influences of TiO_2 and ZnO on mechanical properties of PP reinforced RH composite is investigated. The morphological studies on fractured tensile surfaces were also carried out by using scanning electron microscopes.

2. Experimental details

2.1 Materials

Polypropylene was purchased from Lotte Chemical Titan in the form of pellets with a density of 0.900g/cm^3 and melt flow index of 14g/min. RH was supplied by Bernas rice factory (Besut, Terengganu, Malaysia). TiO_2 and ZnO powder was obtained from Azfa Maju Trading, Terengganu.

2.2 Preparation of the composites and sample

RH was ground and sieved to produce mesh sizes less than 50μm. PP pellets, RH particles and TiO_2 filler were melt-mixed in a Brabender internal mixer. The content of TiO_2 filler was fixed to 3wt% and two levels of RH particle loading (10 and 40wt%) were used. For comparative purposes, ZnO content was also fixed to 3wt% and went through a similar processes. Composite samples were then fed into a Haake injection molding machine to produce specimens in the form of dumbbell shape.

2.3 Characterizations

2.3.1 Mechanical property. The mechanical property of the composites was investigated using an Instron Universal Testing Machine (Model 3366) with maximum load cell of 30kN capacities and cross head speed of 10mm/min.

2.3.2 Thermogravimetric analysis. The measurement of thermal stability was performed using a Thermo Balance SDTA 815E, in the temperature range 30°C to 900°C at heating rate of 10°C /min and under 30ml/min of argon flow.

2.3.3. Scanning electron microscopy. The morphology of the tensile fractured surface of the composite samples was performed using scanning electron microscope (JOEL model JSM 6360LA).

3 Results and Discussion

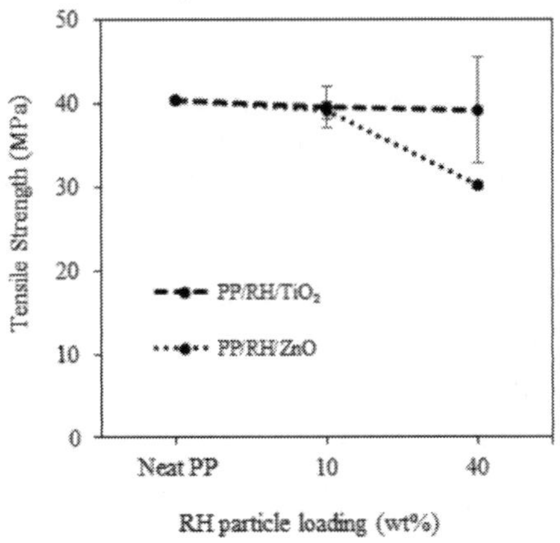

Figure 1. Tensile strength of neat PP, PP/RH/TiO_2 and PP/RH/ZnO

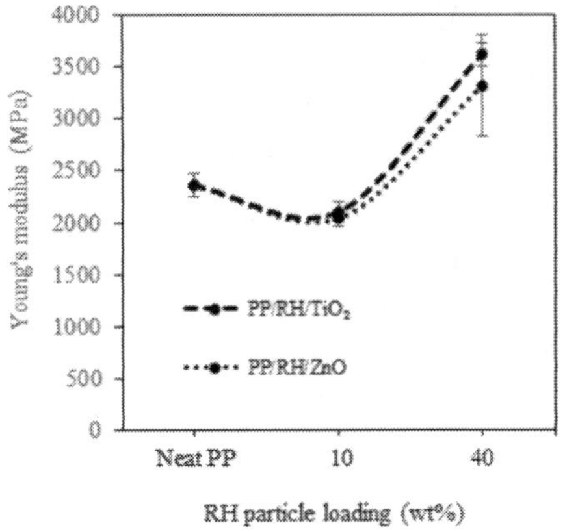

Figure 2. Young's modulus of neat PP, PP/RH/TiO_2 and PP/RH/ZnO

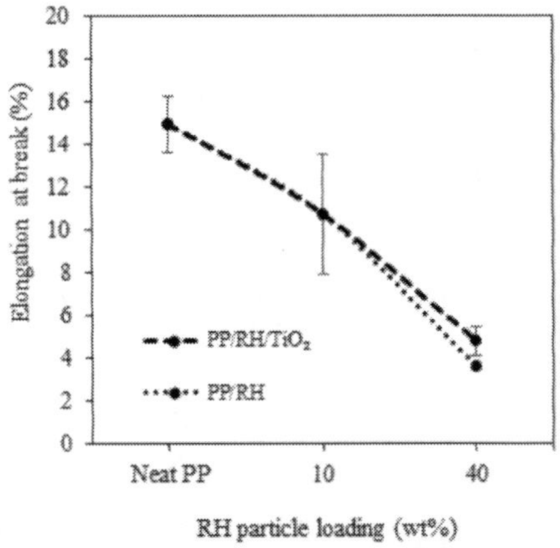

Figure 3. Elongation at break of neat PP, PP/RH/TiO_2 and PP/RH/ZnO

3.1 Mechanical properties

3.1.1 Tensile test. Figure 1 depicts the effect of RH particle loading on the tensile strength of PP/RH/TiO_2 and PP/RH/ZnO. It shows that the value of tensile strength of PP/RH/TiO_2-10wt% and PP/RH/TiO_2-40wt% minimally vary. The value of the tensile strength of these both composites are still in acceptable range and are very close to the value of ultimate tensile strength of neat PP which is 40MPa. It was highlighted that an addition of 3wt% TiO_2 in PP/RH composite has substantially increased tensile strength thus led to excellent mechanical properties. Esthappan et al. also mentioned that an addition of 3wt% of TiO_2, tenacity was significantly increased indicating the uniform distribution of TiO_2 in the PP fiber [22]. As compared to PP/RH/ZnO, by increasing RH particle loading to 40wt%, the tensile strength of the PP/RH/ZnO was slightly decreased. By adding 0.5 to 8wt% of ZnO, the values of tensile strength of PP fiber did not much improve because of poor interaction between filler and matrix [23].

3.1.2 Young's modulus. Overall, all RH particle loading series also indicate that the value of Young's modulus of PP/RH/TiO_2 was higher than PP/RH/ZnO shown in Figure 2. The trends of the plotted graph also increased with increasing of RH particle loading. This phenomenon occurred due to increasing of RH content in composites which directly contributed to an increase in composite stiffness. It was verified that stiffness of the PP/RH/TiO_2 composite increased from 2100MPa to more than 3600MPa while for PP/RH/ZnO composite increased from 2047MPa to more than 3300MPa. Luz et al. reported that natural lignocellulosic materials have higher elastic modulus attribute than PP [24]. A number of researchers also have been found that in the presence of high loading of natural fiber being as important factor in increasing composite's rigidity that causes the reduction of polymer chains mobility thus promote an occurrence of higher stiffness in the composite[25].

3.1.3 Elongation at break. Increasing RH particle loading resulted in an abrupt drop in elongation at break for PP/RH/TiO_2 and PP/RH/ZnO as compared to neat PP in (Figure 3). Generally, an incorporation of inorganic fillers into polymer composites enhances their mechanical properties such as tensile strength, elastic modulus but decreases impact strength and elongation at break [18, 26, 27]. Based on Figure 3, it also can be observed that elongation at break for PP/RH/TiO_2 was higher than PP/RH/ZnO. Altan et al. also found a similar result in PP composites whereby elongation at break of PP/TiO_2 was significantly increased as compared to ZnO reinforced composites especially for the filler loading over 1wt% [28].

3.2 SEM morphology
SEM micrographs of tensile fractured surfaces of neat PP, PP/RH/TiO_2-10wt% and PP/RH/ZnO-10wt% were shown in Figure 4 (a), (b) and (c) respectively. It can be observed that the RH particles were distributed uniformly in Figure 4 (b) unlike Figure 4 (c). It was also can be seen clearly, a poor interaction between polymer matrix and RH particles occurred in Figure 4 (c) as compared to Figure 4 (b). As RH particle loading was increased from 10 to 40wt% as shown in Figure 4(d) and Figure 4(e), there were some agglomerations in morphological structure in PP/RH/TiO_2-40wt%. Aydemir et al. mentioned that the agglomerations occurred due to increasing of filler loading [29]. An increasing RH particle loading also led to phenomenon of fiber pull out due to less interaction between polymer matrix and RH particle as clearly seen in Figure 4(e). Other works reported that when a free space exist between particles and polymer matrix, it is easy to have fiber pull out phenomenon during brittle fracture [30].

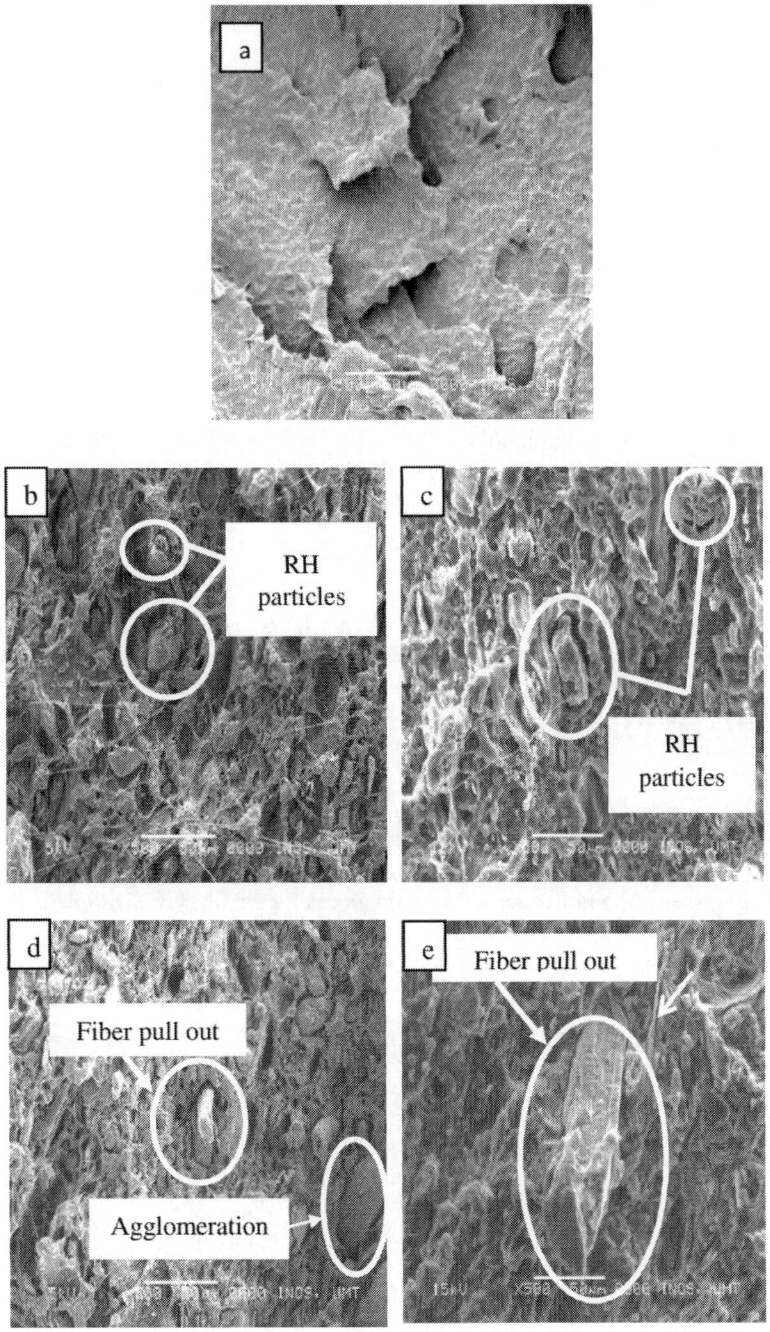

Figure 4. SEM micrographs of tensile fractured surface of (a) Neat PP (b) PP/RH/TiO_2-10wt% (c) PP/RH/ZnO-10wt% (d) PP/RH/TiO_2-40wt% (e) PP/RH/ZnO-40wt% composites

4 Conclusions

Inorganic fillers such as metal oxides behave different according to its physical, mechanical and and chemical properties. In this work, two metal oxides namely TiO_2 and ZnO were used to improve properties of polypropylene filled rice husk composites. An addition of 3wt% of TiO_2 in PP/RH composites imparted an improvement to the mechanical properties. From tensile test, it shows that higher tensile strength and Young's modulus in PP/RH/TiO_2 than that of PP/RH/ZnO composites due to its hardness and stiff structure. Besides, the elongation at break of the PP/RH/TiO_2 was also higher than PP/RH/ZnO composites proven that the compatibility of PP/RH was better with TiO_2 compare than ZnO. From SEM morphology, PP/RH composites filled TiO_2 have a good interaction between polymer matrix and RH particles while fiber pull out appeared in some regions of PP/RH/ZnO led to the reduction of tensile properties of the PP/RH/ZnO composites.

Acknowledgment

The authors would like to thank the School of Ocean Engineering, Universiti Malaysia Terengganu for providing lab's facilities and also to the Ministry of Higher Education (MOHE) for the financial support.

References

[1] Rajendra K and Tejeet S 2014 Rice husk-reinforced composites: A review. Proceedings of the International Conference on Research and Innovations in Mechanical Engineering, pp. 395-405

[2] Majeed K, Jawaid M, Hassan A, Abu Bakar A, Abdul Khalil H P S, Salema A A and Inuwa I 2013 Potential materials for food packaging from nanoclay/natural fibers filled hybrid composites *Mater. Des.* **46** 391-410

[3] Premalal H G B, Ismail H and Baharin A 2002 Comparison of the mechanical properties of rice husk powder filled polypropylene composites with talc filled polypropylene composites Polym. Test. **21** 833-839

[4] Lai S M, Yeh F C, Wang Y, Chan H C and Shen H F 2003 Comparative study of maleated polyolefins as compatibilizers for polyethylene/wood flour composites *J Appl Polym Sci* **87** 487-96

[5] Jiasheng Q, Pingsheng H and Kangming N 2004 Nonisothermal crystallization of PP/nano-SiO_2 composites *J. Appl. Polym. Sci.* **91** 1013

[6] Nicola C and Mahendra 2012 Nano antimicrobials: progress and prospect *Titanium Dioxide–Polymer Nanocomposites with Advanced Properties* ed Kubacka A, Fernández-García M, Cerrada M L and Fernández-García M (Berlin Heidelberg: Springer) pp 119 –149

[7] Altan M and Yildirim H 2012 Mechanical and antibacterial properties of injection molded PP/TiO_2 nano-composites: effects of surface modification *Journal of Material Science and Technology* **28** 686 – 692

[8] Mikesova J, Slouf M, Gohs U, Popelkova D, Vackova T, Hung V N, Kratochvil J and Zhigunov A 2014 Nanocomposites of polypropylene/titanate nanotubes: morphology, nucleation effects of nanoparticles and properties *Polym. Bull.* **71** 795 – 818

[9] Zhao H X and Li R K Y 2006 A study on the photo-degradation of zinc oxide (ZnO) filled polypropylene nanocomposites *Polymer* **47** 3207-3217

[10] Tjong S C, Liang G D and Bao S P 2006 Electrical properties of low-density polyethylene/ZnO nanocomposites: The effect of thermal treatments *J Appl Polym Sci* **102** 1436-1444

[11] Sawai J, Igarashi H, Hashimoto A, Kokugan T and Shimizu M 1995 Evaluation of growth inhibitory effect of ceramics powder slurry on bacteria by conductance method *J Chem Eng Jpn* **28** 288-293

[12] Sawai J, Saito I, Kanou F, Igarashi H, Hashimoto, Kokugan A T and Shimizu M 1995 Mutagenicity test of ceramic powder which have growth inhibitory effect on bacteria *J Chem Eng Jpn* **28** 352-354

[13] Li J H, Hong R Y, Li M Y, Li H Z, Zheng Y and Ding J 2009 Effects of ZnO nanoparticles on the mechanical and antibacterial properties of polyurethane coatings *Prog Org Coat* **64** 504-509

[14] Li S C and Li Y N 2010 Mechanical and antibacterial properties of modified nano- ZnO/high density polyethylene composite films with a low doped content of nano-ZnO *Journal of Applied Polymer Science* **116** 2965-2969
[15] Zeng A, Zheng Y, Guo Y, Qiu S and Cheng Lei 2012 Effect of tetra-neddle- shaped zinc oxide whisker (T-ZnOw) on mechanical properties and crystallization behavior of isostatic polypropylene composites *Mater. Des.* **32** 1477
[16] Yang C, Wu Y and Liu C 2013 Studies on crystallizations and mechanical properties of polypropylene/nano-TiO_2 composites *Material Design, Processing and Applications* **690-693** 494-498
[17] Zohrevand A, Ajji A and Mighri F 2013 Morphology and properties of highly filled PP/TiO_2 nanocomposites *Polymer Engineering and Science* **54** 874-886.
[18] Kruenate J, Tongpool R, Panyathanmaporn T and Kongrat P 2004 Optical and mechanical properties of polypropylene modified by metal oxides *Surf Interf Anal* **36** 1044-7
[19] Li X, Schneider K, Kretzschmar B and Stamm M 2008 Deformation behavior of PP and PP/ZnO nanocomposites as studied by SAXS and WAXS *Macromolecules* **41** 4371-9
[20] Zhou J P, Qiu K Q and Fu W L 2005 The surface modification of ZnOw and its effect on the mechanical properties of filled polypropylene composites *J Compos Mater* **39** 1931-41
[21] Zhang Y C, Xie J F, Wu H Y and Qiu Y P 2009 Crystallization and mechanical properties of nano ZnO/PP/PLA composite filaments *Mater Sci Forum* **620** 485-8
[22] Esthappan S K, Kutappan S K and Joseph R 2012 Thermal and mechanical properties of polypropylene/ titanium dioxide nanocomposite fibers *Mater. Des.* **37** 537-542
[23] Wacharawichanant S and Phutphongsai A 2007 The study of morphology and mechanical propeties of compabitilized polypropylene/zinc oxide composites. *Journal of Solid Mechanics and Materials Engineering* **10** 1231-1237
[24] Luz S M, Del T J, Rocha G J M, Gonçalves A R and Del'Arco Jr A P 2008 Cellulose and cellulignin from sugarcane bagasse reinforced polypropylene composites: effect of acetylation on mechanical and thermal properties *Composites Part A – Applied Science and Manufacturing* **39** 1362-1369
[25] Rana A K, Mandal B C, Mitra R, Jacobson R, Rowell A and Banerjee N 1998 Short jute fiber reinforced polypropylene composites: effect of compatibilizer *J Appl Polym Sci* **69** 329-338
[26] Olmos D, Dominguez C, Castrillo P D and Gonzales-Benito J 2009 Crystallization and final morphology of HDPE: effect of high energy ball milling and the presence of TiO_2 nano particles *Polymer* **50** 1732 – 1742
[27] Altan M and Yildirim H 2010 Mechanical and morphological properties of polypropylene and high density polyethylene matrix composites reinforced with surface modified nano sized TiO_2 particles *International Journal of Chemical, Molecular, Nuclear, Materials and Metallurgical Engineering* **4** 231–7
[28] Altan M, Yildirim H and Uysal A 2011 Tensile properties of polypropylene/metal oxide nano composites *The Online Journal of Science and Technology* **1** 25-30
[29] Aydemir D, Uzun G, Gumus H, Yildiz S, Gumus S, Bardak T and Gunduz G 2016 Nanocomposites of polypropylene/nano titanium dioxide: effect of loading rates of nano titanium dioxide *Materials Science* **22** 364-367
[30] Rosa S M L, Santos E F, Ferreira C A and Nachtigall S M B 2009 Studies on the properties of rice husk filled PP composites- Effect of maleated PP *Materials Research* **12** 333-338

Modelling and control of a nonlinear magnetostrictive actuator system

M H M Ramli[1], A P P Abdul Majeed[2], M A M Anuar[1], and Z Mohamed[1]

[1] Faculty of Mechanical Engineering, Universiti Teknologi MARA, 40450 Shah Alam, Selangor, Malaysia.
[2] Innovative Manufacturing, Mechatronics and Sports Lab (iMAMS), Faculty of Manufacturing Engineering, Universiti Malaysia Pahang, Pekan Campus, 26600 Pekan, Pahang, Malaysia.

E-mail: haniframli@salam.uitm.edu.my

Abstract. This paper explores the implementation of a feedforward control method to a nonlinear control system, in particular, Magnetostrictive Actuators (MA) that has excellent properties of energy conversion between the mechanical and magnetic form through magnetostriction effects which could be used in actuating and sensing application. MA is known to exhibit hysteresis behaviour and it is rate dependent (the level of hysteresis depends closely on the rate of input excitation frequency). This is, nonetheless, an undesirable behaviour and has to be eliminated in realising high precision application. The MA is modelled by a phenomenological modelling approach via Prandtl-Ishlinskii (P-I) operator to characterise the hysteresis nonlinearities. A feedforward control strategy is designed and implemented to linearize and eliminate the hysteresis by model inversion. The results show that the P-I operator has the capability to model the hysteretic nonlinearity of MA with an acceptable accuracy. Furthermore, the proposed control scheme has demonstrated to be effective in providing superior trajectory tracking.

1. Introduction

The Magnetostrictive Actuator (MA) is a solid state magnetic actuator which requires a magnetic field to be driven. It is quite similar to a typical DC motor as it consists of magnets and coils. Nonetheless, it does not need as much power as the conventional DC motor whilst its mechanical strain is typically low. In the recent years, the modelling and control of hysteresis in smart materials have gained considerable attention due to its high precision and nanoscale accuracy. This type of actuator is of particular interest in robotics and new emerging applications that necessitates precision.

Although MA has its inherent advantages namely compact in size, excellent bandwidth (operating frequency) and high stiffness, it also displays a strong hysteresis behavior [1]. Magnetic hysteresis is a condition where there exists a lack of detection property during magnetisation and zero magnetisation process [2]. The presence of hysteresis nonlinearities in such smart materials may leads to uncertainties in its modeling which in turn may also lead to unstable and low performances of the controlled plant. Therefore, it is worth to note that the reduction and elimination the hysteretic behavior is non-trivial in order to make full use of the MA in high precision applications. Typically, there are two approaches of hysteresis modeling i.e. differential based and operator based methods.

Operator based approach is gaining popularity due to its simplicity where it is based on the summation of superposition of elementary relays [3,4]. Moreover, to address the issue of modelling unknown parameters particularly the hysteresis phenomenon, parameter identification or system identification appears to be promising candidate [5,6].

Hitherto, different means of modelling and control of MA have been investigated. Tan et al. utilised a Preisach operator to model the hysteresis of an MA whilst employing inverse compensation algorithms and a robust control framework [3]. The Prandtl-Ishlinskii (P-I) operator has been utilized in modelling the hysteresis of a piezo-actuated nano-positioner [7]. A feed forward compensator and sliding mode controller was used to reduce the hysteresis effect and the inversion error, respectively. The present study aims at adopting the Prandtl-Ishlinskii (P-I) operator as the hysteresis model and to explore feed forward control strategy namely hysteresis linearization and model inversion in compensating of the inherent hysteresis effect in MA.

2. Theoretical Background

The modeling and control an unconventional actuator is often a challenging task, due to its intrinsic hysteresis behaviour. In the present study, a classical Prandtl-Ishlinskii (P-I) hysteresis operator is used to model the MA. This model is selected as it has been demonstrated to characterise hysteretic nonlinearities reasonably well and it is a class of linear model that has invertible properties which is desirable for feedforward control design.

The input-output relationship of the MA based on P-I hysteresis operator is given as

$$y(t) = c_0 u(t) + \int_o^t p(r) F_r[u](t) dr \tag{1}$$

with $u(t)$ is the input function and $y(t)$ is the output response, while c_0 and $p(r)$ are the constants as such that $p(r), c_0 > 0$. While $F_r[u](t)$ is the play operator where the initial condition is given as

$$F_r[u](0) = \max\{u(0) - r, \min\{u(0) + r, 0\}\} \tag{2}$$

and for $\forall t > 0$

$$F_r[u](t) = \max\{u(t) - r, \min\{u(t) + r, F_r[u](0)\}\} \tag{3}$$

where $r_i = \frac{i}{n+1}\|u(k)\|_\infty$ being the threshold vector where its dimension is the same as $p(r)$. In discrete form, the above input-output relationship may be rewritten as follows

$$y(k) = c_0 u(k) + \sum_{i=1}^{n} p_i F_{ri}[u](k) \tag{4}$$

The matrix form of the relationship is given as

$$\begin{bmatrix} y_1 \\ \vdots \\ . \\ y_k \end{bmatrix} = \begin{bmatrix} u_{11} & F_{r12} & \cdots & F_{r1n} \\ u_{21} & \cdot\cdot & . & \vdots \\ \vdots & \cdot\cdot & . & . \\ u_{k1} & F_{rk2} & . & F_{rkn} \end{bmatrix} \begin{bmatrix} c_0 \\ p_1 \\ \vdots \\ p_n \end{bmatrix} \tag{5}$$

From equation (1) to (5), it is apparent that the output response depends on the proper selection of number operators (n) and the corresponding constants c_0 and $p(r) = [p_1 \quad p_2 \quad \ldots \quad p_n]$. Accordingly, there is a need for parameterisation or identification of the system in order to accurately estimate each of the parameter value. The following section discusses the identification method.

3. Parameter identification

The identification of the unknown parameters of the MA's output response is basically attained using least square estimation. For a linear system, the minimisation is quite simple and fairly straightforward. However, for a nonlinear system the minimisation is often carried out via iterative numerical algorithms [6]. The performance index, J utilised in the study is the minimization of the squared error i.e.

$$J = (y - \hat{y})^2 \tag{6}$$

where y is the measured output and the $\hat{y}$ is the estimated output.

4. Controller design

The simplest approach for hysteresis linearisation is the feedforward control and it entails the manipulation of the mathematical model of the actuator itself, mainly the model inversion [8]. However, this scheme is very sensitive, i.e. any changes sensed in the model imposes a strong variation to its inversion (directly depend upon the accuracy of the model). Thus, knowledge of the model is therefore essential to ensure a successful implementation of the feedforward control strategy.

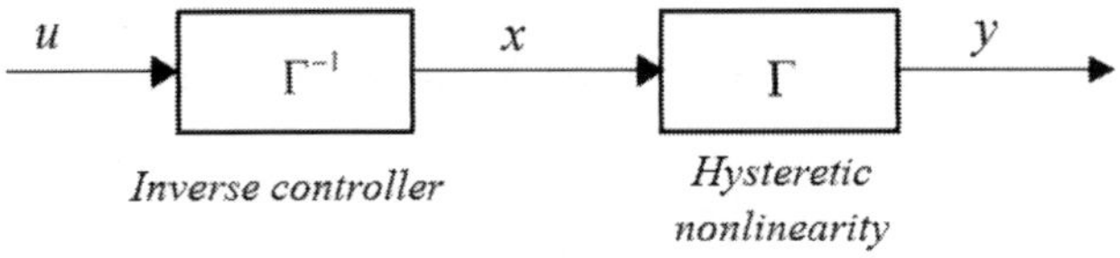

Figure 1. Compensation of the hysteretic nonlinearity by an inverse feedforward controller [9].

The inversion of the P-I model can be derived based on the following relationship shown in figure 1 and through equations (7) to (11) where ***Fs[u](t)***, $\boldsymbol{d_0}$, ***s***, and ***q*** are the inverse play operator, reciprocal of $\boldsymbol{c_0}$, inverse threshold and weight, respectively. As shown in figure 1, the inverse controller is introduced prior to the plant (modelled by a hysteresis operator) to instantaneously cancel the hysteretic nonlinearity effect. The inverse controller acts as if the reciprocal of the forward P-I operator which maps input, u to intermediate output, x before being fed into the plant and later to produce the linearised output, y (at this point y essentially equals to u).

$$x(k) = \Gamma^{-1}[u](t) = d_0 u(k) + \sum_{i=1}^{10} q_i Fs_i[u](k) \tag{7}$$

$$Fs[u](0) = \max\{u(0) - s, \min\{u(0) + s, 0\}\} \tag{8}$$

and for $\forall t > 0$

$$Fs[u](t) = \max\{u(t) - s, \min\{u(t) + s, Fs[u](0)\}\} \tag{9}$$

$$s_j = \sum_{i=1}^{j-1} p_i (r_j - r_i) \tag{10}$$

$$q_i = -\frac{p_i}{\left(d_0 + \sum_{j=1}^{i} p_j\right)\left(d_0 + \sum_{j=1}^{i-1} p_j\right)} \tag{11}$$

IOP Conf. Series: Materials Science and Engineering **342** (2018) 012047 doi:10.1088/1757-899X/342/1/012047

Table 1. Parameters for P-I model and its corresponding inversion.

P-I Model		Inversion of P-I Model	
$c_0 = 1.3169$		$d_0 = 0.7593$	
$p_1 = 1.0679$	$r_1 = 0.0909$	$q_1 = -0.3406$	$r_{e1} = 0.1197$
$p_2 = 1.0087$	$r_2 = 0.1818$	$q_2 = -0.1244$	$r_{e2} = 0.3368$
$p_3 = 1.2515$	$r_3 = 0.2727$	$q_3 = -0.0793$	$r_{e3} = 0.6456$
$p_4 = 1.1736$	$r_4 = 0.3636$	$q_4 = -0.0434$	$r_{e4} = 1.0682$
$p_5 = 0.0100$	$r_5 = 0.4545$	$q_5 = -0.0003$	$r_{e5} = 1.5974$
$p_6 = 2.1807$	$r_6 = 0.5455$	$q_6 = -0.0467$	$r_{e6} = 2.1276$
$p_7 = 0.6657$	$r_7 = 0.6364$	$q_7 = -0.0096$	$r_{e7} = 2.8560$
$p_8 = 0.5863$	$r_8 = 0.7273$	$q_8 = -0.0073$	$r_{e8} = 3.6449$
$p_9 = 1.0490$	$r_9 = 0.8182$	$q_9 = -0.0110$	$r_{e9} = 4.4871$
$p_{10} = 1.1773$	$r_{10} = 0.9091$	$q_{10} = -0.0099$	$r_{e10} = 5.4247$

5. Results and discussion

Figure 2 shows the precision of the P-I model (with identified parameters) compared to the measured experimental data excited at a frequency 1 Hz and 10 Hz using aforementioned estimation method. It is apparent that the classical P-I operator has the capability to model the hysteresis with reasonable accuracy. These results were based on 10 number of play operators (see equation (4)). The corresponding identified parameters are tabulated in table 1.

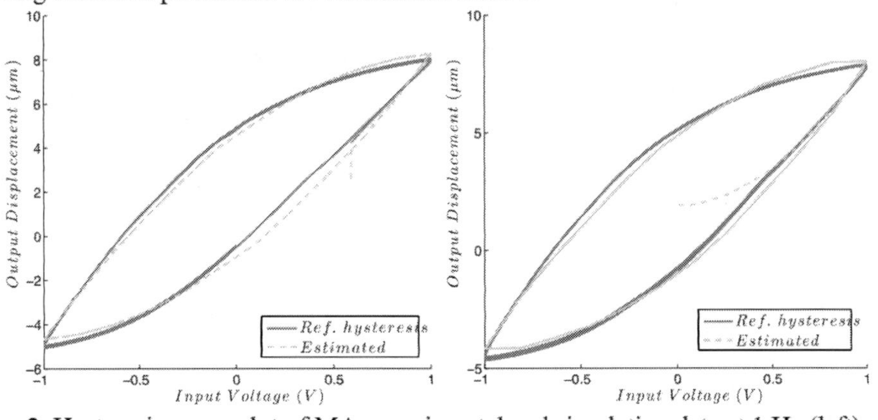

Figure 2. Hysteresis curve plot of MA experimental and simulation data at 1 Hz (left) and 10 Hz (right).

Figure 3 and 4 illustrate and demonstrate the ability of the proposed hysteresis operator, *viz.* Prandtl-Ishlinskii model and its corresponding inversion. The theoretical results expressively show hysteretic behaviour could be characterised by the aforementioned model, and the observation further suggests that the hysteresis curve (traced by blue line in figure 3 and 4) is somewhat symmetric in nature.

IOP Conf. Series: Materials Science and Engineering **342** (2018) 012047 doi:10.1088/1757-899X/342/1/012047

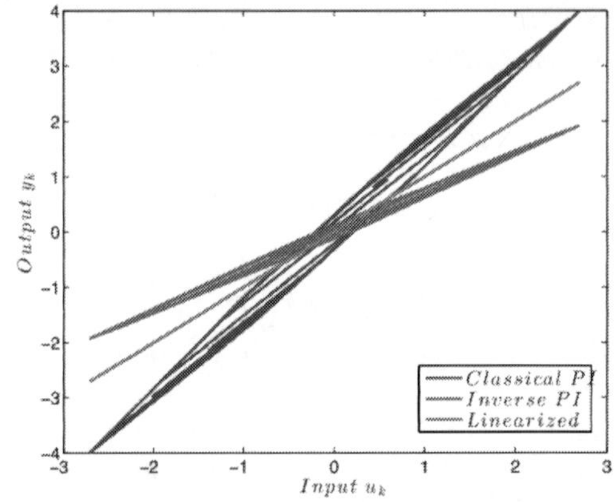

Figure 3. Plot of the hysteresis curve, inversion curve and linearized output based on classical P-I model.

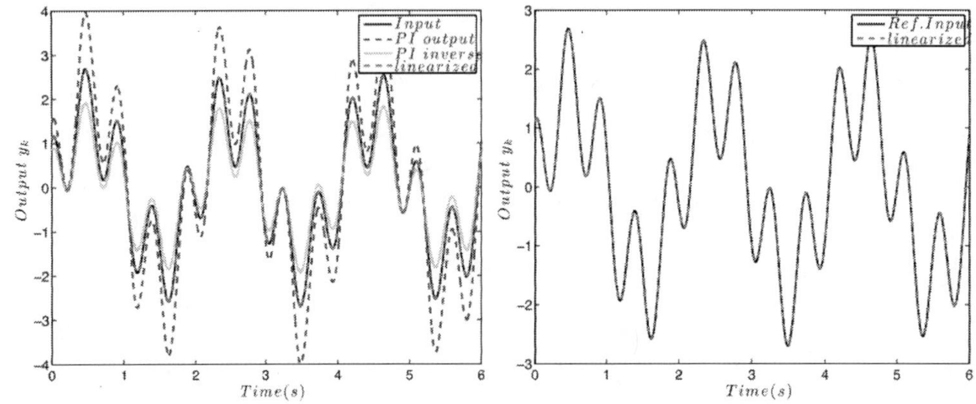

Figure 4. The plot input, P-I (output, inversion) (left) and linearized output (right) with sampling time of 1ms.

It could be observed that the implemented linearisation technique provides exceptional results as shown in figure 4. It could be clearly seen that the linearised output and target (input voltage and output is thought as linear when the hysteresis is eliminated) perfectly overlays on each other, suggesting that the error between them are null (in reality this is not achievable).

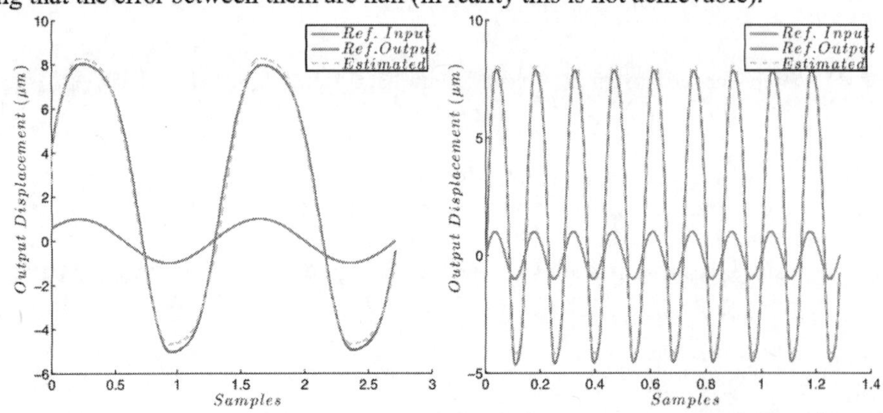

Figure 5. Plot of measured input, output and simulation output at 1 Hz (left) and 10 Hz (right).

Further illustration on hysteresis characterisation is given in figure 5. Figure 5 provides measured output plotted together with the estimated output using P-I. In the figure, it can be clearly noticed the discrepancies between the reference output and the estimated output. These incongruities may be due to asymmetric inherent saturation points of the MA. As a consequence, it appears that the classical P-I model could not to a certain degree precisely characterise the nonlinearity behaviour of the MA. However, the hysteresis compensation through model manipulation is the main interest of the present study and hence, model precision is not of great importance. Furthermore, the deviation between estimated and measured hysteresis curve is not that significant (less than 0.5) as shown in figure 6.

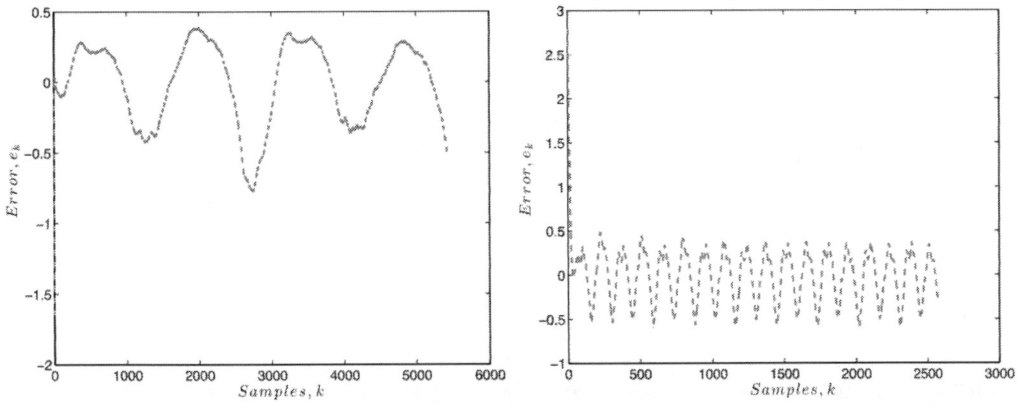

Figure 6. Estimation error between reference hysteresis and estimated ones at 1 Hz (left) and 10 Hz (right).

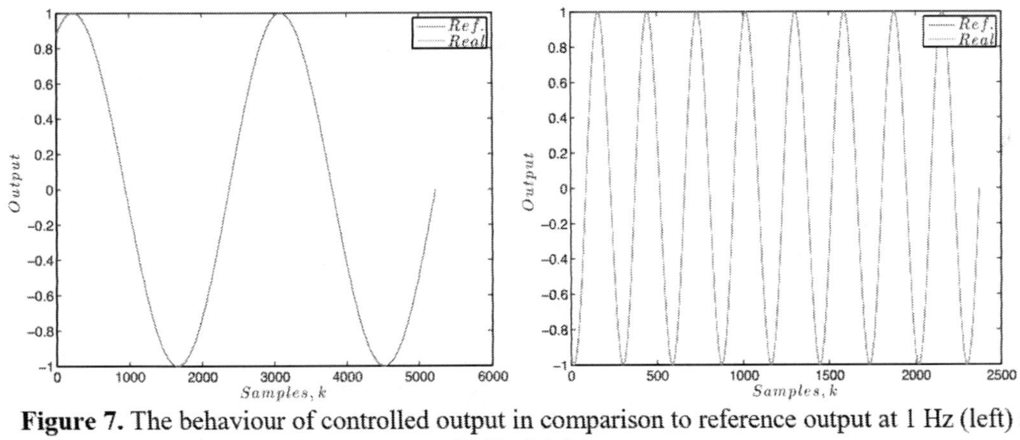

Figure 7. The behaviour of controlled output in comparison to reference output at 1 Hz (left) and 10 Hz (right).

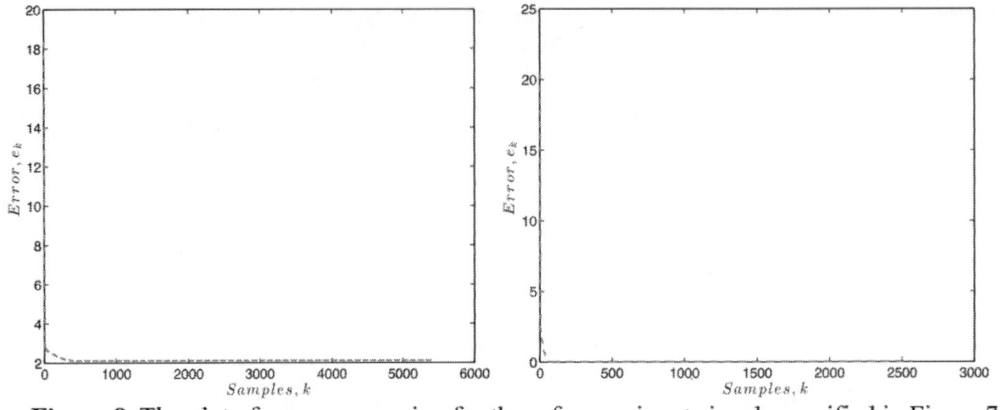

Figure 8. The plot of error progression for the reference input signals specified in Figure 7.

Figure 7 illustrates the trajectory tacking comparison between the linearised and desired output of the system with the inclusion of the feedforward controller. It is evident through figure 8 (tracking error) that the hysteretic nonlinearity effect is effectively compensated by the proposed controller, suggesting the robustness of the controller even in the presence of model uncertainties.

6. Conclusion

In conclusion, the inherent hysteresis behaviour that exist in MA model is successfully characterised via a phenomenological modeling approach i.e. the classical Prandtl-Ishlinskii (P-I) operator. The unknown parameters of the P-I model are addressed by means of parametric identification via the Least Square Estimation. The estimated parameters were observed to be modelled with a certain degree of accuracy, nonetheless, with the incorporation of the proposed control architecture, the estimation accuracy of the model may be neglected. Owing to the linear properties of the P-I model, a feedforward control through model inversion is implemented. The simulation results show the superior tracking are attained upon the inclusion of the proposed controller.

References

[1] Olabi A-G and Grunwald A 2008 Design and application of magnetostrictive materials *Mater. Des.* **29** 469–83

[2] Yamamoto Y, Eda H and Shimizu J 1999 Application of giant magnetostrictive materials to positioning actuators *Advanced Intelligent Mechatronics, 1999. Proceedings. 1999 IEEE/ASME International Conference on* (IEEE) pp 215–20

[3] Tan X and Baras J S 2004 Modeling and control of hysteresis in magnetostrictive actuators *Automatica* **40** 1469–80

[4] Edardar M, Tan X and Khalil H K 2012 Sliding-mode tracking control of piezo-actuated nanopositioners *American Control Conference (ACC), 2012* (IEEE) pp 3825–30

[5] Chen S, Billings S A and Luo W 1989 Orthogonal least squares methods and their application to non-linear system identification *Int. J. Control* **50** 1873–96

[6] Marquardt D W 1963 An algorithm for least-squares estimation of nonlinear parameters *J. Soc. Ind. Appl. Math.* **11** 431–41

[7] Li Z, Aljanaideh O, Rakheja S and Su C 2012 Compensation of play operator-based prandtl-Ishlinskii hysteresis model using a stop operator with application to piezoelectric actuators *Int. J. Adv. Mechatron. Syst.* **4**

[8] Slotine J-J E and Li W 1991 *Applied nonlinear control* vol 199(Prentice Hall Englewood Cliffs)

IOP Conf. Series: Materials Science and Engineering **342** (2018) 012048 doi:10.1088/1757-899X/342/1/012048

Temperature dependence of electrical conduction in PEMA-EMITFSI film

N F Zain[1], M M I Megat Hasnan[1], M F M Sabri[2], S M Said[1], N S Mohamed[3] and F Salleh[1]

[1] Department of Electrical Engineering, Faculty of Engineering, University of Malaya, 50603 Kuala Lumpur, Malaysia
[2] Department of Mechanical Engineering, Faculty of Engineering, University of Malaya, 50603 Kuala Lumpur, Malaysia
[3] Centre for Foundation Studies in Science, University of Malaya, 50603 Kuala Lumpur, Malaysia

E-mail: faizsalleh@um.edu.my

Abstract. Transparent and flexible film of poly (ethyl methacrylate) incorporated with 1-ethyl-3-methyl imidazolium bis(trifluorosulfonyl) imide ionic liquid (PEMA-EMITFSI) with thickness between 100 and 200 μm was fabricated by using solution casting technique. From the ionic transport measurement, it is confirmed that the electrical conduction in PEMA-EMITFSI film is mainly contributed by ionic transport. Moreover, the temperature-dependence of electrical conductivity measurement for 7 days reveals that the electrical properties of PEMA-EMITFSI film could be able to withstand a number of thermal cycles and be lasting for a period of time for potentially used as thermoelectric material through thermal heating.

1. Introduction

Thermoelectric (TE) devices are able to harvest electricity directly from waste heat and vice versa. The performance of TE devices is determined by a dimensionless figure of merit (ZT)

$$ZT = \frac{\sigma S^2 T}{\kappa}, \tag{1}$$

where σ is the electrical conductivity, S is the thermopower, κ is the thermal conductivity and T is the absolute temperature of operation. Based on the expression of ZT, it is known that in order to achieve a high ZT, it is necessary to fabricate TE device by using TE material which has high electrical conductivity and thermopower, and low thermal conductivity. Current high ZT TE materials are mainly inorganic semiconductors like Te based semiconductor material and transition metal oxides [1-5]. However, these materials have facing some issues such as toxicity, highly cost and poor processability, which limit their commercial use. Thus, over the past decade has witnessed remarkable progress in organic TE materials as promising candidates for the development of flexible, cost-effective and scalable TE devices [6-9].

Moreover, the electrical stability of TE material in terms of temperature and time is also important to be considered in practical TE device applications. It is known that the electrical conductivity of organic material such as insulating polymers are mainly governed by ionic transport and strongly depended on the temperature which limits its effectiveness to be applied in wide range temperature TE applications [10]. Moreover, most of the polymer based TE materials are fabricated on a substrate which reduces its flexibility and increases its manufacturing cost [11-13]. In previous study, our group successfully fabricated a free-standing poly (ethyl-methacrylate) incorporated with 1-ethyl-3-methyl imidazolium bis (trifluorosulfonyl) imide ionic (PEMA-EMITFSI) film and found that the film doped with 40 wt.% EMITFSI exhibits the best physicochemical and electrical properties [14]. Therefore, in this study, a same composition of PEMA-EMITFSI film is fabricated and its electrical conductivity is measured. The dependency of electrical conductivity of fabricated PEMA-EMITFSI on temperature and time will be clarified in order to discuss its possibility to be applied as TE material.

2. Experiment

The PEMA-EMITFSI film is fabricated by exactly following the procedure described in our previous report [14]. In the present work, the composition of 40 wt. % 1-ethyl-3-methyl imidazolium bis (trifluorosulfonyl) imide (EMITFSI) was chosen for film fabrication as the film exhibited the highest electrical conductivity at room temperature. 40 wt. % EMITFSI was added into solution containing 1 g PEMA in 20 ml Tetrahydrofuran (THF). The mixture was stirred at 40 °C until it becomes homogeneous. The solution was then casted into teflon petri dish and kept in air to dry. PEMA (Mw ~ 515 000 g mol^{-1}) and EMITFSI were purchased from Aldrich. THF was supplied from R&M Chemicals. All materials were used as received for sample fabrication. The thickness of the samples are between 100 and 200 μm which determined by a digital micrometer (Mitutoyo Corp.).

The electrical conductivity of the film was determined by impedance spectroscopic technique using an impedance analyser (SOLARTRON 1260) over a 100 Hz to 5 MHz frequency range with alternating current (AC) amplitude of 100 mV. The sample was sandwiched vertically between stainless steel blocking electrodes with diameter of 2 cm and placed in a temperature-controlled environmental chamber for measuring temperature-dependence of electrical conductivity from 303 to 373 K. The same procedure was then repeated using the same sample over a period of 7 days. Ionic transport number was measured using direct current (DC) polarization technique in order to confirm the contribution of ionic transport. The sample was sandwiched vertically between stainless steel blocking electrodes with applied potential difference of 1 V. The value of current was monitored until it reached the steady state condition.

3. Results and Discussion

Figure 1 shows the photograph of fabricated PEMA-EMITFSI film with a diameter of ~8 cm. From this figure, the sample which is fabricated without a substrate is seen to be transparent and flexible. This result shows a possibility to obtain TE material based on transparent and flexible material. The time evolution of normalized-measured polarization current of fabricated sample which was cut into 2 cm in diameter is shown in Figure 2. From Figure 2, it is found that the normalized current abruptly decreases with time which is usually observed in characteristic of ionic conducting material [15-16]. Considering the DC polarization technique [17-18], the ionic transport number, T_{ion} of fabricated PEMA-EMITSFI film was determined to be 0.96 by using following equations;

$$T_e = \frac{\sigma_e}{\sigma_i + \sigma_e} = \frac{i_e}{i_T}, \quad (2)$$

$$T_{ion} = 1 - \frac{i_e}{i_T} = 1 - t_e, \quad (3)$$

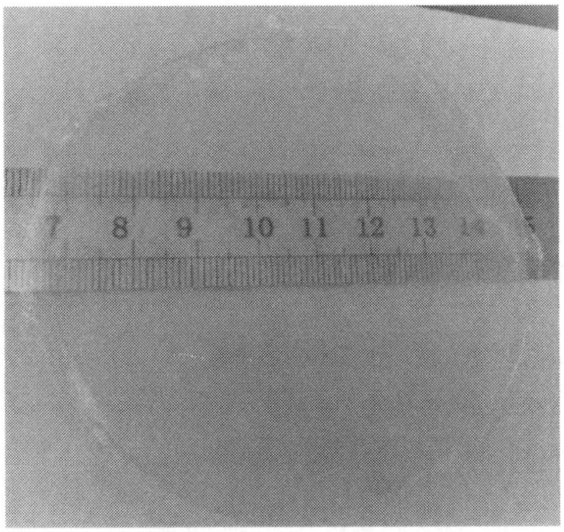

Figure 1. Photograph of free-standing PEMA-EMITFSI film

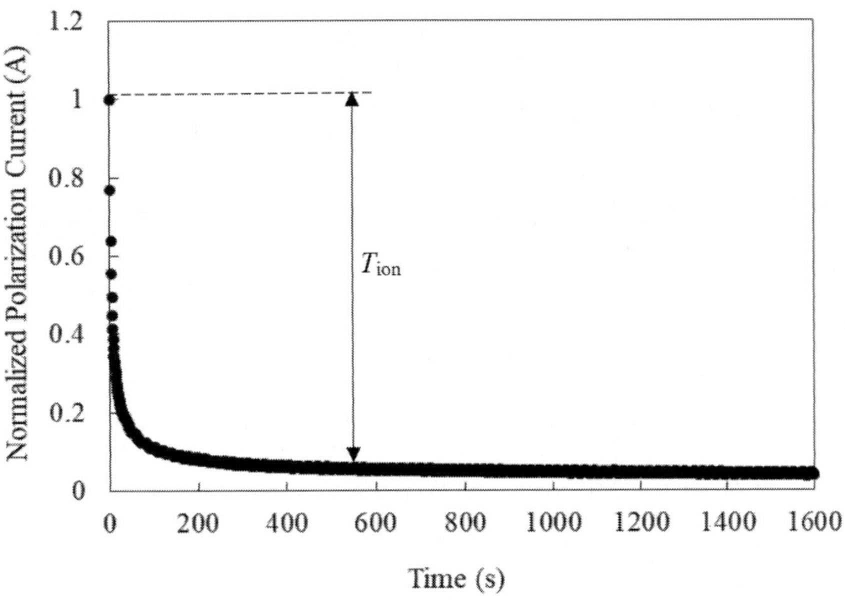

Figure 2. Time evolution of normalized-measured polarization current of PEMA-EMITFSI film at room temperature.

where σ_i is the ionic conductivity, σ_e is the electronic conductivity, i_e is the residual current and i_T is the total current. Therefore, these results indicate that the fabricated free-standing transparent film is an ionic conducting material since the calculated value of T_{ion} is close to 1.

Figure 3 shows the measured electrical conductivity of the fabricated sample in logarithm as a function of inverse temperature for day 1 and 5. The broken lines show the linear regression line. The

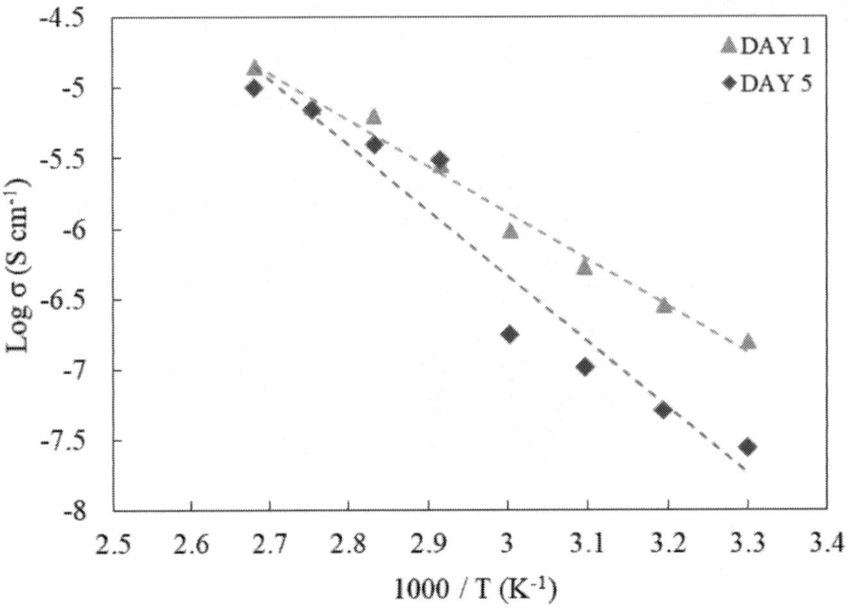

Figure 3. Logarithmic measured electrical conductivity of PEMA-EMITFSI film as a function of inverse temperature for day 1 and 5.

result of day 1 is in good agreement with our previous report which shows a good reproducibility of fabricated sample [14]. Moreover, the electrical conductivity of PEMA-EMITFSI is found to be ~10^{-7} S cm^{-1} which is four orders of magnitude higher compared with reported value of PEMA film which is ~ 10^{-11} S cm^{-1} [14]. From Figure 3, it is found that the logarithmic electrical conductivity linearly increases with increasing the temperature for both cases of day 1 and 5. This strong temperature dependent behavior is consistent with usually observed dependency in polymer materials which can be explained by free-volume model [10]. When temperature is increased as well as the thermal energy, the polymer chain can expand easily and gains higher free volume. This free volume promotes segmental motion of the polymer chains and leads to an increase in ionic mobility. That is, increasing the ionic mobility will increase the electrical conductivity in the polymer network system. Therefore, these results show that the fabricated film is more suitable for higher temperature TE application with low temperature fluctuation.

Moreover, from Figure 3, it is observed that there are differences of electrical conductivity values between day 1 and 5, especially near room temperature region. In order to clarify the influence of time on the electrical conductivity, the logarithmic measured electrical conductivity is replotted in Figure 4 as a function of day for measurement temperature at 303 and 343 K. From Figure 4, it is found that the electrical conductivity decreases with increasing the day for measurement at 303 K. This changes in electrical conductivity of fabricated film is likely due to the physical aging effect which is observed in amorphous polymers [19]. The physical aging effect describes the time-dependent molecular and structural changes which significantly effects on mechanical, structural and electrical properties of the material [19]. However, by heating again the sample around 343 K, the value of electrical conductivity of day 3, 5 and 7 are again identical with the value of day 1, as shown in Figure 4. Thus, it is considered that through physical aging effect the separation of polymeric chains will be reduced and then by thermal heating, a better separation will be obtained again which results pronounced movements of ion through higher free volume. Consequently, since the electrical conductivity is likely can be reset by thermal heating, the PEMA-EMITFSI film is expected can be applied as TE material at higher temperature region.

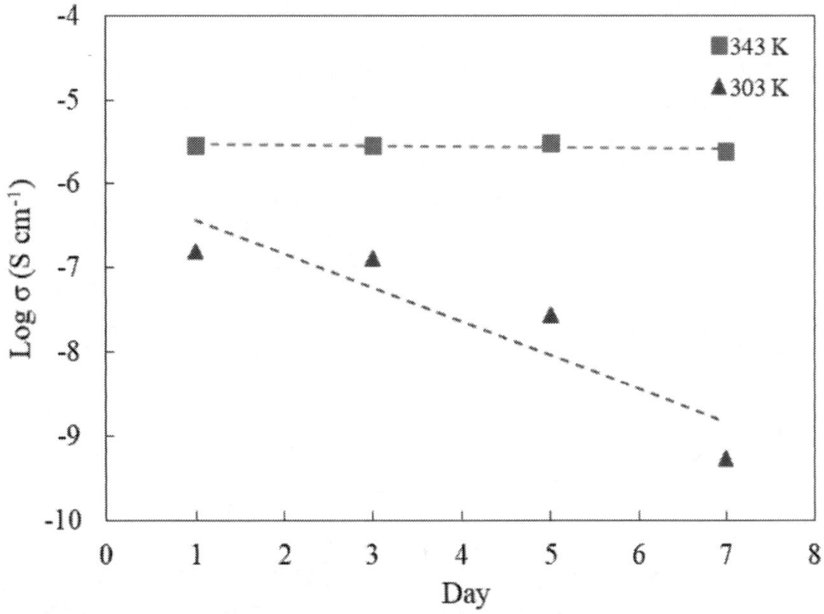

Figure 4. Logarithmic measured electrical conductivity of PEMA-EMITFSI as a function of day for measurement temperature at 303 and 343 K.

4. Conclusion

We have measured the electrical conductivity of PEMA-EMITFSI film doped with 40 wt. % EMITFSI, with thickness between 100 and 200 μm. It is found that the electrical conductivity of PEMA-EMITFSI film increases with increasing the temperature. The increase of electrical conductivity is likely due to the expanding of polymer chain which increases the ionic mobility. In addition, by considering the calculated ionic transport number, it is also confirmed that the electrical conduction in PEMA-EMITFSI film is mainly governed by ionic transport. Moreover, the electrical conductivity decreases with increasing the time which is likely due to the physical aging effect. The electrical conductivity is considered can be increased again by thermal heating. Consequently, PEMA-EMITFSI film may be considered as a good potential for transparent and flexible TE device applications.

Acknowledgments

The work was financially supported by the Fundamental Research Grant Scheme (FP062-2016) from Ministry of Higher Education and University Malaya Research Fund Assistance (BK077-2015) from University of Malaya.

References

[1] Kahalyand M U and Schwingenschlögl U 2014 Thermoelectric performance enhancement of $SrTiO_3$ by Pr doping *J. Mater. Chem. A* **2** 10379

[2] Nethravathi C, Rajamathi C R, Rajamathi M, Maki R, Mori T, Golberg D and Bando Y 2014 Synthesis and thermoelectric behaviour of copper telluride nanosheets *J. Mater. Chem. A* **2** 985

[3] Chatterjee A and Biswas K 2015 Solution-Based Synthesis of Layered Intergrowth Compounds of the Homologous $PbmBi_2nTe_3n^+$ m Series as Nanosheets *Angew Chem. Int. Ed.* **54** 5623

[4] He Q, Hao Q, Chen G, Poudel B, Wang X, Wang D and Ren Z 2007 Thermoelectric property studies on bulk TiO_x with x from 1 to 2 *Appl. Phys. Lett.* **91** 052505

[5] Pu J, Kanahashi K, Cuong N T, Chen C H, Li L J, Okada S, Ohta H and Takenobu T 2016 Enhanced thermoelectric power in two-dimensional transition metal dichalcogenide monolayers *Phys. Rev. B* **94** 014312

[6] Du Y, Cai K F, Chen S, Wang H X, Shen S Z, Donelson R and Lin T 2015 Thermoelectric fabrics: toward power generating clothing *Sci. Rep.* **5** 6411

[7] Russ B, Glaudell A, Urban J J, Chabinyc M L, and Segalman R A 2016 Organic thermoelectric materials for energy harvesting and temperature control *Nat. Rev. Mater.* **1** 16050

[8] McGrail B T, Sehirlioglu A, and Pentzer E, Angewandte 2015 Polymer composites for thermoelectric applications *Chemie International Edition* **54** 1710

[9] Kroon R, Mengistie D A, Kiefer D, Hynynen J, Ryan J D, Yu L, and Müller C 2016 Thermoelectric plastics: from design to synthesis, processing and structure–property relationships *Chem. Soc. Rev.* **45** 6147

[10] Miyamoto T and Shibayama K 1973 Free-volume model for ionic conductivity in polymers *J. Appl. Phys.* **44** 5372

[11] Zhang Q, Sun Y, Qin Y, Xu W and Zhu D 2016 Two soluble polymers with lower ionization potentials: doping and thermoelectric properties *J. Mater. Chem. A* **4** 1432

[12] Fang H, Popere B C, Thomas E M, Mai C K, Chang W B, Bazan G C, Chabinyc M L and Segalman R A 2017 Large-scale integration of flexible materials into rolled and corrugated thermoelectric modules *J. Appl. Polym. Sci.* **134** 44208

[13] Jiao F, Di C A, Sun Y, Sheng P, Xu W and Zhu D 2014 Inkjet-printed flexible organic thin-film thermoelectric devices based on p- and n-type poly (metal 1, 1, 2, 2-ethenetetrathiolate) s/polymer composites through ball-milling *Phil. Trans. R. Soc. A* **372** 20130008

[14] Zain N F, Zainal N and Mohamed N S 2015 The influences of ionic liquid to the properties of poly (ethylmethacrylate) based electrolyte *Phys. Scr.* **90** 15702

[15] Zain N F, Zainal N and Mohamed N S 2016 The effects of MgO nanofiller to the physicochemical and ionic liquid retention properties of PEMA-$MgTf_2$-EMITFSI nanocomposite polymer electrolytes *Polym. Composites*, in press (DOI: 10.1002/pc.24090)

[16] Jibreel U M, Bhattacharya B and Singh P K 2016 Synthesis, characterization, and detailed studies on plasticized poly (ethyl methacrylate): NH_4I polymer electrolyte *Adv. Polym. Tech.*, in press. (DOI: 10.1002/adv.21693)

[17] Bruce P G and Vincent C A 1993 Polymer electrolytes *J. Chem. Soc. Faraday Trans.* **89** 3187

[18] Hashmi S A and Chandra S 1995 Experimental investigations on a sodium-ion-conducting polymer electrolyte based on poly (ethylene oxide) complexed with $NaPF_6$ *Mater. Sci. Eng. B* **34** 18

[19] Kumar B, Koka S, Rodrigues S J and Nookala M 2003 Physical aging effects on conductivity in polymer electrolytes *Solid State Ionics* **156** 163

IOP Conf. Series: Materials Science and Engineering **342** (2018) 012049 doi:10.1088/1757-899X/342/1/012049

Assistive-as-needed strategy for upper-limb robotic systems: A preliminary evaluation of the impedance control architecture

I M Khairuddin[1,2], S N Sidek[1], H Md Yusof[1], A P P Abdul Majeed[2], A Ahmad Puzi[1] and H Mat Rosly[1]

[1] Department of Mechatronics Engineering, International Islamic University Malaysia, Kuala Lumpur, Malaysia
[2] Innovative Manufacturing, Mechatronics, and Sports (iMAMS) Laboratory, Faculty of Manufacturing Engineering, Universiti Malaysia Pahang

Email: ismailkhai@ump.edu.my

Abstract. Rehabilitation is a necessary restoration process of recovering impaired joint motion and muscle strength. Recent trends of rehabilitation have also moved towards providing more participation of the patient in therapy rather than simple passive treatments as it has been demonstrated to be non-trivial in promoting neural plasticity meant to promote motor recovery process. This paper presents an assistive control strategy based on impedance control technique. Dynamic modelling of upper arm is obtained by utilising the Euler-Lagrange formulation. The proportional-derivative (PD), computed torque control (CTC) impedance based framework is applied to examine its effectiveness in performing joint-space control with objectives specified in rehabilitating the elbow joint along the sagittal plane. A feasibility study through simulation was carried out to investigate the efficacy of the proposed controller on acceleration-based impedance model. The results show that impedance controller is more suitable as it allows the cooperative effort of the patient.

1. Introduction

Over the past two decades, the number of live expectancy among the elderly, particularly those in the age group of 60 years and above has increased significantly [1]. In Malaysia, the percentage of elderly in the population is about 8.3% [2]. In addition, the annual report of Malaysian Ministry of Health's in 2011 reported that the Malaysian population between the age group of 0 to 18 years old suffers from both physical disabilities due to traumatic injuries and cerebral palsy at around 11% and 7%, respectively [3]. Furthermore, the report also highlighted that there is a significant growth of stroke patients on the annual basis. These data reflect the size of the segments in the society that are diagnosed with disabilities which in turn depriving them of performing activities of daily living (ADL) [4].

In order to help the patient in regaining and relearning the existing motor skills for ADL, the patients' needs to undergo rehabilitation process. [2]. For the first stage of rehabilitation process, the purpose is to reduce spasticity and muscle tone of impaired limb and increase its movable region or functional range of motion (ROM). Direct involvement is predicted to be sufficient on the condition of recovery.

The involvement of robotic platforms to grant rehabilitation exercise is a comparatively recent area within the field of robotics in healthcare. The primary objective of implementation of robotic platform is to aid the patient to flex out the impaired limb to move along the predefined trajectory. In order to control the robotic platform, many position-based control schemes have been introduced and implemented.

In order to make the patient contributes during the specialized training, an assistance force or resistance force is usually supplied by the robotic platform, notably to rehabilitate the responsive impaired limb. During this stage, the amount of force and/or torque exerted by the robot to the impaired limb is imperative. Therefore, different type of control strategies has been developed to address this issue. For instance, in [5] fuzzy logic is incorporated with hybrid position/force control for the robotic platform to control the elbow and shoulder of the patient. Active force-based control technique has been employed both for upper and lower-limb exoskeletons for passive based rehabilitation therapy [2,6,7]. Impedance control is another type of force control method, which is different from direct force control in a sense that the control law is realised by regulating the dynamic relationship between the robot and the patient [8]. Therefore, various type of control strategies involving impedance controller has been established. Anklebot was reported to utilise adaptive impedance controller [9]. The parameter for impedance controller is obtained by measuring the dynamic contribution from the patient. In [10] the impedance controller is used to control the upper-limb exoskeleton where the input for the controller is taken by measuring the signal from EMG attach to the arm.

This paper aims at benchmarking different existing control strategies namely PD and computed torque control (CTC) based impedance control in order to promote the assist-as-needed strategy for upper limb rehabilitation robot system. The modeling of the robotic platform is described in section 2. It follows in section 3, the results and discussion on simulation study. The conclusion of this paper is described in section 4.

2. Upper-Limb Dynamics and Controller Strategy

2.1. Dynamic Model of the Upper-Limb

The upper-limb dynamics of the human limb and robotic platform are modeled as the motion of an open kinematic chain rigid joint as shown in Figure 1. The single-link model restricted to the sagittal plane, and the human-machine interaction is assumed to be seamless. Furthermore, the frictional elements that act on both the robotic platform and human joints, as well as other unmodelled variables, are also ignored. This model provides movement as in the flexion and extension of the elbow

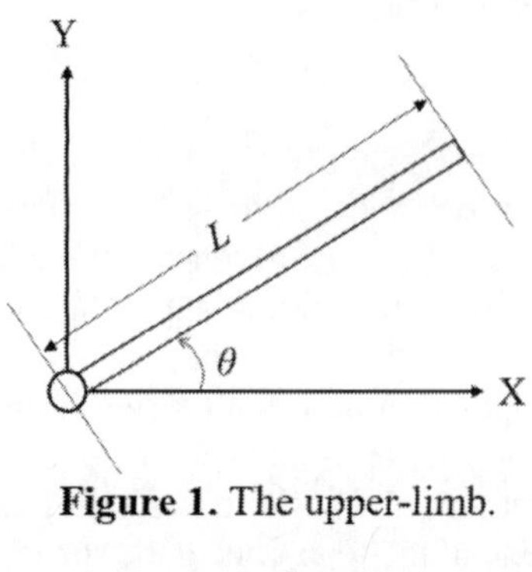

Figure 1. The upper-limb.

In the figure shown, L represent the length segments of the limb and the robotic platform and θ is the angular position of the link. By applying the Newton-Euler formulation, the dynamics of the robotic platform can be written in the following standard form as shown in Equation 1

$$\boldsymbol{\tau} = \mathbf{D}(\theta)\ddot{\theta} + \mathbf{C}(\theta,\dot{\theta}) + \mathbf{G}(\theta) + \boldsymbol{\tau}_d \quad (1)$$

where τ is the input signal represents the actuated torque vector at the joint, D is the uniformly positive definite inertia of the robotic platform and limb, C is the Coriolis and centripetal torque vector, G is the

gravitational torque vector (this term is neglected as the motion of the forearm limb is constrained along the horizontal plane) whilst τ_d is the external disturbance torque vector. The components of following equation that represents the system can be derived as follows
where the component for inertia is

$$D = mL^2 \tag{2}$$

whereas the Coriolis components are

$$C = 0 \tag{3}$$

and for the gravitational terms

$$G = mgL\cos(\theta) = 0 \tag{4}$$

where m is the combination of both masses. The anthropometric parameters of human segments used in this study were adopted from [12]. The remaining parameters are listed in section 3.

2.2. *Controller Strategy*

The primary purpose of robotics based rehabilitation is to assist patients to undergo rehabilitation in addition to complement the dependency on the therapists. However, to complete the prescribed exercise regime demands significant effort from the patient. Therefore, the strategy of assist-as-needed is paramount to contribute the minimal amount of robot assistance to aid the patients in rehabilitation. There are various methods that have been developed by researchers, and the most basic type of controllers found from the literature and often applied in robotic rehabilitation is impedance controller [11]. The objective of utilising impedance controller is to ensure the impaired upper-limb to move on a specific path during the therapy sessions. In this paper, the impedance controller is expressed by equation 5.

$$F = M(\ddot{x} - \ddot{x}_d) + B(\dot{x} - \dot{x}_d) + K(x - x_d) \tag{5}$$

The impedance model represents a mass-spring-damper system, where $M, B,$ and K is the inertia, damping and stiffness factor of the robot respectively. F on the other hand, is the force input from the subjects' upper-limb.

$$\ddot{x} = \ddot{x}_d + M^{-1}\left(B(\dot{x}_d - x) + K(x_d - x) + F\right) \tag{6}$$

Then, the impedance controller has been rearranged as in equation 6 to form the acceleration based impedance controller, in order to describe the model of interaction between the robot and the subjects' upper extremity force input. Conversely, the stiffness, K is the gain factor of displacement, the damping, B is the gain factor of the velocity and the inertia, M is the gain factor of acceleration, respectively. From the equation 6, $x, \dot{x}$ refer to the actual motion and velocity from the robot, while x_d and $\dot{x}_d$ is represented the desired motion and desired velocity of the robot. Whilst, $\ddot{x}$ is the acceleration that used as an input for the inverse dynamics, and $\ddot{x}_d$ is the desired acceleration of the robot.
Figure 2 depicted how the impedance control is being applied to the system.

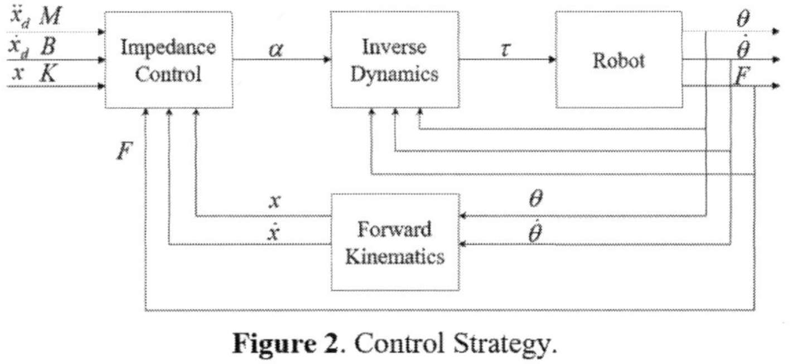

Figure 2. Control Strategy.

3. Result and Discussion

In this study, the simulation works were performed in MATLAB/Simulink computing platform. The simulation parameters employed in this study are tabulated in Table 1 and 2, respectively. The trajectory planner is used as the signal input is derived based on equation 4 [12].

$$a = a_i + \frac{3a_f}{t_f^2}t^2 - \frac{2a_f}{t_f^3}t^2 \tag{7}$$

Where a is the robot joint, a_i represent the initial position of robot joint, while a_f is the final desired of position joint. The time taken for robot motion takes place is represent as t_f while t is the real-time simulation counter.

Table 1. Upper-limb parameters.

Parameters	Values
Limb and robotic platform robot lengths (L)	0.3 m
Upper arm masses ($m_{upperarm}$)	1.91 kg
Robotic platform masses (m_{exo})	0.2 kg
Mass moment of inertia of limb (J_{limb})	0.2374 kg.m^2
Mass moment of inertia of robotic platform (J_{exo})	0.0131 kg.m^2

Table 2. Controller parameters.

Control Parameters	Gains
K_p	360
K_d	60

The simulation time selected for this study is 10 seconds, whilst the fixed-step size of 0.01 is chosen by utilising the ode2 (Heun) solver algorithm.

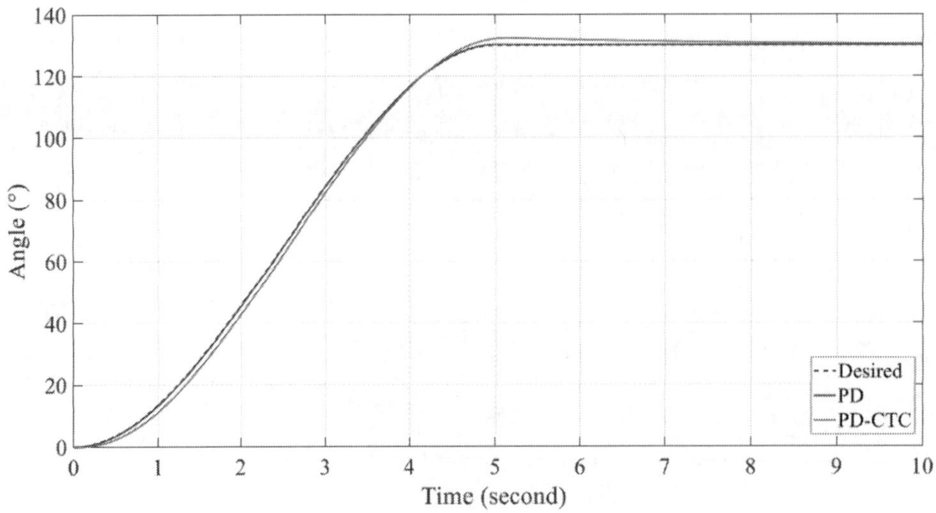

Figure 3. Angle of elbow extension based on Signal Input.

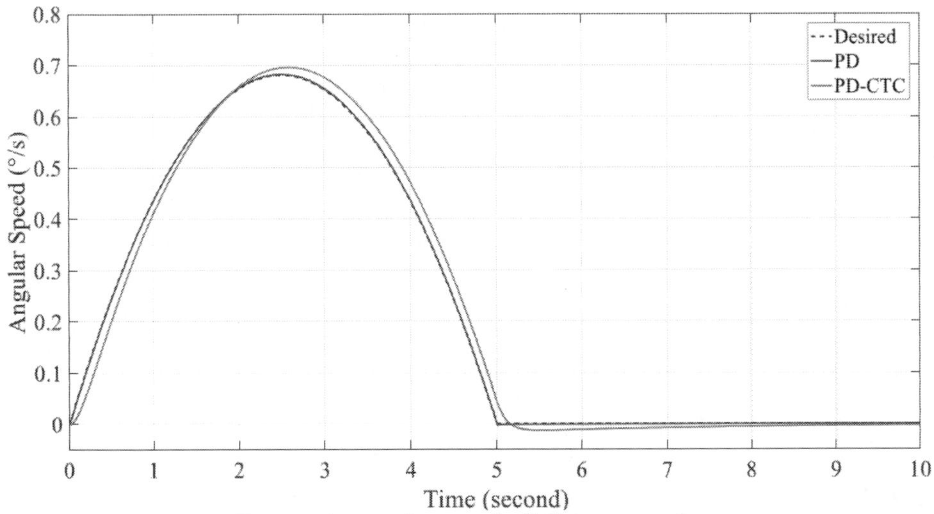

Figure 4. Angular speed of elbow extension.

It could be observed from the preliminary simulation analysis as depicted in Figure 3, that the PD impedance based control scheme tracks well the desired trajectory in comparison to the PD-CTC impedance based scheme. Table 3 tabulates the trajectory error in terms of root-mean-square-error (RMSE) of the compared control schemes. This observation is rather interesting, if passive based rehabilitation (robot perform the rehabilitation without the force/torque from the patient) is concerned, the ability of the PD architecture to track well the desired trajectory is laudable, however, from an active based rehabilitation (the patient will use their strength to complete the rehabilitation) standpoint, the PD-CTC control scheme performs better as it allows or exhibits the natural ability of the impaired limb to perform the task provided. The utilisation of the classical PD impedance scheme somewhat imposes motion of the affected limb or in other words cancels out the influence of the impedance-based controller. Whilst, the whole idea of implementing the impedance controller is to observe the robotic dynamic behaviour according to the natural torque imposed by the patient's limb. It is observed that as the limb is relaxed, it behaves as an admittance and the robotic platform acts as an impedance. As the patient tries to move the limb, the limb can be seen as to have an increase in the impedance but if the robotic platform is operating with a higher impedance, the limb will not respond to the robotic platform. Modulation of impedance allows patient-generated forces to cause a deviation from the joint's target position. This effectively produces an artificial compliance or softness in the interaction between the patient's limb and the robotic platform. A reduced impedance causes the robotic platform to produce a smaller corrective response to position error caused by patient-generated forces. Consequently, there is a trade-off between allowable interaction force and allowable deviation from the desired position which is analogous to the behaviour of a spring.

Table 3. Trajectory error comparison.

Control Scheme	RMSE (°)
PD	107.7
PD-CTC	108.2

4. Conclusion

It is evident from the study that although the conventional PD impedance based control strategy provides excellent joint trajectory tracking, for active based rehabilitation therapy, the PD-CTC impedance based scheme is more suitable as it allows the cooperative effort of the patients rather than solely based the robotic induced torque by the robotic platform. As this work is preliminary in its nature, other forms of task-based trajectories, as well as operating conditions, will be considered in future studies.

Acknowledgment

The work presented was carried out in the Biomechatronics Research Laboratory of International Islamic University Malaysia. The author wishes to gratefully acknowledge the E-Science grant funding (SF15-015-0065) from the Ministry of Science, Technology and Innovation Malaysia

References

[1] Organization W H 2013 The World Health Report 2013: Research for universal health coverage. *Sci. Transl. Med.* **5** 146

[2] Khairuddin I M, Taha Z, Majeed A P P A, Deboucha A H, Razman M A M, Jaafar A A and Mohamed Z 2016 A hybrid joint based controller for an upper extremity exoskeleton *IOP Conf. Ser. Mater. Sci. Eng.* **114** 12133

[3] MOH 2011 Annual Report Ministry of Health 2011 351

[4] Lo H S and Xie S Q 2012 Exoskeleton robots for upper-limb rehabilitation: State of the art and future prospects *Med. Eng. Phys.* **34** 261–8

[5] Ju M-S, Lin C-C K, Lin D-H, Hwang I-S and Chen S-M 2005 A rehabilitation robot with force-position hybrid fuzzy controller: hybrid fuzzy control of rehabilitation robot. *IEEE Trans. Neural Syst. Rehabil. Eng.* **13** 349–58

[6] Majeed A P P A, Taha Z, Khairuddin I M, Wong M Y, Abdullah M A and Razman M A M 2016 The Control of an Upper-Limb Exoskeleton by Means of a Particle Swarm Optimized Active Force Control for Motor Recovery *International Conference on Movement, Health and Exercise* (Springer) pp 56–62

[7] Taha Z, Abdul Majeed A P P, Zainal Abidin A F, Hashem Ali M A, Khairuddin I M, Deboucha A and Wong Paul Tze M Y 2017 A hybrid active force control of a lower limb exoskeleton for gait rehabilitation *Biomed. Eng. / Biomed. Tech.*

[8] Hogan N 1985 Impedance Control: An Approach to Manipulation *Am. Control Conf. 1984 IS - SN - VO* - 304–13

[9] Perez-Ibarra J C, Siqueira A A G and Krebs H I 2015 Assist-As-needed ankle rehabilitation based on adaptive impedance control *IEEE International Conference on Rehabilitation Robotics* vol 2015–Septe pp 723–8

[10] Kiguchi K and Hayashi Y 2012 An EMG-based control for an upper-limb power-assist exoskeleton robot *IEEE Trans. Syst. Man, Cybern. Part B Cybern.* **42** 1064–71

[11] Marchal-Crespo L and Reinkensmeyer D J 2009 Review of control strategies for robotic movement training after neurologic injury. *J. Neuroeng. Rehabil.* **6** 20

[12] Sidek S N, Mat H, Yusof H and Puzi A 2017 Modified Ashworth Scale (MAS) Integrated Adaptive Impedance Control Framework for Upper Extremity Training Platform 893–8

Application of Porous Polydimethylsiloxane (PDMS) in oil absorption

Abdullah Norfatriah, Ahmad Sabli Ahmad Syamaizar and Abu Samah Zuruzi

Faculty of Engineering, Universiti Teknologi Brunei Jalan Tungku Link,Gadong BE1410 Brunei Darussalam

E-mail: zuruzi.abu@utb.edu.bn

Abstract. Porous polydimethysiloxane (PDMS) displays both hydrophobic and oleophilic behaviour which makes it a suitable material to absorb oil in an aqueous stream. Furthermore, its elastomeric nature means that porous PDMS can be a reusable sorbent for oil. For such application, porous PDMS has to (i) absorb oil from aqueous stream quickly and (ii) discharge oil rapidly when compressed. In this study, porous polydimethylsiloxane (PDMS) has been fabricated using sugar templating method. The ability of porous PDMS to absorb olive, sunflower and vegetable oils with and without vibration was investigated. Small amplitude vibration was found to accelerate the oil uptake process and accelerates the absorption of olive and vegetable oil by 2.5 and 3 times, respectively. Compressive stress-strain curves over compression rates between 2 and 100 mm per min are similar and indicate mechanical property of porous PDMS does not vary significantly and can be rapidly compressed.

1. Introduction

Unwanted oil release is a serious environmental and health issue and may result in damages to property, poisoning and death of animals. To mitigate the effects of oil pollution, a wide range of material solutions have been developed that disperse, absorb, solidify (coagulates) and contain organic sources. Absorbent materials are particularly attractive as they promises complete removal of oil pollutants and are may be reusable [1]. Porous materials with unique wettability to water and oil, e.g., porous fibrous mats, superhydrophobic textiles and superhydrophilic / superoleophobic hygro-responsive membranes have recently attracted significant attention for oil/water separation [2]. The collection of oil using materials with both hydrophobic and oleophilic properties is often preferred because it allows for the proper disposal of oil and does not cause secondary pollution [3]. The first example of oil water separation was been reported by Jiang et al. using a mesh coated with hydrophobic and oleophilic materials [4].

Polydimethylsiloxane (PDMS) is an elastomer with desirable properties including clarity, durability, chemically inertness, non-toxicity and non–flammability. It is widely used for biological and medical applications such as in prosthesis, biomedical devices and implants as it is a biocompatible material. Furthermore, PDMS has a low glass transition temperature, low surface tension, no surface viscosity, low boiling point and high compressibility [5]. Unlike bulk polymers, its compressibility is more than 80% without side-wall buckling [6]. The PDMS sponge can completely recover its original shape even after 50 cycles of 90% strain and the elongation at breaking the sponge is as high as 97%. [2].

In this study, porous PDMS was investigated as a medium for oil absorption. Porous PDMS was fabricated using the sugar templating method. It was found that low amplitude vibration increases its oil absorption performance. Its mechanical properties of under compression at various strain rates was investigated. Polydimethylsiloxane has been proven to be safe and present no risk to human and environment.

2. Experimental studies

Chemicals used in this study were purchased from suppliers and used in the as-received condition. Polydimethylsiloxane (PDMS) used in this study was commercial grade Sylgard 184 silicone elastomer (Dow Corning) purchased from a commercial supplier. The elastomer kit consists of a resin and cross-linking agent and used as per manufacturer's instruction. White sugar 'cubes' to be used as templates and three types of oil (olive, sunflower and vegetable) were all purchased from a local supermarket.

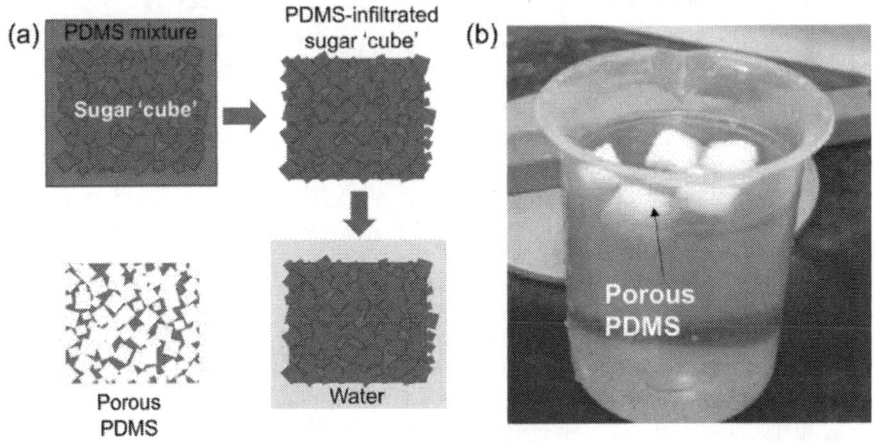

Figure 1. (a) Schematic process flow to fabricate porous PDMS. **(b)** Oil absorption test showing five porous PDMS 'cuboids' in an 2 % (by volume) oil and water suspension.

2.1. Fabrication of porous PDMS

Porous PDMS were manufactured using the templating method with sugar 'cubes' used as templates. Figure 1(a) is a schematic of the process flow. Polydimethylsiloxane resin and crosslinking agent were mixed in the ratio 1 part cross link agent to 10 parts resin as per manufacturer's recommendations. Mixing was done gently to prevent bubbles from forming. The mixture was then poured onto sugar 'cubes' which are then stored under vacuum at room temperature overnight. The PDMS-infiltrated sugar 'cubes' were then placed in boiling water for 15 to 20 mins to dissolve away sugar. After the sugar has been completely removed, porous PDMS foams were dried thoroughly.

2.2. Characterization of structural, mechanical and oil absorption properties

Microstructure of porous PDMS fabricated was investigated using optical microscopy (Nikon). Micrographs obtained were analysed using an open source image analysis software (ImageJ). Average diameter was computed from at least twenty pores observed on the surface of porous PDMS.

Mechanical property of porous PDMS at different strain rates was compared using compression tests using an Instron test machine with a load of cell of 5 kN. Tests were done at constant compression rates by means of Bluehill software. All tests were carried out at room temperature and ambient humidity.

Oil absorption ability was investigated by soaking porous PDMS in a suspension of 2 % oil in water as shown in figure 1(b). Mixtures of oil and water were mixed thoroughly before porous PDMS samples were added. During an oil absorption test five porous PDMS samples were placed in a beaker and the time taken for the PDMS cubes to absorb oil completely was recorded. The average of three closest time durations were computed. Two types of oil absorption experiments were carried out; with and without mechanical agitation. Agitation was induced using a proprietary equipment that applied low amplitude vibration on the base of the beakers.

3. Results and discussion

A typical porous PDMS sample formed is shown in figure 2. Typical porous PDMS formed are cuboids of dimensions 18.5 x 18.5 and 10.5 mm, figure 2(a). Porous PDMS has an open cellular structure and the pores were observed to be cuboids; of similar shape to particles in the sugar 'cube' templates. Each pore connects to adjacent ones to form an interconnected network resulting in the open cell network. Thickness of cell walls range in size between about 25 to 750 μm. Diameter of pores in foams range between about 50 to 500 μm, figure 2(b).

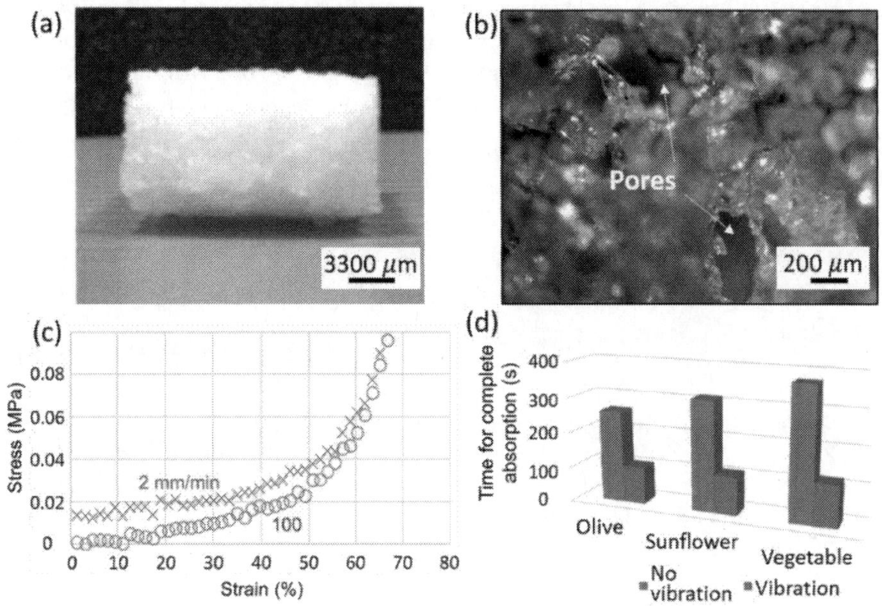

Figure 2. (a) Fabricated porous PDMS cuboid. (b) Optical micrograph showing hexahedral pores which are replicated from particles in the sugar cube. (c) Representative stress-strain curves at 2 and 100 mm/min. (d) Time for complete absorption is reduced significantly with vibration.

Figure 2(c) shows representative stress-strain curves of porous PDMS when compressed at 2 and 100 mm/min. These curves are typical of PDMS during compression with two distinct regions; a gradual linear elastic region followed by another in which stress increases rapidly in a non-linear manner with strain [7]. There is no significant difference between compression curves at various compression rates used in this study although porous PDMS were observed soften at higher compression rates. This observation suggests that the energy cost of compressing oil-laden porous PDMS to discharge the absorbed oil will be lower when compression is carried out at higher compression rates.

During initial stages of compression, ligaments in porous PDMS deforms which leads to pore collapse. Ligament deformation does not require large force which explains the gradual increase in stress over this region. As compressive strain increases, ligament deformation and pore collapse continues until most, if not all, pores in the porous PDMS collapsed. This process occurs between a strain of about 50 - 60 %. As strain increase further, densification of porous PDMS occurred which involved compression of the 'densified' porous PDMS. At this stage, a small increase in strain results in a large increase in stress which explains the rapid increase in stress shown in figure 2(c).

Results of oil absorption tests are shown in figure 2(d). In general, porous PDMS absorbed between 44 % to 96 % by mass of oil relative to its mass. The speed at which oil is absorbed depended on whether

IOP Conf. Series: Materials Science and Engineering **342** (2018) 012050 doi:10.1088/1757-899X/342/1/012050

vibration was used. When vibration was not used, porous PDMS was able to completely absorb olive, sunflower and vegetable oil after 260, 310 and 370 s respectively. When vibration was applied during the oil absorption tests, the corresponding time durations were 105, 110 and 120 s respectively. Hence, vibration shortens duration needed to completely absorb olive, sunflower and vegetable oils by 2.5, 2.9 and 3.1 times, respectively. PDMS is a hydrophobic and oleophilic material. During absorption, PDMS has to uptake oil from the aqueous suspension. Uptake involves mass transfer through flow and diffusion processes. It is postulated that vibration accelerates such mass transfer processes by supplying energy to overcome hindrances during these processes.

4. Conclusion

In conclusion, porous polydimethylsiloxane (PDMS) was fabricated using sugar templating method. The uptake of olive, sunflower and vegetable oils by porous PDMS 'cuboids' with and without vibration was investigated. Applying small amplitude vibration significantly shortens the time duration needed for oil uptake. Time needed to absorb olive, sunflower and vegetable oils was shortened by 2.5, 2.9 and 3.1 times, respectively. Compressive stress-strain curves over compression rates between 2 and 100 mm per min are similar and indicate mechanical property of porous PDMS does not vary significantly and can be rapidly compressed.

References

[1] Adebajo M O, Frost R L, Kloprogge T J, Carmody O and Kokot S 2003 *J. Porous Mater.* **10** pp 159-170

[2] Zhao X, Li L, Li B, Zhang J and Wang A 2014 *J. Mater. Chem. A* **2** pp 18281–18287

[3] Choi S J, Kwon T H, Im H, Moon D I, Baek D J, Seol M L, Duarte J P and Choi Y K 2011 *ACS Appl. Mater. Interfaces* **3** pp 4552–4556

[4] Feng L, Zhong Z, Mai Z, Ma Y, Liu B, Jiang L and Zhu D 2004 *Angew. Chem., Int. Ed.* **43** pp 2012-2014

[5] Paoli F D, 2015 *MethodsX* **2** pp 374-378

[6] Amjadi M, Kim M S and Park I, 2014 *Proc. Int. Conf. on Nanotechnology (Toronto)* (IEEE-NANO) p 764

[7] Johnston I D, McCluskey D K, Tan C K L and Tracey M C, 2014 *J. Micromech. Microeng.* **24** pp 035017

Seebeck coefficient of synthesized Titanium Dioxide thin film on FTO glass substrate

R Usop[1], N K A Hamed[2], M M I Megat Hasnan[1], H Ikeda[3], M F M Sabri[4], M K Ahmad[2], S M Said[1] and F Salleh[1]

[1]Department of Electrical Engineering, Faculty of Engineering, University of Malaya, 50603 Kuala Lumpur, Malaysia
[2]Department of Electronic Engineering, Faculty of Electrical and Electronic Engineering, Universiti Tun Hussein Onn Malaysia, 86400 Johor, Malaysia
[3]Research Institute of Electronics, Shizuoka University, 432-8011 Shizuoka, Japan
[4]Department of Mechanical Engineering, Faculty of Engineering, University of Malaya, 50603 Kuala Lumpur, Malaysia

E-mail: faizsalleh@um.edu.my

Abstract. In order to fabricate a thermoelectric device on glass substrate for harvesting waste heat energy through house appliances, the Seebeck coefficient of translucent TiO_2 thin film was investigated. The TiO_2 thin film was synthesized by using hydrothermal method with F-SnO_2 coated glass as substrate. From scanning electron microscopy analysis, the synthesized TiO_2 thin film was found to be in nanometer-scale rod structure with a thickness of 4 μm. The Seebeck coefficient was measured in the temperature range of 300 – 400 K. The Seebeck coefficient is found to be in negative value which shows that synthesized film is an n-type semiconductor material, and is lower than the value of bulk-size material. This reduction in Seebeck coefficient of TiO_2 thin film is likely due to the low dimensional effect and the difference of carrier concentration.

1. Introduction

Global warming issues such as greenhouse gas emission have been a concern among researchers which encourage them to reuse waste heat as an alternative power generation resource [1, 2]. The ability of thermoelectric devices to directly convert waste heat energy into electrical energy without any moving parts, with high durability, and without maintenance has attracted the significant interest of researchers [3-5]. The thermoelectric conversion efficiency is determined by the figure-of-merit Z, which is proportional to the square of Seebeck coefficient S and electrical conductivity σ and inversely proportional to the thermal conductivity κ [6, 7]. High S and σ are needed to produce a large voltage with a small temperature difference and low κ is needed to reduce heat leakage and maintain temperature difference [8]. Moreover, the enhancement of Z is a difficult task since S, σ and κ are strongly dependent on the carrier concentration and not independent with each other which results a difficulty to find an optimum value [9, 10]. The introduction of nanostructure is theoretically predicted to enhance the S and reduce the κ simultaneously, and this approach has been shown to be one effective way of improving the Z [10, 11].

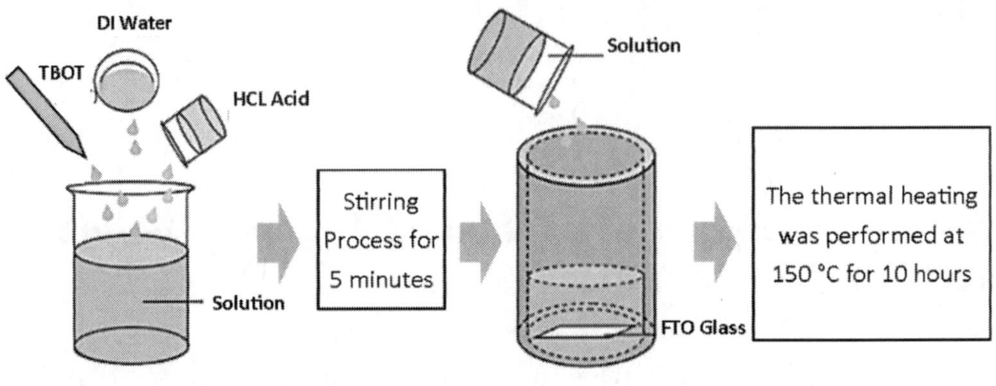

Figure 1. Schematic diagram of synthesis process flow.

Moreover, current thermoelectric material of high performance thermoelectric device is costly and hazardous [12-14]. Thus, material which is less expensive, chemically stable, and non-toxic is considered as an ideal substitution to be used for fabricating the thermoelectric device. In addition, for harvesting waste heat energy through glass based house appliances such as house window, a transparent and manufacturable material on glass substrate is needed. TiO_2 is one of the excellent candidate to full fill the requirement as it provides the desired features [15, 16]. However, its Z is comparatively low with other established thermoelectric materials [17]. It has been reported that the Z of TiO_2 can be enhanced by increasing the σ through impurity doping and by reducing the κ through nanostructuring [17-21]. However, there is no report regarding the enhancement of Z by increasing the S through nanostructuring. Therefore, in this study, we synthesized a translucent nanostructured TiO_2 thin film on F-SnO_2 coated (FTO) glass substrate and its S was measured in order to observe the enhancement of S in nanostructured TiO_2 thin film.

2. Experiment

The TiO_2 thin films discussed in this study was synthesized using the same method as used in our previous investigation and the synthesized procedures is shown in Figure 1 [22]. FTO glass was used as a substrate with a dimension of 10 × 20 mm^2. 120 ml of concentrated Hydrochloric (HCl) acid was dissolved in 120 ml of deionized (DI) water. The mixture was stirred for five minutes until a homogenous solution was obtained. An amount of 5 ml of $C_{16}H_{36}O_4Ti$ (TBOT) was then carefully dripped into the solution by pipette. The solution was inserted into a steel-made autoclave after it was stirred for ten minutes, together with the FTO glass substrate which was placed in the autoclave with the conducting FTO layer surface facing upward. The hydrothermal process was performed at 150 °C for ten hours.

The surface morphology of the samples was investigated by using field-emission scanning electron microscopy (FESEM). The thickness of the sample was estimated from the FESEM image. In this study, the crystallinity of the sample was not analysed since the sample was synthesized in the same manner as in previous study [22]. The thermoelectromotive force and temperature difference of the sample were measured by using conventional S measurement system (ULVAC ZEM-3) in the temperature range of 300 – 400 K. The S was evaluated from the gradient of the linear relation between the thermoelectromotive force and temperature difference which were measured simultaneously.

3. Results and Discussion

Figure 2 a) and b) show the FESEM images of the cross section and the top of sample, respectively. From this figure, it is revealed that the sample is composed with dense TiO_2 nanorod layer which are uniformly deposited on FTO layer. It is also found that the nanorod TiO_2 layer has a thickness of 4 μm. From the inset in Figure 2 b), it is observed that the grown nanorods is in square shape with average

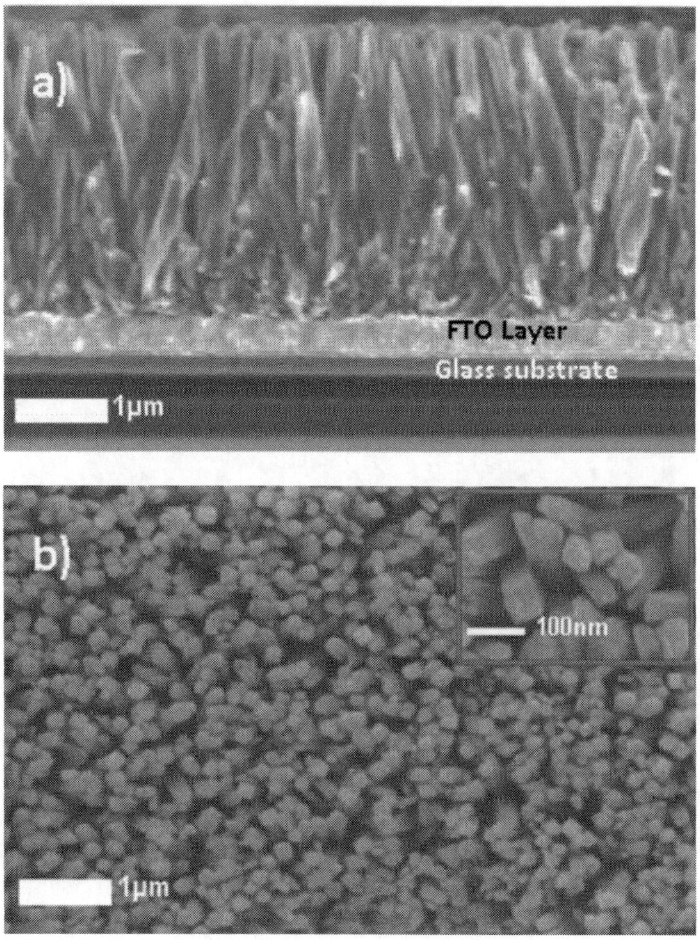

Figure 2. FESEM images of a) cross section and b) top of synthesized sample, and inset shows a magnified image.

width approximately 30 nm. Moreover, from the XRD pattern analysis in previous study, it is confirmed that the synthesized nanorod TiO_2 thin film is rutile since only peaks of the rutile phase were observed [22]. Thus, it is suggested that a uniform and well vertically oriented nanorod TiO_2 thin film can be reproducible by using this synthesis method.

The evaluated S is shown in Figure 3 as a function of average temperature. The filled diamonds represent the reported values of S of bulk TiO_2 and filled triangles represent the evaluated S of FTO glass substrate. From Figure 3, the evaluated S is found to be in negative value which indicates that the synthesized sample is an n-type semiconductor and in good agreement with the polarity of reported bulk TiO_2 [23, 24]. This n-type behaviour of the sample is considered due to the electrons that originating from the Ti and the oxygen vacancies [25]. The evaluated S is also found to be higher than the evaluated S of FTO glass substrate which confirms that the measured S is contributed by nanorod TiO_2 layer. Moreover, the S of the sample is observed to be almost constant in the temperature range of 300 – 370 K and then slightly increases with increasing the temperature. This fact shows that the synthesized TiO_2 thin film on FTO glass substrate is stable for application near room temperature and further investigation is needed for clarifying the increase in S at higher temperature.

However, by comparing the S of synthesized sample and the reported values, it is found that the difference is large compared with the reported value by Kitagawa et al. and small compared with the reported value by Mikami et al. [23, 24]. In addition, the value of S of synthesized sample is close to

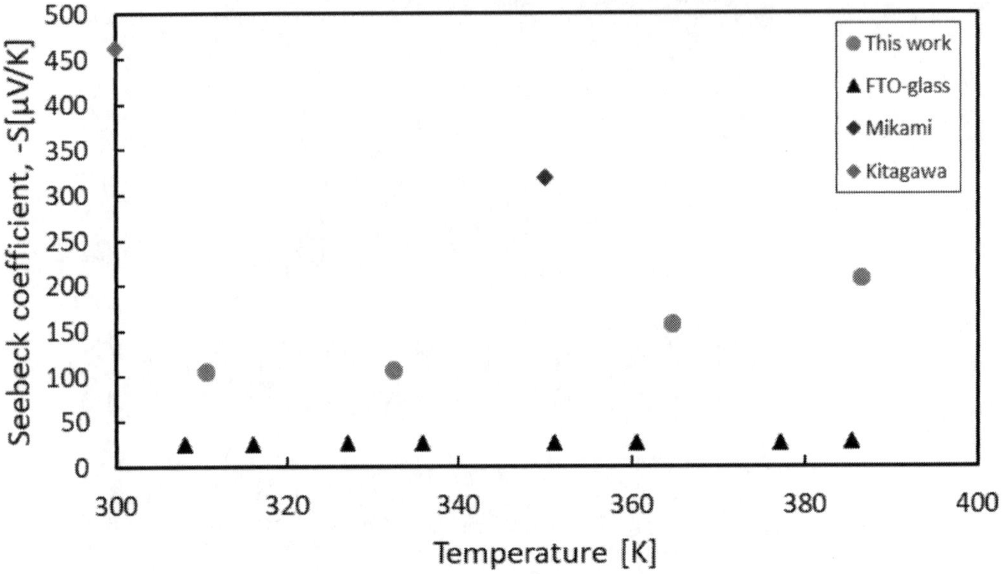

Figure 3. Evaluated S as a function of average temperature. The filled triangle and diamonds represent the evaluated S of FTO glass and reported S of bulk TiO_2, respectively [23, 24].

the calculated value of bulk rutile TiO_2 with carrier concentration of $5x10^{20}$ cm^{-3} by Bayerl et al. using Boltzmann transport formalism which is around -200 μV/K [26]. That is, one of the reasons of the difference in S between the synthesized sample and reported values is likely due to the difference of carrier concentration in the sample. On the other point of views, the dimension of the sample is also different between the samples which is likely influenced the S of the sample. Salleh et al. has reported that by decreasing the cross section of a Si thin film, the contribution of phonon system to S will be eliminated which results a decrease in total S of the sample even at room temperature [27]. Therefore, there is a possibility that the decrease in S of nanorod TiO_2 film is likely due to elimination of phonon contribution part in total S through decreasing the sample dimension, and further investigation is needed.

4. Conclusion

S of nanorod TiO_2 thin films deposited on FTO glass substrate by using hydrothermal method has been investigated. The S is found to be in negative value which shows an n-type semiconductor behaviour and in a good agreement with reported bulk TiO_2. However, the value of S of synthesized sample is found to be much lower than the reported values. This is likely due to the difference in carrier concentration or the elimination of phonon system contribution in S through low dimensional affect, and further investigation is needed. Moreover, the S of the sample is found to be nearly constant around room temperature. Consequently, the TiO_2 thin film deposited on FTO glass substrate is possible to be utilised in near room temperature applications.

Acknowledgments

This work was financially supported by the Fundamental Research Grant Scheme (FP062-2016) from the Ministry of Higher Education and the University Malaya Research Fund (BK077-2015) from University of Malaya.

References

[1] Nam, W. H., Lim, Y. S., Choi, S. M., Seo, W. S., & Lee, J. Y. 2012. High-temperature charge transport and thermoelectric properties of a degenerately Al-doped ZnO nanocomposite. *Journal of Materials Chemistry*, **22**(29), 14633-14638.

[2] DiSalvo, F. J. 1999. Thermoelectric cooling and power generation. *Science*, **285**(5428), 703-706.

[3] Riffat, S. B., & Ma, X. 2003. Thermoelectrics: a review of present and potential applications. *Applied thermal engineering*, **23**(8), 913-935.

[4] Orr, B., Akbarzadeh, A., Mochizuki, M., & Singh, R. 2016. A review of car waste heat recovery systems utilising thermoelectric generators and heat pipes. *Applied Thermal Engineering*, **101**, 490-495.

[5] Agarwal, R. K. 2012. A review on thermocouple for power generation. *International Journal of Applied Science and Engineering Research,* **1**(2), 98-105.

[6] Imai, H., Shimakawa, Y., & Kubo, Y. 2001. Large thermoelectric power factor in TiS_2 crystal with nearly stoichiometric composition. *Physical Review B*, **64**(24), 241104-4.

[7] Snyder, G. J., & Toberer, E. S. 2008. Complex thermoelectric materials. *Nature materials*, **7**(2), 105-114.

[8] Rubenis, K., Populoh, S., Thiel, P., Yoon, S., Müller, U., & Locs, J. 2017. Thermoelectric properties of dense Sb-doped SnO_2 ceramics. *Journal of Alloys and Compounds*, **692**, 515-521.

[9] Ge, Z. H., Zhao, L. D., Wu, D., Liu, X., Zhang, B. P., Li, J. F., & He, J. (2016). Low-cost, abundant binary sulfides as promising thermoelectric materials. *Materials Today*, **19**(4), 227-239.

[10] Rosi, F. D. 1968. Thermoelectricity and thermoelectric power generation. *Solid-State Electronics*, **11**(9), 833IN1849-848IN2868.

[11] Hicks, L. D., & Dresselhaus, M. S. 1993. Effect of quantum-well structures on the thermoelectric figure of merit. *Physical Review B*, **47**(19), 12727.

[12] Vineis, C. J., Shakouri, A., Majumdar, A., & Kanatzidis, M. G. 2010. Nanostructured thermoelectrics: big efficiency gains from small features. *Advanced Materials*, **22**(36), 3970-3980.

[13] Li, F., Li, J. F., Zhao, L. D., Xiang, K., Liu, Y., Zhang, B. P. Lin, Y.H., Nan, C. W. & Zhu, H. M. (2012). Polycrystalline BiCuSeO oxide as a potential thermoelectric material. *Energy & Environmental Science*, **5**(5), 7188-7195.

[14] Chen, R. 2013. Silicide nanopowders as low-cost and high-performance thermoelectric materials. *Jom*, **65**(6), 702-708.

[15] Su, L., & Gan, Y. X. 2012. Experimental study on synthesizing TiO_2 nanotube/polyaniline (PANI) nanocomposites and their thermoelectric and photosensitive property characterization. *Composites Part B: Engineering*, **43**(2), 170-182.

[16] Zaitsev, S. V., Moon, J., Takagi, H., & Awano, M. 2000. Preparation and characterization of nanocrystalline doped TiO_2. *Advanced Powder Technology*, **11**(2), 211-220.

[17] Zi-Hua, W., Hua-Qing, X., Yong-Biao, Z., Liang-Hua, G., & Jun, L. 2015. Enhanced thermoelectric performance of TiO_2-based hybrid materials by incorporating conducting polymer. *Chinese Physics B*, **24**(3), 034402-4.

[18] Kitagawa, H., Kunisada, T., Yamada, Y., & Kubo, S. 2010. Effect of boron-doping on thermoelectric properties of rutile-type titanium dioxide sintered materials. *Journal of Alloys and Compounds*, **508**(2), 582-586.

[19] Xu, L., Garrett, M. P., & Hu, B. 2012. Doping effects on internally coupled seebeck coefficient, electrical, and thermal conductivities in aluminum-doped TiO_2. *The Journal of Physical Chemistry C,* **116**(24), 13020-13025.

[20] Liu, C., Miao, L., Zhou, J., Huang, R., Fisher, C. A., & Tanemura, S. 2013. Chemical Tuning of TiO_2 Nanoparticles and Sintered Compacts for Enhanced Thermoelectric Properties. *The Journal of Physical Chemistry C*, **117**(22), 11487-11497.

[21] Liu, C., Miao, L., Zhou, J., Huang, R., & Tanemura, S. 2012. Bottom-up assembly to Ag

nanoparticles embedded Nb-doped TiO_2 nanobulks with improved n-type thermoelectric properties. *Journal of Materials Chemistry*, **22**(28), 14180-14190.

[22] Ahmad, M. K., Mohan, V. M., & Murakami, K. 2015. Hydrothermal growth of bilayered rutile-phased TiO_2 nanorods/micro-size TiO_2 flower in highly acidic solution for dye-sensitized solar cell. *Journal of Sol-Gel Science and Technology*, **73**(3), 655-659.

[23] Mikami, M., & Ozaki, K. 2012. Thermoelectric properties of nitrogen-doped TiO_{2-x} compounds. In Journal of Physics: Conference Series, **379**(1), 012006-6.

[24] Kitagawa, H., Kunisada, T., Yamada, Y., & Kubo, S. 2010. Effect of boron-doping on thermoelectric properties of rutile-type titanium dioxide sintered materials. *Journal of Alloys and Compounds*, **508**(2), 582-586.

[25] Morgan, B. J., & Watson, G. W. 2010. Intrinsic n-type defect formation in TiO_2: a comparison of rutile and anatase from GGA+ U calculations. *The Journal of Physical Chemistry C*, **114**(5), 2321-2328.

[26] Bayerl, D., & Kioupakis, E. 2015. Theoretical limits of thermoelectric figure of merit in n-type TiO_2 polymorphs. *Physical Review B*, **91**(16), 165104-5.

[27] Salleh, F., Oda, T., Suzuki, Y., Kamakura, Y., & Ikeda, H. 2014. Phonon drag effect on Seebeck coefficient of ultrathin P-doped Si-on-insulator layers. *Applied Physics Letters*, **105**(10), 102104-4.

Development of anthropomorphic robotic hand driven by Pneumatic Artificial Muscles for robotic applications

Mohannad Farag[1], Norsinnira Zainul Azlan[2] and Mohammed Hayyan Alsibai[1]

[1] Faculty of Engineering Technology, Universiti Malaysia Pahang, 26600 Pekan, Pahang, Malaysia.
[2] Department of Mechatronics Engineering, International Islamic University Malaysia, 53100 Kuala Lumpur, Malaysia.

E-mail: mohanadfaraj@gmail.com

Abstract. This paper presents the design and fabrication of a three-fingered anthropomorphic robotic hand. The fingers are driven by tendons and actuated by human muscle-like actuators known as Pneumatic Artificial Muscle (PAM). The proposed design allows the actuators to be mounted outside the hand where each finger can be driven by one PAM actuator and six indirectly interlinked tendons. With this design, the three-fingered hand has a compact size and a lightweight with a mass of 150.25 grams imitating the human being hand in terms of size and weight. The hand also successfully grasped objects with different shapes and weights up to 500 g. Even though the number of PAM actuators equals the number of Degrees of Freedom (DOF), the design guarantees driving of three joints by only one actuator reducing the number of required actuators from 3 to 1. Therefore, this hand is suitable for researches of robotic applications in terms of design, cost and ability to be equipped with several types of sensors.

1. Introduction

In robotics, an end effector is the device at the end of a robotic arm, designed to perform certain tasks depending on the application. A robotic hand refers to a particular kind of end-effector with an anthropomorphic inspiration. Numerous researchers have attempted to create robotic end-effectors that can match the human hand in terms of performance and abilities. [1–3]. Naturally, the human hand contains more than 19 muscles, 17 joints, 19 bones, not counting ligaments, nerves, and thousands of various sensors [4]. Therefore, the development a robotic hand that matches the performance of the human hand is a highly challenging task.

Robotic hand actuators can drive the joints either directly or indirectly. So, the mechanisms of robotic hand can be categorized into direct-driven and wire-driven joint. The former contains an actuator in every active joint which means mismatching with human hand in terms of size and weight. In contrast, the wire-driven structure enables the actuators to be mounted outside of the hand which in its turn reduces the hand's size and weight [2]. By comparing the two mentioned mechanisms, it can be seen that even there are some of limitations in the wire-driven joint mechanism, such as mechanical design and the driving force, it is still the right choice for anthropomorphic hand design in terms of size and weight. For example, Ishihara et al. (2011) developed a motorized anthropomorphic hand using wire-driven joints. However, some problems such as friction between a wire and its guide made its motion unstable [2]. Pneumatic FRH-4 hand developed by Gaiser et al. (2008) is actuated by flexible fluidic actuators (FFA) constitutes a hybrid concept of an anthropomorphic five-fingered hand and a three jaw robotic gripper [5]. Due to the lack of commercial availability of FFAs, other type of actuators should be used in this research. The other example of direct-driven joints is the hand developed by Maeda et al. (2011). This hand is actuated by thirteen miniaturize PAMs actuators [3]. The size of this hand is larger than human hand size due to the length of pneumatic artificial muscles actuators which violates the anthropomorphism concept. By reviewing 25 different robotic hands, the following points can be concluded [4]. First, three fingered hands are most common in robotic community because they can perform different grasping and manipulating tasks. Second, although the development of fully actuated hands has overshadowed underactuated hands, the latter have appeared on recent researches as they reduce the mechanism complexity, hand weight and size and manufacturing cost. Third, even though the fluid actuators have less preferred, but they are still the best choice in terms of safety. The Pneumatic Artificial Muscle (PAM) actuators are widely used as fluidic pneumatic actuators in the robotic and industrial fields due to their characteristics such as high power to weight and power to volume ratios, identical behaviour with human muscles, easy implementation, uncomplicated design, flexible architecture, simple working principle, an efficient performance and safe usage [1,3,6–8]. The PAMs are also known as McKibben actuators that were developed by McKibben and Gaylord in the 1950s [5]. PAM actuator contracts when it is activated by pressurized air like the biological muscle. Typically, PAMs can contract up to 25% of their initial length. Even though some recent prototypes of the robotic hands are overdesigned [4], a trend toward simplifying the hardware design based on application requirements is noted in the art of robot hands [9,10]. In addition, the implementation of simple control approaches causes a lack of robustness in terms of position control. Therefore, the approach of a simple hand design with an advanced control strategy can be adopted where this hand might be used.

A three-fingered pneumatic anthropomorphic hand has been developed in this research. Three wire driven joints are actuated by one PAM actuator for each finger. This design allows the actuators to be mounted outside the hand to achieve the anthropomorphic concept in terms of size, weight and shape. The hand can grasp objects with different shapes and weights up to 500 g and can be used in robotic applications in terms of design, cost and ability to be equipped with several types of sensor

2. Design of Robotic Hand

2.1. Working Principle of the Robotic Finger

The schematic drawing of the robotic finger is shown in Figure 1. The way in which the pulleys and links are attached is based on M. Tatsumi et. al. work where only one finger was driven by PAM actuator using miniaturized unconstrained valves. In this research, the number of fingers is increased to build a full robotic hand to perform grasping tasks of weight-varying objects. The working principle of the robotic finger is described as follows. The free end of the PAM is connected to Pulley 3 in the finger by an inextensible string made up of stainless steel. Since the string can only pull and not push, the PAM actuator is able to bend the finger in a unidirectional motion. The passive extension spring is attached to the opposite side of Pulley 3 allowing it to expand the finger. Link 4 and Pulley 2 are connected by red and blue tendons while Link 3 and Pulley 1 are connected by brown and green tendons. Pulleys 1, 2 and 3 are fixed onto Links 1, 2 and 3 respectively. This mechanism allows the finger to bend under the contraction force F_m produced by PAM actuator and to expand under the spring restoration force F_s. When the PAM actuator is activated, Pulley 3 rotates anticlockwise under the contraction force F_m and Pulleys 1 and 2 rotate in the same direction under pulling forces of brown and red tendons respectively. As a result, the three pulleys revolve with the same angle θ and the finger bends. When the PAM actuator is deactivated, Pulley 3 rotates clockwise under the spring restoration force F_s Pulleys 1 and 2 rotate in the same direction under pulling forces of green and blue tendons respectively, and the finger expands. This mechanism introduces the advantages of a small number of the required actuators, uncomplicated design, compact size and ability to grasp diverse sizes of objects. However, the grasp force produced at the fingertip is smaller than the contraction force produced by PAM actuator.

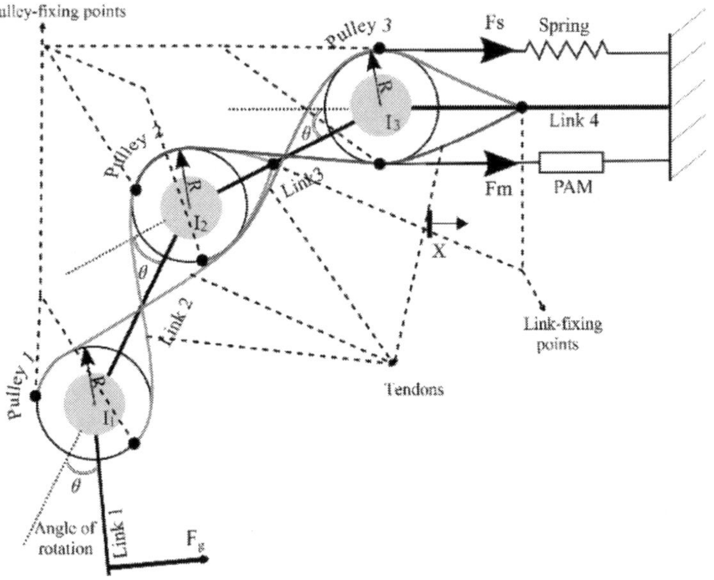

Figure 1. Schematic diagram of the robotic finger.

2.2. Robotic Hand Prototype

The three-fingered robotic hand prototype has been designed using computer aided design software and has been fabricated using 3D print technology and CNC milling machine as shown in Figure 2. The dimensions of the finger and the palm have been chosen to be close to human's as shown in Figures 3 and 4 respectively. The angles between middle, index and thumb fingers are shown in Figure 5. In design perspective, the thumb is given an initial angle of 49.53° with the palm surface to allow it bending across the palm and to give more power grasping. As the object's weight is our concern in this research rather than its shape, this design allows the robotic hand to perform a power grasping of objects with varying weights.

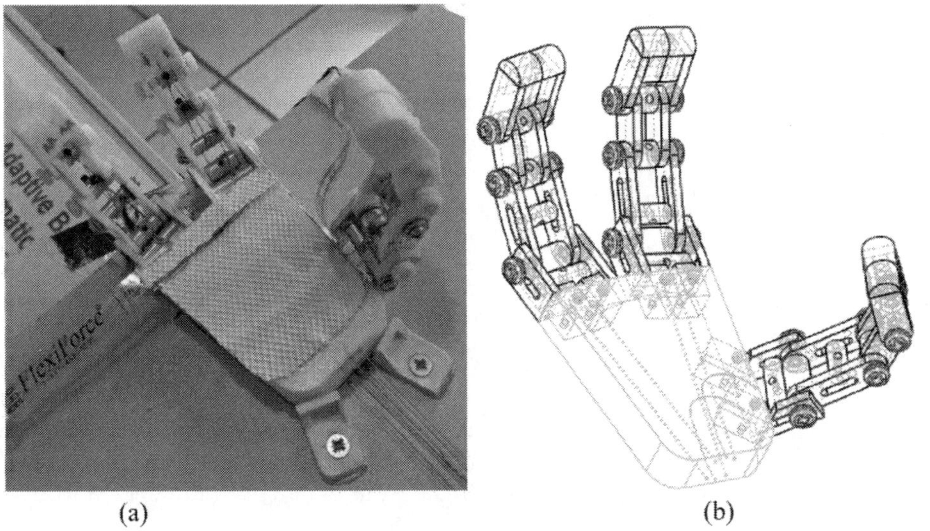

(a) (b)

Figure 2. Robotic hand prototype (a) mechanical drawing, (b) fabricated hand.

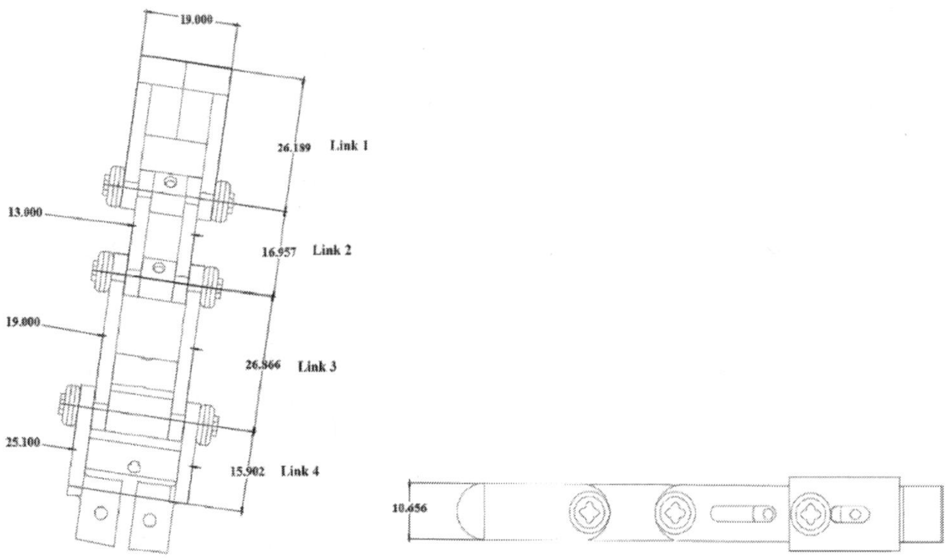

(a) (b)

Figure 3. Mechanical dimensions of the robotic finger, (a) top view, (b) side view.

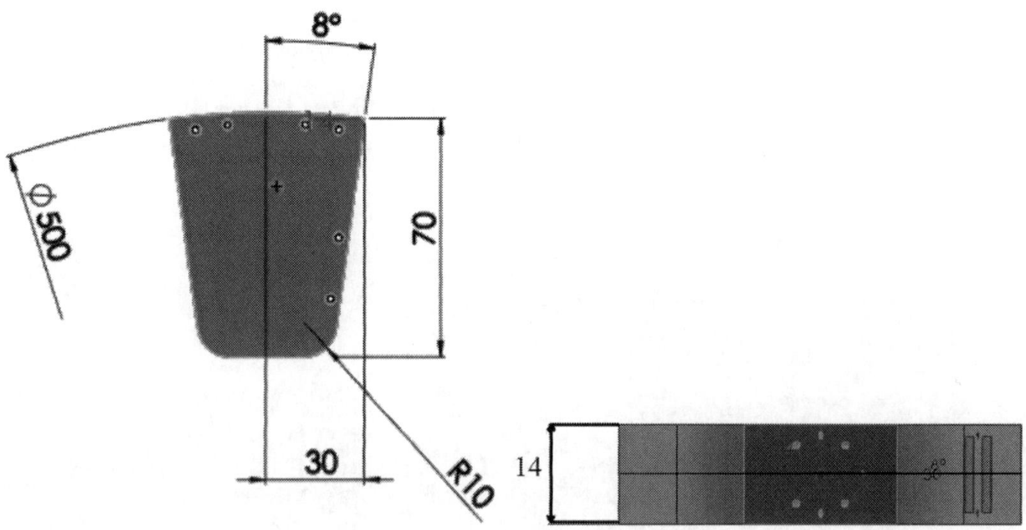

(a) (b)

Figure 4. Mechanical dimensions of the palm, (a) top view, (b) front view.

The finger is composed of 19 parts as shown in Figure 6. This design enables the modification of any hand's part to be made easily during assembling progress. The fingers are installed onto the palm shown in Figure 7. The tendons pass through tiny paths in the palm to be linked to Pulley 3 as illustrated in Figure 1. The pulley fixing point, shown in Figure 8, is located on $\varphi = 111^{\circ}$ and this value has been determined experimentally. This location allows a bigger rotation angle θ for the pulleys producing a power grasping by the robotic hand [11]. Table 1 summarizes the links and pulleys that represent the finger's parts in the schematic diagram shown in Figure 1.

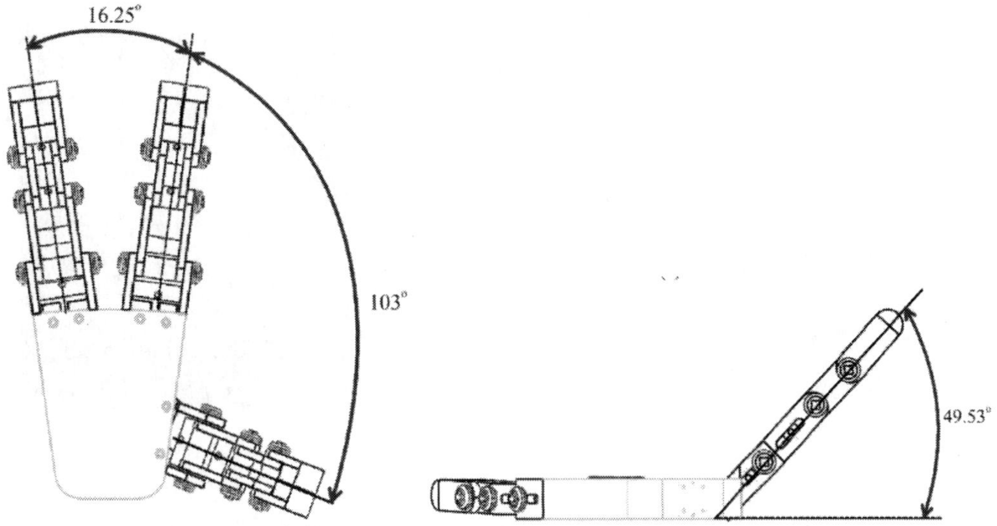

(a) (b)

Figure 5. Angles between the robotic fingers, (a) middle finger, index finger and the thumb (b) the thumb's initial angle.

Two types of material have been used to fabricate the hand in two different manufacturing processes which are the Acrylonitrile Butadiene Styrene (ABS) plastic using 3D printing technology and aluminum using CNC milling machine. Table 2 shows the manufactured material and its quantity. Referring to the table, seventy-six parts should be assembled to build the complete hand. After assembling the three fingers, the pulleys and the pins are linked tightly based on Figure 1. Then, the robotic hand can be actuated by human hand to test the motion of the fingers. Finally, the hand is ready to be installed on the experiment setup and perform grasping task.

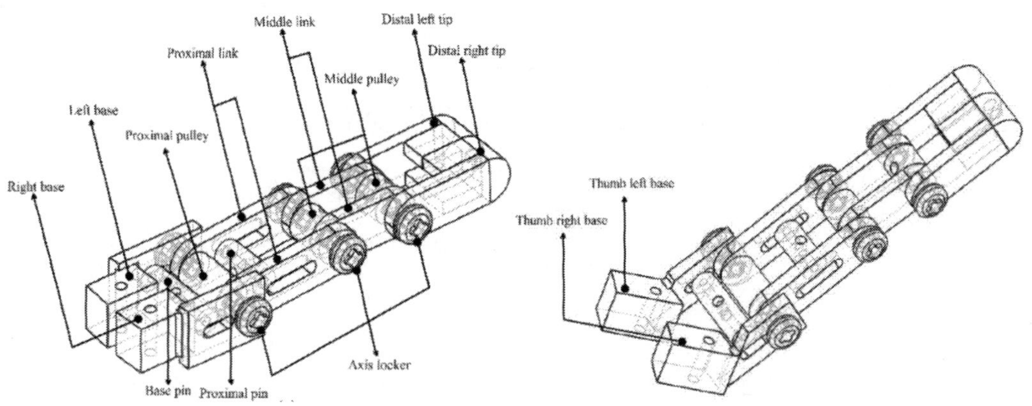

(a) (b)

Figure 6. Robotic fingers parts, (a) middle and index fingers, (b) thumb.

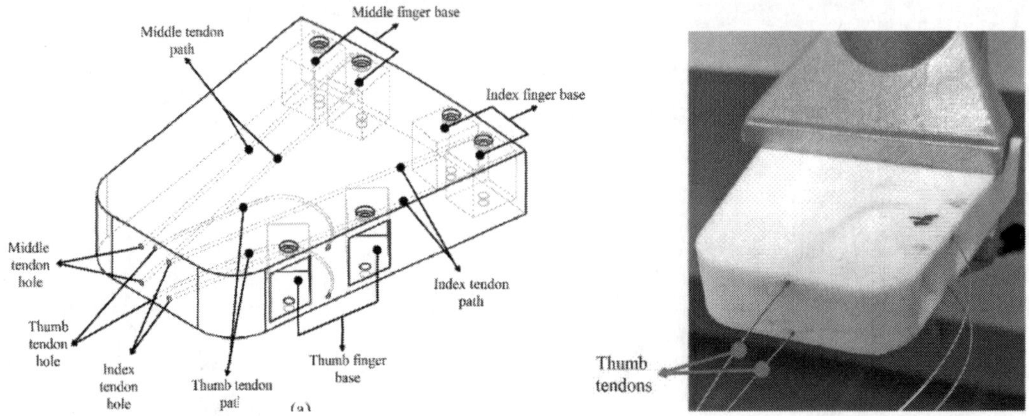

(a) (b)

Figure 7. Robotic palm, (a) tiny paths of the tendons and the fingers bases, (b) thumb tendons passing through the tiny paths.

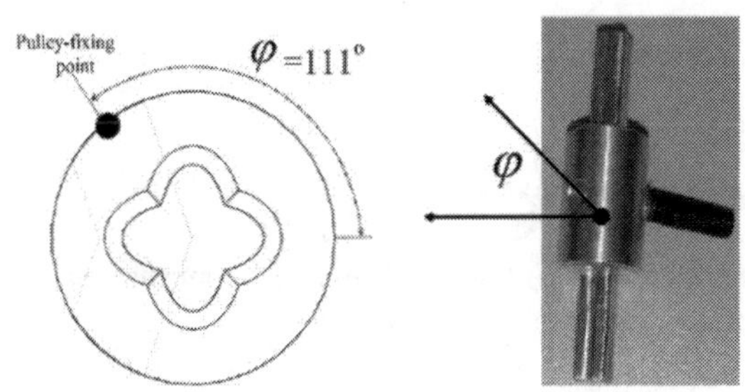

(a) (b)

Figure 8. Pulley-fixing point, (a) back view, (b) real view of the pulley.

Table 1. Finger parts presenting schematic links and pulleys.

Link / Pulley in Figure 1	Finger's part in Figure 6
Link 1	Distal right / left tip
Link 2	Middle link
Link 3	Proximal link
Link 4	Right / Left base
Pulley 1	Middle pulley
Pulley 2	Middle pulley
Pulley 3	Proximal pulley

Table 2. Manufacturing material of hand parts.

ABS		QTY	Aluminium		QTY
Palm	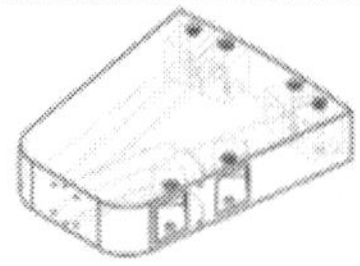	1	Right base	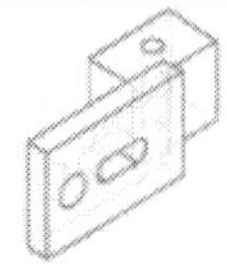	2
Middle Link	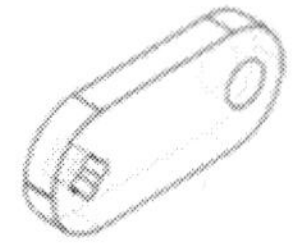	6	Left base	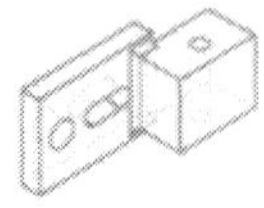	2
Proximal Link	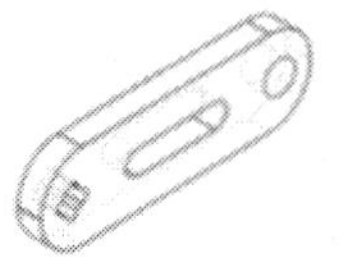	6	Thumb right base	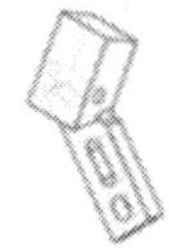	1
Axis locker		18	Thumb left base		1
Distal right tip	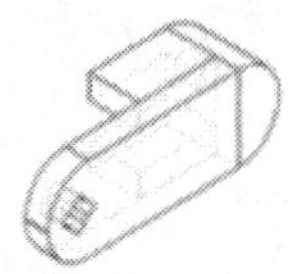	3	Proximal pulley	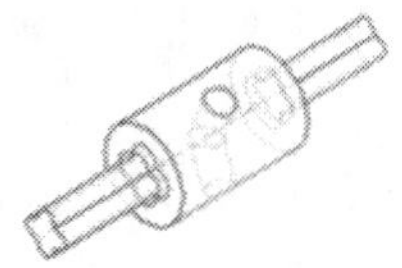	3
Distal left tip	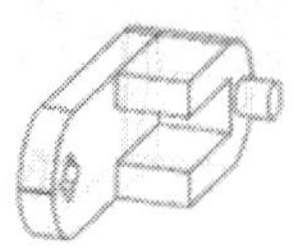	3	Middle pulley	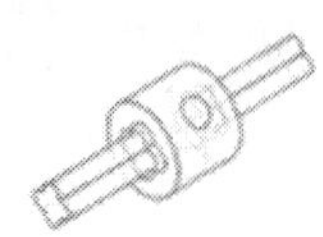	6
Steel		QTY	Base pin	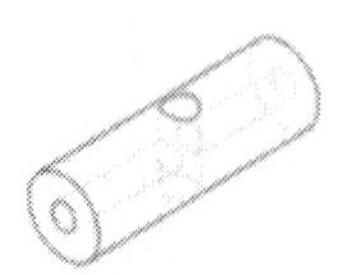	3
Headless screw	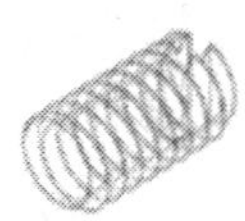	18	Proximal pin	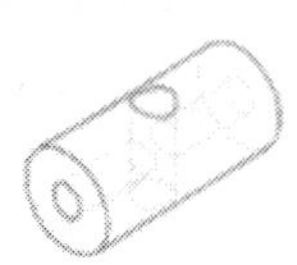	3

The anthropomorphic robotic hand is equipped with two position sensors attached to PAM actuators and two Force Sensing Resistor (FSR) sensors to measure grasping force. The position sensor is a spring return and a linear actuation potentiometer. It converts the mechanical travelling of the PAM actuator into a DC voltage signal in range of 0-5V. First FSR sensor is located on the palm, for middle and index fingers, and the second one is located on the side of the thumb's tip Table 3 summarizes the mechanical properties of the robotic hand and the dimensions of the fingers. The complete set of anthropomorphic robotic hand is shown in Figure 9.

Table 3. Mechanical properties of the robotic hand

Description	Value	unit
Mass of the hand	150.25	g
Volume of the hand	82940.73	mm^3
Finger's Moment of inertia	2.246e-5	$Kg.m^2$
Length of Link 1	26.189	mm
Length of Link 2	16.957	mm
Length of Link 3	26.866	mm
Length of Link 4	15.902	mm
Finger's length	70.012	mm
Finger's width	19.000	mm
Finger's thickness	10.656	mm
Pulley diameter	8.000	mm

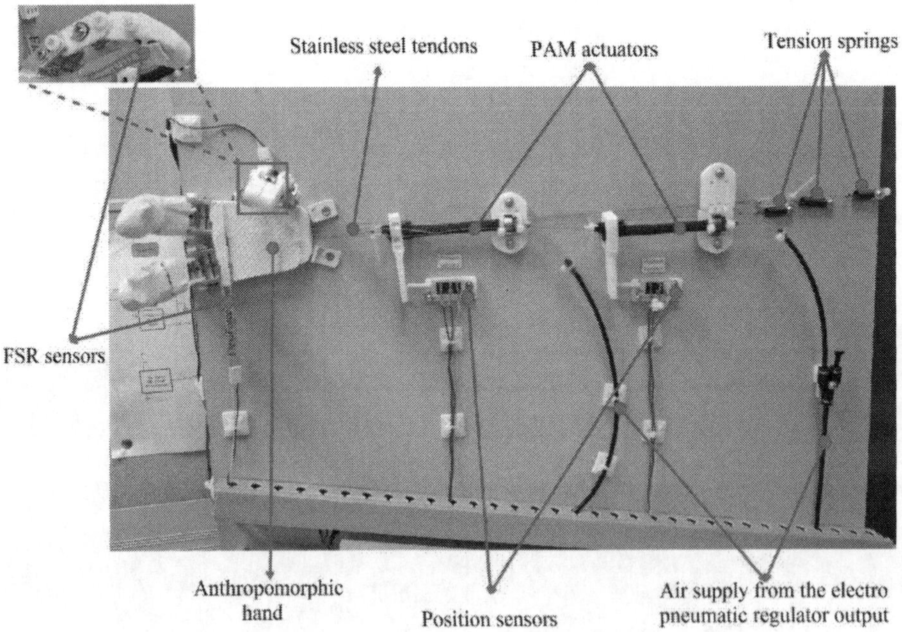

Figure 9. The complete set of anthropomorphic robotic hand.

3. Results and Discussion

Figure 10 shows the experiment results of grasping multiple shape objects with different weight where the force response can be detected using FSR sensors while the position of finger can be detected by position sensors. The maximum weight that the hand can grasp without slipping is 500 g. However, by increasing or decreasing the weight the grasping force can be changed accordingly. The robotic fingers driven by tendons are capable to grasp cylindrical-shaped objects with different diameters. The PAM actuator and the mechanical design have allowed the robotic hand to possess a light weight and a compact size imitating the human being hand and implementing the anthropomorphic concept.

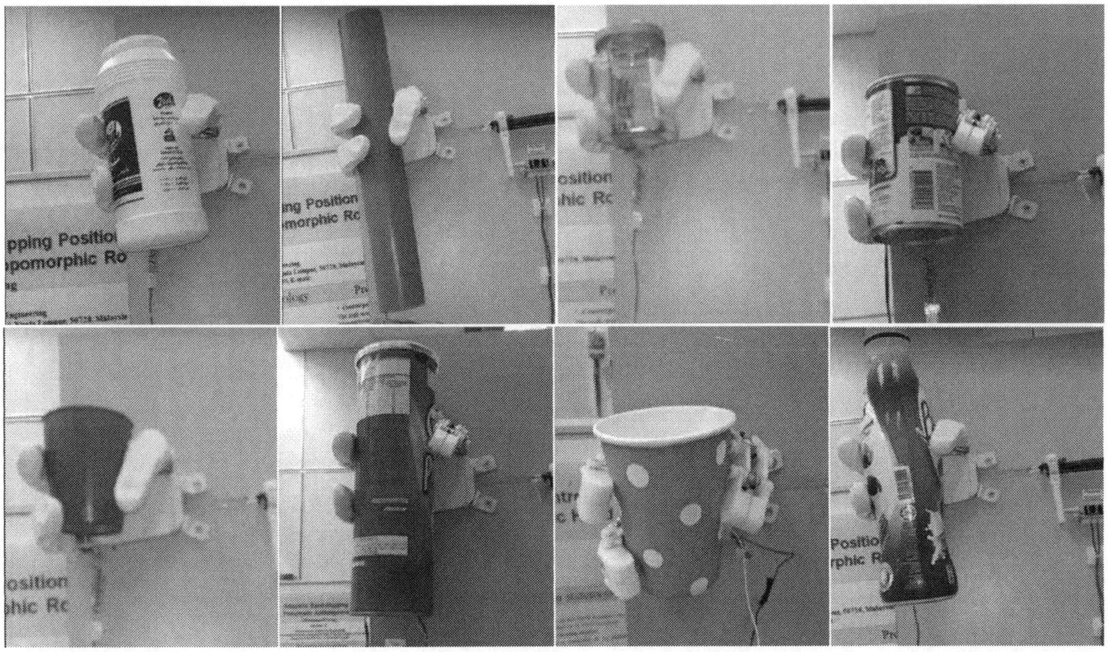

Figure 10. The anthropomorphic robotic hand grasping multiple shape objects with different weight.

In terms of the anthropomorphism concept, the size of the developed hand is closer to human hand size compared to the hand developed by Maeda et al. (2011) where 13 PAMs actuators have been used causing the large size of this hand [3]. In addition, using PAM actuators makes the developed hand more commercialized compared to pneumatic FRH-4 hand developed by Gaiser et al. (2008) where uncommercialized FFA actuators have been used [5].

4. Conclusion

In this work, a three-fingered anthropomorphic robotic hand driven by two PAM actuators has been developed. Based on the results it can be concluded that the proposed design allows the actuators to be mounted outside the hand where each finger can be driven by one PAM actuator and six indirectly interlinked tendons. With this design, the three-fingered hand has a compact size and a lightweight with a mass of 150.25 grams imitating the human being hand in terms of size and weight. The hand also successfully grasped objects with different shapes and weights up to 500 g. In addition, the hand has can avoid the object from dropping by detecting the slip signal generated by FSR sensors when the weight is increased. Therefore, the hand can maintain grasping objects with changing weights up to 500 g. Therefore, this hand is suitable for researches of robotic applications in terms of design, cost and ability to be equipped with several types of sensors.

Acknowledgements

The authors would like to thank Ministry of Education Malaysia for supporting this research under the Research Acculturation Grant Scheme (RAGS13-025-0088).

References

[1] Honda Y, Miyazaki F and Nishikawa A 2010 Control of pneumatic five-fingered robot hand using antagonistic muscle ratio and antagonistic muscle activity *2010 3rd IEEE RAS EMBS Int. Conf. Biomed. Robot. Biomechatronics, BioRob 2010* **400** 337–42

[2] Ishihara H, Hirose N, Funaoka S, Morita K and Miyake T 2011 Proposal of anthropomorphic hand with passive warped fingertip *2011 IEEE International Conference on Mechatronics*

IOP Conf. Series: Materials Science and Engineering **342** (2018) 012052 doi:10.1088/1757-899X/342/1/012052

and Automation (IEEE) pp 1219–24
[3] Maeda S, Tsujiuchi N, Koizumi T, Sugiura M and Kojima H 2011 Development and control of pneumatic robot arm for industrial fields *IECON Proc. (Industrial Electron. Conf.* **9** 86–91
[4] Birglen L, Laliberté T and Gosselin C 2008 *Underactuated Robotic Hands* vol 40 (Springer-Verlag Berlin Heidelberg)
[5] Gaiser I, Schulz S, Kargov A, Klosek H, Bierbaum A, Pylatiuk C, Oberle R, Werner T, Asfour T, Bretthauer G and Dillmann R 2008 A new anthropomorphic robotic hand *2008 8th IEEE-RAS Int. Conf. Humanoid Robot. Humanoids 2008* 418–22
[6] De Volder M, Moers a. J M and Reynaerts D 2011 Fabrication and control of miniature McKibben actuators *Sensors Actuators, A Phys.* **166** 111–6
[7] Tsujiuchi N, Koizumi T, Nishino S, Komatsubara H, Kudawara T and Hirano M 2008 Development of Pneumatic Robot Hand and Construction of Master-Slave System *J. Syst. Des. Dyn.* **2** 1306–15
[8] Tondu B 2012 Modelling of the McKibben artificial muscle: A review *J. Intell. Mater. Syst. Struct.* **23** 225–53
[9] Bicchi A 2000 Hands for dexterous manipulation and robust grasping: A difficult road toward simplicity *IEEE Trans. Robot. Autom.* **16** 652–62
[10] Cutkosky M R and Howe R D 1990 *Human Grasp Choice and Robotic Grasp Analysis* vol 1
[11] Farag M and Azlan N Z 2015 Adaptive Backstepping Position Control of Pneumatic Anthropomorphic Robotic Hand *Procedia Comput. Sci.* **76** 161–7

Effect of fast pyrolysis bio-oil from palm oil empty fruit bunch on bitumen properties

Chia Chin Poh, Norhidayah Abdul Hassan, Noor Azah Abdul Raman, Nurul Athma Mohd Shukry, Muhammad Naqiuddin Mohd Warid, Mohd Khairul Idham Mohd Satar, Che Ros Ismail, Sitti Asmah Hassan and Nordiana Mashros

Faculty of Civil Engineering, Universiti Teknologi Malaysia, Johor, Malaysia

E-mail: hnorhidayah@utm.my

Abstract. Bitumen shortage has triggered the exploration of another alternative waste material that can be blended with conventional bitumen. This study presents the performance of pyrolysis bio-oil from palm oil empty fruit bunch (EFB) as an alternative binder in modified bitumen mixtures. The palm oil EFB was first pyrolyzed using auger pyrolyzer to extract the bio-oil. Conventional bitumen 80/100 penetration grade was used as a control sample and compared with samples that were modified with different percentages, i.e., 5% and 10%, of pyrolysis EFB bio-oil. The physical and rheological properties of the control and modified bitumen samples were investigated using penetration, softening point, viscosity and dynamic shear rheometer (DSR) tests. Results showed that the addition of EFB bio-oil softened the bitumen with high penetration and a reduction in softening point, penetration index, and viscosity. However, the DSR results showed a comparable rutting resistance between the bitumen samples containing EFB bio-oil and virgin bitumen with a failure temperature achieved greater than 64°C.

1. Introduction

The increased demand for crude oil due to the increasing number of automobiles has significantly reduced the crude oil reserves worldwide. This phenomenon resulted in the increased price of bitumen. Thus, many experts attempt to develop economically and environmentally sustainable methods using alternative binders that are suitable for hot mix asphalt. Challenges are also faced by contractors involved in the road construction industry because of the continuous price fluctuation of crude oil due to its low availability and high demand. Bitumen is a viscoelastic material manufactured through the distillation of crude oil in refineries and has been widely used for the construction of asphalt pavement worldwide. The physical properties of bitumen are influenced by its chemical composition, which consists of high-molecular-weight hydrocarbons with some functional groups containing sulfur, nitrogen, and oxygen atoms [1]. In practice, the properties of bitumen change during its service life because of high axle load and extreme environmental effects, such as acid rain, chemical attack, and high surrounding temperatures. Repeated loads from the increased vehicle loads will affect the strength of pavement due to fatigue. Exposure to high temperature will gradually soften the bitumen with low stiffness, thereby becoming susceptible to permanent deformation. These issues have been evident for the last three decades. Pavement research has been facing more demands than before and thus requires the enhancement of the properties of bituminous materials.

Given that the performance of road pavement is mainly affected by the properties of bituminous binder, conventional bitumen often cannot provide the desired resistance to pavement distresses [2]. The use of modified bitumen is an important solution to change the properties of conventional bitumen and sustain the durability of the pavement. Some of these properties include elasticity and adhesive or cohesive strength. Modified bitumen currently has various applications, and bituminous binder has been improved by using various kinds of alternative materials. Polymers are the most commonly used modifying agents that can be effectively used as bitumen modifier. Different types of polymers, including polyethylene, polypropylene, styrene butadiene styrene, styrene butadiene rubber, have been described in previous investigations [3-6]. The potential utilization of various by-products in road construction has been recently investigated. The use of recycled materials, such as tire rubber, plastic wastes, waste engine oil, and cooking oil, as alternative binder modifiers has also been studied [7-11]. Agricultural waste products, such as oil palm fruit ash, palm oil fuel ash, and coconut shell charcoal ash, have also been used as alternative bitumen modifiers by taking advantage of the abundance of waste by-products from palm oil mill industry to help reduce environmental pollution [12-14]. The modification of bitumen with various additives offers several benefits, including improved stiffness at high temperature, cracking resistance at low temperatures, moisture resistance, fatigue life, rutting performance, and decrease in temperature susceptibility. Nanomaterials are widely utilized to improve the performance of asphalt pavement, especially for the modification of bitumen for flexible pavements. Various types of nanomaterials, such as nanoclay, nanosilica, and carbon nanotubes, were studied to reveal their effectiveness in improving the engineering properties of bitumen and mixture [15-17].

The bio-oil extruded from biomass resources, such as oakwood, switchgrass, rice husk, oil palm, and coconut shell, shows significant potential as a bitumen modifier because this bio-oil possesses similar chemical properties with petroleum binders [18]. Among the current technologies, thermochemical conversion processes, such as pyrolysis, gasification, and liquefaction, are the most commonly used to convert biomass materials into sustainable bio-oils [19]. Heating organic materials at high temperature in the absence of oxygen and consequently producing solids, bio-oil, and gases are basically a thermochemical process [20]. Pyrolysis processes are mainly classified into carbonization (very slow), conventional (slow), fast, and flash depending on the operating conditions. Slow pyrolysis is used primarily for charcoal or char production, whereas fast pyrolysis is mainly used to maximize the production of liquid or bio-oil [21]. Fast pyrolysis consists of four main features that help increase the liquid yield; these features include high heating rate, monitoring of reaction temperature in the range of 425 °C to 600 °C, short vapor residence times of less than 3 s, and rapid cooling of the product gas to produce bio oils [22]. These processes yield approximately 75 wt% of liquid bio-oils, 10 wt% to 15 wt% biochar, and 10 wt% to 15 wt% of condensable gases [23]. Bio-oils or pyrolysis oil are dark-brown, free-flowing liquids that are composed of several hundred organic compounds, mainly including acids, alcohols, aldehydes, esters, ketones, phenols, and lignin-derived oligomers [24]. Bio-oil has significantly high moisture content that usually varies from 15% to 30%, thereby contributing to its low viscosity and the difficulty of obtaining homogeneous samples from these bio-oils. The presence of water in biomass reduces the high heating value of bio-oil to below 19 MJ/kg as compared with 42 MJ/kg for conventional petroleum oil. This factor limits the application of bio-oil as a bitumen substitute in paving industry. The high water content of bio-oil can cause spontaneous phase separation, which will then lead to ignition and combustion problems and thus poor quality. This phenomenon will influence the behavior of bio-oil during pyrolysis and affect the physical properties and quality of the pyrolysis liquid [25]. Thus, the phase stability of bio-oil must be improved by reducing the water content and the aging of contributing light components, such as volatile aldehydes and ketones prior to pyrolysis to ensure the high quality and energy value of the bio-oil [26,27].

Yang *et al.* [28] investigated the performance of modified bitumen with bio-oil extracted through pyrolysis from waste wood materials. Three different types of bio-oils, namely, polymer-modified bio-oil (PMB), de-watered bio-oil (DWB) and original bio-oil (OB), were used as additives at 5% and 10% by weight into the virgin bitumen. Their study indicated that the addition of bio-oil improved the high temperature performance and lowered the mixing temperature of asphalt mixtures. The results also showed that OB-modified bitumen binders had the lowest stiffness, whereas the DWB- and PMB-modified bitumen binders had moderate and highest stiffness, respectively. Mohammad *et al.* [29] conducted a comprehensive laboratory investigation of asphalt mixtures containing bio-binder technology. Their findings showed that mixtures modified with bio-binder showed improved rutting resistance performance as compared with conventional mixtures. In addition, the mixtures containing bio-binder exhibited improved low temperature fracture performance and a tensile strength ratio of more than 80% in the moisture susceptibility test. Mills–Beale *et al.* [30] investigated the effect of the addition of bio-binder from swine manure on the rheological properties of bio-bitumen mixtures. Their results showed that bio-modified bitumen had low viscosity and improved thermal cracking performance as compared with conventional bitumen. Fini *et al.* [31] also reported that the swine manure-based bio-bitumen mixtures showed improvements in fatigue cracking resistance, rutting resistance, and low mixing and compacting temperature, which are attributed to the reduction in binder viscosity.

The bio-oil benefits the performance grade of bitumen only when the proper amount is blended or replaced into the conventional bitumen. On the contrary, the utilization of bio-binders derived from the biomass is advantageous because the materials are inexpensive and easily available. These materials can be used by existing asphalt industries to manage problems related to the availability and price of crude oil. The application of bio-oils for the modification of bitumen can improve the performance of asphalt pavements. Therefore, this study aims to investigate the physical and rheological properties of bitumen modified with bio-oils. Fast pyrolysis was conducted to extract the bio-oil from palm oil empty fruit bunch (EFB). The physical and rheological properties of the control and modified bitumen samples were investigated using penetration, softening point, viscosity, and dynamic shear rheometer (DSR) tests.

2. Experimental

2.1. Materials

The bio-oil used in this study was extracted from palm oil EFB through fast pyrolysis. Figure 1 shows the unprocessed and processed palm oil EFB for recycling of palm oil waste. This process involves a series of steps that included the extraction and subsequent purification of the bio-oil. Purification was conducted to remove the water content from the pyrolysis bio-oil, which usually contains a significant amount of condensed water due to the moisture of biomass materials. Elemental investigation of bio-oil was conducted using Automated CHNS Analyzer to determine the chemical composition of carbon (C), hydrogen (H), nitrogen (N), and Sulfur (S). Table 1 shows the comparison of the elemental analysis of pyrolysis EFB bio-oil and of virgin bitumen 80/100. Compared with that for virgin bitumen 80/100, the significantly low percentage by weight of C and H for pyrolysis EFB bio-oil indicates that this bio-oil contains less fraction of long-chain hydrocarbon, which leads to its soft behavior and low stiffness properties. The high sulfur content of bio-oil contributes to the binder viscosity reduction and significantly improves the workability of the mix. This characteristic allows the asphalt mixture to be prepared and compacted at low temperatures, resulting in low energy consumption [32]. For the sample preparation, bitumen with penetration grade 80/100 was used to blend with bio-oils at 140 °C. During asphalt modification, bio-oil was added at 5% and 10% by weight of the control sample. The mixture was blended properly by using high shear mixer at a constant speed of 1000 rpm for 1 h at 140 °C until it was uniformly mixed.

(a) (b)

Figure 1. (a) Unprocessed and (b) processed palm oil empty fruit bunch.

Table 1. Elemental analysis of EFB bio-oil and virgin bitumen 80/100

Elemental composition (wt%)	Bitumen 80/100	EFB Bio-oil
Carbon	81.6	41.9
Nitrogen	0.8	16.3
Hydrogen	11.6	4.68
Sulfur	0.08	18.6

2.2. Laboratory tests

2.2.1. Penetration. Penetration test was performed to determine the consistency of the modified bitumen sample. This test was conducted by determining the distance in tenths of a millimeter using a standard needle penetrated vertically into a sample of the material under fixed conditions of temperature, load, and time. The bitumen sample was heated and stirred until it became sufficiently fluid and then poured into a penetration cup. The sample was cooled at room temperature for an hour and then placed in a water bath for another hour at 25 °C. The penetration equipment was used to perform the test with the applied load of 100 g for 5 s at a temperature of 25 °C. The test was performed by placing the penetrometer outside the bath and in the transfer dish to maintain the temperature. Penetration index (PI) was calculated to measure the temperature sensitivity of bitumen. High PI corresponds to low temperature sensitivity, whereas low PI corresponds to great temperature sensitivity, indicating that the binder tends to change its consistency faster than the temperature change.

2.2.2. Softening point. Softening point test was conducted to determine the softening temperature of bitumen by using ring and ball apparatus. In this test, the heated modified bitumen sample was poured into rings and then cooled at room temperature for 30 min. The excess sample was removed with warm blade to ensure that the test specimens were leveled with the top of the rings. The filled rings were placed in the ring holder, whereas the ball-centering guide was positioned on the rings. In brief, 3.5 g of steel balls were placed on each sample. With the distilled water in the bath at 5 ± 2 °C, the assembled apparatus and steel balls were transferred to the liquid bath. The samples were heated, and the temperature at which the ball touches the base plate was recorded.

2.2.3. Viscosity. This test was conducted using rotational viscometer to determine the resistance to the flow of bitumen. The rotational viscometer determines the torque required to rotate a spindle at constant speed while immersed in the modified bitumen sample. The desired dynamic viscosity value, which is proportional to this measured torque, will eventually be obtained. When conducting the test, the spindle, sample chamber, and viscometer environmental chamber were first preheated to 135 °C. The modified bitumen sample was heated until it became fluid enough to be poured and was then brought to the test temperature of 135 °C within approximately 30 min. Viscosity was reported as the average of three readings. The test was then repeated with temperature of 165 °C.

2.2.4. Dynamic shear rheometer (DSR). DSR involves a parallel plate rheometer that is used to apply shear strain or shear stress under a controlled temperature and frequency. This equipment is used to characterize the viscous and elastic behavior of asphalt at high and intermediate service temperatures. These characteristics can be used to evaluate the rutting and fatigue cracking potential of pavement. In this test, the bitumen sample was preheated until the binder became sufficiently fluid for easy pouring. The bitumen sample was then placed between the test plates of 25 mm diameter with 1 mm height as shown in Figure 2. The test was started after the sample had the desired temperature for at least 10 min. The DSR conditioned the sample for 10 cycles at a frequency of 10 rad/s (1.59 Hz). Thus, the machine gives the value of complex modulus (G*) and phase angle (δ) of the binder at a desired temperature and loading frequency.

Figure 2. Dynamic shear rheometer (DSR) test.

3. Results and discussion

3.1. Penetration and softening point

Figure 3 shows the result of penetration values, softening point, and PI of the bio-oil-modified bitumen. The different bitumen samples had increased penetration values and became soft with the increasing percentage of EFB bio-oil content. The 5% addition of bio-oil resulted in slightly higher penetration value of 90.5 PEN than that of the virgin bitumen at 86.9 PEN. This finding indicated that modification using up to 5% bio-oil on virgin bitumen can still achieve the target binder range in penetration grade of 80/100. For the addition of 10% bio-oil, the penetration value increased up to a maximum of 112.4 PEN. This property would be useful in obtaining soft, modified bitumen by blending the EFB bio-oil into hard-grade bitumen. The soft bitumen is suitable for use in cold climate region. This application agrees with the results obtained by Airey and Mohammed [33], who reported that bio-binders exhibit soft behavior and are more suitable to be used as a modifier for stiffer-grade bitumen rather than as a replacement. Furthermore, the softening temperature of bitumen decreased when the EFB bio-oil content increased, indicating that bitumen became susceptible to temperature change when the EFB bio-oil content increased. The PI variation was reduced from −2.52 for virgin bitumen to −9.86 for 10% of EFB bio-oil-modified bitumen. Therefore, the high content of EFB bio-oil significantly affected the bitumen susceptibility to temperature changes.

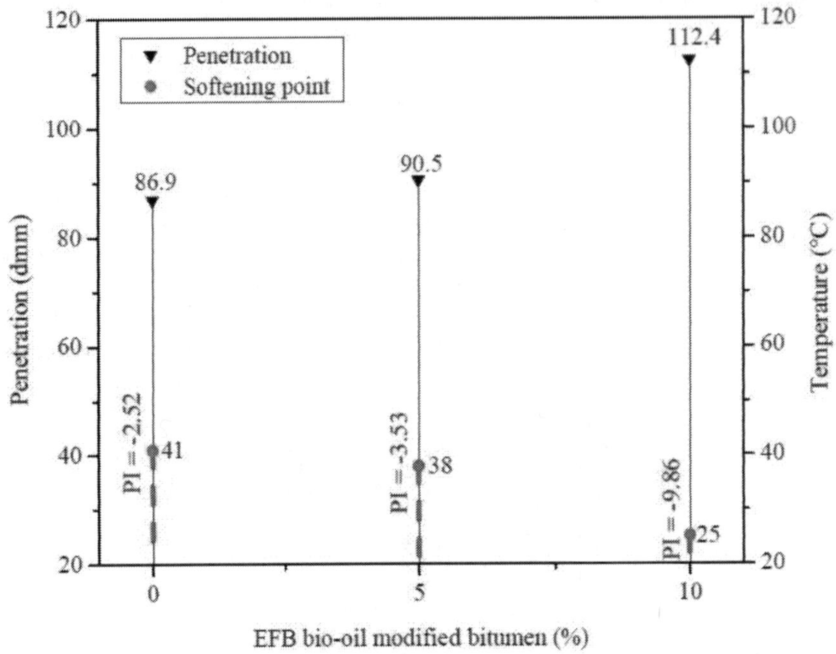

Figure 3. Penetration, softening point and penetration index of EFB bio-oil modified bitumen.

3.2. Viscosity

The viscosity test result of virgin bitumen grade 80/100 PEN and EFB bio-oil-modified bitumen at 135 °C and 165 °C are shown in Figure 4. Modified bitumen with the addition of 5% and 10% of EFB bio-oils achieved lower viscosity than the virgin bitumen. At the temperature of 135 °C, the virgin bitumen showed the viscosity value of 0.47 Pa.s, whereas the modified bitumen with 5% and 10% EFB bio-oils exhibited the same viscosity value of 0.40 Pa.s. This indicates that the viscosity value of bitumen decreases slightly with the increasing bio-oil content, indicating that bio-oil-modified bitumen has lower resistance to flow and small internal friction compared to the virgin bitumen. This phenomenon occurs because the bio-oil is less viscous, thus acting as a lubricant that contributes to the low internal friction. At the temperature of 165 °C, the viscosity of virgin bitumen remained higher but comparable to those with bio-oil at 0.13 Pa.s and 0.1 Pa.s respectively. This finding implies that temperature influences the viscosity reduction for virgin and modified bitumen samples.

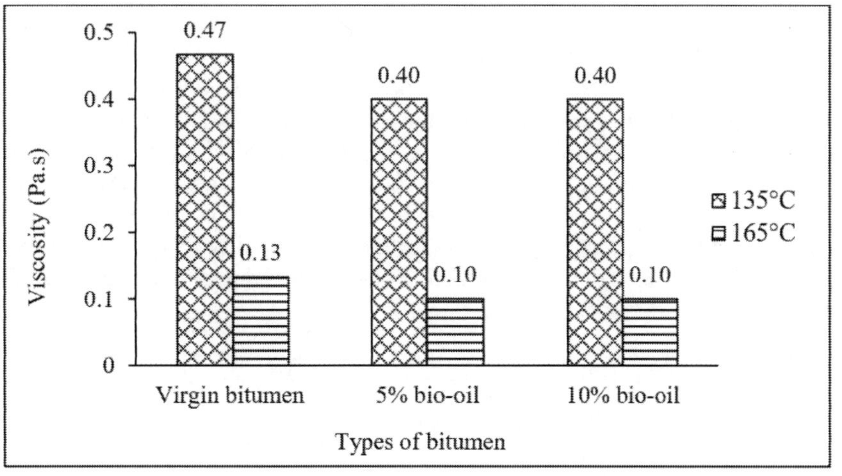

Figure 4. Viscosity at the temperatures of 135°C and 165°C.

3.3. Rutting resistance

DSR test was conducted to measure the rutting resistance of bitumen when $G^*/\sin\delta$ is greater than 1.0 kPa. Based on Superpave specifications, the values of G^* and δ obtained from DSR test were used as predictors of rutting (permanent deformation) temperature. Figure 5 shows the results of rutting resistance for the virgin and bio-oil-modified bitumen samples at five different temperatures (46 °C, 52 °C, 58 °C, 64 °C, and 70 °C). The result shows that the minimum value of $G^*/\sin\delta$ was obtained by the 5% bio-oil, and the virgin bitumen grade 80/100 PEN had the highest value of $G^*/\sin\delta$. The $G^*/\sin\delta$ values were incrementally decreased when the test temperatures increased. Furthermore, the $G^*/\sin\delta$ values increased when the percentage of bio-oil increased from 5% to 10%. The increase in concentration of bio-oil consequently increased the $G^*/\sin\delta$ values, indicating a good rutting resistance.

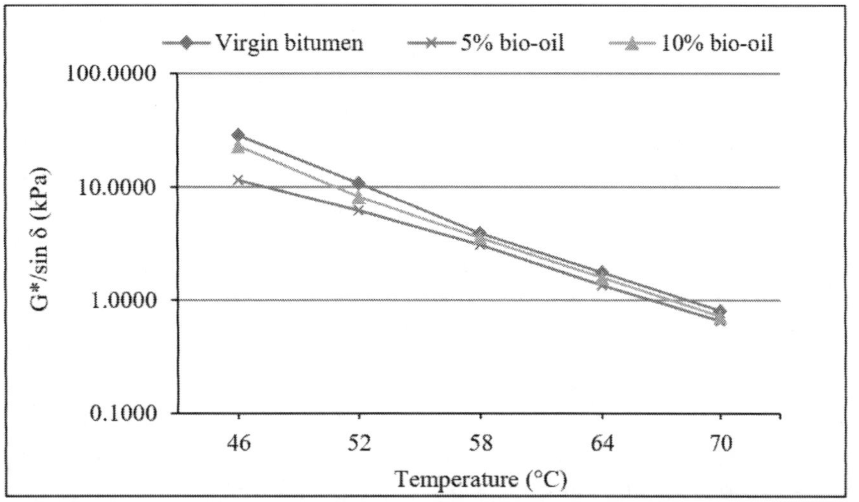

Figure 5. Rutting resistance of virgin and EFB bio-oil modified bitumen

3.4. Failure temperature

According to Superpave specification, the failure temperature of a bitumen is defined as the temperature when the ratio of $G^*/\sin\delta$ is less than 1.0 kPa. High failure temperature values indicate that the binders are less susceptible to permanent deformation at high pavement temperature. From the rheological data plot as in Figure 5, the failure temperature was selected at the rutting resistance of 1.0 kPa. Figure 6 shows the failure temperatures of virgin and bio-oil-modified bitumen samples. It is clearly shown that all bitumen samples have the failure temperatures greater than 64°C and less than 70 °C. The virgin bitumen and modified bitumen with 5% and 10% bio-oil have comparable failure temperatures of around 68 °C, 66 °C, and 67 °C, respectively. This finding indicated that the addition of bio-oil has slightly reduced the failure temperature of the bitumen compared to the virgin bitumen. However, an interesting highlight can be made on the bio-oil's ability to give similar performance as the virgin binder under rutting condition particularly at high temperature even though its physical properties showed bio-oil soften the bitumen.

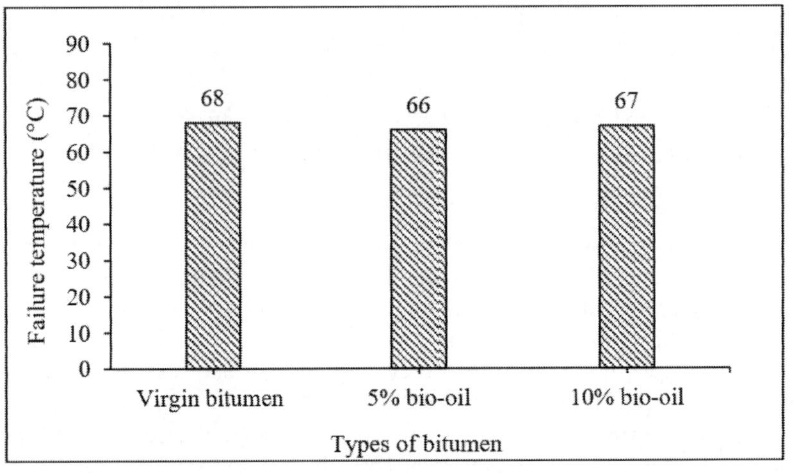

Figure 6. Failure temperature of asphalt binder with different percentages of bio-oil.

4. Conclusion

The findings conclude that adding EFB bio-oil at different percentages affects the temperature susceptibility of the bitumen. When the EFB bio-oil content is increased from 5% to 10%, the penetration value increases significantly, but the softening point and PI values of modified bitumen decrease drastically. The viscosity value of bitumen sample decreases when the EFB bio-oil content is increased, indicating that the modified bitumen has low internal friction and resistance to flow. According to the results of rutting parameter, high content of EFB bio-oil improves the rutting resistance of bitumen at high temperatures and gives a comparable result to the virgin bitumen.

Acknowledgements

The support provided by Malaysian Ministry of Higher Education (MOHE) and Universiti Teknologi Malaysia (research grant no. Q.J130000.2522.11H76 and R.J130000.7822.4F436) for this study is highly appreciated.

References

[1] Weigel S and Stephan D 2017 Relationship between the chemistry and the physical properties of bitumen *Road Materials and Pavement Design* (2017)015

[2] Ali A H, Mashaan N S and Karim M R 2013 Investigations of physical and rheological properties of aged rubberized bitumen *Advances in Materials Science and Engineering* (2013)007

[3] Polacco G, Berlincioni S, Biondi D, Stastna J and Zanzotto L 2005 Asphalt modification with different polyethylene-based polymers *European Polymer Journal* **41** (12) 2831–44

[4] Al-Hadidy A I and Yi-qiu T 2009 Effect of polyethylene on life of flexible pavements *Construction and Building Materials* **23** 1456–64

[5] Baochang Z, Man X, Dewen Z, Huixuan Z and Baoyan Z 2009 The effect of styrene–butadiene– rubber/montmorillonite modification on the characteristics and properties of asphalt *Construction and Building Materials* **23** (10) 3112–17

[6] Kök B V and Çolak H 2011 Laboratory comparison of the crumb-rubber and SBS modified bitumen and hot mix asphalt *Construction and Building Materials* **25** 3204–12

[7] Tuncan M, Tuncan A and Cetin A 2003 The use of waste materials in asphalt concrete mixtures *Waste Management & Research* **21** (2) 83–92

[8] Mahrez A and Karim M R 2010 Rheological evaluation of bituminous binder modified with waste plastic material *5th International Symposium on Hydrocarbons & Chemistry (ISHC5)* May 23–25 Sidi Fredj, Algiers

[9] Kamaruddin N H M, Hainin M R, Hassan N A and Abdullah M E 2014 Rutting evaluation of aged binder containing waste engine oil *Advanced Materials Research* **911** 405–9

[10] Hainin M R, Aziz M M A, Adnan A M, Hassan N A, Jaya R P, Liu H Y 2015 Performance of modified asphalt binder with tire rubber powder *Jurnal Teknologi* **73** (4) 55–60

[11] Azahar W N A W, Jaya R P, Hainin M R, Bujang M and Ngadi N 2016 Chemical modification of waste cooking oil to improve the physical and rheological properties of asphalt binder *Construction and Building Materials* **126** 218–26

[12] Rusbintardjo G, Hainin M R and Yusoff N I M 2013 Fundamental and rheological properties of oil palm fruit ash modified bitumen *Construction and Building Materials* **49** 702–11

[13] Hainin M R, Jaya R P, Ali Akbar N A, Jayanti D S and Yusoff N I M 2014 Influence of palm oil fuel ash as a modifier on bitumen to improve aging resistance *Journal of Engineering Research* **2** (1) 34–46

[14] Abdullah M E, Rosni N N M, Jaya R P, Yaacob H, Hassan N A and Agussabti 2017 Effect of charcoal ash coconut shell from waste material at different size on the physical properties of bitumen *Key Engineering Materials* **744** 121–5

[15] Yang J and Tighe S 2013 A review of advances of nanotechnology in asphalt mixtures *Procedia-Social and Behavioral Sciences* **96** 1269–76

[16] Abdullah M E, Zamhari K A, Buhari R, Kamaruddin N H M, Nayan N, Hainin M R, Hassan N A, Jaya R P and Yusoff N I M 2015 A review on the exploration of nanomaterials application in pavement engineering *Jurnal Teknologi* **73** (4) 69–76

[17] Alhamali D I, Wu J, Liu Q, Hassan N A, Yusoff N I M and Ali S I A 2016 Physical and rheological characteristics of polymer modified bitumen with nanosilica particles *Arabian Journal for Science and Engineering* **41** (4) 1521–30

[18] Mamat R, Hainin M R, Hassan N A, Rahman N A A, Warid M N M and Idham M K 2015 A review of performance asphalt mixtures using bio-binder as alternative binder *Jurnal Teknologi* **77** (23) 17–20

[19] Wiedner K, Rumpel C, Steiner C, Pozzi A, Maas R and Glaser B 2013 Chemical evaluation of chars produced by thermochemical conversion (gasification, pyrolysis and hydrothermal carbonization) of agro-industrial biomass on a commercial scale *Biomass and Bioenergy* **59** 264–78

[20] Raman N A A, Hainin M R, Hassan N A and Ani F A 2015 A review on the application of bio-oil as an additive for asphalt *Jurnal Teknologi* **72** (5) 105–10

[21] Basu P 2010 *Biomass gasification and pyrolysis: Practical design and theory* Technology & Engineering (Academic Press)

[22] Ward J, Rasul M G and Bhuiya M M K 2014 Energy recovery from biomass by fast pyrolysis *Procedia Engineering* **90** 669–74

[23] Bridgwater A V, Meier D and Radlein D 1999 An overview of fast pyrolysis of biomass *Organic Geochemistry* **30** (12) 1479–93

IOP Conf. Series: Materials Science and Engineering **342** (2018) 012053 doi:10.1088/1757-899X/342/1/012053

[24] Xiu S and Shahbazi A 2012 Bio-oil production and upgrading research: A review *Renewable and Sustainable Energy Reviews* **16** (7) 4406–14

[25] Demirbas A 2009 Thermochemical conversion processes *Biofuels* Green Energy and Technology (London: Springer) pp 261–304

[26] Metwally M A R M 2010 *Development of non-petroleum binders derived from fast pyrolysis bio-oils for use in flexible pavement* PhD Thesis Iowa State University

[27] Venderbosch R H and Prins W 2011 Fast pyrolysis *Thermochemical Processing of Biomass: Conversion into Fuels, Chemicals and Power*, ed R C Brown (John Wiley & Sons, Ltd) pp 124–156

[28] Yang X, You Z and Dai Q 2013 Performance evaluation of asphalt binder modified by bio-oil generated from waste wood resources *International Journal of Pavement Research and Technology* **6** (4) 431–9

[29] Mohammad L N, Elseifi M, Cooper III S B, Challa H and Naidoo P 2013 Laboratory evaluation of asphalt mixtures containing bio-binder technologies *Transport Research Record: Journal of the Transportation Research Board.* **2371** 58–65

[30] Mills-Beale J, You Z, Fini E, Zada B, Lee C H and Yap Y K 2014 Aging influence on rheology properties of petroleum-based asphalt modified with bio-binder *Journal of Materials in Civil Engineering* **26** (2) 358–66

[31] Fini E H, Oldham D J and Abu-Lebdeh T 2013 Synthesis and characterization of biomodified rubber asphalt: Sustainable waste management solution for scrap tire and swine manure. *Journal of Environmental Engineering* **139** (12) 1454–61

[32] Gawel I 2000 Sulphur-modified asphalts *Asphaltenes and Asphalts, 2* vol 40B, ed T F Yen and G V Chilingarian (Netherland: Elsevier) pp 515–535

Phase change of iron ore reduction process using EFB as reducing agent at 900-1200°C

H Purwanto[1], H M Salleh[2], A N Rozhan[1], A S Mohamad[1] and A Zakiyuddin[1]

[1] Department of Manufacturing and Materials Engineering, International Islamic University Malaysia (IIUM), Jalan Gombak, 53100 Kuala Lumpur, Malaysia
[2] Department of Biotechnology Engineering, International Islamic University Malaysia (IIUM), Jalan Gombak, 53100 Kuala Lumpur, Malaysia

E-mail: hadi@iium.edu.my

Abstract. Treatment of low grade iron ore involved reduction of oxygen in iron oxide by using reductant such as carbon monoxide or hydrogen gas. Presently, carboneous materials such as coke/coal are widely used as a source to provide reducing gas, but some problem arises from this material as the gas can harm the environments. Therefore, empty fruit bunch biomass from oil palm becomes an alternative to replace the usage of coke/coal as their major composition is carbon and hydrogen. The idea of replacing coke with biomass will reduce the amount of carbon dioxide release as biomass is a carbon neutral and renewable source, and at the same time abundance of waste from oil palm industries can be overcome. Therefore, the aim of this research is to upgrade the low grade iron with reducibility more than 50% being used in iron and steel making. In this research, low grade iron ore are mixed together with EFB then is making into composite pellet before being reduced at certain parameter chosen. The variables involved in this research is composition EFB (10%, 30% and 50%), temperature (1000°C, 1100°C and 1200°C) and reduction time is fixed with 30 minutes. From the experiment conducted, the highest reducibility achieved is 76.37% at temperature 1200°C. While XRD analysis shows the existence of metallic iron phase started to form at 1000°C with composition of 30% of EFB. Meanwhile, from magnetization test show that at 1200°C the highest magnetic susceptibility is achieved as the dominance phase at 1200°C is metallic phase. Therefore it is an interesting alternative to replace coke with biomass for reducing agent in upgrading low grade iron into workable ores.

1. Introduction

Malaysia should upgrade low grade iron as limited sources of workable iron ore is available; hence treatment of low grade iron ore is a good alternative which simultaneously can reduce production cost and also the depletion of high grade iron ore. Malaysia iron ore is impossible to be used in domestic iron and steel making as the ore contain high amount of combined water and impurities [5]. Therefore, the proposed idea to improve the iron content of local iron ore into acceptance level of iron ore is an attractive alternative. The local iron ore is deposited from various mining area in Malaysia and total iron ore deposited is estimated about 9.6 Mt in 2014 as stated in minerals yearbook [1]. The ores are mainly used by the domestic cement, and also iron and steel industries.

Present process of treating the local iron ore with coke have contributed to high amount of carbon dioxide emission and this situation is worrying for the environment. International Energy Statistics estimated about 59 million metric tonnes of CO_2 are given off from the total coal used in Malaysia until 2013 [2]. Since Malaysia is one of the largest palm oil industry in the world, utilizing biomass especially from palm oil industry in the meantime prompt to lower need of fossil fuels which is contributed in reducing carbon dioxide emission to the environment. Hence, alternative ways have

IOP Conf. Series: Materials Science and Engineering **342** (2018) 012054 doi:10.1088/1757-899X/342/1/012054

been taken to replace the coke as reducing agent with biomass waste. As one of the largest palm oil exporters, Malaysia has abundance of oil palm waste which is not being utilized [7]. Therefore, the chosen of EFB from the palm oil waste as a reducing agent is an interesting alternative as if the waste is not being utilized; the cost to eliminate the waste is high.

Hence in this recent study, the oil pam EFB is used as a reductant in order to reduce the iron oxide. Previous research have been focus at low temperature studies, therefore in this study high temperature has been chosen in order to study the effect of the temperature on the composite pellet of iron oxide with EFB. The aim of this research is to determine the suitable and optimum parameters in order to produce metallic iron phase and to study the reduction mechanism by using EFB as the reducing agent.

2. Experimental Procedures

2.1 Raw Materials

The raw materials used in this experiments were obtained from different area which is iron ore was taken from Bahau while the oil palm Empty Fruit Bunch (EFB) is taken from nearby palm oil mill (Mempaga-Bentong,Pahang). Table 1 shown the chemical analysis of original iron ore and both proximate and ultimate analysis of the EFB was provided in Table 2. Before crushing process, the raw materials were dried in the oven to remove moisture for 24h at 110°C. The iron ore was crushed to get coarse powder up to 50 micrometer and then was sieved to get finer iron powder to be observed under XRD, XRF and TG-DTA machine. The fine iron ore and empty fruit bunch was subsequently mixed with chosen mass ratio and was made into pellet by hand rolling. The pellet was compacted to produce pellet with 10mm-20mm in diameter.

Table 1. Chemical analysis of iron ore (Bahau iron ore)

Elements	Fe	Al_2O_3	MgO	SiO_2	K_2O	TiO_2	Cr_2O_3	MnO	Fe_2O_3
Percentage (%)	51.92	9.20	0.44	14.87	0.06	0.52	0.06	0.23	74.23

Table 2. Properties of Empty Fruit Bunch

Proximate Analysis (wt.%)				Ultimate Analysis (wt.%)			
Moisture	Volatile	Fixed	Ash	Carbon	Hydrogen	Oxygen	Nitrogen
6.8	83.94	8.98	7.08	45.64	6.19	48.17	0.35

2.2 Methods

Reduction test is conducted using vertical electric tube furnace with a reactor quartz tube in the middle. The pellets are put in the reactor and are heated at temperature of 1000°C, 1100°C, and 1200°C. Before sample was inserted into the furnace, the argon gas was kept flow into the furnace the make the condition inside in inert state. Heating rate was set at 10°C/min and the temperature was maintained for 30 min for the reduction process to be completed. This experiment was conducted in inert condition with the absence of air or other impurities. During cooling process, argon gas was kept flow in order to avoid oxidation occurred to the composite pellet during reaction inside the furnace.

2.3 Estimation of Reduction Degree

In reduction reaction, the reaction is completed when minerals transformation from hematite had completely changed into metallic iron. Several phase changes during minerals transformation are:

i. **Hematite** to **Magnetite**: $3Fe_2O_3 + CO \longrightarrow 2Fe_3O_4 + CO_2$

ii. **Magnetite** to **Wustite** : $Fe_3O_4 + CO \longrightarrow 3FeO + CO_2$

iii. **Wustite** to **Iron** : $FeO + CO \longrightarrow Fe + CO_2$

By using molecular weight equation, the weight loss for each reduction reaction due to oxygen removal was calculated. From the molecular weight equation, weight for each compound is known. Similar to iron ore from other places, the major elements in Bahau iron ore are geothite and hematite. For every phase transformation involved in reduction reaction the oxygen removed is estimated as shown in Table 3 and calculation of reducibility shown in equation (2.1).

Table 3. Estimation of oxygen removal

Reaction	Reductant	Oxygen removal (%)
$FeO(OH) \rightarrow Fe_2O_3$	CO / H_2	2.5
$Fe_2O_3 \rightarrow Fe_3O_4$		3.33
$Fe_3O_4 \rightarrow FeO$		6.89
$FeO \rightarrow Fe$		22.22

$$\text{Reducibility} = \frac{Reduction\ degree}{Theoretical\ oxygen\ removal} \quad (1)$$

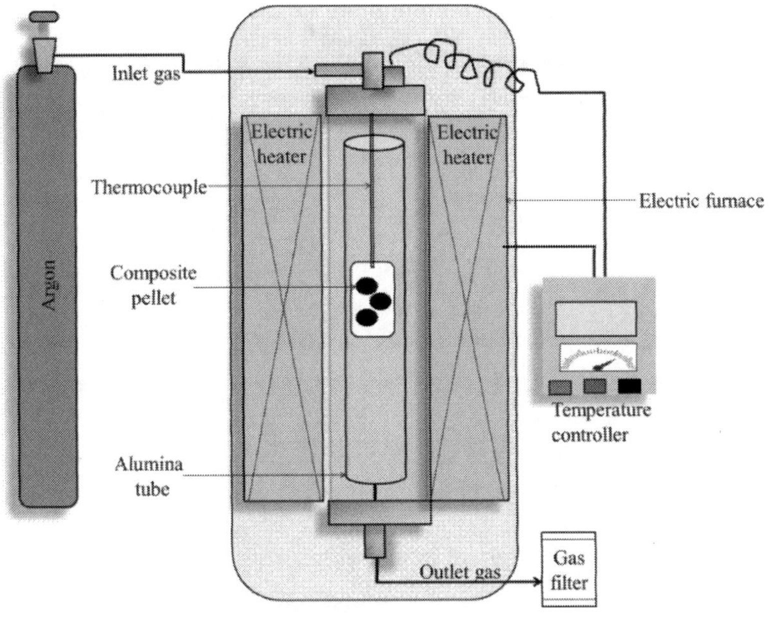

Figure 1. Schematic diagram of experimental apparatus.

3. Results and Discussion

3.1 Reducibility behavior of composite pellet after reduction process

The composite pellet with different composition of EFB was reduced at temperature 1000°C to 1200°C. The reducibility curve of iron ore at different temperature are shown in Figure 2. The graph shows increasing trend as the temperature increase and the reducibility reach the highest value of 78.22% at 1200°C. As there was reaction between the ore particles and the EFB, the reduction degree is corresponding with the increasing temperature. Reduction degree was calculated based on the oxygen removal from iron oxide in the iron ore as shown in equation (2.1). This condition will then contribute to phase change in the iron oxide as the carbon in the EFB act as a solid reductant during reduction reaction.

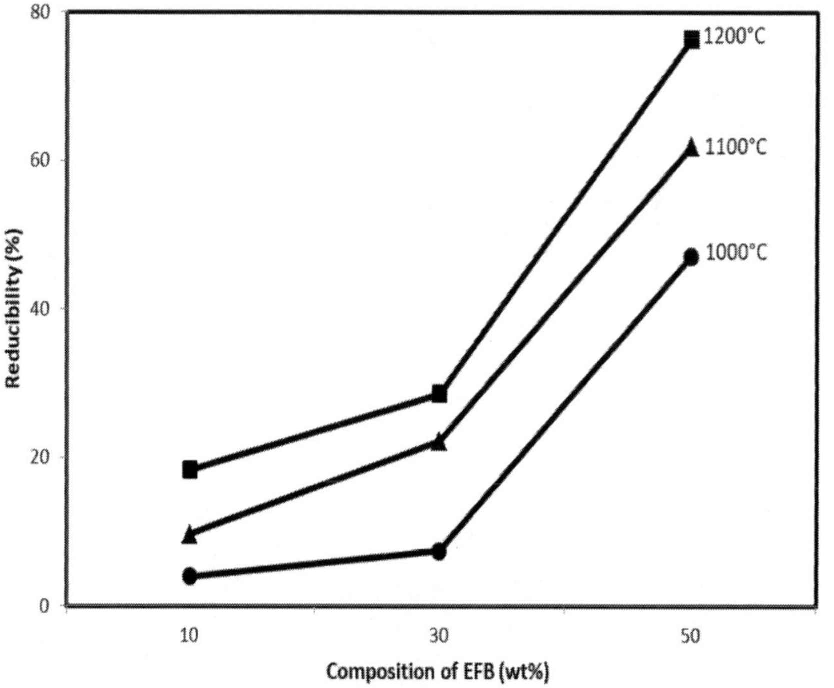

Figure 2. Reduction curves of reduced composite pellet with different composition of biomass and temperature.

3.2 Characterization of reduced iron oxide – EFB composite pellet by XRD

Figure 3 shows XRD pattern of an original iron ore and composite pellet after reduction process at 1200°C with various composition. The pattern of original iron ore is clearly indicated the major compositions are geothite ($Fe_2O_3.H_2O$) and hematite (Fe_2O_3) together with some quartz (SiO_2) as the impurity. The composite pellets of all compositions on reduction at 1000°C shown mainly wustite and iron phase. However, metallic phase started to become a dominance phase when 50% of EFB used compared to a temperature of 1200°C metallic phase is the dominance at lower EFB content. It may be further noted that at higher temperature the reduction is better with lower content of reductant agent. This condition is consistent with the higher reduction degree achieved at 1200°C for all compositions. The reaction between the iron ore and EFB was step up as the temperature increase and the oxygen was removed expeditiously compared to the oxygen released during reduction at lower temperature. Additional heat was required during the reaction since the EFB was in solid form hence more heat was required to vaporize the EFB into gas phase. The EFB then act as reducing agent in the form of carbon monoxide gas to ensure the reaction was proceed.

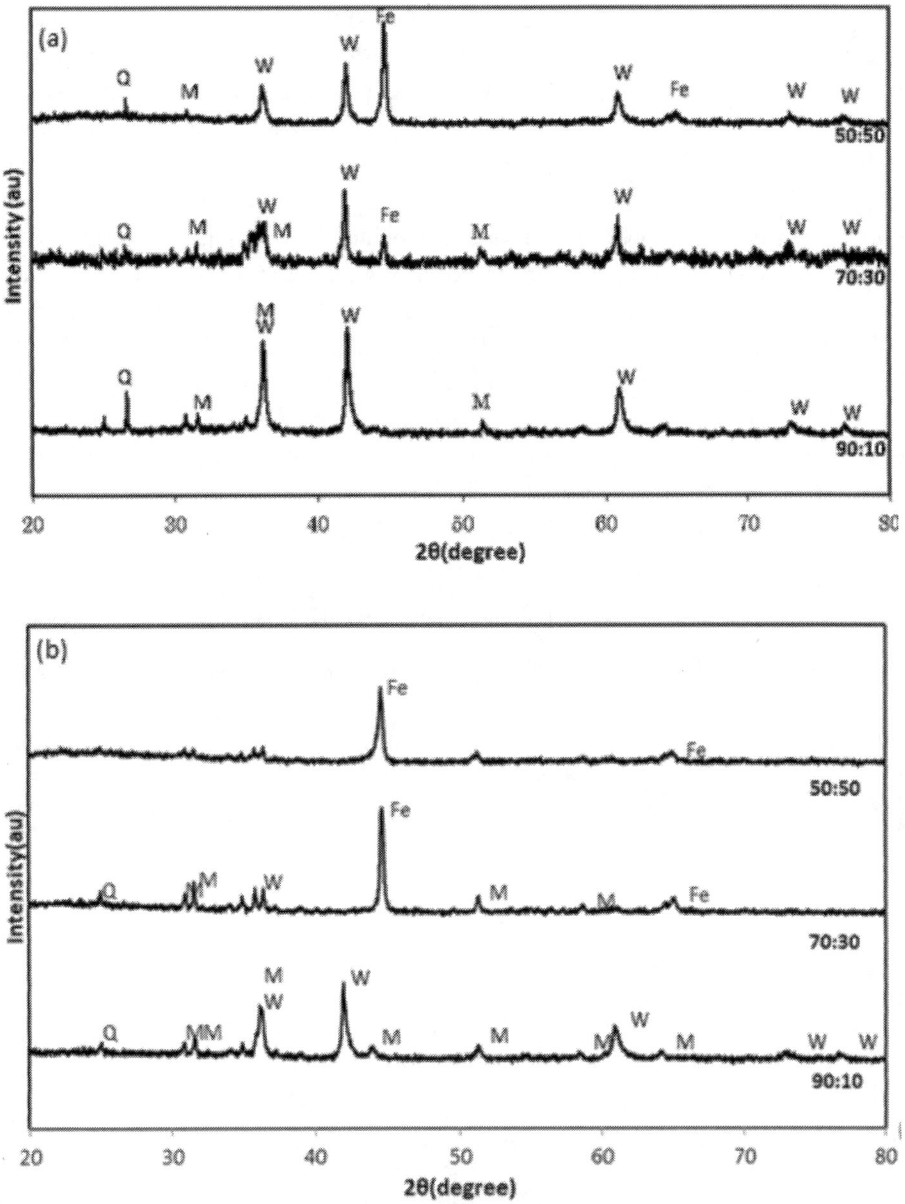

Figure 3. XRD pattern of composite pellet samples after reduction with different EFB composition a. 1000°C; b. 1200°C.

3.3 Analysis of magnetic properties of reduced pellets

Figure 4 shows the magnetization curve of the reduced composite pellet containing 30mass% of EFB at different temperatures. Generally, increasing temperature is enhanced the magnetization of the reduced pellet and this condition can be determined by the magnetic susceptibility and magnetic saturation parameters. Magnetic susceptibility is determined from the slope of the magnetization curve and commonly high in the presence of ferromagnetic materials and low for paramagnetic and diamagnetic materials. The magnetic saturation is measured from the point where the minerals started to reach its saturated state. Purwanto et al.[5] stated that the magnetic saturation for the untreated laterite ore is 3.38emu/g, and the value is quiet low due to the main mineral in the ore are geothite which is considered as weak paramagnetic mineral. Reducing the composite pellet at 1000°C is enhanced both the magnetic saturation and magnetic susceptibility simultaneously. Reducing composite pellet at 1200°C clearly improved magnetic properties as the value for magnetic saturation is comparably higher than original iron ore as reported. At 1000°C, the value of the saturated magnetization is 36.5 emu/g which is relatively higher compared to the original iron ore and this result correspond to the existence of magnetite phase that exhibit ferromagnetic properties or known as strong magnetic minerals. As further reduced up to 1200°C, the magnetization is slightly increased as the iron phase start to become a dominance peak in the reduced composite pellets. The magnetization at 1000°C is lower because of the formation of wustite phase cause the magnetization to decrease because wustite is a weak magnetic phase. As reported, the magnetic susceptibility of original iron ore is low as the main phase in the ore is geothite which attribute properties of weak magnetic phase. Therefore, similar to the magnetic saturation, the susceptibility of the magnetic properties is lower at the presence of wustite or other weak magnetic phase. The magnetic susceptibility at high temperature is in the range of 36-45emu/g, exhibit a degree of susceptibility similar to natural magnetite which is in the range of 35.98-45.14 emu/g [12].

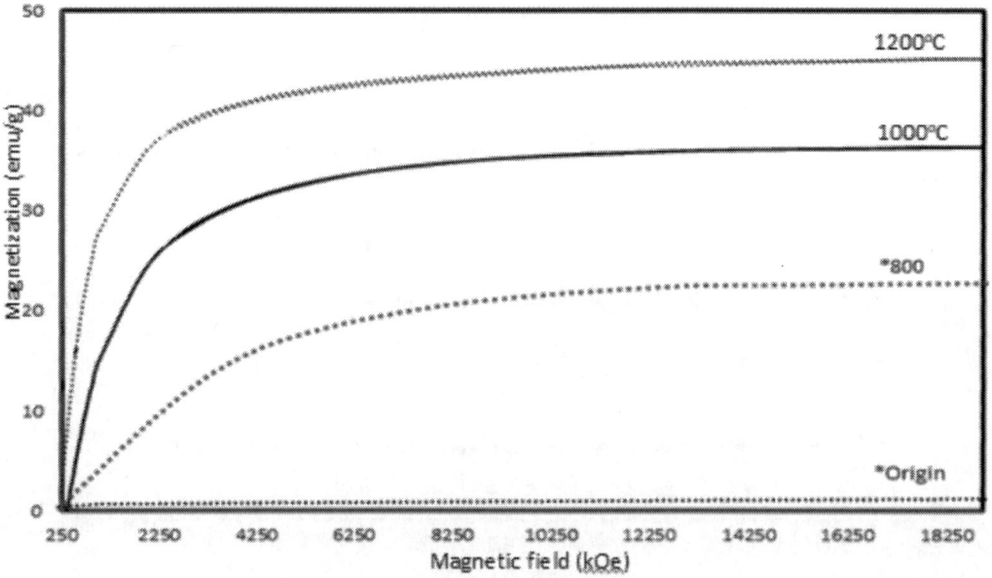

Figure 4. Magnetization curves of composite pellet with 30% EFB content at different temperature.

4. Conclusion

Reduction test of iron ore with EFB at high temperature of 1000°C, 1100°C and 1200°C have been conducted and the following conclusions were achieved. The extent of reduction was higher at maximum temperature compared to a lower temperature. Reduction reaction of iron oxide was rapidly reduced in biomass-iron oxide composite pellet. The highest reduction degree of 38.22% was achieved at 1200°C shown that the reduction degree increase with increasing temperature. Metallic iron phase was existed as the reaction proceeds at higher temperature as indicated in XRD patterns of the composite pellet. At temperature of 1000°C metallic iron phase started to appear in the composite pellet with the right composition of biomass-iron oxide. The magnetic test for different temperature of reduction reaction have revealed that iron phase have higher magnetic properties compared to wustite as iron phase exhibits ferromagnetic properties. The magnetization of reduced pellets at 1200° C was 44.6emu/g which is higher compared to at 1000°C. This can be concluded that iron phase have better magnetic properties compared to other phase and this is one of the preferred condition in reduction reaction process.

References

[1] Takahashi, R., 1990, *Fundamentals of Metallurgical Reaction Kinetics, Group Training Course on Mineral Processing and Metallurgy in the Research Institute of Mineral Dressing and Metallurgy (SENKEN).*

[2] Habermann, A.,Winter, F., Hofbauer, H., Zirngast, J., and Schenk, J. L, 2000, *The Iron and Steel Institute of Japan (ISIJ) International*, **40** 10 p 935-942.

[3] Wei, R., Cang, D., Bai, Y., Huang, D. and Liu, X, 2016, *Ironmaking & Steelmaking*, **43** 2 p 144-152.

[4] Muwanguzi, A., Karasev, A., Byaruhanga, J. and Jönsson, P, 2012, *ISRN Materials Science*, p 1-9.

[5] Abd Rashid, R. Z., Mohd. Salleh, H., Ani, M. H., Yunus, N. A., Akiyama, T., & Purwanto, H, 2013, *Advanced Materials Research*, **701** p 28-31.

[6] Abd Rashid, R. Z., Mohd. Salleh, H., Ani, M. H., Yunus, N. A., Akiyama, T., & Purwanto, H., 2014, *Renewable Energy*, **63** p 617–623.

[7] Choy, Y. K., 2005, *Energy Sources*, **27** 7 p 589–596.

[8] Guo, D., Hu, M., Pu, C., Xiao, B., Hu, Z., Liu, S., Zhu, X, 2015, *International Journal of Hydrogen Energy*, **40** 14 p 4733–4740.

[9] N. F. M. Yunos, N. A. Aziz, A. N. Ismail, M. A. Idris, 2015, *Materials Science Forum*, **819** p 31-36.

[10] Sahajwalla, V., Dubikova, M., & Khanna, R., 2004, *Transformation through Technology*, p 351-362.

[11] Srivastava, U., Kawatra, S. K., & Eisele, T. C., 2013, *International Journal of Mineral Processing*, **119** p 51–57.

[12] Yunus, N. A., Ani, M. H., Salleh, H. M., Rashid, R. Z., Akiyama, T., Purwanto, H., & Othman, N. E., 2014, *International Journal of Minerals, Metallurgy, and Materials*, **21** 4 p 326-330.

[13] Takyu, Y., Murakami, T., Son, S. H., & Kasai, E., 2015, *ISIJ International*, **55** 6 p 1188–1196.

IOP Conf. Series: Materials Science and Engineering **342** (2018) 012055 doi:10.1088/1757-899X/342/1/012055

Characteristics of iron ore sinter with EFB addition

H Purwanto, A N Rozhan, A S Mohamad and A Zakiyuddin

Department of Manufacturing and Materials Engineering, International Islamic University Malaysia (IIUM), Jalan Gombak, 53100 Kuala Lumpur, Malaysia.

E-mail: hadi@iium.edu.my

Abstract. Utilization of EFB-derived biochar in sintering of iron ore has been conducted in this work with regards to the porosity of iron sinter. EFB has been heated up in argon atmosphere to 450°C in order to produce biochar. In the present work, the sintering process was conducted at 1150°C with variations of biochar content from 5% to 10%. In this case, the apparent density for iron sinter shows significant decrease as the biochar addition increase. The porosity of iron sinter showed a gradual increment from 5% to 7.5% and escalated at 10% biochar content. The results of porosity and apparent density were in line with the micrograph of iron sinter.

1. Introduction

The substitution of fossil fuels with renewable biomass is one of the ways to mitigate the CO_2 emissions in the steel industry. Sintering of iron ore produces large amount of flue gas containing pollutant. The combustion of fossil fuel produces main sources like CO_2, SO_2 and NO_2. It is very crucial to find clean and renewable energy sources to replace fossil fuels which make sinter production become eco-friendly. Replacement of fossil carbon by renewable biomass-based carbon is an effective measure to reduce CO_2 emission intensity in the blast furnace ironmaking process [1].

The depletion of fossil fuels resources and problems arising from this process have led to many researches on this matter in order to find the solutions for the problems. Then, biomass has been identified as a potential source for the renewable energy [2-4]. The advantages of using the biomass are that it can scale down the carbon dioxide product and help to alleviate climate change by reducing the greenhouse gases [5]. Besides, utilization of biomass can also decrease the emission of dangerous gases [6-7]. Chean et al., 2008 mentioned that use of sunflower seed husk as a supplementary fuel in iron ore sintering process is feasible to substitute 10% coke breeze [8] as biomass is a potential energy sources that can substitute fossil fuels [9-13]. The usage of biochar can mitigate the emission of hazardous gases such as CO2, NOx, Sox dust etc. [14].

Iron sinter is produced from agglomeration of iron ore fines into compact granules by heating process to the partial fusion and coalescence bonding [15-16]. The structure of iron sinter produced is very important in determining the strength and reducibility of sinter to be charged into blast furnace operation [17]. By increasing the porosity distribution of iron sinter, the contact area for reducing gases are increased, that leads to increase in the efficiency of reducibility for iron sinter [18]. In this research, the effects of oil palm empty fruit bunch (EFB) addition on the production of iron sinter was investigated with regard to the porosity of the product.

2. Experimental procedures

2.1. Materials

Sintering materials which are iron ore, oil palm empty fruit bunch (EFB) and limestone were utilized in this experiment. Iron ore and EFB were taken from mining site in Bahau, Negeri Sembilan and local palm oil mill, respectively. Table 1 shows elemental composition of iron ore (Bahau, Negeri Sembilan). Iron ore was ground and sieved to small particle size of less than 150µm. The shredded EFB was used to produce biochar by heating up to 450°C with heating rate of 10 °C/min. The elemental composition of EFB was given in Table 2. Iron ore, biochar and limestone were mixed together and granulated in a bowl by using water binder. Then, the granulated samples were dried for 24 hours.

2.2. Experimental methods

The granulated samples were put in the electric furnace and set to 1150°C with 10 minutes heating rate and 30 minutes holding time. Figure 1 shows the granulated iron sinter after sintering process. The porosity and apparent density of iron sinter were investigated. Micrograph of iron sinter was examined by using the SEM. Figure 2 shows schematic diagram of furnace for sintering and biochar making (1- Temperature controller, 2- Gas flow controller,3- Gas container,4- Gas filter,5- Thermocouple,6- Electric heater,7- Green sample).

Table 1. Elemental composition of iron ore (Bahau, Negeri Sembilan).

Compound	**Fe**	**Al_2O_3**	**MgO**	**SiO_2**	**K_2O**	**TiO_2**	**CrO_3**	**MnO**	**Fe_2O_3**
Percentage	51.92	9.20	0.44	14.87	0.06	0.52	0.06	0.23	74.23

Table 2. Elemental composition of EFB.

Proximate analysis	wt%	Ultimate analysis	wt%
Moisture	6.80	Carbon	45.64
Volatile matter	83.94	Hydrogen	6.19
Fixed carbon	8.98	Oxygen	48.17
Ash	7.08	Nitrogen	0.35

Figure 1. Granulated iron sinter.

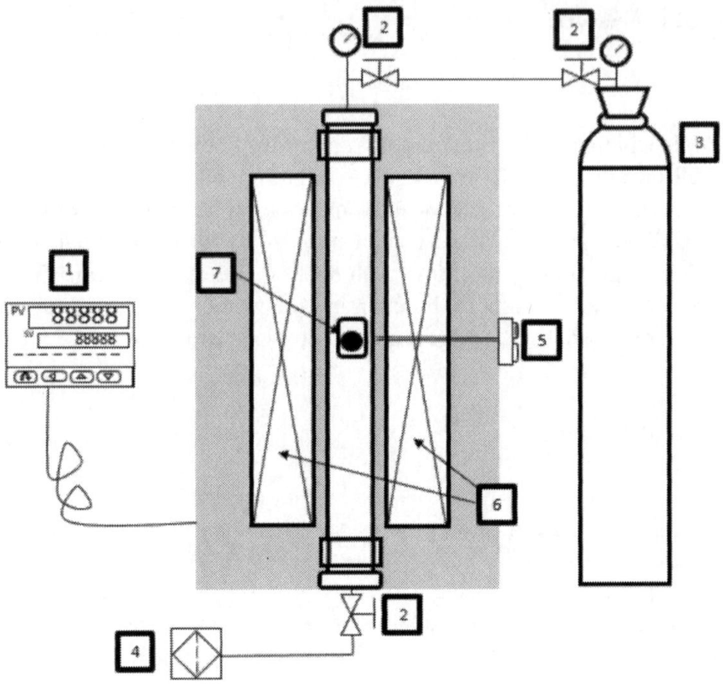

Figure 2. Schematic diagram of furnace for sintering and biochar making (1- Temperature controller, 2- Gas flow controller,3- Gas container,4- Gas filter,5- Thermocouple,6- Electric heater,7- Green sample).

3. Results and Discussion

Figure 3 shows the apparent density and porosity of iron sinter at 1150°C with variations of biochar addition. In this case, the apparent density for iron sinter shows significant decrease as the biochar addition increase from 5% to 10%. The porosity of iron sinter shows a gradual increment from 5% to 7.5% and escalate at 10% biochar content. For iron sinter at 10% biochar content has the highest percentage of porosity because the biochar in the sample was burnt and this left a porous structure without completely coalesce and assimilate with the iron ore.

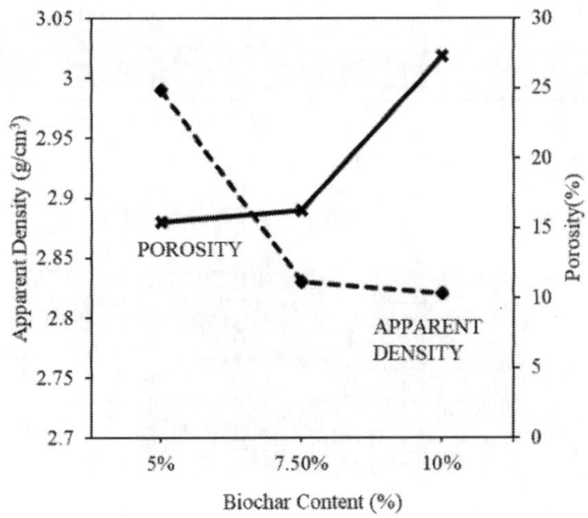

Figure 3. Apparent density and porosity of iron sinter at 1150°C with variations of biochar addition.

Inferring to the Figure 4, it shows the micrograph of iron sinter at 1150°C with variations of EFB addition (a) 5% EFB content, (b) 7.5% EFB content (c) 10% EFB content. It can be seen that at 5% biochar content, a dense product was obtained with complete coalescence of iron sinter. On the other hand, iron sinter produced with 10% biochar content underwent incomplete coalescence and the product was less dense. Results of the porosity and apparent density proved that biochar content have strong effect on the structure and morphology of iron sinter.

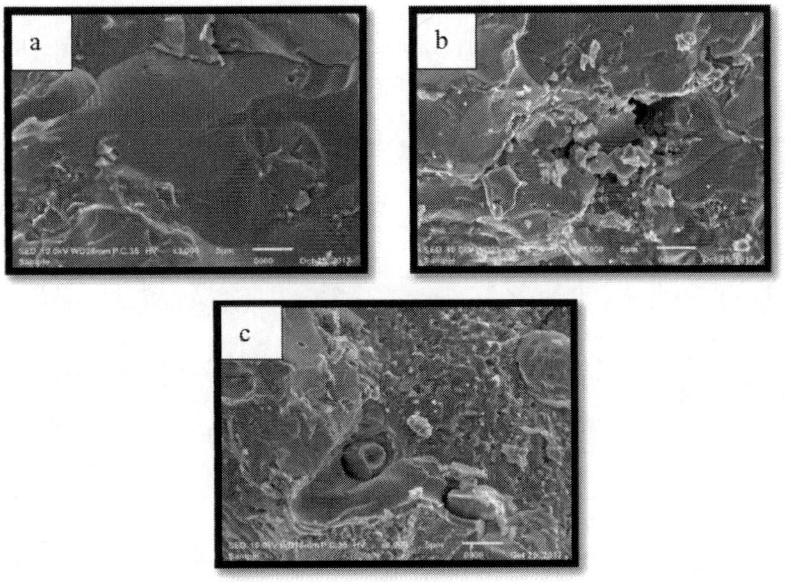

Figure 4. Micrograph of iron sinter at 1150°C with variations of EFB addition (a) 5% EFB content, (b) 7.5% EFB content (c) 10% EFB content

4. Conclusion

In conclusion, the apparent density for iron sinter shows a significant decrease as the biochar addition increased from 5% to 10%. Morphology of the iron sinter revealed that the dense iron sinter has been produced in terms of agglomeration and coalescence of sinter particles at 5% biochar content. At 10% biochar addition, the biochar in the sample was burnt and left a porous structure without completely coalesce and assimilate with the iron ore.

Acknowledgement
This work is partially supported by the International Islamic University Malaysia through IIUM Research Initiative Grant Scheme (RIGS 16-353-0517).

References

[1] Suopajärvi, H., & Fabritius, T. (2013). Towards more sustainable ironmaking-an analysis of energy wood availability in finland and the economics of charcoal production. *Sustainability (Switzerland)*, *5*(3), 1188–1207.
[2] Tobergte, D. R., & Curtis, S. (2013). *Journal of Chemical Information and Modeling* (Vol. 53).
[3] Sulaiman, F., Abdullah, N., Gerhauser, H., & Shariff, A. (2011). An outlook of Malaysian energy, oil palm industry and its utilization of wastes as useful resources. *Biomass and Bioenergy*, *35*(9), 3775–3786.
[4] El-Hussiny, N. A., Khalifa, A. A., El-midany, A. A., Ahmed, A. A., & Shalabi, M. E. H. (2015). Effect of replacement coke breeze by charcoal on technical operation of iron ore sintering.

International Journal of Scientific & Engineering Research, *6*(2).
[5] Shariff, A., Aziz, N. S. M., & Abdullah, N. (2014). Slow Pyrolysis of Oil Palm Empty Fruit Bunches for Biochar Production and Characterisation. *Journal of Physical Science*, *25*(2), 97–112.
[6] Norgate, T., & Langberg, D. (2009). Environmental and Economic Aspects of Charcoal Use in Steelmaking. *ISIJ International*, *49*(4), 587–595.
[7] Gan, M., Fan, X., Ji, Z., Jiang, T., Chen, X., Yu, Z., … Yin, L. (2015). Application of biomass fuel in iron ore sintering : influencing mechanism and emission reduction, *42*(1), 27–34.
[8] Chean, T., Aries, E., Ewan, B. C. R., Thompson, D., Anderson, D. R., Fisher, R., … Tognarelli, D. (2008). The study of sunflower seed husks as a fuel in the iron ore sintering process, *21*, 167–177.
[9] Z. W. Jiang (2010) Study on combustion process of biomass fuels and characteristics of their ash', J. Mater. Rev., 24, (4), 66–80.
[10] J. Y. Li and R. Y. Chen: (2001) 'Direct combustion of biomass fuel characteristics analysis, Northeast Agric. Univ., 32, (2), 290–294.
[11] X. Chen: (2007) Exploitation and utilization of bio-energy, J. Proc. Chem., 19, (7), 1091–1096.
[12] R. C. Guo and J. R. Tang (2007) The statue and vision for utilization of lignified biomass energy, J. Inv. Planning, 32, (1), 90–94.
[13] Y. M. Sun (2006) The status and future of bioenergy and biomass utilization in China, Renewable Energy, , (2), 77–82.
[14] Kawaguchi, T., & Hara, M. (2013). Utilization of Biomass for Iron Ore Sintering, *53*(9), 1599–1606.
[15] Loo, C. E., & Leung, W. (2003). Factors influencing the bonding phase structure of iron ore sinters, *ISIJ International,43*(9), 1393–1402.
[16] Umadevi, T., Bandopadhyay, U. K., Mahapatra, P. C., Prabhu, M., & Ranjan, M. (2010). Influence of limestone particle size on iron ore sinter properties and productivity. *Steel Research International*, *81*(6), 419–425.
[17] Ramos, M. V, Kasai, E., Kano, J., & Nakamura, T. (2000). Numerical simulation model of the iron ore sintering process directly describing the agglomeration phenomenon of granules in the packed bed. *ISIJ International*, *40*(5), 448–454.
[18] Mašlejová, A., Fröhlichová, M., Černík, M., Vlašič, P., & Ivanišin, D. (2014). X-Ray Diffraction analysis of iron sinter, *Acta Metallurgica Slovaca*, 4, 74–81.

IOP Conf. Series: Materials Science and Engineering **342** (2018) 012056 doi:10.1088/1757-899X/342/1/012056

The role of anthropometric, growth and maturity index (AGaMI) influencing youth soccer relative performance

Ahmad Bisyri Husin Musawi Maliki[1], Mohamad Razali Abdullah[1,2], Hafizan Juahir[1], Wan Siti Amalina Wan Muhamad[2], Nur Afiqah Mohamad Nasir[2], Rabiu Muazu Musa[2,3], Siti Musliha Mat-Rasid[1], Aleesha Adnan[2], Norlaila Azura Kosni[2], Farhana Abdullah[2] and Nurul Ain Shahirah Abdullah[2]

[1] East Coast Environmental Research Institute (ESERI), Universiti Sultan Zainal Abidin, 21300, Terengganu, Malaysia.
[2] Faculty of Applied Social Sciences, Universiti Sultan Zainal Abidin, 21300, Terengganu, Malaysia
[3] Innovative Manufacturing, Mechatronics & Sports Lab (iMAMS), Faculty of Manufacturing Engineering, Universiti Malaysia Pahang (Pekan Campus), 26600 Pekan, Pahang, Malaysia.

E-mail: razali896@yahoo.com

Abstract. The main purpose of this study was to develop Anthropometric, Growth and Maturity Index (AGaMI) in soccer and explore its differences to soccer player physical attributes, fitness, motivation and skills. A total 223 adolescent soccer athletes aged 12 to 18 years old were selected as respondent. AGaMI was develop based on anthropometric components (bicep, tricep, subscapular, suprailiac, calf circumference and muac) with growth and maturity component using tanner scale. Meanwhile, relative performance namely physical, fitness, motivation and skills attributes of soccer were measured as dependent variables. The Principal Component Analysis (PCA) and Analysis of Variance (ANOVA) are used to achieve the objective in this study. AGaMI had categorized players into three different groups namely; high (5 players), moderate (88 players) and low (91 players). PCA revealed a moderate to very strong dominant range of 0.69 to 0.90 of factor loading on AGaMI. Further analysis assigned AGaMI groups as treated as independent variables (IV) and physical, fitness, motivation and skills attributes were treated as dependent variables (DV). Finally, ANOVA showed that flexibility, leg power, age, weight, height, sitting height, short and long pass are the most significant parameters statistically differentiate by the groups of AGaMI ($p<0.05$). As a summary, body fat mass, growth and maturity are an essential component differentiating the output of the soccer players relative performance. In future, information of the AGaMI model are useful to the coach and players for identifying the suitable biological and physiological demand reflects more comprehensive means of youth soccer relative performance. This study further highlights the importance of assessing AGaMI when identifying soccer relative performance.

1. Introduction

Anthropometric is a basic research method of somatometric that involves the size of body composition and body part. It is a measurement of the body composition which determines the size of the composition of the body to be tested [1]. The total of the body fat measurement fundamentally in skinfold test has seven sites to be measured namely, biceps, triceps, subscapular, suprailiac, supraspinale, thigh and calf. In the meantime, girth and maturation also contributes to the development of physical characteristic especially in adolescent soccer players. Determination of the optimum physical characteristics integration of the body size, growth and maturation in the early stage of the talent identification and development on the game of soccer are vital.

On the other hand, multilateral factors of anthropometric, physical fitness, soccer specific skill and psychology recently been investigated in predicting soccer players potential [2-4]. Applying multilateral of factors will help coaches, instructors as well as players to focus on the most dominants factors predicting to the soccer player performance. Meanwhile, some of the studies focusing on the variation of the level of the participation in soccer in regard to identify the profiling and distinction of the specific soccer related performance and fitness level [5-7]. It is proven to be variation among soccer players in terms of the position, level of age, level of participation and most of it influence by the physical characteristics, growth and maturation, and also psychosocial of the players. However, most of the studies only consider limit to the anthropometric, physical fitness, physiological, psychological and technical factors [2-7]. In contrast, the variation of the lifestyle by the players influence the body size, somatotype, body fat, growth and maturation. It is fewer study focusing on the variation of the body fat level in contrast to the soccer player performance. Thus, this study aims to develop Anthropometric, Growth and Maturity Index (AGaMI) in soccer and explore its differences to the soccer player physical characteristics, physical fitness, psychological and soccer specific skill attributes.

2. Materials and Method

All the procedures of the current studies such as, anthropometrics, battery test, soccer specific battery test and psychometric related to soccer was undertaken by the following procedures:

2.1 Participants

A random sampling technique were employed to the current study consist of youth soccer players in Malaysia soccer academies by combining ex post facto design. A total of 184 youth soccer players (mean age = 15.2 ± 1.6 years) were voluntary participate in this study which is randomly drawn from soccer academies. Prior of the data collections, all the procedure and protocol were officially approved by the Human Research Ethics Committee of the University (UniSZA/02/1/2016/Jil. 207). Furthermore, aims of the study were informed to all participants and their guidance and were told to signed if they voluntary agreed to be participant of the study.

2.2 Anthropometrics, growth and maturation (AGaMI)

Anthropometric tests were done to measured weight, standing stature, sitting height, and four locales of body fat. Weight is evaluated utilizing a standard electronic advanced scale to the closest of 0.1 kg, in the interim standing stature was measure to the closest 0.5 cm. Meanwhile, chronological age is measured in the month beginning from the date of every player being perceived until the date of test. Moreover, sitting height were measured from the highest point of the head to the base of the sitting and recorded to the closest of 0.5 cm. Level of muscle versus fat is measured through skinfold calliper utilizing four measuring locales to be specific, trisep, biceps, subscapular and suprailiac to the closest of 0.1 mm as indicated by the suggestion from past examinations [6-9]. Furthermore, the body size or girth is measured on two sections of the body in particular, calf circumcision (cc) and middle upper arm perimeter (muac) by non-elastic tape. All procedures were recorded twice and the mean score is recorded for facilitate further analysis. Then again, maturation was measured by applying Tanner scale stages of development as prescribed by prior studies and adopting into current study [3- 4].

2.3 Soccer battery test

Muscle strength assessments are measured in the suggested way for physical fitness assessment. There are seven parameters to be recorded in physical fitness test performance namely, vertical jump, variability of sit up, agility, sit and reach, 5 m, 10 m and 20 m speed. The tape scale is used to measure the range of the flexibility and the long reach of sit and reach and distances of the leg power were measure by using vertical jump. A sit and reach test are undergone to measure the flexibility of the lower back and the hamstring muscles. The player must start in a ready position with both legs are straight. The range is premeditated to the extent that the finger reaches the maximum of the measuring line. Vertical jump tests are doing to measure the distances differences of the starting height and maximum height while jumping. While calculating the height of the jump, the time is also recorded for the process of the jumping routine. The purpose of this test is to measure the horizontal explosive power of the leg muscles [10]. Stopwatch and timing gate were used to compute and recorded the timing of the sit-ups test agility tests. Variability of sit up test is performed to measure the strength and durability of abdominals and hip muscles. The sit up maximum routine of set undertaken for recorded. Meanwhile, agility is the capability of the players to quickly change the body position into the given direction in the minimum requirement time. The timing gate is essential and proven to be valid to record the fastest time from the start line to the finish line as recommended by the previous study [11].

2.4 Soccer technical skills
Soccer specific skill tests were executed to acquire technical skill characteristics by applying from the previous studies namely F-MARC test accompanying to measure skill test namely dribbling with and without ball, ball control using three different part of body, short pass, long pass, shooting with foot and score using head. Time and speed are taken and evaluated during the test process [3].

2.5 Psychometric (Motivation)
Task and Ego Orientation in Sports (TEOSQ) have been implementing in this study. It contains 13 items that assessed the ability of players weather they are task or ego oriented towards the soccer game. The form was translated into Bahasa Melayu using the back-translation method and it was confirmed and added extraordinary reliability in sports. Spectrum for task and ego orientation shows appropriate internal reliability with alpha 0.82 (ego) and 0.71 (task) reliability coefficients respectively [12]. This instrument contains six items that assess the ego components (such as "I can do better than my friends") and seven items assess the task components (such as "I work hard"). Feedback is shown on 5 points of Likert scale where 1=strongly disagrees and 5=strongly agree.

2.6 Data Analysis
A total of 4784 matrices data comprised all 184 observations of players and 26 parameters was calculated to be analyzed. Prior of the main analysis, box plots and Kolmogorov-Smirnov were computed to check the missing, error, typing error and normality of the data. As a result, it is described that the data is clear. Although some of the parameters are not normally distributed, main analysis were computed as recommendation by the prior researches stated that it is normal to have unnormal data to be distributed in human performance because of the individuality variation [6,8,10,11].

To achieve the objective of this study, principal component analysis (PCA) method was conducted a to scrutinize the data. Through the investigation, it can help the researcher to examine the domain components interrelated to the soccer performance. From this current study, seven factors were categorized as an independent component. Nevertheless, only dominant factors will be interpreted after computing varimax rotation [12]. Furthermore, all the expectations of the PCA were calculated prior of the analysis such as sampling adequacy and sphericity test [6,8,10,11]. End of the PCA, an index of anthropometric, growth and maturation (AGAM) will be develop into different categorical of group. It will highlight the range of each groups. Additionally, researcher also applying analysis of variance (ANOVA) in regard to scrutinize group differences of anthropometric, growth and maturation (AGAM) toward soccer performance.

3. Results and Discussion
The output of this study will project in two-fold, 1) examine the domain components interrelated to the soccer performance and 2) scrutinize group differences of anthropometric, growth and maturation (AGAM) toward soccer performance predictors. Prior of the main analysis, PCA were computed to determine the appropriate components with eigenvalue greater than 1 (eigenvalue>1.0). It can be observed from Figure 1 that PCA identified two components as the most essential due to the higher

IOP Conf. Series: Materials Science and Engineering **342** (2018) 012056 doi:10.1088/1757-899X/342/1/012056

eigenvalues (>1). Further analysis performed with varimax rotation by applying two new latent factors in the manner of the appropriate interpretation.

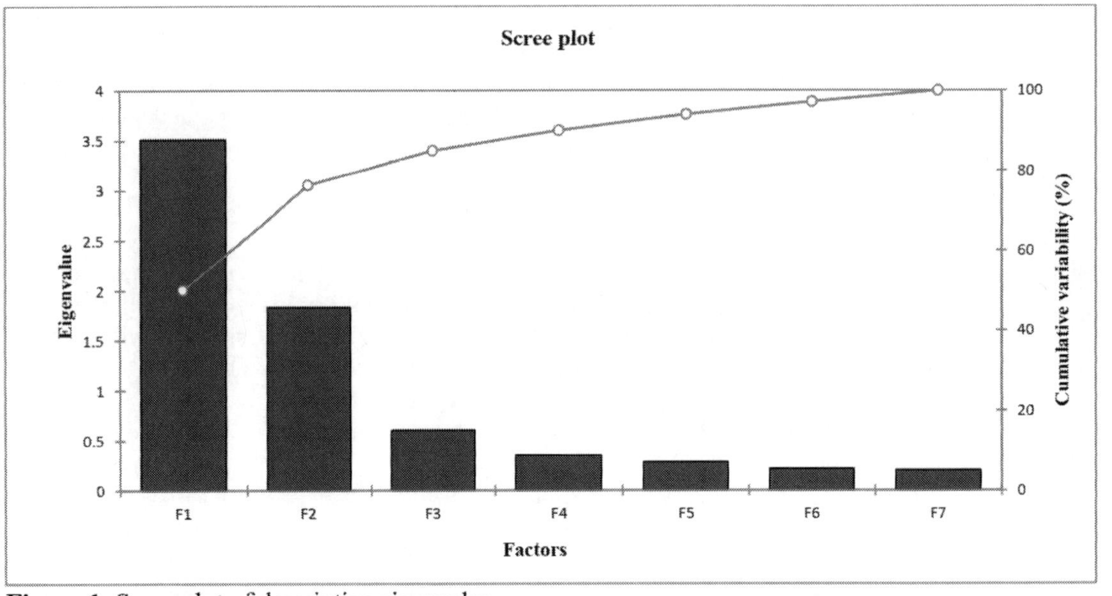

Figure 1: Scree plot of descriptive eigenvalue.

PCA pattern after varimax rotation was disclosed in Table 1. It can be seen the influence of the variance for PCA1 (42.7 %) and PCA2 (33.7 %) with the cumulative of variance is 76.4 %. There are four principal components from first factors (PCA1) satisfied the factor loading threshold (factor loading threshold >0.65) namely bicep, triceps, subscapular, and suprailiac which is reflecting the domain of the body fat. This finding suggesting the important of the body fat needed to be monitor as it have associated with the soccer players performance. According to the previous studies, performance of the soccer (sprint times) players were strongly correlated with the changes in percentage of body fat [13]. Furthermore, percentage of body fat content of elite professional soccer players significantly reduced during the pre-season and seasoning period and increased during the off-season. Thus, physiological operating demanding specific level of body fat, though extra percentages of adipose tissue acts as an additional load in which the body weight must be lifted repeatedly against gravity [14]. Finding of the current study highlighted the important of the domain factor of the percentage of the body fat needed to be monitored as it shown the domain in the first factors which biceps and triceps shows very strong loading, suprailiac part show strong loading and subscapular show moderate loading (see Figure 2).

Meanwhile, second factors explained the domain components from growth and maturation namely muac, cc and maturation. This finding suggesting the domain factors of biological growth and maturity of soccer player. It can be observed that calf circumference (31.9 %) shows the most contributor compared to maturity (28.4 %) and middle upper arm circumference respectively (27.4) (see Figure 2). Finding of the study is in concordance with the previous study stated that biological growth and maturity status contributes positively, although to a small degree, to the dissimilarity in soccer performance [15]. Variation of the anthropometric, growth and maturation by soccer performance were further discuss.

Table 1: Factor loading pattern after Varimax rotation (factor loading set at >0.65).

Components	PCA1	PCA2
Bicep	0.9007	
Tricep	0.9092	
Subscapular	0.6963	
Suprailiac	0.8501	
Muac		0.8039
CC		0.8666
Maturation		0.8119
Eigenvalue	3.5105	1.8354
Variability (%)	42.6542	33.7164
Cumulative %	42.6542	76.3705

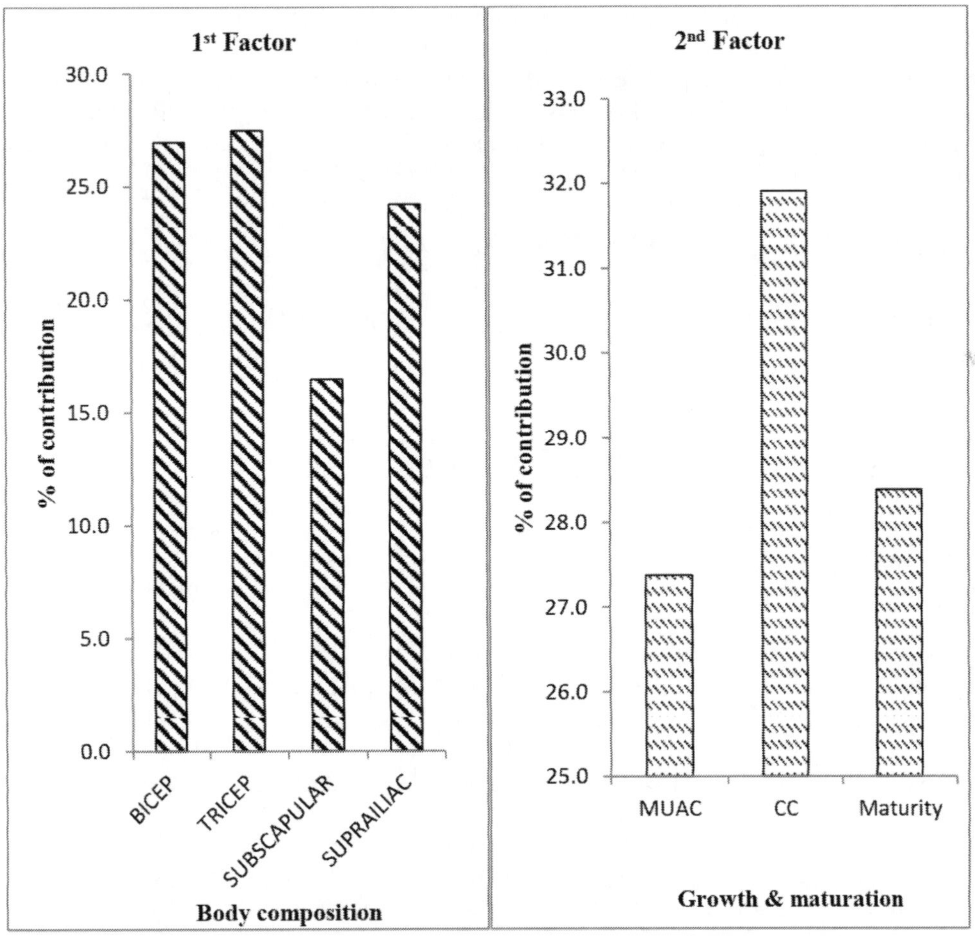

Figure 2. Contribution of The Variables (%) After Varimax Rotation On 1st Factor and 2nd Factor.
*Note: MUAC = middle upper arm circumference; CC = calf circumference.

Furthermore, further analysis was calculated by evolving an index of the anthropometric, growth and maturation index (AGaMI) by applying output of the PCA. Index of the anthropometric, growth and maturation are based on the most dominant components consequential to yield three different categorical sets of anthropometric, growth and maturation namely low, moderate and high AGaMI as shown on Table 2. As shown in Table 2, frequency and cumulative frequency of the players in each group are presented. Based on an index described 5 players consist on high AGaMI group, 88 players on moderate and 91 players on low AGaMI group. Based on the result of AGaMI, each group were further analyzed by seeking the variation on the soccer performance.

Table 2. Index Status of Anthropometric, growth and maturation index (AGaMI).

Status	Frequency	Cumulative Frequency	%	AGaMI
-131.98	91	91	49.46%	Low
-5.34	88	179	47.83%	Moderate
121.31	5	184	2.72%	High

Output of an index were further analyzed by applying ANOVA to determine the differences between groups on soccer performances. Descriptive statistics of mean for functional capacities, chronological age, body size, goal orientation and soccer specific skills for players in three groups of AGaMI are summarized in Table 3. High AGaMI players are significantly lacking of flexibility, leg power, short and long pass, and have equivalently on biological and body size (age, height, weight and sitting height) than other groups ($p<0.05$), but the groups do not differ in muscle strength, agility, speed, aerobic capacity, goal orientation, ball control and shooting ($p>0.05$). Discrepancy of the AGaMI groups based on the significant parameters are summarized in Figure 3.

Table 3. Age, body size, functional capacities, sport-specific skills and goal orientation of youth soccer players classified as high, moderate and low in anthropometric, growth and maturation index and results of ANOVAs.

	Moderate	High	Low	Pr > F	Significant
Sit & Reach (cm)	14.58	13.70	12.20	0.0076	YES
Vertical jump (cm)	66.24	59.60	61.60	0.021	YES
Variability Sit Up (reps)	6.05	6.60	5.90	0.1762	NO
Agility (s)	2.39	2.39	2.35	0.5396	NO
5 m speed	0.81	0.78	0.78	0.3819	NO
10 m speed	1.51	1.49	1.54	0.63	NO
20 m speed	2.83	2.80	2.88	0.4567	NO
VO_2max (ml/kg/min)	48.75	45.18	47.52	0.3747	NO
Chronological age (month)	15.59	15.05	14.74	0.0015	YES
Weight (kg)	60.39	70.12	52.00	< 0.0001	YES
Height (cm)	168.20	166.70	163.87	0.0016	YES
Sitting Height (cm)	88.70	90.06	84.91	< 0.0001	YES
Task	60.31	60.80	59.34	0.685	NO
Ego	32.35	39.00	32.95	0.4262	NO
Ball Control (point)	7.43	6.40	7.11	0.4105	NO
Long Pass (point)	4.31	4.00	3.26	0.037	YES
Short Pass (point)	10.61	8.40	9.20	0.026	YES
Shooting Right top corner (point)	3.44	3.60	3.58	0.9491	NO
Shooting Left top corner (point)	3.05	4.20	3.76	0.2155	NO

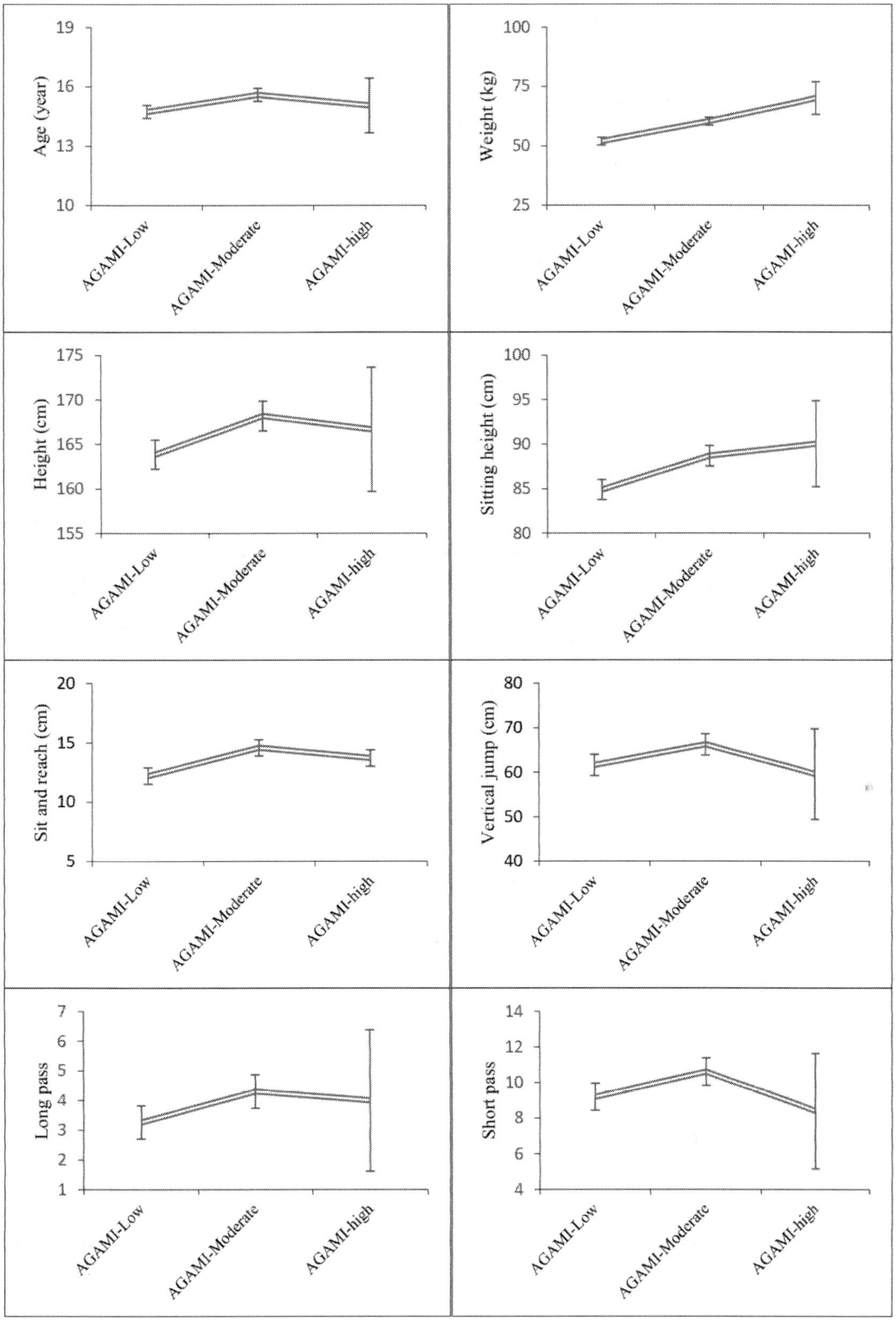

Figure 3: Differences of the most significant variables between AGaMI groups.

Variation in anthropometric, growth and maturity index associated with soccer relative performances in moderate groups of AGaMI was similar in general, between groups advanced in body size, functional capacity and specific soccer attributes than those on low and high groups of AGaMI. Correspondingly, finding of the current study reckon that moderate group of AGaMI advanced their pier groups on the flexibility, vertical jump, short and long pass test. This current finding is similar with previous study stated that body size was significantly vary in maturity status [15] and this discrepancy will lead to the significant differ on the specific soccer performance such as lower leg explosive power test [16]. In contrast to other tests, majority of the assessment of youth soccer players did not differ between groups. It leads to the 11 tested parameters suggesting that there is an equivalent performance between groups that not influence by the variation of the anthropometric, growth and maturation.

Noted that the current study is limited to youth soccer players in Malaysia and an inquisitiveness of potential attention is the anthropometric, growth and maturity status of the players relative to the general population of Malaysia youth. The moderate group of AGaMI tended to be, on average, greater muscle flexibility, more explosive leg power, greater point on short and long pass and taller and aged than low and high of AGaMI. In contrast, the high group of AGaMI tended to be, on average, proportionally taller in sitting height and heavier than other groups. Interestingly, variation of the AGaMI did not affected on the most functional capacity, goal orientation, ball control and shooting test.

4. Conclusion

Model of adaptations to AGaMI include the underlying physical characteristic variations and their explicit indicators. Deliberation of the tested parameters did not consist of perceptual cognitive, tactical, and biomechanical and all these data were not encompassed as it is mostly laboratory-based and not fit for the current study design. Current study summarizes multilaterally the role of AGaMI affecting the variation of the youth soccer performances. This study highlights the soccer players characteristics on the physical characteristic disparities effect. Information of this characteristics on the physical disparities will lead to the effectiveness on the training regime fit with the demand of the soccer game respectively. Specifically, finding of the study suggesting that coaches and instructor need to characterize their player based on the physiological and physical attributes since it proven to influence soccer player performances.

Acknowledgments

The researchers thank to the coaches, players and Innovative Manufacturing, Mechatronics & Sports Lab (iMAMS) for collaborating in this study. No any conflict of interest to be declare by the researchers.

References

[1] Tóth T, Michalíková M, Bednarčíková L, Živčák J, and Kneppo P 2014 Somatotypes in sport *acta mechanica et automatic* **8** 27-32.

[2] Lund S, Söderström T 2017 To See or Not to See: Talent Identification in the Swedish Football Association *Sociology of Sport Journal* 1.

[3] Keller B S, Raynor A J, Bruce L, and Iredale F 2016 Technical attributes of Australian youth soccer players: Implications for talent identification *International journal of Sports Science & Coaching* **11** 819-24.

[4] Wilson R S, James R S, David G, Hermann E, Morgan O J, Niehaus, A C, ... and Smith M D 2016 Multivariate analyses of individual variation in soccer skill as a tool for talent identification and development: utilising evolutionary theory in sports science *Journal of sports sciences* **34** 2074-86.

[5] Gil S M, Zabala-Lili J, Bidaurrazaga-Letona I, Aduna B, Lekue J A, Santos-Concejero J, and Granados C 2014 Talent identification and selection process of outfield players and goalkeepers in a professional soccer club *Journal of sports sciences* **32** 1931-9.

[6] Abdullah M R, Maliki A B H M, Musa R M, Kosni N A, Juahir H, and Mohamed S B 2017 Identification and Comparative Analysis of Essential Performance Indicators in Two Levels of Soccer Expertise *International Journal on Advanced Science, Engineering and Information Technology* **7** 305-14.

IOP Conf. Series: Materials Science and Engineering **342** (2018) 012056 doi:10.1088/1757-899X/342/1/012056

[7] Abdullah M R, Musa R M, Maliki A B H M, Kosni N A, and Aziz M A 2016 The Application of Principle Components Analysis to Identify Essential Performance Parameters in Outfield Soccer Players *Research Journal of Applied Sciences* **11** 1199-205.

[8] Abdullah M R, Eswaramoorthi V, Musa R M, Maliki A B H M, Kosni N A, and Haque M 2016 The Effectiveness of Aerobic Exercises at difference Intensities of Managing Blood Pressure in Essential Hypertensive Information Technology Officers *Journal of Young Pharmacists* **8** 483-6.

[9] Gipit Charles M A, Abdullah M R, Musa R M, Kosni N A, and Maliki A B H M 2017 The Effectiveness of Traditional Games intervention program in the Improvement of Form One School-Age Children's Motor Skills Related Performance Components *Journal of Physical Education and Sport* **7** 925–30.

[10] Johnston K, Wattie N, Schorer J, and Baker J 2018 Talent Identification in Sport: A Systematic Review *Sports Medicine* **48** 97–109.

[11] Woods C T, Raynor A J, Bruce L, McDonald Z, and Robertson S 2016 The application of a multi-dimensional assessment approach to talent identification in Australian football *Journal of sports sciences* **34** 1340-5.

[12] Hirose N, Seki T 2016 Two-year changes in anthropometric and motor ability values as talent identification indexes in youth soccer players *Journal of science and medicine in sport* **19** 158-62.

[13] Ostojic S M 2003 Seasonal alterations in body composition and sprint performance of elite soccer players *Journal of Exercise physiology online* **6** 24-7.

[14] Rico-Sanz J 1998 Body composition and nutritional assessments in soccer *Int J Sport Nutr* **8** 113-23.

[15] Malina R M, Cumming S P, Kontos A P, Eisenmann J C, Ribeiro B, and Aroso J 2005 Maturity-associated variation in sport-specific skills of youth soccer players aged 13–15 years *Journal of sports sciences* **25** 515-22.

[16] Figueiredo A J, Gonçalves C E, Coelho E Silva M J, and Malina R M 2009 Youth soccer players, 11–14 years: maturity, size, function, skill and goal orientation *Annals of human biology* **36** 60-73.

IOP Conf. Series: Materials Science and Engineering **342** (2018) 012057 doi:10.1088/1757-899X/342/1/012057

A multilateral modelling of Youth Soccer Performance Index (YSPI)

Ahmad Bisyri Husin Musawi Maliki[2], Mohamad Razali Abdullah[1,2], Hafizan Juahir[1], Farhana Abdullah[2], Nurul Ain Shahirah Abdullah[2], Rabiu Muazu Musa[2,3], Siti Musliha Mat-Rasid[1], Aleesha Adnan[2], Norlaila Azura Kosni[2], Wan Siti Amalina Wan Muhamad[2] and Nur Afiqah Mohamad Nasir[2]

[1] East Coast Environmental Research Institute (ESERI), Universiti Sultan Zainal Abidin, 21300, Terengganu, Malaysia.
[2] Faculty of Applied Social Sciences, Universiti Sultan Zainal Abidin, 21300, Terengganu, Malaysia
[3] Innovative Manufacturing, Mechatronics & Sports Lab (iMAMS), Faculty of Manufacturing Engineering, Universiti Malaysia Pahang (Pekan Campus), 26600 Pekan, Pahang, Malaysia

E-mail: razali896@yahoo.com

Abstract. This study aims to identify the most dominant factors that influencing performance of soccer player and to predict group performance for soccer players. A total of 184 of youth soccer players from Malaysia sport school and six soccer academy encompasses as respondence of the study. Exploratory factor analysis (EFA) and Confirmatory factor analysis (CFA) were computed to identify the most dominant factors whereas reducing the initial 26 parameters with recommended >0.5 of factor loading. Meanwhile, prediction of the soccer performance was predicted by regression model. CFA revealed that sit and reach, vertical jump, VO_2max, age, weight, height, sitting height, calf circumference (cc), medial upper arm circumference (muac), maturation, bicep, triceps, subscapular, suprailiac, 5M, 10M, and 20M speed were the most dominant factors. Further index analysis forming Youth Soccer Performance Index (YSPI) resulting by categorizing three groups namely, high, moderate, and low. The regression model for this study was significant set as $p < 0.001$ and R^2 is 0.8222 which explained that the model contributed a total of 82% prediction ability to predict the whole set of the variables. The significant parameters in contributing prediction of YSPI are discussed. As a conclusion, the precision of the prediction models by integrating a multilateral factor reflecting for predicting potential soccer player and hopefully can create a competitive soccer games.

1. Introduction

Structured talent detection, identification and development programs have been established for several sports, where success has been associated to anthropometric, physiological and skill characteristics [1-3]. Even though equivalent programs for soccer are less clear, many clubs selectively enrol promising players at a relatively early age and offered specialized programs with the aims of developing and perfecting playing ability [4,5]. The selection, development and professional supervision of young players is thus a priority for many top soccer clubs in order to maintain their game competitiveness. In the world of soccer, physical fitness, anthropometric, technical and physiological is the most important element to improve soccer player performance. Furthermore, this sport can increase the strength and stamina and ensures the health of the players. Therefore, the coach plays a big role in building an active player at the maximum fitness. Based on previous studies, the model that has been used is to find out the most vital variable aims to improve an impact of training [6-8].

Coaches and instructors are often faced with the problem of choosing the appropriate dimensionality of a soccer performance model that will fit a given set of multilateral factors. Construction of soccer performance in adolescents, requires that trainers, teachers or soccer instructors observe carefully and thoroughly in providing guidance to their players. Therefore, football trainers, teachers or trainers must understand and identify the appropriate players' skills and personality at their age [6]. This view is shared by a specialist in youth soccer, who suggest that advanced players in maturity are associated with the advantages of size, fat-free mass, muscle mass, strength, power and speed [7]. In the football team, fulfill players with a very specific role is necessary. Therefore, it is important to measure specific performance by identifying the most important attributes for each player [8]. Thus, this study aims to identifying the most dominants factors affected the performance of soccer players and prediction predictive performance relative to the category of soccer player performance.

2. Materials and Method

The procedures of the soccer battery test and psychometric related to soccer was undertaken the following procedure:

2.1 Participants

A random sampling approaches were applied to this study encompasses of all players in Malaysia soccer academies. Inclusion and exclusion of the players condition were applying whereas only players suit with the ex post facto design. Overall of 184 youth soccer players (mean age = 15.2 ± 1.6 years) were joined to participate in this study which is drawn from two Malaysian youth soccer academies and six from state school centre of soccer. All the process of the research including procedure, protocol and equipment were approved by the university Human Research Ethics Committee (UniSZA/02/1/2016/Jil. 207). Consent forms and objective of the study were informed to the managers of the academies, parents, and guardians of the players with voluntary signed by the players.

2.2 Anthropometrics, growth and maturation

Anthropometric tests were carried out to measured weight, height, sitting height, and four sites of body fat. Weight is assessed using a standard electronic digital scale to the nearest of 0.1 kg, meanwhile height was measure to the nearest 0.5cm. Chronological age is measured in the month starting from the date of each player being born until the date of test. Furthermore, sitting height were measured from the top of the head to the bottom of the sitting and recorded to the nearest 0.5 cm. Percentage of body fat is assessed through skinfold calliper using four measuring sites namely, tricep, biceps, subscapular and suprailiac to the nearest 0.1 mm according to the recommendation from previous studies [9-12]. Additionally, the girth of the body size or circumference is measured on two parts of the body namely, calf circumcision (cc) and medial upper arm circumference (muac) by adhesive and non-elastic tape. All procedures and protocol testing body fat and circumference were obtained twice and the mean score for the final score is recorded for further analysis. On the other hand, the most visible changes during puberty are growth in stature and development of secondary sexual characteristics. Equally profound the achievement of fertility and changes in most body systems especially in soccer. Maturation were measured by applying tanner scale stages of maturity as recommended by previous researchers and adopting the protocol and procedure into current study [10].

2.3 Soccer battery test

Muscle strength tests are measured in the recommended way for physical fitness assessment. There are seven variables to be measured in physical fitness test performance are vertical jump, variability of sit up, agility, sit and reach, 5 m, 10 m and 20 m speed. The tape measurement is used to scale the size of the flexibility and the long reach of sit and reach and measure long jump for vertical jump. A sit and reach test are performed to measure the flexibility of the lower back of the body and the hamstring muscles. The player must start in a ready state with both straight legs. The range is calculated to the extent that the hand reaches the maximum measuring line. Vertical jump tests are performed to measure heights while jumping. While calculating the height of the jump, the time is also taken when the jump is done. The purpose of this test is to measure the strength of the leg muscles [13]. Stopwatch and timing gate were used to record and coordinated the time to test the routine variability of sit-ups and 505 agility tests. Variability of sit up test is performed to measure the strength and durability of abdominals and hip muscles. The sit up maximum routine of set undertaken for recorded. Meanwhile, agility is the ability to quickly change the body position in the minimum requirement time. The timing gate is required to record the fastest time from the start line to the finish line [14].

2.4 Soccer technical skills

Soccer specific battery tests were implemented to acquire technical skill attributes by applying method from the prior studies namely F-MARC test conducting to measured dribbling with and without ball, ball control using three different part of body consecutively, short pass, long pass, shooting with foot and score using head. Time and speed are taken and evaluated during the test process [8].

2.5 Psychometric (Motivation)

Task and Ego Orientation in Sports (TEOSQ) have been employed in this study. It contains 13 items that measure the ability of athletes to task or ego oriented towards the sports. The questionnaire was translated into Bahasa Melayu using the back-translation method and it was confirmed and added extraordinary reliability in sports. Scale for task and ego orientation shows sufficient internal consistency with alpha 0.82 and 0.71 reliability coefficients respectively [15]. This instrument contains six items that assess the ego components (such as "I can do better than my friends") and seven items assess the task components (such as "I work hard"). Feedback is shown on 5 points of the Likert type scale where 1 strongly disagrees and 5 strongly agree.

2.6 Data Analysis

A total 4784 of matrices data that comprised 184 observations of players and 26 parameters was calculated to be analyzed. In this case, box plots and Kolmogorov-Smirnov were used to check the missing data and normality of the data. Based on this result, its described the data do not have any missing although some of the variables are not normal, further analysis were computed as recommendation by the prior researches stated that it is normal to have unnormal data to be distributed in human performance because of the individuality variation [6,8,10,11].

To achieve the objective of this study, the researcher was conducted a factor analysis (FA) method to analyze the data collected. Through the analysis, it can help the researcher to determine the main factors related to the soccer performance. Factor analysis have been used to reduce a large variable into a fewer variable. Hereby, the dominant of factors will be displayed. From this study, 26 of factors were categorized as an independent variable. However, the dominant factors extracted by applying principal components analysis method will only be interpreted when applying varimax rotation [12]. Moreover, all the assumptions of the FA were computed prior of the analysis such as sampling adequacy and sphericity test of the parameters need to be achieve [6,8,10,11]. Furthermore, researcher also applying regression aims to predict the soccer performance by sets of soccer players dominants factors acquire from FA. In specific, regression is used primarily to state the cause and effect. In the simplest form, this formula shows correlation between one independent variable (x) and dependent variable (y), $Y = B_o + B1\ X_1 + B_2 X_2 + B_3 X_3$.

3. Results and Discussion

The output of this study will project in two-fold, 1) identifying the most dominants factors related to the performance of soccer player and 2) predict relative group performance for soccer players based on the most dominants factors related to the performance of soccer player. Initial analysis of FA, exploratory factor analysis (EFA) were computed to determine the latent factors by selecting the appropriate factors with the condition of eigenvalue greater than 1 (>1.0). Based on the Figure 1, scree plot projected the factors that meet the eigenvalue rule which is only four latent factors have the value greater than 1.

Further analysis by applying four new latent factors will further analyzed computing using confirmatory factor analysis (CFA).

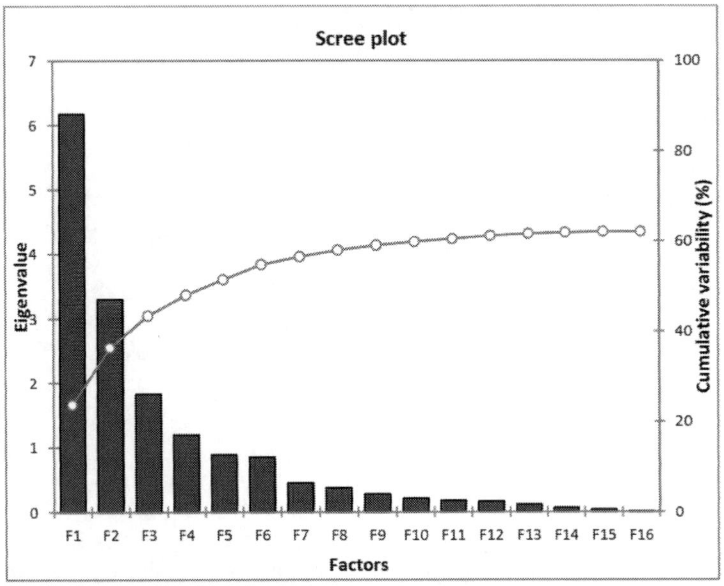

Figure 1: Scree Plot of Descriptive Eigenvalue value.

Factor pattern after varimax rotation was disclosed in Table 1. There are ten components that can be seen from first factor (D1) that are fulfilled the 0.50 factor loading threshold (Sit & Reach, Vertical jump, VO_2max, age, weight, height, sitting height, muac, cc, and maturation). Likewise, second dominant factor (D2) identifies four components with a positive higher factor loading such as bicep, tricep, subscapular and suprailiac. The rest is components from D3 as the third most important factor, had identified three parameters of speed component (5 m, 10 m and 20 m). Meanwhile, there is no dominant factor for D4 that meet the rule of the factor loading setting. Furthermore, it disclosed the most significant component of factor loading after varimax and it can be seen the contribution of the variance for D1 (23.746 %), D2 (12.693 %), D3 (7.045 %) and D4 (4.598 %) with a total of the variance is 48.081 % respectively.

FA by applying a combination of EFA and CFA revealed only seven variables have been dropped out from 26 variables initially. Therefore, 17 variables have been categorized as the dominant factors. From this analysis it has been described from the most dominant factors until the less important factor. In this case, it has proved the most dominant factor were sit & reach, vertical jump, VO_2max, age, weight, height, sitting height, muac, cc, and maturity. It is explaining the important of the coordination, agility, power, and body balance attributes are important component that influencing soccer players performance [16]. The vertical jump influencing of athlete soccer performance due to player's strength in half squat. Hereby, the coach should focus on strength training especially in terms of player movements, and indirectly improve the power of the jump [17].

Table 1. Descriptive of Factor pattern after Varimax rotation.

Variables	D1	D2	D3	D4
Sit & Reach (cm)	**0.514**	-0.067	-0.051	0.072
Vertical jump (cm)	**0.506**	-0.146	-0.155	0.139
Variability Sit Up (reps)	0.218	-0.035	-0.061	-0.175
Agility (s)	0.098	0.089	-0.091	0.213
5 m speed	0.046	0.077	**0.798**	0.150
10 m speed	-0.269	0.053	**0.857**	-0.138
20 m speed	-0.375	0.116	**0.817**	-0.057
VO_2max (ml/kg/min)	**0.516**	-0.322	-0.145	0.179
Chronological age (month)	**0.657**	-0.108	-0.295	0.387
Weight (kg)	**0.847**	0.274	-0.163	-0.117
Height (cm)	**0.807**	-0.057	-0.090	0.026
Sitting Height (cm)	**0.740**	0.026	-0.055	-0.021
Bicep (mm)	0.009	**0.821**	0.244	-0.057
Tricep (mm)	-0.021	**0.844**	0.108	-0.027
Subscapular (mm)	0.397	**0.696**	-0.016	0.094
Suprailiac (mm)	0.124	**0.857**	-0.056	0.025
MUAC (mm)	**0.660**	0.348	-0.117	-0.122
CC(mm)	**0.792**	0.291	-0.169	-0.136
Maturation	**0.719**	-0.064	-0.136	0.272
Task	0.115	-0.053	0.002	-0.121
Ego	0.014	0.040	-0.091	-0.119
Ball Control (point)	0.120	-0.060	0.247	0.281
Long Pass (point)	0.438	-0.044	-0.124	0.384
Short Pass (point)	0.254	-0.085	0.287	0.373
Shooting Right top corner (point)	-0.228	0.103	-0.127	0.463
Shooting Left top corner (point)	-0.216	0.040	-0.117	0.483
Eigenvalue	6.174	3.300	1.8316	1.1953
Variability (%)	23.746	12.693	7.045	4.598
Cumulative %	23.746	36.439	43.483	48.081

*Note: MUAC = middle upper arm circumference; CC = calf circumference; factor loading set at >0.50.

Additionally, further analysis was computed by developing an index of the youth soccer performance by applying output of the CFA. Index of the youth soccer performance are based on the most dominant parameters resulting to produce three different categorical groups of soccer performance namely low, moderate and high soccer performer as shown in Table 2. Categorical groups of performance are define based on the calculated standardization range of all variables and it is tabulate ranked in the index as shown in the group ranged on the Table 2. It also presenting the frequency and cumulative frequency of the players in each group. Based on the result of index, each group were coded as low =1, moderate =2 and high =3 for the analysis of regression. Based on an index described 49 players for high performance, moderate 117 players and low 18 players.

Table 2. Index Status of youth soccer performance.

Status YSP	Freq	Cum. Freq.	%	Cum. %	Group range	Group
-82.86	18	18	9.78%	9.78%	-82.86 ≤ low < -34.39	Low
-34.39	117	135	63.59%	73.37%	-34.39 ≤ < Moderate < 14.08	Moderate
14.08	49	184	26.63%	100.00%	High ≥ 14.08	High

Table 3 shows the goodness of fit statistics for observation of the 184 soccer players in this study. It shows prediction ability of the contributions of the regression model. It is also show that R2 value is 0.822 which explained that the model used in this study has contributed for the total of 82% prediction ability to predict the whole data set of the variables. This result shows that this model is strong enough to measure the set of variables studied and can continue for the further analysis.

Table 3. Goodness of Multiple linear regression fit statistics.

Observations	184
Sum of weights	184
DF	157
R^2	0.8228
Adjusted R^2	0.7934
MSE	0.0697
RMSE	0.2641
DW	1.6996

Further analysis of variance as shown in Table 4, revealed that there is a significant effect on the model ($p<0.001$) explain by the most dominants parameters and it is worth to identify the parameters that contribute to the model.

Table 4. Analysis of Variance.

Source	**DF**	**Sum of Squares**	**Mean Squares**	**F**	**Sig**
Model	26	50.8277	1.9549	28.0307	< 0.0001
Error	157	10.9495	0.0697		
Corrected Total	183	61.7772			

Backward stepwise method was applied to indicate the contribution of the most dominant parameters explaining the youth soccer performance. As disclosed on the Table 4, it is described the most significant parameters statistically contributes to the prediction of the player performance. By applying backward stepwise method, it is show the most significant parameters that explaining the prediction ability of the youth soccer performance namely 5 m speed, age, weight, height, sitting height, bicep, cc, maturity, task, ball control, short pass, and shooting right top corner ($p<0.05$). Meanwhile, 14 parameters were excluded from the prediction model.

The variables included in the prediction model contributes and affects the relative performance of soccer players. This finding is in concordance with the previous studies indicated that to win a game in the soccer match, players must perform an attacking or counter attack to score goals. An attacker must be active in looking for any weakness on the opponent to exploit the opponent properly. Possession of the game is related to the ball control and passing which can dominate the game and producing goal. Both components are related to the performance of the players and it is related in increasing chances of win the game [13].

Table 5. Significant variable resulting from backward regression method.

Source	Value	Standard error	t	Pr > |t|	Lower bound (95%)	Upper bound (95%)
Intercept	7.565	0.763	9.914	< 0.0001**	6.059	9.071
Sit & Reach	0	0				
Vertical jump	0	0				
Var Sit Up	0	0				
AGILITY	0	0				
5m Speed	-0.781	0.142	-5.504	< 0.0001**	-1.061	-0.501
10m Speed	0	0				
20m Speed	0	0				
VO2MAX	0	0				
Age	-0.062	0.017	-3.724	0.0003**	-0.095	-0.029
Weight	-0.032	0.006	-4.929	< 0.0001**	-0.044	-0.019
Height	0.013	0.006	2.368	0.019**	0.002	0.025
Sitting height	-0.021	0.006	-3.410	0.0008**	-0.033	-0.009
BICEP	-0.085	0.024	-3.498	0.0006**	-0.133	-0.037
TRICEP	0	0				
SUBSCAPULAR	0	0				
SUPRAILIAC	0	0				
MUAC	0	0				
CC	-0.038	0.018	-2.121	0.0354**	-0.072	-0.003
Maturity	-0.085	0.038	-2.228	0.0272**	-0.160	-0.010
TASK	-0.005	0.003	-2.058	0.0411**	-0.010	0.000
EGO	0	0				
Ball Control	-0.039	0.010	-3.915	0.0001**	-0.058	-0.019
LONG PASS	0	0				
SHORT PASS	-0.017	0.006	-2.824	0.0053**	-0.028	-0.005
SHOOTING RTC	-0.014	0.007	-1.984	0.0488**	-0.028	0.000
SHOOTING LTC	0	0				

**Note: Significant at p<0.05

In contrast to soccer specific attributes, only the ball control, short pass and shooting at the right top corner test significantly associated among players by level of youth soccer performance as shown in Figure 2. It is interesting that the long pass test showed poor contribute and also cannot predict the performance. Low and moderate performance of the group performed well on this test as did high players. The result suggest that the long pass test may not be a priority attribute in youth soccer performance models. Further information by the figure of the contribution of the parameters suggesting that 5 m speed are more related to the soccer players and also affecting by the physical characteristics as projected in the Figure 2. Meanwhile, maturation and task oriented players were significantly contributing to the youth soccer performance.

Generally, finding of the study are generally consistent with previous researches in light of the fact that the low performance players corresponded to the high performance players in the present design [18]. Consistent with the previous multidimensional investigations of youth soccer and field hockey model, finding of the current study revealed the better prediction models of youth soccer performance by the multilateral set of the most dominant factors in soccer [19-20].

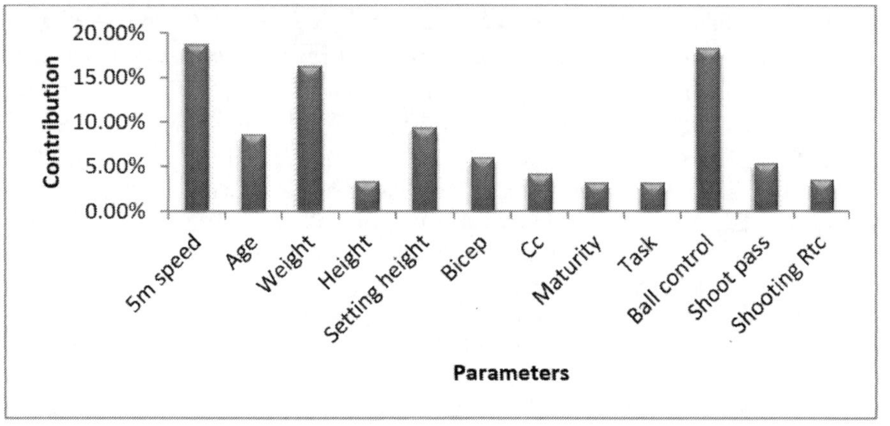

Figure 2. Interface for Camera Live View and Sensors.

4. Conclusion

As noted, the prediction model is limited by its cross-national nature (Malaysia). The consideration of the variables did not include perceptual cognitive, tactical, and biomechanical, all these data were not encompassed because it is generally laboratory-based and not suitable for current studies design. In instant, the present study specifies that prediction model of youth soccer performance was significantly associated with the selected parameters in the current studies and players with different level of performance were statistically influence by the physical characteristics, physiological elements, maturation, soccer specific attributes and task orientation. The finding also highlights the relevance of the contribution on each significant parameter specific on soccer performance. Specifically, contribution of the parameters locus values for the total sample of youth soccer players may be useful for coaches and instructors in both the talent evaluation and development processes.

Acknowledgments

All the researchers thank to the manager, coaches, players and Innovative Manufacturing, Mechatronics & Sports Lab (iMAMS) for collaborating in this study. No any conflict of interest to be declare by all the researchers.

References

[1] Aasa U, Jaric S, Barnekow-bergkvist M, & Johansson H. (2003). Muscle strength assessment from functional performance tests: Role of body size. *Journal of Strength and Conditioning Research*, 17(4), 664-670.

[2] Geladas N D, Nassis G P, & Pavlicevic S. (2005). Somatic and physical traits affecting Sprint swimming performance in young swimmers. *International Journal of Sports Medicine*, *26*(2),139-144.

[3] Girold S, Calmels P, Maurin D, Milhau N, & Chatard J C. (2003). Assisted and resisted sprint training in swimming. *Journal of Strength and Conditioning Research*, *20*(3), 547-554.

[4] Bourgois J, Claessens A L, Vrijens J, Philippaerts R, Van Renterghem B, Thomis M, ... & Lefevre J. (2000). Anthropometric characteristics of elite male junior rowers. *British journal of sports medicine*, *34*(3), 213-216.

[5] Hoare D G, & Warr C R. (2000). Talent identification and women's soccer: an Australian experience. *Journal of sports sciences*, *18*(9), 751-758.

[6] Abdullah M R, Maliki A B H M, Musa R M, Kosni N A, Juahir H, & Haque M. (2016). Multi-Hierarchical Pattern Recognition of Athlete's Relative Performance as A Criterion for Predicting Potential Athletes. *Journal of Young Pharmacists*, *8*(4), 463-470.

[7] Martindale R J, Collins D, & Daubney J. (2005). Talent development: A guide for practice and research within sport. *Quest*, *57*(4), 353-375.

[8] Abdullah M R, Maliki A B H M, Musa R M, Kosni N A, & Juahir H. (2016). Intelligent prediction of soccer technical skill on youth soccer player's relative performance using multivariate analysis and artificial neural network techniques. *International Journal on Advanced Science, Engineering and Information Technology*, *6*(5), 668-674.

[9] Abdullah M R, Maliki A B H M, Musa R M, Kosni N A, Juahir H, & Mohamed S B. (2017). Identification and Comparative Analysis of Essential Performance Indicators in Two Levels of Soccer Expertise. *International Journal on Advanced Science, Engineering and Information Technology*, *7*(1), 305-314.

[10] Maliki A B H M, Abdullah M R, Juahir H, Musa R M, Mat-Rasid S M, Adnan A, ... & Alias N. (2017). Sensitivity pattern recognition and variableness of competitive adolescent soccer relative performance indicators. *Journal of Fundamental and Applied Sciences*, *9*(2S), 539-562.

[11] Abdullah M R, Musa R M, Kosni N A, Maliki A B H M, & Haque M. (2016). Profiling and Distinction of Specific Skills Related Performance and Fitness Level between Senior and Junior Malaysian Youth Soccer Players. *International Journal of Pharmaceutical Research*, *8*(3), 64-71.

[12] Abdullah M R, Musa R M, Maliki A B H M, Kosni N A, & Aziz M A. (2016). The Application of Principle Components Analysis to Identify Essential Performance Parameters in Outfield Soccer Players. *Research Journal of Applied Sciences*, *11*(11), 1199-1205.

[13] Abdullah M R, Musa R M, Maliki A B H M, Kosni N A, & Suppiah P K. (2016). Development of tablet application based notational analysis system and the establishment of its reliability in soccer. *Journal of Physical Education and Sport*, *16*(3), 951-956.

[14] Abdullah M R, Musa R M, Maliki A B H M, Kosni N A, & Suppiah P K. (2016). Role of psychological factors on the performance of elite soccer players. *Journal of Physical Education and Sport*, *16*(1), 170-176.

[15] Maliki A B H M, Abdullah M R, Juahir H, Musa R M, Kosni N A, Mat-Rasid S M, ... & Eswaramoorthi V. (2017). Back translation reliability of TEOSQ in team game, individual game and gender category. *Journal of Fundamental and Applied Sciences*, *9*(2S), 467-484.

[16] Durandt J, Tee J C, Prim S K, & Lambert M I. (2006). Physical fitness components associated with performance in a multiple-sprint test. International journal of sports physiology and performance, 1(2), 150.

[17] Wisløff U, Castagna C, Helgerud J, Jones R, & Hoff J. (2004). Strong correlation of maximal squat strength with sprint performance and vertical jump height in elite soccer players. *British journal of sports medicine, 38*(3), 285-288.

[18] Gil S M, Zabala-Lili J, Bidaurrazaga-Letona I, Aduna B, Lekue J A, Santos-Concejero J, & Granados C. (2014). Talent identification and selection process of outfield players and goalkeepers in a professional soccer club. *Journal of sports sciences*, *32*(20), 1931-1939.

[19] Reilly T, Bangsbo J, Franks A. (2000). Anthropometric and physiological predispositions for elite soccer. *J Sport Sci, 18,* 669–83.

[20] Elferink-Gemser M, Visscher C, Lemmink K, & Mulder T. (2004). Relation between multidimensional performance characteristics and level of performance in talented youth field hockey players. *Journal of Sports Sciences*, *22*(11-12), 1053-1063.

IOP Conf. Series: Materials Science and Engineering **342** (2018) 012058 doi:10.1088/1757-899X/342/1/012058

The axial crushes behaviour on foam-filled round Jute/Polyester composite tubes

A Othman[1,2] and A E Ismail[1]

[1] Department of Engineering Mechanics, Faculty of Mechanical and Manufacturing Engineering, Universiti Tun Hussien Onn Malaysia, Batu Pahat, Johor, Malaysia
[2] Department of Polytechnic Education, Ministry of Higher Education, Putrajaya, Malaysia

E-mail: akbar.othman@gmail.com

Abstract. The present paper investigates the effect of axial loading compression on jute fibre reinforced polyester composite round tubes. The specimen of composite tube was fabricated by hand lay-up method of 120 mm length with fix 50.8 mm inner diameter to determine the behaviour of energy absorption on number of layers of 45^0 angle fibre and internally reinforced with and without foam filler material. The foam filler material used in this studies were polyurethane (PU) and polystyrene (PE) with average of 40 and 45 kg/m^3 densities on the axial crushing load against displacement relations and on the failure modes. The number of layers of on this study were two; three and four were selected to calculate the crush force efficiency (CFE) and the specific energy absorption (SEA) of the composite tubes. Result indicated that the four layers' jute/polyester show significant value in term of crushing load compared to 2 and 3 layers higher 60% for 2 layer and 3% compared to 3 layers. It has been found that the specific energy absorption of the jute/polyester tubes with polystyrene foam-filled is found higher respectively 10% to 12% than empty and polyurethane (PU) foam tubes. The increase in the number of layers from two to four increases the mean axial load from 1.01 KN to 3.60 KN for empty jute/polyester and from 2.11 KN to 4.26 KN for the polyurethane (PU) foam-filled jute/polyester tubes as well as for 3.60 KN to 5.58 KN for the polystyrene (PE) foam-filled jute/polyester. The author's found that the failure of mechanism influence the characteristic of curve load against displacement obtained and conclude that an increasing number of layers and introduce filler material enhance the capability of specific absorbed energy.

1. Introduction

Material of composite have a wide variety of properties related to engineering application because they have their higher stiffness and strength with respect to weight ratio. In addition, composite materials have high corrosion resistance, thermal expansion, thermally resistive and considered as non-conductive materials. Many of crashworthy structures made of metallic materials play important roles in the last decade's dispose of composite materials [1]. The properties of composite materials given more advantageous because their lightweight and cheapest price compared to metal for vehicle applications [2]. Many researcher have been conducted to several studies on crashworthiness capability such as composite tubular cross-section like a tube [1-2], and cone [3] with different applied loading conditions.

Material of composite type of conventional were widely studied previously subjected to axial loading have been carried out many investigator to determine energy absorption performance. Due to the less studied on natural fibre composite, the author need to explore on how the natural fibre can give more impact on crashworthy application. However, several studies on natural fibre can exhibit unstable loaded when subjected to empty case, so that the purpose on this paper is to determine the influence of polyurethane and polystyrene foam as a filler material to strengthen the behaviour of tubular structures when applied to axial loading. Many researcher has been found that the natural fibre composite behaviour achieved fluctuation curve of load against displacement [8-9].

Othman et al. [10] conducted to pultruded foam filled structure subjected to axial loading with different wall-thickness and density foam filled material. They found that the influence of reinforcement internally filler material enhanced the specific absorbed energy when applied axial loading. On those studies, pultruded material was fabricated by pultrusion machine with conventional material like glass, carbon etcetera. In fact, unstable deformation is not found in conventional material. The major objective on this study to determine the failure occurred and specific absorbed energy on round composite jute fiber internally reinforced polymeric foam filler like polyurethane and polystyrene foam. The response behavior of jute/polyester with 2 to 4 layer laminate was investigated in term of load versus displacement curves.

2. Experimental work

2.1 Materials and fabrication

Jute/polyester composite tubes were fabricated using of hand layup methods using woven roving [±45°] jute/epoxy. A stainless steel of tube was use as a mould to fabricate with length of 500 mm. with diameter of 50.8 mm. The plastic bag was use to wrap on surface mould in order to extract specimen from mandrels after curing. Table 1 shows the dimensions of the fabricated composite. Samples of the fabricated specimens are shown in figure 1. The terms J, R, L, PE, and PU are respectively jute, round, layer, polystyrene and polyurethane. The number 2, 3 and 4 referred to the number of layers of the composite laminate structures. The inner diameter of all specimen tube were fixed at 50.8.

Table 1. Dimensions of the composite tubes.

Specimen type				Composite tube	
	Mass (kg)	No of layers	Foam Filler	Wall Thickness t (mm)	Length
JR-2L	0.155	2	-	2.25	120
JR-3L	0.232	3	-	3.21	120
JR-4L	0.310	4	-	4.23	120
JR-2L-PE	0.164	2	Polystyrene	2.21	120
JR-2L-PU	0.166	2	Polyurethane	2.24	120
JR-3L-PE	0.242	3	Polystyrene	3.31	120
JR-3L-PU	0.243	3	Polyurethane	3.28	120
JR-4L-PE	0.320	4	Polystyrene	4.23	120
JR-4L-PU	0.321	4	Polyurethane	4.23	120

Figure 1. Sample of the fabricated composite tubes.

The mechanical properties of the jute and glass/ epoxy materials were obtained from the literatures [13] and listed in Table 2.

Table 2. Mechanical properties of jute/polyester lamina.

Property	Symbol	LS-DYNA parameter	Experimental value
Density	ρ	RO	1350 kg/m^3
Modulus in 1-direction	E_1	EA	4150 MPa
Modulus in 2-direction	E_2	EB	2072 MPa
Shear modulus	G_{12}	GAB	1508 MPa
Major Poisson's ratio	v_{12}	PRAB	0.187
Minor Poisson's ratio	v_{21}	PRAC	0.376
Strength in 1-direction, tension	F_1^{tu}	XT	200 MPa
Strength in 2-direction, tension	F_2^{tu}	YT	200 MPa
Strength in 1-direction, compression	F_1^{cu}	XC	100 MPa
Strength in 2-direction, compression	F_2^{cu}	YC	50 MPa
Shear strength	F_{12}^{su}	SC	100 MPa

2.2. Testing procedure

Quasi-static loading condition were tested on all specimens with constant speed of 5 mm/m compression using a computer-controlled servo-hydraulic Instron machine type GT instruments. The cross-section tube crushed between two parallel steel flat platens, one fix and one moving. The fixed platen was fitted with a load cell from which the load signal is taken directly to the computer. For each test, the axial loading was plotted on the Y-axis and the crosshead displacement on the X-axis. The end front of each specimens was chamfered at 45^0 to ensure that the load fluctuation characteristic were avoided [10].

The determination of crashworthiness mechanism some of following aspect of crush phenomenon is measured during crushes event. The all data were recorded and calculated there are Peak force P_{max}: maximum compressive force. Mean force P_{avg}: average compressive force obtained by the following equation where force and deformation are defined as P and d, respectively and the area of cross section, A, and the density of the material, ρ defined as

$$\bar{P} = \frac{1}{\delta}\int_0^{\delta} P d\delta \tag{1}$$

Energy absorption E: area under the load–displacement curve up to the compaction zone.

$$E = \int_0^{\delta} P d\delta \tag{2}$$

The units of E_s are used to express the crashworthiness parameters, hence they are written in kJ here P is the load acting on the composite specimens. Therefore, the specific energy absorption per unit mass kJ/kg (SEA) where m the crushed mass of the composite is recognized as:

$$SEA = \frac{E}{m} \tag{3}$$

Crash force efficiency (CFE): ratio of the average crushing load P_{avg} to peak load P_{max}.

$$CFE = \frac{P_{avg}}{P_{max}} \tag{4}$$

3. Results and discussion

3.1. Axial loading characteristics

In this section, results of the axial loaded composite jute/polyester were presented. Figures 2, 3 and 4 show the quasi-static load-displacement curves of the composite tubes using jute (J), layer (L), round cross-section (R), polyurethane (PU) and polystyrene (PE) respectively. In figures 2, 3, and 4 are referred to the number of layers with different with and without filler materials. Study indicated that the load progressively increased from the initial failure until final deformed then the load fluctuates with a sharp drop until failure of compaction zone. The foam filler with jute/polyester tubes supported increase load than the empty jute/polyester followed by increasing number of layers of tube composites. Obviously, increase in number of layers from two to four show a significant increase in the maximum load obtained for the composite tubes as shown in figures 2, 3 and 4.

It can be seen from figure 2 also that the increase fluctuation load of jute/polyester composite tubes from empty to polymeric foam profiles until reach the final deformed compaction. The figure 2, 3 and 4 generally indicated that the irregularity fluctuation load observed from up to down loading for empty profile with the internal and external tearing failure of layers of jute/polyester and show drop from maximum load to minimum strength.

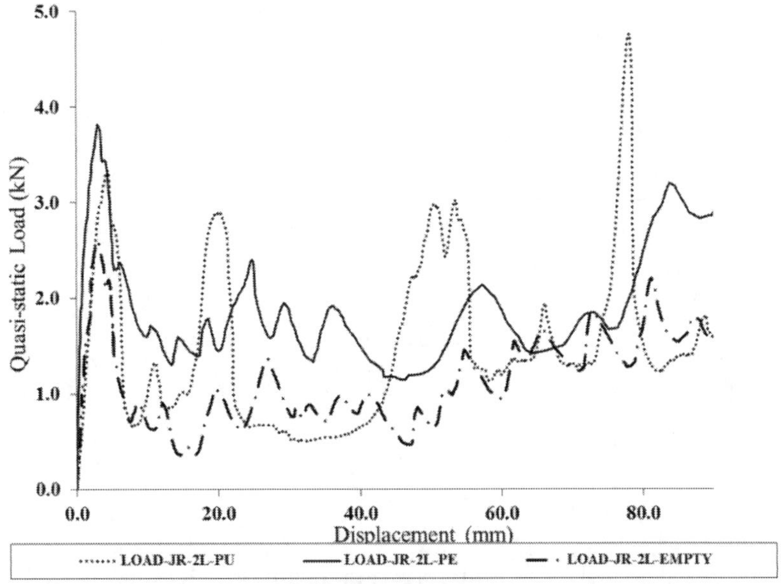

(a) Load against displacement curve of two layers jute/polyester composite tubes.

Unloading 20 mm 40 mm 70 mm 90 mm

(b) Empty profile of jute/polyester of two layers.

Figure 2. Load-displacement relation for jute/polyester tubes for two layers.

Figure 3 and 4 shows the effect number of layer three used on maximum load obtained from the load-displacement graphs. This figure explains that the jute/polyester tubes made of four layers improved crushes energy during quasi-static regime compared to two and three layers. The maximum axial load obtained for the three and four layer jute/polyester tubes with polyurethane and the increase number of layer laminate of the jute significantly affects polystyrene. Composite tubes were fixed between two flat platen as shown in figure 2 to 4 where the load applied axially on the tubes. Picture of each deformation were taken to see the characteristic of each specimen failure. The tested tubes of composite fractured at the surface position where cracks were propagated at the top front-end tube while contact between moving platen and specimen stated until reach the maximum deformation.

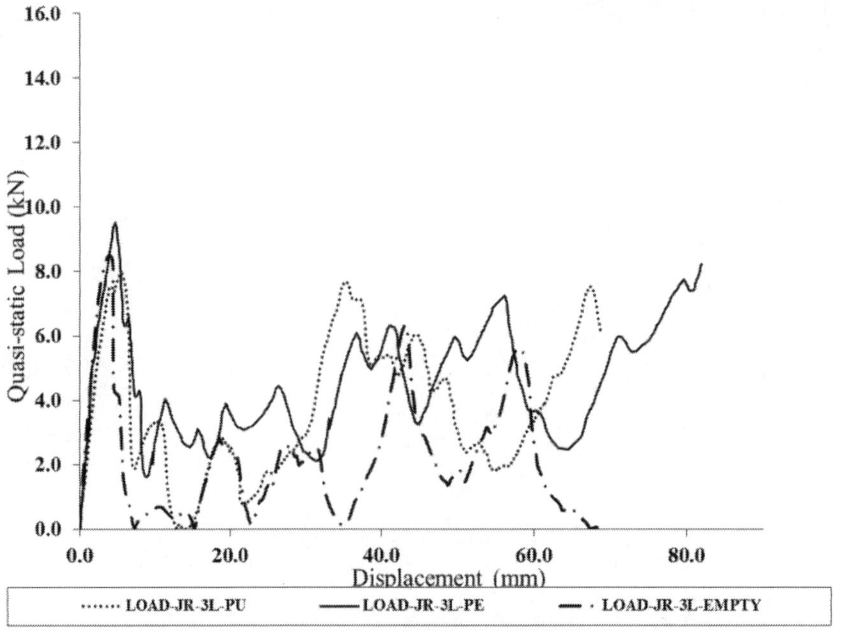

(a) Load against displacement curve of three layer jute/polyester composite tubes.

Unloading 10 mm 50 mm 70 mm 80 mm

(b) Profile of jute/polyester of three layers with polystyrene foam

Figure 3. Load-displacement relation for jute/polyester tubes for three layers.

3.2. Quasi-static loading characteristics

The initial peak and average loads are two characteristics of the failure profile of crashworthy structures. The force needed to crush the surface of the crush box during the initial stages dramatically increased and reached a certain peak force shown in figure 5. Throughout the initial peak force, the influence of failure pattern depended on the initial stages of the force. The load-displacement curve continued with a drop in the force at certain stages, especially during the failure pattern of composite pultruded tubes, whereby the load was decreased with progressive failure.

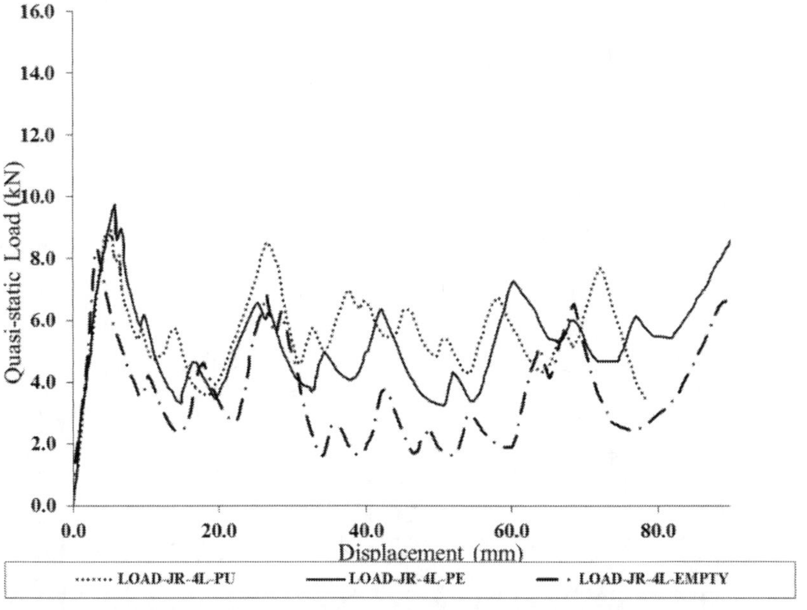

(a) Load against displacement curve of four layers jute/polyester composite tubes.

Unloading 20 mm 50 mm 60 mm 90 mm

(b) Profile of jute/polyester of four layers with polyurethane foam

Figure 4. Load-displacement relation for jute/polyester tubes for four layers.

Figure 5 shows the comparative graph of the initial peak load and the average values of the empty and polyurethane foam-filled specimens. The crush force efficiency (CFE) is an important factor to measure the crush performance and to evaluate the crashworthiness of energy absorber components. The CFE for the composite tubes under quasi-static axial loading is determined using equation 4. Results showed that an almost specimen ranging from 0.4 to 0.6 efficiency.

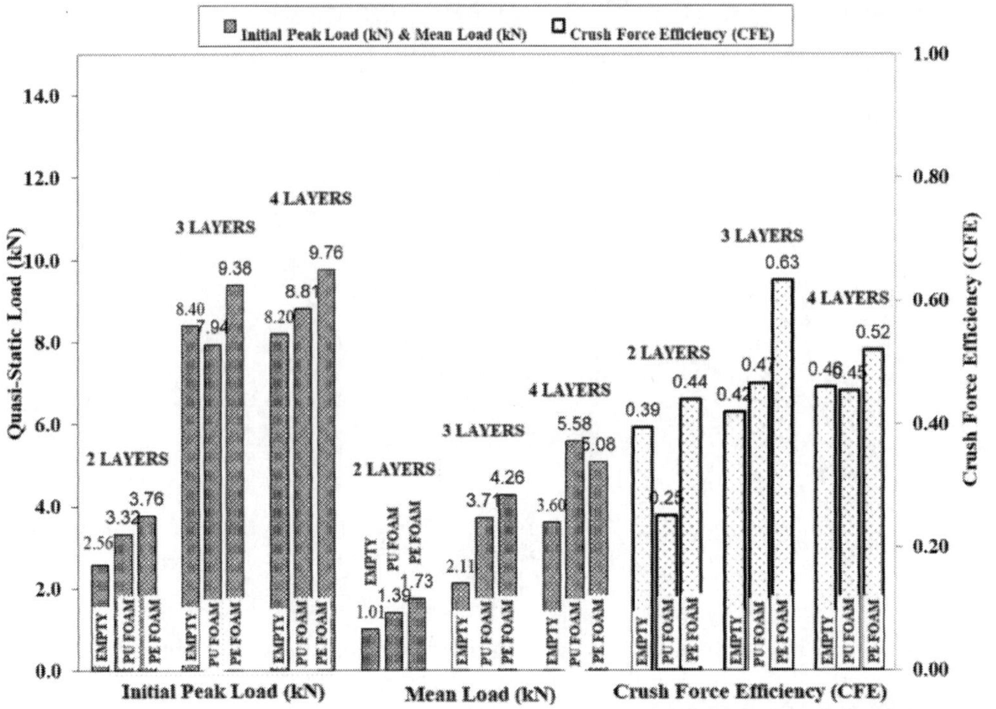

Figure 5. Initial, mean and crush force efficiency relation for jute/polyester tubes for each layers.

3.3. Specific Energy Absorption

Energy absorption is defined as the total area under the load-displacement curve. The absorbed energy is calculated to measure the performance of the crush box as a crashworthy structure. Hence, the total area under the load-displacement curve was measured before the crushing occurred. The capability of the structures and the comparison between each specimen analysis are using equation 1, 2 and 3.

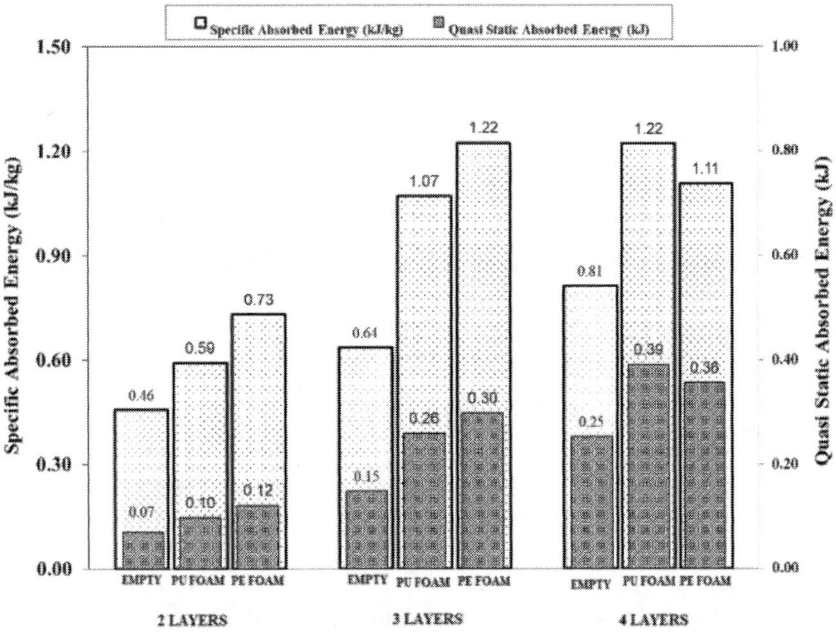

Figure 6. Energy absorption and specific absorbed energy relation for jute/polyester tubes for each layers.

4. Conclusion

In this paper, experimental tests were carried out to investigate the effect of axial loading on the load-displacement curve and specific energy absorption. Hand lay-up method was used to fabricate each specimen of the composite tubes of two, three and four layers, respectively. Results of the load-displacement curve show that the mean axial load obtained for four layered jute/polyester polyurethane is found higher 8.9% than polystyrene. The specific energy absorption of the four-layered jute/polyester polyurethane and three layered jute/polyester polystyrene tubes is found higher respectively 47.5% and 33.6% than empty tubes profile. The increase in the number of layers from two to four increased the mean axial load from 1.01 KN to 3.60 KN for empty jute/polyester tubes. The increase was from 1.39 KN to 5.58 KN for the jute/polyester polyurethane. In general, similar failure mechanism has been obtained from the fractured composite tubes where cracks propagate at the center position of the tube at the top until bottom sides of the tube. The post peak failures showed a significant deformation at the crack line position and then lead to the final fracture.

References

[1] Palanivelu Sivakumar, Paepegem Wim Van, Degrieck Joris, Ackeren Johan Van, Kakogiannis Dimitrios, Hemelrijck Danny Van, et al. 2010 Experimental study on the axial crushing behaviour of pultruded composite tubes. *Polym Test* **29** 2 p 224e34.

[2] Nagel GM, Thambiratnam DP 2004 A numerical study on the impact response and energy absorption of tapered thin-walled tubes. *Int J Mech Sci* **46** 2 p 201e16.

[3] Ahmad Z, Thambiratnam DP, Tan ACC 2010 Dynamic energy absorption characteristics of foam-filled conical tubes under oblique impact loading. *Int J Impact Eng* **37** 5 p 475e88.

[4] Jimenez MA, Miravete A, Larrode E, Revuelta D. 2000 Effect of trigger geometry on energy absorption in composite profiles. *Compos Struct* **48** 1e3 p 107e11.

[5] Othman A, Abdullah S, Ariffin AK, Mohamed NAN 2014 Investigating the quasi-static axial crushing behaviour of polymeric foam-filled composite pultrusion square tubes. Mater Des **63** 44 p 6e59.

[6] Othman A, Abdullah S, Ariffin AK, Mohamed NAN 2013 On the crushing behavior of foam-filled composite tubes under compressive loading. *Adv Mater Res* **626** p 1038e41.

[7] Othman A, Jailani Azrol 2014 Axial crushes simulation of double thin-walled metallic structure inserted with polymeric foam. *Int J Eng Technol* **6** 1 p 43e6.

[8] Jailani Azrol, Othman A 2014 Crush simulation of cross-section conical aluminum alloy 2024-T4 polyurethane foam-filled section. *Int J Eng Technol* **6** 1 p 20e4.

[9] Arifin AA, Othman A 2014 Crush force efficiency of composite square profile under off-axis loading. *Key Eng Mat* 594e595 711e4.

[10] A. Othman, S Abdullah, AK Ariffin, NAN Mohamed, Investigating the crushing behavior of quasi-static oblique loading on polymeric foam filled pultruded composite square tubes, *Composites Part B: Engineering* **95** p 493-514

[11] SNA Khalid, AE Ismail, MH Zainulabidin, Optimization on Impact Strength of Woven Kenaf Reinforced Polyester Composites using Taguchi Method, *IOP Conference Series: Materials Science and Engineering* **165** 1 p 012010

[12] AMT Arifin, M Fahrul Hassan, AE Ismail, M Zulafif Rahim, M Rasidi Ibrahim, RH Abdul Haq, MNA Rahman, MZ Yunos, MHM Amin, Investigation on Suitability of Natural Fibre as Replacement Material for Table Tennis Blade, *IOP Conference Series: Materials Science and Engineering* **226** 1 012037

[13] Anindya Deb, Sumitesh Das, Ashok Mache, Rokesh Laishram 2017 A Study on the Mechanical Behaviors of Jute-polyester Composites, *Procedia Engineering* **173** p 631-638

[14] AE Ismail, Z Noranai, NHM Nor, ALM Tobi, MH Ahmad, Effect of Hybridized Fiber Wrapped Around the Aluminum Tubes on the Crushing Performances, *IOP Conference Series: Materials Science and Engineering* **160** 1 p 012019

[15] AE Ismail, Crushing mechanisms of cylindrical winding kenaf fiber reinforced composites, *Jurnal Teknologi (Sciences & Engineering)* **78** 6 p 45-50

IOP Conf. Series: Materials Science and Engineering **342** (2018) 012059 doi:10.1088/1757-899X/342/1/012059

An overview on real-time control schemes for wheeled mobile robot

M S A Radzak , M A H Ali, S Sha'amri and A R Azwan

Faculty of Manufacturing Engineering, University Malaysia Pahang, 26600 Pekan, Pahang

E-mail: m.sawal.a.r@gmail.com; hashem@ump.edu.my

Abstract. The purpose of this paper is to review real-time control motion algorithms for wheeled mobile robot (WMR) when navigating in environment such as road. Its need a good controller to avoid collision with any disturbance and maintain a track error at zero level. The controllers are used with other aiding sensors to measure the WMR's velocities, posture, and interference to estimate the required torque to be applied on the wheels of mobile robot. Four main categories for wheeled mobile robot control systems have been found in literature which are namely: Kinematic based controller, Dynamic based controllers, artificial intelligence based control system, and Active Force control. A MATLAB/Simulink software is the main software to simulate and implement the control system. The real-time toolbox in MATLAB/SIMULINK are used to receive/send data from sensors/to actuator with presence of disturbances, however others software such C, C++ and visual basic are rare to be used.

1. Introduction

Nowadays, mobile robots are a common use in many applications that represent a hazardous, complex, high accurate or hard tasks in worldwide fields, like aerospace, under-water, military, medicine. Usually the robot is supposed to navigate autonomously in restricted environments; which need high robust control to determine its path within terrain and avoiding obstacles. In some conditions such as road environments, it is indeed required to control robustly the robot in the planned path; otherwise it will crash cars or people.

The wheeled mobile robot usually move using the wheels like two, three and four wheel mobile robot. Two wheel robot or other name is Segway, this type is requires Gyro Sensor to keep it balance in order to remain stable. There are two types of three-wheeled robot. The first one is two differential drive wheel to move the robot plus caster wheel for balancing, others type is a synchronous 3 wheels robot which are both rotating by certain velocity at same time or 3 mecanum Wheels, which have a small cylinder rollers inside and allowing the mobile robot to move in all directions. Lastly, Four Wheeled Mobile Robot or Ackerman driving can be used for driving mobile robot which have two rear wheels that can move continuously, and the front wheel used as ordinary wheels.

A wheeled mobile robot's navigation system controls a task in real-time such as control of motion, planning of the path, and self-localization. The control of motion is the important task to make sure the robot is operating well in restricted environment. The path planning is the process of finding a path for the moving of robot from the start destination to the end destination while avoiding obstacles and disturbances. The self-localization is the robot's skill to specify its position within restricted environment.

In this research, the wheeled mobile robots explores in restricted environment such as factories and road to find the location and then perform any kinds of work. The robot is using laser simulator for navigation system and controlled by active force control. The navigation system is important for generating road path in the whole of the environment and to find the robot position in the restricted environment. It is also to identify all obstacles and disturbance to the path that corresponds to the desired destination in real time. Laser simulator is important to carry out discovery tasks involving the use of sensors.

2. Kinematic Based Controllers

The kinematic based controller have been used to solve obstacle avoidance and tracking problem of wheel mobile robot with nonholonomic [1] and formation control of wheel mobile robot with virtual robot [2]. It is used also for tracking control problem when skipping and slipping for wheeled mobile robot [3], control a wheeled mobile robot using autonomous control in obstacles environment [4] and trajectory tracking for wheeled mobile robot in unknown longitudinal [5, 6] moving in cylindrical environment. It was used for round pipe [7], problem for estimate the slip ratio control of WMR and longitudinal velocity [8]. The PD controller [1] have been resulted big tracking error without PD controller (4.5m – 10m),with PD controller (4.9m-8.5m), then nonlinear control algorithm and cascade system theorem [2] have been resulted error 5 to 10 second to perform the desired formation, GPS-based tracking control and Real- Time Kinematic (RTK) [3] shown the error 0 to -1 and 0 to 4 degree then it stable after 20 second, then lagrange-D'alembert and backstepping-based control law [4], Unscented Kalman Filter and tracking control law [5] the error are close to zero with error 0.2 second. However, the control law and biaxial clinometer method [6] have been resulted much bigger error but remain stable after swinging several times and kinematic controller and Lyapunov stability theory [7] shown the error between actual and desired trajectory tend to zero in a short time. The adaptive unscented Kalman filter [8] have been resulted the big error for wheel angular velocity between measured and AUKF method. The slip ratio error of wheel mobile robot using AUKF is stable from time 2 second until finish and for measured slip ratio of wheel mobile robot, the signal error is oscillate from 0 to -0.2.

3. Artificial Intelligence Based Control System

The artificial intelligence based control system have been used to solve trajectory tracking problem of wheeled mobile robot [9, 10], to control of wheeled mobile robot to reach the target without crash with object and obstacles avoidance [11], the problem unknown disturbance, uncertain parameter and trajectory tracking control of wheeled mobile robot [12], formation control problem of nonholonomic for wheeled mobile robot [13], the basic nonholonomic navigation problem [14], problem to control the movement of outdoor mobile robot [15], the consensus problem for leader-following of wheeled mobile robot [16], The D-type iterative learning control (ILC) [17] and control law method [18] have been resulted the value of tracking error decreases clearly and is close to zero in tenth iteration in 10 second, then fuzzy logic controller and lagrange's equation [11,13] show the small error for mobile robot to avoid the obstacles and the error between leader robot follower robot 1 ,follower robot 2 and follower robot 3 has been resulted with big tracking error (0.1 meter, 0.4 meter, 1 meter) respectively. A adaptive-neural control law and Lyapunov-based stability analysis [12] shown the error between error 1 and error 2 is small error (0.2 meter). A neural network-based adaptive control [14, 18] have been resulted the tracking errors of velocities and posture in graph by using adaptive neural network controller is converge to minimum error and near to zero even if there are external disturbance. The adaptive control method and control law [16] show the result of trajectory tracking error is big error in 2 second and after 2 second the error maintain to zero.

3. Dynamic Based Controllers

The dynamic controllers have been used to solve dynamic design of omni- wheeled robot [19], control the path and motion of wheeled mobile robot [20, 21] , control the external disturbance of WMR [22, 23], trajectory tracking control for two-wheeled mobile robot [24]. The *Computed Torque Control* (Vishal A. 2016)[19] and nonlinear model predictive control (NMPC) [20] have been resulted with big tracking errors (0 -0.5m), (-0.5 - 0.2) respectively , however disturbance observer and dynamic controller [22], Sliding mode dynamic controller (SMDC) [23] and lyapunov theory and nonlinear controller [21] are tended to be zero in a short time. The sliding mode control [24] have been resulted the small error between actual trajectory and reference trajectory. The time response error of simulation 1 and simulation 2 is become stable after 5 second.

4. Active Force Controller

The AFC has generally two loops; first loop is used for kinematic parameters control and another loop for dynamics control of mobile robot. The active force control have been used to solve reject the unwanted disturbance of frictional force in pipe [25], tracking the wheeled mobile robot in pre-planning path in difficult environment [26], the problem to remove the effect of disturbance in the system and any disturbance in the dynamic system [27], problem to reject the disturbance and noise [28,29,30]. The Proportional-integral-derivative (PID), Active Force Control and sliding mode controller (SMC) [25] have been resulted the PID without AFC is the signal error oscillate from 1 second to 10 second and after 10 second the signal become stable at zero level. Resolved Acceleration Control (RAC) with Active force control (AFC) [26] have been resulted no difference between reference path and actual path when there is no disturbance for RACAFC. With constant disturbance, the tracking error has significant difference between reference path and actual path which is small error in power 10^-2. A feedworward active force control control (AFC) with Robust controller [27] and active force control (AFC) with computed torque control [28] show the result tracking error is AFC is perform to oscillate between -0.01 to 0.02 rad. only compare to compute torque that oscillate from -0.04 to 0.03 rad. AFC is perform more better than much lower than compute torque.

5. Conclusion

This review paper shows how to solve the control problem for wheeled mobile robot in many environments using different controller. Most of the controllers that have been used are in the form PD controller, kinematic and dynamic based controllers, Fuzzy logic controller, active force controller. Most of the mobile robot control show good performance to control the robot but it still has some errors and not accurate 100%; however the active force control can guarantee to reach zero level error.

Acknowledgment

Authors would like to thank the Ministry of High Education and Universiti Malaysia Pahang for supporting this project under research grant RDU160131.

References

[1] Hongjiu Y, Xiaozhao F. 2016. Nonlinear Control for Tracking and Obstacle Avoidance of a Wheeled Mobile Robot With Nonholonomic Constraint. *IEEE Transactions On Control Systems Technology, Vol. 2*

[2] LEI, C. and MA, B. 2015. A Nonlinear Formation Control of Wheeled Mobile Robots with Virtual Structure Approach. *Proceedings of the 34th Chinese Control Conference.*

[3] Chang, B.L. and Danwei, W. 2008. GPS-Based Tracking Control for a Car-Like Wheeled Mobile Robot With Skidding and Slipping. *IEEE/ASME transactions on mechatronics, vol. 13, no. 4*

[4] Jaemin, Y., Jong, H.O., Joo, H.P., Suhwan, K. and Dongjun, L. 2014. Autonomous Dynamic Driving Control of Wheeled Mobile Robots. *IEEE International Conference on Robotics & Automation (ICRA).*

[5] Mingyue, C., Wei, L., Hongzhao, L., and Xiaodong, L. 2016. Unscented Kalman Filter-%ased Adaptive Tracking Control for Wheeled Mobile Robots in the Presence of Wheel Slipping. *12th World Congress on Intelligent Control and Automation (WCICA).*

[6] Dawei, H. and Junyong, Z. 2015. Trajectory Tracking Control of Wheeled Mobile Robots Based on Disturbance Observer. *Key Laboratory of Measurement and Control of CSE.*

[7] Zhangjun, S., Hongliang, R., Jianwei, Z., and Shuzhi, S.G. 2016. Kinematic Analysis and Motion Control of Wheeled Mobile Robots in Cylindrical Workspaces. *IEEE transactions on automation science and engineering,vol.13,no.2.*

[8] Maral, P., Guangjun, L. 2015. Slip Ratio Estimation and Control of Wheeled Mobile Robot on Different Terrain. *The 5th Annual IEEE International Conference on Cyber Technology in Automation, Control and Intelligent Systems*

[9] Xiaochun, L., and Jumtao, F. 2015. Tracking Control of Wheeled Mobile Robots

[10] Yasamin, R., Khoshnam, S., and Abbas, C. 2015. Car-like Drive Wheeled Mobile Robot Using RBF Neural Network. *The 6th International Power Electronics Drive Systems and Technologies Conference*

[11] Chian, S.C, Teng, S.C., and Yu, T.Y.2015. Fuzzy Obstacle Avoidance Control of A Two-Wheeled Mobile Robot. *International Automatic Control Conference (CACS).*

[12] Yasamin, R., Khoshnam, S., and Abbas, C. 2015. Car-like Drive Wheeled Mobile Robot Using RBF Neural Network. *The 6th International Power Electronics Drive Systems and Technologies Conference*

[13] Mojtaba, G., and Khalil, A. 2013. Formation Control of Wheeled Mobile Robots based on Fuzzy Logic and System Dynamics. *13th Iranian Conference on Fuzzy Systems (IFSC).*

[14] Zhencai, L., Yang, W. 2015. Neural Adaptive Tracking Control for Wheeled Mobile Robots. *International Conference on Fluid Power and Mechatronics.*

[15] Xiaolong, G., and Peng, Z. 2014. The Adaptive Control for the Outdoor Mobile Robot with Diameter-variable Wheels. *Proceeding of the IEEE International Conference on Information and Automation Hailar, China*

[16] Chunyu, C., Felipe, D.Torre, and Wenjie, D. 2014. Distributed Exponentially Tracking Control of Multiple Wheeled Mobile Robots. *American Control Conference (ACC)*

[17] Xiaochun, L., and Jumtao, F. 2015. Tracking Control of Wheeled Mobile Robots

[18] Xiaochun, L., and Jumtao, F. 2015. Tracking Control of Wheeled Mobile Robots

[19] Vishal, A., Saha, S.K. 2016. Dynamic Identification and Model based Control of an Omni-wheeled Mobile Robot. *Proceedings of the 4th International Conference on Robotics and Mechatronics*

[20] Yang, L., Shuyou, Y., Bingzhao, G., and Hong, C. 2015. Receding Horizon Following Control of Wheeled Mobile Robots: A Case Study. *IEEE International conference on mechatronics and automation*

[21] Ghania, Z., Said, D. , Larbi, C.A. , Djemai, A. , and Pascal, B. 2014. Nonlinear Tracking Control of a Wheeled Mobile Robot. *15th international conference on Sciences and Techniques of Automatic control & computer engineering.*

[22] Dawei, H. and Junyong, Z. 2015. Trajectory Tracking Control of Wheeled Mobile Robots Based on Disturbance Observer. *Key Laboratory of Measurement and Control of CSE.*

[23] Yasmine, K., Mohamed, B., and Tarak, D. 2014. Adaptive Sliding-Mode Control of Nonholonomic Wheeled Mobile Robot. *15th international conference on Sciences and Techniques of Automatic control & computer engineering .*

[24] Jianqiu, M., Xing, G., and Sarah, K. 2015. Trajectory Tracking Control of a Two-Wheeled Mobile Robot Using Sliding Mode Techniques. *Proceedings of the 34th Chinese Control Conference.*

[25] Yaser, S., Musa, M., and Mohamed, H. 2011. Modelling and control of a piezo actuated micro robot with active force control capability for in-pipe application. *Int. J. Modelling, Identification and Control, Vol. 13, No. 4.*

IOP Conf. Series: Materials Science and Engineering **342** (2018) 012059 doi:10.1088/1757-899X/342/1/012059

[26] Mohammed A .H Ali., Musa, M., Wan, A., Zamzuri, H., and Zulkili, Y. 2015. Wheeled Mobile Robot Path Control in a Complex Trajectory using Hybrid Methods. *Recent Advances in Electrical Engineering*

[27] Musa, M., Shariman, A., And Tang, H. 2015. Tracking Performance on Feedfoward Model Base Active Force Control of Mobile Manipulator using Matlab and Adams. *Latest Trends in Circuits, Control and Signal Processing.*

[28] Shariman, A., Musa, M., Collin, T. 2015. Feedforward Model Based Active Force Control of Mobile Manipulator using MATLAB and MD Adams. *WSEAS transactions on systems.*

[29] M. A. H. Ali, Yusoff, W. A. B. W., Hamedon, Z. B., & Yusssof, Z. B. M. (2015). Wheeled Mobile Robot Control in Restricted Environment using PD-Active Force Control Scheme.

[30] Mohammed A. H. Ali, Musa Mailah, Wan Azhar B. Yussof, Zamzuri Bin Hamedon, Zulkili B. M. Yussof. (2017) An Intelligent Robust Control of Wheeled Mobile Robot in Restricted Environment. *International Journal of Control Systems and Robotics,* **2**, 6-11

IOP Conf. Series: Materials Science and Engineering **342** (2018) 012060 doi:10.1088/1757-899X/342/1/012060

Flame analysis using image processing techniques

Albert Chang Her Jie[1], Ahmad Faizal Ahmad Zamli[2], Ahmad Zulazlan Shah Zulkifli[2], Joanne Lim Mun Yee[1] and Mooktzeng Lim[2]

[1] School of Engineering, Monash University Malaysia.
[2] Fuels and Combustion, TNB Research Sdn. Bhd., Malaysia.

E-mail: mook.tzeng@tnb.com.my

Abstract. This paper presents image processing techniques with the use of fuzzy logic and neural network approach to perform flame analysis. Flame diagnostic is important in the industry to extract relevant information from flame images. Experiment test is carried out in a model industrial burner with different flow rates. Flame features such as luminous and spectral parameters are extracted using image processing and Fast Fourier Transform (FFT). Flame images are acquired using FLIR infrared camera. Non-linearities such as thermal acoustic oscillations and background noise affect the stability of flame. Flame velocity is one of the important characteristics that determines stability of flame. In this paper, an image processing method is proposed to determine flame velocity. Power spectral density (PSD) graph is a good tool for vibration analysis where flame stability can be approximated. However, a more intelligent diagnostic system is needed to automatically determine flame stability. In this paper, flame features of different flow rates are compared and analyzed. The selected flame features are used as inputs to the proposed fuzzy inference system to determine flame stability. Neural network is used to test the performance of the fuzzy inference system.

1. Introduction

Flame visualization and characterization using digital image processing is a tool to provide an understanding of flame features related to flame stability. Flame features are extracted in both time and frequency domains of flame with different flow rates to perform flame analysis. The features in time domain are the luminous parameters, while features in frequency domain are the spectral parameters. Flame velocity determines the rate of expansion of flame, while thermal acoustic oscillations cause wear and tear to burners in industry, thus affecting flame stability. Unstable flames also cause harmful effects such as release of nitrogen oxide gas, thermal stress and vibrational phenomena [1].

The flame images' brightness is used to approximate flame velocity [2], while flame visualization and characterization techniques extract luminous and spectral parameters that represents combustion states for monitoring and control purposes [3, 4]. Fuel (or coal) type can be identified through image processing of its flame features under various combustion conditions [5]. Features extracted using digital signal processing are used as inputs to fuzzy inference system for identification of coal type being burnt [6]. Neural networks are used to map flame features to fuel type [7], while power spectral densities (PSD) have been used to provide oscillatory characteristics of flames, which is important to understand the flame structure and stability [8]. Other parameter such as the oscillation frequency has been obtained from image processing to derive a universal index, a simple universal index for flame stability assessment [9].

The above flame visualization methods have been carried out using advanced optical cameras, and are more suitable for laboratory research studies. A more simpler and cost- effective method is required to capture flame images in industrial furnaces. Infrared (IR) cameras are suitable to capture flame images as it is able to capture the flame radiation and movement of flame [2]. Flame images are processed using image processing techniques with Fast Fourier Transform (FFT) to extract flame features.

The main contributions of this work are as follows. First, the mean velocities of the flames for different fuel flow rates are determined using the image processing techniques. The PSD of the flame with different flow rates are then determined for comparison. The second contribution of the work includes a fuzzy logic scheme which uses the extracted luminous features as inputs to predict flame stability. Finally, a neural network approach is used to investigate the performance of the fuzzy inference system and the correlation between the flame features with the flame stability.

2. Methodology

2.1. Combustion Test Facility and Experiments

Flame images are recorded from a lab scale burner in TNB Research Sdn, Bhd. that has a diameter of 50mm as shown in Figure 1. Two videos are recorded for flames with different fuel flow rates of 15 lpm CO2 + 10 lpm CH4 flow rate and 30 lpm CO2 + 10 lpm CH4. The videos are recorded using an IR camera from FLIR (model A615). The videos are used to extract a sequence of images for image processing.

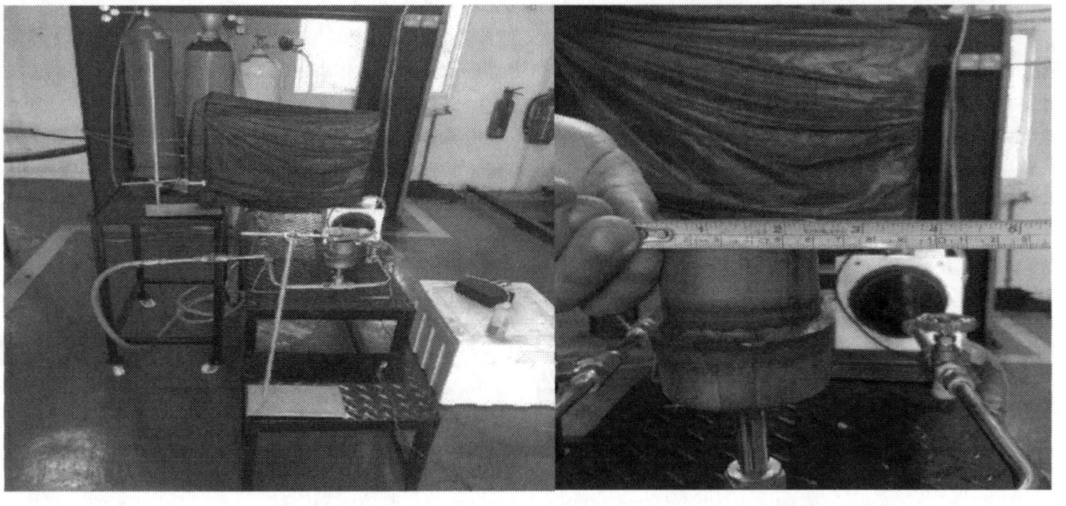

Figure 1. Lab scale (and diameter) of burner at TNB Research Sdn. Bhd.

2.2. Measurement principles

2.2.1. Image processing. The images are first pre-processed to segment the central area of flame and the background noise are filtered. The techniques used are image thresholding, color space transformation and edge detection. The techniques are discussed further in the following sections:

(i) *Image Thresholding.* Otsu's thresholding method is used to select a threshold value of k by computing probabilities of bright and dark pixels based on the histogram. The background noise is separated from the flame central area based on the threshold k value (refer Figure 2) [10].

(ii) *Color space transformation from RGB to HSV.* The hue, saturation, value (HSV) of the images are known to have a better correlation with the optical emission spectra and the flame temperature. This is because the color intensity (as it varies with the flame intensity or brightness) is separately stored in the value component. Thus the RGB images (after the threshold has been applied) are transformed to HSV images for further processing (refer Figure 3).

(iii) *Edge detection using Sobel filter.* Edges of the flame images represent its' shape and structure and therefore needs to be identified as accurate as possible [12]. In this study, a filtering method by Sobel is used to determine the edges of the flame by applying convolving masks to the image in the *x* and *y* directions (refer Figure 4).

2.2.2. Proposed method to determine flame velocity. The area of the flame is filled with white pixels after edge detection is shown in Figure 5. The images are then converted back to grayscale image as shown in Figure 5. A threshold value is applied to the grayscale images based on equation (1). A white pixel is detected if the grayscale value is greater than 1. The mean flame velocity is then approximated by determining the difference in the number of white pixels of the two consecutive image frames, divided by the frame rate of the video as shown in equation (2). The difference in the number of pixels is converted into distance. In this study, 325 pixels represents 50 mm, which is the diameter of the lab scale burner.

$$no.of\ white\ pixels_f = \sum_{i=0}^{row-1} \sum_{j=0}^{column-1} \begin{cases} 1, & I_f(i,j) > 1 \\ 0, & I_f(i,j) < 0 \end{cases} \tag{1}$$

$$V_f = \frac{no.of\ white\ pixels_{f+1} - no.of\ white\ pixels_f}{Frame\ rate\ of\ video} \tag{2}$$

2.2.3 Extraction of flame features. In this study, flame features are divided into luminous parameters and spectral parameters. Luminous parameters such as brightness and its fluctuation provide information on the statistical distribution of the flame's radiation. Spectral parameters such as power spectral density (PSD) are representative of the flame radiation signals' distribution and are related to the flame's stability.

(i) *Luminous parameters.* Each individual pixel of the images represents the grey values in the time domain and is used to determine the brightness and fluctuation. The brightness is determined from equation (3):

$$x = \frac{1}{N} \sum_{i=1}^{N} x_i . \frac{100}{G} \tag{3}$$

where x_i are grey values, *N* is the number of values and *G* is the maximum gray value.

The fluctuation is determined from equation (4):

$$\sigma = \sqrt{\frac{1}{N} \sum\nolimits_{i=1}^{N} (x_i - x)^2} \tag{4}$$

The flame area, A_i is calculated through:

$$A_i = \frac{1}{Width \times Height} \sum_{i=0}^{width-1} \sum_{j=0}^{height-1} \begin{cases} 1, & I_i(i,j) > threshold \\ 0, & other \end{cases} \tag{5}$$

where the threshold is estimated from the maximum background noise of the image [9]. Previous work has shown that the brightness of the image is related to radiation level while the fluctuation is related to non-uniformity of the flame [3]. Thus the flame area indicates the distribution of flame's energy content to the surrounding space.

(ii) *Spectral parameters.* Spectral parameters are frequency domain features, and is an analysis of natural resonance. Spectral parameters are used to monitor vibration signals and ball bearing faults [13]. This study uses power spectral densities (PSD) to determine the fluctuation of the flame. The PSD is given as:

$$P_k(f) = \frac{1}{N}(X_N(f))^2 \tag{6}$$

where P_f is the PSD estimate, N is the number of samples and $X_N(f)$ is obtained via Fast Fourier Transform (FFT) of the N data sequence. Based on the PSD, the flame's flicker F (or the flame's fluctuation) is given as:

$$F = \frac{\sum_{k=0}^{N-1} P_k * f_k}{\sum_{k=0}^{N-1} P_k} \tag{7}$$

where P_k is the PSD of k$_{th}$ frequency and $f_k$is the k$_{th}$ frequency.

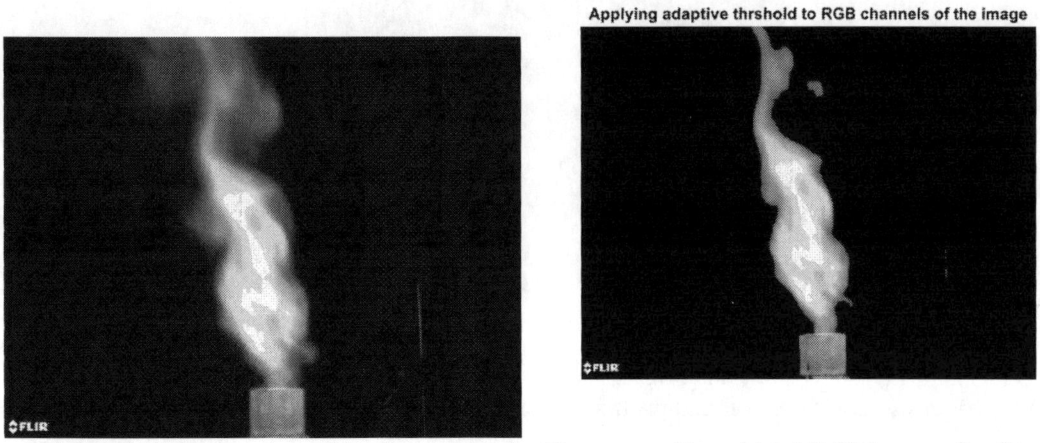

Figure 2. (left) snapshot of the flame from the IR camera videos (right) RGB image after Otsu's thresholding.

Figure 3. (Comparison of RGB and HSV components of flame image.

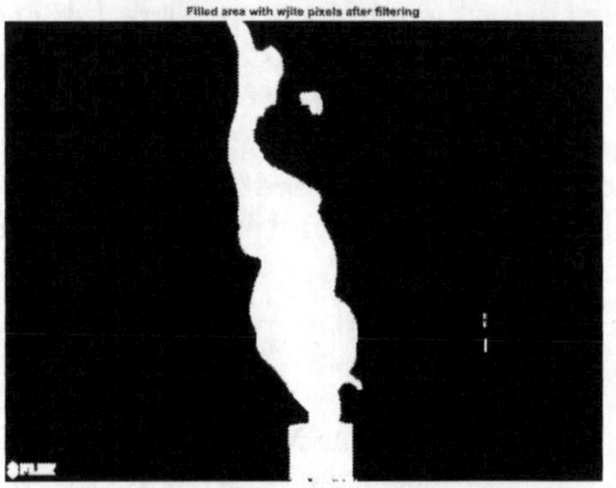

Figure 4. Area filled with white pixels after edge detection.

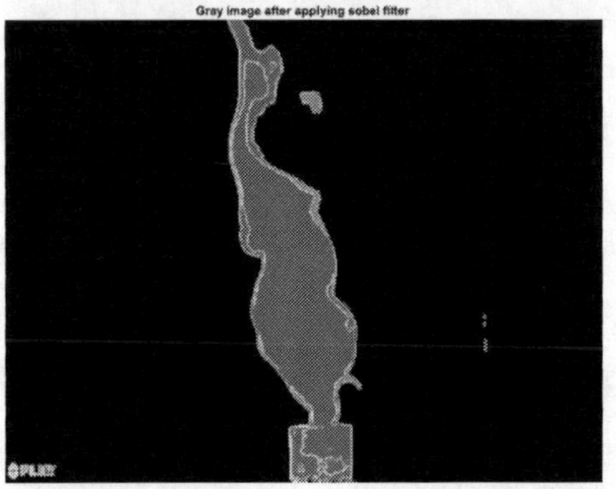

Figure 5. Image converted to grayscale image.

3. Results and discussions

In this section, the mean velocities and PSD of the flame with different flow rates are compared and analyzed. A discussion of the flame features from the proposed fuzzy inference system is presented.

3.1. Mean velocity and PSD of flame with different flow rates

3.1.1 Mean velocities. The flame velocities with respect to time for 15 and 30 lpm CO_2 are shown in Figure 6, while Table 1 shows the mean flame velocities and the standard deviations for both flames. From Table 1, the results show that there is no significant difference in méan velocities and standard deviation for both flames with different CO_2 flow rates. This is because in both cases the flames are opened or exposed to the surroundings, and the amount of oxidant are not in a controlled environment.

3.1.2 Power spectral densities. The PSD of both flames (obtained from FFT) are shown in Figure 7. The histograms for the PSD are shown in Figure 8. Based on the histogram plot in Figure 8, it can be observed that with a higher CO_2 flow rate (30 lpm), there are higher occurrences of oscillations. The dominant PSD range is also higher at 0.8 to 1.0 (refer Table 1). The flame is also observed to be more turbulent. This is because the flame with a higher flow rate (30 lpm) is more affected by non-linearities such as thermal oscillations.

IOP Conf. Series: Materials Science and Engineering **342** (2018) 012060 doi:10.1088/1757-899X/342/1/012060

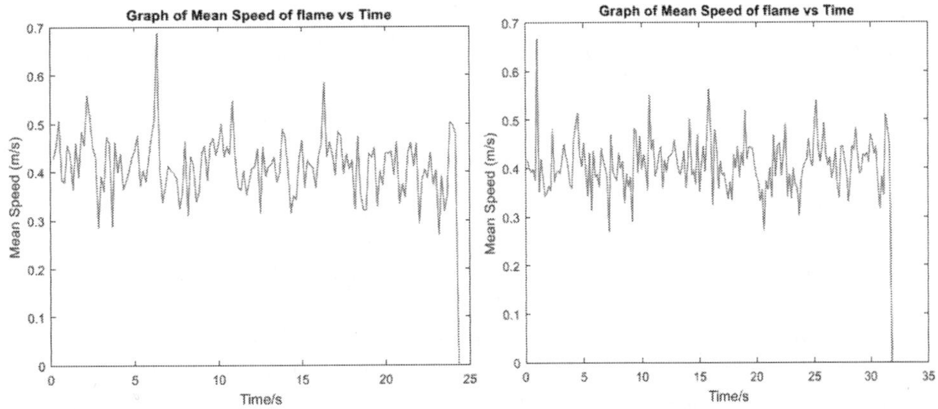

Figure 6. (Graph of mean velocities of 15lpm flame (left) and 30lpm (right).

Table 1. Mean velocity and standard deviation of flame with different CO_2 flow rates and 10 lpm CH_4.

Flow rate	**15 lpm CO_2 + 10 lpm CH_4**	**30 lpm CO_2 + 10 lpm CH_4**
Mean velocity (m/s)	0.4120	0.4074
Standard deviation (m/s)	0.0694	0.0607
PSD	0.6-0.8	0.8-1.0

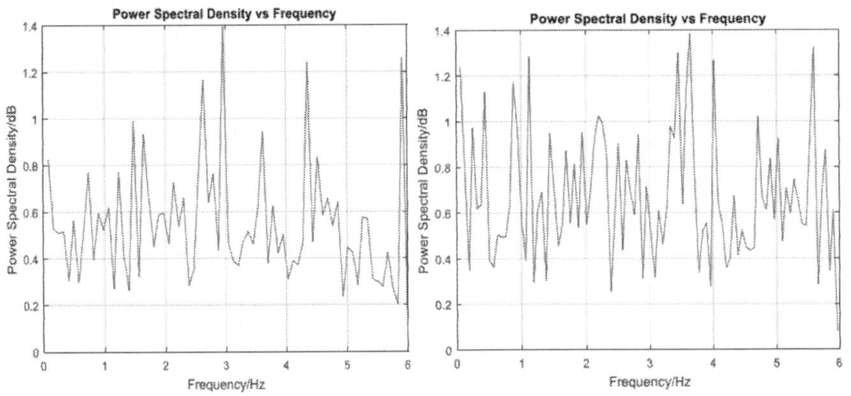

Figure 7. Power spectral density vs Frequency of flame for 15lpm (left) and 30lpm (right).

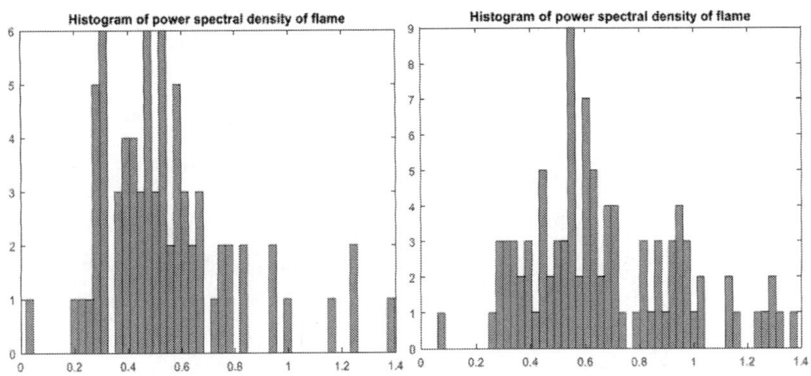

Figure 8. Histogram of PSD of flame for 15lpm (left) and 30lpm (right).

3.2. Features Extraction

The flame features are extracted as inputs to the proposed fuzzy inference system. The extracted features in terms of the brightness, fluctuation amplitude, flame area and flicker are shown in Figure 9 to 11. The mean values of the extracted features are then shown in Table 2.

3.2.1 Brightness. Based on Table 2, the mean brightness of the flame with 30 lpm CO_2 + 10 lpm CH_4 flow rates is lower than the flame with 15 lpm CO_2 + 10 lpm CH_4. This is caused by the higher CO_2 flow rate. Both flames have a constant CH_4 flow rate at 10 lpm and are mixed with CO_2 before it is ignited at the outlet of the burner. At a higher CO_2 flow rate, the total energy content in the gas that comes from CH_4 is diluted. Hence, the flame with a higher CO_2 flow rate is not as intense nor bright. The fluctuation amplitude and flicker for flame with 15 lpm CO_2 + 10 lpm CH_4 is higher because it has a higher energy content compared to the flame with 30 lpm CO_2 + 10 lpm CH_4. The change in the fluctuation amplitude represents a change in the luminous radiation, while the change in flicker reflects a change in the flame geometry. Unlike the mean flame velocities which are similar, these flame features show that both flames are distinctly different.

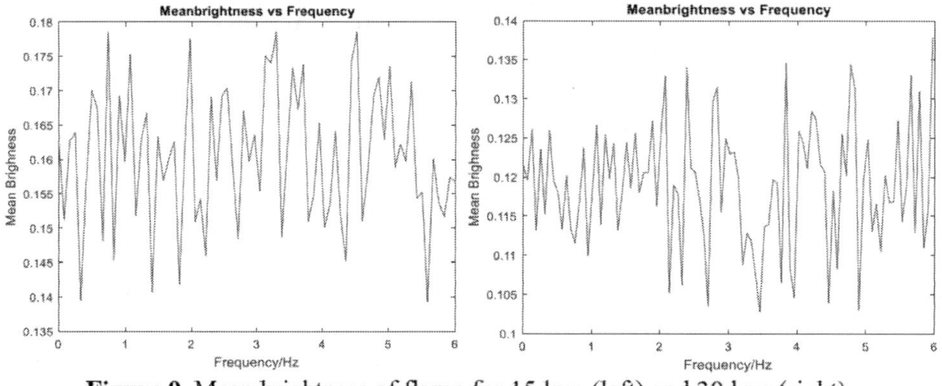

Figure 9. Mean brightness of flame for 15 lpm (left) and 30 lpm (right).

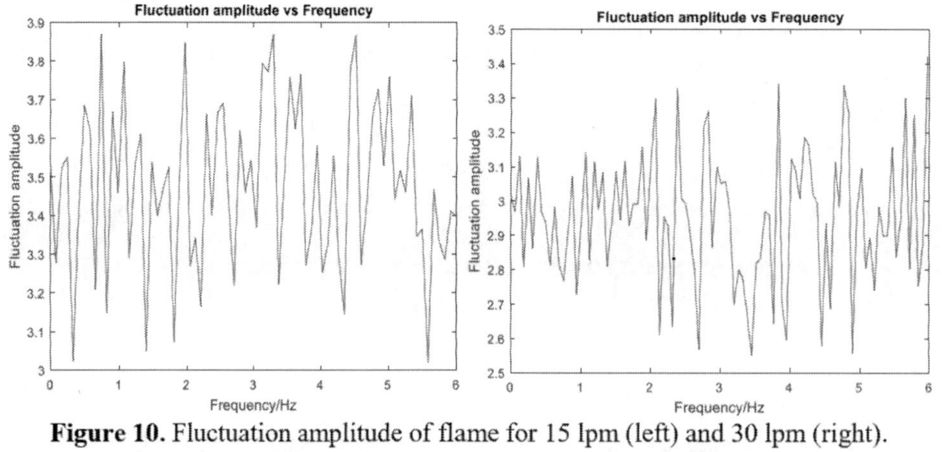

Figure 10. Fluctuation amplitude of flame for 15 lpm (left) and 30 lpm (right).

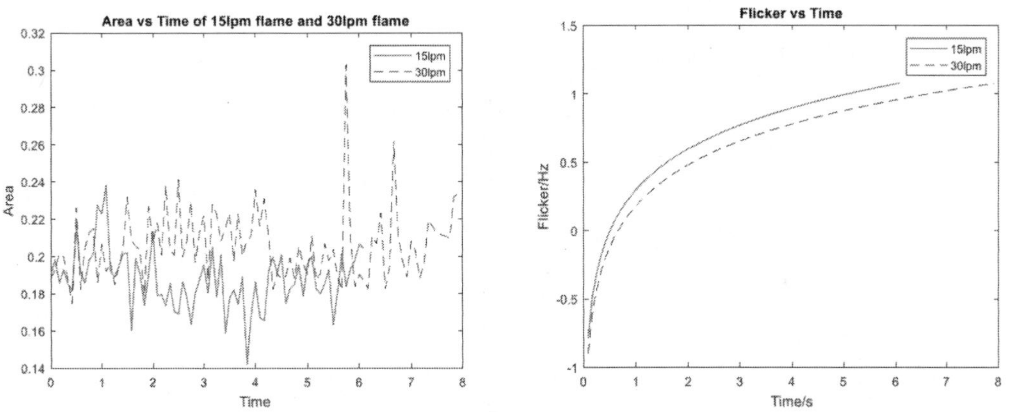

Figure 11. Flame area and flicker for CO_2 flow rates of 15 and 30 lpm.

Table 2. Mean brightness, fluctuation and area of flame with different flow rates.

Flow rate	**15 lpm CO_2 + 10 lpm CH_4**	**30 lpm CO_2 + 10 lpm CH_4**
Mean Brightness	0.16	0.12
Mean Fluctuation	3.48	2.95
Mean Area	0.1855	0.2055

3.3. Implementation and Evaluation of fuzzy inference system

The brightness, fluctuation and flame area are used as inputs for the fuzzy inference system. The membership function for each input is shown from Figures 12 to 13. The performance of the fuzzy inference system is evaluated using a neural network fitting function by plotting the regression and mean square error as shown in Figures 13 and 14.

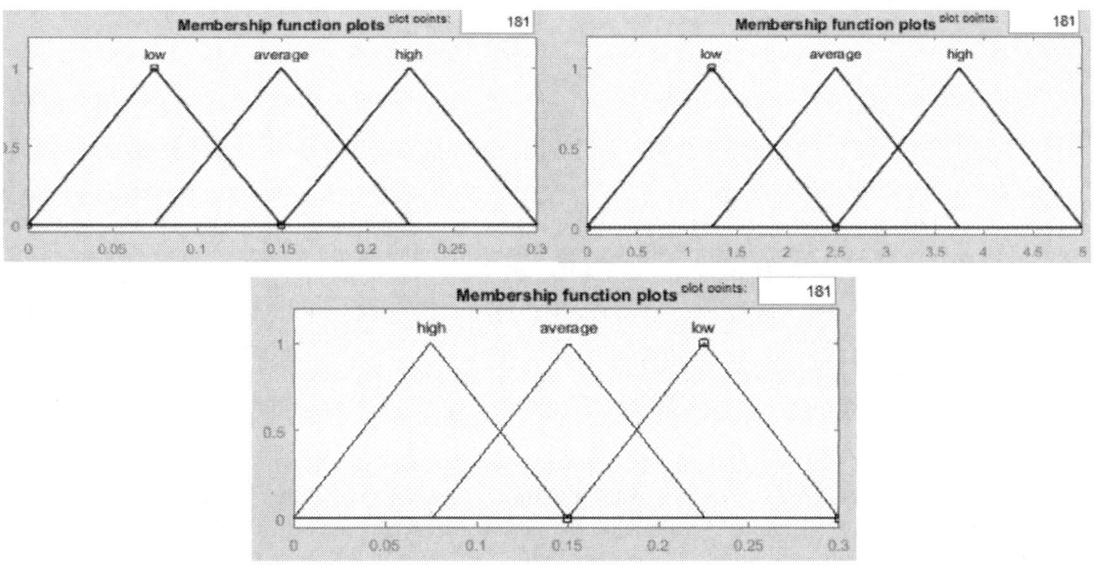

Figure 12. Membership function for (top left) brightness, (top right) fluctuations, and flicker (bottom) area.

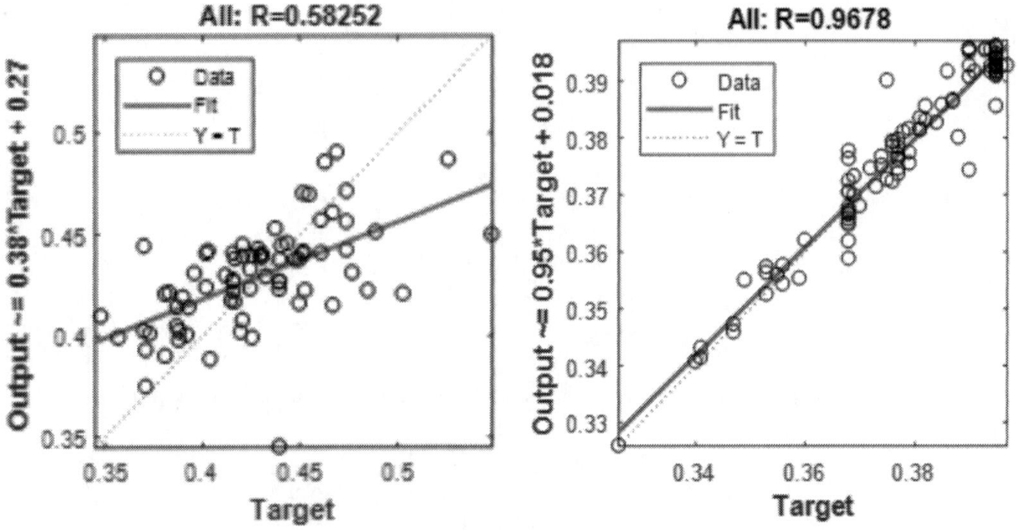

Figure 13. Regression plot of 15lpm flame (left) and 30lpm flame (right).

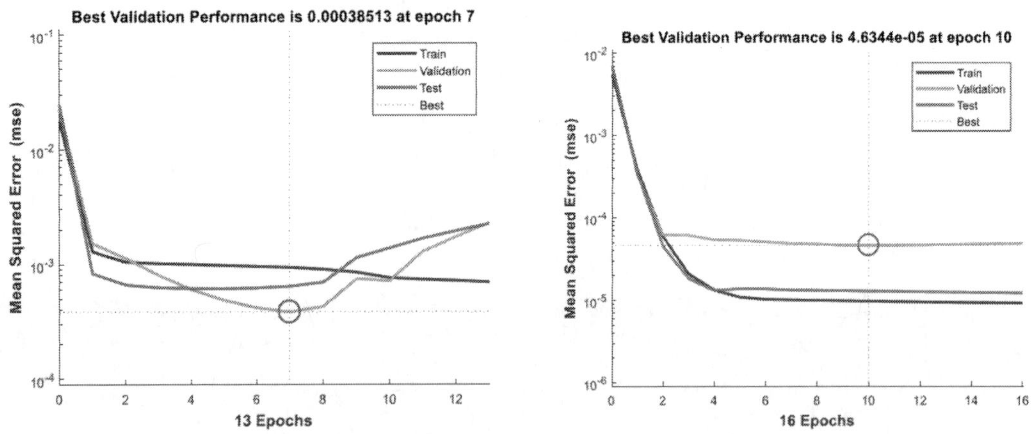

Figure 14. Mean square errors for 15 lpm CO_2 (left) and 30 lpm CO_2 flames (right).

Based on Figure 13, the regression for the 15 lpm CO_2 is 0.58 while the regression for 30 lpm flame is 0.97. The significant difference in the values is due to greater number of samples obtained for the flame with 30 lpm CO2 taken from the neural network. The video recorded for the 30 lpm CO_2 flame was longer (31s) compared to the 15 lpm CO_2 flame (24s). The high regression value for the 30 lpm CO_2 flame shows that the fuzzy inference system has the potential to simplify the non-linear model into a linear model. Furthermore, it shows the potential to derive linear equations using flame features. The low mean square error shown in Figure 14 shows that the fuzzy inference system is reliable in predicting flame stability.

4. Conclusions

An image processing technique is proposed to determine the flame velocities flame for different fuel flow rates. No significant difference in flame velocities is observed for flames with CO_2 flow rates of 15 lpm and 30 lpm. The higher oscillation in flame with flow rate of 30 lpm CO_2 shows that the flame with flow rate of 15 lpm CO_2 is more stable. In addition, flame features such as brightness, fluctuation and flicker show that flame with CO_2 flow rates of 15 and 30 lpm are distinctly different. A proposed fuzzy inference system using extracted features as input to predict flame stability produces good performance from neural network but more experiments are needed to validate the results.

Acknowledgement
The authors would like to thank TNB Research Sdn. Bhd. for providing the facility and the data needed for the experiment.

References

[1] Y. Ren, S. Li, W. Cui, Y. Zhang, and L. Ma, "Low-frequency AC electric field induced thermoacoustic oscillation of a premixed stagnation flame," *Combustion and Flame,* vol. 176, pp. 479-488, 2017.

[2] Mook Tzeng Lim, Ahmad Rashidi, Soong Der Chen, Noor Akma Watie Mohd Noor, Azmi Ahmad, Hamdan Hassan, *et al.*, "Prelimary determination of flame velocities fro infrared images using image processing techniques in a 150 kWth coal-fired combustion test rig," *Advances in Mechanical, Aeronautical and Production Techniques (MAPT),* pp. 1-5, 2015.

[3] A. González-Cencerrado, B. Peña, and A. Gil, "Coal flame characterization by means of digital image processing in a semi-industrial scale PF swirl burner," *Applied energy,* vol. 94, pp. 375-384, 2012.

[4] R. Hernandez and J. Ballester, "Flame imaging as a diagnostic tool for industrial combustion," *Combustion and flame,* vol. 155, pp. 509-528, 2008.

[5] H. Zhou, Q. Tang, L. Yang, Y. Yan, G. Lu, and K. Cen, "Support vector machine based online coal identification through advanced flame monitoring," *Fuel,* vol. 117, pp. 944-951, 2014.

[6] L. Xu, Y. Yan, S. Cornwell, and G. Riley, "On-line fuel identification using digital signal processing and fuzzy inference techniques," *IEEE Transactions on Instrumentation and Measurement,* vol. 53, pp. 1316-1320, 2004.

[7] L. Xu, Y. Yan, S. Cornwell, and G. Riley, "Online fuel tracking by combining principal component analysis and neural network techniques," *IEEE Transactions on Instrumentation and Measurement,* vol. 54, pp. 1640-1645, 2005.

[8] G. Lu, Y. Yan, M. Colechin, and R. Hill, "Monitoring of oscillatory characteristics of pulverized coal flames through image processing and spectral analysis," *IEEE transactions on instrumentation and measurement,* vol. 55, pp. 226-231, 2006.

[9] D. Sun, G. Lu, H. Zhou, Y. Yan, and S. Liu, "Quantitative assessment of flame stability through image processing and spectral analysis," *IEEE Transactions on Instrumentation and Measurement,* vol. 64, pp. 3323-3333, 2015.

[10] N. Otsu, "A threshold selection method from gray-level histograms," *Automatica,* vol. 11, pp. 23-27, 1975.

[11] W.-B. Horng, J.-W. Peng, and C.-Y. Chen, "A new image-based real-time flame detection method using color analysis," in *Networking, Sensing and Control, 2005. Proceedings. 2005 IEEE,* 2005, pp. 100-105.

[12] T. Qiu, Y. Yan, and G. Lu, "An Autoadaptive Edge-Detection Algorithm for Flame and Fire Image Processing," *IEEE Transactions on Instrumentation and Measurement,* vol. 61, pp. 1486-1493, 2012.

[13] I. Attoui, N. Fergani, N. Boutasseta, B. Oudjani, and A. Deliou, "A new time–frequency method for identification and classification of ball bearing faults," *Journal of Sound and Vibration,* vol. 397, pp. 241-265, 2017.

Investigation on heat transfer characteristics and flow performance of Methane at supercritical pressures

Hong Wei Xian[1,2], A N Oumer[1], F Basrawi[1], Rizalman Mamat[1] and A A Abdullah[1]

[1]Faculty of Mechanical Engineering, Universiti Malaysia Pahang, 26600 Pekan, Pahang, Malaysia
[2]Department of Mechanical Precision Engineering, Malaysia – Japan International Institute of Technology (MJIIT)

E-mail: henrywei_xian@hotmail.com

Abstract. The aim of this study is to investigate the heat transfer and flow characteristic of cryogenic methane in regenerative cooling system at supercritical pressures. The thermo-physical properties of supercritical methane were obtained from the National institute of Standards and Technology (NIST) webbook. The numerical model was developed based on the assumptions of steady, turbulent and Newtonian flow. For mesh independence test and model validation, the simulation results were compared with published experimental results. The effect of four different performance parameter ranges namely inlet pressure (5 to 8 MPa), inlet temperature (120 to 150 K), heat flux (2 to 5 MW/m^2) and mass flux (7000 to 15000 kg/m^{2}s) on heat transfer and flow performances were investigated. It was found that the simulation results showed good agreement with experimental data with maximum deviation of 10 % which indicates the validity of the developed model. At low inlet temperature, the change of specific heat capacity at near-wall region along the tube length was not significant while the pressure drop registered was high. However, significant variation was observed for the case of higher inlet temperature. It was also observed that the heat transfer performance and pressure drop penalty increased when the mass flux was increased. Regarding the effect of inlet pressure, the heat transfer performance and pressure drop results decreased when the inlet pressure is increased.

1. Introduction

In this modern age, most of our activities in daily life such as using hand phone, house appliances, and plants in industrial sectors need great amount of electricity. To generate huge amount of electricity, coals and fossil fuels are burnt to provide heat energy in power plants. Burning of these two substances can lead to many environmental problems such as changes of ground surface, emission of carbon dioxide, reduction of renewable energy resources and others. The emission of carbon dioxide leads to greenhouse effect, acid rain and change of climate [1-4]. To reduce these problems, biogas and biofuel can be used to replace coal and fossil fuels as they are more environmental friendly.

Biogas is made up of CH_4 (65-70%), carbon dioxide CO_2 (30-35%) and other gases [5]. It is produced through the breakdown of organic substances without existence of oxygen and this process is known as anaerobic digestion. Anaerobic digestion of wastes, harvests, and residues can reduce the emission of greenhouse gases and develop sustainable energy supplies such as biogas [6]. Biogas is another alternative choice for renewable energy, as methane produced can be used to replace fossil fuels in both heat and power generation and as a vehicle fuel in the form of Liquid Natural Gas (LNG) or Compressed Natural Gas (CNG).

Biogas is mostly used in engine-based combined heat and power (CHP) plants as it can achieve higher efficiency on producing heat and electricity than coal and gas-fired power stations. The electricity produced from power plant can be supplied to public and industrial sectors. The waste heat rejected can be reused in other subsystems for additional power generation hence reduces fossil fuels needed for burning.

Engine can be easily get heated due to long time operation. Methane has good properties as coolant at supercritical pressure and it is normally used in the regenerative cooling system of liquid rocket engine and gas turbine engines. In rocket combustion engines, the combustion chamber wall will undergo high heat flux and to maintain the engine lifetime, it needs to be cooled. It is done by building a cooling jacket around the combustion chamber and putting the supercritical-pressure fluid in the cooling jacket to cool it. Usually, before injecting cryogenic methane into combustion chamber, the working pressure of methane is above its thermodynamic critical pressure and no phase change will occur. At a high working pressure, very high specific heat capacity will occur if temperature near the pseudo-critical temperature. In this condition, the regenerative cooling capability can be enhanced. From studies, methane at supercritical pressure shows significant changes in thermo-physical properties at small temperature and pressure changes. Thus, extensive studies on thermo-physical properties of methane at supercritical pressures are vital. A supercritical status is defined as the temperature and pressure of a working fluid exceeds its critical temperature and critical pressure. Supercritical fluids have few advantages including increased species mixing, heat and mass transfer, fast reaction, environmental friendly, good scalability, as well as being simple and easy for continuous production [7].

Pseudocritical properties are the unusual properties behaviour which near the pseudocritical point. Usually, pseudocritical points are referenced to pseudocritical temperature. Pseudocritical temperature is defined as the temperature where specific heat capacity, C_p reaches a maximum value at a given supercritical pressure [8]. Thermo-physical and transport properties such as density, specific heat, viscosity and thermal conductivity have great variation within a narrow region near the critical point.

Gu et al. [9] in 2013 carried out experimental investigation on convective heat transfer of supercritical methane in a horizontal miniature tube. It is found that heat transfer enhancement could occur at high mass velocity and low heat flux whereas deterioration was observed at high heat flux and low mass velocity in pseudocritical region. Two conventional single phase empirical correlations which are Dittus-Boelter and Gnielinski correlations were used to compare the Nusselt number with experimental data. It is found that both correlations cannot be used for estimating the heat transfer of supercritical methane due to large variation of thermo-physical properties causing overestimation of readings near the pseudocritical temperature region. Recent research on supercritical n-decane in small vertical tubes with 0.95 mm and 2mm inner diameter was carried out by Liu et al. [10] in 2015 experimentally. From their result, high inlet Reynolds number (>7000) will lead to increase of local wall temperature without abnormal temperature distribution for both upward and download flows. On the other hand, for low inlet Reynolds number (2700-4000), buoyancy will cause heat transfer deterioration for upward flow and enhance heat transfer for downward flow.

There are some researchers did their numerical studies on supercritical methane. For instance, Pizzarelli et al. [11] in 2014 analyzed heat transfer behavior of supercritical methane in a asymmetrically heated rectangular-cross-sectional channel (base = 1 mm, height = 3 mm, rib base = 1 mm, internal wall thickness = 1 mm and channel length = 300 mm) using a three-dimensional conjugate heat transfer model. Different coolant pressures and surface roughness were used to determine heat transfer of the cooling system. It was found that at surface roughness = 10 μm, heat transfer deterioration did not occur but coolant pressure drop was increased; at high coolant pressure or rougher surface, heat transfer deterioration is reduced but pressure drop of coolant occurs, same result as Cp et al. [12] obtained in their numerical study in 2010; at lower roughness level, maximum wall temperature is lower but heat deterioration did not vanish. Pizzarelli et al. [13] in 2010 numerically studied the turbulent two-dimensional axisymmetric flow of cryogenic methane at near-critical region in a heated straight tube. Present result and past studies showed that heat transfer deterioration can occur if wall heat flux is high at supercritical pressures [14, 15].

Three-dimensional numerical study on supercritical heat transfer of cryogenic methane in rectangular engine cooling channels with asymmetric heating on top channel surface was conducted by Ruan and Meng [16] in 2012. Their results showed that heat transfer deterioration occurs when supercritical pressure is lowered and wall heat flux is increased. This is due to the abrupt variation of thermo-physical properties near pseudocritical points. Shallow cooling channels showed high Nusselt number and good heat transfer at supercritical pressure but pressure loss is high. A numerical study on cryogenic methane at supercritical pressure of 8 MPa in ribbed circular cooling tube (2mm inner diameter, 3 mm outer diameter and 300 mm heating section) is carried out by Xu et al. [17] in 2015 to investigate the effects of influential parameters on both heat transfer enhancement and pressure loss. Their result showed that rib has no effects on bulk fluid temperature but surface temperature of tube wall is reduced. More recently, Han et al. [18] in 2016 numerically investigated convective heat transfer of supercritical liquefied natural gas (LNG) in a horizontal serpentine tube. Shear-Stress Model (SST) with enhanced wall treatment method worked the most accurately in predicting outlet temperature. At constant mass flux, when bulk LNG temperature is near to pseudocritical point, the surface heat transfer coefficient is maximum. Due to abrupt change of thermo-physical properties, the transport of heat from near wall region to bulk LNG increases which lead to heat transfer enhancement.

Some research works related to heat transfer characteristics of methane have been documented. However, comparing to great amount of studies on supercritical carbon dioxide and water, there are only few papers discussed on heat transfer characteristics of supercritical methane. However, these research works do not extensively investigate the thermal-hydraulic performance of supercritical methane. Therefore, extensive research study is required to find enhanced heat transfer fluids for various engines. The aim of the current work is to investigate the heat transfer characteristics and flow performance of methane at supercritical pressures and temperature.

2. Method of Study

The test section is a horizontal steel miniature circular tube with 200 mm in length, L and 1.6 mm diameter, d as shown in Figure 1. Constant heat fluxes are applied on the outer wall surface (wall boundary) with cryogenic liquid methane injected into the tube from the left side at supercritical pressures.

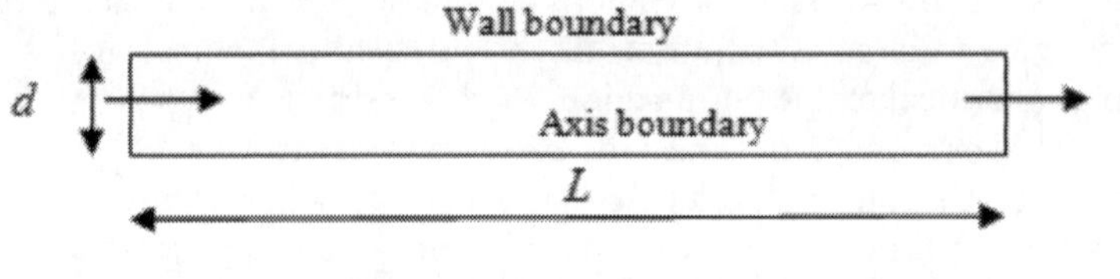

Figure 1. Schematic of the 2D numerical model.

The thermo-physical properties of methane at various supercritical pressures are obtained from NIST database and are entered to the computational fluid dynamics (CFD) software called ANSYS Fluent R17.2. Before analysing the heat transfer performance of supercritical cryogenic methane in circular pipe, mesh independence testing is needed to ensure that the generated results are same for all mesh sizes. Moreover, wall and bulk temperatures are compared to experimental results and it is found that 25,000 mesh size is sufficient to obtain close results respective to experimental data.

Model validation is then carried out to determine whether the developed model is suitable for analysis of heat transfer characteristic and pressure drop in supercritical methane through horizontal tubes. Three different turbulence models namely k-omega SST, k-omega standard and k-epsilon RNG with standard wall function were used. Overall, the k-omega SST showed the lowest discrepancies of results in terms of wall temperature and average fluid bulk temperature for different boundary conditions. Moreover, many research studies have used this turbulence model in their analysis [18, 19]. For model validation, same operating conditions from the experimental work done by Gu et al. [9] are selected.

3. Analysis and Discussion

Effects of inlet pressure, inlet temperature, heat flux and mass flux on heat transfer and flow performance are identified using different boundary conditions.

3.1. Model Validation

Before detail analysis of the heat transfer and pressure drop characteristics of supercritical methane, the developed numerical model should be checked for accuracy. For this, the bulk and wall temperature results from the current model were compared with published experimental data. Figure 2 shows variation of the bulk temperature along the tube length at 7000 kg/m^{2}s, 10 MPa inlet pressure, 130 K inlet temperature and various heat fluxes obtained from the current simulation and experiment. Three different heat fluxes (Q) were applied on the heating section to observe the effect of heat flux on fluid bulk temperature. Similar trend was obtained for each case. As can be seen from the figure, the simulation and experimental results are in good agreement. For 4 MW/m^2 heat flux, the maximum and average deviations between the simulation and experimental results are 2.3 % and 2.2 %, respectively. As the heat flux increases to 5 MW/m2, as expected, the overall fluid bulk temperature increases. Maximum deviation of 3 K between the simulation and experiment results is found at 120 mm from inlet, which is equivalent to 1.6 %. When the heat flux is increased to 6.5 MW/m^2, almost all the simulation and experimental values are close to each other except at the points located at 180 mm onwards. In this case, because of the deviations at the these points, the maximum deviation rose to 2.7%. However, the minimum deviation is only 0.13% and this leads to average deviation of 1% for 6.5 MW/m^2. In all cases, it can be observed that all the average errors are lower than the accepted error of 10% mentioned by most of the previous researchers [18, 19].

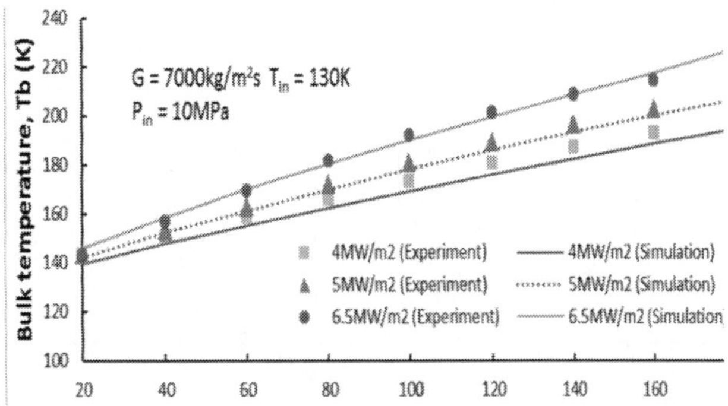

Figure 2. Variation of bulk temperature along the tube length at three different heat fluxes.

Figure 3 shows the effect of three different heat fluxes on wall temperature at the same boundary condition as in Figure 2. All cases show close values between experimental and simulation results. For 4 MW/m^2, the average deviation is found to be 5.9% while the maximum difference is 7.8%. When the heat flux is increased to 5 MW/m^2, both experimental and simulation results show average and maximum deviation of 5.3 % and 8.7% respectively. At 6 MW/m^2, similar trend is shown for both results. There is a sudden increment of wall temperature at 120 mm from inlet. This is due to the supercritical properties of methane at high pressure, in which sudden drop of specific heat capacity leads to sudden change of temperature. For this case, the maximum deviation of wall temperature is 9.6% which is acceptable generally and the average deviation is 5.3%. Overall, the average deviation for all the three cases is about 5 to 6%. This further verifies the validity and accuracy of the developed model used in this study.

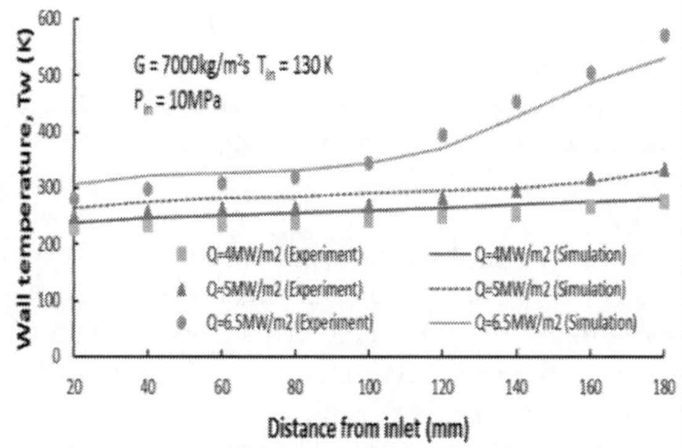

Figure 3. Variation of wall along the tube length at constant mass flux.

Graphs shown in Figures 4a and 4b are variations of bulk and wall temperature along the tube length at various mass fluxes. Both temperatures decreased when the mass flux is increased from 7000 kg/m^{2}s to 7700 kg/m^{2}s for both simulation and experimental data. When the mass flux increases, inlet velocity increases and this leads to more heat at the wall region to be absorbed by the cryogenic methane in the tube. For wall temperature at G = 7000 kg/m^{2}s, there is slight increment of wall temperature at 160 mm from inlet, this is due to the specific heat capacity of methane at this point starts decreasing which leads to heat transfer deterioration and thus wall temperature becomes higher.

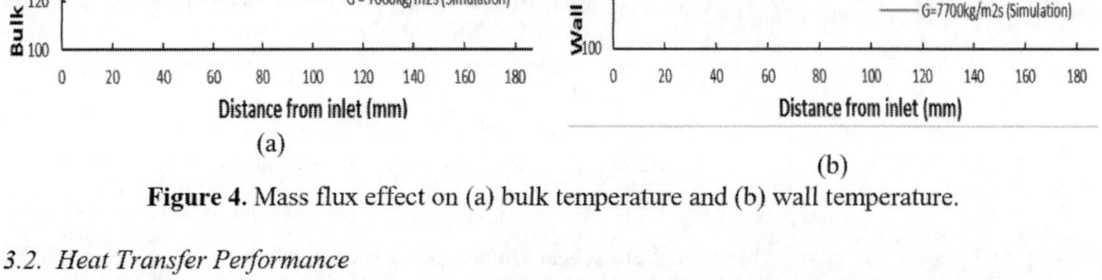

Figure 4. Mass flux effect on (a) bulk temperature and (b) wall temperature.

3.2. Heat Transfer Performance

The performance of heat transfer is proportional to specific heat capacity. Thus, it is important to see how the heat transfer coefficient varies with inlet temperature. Shown in Figure 5 is variation of specific heat capacity at near-wall region along the tube length at various inlet temperatures. At low inlet temperature, the change of specific heat capacity at near-wall region along the tube length is not significant. However, significant variation is observed for the case of higher inlet temperature. It can be clearly seen from the figure that the specific heat capacity at near-wall region increases with inlet temperature at all positions. For instance, at inlet temperature of 150 K, the specific heat capacity starts to increase starting from 120 mm and this leads to enhanced heat transfer performance at the same location as shown in Figure 6. As mentioned earlier, the decrement of specific heat capacity leads to the slight increment of wall temperature and this causes heat transfer performance to drop. For 120 K, 130 K and 140 K, the heat transfer performance is lower due to lower increment of specific heat capacity.

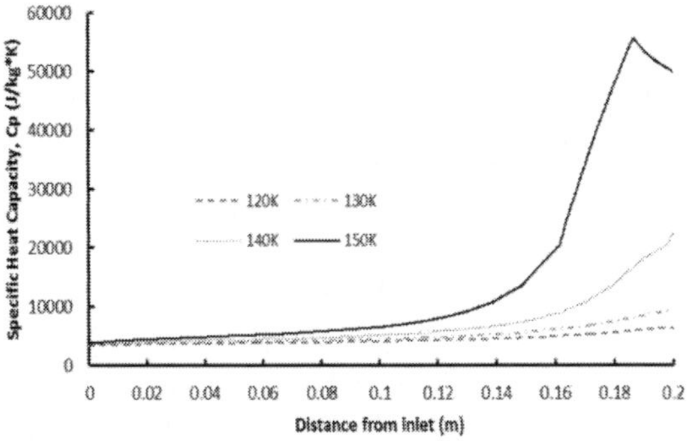

Figure 5. Effect of inlet temperature on specific heat capacity at 5 MPa, 4 MW/m^2 and 10000 kg/m^{2}s. (near-wall region)

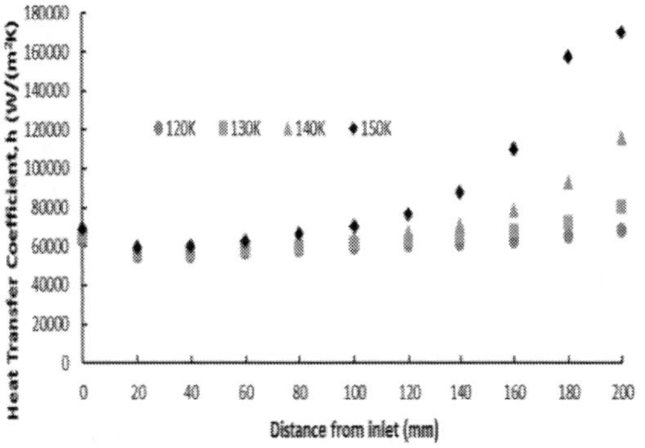

Figure 6. Effect of inlet temperature on heat transfer performance at 5 MPa, 4 MW/m^2, and 10000 kg/m^{2}s.

Shown in Figure 7 is variation of heat transfer coefficient with inlet mass flux. The fixed boundary conditions in this case are 5 MPa inlet pressure, 120 K inlet temperature and 4 MW/m^2 heat flux. The mass fluxes range from 7000 to 15000 kg/m^{2}s. As shown in the figure, the heat transfer performance increases when the mass flux is increased. The same type of result was obtained in another simulation work by Han et al. [18] and it was explained that due to the large variation of thermo-physical properties at near-wall region which lead to enhanced heat transfer performance. However, heat transfer performance increment along the tube length is not significant for G = 10000 and 15000 kg/m^{2}s. This is due to the insignificant change of the specific heat capacity of methane at near-wall region for this two mass fluxes. For near-wall region, the specific heat capacity at G = 7000 kg/m^{2}s is about 12.6 times higher than that of 15000 kg/m^{2}s at outlet. This results in the great difference in variation of heat transfer performance increment. Another point is that the fluid velocity increases when mass flux increases and this will lead to more heat can be transferred away before the fluid temperature becomes higher. Thus, both bulk and wall temperatures are decreasing with increasing mass flux which enhances heat transfer performance.

Another factor that affects the heat transfer performance is inlet pressure. For this, the inlet pressures used are 5 to 8 MPa. The other boundary conditions fixed are mass flux of 10000 kg/m^{2}s, 4 MW/m^2 heat flux and 120 K inlet temperature. The variations of bulk and wall temperature are not shown due to the small difference between each point for every inlet pressure. As shown in Figure 8, the heat transfer performance increases along the tube length for all inlet pressures. For the heat transfer coefficient at 20 mm from inlet, it is higher than that at 40 mm. On the other hand, heat transfer performance decreases when the inlet pressure is increased. This is due to the great difference in thermo-physical properties between these pressures. The same argument was made by Gu et al.[9]. The variation of specific heat capacity at near-wall region for all inlet pressures has the same trend. When it comes to 0.18 m from inlet, the specific heat capacity increment based on 8 MPa is 21.9 %, 12.2 % and 5.2 % for 5 MPa, 6 MPa and 7 MPa respectively. As mentioned previously, heat transfer performance is proportional to specific heat capacity and this section further validates this statement. Thus, for lower supercritical pressures, the variation of thermo-physical properties near pseudocritical point is higher and thus heat transfer performance is enhanced more at lower supercritical pressures before pseudocritical point.

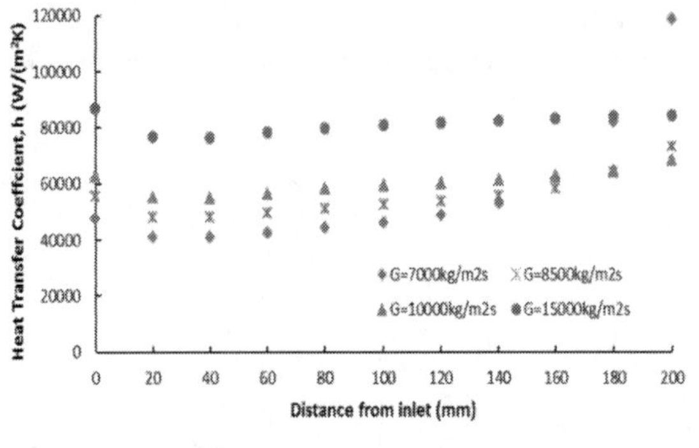

Figure 7. Effect of mass flux on heat transfer performance at 5MPa, 120K and 4 MW/m^2.

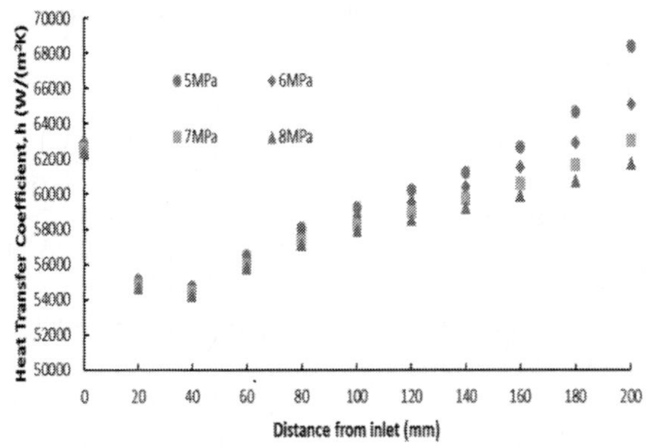

Figure 8. Effect of inlet pressure on heat transfer coefficient at 10000 kg/m^{2}s, 4 MW/m^2 and 120 K.

3.3. Flow Performance

To study the effect of inlet temperature on pressure drop along the tube length, the fixed boundary conditions are mass flux of 10000 kg/m^{2}s and heat flux of 4 MW/m^2. Figure 9 shows the pressure drop of supercritical methane along the tube length at different inlet pressure. It can be clearly seen that the pressure drop decreases as the inlet temperature is lower. Considering the inlet temperature of 120 K at 5 MPa as reference point, the increment of pressure drop is 5.2%, 12.6 % and 22.3% when the inlet temperature is increased to 130 K, 140 K and 150 K respectively. The pressure drop increment is due to higher inlet temperature which causes both wall and bulk temperature increase more than that of lower inlet temperature. Thus, the viscosity of methane decreases more and bulk velocity of methane will increase with increasing inlet temperature. Higher fluid velocity leads to higher pressure drop. On the other hand, increasing inlet pressure with constant heat flux lowers down the pressure drop as well. This is due to the less abrupt variation of thermo-physical properties at higher supercritical pressure. In general, increasing inlet temperature leads to increment of pressure drop.

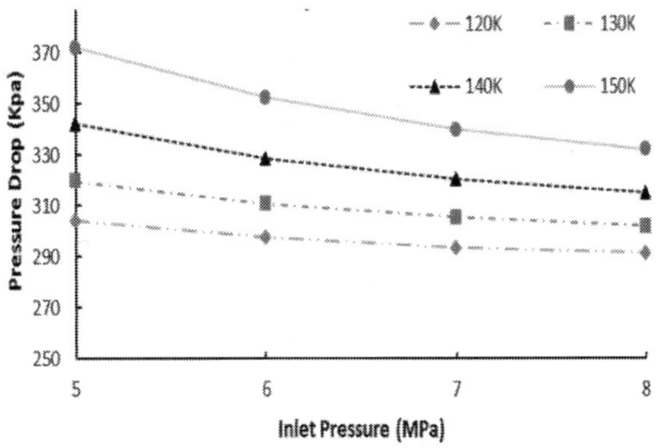

Figure 9. Effect of inlet temperature on pressure drop of methane at 10000kg/m^{2}s, 4 MW/m^2 and various inlet temperatures.

Inlet temperature and heat flux are fixed at 120 K and 4 MW/m^2 respectively. Figure 10 shows the effect of mass flux on pressure drop of methane at different inlet pressure. It is shown that the pressure drop increases when mass flux increases. The huge increment of pressure drop from mass flux of 10000 kg/m^{2}s to 15000 kg/m^{2}s is about 270 kPa for every inlet pressure. This is due to the high increment of average bulk velocity from about 27 m/s to 39 m/s. Since it is found that pressure drop has strong correlation with fluid velocity, it can be concluded that the fluid velocity is always proportional to the pressure drop of supercritical methane in a circular tube. In short, increasing mass flux which increases fluid velocity will lead to increment of pressure drop. On the other hand, the pressure drop decrement is not significant (with maximum deviation of 7 kPa) when the inlet pressure is increased at constant mass flux, this might be due to the low inlet temperature (120 K) and heat flux (4 MW/m^2) used are not sufficient to make the methane bulk temperature near to pseudocritical temperature (which has more abrupt variation of thermo-physical properties) for all inlet pressure.

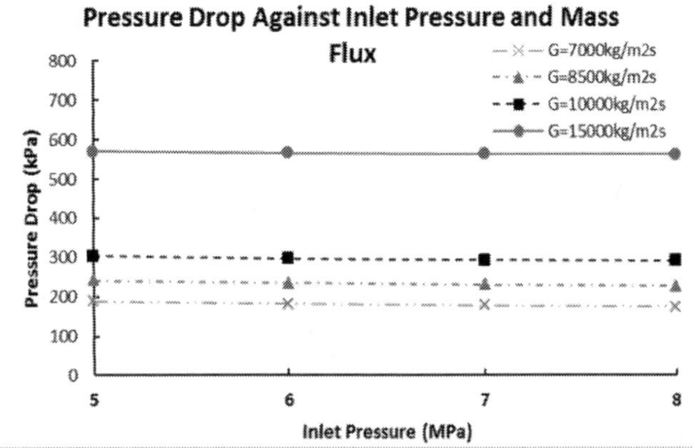

Figure 10. Effect of mass flux on pressure drop of methane at 120 K, 4 MW/m2 and various inlet pressures.

The last parameter examined is inlet pressure. For this, the boundary conditions fixed are inlet temperature of 150 K, mass flux of 10000 kg/m^{2}s and heat flux of 5 MW/m^2. Figure 11 shows the variation of pressure drop and average bulk velocity of supercritical methane. As discussed in the previous section, pressure drop is always proportional to fluid velocity. From this figure, it also shows that pressure drop decreases when average bulk velocity decreases. By comparing with the variation of average bulk velocity through different inlet pressure at G = 10000 kg/m^{2}s it was observed that greater

IOP Conf. Series: Materials Science and Engineering **342** (2018) 012061 doi:10.1088/1757-899X/342/1/012061

variation of the average bulk velocity even though the mass flux and inlet pressures for both cases are the same. However, this great variation of average bulk velocity is due to the difference of inlet temperature and heat flux applied. This can strengthen the statement made earlier which stated that low inlet temperature and heat flux are not enough to make the methane bulk temperature near to pseudocritical temperature. It can be seen from Figure 11 that when inlet pressure is increased, the pressure drop decreases. This is due to the more abrupt variation of thermo-physical properties of methane at lower supercritical pressures. In short, increasing inlet pressure can help to reduce pressure drop across tube length.

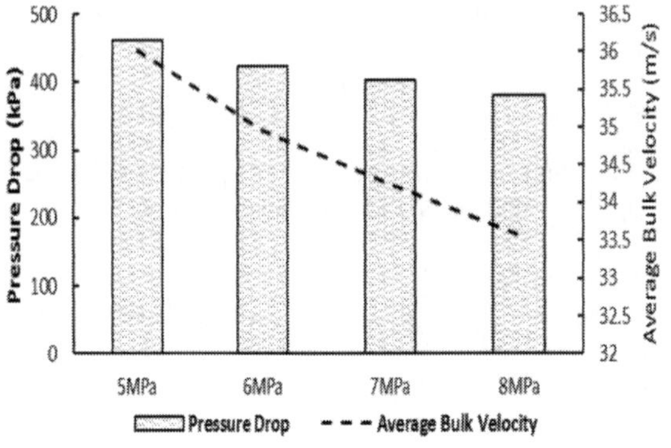

Figure 11. Effect of inlet pressure on pressure drop and average bulk velocity at 150 K, 10000 kg/m^{2}s and 5 MW/m^2.

4. Conclusion

The effect of inlet temperature, mass flux and inlet pressure on heat transfer and flow performance is investigated. High inlet temperature leads to abrupt increment of specific heat capacity which increases heat transfer coefficient after some distance from the inlet of the tube. Next, high mass flux has higher heat transfer coefficient initially but after some distance from inlet, low mass flux which has low flow velocity makes the flowing methane to reach near-pseudocritical temperature which enhances heat transfer performance near the tube outlet due to higher specific heat capacity increment. Lower inlet pressure has higher variation of specific heat capacity and this leads to the increment of heat transfer performance. For pressure drop, it becomes higher when heat flux, inlet temperature and mass flux are increased. Lower inlet pressure leads to higher pressure drop due to the more abrupt variation of thermo-physical properties. Generally, at mass flux of 10,000 kg/m^{2}s and heat flux of 4 MW/m^2, it is found that 150 K inlet temperature with 5 MPa inlet pressure has high heat transfer performance and low pressure drop. Besides that, heat flux of 3 MW/m^2 with mass flux of 15,000 kg/m^{2}s has the highest Goodness factor at 6 MPa inlet pressure and 130 K inlet temperature boundary conditions.

Acknowledgements

The authors would like to thank Universiti Malaysia Pahang for providing necessary assistance for providing financial support under grant number RDU150375.

References

[1] Herzog H, Eliasson B and Kaarstad O 2000 Capturing greenhouse gases *Scientific American* **282** 72-9

[2] Rao N T, Oumer A N and Jamaludin U K 2016 State-of-the-art on flow and heat transfer characteristics of supercritical CO2 in various channels *The Journal of Supercritical Fluids* **116** 132-47

[3] Devandran D, Oumer A N, Rao N T, Basrawi F and Ibrahim H 2017 Investigation on effect of some operating conditions on flow and heat transfer of supercritical CO2 in helical coiled tubes via Multilevel Factorial Methodology *Malaysian Journal of Industrial Technology* **2** 31-8
[4] Yah N F, Oumer A N and Idris M S 2017 Small scale hydro-power as a source of renewable energy in Malaysia: A review *Renewable and Sustainable Energy Reviews* **72** 228-39
[5] Szewczuk S 2015 Biogas as a fuel source for the transport sector. In: *2015 International Conference on the Industrial and Commercial Use of Energy (ICUE)*: IEEE) pp 256-62
[6] Weiland P 2010 Biogas production: current state and perspectives *Applied Microbiology and Biotechnology* **85** 849-60
[7] Wen D, Jiang H and Zhang K 2009 Supercritical fluids technology for clean biofuel production *Progress in Natural Science* **19** 273-84
[8] Hendricks R C, Simoneau R J and Smith R V 1995 *Advances in Cryogenic Engineering*: Springer) pp 197-237
[9] Gu H, Li H, Wang H and Luo Y 2013 Experimental investigation on convective heat transfer from a horizontal miniature tube to methane at supercritical pressures *Applied Thermal Engineering* **58** 490-8
[10] Liu B, Zhu Y, Yan J-J, Lei Y, Zhang B and Jiang P-X 2015 Experimental investigation of convection heat transfer of n-decane at supercritical pressures in small vertical tubes *International Journal of Heat and Mass Transfer* **91** 734-46
[11] Pizzarelli M, Nasuti F, Onofri M, Roncioni P, Votta R and Battista F 2015 Heat transfer modeling for supercritical methane flowing in rocket engine cooling channels *Applied Thermal Engineering* **75** 600-7
[12] Cp C 2010 Numerical Investigation of Supercritical Coolant Flow in Liquid Rocket Engine
[13] Pizzarelli M, Urbano A and Nasuti F 2010 Numerical analysis of deterioration in heat transfer to near-critical rocket propellants *Numerical Heat Transfer, Part A: Applications* **57** 297-314
[14] Jackson J 2013 Fluid flow and convective heat transfer to fluids at supercritical pressure *Nuclear Engineering and Design* **264** 24-40
[15] Dang G, Zhong F, Zhang Y and Zhang X 2015 Numerical study of heat transfer deterioration of turbulent supercritical kerosene flow in heated circular tube *International Journal of Heat and Mass Transfer* **85** 1003-11
[16] Ruan B and Meng H 2012 Supercritical heat transfer of cryogenic-propellant methane in rectangular engine cooling channels *Journal of Thermophysics and Heat Transfer* **26** 313-21
[17] Xu K, Tang L and Meng H 2015 Numerical study of supercritical-pressure fluid flows and heat transfer of methane in ribbed cooling tubes *International Journal of Heat and Mass Transfer* **84** 346-58
[18] Han C-L, Ren J-J, Dong W-P and Bi M-S 2016 Numerical investigation of supercritical LNG convective heat transfer in a horizontal serpentine tube *Cryogenics* **78** 1-13
[19] Wang Y-Z, Hua Y-X and Meng H 2010 Numerical studies of supercritical turbulent convective heat transfer of cryogenic-propellant methane *Journal of Thermophysics and Heat Transfer* **24** 490-500

The use of waste materials for concrete production in construction applications

Ashraf Teara[1,2], Doh Shu Ing[1] and Vivian WY Tam[2]

[1]University Malaysia Pahang, Faculty of Civil Engineering and Earth Resources, Lebuhraya Tun Razak, 26300 Gambang, Kuantan, Pahang.
[2] Western Sydney University, School of Computing, Engineering and Mathematics, Locked Bag 1797, Penrith, NSW 2751, Australia.

E-mail: a.alamen@gmail.com;dohsi@ump.edu.my;V.Tam@westernsydney.edu.au

Abstract: To sustain the environment, it is crucial to find solutions to deal with waste, pollution, depletion and degradation resources. In construction, large amounts of concrete from buildings' demolitions made up 30-40 % of total wastes. Expensive dumping cost, landfill taxes and limited disposal sites give chance to develop recycled concrete. Recycled aggregates were used for reconstructing damaged infrastructures and roads after World War II. However, recycled concrete consists fly ash, slag and recycled aggregate, is not widely used because of its poor quality compared with ordinary concrete. This research investigates the possibility of using recycled concrete in construction applications as normal concrete. Methods include varying proportion of replacing natural aggregate by recycled aggregate, and the substitute of cement by associated slag cement with fly ash. The study reveals that slag and fly ash are effective supplementary elements in improving the properties of the concrete with cement. But, without cement, these two elements do not play an important role in improving the properties. Also, slag is more useful than fly ash if its amount does not go higher than 50%. Moreover, recycled aggregate contributes positively to the concrete mixture, in terms of compression strength. Finally, concrete strength increases when the amount of the RA augments, related to either the high quality of RA or the method of mixing, or both.

1. Introduction

Nowadays, unfriendly environmentally effects of buildings' demolition, and expensive dumping fees of billion tons of concrete and steel has drawn concerns about sustainable materials, such as recycled concrete. In order to ease the environmental worries, landfill taxes, and policy on minimizing disposal places for construction debris have been applied in United Kingdom, Hong Kong, etc. [1]. A number of researches to transform building-waste-materials into "green", ones have been carried out around the world, which shown that recycled concrete has poorer compressive strength than ordinary concrete [1], so that its use has been limited only for reconstructing infrastructures, roads, sidewalk and flooring [2]. Also, recycled concrete derived from many different types of construction waste, in which old concrete constituted from 30 to 40 per cent, could be used in concrete manufacture to save natural resources, and energy consumption [3]. This study investigates the possibility of recycled concrete in construction applications as normal concrete by improving its properties. The investigation and analysis have been conducted to enhance the performance of reclamation materials by adjusting various proportion of RC, for instance, the replacement of natural aggregate by recycled aggregate, and the substitute of cement by associated slag cement with fly ash.

2. Methodology

2.1. Materials and preparation of the materials

Materials have been used to prepare 12 samples of recycled concrete are Portland cement, fly ash, slag, 10mm and 20mm of natural aggregate, 10mm and 20mm of recycled aggregate, water reducer.

• Portland cement, produced in Cement Australia Company, has been recommended to use for construction concrete such as fiber concrete manufacture, building possesses and all projects of civil engineering. Its components included Portland Cement Clinker less than 97 per cent, Gypsum (CaSO4 2H2O) between 2 and 5 per cent, Limestone (CaCO3) from 0 to 7.5 per cent, Calcium Oxide from 0 to 3 per cent, Crystalline Silica (Quartz) under 1 per cent and Hexavalent Chrome (Cr VI) below 20 ppm [4].

• Fly ash is widely trade in Australia, and is natural as industrial mineral with different advantages. As a raw material, fly ash does not have any effect on the chemical composition of mixture because of its 'pozzolanic' attributes. The components of fly ash consist fly ash 100 percent, Crystalline Silica (Quartz) more than 30 percent, Mullite between 5 and 30 percent and Hexavalent Chromium (Chrome VI) less than 1 ppm [4].

• Slag is the waste material removed from the operation of blast furnaces in the smelting of iron ore during steel manufacturing, namely Granulated Blast Furnace Slag (GBFS) and Ground Granulated Blast Furnace Slag (GGBFS). Slag is characterized by its latent hydraulic properties that play important roles in improving the durability properties of cement as mixing together. The information of components for the slag is Granulated Blast Furnace Slag more than 90 per cent, Gypsum (CaSO4 2H20) between 2 and 5 per cent, Hexavalent Chrome Cr (VI) below 1 per cent, Crystalline Silica (Quartz) under 1 per cent [4].

• Natural aggregate, supplied from St Mary's Recycling Facility, about 350 kg size (10 mm) and about 450 kg size (20 mm).

• Recycled aggregate, supplied from St Mary's Recycling Facility about 300 kg size (10 mm) and about 600 kg size (20 mm).

The initial step of preparing concrete was to keep all materials under the laboratory conditions with the temperature between 20 and 26 Celsius degree. Both natural aggregates and recycled ones have been washed carefully to remove impurities, such as dust and dirt. Then the aggregate have been dried by sunlight for seven full sunny days for ensuring no free water is inside the aggregate [5].

2.2. Mixing design

There are 12 different types of concrete has been made in the project experiment, which was divided into 4 groups. Some of them include natural aggregate, while others were made of recycled aggregate, also recycled and natural aggregate were used together in some mixtures. Generally, all the 12 mixtures differed from each other either by the volume of each ingredient, or by using some elements in a few of them, while it is not used in others, for example, recycled aggregate. The processes started with weighting the amount of materials involved in the formation for each type of concrete and placed in a concrete mixer.

In order to comply with the Australian standards [5], at the beginning, the mixer was in charged with approximately half of coarse aggregate was put over the cement or the powdered materials. After that, an appropriate quantity of water was added into the mixer to operate for 30 seconds. Then cement was the next adding to the mixture after the mixing operation was stopped for a while. Before restarting the mixing operation, cement was covered by another half of coarse aggregate to not lose powdered materials. As for admixtures, they were used in two different ways, either before or after adding cement. It has been suggested that admixtures should be distributed separately and uniformly to assure their distribution and to grant their effects [5]. A part of the balanced water should be used to dissolve the soluble admixtures before being added to the mixer. If the powder materials are over 10 percent of the amount of cement, they should be balanced then added to the mixing operation in the same way as cement. Water was added at the end to make a homogeneous mixture. The ingredients were worked out in the mixer until well-mixed. In some cases, the volume of water content of the mixture was not enough, what required the addition of the water reducer to improve the water content and ensure the workability of concrete.

The molds have been prepared and cleaned, and special oil has been sprayed in order to resist the stickiness with the moulds. After filling with concrete, the moulds were put in a covered place in the factory away from rainwater. After 24 hours or may be more (depends on the type of the concrete), the samples were taken and wrapped in plastic cover in order to save the water from evaporation. Table 1 and table 2 shows the quantities of elements that were used in each mixing.

Table 1. Mixing design details.

Group	Group 1			Group 2		
Mix design	NC	S50	FA50	S50+FA50	FA100	S100
Water (kg)	10.50	10.50	10.50	10.50	10.50	10.50
Cement (kg)	23.33	11.67	11.67	-	-	-
Fly ash (kg)	-	-	11.67	11.67	23.33	-
Slag (kg)	-	11.67	-	11.67	-	23.33
Sand (kg)	24.99	24.99	24.99	24.99	24.99	24.99
10mm natural aggregate (kg)	20.39	20.39	20.39	20.39	20.39	20.39
10mm recycled aggregate (kg)	-	-	-	-	-	-
20mm natural aggregate (kg)	40.79	40.79	40.79	40.79	40.79	40.79
20mm recycled aggregate (kg)	-	-	-	-	-	-

Table 2. Mixing design details (continued)

Group	Group 3			Group 4		
Mix design	RAC50	RAC50+S50	RAC50+FA50	RAC50+FA50+S50	RAC50+FA100	RAC50+S100
Water (kg)	10.50	10.50	10.50	10.50	10.50	10.50
Cement (kg)	23.33	11.67	11.67	-	-	-
Fly ash (kg)	-	-	11.67	11.67	23.33	-
Slag (kg)	-	11.67	-	11.67	-	23.33
Sand (kg)	24.99	24.99	24.99	24.99	24.99	24.99
10mm natural aggregate (kg)	10.20	10.20	10.20	10.20	10.20	10.20
10mm recycled aggregate (kg)	10.20	10.20	10.20	10.20	10.20	10.20
20mm natural aggregate (kg)	20.40	20.40	20.40	20.40	20.40	20.40
20mm recycled aggregate (kg)	20.40	20.40	20.40	20.40	20.40	20.40

2.3. Compression test

Compression test is for cylinders which has standard size (height = 2002 mm and diameter = 1002 mm). The samples were tested after 7, 14 and 28 days. Three cylinders were used in each time after the dimensions (high and diameter) have been measured by an electronic ruler. Then caps have been made to the rough poles to make them more regular and straight. Then the cylinders have been tested in the compression device. According to the Australian standards [6], capping is necessary for any rough surface of the cylinder. In addition to that thin, the maximum thickness of each cap can be no more than 6 mm.

3. Results and Discussion

The results of compression test have shown in the table 3 and also it is been shown in figure 1. 100% of slag or fly ash or together 50% slag and 50% fly ash without cement gave a weak concrete as seen in the samples (S50+FA50), (FA100) and (S100). The sample (FA100) which is contained 100% fly ash was so soft and it was broken before testing, therefore, there were no results about compressive strength and the main reason behind that could be the quantity of CaO, since the cement includes about 60% while the volume of this element in the fly ash is less than 3-6% [7] and [8]. The characteristics of these mixtures (S50+FA50, S100) were unsatisfying since compressive strength test was reduced more than 60% and 65% respectively compared to the ideal concrete. According to Cement Australia Company who supplied the fly ash and slag, it was recommended that these materials should be mixed with cement as supplementary elements to improve the concrete properties [4]. As a result, recycled concrete with 100% slag or fly ash are not strong enough in order to be used in structure applications. This result can be also proved by [7] who noticed that, increasing the replacement of fly ash in concrete declined the modulus of rupture and the strength as well.

While a mixture of 50% of slag and 100% of natural aggregate showed more developed results, but it still not as effective as that of the ideal concrete mixture since the strength of concrete was reduced about 15% - 20%, [8] argued that 50% of slag is not only very beneficial but also it gives the best results concerning the strength. The author maintained that the strength showed by the mentioned mixture was 45.7 and 49.8MPa after 28 and 84 days respectively. The difference between Berndt results and the ones concluded from the current study could be based on the quality and the proprieties of the slag.

Moving to the recycled aggregates, the study showed that 50% of recycled aggregate with 100% of cement increased the strength more than 6% compared to that of normal concrete although it has been proved that the RA may be beneficial only in the case when the quantity is less than 30% [9]. So that increase in strength could be due to the methods of mixing and the quality of the RA as it was maintained by [3]. However, 50% of recycled concrete aggregate with 50% of slag give result better than that given by the NA with 50 % of slag, while the results were similar to those of the control one. Therefore, the RA may prove a satisfied result with the slag as long as the quantity of latter does not go more than 50%. In a mixture of 50% of recycled aggregate with 50% of fly ash, the compressive strength declined about 15%.

Table 3. Compressive strength results.

Groups	Label	7 Days (MPa)	14 Days (MPa)	28 Days (MPa)
Group 1	NC	20.6	25.9	31.9
	S50	16.2	21.5	25.7
	FA50	18.7	21.2	31.9
Group 2	S50+FA50	3.2	4	6.9
	FA100	0	0	0
	S100	6.6	8.6	12.7
Group 3	RAC50	25	27.3	34.8
	RAC50+S50	19.8	25.2	29.8
	RAC50+FA50	17.6	20.6	27.3
Group 4	RAC50+S50+FA50	1.9	3.6	6.1
	RAC50+FA100	0	0	0
	RAC50+S100	4.6	9.3	11.1

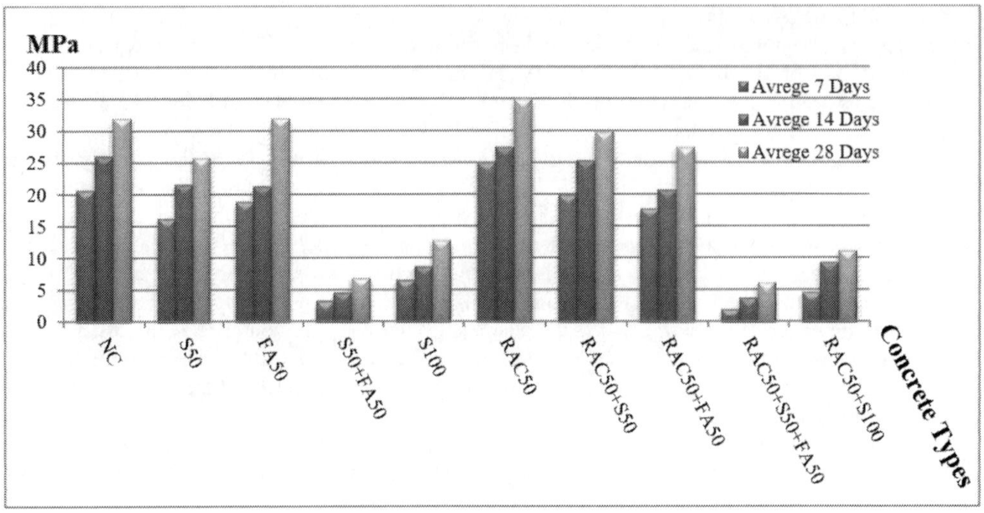

Figure 1. Compression test results.

4. Conclusions

The study reveals that slag and fly ash are effective supplementary elements in improving the properties of the concrete with cement. However, without cement, slag and fly ash together, or slag alone gave weak concrete and unsatisfying compressive strength test, while fly ash alone yielded broken concrete before testing. It was recommended that these materials should be mixed with cement as supplementary elements to improve the concrete properties. This result agreed with previous research that increasing the replacement of fly ash in concrete declined the strength as well [7]. Moreover, the recycled aggregate was slightly beneficial to improve a mixture with fly ash, but not a mixture with slag in terms of compressive strength. Also, fly ash is more useful than slag if its amount does not go higher than 50 per cent. Finally, concrete strength increases when the amount of the RA augments, related to either the high quality of RA or the method of mixing, or both.

References

[1] Tam V, Tam CM and Wang Y 2006 Optimization on proportion for recycled aggregate using two-stage mixing approach *Construction and Building Materials* **21** (1) pp 1928-1939.

[2] Olorunsogo FT and Padayachee N 2002 Performance of recycled aggregate concrete monitored by durability indexes *Cement and Concrete Research* **32** (2) pp 179-185.

[3] Kong D, Lei T, Zheng J, Ma C, Jiang J and Jiang J 2010 Effect and mechanism of surface coating pozzolanics materials around aggregate on properties and ITZ microstructure of recycled aggregate concrete *Construction and Building Materials* **24** (5) pp 701-708.

[4] Cement Australian 2017 Retrieved from *www.cementaustralia.com/ gifriorhi930589.htm*

[5] Standards Australia 1994 AS 1012.2-1994 Methods of testing concrete Preparation of concrete mixes in laboratory *Standards Australia*

[6] Standards Australia 2000 AS 1012.11-R2014 Methods of testing concrete Method 11: Determination of modulus of rupture *Standards Australia*

[7] Kou S, Poon C and Chan D 2007 Influence of fly ash as cement replacement on the properties of recycled aggregate concrete *Materials in Civil Engineering* **19** (9) pp 709-717.

[8] Berndt ML 2009 Properties of sustainable concrete containing fly ash, slag and recycled concrete aggregate *Construction and Building Materials* **23** (7) pp 2606-2613.

[9] Sim, J & Park, C 2011, Compressive strength and resistance to chloride penetration and carbonation of recycled aggregate concrete with varying amount of fly ash and fine recycled aggregate, *Waste Management*, **31**(11), 2352-2360.

IOP Conf. Series: Materials Science and Engineering **342** (2018) 012063 doi:10.1088/1757-899X/342/1/012063

Preparation and characterization of 6-layered functionally graded nickel-alumina (Ni-Al_2O_3) composites

M I A Latiff[1], D M Nuruzzaman[2], S Basri[3], N M Ismail[4], S N S Jamaludin[5] and F F Kamaruzaman[4]

[1,2,4] Faculty of Manufacturing Engineering, University Malaysia Pahang, 26600 Pekan, Pahang Darul Makmur, Malaysia
[3] Office of the Assistant Vice Chancellor (Development), University Malaysia Kelantan, 16100 Kota Bharu, Kelantan Darul Naim, Malaysia
[5] Faculty of Engineering, DRB-HICOM University of Automotive Malaysia, 26607 Pekan, Pahang Darul Makmur, Malaysia

E-mail: dewan052005@yahoo.com

Abstract. The present research study deals with the preparation of 6-layered functionally graded (FG) metal-ceramic composite materials through powder metallurgy technique. Using a cylindrical die-punch set made of steel, the nickel-alumina (Ni-Al_2O_3) graded composite structure was fabricated. The samples consist of four gradual inter layers of varied nickel composition (80wt.%, 60wt.%, 40wt.%, 20wt.%) sandwiched with pure Ni and Al_2O_3 powders at the ends (100wt.% and 0wt.% nickel) were fabricated under 30 ton compaction load using a hydraulic press. After that, two-step sintering was carried out at sintering temperature 1200°C and soaking time 3 hours was maintained in a tube furnace. The properties of the prepared samples were characterized by radial shrinkage, optical microscopy and hardness testing. Results showed that larger shrinkage occurred within the ceramic phase which proves that more porosities were eliminated in the ceramic rich layers. From the microstructural analysis, it was observed that alumina particles are almost uniformly distributed in nickel matrix, so as nickel particles in the ceramic matrix of alumina-dominant layers. From interfacial analyses, it was observed that a smooth transition in microstructure from one layer to the next confirms a good interfacial solid state bonding between metal-ceramic constituents and good compaction process. On the other hand, microhardness test results suggest that there might be increasing percentage of porosities in the graded structure as the ceramic content rises.

1. Introduction

At present, the demand for new materials is growing rapidly for advanced engineering applications. A material of single composition with uniform structure may not satisfy all the requirements simultaneously. Functionally graded material (FGM) is a class of advanced composite which combines both the good properties of pure metal and pure ceramic beside possesses multifunctional properties to accommodate its applications in extreme environments. FGM is a combination of two or more material constituents, gradually changing in composition and properties from one side to the other along the thickness. The graded nature which may be continuous or stepwise (layer by layer) improves the bonding between material constituents and eliminates the stress at the interface. In processing of these graded composite materials various fabrication methods are available namely, thermal spraying, chemical vapour deposition, powder metallurgy and centrifugal casting [1-5]. In recent years, powder metallurgy method is widely used to produce the functional gradients in order to meet the requirements [6,7]. In processing the graded composite system, sintering is the most challenging step. Cracks or defects are very common within the structure during the sintering of multi-layered composite systems due to the mismatch in thermal expansion of the graded layers [8].

Research efforts were carried out in order to investigate the properties of the functionally graded materials of different combinations theoretically and experimentally [9-13]. The obtained results revealed that mechanical responses of the graded composite systems remarkably influenced by material composition, number of layers, sintering temperature, compaction load etc. Mechanical responses of the graded aluminium composites were investigated at the outer, middle and inner regions [14]. The outer regions of the composites showed higher hardness as compared to the middle or inner regions. Moreover, outer region of the composites exhibited higher strength than the inner region. The properties of hydroxyapatite-titanium graded composite structures fabricated under different sintering atmospheres were investigated [15]. From the obtained results it was noticed that the graded structures were significantly influenced by different sintering atmospheres. Stainless steel graded parts were fabricated using laser melting deposition technique [16]. In this investigation, microstructure of different composition gradient, microhardness along the gradient direction were analysed. The obtained results revealed that good bonding was achieved between the deposited material and the substrate.

The design of the graded structure is controlled in terms of the geometrical aspects, such as number of layers and the layer thickness with systematically varied composition. The material variables such as chemical composition of powder compact, particle size distribution and particle shape are also considered as these factors influence the compressibility and sinterability of the fabrication process. In addition, processing parameters such as sintering time, sintering temperature, heating and cooling rate, sintering atmosphere are also important in order to prepare successful specimens. Considering these factors, in this research, Ni-Al_2O_3 functionally graded composite system was chosen since the properties of Ni and Al_2O_3 precursors are well characterized in a wide temperature range [9]. Six-layered nickel-alumina (Ni-Al_2O_3) functionally graded composite specimens were prepared using powder metallurgy and two-step pressureless sintering technique using argon atmosphere. Specimens were designed to have pure constituents (Ni and Al_2O_3) sandwiched the gradual layers which changing uniformly in composition by 20wt.% Ni across each layer. This configuration ultimately formed a six-layered FGM system. In order to characterize and to measure the quality of the graded composite, it is important to analyse the shrinkage, microhardness and microstructural properties. Considering all these issues, the nickel-alumina functionally graded composite specimens were analysed for shrinkage and microhardness. Microstructural and interfacial characteristics were also analysed in this study.

2. Experimental

In processing the Ni-Al_2O_3 functionally graded system, factors such as particle size and shape, layer thickness and number of layers were considered. Ni possesses high ductility and stiffness while being resistive towards corrosive environments. Al_2O_3 on the contrary is low in density and electrical resistivity while having high hardness and stiffness. Besides, low coefficient of thermal expansion (CTE) of Al_2O_3, at the same time having high melting point and highly resistant towards corrosion makes it suitable for combined with metals and applicable as super heat resistance materials [9]. Powder metallurgy (PM) technique was chosen for this research because of the simplicity of the process and it is applicable to a wide range of material combinations including Ni-Al_2O_3 system. Other prominent advantages of PM technique include the possibility of close control over the gradient composition and the net or near net shape forming of the structural materials. Since the Ni and Al_2O_3 precursors are in the form of powders, maximising the exposed surface area for the reaction, thus result in faster reaction and ensure higher percentage of materials to react [1]. The constituent powders nickel (Bendosen) and alumina (Sigma Aldrich) were collected. The size of Ni particles in the range of 10-30 μm and predominantly round in shapes. Al_2O_3 particles have irregular shapes and the size in the range of 50-70 μm. A set of layer mixtures for a 6-layered sample was prepared (100wt.%Ni, 20wt.%Al_2O_3+80wt.%Ni, 40wt.%Al_2O_3+60wt.%Ni, 60wt.%Al_2O_3+40wt.%Ni, 80wt.%Al_2O_3+20wt.%Ni, 100wt.%Al_2O_3) according to the molecular weight of nickel and alumina constituents. After blending, the powder mixtures were sieved through a 100 μm mesh. After that, the mixtures were dried in a vacuum oven. The stacking up process of the powder layers was carried out in a cylindrical steel die of 30 mm diameter according to the arrangement as shown in Figure 1(a).

The powder layers were then compacted using a hydraulic press under a load of 30 ton. After that, the green compact underwent through a heat treatment process to densify the sample using pressureless sintering technique. The sintering was carried out in a tube furnace (Elite Thermal Systems Limited) in argon atmosphere, following the two-step sintering profile as depicted in Figure 1(b) below. The sintering process includes a preheating stage of 30 minutes at 150ºC for the removal of moisture and chemical impurities from the sample. After that, a heating rate of 4°C/min was applied to reach the temperature 1200°C and holding time 3 hours was maintained at this sintering temperature. Finally, a cooling rate of 2°C/min was maintained to allow the specimen to cool down to room temperature.

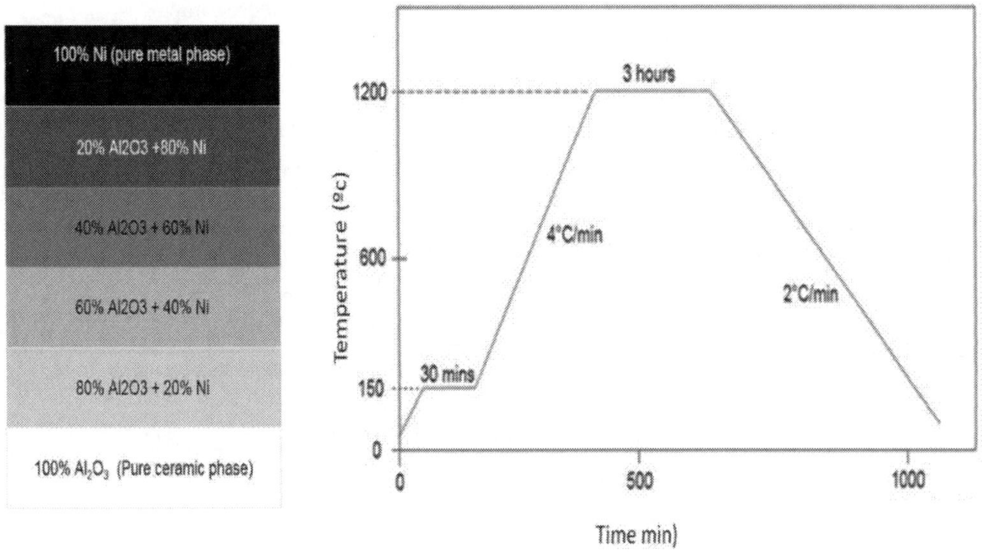

(a) (b)

Figure 1. (a) Illustration of layer arrangement for a 6-layered stepwise-built Ni-Al_2O_3 FGM sample (b) Two-step sintering cycle.

After sintering, the sintered samples were cut, ground and polished to prepare for microstructural characterization and hardness testing. Prepared samples were observed for microstructural characterization under metallurgical microscope (OLYMPUS BX51M). The hardness of individual layer was measured along the sample thickness using Vickers' microhardness tester (Wilson Hardness: Model 402 MD) under a 1 kgf (1 kN) load for dwell time of 15 seconds. Seven measurements of hardness were taken to ensure consistency and the average value was taken into consideration.

3. Results and Discussion

3.1. Densification and shrinkage

After the sintering process of the samples, it was believed that the samples were experienced shrinkage due to the formation of nickel-alumina bonding, hence the decrease of gaps between particles of constituent powders. Figure 2 compares the green compact and sintered samples. The sintered sample was viewed from the front and was seen clearly having a slanted shape along the sides in which the diameter at the top side (nickel) was larger than the diameter at the bottom side (alumina). The radial shrinkage of Ni/Al_2O_3 FGM samples was quantified by measuring the diameter of the green and sintered samples. Three average measurements had been recorded along the thickness for each specimen which were from top (0wt.% alumina), middle (60wt.% alumina) and bottom (100wt.% alumina) position. The measurements were taken from three different samples for consistency of results.

(a) (b)

Figure 2. Samples of 6-layered Ni/Al_2O_3 FGM (a) Green compact and (b) Sintered sample.

Figure 3 compares the diameter of the samples during pre and post sintering process. It can be seen that at pure nickel layer, there was a slight increase in diameter after sintering due to expansion of nickel in the radial direction. From the graph it is apparent that the diameter was shrunk with the increase in alumina wt.%. Larger shrinkage within the ceramic phase proves that during sintering, more porosities in the layer were eliminated.

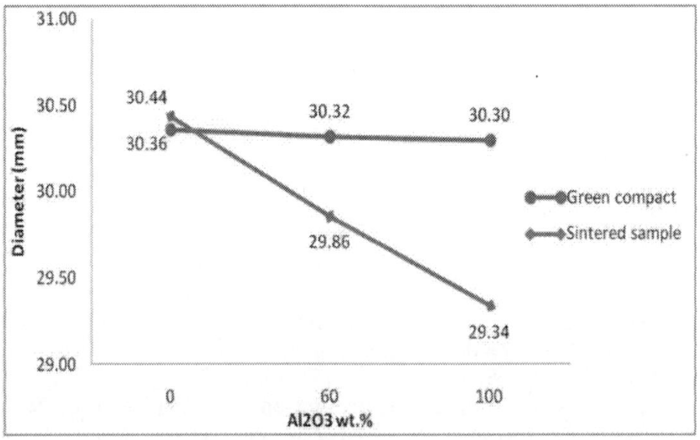

Figure 3. Distribution of diameter along the thickness for green and sintered specimens.

3.2. Microstructural analysis

Figures 4(a)-(f) show the optical micrographs of layers of different compositions for the 6-layered FG samples. In the following micrographs, at pure nickel and higher nickel composition layers, the metal particles (Ni) are revealed by the shiny and bright parts while ceramic constituents are represented by darker regions. On the other hand, at ceramic-dominant layers, Al_2O_3 constituents become more visible due to the lighting adjustment. At 80wt.% Ni composition, the volume part of nickel phase containing interconnected metal particles prevails over the alumina parts and forms the matrix phase. As the composition of alumina is increasing, metal interconnection decreased and the volume of nickel parts isolated between the ceramic constituents become more prevalent and viewed as the reinforcement phases. In the 60wt.% Ni layer, the interconnected nickel phases become isolated whereas alumina phase begin to combine and form clusters. In terms of the particle distribution, it can be referred from the micrographs of 80wt.% Ni layer that alumina particles are almost uniformly distributed in nickel matrix, so as nickel particles in the ceramic matrix of alumina-dominant layers (20wt.% and 40wt.% Ni layers).

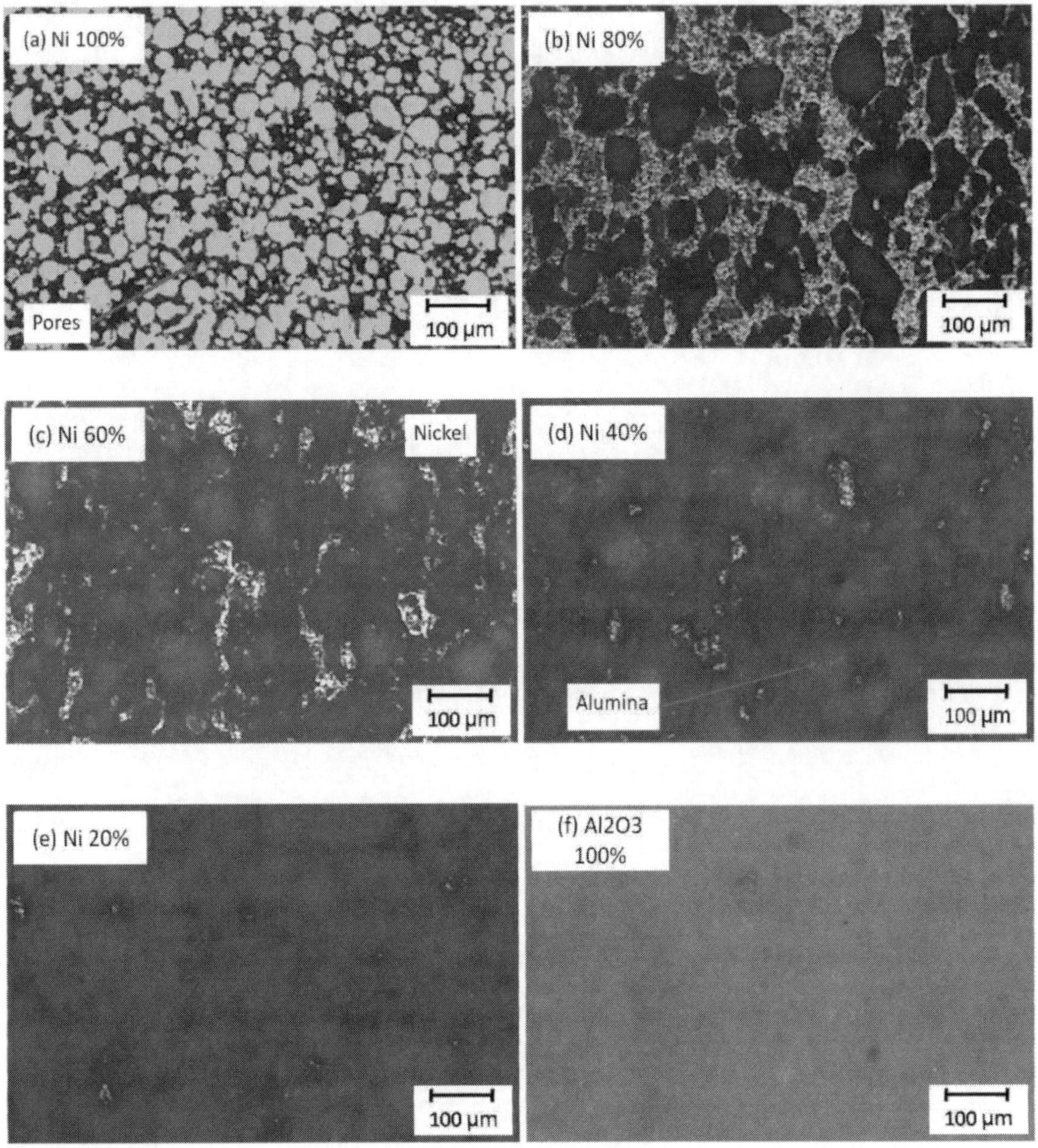

Figure 4. Micrographs of individual layers for 6-layered samples (a) 100wt.% Ni (b) 80wt.% Ni (c) 60wt.% Ni (d) 40wt.% Ni (c) 20wt.% Ni (d) 100wt.% Al_2O_3.

Figures 5(a)-(e) show the interfacial lines of 6-layered Ni/Al_2O_3 FGM. The existences of six layer compositions inside the FGM samples are confirmed from the difference in phase composition of adjacent layers and interfacial barriers. In general, it can be seen that all interfaces show a smooth transition in microstructure from one layer to the next and no crack was found along the interfaces. This confirms a good interfacial solid state bonding between metal-ceramic constituents and proper stacking process. The interface lines represented by dashed lines are almost straight and parallel (judging from the positions of the arrows on both sides of each micrograph which have close gaps) but also wavy at some localized area which indicates non-uniformity in the layer thickness as powders were stacked before compaction. However for 60wt.% Ni and 40wt.% Ni layer combination, the interface is less obvious and little difficult to recognize.

IOP Conf. Series: Materials Science and Engineering **342** (2018) 012063 doi:10.1088/1757-899X/342/1/012063

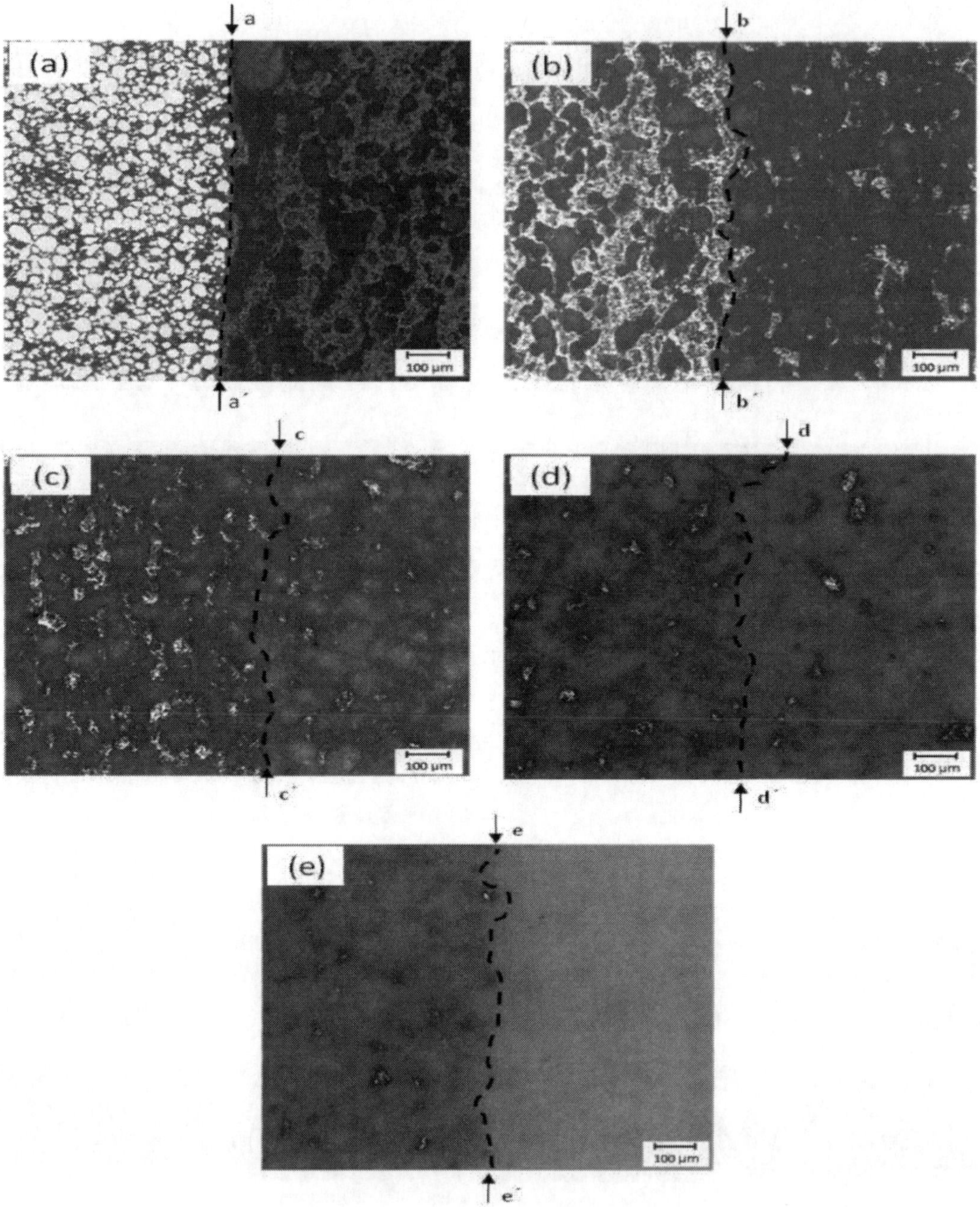

Figure 5. Interface lines of 6-layered Ni/Al_2O_3 FGM. (a) interface aa´ between 100wt.% Ni and 80wt.% Ni layers (b) interface bb´ between 80wt.% Ni and 60wt.% Ni layers (c) interface cc´ between 60wt.% Ni and 40wt.% Ni layers (d) interface dd´ between 40wt.% Ni and 20wt.% Ni layers (e) interface ee´ between 20wt.% Ni and 100wt.% Al_2O_3 layers.

3.3. Microhardness measurement

Figure 6 shows the variation in microhardness values across different layers. The microhardness in the Ni/Al_2O_3 FGM sample was measured at each distinctive layer to observe the hardness variation across the thickness direction. Seven hardness values at different locations were measured for each layer and the average value was taken into consideration. These results are plotted in the figure 6. The highest hardness was recorded on the pure nickel layer which is 150.00 HV while the lowest was at pure Al_2O_3 layer, 12.03 HV. The averaged plotted values show a slight decrease from pure Ni layer to 20 wt.% Al_2O_3 layer. From the obtained hardness values, the trend line implies that as the percentage of

Al_2O_3 composition increases, the microhardness values decrease. The hardness trends are contrary to the expected values and it is believed that sintering temperature and holding time are responsible for the incomplete transformation of Al_2O_3 particles into hard ceramic. This might also relate to the texture of alumina layer after sintering which is not very strong and eroded along the edge. This erosion of alumina is shown in Figure 7. The obtained results suggest that there might be increasing percentage of porosities in the FG structure as the ceramic content rises.

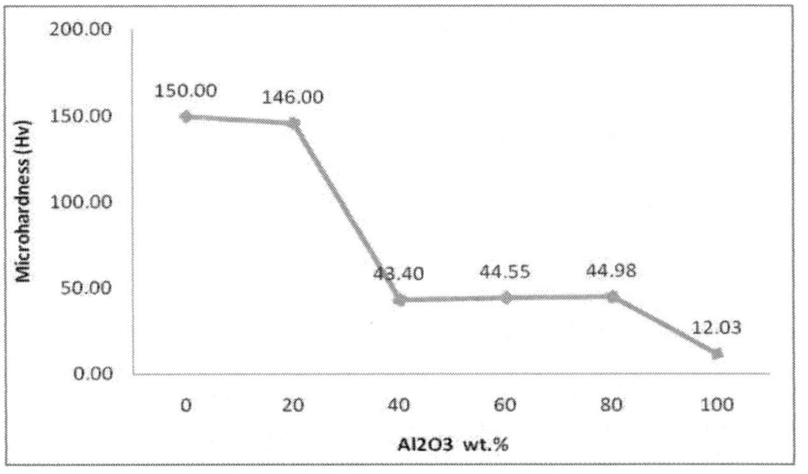

Figure 6. Variation in microhardness values across different layers for sample sintered for 3 hours.

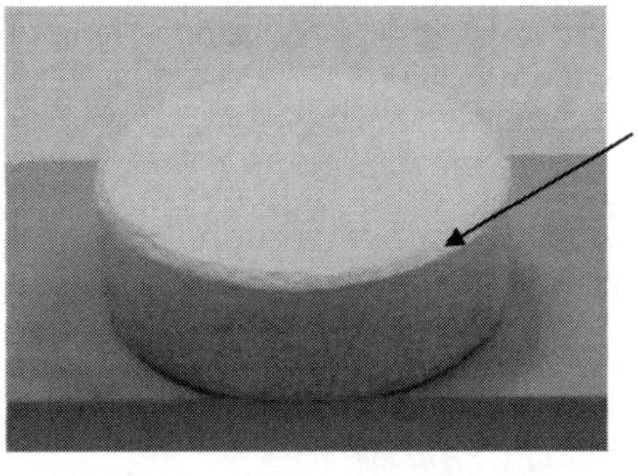

Figure 7. Alumina layer eroded on the edge.

4. Conclusion

Six-layered Ni-Al_2O_3 functionally graded composite materials were prepared through powder metallurgy and pressureless sintering technique. Two-step sintering cycle was applied in a tube furnace at sintering temperature of 1200°C and sintering time of 3 hours. The obtained results revealed that in the ceramic rich layers, larger shrinkage occurred and more porosities were eliminated. Microstructural analyses showed that alumina particles are almost uniformly distributed in nickel matrix and nickel particles are also uniformly distributed in the alumina rich layers. From the interfacial analyses, it was observed that smooth transition of microstructure occurred from one layer to the next layer which confirms a good interfacial solid-state bonding. Microhardness test results suggest that alumina powders might not have sintered completely to transform into hard ceramic and it is believed that there is an increasing amount of micropores/microvoids in the graded structure as the ceramic content rises.

References

[1] EL-Wazery M S, EL-Desouky A R, Hamed O A, Fathy A and Mansour N A 2013 Electrical and mechanical performance of zirconia-nickel functionally graded materials *Int. J. Eng. Trans. A: Basics* **26** 375–82

[2] Khoddami A M, Sabour A and Hadavi S M M 2007 Microstructure formation in thermally sprayed duplex and functionally graded NiCrAlY/Yttria-stabilized zirconia coatings *Surf. Coat. Technol.* **201** 6019–24

[3] Rubio W M, Paulino G H and Silva E C N 2012 Analysis, manufacture and characterization of Ni/Cu functionally graded structures *Mater. Des.* **41** 255-6

[4] Wu K, Scheler S, Park H S and Porada M W 2013 Pressureless sintering of ZrO_2-$ZrSiO_4$/NiCr functionally graded materials with a shrinkage matching process *J. Eur. Ceram. Soc.* **33** 1111–21

[5] Watanabe Y, Hattori Y and Sato H 2015 Distribution of microstructure and cooling rate in Al-Al_2Cu functionally graded materials fabricated by a centrifugal method *J. Mater. Process. Technol.* **221** 197–204

[6] Shahrjerdi A, Mustapha F, Bayat M, Sapuan S M and Majid D L A 2011 Fabrication of functionally graded hydroxyapatite-titanium by applying optimal sintering procedure and powder metallurgy *Int. J. Phys. Sci.* **6** 2258–67

[7] Latiff M I A, Jamaludin S N S, Basri S, Hussain A, Al-Othmany D S, Mustapha F, Nuruzzaman D M, Ismail N M and Ismail I 2014 Effect of sintering temperature on functionally graded nickel/alumina plate *Appl. Mech. Mater.* **629** 437-43

[8] Sun L, Sneller A and Kwon P 2008 Fabrication of alumina/zirconia functionally graded material: from optimization of processing parameters to phenomenological constitutive model *Mater. Sci. Eng. A* **488** 31–8

[9] Bhattacharyya M, Kumar A N and Kapuria S 2008 Synthesis and characterization of Al/SiC and Ni/Al_2O_3 functionally graded materials *Mater. Sci. Eng. A* **487** 524-35

[10] Hosseini S S, Bayesteh H and Mohammadi S 2013 Thermo-mechanical XFEM crack propagation analysis of functionally graded materials, *Mater. Sci. Eng. A* **561** 285-302

[11] Tahvilian L and Fang Z Z 2012 An investigation on thermal residual stresses in a cylindrical functionally graded WC-Co component *Mater. Sci. Eng. A* **557** 106-12

[12] Jamaludin S N S, Basri S, Hussain A, Al-Othmany D S, Mustapha F and Nuruzzaman D M 2014 Three-dimensional finite element modeling of thermomechanical problems in functionally graded hydroxyapatite/titanium plate *Math. Probl. Eng.* **2014** 1-20

[13] Erdemir F, Canakci A and Varol T 2015 Microstructural characterization and mechanical properties of functionally graded Al2024/SiC composites prepared by powder metallurgy techniques *Trans. Nonferrous Met. Soc. China* **25** 3569-77

[14] Radhika N and Raghu R 2016 Development of functionally graded aluminium composites using centrifugal casting and influence of reinforcements on mechanical and wear properties *Trans. Nonferr. Met. Soc. China* **26** 905-16

[15] Jamaludin S N S, Basri S, Mustapha F, Nuruzzaman D M, Latiff M I A and Ismail N M 2017 Phase contamination characterization of stepwise-bulit functionally graded hydroxyapatite/titanium (HA/Ti) sintered under various atmospheres *Mater. Sci. Forum* **889** 90–5

[16] Zhang C H, Zhang H, Wu C L, Zhang S, Sun Z L and Dong S Y 2017 Multi-layer functional graded stainless steel fabricated by laser melting deposition *Vacuum* **141** 181-7

IOP Conf. Series: Materials Science and Engineering **342** (2018) 012064 doi:10.1088/1757-899X/342/1/012064

Quasi-static axial crushes on woven jute/polyester AA6063T52 composite tubes

A Othman[1, 2] and AE Ismail[1]
[1]Department of Engineering Mechanics, Faculty of Mechanical and Manufacturing Engineering, Universiti Tun Hussien Onn Malaysia, Batu Pahat, Johor, Malaysia
[2]Department of Polytechnic Education, Ministry of Higher Education, Putrajaya, Malaysia

E-mail: akbar.othman@gmail.com

Abstract. Quasi-static axial loading have been studied in this paper to determine the behaviour of jute/polyester wrapped on aluminium alloy 6063T52. The filler material also was include into crush box specimen, which is polyurethane (PU) and polystyrene (PE) rigid foam at ranging 40 and 45 kg/m^3 densities. All specimen profile was fabricated using hand layup techniques and the length of each specimen were fixed at 100 mm as well as diameter and width of the tube at 50.8 mm. The two types of tubular cross-section were studied of round and square thin-walled profiles and the angle of fibre at 45^0 were analysed for four layers. Thin walled of aluminium was 1.9 mm and end frontal of each specimen of composite were chamfered at 45^0 to prevent catastrophic failure mode. The specific absorbed energy (SEA) and crush force efficiency (CFE) were analyses for each specimen to see the behaviour on jute/polyester wrapped on metallic structure can give influence the energy management for automotive application. Result show that the four layers' jute/polyester with filler material show significant value in term of specific absorbed energy compared empty and polyurethane profiles higher 26.66% for empty and 15.19% compared to polyurethane profiles. It has been found that the thin walled square profile of the jute/polyester tubes with polystyrene foam-filled is found higher respectively 27.42% to 13.13% than empty and polyurethane (PU) foam tubes. An introduce filler material onto thin walled composite profiles gave major advantage increases the mean axial load of 31.87% from 32.94 kN to 48.35 kN from empty to polystyrene thin walled round jute/polyester profiles and 31.7% from 23.11 KN to 33.84 kN from empty to polystyrene thin walled square jute/polyester profiles. Failure mechanisms of the axially loaded composite tubes were also observed and discussed.

1. Introduction

The tubular cross-section profile of composite have been used to replace the metallic tube as an energy absorber device in automotive application [1]. The type of conventional material of composite such as carbon, kevlar and glass has a major concern for researcher to study the capability of crashworthy structure under dynamic impact and quasi-static loading condition [1-5]. The natural fiber of composite have been popularized in the late of 20 century for do research to improve the capability of their properties under various applied loading condition on benefit for automotive application [1-15]. The mechanical properties of natural fiber reinforced composites highly depend on the interface adhesion property between the fibers and the polymer matrix as have been reported by many researchers [5–9]. Natural fibers contain cellulose, hemicelluloses, pectin's and lignin and are rich in hydroxyl groups, natural fibers tend to be strong polar and hydrophilic materials whilst polymer materials are a polar and exhibit significant hydrophobicity [7]. The problem happen with this phenomenon was the interface between debonding matrix and fiber each layers. The used of thin-walled alloy aluminum 6063-T52 would enhanced and improve the capability of crush box structure as a hybrid material wrapped together with jute/polyester woven composites. On the other hand, surface debonding between interfaces have been modified and cured as well as to reduce air bubble to strengthen material under quasi-static loading condition. The energy management of material have been studied previously with various aspect such as combination of two or more material mentioned in [10-18]. Othman et al. conducted pultruded square cross-section with rigid foam filler material under quasi-static loading condition, they found that an influence cellular material improve the specific absorbed energy with additional density of filler material as well as wall-thickness tube. Owing to the aforementioned observation, this study utilises of using cellular material applied on jute reinforced polyester composite and added filler material to stabilized the energy absorption performance during quasi-static regimes.

The focus on this study is to determine the load-displacement curve and the energy absorption of jute/polyester composite tubes wrapped AA6063T52 under axial quasi-static loading. Furthermore, the study aims to investigate the effect of filler material such as polyurethane and polystyrene rigid foam of the composite tubes.

2. Experimental work

2.1 Materials and fabrication

The natural woven roving jute/polyester composite were fabricated using hand lay-up technique wrapped on aluminium thin walled tube fix with four layers jute of [±45°]. The aluminium alloy 6063T52 was selected as a core material with total length of 500 mm and then cut off for an each 100 mm length for both of round and square cross-section. The diameter and width tube was constant of 50.8 mm and front end of each tube were chamfered end of 45^0 when composite hybrid finished. Each of composite tube were curing with room temperature in order to specimen profile were good enough cure. Table 1 shows the dimensions of the fabricated composite. Samples of the fabricated specimens are shown in figure 1. The terms AL, R, S, L, PE, and PU are respectively aluminium, round, square, layer, polystyrene and polyurethane. The number 4 referred to the number of layers of the composite laminate structures. The inner diameter and width of all specimen tube were fixed at 50.8

Table 1. Dimensions of the composite tubes.

Specimen type				Composite tube	
	Mass (kg)	Profiles	Foam Filler	Wall Thickness Jute/polyester + AA6063 t (mm)	Length
ALR-4L	0.223	Round	-	4.00+1.9	100
ALR-4L-PU	0.230	Round	Polyurethane	4.02+1.9	100
ALR-4L-PE	0.231	Round	Polystyrene	4.02+1.9	100
ALS-4L	0.284	Square	-	4.01+1.9	100
ALS-4L-PU	0.293	Square	Polyurethane	3.98+1.9	100
ALS-4L-PE	0.294	Square	Polystyrene	4.04+1.9	100

Figure 1. Sample of the fabricated composite tubes.

2.2. Testing procedure

The testing procedure of quasi-static loading condition were tested on all specimens with constant speed of 5 mm/m compression using a computer-controlled servo-hydraulic Instron machine type GT instruments. The cross-section tube crushed between two parallel steel flat platens, one fix and one moving. The fixed platen was fitted with a load cell from which the load signal is taken directly to the computer. For each test, the axial loading was plotted on the Y-axis and the crosshead displacement on the X-axis. The end front of each specimens were chamfered at 45^0 to ensure that the load fluctuation characteristic were avoided [15].

Some of following aspect of crush phenomenon was measure during crushes event. The all data were recorded and calculated there are Peak force P_{max}: maximum compressive force. Mean force P_{avg}: average compressive force obtained by the following equation where force and deformation are defined as P and d, respectively and the area of cross section, A, and the density of the material, ρ defined as

$$\bar{P} = \frac{1}{\delta}\int_0^{\delta} P d\delta \tag{1}$$

Energy absorption E: area under the load–displacement curve up to the compaction zone.

$$E = \int_0^{\delta} P d\delta \tag{2}$$

The units of E_s are used to express the crashworthiness parameters, hence they are written in kJ here P is the load acting on the composite specimens. Therefore, the specific energy absorption per unit mass kJ/kg (SEA) where m the crushed mass of the composite is recognized as:

$$SEA = \frac{E}{m} \tag{3}$$

Crash force efficiency (CFE): ratio of the average crushing load P_{avg} to peak load P_{max}.

$$CFE = \frac{P_{avg}}{P_{max}} \tag{4}$$

3. Results and discussion

3.1. Axial loading characteristics

The result of each specimen profile have been shown in figure 2 until 5 below. The letter of AL, R, S, 4L, PE and PU were referred to the aluminium, round, square, four layers, polystyrene and polyurethane, respectively. In figures 2 and 3 indicated that the different between empty and rigid foam-filled both of polystyrene and polyurethane rigid foam filler under quasi-static loading against displacement curves. The behaviour of the quasi-static condition observed that the influence of filler material improved significantly specific energy absorption when introduce foam filler material of the tubes.

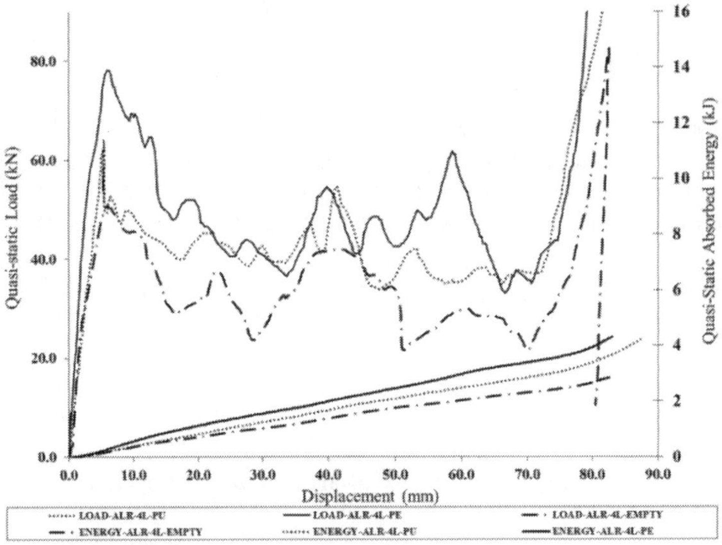

(a) Load against displacement curve of four layers jute/polyester composite tubes.

(b) Profile of jute/polyester of four layers without foam.

(c) Profile of jute/polyester of four layers with polyurethane foam.

Figure 2. Load-displacement relation for jute/polyester tubes for four layers.

The filler material strengthen and sustain force quasi-statically during crush regimes also reported previously done by Othman et al. [10-15]. They observed that the influence of polyurethane with different density of rigid foam bring more additional value on crashworthy structures. The fluctuation curve load against displacement can be seen an increasing force from empty to foam filled profiles. It also can be seen in figure 2 and 3 that the configuration of each behaviour curve and deformation pattern recorded from empty and foam filled tubes.

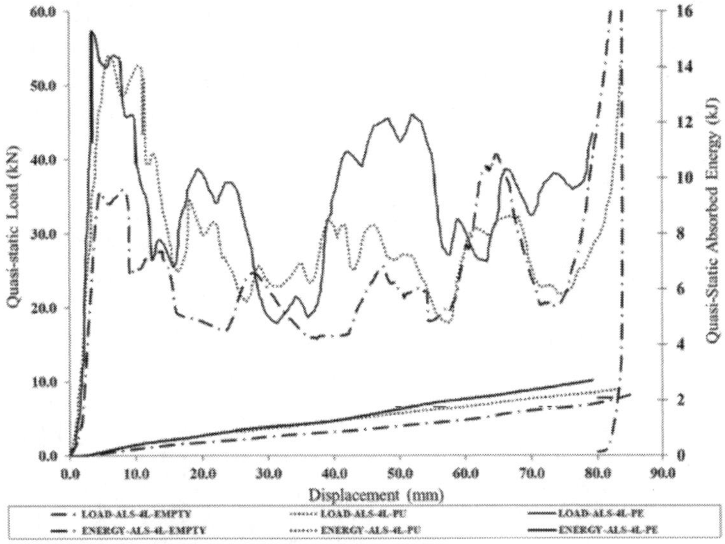

(a) Load against displacement curve of four layer jute/polyester composite tubes.

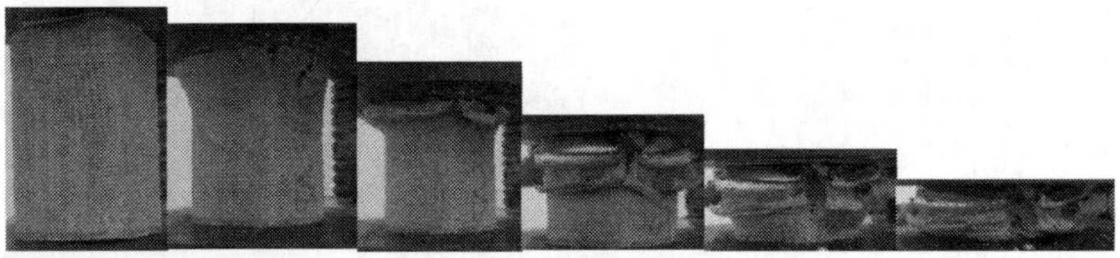

Unloading 15 mm 30 mm 50 mm 70 mm 80 mm

(b) Profile of jute/polyester of four layers with polystyrene foam.

Unloading 20 mm 50 mm 60 mm 80 mm

(c) Profile of jute/polyester of four layers without foam.

Figure 3. Load-displacement relation for jute/polyester tubes for four layers.

Obviously, each curve seen deformed progressively under quasi-static loading condition. The influence of jute/polyester wrapped on aluminium tube has seen less effect when the actual contact occurred, the formation of crack formed and continuously propagated until reach compaction zone (total length deformed). Almost of each specimen of the composite tube fail with the same criteria as well as round and square profiles. Square profile fails with crack formation on the fourth edges estimate around 2 mm of fillet. Different with a round tube profile fail at the front end of the tube with crack formation around eight cracks formed and propagated until final failure.

3.2. Quasi-static loading characteristics

The increasing fluctuation force occurred on the foam-filled profile observed on both round and square composite tubes profiles it can be seen on figure 2 and 3 above. The significant influence of foam-filled as a filler material tremendously attract more advantage when applying on automotive structure.

However, in this study, the main focusing and motivation on how jute/polyester given an impact on increasing specific energy absorption. The crush force efficiency (CFE) of each tube profile have been calculated to determine the mode failure whether specimen fail either under progressive or under catastrophic failure regime. Jimenez et al. [4] state that, either the criteria of failure mode under 0.5 value was fail under catastrophic failure or reach to 1.0 is safe and fail progressively.

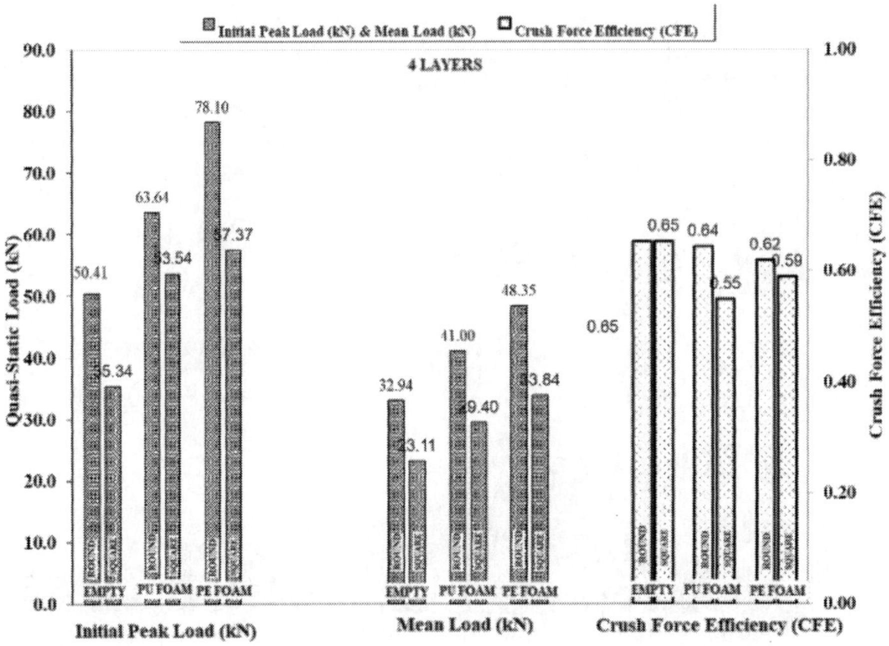

Figure 4. Initial, mean and crush force efficiency relation for jute/polyester tubes for each layers.

Figure 4 below showed all the results that is initial peak, mean load and crush force efficiency (CFE). Result showed that almost specimen profiles value above of 0.5, indicated each specimen profile fail with progressive condition. It also can be seen the value of initial peak load increase when tube profile introduce foam-filled either polyurethane or polystyrene. The mean or average load of each tube have been showed significant increase also, empty than foam-filled profiles. The failure pattern occurred during first contact between platen and specimen, however the crack formation formed on specimen influence the load fluctuation curve and thin walled aluminum tube formed with diamond mode of failure for round tubes. The square tube aluminum formed lobe of 2 stage with total length of 100 mm.

3.3. Specific Energy Absorption

Energy absorption is defined the total area under of load versus displacement curve. Figure 5 showed the different value of square and round tube profile. An increase of energy absorption observed part of foam-filled profiles. The specific energy absorption have been calculated total energy during absorbed quasi-statically divided by weight of each specimen profile. Results showed that the specimen absorbed more energy compared to square profile for value of energy absorption and specific absorbed energy, respectively. This is cause by the easily be fail and deformed by square tube during load compressed. The fourth edge corner of square tube formed crumpled until final deformed. For the square tube deformed with at least two lube and hinge. In addition, the diamond mode failure observed for each specimen of round tube profiles. The specific absorbed energy is a measure to calculate and making comparison to the other tubular profile to evaluate their performance. The capability of the structures and the comparison between each specimen analysis are using equation 1, 2 and 3. The polystyrene foam showed better performance compared to polyurethane foam 15.19% comparison and enhancement for energy absorption. The SEA improve from empty 27.42% to polystyrene and 13.13% improvement compared to polyurethane.

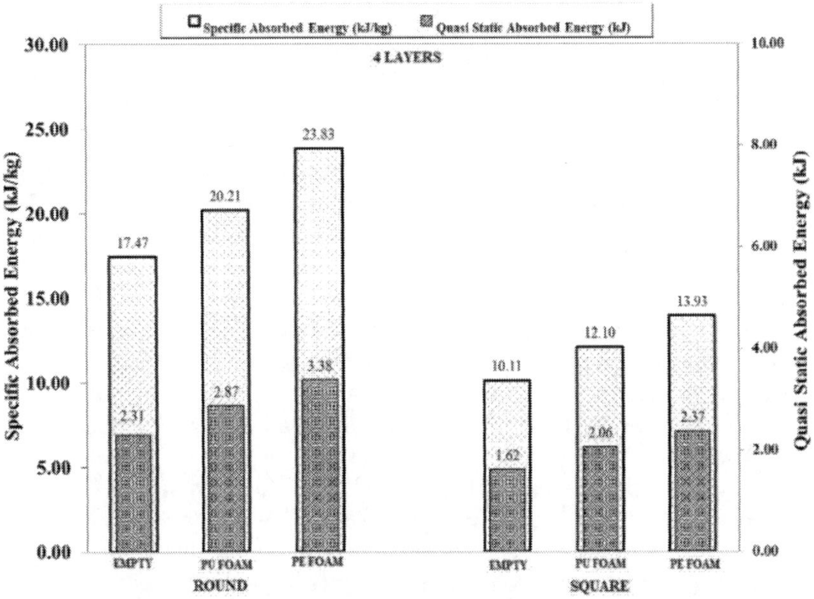

Figure 5. Energy absorption and specific absorbed energy relation for jute/polyester tubes for each layers.

4. Conclusion

The present paper were determine the influence of jute/polyester composite wrapped on AA6063T52 thin walled tube of square and round tube profile under quasi-static loading. The hand layup method were used to fabricate for each specimen profile with fix four layered laminated of tube. Polyester resin and catalyst were used to composite made. The rigid polyurethane and polystyrene were cut off with same dimension cross-section profile as well as inner dimension of both tubes. The rigid foam were provided locally company with estimate of 40 kg/m^3 for polyurethane and 45 kg/m^3 for polystyrene foam. The results of load fluctuation against displacement curve indicate that the 4 layers laminate jute/polyester composite tube bring less improvement compare to foam filler material. Comparison between square and round tube profile in term of quasi-static loading estimate about 15.87 to 26.33% different for initial load contact. The loading average between round and square also estimate between 29.84% to 30%. The polystyrene foam showed better performance compared to polyurethane foam 15.19% comparison and enhancement for energy absorption. The SEA improve from empty 27.42% to polystyrene and 13.13% improvement compared to polyurethane.

References

[1] Sivakumar P, Wim Van P, Joris D, Johan Van A, Dimitrios K and Danny V H 2010 Experimental study on the axial crushing behaviour of pultruded composite tubes *Polym Test* **29** (2) 224e34.
[2] Nagel GM and Thambiratnam DP 2004 A numerical study on the impact response and energy absorption of tapered thin-walled tubes *Int J Mech Sci* **46** (2) 201e16.

[3] Ahmad Z, Thambiratnam DP and Tan ACC 2000 Dynamic energy absorption characteristics of foam-filled conical tubes under oblique impact loading *Int J Impact Eng* **37** (5) 475e88.
[4] Jimenez MA, Miravete A, Larrode E and Revuelta D 2000 Effect of trigger geometry on energy absorption in composite profiles *Compos Struct* **48** (1e3) 107e11.
[5] Khalid SNA, Ismail AE and Zainulabidin MH 2010 Optimization on Impact Strength of Woven Kenaf Reinforced Polyester Composites using Taguchi Method *IOP Conference Series: Materials Science and Engineering* **165** (1)
[6] AMT Arifin, M Fahrul Hassan, AE Ismail, M Zulafif Rahim, M Rasidi Ibrahim, RH Abdul Haq, MNA Rahman, MZ Yunos and MHM Amin Investigation on Suitability of Natural Fibre as Replacement Material for Table Tennis Blade *IOP Conference Series: Materials Science and Engineering* **226** (1)
[7] Anindya Deb, Sumitesh Das, Ashok Mache and Rokesh Laishram 2017 A Study on the Mechanical Behaviors of Jute-polyester Composites *Procedia Engineering* **173** pp 631-638
[8] Ismail AE, Noranai Z, Nor NHM, Tobi ALM and Ahmad MH Effect of Hybridized Fiber Wrapped Around the Aluminum Tubes on the Crushing Performances *IOP Conference Series: Materials Science and Engineering* **160** (1)
[9] Ismail AE Crushing mechanisms of cylindrical winding kenaf fiber reinforced composites *Jurnal Teknologi (Sciences & Engineering)* **78** (6) pp 45-50
[10] Othman A, Abdullah S, Ariffin AK, and Mohamed NAN 2014 Investigating the quasi-static axial crushing behaviour of polymeric foam-filled composite pultrusion square tubes *Mater Des* **63** 446e59.
[11] Othman A, Abdullah S, Ariffin AK and Mohamed NAN 2013 On the crushing behavior of foam-filled composite tubes under compressive loading *Adv Mater Res* **626** 1038e41.
[12] Othman A and Azrol J 2014 Axial crushes simulation of double thin-walled metallic structure inserted with polymeric foam *Int J Eng Technol* **6** (1) 43e6.
[13] Azrol J and Othman A 2014 Crush simulation of cross-section conical aluminum alloy 2024-T4 polyurethane foam-filled section *Int J Eng Technol* **6** (1) 20e4.
[14] Arifin AA and Othman A 2014 Crush force efficiency of composite square profile under off axis loading *Key Eng Mat* **594**
[15] Othman A, Abdullah S, Ariffin AK and Mohamed NAN 2014 Investigating the crushing behavior of quasi-static oblique loading on polymeric foam filled pultruded composite square tubes *Composites Part B: Engineering* **95** pp 493-514
[16] Azrol J and Othman A 2014 A Finite Element Modelling of Polymeric Foam-Filled Aluminium 2024-T4 Alloy Tube under Dynamic Axial Loading, *DOI: 10.4028/www.scientific.net/AMM.315.45*
[17] Othman A and Azrol J 2014 Off-Axis Crushes Simulation of Thin-walled Tapered Tubes Inserted with Foam-filled Structures *DOI: 10.1016/j.apcbee.2014.01.069*
[18] Azrol J and Othman A 2014 Axial Crushes Simulation of Thin-walled Square Aluminum Alloy 6061-T5 Foam-filled Section *DOI: 10.1016/j.apcbee.2014.01.068*

IOP Conf. Series: Materials Science and Engineering **342** (2018) 012065 doi:10.1088/1757-899X/342/1/012065

Finite element modelling of AA6063T52 thin-walled tubes under quasi-static axial loading

A Othman[1, 2] and AE Ismail[1]

[1]Department of Engineering Mechanics, Faculty of Mechanical and Manufacturing Engineering, Universiti Tun Hussien Onn Malaysia, Batu Pahat, Johor, Malaysia
[2]Department of Polytechnic Education, Ministry of Higher Education, Putrajaya, Malaysia

E-mail: akbar.othman@gmail.com

Abstract. The behavior of aluminum alloy 6063T52 thin walled tubes have been present in this paper to determine absorbed energy under quasi-static axial loading. The correlation and comparison have been implemented for each experimental and finite element analysis results, respectively. Wall-thickness of 1.6 and 1.9 mm were selected and all specimen tested under room temperature standard. The length of each specimen were fixed at 125 mm as well as diameter as well as a width and diameter of the tube at 50.8 mm. The two types of tubular cross-section were examined whereas a round and square thin-walled profiles. The specific absorbed energy (SEA) and crush force efficiency (CFE) were analyzed for each specimen and model to see the behavior induced to failure under progressive collapse. Result showed that a correlation less than 5% different between both of comparison experimental and finite element model. It has been found that the thin walled round tube absorbed more energy rather than square profile in term of specific energy with both of either 1.6 or 1.9 of 23.93% and 35.36%, respectively. Overall for crush force efficiency (CFE) of each tube profile around 0.42 to 0.58 value. Indicated that the all specimen profile fail under progressive damage. The calibration between deformed model and experimental specimen were examined and discussed. It was found that the similarity failure mechanism observed for each thin walled profiles.

1. Introduction

Thin walled structure such as tubular form have been used in many engineering application for easy of human being daily life. For crashworthy structure, thin walled tube have been applied as crush box element to protect the compartment occupant during collisions. Metallic material such as steel, mild-steel or aluminum predominantly used to design in structural element such as automotive compartment in order to absorb strain energy during collisions. The peak force would occurred when initial contact between platen and front-end specimen of structure cause by un-deformed of the body and suddenly sudden jerk observed [1-2]. This situation can avoid for applying triggering mechanism to prevent huge peak force induce to catastrophic failure mode [3]. The catastrophic failure is often occurred when the design configuration of tube designed cannot withstand the extreme force and the severity of its deformation [4]. The finite element model of thin walled structure have been more explore by previously by investigators, researchers and engineers [1-18]. This approach were often determine by axial force to develop folding process of the thin walled tubes as well as their properties of material.

The finite element software primary packages in the 21 century provide more option in order to implement develop model and coding to ensure that the numerical model is sufficiently accuracies compared to experimental result or existing analytical and numerical model previously done by investigators. The main objective and propose in this paper is to determine the load-displacement curve and the energy absorption of thin walled tube of AA6063T52 under axial quasi-static loading. Furthermore, the study aims to correlate and calibrate the result done by experimental and numerical model to observe deformation pattern and failure mechanism.

2. Experimental work and modelling technique

2.1 Materials modelling

Thin walled tube of aluminum alloy AA6063T52 were found from Kamco Aluminium Sdn Bhd, Malaysia with the varied cross-section and cut-off for each 120 mm length. The wall thickness of thin walled tube were 1.6 and 1.9 mm with 2 type of cross-section, which is square, and round profile with diameter and width of 50.8 mm. Each of the thin walled tube were not created triggering mechanism in order to find original or actual deformation behavior of tube deformed from undeformed until total displacement onto compaction zone. The finite element software of ls-dyna package was used in this study to model and analyzed results from numerical modeling. The Belytschko–Tsay quadrilateral shell elements was used to model aluminum in the single tube. The element size was 2.0 x 2.0 mm. The following contact algorithms [5] were used between the aluminum shell elements and the rigid top platen solid elements to simulate interaction between them and to prevent interpenetration, contact "automatic surface to surface" to model the interaction between the aluminum tube and rigid top solid element. Contact "automatic single surface" is an option to prevent interpenetration among successive folds as well as to simulate interaction between the aluminum walls and the rigid solid element. The coefficient of friction was assumed equal to 0.3. The loading velocity was assume to be 5.0 mm/minutes. The bottom end of the tube was considered fixed. To carry out the quasi-static analysis condition, some criteria should be implement. In order to reduce the time steps, the mass density of aluminum was scaled up 1000 times the original density. Since the ls-dyna is an explicit finite element code, scaling up the density and the relatively low loading velocity of 5.0 mm/minutes were used to ensure a quasi-static simulation. The interface contact between the loading plate and the tube was plotted as a function of displacement, from which the folding initiation force and energy were obtained. The mean crush force was calculated by dividing the energy with the end displacement.

Table 1. Dimensions of the aluminum.

Specimen/model type				
No. Series	Mass (kg) – (1.6, 1.9 mm)	Profiles	Wall Thickness Jute/polyester + AA6063 t (mm)	Length
ALR	0.0667, 0.07882	Round	1.6, 1.9	125
ALS	0.085, 0.100	Square	1.6, 1.9	125

The MAT24 of ls-dyna was used to model aluminum alloy 6063T52 thin walled tube was used, which is based on piecewise linear isotropic plasticity and uses von Mises flow rule. Fig. 1 shows the stress–strain responses of aluminum under static tensile test. The material has Young's modulus E=68 GPa; Poisson's ratio ν=0.3; yielding stress σ_0=0.117 MPa; plastic modulus E_T=280 MPa; and density $\rho=2.7\times10^3$ kg/m^3. The tension test conducted in accordance with the ASTM: E8M.

Figure 1. Sample of aluminum tube.

2.2. Crashworthiness perimeter

The determination of crashworthiness mechanism some of following aspect of crush phenomenon is measured during crushes event. The all data were recorded and calculated there are Peak force P_{max}: maximum compressive force. Mean force P_{avg}: average compressive force obtained by the following equation where force and deformation are defined as P and d, respectively and the area of cross section, A, and the density of the material, ρ defined as

$$\bar{P} = \frac{1}{\delta}\int_0^{\delta} P d\delta \quad (1)$$

Energy absorption E: area under the load–displacement curve up to the compaction zone.

$$E = \int_0^{\delta} P d\delta \quad (2)$$

The units of E_s are used to express the crashworthiness parameters, hence they are written in kJ here P is the load acting on the composite specimens. Therefore, the specific energy absorption per unit mass kJ/kg (SEA) where m the crushed mass of the composite is recognized as:

$$SEA = \frac{E}{m} \quad (3)$$

Crash force efficiency (CFE): ratio of the average crushing load P_{avg} to peak load P_{max}.

$$CFE = \frac{P_{avg}}{P_{max}} \quad (4)$$

3. Results and discussion

3.1. Axial loading characteristics

The result in Figure 2 show the load against displacement curve subjected to axial loading under quasi-static conditions. The response observed that the characteristic of fluctuation force roughly similar as displacement increased. The experimental test conducted on specimen of 1.6 and 1.9 mm with similar length found similar collapse modes. The collapse pattern of symmetric concertina mode observed for both wall-thickness thin-walled round tube seen in Figure 3 (b). The increasing thickness wall of tube increase absorbed energy when applying quasi-static loading also can be seen in Figure 2. Similar result also found on Othman [13, 16], stated that the when wall-thickness increase, absorbed energy and specific absorbed energy also increase and enhanced the capability crashworthiness performance.

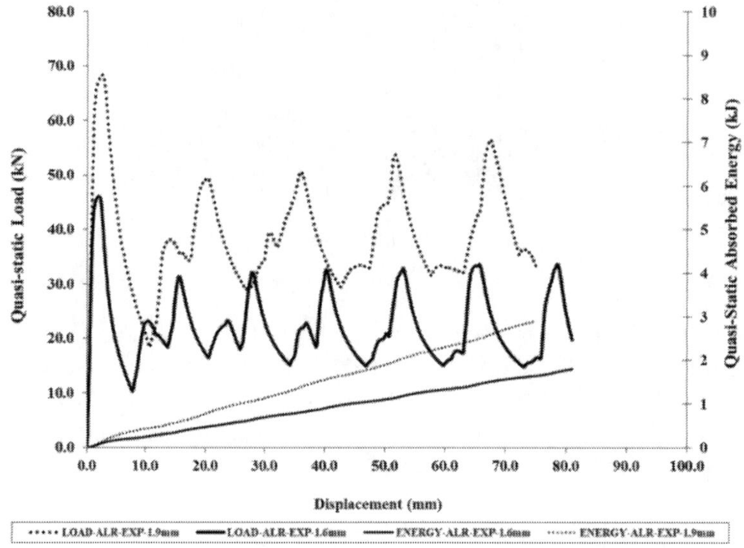

Figure 2. Load-displacement relation for thin-walled AA6063T52 tubes.

The correlation have been studied within the result done by experimental and finite element analysis shown in Figure 3. The result showed that the good agreement between both of result. Similar result also found and reported by [16], showed that similar result observed analysis under finite element simulation. The mode pattern is similar either 1.6 or 1.9 mm wall-thickness as well as wrinkle for each tube for experimental and finite element results. Each results also observed similarity folded pattern for both results, which is 5-fold pattern of 1.6mm, and 5/6 fold pattern on 1.9 mm shown in Figure 3 (b).

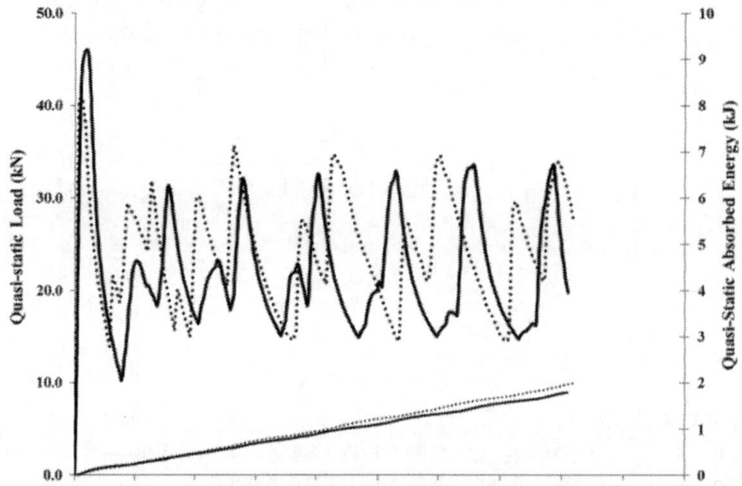

Aluminum alloy 6063T52 (wall-thickness – 1.6 mm)

IOP Conf. Series: Materials Science and Engineering **342** (2018) 012065 doi:10.1088/1757-899X/342/1/012065

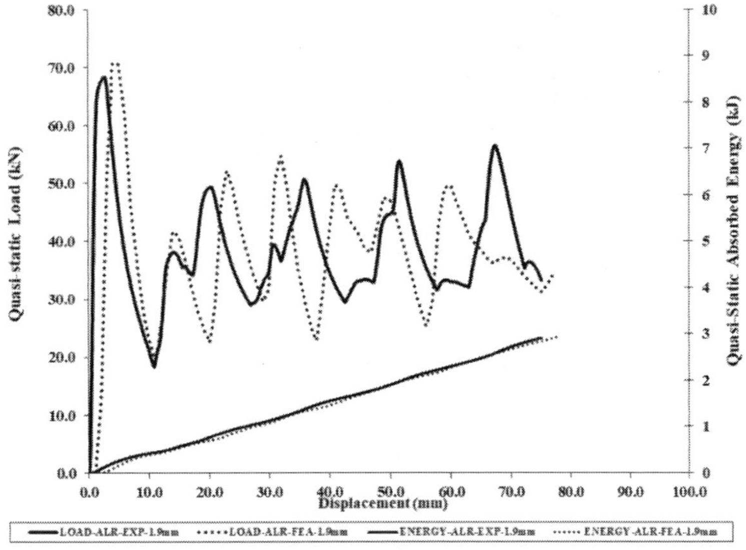

Aluminum alloy 6063T52 (wall-thickness – 1.9 mm)

(a) Load against displacement curve of Aluminum alloy 6063T52 round tubes.

Unloading 20 mm 40 mm 60 mm 80 mm

Aluminum alloy 6063T52 (wall-thickness – 1.6 mm) – Experimental.

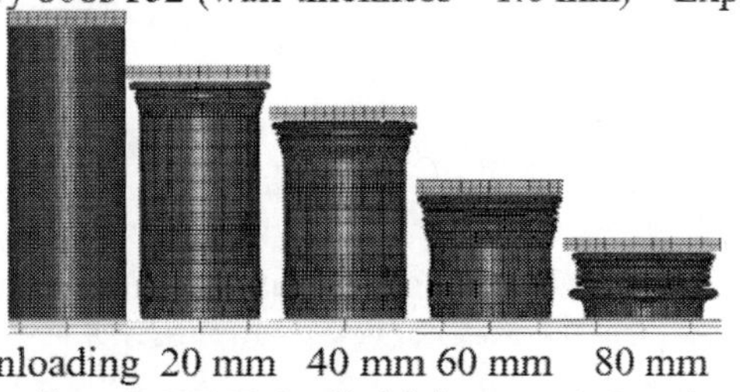

Unloading 20 mm 40 mm 60 mm 80 mm

Aluminum alloy 6063T52 (wall-thickness – 1.6 mm) – FEA.

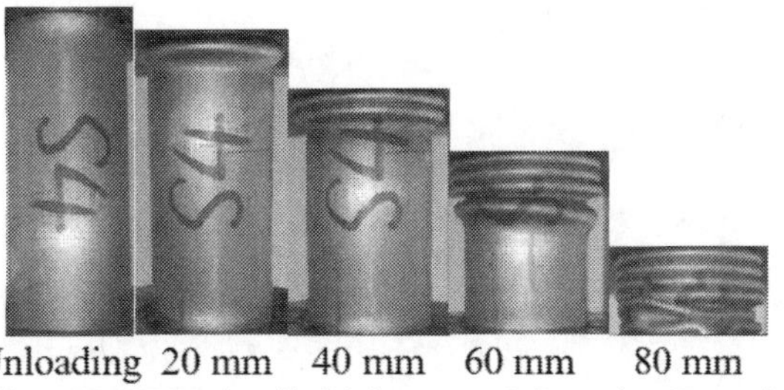

Unloading 20 mm 40 mm 60 mm 80 mm

Aluminum alloy 6063T52 (wall-thickness – 1.9 mm) – Experimental.

Unloading 20 mm 40 mm 60 mm 80 mm
Aluminum alloy 6063T52 (wall-thickness – 1.9 mm) – FEA.
(b) Deformation of Aluminum alloy 6063T52 tubes with experimental and FEA.
Figure 3. Crushing characteristic of Aluminum alloy 6063T52 thin walled round tubes.

The fluctuation force fluctuated with different gab between each result shown in Figure 3, this is because influence by noise and vibration during experimental done. The Figure 3 also showed the deformation pattern observe with simulation and experimental technique. The result indicated that the symmetric concertina mode pattern formed obtained either experimental and simulation results.

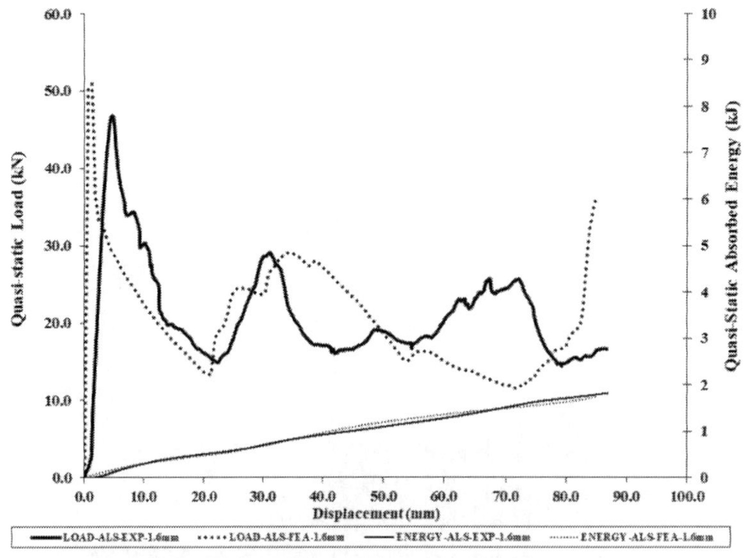

Aluminum alloy 6063T52 (wall-thickness – 1.9 mm)
(a) Load against displacement curve of Aluminum alloy 6063T52 round tubes.

Unloading 20 mm 40 mm 60 mm 80 mm
Aluminum alloy 6063T52 (wall-thickness – 1.9 mm) – Experimental.

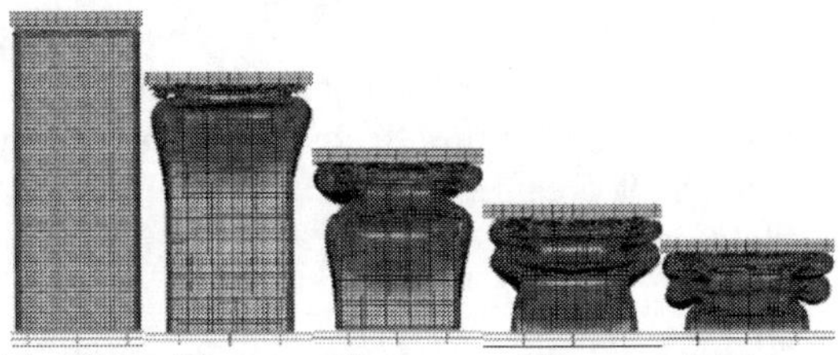

(a) Deformation of Aluminum alloy 6063T52 tubes with experimental and FEA.

Figure 4. Crushing characteristic of Aluminum alloy 6063T52 thin walled square tubes.

The square cross-section profile show efficient crushing mode seen in Figure 4. Definitely, the fluctuation force obtained in this result relatively not the same as a round tube. The initial contact strongly huge at the higher peak maximum one compared to round tube, but the squared formed the wrinkle mode started on the four corner edges. On the other hand, the progressive collapse occurred after the peak force begin and the force fluctuated throughout entirely the wrinkle process. The fluctuation force uniformly recorded during that process and the specimen and numerical model crumpled until compaction zone area. The mode of pattern wrinkled observed showed in Figure (b), and indicate that the similar lobe of failure found from the results.

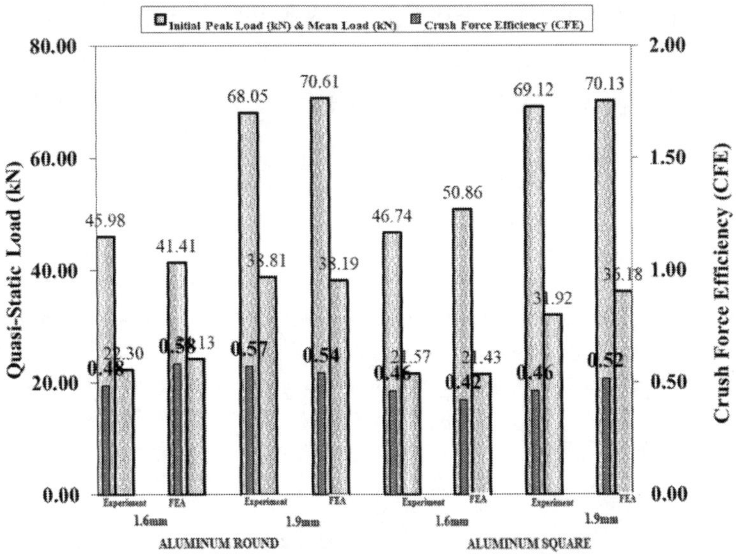

Figure 5. Initial, mean and crush force efficiency relation for round and square of aluminum.

Figure 5 show a systematic comparison obtained by experimental and finite element results. The crush force efficiency obtained by both result around 0.42 to 0.57, it is indicate that the almost of specimen and model fail under progressive folding. For both round and square cross-section obtained increasing load when wall-thickness increase.

3.2. Specific Energy Absorption

To specify the energy absorbed for each profile, the sum of area under load against displacement curve was calculated. The result found that the round cross-section tube absorbed more energy compare to square tube profile; this is because the restriction for round tube to wrinkle on failure deformed than square tube profile. Result also indicate that the specific energy absorption (SEA) for 1.9 mm wall-thickness, improvement around of 25.66 - 35.36% different between round and square thin walled profile and around a 23.93 - 30.43% for 1.6 mm. The result also supported from previous study found in literature [16-18] indicated that a round tube profile absorbed more energy compared to square tube profile. This is an evident for the result obtained and justification on how for a selection of material structure on better design in developing vehicle structure.

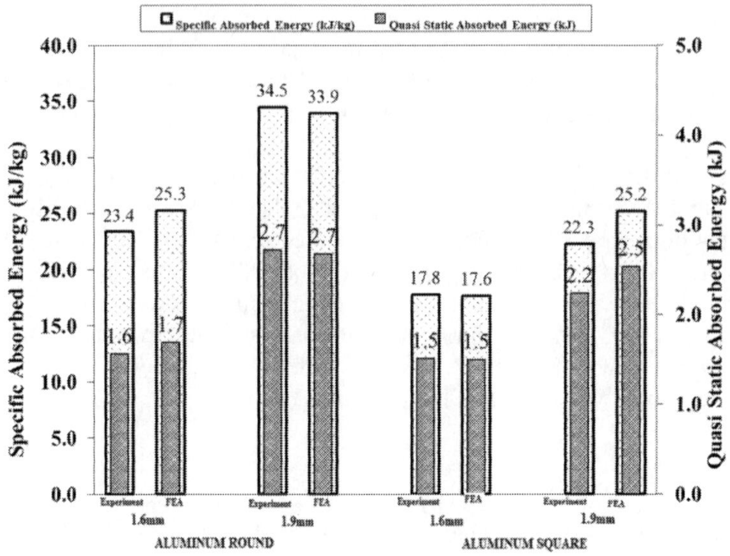

Figure 6. Energy absorption and specific absorbed energy relation for round and square of aluminum.

4. Conclusion

The crushing behavior on thin walled profile of round and square tubes have been understood for empty case. The concertina mode failure obtained on each tube of round given most predominantly absorbed more energy compared to square profile fail under wrinkle and crumpling on four round corners. The buckling mode of wrinkle found by round tube profile created a fluctuate force throughout crushing proses under quasi-static condition. The studies found that the good agreement obtained by of both result, which is a 5.0 to 10.0% different between result found by experimental and finite element analysis. The correlation have been discussed both of result tailored by existing analysis worked by previous researcher in literature.

References

[1] Palanivelu Sivakumar, Paepegem Wim Van, Degrieck Joris, Ackeren Johan Van, Kakogiannis Dimitrios, Hemelrijck Danny Van, et al 2010 Experimental study on the axial crushing behaviour of pultruded composite tubes *Polym Test* **29** (2) 224e34.

[2] Nagel GM and Thambiratnam DP 2004 A numerical study on the impact response and energy absorption of tapered thin-walled tubes *Int J Mech Sci* **46** (2) 201e16.

[3] Ahmad Z, Thambiratnam DP and Tan ACC 2010 Dynamic energy absorption characteristics of Foam filled conical tubes under oblique impact loading *Int J Impact Eng* **37** (5) 475e88.

[4] Jimenez MA, Miravete A, Larrode E and Revuelta D 2000 Effect of trigger geometry on energy absorption in composite profiles *Compos Struct* **48** (1e3) 107e11.

[5] SNA Khalid, AE Ismail and MH Zainulabidin 2010 Optimization on Impact Strength of Woven Kenaf Reinforced Polyester Composites using Taguchi Method *IOP Conference Series: Materials Science and Engineering* **165** (1) 012010

[6] AMT Arifin, M Fahrul Hassan, AE Ismail, M Zulafif Rahim, M Rasidi Ibrahim, RH Abdul Haq,

MNA Rahman, MZ Yunos and MHM Amin Investigation on Suitability of Natural Fibre as Replacement Material for Table Tennis Blade *IOP Conference Series: Materials Science and Engineering* **226** (1) 012037
[7] Anindya Deb, Sumitesh Das, Ashok Mache and Rokesh Laishram 2017 A Study on the Mechanical Behaviors of Jute-polyester Composites *Procedia Engineering* **173** pp 631-638
[8] AE Ismail, Z Noranai, NHM Nor, ALM Tobi and MH Ahmad Effect of Hybridized Fiber Wrapped Around the Aluminum Tubes on the Crushing Performances *IOP Conference Series: Materials Science and Engineering* **160** (1) 012019
[9] AE Ismail Crushing mechanisms of cylindrical winding kenaf fiber reinforced composites, *Jurnal Teknologi (Sciences & Engineering)* **78** (6) pp 45-50
[10] Othman A, Abdullah S, Ariffin AK and Mohamed NAN 2014 Investigating the quasi-static axial crushing behaviour of polymeric foam-filled composite pultrusion square tubes *Mater Des* **63** 446e59.
[11] Othman A, Abdullah S, Ariffin AK and Mohamed NAN 2013 On the crushing behavior of foam-filled composite tubes under compressive loading *Adv Mater Res* **626** 1038e41.
[12] Othman A and Jailani Azrol 2014 Axial crushes simulation of double thin-walled metallic structure inserted with polymeric foam *Int J Eng Technol* **6** (1) 43e6.
[13] Jailani Azrol and Othman A 2014 Crush simulation of cross-section conical aluminum alloy 2024-T4 polyurethane foam-filled section *Int J Eng Technol* **6** (1) 20e4.
[14] Arifin AA and Othman A 2014 Crush force efficiency of composite square profile under off-axis loading *Key Eng Mat* 594e595 711e4.
[15] A. Othman, S Abdullah, AK Ariffin and NAN Mohamed Investigating the crushing behavior of quasi-static oblique loading on polymeric foam filled pultruded composite square tubes *Composites Part B: Engineering* **95** pp 493-514
[16] Jailani Azrol and Othman A Finite Element Modelling of Polymeric Foam-Filled Aluminium 2024-T4 All6oy Tube under Dynamic Axial Loading *DOI: 10.4028/www.scientific.net/AMM.315.45*
[17] Othman A and Jailani Azrol 2014 Off-Axis Crushes Simulation of Thin-walled Tapered Tubes Inserted with Foam-filled Structures *DOI: 10.1016/j.apcbee.2014.01.069*
[18] Azrol Jailani and Othman A 2014 Axial Crushes Simulation of Thin-walled Square Aluminum Alloy 6061-T5 Foam-filled Section *DOI: 10.1016/j.apcbee.2014.01.068*

Numerical investigation on effect of blade shape for stream water wheel performance.

N F Yah, A N Oumer, A A Aziz and I M Sahat

Faculty of Mechanical Engineering, Universiti Malaysia Pahang, 26600 Pekan Pahang.

E-mail: norfadilahyah@gmail.com

Abstract. Stream water wheels are one of the oldest and commonly used types of wheels for the production of energy. Moreover, they are economical, efficient and sustainable. However, few amounts of research works are available in the open literature. This paper aims to develop numerical model for investigation of the effect of blade shape on the performance of stream water wheel. The numerical model was simulated using Computational Fluid Dynamics (CFD) method and the developed model was validated by comparing the simulation results with experimental data obtained from literature. The performance of straight, curved type 1 and curved type 2 was observed and the power generated by each blade design was identified. The inlet velocity was set to 0.3 m/s static pressure outlet. The obtained results indicate that the highest power was generated by the Curved type 2 compared to straight blade and curved type 1. From the CFD result, Curved type 1 was able to generate 0.073 Watt while Curved type 2 generate 0.064 Watt. The result obtained were consistent with the experiment result hence can be used the numerical model as a guide to numerically predict the water wheel performance

.

1. Introduction

Renewable energy is the cleanest energy resources compared to fossil fuel and nuclear energy. Hydro power is one of the renewable energy resources which is expected to play a prominent role in power generation due the decreasing of global fossil fuels sources and supplying about 19% of electricity worldwide. Hydropower energy sources consists of waves, tides, ocean currents, natural flow of water in rivers, or marine thermal gradients [1]. Where it is not feasible to extend the main grid to Malaysia's remote rural communities, the decentralized generation of hydroelectricity can contribute significantly to improving the economic conditions of the rural populations of Malaysia. Hence, small scale hydro power plants may provide cost-effective energy alternatives to grid extension or isolated diesel mini-grids in rural areas of Malaysia. With hilly topography running almost the entire length and width of the country, and abundant number of streams flowing to foothills, Malaysia has a lot of small-hydro potential [2]. Basically, Small scale hydro power plant mainly equipped with reservoir or irrigation canal, governor, turbine and generator. Turbine is the heart of the system since it becomes the part which hydraulic energy will be converted to the mechanical energy.

Turbine broadly classified into positive displacement and dynamic categories. Basically, positive displacement turbines consist of small devices especially used for volume flow measurement, while dynamic turbine range from tiny to huge and used for flow measurement and power production. Dynamic turbines consist of impulse turbines and reaction turbines. Impulse turbines produced power by the change in momentum due to the change of velocity direction of fluid flow. Impulsive force generated from the change of momentum will rotate the turbine runner. Therefore, the turbine are not necessarily to be immersed in the fluid since it is not driven by the pressure change in the fluid surrounding the runner. Reaction turbines are driven by the pressure change of fluid surrounding the runner that will cause it to rotate. Since the power generation depends on the pressure change of fluid, so the turbine need to be immersed in the fluid to make it functioning. Turbine selection process will consider the available head and flow on the site. For this study, the water wheel will be considered since the application is for the ultra-low head.

Water wheel are preferable energy converter for the small hydropower generation especially pico hydropower due to its ability to extract energy from low head water resources. The water wheel are believed to be originated in someplace at Mediterranean around 50 to 100 B.C [3]. It is classified as one of the oldest water turbine that has been used since antiquity until the introduction of high-pressure steam engines at the end of 18th century. Although this technology are invented since pre steam era, but the researcher and engineer nowadays are interested to study this hydraulic energy converter. This simple ancient machine that can be made of wood or steel with equally spaced blades or paddles around its circumference. It uses the falling or flowing water which flow and strikes surfaces of the blades. The impulsive force of water will rotates the wheel and generated torque will be transmitted to machinery via the shaft of the wheel [4]. Water wheels are widely used not only for mechanical power sources in flour and mineral mills, textile and tool making devices, but also used to generate electricity. Main advantages of water wheel compared to steam engines and turbines were their comparatively low cost and high efficiencies for a wide range of flow rates respectively [5].

There are four main type of water wheel which are undershot, breastshot, overshot and stream water wheel. They are characterized based on different level of water flow through the wheel and head required to operate. The first and believed as an oldest type is undershot water wheel. Some undershot wheel such as Zuppinger wheel employs only the potential energy of the water as the driving force. The idea of utilizing potential energy instead of kinetic energy of the low head water in rivers was noticed by a French engineer, Poncelet [5]. Undershot water wheel could operate in very small head differences of 0.5m to 2.5m and large flow volume ranging from 0.5 to 0.95 m^3/s per m width. The water enters the wheel's blade below its axis.This water wheel utilized wherever a swiftly flowing river is available. In 19th century, the efficiency is around 25 % and was further developed until the efficiency can reached 70 % by Poncelet [1]. Figure 1 (a) below shows the undershot water wheel.

Breastshot water wheels were developed for head differences of 1.5 to 4m. Water enters the wheel at the level approximately to the level of wheel axis. The principal driving force of this water wheel is the weight of the water enclosed in the blade cell [2]. Figure 1 (b) below shows the breasthot water wheel. Overshot water wheel also known as the backshot water wheel was constructed to utilize head differences of 2.5m to 10m and flow rates of 0.1 to 0.2 m^3/s per m width. As can be seen from Figure 1(c), water enters the wheel bucket from the top and the different weight of the water on other side causing the rotation of the vertically mounted wheel. Therefore, the buckets are designed in a way so that the water from above can enter each bucket at its natural angle of fall. The wheel design have utilized the force from falling water and weight of water collected in the bucket in order to rotate the wheel. But, the limitation is the head of water must be greater than the wheel diameter so that the water can be channelled from the top. This type of water wheel is fit for mountainous and hilly countries because its mechanical power is based on the wheel physical size and the available head [3]. Therefore, it requires the construction of dam, millpond and waterways. Stream wheels only utilizes kinetic energy of flowing water into mechanical energy. They are cost effective since only small construction works required to install the water wheel. Figure 1 (d) shows the stream water wheels. This paper aimed to study the effect of blade design and immersed depth ratio on the performance of the stream water wheel.

IOP Conf. Series: Materials Science and Engineering **342** (2018) 012066 doi:10.1088/1757-899X/342/1/012066

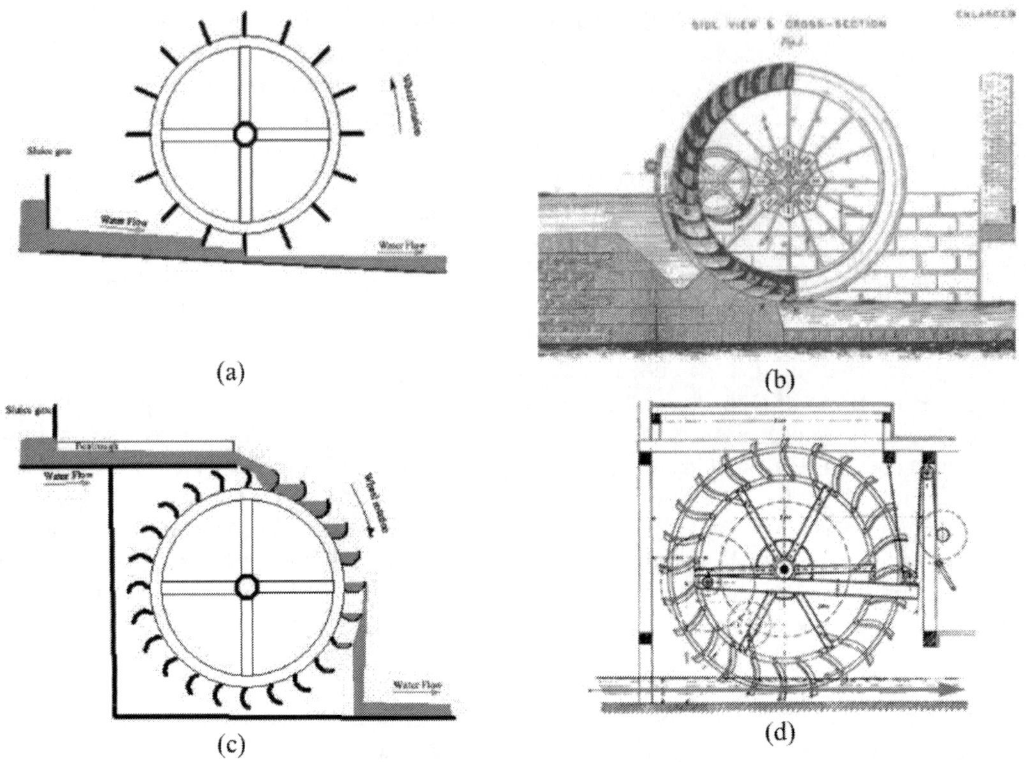

(a) (b) (c) (d)

Figure 1. (a) Undershot water wheel (b) Breasthot water wheel (c) Overshot water wheel (d) Stream water wheel.

This paper aimed to study the effect of paddle number and immersed radius ratio on the stream wheel performance. The maximum powers of each paddle number and each immersed radius ratio were compared.

2. Physical Domain

The blade shape included straight, curved type 1 and curved type 2 with six number of blades. The schematic diagram of each blade design are presented in Figure 1 below.

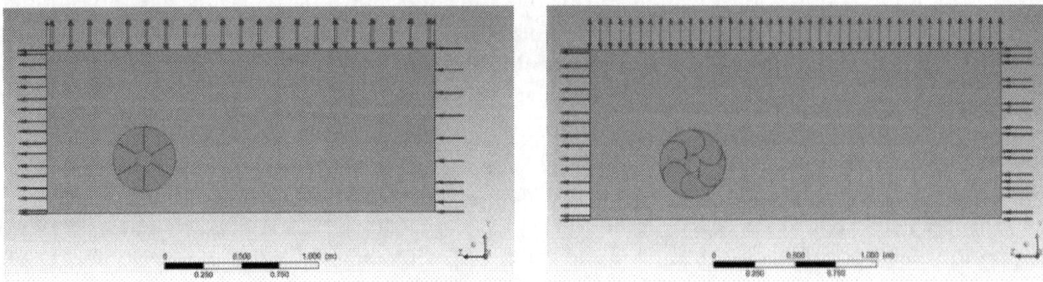

(a) (b)

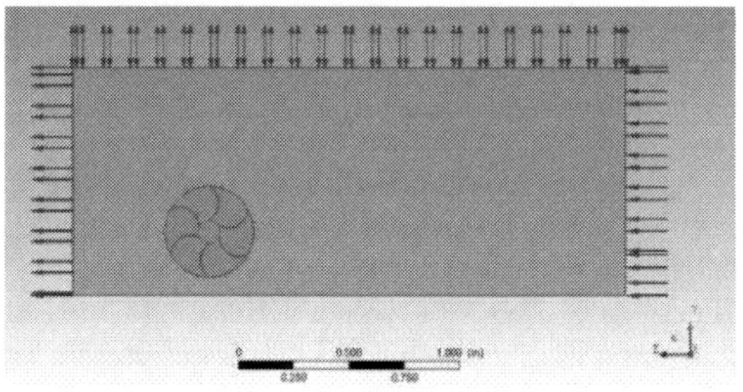

(c)

Figure 2. Schematic diagram of the (a) Straight blade (b) Curved type 1 (c) Curved type 2.

The turbine is immersed in the flowing fluid (river) which is assumed to have constant flow velocity. The river flow field is considered to be incompressible, non-isothermal and behaves as Newtonian fluid (neglecting the effects of slurries inside the river). Therefore, the general forms of the continuity and momentum equations can be written as:

$$\frac{\partial \rho}{\partial t}+\frac{\partial}{\partial x_j}\left(\rho U_j\right)=0 \tag{1}$$

$$\frac{\partial \rho U_i}{\partial t}+\frac{\partial}{\partial x_j}\left(\rho U_i U_j\right)=-\frac{\partial p'}{\partial x_i}+\frac{\partial}{\partial x_j}\left[\mu_{eff}\left(\frac{\partial U_i}{\partial x_j}+\frac{\partial U_j}{\partial x_i}\right)\right]+S_M \tag{2}$$

Where ρ is the density of water, U_i and U_j is the velocity vectors in direction of i and j, p' is the modified pressure, μ_{eff} is the effective viscosity accounting for turbulence. The term S_M represents the sum of body forces. The k-ε model, like the zero equation model, is based on the eddy viscosity concept, so that:

$$\mu_{eff}=\mu+\mu_t \tag{3}$$

Where is the turbulence viscosity. The k-ε model assumes that the turbulence viscosity is linked to the turbulence kinetic energy and dissipation via the relation:

$$\mu_t=C_\mu \rho \frac{k^2}{\varepsilon} \tag{4}$$

Where μ_t is a constant.

Torque and power is the essential parameter which need to be identified in order to evaluate the performance of the water wheel. From the experiment and numerical analysis, torque will be identified based on the different approach. During experiment, torque will be identified through the value of rotational speed, n. While, in numerical analysis the value of torque will be calculated by:

$$T=F_z \times r \tag{5}$$

$$P=T\times\omega \tag{6}$$

$$\omega=2\pi n/60 \tag{7}$$

Where T is the Torque in x axis direction of rotation in Nm, F_z is the force acted on the blade in z direction in N, r is the radius of the water wheel in m, P is the power generated by the water wheel in Watt and ω is rotational speed in rev/s. The analysis of change in performance at different configuration is performed by calculation the relative percentage in the interested variables. Percentage change of the variable on interest is calculated by Eq 8.

$$\text{Percentage change} = \frac{\text{Modified case - Reference case}}{\text{Reference case}} \times 100\,\% \tag{8}$$

In this study, a three dimensional with transient flow analysis was established by using Ansys CFX 15.0. In order to simulate the flow of the open channel flow, the uniform model of an Euler-Euler simulation was selected. In Ansys CFX, multiphase flow model can be applied with combination of Volume Fraction method. The basic equation used are based on mass, momentum and volume conservation. The standard k-epsilon model was adopted as the turbulent flow model, and the standard wall function was used to handle regions near wall surfaces. k is the turbulence kinetic energy and is defined as the variance of the fluctuations in velocity [10].

2.1 Geometry

The geometry of the water wheel was modelled in CAD software (Solidworks) and hence converted into readable file in Ansys such as Parasolid and STEP files. The dimension of the study from Anurat et al was used hence the water wheel is 0.2 m radius and each paddle is 0.1 m (W) x 0.15 m (H). The model was tested in 0.3 m (W) x 0.35 m (H) x 2.5 m (L) water channel with 0.3 m/s of water velocity [9].

3. Boundary Conditions and Meshing

The mesh generation was done by using Ansys-ICEM CFD. This software is not only effectively able to mesh large and complex models but also can provide advanced size function of the grid. Basically, computational mesh are classified into structured mesh and unstructured mesh [11]. Structured meshes are not preferred for complex geometries but it is more efficient in terms of accuracy, CPU time and in most cases preferable. Unstructured mesh are preferable for complex geometries but the size difference between two adjacent elements can be large thus affect its numerical accuracy.

In this study, the simulation domain was divided into two parts; rotating and stationary parts. The stationary domain represents the domain for control surfaces of upstream and downstream region of the water channel. The rotating (turbine blade) domain meshed with advanced size function "proximity and curvature" and consist mainly of tetrahedral elements. The stationary domain which represent the upstream and downstream regions were meshed with advanced size function turned off and consist mostly hexahedral elements with some tetrahedral, prismatic and wedges element type. *Figure 3* represent the mesh for both stationary and rotating domain.

Analysis type was selected as transient analysis. There are two fluid domains were created, which is representing the rotating and stationary domains. The region of the turbine and water was separated into two different domain and mesh types to enable the rotating frame of reference used in the CFX [12][8]. The pressure around the turbine runner is the atmospheric pressure which is equal to 1 atm. The k-epsilon model was adopted since it has proven to be stable and numerically robust and has a well-established regime of predictive capability. This model are suitable for flows with boundary layer separation, flows with sudden changes in the mean strain rate, flows in rotating fluids, and flows over curved surfaces [13].

The inlet of water flow was set at the level depth of 0.18 m with the velocity of 0.3 m/s. In order to separate the water and air interface, volume fraction of the water was set to 1 and 0 for air. Water wheel was immersed in the water region with two different depth ratio which is 0.5. The complete setup was represented in the **Figure *4***.

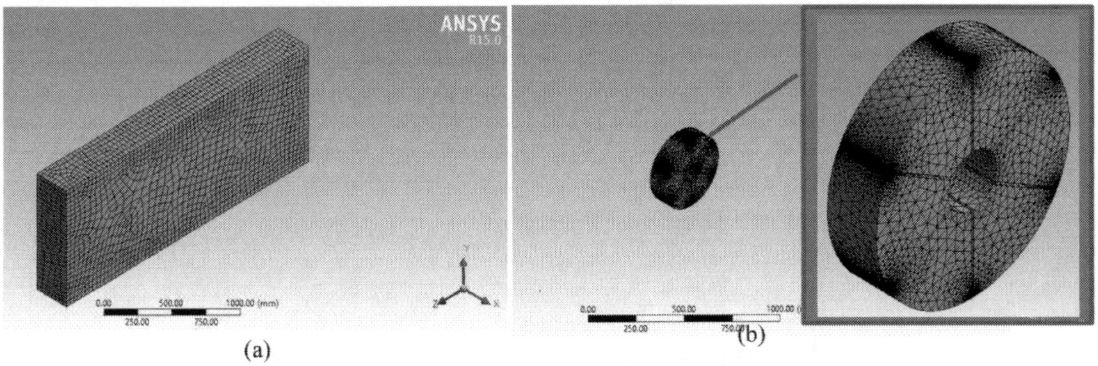

Figure 3. Meshing for (a) stationary and (b) rotating domain.

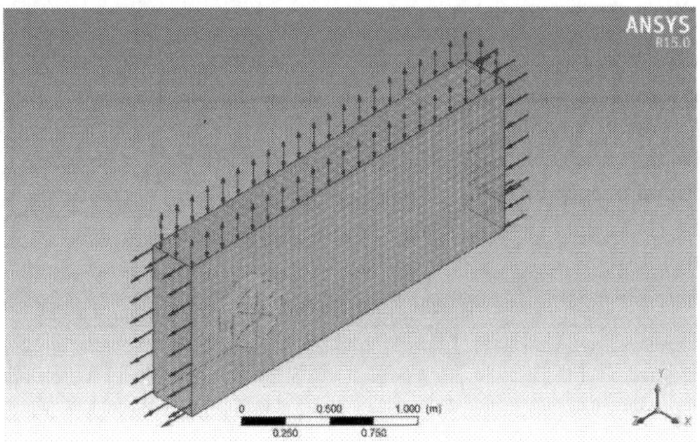

Figure 4. Complete setup for stationary and rotating domain.

Figure 5 below was plotted to represent the mesh independency test based on the mesh size of the both rotating and stationary domain. The force acted on the blade was numerically calculated based on the changes of the mesh sizes. It is observed that the mesh size at 0.05 and 0.06 have no changes on the force acted on the blade and the value of force is closer to experimental value. Therefore, the mesh size of 0.06 was predicted as a stable mesh size for the numerical study.

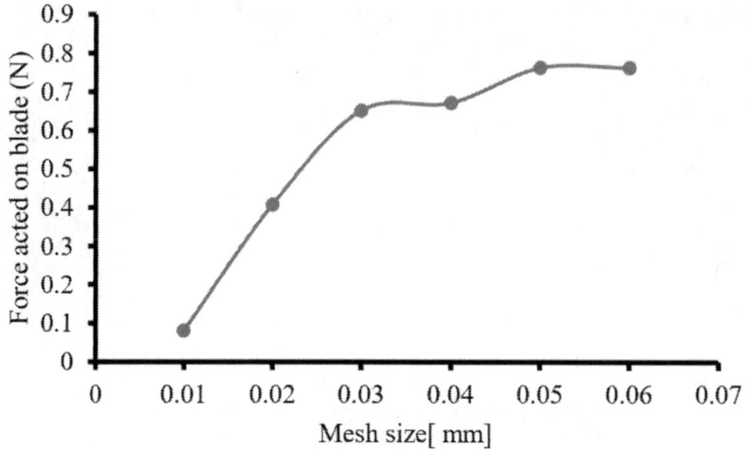

Figure 5. Mesh independency test for rotating and stationary domain.

The model was validated through comparison with the experimental result. Hence, the numerical study on the water wheel with different blade shape was conducted by applying the same numerical model. From the study, the torque and power generated from the flow was numerically calculated. Both of the model was tested with 0.5 immersed depth as this ratio was experimentally and numerically proven to be the optimum immersed depth ratio.

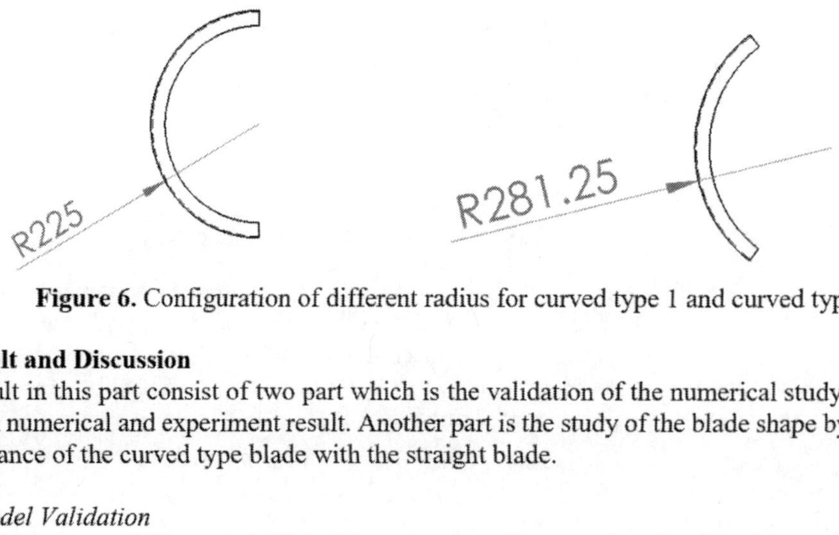

Figure 6. Configuration of different radius for curved type 1 and curved type 2.

4. Result and Discussion

The result in this part consist of two part which is the validation of the numerical study by comparison between numerical and experiment result. Another part is the study of the blade shape by comparing the performance of the curved type blade with the straight blade.

4.1. Model Validation

Validation process of the numerical model was conducted in order to validate the accuracy of the model used. Therefore, a numerical model was constructed based on the study proposed by the Nishi et al [14]. In this simulation, rotational speed was varied as experiment and force acted on the blade was numerically calculated. Immersed depth ratio of the water wheel was set to 0.5 and 0.75 and performance was observed [9]. The force acted in z-directions was plotted in the figures below. As can be seen in the Figure 7 (a), force generated on the blade surface at the immersed depth ratio of 0.5 was decreasing with the increasing of the rotational speed. While Figure 10 (b) plots the force generated on the blade for 0.75 immersed depth ratio. The force was responsible to generate the torque and power.

In experiment, the power generated from the water wheel at the flow velocity of 0.3 m/s was increased up to a maximum and then it decreases. The highest value of power generated at the 2.797

rpm and torque load of 0.14 Nm. After this value, which is equivalent to 0.041 Watt the power generated was decreased slightly although with increasing of rotational speed.

In CFD study, the inlet flow velocity was set to 0.3 m/s and the rotational speed of the water wheel was changed as same as in the rotational speed varied in the experiment. During 0.5 immersed depth ratio, power generated from the water wheel initially increase until a maximum value of 0.045 Watt at the 2.797 rpm and was slightly decreases although with the increasing of rotational speed. The result of generated power for both experiment and CFD was compared and presented in Figure 11. Therefore, the 0.5 immersed depth ratio performance predicted in the CFD study are likely same as the experimental results obtained in the literature and the highest power generated at the same rotational speed.
The normalized power generated from the numerical calculation and experiment for 0.75 immersed depth ratio was plotted in the Figure 8 (b). From the previous experiment result, the value of normalized power initially increase to maximum value and slightly decrease to minimum. Maximum power was generated at the 1.91 rpm and torque load of 0.16 Nm. Maximum power is 0.03 Watt and the generated power start to decrease after this maximum value. In CFD study, power generated also slightly increase to 0.034 Watt at 1.91 rpm and rapidly decrease to minimum value.

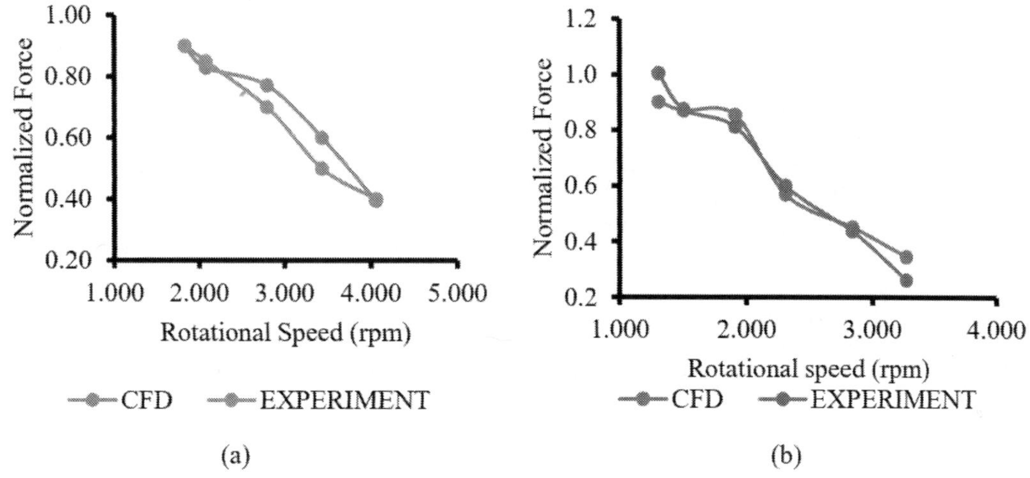

Figure 7. Generated force for (a) 0.5 immersed depth ratio (b) 0.75 immersed depth ratio.

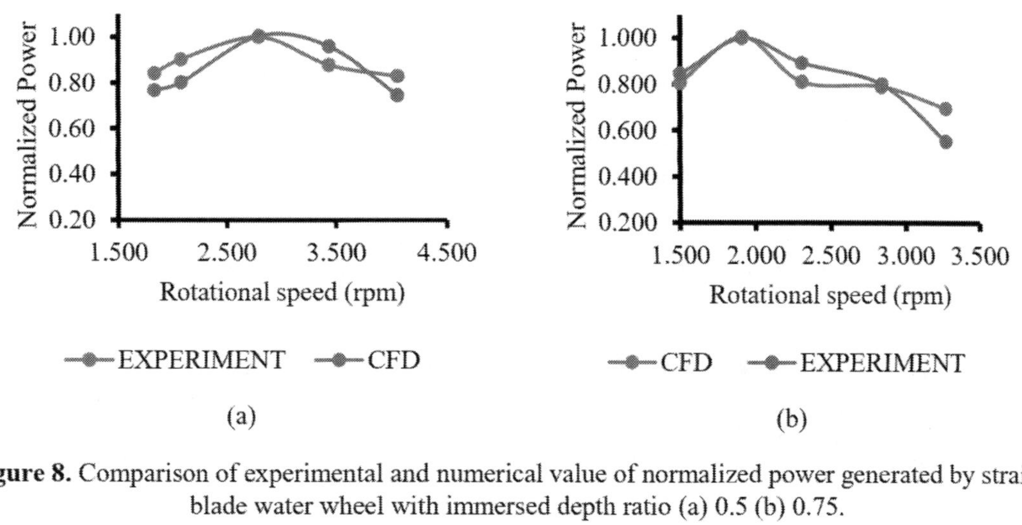

Figure 8. Comparison of experimental and numerical value of normalized power generated by straight blade water wheel with immersed depth ratio (a) 0.5 (b) 0.75.

Hence, the numerical model proposed in this study are acceptable since the performance of the water wheel in the experimental and numerical study are presenting the same trends. The average percentage of the error are around 9.3 %. Therefore, the model are validated since it able to predict the maximum power generated by the water wheel with acceptable percentage of error.

4.2. Effect of Blade Shape

The performance of the water wheel with the straight and curved blade was studied by performing computational fluid dynamics with the same numerical model. The rotational speed was changed and the generated torque and power of the models was observed. Figure represents the changes of power generated by the water wheel based on the different rotational speed at the optimum immersed depth ratio of 0.5 [9]. It can be observed that the performance of each of the water wheel with curved blade type 1 and type 2 have increasing compared to the straight blade.

Previously, straight blade water wheel produces 0.041 Watt experimentally and 0.045 Watt numerically. From the study, it is observed that the curved type 2 has generate highest power output which is 0.073 Watt compared to 0.064 Watt generated by curved type 1.

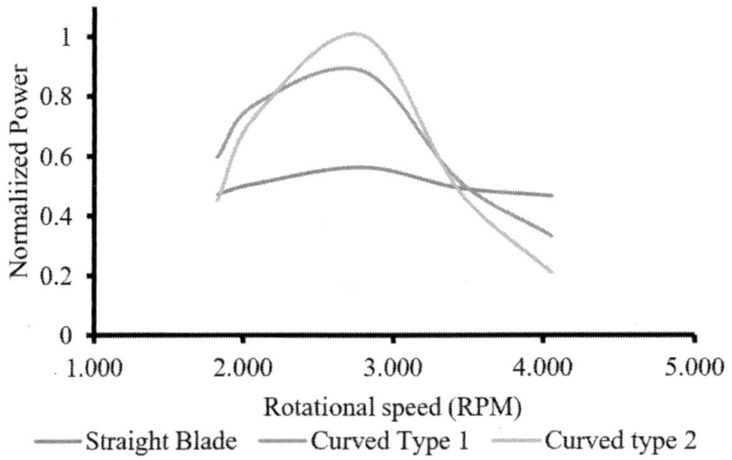

Figure 9. The performance of straight, curved type 1 and curved type 2 based on the different rotational speed (rpm).

5. Conclusion

The performance of straight blade water wheel was numerically and experimentally investigated. From the simulation, result obtained was compared with the previously conducted experiment in order to validate the computational model used. The maximum power generated was 0.045 Watt which is small error of 9.3 % compared to the power generated during experiment. Therefore, the computational model used are validated to be appropriate prediction of the water wheel performance with different blade shape. Another two blade shape was numerically investigated which is curved blade with different curvature. From the simulation, curved blade type 2 has numerically generate highest power which is 0.0.73 Watt. Therefore, it can be concluded that curved blade are better than straight blade in terms of power generation.

This research examined the potential of utilizing very low head sites for small hydropower technology. Stream water wheel is a suitable water turbine for utilizing the water power from open channel flow. The outcome of this study contribute to the existing knowledge on very low head hydropower and the performance characteristics of hydraulic turbines with similar working principle.

IOP Conf. Series: Materials Science and Engineering **342** (2018) 012066 doi:10.1088/1757-899X/342/1/012066

Acknowledgement

The authors would like to thank Universiti Malaysia Pahang for providing necessary assistance and Ministry of Higher Education for providing financial support under FRGS project number RDU140127.

References

[1] Güney MS and Kaygusuz K 2010 Hydrokinetic energy conversion systems: A technology status review *Renew. Sustain. Energy Rev.* **14** (9) pp. 2996–3004

[2] Liu L, Masera H and Esser D 2013 World Small Hydropower Development Report

[3] Reynolds TS 2002 *Stronger than a hundred men: a history of the vertical water wheel*

[4] Farret FA 2006 *Integration of alternative sources of energy* John Wiley & Sons.

[5] Müller G 1983 Water Wheels As a Power Source pp. 1–9

[6] Turnock S, Muller G and Nicholls-Lee R 2007 Development of a floating tidal energy system suitable for use in shallow water

[7] Cengel JM and Cimbala YA 20016 *Fluid Mechanics Fundamentals and Applications* The McGraw Hill Companies

[8] Barstadm LF 2012 CFD Analysis of a Pelton Turbine

[9] Tevata A and Inprasit C 2011 The Effect of Paddle Number and Immersed Radius Ratio on Water Wheel Performance *Energy Procedia* **9** pp 359–365

[10] ANSYS CFX-Solver Theory Guide 2013 **15317** pp 724–746

[11] Al Sam A 2010 Water wheel CFD simulations

[12] Nishi Y, Inagaki T, Li T, Omiya R and Hatano K 2014 Research on the flow field of undershot cross-flow water turbines using experiments and numerical analysis *IOP Conf. Ser. Earth Environ. Sci.* **22** (6) p 062006

[13] Canonsburg TD 2009 ANSYS CFX-Solver Modeling Guide **15317** pp 724–746

[14] Nishi Y, Inagaki T, Li Y and Hatano K 2015 Study on an Undershot Cross-Flow Water Turbine with Straight Blades

IOP Conf. Series: Materials Science and Engineering **342** (2018) 012067 doi:10.1088/1757-899X/342/1/012067

Assessing self-efficacy and college readiness level among new undergraduate students in computer science using metacognitive awareness inventory (MAI)

Wan Nor Afiqah Wan Othman and Aziman Abdullah

Faculty of Computer Systems & Software Engineering, Universiti Malaysia Pahang, Lebuh Raya Tun Razak, 26300 Kuantan, Pahang, Malaysia.

Email: aziman@ump.edu.my

Abstract. This preliminary study was conducted to address the issue of academic planning skills among new university student. Due to lack of proper measurement mechanism for awareness and readiness among students, this study proposes Metacognitive Awareness Inventory (MAI) to assess the connection between student self-efficacy and college readiness. Qualitative and quantitative approach were used by provide an online self-assessment for new student of Faculty of Computer Systems & Software Engineering (FSKKP) and analyse the data respectively. The possible relationships between MAI and College Readiness Item (CRI) in self-assessment has been evaluated. The sample size of 368 respondents from UMP are responding to the online self-assessment. The initial finding shows most student (71%) of the respondent lack of skills in planning. We manage to use *Pearson Product-moment* correlation coefficient to find the significant relationship between MAI and CRI. Thus, we found that College Readiness provide sufficient evidence that there is a significant correlation with most of MAI items. The findings also indicated not much difference was found between gender in terms of self-efficacy level. This paper suggests the MAI and CRI is a reliable and valid scale to respond the planning skills issues among new university students.

1. Introduction

Metacognitive Awareness Inventory (MAI) has been developed by [1] to evaluate metacognition as comprised of two major components: metacognitive knowledge and metacognitive regulation. Metacognition plays an important role in helping learners in learning process [2–4]. The MAI consists of 52 statements relate to the both metacognition components which assign the ability to monitor and control one's learning [1,5,6]. Metacognitive knowledge includes three sub-processes: declarative, procedural, and conditional. Declarative knowledge means knowledge about self and the strategies. Procedural knowledge is to know how to handle the procedures in study strategies. Conditional knowledge refers to the knowledge about *when* and *why* to use the strategies [5,7]. Whereas, metacognitive regulation also includes three sub-processes: planning, monitoring, and evaluation. This sub-processes more to the steps which an individual take to solve problems, set goals and analyse what they are studied about [5,7]. The findings of MAI in these study can be used [1,7,8] to assess the self-efficacy and college readiness among new students.

IOP Conf. Series: Materials Science and Engineering **342** (2018) 012067 doi:10.1088/1757-899X/342/1/012067

As mentioned by [9], self-efficacy is capabilities of one's beliefs to produce the required level of performance. There is an evidence [10] indicates students who are performed in study posse a high level of self-efficacy. Furthermore, there are other studies [11–13] recommended that self-efficacy correlates highly with college achievement. Besides, it is also affected to academic achievement and motivation [13], students with high level of self-efficacy are more prepared, hardworking, pursue challenging goals, increase effort to perform the goals, and able to deal with problems [9,10]. These findings supported by [10] defined that students need to build a strong belief that they can complete the task successfully, rather than relying on the ability and skills to perform well on academic tasks. Furthermore, in following research, the result shown that self-efficacy beliefs be related to decisional behaviors and various indices of career choice [13].

The lack of college readiness skills is a national issue in certain countries as addressed in many previous studies [14–16]. Malaysia is also not exempted in facing this challenging issue in higher education particularly for the new higher education system defined in Malaysia Education Blueprint 2015-2025 (Higher Education) [17]. In a study conducted by [18], lack of college preparation can be caused by the gap that exists in high school learning and what is expected when students reach college. College-ready student is one that capable to understand what is college expected, know the content knowledge, and can adapt the key intellectual lessons as defined by [16]. There is very little research that address college readiness in education area. The development and employ an instrument that assess the college readiness will be constructed in this study.

This paper presents the investigation on self-efficacy and college readiness among new undergraduate students in computer science discipline. This study is important in emphasizing the issue of planning skills in academic among new university student, since there is no proper mechanism to measure student awareness and readiness to venture their life in university environment. The following research questions have been searched in this designated study:

1. Is there a significant difference among new students in term of degree of awareness studying at university?
2. Is there a significant relationship between MAI and college readiness instrument?
3. Is there a significant difference between gender in term of self-efficacy level based on MAI instrument?

Hence, the aim of the study is to assess the connection between new students' self-efficacy and college readiness by analyse data using MAI instrument.

2. Literature Review

2.1. Self-efficacy
This preliminary study has reviewed self-efficacy in education area that address student self-efficacy and academic performance. [19] mentioned that a deep or strategic approach to general learning are being adopted for students with a high self-efficacy relating their reading and writing abilities, others with a low self-efficacy (reading and writing) adopt a surface approach. Students with a high self-efficacy adapted better in college [20]. The high school grade point average and student self-efficacy were strongly correlated in college setting with academic performance measured by course grades [21]. Literatures shown that [10,11,22,23] student self-efficacy and academic performance have a strong positive correlation. However, there is a need to develop a general scale for students, thus MAI instrument is the best option to be implement as measurement for self-efficacy among students.

2.2. College readiness
[16] mentioned college readiness as preparation level for student to enrol and succeed without remediation in a credit-bearing general education course at a postsecondary institution. "Succeed" is determined as finishing entry-level courses at a level of understanding and skill that makes it possible

for the student to consider taking the next course in the sequence or the next level of course in the subject area [16]. College readiness skills are categories into four element included content knowledge, contextual skills and knowledge, cognitive strategies and academic behaviors, which would familiarize the student with a college environment [24]. From this review, it come out with those approaches. However, it seems there are few sources that study about the relationship of college readiness with other instruments to measure the confidence in student awareness in their study. Thus, this paper will evaluate the relationship between college readiness and MAI instrument.

2.3. Metacognitive Awareness Inventory (MAI)

This study has examined MAI in relationship to the academic achievement, learning skills and self-regulation of students in academic area [18]–[21]. [25] investigated the MAI intent on how it relates to broad and single measures of academic achievement in college students. As a result, a significant correlation between the MAI and cumulative GPA as well as end of course grades were found. Based on [26] the findings show significant correlations between metacognition and academic monitoring. The result on [27] indicated that metacognition level and self-efficacy has a weak relationship based on *Pearson* correlation coefficient and *linear regression* test indicated a positive and significant relationships.

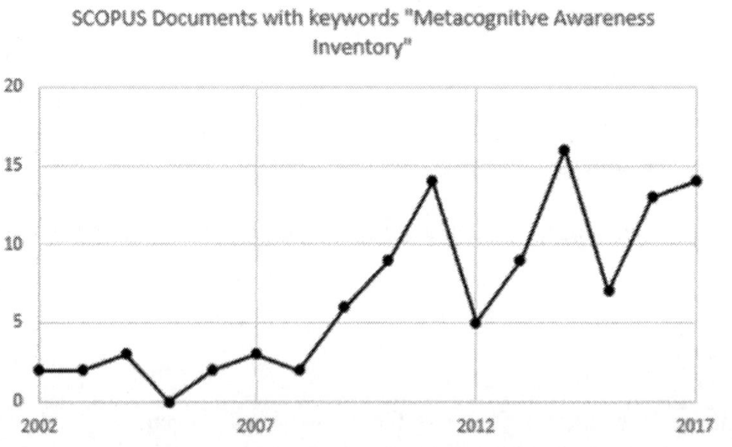

Figure 1. Search Result from SCOPUS database with keywords "Metacognitive Awareness Inventory" retrieved on 8 November 2017.

Based on our search in SCOPUS database, the use of MAI is to evaluate or assess students' skills for learning purpose in various subject area is still in infancy stage with 112 documents retrieved globally, although we can still take this as a reference that the MAI instrument can be used for this preliminary study (Figure 1).

3. Methodology

3.1. Study subjects

This preliminary study is conducted in qualitative and quantitative approaches by analysing data from MAI and CRI. The CRI is an instrument devoted in UMP context and includes 4 items regarding students' preparation at university. Although there is a specific instrument for measuring the self-efficacy, there also other studies used MAI to measure or associate with self-efficacy[8,28]. The MAI items shown in Table 1 [25]. These two instruments, were used by respondents in the self-assessment with a true or false choice and 5-point Likert scale (strongly agree, agree, neutral, disagree and strongly disagree) respectively. The qualitative approach has been used in detailed description of observations and individual insights of the respondents [22]. While the quantitative approach to quantify the self-efficacy and college readiness among UMP students by applies statistical techniques to recognize patterns [22] whether this preliminary study is applicable or not. The MAI and CPI is a general self-assessment has been used in an online environment targeting respondent among new FSKKP students. The rational online self-assessment has been used is, data collection easy to manage within the time, it is well accepted especially relate with this kind of self-assessment method, and in this era of technology, students have their own smartphone to access it on time. The session was conducted on 23rd September 2017 with 368 respondents by distributing the Google form link through First Year Seminar (FYS) held at Universiti Malaysia Pahang (UMP).

Table 1. MAI items.

Components	Items	Number of items
Knowledge about Cognition	Declarative Knowledge	8
	Procedural Knowledge	4
	Conditional Knowledge	5
Regulation of Cognition	Planning	7
	Information Management Strategies	10
	Comprehension Monitoring	7
	Debugging Strategies	5
	Evaluation	6

3.2. Data analysis

Two dependent variables regarding students' self-assessment, MAI (Dependent Variable) and CPI (Independent Variable), were analysed in this study. *Pearson Product-moment* correlation coefficient was used to evaluate possible relationships between MAI and CPI items respectively. A factor of α=0.05 and p value less than 0.05 was considered statistically significant. A common spreadsheet application (Microsoft Excel 2016) was employ as a statistical tool for data analysis purpose.

4. Results and Discussion

This section provides the relevant results for data analysis from MAI and CRI. The percentage of students' preparation is summarized in Table 2.

Table 2. Percentage of students' preparation (n=368).

Concept	**Strongly Agree**	**Agree**	**Neutral**	**Disagree**	**Strongly Disagree**
College Readiness					
CR1: Actually, I don't fully understand how UMP grading system Cumulative Grade Point Average (CGPA) works.	6(2%)	58(16%)	**159(43%)**	111(30%)	34(9%)
CR2: Actually, I have no idea what is new UMP grading system Integrated Cummulative Grade Point Average (iCGPA) works.	11(3%)	95(26%)	**157(43%)**	82(22%)	23(6%)
CR3: At this moment, I have no idea how to get first class award upon my graduation.	20(5%)	66(18%)	**124(34%)**	107(29%)	51(14%)
CR4: At this moment, I never thought what kind of career I would do after completing my study.	22(6%)	61(17%)	110(30%)	**123(33%)**	52(14%)

IOP Conf. Series: Materials Science and Engineering **342** (2018) 012067 doi:10.1088/1757-899X/342/1/012067

Table 2 show that new students are mostly in balance criteria for the college readiness. Due to most of feedback dominated at neutral scale, we found it is difficult for us to distinguish either students were having readiness or not. Thus, Case 1 and Case 2 are being developed to provide a clear picture for the scenarios analysis as shown in Figure 2. Case 1 is a scenario with assumption to those who choose neutral scale are tend to not ready or grouped with the scale strongly agree and agree, while Case 2 refers to those who choose at neutral scale tend to have some readiness. As illustrated in Figure 2, the career scenario in Case 1 shows (52%) students are not concerned about their future career. Whereas, Case 2 shows mostly students are well prepared in all items.

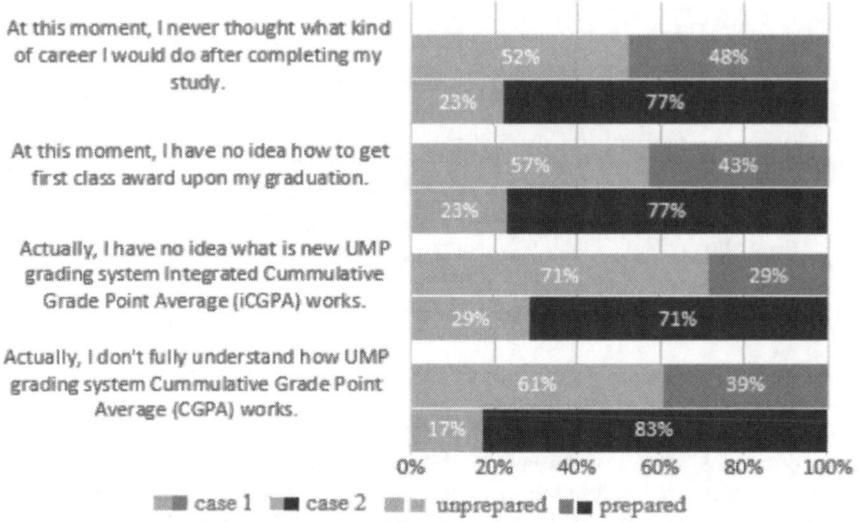

Figure 2. Percentage of student readiness based on case 1 and case 2 (n=368).

Based on Table 3, it indicates the percentage of students' self-efficacy using MAI instrument. Majority (90%) of respondents are not competent in declarative knowledge and (71%) of the respondent lack of planning skills.

Table 3. Percentage of students' self-efficacy using MAI (*n*=368)

MAI	Competent	Not Competent
Knowledge about Cognition		
Declarative Knowledge	10%	90%
Procedural Knowledge	42%	58%
Conditional Knowledge	38%	62%
Regulation of Cognition		
Planning	29%	71%
Information Management Strategies	23%	77%
Comprehension Monitoring	30%	70%
Debugging Strategies	62%	38%
Evaluation	24%	76%

Table 4 depicts the possible relationships between items related to MAI and CRI. The MAI divided by 8 categories and each category will be associated with CRI to find the statistically significant between them. Among the 8 categories, only Information Management System and Debugging Strategies not have significant relationship with CRI. The relationship between Conditional Knowledge and CRI was found as a highest significant positive correlation of $r(366) = 0.2458$ (n=368, α=0.05, p<0.001), but it still showing a weak positive correlation, followed by Procedural Knowledge and Declarative Knowledge (**Table 4**).

Table 4. Relationship between MAI and CRI (n=368, $\alpha = 0.05$, $p < 0.05$).

MAI	College Readiness			
	r	t	df	p
Declarative Knowledge (DK)	0.2172	4.2575	366	2.63×10^{-5}
Procedural Knowledge (PK)	0.2222	4.3599	366	1.69×10^{-5}
Conditional Knowledge (CK)	0.2458	4.8512	366	1.82×10^{-6}
Planning (P)	0.1677	3.2545	366	1.24×10^{-3}
Information Management Strategies (IMS)	0.0440	0.8427	366	3.99×10^{-1} *
Comprehension Monitoring (CM)	0.1496	2.8950	366	4.02×10^{-3}
Debugging Strategies (DS)	0.0919	1.7665	366	7.82×10^{-2}*
Evaluation (E)	0.1433	2.7696	366	5.90×10^{-3}

**not significant*

Figure 3 shows overview of how strong the relationship between the items of MAI and CRI. It shows that the green line is the strongest relationship between them, but the red line indicates there is no significant relationship between CRI with Information Management Strategies and Debugging Strategies respectively.

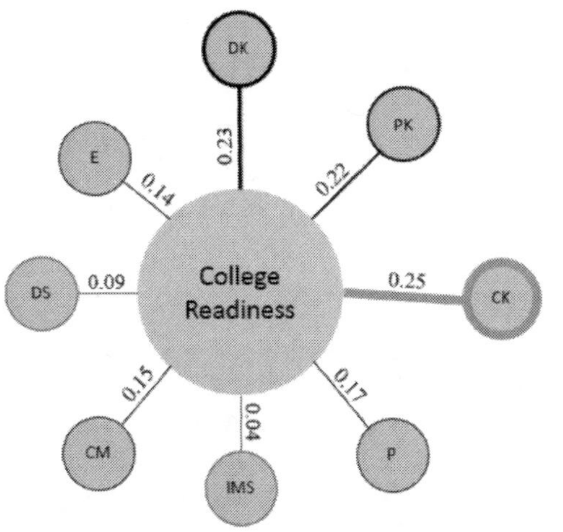

Figure 3. Relationship between MAI and CRI.

Table 5 shows the results of the MAI comparison, based on whether the student was male or female. The MAI is divided into 8 categories and we compare how many respondents feedback for each MAI items through status (1 = Good, <1 = Need Improvement) on both gender. The good status means that the students already prepare for their study in university level and vice versa. The result shows the status "1" and "<1" of male and female students was examined and found there is not much difference between both gender, average difference is between 1 to 17 percent. Table 5 shows male DS is 51% and 49% and female DS is 68% and 32% respectively.

Table 5. Self-efficacy differences between gender through MAI instrument

Gender	Status (1 = Good, <1 = Need Improvement)	DK	PK	CK	P	IMS	CM	DS	E
Male (130)	1	16 (12%)	53 (41%)	45 (35%)	28 (22%)	20 (15%)	33 (25%)	66 (51%)	27 (21%)
	<1	114 (88%)	77 (59%)	85 (65%)	102 (78%)	110 (85%)	97 (75%)	64 (49%)	103 (79%)
Female (238)	1	20 (8%)	100 (42%)	96 (40%)	78 (33%)	64 (27%)	78 (33%)	163 (68%)	63 (26%)
	<1	218 (92%)	138 (58%)	142 (60%)	160 (67%)	174 (73%)	160 (67%)	75 (32%)	175 (74%)

5. Conclusion

According to the findings of this preliminary study, self-efficacy and college readiness level among new students using MAI found that majority students have low awareness about studying at university. This finding indicates that students are not well planned about their future careers and potential for not having initial plans to address the changing nature of works in future due to Industry Revolution 4.0. The differences found between male and female students illustrated in Figure 3 can be considered as no different gender perception regarding to the self-efficacy level. Although the significant relationship between MAI and CRI were found, the associations is still unclear since there are two MAI items; Information Management Strategies and Debugging Strategies are not significant to CRI. Based on this finding, we found that the neutral scale might not be included in the self-assessment since it potentially gives unprecise analysis for evaluation. Besides, the proper instrument need to be further developed and how it can be officially 'compulsory' subscribed by all new students in a more engaging gamification approach fitting the nature of millennials through interactive analytics or personalized dashboard. Therefore, more research is needed to investigate the theoretical and application in measuring self-efficacy and college readiness among new students.

Acknowledgement

We would like to thank to all participated respondents in this study and volunteer reviewers for their comments that greatly improved the quality of this paper. This work is supported by Universiti Malaysia Pahang internal grant **RDU170317.**

References

[1] Schraw G and Dennison R S 1994 Assessing Metacognitive Awareness *Contemp. Educ. Psychol.* **19** 460–75

[2] Pintrich P R 2002 The Role of Metacognitive Knowledge in Learning, Teaching, and Assessing *Theory Pract.* **41** 219–25

[3] Efklides A 2006 Metacognition and affect: What can metacognitive experiences tell us about the learning process? *Educ. Res. Rev.* **1** 3–14

[4] Reza Ahmadi M, Nizam Ismail H and Kamarul Kabilan Abdullah M 2013 The importance of metacognitive reading strategy awareness in reading comprehension *English Lang. Teach.* **6** 235–44

[5] Kincannon J, Gleber C and Kim J 1999 The Effects of Metacognitive Training on Performance and Use of Metacognitive Skills in Self-Directed Learning Situations *Proc. Sel. Res. Dev. Pap. Present. Natl. Conv. Assoc. Educ. Commun. Technol.* 171–86

[6] Tok H, Özgan H and Döş B 2010 Assessing Metacognitive Awareness and Learning Strategies As Positive Predictors for Success in a Distance Learning Class *Mustafa Kemal Üniversitesi Sos. Bilim. Enstitüsü Derg.* **7** 123–134

[7] Cera R, Mancini M and Antonietti A 2013 Relationships between Metacognition, Self-efficacy and Self-regulation in Learning *ECPS - Educ. Cult. Psychol. Stud.* 115–41

[8] Nosratinia M, Saveiy M and Zaker A 2014 EFL Learners' Self-efficacy, Metacognitive Awareness, and Use of Language Learning Strategies: How Are They Associated? *Theory*

Pract. Lang. Stud. **4** 1080–92
[9] Bandura A 1994 Self-Efficacy *Encycl. Hum. Behav.* **4** 71–81
[10] Hsieh P, Sullivan J R and Guerra N S 2007 A closer look at college students: Self-efficacy and goal orientation *J. Adv. Acad.* **18** 454–76
[11] Bong M 2001 Role of Self-Efficacy and Task-Value in Predicting College Students' Course Performance and Future Enrollment Intentions *Contemp. Educ. Psychol.* **26** 553–70
[12] Chemers M M, Hu L and Garcia B F 2001 Academic self-efficacy and first year college student performance and adjustment. *J. Educ. Psychol.* **93** 55–64
[13] Multon K D, Brown S D and Lent R W 1991 Relation of self-efficacy beliefs to academic outcomes: A meta-analytic investigation. *J. Couns. Psychol.* **38** 30–8
[14] Komarraju M, Ramsey A and Rinella V 2013 Cognitive and non-cognitive predictors of college readiness and performance: Role of academic discipline *Learn. Individ. Differ.* **24** 103–9
[15] Tierney W G and Sablan J R 2014 Examining College Readiness *Am. Behav. Sci.* **58** 943–6
[16] Conley D T 2011 Redefining college readiness *Educ. Policy* **5** 1–6
[17] Malaysia M of E 2015 *Malaysia Education Blueprint 2015-2025 (Higher Education)* vol 2025
[18] Journal T O and Horizons N 2007 College Readiness and Black Student Performance : Disaffirmed Equity **3** 23–44
[19] Prat-Sala M and Redford P 2010 The interplay between motivation, self-efficacy, and approaches to studying *Br. J. Educ. Psychol.* **80** 283–305
[20] Ramos-Sánchez L and Nichols L 2007 Self-Efficacy of First-Generation and Non-First Generation College Students: The Relationship With Academic Performance and College Adjustment *J. Coll. Couns.* **10** 6–18
[21] Klomegah R Y A O 2005 Predictors of Academic Performance of University Students: An Application of the Goal Efficacy Model *Coll. Stud. J.* **44** 407–16
[22] Lane J, Lane A M and Kyprianou A 2004 Self-efficacy, self-esteem and their impact on academic performance *Soc. Behav. Pers.* **32** 247–56
[23] Zajacova A, Lynch S M and Espenshade T J 2005 Self-efficacy, stress, and academic success in college *Res. High. Educ.* **46** 677–706
[24] Conley D T 2008 Rethinking College Readiness *New Engl. J. High. Educ.* 24–6
[25] Young A and Fry J D 2008 Metacognitive Awareness and Academic Achievement in College Students *J. Scholarsh. Teach. Learn.* **8** 1–10
[26] Rayne A. Sperling, Bruce C. Howard, Richard Staley N D 2004 Metacognition and Self Regulated Learnin Constructs *Educ. Res. Eval.* **10** 117–39
[27] Ridlo S and Lutfiya F 2017 The Correlation Between Metacognition Level with Self-Efficacy of Biology Education College Students *J. Phys. Conf. Ser.* **824** 12067
[28] Alkan F and Erdem E 2014 The Relationship between Metacognitive Awareness, Teacher Self efficacy and Chemistry Competency Perceptions *Procedia - Soc. Behav. Sci.* **143** 778–83

IOP Conf. Series: Materials Science and Engineering **342** (2018) 012068 doi:10.1088/1757-899X/342/1/012068

Applications of Nano palm oil fuel ash and Nano fly ash in concrete

Hussein M Hamada[1], Gul Ahmed Jokhio[1], Fadzil Mat Yahaya[1] and Ali M Humada[2]

[1]Faculty of Civil Engineering and Earth Resources, University Malaysia Pahang, 26300, Gambang, Malaysia

[2]Electricity Production Directorate of Salahaldeen, Ministry of Electricity, 34007, Baiji, Iraq

E-mail: Enghu76@gmail.com

Abstract: This paper discusses the applications of Nano waste materials including palm oil fuel ash and fly ash in the concrete production. The implementation of nanotechnology has been instrumental in the development of significant interest among the stakeholders to improve the mechanical and chemical properties of materials involved in the production of concrete. Although many researchers have shown the potential of nanomaterials to increase strength and durability of concrete and improve its physical and chemical properties, there is still a knowledge gap regarding the preparation of Nano waste materials from agricultural waste to use as cement replacement instead of non-renewable materials. Therefore, it should be focused on to study Nano- waste materials to benefit from these characteristics during preparation of concrete mixtures. Therefore, this paper highlights the potential of waste materials in the Nano size to partially replace cement in concrete and achieve the same or better result than the traditional concrete. This paper recommends to conduct further experimental works to improve the concrete material properties by investigating the properties of waste materials in Nano size.

1. Introduction and background

Nanotechnology is the re-engineering of materials by manipulation at the atomic level to get specific properties of nanomaterials [1]. It is concerned with the adjustment and understanding of materials on the nanoscale (<100 nm) which have chemical properties that are to be determined by quality technicians. In case of nanomaterials, the effects of the surface are dominate instead of bulk properties [2]. Many researchers have examined the nanomaterials in biology, chemistry, and physics; and have investigated Feynman's ideas regarding manipulation and adjustment of the small-size materials similar with nanoscale and atomic scale [3]. The word 'nanotechnology' is related to the small size materials with the specific definition differing from field to field. In general, it can be said that these materials are not more than 100 nm in size, which can be manipulated and their characteristics changed to produce new features and specifications that can be benefitted from [4]. Nanotechnology has been explored to a considerable degree to target the problems in the design and construction of practical structures with one or more dimensions;

to benefit from the characteristic of nanomaterials [5]. Nano particles or ultrafine particles can increase the density due to filler effect and thus enhance the durability and compressive strength of concrete produced. Besides that, these also decrease the absorption of water and improve the microstructure properties to protect against sulphate and acid attacks. Due to the damaging effects of cement on the environment and its high financial cost, Nano particle additives have been used in wide variety in a range of proportions to reduce the quantity of cement in concrete. However, the high cost of Nano particles additives such as Nano silica made the researchers consider alternative methods through manipulation of the materials that have same chemical compositions as cement such as palm oil fuel ash POFA and fly ash FA to replace cement by different percentages. Therefore, this paper reviews the effects of cement replacement by Nano POFA (NPOFA) and Nano FA (NFA) on the concrete properties.

2. Production of Nano Materials

Nano materials can be obtained from various sources by different methods to give unique properties not available in the original particles [6]. There are two approaches to prepare Nano particle size from the original materials, the first approach is top-down [7], while bottom-up is the second approach [8]. Selection of a suitable approach depends on many factors such as cost, suitability, and behaviour of Nano particles [3, 9]. Milling by the grinding machine is one of the applications of the top-down approach [10]. Due to the ease of this approach and availability of the machine, many researchers select this method to prepare Nano particles from the waste materials without any chemical additions [10], which would be bottom-up approach. Despite a few disadvantages of the top-down approach due to many factors such as limitations on types and number of milling balls and the speed of milling, the quality of Nano particles can be improved [7, 11, 12]. High energy ball milling is one of the machines developed to grind micro particles to the Nano size.

3. Nano waste materials

Many researchers have investigated the waste materials in Nano size, which are prepared by grinding in Laos Angeles machine and/or high energy ball mill [13, 14]. The presence of significant amount of silica oxide content and other materials in some of waste materials in fine particle size has encouraged academics and researchers to grind the particles into micro and Nano particle size. In the following subsections, the agro-waste materials (POFA and FA) in Nano size used as replacement cement in ultrafine and Nano particle size have been discussed.

3.1 Palm Oil Fuel Ash (POFA)

Palm oil fuel ash is a by-product obtained from the burning of palm oil shell and husk in palm oil mills to generate electricity. Many researchers have studied the effects of Nano POFA on the concrete properties as shown in Table 1.

Table 1. Effects of nano particle size POFA on concrete properties.

Ref.	Particle size	Field of study	Results
[13]	less than 1 μm	heat of hydration and compressive strength of cement mortar	higher strength than cement mortar and reduces hydration heat
[14]	20 – 90 nm	study morphological analysis of hardened cement paste	NPOFA particles have high density due to the filling effect.
[17]	50 nm	study the shape and texture of raw and Nano-structured POFA	The raw POFA had an irregular spongy shape, while Nano POFA had irregular and crushed shape

POFA has a high content of silica ranging from 43 to 66%, as can be seen in Table 2. The grinding of POFA by the grinding machine increases the amount of silica oxide SiO_2. The loss of ignition LOI is achieved due to burning of the carbon in the heating process by gas or electric furnaces, as given in Table 2 [15, 16].

Lim et al. [13] studied the effects of cement replacement by high volume Nano palm oil fuel ash NPOFA (up to 80%) to produce concrete with high performance quality. The study investigated the composite morphology, strength activity index, the hydration temperature, microstructure properties, and thermal conductivity. The POFA was ground many times to obtain particle size less than 1 µm. The colour of POFA collected was dark in the first round, which was then dried in electrical oven for 24 hours at 110°C. The dried POFA was sieved by 300 mm, and the passing quantity was again ground by Los Angeles machine. The POFA thus produced in micro particle size was heated in a gas or electric furnace at 500°C to burn the carbon particles and enhance the microstructure properties and pozzolanic reaction. The treated POFA by furnace was ground once again by high energy ball milling for 30 hours to get particles size less than 1 µm. In this case ultrafine POFA [13] or Nano POFA [17] was produced. The main composite in the Nano POFA is SiO_2 which ranges from 55 to 72% of total. The results of the study indicated that Nano POFA can be used as cement replacement up to 80% to achieve higher strength than normal concrete at a later age. In addition, it also reduces the hydration heat of concrete.

Table 2. Chemical composition of ground POFA and ultrafine/Nano POFA.

Chemical composition	Lim et al. [13]		Johari et al. [16]	
	Ground POFA	Ultrafine POFA	Ground POFA	Ultrafine POFA
SiO_2	35.5	69.3	51.18	65.01
Al_2O_3	1.9	5.3	4.61	5.72
Fe_2O_3	1.1	5.10	3.42	4.41
CaO	8.3	9.15	6.93	8.19
K_2O	6.5	11.10	5.52	6.48
MgO	4.1	4.10	4.02	4.58
SO_3	2.36	1.59	0.36	0.33
LOI	20.9	1.3	21.6	2.53

(a) before treatment (GPOFA) (b) after treatment (UPOFA)

Figure 1. Colour of POFA before and after treatment [13].

Rajak et al. [14] studied the effects of Nano POFA and micro POFA on cement mortar and the comparison between the two. The NPOFA was prepared with average particle size ranging between 20 nm to 90 nm. The range of replacement of 10 – 50% of cement by NPOFA was used to prepare NPOFA mortar. The authors performed the morphological analysis of hardened cement paste (HCP) containing Nano POFA (NPOFA) and Micro POFA (MPOFA) at curing ages of 7, 28, and 90 days. The authors prepared the Nano POFA with high specific surface area of 145.35 m^2/g by milling the micro POFA through high energy ball milling machine for 30 hours. The average particle size of Nano POFA obtained from grinding MPOFA ranged between 20 to 90 nm as shown in Figure 2. The authors concluded that NPOFA has filler actions and pozzolanic behaviours more than MPOFA particles due to the production of further C-S-H gels. This behaviour leads to improve the durability and mechanic properties of hardened cement.

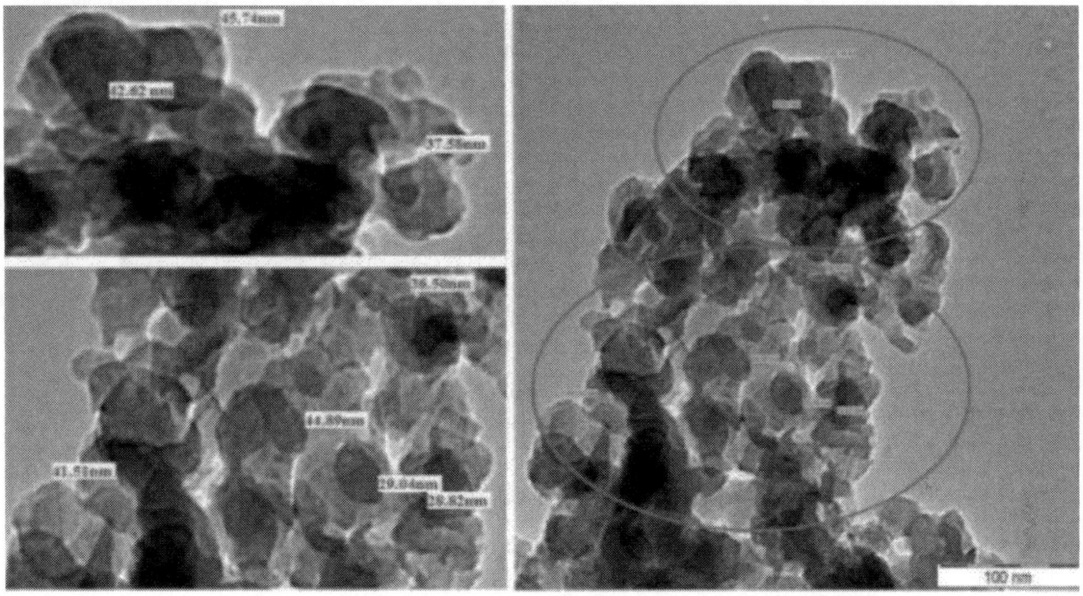

Figure 2. Particle sizes of NPOFA by TEM test analysis [14].

3.2 Fly Ash (FA)

Prince et al. [18] studied the effects of Nano FA on the compressive strength and workability of concrete grades 20, 30, 40, and 50. The results of the study indicate increase in the workability value when the Nano FA amount is increased, which ranged between 90 to 140% of that of the normal concrete. The compressive strength also increases due to addition of 10% of Nano FA to the concrete mix, which was recorded between 17 to 46% more than normal concrete at 28 days. A study was conducted by Raghavendra et al. [19] to prepare the Nano FA by utilization a planetary ball mill in a stainless steel chamber with zirconia balls with diameter of 5 and 10 mm. 6 hours was the duration of milling and the rotation speed of the planet carrier was 400 rpm. The results show increase in the surface area of Nano fly ash from 0.31 to 24.65m^2/g and improvement in the chemical composition of Nano FA. Tudjono et al. [20] studied the effects of Nano FA and Nano lime on the compressive strength of concrete through replacement of cement by different dosages (of Nano FA and Nano lime) from 0% to 15% at intervals of 2.5%. To prepare Nano FA and Nano lime into Nano particles, the researchers used a Planetary Ball Mill (PBM). XRD test was used to analyse Nano FA and Nano lime size required to ensure the particles were within Nano size after grinding process. The researchers found a relation between 2 theta and intensity as illustrated in Figure 3. Scherer's formula has been used to analyse the numbers to get the particle size of FA and lime, before and after grinding process, whereas the particle size was 44.39 nm, 21.03 nm for Nano FA and Nano lime, respectively. The compressive strength of mortar at 7 days was obtained as in Table 2. The compressive strength of control sample was 26.47 MPa.

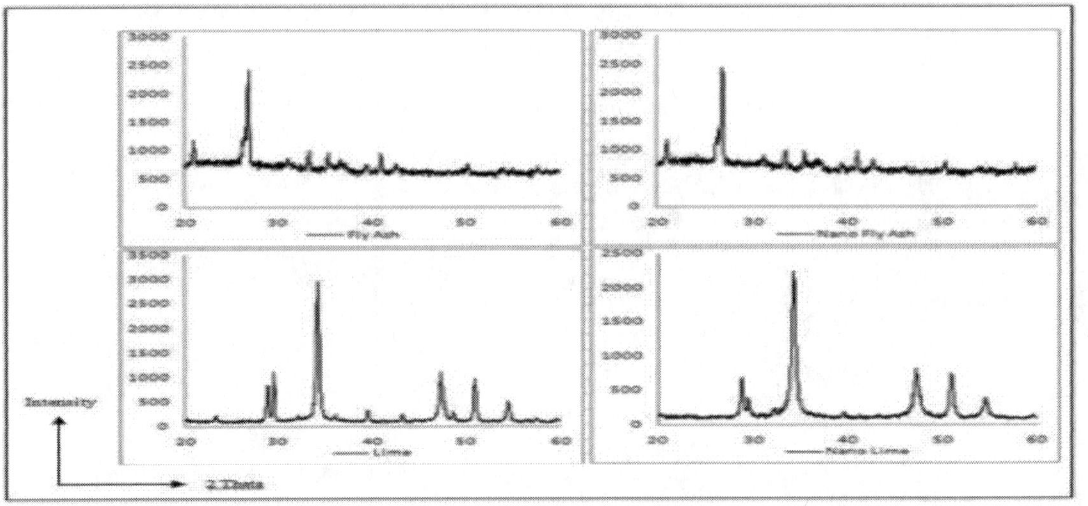

Figure 3. Test of XRD for (a) Fly Ash (b) Nano Fly Ash (c) Lime (d) Nano Lime [20].

Murthy et al. [21] conducted their study by grinding micro fly ash MFA with Al into Nano particle fly ash NFA by high energy ball milling machine for 30 hours. The Nano FA particles of 23 nm size were obtained. Figure 4 shows the Nano FA mix before and after grinding process.

(a) (b)

Figure 4. Al/Nano fly ash (a) before ball milling, (b) after ball milling for 20 min [21].

Jemimah et al. [22] studied the effects of replacement of cement by Nano FA on the compressive strength at different ages of 3, 7, and 21 days. The researchers showed that the compressive strength increases with the addition of more quantity of Nano FA as cement replacement. It was found that 30% of Nano FA as cement replacement gave the optimum value of compressive strength and 20% was the next optimum value. Besides that, the authors also studied the effects of the Nano FA on the initial and final setting times of cement mortars due to replacement of cement weight by 0%, 10%, 20%, 30%, 40%, and 50%. The results illustrated that increasing dosage of Nano FA lead to delay of initial and final setting time, whereas the final setting time was 7.84%, 21.57%, 30%, 53.92%, and 80.39% due to the replacement of cement by POFA with 10%, 20%, 30%, 40% and 50%, respectively. Roychand et al. [23] studied the effects of Nano silica and silica fume individually and in combination with the accelerator and hydrated lime, on the ultra-fine fly ash to replacement cement by 80 % of total weight. The authors concluded that Nano fly ash when combined with Nano silica can replace cement by 80 % of while maintaining the same properties obtained when the ordinary cement was used.

4. Conclusion

The NPOFA and NFA particles from waste materials show good pozzolanic behaviour due to high silica content and high specific surface area in addition to being lower in the Los of Ignition (LOI). The pozzolanic behaviour of NPOFA particles leads to the production of secondary C-S-H gels via pozzolanic reactions with free lime. Using these wastes in Nano particle size can reduce the cement quantity in concrete production up to 80% and thus reduce the environmental impact due to generation of CO_2 resulting from cement production as well as lower the cost. This study recommends conducting further experimental works to prepare Nano particles from other waste materials to improve the characteristics of concrete to make it more environmentally friendly and economic material.

References

[1] Roco MC, Williams R, and Alivisatos P 1999 Nanotechnology Research Directions: IWGN Workshop Report *Vision for Nanotechnology R&D in the Next Decade DTIC Document*
[2] Guo D, Xie G and Luo J 2013 Mechanical properties of nanoparticles: basics and applications *Journal of Physics D: Applied Physics* **47** p 013001
[3] Sanchez F and Sobolev K 2010 Nanotechnology in concrete–a review *Construction and building materials* **24** pp 2060-2071
[4] Kaiser DL, Standridge S, Friedersdorf L, Geraci CL, Kronz F, Meador M, *et al.* 2014 National Nanotechnology Initiative Strategic Plan
[5] Kelsall RW, Hamley IW and Geoghegan M 2005 *Nanoscale science and technology*: Wiley Online Library
[6] Sobolev K and Gutiérrez MF 2005 How nanotechnology can change the concrete world *American Ceramic Society Bulletin* **84** p 14
[7] Abdoli H, Farnoush H, Asgharzadeh H and Sadrnezhaad S 2011 Effect of high energy ball milling on compressibility of nanostructured composite powder *Powder Metallurgy* **54** pp 24-29
[8] Jankowska E and Zatorski W 2009 Emission of nanosize particles in the process of nanoclay blending *Quantum, Nano and Micro Technologies (ICQNM'09) Third International Conference pp* 147-151.
[9] Sobolev K and Shah S 2008 Nanotechnology of concrete: recent developments and future perspectives: *ACI*
[10] Norhasri MM, Hamidah M and Fadzil AM 2017 Applications of using nano material in concrete: A review *Construction and Building Materials* **133** pp 91-97
[11] Yadav TP, Yadav RM and Singh DP 2012 Mechanical milling: a top down approach for the synthesis

of nanomaterials and nanocomposites *Nanoscience and Nanotechnology* **2** pp 22-48
[12] Saleh NJ, Ibrahim RI and Salman AD 2015 Characterization of nano-silica prepared from local silica sand and its application in cement mortar using optimization technique *Advanced Powder Technology* **26** pp 1123-1133
[13] Lim NHAS, Ismail MA, Lee HS, Hussin MW, Sam ARM and Samadi M 2015 The effects of high volume nano palm oil fuel ash on microstructure properties and hydration temperature of mortar *Construction and Building Materials* **93** pp 29-34
[14] Rajak MAA, Majid ZA and Ismail M 2015 Morphological characteristics of hardened cement pastes incorporating nano-palm oil fuel ash *Procedia Manufacturing* **2** pp 512-518
[15] Zeyad AM, Johari MAM, Tayeh BA and Yusuf MO 2017 Pozzolanic reactivity of ultrafine palm oil fuel ash waste on strength and durability performances of high strength concrete *Journal of Cleaner Production* **144** pp 511-522
[16] Johari MM, Zeyad A, Bunnori NM and Ariffin K 2012 Engineering and transport properties of high-strength green concrete containing high volume of ultrafine palm oil fuel ash *Construction and Building Materials* **30** pp 281-288
[17] Khalil HA, Fizree H, Jawaid M and Alattas OS 2011 Preparation and characterization of nano structured materials from oil palm ash: a bio-agricultural waste from oil palm mill *Bio Resources* **6** pp 4537-4546
[18] Arulraj PG and Carmichael JM 2011 Effect of nano-flyash on strength of concrete *International journal of civil and structural Engineering* **2** p 475
[19] Raghavendra G, Ojha S, Acharya SK and Pal S 2014 Fabrication and characterization of nano Fly ash by planetary ball Milling *International Journal of Material Science Innovations* **2** pp 59-68
[20] Tudjono S, Purwanto X and Apsari KT 2014 Study the effect of adding nano fly ash and nano lime to compressive strength of mortar *Procedia Engineering* **95** pp 426-432
[21] Murthy IN, Rao DV and Rao JB 2012 Microstructure and mechanical properties of aluminum–fly ash nano composites made by ultrasonic method *Materials & Design* **35** pp 55-65
[22] Jemimah C and Arulraj P 2012 Influence of nano materials on consistency, setting time and compressive strength of cement mortar *Engineering Science and Technology* pp 2250-3498
[23] Roychand R, De Silva S, Law D and Setunge S 2016 Micro and nano engineered high volume ultrafine fly ash cement composite with and without additives *International Journal of Concrete Structures and Materials* **10** pp 113-124

Two-fold sustainability – Adobe with sawdust as partial sand replacement

Gul A Jokhio[1], Sharifah M Syed Mohsin[1] and Yasmeen Gul[2]

[1]Faculty of Civil Engineering and Earth Resources, University Malaysia Pahang, Malaysia
[2]Faculty of Built Environment, University Technology Malaysia, Malaysia

E-mail: johkio@ump.edu.my

Abstract. Adobe is a material that is economic, environment friendly, and provides better indoor air quality. The material required for the preparation of adobe include clay, sand, and sometimes straw or other organic materials. These materials do not require industrial processing or transportation, however, sand mining has been recently posing a threat to the environment. Therefore, to enhance the existing sustainability of adobe, sand can be partially or fully replaced by other waste materials. This approach will not only solve the problem of excessive sand mining, it will also address the issue of waste management. Sawdust is one such waste material that can be used to partially replace sand in Adobe. This paper presents the results of compressive and flexural test carried out on Adobe samples with partial sand replacement by sawdust. The results show that about 4% sand replacement by volume produces higher compressive strength, whereas the flexural strength reduces with the use of sawdust. However, since flexural strength is not a critical property for adobe, it is concluded that replacing sand with sawdust by about 4% of volume will be beneficial.

1. Introduction

When it comes to low-cost sustainable construction, adobe is an ideal material [1]. In the regions of the world where it remains dry during most of the year, the legacy of the use of this material goes as far back as 10,000 years [2-4]. Adobe structures have been found in South Asia, Africa, Europe, and Central Asia with the types of structures ranging from important religious buildings to ordinary houses [5-11]. As well as the historical use of adobe [12], it is estimated that a considerable proportion (about 30%) of the current world population lives in houses made of adobe [13]. In the present day, adobe structures can be found in Africa, Iran, Afghanistan, Yemen, Iraq, Syria, Spain, Germany, England, France, Portugal, Italy, Denmark, and Sweden [14].

Despite its economy and sustainability, adobe suffers from challenges such as weakened tensile strength, which results in faulty behaviour when subjected to adverse conditions, such as earthquake, for e.g. [15]. The interest of the scientific and engineering community to increase the mechanical properties of this material has increased over the past few years [16]. Some of the efforts at increasing the mechanical properties have been directed at including some kind of reinforcement, for example, wire mesh [17]. Other researchers have focused on addition of materials to reduce the brittle behaviour and cracking due to shrinkage [3]. The additional materials sometimes consist of processed and synthetic materials such as cement or plastic fibres; however, most of the time in order to maintain the sustainable nature of this material, natural fibres such as kenaf or waste materials such as fuel ash have been used [6]. Saw dust is another such waste material that is produced out of woodwork. The conventional uses of sawdust produced at sawmills include preparation of feed for livestock or re-use for making plywood etc., and such other uses. However, most of the times, the disposal of excess sawdust poses a challenge. The timber industry plays a vital role in Malaysian economy [18] and large quantities of sawdust waste materials are produced as a result. The disposal of this material, such as in landfills, poses a great challenge [19]. Moreover, it has been reported that sand mining for use as aggregate in concrete is the cause of many environmental challenges such as degradation of rivers, rising sea levels, destruction of aquatic habitats, destruction of ecosystems and biodiversity, etc. [20].

To address the above mentioned challenges that arise from excessive sand mining and disposal of waste sawdust into landfills, it is a good idea to use sawdust as partial sand replacement during the production of concrete as well as adobe. If added to adobe, it is expected that sawdust will positively affect the mechanical properties of adobe by reducing the dry shrinkage and increasing the ductility of the material. Such a use can be referred to as two-fold sustainability as it is aimed at addressing the environmental issues of sand mining and sawdust disposal by adding it to an already sustainable and environment friendly material, adobe. Therefore, the present study is aimed at exploring the use of adobe as partial sand replacement during the preparation of adobe.

2. Methodology

In order to achieve the objectives set out for this study, a total of 36 specimens were tests, 18 out of which were cubes tested for compressive strength and the remaining 30 were prismatic specimens tested for flexural strength. The cubes were 150 mm × 150 mm × 150 mm in size, whereas the prismatic specimens were 150 mm × 150 mm × 500 mm. The materials used for the preparation of these specimens included clay, sand, and sawdust. The clay proportion consisted of 30%. The details of specimens along with mix proportions used are provided in Table 1.

Table 1. Details of specimens and mix proportions used.

S. No.	Detail of Specimen	Clay Proportion	Sand Proportion	Sawdust Proportion
1	3 cubic specimens named C0-1, C0-2, and C0-3; and 3 prismatic specimens named P0-1, P0-2, and P0-3	30%	70%	0%
2	3 cubic specimens named C1-1, C1-2, and C1-3; and 3 prismatic specimens named P1-1, P1-2, and P1-3	30%	60%	10%
3	3 cubic specimens named C2-1, C2-2, and C2-3; and 3 prismatic specimens named P2-1, P2-2, and P2-3	30%	50%	20%
4	3 cubic specimens named C3-1, C3-2, and C3-3; and 3 prismatic specimens named P3-1, P3-2, and P3-3	30%%	40%	30%
5	3 cubic specimens named C4-1, C4-2, and C4-3 and 3 prismatic specimens named P4-1, P4-2, and P4-3	30%%	30%	40%

The mixing procedure involved adding all the ingredients first to the concrete mixer followed by addition of water in small amounts until a malleable adobe mass was obtained. The adobe mix was then poured into moulds for the preparation of test specimens. The test set up for the compressive testing is shown in Figure 1, whereas the test set up for flexural testing is shown in Figure 2.

Figure 1. Compressive strength test setup.

Figure 2. Flexural strength test setup.

3. Compressive Strength Test Results

The results of the compressive strength test have been summarized in Table 2 and illustrated in Figure 3. It can be seen that while the average compressive strength for the three specimens with no sawdust is 2.1 MPa, it increases when sand is partially replaced by sawdust. For sawdust comprising 10% of the total mix, the average compressive strength increases to 2.19 MPa. This keeps on increasing to 2.43 MPa for 20% of the mix comprising of sawdust. However, when the proportion of sawdust is increased beyond 20%, the compressive strength starts decreasing. It drops back to an average value of 2.19 MPa for 30% and 2.05 MPa for 40%, which is even less than when no sawdust was used at all. Therefore, it can be concluded that the addition of sawdust increases the compressive strength of adobe with the optimum proportion of sawdust being around 20% when 30% clay is used. A look at the standard deviation values indicates that all the results are closely grouped together, which adds to the reliability of these results.

Table 2. Results of the compressive strength test

Sawdust Proportion	Specimen	Compressive Strength (MPa)	Average (MPa)	Standard Deviation
Nil	C0-1	2.16	2.1	0.06
	C0-2	2.09		
	C0-3	2.05		
10%	C1-1	2.24	2.19	0.04
	C1-2	2.18		
	C1-3	2.16		
20%	C2-1	2.47	2.43	0.11
	C2-2	2.31		
	C2-3	2.51		
30%	C3-1	2.17	2.19	0.02
	C3-2	2.2		
	C3-3	2.21		
40%	C4-1	2.14	2.05	0.08
	C4-2	1.98		

	C4-3	2.03		

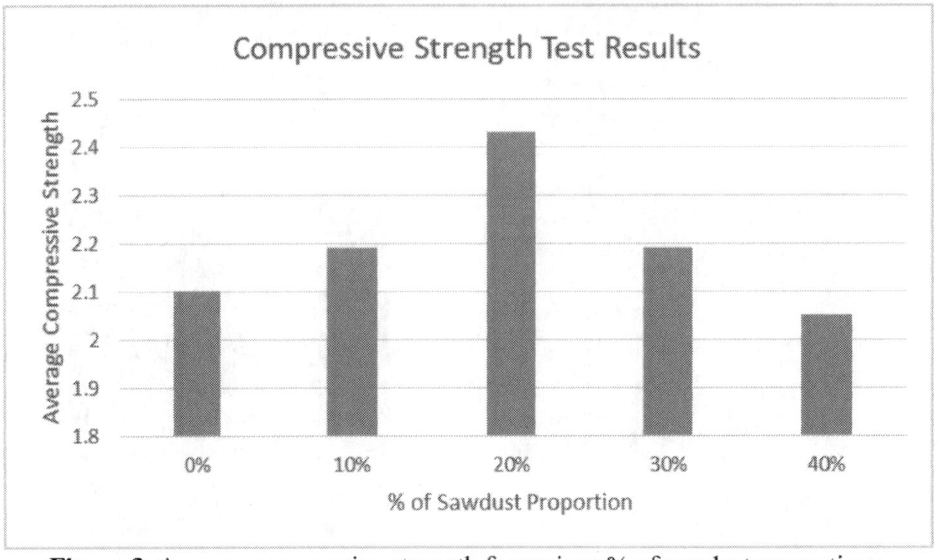

Figure 3. Average compressive strength for various % of sawdust proportions.

4. Flexural Strength Test Results

The results of the flexural strength test have been summarized in Table 3 and illustrated in Figure 4. It can be seen that while the standard remains low indicating close grouping of the results, the flexural strength actually reduces when sawdust is added to adobe mixture. The average flexural strength when no sawdust was used is 0.51 MPa, which reduces to 0.45 MPa with 10% of mix proportion comprising of sawdust. The flexural strength keeps reducing as the proportion of sawdust is increased to 20%, 30%, and 40% with the values of 0.42 MPa, 0.39 MPa, and 0.34 MPa, respectively. Although this reduction in the flexural strength of adobe with the addition of sawdust suggests that it may be optimal not to use sawdust as partial sand replacement. However, it should be noted that the flexural strength of adobe is very low to start with. The flexural strength of 0.51 MPa without any sawdust is also negligible and it is not possible to design any flexural structural members using adobe only. Adobe structures usually consist of load bearing walls that are primarily loaded in compression. The small openings such as door and windows are usually supported by the help of other materials such as wooden boards etc. Therefore, the negative effects of sawdust on the flexural strength of adobe can be discounted as insignificant.

Table 3. Results of the compressive strength test

Sawdust Proportion	Specimen	Compressive Strength (MPa)	Average (MPa)	Standard Deviation
Nil	P0-1	0.51	0.51	0.01
	P0-2	0.5		
	P0-3	0.52		
10%	P1-1	0.44	0.45	0.01
	P1-2	0.44		
	P1-3	0.46		
20%	P2-1	0.41	0.42	0.02
	P2-2	0.42		
	P2-3	0.44		
30%	P3-1	0.35	0.39	0.03
	P3-2	0.42		
	P3-3	0.39		
40%	P4-1	0.3	0.34	0.06

	P4-2	0.33		
	P4-3	0.41		

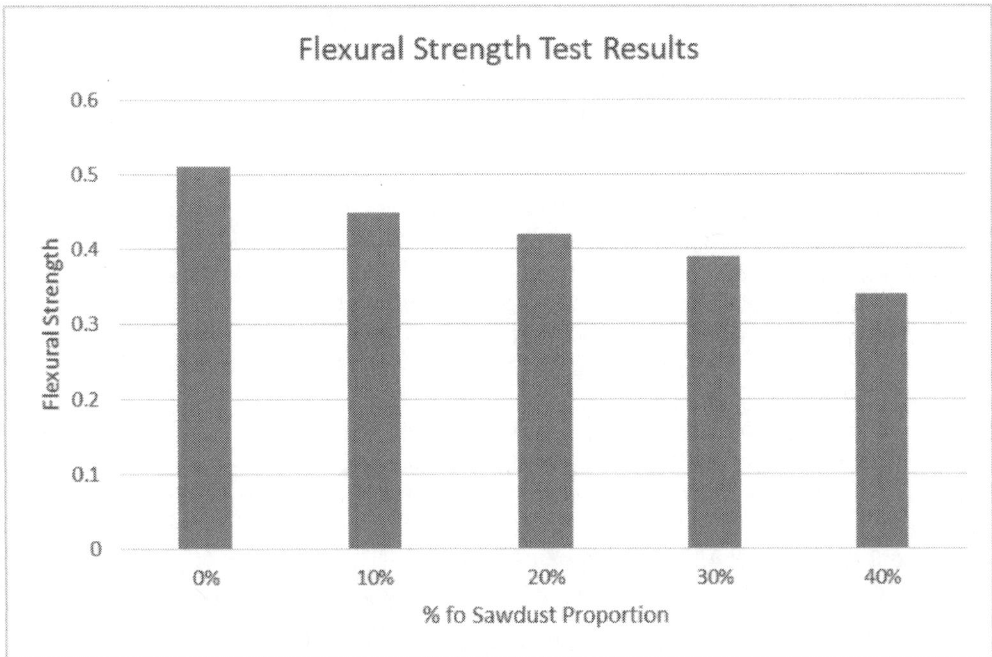

Figure 4. Average compressive strength for various % of sawdust proportions.

5. Conclusion and Recommendations

The results of the compressive strength and flexural strength tests have been presented when sawdust was used as partial sand replacement in adobe. The proportions of sawdust were varied from 0% to 40% at 10% intervals. From the results obtained, the following 2 conclusions can be drawn:

- Addition of sawdust increases the compressive strength of adobe. The ideal proportion of sawdust lies at 20%, which adds about 15% to the compressive strength.
- The addition of sawdust reduces the flexural strength of adobe, however, because of the fact that flexural strength is not a critical parameter for adobe, this effect is detrimental to the use of sawdust as partial sand replacement.

It is worth noting here that the clay proportion used for all the specimens in this study was fixed at 30%. It would be interesting to explore how the addition of sawdust influences both the compressive and flexural strength with varying proportion of clay in the mixture. Furthermore, it is noted that the generalizability of these results may be limited by the number of specimens used in the present study. However, these results may be treated as a starting point in order to set a direction for the future research in this field.

Acknowledgements

The authors acknowledge the research environment and infrastructure provided by University Malaysia Pahang. The research funding for the present study was also provided by University Malaysia Pahang through research grant number RDU160370, for which the authors are highly grateful.

References

[1] Blondet M and Garcia GV 2003 Adobe Construction Tech. Rep.
[2] Angulo-Ibanez Q, Mas-Tomas A, Galvan-LLopis V and Santolaria-Montesinos JL2012 *Const & Bldg Mat* **30** 389-399
[3] Keefe L 2005 Earth Buildings - Methods, materials, repair and conservation Taylor & Francis
[4] Morton T 2008 Earth masony - design and construction guidelines HIS BRE Press

[5] Fodde E 2009 *Int J of Arch Heritage* **3** 2 145-68
[6] Adorni E, Coisson E and Ferretti D 2013 *Const & Bldg Mat* **40** 1-9
[7] Chasagnoux A1996 L'architecture voute iranienne-modelisation et simulation par elements finis. PhD thesis (in French) Universite de Nantes
[8] Ghirshman R, Gilbert S and Emmons J 1962 Persian Art, The Parthian and Sassanian Dynasties, 249{BC}-{AD}651. Golden Press
[9] Hejazi M 1997 Historical buildings of Iran: their architecture and structure Queen Mary and Westfield College University of London
[10] Pope AU 1965 Persian Architecture George Braziller
[11] Wulff HE 1966 The traditional crafts of Persia MIT Press
[12] Binici H, Aksogan O, Bakbak D, Kaplan H and Isik B 2009 *Const & Bldg Mat* **23** 2 1035-1041
[13] Fratini F, Pecchioni E, Rovero L and Tonietti U 2010 *App Clay Sc* **53** 3 509-516
[14] Delgado MCJ and Guerrero IC 2006 *Const & Bldg Mat* **20** 9 679-690
[15] Gul Y, Jokhio GA and Kakar E 2013 6th Int Civ Engg Cong
[16] Illampas R, Ioannou I and Charmpis DC 2013 *Int J of Arch Heritage* **7** 2 165-188
[17] Jokhio GA, Al-Tawil YMY, Syed Mohsin SM, Gul Y, and Ramli NI 2017 MUCET Penang
[18] Ministry of Plantation Industries and Commodities 2009 National Timber Industry Policy 2009-2020, http://www.mtib.gov.my/natip/Natip3.pdf [Accessed: December 21, 2017].
[19] Elinwa AU and Mahmood YA 2002 *Cement and Concrete Composites* **22** 219-222
[20] Desprez M, Pearce B and Le Bot S 2009 International Council for Exploration of the Sea Oxford Journals 271-277, doi:10.1093/icesjms/fsp245

IOP Conf. Series: Materials Science and Engineering **342** (2018) 012070 doi:10.1088/1757-899X/342/1/012070

The efficiency of average linkage hierarchical clustering algorithm associated multi-scale bootstrap resampling in identifying homogeneous precipitation catchments

Zun Liang Chuan[1], Noriszura Ismail[2], Wendy Ling Shinyie[3], Tan Lit Ken[4], Soo-Fen Fam[5], Azlyna Senawi[1] and Wan Nur Syahidah Wan Yusoff[1]

[1]Faculty Industrial Sciences & Technology, Universiti Malaysia Pahang, Lebuhraya Tun Razak, 26300 Gambang, Kuantan, Pahang
[2]School of Mathematical Sciences, Faculty Science and Technology, Universiti Kebangsaan Malaysia, 43600 UKM Bangi, Selangor
[3] Department of Mathematics, Faculty of Science, Universiti Putra Malaysia, 43400 UPM Serdang, Selangor
[4]Malaysia-Japan International Institute of Technology (MJIIT), Universiti Teknologi Malaysia Kuala Lumpur, Jalan Sultan Yahya Petra (Jalan Semarak), 54100 Kuala Lumpur
[5]Centre of Technopreneurship Development, Universiti Teknikal Malaysia Melaka, Hang Tuah Jaya, 76100, Durian Tunggal, Melaka

Email: chuanzunliang@ump.edu.my

Abstract. Due to the limited of historical precipitation records, agglomerative hierarchical clustering algorithms widely used to extrapolate information from gauged to ungauged precipitation catchments in yielding a more reliable projection of extreme hydro-meteorological events such as extreme precipitation events. However, identifying the optimum number of homogeneous precipitation catchments accurately based on the dendrogram resulted using agglomerative hierarchical algorithms are very subjective. The main objective of this study is to propose an efficient regionalized algorithm to identify the homogeneous precipitation catchments for non-stationary precipitation time series. The homogeneous precipitation catchments are identified using average linkage hierarchical clustering algorithm associated multi-scale bootstrap resampling, while uncentered correlation coefficient as the similarity measure. The regionalized homogeneous precipitation is consolidated using K-sample Anderson Darling non-parametric test. The analysis result shows the proposed regionalized algorithm performed more better compared to the proposed agglomerative hierarchical clustering algorithm in previous studies.

1. Introduction

The projection accuracy of the magnitudes and frequency of extreme hydro-meteorological such as extreme precipitation events is highly dependent on the availability and quality of the historical precipitation records. However, the accuracy of the projection for extreme hydro-meteorological at East-

Coast region of Peninsular Malaysia based on at-site frequency analysis is always uncertain due to the limitation of historical precipitation records. In order to reduce the uncertainty of projection, extrapolation information from gauged to ungauged precipitation catchments based on regionalization is indeed much needed. For the past decades, several regionalization algorithms have been widely applied in hydro-meteorological studies, including the agglomerative hierarchical clustering algorithm [1-7], the canonical correlation algorithm [8], the neural network algorithm [9,10] and the principal component algorithm [11,12].

For the regional frequency analysis of hydro-meteorological studies carried out in Malaysia, the agglomerative hierarchical clustering are frequently applied as the regionalization algorithm. As examples, Ahmad et al. [5] studied the efficiency of 77 regionalized algorithms in identifying the homogenous precipitation catchments based on precipitation amount, which the 77 regionalized algorithms formed from all possible combinations of 7 agglomerative hierarchical clustering algorithms and 11 similarity measures. They concluded that the combination of complete linkage hierarchical clustering algorithm and correlation similarity measure is the most appropriate algorithm to identify the homogeneous precipitation catchments in Peninsular Malaysia. They also proposed to determine the optimum number of homogeneous regions when most of the internal clustering validation indices indicated the similar results.

Zakaria et al. [7] carried out another regional frequency analysis focused on East-Coast region, Malaysia. The Ward's minimum variance hierarchical clustering applied in their studies in identifying the homogenous precipitation catchments, which the topographic characteristics of precipitation catchments such as elevation, latitude and longitude are used as attributes of regionalization. Meanwhile, the regionalized homogenous precipitation catchments consolidated using discordant and heterogeneity tests.

Most of the regional frequency analyses carried out in previous studies are mainly focused on the stationary precipitation time series. However, past studies showed that the regional phenomenon, such as the monsoon, El Nino-Southern Oscillation, Indian Ocean Dipole and Madden-Julian Oscillation has created non-stationary components in climate variability [13,14]. In addition, the discordant and heterogeneity tests more appropriate for low skewed data [15]. Therefore, an efficient regionalization algorithm which more suitable for non-stationary precipitation time series is absolutely necessary to result more trustworthy information for projection of extreme hydro-meteorological events.

The main objective of this study is to propose an efficient regionalization algorithm to identify the homogeneous precipitation catchments for non-stationary precipitation time series, which unify the average linkage hierarchical clustering algorithm and multi-scale bootstrap resampling. On the other hands, K-sample Anderson Darling non-parametric test is also proposed to consolidate the results of the proposed regionalization algorithm.

2. Study Area

Agriculture sector plays a substantial role in ensuring food security, economic growth, socioeconomic improvement, employment generation and poverty reduction of the nation in Malaysia [16,17]. Therefore, the 20 precipitation catchments (figure 1) located at Kuantan River Basin, Pahang are selected as the study area. This is due to the Kuantan River Basin is one of the significant tributaries that irrigates the majority of the rural, urban, agriculture and industrial areas of Kuantan District [18], and frequently exposed to risks of flood occurrence during the Northeast Monsoon. The topography characteristics and descriptive statistics of monthly precipitation amount for 20 precipitation catchments are illustrated in table 1. The monthly precipitation amount applied this study covers the period February 2010 until November 2014.

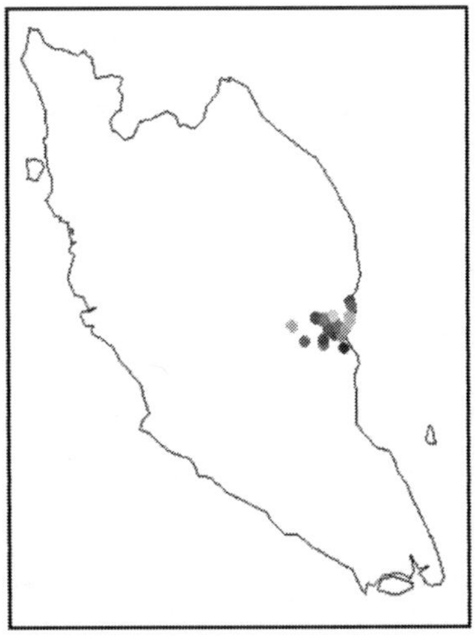

Figure 1. Location of 20 precipitation catchments in the Kuantan River Basin, Pahang.

Table 1. The topography and descriptive statistics of the monthly precipitation amount for 20 precipitation catchments in Kuantan River Basin, Pahang.

Catchment	Catchment Name	Elevation(m)	Latitude(N)	Longitude(E)	MN(mm)	CV(%)	SKEW	KURT
01	Sri Damai	14.9	03°44'47''	103°13'20''	90.5552	165.6909	3.3014	14.5292
02	Paya Bungor	34.7	03°41'30''	102°56'00''	158.1138	77.1298	1.0290	1.3189
03	Kampung Pulau Manis	37.4	03°39'10''	103°07'10''	181.5224	71.5749	2.0425	7.3836
04	Kampung Bahru	7.6	03°37'50''	103°18'55''	179.6224	92.8763	2.3708	6.9554
05	JKR Gambang	41.3	03°42'20''	103°07'00''	234.0069	66.1214	1.5544	3.3734
06	Paya Besar	6.0	03°46'20''	103°16'50''	162.8810	93.1208	2.2039	8.2500
07	Kampung Sungai Soi	11.9	03°43'50''	103°18'00''	210.9621	94.1730	2.6879	8.4473
08	Ladang Ulu Lepar	91.7	03°50'25''	102°48'00''	167.2017	65.7289	0.9717	1.0498
09	Ladang Mentiga	9.4	03°48'58''	103°19'30''	199.1931	73.1688	1.1191	2.0870
10	Panching	71.4	03°48'53''	103°09'38''	234.0707	86.9853	2.0654	6.9068
11	Paya Pinang	6.7	03°50'30''	103°15'30''	209.7328	93.7746	3.0494	13.5455
12	JPS Pahang	10.3	03°48'30''	103°19'45''	180.8603	97.8549	2.7916	12.2130
13	Ladang Jeram	-1.4	03°53'40''	103°23'00''	210.9414	119.548	3.3951	15.1796
14	Sungai Lembing	33.1	03°55'00''	103°02'10''	245.6690	62.2638	0.3866	-0.4114
15	Ladang Nada	16.9	03°54'30''	103°06'20''	227.9431	73.9733	1.6809	4.9231
16	Ladang Kuala Reman	29.9	03°54'00''	103°08'00''	201.8638	74.4579	0.7849	1.0727
17	Balok	4.1	03°56'40''	103°23'00''	220.8241	110.4599	2.6069	8.4019
18	Bukit Sagu	20.9	03°56'14''	103°12'52''	511.7517	79.6973	2.7206	12.0137
19	Kampung Cherating	9.0	04°05'35''	103°22'50''	221.2155	108.1321	2.5446	8.0246
20	Kampung Sungai Ular	58.5	04°30'00''	103°23'40''	228.7362	108.5308	3.0108	11.2957

***Note:** **MN**: Monthly average precipitation amount; **CV**: Coefficient of variation; **SKEW**: Skewness; **KURT**: Kurtosis

3. Methodology

Figure 2 illustrated the overview of the proposed regionalization in identifying the homogenous precipitation catchments for non-stationary precipitation time series.

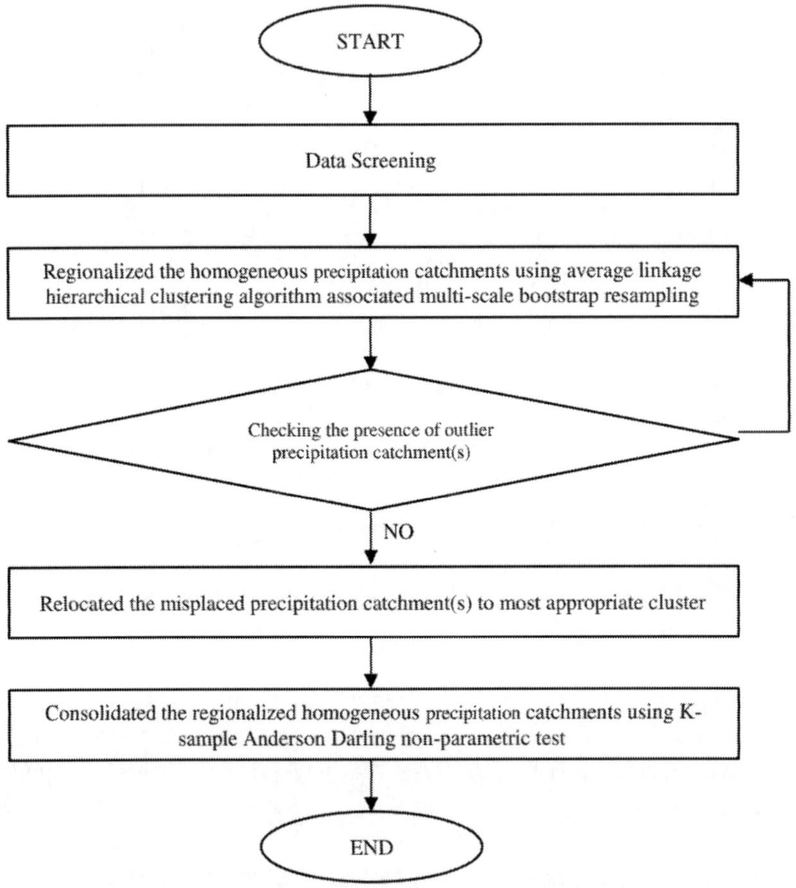

Figure 2. The overview of the proposed regionalized algorithm for non-stationary time series.

3.1 Data screening

Suppose that $\mathbf{Y} = \left[y_{ij} \right]_{MN}$ represents the *ith* month precipitation amount for *jth* precipitation catchments, such that $i,(j) = 1,\ 2,\ \ldots,\ M,(N)$, and $y_{ij} \in \mathbf{Y}$, the available monthly precipitation amount, $y_{ij}^{miss} \in \mathbf{Y}$, the missing precipitation amount. In this study, the y_{ij}^{miss} can be imputed based on the algorithm as below:

$$y_{ij}^{miss} = \left(\prod_{i=1}^{m_i} y_{ij} \right)^{\frac{1}{m_i}} \quad (1)$$

which is the superior imputation missing precipitation amount proposed by Saeed et al. [19]. Therefore, a complete $\hat{\mathbf{Y}} = \left[\hat{y}_{ij} \right]_{MN}$ is resulted. On the other hands, it should be highlighted that $\hat{\mathbf{Y}}$ is independent of the inherent assumptions of stationary time series after testing using Mann-Kendall test [20]. This is due to the Mann-Kendall test show that there is significant serial correlation over time presented in $\hat{\mathbf{Y}}$ although pre-whitening filter is applied.

3.2 Regionalized homogeneous precipitation catchments

The average linkage agglomerative hierarchical clustering algorithm is selected in this study due to this algorithm yielded more reasonable results compared to the single linkage and Ward's minimum variance agglomerative hierarchical algorithm, which the topography and descriptive statistics of monthly precipitation amount (table 1) as attributes of regionalization, and uncentered correlation coefficient as the similarity measures. In principle, the average linkage is performed $N-1$ successive fusion by agglomerating the closest pair of precipitation catchments based on the predetermined similarity measure, until N precipitation catchments fused as a single dendrogram.

Suppose that $\theta\left(\hat{\mathbf{Y}}_{a_0}, \hat{\mathbf{Y}}_{b_0}\right)$ represents the minimum value of predetermined similarity measure for a single cluster, $\hat{\mathbf{Y}}_a$, constitutes the fused pair of $\hat{\mathbf{Y}}_{a_0}$ and $\hat{\mathbf{Y}}_{b_0}$ clusters. A new dendrogrammatic similarity measure, $\tilde{\theta}\left(\hat{\mathbf{Y}}_a, \hat{\mathbf{Y}}_b\right)$, between a new single cluster, $\hat{\mathbf{Y}}_a$, and remaining infused clusters, $\hat{\mathbf{Y}}_b$ is yielded by updating the dendrogrammatic similarity function [21] given as

$$\tilde{\theta}\left(\hat{\mathbf{Y}}_a, \hat{\mathbf{Y}}_b\right) = \frac{n_{a_0}}{n_{a_0} + n_{b_0}} \tilde{\theta}\left(\hat{\mathbf{Y}}_{a_0}, \hat{\mathbf{Y}}_a\right) + \frac{n_{b_0}}{n_{a_0} + n_{b_0}} \tilde{\theta}\left(\hat{\mathbf{Y}}_{b_0}, \hat{\mathbf{Y}}_b\right) \tag{2}$$

where n_{a_0} and n_{b_0} represents the number of precipitation catchments in the cluster $\hat{\mathbf{Y}}_{a_0}$ and $\hat{\mathbf{Y}}_{b_0}$, and the uncentered correlation coefficient as the similarity measure, $\bar{R}$, are given as

$$\bar{R} = 1 - \frac{\sum_{i=1}^{M}\left(\hat{y}_{ij}\hat{y}_{ik}\right)}{\sqrt{\sum_{i=1}^{M}\left(\hat{y}_{ij}\right)^2 \sum_{i=1}^{M}\left(\hat{y}_{ik}\right)^2}}; \; \hat{y}_{ij} \in \hat{\mathbf{Y}}_a, \hat{y}_{ik} \in \hat{\mathbf{Y}}_b \tag{3}$$

Since identifying the homogeneous precipitation catchments based on a single dendrogram is very uncertain, therefore this study suggested the use of multi-scale bootstrap to reduce the uncertainty in identifying the homogeneous precipitation catchments and outlier precipitation catchment(s). The multi-scale bootstrap resampling is an approximately unbiased test based on the bootstrap probabilities.

3.3 Consolidated the regionalized precipitation catchments

The K-sample Anderson Darling non-parametric test, λ_{AD}^2, is a rank test, which independent on any assumption of statistical assumption [15]. In this study, the K-sample Anderson Darling non-parametric test is selected due to the distribution for most of precipitation amount (table 1 and figure 3) is highly positively skewed, therefore the discordant and heterogeneity tests might be yielded a misleading result.

Suppose that $\hat{\mathbf{Y}}$ is regionalized into Q regions of homogeneous precipitation catchments, which each region can be represented by $\mathbf{C}_q = [\hat{y}_{il}]_{MQ}$; $i,(l) = 1,2,3,\ldots,\mathrm{M},(Q)$ and $\mathbf{V} = [v_r]_{(MQ)} = vec(\mathbf{C}_q)$; $r = 1, 2, \ldots, MQ$ is a vector of the pooled precipitation amount, such that $v_1 < v_2 < \ldots < v_{MQ}$, which $vec(\cdot)$ is vectorization. The $\mathbf{C}_q$ is considerable statistical heterogeneity if

$$\left(\phi_{\mathbf{C}_q} = \frac{\lambda_{AD}^2 - MQ + 1}{\sqrt{\sigma_{\lambda_{AD}^2}^2}} \right) \geq \phi_{(MQ-1,\alpha)} \tag{4}$$

where λ_{AD}^2 and its variance, $\sigma_{\lambda_{AD}^2}^2$, is given in equations (5) and (6), respectively.

$$\lambda_{AD}^2 = \frac{1}{(MQ)^2} \sum_{r=1}^{MQ-1} \left(\frac{\left(\tilde{n}(MQ) - rM\right)^2}{r(MQ-r)} \right) \tag{5}$$

$$\sigma_{\lambda_{AD}^2}^2 = \frac{\Gamma(MQ-3)}{\Gamma(MQ)} \left(\omega_1 + \omega_2 (MQ) + \omega_3 (MQ)^2 + \omega_4 (MQ)^3 \right) \tag{6}$$

The $\tilde{n}$ represents the number of precipitation amount in the lth precipitation catchments that are not greater than v_r, and ω_1, ω_2, ω_3 and ω_4 are defined in equations (7), (8), (9) and (10), respectively.

$$\omega_1 = 2Q(\gamma Q - 2\gamma + 3Q) \tag{7}$$

$$\omega_2 = 2\gamma \left(\frac{Q^2(4\delta - 1) + 3Q + 2\delta}{\delta} \right) + 2\beta(\gamma - 3) \tag{8}$$

$$\omega_3 = 2\delta\left(Q^2 + \beta + 2\right) + 2\left(4\gamma(Q-1) - \beta(7\gamma - 2) + 3 - 2Q^2\right) \tag{9}$$

$$\omega_4 = 2\delta(2Q - 3\beta - 2) - 2(3Q + 2\gamma - 3) \tag{10}$$

where $\beta = \frac{1}{QM}$, $\gamma = \sum_{r=1}^{QM-1} \frac{1}{r}$, and $\delta = \sum_{i=1}^{QM-2} \sum_{l=1+1}^{QM-1} \left(\frac{1}{l(QM-i)} \right)$.

4. Analysis Results

The five-number summary of monthly precipitation for 20 precipitation catchments located Kuantan River Basin, Pahang are illustrated as figure 3. On average, it can be observed (figure 3) that Catchments 1 and 18 respectively received the lowest and highest monthly precipitation compared to other 18 precipitation catchments. The description also can be consolidated by the result of nonparametric pairwise comparison, which indicates that the average monthly precipitation amount of Catchment 18 is significantly higher, while the average monthly precipitation amount of Catchment 01 is significantly lower compared to other precipitation catchments. On the other hands, the non-parametric multiple comparison tests also indicates that there are significant differences between Catchment 05 with Catchments 02 and 08 and Catchment 14 with Catchments 02, 03, 06, 08, and 18 at significant level of 0.05.

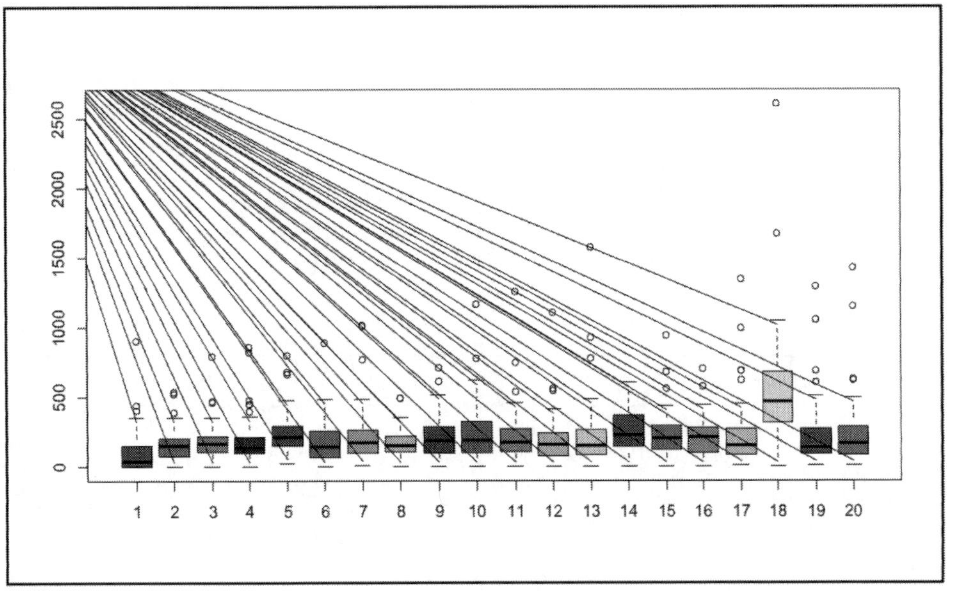

Figure 3. The five-number summary of precipitation amount for 20 precipitation catchments located Kuantan River Basin, Pahang.

In order to authenticate the efficiency of the proposed algorithm, Ward^{LR} and Complete^{LR}, algorithms respectively proposed by Ahmad et al. [5] Zakaria et al. [7] are applied as comparisons. Figure 4 illustrated the regionalized homogeneous precipitation catchments based on Ward^{LR}, Complete^{LR}, Complete^{MLR} and the proposed algorithm (Average^{P}), while table 2 illustrated the resulted p-value of K-sample Anderson Darling non-parametric test for various sample sizes. It should be highlighted that this study replaced Euclidean distance as similarity measure for Complete^{MLR} algorithm instead of correlation coefficient (Complete^{LR}). This is due to regionalize the precipitation catchments using Complete^{LR} (figure 4(b)) yielded an unreasonable result, which the 16 precipitation catchments classify as outlier catchments (orange color points).

In principle, the distribution of spatial and temporal is varied between the coastal and inland region [22,23]. Therefore, this study concluded the proposed regionalized algorithm performed more better compared to Ward^{LR}, Complete^{LR}, and Complete^{MLR} although K-sample Anderson Darling non-parametric test shows that all of the regionalized regions using Ward^{LR}, Complete^{LR}, Complete^{MLR} and Average^{LR} algorithms are significantly homogeneous. The proposed algorithm in this study isolated the precipitation catchments located Kuantan River Basin, Pahang into 2 homogeneous regions, namely inland and coastal regions, where 3 out of 20 precipitation catchments classify as outlier precipitation catchments. This study also relocated the misplaced of Catchment 20 as the proposed regionalized algorithm misplaced Catchment 20 in the inland region.

IOP Conf. Series: Materials Science and Engineering **342** (2018) 012070 doi:10.1088/1757-899X/342/1/012070

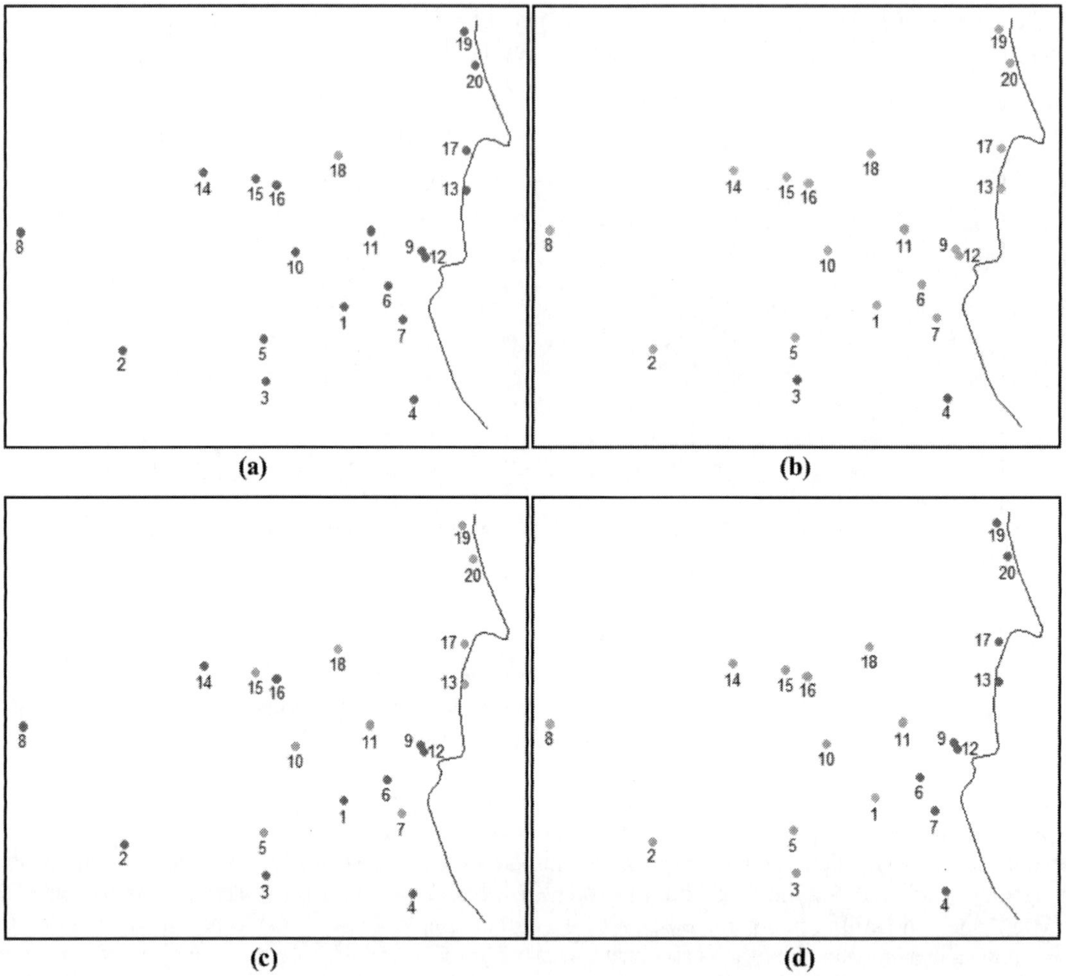

Figure 4. (a), (b), (c) and (d) are the regionalized homogeneous precipitation catchments based on Ward^{LR}, Complete^{LR}, Complete^{MLR} and Average^{P}, respectively.

Table 2. The p-value of K-sample Anderson Darling nonparametric test based on $Ward^{LR}$,$Complete^{LR}$, $Complete^{MLR}$, and $Average^{P}$ algorithms, respectively.

Algorithm	Cluster	Catchment	P-value					
			100	500	1000	5000	10000	50000
$Ward^{LR}$	1	01, 02, 03, 04, 05, 06, 07, 08, 09, 10, 11, 12, 13, 14, 15, 16, 17, 19, 20	1.0000	1.0000	1.0000	1.0000	1.0000	1.0000
$Complete^{LR}$	1	03, 04	0.6500	0.5600	0.5390	0.5178	0.5170	0.5155
	2	06, 11	0.5800	0.5740	0.5750	0.5708	0.5668	0.5661
$Complete^{MLR}$	1	01, 02, 03, 04, 06, 08, 09, 12, 14, 16	1.0000	1.0000	1.0000	1.0000	1.0000	1.0000
	2	05, 07, 10, 11, 13, 15, 17, 19, 20	0.6100	0.5440	0.5500	0.5836	0.5804	0.5839
$Average^{P}$	1	04, 06, 07, 09, 12, 13, 17, 19, 20	0.6200	0.5980	0.5700	0.5558	0.5648	0.5699
	2	02, 03, 05, 10, 11, 14, 15, 16	0.5000	0.3840	0.3930	0.4256	0.4246	0.4266

5. Conclusion

This study proposed an efficient regionalized algorithm in identifying homogeneous precipitation catchments for non-stationary time series. The analysis result shows that the proposed regionalized algorithm successfully isolated 20 precipitation catchments located Kuantan River Basin, Pahang into 2 homogenous regions, namely inland and coastal regions. In addition, the analysis result shows that the proposed regionalized algorithm performed more better compared to the proposed regionalized algorithms in previous studies. For future works, this study suggested to further investigation of efficiency for other hierarchical clustering algorithms associated multi-scale bootstrap resampling, which correlation coefficient and absolute correlation coefficient as similarity measures.

Acknowledgement

The authors would like to express appreciation to Department of Irrigation and Drainage for providing the precipitation data used in this study. A word of appreciation also goes to the Universiti Pahang Malaysia for providing the flagship research grant RDU150393 and the internal research grant RDU1703184. The authors also extend appreciation to all reviewers who providing valuable comments.

References

[1] Guttman NB 1993 The use of L-moments in the determination of regional precipitation climates *Journal of Climate* **6** (12) 2309-2325.

[2] Burn DH, Zrinji Z and Kowalchuk M 1997 Regionalization of catchments for regional flood frequency analysis *Journal Hydrologic Engineering* **2** (2) 76-82.

[3] Venkatesh B and Jose MK 2007 Identification of homogeneous rainfall regimes in parts of Western Ghats region of Karnataka *Journal of Earth System Science* **116** (4) 321-329.

[4] Ngongondo CS, Xu C-Y, Tallaksen LM, Alemaw B and Chirwa T 2001 Regional frequency analysis of rainfall extremes in Southern Malawi using the index rainfall and L-moments approaches *Stochastic Environmental Research and Risk Assessment* **25** (7) 939-955.

[5] Ahmad NH, Othman IR and Deni SM 2013 Hierarchical cluster approach for regionalization of Peninsular Malaysia based on the precipitation amount *Journal of Physics: Conference Series* **423** 012018.

[6] Pansera WA, Gomes BM, Boas MAV and Mello ELd 2013 Clustering rainfall stations aiming regional frequency analysis *Journal of Food, Agriculture & Environment* **11** (2) 877-885.

[7] Zakaria R, Radi NFA, Ngongondo CS and Azman MA 2014 A regional frequency analysis of annual maximum rainfall in Terengganu using the index rainfall and L-moments approaches *Proc. of the 10th Indonesia-Malaysia-Thailand Growth Triangle Int. Conf. on Mathematics, Statistics and Its Applications, 14th-16th October 2014, Kuala Terengganu* pp 667-674.
[8] Cavadias GS, Ouarda TBMJ, Bobée B and Girard C 2001 A canonical correlation approach to the determination of homogeneous regions for regional flood estimation of ungauged basins *Hydrological Sciences Journal* **46** (4) 499-512.
[9] Hall MJ, Minns AW and Ashrafuzzaman AKM 2002 The application of data mining techniques for the regionalisation of hydrological variables *Hydrology and Earth System Sciences* **6**(4) 685-694.
[10] Lin G-F and Chen L-H 2006 Identification of homogeneous regions for regional frequency analysis using the self-organizing map *Journal of Hydrology* **324** (1-4) 1-9.
[11] Baeriswyl P-A and Rebetez M 1997 Regionalisation of precipitation in Switzerland by means of principal component analysis *Theoretical and Applied Climatology* **58** (1-2) 31-41.
[12] Singh KK and Singh SV 1996 Space-time variation and regionalization of seasonal and monthly summer monsoon rainfall on sub-Himalayan region and Gangetic plains of India *Climate Research* **6**(3) 251-262.
[13] Tangang FT, Juneng L, Salimun E, Sei KM, Le LJ and Muhamad H 2012 Climate change and variability over Malaysia: Gaps in sciences and research information *Sains Malaysiana* **41** (11) 1355-1366.
[14] Agilan V and Umamahesh NV 2016 Is the covariate based non-stationary rainfall IDF curve capable of encompassing future rainfall changes? *Journal of Hydrology* **541** (Part B) 1441-1455.
[15] Viglione A, Laio F and Claps P 2007 A comparison of homogeneity tests for regional frequency analysis *Water Resources Research* **43** W03428.
[16] Alam MM, Morshed MG, Siwar C and Murad MW 2012 Initiatives and challenges of agricultural crop sector in East Coast Economic Region (ECER) development projects in Malaysia *American-Eurasian Journal Agriculture & Environmental Sciences* **12** (7) 922-931.
[17] Jackson R 2012 *Occupy World Street: A Global Roadmap for Radical Economic and Political Reform* (Hartford: Chelsea green).
[18] Win NL and Win KN 2014 The probability distributions of daily rainfall for Kuantan Basin in Malaysia *Int. Journal of Science and Research* **3**(8) 977-983.
[19] Saeed GAA, Chuan ZL, Zakaria R, Yusoff WNSW and Salleh MZ 2016 Determination of the best single imputation algorithm for missing rainfall data treatment *Journal of Quality Measurement and Analysis* **12** (1-2) 79-87.
[20] Mann HB 1945 Nonparametric tests against trend *Econometrica* **13**(3) 245-259.
[21] Lance GN and Williams WT 1967 A general theory of classificatory sorting strategies 1. hierarchical systems *The Computer Journal* **9**(4) 373-380.
[22] Hamada J-I, Yamanaka MD and Mori S 2008 Differences of rainfall characteristics between coastal and interior areas of central Western Sumatera, Indonesia *Journal of the Meteorological Society of Japan* **86** (5) 593-611.
[23] Azman MA, Zakaria R, Satari SZ and Radi NFA 2017 Entropy of stable seasonal rainfall distribution in Kelantan *AIP Conf. Proc.* **1842** 020002.

Nano-silica as the go material on heat resistant tunnel lining

Faizah Omar, S A Osman and A Mutalib

Department of Civil & Structural Engineering, Faculty of Engineering and Built Environment, Universiti Kebangsaan Malaysia, 43600, Bangi, Selangor Darul Ehsan, Malaysia.

E-mail: faizah.omar@siswa.ukm.edu.my, saminah@ukm.edu.my

Abstract. This paper is concerned with passive fire protection method of protective concrete mix that is made up of fly ash, polypropylene fibre, and nano-silica. Nano-silica is focused on as the innovative material to be used in the composition of the protective concrete mix. The previous experimental studies which analyse the performance of passive fire protection on tunnels are discussed. This paper also discusses passive fire protection. The fire protection materials and behaviour analyses of tunnel structure are also presented. At the end of the paper, the recommendation of the optimum composition concrete material with fly ash, polypropylene fibre and nano-silica as tunnel lining fire protective materials is proposed.

1. Introduction

Since the early 1970s, there have been numerous minor, and several major fire incidents in underground transportation systems are increasing the interest in structural fire safety of underground facilities. The high temperature in a very confined space is able to damage the concrete lining and threaten the safe operation of the tunnel. In a major fire, the high temperature may also cause the tunnel lining to collapse.

When structures such as tunnels are exposed to a high temperature of 1000°C in a short span of time the structural, physical, and mechanical properties of the structure could be affected and consequently leads to irreversible loss of strength and instability of the tunnel structure. These effects could be due to the micro-cracking and spall generated by high vapour pressure from the temperature of the smoke and the radiation from the fire. These effects can be prevented, or at least minimised, by choosing the right material to provide fire protection. Massive amounts of wastes, such as fly ash and silica fume produced by the industries daily, can be added to or replaced a certain percentage of cement in a concrete mix.

2. Protective Concrete Mix

The three primary failures in tunnel fire are a failure of segments and segmental joints and yielding of the joint bolts [1]. These failures are mostly caused by the explosive spalling that takes place during sudden elevating temperature of fire in the tunnel. Few methods can be used to reduce the chances of explosive spalling such as protective concrete mix, boards and heat resistant segments. Bottomline is that focus should be given to the protective concrete mix. The right ratio of materials in the concrete mix will be able to suppress the possibility of spalling and subsequently avoid the failure aforementioned. The usage of supplementary cement replacements such as fly ash, polypropylene fibres, air-entraining agent, admixtures, silica fumes, and nano-silica is highly recommended while to minimise the effect of elevated temperature, the correct preventative measures with suitable material and proper insulation methods should be carried out.

Concrete is a porous material formed by combining cement, aggregates & steel reinforcements. Compared with steel structures, concrete structures have excellent fire resistance characteristics and are often used to protect steel from fire effects. Concrete has a low heat flow rate and high specific latent length compared to steel. The conductivity values of iron and steel are between 16 W / mK and 43 W / mK while the conductivity values of concrete are 0.62 W / mK to 3.3 W / mK [2]. The advantage of concrete when exposed to fire is to store strength if the exposure is not too long, and does not produce smoke and toxic waste at the end of the combustion. High-temperature exposure affects all concrete components which ultimately result in severe concrete loss strength. Detailed studies showed that the strength of the concrete is almost unaffected until the temperature is 200 ° C to 300 ° C, but starts to decrease at a higher temperature [3].

Fly ash is a waste material from coal-burning power plants. It has been used extensively as supplementary material in high strength concrete due to its exceptional performance as a pozzolanic material. When 50% of the cement is replaced with fly ash, the fly ash will act as a pozzolanic material and micro filter in concrete [4]. Adding 15 to 25% fly ash in a concrete mix help to increase the tensile strength of the concretes. However, adding a higher percentage of fly ash reduces the tensile strength of concrete mix slightly [5]. Regarding compressive strength, however, compressive strength decreases with the increase of fly ash content [4]. To show that fly ash's strength increases with age, the highest increase in strength from 28 to 365 days was a specimen with 40% fly ash by 45% and the lowest increase in strength from 28 to 365 days was a specimen with 45% fly ash with only 39.27% [4]. This is due to the cement that continues to hydrate. Under the high heating condition, fly ash shows superior performance than pure cement concrete at temperatures below 600°C [6]. After being exposed to heat, the specimen containing fly ash only produced fine cracks and did not spall. To maintain maximum strength and durability after exposure, replacing cement with 30% fly ash is recommended [6].

Polypropylene (PP) fibre is famous for its fire-resistant properties. In a normal condition, adding PP fibre to the mix does not affect the concrete's compressive strength positively because of the weak interfacial bond between the fibre and cement fly ash grains [7]. Regarding indirect tensile strength, on the other hand, the inclusion of a high volume of fly ash and 0.75% PP fibre resulted in an increase of 53% at 28 days of curing [8]. This is due to the microstructural densification in the transition zone between the matrix and the fibre [8]. In elevated temperatures, PP fibres create pores in concretes when they disintegrate at approximately 160 °C [9]. The powerful vapour pressure will then be released into the additional pores or voids formed by the PP fibre, thereby reducing cracking and spalling. PP fibre is added into the concrete mixture with quantity between 1 to 2 kg / m^3 of concrete. However, for real tunnel development projects, a 1.5 kg / m^3 PP fibre is recommended [10]. Note that rates greater than 1.5 (kg of fibre PP) / (m^3 concrete), however, result in severe reduction of workability and densification ability and, therefore, increase in pore space[11]. Studies found that although fibre enhancement improves concrete performance, it also reduces workability by 2-8% [12]. This problem can be avoided by adding fly ash. Figure 1 shows the image of scanning electron microscopy (SEM) of control specimen with significant cracks after heated to 750°C. Figure 2 then shows that after the PP fibre specimen being exposed to 500°C however, only microcrack can be seen as the fibres have completely melted and vaporised.

Nano-silica (NS), the focus of this paper, is a material which exhibits advanced pozzolanic properties that enhance the microstructure of cement. During cement hydration, nano-silica reacts with calcium hydroxide ($Ca(OH)_2$) to form calcium silicate hydrate (CSH) gel [13][14]. CSH gel is the main component responsible for the high strength of siliceous mixed specimens. The primary purpose of using nano-silica is to enhance a material's hydration, thus raising pozzolanic activity, and to act as a micro-filler in cement [14][15]. Since nano-silica can produce high-density calcium silicate, it can accelerate hydration process and thus increase the initial strength of fly ash and nano-silica mix. Since there is a risk of the mixture undergoing exaggerated self-desiccation and cracks, high amounts of nano-silica compensated with the right water and superplasticisers ratio are recommended [14].

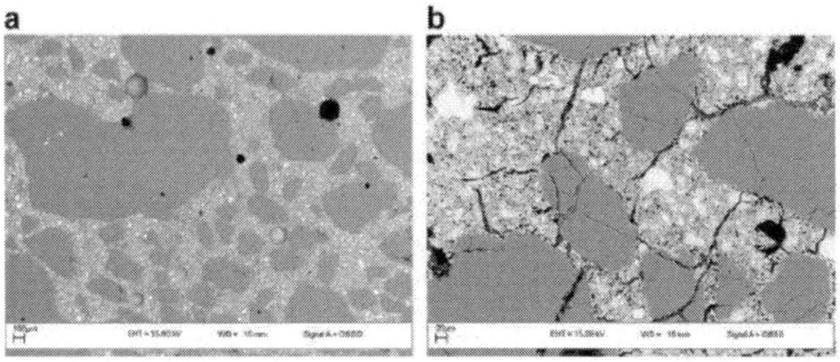

Figure 1. SEM images of control specimen at (a) 20°C and (b) 750°C [9].

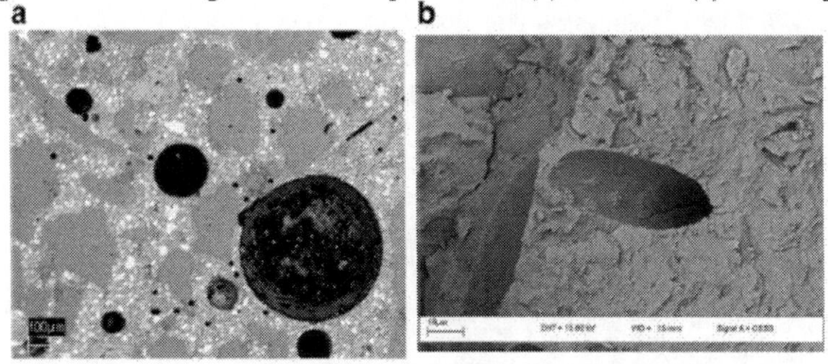

Figure 2. SEM Images of PP fibres RC (a) 20°C and (b) after exposure to 500°C [9].

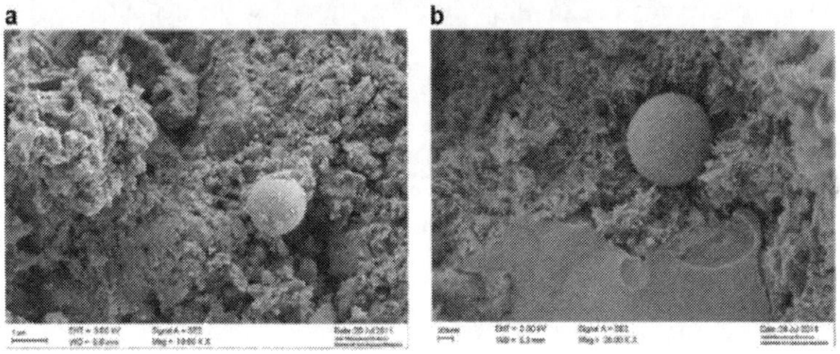

Figure 3. SEM images of fly ash and nano-silica specimen a)27°C and b) after exposure to 700°C [16].

An experiment was carried out to investigate the behaviour of nano-silica when added in a high volume of fly ash specimen and resulted in no improvement in strength when only 60 and 70% fly ash was used [15]. However, when 2% nano-silica was added to the mixture, the 28-day compressive strength of the specimen increased to 33 and 48%, respectively. This shows that the correct ratio of all substances is crucial to ensure that maximum strength is obtained. As a result of these observations, the addition of 2% nano-silica is recommended as the optimum value to achieve the highest compressive strength.

As a result of the better particle packing of the extremely small-sized nano-silica and the high pozzolanic activity of nano-silica reacting with fly ash as shown in Figure 3, the nano-silica and fly ash can enhance the strength of the calcium silicate hydrate structure, the specimens containing 37.5% fly ash and 7.5% nano-silica exhibited a maximum increase in strength [16]. When heated, all specimen containing nano-silica showed an obvious increase in compressive strength because of the filler effect of nano-silica and the enhancement of hydration process which then produced more high-density calcium silicate hydrate that will increase the reactivity of nano-silica and subsequently boost the strength [16]. Nano-silica, however, has a more negative influence on flexural strength when exposed to higher temperature [16]. High temperature causes the specimen to crack due to the vapour pressure. The addition of tensile stresses increased the expansion of cracks thus reducing the strength. These effects are believed can be improved with the addition of PP fibre to increase the tensile strength of the specimen thus reducing cracks and spalling. Therefore a combination of fly ash, PP fibre and nano-silica could provide the best protection method for concrete exposed to high temperature.

3. Discussion

Due to all the advantages that siliceous materials have to offer, this paper proposes using nano-silica together with fibre and fly ash to ensure the fire resistance of cementitious concrete mix. The composition of 30% of fly ash, 1.5 kg / m^3 of PP fibre and 5-10% of nano-silica is suggested to replace the cement composition. This can be used as a basis for investigating the performance of the concrete mix in tunnel lining.

The previous studies that were explained previously were from experimental studies on normal structural specimens. However, note should be taken that there is yet an experiment, reduced-scale or full-scale that investigates the usage of nano-silica in a heat-resistant tunnel segments.

Due to all the advantages that siliceous materials have to offer, this paper proposes using nano-silica together with fibre and fly ash to ensure the fire resistance of cementitious concrete mix. The effect of thermal, axial and combined loading should be investigated further to ensure that the chosen material is the best for the safety of the tunnels and tunnels users. This paper also recommends that more study should be carried out on the strength and mechanical, thermal and structural properties of nano-silica as it is believed that nano-silica can be an excellent fire protection instrument. A full-scale fire test should be done on new materials containing fly ash, fibre and nano-silica along with numerical simulation using CFD software and parametric study as to investigate the properties of the materials. A full-scale structural analysis should not only to examines the properties of segments but also the properties of the joint between segments and the segments with surrounding soils by taking into account initial load, construction load, service load, and thermal load.

4. Conclusion

This paper presented a collection of a literature review of passive fire protection materials to be used as tunnel lining fire protection. Gaps and recommendations are highlighted for future research. Since very few studies have been done on the benefits of nano-silica as a tunnel fire protection instrument, research on this topic could be very helpful to tunnel designers since the use of nano-silica has been proven to strengthen structures. To further understand the behaviour of nano-silica as a tunnel lining subjected to thermal loading, a very detailed study which involves analytical, experimental, and numerical simulation needs to be carried out.

Acknowledgement

The authors acknowledge the support of the Universiti Kebangsaan Malaysia and the financial support of the Ministry of Higher Learning Malaysia under grant AP-2015-011, DIP-2014-019 and FGRS/1/2015/TK01/UKM/02/4.

References

[1] Yan Z, Shen Y, Zhu H, and Lu Y 2016 *Tunn. Undergr. Sp. Tech.,* **53** 46-60.
[2] Naus N 2006 US Nuclear Regulatory Commission, Office of Nuclear Regulatory Research, Washington, DC.
[3] Khoury G 2008 *Proceedings of the Institution of Civil Engineers-Structures & Buildings,* **161** 135-45.
[4] Siddique R 2004 *Cement Concrete Res.* **34** 487-93.
[5] Lam L, Wong Y, and Poon C 1998 *Cement Concrete Res.* **28** 271-83.
[6] Poon C-S, Azhar S, Anson M, and Wong Y-L 2001 *Cement Concrete Res.* **31** 1291-300.
[7] Haque M, Langan B, & Ward M. 1984 *ACI Mater. J.* **81** 54-60.

[8] Kayali O 2004 *Mater. Struct.* **37** 318-27
[9] Biolzi L, Cattaneo S, and Rosati G 2008 *Cement Concrete Comp.* **30** 907-16.
[10] Smith K and Atkinson T 2010 *Propex Concrete Systems (International)*
[11] Pichler C, Lackner R, Mang H, and ASCE F 2006 *J. Struc. Eng.* **132** 961-9.
[12] Topcu I and Canbaz M 2007 *Constr. Build. Mater.* 21 1486-91.
[13] Li H, Xiao H, Yuan J, and Ou J 2004 *Composites Part B: Engineering,* **35** 185-9.
[14] Shaikh F, Shafaei Y, and Sarker P 2016 *Constr. Build. Mater.* **115** 690-98.
[15] Shaikh, F, Supit S, and Sarker P 2014 *Materials and Design* **60** 433-42.
[16] Ibrahim R, Hamid R, and Taha M 2012 *Constr. Build. Mater.* **36** 779-86.

Characterization of industrial wastes as raw materials for Emulsified Modified Bitumen (EMB) formulation

Mohd Najib Razali, Syarifah Nur Ezatie Mohd Isa, Noor Adilah Md Salehan, Musfafikri Musa, Mohd Aizudin Abd Aziz, Abdurahman Hamid Nour and Rosli Mohd Yunus

Faculty of Chemical and Natural Resources Engineering, Universiti Malaysia Pahang, Malaysia .

Email: najibrazali@ump.edu.my

Abstract. This study was conducted to characterize industrial wastes for formulation of emulsified modified bitumen (EMB) in relation to their physical characteristic and elemental composition. This analysis will give information either raw materials from industrial wastes can be used for EMB formulation. Bitumen is produced from crude oil that is extracted from the ground which categorizes the crude oil as one of the non-renewable form of product. A vast environmental problem issues arises in Malaysia cause by the excessive manufacturing activity that lead to a miss-management of industrial waste has leads to the used of industrial waste in the EMB formulation. Industrial waste such as polystyrene, polyethylene and used automotive oil can be used as alternative to formulate bitumen. Then a suitable emulsifier needs to be added to produce the final product which is EMB. The emulsifier will yield a charge depends on its properties to bind the oily bitumen with water. Physical characteristic studies were performed by thermogravimetric Analysis (TGA), differential scanning calorimetry (DSC), flash point test, density rest and moisture content test. Fourier Transform Infrared Spectroscopy (FTIR) analysis was measured to determine the material's molecular composition and structure.

1. Introduction

Bitumen is used mainly in road paving, roofing application, roads construction, waterproofing products, building materials and industrial coatings [1]. It is estimated that the current world consumption of bitumen is 102 million tonnes per year which 85% from it is used as binder for pavements, 10% for roofing application and the rest is used for other variety kind of purposes [2]. In Malaysia, the urge to solve environmental issue is getting crucial. Significantly, the industrial waste such as crumb rubber, polystyrene, plastics and used oil has been contributed to this problem. These wastes can be formulate to become modified bitumen and essentially with the help of suitable emulsifier, it can become emulsified modification bitumen (EMB).

Bitumen is considered as one of the non-renewable product when it is produced from crude oil. In order to prevent this renewable source replenished from earth as well as to maximize the use of waste produce by the industry, EMB can be formulated from industrial waste by using crumb rubber, polystyrene, plastics and used oil to replace the existed bitumen from crude oil. The EMB formulated will solve the dependency to the natural resources as well as the vast environmental problem that

IOP Conf. Series: Materials Science and Engineering **342** (2018) 012072 doi:10.1088/1757-899X/342/1/012072

consequently happen because of the overused of natural resources. The EMB formulated will be used for insulation and coating of building. Not only that, most EMB shows a better performance than unmodified one [3].

The increasing quantities of industrial waste in Malaysia have led to many serious environmental problems which require a better solution to handle it. The solid waste in Malaysia increased from 16,200 tonnes/day in 2001 to 19,100 tonnes/day in 2005 or an average of 0.8 kilogram per capita/ day [4]. In 2014, the Environmental Protection Expenditure (EPE) was reported at a total of RM2.244 billion where the manufacturing sector was the highest contributor to this EPE at RM1.619 billion [5] and since the solid wastes generated are exceeding the maximum amount, these generated wastes are finally disposed in landfills as the current implemented waste management is poorly managed. This method is not sustainable and need to be improved or else it would affect our environment, social and lead to economic lose [6]. Moreover, with the growing demand of crude oil at a rate of 4.5% annually during 2000-2005 to 6.2% per annum during 2006-2010 [7], it is consequently decrease the rate of natural resources in Malaysia. Thus, the development of emulsified modification bitumen from industrial waste is one of the alternative ways to reduce this problem.

As the EMB will be formulated through recycling technology, many environmental aspects that will lead to serious effect can be decreased such as energy consumption, high level of CO_2 emission to the environment during work construction, maintenance operation and worker's health [8]. Besides, bitumen emulsion presents much lower viscosity as it is diluted with water and can be applied at lower temperature as low as 80°C. Conventional binders are sprayed at 160-200°C so there is much more additional risk of fire and explosion [1]. Physical characteristic studies were performed by thermogravimetric Analysis (TGA), differential scanning calorimetry (DSC), flash point test, density rest and moisture content test. Fourier Transform Infrared Spectroscopy (FTIR) analysis was measured to determine the material's molecular composition and structure. The suitable raw material for formulation of emulsified modified bitumen will be choose based on six analysis conducted.

2. Materials and Methods

2.1. Materials

The raw materials used for this research are bitumen grade 80/100 from Kemaman Bitumen Company (KBC), high density polyethylene and polystyrene from landfill. Three type of recycled base oil collected from Kemaman, Banting and Kuantan Lastly, the waste sludge is from Kemaman.

2.2. Thermogravimetric Analysis (TGA)

TGA was performed to determine the decompose temperature and the composition in a sample. This analysis measures the percent weight loss of a test sample while it is heated at 10°C/min until approximately 900 °C under inert air. Firstly, 2.5 mg pf test specimen is weight and placed in an aluminium pan and sits upon a constantan disc on a platform in the TGA cell and it is hangs off a hook which is connected by a microgram arm to a tare pan. The loss in weight over specific temperature ranges provides an indication of the composition of the sample.

2.3. Differential Scanning Calorimetry (DSC)

This analysis was done to determine its melting temperature (Tm). Approximately 2.5 mg of test specimen is weight and placed in a aluminium pan and sits upon a constantan disc on a platform in the DSC cell with a chromel wafer immediately underneath. A chromel-alumel thermocouple under the constantan disc measures the sample temperature. An empty reference pan sits on a symmetric platform with its own underlying chromel wafer and chromel-alumel thermocouple. Heat flow is measured by comparing the difference in temperature across the sample and the reference chromel wafers. The sample was running for about 33 minute by specifying the final temperature of the sample at 350°C.

2.4. Fourier-transform infrared spectroscopy (FTIR)
FTIR analysis was measured to determine the material's molecular composition and structure. It measures the range of wavelengths in the infrared region that are absorbed by a material. This is accomplished through the application of infrared radiation (IR) to samples of a material. The sample's absorbance of the infrared light's energy at various wavelengths. The first step is to collect a background spectra to subtract from the test spectra to ensure the actual sample is all that is analyzed. A simple device called an interferometer is used to identify samples by producing an optical signal with all the IR frequencies encoded into it. Then, the signal is decoded by applying a mathematical technique known as Fourier transformation which generates the absorbance spectra showing the unique chemical bonds and the molecular structure of the sample material.

2.5. Flash Point Test
A flash point test was performed. Approximately 70 ml test specimen is placed into a test cup. The temperature of the test specimen is increased rapidly at first and then at a slower constant rate as the flash point was approached. At specified intervals a test flame was passed across the cup. The flash point is the lowest liquid temperature at which application of the test flame causes the vapors of the test specimen to ignite. To determine the fire point, the test is continued until the application of the test flame causes the test specimen to ignite and sustain burning for a minimum of 5s.

2.6. Density Test
Density test was performed by using gas pycnometer device to determine the density of each samples. To conduct the testing, the liquid sample is filled in a steel cup and then placed in the device. The gas pycnometer works by employing Archimedes' principle of fluid displacement, and Boyle's Law of gas expansion. A sealed sample chamber of known volume is pressurized to a target pressure with the displacement gas. Once stabilized, this pressure is recorded. A valve is then opened allowing the gas to expand into a reference chamber whose volume is also known. Once stabilized, this second pressure is recorded. This pressure drop ratio is then compared to the behavior of the system when a known volume standard underwent the same process.

2.7. Moisture Content Test
Moisture content analysis was performed to check on any moisture in each samples. About 2g of the sample is filled in its pan evenly and placed on the platform. The drying temperature was set to standard drying temperature which is 105 °C and the lid was closed.

3. Results and Discussions

3.1. Thermogravimetric Analysis (TGA)
Based on figure 1, the curve shows a continuous mass loss steps relating to the loss of volatile components such as moisture and additives, decomposition temperature of the polymer and combustion of final residue such as ash and filler [9]. According to [10], the multicomponent material in one polymer which are low-molecular mass compounds, polymeric material and inorganic additives can be divided by temperature. Thus, each composition and degradation temperature can be known.

By referring to figure 1, all of the sample shows the same degradation temperature but different weight loss for three existing component which for polystyrene sample, 0.4476% of its component degrade at temperature below 200°C, 97.16% at 402.9°C and 2.355% at 900°C while for high density polyethylene 2 (HDPE 2), 0.0852% of its component degrade at temperature below 200°C, 97.33% at 402.9°C, 2.606% at 900°C and lastly for high density polyethylene 1 (HDPE 1), 0.3235% degrade at temperature below 200°C, 98.17% at 402.9°C, 1.510% at 900°C. These results show that all of the sample consist of low-molecular-mass components such as water and additives as it completely losses its mass at low temperature. Furthermore, the major weight loss which is approximately 80% of each samples occur at 402.9°C. At temperature around 350-500°C, all carbon-carbon bonds is typically rupture by undergoing three possible mechanism; random scission, depolymerisation and side group elimination (Crompton, 2012). Besides that, the residue left which about 3% of the samples at the end of analysis can be classified as additives or fillers because inorganic additives are usually stable in an inert atmosphere up to 900°C or higher. The residue that remain above 600°C are normally associated with inorganic compounds such as silica particles, glass fibers or calcium carbonate [10].

Consequently, based on the result above, HDPE 1 sample should be able to tolerate with the bitumen formulation the most as it contains the least amount of filler and the highest purity of polymer with 98.17% of the total weight is the hydrocarbon component.

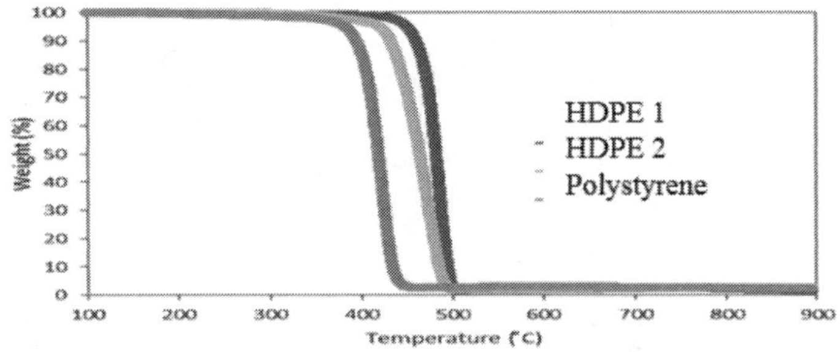

Figure 1. Graph of Weight Loss (%) against Temperature (°C).

3.2. Differential Scanning Calorimetry (DSC)

From figure 2, the HDPE 1 and 2 sample shows approximately the same trend which deviate from the polystyrene sample. This is because both sample is come from the same group of polymer. Graph form DSC will provide 3 different value of melting for each material which from the graph, it is denotes as A, B and C. At point A, it is the point where the material starts to melt. Point B depicts the peak temperature of melting where it indicates the maximum rate of melting. Lastly, the real melting point value is at point C when the graph reach steady state before it is going down the endotherm [10]. Therefore, in order to formulate a bitumen, the melting point of each sample need to determine so that the method developed for this research will ensure all of the material use is melted accordingly based on the temperature. In semicrystalline polymers such as polyethylene, the degree of crystallinity (% crystallinity) influences the degree of stiffness, hardness and heat resistance. The ratio of meltiing heat of the samples and the melting heat of the completely crystallized polyethylene was used to calculate the degree of crystallization. From table 2, it is clearly shown that polystyrene sample exhibit much deviate value from HDPE sample at 101.97% of crystallinity. This is because polystyrene is an amarphous polymer while HDPE is a type of semicrystalline polymer. Semi-crystalline polymers exhibit organized and tightly packed molecular chains also consist of at least some disorganized region whereas the polymer chains for amorphous plastics are more disorganized. When heat is applied on semicrystalline polymer, weaker bonds in polymer chains which is in disorganized areas break at Glass Transition Temperature. However, the bonds at organized areas do not break until the Melting Temperature. At this point the material changes from a solid-like to a liquid-like state [11].

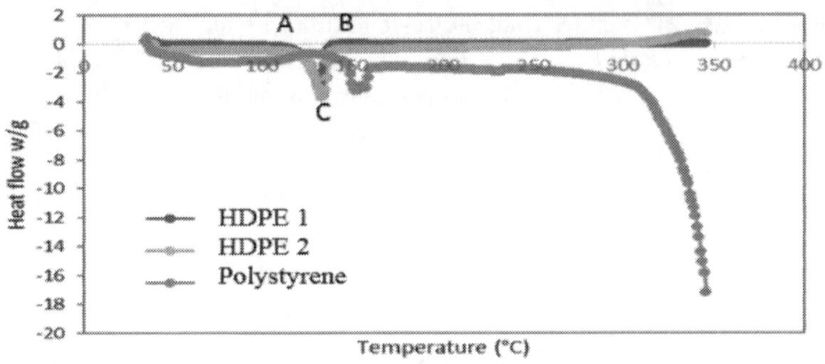

Figure 2. Graph of Heat Flow (w/g) against Temperature (°C).

Table 1. Degree of Crystallinity of Polymer Sample.

Sample	Melt Peak Temperature (°C)	Enthalpy (J/g)	Degree of Crystallinity (%)
HDPE1	125.72	172.2	58.65
HDPE2	123.75	166.4	56.67
Polystyrene	147.79	98.69	101.97

3.3. Fourier Transform Infrared Spectroscopy (FTIR)

Figure 3 shows the spectra of three types of polymer which are HDPE 1, HDPE 2 and polystyrene. FTIR is a technique based on the vibrations of atoms of a molecule. An IR spectrum is commonly obtained by passing IR radiation through a sample and determining what fraction of the incident radiation is absorbed at a particular energy. The energy at which any peak in an absorption spectrum appears corresponds to the frequency of a vibration of a part of a sample molecule. This test was conducted to know the actual group of the three samples especially HDPE 1 and 2 sample as they were used as the same application which is as detergent bottle. From figure 3, it is noted that all present the same absorptions, varying only in terms of intensity which can be attributed to their degree of branching, that is, number and size of ramifications. Farm fresh bottle and dynamo bottle sample shows the exact same trend as both of it comes from the same group. For HDPE or LDPE group, the significant band are in the regions of 3000–2800, 1550– 1400 and 750–650 cm^{-1} [12] because they have the nearly same spectra. When one compares the HDPE-LDPE spectra, it lacks any noticeable difference in structure. The only exception would be the inclusion of additional thickness or length to peaks due to additional branching in some sample [13]. FTIR also allow the determination if any water molecule exist in a sample. If the sample is having water molecule, it will resonate around 3450 cm^{-1} region. Amongst the three samples, none of them shows significant resonance around 3450 cm^{-1} so it can be said that the water content in all of the sample are so small. So, the moisture content analysis need to be done in order to know the exact amount of moisture content in a sample.

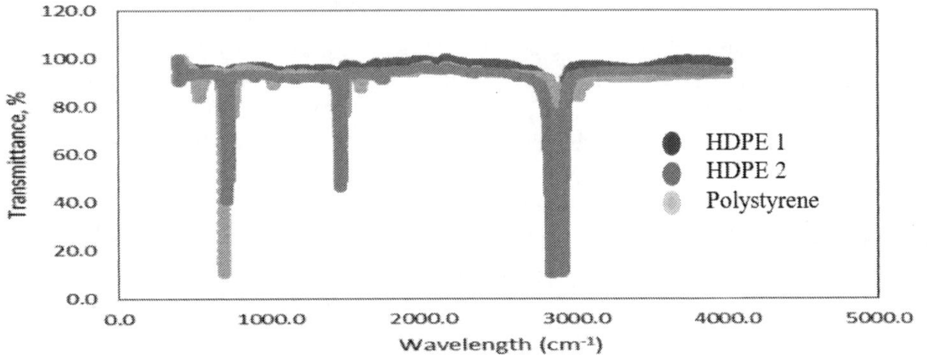

Figure 3. Graph of Transmittance (%) against Wavelength (cm^{-1}).

3.4. Flash Point (ASTM D93-08)

Table 2 shows the flash point temperature result for four different samples mainly contain different blend of oil which were collected from different sources. Flash point is defined as the lowest temperature of a liquid at which its vapors will form a combustible mixture with air [14]. Generally, the more viscous and more contained of additives in the oil, the higher the flash point. From table 1, base oil 1 have the highest flash point with 132.5°C while base oil 2 is having the lowest flash point with 20° C. Physically, the base oil 2 is much more watery-like compare to others. This might be affected by the addition of other component that has diluted the oil. In order to formulate a bitumen, the oil needs to have a high flash point as this experiment is conducted at maximum temperature of 180°C and the oil was used as the medium of the polymer to melt.

Table 2. Flash Point Temperature for liquid samples.

Sample Name	Flash Point Temperature (°C)
Base oil 1	132.5
Base oil 2	20.0
Base oil 3	65.0
Sludge	145.5

3.5. Density

Table 3 shows the density result for four different sources of oil sample mainly contain different blend of oil. This test was conducted to check on the physical behavior of the oils. Generally, in order to mix a different material in one mixture, the density of all the materials involved need to be closed enough to each other to make the mixture homogeneously mixed. To compare the density amongst the oil, base oil 1 has the highest density which is 0.9304 kg/m^3 whereas the lowest density is base oil 2 with a density of 0.8070 kg/m^3. In this case, base oil 3 was chosen because the supplier for base oil 1 have no more production of base oil to supply and base oil 3 has the higher density compare to base oil 2.

Table 3. Density for liquid samples.

Sample Name	Density (kg/m^3)
Base oil 1	0.9304
Base oil 2	0.8070
Base oil 3	0.8720
Sludge	1.1831

3.6. Moisture Content

Table 4 shows the moisture content result for all raw material which consist of oil and polymer. Moisture content analyzer test was run to check on any moisture existed in the sample because one standard of moisture content in raw material need to develop in order to control the quality of the final products. Amongst three polymeric material, HDPE 1 sample has the least moisture content of 0.266% followed by HDPE 2 sample at 0.389% and polystyrene sample at 0.757%. Whereas for the base oil samples, base oil 1 has the highest moisture content at 1.02% and the lowest content is base oil 3 which at 0.311%. Since the formulated bitumen was then being emulsified by using a suitable emulsifier by means, to dilute the bitumen into liquid form, so the bitumen sample that was formulated is not necessarily has the exact same performance as the original pavement bitumen which is more viscous compare to the bitumen formulated in this research.

Table 4. Moisture content for all solid and liquid samples.

Sample Name	Moisture Content (%)
Base oil 1	1.019
Base oil 2	0.504
Base oil 3	0.311
Sludge	14.479
HDPE 1	0.266
HDPE 2	0.389
Polystyrene	0.757

Table 5 summarize all the results for all the testing that already conducted. TGA and DSC only allow solid sample, flash point and density testing only allow liquid sample whereas moisture content analysis allow both liquid and solid sample to be run on the analyzer. From all of these analysis, TGA measures the decomposition temperature and composition in one sample so the exact amount of filler content can be known throughout this analysis. Furthermore, DSC determine the melting point of each samples as it is an important parameter we need to know before establish a method for bitumen formulation. In addition, it is important for the oil that used in this formulation to have a high flash point.

Table 5. Summary and result of all the test conducted.

Sample Test	HDPE 1	HDPE 2	Polystyrene	Base oil 1	Base oil 2	Base oil 3	Sludge
TGA (°C)	468.68	439.54	402.90	-	-	-	-
	Solid sample only						
DSC (°C)	131.24	133.19	151.90	-	-	-	-
	Solid sample only						
Flash Point (°C)	-	-	-	20	132.5	65	145.5
	Liquid sample only						
Density (kg/m³)	-	-	-	0.807	0.9304	0.872	1.1831
	Liquid sample only						
Moisture Content (%)	0.266	0.389	0.757	0.504	1.019	0.311	14.48

4. Conclusion

HDPE 1 is chose above other polymer as it contain less amount of filler so it is easier to undergo depolymerisation process while being heated and mix with the bitumen. Secondly, it is important to know the melting temperature of the material through DSC analysis as the material need to be melt to mix it in the formulation homogenously. Thirdly, automotive oil is used as the medium of the polymer to melt so an oil that having a high flash point is necessary for this experiment. Thus by comparing all type of base oil, there are two supplier that can supply the base oil for long term which are supplier for base oil 2 and base oil 3. Hence, base oil 3 is selected since it has higher flash point compared to base oil 2. Also, to compare in term of density, the higher density which is base oil 3 is chosen as it will tolerate with the polymer in the mixture the most.

Acknowledgments

The authors would like to thank the reviewers for their constructive comments. The authors also wish to express their appreciation for the financial support from Ministry of Higher Education (MOHE), Malaysia for Fundamental Research Grant Scheme (FRGS – RDU160129) and Universiti Malaysia Pahang Internal Grant (RDU160324). The support of Faculty of Chemical and Natural Resources Engineering and Universiti Malaysia Pahang, Malaysia are also acknowledged.

References

[1] Charles E and Carraher Jr 2010 *Introduction to Polymer Chemistry* (Florida: Taylor & Francis Group) p 360
[2] Asphalt Institute and Eurobitume 2011 *The Bitumen Industry – A Global Perspective: Production, Chemistry, Use, Specification and Occupational Exposure* (Kentucky: Lexington)
[3] Carrera V, Cuadri A, A and Partal P 2015 The development of polyurethane modified bitumen emulsions for cold mix applications pp 3407–3414
[4] Tarmudi Z, Abdullah M and Tap A M 2009 An Overview of Municipal Solid Wastes Generation in Malaysia *Jurnal Teknologi.* **51** *(F)*(November 2016) 1–15
[5] Department of Statistics Malaysia 2016 *Survey of Manufacturing Industries 2015* (Department of Statistics Malaysia) p 2–6
[6] Sun Y, Kwok Y C and Nguyen N T 2006 Low-pressure, high-temperature thermal bonding of polymeric microfluidic devices and their applications for electrophoretic separation *Journal of Micromechanics and Microengineering.* **16**(8) 1681–1688
[7] Rahim K A and Liwan A 2012 Oil and gas trends and implications in Malaysia. *Energy Policy.* **50** 262–271
[8] Yaacob H, Hainin M R, Aziz M M A, Warid M N M, Chang F L, Ismail C R and Hassan N A 2013 Bitumen emulsion in malaysia-a conspectus *Jurnal Teknologi (Sciences and Engineering).* **65** (3) 97–104
[9] Turi A 1997 *Thermal Characterization of Polymeric Materials* (San Diego: Academic Press)
[10] Joseph W 2009 *Characterization and Analysis of Polymers* (New Jersey: John Wiley & sons) 230
[11] Raymond F Boyer 1976 Mechanical Motions in Amarphous and Semi-crystalline Polymer *Polymer.* **17** 996-1008
[12] Gulmine J, Janissek P R 2002 Polyethylene Characterization by FTIR **21** 557-563
[13] Petrovich J 2015 *FTIR and DSC of polymers films used for packaging: LLDPE, PP, PVDC* (Casteau: SHAPE American High School) p 1-13
[14] Gharagheizi F and Alamdari R F 2008 Prediction of Flash Point Temperature of Pure Components Using a Quantitative Structure–Property Relationship Model. *QSAR & Combinatorial Science* **27**(6) 679–683

Preliminary study on gas separation performance of flat sheet mixed matrix (PVDF/Zeolite)

Sunarti Abd Rahman[1], Gamal Abdalla Suliman Haron[1], Raj Krishna Roshan Kanasan[1] and Hasrinah Hasbullah[2]

[1]Faculty of Chemical and Natural Resources Engineering, Universiti Malaysia Pahang, 26300 Gambang, Pahang, Malaysia
[2]Advanced Membrane Technology Research Centre (AMTEC), Universiti Teknologi Malaysia, 81310 Skudai, Johor Darul Ta'zim, Malaysia

E-mail: sunarti@ump.edu.my

Abstract. Membrane separation has attracted a lot of attention over the last years mainly due to its separation ability, operational capability and economical viability. Mixed matrix membrane (MMM) combines the superior transport and selectivity properties of inorganic membrane materials and the excellent fabrication properties of organic polymers. This emerging technology can be utilized to purify biogas which can be used in a variety of applications. In this study, flat sheet mixed matrix membranes were synthesized with different percentages of N-Mehtyl-2-pyrrolidone (NMP) as solvent, Polyvinylidene Fluoride (PVDF) as the polymer matrix and zeolite 4A as the dispersed fine particles, membrane A (80: 20: 0), membrane B (80: 18: 2), membrane C (80: 15: 5), and membrane D (75: 15: 10) respectively. The membranes were fabricated using dry/wet phase inversion method. The membrane's performance in terms of permeability and selectivity was examined using the single gas permeation device. The general trend was that, the permeability of the two gases (CO2/CH4) decreased with the increase of the pressure (0.5, 1, 1.5) bar. Membrane D was found to be suitable to separate the pair gas (CO2/CH4) as the permeability was 65623.412, Barrer and 15587.508, Barrer respectively, and its selectivity for was 4.21 at 0.5 bar.

1. Introduction

Generally, membranes for gas separation can be categorized as inorganic membranes and organic membranes depending on the constituent materials. Inorganic membranes exhibit excellent selectivity because of their regular pore structures[1]. On the other hand, academic investigators and industrial suppliers have employed organic polymers as asymmetric nonporous membranes that offer many desired properties including low operating cost and ease of construction and excellent processability [2]. And the latest emerging membrane with the potential for many applications involves mixed matrix membranes (MMM). The concept of mixed matrix membrane is the combination of polymeric membranes and inorganic membranes such as zeolite, activated carbon, and carbon nanotube. Inorganic molecular sieving materials have high separation performance properties and have the potential to surpass Robeson 2008 trade-off limit compared to polymeric membranes which are bounded by the trade-off limit [3]. The Polymer chose and the structure of the membrane are key aspects in selecting the appropriate membrane fabrication method. Phase inversion, interfacial polymerization, stretching,

IOP Conf. Series: Materials Science and Engineering **342** (2018) 012073 doi:10.1088/1757-899X/342/1/012073

and track-itching, are by far the most used techniques for membrane preparation. Phase inversion is defined as a process where a homogeneous polymer can be transformed from a liquid state to a soled state [4]. On the other hand, interfacial polymerization is One the most important techniques for commercial preparation of thin-film composite (TFC), (NF), and (RO) membranes [5].

Nowadays gas separation membranes find many applications such as hydrogen separation, oxygen–nitrogen separation, natural gas separation (carbon dioxide separation, dehydration, and dew point adjustment), vapor–vapor separation, and dehydration of air [6]. Membrane is one of the main methods of treatment of landfill gas mainly because of its lower energy requirements and capital investment cost as compared with the traditional separation techniques such as cryogenic distillation, pressure swing adsorption, and chemical absorption[7]. For an effective use of biogas as potential vehicle fuel, it has to be enriched in methane and this primarily achieved by carbon dioxide removal which then enhances the energy value of the gas to give longer driving distances with a fixed gas storage volume [8].

In this research, based on its unique features that the mixed matrix membrane (MMM) provides, it has been selected as a feasible method for CO2/CH4 separation. MMM combines the superior transport and selectivity properties of inorganic membrane materials (zeolite) and the excellent fabrication properties of organic polymers (PVDF). The aim of the study is to fabricate flat sheet mixed matrix membrane for CO2/CH4 biogas separation and test its permeability and selectivity. As well, it aims to characterize the membrane physically by scanning electron microscopy (SEM).

2. Experimental Set up

2.1. Material

Pallet of analytic grade Polyvinylidene fluoride (PDVF) with 177°C melting temperature, the inorganic filler material, Zeolite 4A with particle size 8-12mesh, and biotech grade N-Mehtyl-2-pyrrolidone, NMP with 99.5% purity, were bought from Sigma Aldrich. The two gases CO2 and CH4 which were utilized for gas permeation test were provided by Chemical Engineering Laboratory, Universiti Malaysia Pahang, which were utilized for gas permeation.

2.2. Preparation of Dope Solution

The dope solution was prepared be mixing the PVDF solution with the zeolite particles. The PVDF was dissolved in the NMP solvent and stirred with magnetic stirrer at 180°C for 6 hours. Then the zeolite was added to the PVDF solution according to the ratio given in the Table 1. Then continuously the dope was stirred for almost 3 hours at 180°C to ensure its homogeneity. Finally, the stirring was stopped and the gas bubbles in the dope were released by leaving the solution at room temperature for at least 72h [9]. Figure 1. illustrates the dope solution preparation steps.

Table 1. The component ratio of dope solution.

Membrane type	PVDF (wt.%)	Zeolite (wt.%)	NMP (wt.%)
A	20	0	80
B	8	2	80
C	15	5	80
D	15	10	75

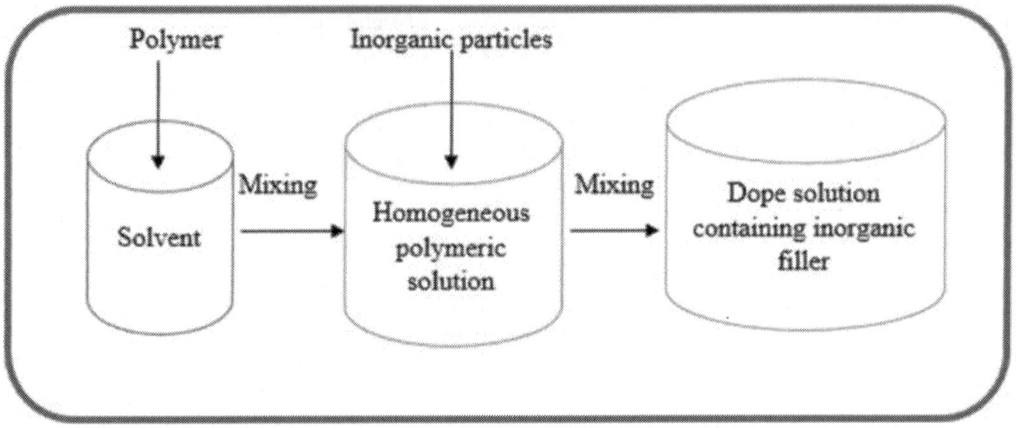

Figure 1. Method to produce dope solution for mixed matrix membrane [6].

2.3. Membrane Casting

The casting solutions were left at the room temperature for 24 h to de-gas to avoid bubbles. Then they were casted uniformly on a glass plate at room temperature with a knife gap of 0.2 mm. After it was subjected to air for 30s, the casted film was immersed into coagulation bath. The formed membrane was taken out of the coagulation bath to a container containing fresh running water to remove the excessive (NMP) solvent. Last, the wetting membrane was dried in air at room temperature until a dry flat sheet membrane was achieved.

2.4. Permeability Test

The single gas permeation device was used to test the membrane performance. CO2 and CH4 were tested separately by using different pressure set (0.5, 1, 1.5) bar, and the permeability's of the two gases were measured by the bubble flow meter as illustrated in Figure 2. The gas permeability was calculated using equation 1. and it was expressed in Barrer [1 barrer = 1*10^-10 cm3 (STP) cm/ (cm2scm Hg)]. The gas permeability in Barrer can be found by utilizing Equation 2, but the conversion factor has to be considered.

$$P = \frac{LV}{At\Delta p} \quad (1)$$

$$P(\text{Barrer}) = \frac{1*10^{10}*LV}{At\Delta P} \quad (2)$$

Where P is permeability, L is membrane thickness, V is gas volume in cm3 displaced in t(s), A is effective membrane area, ΔP is pressure difference across the membrane. Equation 3. is membrane selectivity which is the membrane ability to separate the pair gas and it is the ratio of CO2 and CH4 permeabilities.

$$\alpha_{CO2/CH4} = \frac{P_{CO2}}{P_{CH4}} \quad (3)$$

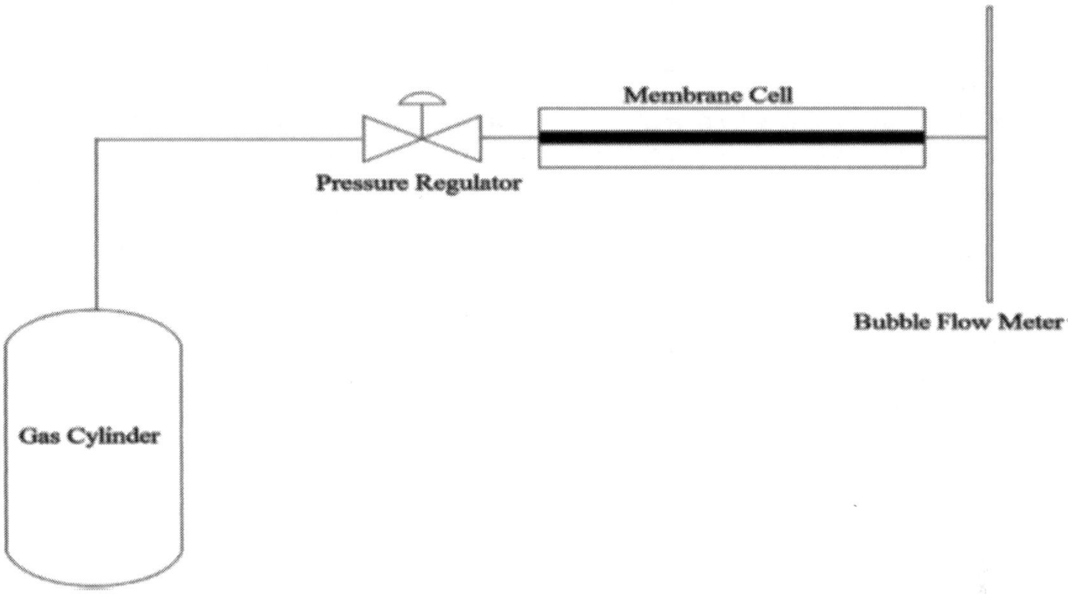

Figure 2. Single gas permeation device [10].

2.5. Membrane Characterization

A Scanning Electron Microscopy, SEM is used for the characterization of the membrane. SEM device is to examine the membrane morphology. The sample was fractured in liquid nitrogen and then inserted into the SEM device for cross sectional view of membrane. The size of pores on SEM micrograph that is scanned by a computer is calculated using the Sigmascan software.

3. Result and discussion

3.1 Performance of the membrane

The single gas permeation device was utilized to carry out the test. Table 2 shows the data that was obtained from the test to calculate the permeability and the selectivity of the two gases. The general trend is that, all four membranes showed an increase in permeability when there was a decrease in pressure. It is obvious that, membrane D at pressure 1.5 bar exhibits the highest selectivity of CO_2 among all other membranes, and that believed to be due to the loading rate of the fine zeolite particles which decreased the free volume of the membrane, thus influencing the gas transport behaviour. Then followed by membrane C, B and A at the same pressure 1.5 bar.

In terms of permeability, membrane D was the highest permeable to CO_2 65623.412 Barrer followed by membrane C, B and A 57980.535, 41725.869 and 30415.773 Barrer respectively. For CH_4 permeability, it very clear that membrane B was the highest permeable 36092.877 Barrer followed by membrane C, A and D, 32082.557, 26538.880, and 15587.508 Barrer respectively. On the other hand, the least permeable membrane for both gases CO_2 and CH_4 were membrane A with permeability 11577.692 Barrer and membrane D 7573.786 Barrer respectively, and this gas transport behavior could be attributed the slow gas diffusion through the membranes. plot of the membrane's permeability results that obtained from the test verses different set of pressure showed in Figure 3.

Table 2. The Performance of the Membranes.

NMP: ZEOLITE	PVDF:	Pressure, Bar	PCO_2 (Barrer)	PCH_4 (Barrer)	Selectivity CO_2/CH_4
80: 20: 0 A		0.5	30415.773	26538.880	1.15
		1	15154.568	12381.776	1.22
		1.5	11577.692	8797.776	1.32
80: 18: 2 B		0.5	41725.869	36092.877	1.16
		1	18782.232	17394.157	1.08
		1.5	18556.987	9663.420	1.92
80: 15: 5 C		0.5	57980.535	32082.557	1.80
		1	23590.115	13594.303	1.70
		1.5	19996.607	9238.820	2.16
75: 15: 10 D		0.5	65623.412	15587.508	4.21
		1	26735.464	7573.786	3.53
		1.5	16113.251	6146.083	2.62

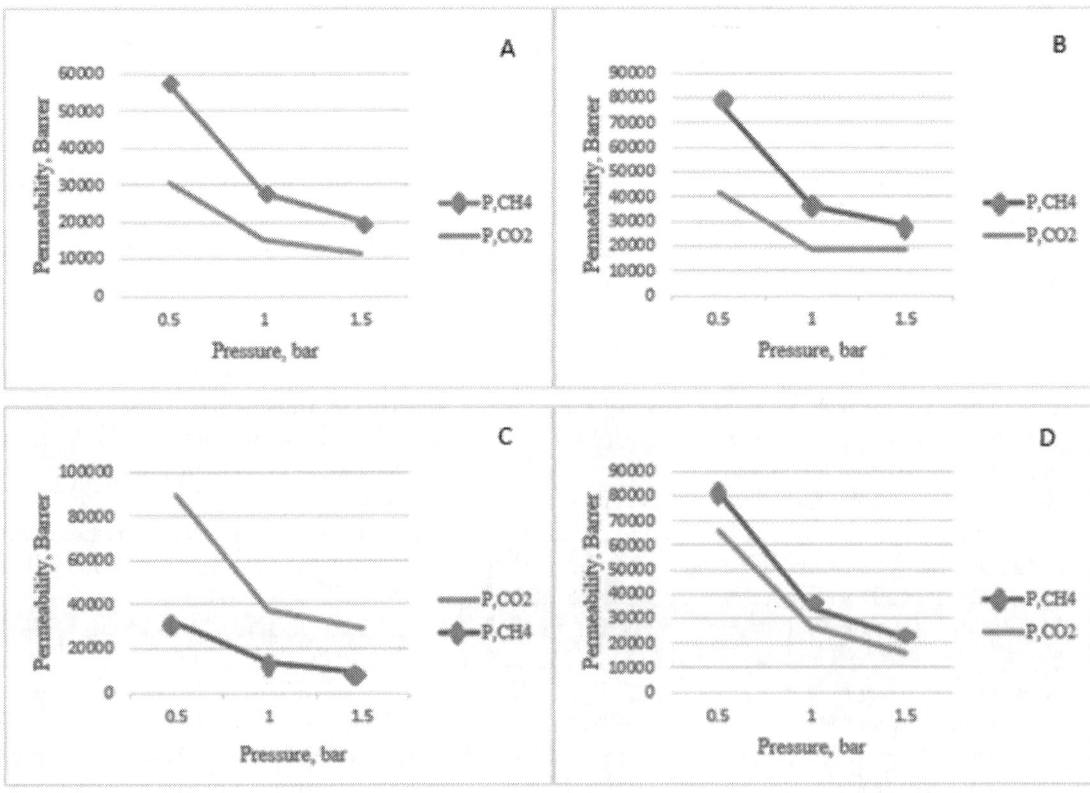

Figure 3. Permeability, Barrer of Membranes: A (80: 20: 0), B(80: 18: 2), C (80: 15: 5) and D (75: 15: 10) VS Pressure, bar.

The influence of the pressure on the membrane's selectivity was also obvious as the pressure increased there was a decrease in selectivity for membrane C and D, however, the result was a bit different for membrane A and B as the pressure increased there was a decrease in selectivity. Thus, membrane D which consist of 80%, 15%, 10% NMP, PVDF, and Zeolite respectively was regarded as the suitable membrane to separate CO_2/CH_4 biogas as the selectivity was 4.21 at 0.5 bar. Figure 4 illustrates the selectivity of the membranes versus pressure.

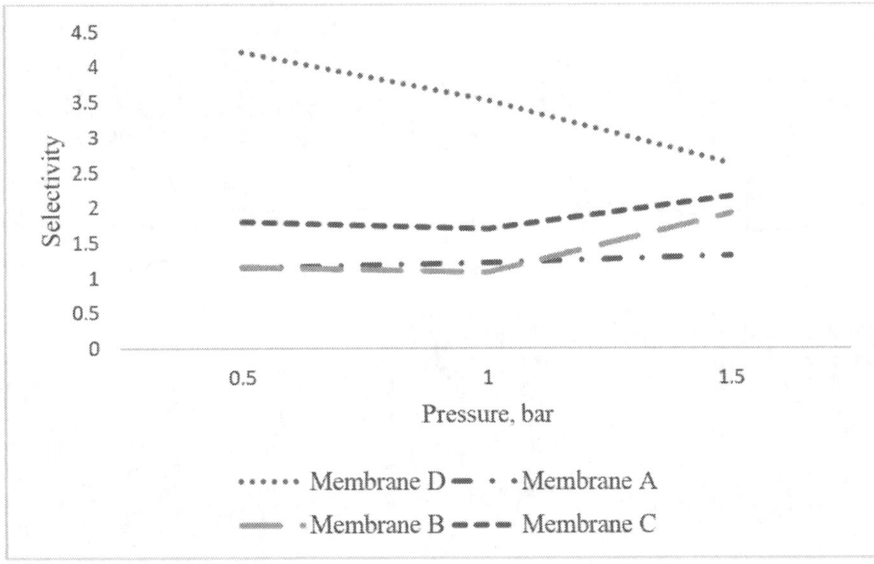

Figure 4. Selectivity of all membranes vs Pressure.

3.2. Physical characterization by SEM

After the fabrication of the membranes, they were cut into small pieces and dipped into liquid nitrogen to get a clear cross section. For the coating purpose, the membranes then placed on a metal holder and platinum. The surface morphology of the membranes is illustrated in Figure 5 and Figure 6 In general, the long finger-like structure and the main structures of the four membranes is almost the same. However, the dense membrane which has no zeolite added to it shown in Figure 5A and Figure 6A has a slightly different surface morphology in comparison with the rest of the membranes. Beside finger-like structure, Figure 6A clearly shows a consistent pores diameter. For membranes B, C and D which consist of 2%, 5% and 10% zeolite respectively, illustrate different degrees of disruption of the surface morphology made by the addition of the fine zeolite particles.

As could be seen from Figure 6 with 1000x magnification, the long finger-like structure of the dense membrane with pore diameter of 2.9 µm- 3.4µm, is attributed to the fast coagulation of the solvent NMP and the non-solvent water [11]. This range of pore diameter is not viable for CO2/CH4 separation as the pore diameter of these gases is 3.3x10-4, 3.8x10-4 respectively. There is unbalanced dispersion of the inorganic particles in Figure 6B. Therefore, this membrane displayed many cavities with pores diameter of 1.9µm-2.6µm which were formed during sample preparation by the cleavage of particles. [7] attributed the presence of the voids between the zeolite particles and the polymer chain to the existence of the fine particles in different quantities so more content of the zeolite more increase of the voids size. But unexpectedly membranes C, D with 5%, 10% zeolite content respectively showed smaller pores size of 1.7µm-2.2µm, 0.62µm-1.1µm respectively. And may be attributed to the good homogeneity of the solution which increased the adhesion of the zeolite particles with the polymer structure resulting in perfectly smaller pore sizes.

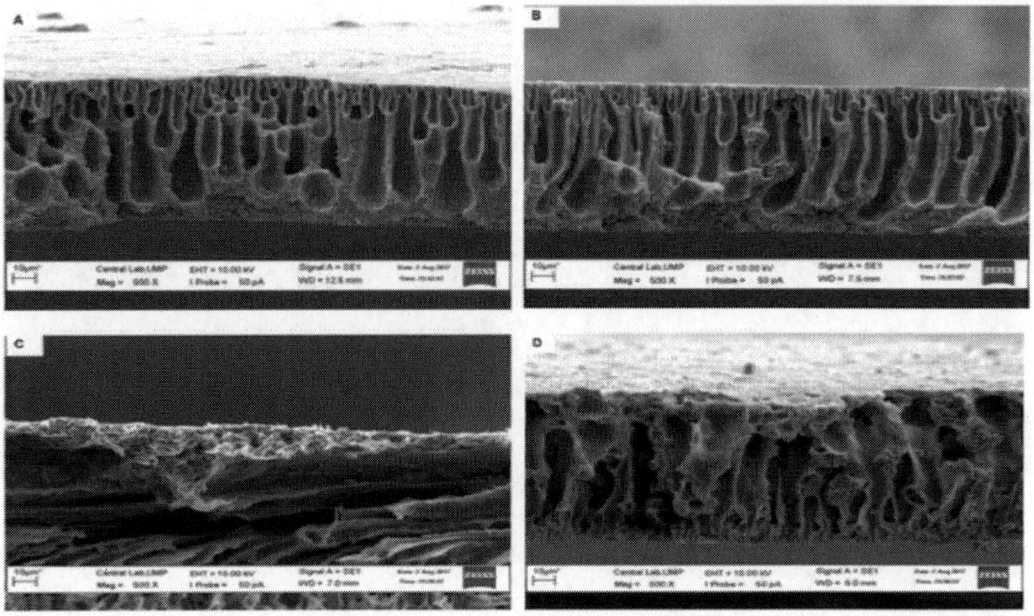

Figure 5. 500X magnification SEM image of NMP, PVDF, Zeolite: A (80: 20: 0), B (80: 18: 2), C (80: 15: 5), D (75: 15: 10).

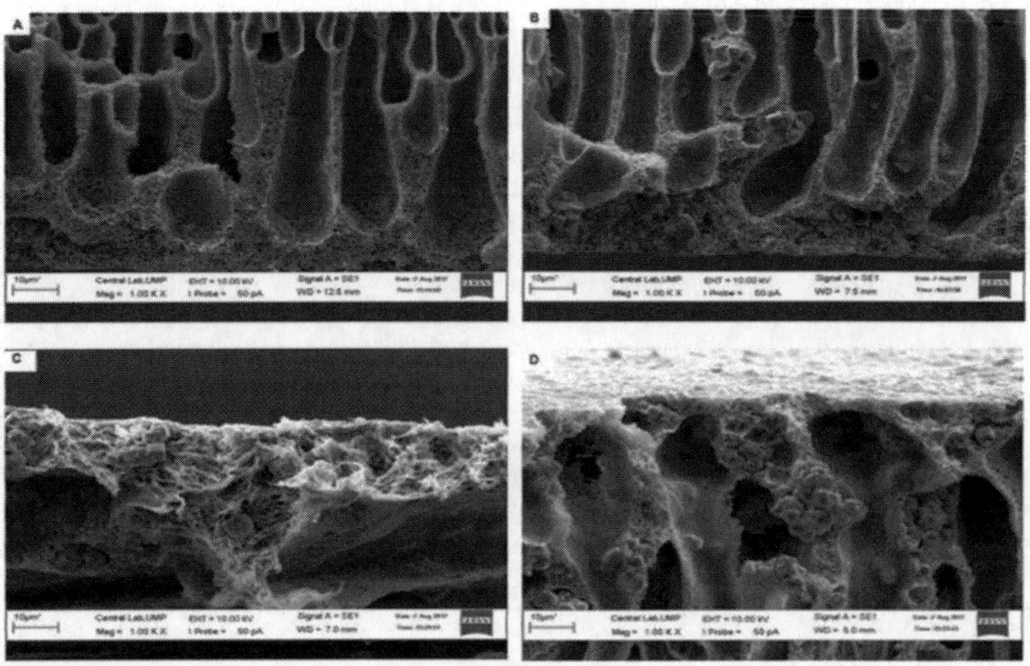

Figure 6.1000X Magnification SEM image of NMP: PVDF: Zeolite: A (80: 20: 0), B (80: 18: 2), C (80: 15: 5), D (75: 15: 5).

4. Conclusion

In this paper, flat sheet mixed matrix membranes were synthesized with different percentages of N-Mehtyl-2-pyrrolidone (NMP) as solvent, Polyvinylidene Fluoride (PVDF) as the polymer matrix and zeolite 4A as the dispersed fine particles, membrane A (80: 20: 0), membrane B (80: 18: 2), membrane C (80: 15: 5), and membrane D (75: 15: 10) respectively. The membranes were fabricated using dry/wet phase inversion method. The membrane's performance in terms of permeability and selectivity was examined using the single gas permeation device. The general trend was that, the permeability of the two gases (CO2/CH4) decreased with the increase of the pressure (0.5, 1, 1.5) bar. Membrane D was found to be suitable to separate the pair gas (CO2/CH4) as the permeability was 65623.412, Barrer and 15587.508, Barrer respectively, and its selectivity for CO2 was 4.21.

Acknowledgement

The authors would like to thank Universiti Malaysia Pahang for providing financial assistance (RDU 1702203), and Faculty of Chemical and Natural Resources Engineering for the gas engineering lab facilities.

References

[1] Yeung K 2011 *Chem. Commun.* **47** 8085
[2] Abraham J, Jose T, George S C, Kalarikkal N, and S Thomas 2015 *Int. J. of Memb. Sci and Technol* **2** 31
[3] Nour M K 2014 *PhD Thesis* RMIT Univ
[4] Mahajan R and Koros W J, 2000 *Ind. Eng. Chem. Res.* **39** 2692
[5] Lalia B S, Kochkodan V, Hashaikeh R, and Hilal N 2013 *Desalination* **326** 77
[6] Aroon M A, Ismail A F, Matsuura T, and Montazer-Rahmati M M 2010 *Sep. Purif. Technol.* **75** 229
[7] Shen Y and Lua A C 2012 *Chem. Eng. J.* **192** 201
[8] Eze J I and Agbo K E 2010 *Am J Sci Ind Res* **1** 604
[9] Hou D, Fan H, Jiang Q, Wang J and Zhang X 2014 *Sep. Purif. Technol.* **135** 211
[10] Ahmad A L, Sunarti A R, Teong L K, and Fernando W J N 2009 *Asia-Pacific J. Chem. Eng.* **4** 787
[11] Shi H, Liu F and Xue L 2013 *J. Memb. Sci.* **437** 205

Development of wireless vehicle remote control for fuel lid operation

N Sulaiman, M S Jadin, M S Najib, M. Mustafa and S N F Azmi

Faculty of Electrical & Electronics Engineering, Universiti Malaysia Pahang, Pahang, Malaysia

E-mail: norizam@ump.edu.my

Abstract. Nowadays, the evolution of the vehicle technology had made the vehicle especially car to be equipped with a remote control to control the operation of the locking and unlocking system of the car's door and rear's bonnet. However, for the fuel or petrol lid, it merely can be opened from inside the car's cabin by handling the fuel level inside the car's cabin to open the fuel lid. The petrol lid can be closed by pushing the lid by hand. Due to the high usage of using fuel lever to open the fuel lid when refilling the fuel, the car driver might encounter the malfunction of fuel lid (fail to open) when pushing or pulling the fuel lever. Thus, the main aim of the research is to enhance the operation of an existing car remote control where the car fuel lid can be controlled using two techniques; remote control-based and smartphone-based. The remote control is constructed using Arduino microcontroller, wireless sensors and XCTU software to set the transmitting and receiving parameters. Meanwhile, the smartphone can control the operation of the fuel lid by communicating with Arduino microcontroller which is attached to the fuel lid using Bluetooth sensor to open the petrol lid. In order to avoid the conflict of instruction between wireless systems with the existing mechanical-based system, the servo motor will be employed to release the fuel lid merely after receiving the instruction from Arduino microcontroller and smartphone. As a conclusion, the prototype of the multipurpose vehicle remote control is successfully invented, constructed and tested. The car fuel lid can be opened either using remote control or smartphone in a sequential manner. Therefore, the outcome of the project can be used to serve as an alternative solution to solve the car fuel lid problem even though the problem rarely occurred.

1. Introduction

Nowadays, the vehicle technology in this era involved the development of the remote control to control the operation of the vehicle especially car's door and rear bonnet locking system which was powered up by the battery. In addition, this type of remote control can be used to handle the operation locking and unlocking system from outside of the car . For the current vehicle technology, in order to open the fuel lid of the car, the operation merely can be done inside the car's cabin. Here, the car driver is required to push or pull up the fuel lever to release the car fuel lid. The existing configuration of car fuel lid is illustrated by figure 1 and figure 2 which involves the mechanical-based operation of car fuel lid.

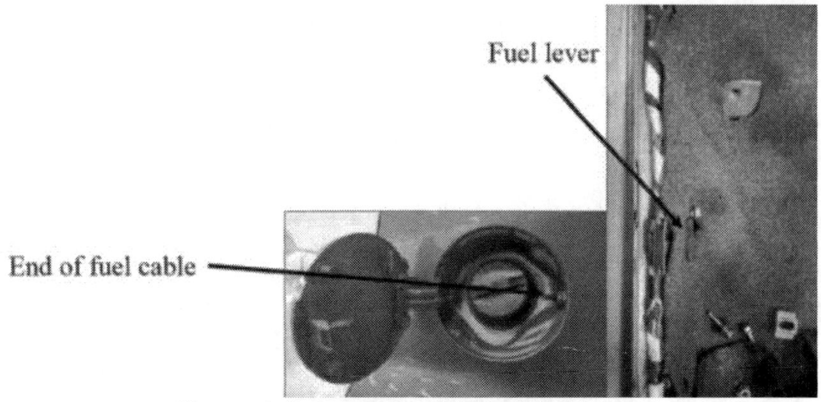

Figure 1. Existing petrol lid and fuel lever configuration.

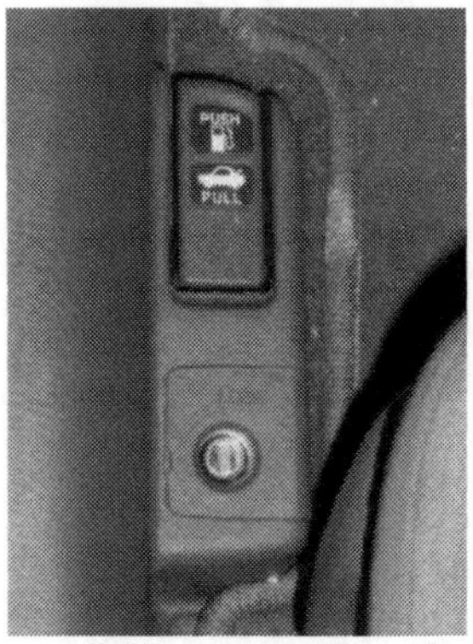

Figure 2. Example of car fuel lever.

As depicted in figure 1 and figure 2, the assembly of fuel cable is used to lock and release the hook that located inside the car's fuel cap compartment. The mechanical configuration of the fuel cable among the vehicle manufacturer or type of vehicle might vary which depends on the design of fuel cable assembly for each car. The end of fuel cable will pull in when the fuel lever is being pushed or pull up which resulting the fuel lid to open. Even though the current operation of the fuel lid is easy and the malfunction of the fuel lid seldom occurred, the vehicle manufacturer still need to consider the operation of the fuel lid seriously. It is because if the malfunction of fuel led occurred, the car driver absolutely cannot open the petrol lid from inside the car to refill the petrol. Figure 3 illustrates the example of the damaged assembly of the fuel cable which might occur due to the high usage of the fuel lever by the car user to re-fill the petrol.

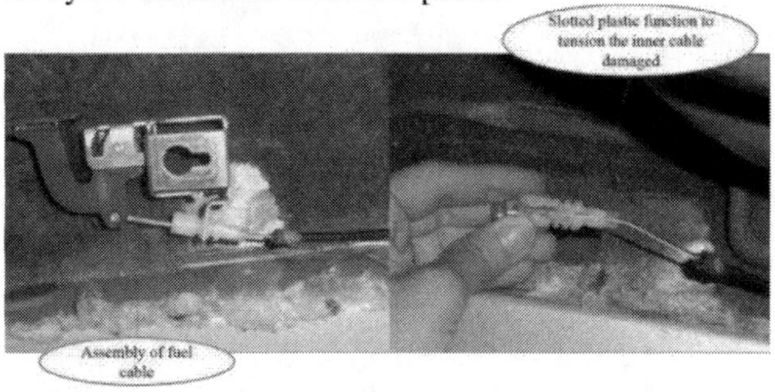

Figure 3. Example of damage assembly of fuel cable.

IOP Conf. Series: Materials Science and Engineering **342** (2018) 012074 doi:10.1088/1757-899X/342/1/012074

Since the operation of fuel lever is mechanical-based and might subject to malfunction due to the high usage of the fuel level, it is a must to have another alternative method to open the fuel lid. Therefore, this project is implemented to have wireless remote control to open the fuel lid from outside of the car which can be integrated with the existing car remote control. Here, the car remote control applied wireless communication to control the operation of the car door and rear bonnet. Therefore, to equip the existing car remote control with another function to open the car fuel lid, the type of the wireless sensor network needs to be selected. In this project, the *XBee* wireless sensor network based on IEEE 802.15.4 is selected due to its wide range coverage up to 100 meters in distance and apply low power consumption [1]. In order to process the wireless data more precisely, Arduino Bluetooth microcontroller is selected since Bluetooth is one of the wireless technologies to exchange the data over the short distances. Here, the Bluetooth device for locking system is very practical as the users can store and manage lock information in real-time [2]. The previous researchers have applied Bluetooth in many wireless communication applications. One of them, Srikanth, *et. al* had proposed *XBee* wireless sensor integrated with the Arduino to develop the remote access for streetlight maintenance and control [3-4]. As the results, the streetlight condition can be monitored via the wireless monitoring system for the maintenance and control works. *XBee* is the family of ZigBee where the USB cable can be easily connected and enable the microcontroller to use the services of ZigBee protocol [5-7]. Hence, this study is carried out to enhance the existing vehicle remote control where the car driver can control the operation of the fuel lid outside of the vehicle.

2. Methodology

In this project, the operation of fuel lid is stimulated with two different techniques which are remote control and smartphone. The implementation of the operation can be done in the sequence, one by one. The car driver can start the operation of the fuel lid first and then followed by the smartphone or vice=versa. Figure 4 illustrate the graphical block diagram of the project implementation which requires the use of Arduino UNO Rev3 as the main microcontroller, Arduino shield, servo motor, *XBee*, push button, Bluetooth and smartphone. To provide better communication between *XBee* transmitter and receiver and also to process the data efficiently, two Arduino microcontrollers are required.

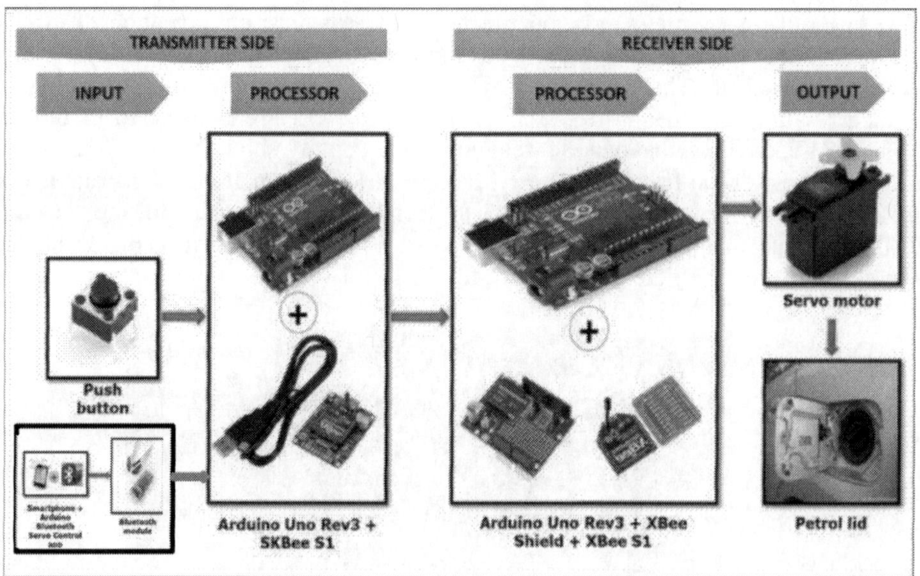

Figure 4. Graphical block diagram for remote control-based and smartphone-based technique.

The overall process flow to implement the project is depicted by figure 5. As illustrated by figure 5, the project involved the construct of the circuit and simulation for both remote control and mobile phone techniques. The communication for both techniques is implemented using XCTU software. It is simulated first using Proteus software. Once the simulation is successfully done, both the remote control will be tested to run the servo motor attached to the fuel lid. If the servo motor fail to open the fuel lid, need to trouble shoot the system at the circuit parameters and Arduino coding.

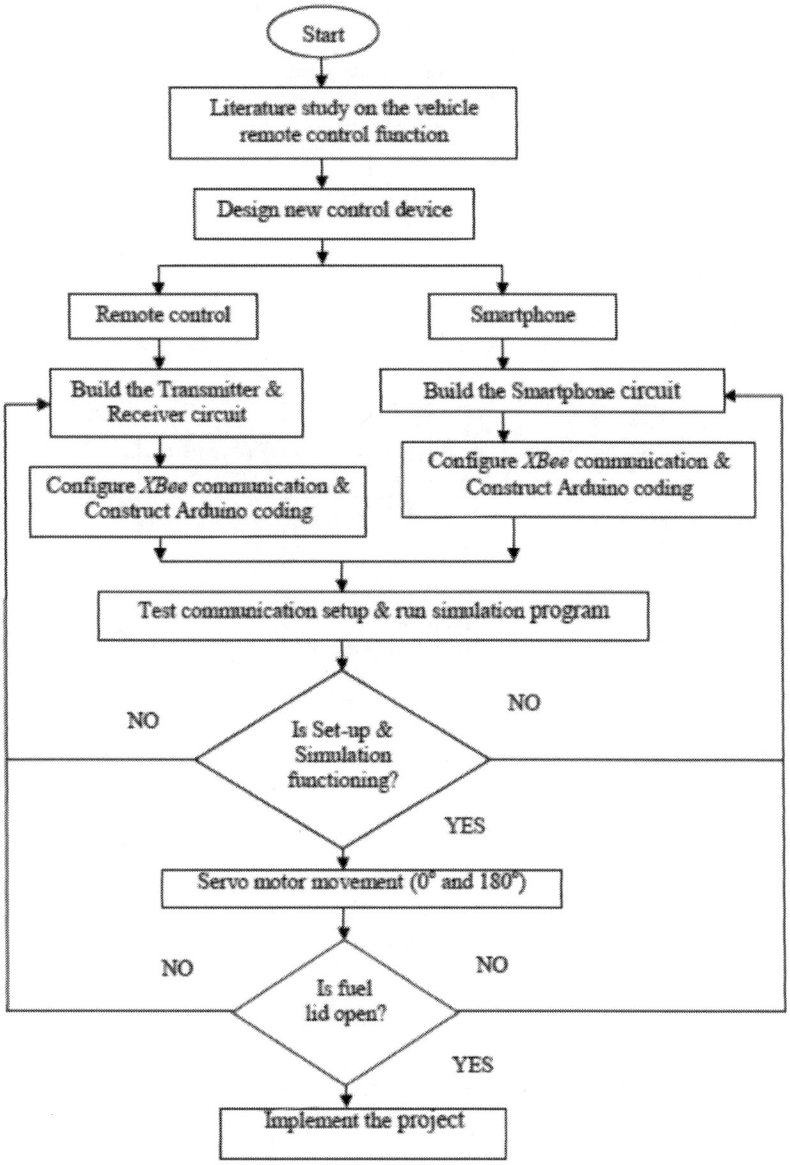

Figure 5. Overall process flow for remote control-based & smartphone-based technique.

2.1. Communication process flow

For the communication process, figure 6 depicts the process flow on the communication process between *XBee* at the transmitter and receiver part with the Arduino microcontroller. Here, when the push button at the remote control is pressed, *XBee's* transmitter will send value '1' or '0' to the

XBee's receiver. If the *XBee*'s received value '1', the Arduino will instruct the servo motor to rotate to180 degree. As a result, the fuel cable will be stretched inward, and the fuel lid will be released. On the other hand, if the *XBee's* received value '0', the Arduino will instruct the servo motor to rotate to 0 degree. As a result, the fuel cable will be stretched out, and the fuel lid will remain closed.

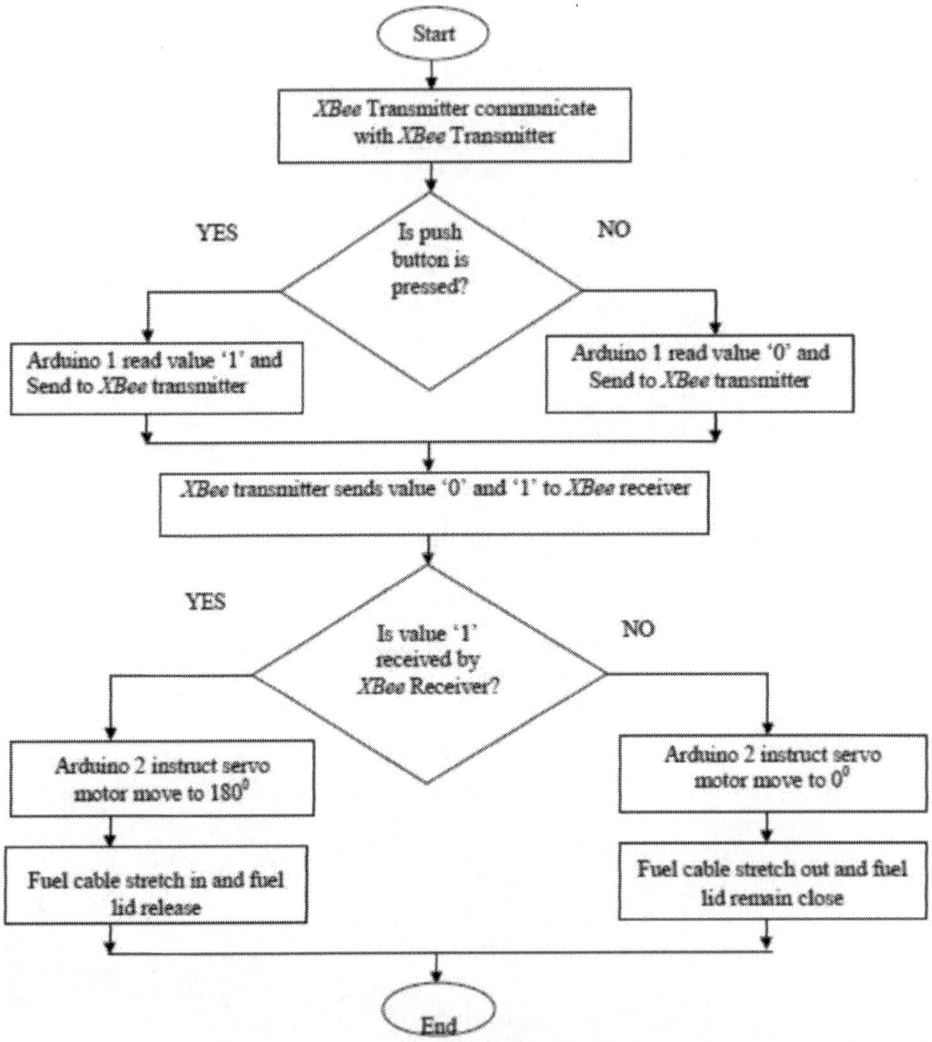

Figure 6. Process flow for remote control-based technique.

The communication process between the smartphone with the Arduino microcontroller to control the operation of the fuel lid is elucidated by figure 7. Here, the car driver need to use servo motor apps in the smartphone to give instruction to Arduino to rotate the servo motor to servo motor to rotate to 180 (to open the fuel lid) or 0 (to let the fuel lid remain closed).

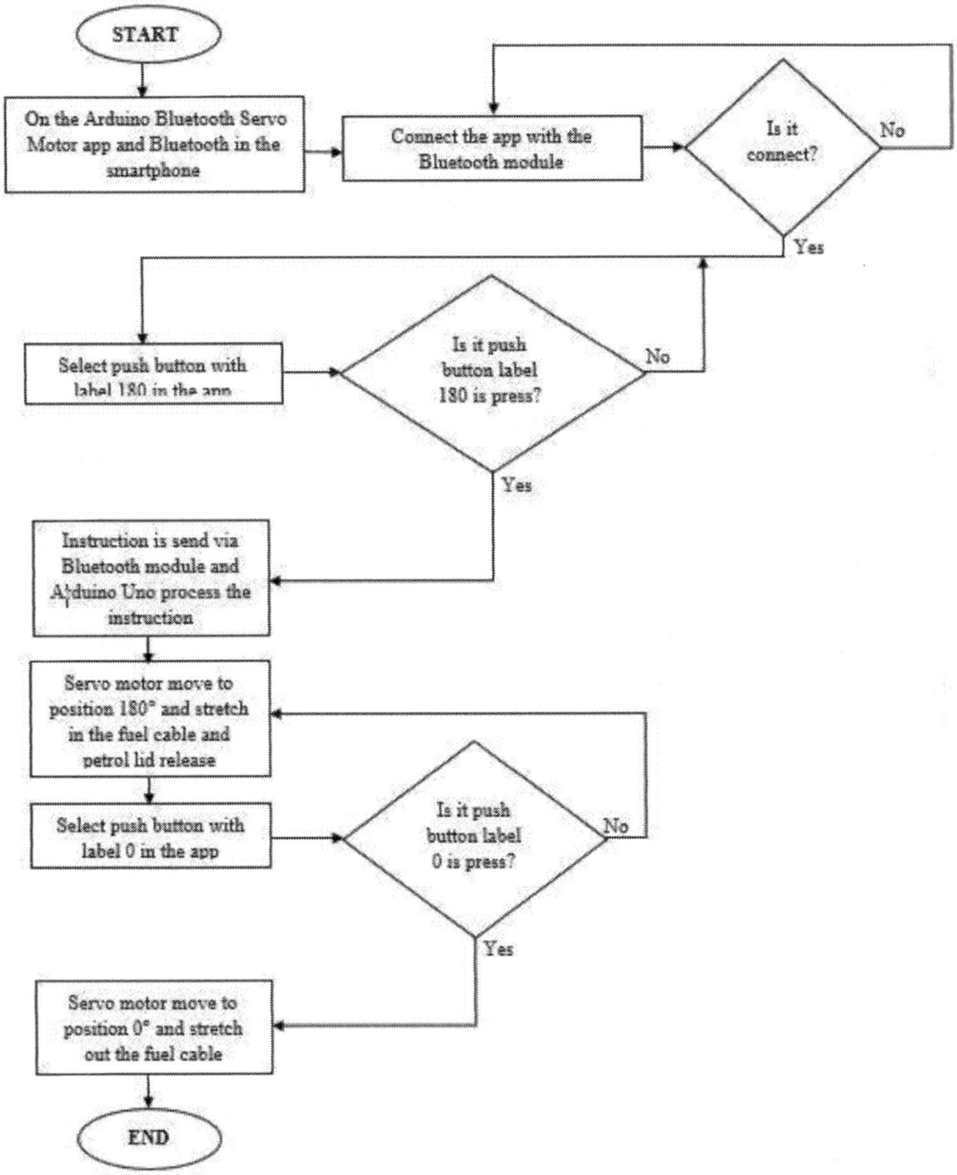

Figure 7. Process flow for smartphone-based technique.

2.2 Setup of XCTU Software and XBee

In order to run *XBee*, the installation of the Future Technology Devices International (FTDI) and Combined Driver Model (CMD) in the computer is required as shown in figure 8. Here, each *XBee* is assigned to different port to communicate with each other using *XCTU* software. There are

IOP Conf. Series: Materials Science and Engineering **342** (2018) 012074 doi:10.1088/1757-899X/342/1/012074

several steps in order to configure *XBee* as illustrated in figure 8. *XBee* must be configured by *XCTU* software before the device be used to transmit or receive signals.

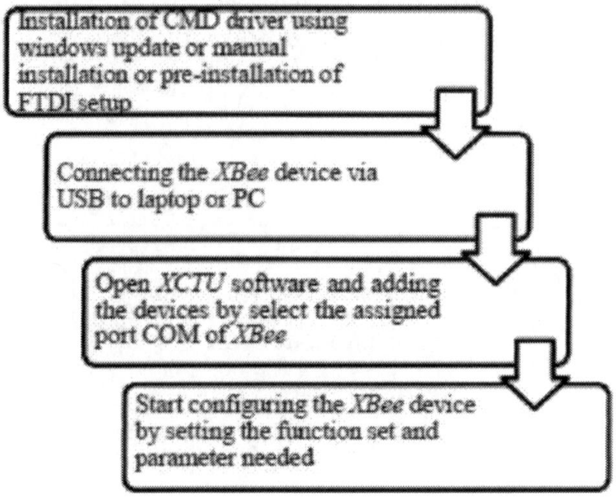

Figure 8. Steps for *XBee* configuration.

The configuration parameter for *XBee* is shown in Table 1. According to table 1, the configuration parameters for *XBee* are set using *XCTU* software. Here, both *XBee* transmitter and receiver are set with the same Network ID (3110) and Baud rate (9600). In order to allow the communication between 2 *XBee* devices, both devices must be in the same network which is defined by Personal Area Network (PAN). Then, the source address (MY) and the destination address is defined for each to recognise each other. The lower byte of the destination address (DL) is enough to enter the *XBee* S1 into the broadcast mode by setting the value of 0xFFFF. The entry channel (CH) is in the range of 0B until 1A but it will be easier to set it to the default channel as 0C. The baud rate is set to the default rate, 9600.

Table 1. *XBee* configuration parameters for *XCTU* software.

Parameters	Transmitter	Receiver
Network ID	3110	3110
Channel(CH)	0C	0C
Baud Rate (BR)	9600	9600
Source Address (MY)	0010	0011
Destination Address (DL)	0011	0010

Meanwhile, the configuration of the communication port to communicate between *XBee*, *XBee* shield and *XCTU* software is illustrated by figure 9. As illustrated by figure 9, Com 3 port is used to communicate with *XBee*. Meanwhile, Com 4 port is used to communicate with *XBee* shield and Arduino microcontroller.

IOP Conf. Series: Materials Science and Engineering **342** (2018) 012074 doi:10.1088/1757-899X/342/1/012074

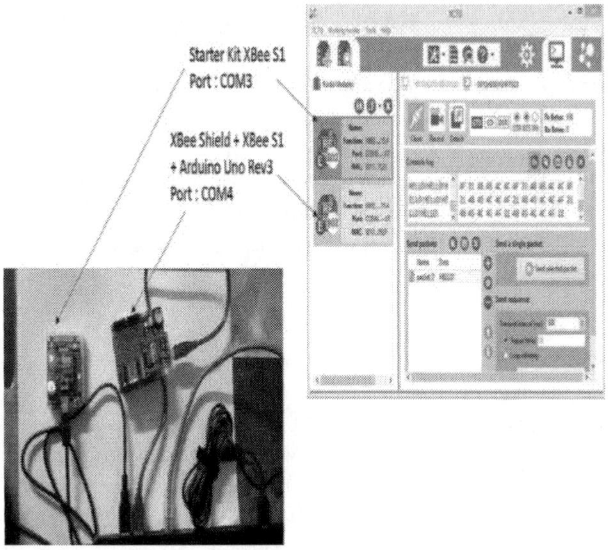

Figure 9. XCTU port setting for *XBee*, *XBee* shield and Arduino microcontroller.

3. Results and Discussion

In this section, the results of the development of the remote control, smartphone mode and system prototyping and model will be discussed.

3.1. Development of remote control technique

The remote control technique is constructed using 2 wireless *XBee* S1 devices to act as transmitter and receiver. Figure 10 shows the overall schematic diagram for fuel lid operation in remote control mode which is built and analysed using Proteus software before set-up the hardware components. According to figure 10, the schematic diagram consists of 2 *XBee* S1 and each *XBee* S1 is connected to one Arduino microcontroller to process the radio signals that are received or transmitted by the *XBee* S1. The simulation in Proteus software is conducted to test the communication between 2 *XBee* S1. The Proteus software is also used to calculate the required power supply by the *XBee* S1. Once the push button of remote control is pressed, the communication between *XBee* transmitter and receiver occurred. It is indicated by the change of serial monitor in Arduino microcontroller from '0' to '1'. Besides, the simulation results also show that the required power supply by the *XBee* is merely 3.31V.

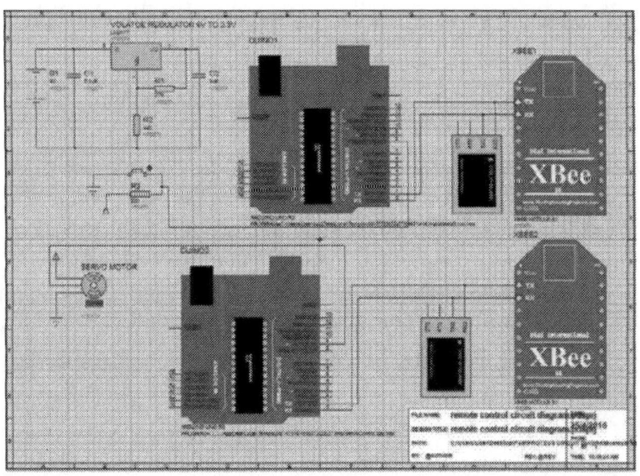

Figure 10. Schematic of remote control mode in Proteus software.

In this project, the 9V battery is used to power up the remote control. The voltage calculation is required in order to make voltage regulator LM317 to produce output voltage about 3.3 V. Equation (1) is used to decide the resistor values to be used in the schematic in order to get the desired output voltage which is 3.3V. Thus, it is required to calculate the value of *R*1, *R*2 and *R*3 based on the desired output voltage and adjustment of the current (*Iadj*) which can be obtained from the analysis of the circuit in Proteus software. Meanwhile, in constructing remote control technique, the push button is used as a switch for the remote control to open the fuel lid. Here, the push button is connected to the pin 8 of the input port of the Arduino microcontroller.

$$V_o = 1.25(1 + (R1 + R2) / R3) + (Iadj)R2 \qquad (1)$$

3.2. Development of smartphone technique

In the smartphone technique, the circuit consists of Bluetooth module HC-05 as a wireless communication device to the smartphone in order to control the operation of fuel lid. Figure 11 elucidates the schematic diagram for the operation of fuel lid using smartphone. The circuitry is analysed using Proteus software. Here, the Bluetooth module HC-05 is connected to the "TX" and "RX" port of Arduino microcontroller to receive the instruction from the smartphone. Then, the "TXD" pin at the Bluetooth module HC-05 is connected directly to the "RX" pin 0 of the Arduino microcontroller. However, in order to avoid the Bluetooth module HC-05 to get burnt out due to the high voltage in the circuit, the voltage divider circuit is applied to regulate voltage of 5 V from "TX" Arduino microcontroller in pin 1 to become 3.3 V. It is because the Bluetooth module HC-05 can be damaged easily if the "RXD" pin at Bluetooth module HC-05 received voltage more than 3.3V. Hence, the simulation of the smartphone mode must be run and analyzed in Proteus software to ensure the correct voltage is supplied to the Bluetooth module. Besides, the software simulation will ensure the correct parameters are selected for voltage divider circuit.

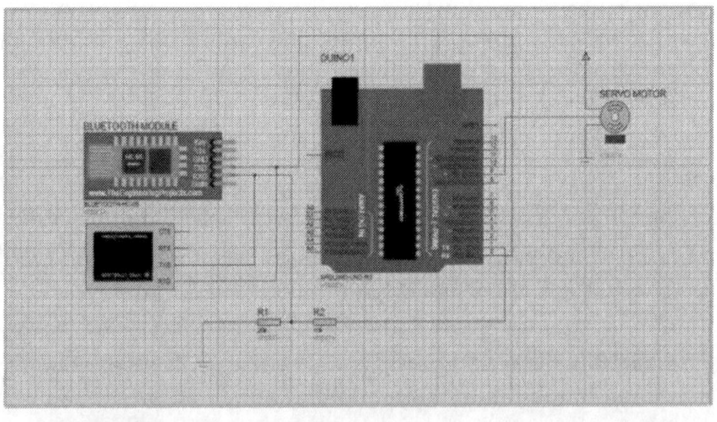

Figure 11. Schematic of smartphone mode in Proteus software.

Furthermore, the servo motor which is connected to the pin 9 of Arduino microcontroller will move according to the instruction received from the smartphone. The rotation of the servo motor will release the hook of fuel cable and then, open the petrol lid. In addition, the movement of the servo motor is controlled by an Android application that can be downloaded from the Android Play Store. The name of the application is Arduino Bluetooth Servo Motor. Here, the movement of the servo motor can be controlled by this application. Figure 12 depicts the diagram of the Arduino Bluetooth Servo Motor application which can be used to communicate between Arduino, servo motor and smartphone.

IOP Conf. Series: Materials Science and Engineering **342** (2018) 012074 doi:10.1088/1757-899X/342/1/012074

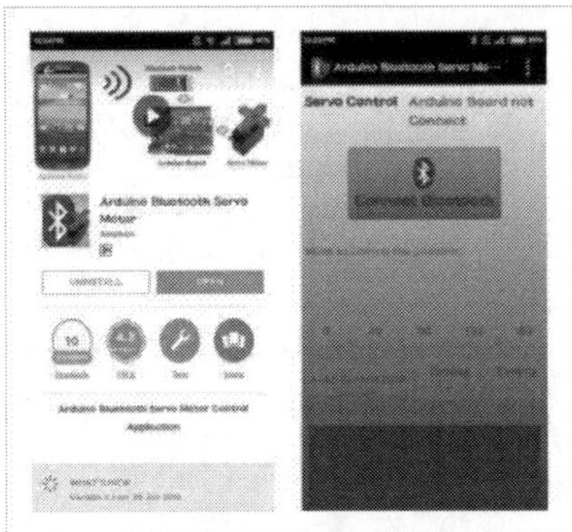

Figure 12. Application of Arduino Bluetooth Servo Motor for smartphone.

3.3. Development of the prototype of remote control technique

Once the remote control and smartphone techniques are successfully simulated, a prototype of the remote control mode is constructed as shown in figure 13 and figure 14 respectively. The prototype is built according to the schematic shown in figure 10. Meanwhile, the prototype of smartphone technique is elucidated by figure 15 and it is developed according to the schematic shown in figure 11.

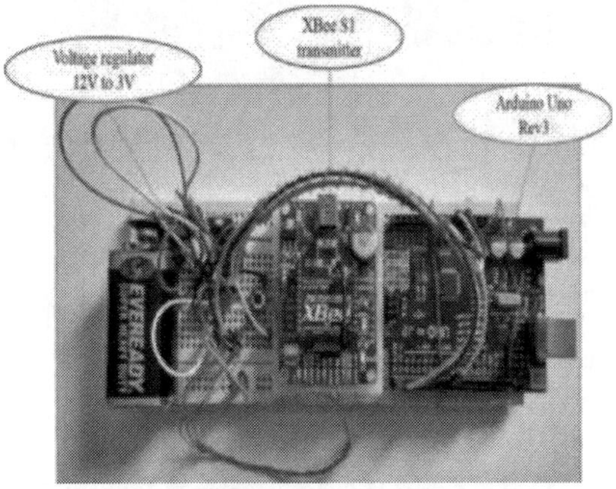

Figure 13. Transmitter part of remote control prototype.

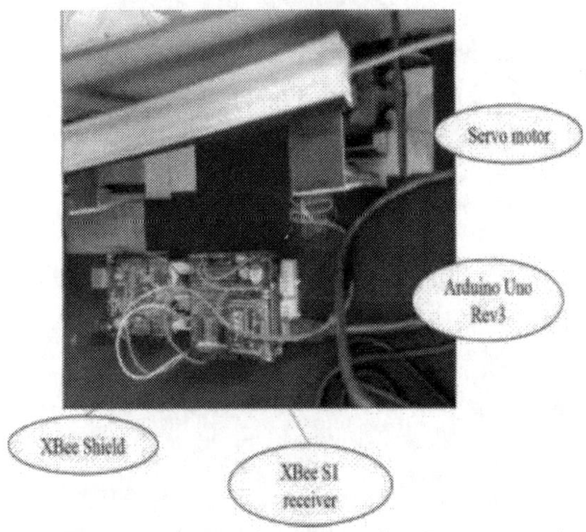

Figure 14. Receiver part of remote control prototype.

The smartphone technique for the operation of the petrol lid consist of three major part of circuit diagram as shown in figure 15. For the first part, the Bluetooth module is employed as an input device that received the instruction from the Android application in smartphone. Meanwhile, for the second part, Arduino microcontroller is employed to process the value send by input device. The final part is the movement of servo motor which acts as an output device that moves according to the direction given by the input device.

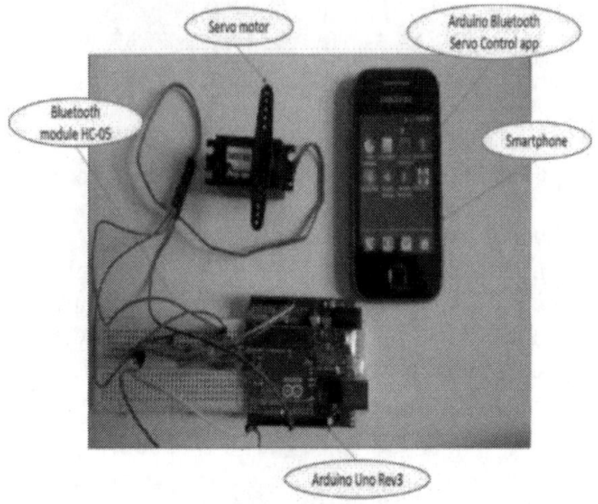

Figure 15. Smartphone prototype to control servo motor movement.

3.4. Implementation of the remote control and smartphone prototyping

Before implementing the proposed design to the real car, the model of car is built in order to demonstrate the operation of car fuel lid using remote control and smartphone techniques. The model of car is constructed using cardboard paper and the plastic cable to represent the cable fuel at the petrol lid mechanical part as shown in figure 16. Once the remote control or smartphone is activated, it will send instruction to servo motor attached to petrol lid to release the cable fuel.

Figure 16. Model of car for operation of fuel lid and fuel cable.

IOP Conf. Series: Materials Science and Engineering **342** (2018) 012074 doi:10.1088/1757-899X/342/1/012074

Figure 17 shows the placement of the servo motor, remote control, microcontroller and Bluetooth circuit inside the compartment of the model of the car. Finally, the prototyping model of car is built and produced to demonstrate the operation of car fuel lid by using remote control and smartphone. Both remote control and smartphone mode are functioning and successfully tested to release the fuel cable hook and open the fuel lid. The testing and verification is done several time to test the robustness of the system. The testing is implemented in the sequence manner; either the system is started using remote control first and then follow with smartphone or vice versa.

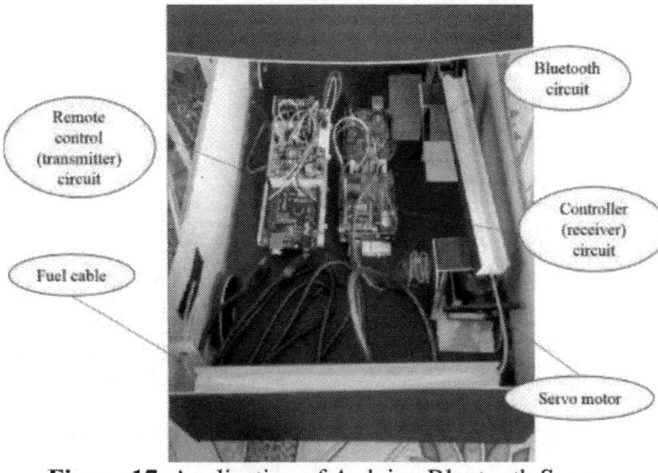

Figure 17. Application of Arduino Bluetooth Servo Motor.

3.5 Testing of the Servo Motor movement

The Servo Motor is tested using Pulse Width Modulation (PWM) technique. The electrical pulse is given to the servo motor with the variable pulse to determine how far the servo motor will move. Figure 18 illustrates the basic movement of servo motor by the given PWM which can be used for micro-controller. It indicates that the servo motor will take about less than or equal 1.5ms to move from 0 degree direction to 90 degree direction. Meanwhile, it will take about less than or equal 2ms to move from 90 degree to 180 degree direction. The plot of the duration of the pulse versus direction of servo motor is shown in Figure 19. It shows that the 1.86ms or about 2ms of the pulse is required to move the servo motor from 0 degree to 190 degree direction which provides enough time to control the fuel lid operation if the operation is done using smartphone. The graph also shows that the car driver has enough time to operate the fuel lid using remote control as well.

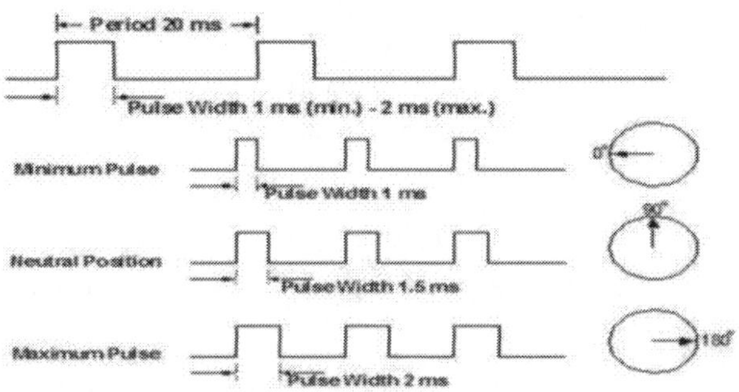

Figure 18. Servo motor movement according to PWM which is applied to microcontroller [5].

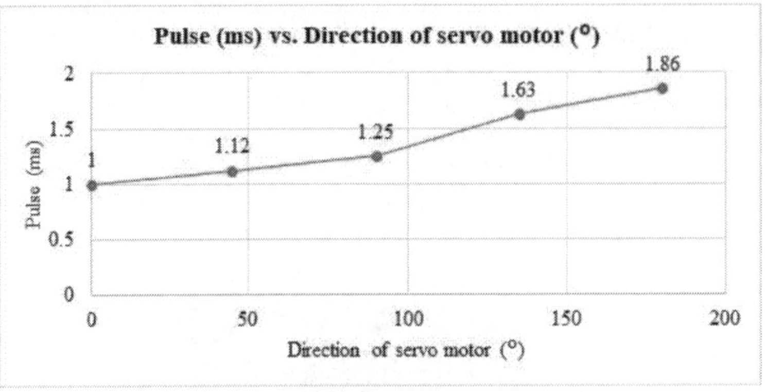

Figure 19. Servo motor movement according to PWM.

4. Conclusion

This project is implemented to provide an alternative method to open the car fuel lid from outside of car's cabin. Even though there is seldom case of malfunction of car fuel lid, when the malfunction of fuel cable occurred, the fuel lid absolutely cannot be opened. Thus, it will create troublesome to car driver. The project output shows that both remote control and smartphone technique can control the operation of the fuel lid with duration less than 2ms. Hence, the project succeed in providing multi- function of the car remote control where the existing vehicle just capable to control merely the car door and rear bonnet. In the case that if the proposed car remote unable to operate the car fuel lid, the car driver still has another option to open the fuel lid using smartphone beside using the existing car fuel lever. The project output proved that it is very beneficial to develop the vehicle remote control to operate car fuel lid wirelessly via remote control and smartphone techniques. This project is developed to overcome the malfunction of fuel lid which the car driver can open the fuel lid in wireless manner. Hence, for the future work, it will include the implementation of the project at the real vehicle and communication performance between the 2 techniques will be studied and improvised. Besides, the appropriate sensor will be allocated either at the fuel lever or fuel lid to alert the car driver about the malfunction of the fuel lid.

Acknowledgment

The author would like to acknowledge the great supports and contributions given by the final year Undergraduate student and project team members. High appreciation also goes to the faculty of Electrical & Electronics Engineering and University Malaysia Pahang by providing financial support through research grant (RDU170365) to make the project succeed.

References

[1] N. Patel, H. Kathiriya, and A. Bavarva 2013 Wireless sensor network using ZigBee, *Int. J. Res.Eng. Tech.* **2** 1038–1042

[2] H.D.J. Jeong, W. Lee, J. Lim, and W. Hyun 2015 Utilizing Bluetooth remote lock system for a smartphone *Pervasive Mob. Comput.* **24** 150-165

[3] D. Anusha, P. Sarma, and M. N. Sandhya Rani 2013 Appliance remote control using Arduino *Int. J. Latest Trends Eng. Tech.* **2**

[4] M. Srikanth and K. N. Sudhakar 2014 ZigBee based remote control automatic streetlight system Int. *J. Eng. Sc. Comp.* 639-642

[5] A. M. A. Haidar, C. Benachaiba, and M. Zahir 2013 Software interfacing of servo motor with microcontroller **1** 84-99

[6] B. Aralov, M. Sarsengeldin, O. Baimuratov, O. Ulbossyn, and A. Ulzada 2015 *Recent Advances in Electrical Engineering*, ed V. Mladenov (Kuala Lumpur: WSEAS Press) p 134-138

[7] J. Abdullah 2011 The Design of Mobile Control Car Security System *Int. J. Eng. Tech.* **3** 254-262

Mechanical properties of steel/kenaf (hybrid) fibers added into concrete mixtures

A O Baarimah and S M Syed Mohsin

Faculty of Civil Engineering and Earth Resources, Universiti Malaysia Pahang, Lebuhraya Tun Razak, 26300 Gambang, Kuantan, Pahang, Malaysia

E-mail: eng.baarmh@gmail.com

Abstract. This paper investigates the potential advantages of adding hybrid steel-kenaf fibers to concrete mixtures. Compression and flexural test were conducted on six concrete mixtures at 28 days to investigate the mechanical properties of the concrete. The experimental work consists of six concrete mixtures, in which the first mixture was a control mixture without adding any fiber. The following five concrete mixtures contain a total of 1% of volume fraction for steel, kenaf and a mixture of steel-kenaf (hybrid) fibers. Three ratios were considered for hybrid fibers with the ratios of 0.25/0.75, 0.5/0.5 and 0.75/0.25 for steel and kenaf fibers, respectively. From the investigation, it was observed that fibers have minimal effect on compressive strength of the concrete. However, the findings suggest promising improvement on the flexural strength of the concrete added with hybrid fiber (up to 86%) as well as manages to change the mode of failure of the beam from brittle to a more ductile manner.

1. Introduction

Concrete has become one of the most important construction materials commonly used in many types of engineering structures. Concrete is a material which is very strong in compression and weak in tension, thus causes cracking in the tension zone. In the past decades, fiber reinforced concrete (FRC) has been gaining more attention in the concrete research development and were used in numerous types of civil engineering application such as pavement slabs, precast products, shotcrete, tunnel linings, seismic structures, marine and refractory applications [1-4]. There are many advantages of inclusion of fiber into reinforced concrete structures, such as improving the load carrying capacity and ductility of the structural members, controlling crack propagation, increasing energy absorption and altering the mode of failure [5-11].

Steel fiber has been demonstrated its efficiency in enhancing the structural behaviour of reinforced concrete structural members [11-16]. The addition of kenaf fibers into concrete mixtures as this type of fiber reinforced concrete was investigated as a potentially promising "green" construction material to be used in many types of structural engineering [17,18]. However, since kenaf fiber is a natural fiber, some treatment is needed to reduce the hydrophobic characteristic of the fiber in order to maintain water/cement ratio and ensure sufficient workability of the concrete mixtures [17,19]. One of the treatments recommended is to use a chemical such as sodium hydroxide (NaOH) to reduce the hydrophobic characteristic, thus enhancing the adhesion between the fiber surface and the matrix, by removing the hydroxyl group in cellulose and increased the surface roughness which resulted in the improvement of the tensile properties of kenaf fibers as compared to untreated kenaf fibers [17,20-22].

Previous works reported in literature have shown that promising results were observed in the potential of steel fibers and kenaf fiber to improve the structural behaviour of reinforced concrete members. Present work aims to study the behaviour of single and hybrid (steel-kenaf) fibers and its ratio in the mechanical properties of concrete through compressive strength, flexural strength, cracking pattern and mode of failure of concrete.

2. Materials and Methods

The control concrete mixture is designed in accordance to British Standards (BS EN 206-1, 2000) for 20 MPa of concrete compressive strength. In order to investigate the mechanical behaviour of single fibers and hybrid fibers added into the concrete mixture, three types of fibers, namely steel fiber, kenaf fiber, and hybrid steel-kenaf fiber were added in the concrete. The fiber volume fraction (V_f) considered was 1 % for single and hybrid fiber, where the hybrid fiber divided between steel-kenaf into ratios of 0.25–0.75%, 0.5–0.5% and 0.75–0.25%, respectively.

Six concrete mixtures were produced for single and hybrid fibers using different fiber volume fraction as listed in Table 1. The first mixture (Mix 1) is a control mixture without adding any fiber, V_f = 0%. The following five concrete mixtures contain V_f = 1%, 1%, 0.25/0.75%, 0.5/0.5% and 0.75/0.25% and represent in Mix 2, Mix 3, Mix 4, Mix 5 and Mix 6, respectively. Hooked-end steel fiber and kenaf fiber were considered for this study. The properties of the steel fiber and kenaf fiber are presented in Table 2 while the shape of steel and kenaf fibers are shown in Figure 1 and 2, respectively. In addition, the kenaf fiber was treated by 1% of NaOH in order to clean and remove impurities from the fiber surface as well as enhancing the tensile properties of kenaf fiber. Moreover, super-plasticizer (SP) is added into the mixtures to improve workability and achieve the required slump.

Table 1. Concrete mix design.

Properties	Mix 1 C0	Mix 2 S1	Mix 3 K1	Mix 4 H25/75	Mix 5 H50/50	Mix 6 H75/25
Type of fiber	Control	Steel	Kenaf	Hybrid	Hybrid	Hybrid
V_f (%)	0	1	1	0.25-0.75	0.5-0.5	0.75-0.25
Cement (kg/m^3)	280	280	280	280	280	280
Fine Aggregate (kg/m^3)	556	556	556	556	556	556
Coarse Aggregate (kg/m^3)	1292	1292	1292	1292	1292	1292
Water (Liter/m^3)	162	162	162	162	162	162
W/C ratio	0.58	0.58	0.58	0.58	0.58	0.58
SP (Liter/m^3)	5.6	5.6	5.6	5.6	5.6	5.6
Steel fiber (kg/m^3)	0	75	0	18.8	37.5	56.3
Kenaf fiber (kg/m^3)	0	0	6.5	4.9	3.3	1.7

Table 2. Properties of steel and kenaf fibers.

Properties	Steel Fiber	Kenaf Fiber

Length, L (mm)	60	30
Diameter, D (mm)	0.75	0.1-0.5
Aspect Ratio, L/D	80	-
Tensile Strength (MPa)	1100	400-550
Unit Weight (kg/m^3)	7500	650

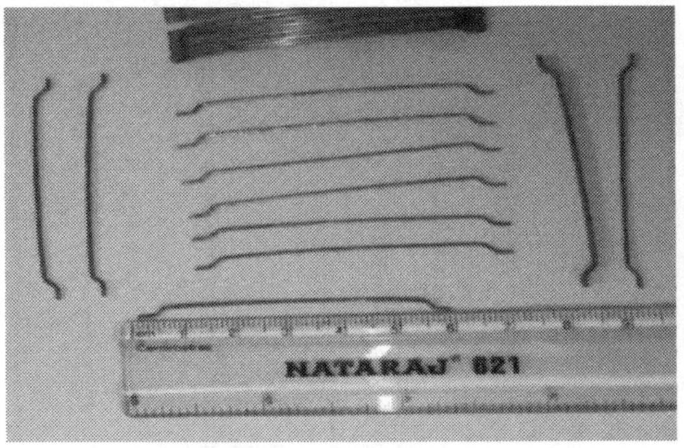

Figure 1. Steel fiber.

Figure 2. Kenaf fiber.

In order to measure the compressive and flexural strength of the concrete mixtures, compression test and flexural test were conducted in this study using cube compression test machine MATEST and Universal Testing Machine under four-point bending test, respectively. Three cubes, as well as three beams were prepared for each mixture. A total number of eighteen cubes with a standard size of 150 x 150 x 150 mm as well as eighteen beams with a dimension of 100 x 100 x 500 mm were tested on the 28th day as recommended in British Standards BS EN 12390-3, 2009 and BS EN 12390-5, 2009, respectively. Figures 3 and 4 show the machine used for compression test and flexural test, respectively.

Figure 3. Compression Machine MATEST.

Figure 4. Universal Testing Machine.

3. Results and Discussion
Present work discusses the mechanical properties of concrete mixtures added with fibers, which includes the compression and flexural strength of the concrete and its cracking pattern.

3.1 Compressive Strength Test
Figure 5 shows the average compressive strength of fiber reinforced concrete cube specimens tested at 28th day. It is noteworthy that all concrete mixtures achieved more than its target compressive strength (20 MPa) except the kenaf fiber reinforced concrete mixture which was due to the low density of kenaf fiber concrete mixture compared to the control mixture. In comparison to the control cubes (C0),

only steel fiber reinforced concrete (S1) and hybrid fiber reinforced concrete (H75/25) show higher compressive strength with increasing up to 6% and 11%, respectively. This is due to the increase in the amount of steel fiber in the concrete mixture that makes the specimen denser and stronger. However, the compressive strength of cube specimens K1, H25/75 and H50/50 that contain kenaf fiber decrease up to 27%, 7% and 4%, respectively, in comparison to the C0, as the amount of kenaf fibers increases. This observation is in agreement with the previous study [17]. It was reported that the compressive strength of the concrete mixture decreases with an increase of kenaf fiber volume fraction. Moreover, this can be explained due to the decrease in the density of the mixture as well as the specimens with kenaf fiber were almost not thoroughly dried and hardened as reported by [19].

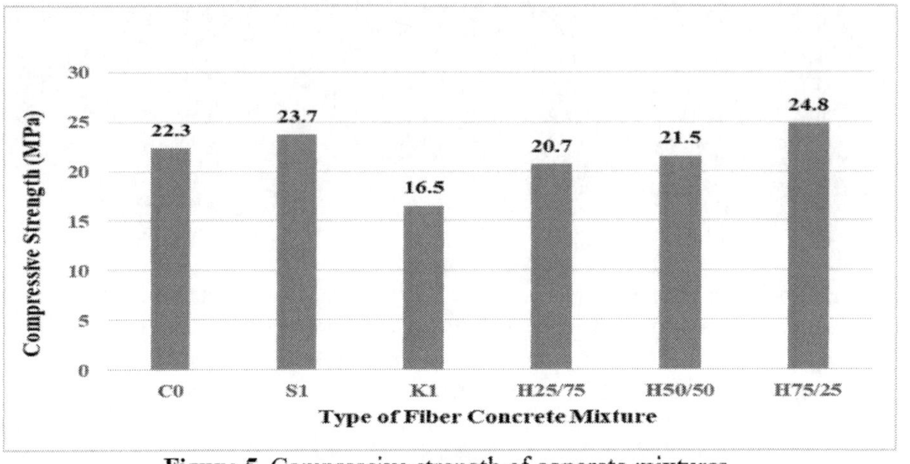

Figure 5. Compressive strength of concrete mixtures.

3.2 Flexural Strength Test

The results of the flexural strength of the fiber concrete mixtures beam specimens tested on the 28th day is illustrated in Figure 6. The results show that the inclusion of single and hybrid fibers into concrete mixture fibers has a moderate influence on the flexural strength of the concrete mixtures. The findings showed that the beams specimens that contain steel fiber and hybrid fiber have a higher flexural strength compare to control specimen. However, kenaf fiber specimen and control specimen have almost same flexural strength.

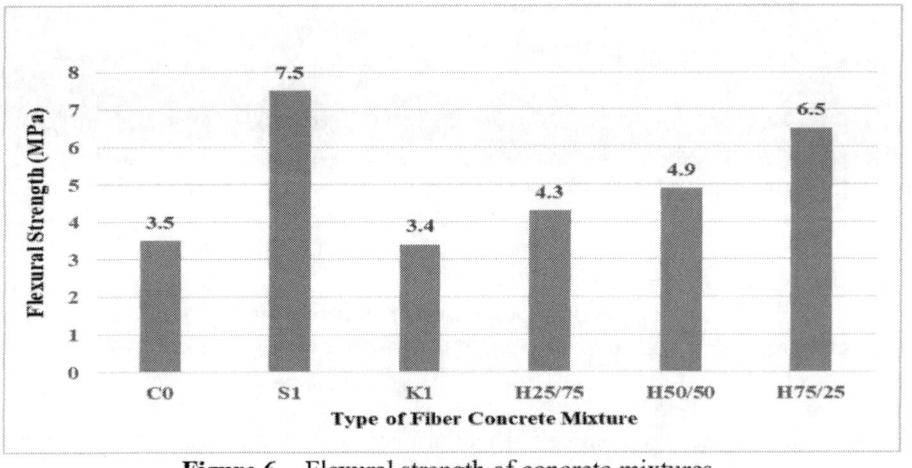

Figure 6. Flexural strength of concrete mixtures.

The highest flexural strength of the beam specimen was observed in beam S1 with 7.5 MPa, which is 114% higher than beam C0. In addition, the flexural strength of hybrid concrete beam specimens H75/25, H50/50 and H25/75, show improvement up to 23 %, 40 %, and 86 %, respectively, in comparison to the beam C0. The best ratio of hybrid fibers can be seen from beam H75/25. As the amount of steel fibers in the hybrid fibers' ratio increases, the flexural strength of the concrete also increases. This observation can be explained in terms of the synergy between steel and kenaf fiber which implies that the comprehensive performance in the flexural strength would exceed that produced by steel fibers or kenaf fibers separately at the same amount of fiber.

3.3 Cracking pattern and mode of failure

Looking at the cracking pattern of the beams at failure shown in Figure 7, most of the beams show cracking propagation along the mid-span between the two-point of loading. During testing, it could be observed that the control beam (C0) has failed in a brittle manner which occurred suddenly without any warning. However, the beams S1, K1, H25/75, H50/50 and H75/25 failed in the ductile manner, in which the cracking is propagated and extended without breaking the beam into half as shown in Figure 7. This confirms the fact that adding fibers to concrete has potential to delay the cracking propagation and change the mode of failure from brittle to ductile manner.

Beam	Cracking pattern	Failure mode
C0		Brittle
S1		Ductile
K1		Ductile
H25/75		Ductile
H50/50		Ductile
H75/25	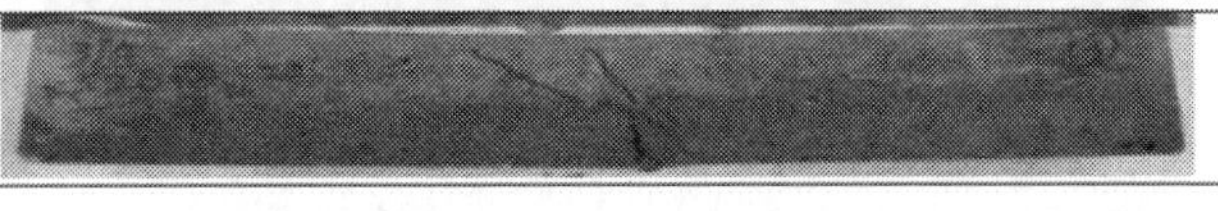	Ductile

Figure 7. Cracking pattern and failure mode of beam.

4. Conclusion

Based on the findings, it can be seen that the inclusion of steel fiber and a higher ratio of steel fiber in hybrid fiber slightly improve the compressive strength of the concrete. In term of flexural strength, significant enhancement was observed in the concrete beam added with fibers. With a minimum ratio of steel fiber in hybrid fiber (H25/75), a notable improvement on the flexural strength can be observed. Furthermore, the cracking propagation was delayed and the mode of failure of the beam also change to a ductile manner.

Acknowledgement
The authors would like to acknowledge the support and kind assistance received from the technicians in the Concrete Structural Laboratory of Faculty of Civil Engineering and Earth Resources, Universiti Malaysia Pahang (UMP) in conducting the experimental work. This work was financially supported by UMP through Research Grant Scheme (RDU 170340).

References
[1] Swamy R N, Lankard D R 1974 Some practical applications of steel fibre reinforced Concrete, *Proc. of the Instit. of Civil Eng.* **56** 235-256
[2] Brandt A M 2008 Fibre reinforced cement-based (FRC) composites after over 40 years of development in building and civil engineering, *Compos. Struct.* **86** 3-9
[3] Syed Mohsin S M 2012 *Behaviour of fibre-reinforced concrete structures under seismic loading* Doctoral dissertation (London: Imperial College London)
[4] Mansur M A, Aziz M A 1982 A study of jute fibre reinforced cement composites, *Int. J. of Cem. Compos. and Lightweight Concrete* **4** 75-82
[5] Hannant D J 2003 Fibre-reinforced concrete *Advanced Concrete Technology–processes* ed. J Newman and B S Choo (Oxford: Butterworth-Heinemann, Elsevier) chapter 6 pp 1 - 17
[6] Cucchiara C, La Mendola L, Papia M 2004 Effectiveness of stirrups and steel fibres as shear reinforcement, *Cem. and Concr. Compos.* **26** 777-786
[7] Khaloo A R, Afshari M 2005 Flexural behaviour of small steel fibre reinforced concrete slabs, *Cem. and Concr. Compos.* **27** 141-149
[8] Juárez C, Valdez P, Durán, A, Sobolev K 2007 The diagonal tension behavior of fiber reinforced concrete beams, *Cem. and Concr. Compos.* **29** 402-408
[9] Ding Y, You Z, Jalali S 2011 The composite effect of steel fibres and stirrups on the shear behaviour of beams using self-consolidating concrete, *Eng. Struct.* **33** 107-117
[10] Minelli F, Conforti A, Cuenca E, Plizzari G 2014 Are steel fibres able to mitigate or eliminate size effect in shear?, *Mater. Struct.* **47** 459-473
[11] Baarimah A O, Syed Mohsin S M 2017 Behaviour of reinforced concrete slabs with steel fibers *IOP Conf. Ser.: Mater. Sci. Eng.* **271** 012099
[12] Conforti A, Minelli F, Plizzari G A 2013 Wide-shallow beams with and without steel fibres: a peculiar behaviour in shear and flexure, *Compos. Part B: Eng.* **51** 282-290
[13] Abbas A A, Syed Mohsin S M, Cotsovos D M, Ruiz-Teran A M 2014 Shear behaviour of steel-fibre-reinforced concrete simply supported beams, *Struct. Build.* **167** 544-558
[14] Gouveia N D, Fernandes N A, Faria D M, Ramos A M, Lúcio V J 2014 SFRC flat slabs punching behaviour–Experimental research, *Compos. Part B: Eng.* **63** 161-171
[15] Baarimah A O, Syed Mohsin S M 2016 An overview of using steel fibers in reinforced concrete structural elements to improve shear reinforcement *In: Proc. of the Natl. Conf. for Postgraduate Research 2016 (Malaysia: Pahang)* (Malaysia: Universiti Malaysia Pahang) pp 260-265
[16] Marar K, Eren Ö, Roughani H 2017 The influence of amount and aspect ratio of fibers on shear behaviour of steel fiber reinforced concrete, *KSCE J. Civil Eng.* **21** 1393-1399
[17] Elsaid A, Dawood M, Seracino R, Bobko C 2011 Mechanical properties of kenaf fiber reinforced concrete *Constr. and Build. Mater.* **25** 1991-2001
[18] Syed Mohsin S M, Azimi S J, Namdar A 2014 Behaviour of Oil Palm Shell Reinforced Concrete Beams Added with Kenaf Fibres. *In Appl. Mech. and Mater.* **567** 351-355
[19] Syed Mohsin S M, Manaf M F, Sarbini N N, Muthusamy K 2016 Behaviour of reinforced concrete beams with kenaf and steel hybrid fibre, *ARPN J. of Eng. and Appl. Sci.* **11** 5385-5390
[20] Li X, Tabil L G, Panigrahi S 2007 Chemical treatments of natural fiber for use in natural fiber-reinforced composites: a review *J. of Polym. and the Environ.* **15** 25-33

[21] Yousif B F, Shalwan A, Chin C W, Ming K C 2012 Flexural properties of treated and untreated kenaf/epoxy composites *Mater.and Des.* **40** 378-385

[22] Mahjoub R, Yatim J M, Sam A R M, Hashemi S H 2014 Tensile properties of kenaf fiber due to various conditions of chemical fiber surface modifications *Constr. and Build. Mater.* **55** 103-113

Incorporating attitudinal parameter in assessing sustainability of Malaysia manufacturing industry

Wan Nurul Syahirah Wan Lanang, Faiz Mohd Turan and Kartina Johan

Faculty of Manufacturing Engineering, Universiti Malaysia Pahang, 26600 Pekan, Pahang.

E-mail: w.n.syahirah.w.lanang@gmail.com

Abstract. Numerous companies are accepting sustainability as an organizational peremptory. There is, however, little convergence on how organizations become sustainable. The previous study suggests that a paradigm shift is necessary to incorporate more sustainable ways of thinking, while others advocate that sustainability requires only moderate behavioural changes as in attitude. In addition, it is also suggesting that sustainability develops most effectively when a singular view of sustainability is applied throughout the company; others contend that differentiated views of sustainability emerge within the various subcultures of an organization. The aim of this paper is to analyse the aftermath of considering attitudinal parameter into the initial data, portraying the true nature of personality during the survey. The research presented was carried out with employees from various manufacturing companies with different branches of knowledge and attitude. Survey methodology was employed by building a questionnaire combining Likert-type items, Sustainable Development Goals (SDGs), Green Project Management (GPM) P5 Integration and multiple-option items. The results allow identifying the knowledges and attitudes of the employees in Malaysia context, contributing relevant data in regard to future engagement relating to sustainability and attitudinal parameter.

1. Introduction

Sustainability has commonly been regarded as a crucial goal for organizations. Mostly refers to longevity, continuity, and capability to be maintained. According to the World Commission on Environment and Development, sustainability means to 'meet the needs of present without compromising the ability of future generations to meet their own need' (United Nations Documents 1987, no page) [1]. Sustainability seems to be agreeable proposal because of its meeting points among environmental concerns, manufacturing, and product design activities.

Over the last decade, the number of sustainability indicators and their use in decision-making has greatly increased [2]. However, the existing sustainability evaluation still do not integrate a nature-economic-society aspect, some of these tools are focused on just one or two dimension(s) of sustainability, product sustainability perspective [3], environmental aspect [4-5]. Moreover, some others focused on all three dimensions [6-8], but there is a same gap in all of these methods which is limited attempts at bringing Green Project Management (GPM) P5 method to use in sustainability practices. Besides that, there is no present research that has been attempted from the viewpoint of focusing on sustainable parameters toward achieving a more systematic assessment model which can contribute to sustainability reporting.

Encouraged by Bursa Malaysia, sustainability view can be referring as crucial point to a successful business in this present-day. Every companies in the auspices of Bursa Malaysia also required to embed the sustainability concept as a vanguard of their business [9]. Besides, each company also need to provide a sustainability report as a requirement of Bursa Malaysia Securities Berhad while currently, every single company that under the auspices of BURSA Malaysia is required to yield the sustainability reporting. The concept of P5 Integration matrix will conclude all major sustainability standard refer to the Triple Bottom Line of sustainability: people (society), profit (financial), and planet (environment), whereas another two are process and product [10].

In principle, attitudes refer to a person's feeling, opinions and general approach towards a person or object [11]. By contrast to personality, attitudes are often influenced by situational and circumstantial factors and hence, they are believed to be less stable than personality traits [12]. Technically, the word risk refers to situations in which a decision is made whose consequences depend on the outcomes of future events having known probabilities. When it comes to the psychology of risk, "there is still a lot of room to go and businesses are only at the beginning in terms of understanding that there are coherent frameworks available that could help them improve the practice of risk management," aforesaid by Hersh Shefrin [13]. Besides, psychology has likely played a larger role in crisis response than many people realize. "Every single risk management disaster in the last 15 years, including financial disasters, has had psychological issues at the root." "Whether it's an earthquake, natural catastrophe or a financial disaster, it is often compounded by our psychological imperfections."

This study aimed to evaluate the influences of a persons' attitude affected their every aspect of risk management in making the decision regarding sustainability practice of their company. The results allow us to identify and analyze the impact of using attitudinal parameter into the data also affected the nature of someone's personality during the survey.

2. Methodology

The proposed research has six steps as follows:

- Step 1: Data sources – case company, existing research
- Step 2: Criteria selection – Weighing criteria
- Step 3: Data collection – analysis
- Step 4: Evaluating attitudinal rating (resolution)

2.1. "Sampling" / Data sources.

The boundary of this study is limited to the manufacturing industry, and the target population was employees who work in the top management of the industry. Hence, the studies from the existing research is done to analyse the frequently used parameters in sustainable assessment and the tools used to measure the sustainability. The case study will be conducted in any company in Malaysia. Thus, five companies from the manufacturing sector that covers in Nilai, Shah Alam, Pekan, Kuantan and Port Klang is selected during the data collection.

2.2. Criteria selection – weighing criteria.

The scale between +3 – -3 as depicted in Table 1 was developed to ease the respondents' group for rating the evaluation criteria, which initially selected by the design engineers based on technical documents and the results of a prior survey.

Table 1. Scale of "Weighting criteria".

Numerical rating	Description
3	Negative Impact High
2	Negative Impact Medium
1	Negative Impact Low
0	Neutral
-1	Positive Impact Low
-2	Positive Impact Medium
-3	Positive Impact High

The other fairly important in this method is where the questions are generated by using the green project management (GPM) that serves as a reference. The green project management (GPM) concept integration matrix is describing below:

a) People – labor practices and decent work, society and customers, human rights, ethical behaviour
b) Planet – material and procurement, energy, water, transport, waste
c) Profit – return on investment, business agility, economic simulation.

2.3. Data collection – analysis.

In this case study, owner of the company, chief executive officer, general manager and a system manager were selected as the expert decision makers. This will be assumed as an act of a field study including in-depth interviews with selected experts. The experts' opinions are used for providing a sufficient data to fulfil the research objectives. The GPM P5 checklist is being modified with respect of Malaysia industry context for each parameter before the respective checklist is used in the interview (data collection).

The process of gathering the data has been conducted in several departments of the companies using the same research questions which are: Department of Production (Parts), Department of Production (Assembly), Department of Engineering, Department of Environmental Quality, Department of Quality Control. From here, the questionnaire has been assigned at the stated departments where the results are then transmitted into the scoring board that has used the green project management as the guideline.

3. Result and Discussion

Below is the initial result concluded from the data analysis that have been done. The graph consists of three different parameters: people, planet and profit with their respected sub-parameter from operational management, showing the relationship between each parameter with sustainability compliance index as stated in scale of 'weighing criteria', Table 1. The data (value) from the graph is already been calculated using min formula in the excel and the equation is developed using polynomial equation: second order in Matlab, also considering 5% error in plotting the graph (95% confidence level). Figure 1 and 2 show the pattern of each sub-parameter before considering the attitudinal parameter revolving the raw data only, which divided into risk averse and risk seeker personality respectively.

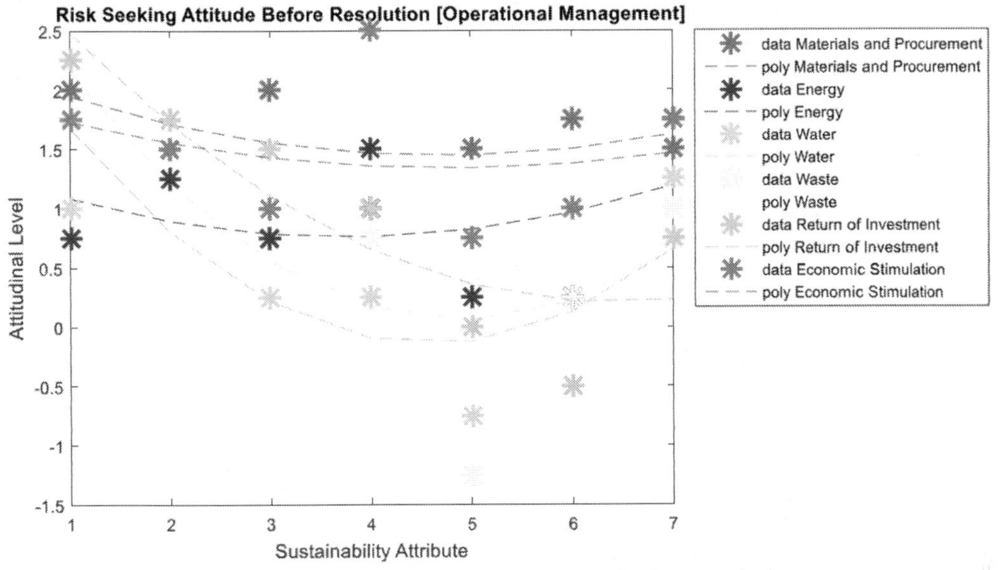

Figure 1. Risk seeking's parameter before resolution.

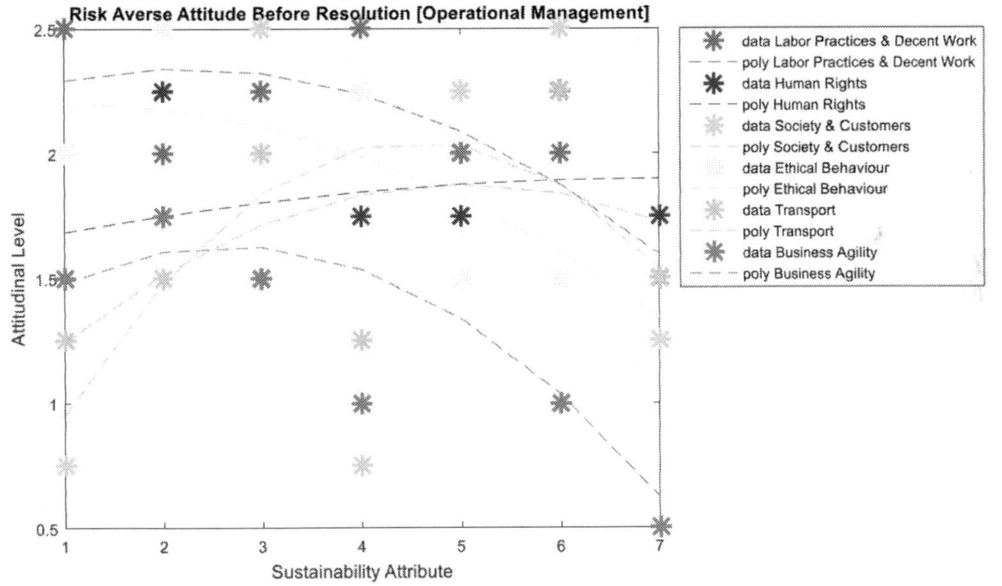

Figure 2. Risk averse's parameter before resolution.

3.1. Evaluating attitudinal rating (resolution)

In this step, a new resolution based on attitudinal parameter is proposed. All crisp data that are gathered in step 3 are being multiply by the proposed resolution. By principle, there is a point of neutral sustainability for every measured aspect of a system. In the measurement of sustainability, there is a tipping point where a system is either sustainable or unsustainable, which we concluded it as impact high, medium and low. Table 2 below presented the proposed resolution based on attitudinal parameter [14], which the author believe could be effective in tackling the complex; ill-defined and human-oriented decision problems in the assessment of product reliability. The attitudinal parameter (λ) is within the range [-0.9,0.9], hence, if $0.1 < \lambda \leq 0.9$, then it's said to be risk-averse. If $\lambda = 0$, it is risk neutral and finally, if $-0.1 < \lambda \leq -0.9$, then it is risk-seeking.

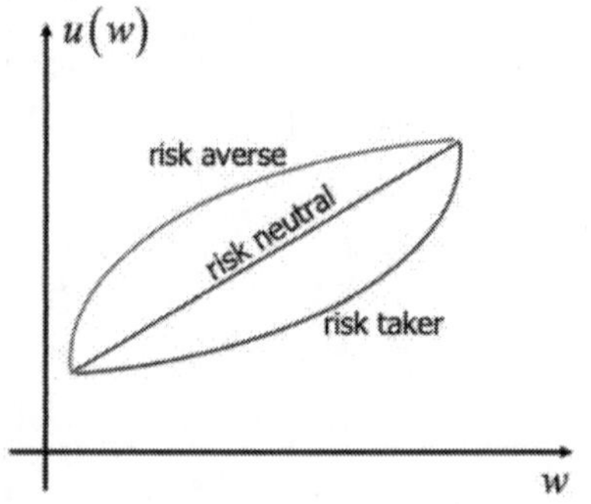

Figure 3. Risk-aversion (green) contrasted to risk-neutrality (blue) and risk loving (red).

Table 2. Proposed resolution based on attitudinal parameter.

Numerical rating	Resolution
-3	0.8
-2	0.4
-1	0.2
0	0
1	-0.2
2	-0.4
3	-0.8

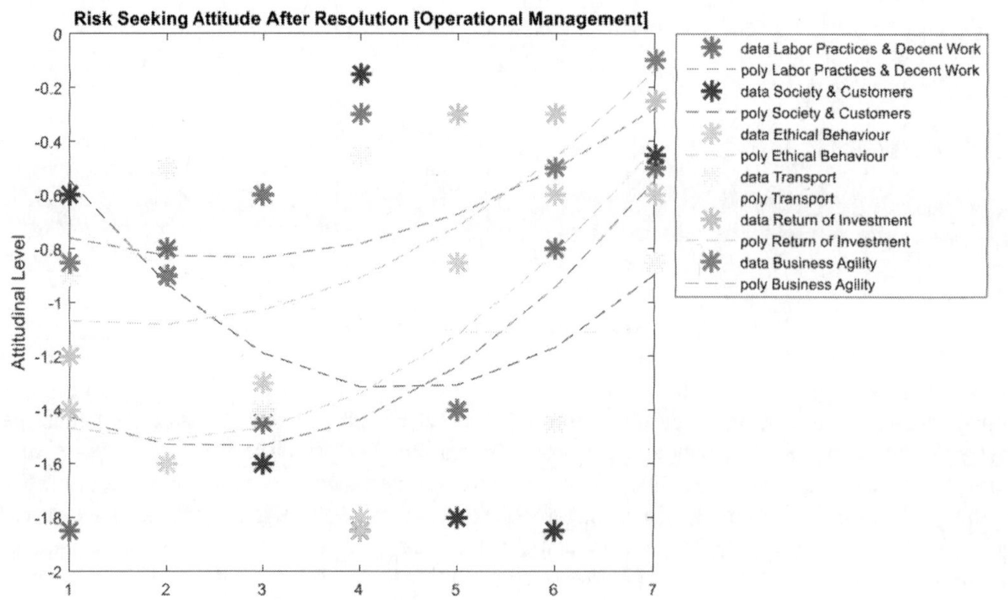

Figure 4. Risk seeking's parameter after resolution.

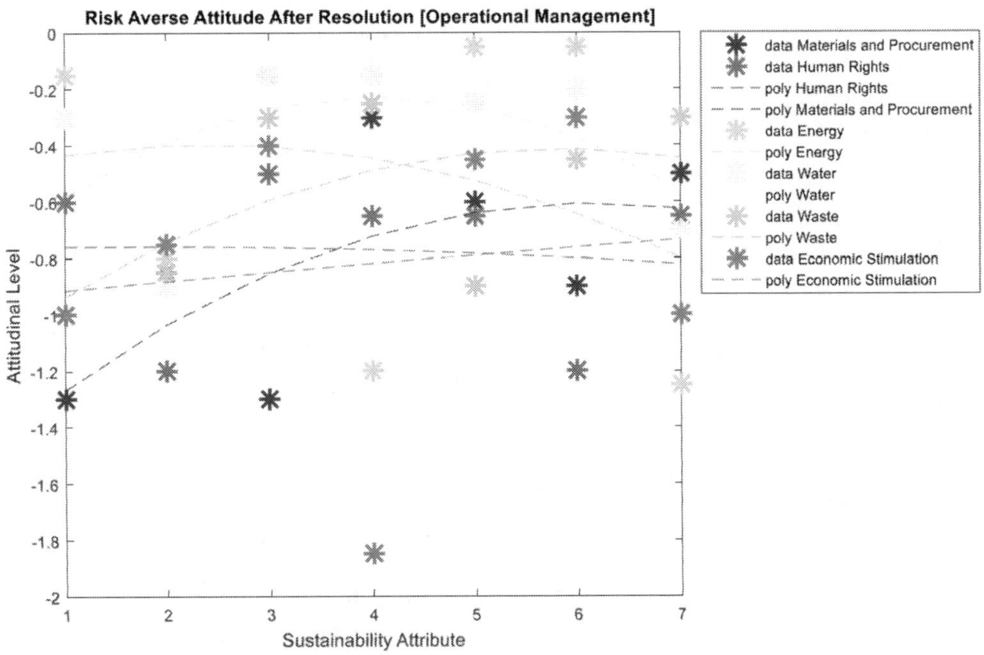

Figure 5. Risk averse's parameter after resolution.

Table 3. Type of risk for each parameter before resolution.

TYPE OF RISK	PARAMETERS
RISK SEEKING	Materials and Procurement, Energy, Water, Waste, Return of Investment, Economic Stimulation
RISK NEUTRAL	-
RISK AVERSE	Labor Practices & Decent work, Human Rights, Society & Customers, Ethical behaviour, Transport, Business Agility

Table 4. Type of risk for each parameter after resolution.

TYPE OF RISK	PARAMETERS
RISK SEEKING	Labor Practices & Decent work, Society & Customers, Ethical behaviour, Transport, Return of Investment
RISK NEUTRAL	-
RISK AVERSE	Human Rights, Materials and Procurement, Energy, Water, Waste, Economic Stimulation, Business Agility

Table 3 and 4 show the clearer comparison of all parameters between before and after considering attitudinal parameter's resolution as suggested in Table 2. As presented in Table 3, 6 over 12 parameters show the trait of risk seeking personality, while another half parameters lead to risk averse personality. However, as far as attitudinal parameter was concerned; after multiplying by the resolution, most of the employees seemed to be more on risk averse personality, where 7 out of 12 parameters indicate the trait by considering the risk utility function as shown in Figure 4 (thus indicated by 58.33%), and each of the parameter under the respective trait is totally different compare with the trait before undergoing the resolution. In economic theories, it is assumed that risk aversion is a typical human attitude toward risk, where someone will find themselves immediately leaning toward the guaranteed results/success, and the differences between risk lover and risk averse are determined by the curvature of the utility function, Figure 3. Moreover, another 5 over 12 parameters (indicated 41.67%) show the trait of risk seeking personality. Risk seeker is assumed that someone who is naturally find themselves drawn to situations where they could win or lose in the end. The results in table 3 and 4 also showed that before considering attitudinal parameter, the employee who is said to be a risk seeker, in reality is not the same as the mentioned personality, where it is originally risk averse and vice-versa.

Nevertheless, these trait of personality as seen in Figure 4 and 5 can also fit roughly in the risk-neutral area while still being slightly averse or slightly risk-seeking if and only if the pattern of graph is to be considered. Likewise, they could be a risk seeker who leans toward neutrality, or a risk-avoider who tends toward neutrality as well. In addition, they also could have a measurable difference between instinct and behavior where he/she for example be a natural risk-seeker who forces himself or herself to make more neutral decisions in order to live responsibly. Or they could be a natural risk-avoider who trains himself or herself to take more risks to get more out of life. Lastly, the results of psychological studies have indicated, however, that people differ in how they make decisions under uncertainty and what motivates them to take economic risks.

4. Conclusion

Once and for all, this study is expected to assist any organizations including an engineer or project managers in producing a better sustainability reporting while incorporating the attitudinal parameter based on GPM P5 integration without neglecting the major pillar of sustainability standard.

In addition, it portrays on how a company need to consider the attitude of the employee to determine the level of sustainability's practice for their organization. From this research, the expectation of the employees' personality from the first place is not true after attitudinal parameter is considered into the raw data.

Meanwhile, due to the diverging comprehension about the sustainability compliance gained from the feedback including their diametrical attitude during the survey, their ideas regarding this assessment are restricted on the existed sustainability tool. Thence, since this assessment method not only based on triple bottom line principle, but also include process and product elements, the outcomes of the research certainly sweeping of the current assessment in the sustainability practices.

Acknowledgments

The authors would like to give special thanks to Research & Innovation Department, Universiti Malaysia Pahang for funding this research project.

References

[1] Florea, L., Cheung, Y. H., & Herndon, N. C. (2013). For all good reasons: Role of values in organizational sustainability. *Journal of Business Ethics, 114(3), 393-408.*

[2] Zamagni, A., Pesonen, H. L., & Swarr, T. (2013). From LCA to Life Cycle Sustainability Assessment: concept, practice and future directions. *The International Journal of Life Cycle Assessment, 18*(9), 1637-1641.

[3] Moldan, B., Janoušková, S., & Hák, T. (2012). How to understand and measure environmental sustainability: Indicators and targets. *Ecological Indicators, 17*, 4-13.

[4] Ghadimi, P., Azadnia, A. H., Yusof, N. M., & Saman, M. Z. M. (2012). A weighted fuzzy approach for product sustainability assessment: a case study in automotive industry. *Journal*

of Cleaner Production, *33*, 10-21.

[5] Bjørn, A., Margni, M., Roy, P. O., Bulle, C., & Hauschild, M. Z. (2016). A proposal to measure absolute environmental sustainability in life cycle assessment. *Ecological Indicators*, *63*, 1-13.

[6] Strazza, C., Magrassi, F., Gallo, M., & Del Borghi, A. (2015). Life Cycle Assessment from food to food: A case study of circular economy from cruise ships to aquaculture. *Sustainable Production and Consumption*, *2*, 40-51.

[7] Chong, Y. T., Teo, K. M., & Tang, L. C. (2016). A lifecycle-based sustainability indicator framework for waste-to-energy systems and a proposed metric of sustainability. *Renewable and Sustainable Energy Reviews*, *56*, 797-809.

[8] Scandelius, C., & Cohen, G. (2016). Sustainability program brands: Platforms for collaboration and co-creation. *Industrial Marketing Management*, *57*, 166-176.

[9] B. Malaysia. (2015). Sustainability Reporting Guide. Available: http://www.bursamalaysia.com/

[10] G. Global. (2015). The GPM® P5™ *Standard for Sustainability in Project Management*.

[11] Funder, J. W. (2012). Aldosterone and mineralocorticoid receptors: a personal reflection. *Molecular and cellular endocrinology*, 350(2), 146-150.

[12] Armstrong, A. C., Wakefield, A., Foy, S., Howell, S. J., & Lee, R. J. (2010). 64 The conflict between survival and fertility post breast cancer treatment–attitudes and beliefs of young women with breast cancer. *European Journal of Cancer Supplements*, *8*(3), 77.

[13] Shefrin, Hersh. *Behavioral Risk Management: Managing the Psychology That Drives Decisions and Influences Operational Risk*. Palgrave Macmillan US: Springer, 2016. Print

[14] Wu, J., & Chiclana, F. (2014). A risk attitudinal ranking method for interval-valued intuitionistic fuzzy numbers based on novel attitudinal expected score and accuracy functions. *Applied Soft Computing*, *22*, 272-286.

Microstructure and hardness performance of AA6061 aluminium composite using friction stir processing

C D Marini and N Fatchurrohman

Faculty of Manufacturing Engineering, Universiti Malaysia Pahang, Malaysia

E-mail: chedinimarini@gmail.com; fatchurrohman@ump.edu.my

Abstract. Rice husk ash (RHA) is an industrial waste that has become a potential reinforced material for aluminium matrix composite (AMCs) due to low cost and abundantly available resources. Friction stir processing (FSP) has been introduced as a method to modify surface properties of the metal and alloy including theirs composite as well. The present work reports the production and characterization of AA6061 and AA6061/5 vol% RHA using FSP using parameters rotation speed 1000 rpm and traversed speed 25 mm/min. The microstructure was studied using optical microscopy (OM). A homogenous dispersion of RHA particles was obtained in the composite. No agglomeration or segregation was observed. The produced composite exhibited a fine grain structure. An improvement in hardness profile was observed as AA6061/5 vol% RHA improves in hardness compared to FSPed of AA6061 without reinforcement.

1. Introduction

Surface modification technique such as plasma spray, chemical vapor deposition (CVD), physical vapor deposition (PVD), electron beam welding (EBM), plasma transferred arc welding (PTAW), thermal spraying and laser beam techniques will generate the formation of unfavorable phase which degrades the properties of the composite. Consequently, all of these techniques are conducted in high temperature which they will produce a detrimental phase meaning difficult to avoid reaction between matrix and reinforcement [1]. In recent studies, friction stir processing (FSP) has been introduced to modify the upper surface of metallic materials [2]. FSP has been developed for use with aluminium alloys, with the goal of high-strain-rate processing [3]. FSP is a fast method with simple processing and low cost to in situ synthesizes particulate-reinforced composites especially in aluminium alloy [4-5] as the composite materials promote better properties including mechanical properties [6]. FSP enhanced the distribution of the reinforced particles which resulting more homogeneous and thermodynamically more stable in metallurgical properties as the matrices and reinforcement have a strong interfacial bonding [7].

Aluminium matrix composites (AMCs) become potential materials for various applications including in aerospace and automotive as they required a high strength and lightweight components [8-9]. Rice husk ash (RHA) has been introduced as reinforced material in AMCs. RHA offered such as high silica contents (87– 97 wt% SiO_2), high porosity, lightweight and very high external surface area which is a valuable material for industrial applications. Rice husks contain high silica content as potential reinforcement for metal matrix composites (MMCs) and ceramic matrix composites (CMCs) [10].

Few studies were reported on fabrication of AMCs with RHA as the reinforced material. A study done by Dinaharan et al. on AA6061 reinforced with 18 vol% of RHA using FSP resulting a homogenous dispersion with no agglomeration and segregation of RHA particles in the composite. AA6061/18 vol% RHA composite demonstrated a fine and equiaxed grain structure. RHA particles fragmented during FSP. An improvement in the tensile strength was observed subsequent to reinforcement of RHA particles. The UTS of AA6061 and AA6061/18 vol% RHA were found to be 220 MPa and 285 MPa, respectively. The fracture surface was dispersed with fractured RHA particles confirming excellent interfacial bonding with the aluminum matrix [11]. A work done by Zuhailawati et al. on aluminium 1100 reinforced with RHA reported that FSP has refined the silica particle associated with excellent distribution in the matrix. Reduction in wear rate of FSPed Al1100 with improved hardness was believed to be due to the presence of hard silica with high interfacial strength and high hardness of recrystallized aluminum grains in the stir zone. The SEM observation of the worn surface of the alloy with added amorphous silica particles showed no deboning between the particles and the matrix [12].

In the present work, a challenge is made to introduce FSP for surface modification of aluminium AA6061 and AA6061 reinforced with RHA. Further studies were conducted in order to investigate the effect on their microstructure and hardness behavior.

2. Experimental procedure

The plates of aluminium alloy AA6061 with dimension of length 100 mm, width 50 mm and thickness 10 mm were used for this experiment. A net of holes were designed on the surface of the plates which corresponding to 5 vol% of RHA particle (40 µm).

Rice husk (RH) was collected from rice mill, BERNAS Rompin. The RH then were washed with distilled water and dried at room temperature for 1 day. Dried RH was heated to 200 °C for 1 hour in order to remove the moisture and organic matter. The color of RH was changed from yellowish to black. Then RH was continued heated at 600 °C for 12 hours to remove carbonaceous matter. During this operation the color of black RH has been change to greyish white ash [13-15]. Table 1 represents the composition of RHA.

Table 1. The composition of RHA.

Constituent	SiO_2	K_2O	P_2O_5	CaO	SO_3	MgO	MnO	Fe_2O_3	ZnO	Rb_2O	MoO_3	SrO
%	43.87	2.18	2.00	1.49	0.61	0.51	0.20	0.10	0.06	0.03	0.02	0.01

The tool was made of H13 tool steel with a shoulder diameter of 16 mm, probe length of 4 mm, probe height 5 mm and probe diameter of 6.25 mm. All samples were subjected to single pass of FSP either with or without reinforced with parameter of rotation speed of 1000 rpm and traversed speed of 25 mm/min. FSP was accomplished by using vertical milling machine (MAKINO).

All samples were prepared for microstructural and mechanical characterization. All the specimens were cut from the cross section, grinded together with polished using standard metallographic technique and etched with Weck reagent. The hardness profile of the thickness cross section samples was determined by using a Vickers indenter at load of 1000 g and dwell time of 15 s.

3. Results and discussion

Aluminium alloy AA6061 with and without reinforced with RHA particulates were successfully produced using friction stir processing. The microstructure and mechanical characterization of both FSPed AA6061 and FSPed AA6061/5 vol% RHA are discussed below.

3.1. Microstructure of FSPed AA6061 and FSPed AA6061/RHA

The micrographs were recorded within the stir zone of FSP. The optical micrograph of the produced AA6061 and AA6061/5 vol% RHA are shown in Figure 1. Figure 1(a) represents the microstructure of FSPed AA6061. Figure 1(b) show clearly the uniform distribution of RHA in the matrix, and no void and discontinuities were observed. There was a good interfacial bonding between the particles and matrix material. Furthermore, in Figure 1(b), the stirred zone showed fine grain compared to FSPed AA6061. This fine structure was produced by dynamic recrystallization and static grain growth after the stirring process, caused by the influence of frictional heat and plastic deformation. The alloy matrix grain size decreased slightly in the presence of silica particles, which may have limited the extent of grain growth during FSP [12].

There are no agglomerations and segregations of RHA particles in the stirred zone. In order to get isotropic properties, the composite need to have a homogeneous dispersion between matrix and reinforcement. The rubbing tool shoulder and shearing action of the pin surface will produced the fractional heat which leads the aluminum matrix to plasticize. The holes are sheared off during material flow. The composite will be fabricated as the rotating action of the tool enhanced the compacted particles to mix with the plasticize aluminium. Since FSP is a solid state process, thus the aluminum matrix does not melt during the process. After the composite is formed in the solid state, the free movement of reinforcement particle due to density gradient with the aluminum matrix is restricted. On the other hand, due to density gradient in molten aluminum, the reinforcement particle will tend to float up or sink down. This has been the major obstacle of stir casting which is overcome by FSP. Both of the matrix and reinforcement, aluminium and RHA particles are subjected to the severe plastic strain induced by the FSP process. The strain developed during FSP is much higher compared to conventional severe plastic deformation processes. Reinforcement particles commonly breakdown and undergo a change in shape and size [11].

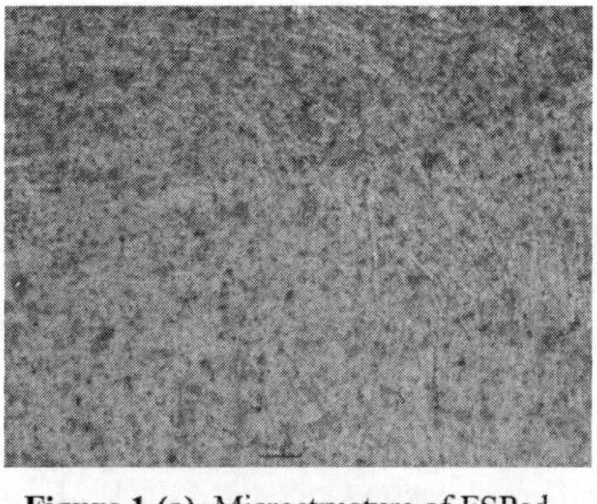

Figure 1 (a). Microstructure of FSPed AA6061 (5X).

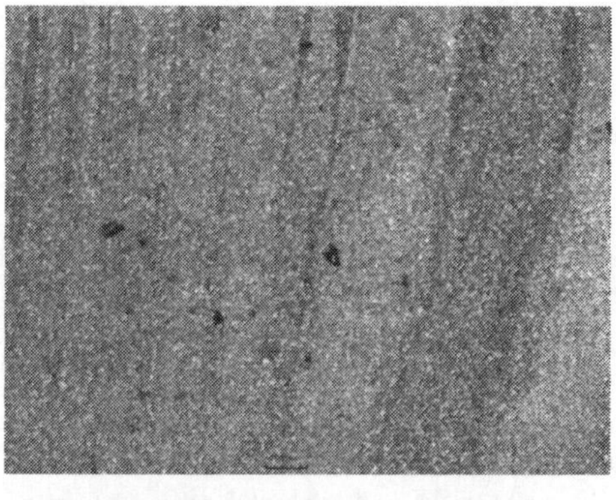

Figure 1(b). Microstructure of FSPed AA6061/5 vol% RHA (5X)

3.2. Hardness profile of FSPed AA6061 and FSPed AA6061/RHA

Figure 2 shows that the harness value for within stirred zone for FSPed of AA6061 and AA6061/5 vol% RHA at 1000 rpm at different point. The harness value for FSPed AA6061 is lower compared to AA6061/5 vol% RHA, as in average of 114.1 Hv and 131.7 Hv, respectively. As the RHA is incorporate within AA6061 matrix, these will influenced the formation of new recrystallization of aluminium grains which lead to refine the grain in the stirred zone [12].

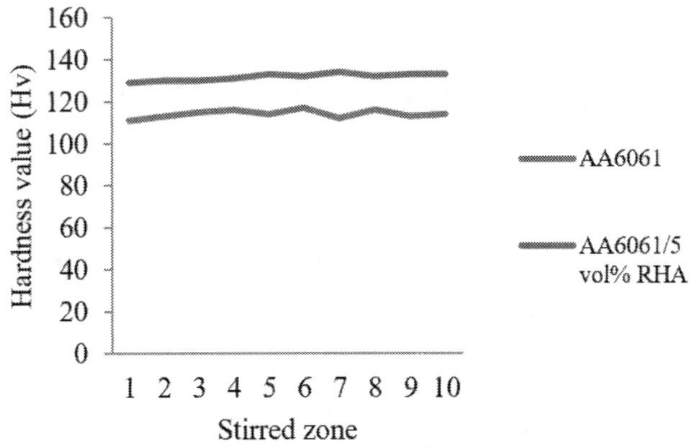

Figure 2. Hardness value in stirred zone.

4. Conclusions

RHA has been introduced in aluminium matrix AA6061 using FSP. RHA particles were homogenously distributed within the AA6061 matrix across the stirred zone. There was no agglomerations and segregation of the reinforced particles within the composite. A homogeneous dispersion lead to grain refinement within the composite. RHA particles improved the harness profile of the composite. In future work, FSP is suitable process to produce RHA reinforced AMCs by altering the processing parameters and compositions.

Acknowledgement

The authors gratefully appreciate Universiti Malaysia Pahang for the financial support through Postgraduate Research Scheme PGRS160343 and Internal Research Grant RDU150371.

References

[1] Devaraju A, Kumar A, Kumaraswamy A and Kotiveerachari B 2013 *Mater. Des.* **51** 331–41
[2] Wang W, Shi Q, Liu P, Li H and Li T 2009 *J. Mater. Process. Technol.* **209** 2099–103
[3] Morishige T, Hirata T, Tsujikawa M and Higashi K 2010 *Mater. Lett.* **64** 1905–08
[4] Abdollahi S H, Karimzadeh F and Enayati M H 2015 *J. Alloys Compd.* **623** 335–41
[5] Qian J, Li J, Xiong J, Zhang F and Lin X 2012 *Mater. Sci. Eng. A* **550** 279–85
[6] Marini C D and Fatchurrohman N 2015 *Jurnal Teknologi (Sci & Eng.)* **74** 103-109
[7] Hsu C J, Kao P W and Ho N J 2007 *Mater. Lett.* **61** 1315–18
[8] Fatchurrohman N, Iskandar I, Suraya S, Johan K 2015 *App Mechanics & Mater.* **695** 32-35
[9] Fatchurrohman N, Sulaiman S, Sapuan S M, Ariffin M K A and Baharuddin B T H T 2015 *Int. J. Automot. Mech. Eng.* **11** 2531–40
[10] Soltani N, Bahrami A, Pech-Canul M I, Gonzlez L A and González L A 2015 *Chem. Eng. J.* **264** 899–935
[11] Dinaharan I, Kalaiselvan K and Murugan N 2017 *Compos. Commun.* **3** 42–6
[12] Zuhailawati H, Halmy M N, Almanar I P, Seman A A and Dhindaw B K 2016 *Int. J. Metall. & Mater. Eng.* **2** 2–7
[13] Saravanan S D and Kumar M S 2013 *Procedia Eng.* **64** 1505–13
[14] Narasaraju G and Raju D L 2015 *Mater. Today Proc.* **2** 3056–64
[15] Prasad D S, Shoba C and Ramanaiah N 2014 *J. Mater. Res. Technol.* **3** 79–85

Nanofluid as coolant for grinding process: An overview

J Kananathan[1], M Samykano[1], K Sudhakar[1,3], S R Subramaniam[2], S K Selavamani[2], Nallapaneni Manoj Kumar[1], Ngui Wai Keng[1], K Kadirgama[1], W A W Hamzah[1] and W S W Harun[1]

[1] Faculty of Mechanical Engineering, Universiti Malaysia Pahang, 26600 Pekan, Pahang, Malaysia.
[2] Faculty of Electrical and Electronics Engineering, Universiti Malaysia Pahang, 26600 Pekan, Pahang, Malaysia.
[3]Energy Centre, Maulana Azad National Institute of Technology Bhopal, M.P., India

E-mail: mahendran@ump.edu.my

Abstract. This paper reviews the recent progress and applications of nanoparticles in lubricants as a coolant (cutting fluid) for grinding process. The role of grinding machining in manufacturing and the importance of lubrication fluids during material removal are discussed. In grinding process, coolants are used to improve the surface finish, wheel wear, flush the chips and to reduce the work-piece thermal deformation. The conventional cooling technique, i.e., flood cooling delivers a large amount of fluid and mist which hazardous to the environment and humans. Industries are actively looking for possible ways to reduce the volume of coolants used in metal removing operations due to the economical and ecological impacts. Thus as an alternative, an advanced cooling technique known as *Minimum Quantity Lubrication (MQL)* has been introduced to the enhance the surface finish, minimize the cost, to reduce the environmental impacts and to reduce the metal cutting fluid consumptions. Nanofluid is a new-fangled class of fluids engineered by dispersing nanometre-size solid particles into base fluids such as water, lubrication oils to further improve the properties of the lubricant or coolant. In addition to advanced cooling technique review, this paper also reviews the application of various nanoparticles and their performance in grinding operations. The performance of nanoparticles related to the cutting forces, surface finish, tool wear, and temperature at the cutting zone are briefly reviewed. The study reveals that the excellent properties of the nanofluid can be beneficial in cooling and lubricating application in the manufacturing process.

1. Background

In recent years, the increased requirement for product quality and precision has demanded the use of high-performance grinding technology, particularly in automotive and semiconductor industries. These have also necessitated an enhanced grinding process which capable of providing near optimal yield with respect to productivity, precision and cost. In general, the goal of manufacturing processes is to maximize the production rate to the specified product quality bounds and simultaneously reduce the production costs. In precision manufacturing industries, components that require precise tolerances and smooth finishes are typically manufactured by the grinding process. Figure 1 shows the representative of the grinding machine setup. Grinding process guarantees the final accuracy and surface quality of any produced parts [65,66]. However, the energy consumption for the removal of a unit volume of material in grinding is typically higher than other machining processes. One of the major limiting factors in grinding production rates are the thermal damage [31]. The most severe damage that usually occurs on the grinded work-piece is the 'work-piece burn,' where discoloration and blemishes can be observed on the work-piece [37]. Also, during the grinding process the surface microstructure of the material expected to change due to the increase in the working temperature. These microstructure changes eventually vary the hardness of the material and subsequently results in detrimental internal stresses. This internal stress creates a higher tensile stress on the workpiece surface that leads to a reduced fatigue life [12]. As a result, it decreases the service life, reliability of the produced parts and the precision of the grinding wheels [66]. Also, while machining a high strength material, the working temperature rises with the speed and load of the cutting tool used, this decreases the tool strength leading to a quicker tool wear [11]. Besides, the dulling (removal of the worn grains to develop the new cutting points) drives the formation of scattering marks on the work-piece surface upon machining which later affects the surface quality of the workpiece. However, this damage can be reduced by the application of a coolant to aid in removing the heat created by the work piece-tool interaction. The primary objective of lubricators is to decrease the amount of friction between two sliding surfaces and to maintain the work piece surface property without any alteration or damages [31,47].

The suitability of using coolants in abundance to cool and lubricate machining processes become one of the most crucial subjects in the manufacturing industry. In metalworking operations, application of the coolants is involved in processes such as cutting (separation of the metal from the workpiece in the form of chips) and abrading (rubbing away the surface by friction). The coolants also used to protect the machined surface from corrosion. Usually, coolant is applied directly to the grinding zone to limit the heat generation. The fluid accomplishes this by reducing the amount of friction in the grinding zone through its lubrication properties which minimize the cutting forces thereby saving the energy [58]. These results in an increased of cutting tool life (grinding wheel) and capacity utilization. Thus the production is kept consistently at a high level in terms of quality and efficiency [3]. Thus, a proper selection of coolant is crucial to eliminate the heat effectively and efficiently [23]. In addition, the cutting fluid also helps to flush the produced chips from the grinding process [18]. Whereby improper chip removal could clog and damage the wheel. Also, the forces and energy input would substantially increase due to clogging and would result in heat input to the work-piece [22].

The conventional mechanism of applying the coolant is known as flood coolant system, where a large quantity of coolant continuously impinges on the rake face of the tool [28]. The conventional flood delivery method is as shown in Figure 2. This mechanism has several drawbacks such as high cost, high pumping power requirement, and limited practical usefulness. Due to the large fluid delivery, an extensive amount of hazardous mist is generated during grinding process which contributes to health and environmental hazards. However total elimination of this conventional coolant delivery is not recommended because for high depth machining it would shorten the tool life [16]. Another drawback that is associated with conventional flood coolant mechanism is that the hefty amount of fluid would bound its way to the machining surface owing to the chips and swarf obstruction [59].

Figure 1. Illustration of whole grinding machine set-up.

Figure 2. Conventional fluid delivery system (Flood Coolant system).

There are two alternatives available for flood coolant mechanism. One is known as dry grinding and another is near dry grinding (minimum quantity lubrication, MQL). Dry grinding utilizes no cutting fluid, as such, it is not recommended for machining hard materials due to the high heat generation. Since there is no coolant used, the transfer the heat from the contact zone is disrupted resulting in severe thermal damages on the ground surface of the material. These also increase the grinding energy and grinding forces, wear of the grinding wheel, low material removal rate (low depths of cut) and reduced surface integrity compared to conventional flood delivery mechanism [27].

Whereas, MQL is a technique of supplying a minute quantity of cooling lubricant to the zone or contact point so that the applied amount of coolant can be reduced immensely while retaining the cooling and lubrication effects that are absent in dry machining [56]. MQL helps in reducing the friction at the tool-workpiece interface. MQL technique produces the best surface finish, and improved tool life compared to dry grinding [36]. MQL with oil shown to produce better surface finish than MQL-water grinding [63].

As to date, many researches have reported on the effect of different cooling techniques using conventional cutting fluid in the different machining process. However, there is limited or no comprehensive review available on the effect of MQL using conventional and nanoparticle enhanced cutting fluid in grinding machining. Thus, this paper targets to provide a review focusing on the use of Nano fluids in grinding applications. Present work aims to review the effect of MQL using different Nano fluids and conventional cutting fluid on grinding machining operation.

2. Minimum Quantity Lubrication (MQL)

MQL technique implicates the application of a small quantity coolant dispensed to the tool-work piece interface by compressed air flow. MQL grinding refers to the delivery of a minute quantity of fluid following an aerosol mechanism to the grinding zone for cooling and lubricating the work piece. The MQL lubrication system is as shown in figure 3.

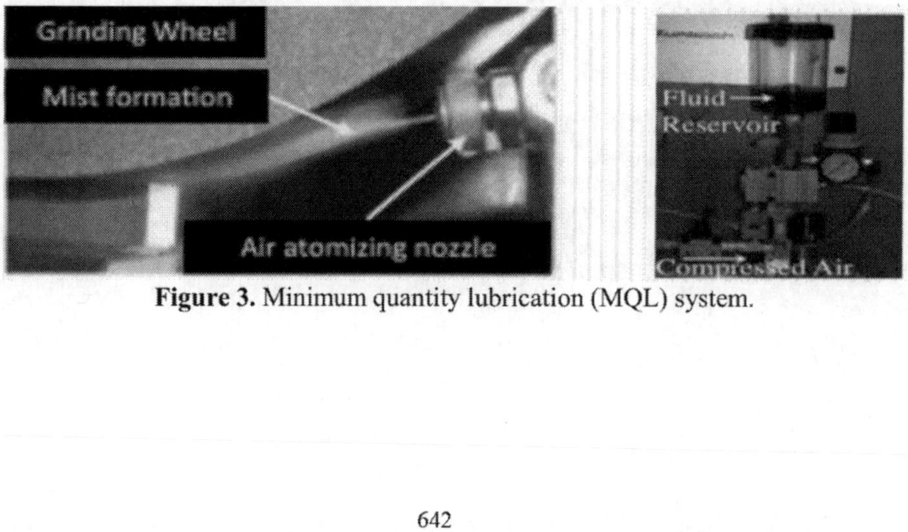

Figure 3. Minimum quantity lubrication (MQL) system.

In MQL system, a compressor is used to supply the air at high pressure. The fluid chamber is connected to a small diameter flexible tube. This tube is made to pass through a roller type flow controller which permits small amount of fluid to flow at high pressure and velocity. The compressed air entering into the inlet port produces pressure to cause the fluid to flow continuously to the mixing chamber through the controller at a constant rate. The air and the cutting fluid then mixed in the mixing chamber so that the mixture of cutting fluid and air impinged at a high velocity through the nozzle on the tool-work piece interface. The schematic and working mechanism of the MQL set-up is shown in figure 4. In MQL technique, the typical flow rate for the coolant is 10–100 ml/h [33] compared to flood machining which is 30000–60000 ml/h [17].

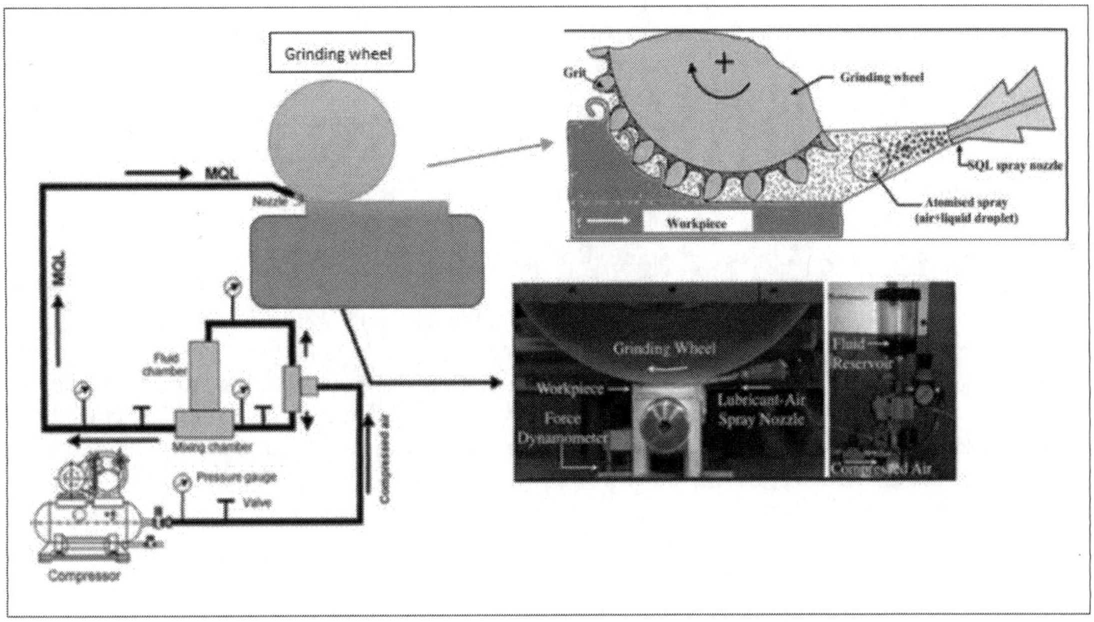

Figure 4. Schematic of MQL Set-Up in Grinding,

Numerous advantages have been reported for the MQL grinding compared to the conventional technique and dry grinding. The cooling and lubrication performance of the grinding fluid is the key to the success of MQL grinding process. As to date, numerous studies related to MQL have been carried out to investigate the effectiveness of this technique. Most of the investigators have indicated that minimum quantity lubrication (MQL) technique is the best alternative to flood cooling in machining. Moreover, the MQL technique also enhances the surface finish, reduces the cost, reduces the impact on the environment and the health of the operating persons. In grinding process, it was shown that MQL provides low friction with the development of thin-film layer lubrication between the tool and workpiece which significantly reduces the tool wear and cutting forces.

Da Silva et al. [13] have studied the effect of surface integrity for minimum quantity lubricant (MQL). This technique showed superior effect in the grinding process in terms of surface integrity (roughness, residual stress, microstructure and micro hardness). It was observed that surface roughness values substantially reduced with the use of the MQL technique, probably due to excellent properties of lubricity and also no significant clogging of the grinding wheel pores was found. Despite of high residual stress value obtained, MQL technique did not negatively affect the surface integrity after Grinding.

Tawakoli et al. [62] studied the role of MQL in grinding soft and hardened steel. It was observed surface roughness was higher in grinding the soft steel. Whereas improved surface roughness in grinding hardened steel. MQL technique shows reduction in the cutting force in both cases which has led to lesser wheel warn out, improved wheel life.

3. MQL Techniques Using Conventional Cutting Fluid

Application of MQL in machining has emerged as an alternative for reducing excessive use of cutting fluid and cost saving with in a more economical way. This section discusses use of conventional cutting fluid with MQL and its influence on machining parameters in grinding machining process based on available research works done. [63] have investigated an MQL grinding or near dry grinding (NDG) system. In this system, an air–oil mixture called an aerosol is fed into the wheel-work zone. Compared to dry grinding, MQL grinding substantially enhances cutting performance in terms of increasing wheel life and improving the quality of the ground part.

In general, the use of synthetic oil is used widely been used in researches done and gives better results than the vegetable oil [24]. [51] studies the effect of grinding 17CrNi6-6 steel.it seen that MQL shows better residual stress when CBN wheel is used. It was shown that MQL does not affect the surface integrity. Anand et al. [4], investigated grinding of Pre-Sintered zirconia using synthetic oil with water soluble oil. It was reported that, MQL showed reduced grinding force and lesser wheel warn out. However, the G-ratio was better in wet condition. Tawakoli et al. [61] found that MQL with oil produced better surface than the MQL-water grinding. Rodriguez, Hildebrandt and Lopes [45] in their study noticed that MQL using lower flow rate less than 120ml/h exhibit bad surfaces and higher wheel wear.

MQL reduces specific grinding energy thus also providing efficient lubrication in able to control the thermal deformation occurring on the work piece and able to give good surface quality then the flood grinding [8]. Belentani et al. [10] found out that MQL not only improves surface roughness but also helps in disrupting the air barrier formed around the rotating wheel thus lower grinding power obtained. Numerous advantages were found for MQL grinding compared with the conventional technique and dry grinding. The cutting fluid parameters along with the summary of MQL influence on cutting parameter performances is briefed in Table 1.

Table 1. Summary of cutting fluids, MQL and machining parameters, mode of lubrication for grinding and machining performance findings.

Authors	Cutting Fluid	Work-Piece material	Lubrication	MQL flow rate	MQL air/fluid pressure	Depth of Cut (μm)	Wheel speed (m/s)	Table speed (mm/min)	Findings
Guo, Li, Zhang, & Wang, 2017 [24]	Castor oil	nickel-based alloy named GH4169	Mode	50 ml/h	0.6 Mpa	10	30	3000	Use of Mixed base oils shows lesser Specific Grinding energy and better surface roughness compared to pure castor oil.
De Mello, de Silva, Machado, Gelamo, & Diniz, 2017[15]	Water soluble synthetic oil	Ti-6Al-4V alloy	MQL	240 ml/h	0.3 MPa	10, 20	32	6.8 m/min	MQL shows lesser surface roughness and better surface quality in lesser depth of cut. At higher depth of cut, flood outperformed MQL.
Anand, Arunachalam, & Vijayaraghavan, 2017[4]	Synthetic Oil and Water soluble oil of 5% concentration	Pre-Sintered zirconia	MQL and flood	250 ml/h	6 bar	10, 15, 20, 25 and 30	22, 26, 32, 36 and 42	2, 4, 6, 8 and 10 (m/min)	MQL showed reduced grinding force and lesser wheel warn out. The surface finish and G-ratio was better in wet condition.
Huang, Ren, Li, Zhou, & Zhang, 2017[30]	MB-103A coolant based on synthetic oil	AISI5140 annealed steel	MQL and wet cooling	120, 240, 360, 480 ml/h	3, 4, 5, 6, 7 bar	100	45	0.5 m/min	MQL produced better surface quality with reduced harden layer depth with increased fluid rate and pressure .MQL has higher cooling rate.
Rodriguez, Hildebrandt, & Lopes, 2017[45]	Vegetable Oil, Water miscible – Semi-synthetic vegetable oil	AISI 4340 steel	MQL	30, 60 and 120 ml/h	0.6 Mpa	1.2, 2.5, 3.7	30	0.5 m/min	MQL with lower flow rate than 120ml/h exhibited inferior performances in terms of bad surface roughness, higher wheel wear and more power consumption.
Sawicki, Kruszyński, & Wójcik, 2017[47]	Micro5000 oil and oil emulsion 5%	17CrNi6-6 steel	MQL and flood	25 ml/h	-	25	30	-	Flood cooling with oil emulsion produces better residual stress than MQL. MQL shows better residual stress when CBN wheel in use based on grinding condition. Use of MQL did not negatively affect the surface integrity
Saber, Rahimi, Parsa, & Ashrafijou, 2016[46]	RS1642 Behran Oil, Water-miscible coolant lubricant at 5% concentration	CK45, 90 ± 3 HRB (mild carbon steel)	MQL and flood	120 ml/h	0.7 Mpa	5, 10	20	4000	CAMQL shows reduction of power consumption than to dry grinding. Better surface Quality using MQL.CAMQL is used for fine grinding of soft steels.
Balan, Vijayaraghavan, & Krishnamurthy, 2016[5]	Cimtech D14 MQL oil	Inconel 751 super alloy	MQL and dry-grinding	60 ml/h	6 bar	10, 20, 30		0.4, 0.6, 0.9 m/min	MQL produces better performances than dry grinding. Thin flaked and elongated chips produced by MQL.
Wójcik & Wejman, 2016[67]	(MC) - Micro 3000, (EC) – Eco Cut micro 82, (BO) 3000	Titanium Alloy(Ti-6Al-4V and TIGER 5)	MQL and dry-grinding	50 ml/h	0.6 Mpa	5	22.5-26.5	0.2 - 0.5 m/s	MQL has contributed in reduced cutting force and thermal deformation at top layer of object minimised.

Rabiei, Rahimi,, & Hadad, 2015[44]	RS1642 Behran oil	Mild carbon, bearing, & tool steels, raw HSS,	MQL and WET cooling	120 ml/h	4 bar	5, 20, 35, 50	30	3	MQL reduced the friction coefficient and reduced the tangential forces. MQL was more useful for grinding of hard steels
Hadad, 2015[25]	Syntilo XPS Castrol in a 5% concentration	hardened 100Cr6	MQL,DRY,F LOOD	20, 50, 100 ml/h	2, 3, 4, and 7 bar	20	30	1000	Surface Roughness improved better using MQL than in dry condition. Moreover MQL reduces grinding force which lowers heat generation in contact zone. The nozzle distance from the grinding zone plays an important role. The lesser the distance of MQL spray from grinding zone, the lubrication effect reduces
Batako & Tsiakoumis, 2015[9]	Castrol "Carecut ES1" oil	nickel alloy and BS-534A99 steel	MQL,DRY,F LOOD	30 ml/h	0.45 MPa	10-30	30	200	MQL also showed reduced grinding forces and better surface finish than dry grinding. The relative improvement in surface finish (lower roughness) is explained by the semi-lapping effect induced by the superimposed oscillations to the cutting process.
Emami, Sadeghi, & Sarhan, 2014[20]	Synthetic oil, hydrocracked oil, vegetable oil, and (conventional) mineral oil.	alumina (Al2O3) ceramic	MQL and Dry condition	150 ml/h	-	20	30	200	Use of MQL produce higher surface quality and preferred in finish grinding. Use of MQL retains its lubricity ability property at high temperature and wheel-work piece contact.
Barros, Silva, & Canarim, 2014[8]	Quimatic ME-I semi-synthetic soluble oil, with 2.5 % concentration	AISI 4340 steel rings	MQL	2.7 × 10–8 m^3 /s	0.6 Mpa	-	30	0.25, 0.50, and 0.75 mm/min	Flood coolant produces better surface roughness. The high rate of flow if flood coolant is responsible for removing chips from the cutting zone. MQL reduces specific grinding energy thus also providing efficient lubrication in able to control the thermal deformation occurring on the work piece and able to give good surface quality.
Belentani, Funes Júnior, Canarim, & Hassui, 2014[10]	MQL with water, vegetable oil in the proportions of: 1:1, 1:3 and 1:5 parts of oil	SAE/AISI 4340	MQL and WET cooling	-	0.8 Mpa	-	30	0.25, 0.5 and 0.75 mm/min	MQL improves surface Roughness, with addition of water in decreases wheel wear than the conventional cooling lubrication. It helps in disrupting air barrier formed around the rotating wheel thus lower grinding power obtained.
Abdul Rahim, Ibrahim, Mohid, & Ahmad, 2014[2]	-	AISI 1020	MQL and Flood cooling	80 l/h	-	0.2 mm	-	400, 500, 600 mm/min	Use of MQL reduces surface temperature than to flood coolant. The grinding forces are reduced with increased cutting speed. MQL also helps in reduction of temperature which attributed to the capability of the tiny mist particles of MQL penetrating into the grinding zone, subsequently reduces the grinding friction.

Hadad & Hadi, 2013[26]	Vegetable oil and synthetic ester oil	Aluminium alloy 6061 & hardened steel	MQL and FLOOD cooling	4 ml/h	-	5, 15, 25	30	1.5	MQL reduced the surface roughness and grinding forces as compared to dry and wet grinding and provided better cooling than dry grinding though not better than conventional wet grinding.
Balan, Vijayaraghavan, & Krishnamurth, 2013[6]	Cimtech D14 MQL oil	Inconel 751 super alloy	MQL, dry, flood	60, 80, 100 ml/h	2, 4, and 6 bar	30	2826 m/min	0.9	Minimum grinding force and surface roughness could be achieved by increasing the MQL oil flow rate and air pressure. MQL considerably reduced the grinding forces, temperature and surface roughness.
Emami, Sadeghi, & Sarhan, 2013[19]	Mineral oil	Al2O3 engineering ceramic	MQL	25-400 ml/h	-	18	30	10	Use of optimally designed MQL in grinding reduced the tangential as well as normal grinding force and surface roughness.
Li & Lin, 2012[36]	Vegetable oil (Bluebe lubricant LB-1)	SK3 steel (HRC 18)	MQL	1.88 and 0.63 ml/h	0.5 MPa	50	30000, 39000, 48000 RPM		-
Barczak, & Batako, 2012[7]	Synthetic oil Castrol Carecut ES1	Common steels EN8, M2 & EN31		33 ml/h	0.4 Mpa	5,15	25, 45	6.5, 15	Low grinding temperature and lower surface roughness values were recorded with MQL.
Morgan, Barczak, & Batako, 2012[40]	Castrol Carecut ES1 oil	Common steels EN8, M2 and EN31	MQL, dry, wet	33 ml/h	0.4 Mpa	5.15	25, 45	6.5, 15	More favourable for softer materials.
Tawakoli, Hadad, & Sadeghi, 2011[63]	Mineral oil, hydrocracked oil, synthetic oil, ester oil	Hardened steel (100Cr6)	MQL, dry, wet	100 ml/h	4 bar	30	30	3	Reduced the grinding forces and surface damages Among many oils, MQL produced best surface quality with mineral hydrocracked oil
Tawakoli, Hadad, & Sadeghi, 2010[61]	HAKUFORM 20e30 MQL oil	Hardened steel (100Cr6)	MQL, dry, flood,	20, 50, 100 ml/h	2, 3, 4, and 7 bar	20	30, 45	1, 2	MQL, dry, flood MQL nozzle, while set angular toward the wheels, produced lowest surface roughness and tangential grinding and normal forces. Increase of nozzle distance improved the surface quality.

IOP Conf. Series: Materials Science and Engineering **342** (2018) 012078 doi:10.1088/1757-899X/342/1/012078

4. Role of Nanoparticles in Nano-fluid Machining

Recently, the inclusive of nanoparticles in conventional lubricants has shown remarkable improvement in thermo-physical, heat transfer capabilities, reduction in the coefficient of friction, wear effect to enhance the efficiency, and reliability of machined part [47,48]. The friction force at the tool chip interface normally increases due to tool wear which leads to increment in the cutting force and friction force. It is noted that by adopting to the Nano lubricants at the tool chip interface, the coefficient of friction can be reduced. It is believed that nanoparticles deposits on the friction surface and compensate for loss of mass, which called 'mending effect'. Due to porous nature of spherical nanoparticle, it could impart high elasticity and enhances the gap at tool to work-piece interface. Some particles have rolling effects and some are sheared due to high pressure at the cutting zone. The shape of particles changes due to high compression. With increasing the concentration of nanoparticles the degree of shaping and shearing increases. For this reason, Nano lubricants are able to reduce the cutting force with less power consumption [49,50]. Nanoparticles in the mineral oil impart their effect by combined effect of rolling and sliding at the tool-chip interface. This effect reduces the coefficient of friction. The lubrication mechanism is shown in figure 5.

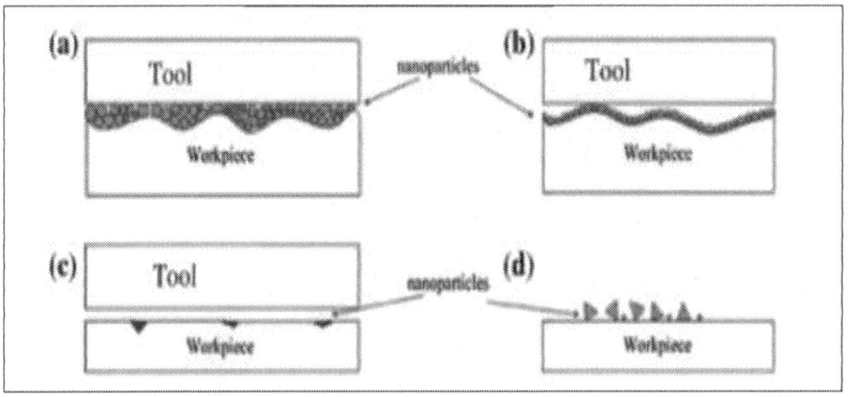

Figure 5. (a). Rolling effect. (b) Protective film (c) Mending effect (d) Polishing effect [34].

Nano-fluid is a new class of fluids engineered by dispersing nanometer size solid particles into base fluids such as water, lubrication oils. Nanoparticles have much higher and stronger temperature dependent thermal conductivity, and enhanced heat transfer coefficients at very low particle concentration. These are the key parameters for their enhanced performance in many of the machining applications. Their ball bearing effect lubrication at the cutting zone through formation of film layer which reduces friction between the contact surfaces thereby reducing cutting force, temperature and tool wear [41,49]. Also, the addition of nanoparticles into base fluid can enhance their convective heat transfer coefficient, lubricating properties, and wettability apart from their tribology, and wear characteristics. The excellent properties of the Nano-fluids makes them very attractive in cooling and lubricating application in manufacturing. The application of various nanoparticles and its performances in metal cutting operations with respect to the cutting forces, surface finish, tool wear and temperature at the cutting zone are evaluated and highlighted.

5. MQL Using Nano-fluid in Grinding Process

In the MQL lubrication process, the coolant or lubrication medium used is posed certain restrictions especially at very high cutting speeds where the lubricating oil tends to evaporates as it strikes the already heated cutting tool at elevated temperature. Desire to compensate for the shortcomings of the lubricating medium in the MQL technique led to the introduction of nanoparticles in the cutting fluids for use in the MQL lubrication process. Nano-fluids has emerged as a promising solution to this problem.

Friction, and wear result in the increase of energy consumption, and reduction in the life of mechanical parts [54,50]. Friction is a principal cause of wear and energy dissipation. Moreover high temperature generations at cutting zone affects the work-piece dimensional accuracy and surface quality [72,73]. It has been reported in various studies that nanoparticles introduction in cutting fluids led to excellent machining performance in reduction of cutting forces, reduced tool wear, reduced cutting temperature and improved surface finish of the work piece thereby increasing productivity and reduction of hazards to the health of personnel and the environment better than the pure or conventional MQL process [57]. Nano lubricant confers rolling action of billions of nanoparticles at the tool chip interface thereby reducing friction and thermal deformation of the work piece in addition to less consumption of lubricating oil. Hisakado et al. [29] added copper and nickel nanoparticle into paraffin and to investigate its frictional properties. It showed that coefficient of friction reduced by 18%. This finding shows that nanoparticle improves the anti-wear and friction reducing performance of lubrication oil. Nanoparticle size and concentration are two important parameters that affect grinding performance.

Lee et al. [34] studied micro grinding on of a tool steel under four lubrication media (Nano-diamond of 2 level concentration, Nano-Al_2O_3 of 2-level concentration, dry air and pure MQL of the paraffin based oil. Nanoparticle concentration considered are 2% vol. and 4% vol. while particle size of 30nm and 150nm selected for both Nano-diamond and Nano-Al_2O_3, shows that the smaller particle size surface roughness enhanced producing smoother surface than the larger particle. Optimizing the nanoparticles concentration leads to better surface roughness and reduced grinding force. [52] studied grinding of titanium alloy using silicon carbide wheel with different lubrication media and with water based Al_2O_3 nano-fluid of 1% and 4% concentration. It was reported that grinding forces reduced using Nano cutting fluid with lower nanoparticle concentration. However, the surface finish was improved with increased nanoparticles concentration. [64] compared the performance of MQL with vegetable oil with MQL+Vegetable oil+ Al_2O_3 nanoparticle in machining Inconel 600alloy.The experiment resulted temperature, surface roughness and tool wear were reduced under the MQL with 6% vol. of Al_2O_3 particle). Mao, Huang, Zhou, Gan and Zhang [38] investigated study of Nano-fluid (Al_2O_3/water) in grinding of hardened AISI 52100 steel. It was found that addition of Al_2O_3 nanoparticle into deionized water helps in reducing grinding force. Setti et al. [53] continued the research comparing the performance of two different Nano-fluid (Al_2O_3/water, and CuO/water) in grinding of Ti–6Al–4V using MQL system. It showed that Al_2O_3/water reduction in friction coefficient due to its ability to form and prevent tribo films on the work-piece surface. The lotus effect of titanium and Al_2O_3 lead to effective flushing of debris thus improved the grind ability by reduction friction, temperature from grinding zone.

Kalita et al. [32] studied that surface-grinding test were performed on ductile iron work pieces using an aluminum oxide wheel and impingement of advanced Nano lubricants, consisting of organic molecules with phosphide intercalated-MoS_2 Nano particles (<100nm), was observed the effectiveness of Nano lubricants measuring reduction of 45-50% in the grinding force and 48-55% in the reduction in the abrasive wheel wear than MQL with paraffin lubricant. Zhang et al. [69] evaluated MoS_2 nanoparticles in three different oils (soybean, palm and rapeseed oil) as base fluid.it showed that the optimal concentration range of MoS_2 nanoparticle is 6% which showed excellent lubricating property which have reduce specific grinding energy and lower coefficient of friction. Zhang et al. [68] extended the research by investigating surface quality under MQL grinding of Ni-based alloy with MoS_2, CNT and Mix (MoS_2+CNT). It showed that Mix MQL showed the best surface quality followed by MoS_2, and CNT. It is also observed increase of Nano concentration lowers grinding force ratio (G) which improves the lubrication and surface quality.

Prabu and Vinayagam [43] carried out experimental studies to evaluate the impact of using Nano lubricants containing MWCNT and SAE20W40 oil with MQL technique during grinding process on work piece (AISI D2 tool steel) for surface smoothness. It was observed specimen sparked using MWCNTs have better surface finish, reduced micro cracks and better surface morphology [70]. The surface smoothness was evaluated using surface roughness meter and atomic force microscope (AFM). The MWCNT and SAE20W40 oil improved its surface quality to Nano level from micro level. [42] extended the study on surface characteristic of AISI D2 tool steel in grinding process but using SWCNTs in SAE20W40 oil base fluid as the Nano lubricant. The fractal dimension and surface roughness were measured. It was found that under the fractal dimension and the roughness are decreased when without using Nano fluid. However, under presence of Nano fluid with optimum speed of the grinding wheel, the surface was even and smooth. CNTs increases the heat carrying capacity and thermal conductivity of the lubricating oil and thus prevents any damage to the work piece. [53] studied the effect of CuO, and Al_2O_3 Nano-fluids with different concentration in MQL grinding of Ti–6Al–4V on grinding force, surface roughness and wheel morphology. It is observed that Al_2O_3 offered significant reduction in the coefficient of friction due to its ability to form and prevent tribo film in ground surface. Al_2O_3 Nano-fluid improved the grindability of Ti–6Al–4V with conventional abrasive wheel by reducing friction, temperature, favoring effectively grain fracturing, ideal chip formation and effective flushing out of debris from grinding zone.

[39] studied the effects of Nano-fluid MQL combined with ultrasonic vibration on surface grinding. Seen that the lubricants with MoS_2 additives show significant improvement in grinding performance by lowering the grinding forces and the force ratio. Normal force is reduced by around 25% and tangential force is reduced by around 50% compared to dry grinding, and it also produces smaller forces than flood cooling. The reduction can be attributed to penetration MoS_2 NP into the grinding interface. It is found that using UAG with MoS_2 Nano-fluid MQL lead to significant improvement in the surface roughness. Improvement in surface is due to the fact that the grit motion with low friction in horizontal vibration creates a higher chance of cutting the peaks of the surface. [60,70] investigated grinding of Inconel 718 of two different Nano particle (Ag/water and ZnO/water) under SQL mode. It is observed that addition of NPS in DI water have further improved in further convection in the higher grinding zone temperature. The thermal load on the abrasive grits is improved by the high heat carrying capacity and improved lubricity of the Ag and ZnO Nano-fluids. Low grinding forces, smaller friction coefficient, minimal sticking of debris over alumina grits, and better ground surface have been observed in the case of grinding with ZnO Nano-fluids. This is due to the spreading nature of the ZnO. The wheel wear is almost invisible when ZnO Nano-fluid is used. This is due to better wettability of the ZnO NF. [35] investigated heat transfer performance of MQL grinding of Ni-based alloy GH4169 with different Nano-fluid (MoS_2, SiO, PCD, CNT, Al_2O_3 and ZrO_2,) which palm oil as MQL base fluid. Nano-fluid with high thermal conductivity is believed to have good heat transfer performance. CNT have stated the highest thermal conductivity when added with to base fluid. It is observed that SiO have the lowest grinding force followed by MoS_2 and PCD Nano-fluids.

However, CNT have the lowest grinding temperature, thus CNT has the best heat transfer performance compared to other Nano-fluids analyzed. [66,71] further investigated the lubricating property of different Nano-fluid (MoS_2, SiO_2, ND, CNTs, α- Al_2O_3, and ZrO_2) in MQL grinding of 440C steel ball. It showed that Al_2O_3 Nano-fluid has the best antifriction property contributing the best surface morphology. Based on the frictional test and grinding experiment the Diamond, ND nanoparticle can cause grinding and polishing effect on the work piece surface during plane grinding. The Al_2O_3, MoS_2 and SiO_2 nanoparticles however are suitable as friendly addictive's of base oil. Wang et al. [65] studied on the effect of using different concentration of Al_2O_3 on surface quality specific energy and wheel morphology in MQL grinding Ni-based alloy GH4169. Different volume concentrations (0.5%, 1.0%, 1.5%, 2.0%, 2.5%, 3.0%, 3.5%, and 4.0%) of Al_2O_3 nanoparticles were added into palm oil. Al_2O_3 nanoparticles can form thin protective oil films on the grinding wheel and work piece surface. These films improve the tribological performance of grinding wheel/work piece interfaces significantly.it is observed that as Nano-fluid concentration in base oil increases, the force ratio and specific grinding energy decrease first and then increase, reaching the minimum at 1.5% vol. At 2.5% vol., the wearing loss of the grinding wheel is the lowest. Best tribological performance is gained when the Nano-fluid concentration is 2% vol. However further increase of vol. % than 2%, could have possibly increase the thickness of lubricating film but it will not reduce the friction and wearing loss. At 2% vol. is where the largest wetting area and lubricating effect occurs.

6. Conclusion

This paper has presented a review of research activities carried in the application different types of cutting fluid and Nano-fluid based cutting fluid with the help of the MQL lubrication technique. This paper reviews MQL lubrication technique in the grinding machining process. A brief description and mechanism of the MQL technique and systematically discussing its effect on performance in the grinding process are also presented. There is enough literature which clearly reveals that MQL system provides better performance than dry machining. So MQL is found to be promising alternative for flooded and dry machining due to their environmental friendly characteristics. Researcher have done inclusive review on the application of Nano-fluid in grinding process. Different type of nanoparticles which benefits in grinding were reviewed. The nano fluid has gained significant attention due to its capability to enhance the heat transfer and lubrication performance. Nano fluid MQL was proven to reduce the coefficient of friction and wear effect by enhancing the efficiency and reliability of machine tools. MQL gave the best performance in terms of the surface roughness, feed forces, tool wears and surface quality. Very few records can be seen in the literature indicating more experimental research that is needed to ensure maximum performance of using nano fluids as a coolant in grinding process.

Acknowledgement

Universiti Malaysia Pahang fully supports the facilities and resources for this research. The authors would like to acknowledge the support from the Universiti Malaysia Pahang internal grant RDU1703150, RDU170320, and Ministry of Higher Education, Malaysia (FRGS Grant) RDU160119.

References

[1] Sridharan, u., & Malkin, s. 2009. Effect of minimum quantity lubrication (MQL) with nanofluid on grinding behavior and thermal distortion. Transactions of NAMRI/SME, 37, pp.629-636.

[2] Abdul Rahim, E., Ibrahim, R., Mohid, Z., & Ahmad, M. 2014. Study on temperature, force and specific energy of AlSl 1020 under MQL grinding process. Applied Mechanics and Materials, 465, pp.1119-1123.

[3] Adedayo, S., & Momoh, S. 2010. Effect of Initial Elevated Metal Temperature on Mechanical Properties of an ARC-Welded Mild Steel Plate. Indian Journal of Science and Technology, 3(12), pp.1224-1228.

[4] Anand, P., Arunachalam, N., & Vijayaraghavan, L. 2017. Investigation on grindability of medical implant material using a silicon carbide wheel with different cooling conditions. Procedia Manufacturing,10, pp.417-428.

[5] Balan, A., Vijayaraghavan, L., & Krishnamurthy, R. 2016. An experimental assessment on the performance of different lubrication techniques in grinding of Inconel 751. Journal of advanced research, 7(5), pp.709-718.

[6] Balan, A., Vijayaraghavan, L., & Krishnamurth, R. 2013. Minimum quantity lubricated grinding of Inconel 751 alloy. Materials and Manufacturing Processes, 28(4). Materials and Manufacturing Processes, 28(4), pp.430-435.

[7] Barczak,, L., & Batako, A. 2012. Application of minimum quantity lubrication in grinding. Materials and Manufacturing Processes, 27(4), pp.406-411.

[8] Barros, B., Silva, T., & Canarim, R. 2014. Utilization of teflon and aluminum oxide for wheel cleaning in Minimum Quantity Lubrication (MQL) grinding. Materials Research, 17(1), pp.23-32.

[9] Batako, A., Tsiakoumis, V. 2015. Experimental Investigation into Resonance Grinding of Hardened Steel & Nickel Alloys with MQL. International Advance Manufacuring Technology, pp.1-26.

[10] Belentani, R., Funes Júnior, H., Canarim, R., & Hassui, A. 2014. Utilization of minimum quantity lubrication (MQL) with water in CBN grinding of steel. Materials Research, 17(1), pp.88-96.

[11] Bruni, C., Forcellese, A., & Gabrielli, F. 2006. Effect of the lubrication-cooling technique, insert technology and machine bed material on the workpart surface finish and tool wear in finish

turning of AISI 420B. International Journal of Machine Tools and Manufacture, 46(12),, pp.1547-1554.
[12] Chen, X., Rowe, W., & McCormack, D. 2000. Analysis of the transitional temperature for tensile residual stress in grinding. . Journal of Materials Processing Technology, 107(1), pp.216-221.
[13] Da Silva, L., Bianchi, E., & Fusse, R. 2007. Analysis of surface integrity for minimum quantity lubricant—MQL in grinding. International Journal of Machine Tools and Manufacture, 47(2), pp.412-418.
[14] Davis, B., Schueller, J., & Huang, Y. 2015. Study of ionic liquid as effective additive for minimum quantity lubrication during titanium machining. Manufacturing Letters, 5, pp 1-6.
[15] De Mello, A., de Silva, R., Machado, A., Gelamo, R., & Diniz, A. 2017. Surface Grinding of Ti-6Al-4V Alloy with SiC Abrasive Wheel at Various Cutting Conditions. Procedia Manufacturing, 10, , pp.590-600.
[16] Diniz, A., & Micaroni, R. 2002. Cutting conditions for finish turning process aiming: the use of dry cutting. International Journal of Machine Tools and Manufacture, 42(8),, pp.899-904.
[17] Dixit , U., Sarma, D., & Davim, J. 2012. 'Environmentally friendly machining'. Briefs in Applied Sciences and Technology,Springer,New York,USA.
[18] Ebbrell, S., Woolley, N., & Tridimas, Y. 2000. The effects of cutting fluid application methods on the grinding process. International Journal of Machine Tools and Manufacture, 40(2), pp.209-223.
[19] Emami, M., Sadeghi, M., & Sarhan, A. 2013. Investigating the effects of liquid atomization and delivery parameters of minimum quantity lubrication on the grinding process of Al 2 O 3 engineering ceramics. Journal of Manufacturing Processes, 15(3), pp.374-388.
[20] Emami, M., Sadeghi, M., & Sarhan, A. 2014. Investigating the minimum quantity lubrication in grinding of Al 2 O 3 engineering ceramic. ournal of cleaner production, 66,, pp 632-643.
[21] Fritsche, A., & Bleicher, F. 2015. Experimental Investigation of the Heat Flux Distribution in Grinding of Titanium Alloy. Procedia Engineering, 100, pp.987-993.
[22] Ge, P., Li, G., Lu, C., & Liu, Z. 2003. Performance Evaluation & Action Mechanism Analysis of Extreme Pressure Additives Used for Oil-Based Cutting Fluids. In Key Engineering Materials, Vol. 250, pp. 281-286.
[23] Guo, C., & Malkin, S. 2000. Energy partition and cooling during grinding. Journal of Manufacturing Processes, 2(3), pp.151-157.
[24] Guo, S., Li, C., Zhang, Y., & Wang, Y. 2017. Experimental evaluation of the lubrication performance of mixtures of castor oil with other vegetable oils in MQL grinding of nickel-based alloy. Journal of Cleaner Production, 140, pp.1060-1076.
[25] Hadad, M. 2015. An experimental investigation of the effects of machining parameters on environmentally friendly grinding process. Journal of Cleaner Production, 108, pp.217-231.
[26] Hadad, M., & Hadi, M. 2013. An investigation on surface grinding of hardened stainless steel S34700 and aluminum alloy AA6061 using minimum quantity of lubrication (MQL) technique. . The International Journal of Advanced Manufacturing Technology, 68(9-12), pp.2145-2158.
[27] Hadad, M., & Sharbati, A. 2016. Thermal aspects of environmentally friendly-MQL grinding process. Procedia Cirp, 40, pp.509-515.
[28] Heisel, U., Lutz, M., & Spath, D. 1994. Application of minimum quantity cooling lubrication technology in cutting processes.
[29] Hisakado, T., Tsukizoe, T., & Yoshikawa, H. 1983. Lubrication mechanism of solid lubricants in oils. ASME, Transactions. Journal of Lubrication Technology (ISSN 0022-2305), 105, pp.245-252.
[30] Huang, X., Ren, Y., Li, T., Zhou, Z., & Zhang, G. 2017. Influence of minimum quantity lubrication parameters on grind-hardening process. Materials and Manufacturing Processes, , pp.1-8.
[31] Irani, R., Bauer, R., & Warkentin, A. 2005. A review of cutting fluid application in the grinding process. International Journal of Machine Tools and Manufacture, 45(15), pp.1696-1705.

[32] Kalita, P., Malshe, A., & Rajurkar, K. 2012. Study of tribo-chemical lubricant film formation during application of nanolubricants in minimum quantity lubrication (MQL) grinding. CIRP Annals-Manufacturing Technology, 61(1), , pp.327-330.

[33] Khan, M., Mithu, M., & Dhar, N. 2009. 'Effects of minimum quantity lubrication on turning AISI 9310 aloy steel using vegetable oil-based cuttig fluid . Journal of Materials Processing Technology,Vol.209, pp.5573-5583.

[34] Lee, P., Nam, J., Li, C., & Lee, S. 2012. An experimental study on micro-grinding process with nanofluid minimum quantity lubrication (MQL). . International Journal of Precision Engineering and Manufacturing, 13(3), pp.331-338.

[35] Li, B., Li, C., Zhang, Y., Wang, Y., & Jia, D. 2017. eat transfer performance of MQL grinding with different nanofluids for Ni-based alloys using vegetable oil. Journal of Cleaner Production, 154, pp.1-11.

[36] Li, K., & Lin, C. 2012. Study on minimum quantity lubrication in micro-grinding. Int. J. Adv. Manuf. Technol. 62, 99-105.

[37] Liu, Z., Abe, S., & Noda, M. 1995. he influence of grinding oil viscosity on grinding heat and burn damage in creep-feed grinding . Lubrication engineering, 51(8).

[38] Mao, C., Huang, Y., Zhou, X., Gan, H., & Zhang, J. 2014. The tribological properties of nanofluid used in minimum quantity lubrication grinding. he International Journal of Advanced Manufacturing Technology, 71(5-8, pp.1221-1228.

[39] Molaie, M., Akbari, J., & Movahhedy, M. 2016. Ultrasonic assisted grinding process with minimum quantity lubrication using oil-based nanofluids. Journal of Cleaner Production, 129, pp.212-222.

[40] Morgan, M., Barczak, L., Batako, A. 2012. Emperatures in fine grinding with minimum quantity lubrication (MQL). International Journal of Advanced Manufacturing Technology, 60(9), pp.951-958.

[41] Park, K., Ewald, B., & Kwon, P. 2011. Effect of nano-enhanced lubricant in minimum quantity lubrication balling milling. Journal of tribology, 133(3).

[42] Prabhu, S., & Vinayagam, B. 2011. Fractal dimensional surface analysis of AISI D2 Tool steel material with nanofluids in grinding process using atomic force microscopy. ournal of the Brazilian Society of Mechanical Sciences and Engineering, 33(4),, pp.459-466.

[43] Prabu, S., & Vinayagam,, B. 2010. Nano surface generation of grinding process using carbon nano tubes. . Sadhana, 35(6), pp.747-760.

[44] Rabiei, F., Rahimi,, A., & Hadad, M. 2015. Performance improvement of minimum quantity lubrication (MQL) technique in surface grinding by modeling and optimization. Journal of Cleaner Production, 86, pp.447-460.

[45] Rodriguez, R., Hildebrandt, R., & Lopes, J. 2017. Application viability evaluation of the Minimum Quantity Lubrication coolant technique under different flow rates in Plunge Cylindrical Grinding of the ABNT 4340 steel with aluminum oxide wheel. REM-International Engineering Journal, 70(4), pp.429-436.

[46] Saber, A., Rahimi, A., Parsa, H., & Ashrafijou, M. 2016. Improvement of surface grinding process performance of CK45 soft steel by minimum quantity lubrication (MQL) technique using compressed cold air jet from vortex tube. Journal of Cleaner Production, 131, pp.728-738.

[47] Samykano, M. (2015). Physical and Mechanical Characterization of Electrodeposited Nickel Nanowires–Influence of Current Density and External Magnetic Field. . *North Carolina Agricultural and Technical State University.*

[48] Samykano, M., Mohan, R., & Aravamudhan, S. (2014). Morphology and Crystallographic Characterization of Nickel Nanowires—Influence of Magnetic Field and Current Density During Synthesis. *Journal of Nanotechnology in Engineering and Medicine, 5(2)*, p.021005.

[49] Samykano, M., Mohan, R., & Aravamudhan, S. (2015). Structure and Morphology of Electrodeposited Nickel Nanowires at an Electrode Distance of 20mm. *Structure, 1*, p.23704.

[50] Samykano, M., Mohan, R., & Aravamudhan, S. (2015). Effect of current density and magnetic field on the growth and morphology of nickel nanowires. *In MEMS and Nanotechnology, Volume 8* , pp. 75-83.

[51] Sawicki, j., Kruszyński, b., & Wójcik, R. 2017. He influence of grinding conditions on the distribution of residual stress in the surface layer of 17crni6-6 steel after carburizing. Advances in Science and Technology. Research Journal, 11(2), pp.17-22.//47

[52] Setti, D., Ghosh, S., & Rao, P. 2012. Application of nano cutting fluid under minimum quantity lubrication (MQL) technique to improve grinding of Ti–6Al–4V alloy. In Proceedings of World Academy of Science, Engineering and Technology No. 70, pp. 512-516.//48

[53] Setti, D., Sinha, M., Ghosh, S., & Rao, P. 2015. Performance evaluation of Ti–6Al–4V grinding using chip formation and coefficient of friction under the influence of nanofluids. nternational Journal of Machine Tools and Manufacture, 88, pp.237-248.//49

[54] Sharif , S., Sadiq, I., & Yusof, N. 2017. A Review of Minimum Quantity Lubrication Technique with Nanofluids Application in Metal Cutting Operations. International Journal on Advanced Science, Engineering and Information Technology, 7(2), pp.587-593.//50

[55] Sharma,, J., & Sidhu, B. 2014. Investigation of effects of dry and near dry machining on AISI D2 steel using vegetable oil. Journal of cleaner production, 66, pp 619-623.//59

[56] Shen, B., Shih, A., & Tung, S. 2009. Application of nanofluids in minimum quantity lubrication grinding. Tribology & Lubrication Technology, 65(3), p.73.

[57] Shokoohi, Y., & Shekarian, E. 2016. 'Application of nanofluids in Machining processes-A review". Journal of Nanoscience and Technology, vol 2(1), pp.59-63.

[58] Shokrani, A., Dhokia, V., & Newman, S. 2012. Environmentally conscious machining of difficult-to-machine materials with regard to cutting fluids. International Journal of Machine Tools and Manufacture, 57, pp.83-101.

[59] Singh, T., Singh, P., Dureja, J., & Dogra, M. 2016. A review of near dry machining/minimum quantity lubrication machining of difficult to machine alloys. International Journal of Machining and Machinability of Materials, 18(3), pp.213-251.

[60] Sinha, M., Madarkar, R., Ghosh, S., & Rao, P. 2017. Application of eco-friendly nanofluids during grinding of Inconel 718 through small quantity lubrication. ournal of Cleaner Production, 141, pp.1359-1375.

[61] Tawakoli, T., Hadad, M., & Sadeghi, M. 2010. Influence of oil mist parameters on minimum quantity lubrication–MQL grinding process. International Journal of Machine Tools and Manufacture, 50(6), pp.521-531.

[62] Tawakoli, T., Hadad, M., & Sadeghi, M. 2009. An experimental investigation of the effects of workpiece and grinding parameters on minimum quantity lubrication—MQL grinding. International Journal of Machine Tools and Manufacture, 49(12), pp.924-932.

[63] Tawakoli, T., Hadad, M., Sadeghi, M., & Daneshi, A. 2011. Minimum quantity lubrication in grinding: effects of abrasive and coolant–lubricant types. ournal of Cleaner Production, 19(17), pp.2088-2099.

[64] Vasu, V., & Kumar, K. 2011. Analysis of nanofluids as cutting fluid in grinding EN-31 steel. Nano-Micro Letters, 3(4), , pp.209-214.

[65] Wang, Y., Li, C., Zhang, Y., & Yang, M. 2017. Experimental evaluation on tribological performance of the wheel/workpiece interface in minimum quantity lubrication grinding with different concentrations of Al 2 O 3 nanofluids. Journal of Cleaner Production, 142,, pp.3571-3583.

[66] Wang, Y., Li, C., Zhang, Y., & Yang , M. 2017. Comparative evaluation of the lubricating properties of vegetable-oil-based nanofluids between frictional test and grinding experiment. Journal of Manufacturing Processes. 2017 Apr 30, 94-104.

[67] Wójcik, R., & Wejman, P. 2016. Grinding Process Results with Oil Mist. Mechanics and Mechanical Engineering Vol. 20, 255-261.

IOP Conf. Series: Materials Science and Engineering **342** (2018) 012078 doi:10.1088/1757-899X/342/1/012078

[68] Zhang, Y., Li, C., Jia, D., & Wang, Y. 2016. Experimental study on the effect of nanoparticle concentration on the lubricating property of nanofluids for MQL grinding of Ni-based alloy. *Journal of Materials Processing Technology, 232,*, pp.100-115.

[69] Zhang, Y., Li, C., Jia, D., & Zhang, D. 2015. Experimental evaluation of MoS 2 nanoparticles in jet MQL grinding with different types of vegetable oil as base oil. *Journal of Cleaner Production, 87,*, pp.930-940

[70] Sofiah, A., Kananathan, J., & Samykano, M. (2017). Electrochemical deposited nickel nanowires: influence of deposition bath temperature on the morphology and physical properties. *Materials Science and Engineering Vol. 257*, p. 012032.

[71] Lah, N., Samykano, M., Johan, M., & Othman, N. (2017). Plasmonic behaviour of phenylenediamine functionalised silver nanoparticles. *Materials Research Express, 4(9)*, p.095018.

[72] Hamidi, M., Harun, W., & Samykano, M. (2017). A review of biocompatible metal injection moulding process parameters for biomedical applications. *Materials Science and Engineering*.

[73] Asri, R., Harun, W., Samykano, M., & Lah, N. (2017). Corrosion and surface modification on biocompatible metals:A Review. *Materials Science and Engineering*.

IOP Conf. Series: Materials Science and Engineering **342** (2018) 012079 doi:10.1088/1757-899X/342/1/012079

Framework of Sustainability Assessment (FSA) method for manufacturing industry in Malaysia

Nur Soliha Sahimi, Faiz Mohd Turan and Kartina Johan

Faculty of Manufacturing Engineering, Universiti Malaysia Pahang, 26600 Pekan, Pahang, Malaysia

E-mail: solihasahimi.ump@gmail.com

Abstract. In recent years, our planet faced massive social, environmental and economic challenges. Moreover, Environmental Democracy Index (EDI) published that Malaysia has been ranked as the second worst environmental democracy among 70 countries in May 2015. This is due to lack of participation in sustainability compliance and less awareness on sustainability by Malaysia's business partners. By combating and achieving these, the Sustainable Development Goals (SDGs) defines global priorities and inspirations through Agenda 2030 that recognize by United Nation. This paper attempts to propose Framework Sustainability Assessment (FSA) to assess sustainability for the various sector in Malaysia. This framework will give an overview to reduce the noise from wide data collection to a single element. Lastly, it will integrate the current performance with the linkage of the SDGs and the sustainability elements by dividing into three phase which is baselining, clustering and correlation. Now it is time for business to take action especially every sector or industry in Malaysia and it will improve the awareness and enhance Environmental Democracy Index to a better rank in future.

1. Introduction

All sectors have significantly contributed towards strengthening the economy of many countries especially in Malaysia. From Gross Domestic Product (GDP), economy performances in Malaysia are depending on five main sectors which are services, manufacturing, agriculture, construction and mining & quarrying but exclude on import duties [1]. The various industry creates job opportunities, alleviates poverty and improves healthcare as well as education especially in rural areas [2]. In fact, based on the previous statistics from Department of Statistic Malaysia (2017), Malaysia's economy remained its uptrend momentum since second quarter of 2016 up to now [3].

From early 1990s, sustainability assessment has gradually become a new important appraisal method for supporting decision making and policy in a broad of sustainability elements in many sectors all over the world. The first serious discussions and analyses of Sustainability emerged during 1983 at new World Commission on Environment and Development (WCED)[4][5] by Norwegian prime minister Gro Harlem Brundtland. In the report titled "Our Common Future", sustainable development is define as, "...the development that meets the needs of the presents without compromising the ability of future generations to meet their own need" (WCED,1987)[6]. The persistence of this report is to provide long-term environmental strategies, to define shared perceptions of long-term environmental issues and to concern into greater co-operation among developing countries and between countries at different stages of economic and social development that interrelationship between people, resources, environment and development [7]. After recent year, sustainability always be a hot debatable issue over the world, constantly being discussed in huge number of professional and growing realization among leading scientists, the public and politicians. Thus, it is growing concern about the long-term future in many industries in Malaysia.

Although sustainability is drastically developed since 1987 all over the world up until now but from World Resources Institute (WRI) with the Access Initiative, Malaysia has been ranked as the second worst environmental democracy among 70 countries in Environmental Democracy Index (EDI) evaluation. Therefore, government, business and civil society together with United Nation have started to mobilize efforts to achieve the sustainable development agenda by 2030 in every country. The agenda calls for universal and inclusive action by all countries to improve the lives of people everywhere [8]. In 2015, countries adopted the 2030 Agenda for sustainable Development and its 17 Sustainable Development Goal (SDG). On 1 January 2016, the 17 goals of the agenda 2030 agenda for sustainable development officially came into force. These goals universally apply to all and demand countries to mobilize efforts to end all forms of poverty fight inequalities and tackle climate change and ensuring no one left behind the process. By growing the SDGs, it is an opportunity for business-led solutions and technologies to be developed and implemented to address the world's biggest sustainable development challenges. The SDGs will help to connect business strategies with global priorities especially to industry in Malaysia.

Since every company in Malaysia was registered under Bursa Malaysia, it is compulsory for them to return the sustainability reporting of their business to Bursa Malaysia [9]. A key of Bursa Malaysia's success today is by advocating their business with sustainability concept. Sustainability is a holistic approach to business management, incorporating economic, environmental, social, and governance considerations. Thus, it can support business continuity and long-term value for stakeholders and society at large. From Stubbs et al. [10], they measured the performance of sustainability at Malaysia. The majority companies in Malaysia which 77% is aware about the sustainability but only 40% embed a sustainability concept on their overall project or business. Thai et al. with their research on companies that are listed in Bursa Malaysia review that predicting the business can help corporation or investors in prudent decision-making either in terms of financial, management and others [11].

To remain competitive among other countries around the world, Malaysia was prioritized and escalates the number of key growth alternative to the support policy. Thus, the Economic Transformation Programme (ETP) was announced in Tenth Malaysia Plan or "*Rancangan Malaysia Ke-10*" (RMK 10) that focus on twelve National Key Economic Area (NKEAs) such in healthcare, greater Kuala Lumpur/Klang Valley, wholesale, financial services, education, palm oil & gas, business services, communication, content & infrastructure, oil & gas (energy), agriculture, tourism and electrical & electronics [12]. Therefore, this paper propose a framework of sustainability assessment in Malaysia's Industry by linking the SDGs and sustainability elements. As a result, develop a sustainability assessment's framework and tool to indicate the level of sustainability compliances with current world development using baselining, clustering and correlation method. At the same time, to promote sustainability practice as the effort have been made by Department of Statistic Malaysia specifically to Malaysia's Industry.

2. Methodology

The development of Sustainability Assessment Framework is to quantify the current sustainability performance industries in Malaysia. This methodology based on the real data from the survey by considering any related criteria that assume to improve the framework of Sustainability Assessment. Thus, it will presenting the best sustainability concept for industry in Malaysia. The Framework of Sustainability Assessment (FSA) will amend to the performance of sustainability business in Malaysia's Industry. Table 1 below, show the general framework of Sustainability Assessment.

Table 1. Proposed Framework of Sustainability Assessment (FSA) in Malaysia's Industry.

Phase	Tool	Purpose	Deliverables
Baselining	Survey	• Collecting the respondent information	Variety of respondent from different sector
Clustering	Ward's algorithm agglomerative and Pairwise linkage ration	• Minimize the variance data collection • Fit the hierarchical clustering model due to the data collection from survey	Form a single element in data respondent for correlate to the policy
Correlation	SDG Compass	To link the data respondent to the business action and policy.	Epistemology the current practice to an existing Malaysia's policies.

2.1. Baselining phase

In this proposed framework, the baselining is the initial phase of the entire framework. This study made use of a survey research based on the quantitative approach using industries exist in Malaysia. The population in the baselining phase was referring to Gross Domestic Product (GDP) which is using random sampling technique for collecting population with different sectors in Malaysia. Based on the previous researcher, the good practice in conducting survey will give the best result in related case study. Therefore, there are three elements should be considered in this survey such descriptive research, analytical studies and evaluation research. From that three elements it can make the survey on track by giving the estimation on specific parameters in the targeted population and to describe the association. In addition, the illumination for a specific problem will be solved through focused data and the effect of one set of variable can be analysed. Lastly, the data collection will ascertain the effects of a planned change onward [13].

In the baselining phase it will more focus on constructing a survey. The survey's questions consist of three parts: (1) the demographic or respondent background, (2) the employees, attitude or concern about the organizational effort with the current practice of the company, and (3) the employees' attitude or opinion about the situation that related to the GPM and the SDGs. For part 1, there are several industries in Malaysia that recognized by Department of Statistic Malaysia in Gross Domestic Product (GDP). Baselining phase is to obtain as much as possible the feedback answers from the survey by targeted population of respondents. The structure of surveys made by concerning the objectives and the related factor of the study. In addition, there are several question touch on the linkage of the SDGs with elements in the Green Project Management (GPM) P5 Standard for sustainability. These linkage are based on a more detailed analysis available on the SDG Compass [14] and all twelve elements in Green Project Management (P5) were map to the 17 SDGs. The performance was measured by providing several situations that related to those linkages.

The outcome of this phase is to identify the awareness level of respondent about sustainability according to various sector in Malaysia. Since random sampling are chosen while conducting this survey. Thus, the random sampling allows the results to be generalize to the larger population and statistical analysis will be perform if appropriate [13] and to make reduce the noise. A clustering phase will be prepared in the next sub section.

2.2. Clustering

After baselining process which is data collection phase, an analysis of data should be done for the complete study. The data analysis has to identify the homogenous groups of respondent based on the deliverables from agglomerative hierarchical cluster analysis and pairwise linkage ration. This analysis, generally will reduce the dimensions' number of variables and minimize the multi collinearity effects by determine the distance between the sets. The distance between a pair of clusters or set is equal to how much will the error sum of square (ESS) increase by merging them in grouping process identifying. This step will use pair wise linkage ration. Lastly, the significant test is applied to check the fit goodness of the model by delivered the variances value[15]. The hierarchy within the final cluster has the properties as follow:

i. Clusters generated in early stages are nested in those generated at later stages.
ii. Clusters with different sizes in the tree can be valuable for discovery.

2.2.1. Ward's algorithm agglomerative

This is a tool for clustering method by using variance by determine the distance between set of cluster. The ward error sum of square hierarchical clustering method has been very widely used since 1963 on its publication by Ward [16]. There are some process involve in Agglomerative Ward's hierarchical clustering as in figure 1 below:

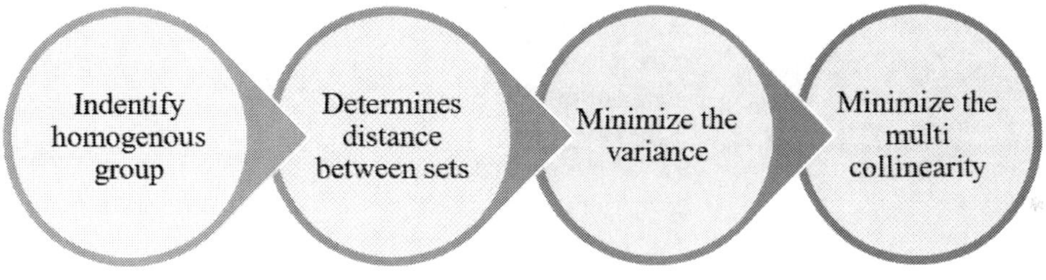

Figure 1. Process involves in Agglomerative Ward's hierarchical clustering phase

Lastly, there are some processes by performing the clustering method. Figure 2 below show the flow chart on performing a single element from the data respondent as the current practice that can be incorporate with the existing policy:

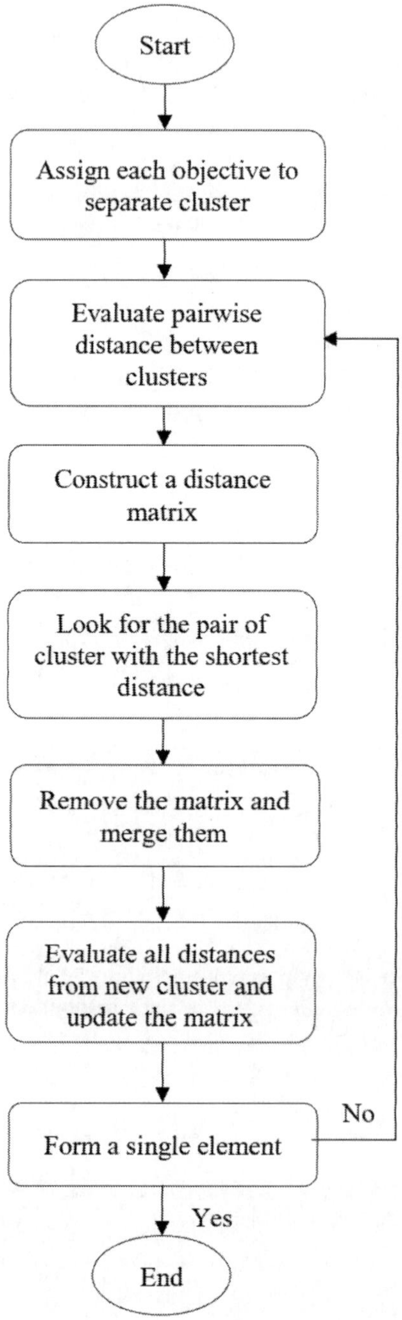

Figure 2. Flow chart for the clustering method.

2.3. Correlation

Since, a single element from data respondent is collected, the correlation process between that element with the Sustainable Development goals (SDG) and Sustainability element are determined. From SDG Compass, it will guide business especially industry in Malaysia on how to align their business strategies as well as measure and manage their contribution to the sustainable development goals.

There are five steps of SDG Compass rest on the recognition of the responsibility of all companies to comply with all relevant legislation, respect international minimum standards and address as a priority all negative human rights impacts. From this correlation, it will guide an industry to organize their business and evaluate their own level in sustainability with comply the policy indirectly. Figure 3 shows the 5 steps that can be apply to set or align their course, depending on their current level of ensuring that sustainability is an outcome of core business strategy.

Figure 3. Steps that assist bussiness contribute to the SDGs.

Lastly, the baselining, clustering and correlation phases in this Framework of Sustainability Assessment (FSA) will be as a catalyst on sustainability practice in Malaysia's industry. It is also recommended for the document to be made as a strategic tool in business. Therefore, it has high priority with respect to the business objectives of the company and the policies in Malaysia to be integrate by referring this framework.

3. Conclusion

In order to encourage the sustainability practices for every sector in Malaysia, incentives and reward should be awarded to the rightful party who practices sustainability. Since, government is an important role in solving sustainable development challenges. The awareness and the enforcement of the rules and laws must be abided by the end user which are all sector that contribute to the destruction of environment, economic and society directly or indirectly. But, at the same time they are an agent to uptrend the momentum in Malaysia's economic. Because of that, the best initiative should be proposed to avoid terrible causes and effects in future but will enhance a good benefit to the country and also the future generations.

This paper proposes the framework of sustainability assessment (FSA) to the Malaysia Industry by integrating the current level of sustainability to the linkage between SDGs and the sustainability elements. A questionnaire survey is conducted to collect a data to be analysed. Three factor are investigated; Factor (1) the working area, job designation and the working sector, Factor (2) current level of awareness on sustainability, and Factor (3) their perception on the sustainability according to the attitudinal theory. It was found that since the framework on sustainability assessment is developed, the correlation between the current business practice and the existing policies are not aligned together. Therefore, there are several phase to correlate the current practice in different sector in Malaysia to the sustainable development goal and sustainability element such as baselining, clustering and correlating phases. The integration between them were based on the business core value, the current policy in Malaysia and the world demand.

As the recommendation for the future study, the indexes for every sector would be generated by focusing the sampling population in every sector in Malaysia. In addition, the indexes will give a number that will be measured and compared to the current business practice due to the sustainability development practice. Subsequently, an improvement of the business especially industry's performances in Malaysia are growing up and able to compete with other developed countries.

Acknowledgement
The authors would like to give special thanks to Research & Innovation Department, Universiti Malaysia Pahang for funding this research project.

References
[1] J. Perangkaan Malaysia "2 . 1 — keluaran dalam negeri kasar mengikut jenis aktiviti ekonomi pada harga malar 2010 , Malaysia RM juta," p. 2016, 2016.
[2] B. D. N. Awang Ali, R. Kunjappan, M. Chin, G. Schoneveld, L. Potter, and R. Andriani, "The local impacts of oil palm expansion in Malaysia: an assessment based on a case study in Sabah State.," *CIFOR Work. Pap.*, no. 78, p. 17 pp., 2011.
[3] D. of S. M. (DOSM), "Press Release Gross Domestic Product Second Quarter 2017," *Stat. Malaysia 2015*, no. August, pp. 7–9, 2017.
[4] F. M. Turan, K. Johan, W. N. S. Wan Lanang, and A. Asmanizam, "Assessing Sustainability in Environmental Management: A Case Study in Malaysia Industry," *IOP Conf. Ser. Mater. Sci. Eng.*, vol. 226, no. 1, 2017.
[5] N. S. Sahimi, F. M. Turan, and K. Johan, "Development of Sustainability Assessment Framework in Hydropower sector," *IOP Conf. Ser. Mater. Sci. Eng.*, vol. 226, no. 1, 2017.
[6] G. H. Brundtland, "Our Common Future: Report of the World Commission on Environment and Development," *United Nations Comm.*, vol. 4, no. 1, p. 300, 1987.
[7] WCED, "Report of the World Commission on Environment and Development: Our Common Future," *Sustain. Dev.*, vol. 154, pp. 1–374, 1987.
[8] K. Johan and F. M. Turan, "Framework of systematic sustainability assessment strategy (FSSAS) for hydroelectric power industry in Malaysia," *IOP Conf. Ser. Mater. Sci. Eng.*, vol. 226, no. 1, 2017.
[9] A. Today and E. Team, "Integrated Reporting in Malaysia Where do Malaysian," 2014.
[10] T. Thijssens, L. Bollen, and H. Hassink, "Managing sustainability reporting : many ways to publish exemplary reports," *J. Clean. Prod.*, vol. 136, pp. 86–101, 2016.
[11] S. B. Thai, H. H. Goh, B. H. Teh, J. C. Wong, and T. S. Ong, "A Revisited of Altman Z- Score Model for Companies Listed in Bursa Malaysia," *Int. J. Bus. Soc. Sci.*, vol. 5, no. 12, pp. 197–207, 2014.
[12] J. Perdana Menteri, "Government Transformation Programme," *Report*, pp. 1–264, 2010.
[13] K. Kelley, B. Clark, V. Brown, and J. Sitzia, "Good practice in the conduct and reporting of survey research," *Int. J. Qual. Heal. Care*, vol. 15, no. 3, pp. 261–266, 2003.
[14] Global Reporting Initiative, "Linking the SDGs and GRI," p. 6, 2015.
[15] N. Shuja and Y. B. Wah, "Functional Analysis of Industrial Clusters in Malaysia," *Proc. 2nd ISI Reg. Stat. Conf.*, no. March, pp. 20–24, 2017.
[16] F. Murtagh and P. Legendre, "Ward ' s Hierarchical Agglomerative Clustering Method : Which Algorithms Implement Ward ' s Criterion ?," *J. Classif.*, vol. 31, no. October, pp. 274–295, 2014.

IOP Conf. Series: Materials Science and Engineering **342** (2018) 012080 doi:10.1088/1757-899X/342/1/012080

Effect on Ammonium Bromide in dielectric behavior based Alginate Solid Biopolymer electrolytes

A F Fuzlin, N M J Rasali and A S Samsudin

Advanced Material, Faculty of Industrial Science & Technology, Universiti Malaysia Pahang Lebuhraya Tun Razak, Gambang 26300, Kuantan, Pahang, Malaysia

E-mail: ahmadsalihin@ump.edu.my

Abstract. This paper present the development of solid biopolymer electrolytes (SBEs) system which has been accomplished by incorporating various composition of ionic dopant namely ammonium bromide (NH_4Br) with alginate solution casting method. The prepared sample of SBEs has been analyzed via electrical impedance spectroscopy (EIS) showed that the ionic conductivity at room temperature was increased from 4.67 x 10^{-7} S cm^{-1} for un-doped sample to optimum value at 4.41 x 10^{-5} S cm^{-1} for composition of 20 wt. % NH_4Br. The SBEs system was found to obey the Arrhenius characteristics with R^2~1where all sample is thermally activated when increasing temperature. The dielectric behavior of the alginate-NH_4Br SBEs system were measured using complex permittivity (ε*) and complex electrical modulus (M*) and shown the non-debye behavior where no single relaxation was found for present SBEs system.

1. Introduction

Nowadays, solid biopolymer electrolytes (SBEs) system already been used in battery industry to avoid hazardous in environment. Biopolymer (natural polymer) is one of the biodegradable product which made from waste material that been used in electrolyte study. Biopolymer is the host polymer in the proton-conducting solid biopolymer electrolytes (SBEs) system because of it safety issues and low-cost in battery production [1]. The examples of host polymer that been study are carboxymethyl cellulose (CMC) [2-4], chitosan [5-7], cornstarch [8], and 2-hydroxyethyl cellulose (2-HEC) [9].

Alginic acid, or called alginate, is the cell wall extraction of brown seaweed, which considered being biocompatible, non-immunogenic, biodegradability and having low toxicity [10, 11]. The pure alginate shows emphirical formula, which is $NaC_6H_7O_7$ [12]. Alginate is highly water absorbing that consist of (1-4)-linked β-d-mannuronic acid (M) and α-l-gluronic acid (G) units [11]. Alginate can be host polymer in SBE system but its conductivity is low. Therefore, the addition of dopant system was found to be a suitable way to enhance the ionic transports and conductivity performances.

In proton-conducting solid biopolymer electrolyte (SBEs) system, ammonium bromide (NH_4Br) is good proton donor which could approve the proton conduction through solid polymer electrolyte matrix as well as could enhance the ionic conductivity to the higher value [9] [13-16]. NH_4Br is ammonium halide type that consists of NH_4^+ ion and Br^- ion. Ionic salt will doped with biopolymer to increase the ionic conductivity. There are many research proved that used of ammonium salt can increase conductivity such as carboxymethyl cellulose (CMC) doped with ammonium thiocyanate (NH4SCN) [17].

In the present study, the proton conducting SBEs films based on Alginate incorporating with NH_4Br have been prepared via solution casting technique with various compositions. The ionic conductivity and dielectric behaviour of alginate-NH_4Br based biopolymer electrolytes (BEs) system was observed by using electrical impedance spectroscopy (EIS).

2. Material and methods

2.1. Preparation of SBEs thin films

The preparation of the sample was started by dissolving 2 g of alginate into 98 ml of distilled water. Then, different amount of NH_4Br (5 – 30 wt. %) were added into the alginate solution. The alginate-NH_4Br solution is then was stirred slowly until a homogeneous solution is obtained. The mixture of alginate-NH_4Br was casted into different petri dishes and left to dry until film completely formed. The film is then transferred into a desiccator filled with silica gel to further up drying and also prevent water content appear in the film [18] [19].

2.2. Preparation of SBEs thin films

The alginate-NH_4Br electrolytes will be press and sandwiched between two stainless steel electrodes of a conductivity holder. The dielectric properties of the proton conducting SBEs system were tested using HIOKI 3532–50 LCR Hi-tester interfaced to a computer in a frequency range of 50 Hz to 1 MHz and were tested with different temperature from 303K to 353K. The value of bulk resistance (R_b) was determined from the Cole-Cole plots imaginary impedance (Z_i) versus real impedance (Z_r) which obtained from the impedance measurement. The ionic conductivity, σ of the electrolyte was measured using the following equation:

$$\sigma = \frac{t}{R_b A} \tag{1}$$

where t is the thickness of the electrolytes and A is the electrode–electrolyte contact area (cm^2).

Dielectric constant also called as amount of charge stored in material. The loss of energy represents the dielectric loss, εi that normally produces a rise in temperature of a dielectric placed in an alternating electrical field. The amount of energy losses represents to the dielectric loss, εi where the movement of ions began when the polarity of electric field reverses rapidly during increasing temperature [20,21].

The dielectric constant (the real part of modulus, εr) and dielectric loss (the imaginary part of modulus, εi) is calculated using the following equation:

$$\varepsilon_r = \frac{Z_i}{\omega C_o (Z_r^2 + Z_i^2)} \tag{2}$$

$$\varepsilon_i = \frac{Z_r}{\omega C_o (Z_r^2 + Z_i^2)} \tag{3}$$

The complex electric modulus (M*) is used to investigate the conductivity relaxation phenomena. It also needed to suppress the effects of electrode polarization and bulk dielectric properties of the SBE system. The real modulus, Mr and imaginary modulus, Mi was calculated using the following equation:

$$M_r = \frac{\varepsilon_r}{(\varepsilon_r^2 + \varepsilon_i^2)} \tag{4}$$

$$M_r = \frac{\varepsilon_r}{(\varepsilon_r^2 + \varepsilon_i^2)} \tag{5}$$

3. Results and discussion

3.1. Conductivity studies

The variation of ionic conductivity values of alginate samples containing with different amounts of NH_4Br was shown in Figure 1. The ionic conductivity at ambient temperature was found to increase from 4.67×10^{-7} S cm^{-1} for un-doped sample towards optimum value with six magnitudes order at 4.41×10^{-5} S cm^{-1} for composition of 20 wt. % NH_4Br. The increment of ionic conductivity with increasing of NH_4Br concentration can be related to the increment of number of mobile charge carriers [21]. This is because the complexation between alginate and NH_4Br gives to the high dispersion of H^+, therefore will effect towards the increment of ionic conductivity [22, 23]. It clearly can be seen that with the addition above than 20 wt. % NH_4Br, the ionic conductivity start to decreased and this can be attributed due to neutral aggregation of the ions re-associated and also lead to formation of ion cluster as reported by other researcher [14, 24].

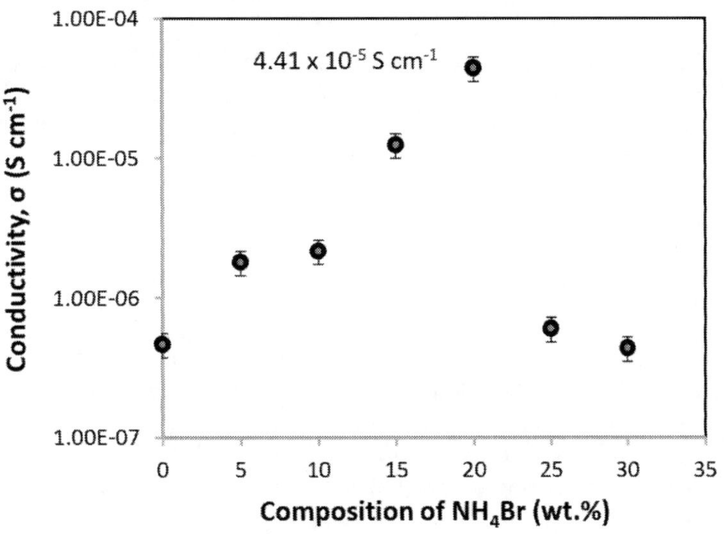

Figure 1. Ionic conductivity of alginate-NH_4Br SBEs system at room temperature.

The effect of temperature on the properties of the ionic conductivity in alginate-NH_4Br SBE system can be analysed using temperature dependence study of ionic conductivity. The log conductivity versus 1000/T for various samples at the temperature range from 303 K to 353 K are shown in Figure 2. It shows that the ionic conductivity increased linearly with temperature, which showing that the conductivity is thermally assisted for the present system. The regression value, R^2 was found close to unity (R^2~1) suggesting that all composition for alginate-NH_4Br SBE system in ionic study of temperature-dependence obey Arrhenius behaviour [14]. From the plot of log conductivity versus 1000/T, the activation energy, E_a was calculated using the Arrhenius relationships equation:

$$\sigma = \sigma_0 \, exp\left(-\frac{E_a}{kT}\right) \tag{6}$$

where E_a is the activation energy, σ is the conductivity of SBEs system, σ_0 is the pre-exponential factor, k is the Boltzmann constant, and T is the temperature in Kelvin.

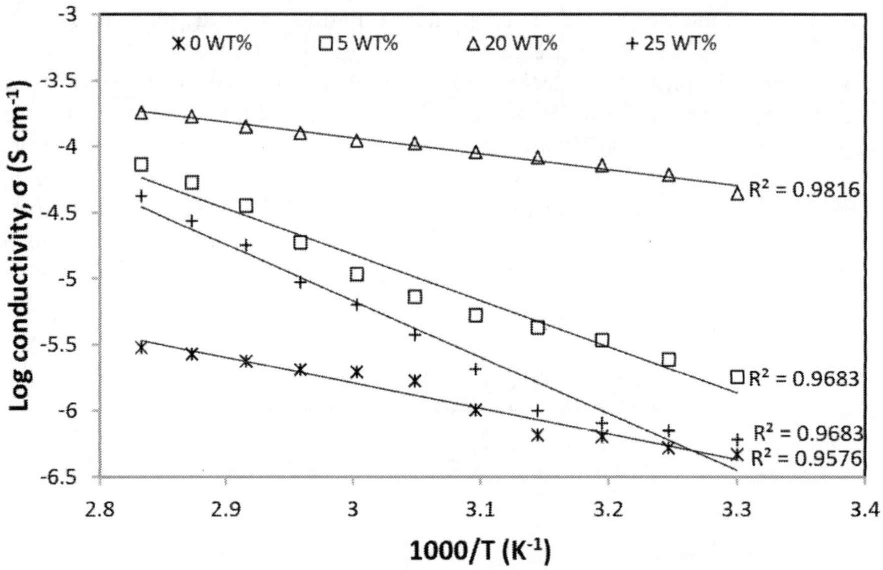

Figure 2. Temperature dependence of alginate-NH_4Br SBEs system.

The activation energy, E_a is the energy needed an ion to migrate from one site to another and can be due to the defect formation [18, 25]. The activation energy, E_a for selected sample of the alginate-NH_4Br of SBE system is shown in Figure 3. It can be observed that the increment of ionic conductivity SBE system lead to decrease the E_a value and this suggesting the ions which correspond to H^+ only require lower energy to migrate to polymer backbone thus will enhanced the ionic mobility.

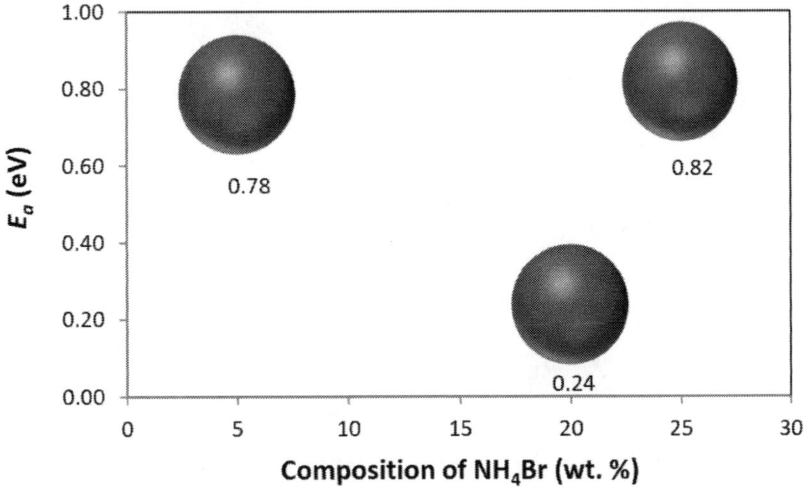

Figure 3. Activation energy for selected sample of alginate-NH_4Br SBEs system.

3.2. Dielectric studies

The dielectric constant is the amount of charge stored in material. In polymer electrolyte study, the charge carrier is ions. Figure 4 shows the study of the dielectric constant for different concentration at selected frequency. It can be observed that increased in concentration of salt, the dielectric constant, ε_r also increased for every frequency until 20 wt. % when the ε_r value decreased rapidly. This implies that as the concentration of salt is increased, the stored charge also increased which reflected the increasing number of ions in SBE system [26]. Therefore, the reducing of dielectric constant at 25 wt. % NH_4Br is due to decrement in number of density of charge carriers which attributed to the re-association of ions [14]. In Figure 5, the result shown that rising in temperature also increased the dielectric constant value. This is because the re-dissociation of ion starts to aggregates starting to grow resulting in the increment of number of charge carrier density [14].

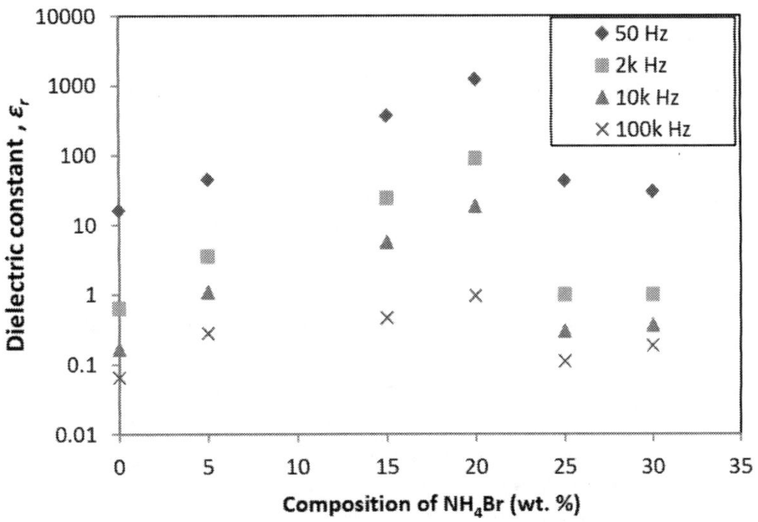

Figure 4. Dielectric constant of alginate-NH_4Br SBEs system.

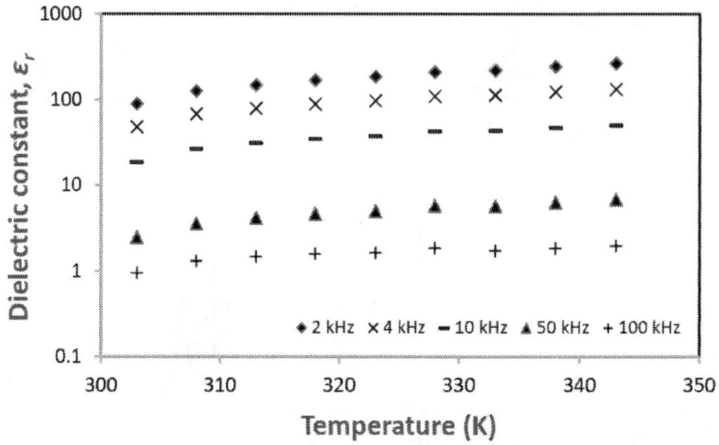

Figure 5. Dielectric constant for the highest sample of SBEs system.

Figure 6 shows that the dielectric loss, ε_i for SBEs system. Dielectric loss, ε_i is the measure of dispersed energy and mostly contributes in the ionic transport and the polarization of the charge [17, 27]. Based on Figure 6, it can found that the value of dielectric loss increasing rapidly at low frequencies and gradually dropped at higher frequencies for all sample. There were no appreciable relaxation peaks observed in the frequency range that been used in this study. The value of ε_i have increase sharply at the low frequencies due enhancement in mobility of charge carrier and can be related to the free charge motion within the electrolytes [17]. Meanwhile, the value of ε_i decreases at higher frequencies because of the lesser ions that can pile up at the interface, which have faster rate in direction of the electric field change [17, 29]. The value of ε_i reduce at higher frequencies may due to periodic reversal of the electric field occurs so fast that there is no excess ion diffusion in the direction of the field [19, 21].

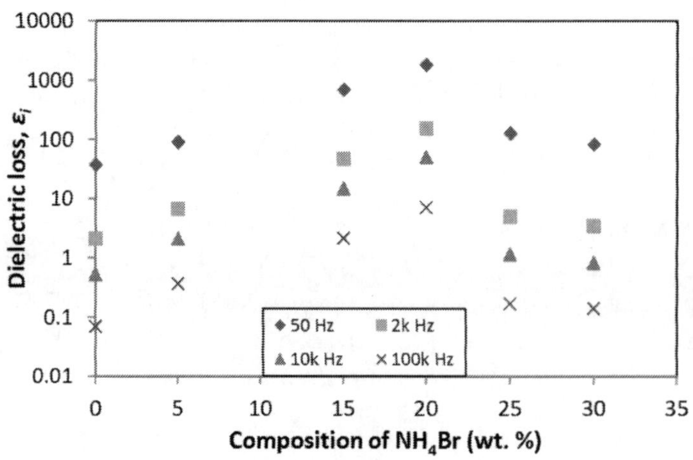

Figure 6. Dielectric loss of alginate-NH_4Br SBEs system at room temperature.

3.3. Modulus studies

Electrical modulus study is a study of the dielectric behaviour of alginate-NH_4Br SBEs system that suppressed the effect of the electrode polarization [25]. Figure 7 show the frequency dependence of the real and imaginary parts modulus formalism respectively, as a function of frequency for all composition of alginate-NH_4Br SBEs system at room temperature. Figure 7 (a) and (b) show both M_r and M_i are approaches to zero at low frequency but it begin to rise sharply as frequency increased [17, 27, 28]. At highest conductivity, the real and imaginary part modulus becomes lowest at higher frequency. The presence of peaks in the modulus formalism for all the polymer system shows that the polymer electrolyte films are ionic conductor at higher frequencies [18]. Figure 7 show long tail at low frequencies due to the large capacitance applied with the electrodes and also removal of electrode polarization [18, 29-32]. This confirmed the non-Debye behaviour in the alginate-NH_4Br SBE system.

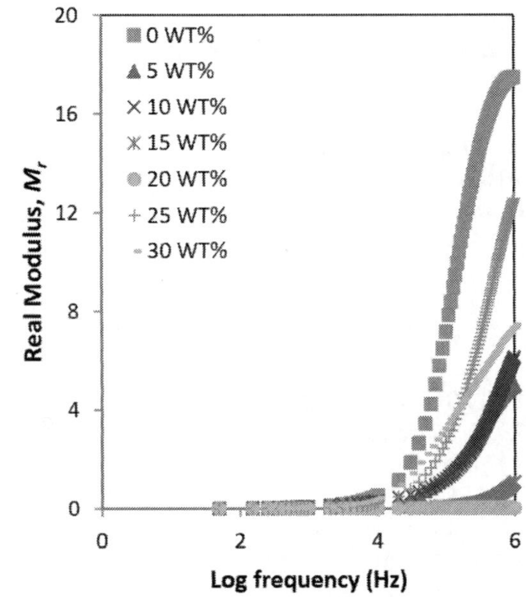

Figure 7 (a). Frequency dependence of real part modulus of alginate-NH_4Br SBEs system at room temperature.

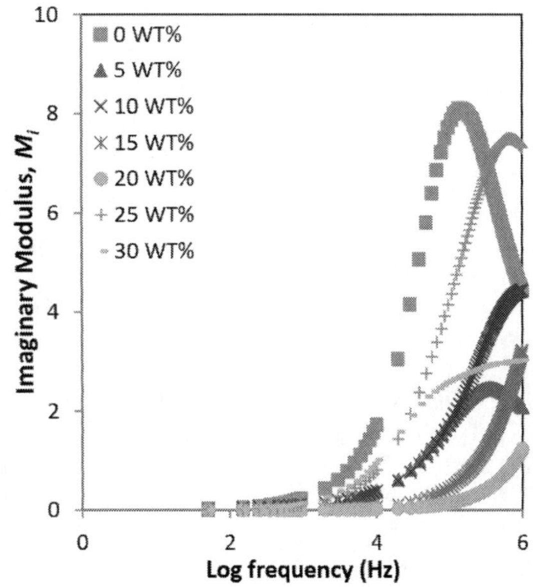

Figure 7 (b). Frequency dependence of imaginary part modulus of alginate-NH_4Br SBEs system at room temperature.

4. Conclusion

Solid biopolymer electrolytes (SBEs) system based on alginate as host and doped with NH_4Br at various salt composition were successfully prepared via solution cast technique. The highest ionic conductivity at ambient temperature was obtained at 4.41×10^{-5} S cm^{-1} for sample containing 20 wt. % NH_4Br. Temperature dependence shows that the SBEs system in the present work following Arrhenius behaviour where the R^2 were found closed to unity. The calculated activation energy were found to be inversely proportional with ionic conductivity of SBEs system where only require small amount of energy for ions migration. The dielectric and modulus properties shown that the alginate-NH_4Br SBEs system can be described with non-Debye behaviour where non-single relaxation of ions were observed in the present work. Based on the results, it shows that the biopolymer namely alginate has a good potential to be use as an electrolytes system for application in electrochemical devices.

Acknowledgement

The authors would like to thank Universiti Malaysia Pahang for the financial via internal grant (RDU1703189), Faculty of Industrial Sciences and Technology, University Malaysia Pahang for the support given in this work.

References

[1] Ramlli M A and Isa M I N. 2016 *Phys. Chem. B*, 2016 **120**,11567–11573
[2] Isa M I N and Samsudin A S 2017 *Int. J. Polym. Anal. Charact.* .**22**, 447-454
[3] Ahmad N H and Isa M I N 2015. *Adv Mat Res* **1107**, 247-252
[4] Chai M N and Isa M I N 2013 Int J Polym Anal **18**, 280–286
[5] Fadzallah I A. Noor I M., Careem M A, and Arof A K.. 2016 *Ionics* . **22**, 1635–1645
[6] Buraidah M H. Teo L P, Majid S R, Yahya R, Taha R M, and Arof A K.2010. *Int. J. Photoenergy* **2010**, 7
[7] Yahya M Z A, Harun M K, Ali A M M, Mohammat M F, Hanafiah M A K M , Ibrahim S C , Mustaffa M , Darus Z M and Latif F. 2006. *J Appl Sci* **6**, 3510–3154
[8] Shukur M F and Kadir M F Z. *Ionics* 2015 **21**,111-124
[9] Zainuddin N K and Samsudin A S. 2017 *Makara J. Technol.* **21/1**, 37-42
[10] Kosik A, Luchowska U and Święszkowski W 2016 *Mater. Lett.* **184**, 104–107
[11] Syed Junaid Mahmood , Azhar Siddique.2010 *J. Saudi Chem. Soc.* **14**,117–123
[12] Jen-Ming Yan , Nian-Ci Wang, Hsien-Chih Chiu 2014 *J. Membr. Sci.* **457**, 139–148
[13] Zainuddin N K and Samsudin A S. 2017 *Makara J. Technol.* **21/1**, 37-42
[14] Samsudin A S and Isa M I N 2012 *J. Appl. Sci.* **12**, 174-179
[15] Radha, K P, Selvasekarapandian S, Karthikeyan S, Hema M., and Sanjeeviraja C. 2013. *Ionics* **19**, 1437–1447
[16] Salleh N S, Aziz S B, Aspanut Z and Kadir M F Z. 2016 *Ionics*, **22**, 2157–2167
[17] Ramlli M A and Isa M I N 2014 *Res.J.Recent Sci.* **3**, 59-66
[18] N A M Noor and Isa M I N 2015 *Adv Mat Res* **1107**, 230-235
[19] Isa M I N. and Noor N A M 2015 *AMM*, 719-720, 114-118
[20] Salleh N S, Aziz S B , Aspanut Z and Kadir M F Z. 2016 *Ionics* **22**, 2157–2167
[21] Abu Bakar N Y, Muhamaruesa N H, Bashirah Aniskari N A, Mohd Isa M I N 2015 *Am J Appl Sci* **12**,40–46
[22] Padmasree K P, Kanchan D K and Kulkarni A R. 2006 *Solid State Ionics*, **177**, 475-482
[23] Nik Aziz N A, Idris N K and Isa M I N. 2010. *Int. J. Phys. Sci.* **5**,748-752
[24] Othman L, Md Isa K B, Osman Z, Yahyaa R. 2017 *Materials Today: Proceedings*. **4**, 5122– 5129
[25] Rani M S A, Dzulkurnain N A, Ahmad A and Mohamed N S 2015 *Int. J. Polym. Anal. Charact.* **20**, 250-260
[26] Buraidah M H , Teo L P, Majid S R, Arof A K 2009 *Physica B* **404**,1373–1379
[27] Rozali M L H, Samsudin A S and Isa M I N. 2012 *J Appl Sci Technol, 2*, 113-121
[28] Salleh N S, Aziz, S B, Aspanut Z, and Kadir M F Z. 2016 *Ionics*, **22**, 2157–2167
[29] Majid S R and Arof A K. 2007 *Pysica B*. **390** 209-215
[30] Dasari M P, Rao K S, Krishna P M, and Krishna G G. 2011 *Acta Phys. Pol. A* **119**, 387–394.
[31] Muralidharan, P., Venkateswarlu M., and Satyanarayana N. 2004 *Mater. Res. Bull.* **39**, 1753–1762
[32] Baskaran R, Selvasekarapandian S, Kuwata N, Kawamura J, and Hattori T. 2006 *Solid State Ionics* **177**, 2679–2682

IOP Conf. Series: Materials Science and Engineering **342** (2018) 012081 doi:10.1088/1757-899X/342/1/012081

Mechanical properties of polymer-modified porous concrete

N F Ariffin[1], M F Md. Jaafar[1], N H Abdul Shukor Lim[2], M A R Bhutta[3] and M W Hussin[2]

[1]Faculty of Civil Engineering & Earth Resources, Universiti Malaysia Pahang, Lebuhraya Tun Razak, Gambang, 26300, Pahang, Malaysia
[2]Department of Structure and Material, Faculty of Civil Engineering, Universiti Teknologi Malaysia, 81310, Johor, Malaysia
[3] Department of Civil Engineering, Faculty of Applied Sciences, University of British Columbia, Vancouver, B. C., Canada V6T 1Z4, Canada.

E-mail: nfaump@gmail.com

Abstract. In this research work, polymer-modified porous concretes (permeable concretes) using polymer latex and redispersible polymer powder with water-cement ratio of 30 %, polymer-cement ratios of 0 to 10 % and cement content of 300 kg/m^3 are prepared. The porous concrete was tested for compressive strength, flexural strength, water permeability and void ratio. The cubes size of specimen is 100 mm ×100 mm × 100 mm and 150 mm × 150 mm × 150 mm while the beam size is 100 mm × 100 mm × 500 mm was prepared for particular tests. The tests results show that the addition of polymer as a binder to porous concrete gives an improvement on the strength properties and coefficient of water permeability of polymer-modified porous concrete. It is concluded from the test results that increase in compressive and flexural strengths and decrease in the coefficient of water permeability of the polymer-modified porous concrete are clearly observed with increasing of polymer-cement ratio.

1. Introduction

Porous concrete made from 80% of course aggregates which contains little or no fine aggregate. Therefore, an adequate amount of cement paste was used to coat and bind the aggregate particles to create a system of high porosity of the concrete. The porosity is depends upon the application of the concrete and size of aggregate used. Porous concrete is concrete which is designed to have many voids according to it applications such as to reduce the amount of runoff water and improve the water quality near pavements and parking lot [1–6]. There are a numbers of alternate names for porous concrete including permeable concrete, porous concrete pavement and pervious concrete. All names basically giving the same meaning which is the form of concrete is permeable rather than solid state. Porous concrete has been developing as an environmentally friendly material in in North America, Europe and Japan since1980's. Due to water-permeating, water-draining, and water-retaining performances of porous concrete, it has been utilized in road pavement, sidewalks, parks and building extension, as well as for plant bedding and permeable gutters [5, 7-11]. Unfortunately, limited technical data and information are available on development activities and applications of porous concrete in South East Asia particularly, Malaysia. Located near the equator, Malaysia's climate is categorized as equatorial being hot and humid throughout the year. The average rainfall is 2500 mm a year [12], necessitating the introduction and popularization of porous concrete technology in Malaysia in order to minimize the adverse effects of heavy rainfall as floods and mudslides. The issues rise during the end of year in Malaysia is heavy raining and flooding which cause a big chaos to Malaysian citizen. Porous concrete pavement can potentially infiltrate storm water at the source which will allow the oils from cars and trucks to biodegrade safely, improve driving safety, reduce traffic noise and also reduce urban temperatures.

2. Experimental Program

2.1 Materials

The cement used in the study was ordinary Portland cement (OPC) obtained from Holcim Cement Manufacturing Company of Malaysia, conforming to ASTM C150 standard. Crushed granite aggregate with the size between 5 mm to 10 mm was used. The aggregate occupied 60 to 80 percent of the concrete volume. The coarse aggregate is air dried to obtain saturated surfaces dry condition to ensure the water cement ratio is not affected. As for the binder, two types of commercially available styrene butadiene rubber (SBR) based redispersible polymer powder (RPP) and latex emulsion (Latex) were used as polymeric admixtures into the porous concrete. The aim was to enhance the strength of porous concrete. The percentage of polymer added is 0 %, 3 % and 5 % over percentage of cement. The properties of polymeric admixtures are given in Table 1.

Table 1. Properties of redispersible polymer powder (RPP) and latex.

Type of polymer	Appearance	Particle size (μm)	Density (g/cm^3)	Solid content (%)	pH
RPP	Grayish-white powder	0.15	0.50	50	7–8
Latex	Milky white	200	0.98	49	10–11

2.2 Mix Proportion

The porous concrete mixtures comprised of Portland cement, water and coarse aggregate. To improve the strength of porous concrete, latex polymer in powder and liquid forms were used. Various preliminary mixes were prepared to obtain the optimum mix proportion of porous concrete. Polymer was first mixed with mix design water and no extra water was added. The optimum mix proportions are presented in Table 2. The polymer–cement ratios (P/C) of 0 %, 3 % and 5 % by mass of cement were used. Porous concrete has low water–cement ratio. Therefore, it is important to maintain the mix design water content in porous concrete for proper mixing.

Table 2. Mix proportions of porous concrete.

Type of concrete	W/C (%)	Void ratio (%)	Aggregate content (%)	Mix proportions (kg/m^3)			
				Water	Cement	Aggregate	Polymer
Normal							-
RPP	30		80	89	300	1640	3%
							5%
Latex							3%
							5%

2.3 Specimen Preparation

The specimens were prepared at ambient laboratory temperatures according to the JCI Standard – "Method of Making Porous Concrete Specimens (draft), Report on Eco-Concrete Committee on

Design, Construction and Recent Applications of Porous Concrete''. The specimens were then cured at room temperature for 1-day. According to this method, mixed porous concrete was filled into moulds up to one third of its height and subjected to 25 times hand tamping. The same procedure was applied for the two third and complete fill-ups. Placing and consolidating the porous concrete in moulds can cause excessive void and strength losses, thus adversely affecting the test results. Porous concretes contained SBR-based latex and RPP is well known for an optimum film formation, thus, a dry cure is required [13]. On the other hand, cement hydration only takes place in the presence of sufficient water. Since porous concrete has low W/C, therefore, both the specimens were placed in water for 5-d and then subjected to 25 ± 1°C and 60% R. H. for 22-d. The cube specimens size is 100×100×100 mm for compressive strength, beam specimens sized (100×100×400 mm) for flexural strength and cylinder specimens sized (ø100×200 mm) for total void ratio and water permeability tests. To obtain mean value, three identical concrete specimens were prepared for each test.

3. Experimental Tests

3.1 Total void ratio test

JCI Test Method (Report on Eco-Concrete Committee for Void Ratio of Porous Concrete (draft) was employed to determine the total void ratio of porous concrete cylinders (ø10 × 200 mm) [14]. The total void ratio was obtained by dividing the difference between the initial mass (M1) of the cylinder specimen in the water and the final mass (M2) measured following air drying for 24 h with the specimen volume (V), where ρM is the density of water. The specimens were placed inside the cage immersed in clean water at room temperature. The mass of the specimen in water (M1) was determined after removing the air bubbles on the particle surfaces and between the particles of specimen. The equation used to obtain total void ratio (A) is as follows:

$$A\ (\%) = 1 - [(M2 - M1)/\rho w] / V1 \times 100 \qquad (1)$$

3.2 Coefficient of Water permeability

Permeability is an important parameter of porous concrete since the material is designed to perform as drainage layer in pavement structures. Due to the high porosity and the interconnected air voids path, Darcy's law for laminar flow is no longer applicable for porous concrete. In this study, a permeability measurement device and method developed by Huang et al. for drainable asphalt mixture similar to porous concrete in function were used [5]. This test is done according to JCI-SPO3-1 (Japan Concrete Institute) standard.

3.3 Compressive and flexural strengths

The compressive strength was performed according to JIS (Japanese Industrial Standard) A 1108 (Method of Test for Compressive Strength of Concrete). The flexural strength test was conducted in accordance with JIS A 1106 (Method of Test for Flexural Strength of Concrete) which is similar to ASTM C 78/C 78M-10e1 ''Standard Test Method for Flexural Strength of Concrete''(Using Simple Beam with Third-Point Loading).

4. Results and Discussions

4.1 Effect of polymer-modified porous concrete on total void ratio

Total voids ratio is defined as the percentage of the total volume of voids to the total volume of specimen. Void ratio is very important in porous concrete because more void present, it will gives lower strength to porous concrete. Same as conventional concrete, present of void will effect the strength of the concrete. Figure 1 shows the relationship between total void ratio and type of porous concrete. The result shows that the type admixture added in porous concrete effect the void content. Normal porous concrete without polymer shows high void ratio than polymer-modified porous concrete. By adding polymer in porous concrete mixture, the void ratio can be decrease and as a results, it will produce high strength of porous concrete. The redispersible polymer in powder form has somehow lower the total void ratio due to its properties that bind the cement paste well and also produce a sticky mixture. Thus, it lower the total void percentage of polymer-modified porous concrete.

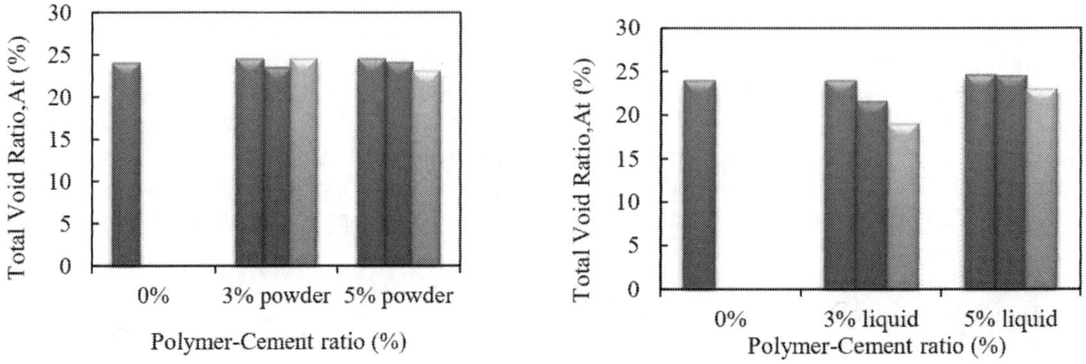

Figure 1. Relationship between total void ratio and polymer-cement ratio.

4.2 Compressive strength

Figure 2 shows the compressive strength of the various polymer-modified porous concrete. Figure shows that by adding polymer admixture into the mixture, the strength of porous concrete is increase. Powder type of polymer gives higher strength compare to liquid type polymer. The additional of 5% polymer powder into the mixture was resulted in higher compressive strength compared to 3% of polymer powder content. This is due to the adequate amount of polymer that binds the paste together to produce stronger bonding in the concrete. A polymer modification is considered and incorporated into the mixtures to improve the strength of porous concrete. It is evident that the addition of both RPP in powder form and liquid latex polymer could increase the compressive strength of porous concretes as well as increase the contact area between neighbouring aggregate particles. Moreover these polymers react along with cement hydration products and create two interpenetrating matrices which work together, resulting in improvement of compressive strength [2,13].

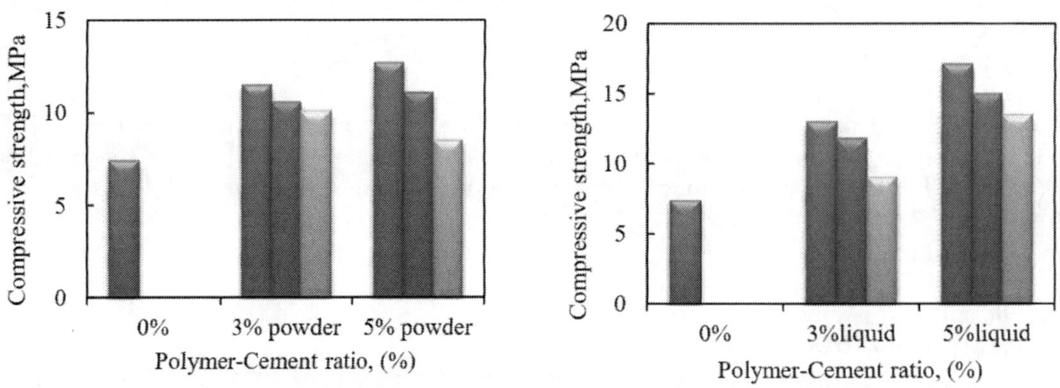

Figure 2. Relationship between compressive strength and polymer-cement ratio

4.3 Flexural strength

Figure 3 exhibits the effect of polymer addition on the flexural strength of porous concretes. The introduction of polymer modification enhanced the flexural strength in all porous concretes. As mentioned, this is attributed to the polymer network formed during the commingling and inter-penetration of the polymer and cement hydra tion products [13]. Unlike the brittle cement paste, the polymer network is relatively strong in tension zone, which contributes significantly to the flexural strength of porous concrete.

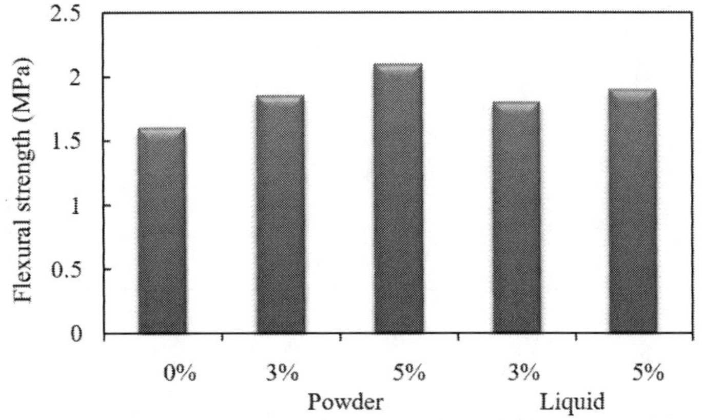

Figure 3. Relationship between flexural strength and polymer-cement ratio.

4.4 Coefficient of Permeability

Figure 4 represents the coefficient of permeability of porous concrete with different percentage of polymer admixtures. Porous concrete shows water permeability values between 2.4 to 3.7 cm/s, which is adequate to be used for drainage purpose in pavement or precast product application [15]. The addition of powder and liquid polymer could lead to a further reduction in permeability as shown by 5% of liquid addition, the coefficient of permeability is the higest compared to others porous concrete.

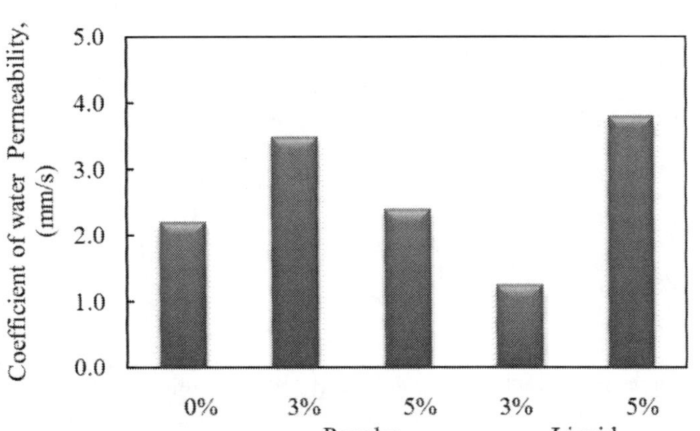

Figure 4. Relationship between coefficient of permeability and polymer-cement ratio.

Figure 5 exhibits the relationship between total void ratio and coefficient of water permeability of porous concretes. Regardless of polymer types, the coefficient of water permeability for all porous concretes became larger as the total void ratio increased. In the range of 23–28 % total void ratio, the coefficient of water permeability of all porous concretes increased linearly.

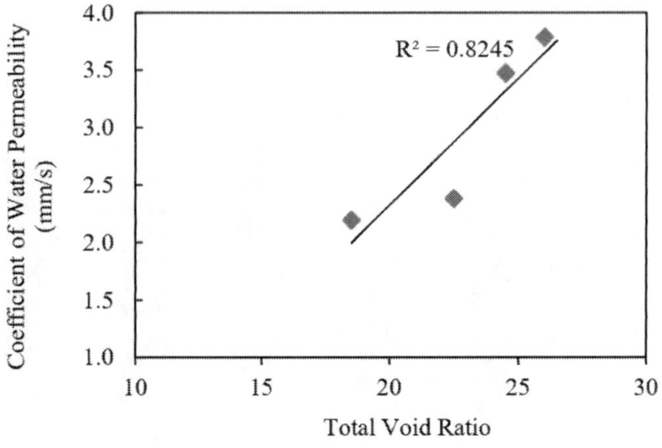

Figure 5. Relationship between coefficient of permeability and total void ratio.

5. Conclusions

An experimental study was conducted to investigate the mechanical properties of polymer-modified porous concretes. The effects of using different types of polymer on total void ratio, permeability and strengths were evaluated. Based on this study, the following conclusions can be drawn:

1. Use of the polymer latex and redispersible polymer powder could produce acceptable porous concrete with adequate strength properties and water permeability.
2. Strength properties of polymer-modified porous concretes are significantly improved by the polymer modification.
3. There is no significant change in void ratio of polymer-modified porous concretes compared to unmodified (normal) porous concrete. That means without changing void ratio of porous concrete, the strength can be improved by the polymer-modification.
4. Polymer powder show high compressive strength than liquid polymer.
5. Coefficient of permeability is directly proportional to void ratio.

Acknowledgments

The authors greatly appreciate Ministry of Education (MOE), Universiti Malaysia Pahang (UMP) and Research, Universiti Teknologi Malaysia and Innovation (PnI) UMP for financial aid that helped out through this project under RDU 170341.

References

[1] Tamai M, Mizuguchi H, Hatanaka S, Katahira H, Nakazawa T, Yanagibashi K and Kunieda M 2004 *Proc. JCI symp. on design, construction and recent applications of porous concrete* (Tokyo) p 1
[2] Huang B, Wu H, Shu X and Burdette E G 2010 *Construction Building Material* **24** 818
[3] Shu X, Huang B, Wu H, Dong Q and Burdette E G 2011 *Construction Building Material* **25** 3187
[4] Wu H, Huang B, Shu X and Dong Q 2010 *Journal Material Civil Engineering* **23** 697
[5] Huang B, Dong Q and Burdette E G 2009 *Construction Building Material* **23** 3451
[6] Aamer Rafique Bhutta M, Tsuruta K and Mirza J 2012 *Construction Building Material* **31** 67

IOP Conf. Series: Materials Science and Engineering **342** (2018) 012081 doi:10.1088/1757-899X/342/1/012081

[7] Beeldens A 2002 *Influence of Polymer Modification on the Behavior of Concrete under Severe Conditions*. Ph. D. dissertation (Belgium)
[8] Beeldens A, Van Gemert D, Caestecker C and Van Messem M 1998 *Proc. 8th International Symp .on Concrete Roads* (Lisbon) p 129
[9] Monters F 2006 *Pervious Concrete: Characterization of Fundamental Properties and Simulation of Microstructure* PhD dissertation (South Carolina)
[10] Kajio S, Tanaka S, Tomita R, Noda E and Hashimoto S 1998 *Proc. 8th Int. Symp. on Concrete Roads* (Lisbon) p 171
[11] Wang W 1997 *Science Technology Building Material* **6** 25
[12] Swee-Hock S 2007 *The population of peninsular Malaysia* (Singapore Institute of Southeast Asian Studies ISBN 978-981-230-730-9)
[13] Ohama Y 1995 *Handbook of Polymer-Modified Concrete and Mortars: Properties and Process Technology* (New Jersey: Noyes Publications)
[14] Japan Concrete Institute 2004 *Proc. JCI Symp. On Design, Construction and Recent Applications of Porous Concrete* (Tokyo) p 1
[15] Kardon J B. 1997 *ASCE Journal Material Civil Engineering* **9** 85

Compressibility characteristics of Sabak Bernam Marine Clay

D C Lat[1], N Ali[2], I B M Jais[3], B Baharom[1], N Z M Yunus[2], S M Salleh[1] and N A C Azmi[1]

[1]Universiti Teknologi MARA Johor Kampus Pasir Gudang, 81750, Masai Johor
[2]Universiti Teknologi Malaysia, Skudai, 81310, Johor Bahru
[3]Universiti Teknologi MARA Shah Alam, 40450, Selangor

E-mail: dianacl@johor.uitm.edu.my

Abstract. This study is carried out to determine the geotechnical properties and compressibility characteristics of marine clay collected at Sabak Bernam. The compressibility characteristics of this soil are determined from 1-D consolidation test and verified by existing correlations by other researchers. No literature has been found on the compressibility characteristics of Sabak Bernam Marine Clay. It is important to carry out this study since this type of marine clay covers large coastal area of west coast Malaysia. This type of marine clay was found on the main road connecting Klang to Perak and the road keeps experiencing undulation and uneven settlement which jeopardise the safety of the road users. The soil is indicated in the Generalised Soil Map of Peninsular Malaysia as a CLAY with alluvial soil on recent marine and riverine alluvium. Based on the British Standard Soil Classification and Plasticity Chart, the soil is classified as a CLAY with very high plasticity (CV). Results from laboratory test on physical properties and compressibility parameters show that Sabak Bernam Marine Clay (SBMC) is highly compressible, has low permeability and poor drainage characteristics. The compressibility parameters obtained for SBMC is in a good agreement with other researchers in the same field.

1. Introduction

Due to the rapid urbanization of Malaysia's city centre and around the world, there has been an increase in demand for developing areas nearby coastal plain. The development of housing and infrastructure has led to the discovery of new land and existing land reclamation. In conjunctions with rapid housing development, roads to connect remote areas with cities concurrently increase in demands. Major concern for construction nearby coastal plain is the problem with soft ground. Soft ground has caused many disasters due to high compressibility, low shear strength and poor drainage characteristics of the soft soil [1,2]. There are always two main problems regarding soft soil, namely stability and settlement. However, many practicing engineers have always forgotten the importance of solving settlement problems. Since 2010, statistics of geotechnical forensic investigation on problematic projects carried out by JKR summarized that 182 cases out of 252 forensic cases (approximately 72%) are related to the issue of ground settlement, while the remaining 28% are caused by other factors such as vibration, erosion, foundation failures and so forth [3]. Problems of compressibility occurred when a change in volume is induced in all soils when the external loads increase in soft soil which produce serious settlement problem. Therefore, pre-consolidation works are necessary to be carried out to accelerate the consolidation settlement thus to prevent future excessive post construction settlement and differential settlement that will lead to failure and instability of structure. Ground improvement must be performed before any construction is carried out to ensure that the soil has enough strength and bearing capacity to resist the load from structure. Various methods of ground improvement have been implemented nowadays such as preloading with surcharge [4], prefabricated vertical drain (PVD) coal bottom ash from power plants, bottom ash column [5] and lime and silica fume [6,7,8].

To have an efficient method of ground improvement, determination and interpretation of geotechnical properties of the soil are very crucial. Sufficient subsurface investigation with thorough monitoring by consultant engineer is crucial to ensure the SI carried out is convincing for further design of ground improvement [9]. The challenges in predicting settlements in highway embankments and reclamation works in marine, deltaic and estuarine type of deposits have been previously discussed [10]. The emphasis is on practical aspects and the difficulties experienced in confidently estimating settlements even after a century of developments and contributions. Interpreting soil parameters from laboratory and field test is important for settlement analysis. Some of the compressibility parameters are natural void ratio (e_o), preconsolidation pressure (P'c), overconsolidation ratio (OCR), compression ratio (CR) and recompression ratio (RR). All the abovementioned parameters always play major role in the settlement analyses. Many researches have been carried out in order to have more understanding in these parameters. However, the reported literature is different for all types of clay around the world [11,12]. To overcome the problem encountered in soft clay, knowledge and deep understanding about the engineering characteristics of soft clay are very important. The data will hopefully become a part of soft clay database in Malaysia. The result from this study can be referred by engineers as a useful guidance for construction on soft clay soil. The correlations produced could be used as a preliminary design for structure on soft clay soil.

2. Objective of Study

Objectives of this study are to determine the geotechnical properties and compressibility parameters of Sabak Bernam Marine Clay and carry out verifications of the compressibility parameters obtained for SBMC with existing correlation by other researchers.

3. Alluvial Soil Formation in Malaysia

A large acreage of soils in Peninsular Malaysia is derived from riverine alluvial deposits. The subrecent alluvial deposits of peninsular Malaysia consist of kaolinite, gibbsite, goethice, anatase, mica, chlorite and mixed layer. The most important mineral is kaolinite which contains more than 50% of the clay friction [13]. In the lowlands where alluvial soils dominate, the geomorphic surface and the groundwater table determine the soil type. The alluvial soil can be found along the east and west coast of Peninsular Malaysia. It is divided into two broad categories which are riverine deposits and marine, estuarine and brackish deposits [14]. The riverine deposits including colluvial and hill wash sediments are further divided into three categories, which are older, sub-recent and recent alluvia depending on the morphological and ages of the formation as shown in Figure 1.

IOP Conf. Series: Materials Science and Engineering **342** (2018) 012082 doi:10.1088/1757-899X/342/1/012082

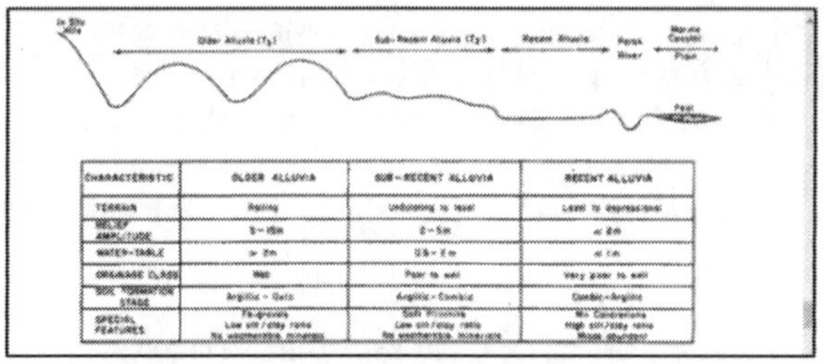

Figure 1. Typical soil landscape of riverine soil deposits [14].

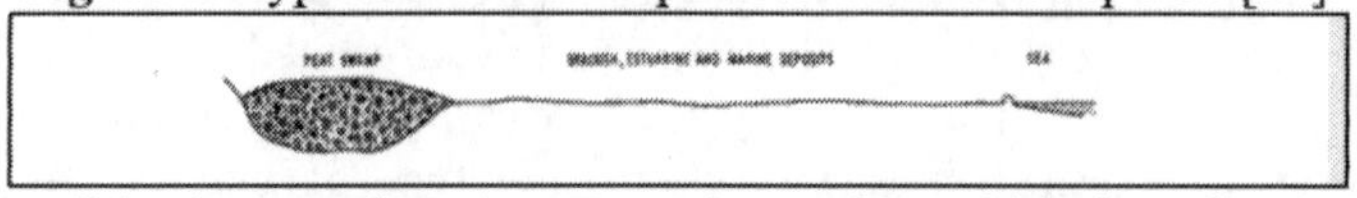

Figure 2. Typical soil landscape of Marine, Estuarine and Brackish deposits along coastal plain of west coast of Peninsular Malaysia [14].

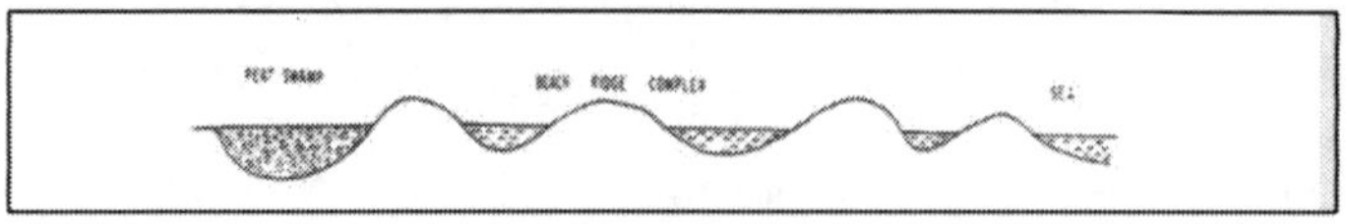

Figure 3. Typical soil landscape of Marine, Estuarine and Brackish deposits along coastal plain of east coast of Peninsular Malaysia [14].

Marine, estuarine and brackish deposits occur along the coastal plain of Peninsular Malaysia. At the west coast of Malaysia, due to present of Sumatera, calm seas condition causes the clayey soil to deposit along the coastal plain as shown in Figure 2. On the other hand, at the east coast of Peninsular Malaysia, due to strong waves of the open sea, the soil forms a series of beach ridges along the coastal plain as shown in Figure 3. Study by [15] on humus forms in alluvial soils, considered as young and heterogeneous environments. The soil structure formation is determined by both the nature of the recent alluvial deposits and the soil fauna. Paleosols formed from weathering of alluvial mudstones in the Late Cretaceous (Maastrichtian) Prince Creek Formation, North Slope Alaska, are dominated by detrital smectite, discrete illite, kaolinite, chlorite, quartz, and pedogenicillite–smectite mixed-layer clays [16].

3.1 Geological Formation of Study Area

The soil sample was obtained at Sungai Besar, Sabak Bernam, Selangor where the area is about 2 km from coastal plain of Malacca Straits as indicated in the Generalised Soil Map of Peninsular Malaysia, 1970 [17] in Figure 4. Based on the map, the area whereby the soil sample was obtained is classified as a clay soil with alluvial soil on recent marine and riverine alluvium as shown in Figure 5 [17,18]. The verification of the sampling location has been indicated in the map as shown in Figure 5. The physical appearance of the marine clay sample is as shown in Figure 6 which is greyish in colour. Study conducted by [19] revealed that Sabak Bernam Marine clay (SBMC) soils have high pH, high cation exchange capacity (CEC) – specific surface area (SSA) values, and high clay content with the presence of montmorillonite. SBMC soils contain montmorillonite, which is an active clay mineral that increases the CEC and SSA values. These values have contributed to maximum sorption capacity for SBMC soils.

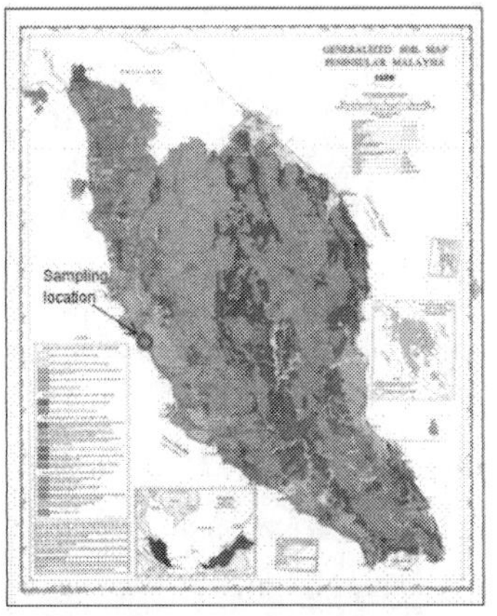

Figure 4. Generalised Soil Map of Peninsular Malaysia, 1970 [17].

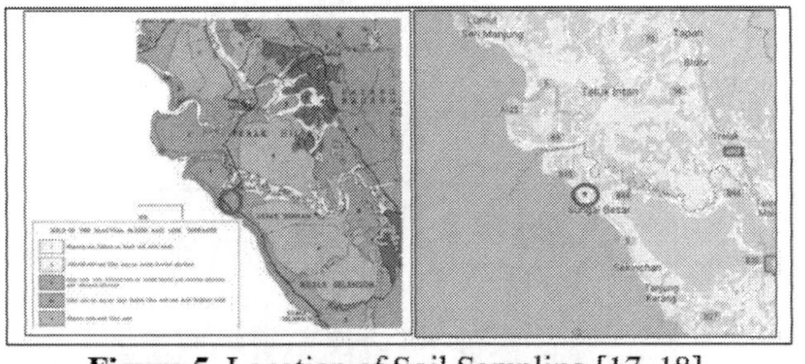

Figure 5. Location of Soil Sampling [17, 18].

4. Geotechnical Properties and Laboratory Compressibility Test

Determination of geotechnical properties of the SBMC is undertaken in this study. Laboratory test were carried out to determine the physical and compressibility properties of the SBMC as shown in Figures 7 to 9. Standard test method BS 1377:1990 is referred. Moisture content test was carried out to determine the natural moisture content of the soil taken from the site location. To prevent the loss of moisture, the soil was kept in a sealed container to the laboratory test for moisture content test. Sieving test was carried out to determine the particle size of soil passes the 0.063mm sieve size which is classified as a fine soil. Hydrometer test was performed on the soil particle passing 0.063mm sieve size to determine particle size distribution of fine-grained soil. Atterberg limit test was performed to determine the plasticity index and liquid limit of the soil. Two methods have been executed for determination of liquid limit of the soil namely cone penetrometer test and Casagrande test. Particle density test has been performed to determine the specific gravity of soil by using Pycnometer apparatus. On the other hand, for compression characterization of the soft soil, 1-D consolidation test was performed using Oedometer apparatus as shown in Figure 10.

Figure 6. Marine clay soil taken at Sabak Bernam Selangor.

Figure 7. Some of physical properties test setup.

Figure 8. Atterberg limit test.

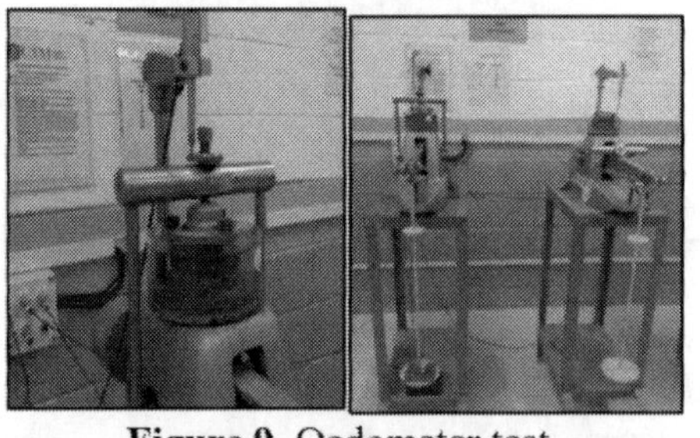

Figure 9. Oedometer test.

5. Results and Discussion

5.1 Physical and Chemical Properties of Sabak Bernam Marine Clay

Particle size distribution of fine soil and plasticity chart as shown in Figure 10 indicated that the soil is classified as CLAY with Very High Plasticity, CV. From the plasticity chart, the test results are recorded to be above the A-Line. The summary of soil parameters from physical properties tests are shown in Table 1. Moisture content and liquid limit of the soil is high which are 80% and 79% respectively indicated that the soil is highly compressible. The plasticity index is 48%. Results obtained were compared with other marine clay properties in South East Asia [20] as shown in Figure 11. It is shown that the physical properties of SBMC sample in this study is in the range of other marine clays in South East Asia.

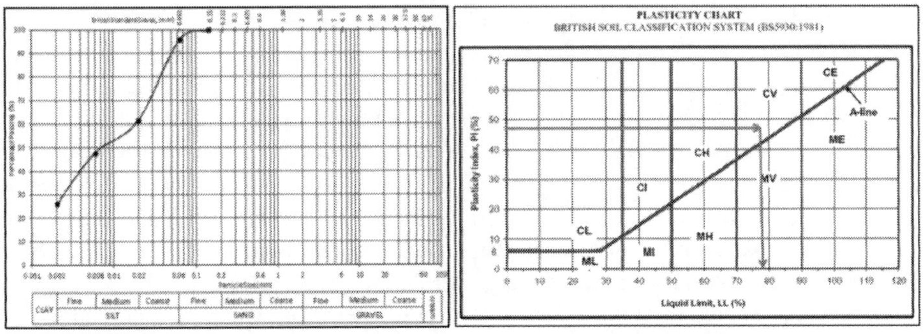

Figure 10. Particle size distribution and plasticity chart of fine soil.

Table 1. Physical and Chemical Properties of Sabak Bernam Marine Clay

Soil Parameters	Value
Moisture content, Mc (%)	80
Bulk density (Mg/mm^3)	1.474
Dry density (Mg/mm^3)	0.819
Liquid limit, LL (%)	79
Plastic limit, PL (%)	31
Plasticity index, PI	48
Linear shrinkage (%)	17
Specific gravity	2.352
pH value	5
Chloride content, CL(%)	0.01
Sulphate content (SO^3) (mg/L)	<200

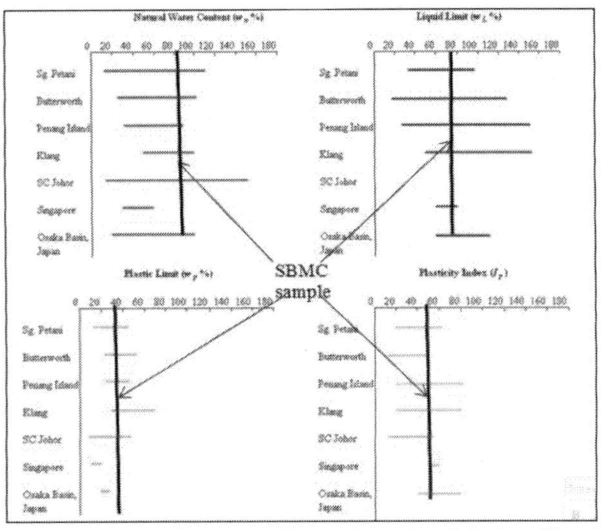

Figure 11. Atterberg Limits of SBMC in comparison with other Marine Clay in South East Asia [20].

5.2 Compression Characteristics of Sabak Bernam Marine Clay

Compression index, C_c is the slope of the linear portion of the e – log σ′ plot and is dimensionless. The recompression index, C_r is also an important parameter for soft compressible ground settlement analysis. C_r is defined in the same way as C_c except that it applies to the unloading reloading phase of the oedometer test. Figures 12 to 14 show void ratio and log effective stress relationship curve (e log σ') for undisturbed and remoulded sample of Sabak Bernam Clay soil from oedometer laboratory test. From the e log σ' curve, the average compression index is 0.614 whereas the average recompression ratio is 0.169. The compression parameters for the tested soil are tabulated in Table 2. The results of compression ratio are compared with the existing correlations from established researchers. Values of C_c by the researchers as shown in Table 3 is in a good agreement with C_c value obtained in this study.

The results of $C_c/(1+e_o)$ in this study is 0.213 and is applicable for normally consolidated soil and these results show that the soil is classified as highly compressible according to Table 4 [20]. According to Neoh [11], the normal compression ratio ($Cc/[1+e_o]$) for soft clay is from 0.2 to 0.4 which is within the range of ($Cc/[1+e_o]$) obtained in this study. The over consolidated soil is slightly compressible as shown by the shallow slope whilst the normal consolidated soil is highly compressible as shown by the steep slope. The volume of compressibility, m_v for SBMC soil is 0.571 as shown in Figures 15 to 17 and the coefficient of permeability is 1.68×10^{-10} m/s indicated that the drainage properties of this soil is very poor as shown in Figure 19. C_v value obtained in this study is 1.8mm²/min (0.93m²/yr) as shown in Figure 18 which is in the range of C_v for soft clay according to Neoh [12] those are 0.5 to 1.5 m²/ year.

The response of C_v to pressure increase in clays is governed by the mechanical and physicochemical factors that govern the compressibility [23]. The comparison between in situ and laboratory C_v values show that the macrostructure of the soil layer influences the behavior of the whole layer [24].

Table 2. Compressibility Parameters of Sabak Bernam Marine Clay.

Compression Parameters	**Value**
Compression index, C_c	0.614
Recompression index, C_r	0.169
Preconsolidation pressure, P_c	60kPa
Coefficient of consolidation, C_v	1.8mm²/min(0.93m²/yr)
$C_c/(1+e_o)$	0.213
$C_r/(1+e_o)$	0.059
C_c/C_r	3.63
m_v (m²/MN)	0.571
k (m/s)	1.68×10^{-10}

Table 3. Compression index, C_c for soft clay by other researchers.

Equation	C_c (with respect to LL=79%, PI= 48)
Terzaghi and Peck, 1967 For undisturbed clay: C_c= 0.009 (LL – 10) For remolded clay C_c= 0.007 (LL – 10)	0.62 0.483
Schofield and Wroth, 1968 C_c =1.35 PI	0.65
Tan et al (2000) C_c= 0.02LL – 0.87 (Klang Marine Clay)	0.71
Kulhawy and Mayne (1990) C_c=PI/74	0.65

Table 4. Classification of soil compressibility[a] [21].

$C_c/(1+e_o)$ or $C_r/(1+e_o)$	Classification
0 – 0.05	Very slightly compressible
0.05 – 0.10	Slightly compressible
0.10 - 0.20	Moderately compressible
0.20 – 0.35	Highly compressible
>0.35	Very highly compressible

[a]For soils that are normally consolidated, base the classification on $C_c/(1+e_o)$.
For soils that are overconsolidated, base the classification on $C_r/(1+e_o)$

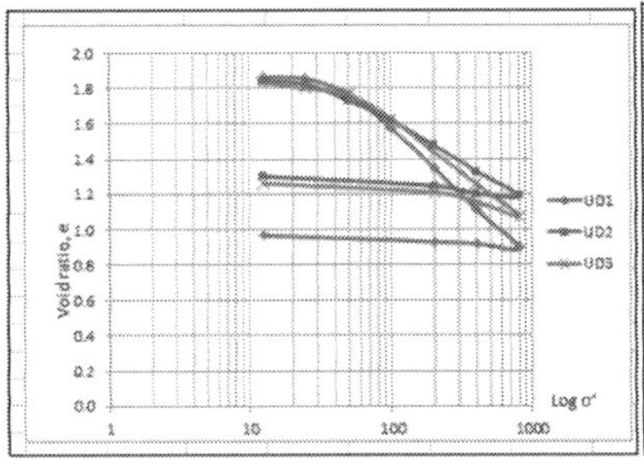

Figure 12. e log σ' curve for undisturbed soil .

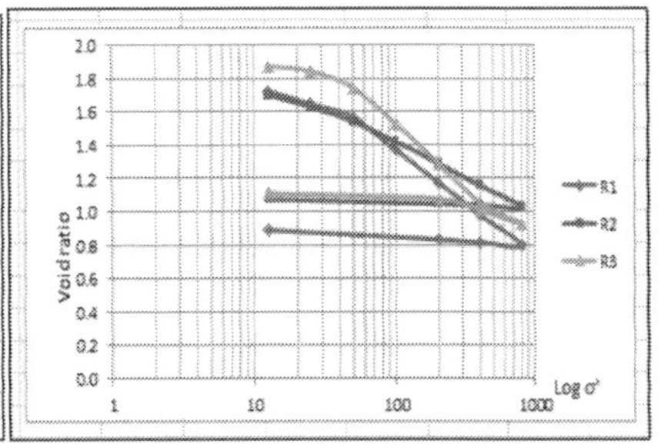

Figure 13. e log σ' curve for remoulded soil.

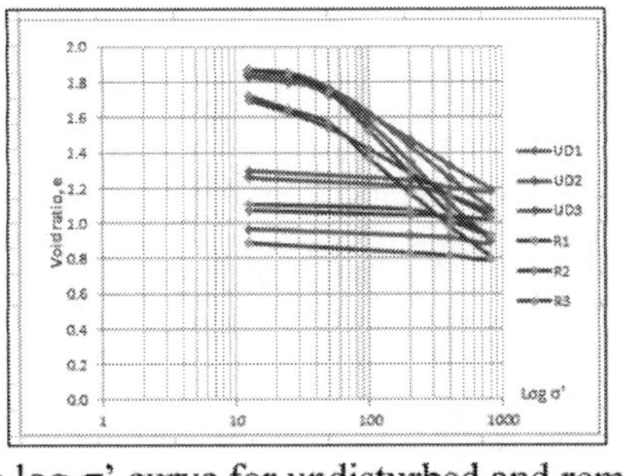

Figure 14. e log σ' curve for undisturbed and remoulded soil.

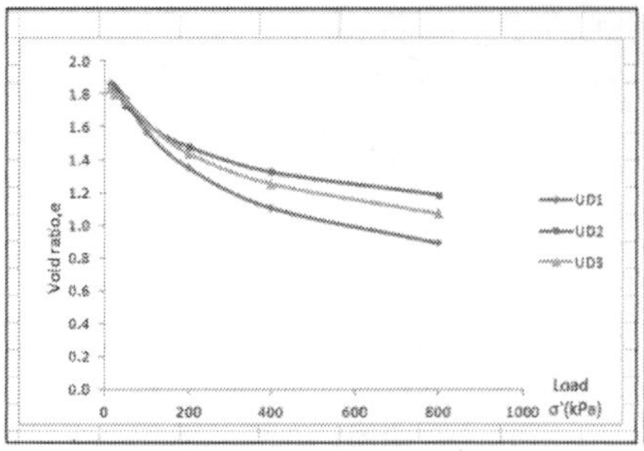

Figure 15. e vs σ' curve for undisturbed soil.

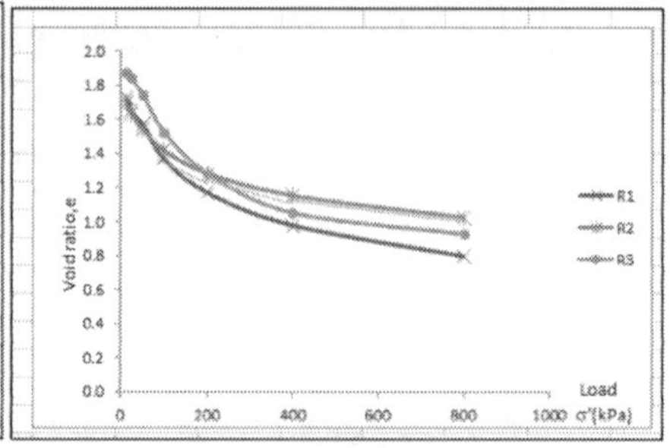

Figure 16. e vs σ' curve for remoulded soil.

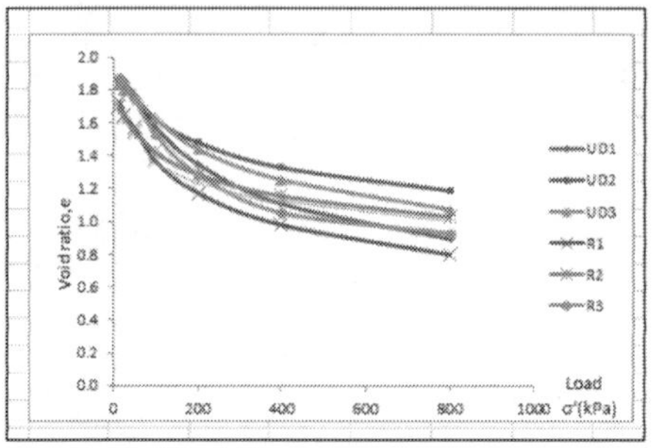

Figure 17. e vs σ' curve for undisturbed and remoulded soil.

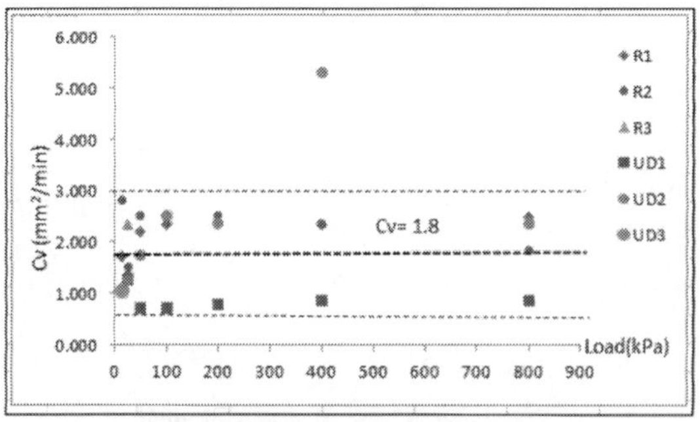

Figure 18. C_v for SBMC.

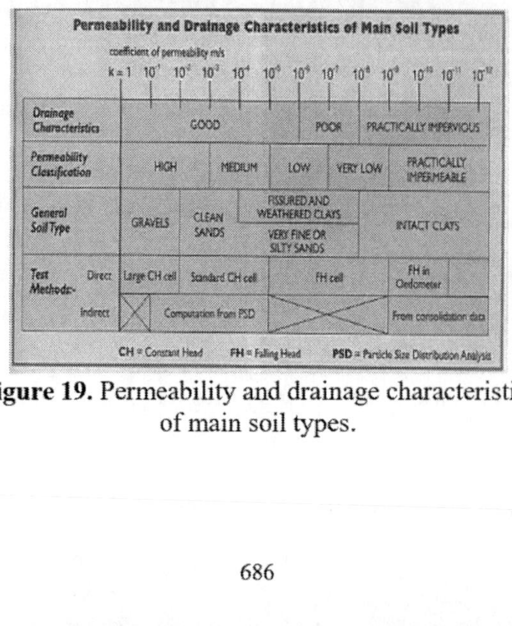

Figure 19. Permeability and drainage characteristics of main soil types.

6. Conclusion

Soil sample collected at Sabak Bernam is an alluvial deposit of marine clay. Based on British Standard Soil Classification and Plasticity Chart, the soil is classified as CLAY with very high plasticity (CV). The moisture content of SBMC is high with 80% and the coefficient of permeability is very low which denote the poor drainage characteristics of the soil. The physical properties of SBMC sample is in the range value of other marine clay in Malaysia and South East Asia. Compressibility parameters show that SBMC is highly compressible, has low permeability and poor drainage characteristics. The compressibility parameters obtained for SBMC are in good agreement with other researchers in the same field.

References

[1] Huat B B K, Othman K and Jaafar A A, 1995 Geotechnical Properties of Malaysia Marine Clays, Journal – Institution of Engineers Malaysia.

[2] Huat B B K, Maail S and Mohamed T A, 2005 Effect of Chemical on the Engineering Properties Of Tropical Peat Soil, American Journal of Applied Sciences, Science Publication.

[3] N O Mohamad, C E Razali, A A A Hadi, P P Som, B C Eng, M B Rusli and F R Mohamad, 2016 Challenges in Construction Over Soft Soil - Case Studies in Malaysia Soft Soil Engineering International Conference 2015 (SEIC2015) IOP Publishing IOP Conf. Series: Materials Science and Engineering 136 (2016) 012002 doi:10.1088/1757-899X/136/1/012002

[4] Soemitro R, Leong E C and Rahardjo H, 2000 Soil improvement by surcharge and vacuum preloadings, Géotechnique, *50*(5), 601–605 https://doi.org/10.1680/geot.2000.50.5.601

[5] Marto A, and Tan C S, 2016 Properties of Coal Bottom Ash From Power Plants in Malaysia and Its Suitability As Geotechnical Engineering Material, JurnalTeknologi, 78(8–5), 1–10 https://doi.org/10.11113/jt.v78.9603

[6] Alrubaye A J, Hasan M and Fattah M Y, 2017 Stabilization of Soft Kaolin Clay with Silica Fume and Lime, International Journal of Geotechnical Engineering, 11(1), 90–96 https://doi.org/10.1080/19386362.2016.1187884

[7] Fattah M Y, Al-Saidi A A and Jaber M M, 2015 Improvement of Bearing Capacity of Footing on Soft Clay Grouted with Lime-Silica Fume Mix, Geomechanics and Engineering, 8(1), 113–132 https://doi.org/10.12989/gae.2015.8.1.113

[8] Hasan M Bin, Marto A B, Hyodo M and Makhtar A M Bin, 2011 The Strength of Soft Clay Reinforced with Singular and Group Bottom Ash Columns, Electronic Journal of Geotechnical Engineering, 16 N, 1215–1227.

[9] D C Lat, IB Jais, Khatif T, Razali R, 2015 Evaluation of PU Injection Pile and Slab in Comparison with Lighweight Concrete for Ground Improvement using PLAXIS, Journal of Applied Science Research, 11(20), special, pages 11-16.

[10] Balasubramaniam A S, Cai H, Zhu D, Surarak C and Oh E Y N, 2010 Settlements of Embankments in Soft Soils, Geotechnical Engineering, 41(2).

[11] Tan Y C, Gue S S, Ng H B and Lee P T, 2007 Some Geotechnical Properties of Klang Clay. Jurutera, 32–33 https://doi.org/https://doi.org/10.3311/PPci.9350

[12] Tan Y C, Gue S S, Ng H B, and Lee P T, 2000 Some Geotechnical Properties of Klang Clay Gue & Partners Sdn Bhd, Malaysia

[13] Shamshuddin J, 1986 Clay Mineralogy of Selected Alluvial Soils from Peninsular Malaysia GEOSEA V Proceedings Vol. /. Geol. Soc. Malaysia. Bulletin /9. April 1986; pp. 443-451

[14] Paramananthan S and Zauyah S, 1986 Soil Landscapes in Peninsular Malaysia, GEOSEA V P;oceedings Vol. /, Geol. Soc. Malaysia, Bulletin /9, April 1986; pp. 565-583

[15] Bullinger-Weber G, Le Bayon R C, Guenat C, and Gobat J M, 2007 Influence of Some Physicochemical and Biological Parameters on Soil Structure Formation in Alluvial Soils. European Journal of Soil Biology, 43(1), 57–70, https://doi.org/10.1016/j.ejsobi.2006.05.003

IOP Conf. Series: Materials Science and Engineering **342** (2018) 012082 doi:10.1088/1757-899X/342/1/012082

[16] Jaramillo S S, McCarthy P J, Trainor T, Fowell S J and Fiorillo A R, 2015 Origin of Clay Minerals in Alluvial Paleosols, Prince Creek Formation, North Slope, Alaska, U.S.A Influence of Volcanic Ash on Pedogenesis in the Late Cretaceous Arctic, Journal of Sedimentary Research, 85(2), 192–208, https://doi.org/10.2110/jsr.2015.12
[17] Director of National Mapping Malaysia, 1970 Generalised Soil Map of Peninsular Malaysia 1970
[18] Google Map Malaysia assessed on 10th November 2017
[19] Wan Yaacob W Z, Samsudin A R, Ramziemran M and Loon C Y, 2004 Natural Sorption Capability of Heavy Metals Granitic Residual Soil from Broga and Marine Clay from Sg. Besar Geological Society of Malaysia, Bulletin 48 June 2004, p. 13 -16.
[20] Suriawati A/P Ramamoorthy, 2007 Correlation Of Engineering Characteristics of Marine Clay from Central West Coast Of Malaysia A Thesis of Master of Engineering (Civil – Geotechnics)
[21] Coduto D P, Yeung M R and Kitch W A, 2011 Geotechnical Engineering Principles and Practices (2nd ed.) Pearson Education Inc.Pearson Prentice Hall.
[22] Budhu M, 2015 Soil Mechanics Fundamentals Nature (Vol. 143) https://doi.org/10.1038/143865a0
[23] Robinson R G and Allam M M, 1998 Effect of Clay Mineralogy on Coefficient of Consolidation, Clays and Clay Minerals 46(5) 596–600 https://doi.org/10.1346/CCMN.1998.0460514
[24] Cortellazzo G, 2002 Comparison between Laboratory and In Situ Values of the Coefficient of Primary Consolidation Cv, Canadian Geotechnical Journal, 39(1), 103–110 https://doi.org/10.1139/t01-080

IOP Conf. Series: Materials Science and Engineering **342** (2018) 012083 doi:10.1088/1757-899X/342/1/012083

Investigating the ability of shunt hybrid power filter based on SRF method under non-ideal supply voltage

Mohamed A Omran[1, 2], Izzeldin I Mohd[1], Mohamad M Almelian[1,3], Usman Ullah Sheikh[4] and Mustafa E A A Bofares[5]

[1]Faculty of Electrical and Electronics Engineering, University Malaysia Pahang (UMP), 26600 Pekan, Pahang, Malaysia
[2]College Of Technical Sciences – Bani-Walid, Libya
[3]Higher Institute Of Sciences And Technology, Mezdah, Libya
[4]Faculty of Electrical Engineering, Universiti Teknologi Malaysia, 83100 UTM Skudai, Johor Bahru, Johor, Malaysia
[5]Higher and Intermediate Institute of Engineering Professions-Bani-Walid, Libya

E-mail: omranmohamed346@yahoo.com

Abstract. This study presents the capacity of a self-tuning filter based on the synchronous reference frame method with a fuzzy logic controller for the improvement of the efficiency of harmonic suppression of a shunt hybrid active power filter in an unbalanced distorted and un-distorted voltage supply conditions. The simulation results indicated that the filter with a fuzzy logic controller had a good filtering performance in steady and transient states, irrespective of whether the voltage supply is distorted or unbalanced.

1. Introduction

The issues with power quality have become critical in recent times due to the use of electronic circuits with non-linear behaviors in industrial and domestic devices such as inverters, power supply switches, single/ three phases of rectifiers. These circuits generate non-sinusoidal currents in the distribution power networks. The issues of reactive and active power occur usually due to the connection of nonlinear single-phase loads with either three-phase or four-wire structures. Even when these loads are stable, harmonic currents circulate through the neutral conductor, and this is bound to happen as a result of the presence of zero sequence elements [1-2]. These currents might have higher amplitudes compared to those of the component phase current. This difference of current amplitudes can lead to damage the transformers and neutral conductor which bear these loads [3-5].

To disband this problem of current harmonics in the power distributed network, several solutions have been proposed. These solutions can be grouped into two: the first group is the conditioning loads group which provides confer equipment with less sensitivity to energy instabilities, thereby, allowing service continuity even when there is a significant distortion in the voltage [6]. The other technique is the installation of conditioning structures which minimizes harmonic disturbances.

Harmonic currents can be minimized with passive filters (PFs) to compensate for reactive power. However, there are some issues with these PFs, such as heavy and bulky devices, parallel and series resonance with load and source impedances. The active filters were established to reduce the issues with the passive filters [4, 7, and 8] but they have high operational costs and also not appropriate for moderate or high voltage conditions. A synergistic combination of the passive and active filters in the form of a hybrid power filter (HPF) is appealing [7]. Various forms of hybrid filters with varying control systems have been suggested for load harmonics compensation.

These structures allow the separate use of parallel/series active filters to design small-rated active filters. The harmonic extraction method for the detection of voltage and current harmonic as well as sub-harmonics and components are mainly according to either direct or indirect control technique which is dependent upon the theory of instantaneous reactive power [9]. The efficiency of the active power filters has also been improved using adaptive notch filters, synchronous rotational reference frame, flux-based controllers, and nonlinear controllers. However, most of these strategies are difficult to implement as they contain a number of transformations [10].

The fuzzy logic controllers [4,6] have recently gained attention in terms of application to APFs. A fuzzy logic controller is an ideal approach compared to the conventional methods because they do require any accurate mathematical model, can hand non-linearity, can apply inaccurate inputs, and are more vigorous to implement [11].

In this paper, a synchronous reference frame (SRF) with a self-tuning filter (STF) was used to shape ac current and regulate dc voltage intended for a-three-phase shunt HAPF under ideal and non-ideal power source conditions.

2. Control algorithms of the SHAPF

Figure 1 depicts the controlling algorithm block diagram. The signals of feedback are detected from the load currents, as well as AC source and DC bus voltages of SAPF [12]. The three-phase currents (i_L^a , i_L^b and i_L^c) were converted from from (a-b-c) phases reference frame to (α-β) phases static reference frame currents iα and iβ using

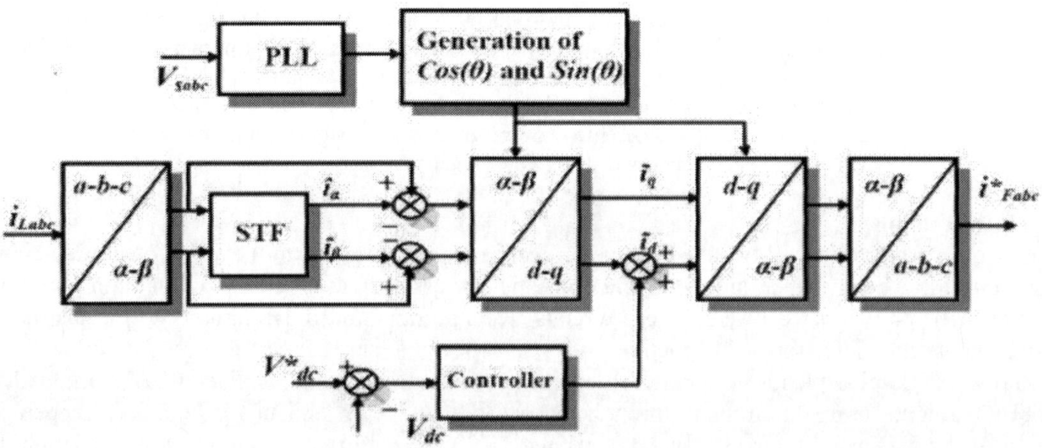

Figure 1. SRF control strategy with STF based on SHAPF.

$$\begin{bmatrix} I_\alpha \\ I_\beta \end{bmatrix} = \sqrt{2/3} \begin{bmatrix} 1 & -1/2 & -1/2 \\ 0 & \sqrt{3}/2 & -\sqrt{3}/2 \end{bmatrix} \begin{bmatrix} I_L^a \\ I_L^b \\ I_L^c \end{bmatrix} \tag{1}$$

The instantaneous current could be represented as the total fundamental and alternative components:

$$\begin{cases} I_\alpha = \bar{I}_\alpha + \tilde{I}_\alpha \\ I_\beta = \bar{I}_\beta + \tilde{I}_\beta \end{cases} \tag{2}$$

Then the STF directly extracts the fundamental components from the currents in the α-β axis. Then, the α-β harmonic current components are generated by deducting the signals of STF input from the resultant outputs (Figure 1). The signals from this computation are the AC elements (ĩα and ĩβ) which are the corresponding harmonic elements generated in the stationary reference frame by the load currents for all phases i_L^a , i_L^b and i_L^c [13-14]. Cos(θ) and sin(θ) can be generated from the phase voltage source (Vsa, Vsb and Vsc) with the help of Phase Locked Loop (PLL). The currents ĩα and ĩβ within the d-q reference frame can be expressed as

$$\begin{bmatrix} \tilde{I_d} \\ \tilde{I_q} \end{bmatrix} = \begin{bmatrix} sin(\theta) & -cos(\theta) \\ cos(\theta) & sin(\theta) \end{bmatrix} = \begin{bmatrix} \tilde{I_\alpha} \\ \tilde{I_\beta} \end{bmatrix} \quad (3)$$

The reference currents (I_F^*) in the a-b-c frame are expressed as

$$\begin{bmatrix} I_{Fa}^* \\ I_{Fb}^* \\ I_{Fc}^* \end{bmatrix} = \sqrt{2}/_3 \begin{bmatrix} 1 & 0 \\ -1/_2 & \sqrt{3}/_2 \\ -1/_2 & -\sqrt{3}/_2 \end{bmatrix} \begin{bmatrix} I_\alpha^* \\ I_\beta^* \end{bmatrix} \quad (4)$$

The STF is used in Figure 2 as a harmonic isolator rather than filters which based on classical extraction methods. Hong-sock Song in [5] had earlier justified the inclusion of the SRF method

$$V_{XY}(t) = e^{j\omega t} \int e^{-j\omega t} V_{XY}(t)dt \quad (5)$$

By transfer function of filter, the equation (5) can be expressed: [15]

$$\tilde{I_\alpha}(s) = \frac{k}{s}\left[I_\alpha(s) - \tilde{I_\alpha}(s)\right] - \frac{wc}{s}\tilde{I_\beta}(s) \quad (6)$$

$$\tilde{I_\beta}(s) = \frac{k}{s}\left[I_\beta(s) - \tilde{I_\beta}(s)\right] + \frac{wc}{s}\tilde{I_\alpha}(s) \quad (7)$$

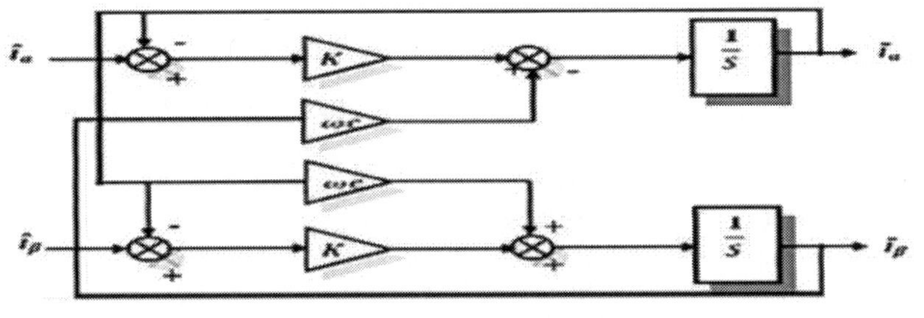

Figure 2. The structure of self-tuning filter tuned STF.

3. DC capacitor voltage control

The fuzzy control (FL) algorithm was used for the control of the DC capacitor voltage according to the processing of DC voltage error as well asits variation to maintain the dc voltage as a reference value, improve the SHAPF performance, and minimize the THD as shown in figure3.While designing a fuzzy control system [16]. FL was described by seven fuzzy sets entailing NM, NB, ZE, NS, PM, PS, and PB for its input and output variables, function oftriangular membership utilized for thesimplicity, defuzzification by means of the height technique and implication by means of Mamdani-type min-operator.

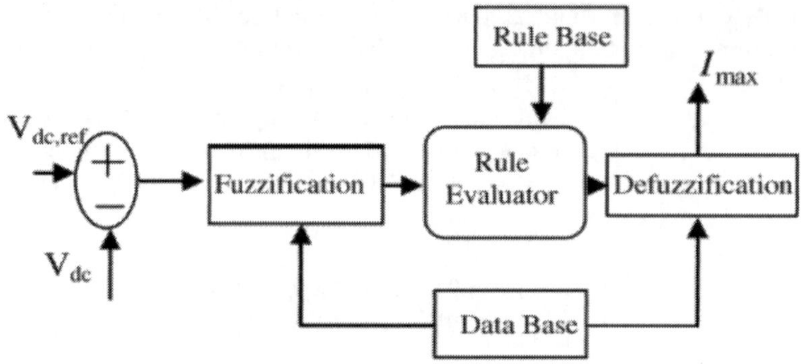

Figure 3. The structure of FLC.

4. Results and discussion

The source voltage, source current with filter, compensation current, compensation current, compensation current, dc link voltage and THD under conditions of balanced un-balanced and non-sinusoidal supply voltage with FLC are detailed in Figures 4, 5,6 and 7. The specification and circuit parameters are used in the simulations are written in table 1 below:

Table 1. presented system parameters.

PARAMETERS	VALUE
Source voltage	240V
voltage frequency	50 HZ
Inductor Ls	0. 0001 H
Resistor Rs	0. 0005 Ω
Inductor LL 1	0.05 H
Resistor RL 1	80 Ω
Inductor LL 2	0.5 H
Resistor RL 2	120 Ω
Inductor Lf	0. 000075 H
Resistor Rf	0.00050 Ω
DC bus voltage	450V
DC capacitor	350 μF

The system performance was initially investigated with nonlinear (0 to 0.25s) and additional nonlinear loads (0.25to 0.4 s) under balanced sinusoidal conditions and found to be effective at harmonics suppression (Figures 4 and 5). The THDs of the currents were reduced from 28.45% and 29.33% to 2.08% and 1.78% respectively. Meanwhile, there was an increase of 30% in load at t = 0.25s in the case of changing loads. As evidenced, the filter current was relational to the changes in the load as no significant current transition was observed. The line current remained sinusoidal and attained the steady value. Besides, the DC-link capacitor was maintained with very small ripple by the FLC at the required range (450v) during load changes.

The filter capability to compensate harmonicloads under unbalanced distorted ac source (includes a 5th and 7th harmonic components (THD= 14.67%) is shown in Figure 6. There was a decrease in the measured THD of the source current from 29.31% and 31.25% prior to compensation to 2.23% and 2.71% post-compensation for the non-linear loads as in (Figure 7). The voltage of the dc capacitors dc was within the reference range and the voltage variation was within the limit in the controller under distorted unbalanced supply.

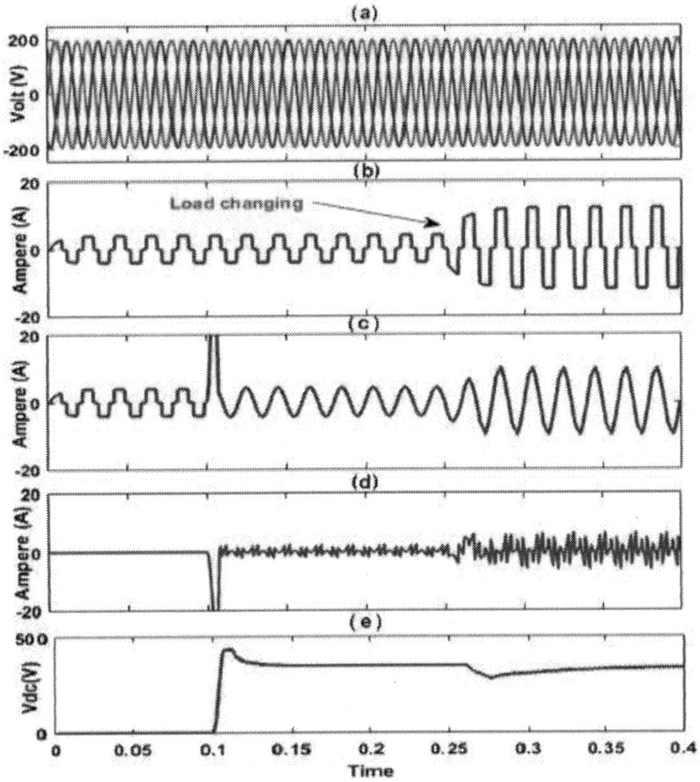

Figure 4. SHAF response under ideal voltage situation (a) Source voltage (Vs), (b) Load current (I_L), (c) load current with filter, (d) Filter compensation current (If) , and (e) DC link voltageof system (Vdc).

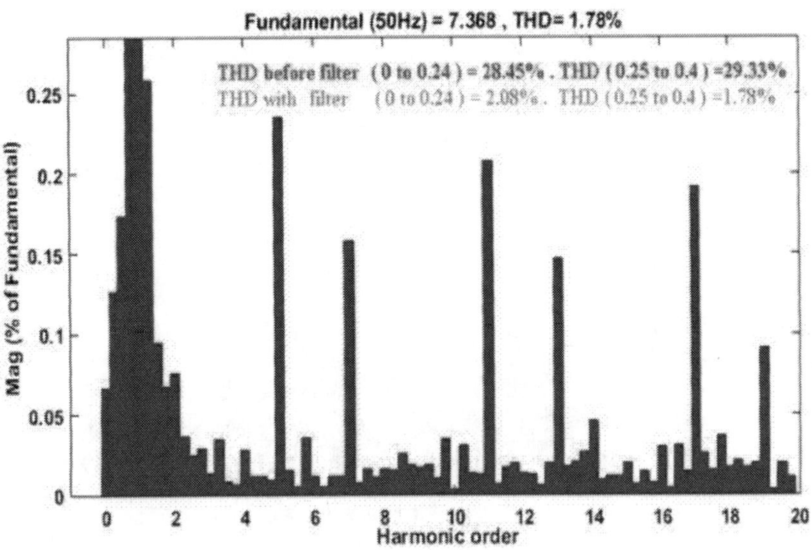

Figure 5. THD of current (Is)under ideal voltage supply.

IOP Conf. Series: Materials Science and Engineering **342** (2018) 012083 doi:10.1088/1757-899X/342/1/012083

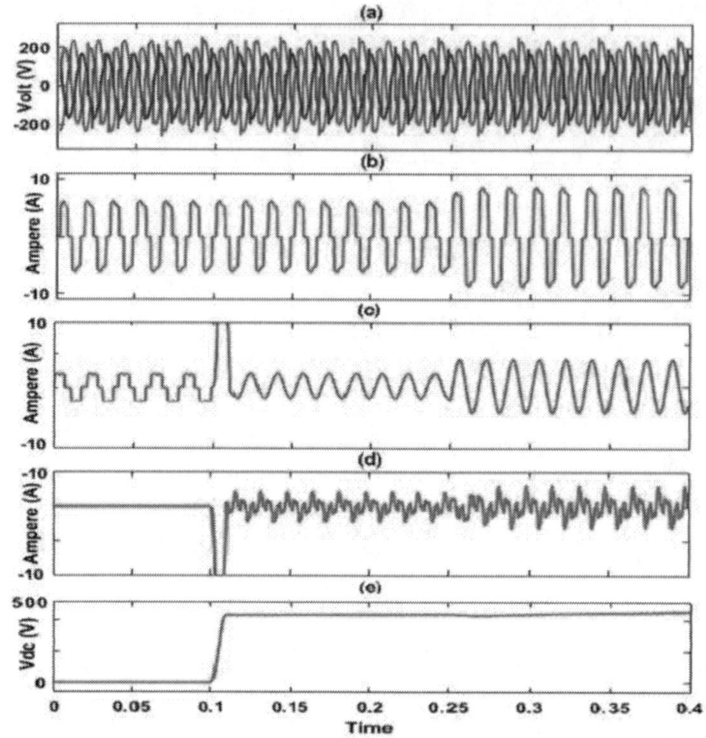

Figure 6. SHAF response under non ideal voltage situation (a) Source voltage (Vs), (b) Load current (Il), (c) load current with filter, (d)Filter compensation current (If), and (e) DC link voltageof system (Vdc).

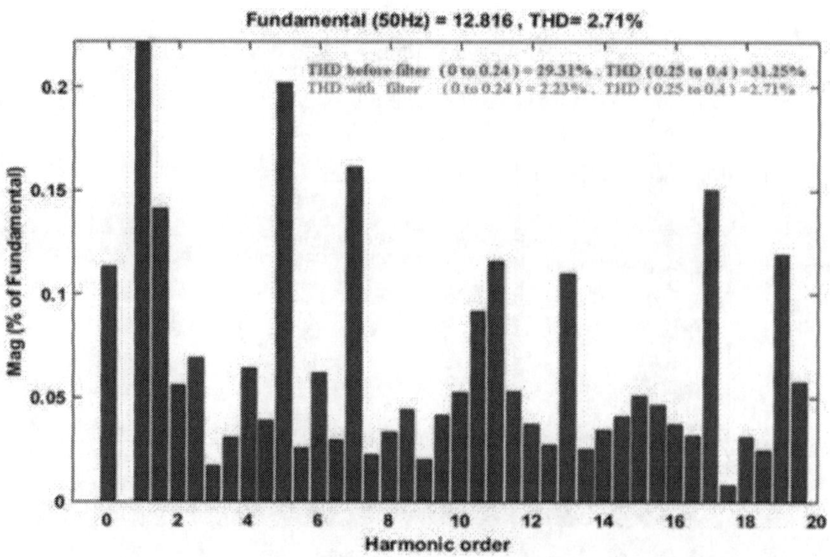

Figure 7. THD of current (Is) under non ideal source voltage.

5. Conclusion

This paper investigated the effectiveness of a synchronous reference frame (SRF) with a self-tuning filter (STF) control strategy in controlling the performance of a three-phase SHAPF system under conditions of non-ideal and balanced supply voltage. The fuzzy logic controller was utilized for the adjustment of the DC voltage. The performance of the SHAPF system was investigated under a dynamic and steady state and under different load operating conditions. The simulation results showed the SAPF to have successfully reduced current harmonics to about 1.7 and 2.7 % for both cases of source voltages.

Acknowledgments

The authors thank the Ministry of Education Malaysia and Universiti Teknologi Malaysia (UTM) for their support under grant number Q.J130000.2623.14J48.

References

[1] Eltawil, M.A. and Zhao, Z., 2010. Grid-connected photovoltaic power systems: Technical and potential problems—A review. *Renewable and Sustainable Energy Reviews*, *14*(1), pp.112-129.

[2] Bhat, A.H. and Agarwal, P., 2008. Three-phase power quality improvement ac/dc converters. *Electric Power Systems Research*, *78*(2), pp.276-289.

[3] Wagner, V.E., Balda, J.C., Griffith, D.C., Mceachern, A., Barnes, T.M., Hartmann, D.P., Phileggi, D.J., Emannuel, A.E., Horton, W.F., Reid, W.E. and Ferraro, R.J., 1993. Effects of harmonics on equipment. *IEEE Transactions on Power Delivery*, *8*(2), pp.672-680.

[4] Mohamed, M.A., 2015. Design of shunt active power filter to mitigate the harmonics caused by nonlinear loads (Doctoral dissertation, *Universiti Tun Hussein Onn Malaysia*).

[5] Soomro, D.M. and Almelian, M.M., 2015. Optimal design of a single tuned passive filter to mitigate harmonics in power frequency.

[6] Ghosh, A. and Ledwich, G., 2012. Power quality enhancement using custom power devices. *Springer Science & Business Media*.

[7] Rahmani, S., Hamadi, A. and Al-Haddad, K., 2012, October. A comprehensive analysis of hybrid active power filter for power quality enhancement. In *IECON 2012-38th Annual Conference on IEEE Industrial Electronics Society* (pp. 6258-6267). IEEE.

[8] Wang, Y.F. and Li, Y.W., 2013, September. An overview of grid fundamental and harmonic components detection techniques. In *Energy Conversion Congress and Exposition (ECCE), 2013 IEEE* (pp. 5185-5192). IEEE.

[9] Rahmani, S., Hamadi, A., Al-Haddad, K. and Alolah, A.I., 2013. A DSP-based implementation of an instantaneous current control for a three-phase shunt hybrid power filter. *Mathematics and Computers in Simulation*, *91*, pp.229-248.

[10] Green, T.C. and Marks, J.H., 2005. Control techniques for active power filters. *IEE Proceedings-Electric Power Applications*, *152*(2), pp.369-381.

[11] Mikkili, S. and Panda, A.K., 2012, March. Mitigation of harmonics using fuzzy logic controlled shunt active power filter with different membership functions by instantaneous power theory. In *Engineering and Systems (SCES), 2012 Students Conference on* (pp. 1-6). IEEE.

[12] Shahbaz, M., 2012. *Active Harmonics Filtering of Distributed AC System* (Master's thesis, Institutt for elkraftteknikk).

[13] Abdusalam, M., Poure, P. and Saadate, S., 2008, June. Hardware implementation of a three-phase active filter system with harmonic isolation based on self-tuning-filter. In *Power Electronics Specialists Conference, 2008. PESC 2008. IEEE*(pp. 2875-2881). IEEE.

[14] Abdusalam, M., Poure, P., Karimi, S. and Saadate, S., 2009. New digital reference current generation for shunt active power filter under distorted voltage conditions. *Electric Power Systems Research*, *79*(5), pp.759-765.

[15] Ghamri, A., Benchouia, M.T. and Golea, A., 2012. Sliding-mode control based three-phase shunt active power filter: Simulation and experimentation. *Electric Power Components and Systems*, *40*(4), pp.383-398.
[16] Mahajan, V., Agarwal, P. and Gupta, H.O., 2014. An artificial intelligence based controller for multilevel harmonic filter. *International Journal of Electrical Power & Energy Systems*, *58*, pp.170-180.

IOP Conf. Series: Materials Science and Engineering **342** (2018) 012084 doi:10.1088/1757-899X/342/1/012084

Performance of unified power quality conditioner (UPQC) based on fuzzy controller for attenuating of voltage and current harmonics

Mohamad Milood Almelian[1,2], Izzeldin I Mohd[1], Mohamed Asghaiyer Omran[1,3] and Usman Ullah Sheikh[4]

[1]Faculty of Electrical and Electronics Engineering, University Malaysia Pahang (UMP), 26600 Pekan, Pahang, Malaysia.
[2]Department of Electrical and Electronics Engineering, Higher Institute of Sciences and Technology, Mezdah, Libya.
[3]Faculty of Electrical and Electronics Engineering, Bani Walid, Libya.
[4]Faculty of Electrical Engineering, Universiti Teknologi Malaysia, 83100 UTM Skudai, Johor Bahru, Johor, Malaysia

E-mail: almalyan1984@yahoo.com; izzeldin@ump.edu.my

Abstract. Power quality-related issues such as current and voltage distortions can adversely affect home and industrial appliances. Although several conventional techniques such as the use of passive and active filters have been developed to increase power quality standards, these methods have challenges and are inadequate due to the increasing number of applications. The Unified Power Quality Conditioner (UPQC) is a modern strategy towards correcting the imperfections of voltage and load current supply. A UPQC is a combination of both series and shunt active power filters in a back-to-back manner with a common DC link capacitor. The control of the voltage of the DC link capacitor is important in achieving a desired UPQC performance. In this paper, the UPQC with a Fuzzy logic controller (FLC) was used to precisely eliminate the imperfections of voltage and current harmonics. The results of the simulation studies using MATLAB/Simulink and Simpower system programming for R-L load associated through an uncontrolled bridge rectifier was used to assess the execution process. The UPQC with FLC was simulated for a system with distorted load current and a system with distorted source voltage and load current. The outcome of the comparison of %THD in the load current and source voltage before and after using UPQC for the two cases was presented.

1. Introduction

One of the issues that is increasingly bordering electricity consumers is power quality. This is due to the sensitivity of the equipment and the non-linear nature of loads in both domestic environments and the power sectors. Power quality is generating more concerns among electricity users. Electricity supplies that were once acceptable to the users are now considered as problematic [1,2] and the solution to issues of power quality is within the reach of the users and the distribution operators. The active power filters are connected in shunt, series and vice versa to compensate for the distortions in

current and voltage and to provide quality solutions. The UPQC has been proposed as a powerful compensating device for simultaneously dealing with the issues of current and voltage qualities [3-5]. The UPQC which is made up of shunt and series active filters can compensate for distorted voltage at the source side and current at the load side to ensure the load voltage and supply currents are sinusoidal. Although the effectiveness of UPQC in the improvement of power quality, its application is still limited. The passive component of the proposed system was designed to achieve a good performance [6-9]. The UPQC was used in this study as a universal active power conditioning device for the mitigation of both current and voltage harmonics at a power distribution network.
The UPQC performance rely mainly on a quick and accurate compensation of the derived signals. The UPQC is designed based on FLC strategy to actively and reactively compensate for power. The FLC is used to overcome the nonlinear problems efficiently. The schematic representation of the UPQC is presented in Figure 1.

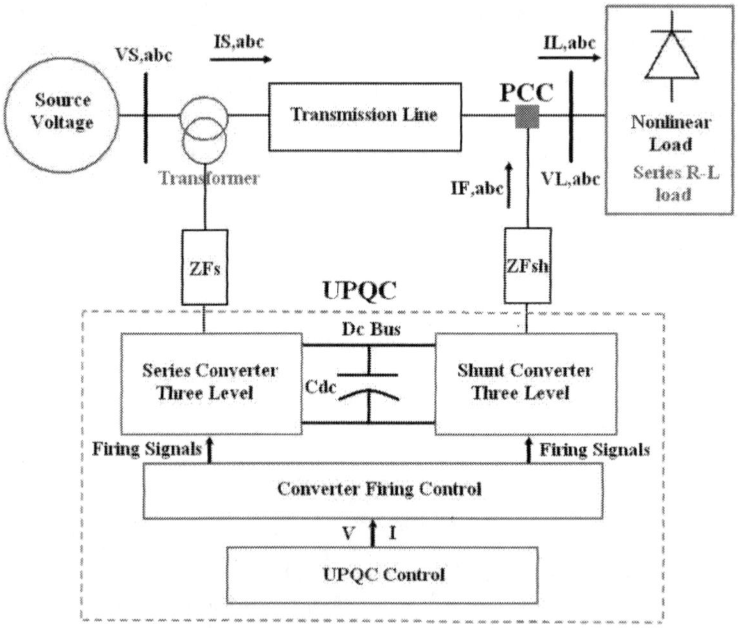

Figure 1. A representation of a UPQC.

Series filters are active filters or series converter which are serially connected through an injection transformer and acts as a controlled voltage generator. Series filters can compensate for voltage imbalance, harmonics and voltage regulation at the point of common coupling (PCC). Additionally, it can provide the harmonic isolation between a distribution system and a sub-transmission system. The shunt active filter is the second unit which is parallel connected with the load and it also generates controlled current. The current harmonics are absorbed by the shunt active filter, which similarly compensates for the current harmonics introduced by the load. Additionally, it maintains the DC current to a desired value. Another element in the power line conditioner is a DC link inductor which stores energy. The active power filter requires a minute amount of power to compensate for harmonics. The DC inductor serves as a source of DC power and therefore, does not require external power source. Meanwhile, the maintenance of a steady DC in the energy storage component requires drawing a small fundamental current to compensate for the active filter losses [10-12].

2. Control algorithm

FLC is a systematic approach for the control of non-linear-based procedures which depends on human knowledge and experience. A fuzzy controller can utilize multiple inputs and output variables. Inverter convert either DC voltage or AC currents into AC current or voltage. Hysteresis voltage and current control is used to generate to make a switching signal for operating the inverter. RL Load is connected. The overall function of UPQC mainly depends on the series and shunt active power filter controller.

2.1. Series converter control

In the voltage-source-inverter-based UPQC, hysteresis voltage control (HVC) is used for the control of the series converter (series filter). Series active filters provide a high impedance to current harmonics and obstructs their load to source streaming and source to load headings, and in like manner, presents as a regulated voltage supply. The elimination of the harmonic components in voltage supply is the major objective of series compensators; the voltage from the series filter is given by Equation 1 [13], where Figure 2 showed the series converter in terms of a control block diagram.

$$V_{ah} = V_{1na} + V_{1pa} + \sum_{k=2}^{\infty} V_{kn} \sin(k\omega t + \theta_{kn}) \quad (1)$$

Where: V_{ah} is the voltage which created by series filter, V_{1na} is the negative sequence components, and V_{1pa} is the positive sequence components.
To ensure the load voltage is sinusoidal, the phase-locked loop (PLL) based unit vector template (UVT) is multiplied with a constant of equal strength to the peak amplitude of the fundamental input voltage so that the voltage of the reference loads is obtained. The UVTs with the respective phase delays for individual phases are achieved and are stated below:

$$U_a = \sin(\omega t) \quad (2)$$

$$U_b = \sin(\omega t + 120) \quad (3)$$

$$U_c = \sin(\omega t - 120) \quad (4)$$

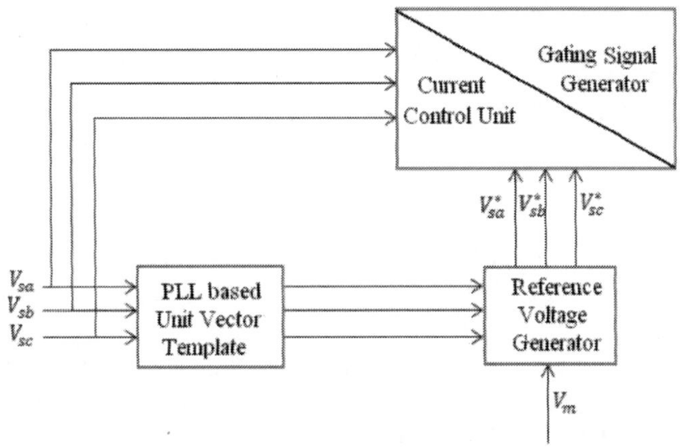

Figure 2. Block diagram of series converter controller.

The generation of the compensation signals is as follows:

$$V_{fa} = V_{sa} - V_m . U_a \quad (5)$$

$$V_{fb} = V_{sb} - V_m.U_b \quad (6)$$

$$V_{fc} = V_{sc} - V_m.U_c \quad (7)$$

Where: (U_a, U_b, U_c) and (V_{sa}, V_{sb}, V_{sc}) represent 3-phase unit current vectors multiplication and the 3-phase supply voltage respectively, and V_m is the supplied voltage amplitude.
The gating signal needed for the series converter is obtained by comparing the actual and compensation signals at the terminals of the series converter. The generated error is passed to the hysteresis controller.

2.2. Shunt converter control
The increasing nonlinearity of loads and power in electronic equipment is causing harmonics in the distribution systems. These harmonics can be compensated by a shunt converter (shunt filter) where the DC voltage is detected and benchmarked to a reference voltage and the signal error processed and seen as a magnitude of the reference 3-phased supply current. The phase unit vector is used to calculate the reference current. The 3-ph shunt current for harmonics compensation is presented in equation 8 [14].

$$I_{sha} = \frac{V_{ia}}{Z_{sh}};\ I_{shb} = \frac{V_{ib}}{Z_{sh}};\ I_{shc} = \frac{V_{ic}}{Z_{sh}} \quad (8)$$

Where: I_{sha}, I_{shb} and I_{shc} are the shunt currents of phase a, b and c , V_{ia}, V_{ib}, V_{ic} are the injected voltage in phase a, b and c, and Z_{sh} is the shunt impedance. A block diagram of the shunt converter is shown in Figure 3.

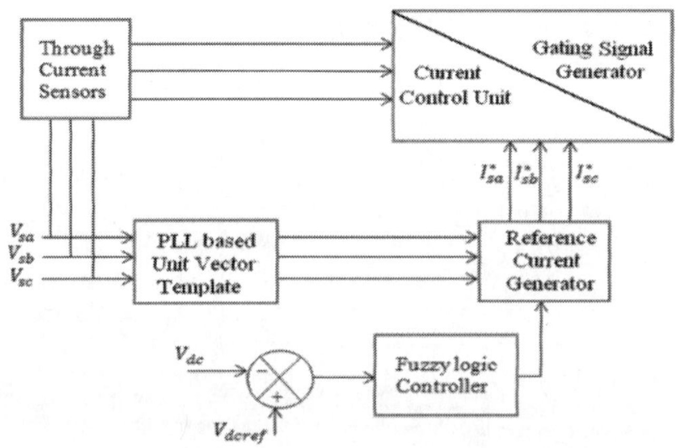

Figure 3. A shunt converter controller.

3. Fuzzy logic control
The FLC is based on the fuzzy set theory in which there is an inevitable transition between membership and non-membership function, which results in fuzzy set ambiguity. When there are lengthy mathematical calculations, the best approach is to deploy the FLC. The fuzzy logic theory can be regarded as a qualitatively-defined mathematical multi-valued logic that involves artificial intelligence and probability theory to proffer solution to problems since it mimics the human brain and uses approximate reasoning when relating different data sets and making decisions. Figure 4 show the three basic components (fuzzification, decision making and de-fuzzification) of fuzzy logic-based control schemes.

The FLC input is the sensed DC link voltage value. The voltage value is through fuzzification to a fuzzy value. The input values (in the fuzzification method) are converted to linguistic values such as positive-small (PS), positive-medium (PM) positive-big (PB) and negative-big (NB), negative-medium (NM), negative-small (NS), very-small (VS). These values are characterized by memberships.

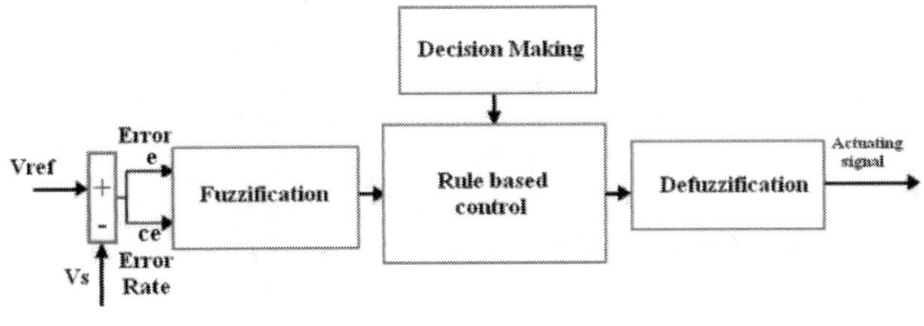

Figure 4. Fuzzy logic control scheme.

The input signal measurement in a fuzzy control system is interpreted as a fuzzy singleton using the Mamdani-type-inference system in which the linguistic variables (input and output) assume are presented as fuzzy variables. Figure 5(a) and (b) show the input variable of fuzzification while Figure 5(c) show the output variable of de-fuzzification. The fuzzy set membership function is an assembly of the indicator functions in classical sets. The fuzzy variables are determined by membership functions and typified by position, width, shape or whole overlap. In the controller, the triangular membership form is used.

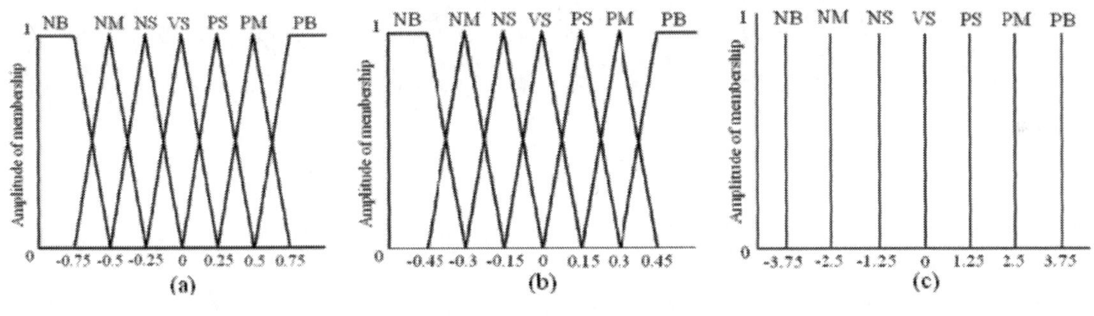

Figure 5. Variable of fuzzy logic (a) Input variable of fuzzification (error), (b) Input variable of fuzzification (error rate), (c) Output of defuzzification (Actuating signal) .

The fuzzy rule base gathers the information of a given problem and employ human control. The fuzzy rule base is developed using 7 x 7 matrix, where the central area of the method is for the de-fuzzification method of getting the crisp function from the fuzzy set.

4. Results and discussion

A three-phase UPQC with fuzzy logic controller is connected to supply voltage 400V, 50Hz and with a nonlinear load comprised of 3-phase bridge rectifier feeding an RL load. This configuration is modeled in MATLAB/Simulink and the performance of this UPQC is estimated in terms of mitigation of current and voltage harmonics, where the simulation circuit is shown in Figure 6 and the parameters circuit states as following:

Grid line voltage is 400V (peak), grid resistance is 0.001 Ω, grid impedance is 10 mH, grid frequency is 50 HZ, load inductance is 0.05 H, load resistance is 100 Ω, filter resistance is 10 mΩ, filter inductance is 3.6 mH, DC link voltage is 680 V and DC link capacitor is 750 μF.

IOP Conf. Series: Materials Science and Engineering **342** (2018) 012084 doi:10.1088/1757-899X/342/1/012084

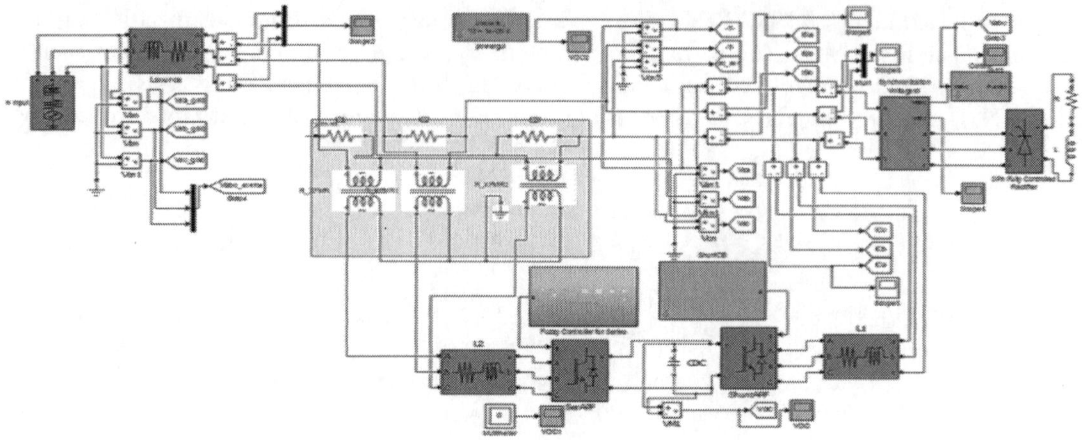

Figure 6. Matlab simulation circuit.

4.1. Case study one :System with distorted load current

A nonlinear load is a diode bridge rectifier connected to the series RL load to create the harmonics inside the system. UPQC is connected to operate for compensate the load current. FLC is employed to regulate the DC link voltage. UPQC steady state performance is examined and the simulation results are shown in the Figure 7.

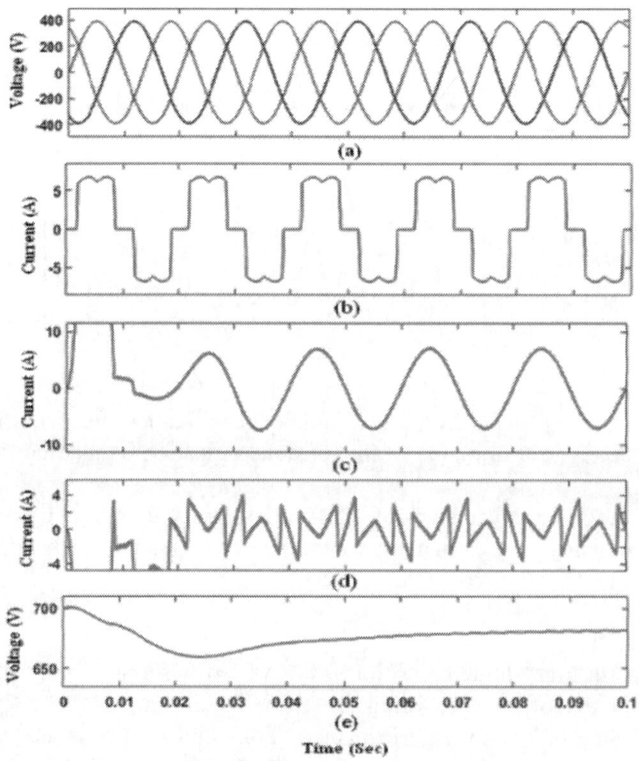

Figure 7. Simulation circuit (a) Source voltage without UPQC, (b) Load current without UPQC, (c) Load current with UPQC, (d) Filter current, and (e) DC link voltage.

IOP Conf. Series: Materials Science and Engineering **342** (2018) 012084 doi:10.1088/1757-899X/342/1/012084

It can be clearly seen that the source voltage waveform is pure sinusoidal and the load current waveform has some harmonic components as shown in figure 7(a) and 7(b) respectively. Figure 7(c) and (d) illustrate the operation UPQC has cancelled the effect of distorted load current by injecting the current is equal to harmonic current magnitude. The THD of load current is reduced from 36.37% to 2.10% as in Figure 8. Also UPQC is able to regulate the DC link voltage to its reference value (680v) as shown in Figure 7(e).

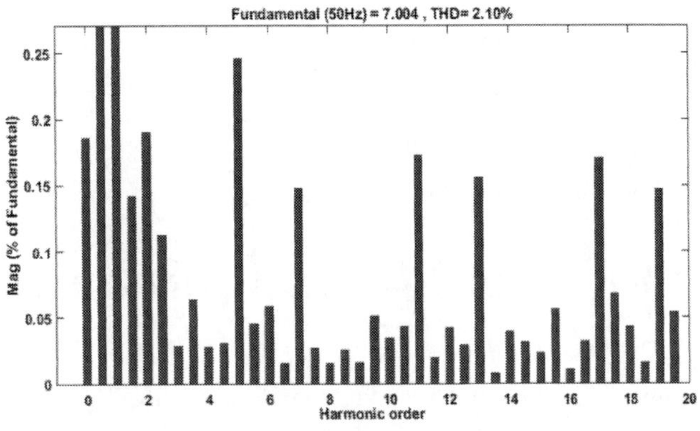

Figure 8. THD of load current after connecting UPQC filter.

4.2. Case study two: System with distorted source voltage and load current

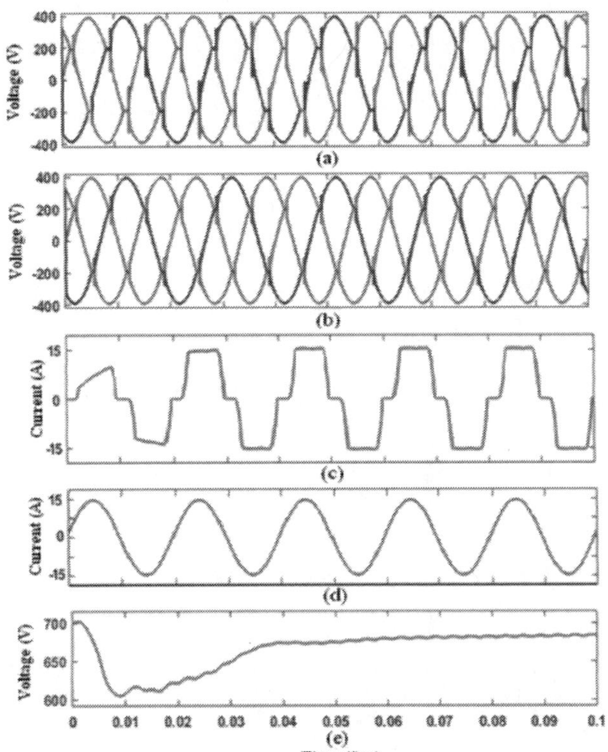

Figure 9. Simulation circuit (a) Source voltage without UPQC, (b) Source voltage without UPQC, (c) Load current without UPQC, (d) Load current with UPQC, and (e) DC link voltage.

IOP Conf. Series: Materials Science and Engineering **342** (2018) 012084 doi:10.1088/1757-899X/342/1/012084

Source voltages with 5th, 7th and 11th harmonics order are shown in Figure 9(a), as well the current waveform as displayed in Figure 9(c) which has high harmonics components. The compensated supply voltage and current waveform are depicted in Figure 9(b) and (d) when the UPQC becomes active. Also FLC has been maintained the DC link voltage to its reference value (680v) as in Figure 9(e).

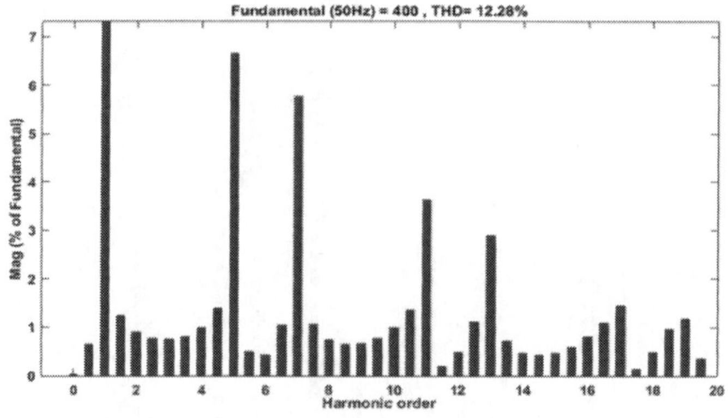

Figure 10.THD of source voltage without UPQC filter.

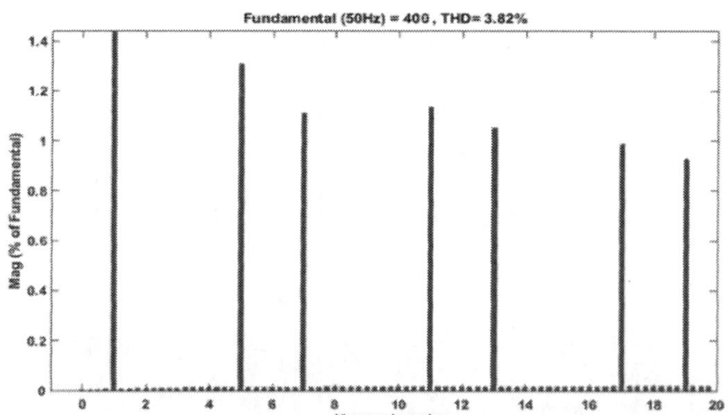

Figure 11 . THD of source voltage with UPQC filter.

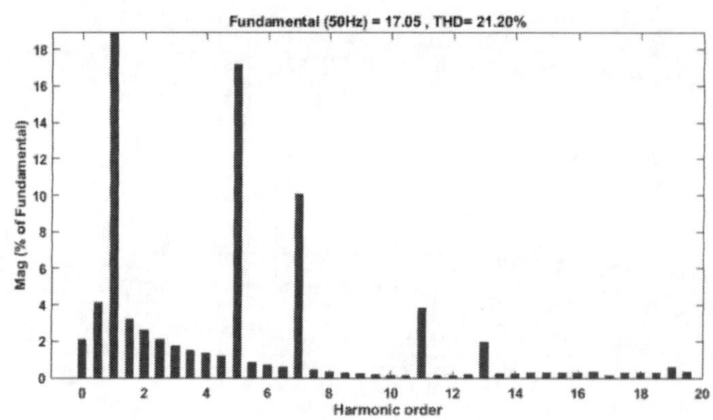

Figure 12 . THD of load current without UPQC filter.

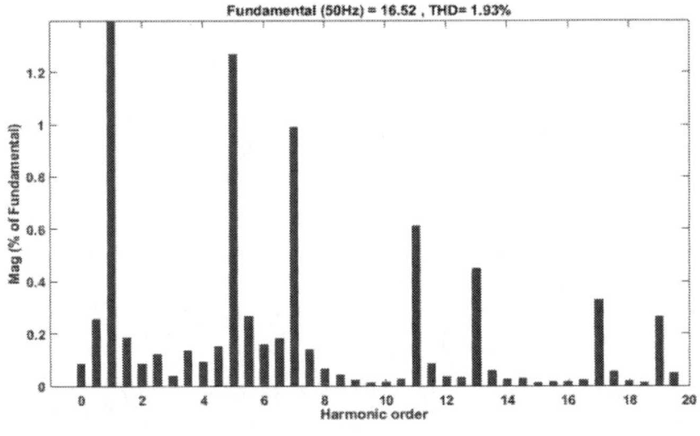

Figure 13 . THD of load current with UPQC filter.

The harmonics in the model system is analyzed in the source voltage and load current, where THD of the voltage before being connected to the UPQC was 12.28% as in Figure 10 and 3.82% after being connected to the UPQC with FLC, as shown in Figure 11.The THD of the load current before applying the UPQC was 21.20% as shown in Figure 12 and 1.93% after connection to the UPQC with FLC, as shown in Figure 13.

Table 1. THD values of load current and source voltage (with and without UPQC filter).

Harmonic order	Performance system					
	Case one		Case Two			
	THD of load current before filter	THD of load current with filter	THD of voltage source before filter	THD of voltage source with filter	THD of load current before filter	THD of load current with filter
3th	2.52	0.12	0.74	0.01	1.76	0.04
5th	11.07	2.02	6.64	1.31	17.18	1.27
7th	4.03	0.9	5.77	1.11	10.4	0.99
9th	1.20	0.01	0.66	0.01	0.21	0.02
11th	4.67	1.33	3.63	1.13	3.78	0.61
13th	2.58	1.12	2.88	1.05	1.95	0.45
15th	1.32	0.01	0.46	0.02	0.26	0.01
17th	3.55	1.11	1.64	0.99	0.11	0.33
19th	1.54	0.95	1.17	0.39	0.56	0.26
THD %	36.37	2.10	12.28	3.82	21.20	1.93

Table 1 depicts the harmonics values for each harmonic order before and after using UPQC with FLC for voltage and current waveforms. Based on the Table 3, the high harmonics value were the 5th and 7th for voltage, which were 6.64% and 5.77% respectively, while the values were 17.18% and 10.04% for current respectively. However, the 5th and 7th harmonics value have been reduced by UPQC to 1.31% and 1.11% in voltage waveform and to 1.27% and 0.99% in current waveform.

5. Conclusions

This paper presents the proposed scheme of UPQC for harmonics compensation of load current and source voltage in distribution systems using FLC which regulates the DC voltage. The performance of proposed UPQC for power quality improvement was extensively investigated under two simulation case studies and from the studies, it was noted that the proposed scheme adequately compensated the load current and source voltage harmonics to meet the regulations of IEEE Standard 519.

Acknowledgments

The authors thank the Ministry of Education Malaysia and Universiti Teknologi Malaysia (UTM) for their support under grant number Q.J130000.2623.14J48.

References

[1] Soomro, D.M. and Almelian, M.M., 2015. Optimal design of a single tuned passive filter to mitigate harmonics in power frequency.
[2] Soomro, D.M., Omran, M.A. and Alswed, S.K., 2006. Design of a Shunt Active Power Filter to Mitigate the Harmonics Caused by Nonlinear Loads.
[3] Otto, R.A., Putman, T.H. and Gyugyi, L., 1978. Principles and applications of static, thyristor-controlled shunt compensators. *IEEE Transactions on Power apparatus and Systems*, (5), pp.1935-1945.
[4] Gyugyi, L., 1976. Active ac power filters. In *IEEE/IAS Annual Meeting, 1976* (Vol. **529**).
[5] Gyugyi, L., 1992, July. Unified power-flow control concept for flexible AC transmission systems. In *IEE proceedings C (generation, transmission and distribution)* (Vol. **139**, No. 4, pp. 323-331). IET Digital Library.
[6] Chaithanyakumar, P. and Varaprasad, T., 2012. Design of Unified Power Quality Conditioner (UPQC) to Improve the Power Quality Problems by Using PQ Theory. *Power*, **2**(3).
[7] Aredes, M. and Fernandes, R.M., 2009, September. A unified power quality conditioner with voltage sag/swell compensation capability. In *Power Electronics Conference, 2009. COBEP'09. Brazilian* (pp. 218-224). IEEE.
[8] Han, B., Bae, B., Kim, H. and Baek, S., 2006. Combined operation of unified power-quality conditioner with distributed generation. *IEEE Transactions on Power Delivery*, **21**(1), pp.330-338.
[9] Fujita, H. and Akagi, H., 1998. The unified power quality conditioner: the integration of series- and shunt-active filters. *IEEE transactions on power electronics*, **13**(2), pp.315-322.
[10] Kesler, M. and Özdemir, E., 2009, November. Operation of shunt active power filter under unbalanced and distorted load conditions. In *Electrical and Electronics Engineering, 2009. ELECO 2009. International Conference on* (pp. I-92). IEEE.
[11] Shankar, S., Kumar, A. and Gao, W., 2011, July. Operation of unified power quality conditioner under different situations. In *Power and Energy Society General Meeting, 2011 IEEE* (pp. 1-10). IEEE.
[12] Monteiro, L.F., Aredes, M. and Neto, J.M., 2003, June. A control strategy for unified power quality conditioner. In *Industrial Electronics, 2003. ISIE'03. 2003 IEEE International Symposium on* (Vol. **1**, pp. 391-396). IEEE.
[13] Kumar, A., Bhat, A.H. and Singh, S.P., 2016, September. Performance evaluation of fuzzy logic controlled voltage source inverter based unified power quality conditioner for mitigation of voltage and current harmonics. In *Advances in Computing, Communications and Informatics (ICACCI), 2016 International Conference on* (pp. 1799-1804). IEEE.
[14] Elango, M.K. and Tamilarasi, T., 2016, March. Improvement of power quality using a hybrid UPQC with distributed generator. In *Circuit, Power and Computing Technologies (ICCPCT), 2016 International Conference on* (pp. 1-8). IEEE.

IOP Conf. Series: Materials Science and Engineering **342** (2018) 012085 doi:10.1088/1757-899X/342/1/012085

Unsteady free convection flow past a semi-infinite vertical plate with constant heat flux in water based nanofluids

Marneni Narahari

Fundamental and Applied Sciences Department, Universiti Teknologi PETRONAS, 32610 Seri Iskandar, Malaysia

E-mail: marneni@utp.edu.my

Abstract. The unsteady free convective flow of nanofluids past a semi-infinite vertical plate with uniform heat flux has been investigated numerically. An implicit finite difference technique of Crank-Nicolson scheme has been employed to solve the governing partial differential equations. Five different types of water based nanofluids containing Cu, Ag, Al_2O_3, CuO and TiO_2 nanoparticles are considered to study the fluid flow characteristics with various time and solid volume fraction parameters. It is found that the local as well as the average Nusselt number for nanofluids is higher than the pure fluid (water). The local skin-friction is higher for pure fluid as compared to the nanofluids. The present numerical results obtained for local Nusselt number are validated with the previously published correlation results for a limiting case and it is found that the results are in good agreement.

1. Introduction

Nanofluids have received great attention lately due to its practical applications in many engineering and industrial processes. In 1995, Choi pioneered the concept of suspending ultra-fine particles in a base fluid and name it as nanofluid [1]. The promising outcome was obtained where the nanofluids exhibit higher thermal conductivity properties than pure fluid such as water, oil and ethylene glycol. According to Choi et al. [2], the thermal conductivity of nanofluid that consists of nanoparticles with total volume fraction less than 1% is approximately 2 times greater than pure fluid. Due to its enhanced thermal conductivity properties, nanofluids have wide applications especially in convective heat transfer which plays an important role in engineering processes such as the cooling system of electronic components, advanced nuclear system, solar collector, thermal insulation systems, food processes etc.

Convective heat transfer using nanofluid is an area that receives most scientific attention in the heat transfer research community. Most of the studies on convective heat transfer considered the steady state flow. Khan and Aziz [3] studied the steady free convective flow past a vertical plate with uniform surface heat flux in nanofluids. They highlighted that the reduced Nusselt number increased with the increase of Prandtl number, but decreased with the increase of Brownian motion, thermophoresis and the buoyancy ratio parameters. Chamkha and Rashad [4] investigated the steady state free convection flow past a permeable vertical cone embedded in a nanofluid filled porous medium under the uniform lateral heat and mass fluxes. It was concluded that an increase in the Lewis number increased both local Nusselt and Sherwood numbers. Tham et al. [5] investigated the steady mixed convection flow over a horizontal circular cylinder with constant heat flux in a porous medium filled by a nanofluid. They observed that the Brownian motion and buoyancy ratio parameters affected the fluid flow and heat transfer profiles.

Noghrehabadi et al. [6] numerically analyzed the natural convection flow of nanofluid over a vertical plate embedded in a Darcy porous medium subjected to surface heat and nanoparticle fluxes. They found that both reduced Nusselt and Sherwood numbers increased with the increase of Lewis number. Sheikholeslami et al. [7] presented the free convection flow of nanofluid in an enclosure with magnetic field. The control volume based finite element method was used to solve the governing equations and they found that the increase of Hartmann number reduced the rate of heat transfer while the augmentation of aspect ratio, nanoparticle volume fraction and Rayleigh number enhanced the rate of heat transfer. Jalilpour et al. [8] performed an investigation on the MHD stagnation-point nanofluid flow over a heated porous stretching surface with suction or blowing conditions and surface heat flux. The governing equations were solved through homotopy perturbation method (HPM). They concluded that an increase in the magnetic and thermophoresis parameters as well as Lewis number decreased the magnitude of the reduced Nusselt number. Reddy [9] performed an analysis of the natural convection flow of nanofluids over a vertical plate with uniform heat flux in the presence of thermal radiation. Similarity solutions were obtained using the fourth order Runge-Kutta method together with shooting technique and concluded that an increase in Prandtl number and radiation parameter increased the reduced Nusselt number. The natural convection flow of nanofluid past an inclined wavy surface with heat flux in a porous medium was analyzed by Srinivasacharya and Kumar [10] using the successive linearization method. They concluded that the local nanoparticle mass transfer coefficient increased and the local heat transfer coefficient decreased by increasing thermophoresis parameter. Trimbitas et al. [11] carried out an investigation on the steady mixed convection flow past a vertical semi-infinite flat plate in nanofluids. The governing equations were transformed to a system of coupled nonlinear ordinary differential equations using an appropriate similarity transformation and then solved numerically. They concluded that adding the nanoparticles into the based fluid (water) enhanced the skin-friction and heat transfer coefficient. All the above mentioned studies are confined to only convective heat transfer of steady state nanofluid flows.

The main purpose of the present paper is to investigate the unsteady free convective flow of water based nanofluid past a semi-infinite vertical plate with constant heat flux by using implicit finite difference technique of Crank-Nicolson scheme. This study is motivated by the need to explore the thermal performance of nanofluids in an unsteady state flow. A total of five different types of nanofluids, namely, Cu-water, Ag-water, CuO-water, TiO_2-water and Al_2O_3-water are taken into consideration. The influences of time and solid volume fraction on the fluid flow and heat transfer characteristics are discussed through graphs.

2. Mathematical analysis

Consider the unsteady two-dimensional free convective flow and heat transfer of a nanofluid past a semi-infinite vertical plate. The x' - axis is taken along the plate in the upward direction and the y' - axis is taken perpendicular to the plate into the fluid. Initially, the plate and the fluid are at the same temperature T'_∞. At time $t' > 0$, the surface of the plate is subject to a constant heat flux q'_w. The governing boundary layer equations of mass, momentum and energy for the free convective flow of nanofluid under the Boussinesq's approximation are as follows:

$$\frac{\partial u'}{\partial x'} + \frac{\partial v'}{\partial y'} = 0 \tag{1}$$

$$\rho_{nf}\left(\frac{\partial u'}{\partial t'} + u'\frac{\partial u'}{\partial x'} + v'\frac{\partial u'}{\partial y'}\right) = \mu_{nf}\frac{\partial^2 u'}{\partial y'^2} + g(\rho\beta)_{nf}(T' - T'_\infty) \tag{2}$$

$$(\rho C_P)_{nf}\left(\frac{\partial T'}{\partial t'} + u'\frac{\partial T'}{\partial x'} + v'\frac{\partial T'}{\partial y'}\right) = k_{nf}\frac{\partial^2 T'}{\partial y'^2} \tag{3}$$

The appropriate initial and boundary conditions are:

$$\left.\begin{array}{ll} t' \le 0: & u'=0,\ v'=0,\ T'=T'_\infty \quad \text{for all} \quad x',y' \\ t'>0: & \begin{cases} u'=0, v'=0, \dfrac{\partial T'}{\partial y'} = -\dfrac{q'_w}{k_{nf}} & \text{at} \quad y'=0 \\ u'=0, v'=0,\ T'=T'_\infty & \text{at} \quad x'=0 \\ u'\to 0, v'\to 0,\ T'\to T'_\infty & \text{as} \quad y'\to\infty \end{cases} \end{array}\right\} \tag{4}$$

Where u' and v' are the velocity components in the x' and y' directions, respectively, T' is temperature of the fluid near the plate, T'_∞ is the temperature of the fluid far away from the plate, t' is the time, g is the acceleration due to gravity. μ_{nf} is the dynamic viscosity of the nanofluid, β_{nf} is the thermal expansion coefficient of the nanofluid, ρ_{nf} is the density of the nanofluid, k_{nf} is the thermal conductivity of the nanofluid, q'_w is the constant heat flux and $(C_p)_{nf}$ is the specific heat of the nanofluid which are given by:

$$\left.\begin{array}{l} \rho_{nf} = (1-\phi)\rho_f + \phi\rho_s \\ \mu_{nf} = \dfrac{\mu_f}{(1-\phi)^{2.5}} \\ (\rho\beta)_{nf} = (1-\phi)(\rho\beta)_f + \phi(\rho\beta)_s \\ (\rho C_P)_{nf} = (1-\phi)(\rho C_P)_f + \phi(\rho C_P)_s \\ k_{nf} = k_f\left[\dfrac{k_s + 2k_f - 2\phi(k_f - k_s)}{k_s + 2k_f + \phi(k_f - k_s)}\right] \end{array}\right\} \tag{5}$$

where ϕ is the solid volume fraction of the nanoparticles, ρ_f is the density of the base fluid, ρ_s is the density of the nanoparticles, μ_f is the viscosity of the base fluid, $(C_P)_f$ is the specific heat of the base fluid and $(C_P)_s$ is the specific heat of the nanoparticles. k_f is the thermal conductivity of the base fluid and k_s is the thermal conductivity of the nanoparticles.

The local as well as the average skin-friction and Nusselt number can be computed as follows:

$$\tau_{x'} = \mu_{nf}\left(\frac{\partial u'}{\partial y'}\right)_{y'=0} \tag{6}$$

$$Nu_{x'} = \frac{q'_w x'}{k_f (T' - T'_\infty)_{y'=0}}, \quad \text{where } q'_w = -k_{nf}\left(\frac{\partial T'}{\partial y'}\right)_{y'=0} \tag{7}$$

$$\tau_L = \frac{1}{L}\int_0^L \mu_{nf}\left(\frac{\partial u'}{\partial y'}\right)_{y'=0} dx' \tag{8}$$

$$Nu_L = \int_0^L \frac{q'_w}{k_f (T' - T'_\infty)_{y'=0}} dx' \tag{9}$$

Where $\tau_{x'}$ is the local skin-friction, $Nu_{x'}$ is the local Nusselt number, τ_L is the average skin-friction and Nu_L is the average Nusselt number.

On introducing the following non-dimensional quantities:

$$\left.\begin{aligned} & x=\frac{x'}{L},\ y=\frac{y'}{L}Gr^{1/4},\ t=\frac{t'\nu_f}{L^2}Gr^{1/2},\ u=\frac{u'L}{\nu_f}Gr^{-1/2},\ v=\frac{v'L}{\nu_f}Gr^{-1/4}, \\ & \theta=\frac{T'-T'_\infty}{T^*},\ Gr=\frac{g\beta_f L^3 T^*}{\nu_f^{\ 2}},\ \Pr=\frac{\nu_f(\rho C_p)_f}{k_f}. \end{aligned}\right\} \tag{10}$$

Where $T^* = \dfrac{q'_w L}{k_f} Gr^{-1/4}$, x and y are the dimensionless coordinate axes, t is the dimensionless time, θ is the non-dimensional temperature, u and v are the non-dimensional velocity components in x and y directions, respectively, Gr and $\Pr$ are the Grashof and Prandtl numbers, respectively.

The non-dimensional governing equations are:

$$\frac{\partial u}{\partial x}+\frac{\partial v}{\partial y}=0 \tag{11}$$

$$\frac{\partial u}{\partial t}+u\frac{\partial u}{\partial x}+v\frac{\partial u}{\partial y}=a_1\frac{\partial^2 u}{\partial y^2}+a_2\theta \tag{12}$$

$$\left(\frac{\partial \theta}{\partial t}+u\frac{\partial \theta}{\partial x}+v\frac{\partial \theta}{\partial y}\right)=\frac{a_4}{\Pr a_3}\frac{\partial^2 \theta}{\partial y^2} \tag{13}$$

The initial and boundary conditions are:

$$\left.\begin{aligned} & t\le 0: \quad u=0,\ v=0,\ \theta=0 \quad \text{for all} \quad x, y\ge 0 \\ & t>0: \begin{cases} u=0, v=0,\ \dfrac{k_{nf}}{k_f}\dfrac{\partial\theta}{\partial y}=-1 & \text{at} \quad y=0 \\ u=0, v=0,\ \theta=0 & \text{at} \quad x=0 \\ u\to 0, v\to 0,\ \theta\to 0 & \text{as} \quad y\to\infty \end{cases} \end{aligned}\right\} \tag{14}$$

Where,

$$a_1=\frac{1}{(1-\phi)^{2.5}}\left[1/\{1-\phi+\phi(\rho_s/\rho_f)\}\right],\ a_2=\frac{1-\phi+\phi[(\rho\beta)_s/(\rho\beta)_f]}{1-\phi+\phi[\rho_s/\rho_f]},$$

$$a_3=1-\phi+\phi[(\rho C_P)_s/(\rho C_P)_f],\ a_4=\frac{k_{nf}}{k_f}=\left[\frac{k_s+2k_f-2\phi(k_f-k_s)}{k_s+2k_f+\phi(k_f-k_s)}\right].$$

Using the non-dimensional quantities specified in (10), the local as well as the average skin-friction and Nusselt number are as follows:

$$\tau_x/Gr^{3/4}=\frac{\tau_{x'}L^2}{\rho_f\nu_f^{\ 2}}=\frac{1}{(1-\phi)^{2.5}}\left(\frac{\partial u}{\partial y}\right)_{y=0} \tag{15}$$

$$Nu_x/Gr^{1/4}=-a_4 x\left[\left(\frac{\partial\theta}{\partial y}\right)_{y=0}\Big/(\theta)_{y=0}\right] \tag{16}$$

$$\tau/Gr^{3/4}=\frac{\tau_L L^2}{\rho_f\nu_f^{\ 2}}=\frac{1}{(1-\phi)^{2.5}}\int_0^1\left(\frac{\partial u}{\partial y}\right)_{y=0}dx \tag{17}$$

$$Nu/Gr^{1/4} = -a_4 \int_0^1 \left[\left(\frac{\partial \theta}{\partial y} \right)_{y=0} \Big/ (\theta)_{y=0} \right] dx \tag{18}$$

3. Numerical technique

In order to solve the unsteady, non-linear coupled partial differential equations (11) to (13) under the conditions (14), an implicit finite difference scheme of Crank-Nicolson type has been employed. The finite difference equations corresponding to equations (11) to (13) are as follows:

$$\frac{u_{i,j-1}^{k+1} - u_{i-1,j-1}^{k+1} + u_{i,j-1}^{k} - u_{i-1,j-1}^{k} + u_{i,j}^{k+1} - u_{i-1,j}^{k+1} + u_{i,j}^{k} - u_{i-1,j}^{k}}{4\Delta x} + \frac{v_{i,j}^{k+1} - v_{i,j-1}^{k+1} + v_{i,j}^{k} - v_{i,j-1}^{k}}{2\Delta y} = 0 \tag{19}$$

$$\begin{aligned} &\frac{u_{i,j}^{k+1} - u_{i,j}^{k}}{\Delta t} + u_{i,j}^{k} \frac{\left[u_{i,j}^{k+1} - u_{i-1,j}^{k+1} + u_{i,j}^{k} - u_{i-1,j}^{k}\right]}{2\Delta x} + v_{i,j}^{k} \frac{\left[u_{i,j+1}^{k+1} - u_{i,j-1}^{k+1} + u_{i,j+1}^{k} - u_{i,j-1}^{k}\right]}{4\Delta y} \\ &= a_1 \frac{\left[u_{i,j-1}^{k+1} - 2u_{i,j}^{k+1} + u_{i,j+1}^{k+1} + u_{i,j-1}^{k} - 2u_{i,j}^{k} + u_{i,j+1}^{k}\right]}{2(\Delta y)^2} + a_2 \frac{\left[\theta_{i,j}^{k+1} + \theta_{i,j}^{k}\right]}{2} \end{aligned} \tag{20}$$

$$\begin{aligned} &\frac{\theta_{i,j}^{k+1} - \theta_{i,j}^{k}}{\Delta t} + u_{i,j}^{k} \frac{\left[\theta_{i,j}^{k+1} - \theta_{i-1,j}^{k+1} + \theta_{i,j}^{k} - \theta_{i-1,j}^{k}\right]}{2\Delta x} + v_{i,j}^{k} \frac{\left[\theta_{i,j+1}^{k+1} - \theta_{i,j-1}^{k+1} + \theta_{i,j+1}^{k} - \theta_{i,j-1}^{k}\right]}{4\Delta y} \\ &= \frac{a_4}{\Pr a_3} \frac{\left[\theta_{i,j-1}^{k+1} - 2\theta_{i,j}^{k+1} + \theta_{i,j+1}^{k+1} + \theta_{i,j-1}^{k} - 2\theta_{i,j}^{k} + \theta_{i,j+1}^{k}\right]}{2(\Delta y)^2} \end{aligned} \tag{21}$$

The finite-difference form of the thermal boundary condition at $y=0$ can be written as

$$a_4 \left[\frac{\theta_{i,1}^{k+1} - \theta_{i,-1}^{k+1} + \theta_{i,1}^{k} - \theta_{i,-1}^{k}}{4\Delta y} \right] = -1 \tag{22}$$

At $y=0$ (i.e., $j=0$), the finite-difference equation (21) for energy equation becomes

$$\begin{aligned} &\frac{\theta_{i,0}^{k+1} - \theta_{i,0}^{k}}{\Delta t} + u_{i,0}^{k} \frac{\left[\theta_{i,0}^{k+1} - \theta_{i-1,0}^{k+1} + \theta_{i,0}^{k} - \theta_{i-1,0}^{k}\right]}{2\Delta x} \\ &= \frac{a_4}{\Pr a_3} \frac{\left[\theta_{i,-1}^{k+1} - 2\theta_{i,0}^{k+1} + \theta_{i,1}^{k+1} + \theta_{i,-1}^{k} - 2\theta_{i,0}^{k} + \theta_{i,1}^{k}\right]}{2(\Delta y)^2} \end{aligned} \tag{23}$$

After eliminating $\theta_{i,-1}^{k+1} + \theta_{i,-1}^{k}$ from Eq. (23) using Eq. (22), the finite-difference equation for the energy equation (13) at the boundary reduces to the form

$$\begin{aligned} &\frac{\theta_{i,0}^{k+1} - \theta_{i,0}^{k}}{\Delta t} + u_{i,0}^{k} \frac{\left[\theta_{i,0}^{k+1} - \theta_{i-1,0}^{k+1} + \theta_{i,0}^{k} - \theta_{i-1,0}^{k}\right]}{2\Delta x} \\ &= \frac{a_4}{\Pr a_3} \frac{\left[-\theta_{i,0}^{k+1} + \theta_{i,1}^{k+1} - \theta_{i,0}^{k} + \theta_{i,1}^{k} + 2(\Delta y/a_4)\right]}{(\Delta y)^2} \end{aligned} \tag{24}$$

To solve the above equations, the region of integration is considered as a rectangle with sides $x_{max}(=1)$ and $y_{max}(=7)$, where y_{max} corresponds to $y \to \infty$ which lies very well outside the momentum and energy boundary layers. The maximum value of *y* is chosen as 7 after some preliminary investigations so that the last two of the boundary conditions in (14) are satisfied. Here, the subscript *i*-designates the grid point along the *x*-direction, *j* along the *y*-direction and the superscript *k* along the *t*-direction. An appropriate mesh size considered for the calculation is $\Delta x = 0.01$, $\Delta y = 0.05$, and time step $\Delta t = 0.01$.

During any one time step, coefficients $u_{i,j}^n$ and $v_{i,j}^n$ appearing in the difference equations are treated as constants. The values of u, v and θ are known at all grid points at $t = 0$ from the initial conditions. The computations of u, v and θ at time level $(k+1)$ using the known values at previous time level (k) are calculated as follows: The finite difference equation (21) at every internal nodal point on a particular i-level constitute a tri-diagonal system of equations. Such a system of equations is solved by Thomas algorithm as described in Carnahan et al. [12]. The values of θ are calculated from equation (21). Using the values of θ at $(k+1)^{th}$ time level in the equation (20), the values of u at $(k+1)^{th}$ time level are found in a similar manner. Thus the values of θ and u are known on a particular i-level. The values of v are calculated explicitly using the equation (19) at every nodal point on a particular i-level at $(k+1)^{th}$ time level. This process is repeated for various i-levels. Thus the values of θ, u and v are known at all grid points in the rectangular region at $(k+1)^{th}$ time level.

4. Results and discussion

The unsteady free convection flow past a semi-infinite vertical plate with constant heat flux in water based nanofluids such as Cu-water, Ag-water, CuO-water, TiO_2-water and Al_2O_3-water are investigated numerically by employing an implicit finite difference technique of Crank-Nicolson scheme. The thermo-physical properties of water and nanoparticles are shown in Table 1. The numerical results are illustrated graphically and discussed thoroughly in order to get an in depth study on the relationship between various factors. Table 2 illustrates the comparison of the current numerical results of steady state local Nusselt number with pure fluid correlation results of Chen et al. [13] when $\phi = 0$. It is observed that the comparison outcome is in good agreement and this justifies the accuracy of the present numerical technique.

Table 1. Thermo-physical properties of water and nanoparticles.

	ρ (Kg/m^3)	C_p (J/Kg.K)	k (W/m.K)	$\beta \times 10^5$ (K^{-1})
Water	997.1	4179	0.613	21
Ag	10500	235	429	1.89
Cu	8933	385	401	1.67
Al_2O_3	3970	765	40	0.85
CuO	6320	531.8	76.5	1.80
TiO_2	4250	686.2	8.9538	0.9

Table 2. Comparison of the local Nusselt numbers when $\phi = 0$ with the correlation results of Chen et al. [13] for regular fluids.

Pr	Local Nusselt Number	
	Chen et al. [13]	Present
1	0.5341	0.5344
7	0.8709	0.8704
10	0.9453	0.9447
15	1.0356	1.0348
20	1.1037	1.1027
50	1.3452	1.3437
100	1.5567	1.5547
500	2.1692	2.1655
1000	2.4976	2.4929

The effect of time on both velocity and temperature profiles of Al_2O_3-water nanofluid are depicted in Figures 1 and 2. It is noticed that both velocity and temperature profiles increase prominently with the increase of time until the fluid achieves steady state conditions at $t = 9.5$ when $\Pr = 6.2$ and $\phi = 0.04$. The prolonged time enables more heat to transfer from the plate to the fluid, causing an increase in the fluid temperature and buoyancy effect, and hence, the fluid velocity increases as well. Figure 1 shows that the velocity increases to a maximum value near the vertical plate, then it decays as the distance from the vertical plate increases till it satisfies the free stream condition. The maximum velocity increases significantly and formed further away from the plate with the increase of time. Figure 2 demonstrates that the temperature declines gradually to a free stream temperature as the distance from the vertical plate increases.

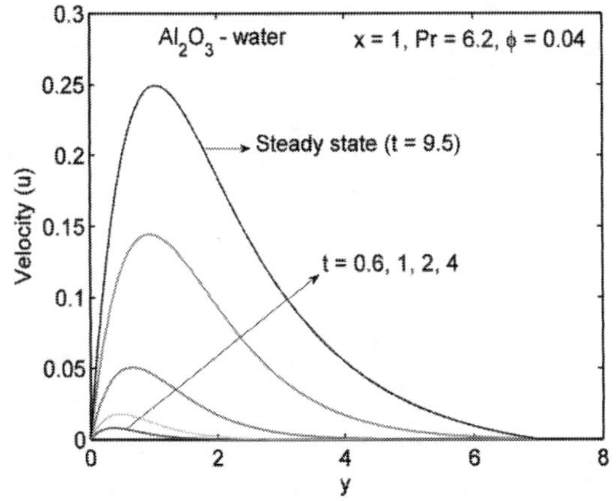

Figure 1. Velocity profiles at different t for Al_2O_3-water nanofluid.

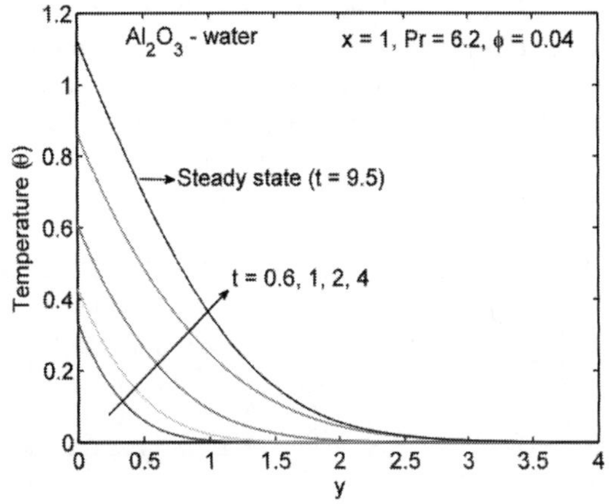

Figure 2. Temperature profiles at different t for Al_2O_3-water nanofluid.

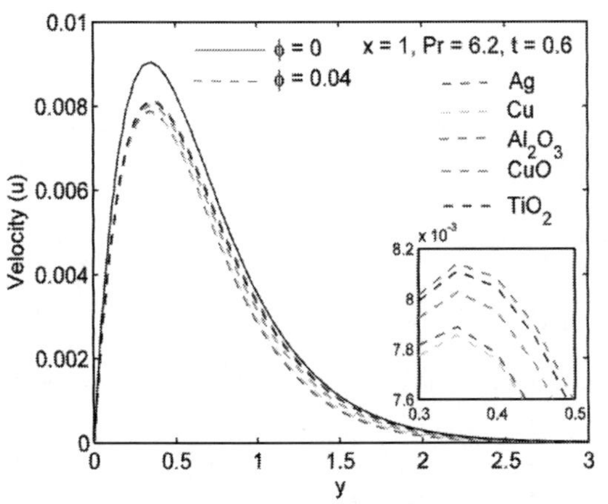

Figure 3. Velocity profiles of nanofluids at different ϕ.

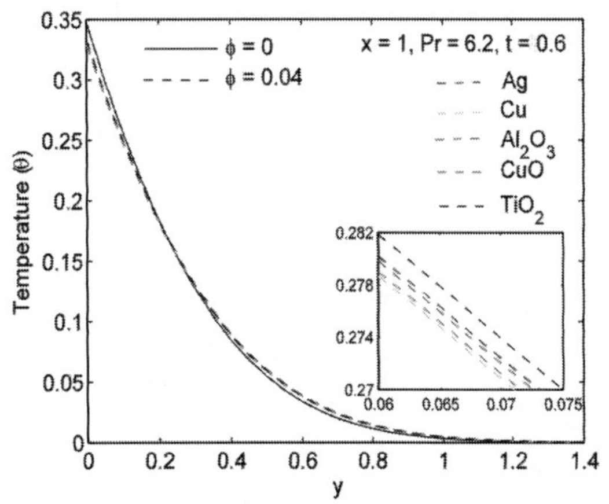

Figure 4. Temperature profiles of nanofluids at different ϕ.

The velocity and temperature profiles of pure fluid and nanofluids with the volume fraction of 0.04 are shown in Figures 3 and 4, respectively. The velocity rises rapidly to a peak value near the vertical plate, then it drops as the distance from the vertical plate increases until the free stream boundary condition is satisfied. The maximum velocity of the pure fluid and the nanofluids are formed in the vicinity of the plate and at the same distance from the vertical plate. The pure fluid exhibits higher velocity profile compared to nanofluids while Al_2O_3-water has the highest velocity profile among the nanofluids and Cu-water has the lowest. It is seen from Figure 4 that the temperature decays gradually to a free stream temperature as the distance from the vertical plate increases. The pure fluid has higher temperature

IOP Conf. Series: Materials Science and Engineering **342** (2018) 012085 doi:10.1088/1757-899X/342/1/012085

profile than nanofluids in close proximity to the vertical plate but the temperature profile of pure fluid is lower than that of nanofluids after a certain distance from the plate. Figure 4 also shows that nanofluids exhibit minor variations in temperature profiles. A further scrutiny reveals that TiO_2-water has the highest temperature profile while Cu-water has the lowest among the nanofluids. However, the trend changes after a certain distance from the vertical plate where Ag-water has the highest temperature profile whilst TiO_2-water has the lowest.

The local skin-friction and local Nusselt number variations for various nanofluids with the volume fraction of 0.04 and pure fluid are depicted in Figures 5 and 6, respectively. It is observed that the local skin-friction escalates drastically and reaches steady state near the leading edge. A prominent variation in the local skin-friction can be seen once the fluids achieve steady state along the vertical plate. Compared to nanofluids, the pure fluid has higher local skin-friction while Ag-water possesses slightly maximum and Cu-water possesses the minimum local skin-friction among the nanofluids. It is seen from Figure 6 that the local Nusselt number increases linearly with the distance along the vertical plate. Nanofluids have the higher local Nusselt number than pure fluid due to the addition of nanoparticles that enhances the fluid thermal conductivity. Meanwhile, Cu-water possesses the maximum and TiO_2-water possesses the minimum local Nusselt number among the nanofluids.

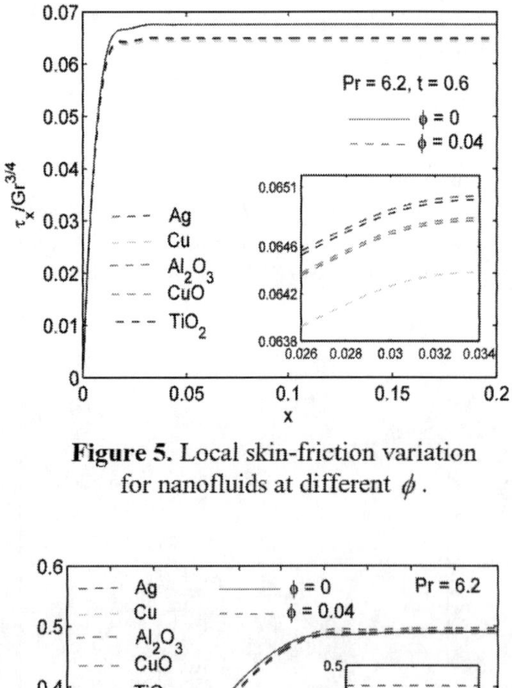

Figure 5. Local skin-friction variation for nanofluids at different ϕ.

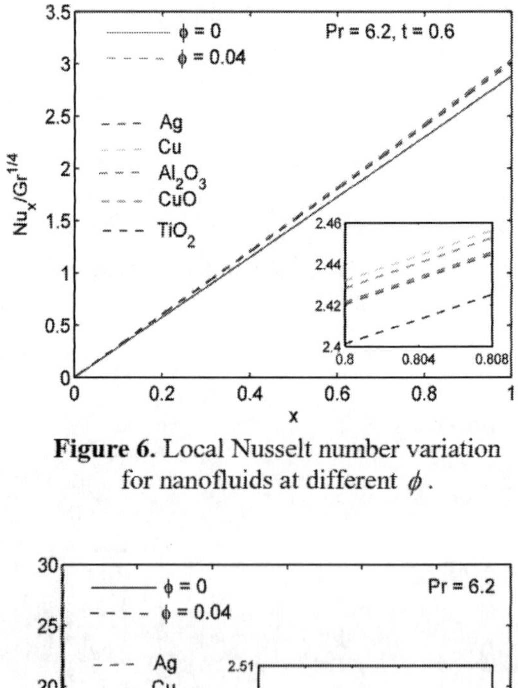

Figure 6. Local Nusselt number variation for nanofluids at different ϕ.

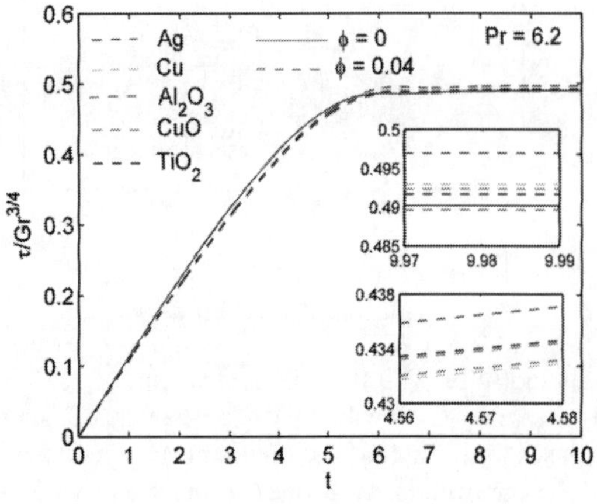

Figure 7. Average skin-friction variation for nanofluids at different ϕ.

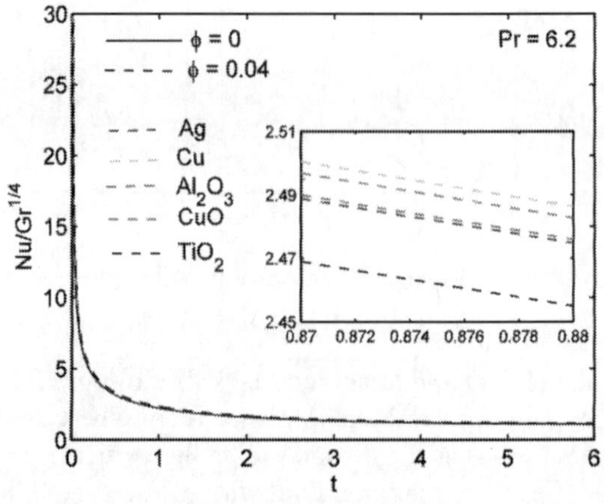

Figure 8. Average Nusselt number variation for nanofluids at different ϕ.

The average skin-friction and Nusselt number variations with time for nanofluids and pure fluid are illustrated in Figures 7 and 8, respectively. The average skin-friction increases as time progress until it reaches the steady state. Pure fluid exhibits higher average skin-friction compared to nanofluids during the transient state, thereafter the average skin-friction for pure fluid is lower than nanofluids instead. Among the nanofluids, Ag-water causes the most average skin-friction while Al_2O_3-water causes the least in the steady state. From Figure 8, it is observed that the average Nusselt number experiences a sudden drop at small values of time till it reaches steady state as time progresses. However, the nanofluids with volume fraction of 0.04 and pure fluid exhibit meagre variations of average Nusselt numbers. A close observation on the graph reveals that pure fluid has lower average Nusselt number compared to nanofluids while Cu-water has the maximum and TiO_2-water has the minimum average Nusselt number among the nanofluids.

5. Conclusions

In this paper, the unsteady free convection flow past a semi-infinite vertical plate with constant heat flux in water based nanofluids has been investigated numerically using an implicit finite difference technique of Crank-Nicolson type. The accuracy of the present numerical scheme is verified by comparing the local Nusselt number with the well-known pure fluid correlation results and the outcome is excellent. The flow characteristics with different time and solid volume fraction are discussed and the important conclusions are listed below:

1. An increase in time leads to increase in temperature and velocity profiles of nanofluid until they reach steady state profiles.
2. Al_2O_3-water nanofluid has the highest velocity profile while Cu-water nanofluid has the lowest among the nanofluids chosen.
3. TiO_2-water and Cu-water have the highest and lowest temperature profiles, respectively, near to the vertical plate. However, Ag-water has the highest temperature profile whilst TiO_2-water has the lowest after a certain distance from the plate.
4. The pure fluid causes higher local and average skin-frictions compared to the nanofluids.
5. The pure fluid has lower local and average Nusselt numbers compared to the nanofluids.
6. Ag-water nanofluid has the maximum and Cu-water nanofluid has the minimum local skin-friction.
7. Cu-water nanofluid has the maximum and TiO_2-water nanofluid has the minimum local Nusselt number.
8. Ag-water nanofluid has the maximum and Al_2O_3-water nanofluid has the minimum average skin-friction.
9. Cu-water nanofluid has the maximum and TiO_2-water nanofluid has the minimum average Nusselt number.

Acknowledgement

The author would like to thank Universiti Teknologi PETRONAS for the financial assistance provided by the research grant YUTP 0153AA-E28.

References

[1] Choi S U S and Eastman J 1995 Enhancing thermal conductivity of fluids with nanoparticles, The Proceedings of the 1995 ASME International Mechanical Engineering Congress and Exposition, San Francisco, USA, ASME, FED 231/MD66. 99-105.

[2] Choi S, Zhang Z, Yu W, Lockwood F and Grulke E 2001 Anomalously thermal conductivity enhancement in nanotube suspensions *Appl. Phys. Lett.* **79** 2252

[3] Khan W A and Aziz A 2011 Natural convection flow of a nanofluid over a vertical plate with uniform surface heat flux *Int. J. Therm. Sci.* **50** 1207

[4] Chamkha A J and Rashad A M 2012 Natural convection from a vertical permeable cone in a nanofluid saturated porous media for uniform heat and nanoparticles volume fraction fluxes *Int. J. Numer. Methods Heat Fluid Flow* **22** 1073

IOP Conf. Series: Materials Science and Engineering **342** (2018) 012085 doi:10.1088/1757-899X/342/1/012085

[5] Tham L, Nazar R and Pop I 2016 Mixed convection flow over a horizontal circular cylinder with constant heat flux embedded in a porous medium filled by a nanofluid: Buongiorno-Darcy model *Heat Mass Transf.* **52** 1983

[6] Noghrehabadi A, Behrseresht A and Ghalambaz M 2013 Natural convection of nanofluid over vertical plate embedded in porous medium: Prescribed surface heat flux *Appl. Math. Mech. – Engl. Ed.* **34** 669

[7] Sheikholeslami M, Gorji-Bandpy M, Ganji D D and Soleimani S 2014 Heat flux boundary condition for nanofluid filled enclosure in presence of magnetic field *J. Mol. Liq.* **193** 174

[8] Jalilpour B, Jafarmadar S, Ganji D D, Shotorban A B and Taghavifar H 2014 Heat generation/absorption on MHD stagnation flow of nanofluid towards a porous stretching sheet with prescribed surface heat flux *J. Mol. Liq.* **195** 194

[9] Reddy M G 2014 Influence of thermal radiation on natural convection boundary layer flow of a nanofluid past a vertical plate with uniform heat flux *International Journal of Heat and Technology* **32** 1

[10] Srinivasacharya D and Kumar P V 2015 Free convection of a nanofluid over an inclined wavy surface embedded in a porous medium with wall heat flux *Procedia Engineering* **127** 40

[11] Trimbitas R, Grosan T and Pop I 2015 Mixed convection boundary layer flow past vertical flat plate in nanofluid: case of prescribed wall heat flux *Appl. Math. Mech. -Engl. Ed.* **36** 1091

[12] Carnahan B, Luther H A and Wilkes J O 1969 Applied Numerical Methods, Jhon Wiley & Sons, New York

[13] Chen T S, Tien H C and Armaly B F 1986 Natural convection on horizontal, inclined, and vertical plates with variable surface temperature or heat flux *Int. J. Heat Mass Transf.* **19** 1465

Extreme Learning Machine and Particle Swarm Optimization in optimizing CNC turning operation

Tiagrajah V Janahiraman, Nooraziah Ahmad and Farah Hani Nordin

Centre for Signal Processing and Control Systems, Dept. of Electronics and Communication Engineering, College of Engineering, Universiti Tenaga Nasional, Jalan IKRAM-UNITEN, 43000 Kajang, Selangor, Malaysia.

E-mail: tiagrajah@uniten.edu.my

Abstract. The CNC machine is controlled by manipulating cutting parameters that could directly influence the process performance. Many optimization methods has been applied to obtain the optimal cutting parameters for the desired performance function. Nonetheless, the industry still uses the traditional technique to obtain those values. Lack of knowledge on optimization techniques is the main reason for this issue to be prolonged. Therefore, the simple yet easy to implement, Optimal Cutting Parameters Selection System is introduced to help the manufacturer to easily understand and determine the best optimal parameters for their turning operation. This new system consists of two stages which are modelling and optimization. In modelling of input-output and in-process parameters, the hybrid of Extreme Learning Machine and Particle Swarm Optimization is applied. This modelling technique tend to converge faster than other artificial intelligent technique and give accurate result. For the optimization stage, again the Particle Swarm Optimization is used to get the optimal cutting parameters based on the performance function preferred by the manufacturer. Overall, the system can reduce the gap between academic world and the industry by introducing a simple yet easy to implement optimization technique. This novel optimization technique can give accurate result besides being the fastest technique.

1. Introduction

Currently, the metal cutting sector has been expanding as it is widely used for the production of home appliance, automotive industry, building construction, shipbuilding and aerospace industry; these industries are well-equipped with large machine for the use of their employees in metal cutting activities. Out of all the processes, turning operation has been commonly used in the experimental work of metal cutting [1].

There are two types of lathe machine available and still in use: conventional lathe and computer numerical control (CNC) machine. The manufacturers for machine tool produced computer numerical control (CNC) machine for the purpose of maximizing the capability of the conventional lathe that has the capability of reducing cost, producing highly accurate part and increasing the efficiency of the machine tool.

The CNC machine is controlled by manipulating cutting parameters that could directly influence the performance of the process. This concern is also mentioned by Park and Kim in their review that changing the parameters provides better quality as well as accommodating the machinists in the increase of profit and productivity [2].

In the industry, the machinists have to handle the parameters according to their experiences or by the help of the manual handbook that is suggested by the tool supplier. It is possible for people to learn from their experiences and they can also easily learn from a manual handbook. However, the selected cutting parameters are barely at their optimal value. Therefore, many studies have been done by researchers for the introduction of optimization techniques in the effort to solve this problem.

Many optimization techniques such as statistical regression technique and Artificial Intelligence (AI) were applied to obtain the optimal parameters for the desired performance function. Nonetheless, the industry still uses traditional technique to obtain the optimal value. The lack of knowledge on optimization techniques is the main reason for this issue to be prolonged. In this paper, a system has been made to reduce the human intervention in selection of optimal cutting parameters for CNC turning operation. This system need to be fast and effective technique for the machinists to select the right cutting parameters to improve the production.

The paper is organised as follows: Section 1 is the introduction; Section 2 is the literature review of optimization system; Section 3 explains the modelling and optimization technique in determining the cutting parameters; Section 4 describes the proposed graphical user interface; Section 5 discusses the optimization results; and the last section presents the conclusion of this study.

2. Literature Review

There are many computer aided manufacturing(CAM) software available for CNC turning process such as MasterCAM and Vericut. These software offer tool selection, CNC code generation, and process simulation. However, the optimization tool that is available in certain software is limited to minimization of time. Some of the software provide suggestion for the cutting conditions, but these parameters only consider the type of tool and material that are used in the process, for example, MasterCAM. The available software also requires too many variables and it is difficult to be used [3].

There are research done in the effort to develop a system that could optimize the manufacturing process. However, most of the system only focus on the optimal tool for the process. In 2001, Mookherjee has developed a tool selection software called EXTOOL that could help in choosing the right tool and material, as well as inserting geometry based on the requirement by the user [4]. There are also research done in computer aided process planning system in identifying the best cutting tool for the machining operation [5, 6, 7, 8].

The expert computer aided cutting tool selection (EXCATS) was developed to select cutting tools and conditions for major turning operations [9]. Gecevska has introduced optimization of milling and drilling (OPTIMAD) and Genetic Algorithm for machining optimization (GAMO) to optimize and analyse machining conditions in multipass NC machining operation, specifically in milling and drilling operation [10].

One of the major drawbacks of the existing system is that it does not consider numerous parameters of influential factor. According to Kalpakjian, there are many factors that could influence the turning process, including tool (its material, geometry, and shape), work piece (material, condition, and temperature), cutting parameters (speed, feed rate, and depth of cut), the use of cutting fluid, chip formation, the characteristics of the machine tool (feed and speed ranges), work-holding devices, and part characteristic (geometry, accuracy and finish) [11]. It is important to consider all these factors in order to obtain the optimal parameters. Therefore, the experimental design based optimization is preferred.

Numerous studies have identified that the experimental design based optimization techniques can achieve the desired performances for not only in turning operations [12] but also in many manufacturing operation such as drilling [13, 14], milling [15], injection moulding [16, 17], and electrical discharge machining [18, 19]. According to the literature, the optimal process parameters that were obtained by the researchers through the optimization techniques have proven to provide the desired performance for that particular process.

Nalbant et al. has used Taguchi method, the smaller-the-better characteristic, in optimizing the parameters for surface roughness while Mandal et al. evaluate the influence of cutting parameters using combination of response surface methodology (RSM) and Taguchi as the optimization [20, 21]. Since most of the conventional optimization often stuck at local minima, the researchers move to nonconventional or evolutionary optimization to determine the optimal parameters. Zuperl and Cus in their research using genetic algorithm (GA) to determine the optimal cutting parameters for turning cast steel material [22].

Among the artificial intelligent system, Huang et al. has introduced extreme learning machine (ELM), a new technique which simplify single layer feed forward neural network (SLFN) technique and requires less user-defined parameters which lead to fastest technique [23]. It also has a structure that is similar to neural network, which is equipped with an input layer, a hidden layer, and an output layer as shown in Figure 1.

Since the introduction of ELM, numerous researchers have investigated the technique for regression or classification problem. The results show that the ELM performed better than other modelling technique [24, 25, 26, 27]. ELM is identified as an extraordinarily fast learning algorithm. Unlike the traditional SLFN, the weight in ELM does not need to be tuned and it can perform well even with insufficient data [28]. Figure 2 shows the steps of the algorithm.

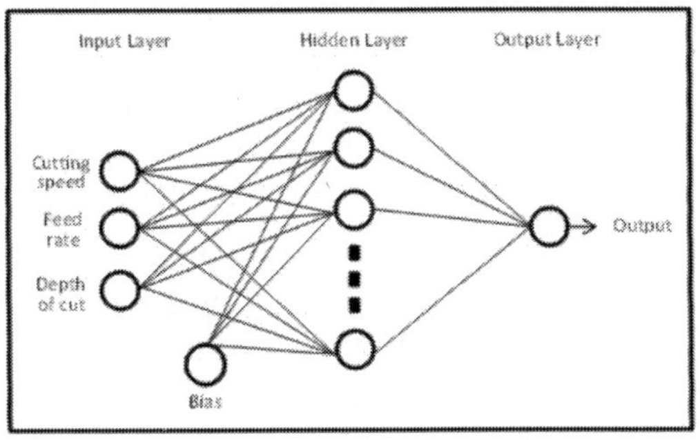

Figure 1. Extreme Learning Machine structure.

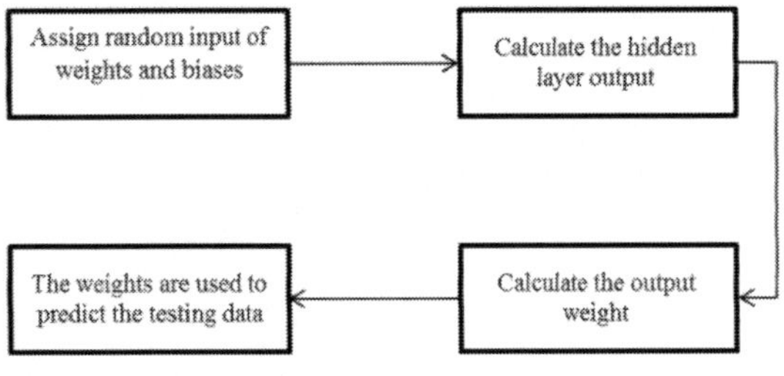

Figure 2. Steps in ELM algorithm.

Since input weight and hidden biases for ELM is assigned randomly, the algorithm is often ends up being unsatisfactory. Xu and Shu has introduced a combination of ELM with PSO to solve the issue [29]. This idea can achieve good generalization performance compared to traditional ELM. The ELM with PSO algorithm can effectively model the process in fast and efficient process [30]. Therefore, the combination of ELM and PSO is used as the modelling technique.

The optimization technique can be applied to obtain the solution for many mathematical problems under the given constraint. From the literature, there have been many articles published on optimization in the effort to prove the efficiency and accuracy of the existing optimization techniques [22, 31, 32, 33]. The Genetic Algorithm is widely used in many applications because it was introduced earlier than the other techniques. However, the comparison of the literature on the optimization technique has discovered that an optimization technique called Particle Swarm Optimization (PSO) has the ability to outperform other techniques [34, 35, 36, 37]. The Particle Swarm Optimization is simulated based on the bird flocking and fish schooling model. According to Chavan and Adgokar and Kim et al., PSO is simple and easy to be implemented [38, 39]. Besides that, Boudjelaba et al. mentioned that PSO is less complex and requires less parameter to adjust as this technique has no mutation and crossover operator compared to GA. Moreover, PSO tends to converge faster than GA [40]. Therefore in this study, PSO is chosen as the optimization tool and the improver to converge ELM.

However, in order to implement this technique in industry, the machinist need to understand throughly the optimization technique. Therefore, a graphic user interface of optimization technique is needed to reduce the human intervention. This intelligent system will require less time to operate, provide accurate results, and easy to understand.

3. Optimization System

There are two stages in the optimization system. First, the modelling of input-output and in-process parameters will produce a mathematical input for the formulation of the process objective function. The second stage is the optimal or near optimal conditions calculation for the turning process. Both stages, as depicted in Figure 3, require either mathematical approaches or artificial intelligent techniques to achieve the goal of optimization.

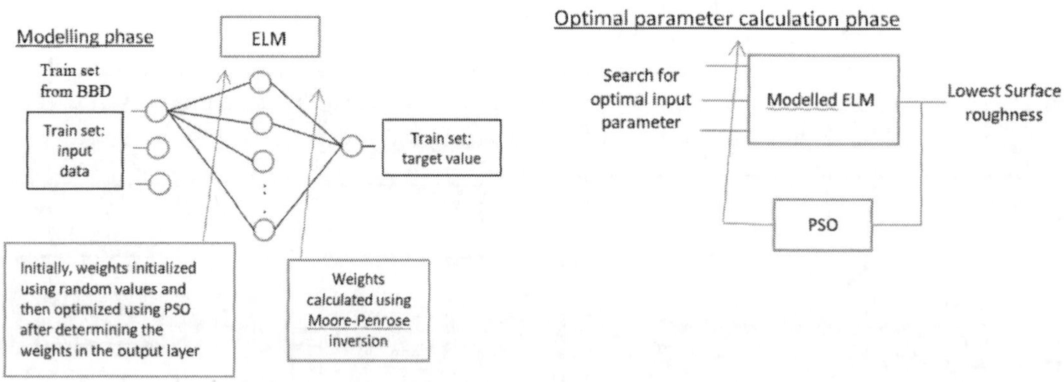

Figure 3. Diagram describing the overall system.

The relationship of input-output and in-process parameters is important to develop the model. This relationship can be determine based on the experimental works done in the manufacturing process. In this study, the Box Behnken Design (BBD) was selected as the design of experiment. BBD is also known as the statistical technique that has the lowest number of experimental run, and it is reliable and efficient in developing a model.

For the three level factors, BBD only requires 15 number of runs including two repetitions. It is the best design for fitting a quadratic model. This is because, the geometry of this design suggests a sphere within the process space. Therefore, the prediction model can be created by using the BBD data, which is much better than other design of experiment technique. Table 1 shows the experimental investigation that uses BBD.

The experimental investigation approach usually employs the regression analysis to build the modelling function. Even though it is a straight forward approach, the result is unreliable. Therefore, this study employed the experiment technique. The industries require less number of runs as the amount of experimental runs could affect the company's production time and cost.

The BBD data will be key into the extreme learning machine with particle swarm optimization (ELMPSO) algorithm in order to train and develop the accurate model.

Consider N data samples of (x_i, t_i), where x_i is an element of $R_n{\times}R_m$. The output function for generalized SLFN with L hidden nodes and activation function $g(x)$ is given in Equation 1.

$$f(x) = \sum_{i=1}^{L} \beta_i g(w_i x_j + b_i), j = 1,2,\cdots,N \tag{1}$$

Table 1. The experimental design using BBD.

Experiment	Parameter 1	Parameter 2	Parameter 3
1	-1	-1	0
2	1	-1	0
3	-1	1	0
4	1	1	0
5	-1	0	-1
6	1	0	-1
7	-1	0	1
8	1	0	1
9	0	-1	-1
10	0	1	-1
11	0	-1	1
12	0	1	1
13	0	0	0
14	0	0	0
15	0	0	0

β_i is the output weight, w_i is the input weight vector and b_i is the bias of the i-th hidden node.

The activation functions are the sigmoid in the hidden layer and a linear function for the output layer. The β_i is determined using Moore-Penrose inverse that is shown in Equation 2.

$$\hat{\beta} = H^{\dagger}T \tag{2}$$

where H stands for Moore-Penrose inverse of hidden layer output matrix and T is the target. The w_i and b_i is determined using PSO. The particle,x_i^k in the PSO is refer to the input weight and bias. The movement of each particle depends on the local best position, p_i and also guides the best known position found in the search space by other particle, p_g.

$$v_i^{k+1} = wv_i^k + c_1R_1(p_i - x_i^k) + c_2R_2(p_g - x_i^k) \tag{3}$$

$$x_i^{k+1} = x_i^k + v_i^{k+1} \tag{4}$$

v_i^k is the velocity of i-th particle at k-th iteration in the swarm while x_i^k is the current position of the particle. c_1 and c_2 are the acceleration coefficients which are 2, R_1 and R_2 are two different random numbers between 0 and 1, p_i is the Pbest and p_g is the Gbest that was achieved by the particle in its neighbourhood. The movement of these particles depend on the fitness value calculated using ELM algorithm. The fitness value for each particle refers to the prediction error of the validation data set. The best particle that has the best quality measurement will be updated in every iteration for 100 iterations using the velocity and particle equation until the best solution for the problem is reached.

In order to determine the optimal cutting parameters, the PSO is used again. In this stage, the random value of cutting parameters acts as a particle, adjusts the velocity vector based on the best position (p_b) as well as the best position of the neighbours (p_g) by referring to the best selected performance function at every iteration. Then, the new position can be obtained. In this optimization technique, the ELM acts as an objective function. The best result is the cutting parameters that can satisfy the objective function.

4. Implementation in Matlab

The graphical user interface (GUI) is applied in the development of the system based on the proposed optimization technique. The development of GUI consists of two phases: GUI design and GUI implementation. In the design phase, the overall interface layout can be set up; while in the implementation phase, the interface can invoke modularly designed Matlab source code.

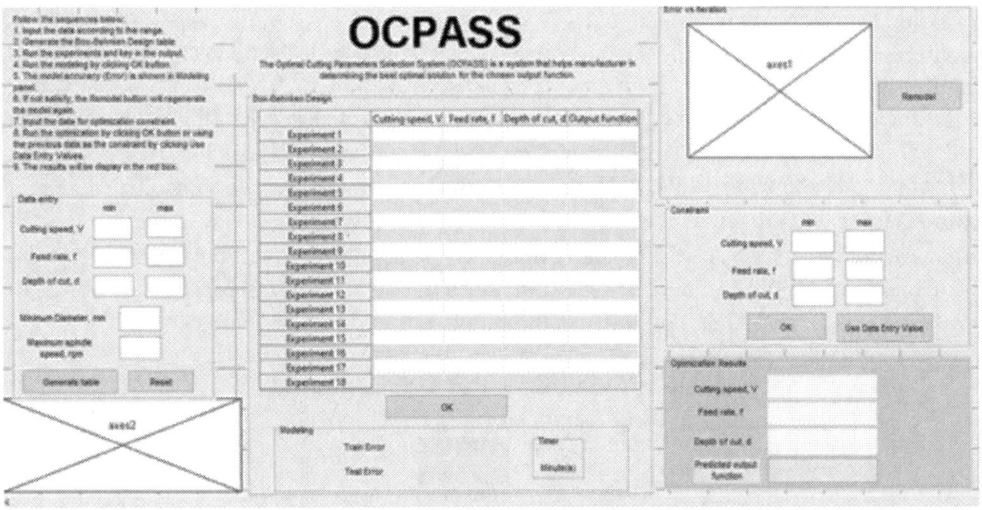

Figure 4. GUI layout in the design area.

The graphical user interface development environment toolbox (GUIDE) that is provided by Matlab is utilized for the development of GUI in an easy and straight forward manner. The main aim of the GUI is to assist the CNC turning operator in selecting the optimal cutting parameters. The GUI components are placed inside the design area using drag and drop technique. Then, the selected component is aligned in its place. Based on the framework, the final interface for the Optimal Cutting Parameters Selection System (OCPASS) is shown in Figure 4.

5. Optimization Result

An experiment is done using AISI 1045 cylindrical bar with 100mm in length and diameter 20mm to validate the system using cutting parameters in Table 2. The objective of the performance that is required by the machinist is that the surface roughness must be lower than 3.2 μm. In manufacturing industry, the surface roughness value should be 3.2 μm and below for jointing and bearing purposes. In order to achieve the objective function of surface roughness, the new cutting edge of CNMG 120404N-GU AC603 is used in each experiment to ensure the accuracy of the reading for the finished surface.

Table 2. The cutting parameters for lower and upper boundary.

Cutting parameters	Lower Boundary	Upper Boundary
Cutting speed (m/min)	80	100
Feed rate (mm/rev)	0.1	0.2
Depth of cut (mm)	0.8	1

Each experiment was carried out until reaching the 40mm cutting length and 17.7mm diameter. The surface roughness tester, Rugosurf 100, is used to record the surface roughness. The samples were measured three times at different locations and the average surface roughness for each sample is recorded. The first step is to input the data entry according to the requirement in Optimal Cutting Parameters Selection System (OCPASS). Then, the machinist is required to generate the BBD table.

The machinist is required to follow the BBD table when running the 18 experiments and recording the surface roughness value in the output function column. Next, the model for the system can be developed and the OCPASS will display the training and testing error in the graph. During modelling, 15 data will be used as train set while the other three data will be used as test set. The modelling technique is referred as ELM and PSO.

Finally, the optimal cutting parameters are determined after clicking the OK button. The suggested parameters (Figure 5) by the OCPASS are applied in the manufacturing area for AISI 1045. The result for the machining is shown in Table 3. The overall result shows that the difference between prediction and real machining is 24%. This result reveals that the percentage for prediction is quite low although the surface roughness complies with the experimental objective. Moreover, the time taken to complete the work using OCPASS is 36 minutes, which includes the manufacturing process. However, the total time for the system to model and optimize is 6 minutes.

Table 3. The validation experiment's result.

	Cutting speed (m/min)	Feed rate (mm/rev)	Cut depth (mm)	Surface roughness (μm)
OCPASS parameters	86	0.1	0.9	1.91
Real machining	86	0.1	0.9	2.51

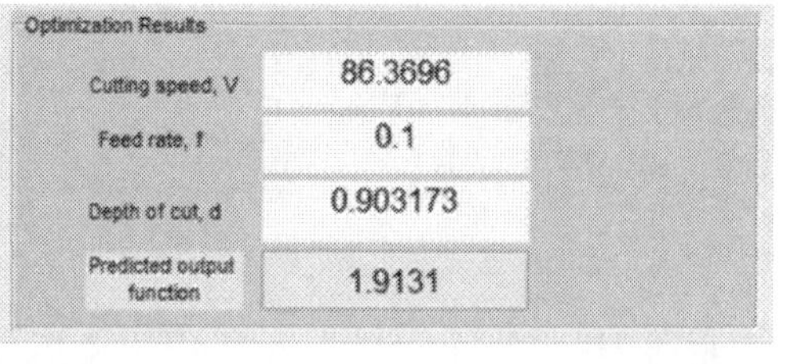

Figure 5. The result of the optimization process using OCPASS.

6. Conclusion

It is proven that optimization technique has been successfully used in several engineering application. However, this technique could not applied in the industry due to time constraint and lack of knowledge. This paper has demonstrated the development of the fastest optimization system together with the graphical user interface for the optimization of cutting parameters in turning operation. This system has three steps, which are design of the experiment using Box Behnken Design, modelling using Extreme Learning Machine with Particle Swarm Optimization, and optimization using Particle Swarm Optimization. This is a novel optimization technique that can accurately optimize besides being the fastest optimization technique. The developed graphical user interface is easy to be understood by all level of users. For future work, this approach can be applied for the optimization of other machining process and extend to multi-objective problems.

Acknowledgement

The authors would like to be obliged to Kulim Techno Centre for providing laboratory facilities for this project.

References

[1] Trent E M 2010 *Metal cutting* Woburn, Massachusetts: Butterworth-Heinemann Fourth edition.
[2] Park K S and Kim S H 1997 Artificial intelligent approaches to determination of CNC machining parameters in manufacturing: a review *Artificial Intelligence in engineering* **12** 127-134
[3] Pena A E, Anania F D, and Zapciu M 2014 Research regarding software developing for machining parameters optimization calculus *Applied Mechanics and Materials* **656** 192-199
[4] Mookherjee R and Bhattacharyya B 2001 Development of an expert system for turning and rotating tool selection in a dynamic environment. *Journal of Materials Processing Technology* **113 (1-3)** 306–311
[5] Giusti F, Santochi M and Dini G 1986 COATS: an expert module for optimal tool selection *CIRP Annals-Manufacturing Technology* **35(1)** 337-340
[6] Oral A and Cakir M C 2004 Automated cutting tool selection and cutting tool sequence optimisation for rotational parts *Robotics and Computer-Integrated Manufacturing* **20(2)** 127-141
[7] Rho H M, Geelink R, Van't Erve A H and Kals H J J 1992 An integrated cutting tool selection and operation sequencing method *CIRP Annals-Manufacturing Technology* **41(1)** 517-520
[8] Fernandes K J and Raja V H 2000 Incorporated tool selection system using object technology *International Journal of Machine Tools and Manufacture* **40(11)** 1547-55
[9] Arezoo B Ridgway K and Al-Ahmari A M A 2000 Selection of cutting tools and conditions of machining operations using an expert system *Computers in Industry* **42(1)** 43-58
[10] Gecevska V Cus M, Lombardi F Dukovski V and Kuzinovski M 2006 Intelligent approach for optimal modeling of manufacturing systems *Journal of Achievements in Materials and Manufacturing Engineering* **14(1-2)** 97-103
[11] Kalpakjian S Schmid S R and Hamidon M 2006 *Manufacturing engineering and technology*. Jurong: Prentice Hall
[12] Sardinas R Q, Santana M R and Brindis E A 2006 Genetic algorithm-based multi-objective optimization of cutting parameters in turning processes *Engineering Applications of Artificial Intelligence* **19(2)** 127-133
[13] Kurt M, Bagci E and Kaynak Y 2009 Application of Taguchi methods in the optimization of cutting parameters for surface finish and hole diameter accuracy in dry drilling processes *The International Journal of Advanced Manufacturing Technology* **40(5-6)** 458-469
[14] Haq A N, Marimuthu P and Jeyapaul R 2008 Multi response optimization of machining parameters of drilling Al/SiC metal matrix composite using grey relational analysis in the Taguchi method *The International Journal of Advanced Manufacturing Technology* **37(3-4)** 250-255
[15] Moshat S, Datta S, Bandyopadhyay A and Pal P 2010 Optimization of CNC end milling process parameters using PCA-based Taguchi method *International Journal of Engineering, Science and Technology* **2(1)** 95-102
[16] Oktem H, Erzurumlu T and Uzman I 2007 Application of Taguchi optimization technique in determining plastic injection molding process parameters for a thin-shell part. Materials & design **28(4)** 1271-78
[17] Shen C Wang L and Li Q 2007 Optimization of injection molding process parameters using combination of artificial neural network and genetic algorithm method *Journal of Materials Processing Technology* **183(2)** 412-418
[18] Singh P N, Raghukandan K and Pai B C 2004. Optimization by grey relational analysis of EDM

parameters on machining Al–10% SiC P composites *Journal of Materials Processing Technology* **155** 1658-61

[19] Lin J L and Lin C L 2002 The use of the orthogonal array with grey relational analysis to optimize the electrical discharge machining process with multiple performance characteristics *International Journal of Machine Tools and Manufacture* **42(2)** 237-244

[20] Nalbant M Gökkaya H & Sur G 2007 Application of Taguchi method in the optimization of cutting parameters for surface roughness in turning *Materials and Design* **28(4)** 1379-85

[21] Mandal N Doloi B Mondal B & Das R 2011 Optimization of flank wear using Zirconia Toughned Alumina (ZTA) cutting tool: Taguchi method and regression analysis. *Measurement* **44** 2149-55

[22] Zuperl U & Cus F 2003 Optimization of cutting process by GA approach *Robotics and Computer-Integrated Manufacturing* **19** 113-121

[23] Huang G B, Zhu Q Y and Siew C K 2006 Extreme learning machine: theory and applications Neurocomputing **70(1)** 489-501

[24] Tang X and Han M 2009 Partial Lanczos extreme learning machine for single-output regression problems *Neurocomputing* **72(13)** 3066-3076

[25] Cheng G J, Cai L and Pan H X 2009 December Comparison of extreme learning machine with support vector regression for reservoir permeability prediction *IEEE International Conference on In Computational Intelligence and Security* vol 2 pp. 173-6

[26] He Q Shang T Zhuang F and Shi Z 2013 Parallel extreme learning machine for regression based on MapReduce *Neurocomputing* **102** 52-58

[27] Javed K Gouriveau R Zerhouni N Zemouri R and Li X 2012 June Robust, reliable and applicable tool wear monitoring and prognostic: approach based on an improved-extreme learning machine. *IEEE Conference on Prognostics and Health Management (PHM)* pp. 1-9

[28] Natarajan C, Muthu S and Karuppuswamy P 2011 Prediction and analysis of surface roughness characteristics of a non-ferrous material using ANN in CNC turning *The International Journal of Advanced Manufacturing Technology* **57(9-12)** 1043-51

[29] Xu Y & Shu Y 2006 *Evolutionary extreme learning machine–based on particle swarm optimization* Advances in Neural Networks pp 644-652

[30] Ahmad N and Janahiraman T V 2015 Modelling and Prediction of Surface Roughness and Power Consumption Using Parallel Extreme Learning Machine Based Particle Swarm Optimization *In Proceedings of ELM-2014* Vol 2 Springer International Publishing pp. 321-329

[31] Chandrasekaran M Muralidhar M Krishna C M and Dixit U S 2010 Application of soft computing techniques in machining performance prediction and optimization: a literature review. *The International Journal of Advanced Manufacturing Technology* **46(5-8)** 445-64

[32] Neseli S, Yaldiz S and Türkes E 2011 Optimization of tool geometry parameters for turning operations based on the response surface methodology *Measurement* **44(3)** 580-587

[33] Karpat Y and Özel T 2007 Multi-objective optimization for turning processes using neural network modeling and dynamic-neighborhood particle swarm optimization *The International Journal of Advanced Manufacturing Technology* **35(3-4)** 234-247

[34] Aggarwal A, Singh H, Kumar P and Singh M 2008 Optimizing power consumption for CNC turned parts using response surface methodology and Taguchi's technique—a comparative analysis *Journal of materials processing technology* **200(1)**, 373-384.

[35] Subramonian S and Abdul Rahim S 2014 Comparison between Taguchi Method and Response

Surface Methodology (RSM) in modelling CO2 laser machining *Jordan Journal of Mechanical and Industrial Engineering* **8(1)** 35-42

[36] Elbeltagi E, Hegazy T and Grierson D 2005 Comparison among five evolutionary-based optimization algorithms *Advanced engineering informatics* **19(1)** 43-53

[37] Eberhart R C and Shi Y 1998 March Comparison between genetic algorithms and particle swarm optimization *In International Conference on Evolutionary Programming* Springer Berlin Heidelberg pp 611-616

[38] Chavan S D and Adgokar N P 2015 An Overview on Particle Swarm Optimization: Basic Concepts and Modified Variants *International Journal of Science and Research* 4(5) 255-260

[39] Kim T H, Maruta I and Sugie T 2010 A simple and efficient constrained particle swarm optimization and its application to engineering design problems *Proceedings of the Institution of Mechanical Engineers Part C: Journal of Mechanical Engineering Science* **224(2)** 389-400

[40] Boudjelaba K, Ros F and Chikouche D 2014 Potential of particle swarm optimization and genetic algorithms for FIR filter design *Circuits, Systems, and Signal Processing* **33(10)** 3195-3222

IOP Conf. Series: Materials Science and Engineering **342** (2018) 012087 doi:10.1088/1757-899X/342/1/012087

Application of Response Surface Methodology for characterization of ozone production from Multi-Cylinder Reactor in non-thermal plasma device

Tan Lian See[1], Ahmad Zulazlan Shah Zulkifli[2], Lim Mook Tzeng[2]

[1]Malaysia - Japan International Institute of Technology (MJIIT), Universiti Teknologi Malaysia (UTM), 54100 Kuala Lumpur, Malaysia.
[2]TNB Research Sdn. Bhd., 43000 Kajang, Selangor, Malaysia.

E-mail: mook.tzeng@tnb.com.my; mooktzeng.lim@live.com

Abstract. Ozone is a reactant which can be applied for various environmental treatment processes. It can be generated via atmospheric air non-thermal plasmas when sufficient voltages are applied through a combination of electrodes and dielectric materials. In this study, the concentration of ozone generated via two different configurations of multi-cylinder dielectric barrier discharge (DBD) reactor (3 x 40 mm and 10 x 10 mm) was investigated. The influence of the voltage and the duty cycle to the concentration of ozone generated by each configuration was analysed using response surface methodology. Voltage was identified as significant factor to the ozone production process. However, the regressed model was biased towards one of the configuration, leaving the predicted results of another configuration to be out of range.

1. Introduction

Ozonation is gaining attention as an environmental remediation process due to its ability to decompose contaminants into safer compound compared to the conventional chemical treatments [1, 2]. Ozone (O_3) can be used for drinking water treatment as an alternative to chlorination [3-5]. It can be used to degrade antibiotics and organic compounds in water [6-11].

Ozone is also a powerful antimicrobial agent which can be applied in processing of food. It is capable of inactivating microorganisms rapidly by reacting with intracellular enzymes, nucleic material and components of their cell envelope, spore coats, or viral capsids [12]. This is an important aspect in preservation of food.

Non-thermal plasma discharge is one of the popular and cost effective method to generate ozone artificially [1, 13-16] and is synthesized from corona discharges [17]. Non-thermal plasma (NTP) reactor can be operated at ambient temperature and atmospheric pressure condition [3, 18]. It is also more energy saving compared to thermal plasma [19]. There are various types of NTP discharge and they are usually categorized based on their electrode geometry, breakdown mechanism, and the presence (or absence) of dielectric materials between the gas flow and the electrodes [13].

One of the common types of NTP reactor is based on dielectric barrier discharge (DBD) [13, 16]. The electrode configuration is characterized by the presence of at least one dielectric barrier (insulator) in the current path in addition to the gas gap used for discharge initiation [16]. It has a relatively quiet discharge operation when the narrow discharge gap of a few millimeters in the reactor is fed with low flow rate of oxygen or air for ozone synthesis [15].

The reactor is able to convert majority of the electrical energy into producing high temperature electrons, rather than directly into gas heating [14, 16]. The energetic electrons rarely collide directly with the pollutants. Rather, the electrons tend to collide with the dominant nitrogen (N_2) and O_2 molecules and produce radicals such as oxygen atom which subsequently lead to the generation of O_3 [14, 16]. O_3 has relatively longer lifetime than electrons. It is well-known for its strong oxidizing capability [15, 16]. It decays without residues that are harmful to the environment [12, 14]. Since the production of ozone requires only air/oxygen (O_2) and electricity, the need for material handling of hazardous chemicals (for decontamination) is eliminated [14]. Therefore, ozonation process can be considered as a treatment process which is benign to the environment.

Ozone generation is expected to increase with the increasing activities of a corona discharge on the surface of dielectric barrier. Specifically, the generated ozone concentration would increase with the increasing voltage applied to the system [14]. However, after optimum condition of the corona discharge, the generated ozone concentration could decrease with increasing applied voltage [14]. This is because the heat from the discharges in the small discharge gap promotes ozone disassociation [20]. Hence, it is important to identify the optimum voltage level to be applied to a NTP reactor. The ozone generation rate is also associated with the duty cycle, which controls the amount of current injected into the DBD. Higher duty cycles (and amperes) could cause localized heating at the location of the discharges, dissociating ozone.

In this work, a multi-cylinder in an enveloping cylinder was developed in order to minimize the possibility of fine particle clogging in the reactor. It is an alternative to the conventional plate-type reactor. A multipoint configuration electrode is also expected to present the advantage of low voltage operation [21]. In this paper, specifically, the ozone generation performance with respect to the voltage and the duty cycle for two configurations of multi-cylinder DBD was investigated using response surface methodology. This is important to get insight of influencing parameters for characterization of ozone production from the developed device.

2. Materials and methods

2.1 Experimental setup

A schematic flow of the experimental setup used in this study is shown in Figure 1. The NTP rig had an enveloping cylinder with inner diameter of 110 mm which acts as the ground electrode. The dielectric barrier was silica cylinders with a thickness of 2 mm and held in place by a Teflon holder. Two different configurations of dielectric barrier discharge (DBD) reactors were tested in this study. In the first configuration, the arrangement consisted of three silica cylinders with 40 mm outer diameter (3 x 40 mm). In the second configuration, the arrangement consisted of ten silica cylinders with 10 mm outer diameter (10 x 10 mm). The inner surface of the silica cylinders were lined with copper sheets that are connected to the high voltage supply. The smallest air gap formed between the silica cylinders and the outer enveloping shell was 1 mm.

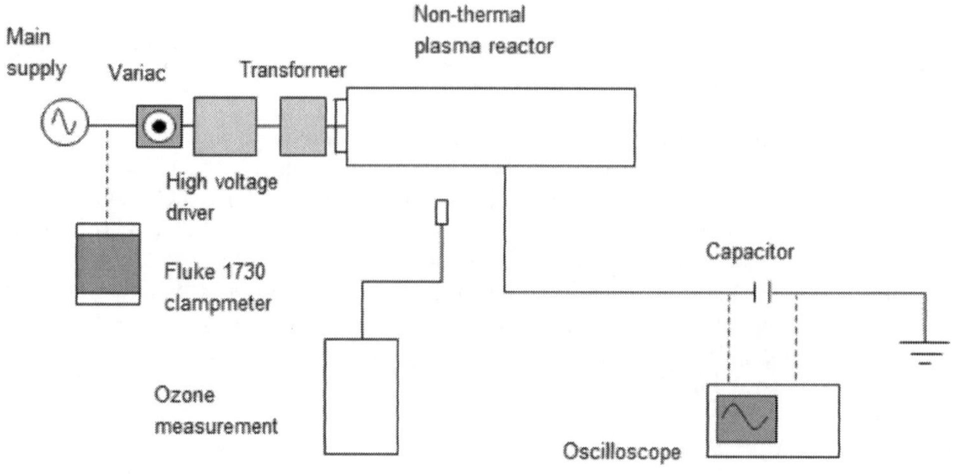

Figure 1. Schematic diagram of the experimental setup.

The input power into the NTP rig was supplied by a high voltage (HV) power supply (PVM 2000). The HV source included a 2 kW VARIAC which controlled the total power to the system. The current was controlled by the duty cycle which provided the current pulses. A circuit breaker was placed before the NTP reactor as safety device to protect the reactor from damage. It would cut off voltage supply if the voltage or current induced from the HVT was too high.

All the experiments were conducted at atmospheric pressure and room temperature. The concentration of ozone produced from the NTP reactor was measured using Aeroqual Series 500 Portable Ozone Analyzer.

2.2 Design of experiment

The variables in this experiments were voltage, duty cycle and configuration of the multi-cylinder in the reactor. The numerical variables are voltage and duty cycle while the categorical variable is the configuration of the cylinders. Experimental runs were conducted based on the combinations of the variables were generated using response surface methodology (RSM) provided by Design-Expert software 6.0.6 (Stat-Ease Inc., Minneapolis, USA). This is done to investigate the influence of variables selected to the amount of ozone synthesized

Central Composite Design (CCD), with an alpha value of ±1.00 was applied to study the experimental variables. The range for voltage was 12 kV to 16 kV while the range for duty cycle was 10–30 based on the equipment specifications. The level of condition for the numerical and categorical variables are given in Table 1 and Table 2, respectively.

Table 3 shows the twenty six sets of experimental condition generated by Design Expert software. Evaluation on the effect of voltage, duty cycle and electrode configuration to the ozone concentration synthesized from the multi-cylinder reactor was subsequently conducted using analysis of variance, ANOVA.

IOP Conf. Series: Materials Science and Engineering **342** (2018) 012087 doi:10.1088/1757-899X/342/1/012087

Table 1. Levels of the condition for numerical variables.

Numerical Factors	Coding	Unit	Levels		
			-1	0	1
Voltage	A	kV	12	14	16
Duty cycle	B	-	10	20	30

Table 2. Levels of the condition for categorical variables.

Categorical Factor	Coding	Level 1	Level 2
Configuration	C	3 x 40 mm	10 x 10 mm

Table 3. Central composite experimental design matrix.

Run	Voltage (kV)	Duty cycle	Configuration
1	16	20	10x10 mm
2	16	10	3x40 mm
3	14	20	10x10 mm
4	12	10	3x40 mm
5	12	30	3x40 mm
6	16	30	10x10 mm
7	14	30	10x10 mm
8	12	30	10x10 mm
9	14	20	3x40 mm
10	14	20	3x40 mm
11	12	10	10x10 mm
12	16	20	3x40 mm
13	14	20	10x10 mm
14	12	20	10x10 mm
15	16	10	10x10 mm
16	14	10	3x40 mm
17	16	30	3x40 mm
18	14	20	3x40 mm
19	14	30	3x40 mm
20	14	10	10x10 mm
21	14	20	10x10 mm
22	14	20	10x10 mm
23	14	20	10x10 mm
24	14	20	3x40 mm
25	14	20	3x40 mm
26	12	20	3x40 mm

3. Results and Discussion

3.1 Ozone Production Model based on RSM

Based on the response (ozone concentration) obtained for the designated set of experiments in Table 3, linear model was analyzed by the Design Expert software as the suggested model. It was the best model with highest order polynomial whereby the additional terms were significant and the model was not aliased.

Table 4. Fit summary based on sequential model sum of squares

Source	Sum of Squares	Mean Square	F Value	Prob > F	Remarks
Mean	70.95	70.95			
Linear	2.39	0.80	5.50	0.0057	Suggested
2FI	0.27	0.091	0.59	0.6272	
Quadratic	0.59	0.29	2.16	0.1460	
Cubic	1.13	0.23	2.28	0.1119	Aliased
Residual	1.19	0.099			
Total	76.52	2.94			

The significance and the fitness of the model was also verified by using Analysis of Variance (ANOVA) in the Design Expert software. The results are presented in Table 5. The model F-value of 5.50 implies the model generated was significant. Based on the analysis by ANOVA, there is a 0.57% chance that a "Model F-Value" this large could occur due to noise. The model terms with "Prob > F" less than 0.0500 indicates that it contributed significantly to the model.

Based on statistical analysis, voltage was the only significant factor in influencing the amount of ozone generated. Duty cycle was deemed as not significant in influencing the model. The configuration of cylinders was also deemed as not significant in affecting the results of the ozone production, based on ANOVA results.

The increasing trend of ozone concentration with the increase of voltage was also observed by Chae et al. [13] and Fang et al. [7]. However, if it passed an optimum point, a drastic decrease of ozone concentration could occur with increased voltage applied to the NTP reactor [14]. Increasing voltage would lead to the increased activities of corona discharge. However, the temperature would also increase due to the heat coming from the intense discharges in the air gap. This would lead to the dissociation of the generated ozone, hence, it reduced the generated ozone concentration [14]. Therefore, careful consideration must be taken while increasing the voltage as total net ozone concentration would decrease once it passed the optimum condition.

Table 5. ANOVA for the significance of model and model terms

Source	Sum of Squares	DF	Mean Square	F Value	Prob > F	Remarks
Model	2.39	3	0.80	5.50	0.0057	Significant
Voltage	2.24	1	2.24	15.44	0.0007	Significant
Duty cycle	0.007	1	0.007	0.048	0.8534	Not significant
Configuration	0.15	1	0.15	1.01	0.3258	Not significant
Residual	3.19	22	0.14			
Lack of Fit	2.36	14	0.17	1.63	0.2477	Not significant
Pure Error	0.83	8	0.10			
Cor Total	5.57	25				

Since configuration of the electrodes was considered as categorical factor while the voltage and duty cycle was the numerical factors in this analysis, the correlations for each categorical factor could be generated by RSM accordingly. The corresponding coefficients of the model obtained by regression analysis of the experimental data are shown in Equation (1) and (2) for configuration of 3 x 40 mm and 10 x 10 mm, respectively.

$$\text{Ozone concentration} = -0.2464 + 0.2158*\text{Voltage} - 0.00242*\text{Duty cycle} \quad (1)$$

$$\text{Ozone concentration} = -1.3964 + 0.2158*\text{Voltage} - 0.00242*\text{Duty cycle} \quad (2)$$

where the units for ozone concentration, voltage, and duty cycle are ppm, kV and percentage respectively.

Based on the correlations generated by RSM, the predicted versus actual data for configuration of 3 x 40 mm and 10 x 10 mm were plotted and presented in Figure 2 and Figure 3 respectively. It is observed that the correlation generated for 3 x 40 mm, as in Figure 2, consistently over-predict the ozone concentration. Meanwhile, the ozone concentration was quite well predicted for configuration of 10 x 10 mm as almost all data fell within the 25% range of prediction target. Therefore, although the model generated by RSM was analysed to significant model based on ANOVA, it appeared that the model was biased toward cylinders configuration of 10 x 10 mm.

RSM is able to analyse the significance and conformation of both numerical and categorical factors based on statistical analysis. However, the resulting model from simultaneous numerical and categorical factor needs to be graphically verified based on set deviation target such as in Figure 2 and Figure 3. Otherwise, the biasness of the model towards a certain categorical factor might be undetected.

It should be noted here that the trend of ozone production depends on the parameter or combination of parameters varied and it is not always in polynomial form. Under this circumstance, an alternative method needs to be used to correlate the variables with the response. This method is under development and would be presented in an upcoming manuscript.

IOP Conf. Series: Materials Science and Engineering **342** (2018) 012087 doi:10.1088/1757-899X/342/1/012087

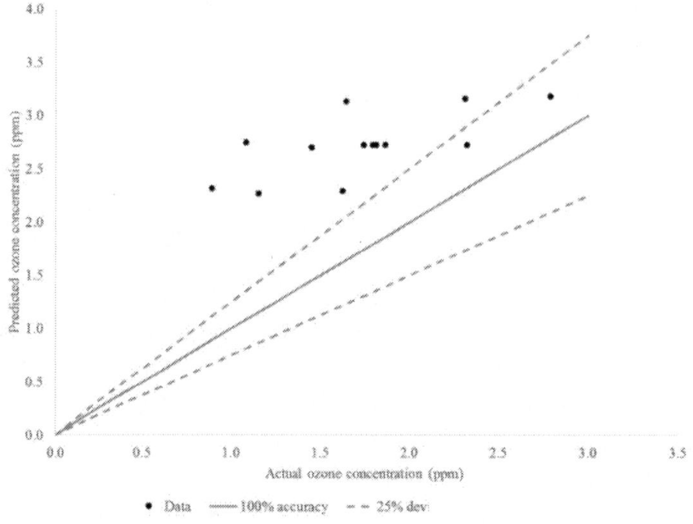

Figure 2. Predicted versus actual ozone concentration for 3 x 40 mm configuration based on correlation by RSM .

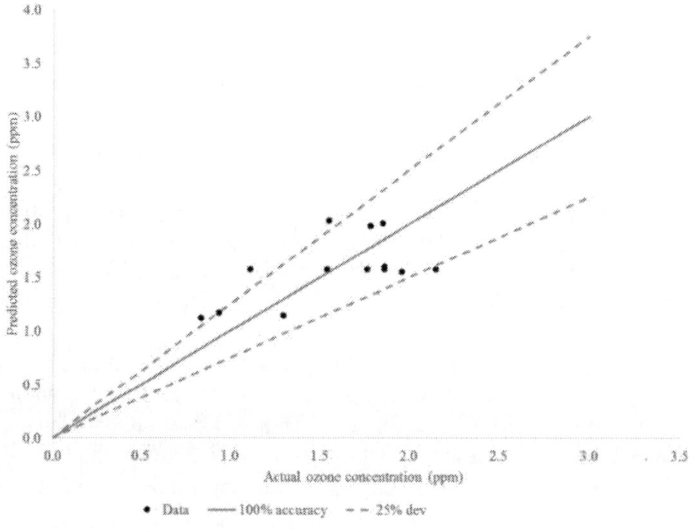

Figure 3. Predicted versus actual ozone concentration for 10 x 10 mm configuration based on correlation by RSM .

4. Conclusion

The effect of voltage and duty cycle for two different configuration of dielectric barrier on ozone concentration in multi-cylinders reactor in NTP was investigated using RSM. Based on analysis using RSM, voltage was identified as significant factor to the ozone production process. Duty cycle was found to be insignificant in influencing the concentration of ozone generated. However, the regressed model was biased towards one of the configuration, leaving the predicted results of another configuration to be out of range.

Acknowledgement

The authors would like to thank TNB Research Sdn. Bhd. (TNBR/SF 195/2015) for providing the experimental facility for this manuscript.

References

[1] Echeverry Ibarra D F, Cadavid Ramírez H, Alonso J M, Aponte Mayor G and Gálvis Castaño A 2008 *Ozone: Sci. Eng.* **30** 202-9

[2] Hashim S A, Samsudin F N D b, Wong C S, Abu Bakar K, Yap S L and Mohd. Zin M F 2016 *Archives Biochem. Biophys.* **605** 34-40

[3] Goldman M, Goldman A and Sigmond R S 1985 *Pure Appl. Chem.* **57** 1353-62

[4] Muhammad Arif M, Abdul G and Salman Akbar M 2001 *Plasma Sources Sci. Tech.* **10** 82

[5] Park S-L, Moon J-D, Lee S-H and Shin S-Y 2006 *J. Electrost.* **64** 275-82

[6] Petr L, Martin C, Vaclav B, Vaclav J and Pavel S 2005 *J. Phys. D: Appl. Phys.* **38** 409

[7] Ikehata K, Gamal El-Din M and Snyder S A 2008 *Ozone: Sci. Eng.* **30** 21-6

[8] Yargeau V and Leclair C 2008 *Ozone: Sci. Eng.* **30** 175-88

[9] Zeng Z, Zou H, Li X, Arowo M, Sun B, Chen J, Chu G and Shao L 2013 *Chem. Eng. J.* **229** 404-11

[10] Zhou S, Bu L, Shi Z, Bi C and Yi Q 2016 *Chem. Eng. J.* **306** 719-25

[11] Dudasova H, Derco J, Sumegova L, Dercova K and Laszlova K 2017 *J. Hazard. Mat.* **321** 54-61

[12] Khadre M A, Yousef A E and Kim J G 2001 Microbiological Aspects of Ozone Applications in Food: A Review *J. Food Sci.* **66** 1242-52

[13] Chae J-o, Moon S-i, Sun H-s, Kim K-y, Vassiliev V A and Mikholap E M 1999 *KSME Int. J.* **13** 647-55

[14] Moon J-D and Geum S-T 1998 *IEEE Trans. Ind. Appl.* **34** 1206 - 11

[15] Ma H and Qiu Y 2003 *Ozone: Sci. Eng.* **25** 127-35

[16] Kogelschatz U 2004 *Plasma Phys. Controlled Fusion* **46** B63

[17] Yehia A and Mizuno A 2012 *Phys. Plasmas* **19** 033513-1--6

[18] Sung Y-M and Sakoda T 2005 *Surface Coatings Tech.* **197** 148-53

IOP Conf. Series: Materials Science and Engineering **342** (2018) 012087 doi:10.1088/1757-899X/342/1/012087

[19] Wang Z H, Yang L, Li B, Li Z S, Sun Z W, Aldén M, Cen K F and Konnov A A 2012 *Combustion Flame* **159** 120-9

[20] Moon J-D and Jung J-S 2007 *J. Electrost.* **65** 660-6

[21] Takaki K, Hatanaka Y, Arima K, Mukaigawa S and Fujiwara T 2008 *Vacuum* **83** 128-32

IOP Conf. Series: Materials Science and Engineering **342** (2018) 012088 doi:10.1088/1757-899X/342/1/012088

The comparison respond of braking torque control between PID and SMC controller for electric powered wheelchair descending on slope condition

S M Asyraf, P M Heerwan and I M Izhar

Automotive Engineering Centre (AEC), Universiti Malaysia Pahang, Pekan Campus, 26600 Pekan, Pahang, Malaysia.

E-mail: mheerwan@ump.edu.my

Abstract. During descending on a slope, the speed of Electric Powered Wheelchair (EPW) tends to changed rapidly. Normally, most EPW is provided with mechanical braking system which transfers human pulling force of the lever creating friction at the tire. However, the task is difficult for the users are elderly or paralyses. However, even for normal user with good strength, in fear condition they tend to give sudden braking which leads to tire locking up and skidding, eventually EPW unstable. These problems will cause accident and injuries to the users if speed does not properly control. In this paper, the automated braking torque control method was proposed in EPW as alternative to solve this problem and increase the mobility and stability especially during descending on slope in other to help the user of the EPW as their daily transportation. In this research, Proportional-Integral-Derivative and Sliding Mode Control controller are compared to determine the best response for torque braking control. The rapid change of speed can be controlled by the braking torque using proposed controllers based on the desired constant speed set by the control designer. Moreover, the sudden braking that caused tire to lock up and skid can be avoided. Furthermore, result from SMC shows this controller have good time respond to maintain the speed based on desired value when descending at slope condition by controlling the braking torque compared to the PID controller.

1. Introduction

The usage of wheelchairs is increasing every year due to the increase of disabled people and elderly. The statement of "World Population Prospects: The 2012 Revision" conducted by the UN estimates that the population of 65 years old peoples and above is 16.5% by 2060 [1]. Moreover, there are almost 200 000 disabled people use wheelchairs as transportation in daily [2]. Meanwhile in Japan, EPW sales at the end of 2008 reached 530 00 units and continues to rise due to the age and ability factor [3]. From the statement of the use of wheelchairs, elderly and disable people depend on EPW in their daily life. Therefore, the research on wheelchairs is also important in other to improve performance and safety factors.

The EPW was introduced during World War II by George Klein who worked for the National Research Council of Canada. At that time, the EPW is just only equipped with electric motors, battery and controllers [4]. Currently, EPW had been integrated with a microcontroller system to make EPW performance more stable and user-friendly since the electric motor such as fast and precise torque respond [5]. Because of structural construction such as staircase, the EPW it is difficult to move. Nowadays, EPW has become intelligent that able to move even in a critical condition [6].

Furthermore, the EPW performance is further enhanced by introducing a stable system that is able to prevent the EPW from falling backward when passing a slope [7].

As we know, Ramps or slopes are provided when in the existence of a low staircase in most of new buildings to assist wheelchair users. However, when moving down the slope, it becomes difficult for users especially elderly and disabled because they have not enough strength to pull the braking lever in order to control the speed of EPW [8]. Even for user with good strength, during descending on slope in fear condition, they tend to give exceed force which leads to sudden braking. Effect from that, the tire will lock up and cause EPW to skid. Moreover, other condition such as slippery road and small contact patch of tire can also lead to this problem.

To avoid this problem, in this paper, we proposed the automated braking control during moving down on the slope instead using human braking inputs that caused sudden braking. We also compared the output response of speed between PID and SMC controller to identify the best controller to control the braking torque.

2. Analysis of EPW model

2.1. Specification of EPW

In this analysis, an EPW was used as shown in Figure 1 and the physical specifications of the EPW are shown in table 1 below. EPW is driven by in-wheel electric motor which is the each tire can be controlled independently. This EPW also be equipped with mechanical braking system for the manual braking form user.

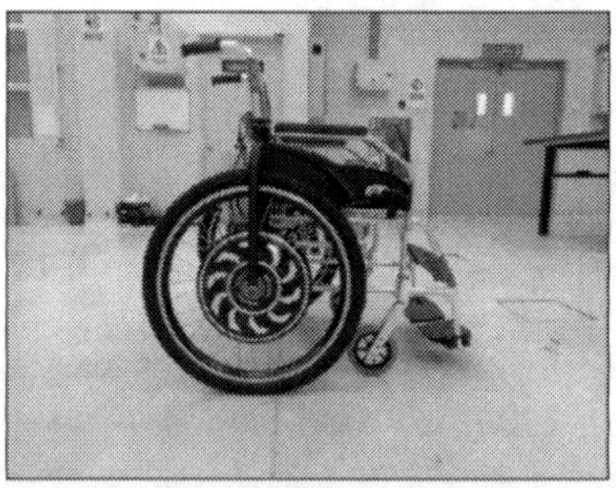

Figure 1. Example Electric Powered Wheelchair for analysis.

Table 1. List of specification for EPW.

Parameter	Symbol	Unit
Maximum Speed	v_t	20 km/h
EPW mass	m_{EPW}	24 kg
Tire mass	m_t	12 kg
Tire inertia	I_t	0.635 kg.m^2

IOP Conf. Series: Materials Science and Engineering **342** (2018) 012088 doi:10.1088/1757-899X/342/1/012088

2.2. Mathematical modeling of EPW

2.2.1 Modeling of tire speed

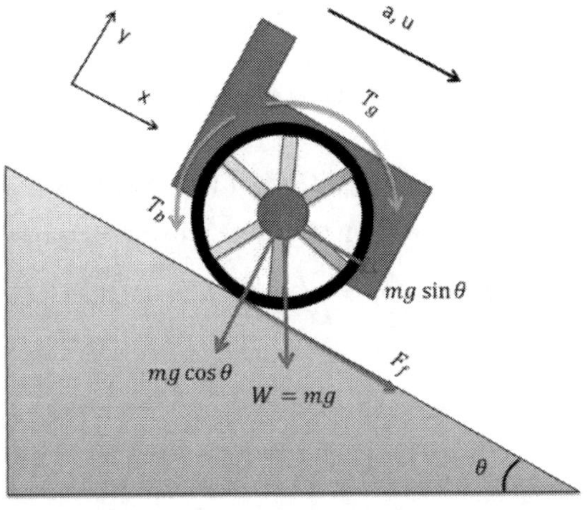

Figure 2. Dynamic body diagram of EPW during descending on the slope.

Figure 2 above shows the free rolling dynamic motion of EPW. Free rolling mean the EPW is moving without giving any external torque or forces. During descending on slope, EPW tends to move forward. This is because the existence of gravity weight ($mg\sin\theta$) at x-component. The gravity weight at x-component will pull the EPW naturally. As result, the speed of EPW will increase. In this analysis, Castrol tire is neglected. When the EPW is pulled naturally by weight gravity at x-component, the torque is generated, T_g as derived as equation (1). The speed of tire is determined by using equation (2) and (3).

$$T_g = mg\sin(\theta)R \tag{1}$$

$$\dot{\omega} = \left(\frac{T_g - F_r R - T_b}{I_t} \right) \tag{2}$$

$$V_t = R\omega \tag{3}$$

Where ω is the angular speed of the tire, F_f is friction force, T_b is the applied braking torque, I_t and R is tire inertia and radius respectively. Since EPW moved at slope condition, F_f can be calculated by equation (4) where m is the tire mass and μ is the friction coefficient can be determine in equation (5) where k is depend on surface condition and ρ is slip ratio [9].

$$F_f = mg\cos(\theta)\mu \tag{4}$$

$$\begin{aligned} \mu_{driving} &= -1.05k(e^{-45\rho} - e^{-0.45\rho}) \\ \mu_{braking} &= 1.1k(e^{35\rho} - e^{35\rho}) \\ k &= 0.8(dry) \\ k &= 0.12(icy) \end{aligned} \tag{5}$$

Slip ratio can be determined by using equation (5) and (6). When the $v_t >$ v known as driving and equation (6) is used. When $v_t <$ v this condition known as braking and equation (7) is used.

$$\rho_{driving} = \frac{v_t - v}{v_t} \tag{6}$$

$$\rho_{braking} = \frac{v - v_t}{v} \tag{7}$$

2.3. Calculating speed of EPW

Determine the speed of EPW very important in order to know the behavior of EPW when moving in different surface condition. The EPW speed can be calculated by using Newton's second law as shown in equation (8) where F_{fi} is force friction and subscript i to indicate left or right tire.

$$\dot{v} = \frac{\sum F_{fi}}{m_{EPW}} \tag{8}$$

3. Controller design

3.1 PID

Proportional, integral and derivative are the basic modes of PID controller, proportional mode provides a rapid adjustment of the manipulating variable, reducing error and speeds up dynamic response. Integral mode achieves zero offset. The derivative mode provides rapid correction based on the rate change of the controlled variable. The controller equation is given by equation (9).

$$u(t) = k_p.e + k_i.\int e + k_d.\frac{de}{dt} \tag{9}$$

Where, e is the error between desired speed and actual speed. k_p, k_i, and k_d is the gain for proportional, integral and derivative respectively that need to be tuned in other to find the best speed response [10].

3.2 Sliding Mode Control

It is well known that SMC is one of the robust control methods to stabilize the disturbances, linear and non-linear system [11]. Figure 3 shows the block diagram of SMC. The desired angular speed in this simulation is set as 5 rad/s. This angular speed can be adjusted by user itself. Different from desired and actual angular speed know as error that used for SMC controller information.

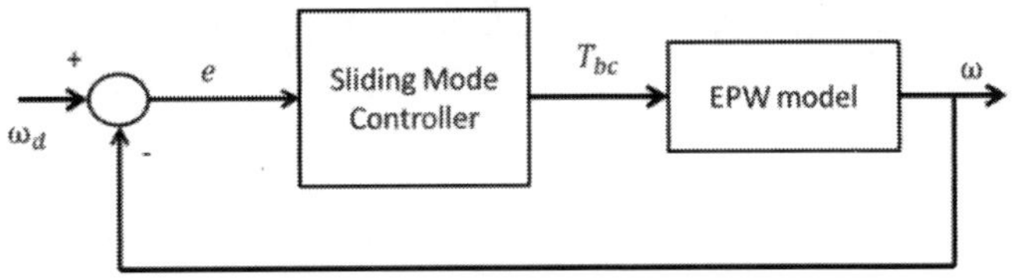

Figure 3. Block diagram of SMC.

The main objective of torque braking controller in this paper is to maintain the speed of the wheel during descending on the slope. By descending slowly, the stability of EPW will not lose. Therefore, the control objective becomes a tracking problem, and the tracking error can be defined as equation (10) where ω is the actual angular speed of tire, and ω_d is the desired angular speed of tire.

$$\omega_e = \omega - \omega_d \tag{10}$$

It is by designing the control law, the torque braking controller make its tracking error tend to zero in a finite time. Based on the above analysis, the sliding surface is taken as equation (11) where c must satisfy the Hurwitz condition, $c > 0$.

$$s = c(\omega_e = c(\ \omega - \omega_d), \text{sgn}\ (s) = \begin{cases} 1, & s > 0 \\ 0, & s = 0 \\ -1, & s < 0 \end{cases} \tag{11}$$

Differentiating the sliding surface, S with respect to time and taking the general reaching law into consideration. The purpose of reaching law is how faster the error to reach in sliding surface. It also will effect to the system responses. The equation obtained can represent as:

$$\dot{s} = ce \doteq c(\dot{\omega} - \dot{\omega}_d) \leq -k\,.sgn\ (s) \tag{12}$$

where $k > 0$ is the control parameter which should be adjusted according to the number of uncertainties and disturbance. Recalling back from tire dynamic rotation in equation (1), then substitute into equation (12). The torque braking control to control the angular speed of the tire by using SMC is derived as equation (13) below.

$$T_{bc} = T_g - I\left(\frac{\dot{S}}{c} + \dot{\omega}_d\right) = T_g - T_f - \left(\frac{-k\ sgn\ (s)}{c} + \dot{\omega}_d\right) I_t \tag{13}$$

4. Result and discussion

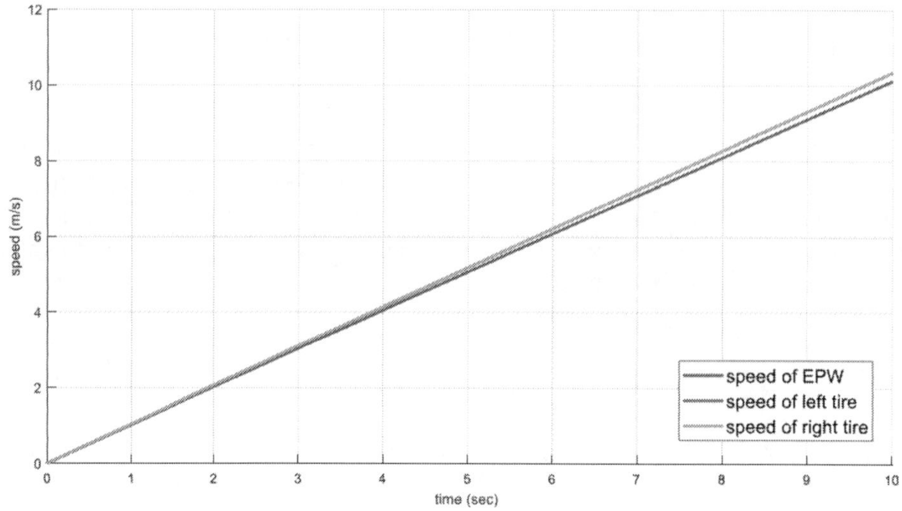

Figure 4. Speed behavior of a wheelchair during descending on slope 15°.

Figure 4 shown the changes of speed when the EPW is release from the top of slope. In this analysis, the slope is set as 15° more than standard slope recommended for building slope[12]. These conditions also known as free rolling. From the plot shows the speed of left and right tire and also speed of EPW which indicate red, yellow and blue respectively gradually increase. At 10 second, the speed of EPW was recorded almost 10 m/s.

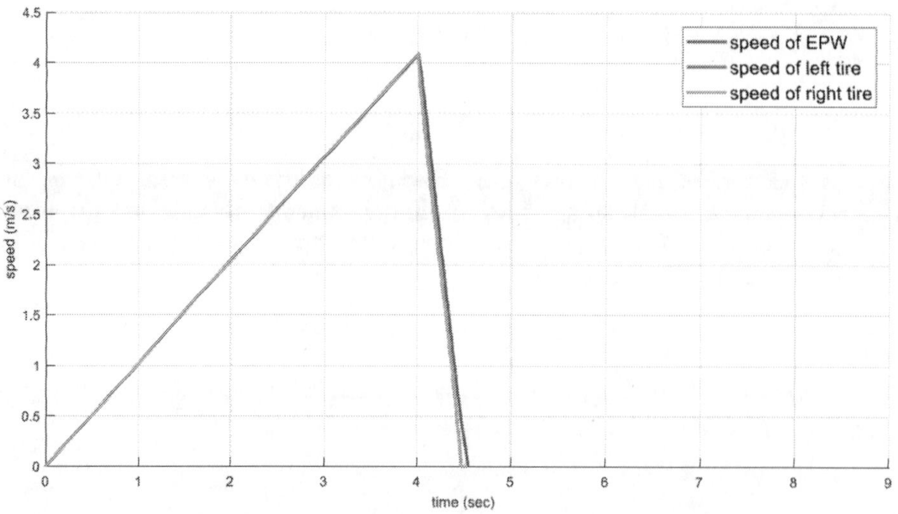

Figure 5. Applied sudden braking at dry condition.

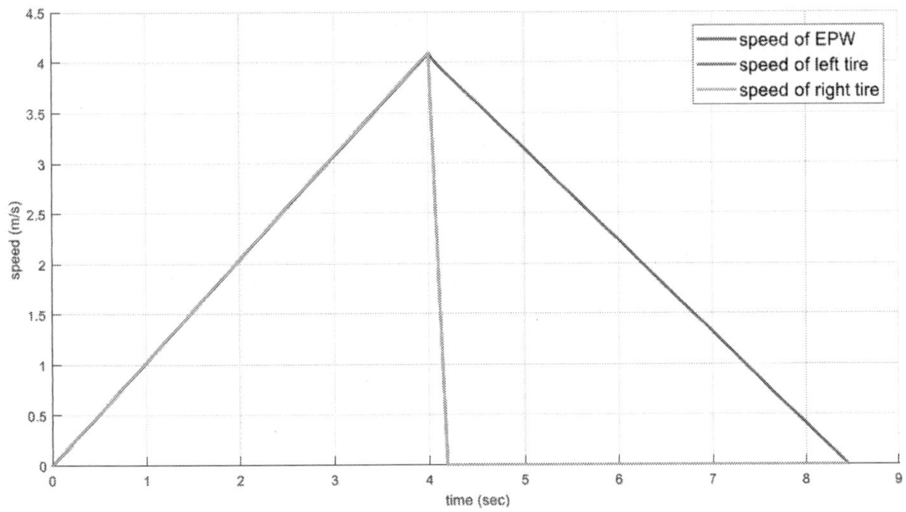

Figure 6. Applied sudden braking at icy condition.

When sudden braking was applied at t=4 sec second in dry condition ($\boldsymbol{k} = \mathbf{0.8}$), from the Figure 5, the speed of left and right tire which indicated with red and yellow line respectively are stopped abruptly at 4.5 seconds and EPW still moved slightly and stopped at 4.7 second. Next, when sudden braking was applied at the low adhesive surface condition ($\boldsymbol{k} = \mathbf{0.12}$), the tires are suddenly locked at 4.2 second. However, EPW kept sliding ad stopped at 8.4 seconds as shown in Figure 6. This is shown that by when sudden braking is dangerous to apply especially at slippery condition.

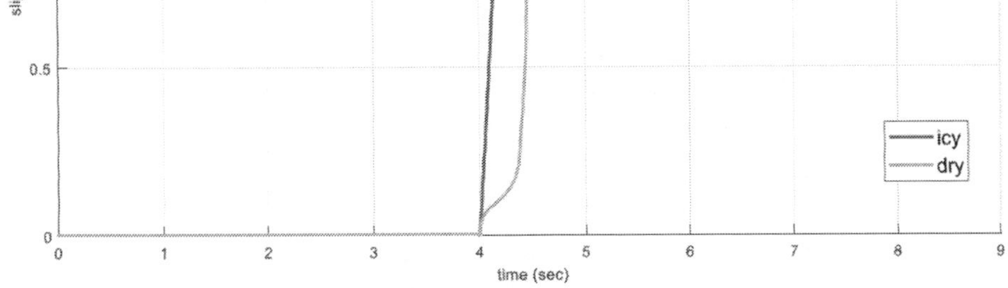

Figure 7. Slip ratio during braking in dry condition.

Slip ratio plotted in Figure 7 also can be used to identify the slip behavior for the EPW. The yellow and blue plot indicted as the slip ratio during braking in dry and icy condition respectively. When the EPW is acceleration (t < 4 seconds) the slip ratio was recorded as 0.08. That means the EPW does not slip. After a sudden braking is given at t = 4 for dry condition, the slip ratio became 1 at $t = 4.5$ seconds and EPW stopped at t = 4.7 seconds. When sudden braking was applied at icy condition, the slip ratio became 1 at t = 4.2 seconds and EPW stopped at t = 8.4 seconds.

To avoid from speed became rapidly change and EPW loose stability, the speed need to be maintain in less fear condition speed for wheelchair user which follow the jerk based $v = 0.6\ m/s$ [13]. In other word, the EPW need to descend on slope slowly. In Figure 8 shows the blue line is speed responds when using PID controller. From the plotted graph, the speed is maintained as desired which is 0.6 m/s and settling time, t_s is 1.8 second. In order to archive the best time responses speed, it seems 13.9 % of overshoot was happening and this is the limitation of PID.

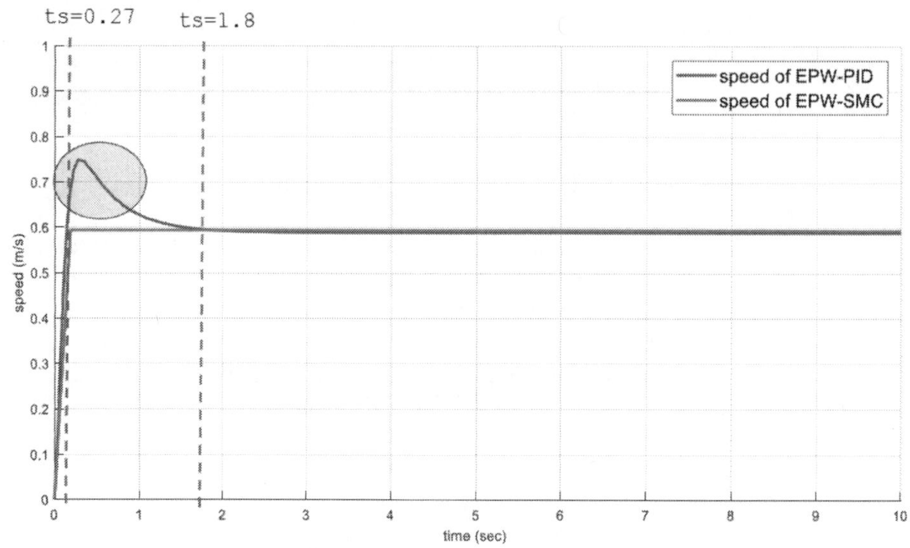

Figure 8. Time respond of Controller by using PID and SMC controller.

By using Sliding Mode Control (SMC) method control, the speed also can be maintained as desired 0.6 m/s as shown a red plot. Even the time rise for SMC and PID almost same, the different of PID and SMC can be identified in settling time and overshoot percentage. SMC gives the much faster speed response. The settling time, t_s was recorded at 0.27 second, but there is no overshoot was recorded.

Table 2. Comparison parameters between PID and SMC.

Parameter	PID	SMC
Gain	$k_p = 5.22$	$k = 5$
	$k_i = 1.01$	$c = 0.5$
	$k_p = 1.43$	
Rise time, t_r (seconds)	0.25	0.27
Settling time, t_s (seconds)	1.8	0.27
Overshoot (%)	13.9	No

5. Conclusion

In conclusion, the proposed automated torque braking control gives the best braking control during descending on a slope by controlling the speed as desired instead using human strength that caused sudden braking and loss stability of EPW. Moreover, by using Sliding Mode Control (SMC) control technique is more effective rather than using PID controller. SMC gives the fastest speed responses without giving any overshoot value than PID. By maintaining the speed of the tire, the slip ratio also became smaller. In other words, the sudden braking problem that caused tire to lock can be eliminated and the stability of EPW during descending on the slope also enhanced.

Acknowledgements

The authors would like to thank Ministry of Higher Education Malaysia for providing Research Acculturation Grant Schemt (RAGS) under project number RDU151403. Special thanks to Universiti Malaysia Pahang for providing laboratory facilities and technical support for this research.

References

[1] Hamatani S 2015 A Novel Steering Mechanism of Two-Wheeled Wheel Chair for Stability Improvement 2154–9

[2] Pierret B, Desbrosses K, Paysant J and Meyer J 2014 Cardio-respiratory and subjective strains sustained by paraplegic subjects , when travelling on a cross slope in a manual wheelchair (MWC) *Appl. Ergon.* **45** 1056–62

[3] Takei T, Suzuki Y, Matsumoto O, Adachi Y, Sasaki Y and Kamo M 2010 Development of assistive technologies for safe operation of electric wheelchairs on sloping sidewalks and grade height differences 43–8

[4] Ba B W, Watson N and Watson N 2017 A Short History of Powered Wheelchairs A Short History of Powered Wheelchairs **435**

[5] Peeie M H, Ogino H and Oshinoya Y 2014 Skid control of small electric vehicle with hydraulic-mechanical hybrid brake system (Effect of ABS and regenerative brake control on an icy road) **1** 1–15

[6] Morales R, Feliu V, González A and Pintado P 2006 Coordinated Motion of a New Staircase Climbing Wheelchair with Increased Passenger Comfort 3995–4001

[7] Hamatani S 2016 Steering Control in Multi-degrees-of-freedom Two-wheeled Wheel Chair on Slope Environment 6181–6

[8] Massy-westropp N M, Gill T K, Taylor A W, Bohannon R W and Hill C L 2011 Hand Grip Strength : age and gender stratified normative data in a population-based study *BMC Res. Notes* **4** 127

[9] Sakai S-I, Sado H and Hori Y 2000 Anti-skid control with motor in electric vehicle *6th International Workshop on Advanced Motion Control Proceedings* pp 317–22

[10] Munje R K, Roda M R and Kushare B E 2010 Speed Control of DC Motor Using PI and SMC *Conf. Proc. IPEC* 945–50

[11] Nam K, Hori Y and Lee C 2015 Wheel Slip Control for Improving Traction-Ability and Energy Efficiency of a Personal Electric Vehicle 6820–40

[12] Blanck P D 1996 Economics of the Employment Provisions of the Americans with Disabilities Act Part I - Workplace Accommodations, The *DePaul Law Rev.* **46**

[13] Seki H, Ishihara K and Tadakuma S 2009 Novel Regenerative Braking Control of Electric Power-Assisted Wheelchair for Safety Downhill Road Driving **56** 1393–400

IOP Conf. Series: Materials Science and Engineering **342** (2018) 012089 doi:10.1088/1757-899X/342/1/012089

Understanding behavioural intention to play online game: The case of VocBlast

Z Ali

Centre for Modern Languages & Human Sciences, Universiti Malaysia Pahang 26600 Pekan, Pahang

E-mail: zuraina@ump.edu.my

Abstract. Research has shown that mobile learning enables its users to learn at any time and place. The current study investigates the use of VocBlast; an app that integrates technical and engineering vocabulary, in terms of understanding the behavioural intention of its players. The study employs 129 engineering and technical students from Universiti Malaysia Pahang (UMP). Online survey was used to collect their opinions; in particular male and female students' opinions on the use of the app in the future. The results of the study indicated that there was no significant difference pertaining to their behavioural intention using VocBlast in the course of time. The study implies that more time needs to be given to the students in playing VocBlast as it is believed that playing the game repetitively would promote positive perceptions among its players.

1. Introduction

Mobile learning has emerged as a powerful platform for learning. A report prepared by the Educause Center for Analysis and Research in 2013 revealed that 45 percent of the students at tertiary level education used mobile devices while 37 percent of them used smartphones in their daily routine as students. With this statistic in mind, it brings a major challenge for educators to recognize the benefits that the devices can bring to the students. Numerous studies have been in conducted in the use of m-learning particularly in language learning. An app namely "Remword", a free mobile English-Chinese dictionary app, integrates m-learning to enable its users to remember words they learnt [1]. All the words that have been remembered can easily be retrieved by the users if they want to revise them. This is done by clicking the search engine that is available in "New Word Book" in the app. By using the app they can search for words they have remembered and then identify new words they need to memorise. Meanwhile, the use of "English Vocabulary Crosswords" revealed that the app is useful in assisting students to know definition, terminology and lexis that consequently enables them to retain and memorise words learnt [2]. The app provides a word-rich learning for students as they can practice and review vocabulary in the app. Meanwhile, it is shown that students had positive attitudes towards mobile-assisted vocabulary exercises. More specifically they were satisfied when the teachers gave those exercises via Short Messaging System (SMS) [3]. Likewise, an app namely Kadazandusun to learn Kandazan and Dusun language; a Malayo-Polynesian language family, proved to be useful for those who were interested in learning the language [4].

On a different note, previous research comparing males and females in their behavioural intention to play online games showed that gender; as a variable, had and had not play significant roles in giving decision to its players to play or otherwise in the future. For instance, in Malaysia, a study on consumers' behavioural intention was investigated in its relation to adopting mobile entertainment (m-entertainment) between gender. The study suggests that users' behavioural intention is related to their perceived ease of use and social influence instead of gender issues [5]. The researcher found that male and female consumers used mobile phones, tablet PCs and PDAs if the cost of adopting the services were reasonable. It seemed that there was no gender difference in the adoption of m-entertainment since Malaysian government gave equal opportunity for its citizen to use or not to use the services for entertainment if they were able to afford it. On the contrary, it was found that gender influenced players in their behavioral intention to play Social Network Game (SNG) among its players in China [6]. The variable had a significant effect in moderating other variables namely achievement, social influence, perceived enjoyment, fantasy, price value; that were experimented in the study. Interestingly, females showed stronger intention to play the game in the future than males. Such reported was due to the entertainment they obtained from the online games. Moreover, the less violent and less competitive features were the reason that SNG became popular among females. In another study, albeit comparing gender was not directly investigated in the adoption of m-learning, it was found that younger female academicians were increasingly engaging in m-learning resources compared to the older female staff [7]. Such was due to their awareness in using technology for their educational methodologies. In fact, they needed to know the nuts and bolts of using the tool in delivering contents in the 21st classroom. In addition, it was found that investigating gender differences was significant in relation to playing online games [8]. Such was the limitation in Park et. Al.'s study [8] as the researchers recommended that comparing the demographic information might enable them to predict more salient outcome of their study. The study that employed approximately equal number of male and female amounting to the total number of 1049 students did not investigate the extent as to whether gender played any roles in their study. Nevertheless, one major issue in the previous research concerned with the contents of m-learning in that researchers investigated the use of m-learning as resources, social networking and entertainment between gender. No attention has been paid to the use of m-learning in relation to learning engineering and technical vocabulary between male and female students as players. Thus, the nature of the use of m-learning between gender remains unclear. This paper, therefore, attempts to understand the behavioural intention of playing VocBlast by comparing male and female students' opinions in their adoption to use the app in the future.

2. Methodology

The current study involved 129 engineering and technical students from Universiti Malaysia Pahang (UMP) as its participants. The study employed VocBlast as a research material in investigating its adoption in the future between male and female students. The app consists of ten (10) games; crossword puzzle, matching, word search; to name a few, that are arranged according to its level of difficulty. The target vocabularies selected in the app are derived from students' course books. Meanwhile, an online questionnaire was prepared for the purpose of collecting data for the study. In designing the tool, Likert scales from 1 (strongly disagree) to 5 (strongly agree) were formulated. Participants were required to give their opinions on their behavioural intention when using VocBlast. It is worth to report that during the collection of data, it was found that majority of them used Android and due to the limitation in having only two (2) iPads, they were required to share the gadgets and play in groups of five members. They took about 30 minutes to play all the games in VocBlast. Once they completed playing, the were required to answer the online questionnaire. After completing the game, they were informed to answer 24 items in the questionnaire. They took about 10-15 minutes to answer the items in the questionnaire. In analysing the data for the study, t- test was used to compare male and female students' behaviour intention in using the app; VocBlast.

3. Results and Discussion

An independent sample t-test was conducted to compare students' behavioural intention scores in using VocBlast between male and female students. Results from Table 3.1 and Table 3.2 show that there were no significant differences in scores for males (M=3.81, SD=.83) and females (M=3.78, SD=.62, t (127) = .22, p.82<.0005).

IOP Conf. Series: Materials Science and Engineering **342** (2018) 012089 doi:10.1088/1757-899X/342/1/012089

Table 3.1. Mean and Standard Deviation of students' behavioural intention of using VocBlast.

Variables	Gender	N	Mean	Std. Deviation	Std. Error Mean
Mean Behavioural Intention		81	3.8086	.82585	.09176
	Female	48	3.7778	.61917	.08937

Table 3.2. Independent Sample t-test pertaining to students' behavioural intention in using VocBlast.

	Levene's Test for Equality of Variances		t-test for Equality of Means				
	F	Sig.	t	df	Sig. (2-tailed)	Mean Difference	Std. Error Difference
Mean Behavioural Intention	4.614		.224	127	.823	.03	.14

From the results of the study, it was found that male and female students did not show positive remarks in their behavioural intention to use VocBlast. This may be due to the fact that they played VocBlast in the class. Grouping them was another barrier in getting their positive intention when they were asked of the possibility of playing VocBlast in the future. In the online questionnaire, many of the students chose scales from one (1) to three (3); having 1 as Disagree Strongly, 2 as Disagree and 3 as Neutral; when they were asked pertaining to their opinions on their intention to use VocBlast to improve technical and engineering vocabulary. This result may be explained by the fact that they need to complete each level of the game before they get to another level. In other words, they were required to complete playing ten (10) levels of the game in VocBlast. This may cause a hindrance for them to enjoy playing the game at the time that was requested by the researchers. This result is consistent with data obtained in a study that showed that behaviour intention of purchasing an app is due to its playfulness; among others [9]. Enjoyment serves as an emotional value when players play games. And as such, it is a contributing factor that makes an app to be purchased or otherwise by its players and non-players. To ensure enjoyment in playing VocBlast, therefore, it can thus be suggested that students need to play the game according to their free time. Moreover, they need to be playing the game alone as they may have the chance to think of the answers in each of the level they play in VocBlast.

The current study also found that participants gave low consideration in downloading the application in their devices. This result matched with an earlier study showing that smartphone users' in-game purchase intention was due to their perceived added values on the game itself [10]. Nevertheless, habit in the first place, is the determiner that makes ones to use in-app purchase in mobile games [11]. In fact, it is argued that habit needs to be considered when conducting research concerning online games [10]. In the previous study conducted, it was found that players need to play a particular game repetitively to promote positive perceptions with regard to the value that a particular game has to offer [10]. In the current study, being not given the chance to play all levels of games in VocBlast continually was the factor that contributed to the non-significant result of male and female students' behavioural intention to use VocBlast in the future. However, the aforesaid decision was due to it was afraid that it would be difficult to gather all participants for the slot to play the app in the next time around. An implication of this is the possibility that there is a necessity to plan for sufficient slots for the students to play VocBlast considering that majority of them studying in the university are not Apple users.

4. Conclusion

This paper has argued that there was no significant result in gender difference pertaining to their behavioural intention using VocBlast in the future. An implication of this is the possibility that in the future there is a need to require students to play VocBlast on their own and according to their free time. Such was necessary since they would be able to complete all the levels of games in VocBlast on their own. Moreover, VocBlast does not offer social relationship as it does not involve multiplayers. Therefore, students as players are required to figure out the answers in each level of the games they are playing by themselves before they move on to the other level. Nevertheless, this research extends our knowledge of online games in that players need to enjoy playing VocBlast, and requiring them to play in groups seemed to hinder their excitement. As a result of not being able to play VocBlast repetitively due to they played the game in class only, they felt that it was not worth to play it in the future. Yet, this discrepancy could be attributed to many of the students used Android as its operating system. As VocBlast operates on iOS, many students who were not familiar with the system could not enjoy the game. The evidence from this study suggests that teachers may use online games to assist students in learning vocabulary. The use of technology is convenient due to students, in particular, as it is 'part and parcel' of their lives [12]. However, for teachers, the conventional method of teaching vocabulary as requiring students to check the meaning of a particular vocabulary in dictionary; whether online or paper dictionary, is still relevant. Greater efforts for further research are needed to understand the potential use of VocBlast in the future in terms of its players' behavioural intention of playing the game to assist them in learning vocabulary.

Acknowledgement
This research was supported by Universiti Malaysia Pahang (UMP) Project ID RDU141310.

References

[1] Deng H, Shao Y. Self-directed English vocabulary learning with a mobile application in everyday context. In10th World Conference on Mobile and Contextual Learning: mLearn 2011 Conference Proceedings, Beijing, China. mLearn 2011 Oct 18 (pp. 24-31).
[2] Keyes C, Shroff RH, Chow E. Pedagogical foundations of a mobile application for language acquisition. Ubiquitous Learning: An International Journal. 2016;9(2):1.
[3] Suwantarathip O, Orawiwatnakul W. Using mobile-assisted exercises to support students' vocabulary skill development. TOJET: The Turkish Online Journal of Educational Technology. 2015 Jan 1;14(1).
[4] Pindeh N, Suki NM, Suki NM. User acceptance on mobile apps as an effective medium to learn Kadazandusun language. Procedia Economics and Finance. 2016 Jan 1;37:372-8.
[5] Leong LY, Ooi KB, Chong AY, Lin B. Modeling the stimulators of the behavioral intention to use mobile entertainment: does gender really matter?. Computers in Human Behavior. 2013 Sep 30;29(5):2109-21. [1] Kuhn E, Eftekhari A, Hoffman JE, Crowley JJ, Ramsey KM, Reger GM, Ruzek
[6] Xu X. Understanding users' continued use of online games: An application of UTAUT2 in social network games. MMEDIA 2014. 2014.
[7] Camilleri MA, Camilleri AC. Measuring the Educators' Behavioural Intention, Perceived Use and Ease of Use of Mobile Technologies.
[8] Park E, Baek S, Ohm J, Chang HJ. Determinants of player acceptance of mobile social network games: An application of extended technology acceptance model. Telematics and Informatics. 2014 Feb 28;31(1):3-15.
[9] Mulcahy R, Russell-Bennett R, Kuhn KA. Balancing entertainment and behaviour value: M-games as a social marketing agent of change. InProceedings of the 2014 Australian and New Zealand Marketing Academy Conference: Agents of Change 2014 (p. 627). ANZMAC.
[10] Han B, Windsor J. An investigation of the smartphone user's in–game purchase intention. International Journal of Mobile Communications. 2013 Jan 1;11(6):617-35.
[11] Cho D. What influences people to purchase in-game mobile items?: Analysis of motivational drivers to use ingame mobile game items in the US. Michigan State University; 2015.

IOP Conf. Series: Materials Science and Engineering **342** (2018) 012089 doi:10.1088/1757-899X/342/1/012089

[12] Ali Z, Mukundan J, Baki R, Ayub AF. Second language learners' attitudes towards the methods of learning vocabulary. English Language Teaching. 2012;5(4):24.

IOP Conf. Series: Materials Science and Engineering **342** (2018) 012090 doi:10.1088/1757-899X/342/1/012090

A study of waste and delivery valve design modification to the pump performance

M N Harith, R A Bakar, D Ramasamy, K Kardigama and Ma Quanjin
Faculty of Mechanical Engineering, Universiti Malaysia Pahang. 26600 Pekan Pahang Darul Makmur. Malaysia

E-mail: najmiharith@gmail.com

Abstract. This paper objective is to share design revolution of waste and delivery valve that contribute to the overall pump performance. In this paper, 3 new designs of waste and delivery valve pump are presented with comprehensive internal flow analysis using computational fluid dynamics (CFD) simulation over 4 cases that have been deeply study for one of the design chosen. 4 cases involving opening and closing both valve or either one. 0.265m height size of customized waste valve with an opening limiter and spring was used to demonstrate cyclic closing and opening valve operation extended up to 0.164m gap. Based on result, this characteristics contribute to 10-20% waste water reduction and enhancement of flow rate height up to 80m. Apart from that this paper also share some of pressure (dynamic, total, static), velocity (x, y, z axis) simulation including the vector flow were under different flow cases.

1. Introduction

Research with regards to hydraulic ram pump has been conducted extensively. Over the past 30 years, much more information has become available on new design approach and enhance performance of hydraulic ram pump. The first ever hydraulic ram had been constructed by Whitehurst in 1775, which had to be operated manually the opening and closing of the stopcock. This pump capable to raised water up to a height of 4.9m. Meanwhile, in 1796 the first automatic or self-acting ram pump was invented by the Frenchman Joseph Michel Montgolfier in order to raise water in his paper mill at Voiron [1]. Furthermore, despite many advantages such as free energy, self-maintenance and long service life, hydraulic ram pump also has been intensively investigated especially at the waste valve area recently due to high water losses and pressure [2].

In an analysis and enhancement of hydraulic ram pump using computational fluid dynamics (CFD), Piyush et al. [3] specifies that mass flow through the waste valve of hydram is excessive, so to reduce it, several enhancement is done and a new design is created. These are based on a systematic study of hydraulic ram pump and testing of on hydraulic ramp pump model. In another major study [4], established a suitable methodology including turbulence model selection for simulating both water and solid distribution data on different geometries using commercially available CFD software Fluent 6.1.22. The simulation results are validated with experimental data generated in the laboratory. Apart from that [5] describes the techniques and guidelines to successfully install the modern customized hydraulic ram pump. The proposed technique illustrates the methodology that can be used for the primary design considerations and applications in various ways. For this experiments, waste valve has been identified as a major contributing factor to the decline of pump performance, resulting elimination of retainer spring and additional of weighted valve. These research are supported by the findings of Awari et al. [6] which evaluate the performance of a pump under predetermined operating conditions in order to optimize the related parameters for the use of the pump under two-phase mixtures of air and water.

Surprisingly, internal flow of any hydraulic ram pump new design still not yet been closely investigated. There is a relative paucity of high quality research describing the impact of design improvement on the overall pump internal flow simulation and performance. Although studies have recognized waste valve as the main water losses area, research has yet to systematically investigate flow condition around the area. For this purpose, the aim of this work to improve the design by the additional waste control mechanism and also undergo CFD simulation classified into 4 different cycles.

The present study will demonstrate design improvements that have been made on both valve (delivery and waste valve) that will justify every cycles and stages including during acceleration, delivery and recoil period in the aspects of pressure and velocity within same new design pump. The contribution of this study is obvious as the resulting outcomes can be capitalized as guidelines to study internal flow simulation for another well improve designs and become remarkable design information to engineers for actual prototype and fabrication.

2. Method

In this work, the novel design idea and arrangement are adapted from [7], where this research managed to obtained 70% overall efficiency of delivery heads control and regulation of the hydraulic ram pump by adding adjustable local resistance components in parallel with the delivery pipes. Basically for this research, 3 designs as shown in Figure 1 were presented, but for analysis only 1 design will be selected for this paper. Basically, a customized dimension of 1.064 m length, 1.106 m height and 0.362 m width of hydraulic ram pump as shown in Figure1 was sketched by using Solid Work software. Next Figure2 shows a dedicated spring attached to the delivery instead of conventional loaded spring type valve to eliminate unnecessary mechanical movement. Waste valve in Figure3 uniquely customized with opening/closing regulator which explain the significant improvement and pressure chamber was designed longer height than usual to provide significant effect on the delivery flow rate [8].

Figure 1. Isometric View for 3 Propose Designs.

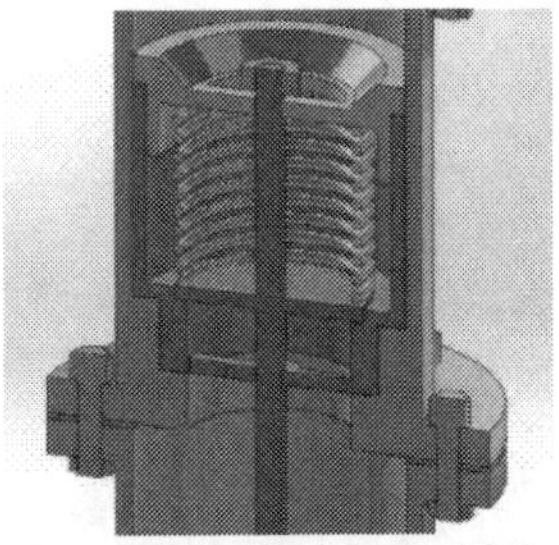

Figure 2. One of the Delivery Valve Design.

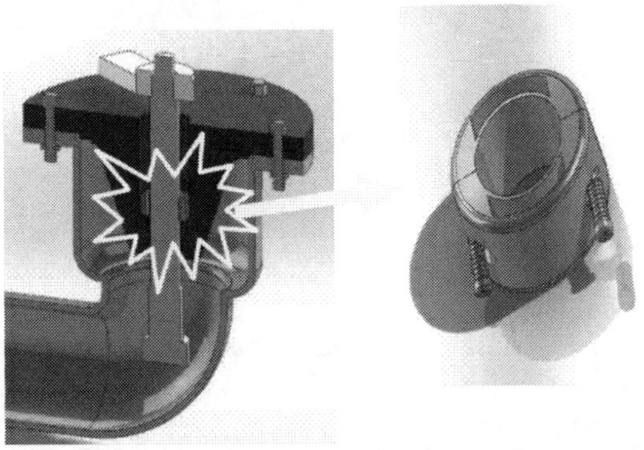

Figure 3. One of the Waste Valve Design with Customized Opening Regulator.

2.1. Working Principle of Design Selected

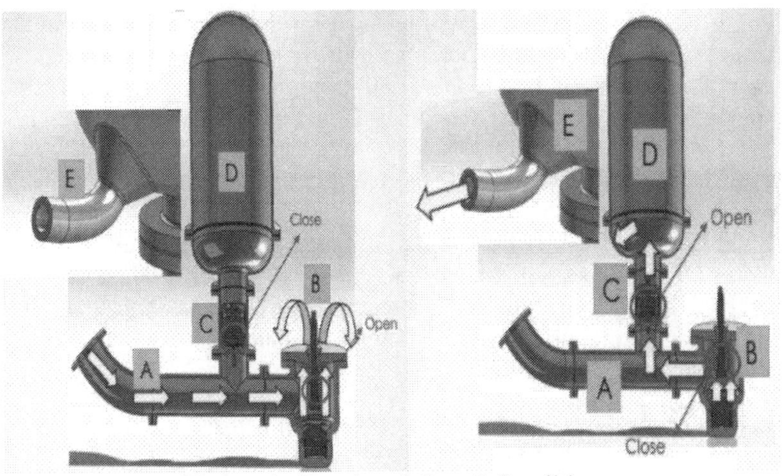

Figure 4. Case 1 and Case 2 Condition.

IOP Conf. Series: Materials Science and Engineering **342** (2018) 012090 doi:10.1088/1757-899X/342/1/012090

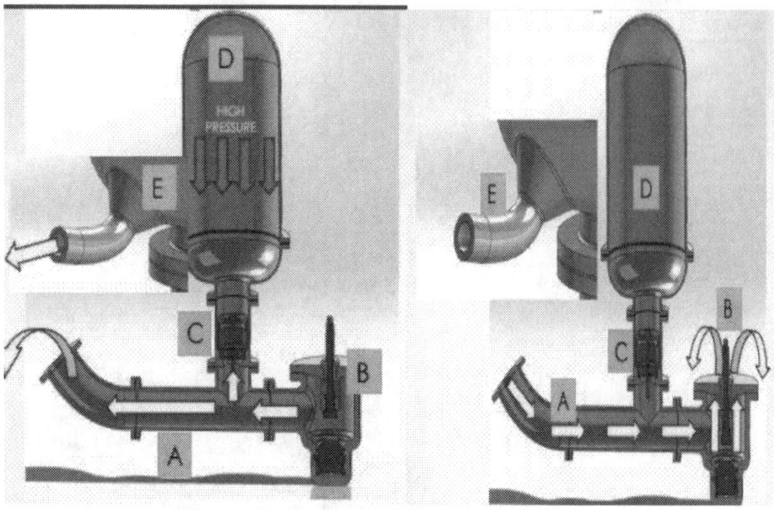

Figure 5. Case 3 and Case 4 Condition.

Figure 4 case 1 and 2 shown the initial process of pump operation by which the waste valve (B) in open condition whereas delivery valve (C) in close condition state. At a consistent flow, water flow from inlet (A) towards (B) pass through (C). When it reach (B), the water gain enough momentum to close (B). At the same time water continue flow in and create pressure to the (B) surface. As a result vigorous water flow that collide with valve surface, hammering effect is created that lead to surge pressure. This caused water to recoil back towards (C) and provide significant pressure to the chamber. During this process pressure along pipeline (A) and (B) dropped drastically.

Figure 5 case 3 illustrates how accumulation of pressure inside the chamber caused the delivery valve to close and water to be pump out towards greater heights. After the water flow out, the pressure in the chamber become low. Case 4 where all the process will start again in cycles. This processes will continue to occur with condition of consistent low viscosity clean water.

2.2. Application of ICEM and CFD

Increasing interest in the performance characteristics of hydraulic rams made clear that there is a lack of simulation skills [9]. In this research, 2 engineering analysis software being used. One of them was the ICEM computational fluid dynamics (CFD) which function as to initiate mesh preparation and generator settings as shown in Figure 6. Meanwhile, Ansys Fluent software specifically to solve flow problem and result simulation over required stages.

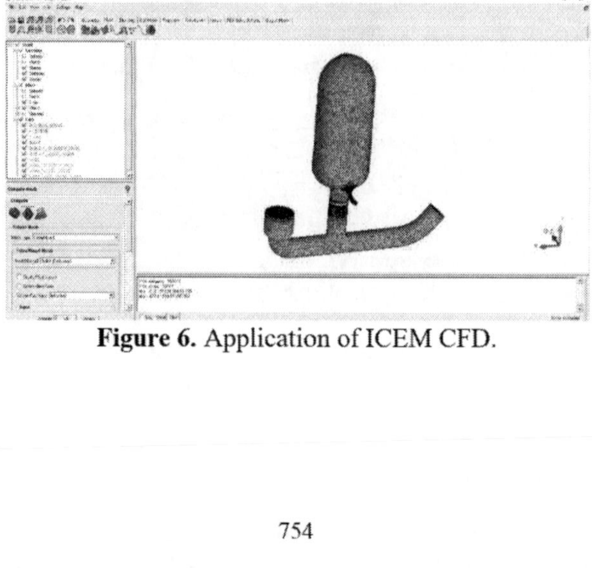
Figure 6. Application of ICEM CFD.

IOP Conf. Series: Materials Science and Engineering **342** (2018) 012090 doi:10.1088/1757-899X/342/1/012090

Before run the simulation, good meshing quality process as shown in Figure 7 is required to ensure accurate determination of numerical study and give more accurate insight into the physics of complex problem that would very impractical to study experimentally. Thus a grid and mesh independence study must be conducted to ensure good quality of mesh selection and result consistency over a selected period of iterations. For this research, the overall skewness value generated is 0.7 for surface and volume mesh as shown in Table 1 that clearly good enough for accuracy and convergence.

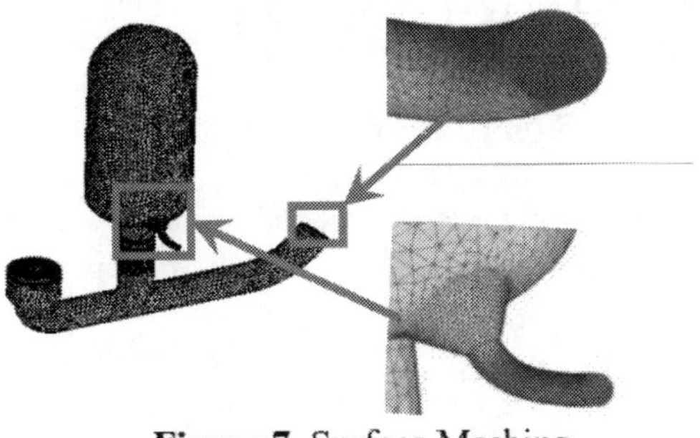

Figure 7. Surface Meshing.

Table 1. Grid Independence Specification.

Condition	Criteria	Value
Waste Valve Open, Delivery Valve Close	Mesh Method	Tetrahedral
	Quality	0.25
	Skewness	0.7
	Elements	557024
Waste Valve Close, Delivery Valve Open	Mesh Method	Tetrahedral
	Quality	0.25
	Skewness	0.7
	Elements	499094

Table 2. Mesh Independence Specification.

Type of Case	Criteria	Value
Case 1	Total Elements	560249
	Total Nodes	94275
Case 2	Total Elements	431892
	Total Nodes	73264
Case 3	Total Elements	644556
	Total Nodes	108683
Case 4	Total Elements	560249
	Total Nodes	94275

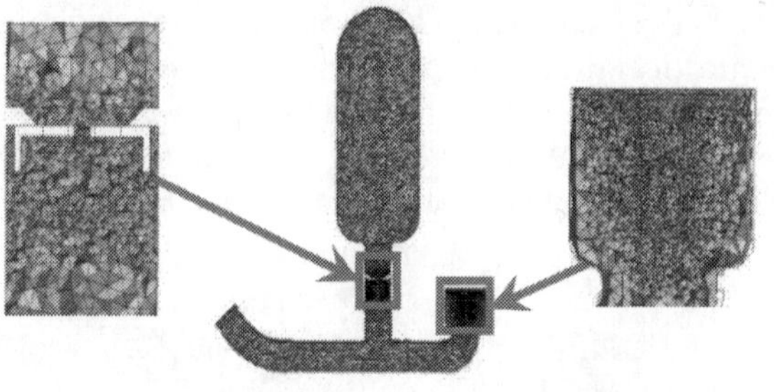

Figure 8. Volume Meshing.

Table 2 explains different geometrical over 4 sequential pump operation as explain previously that affected the simulated geometry result. Then the final stage before simulation is the boundary condition setting as being shared in Table. 3 below. Moreover, based on Figure 7 this new hydraulic ram pump design will have a fixed 1 inlet and 2 outlets for simulation purpose. Other than that, the simulation result is being illustrated in 4 cases as in Table 4 that demonstrate the cyclic opening/closing of both valve over a period of 2 seconds under transient analysis [10]

Table 3. Boundary Condition.

Condition	Setting Data
Inlet flow rate	10 m^3/h
outlet	Mass flow outlet
Environment pressure	101326 KPa
Analysis type	Transient
Turbulence model	Realizable-K epsilon turbulence model
Solver set-up	Second order
Other assumption	Wall as stationary and 4 different cases used for this simulation

Table 4. 4 Cases Description.

Type of Case	Waste Valve Condition	Delivery Valve Condition
Case 1	Open	Open
Case 2	Close	Open
Case 3	Open	Close
Case 4	Open	Open

2.3. Analysis of Convergence

During simulation, convergence is one of the method to check validity of solution or result [11]. Figure 9 provided a fluctuation downwards trend along the analysis, this is normal for the unsteady calculation. Anyhow, by increasing the number of iterations the fluctuations will still be there but can be ignored for a huge number of iterations. Alternatively, in other word increase the scope of analysis.

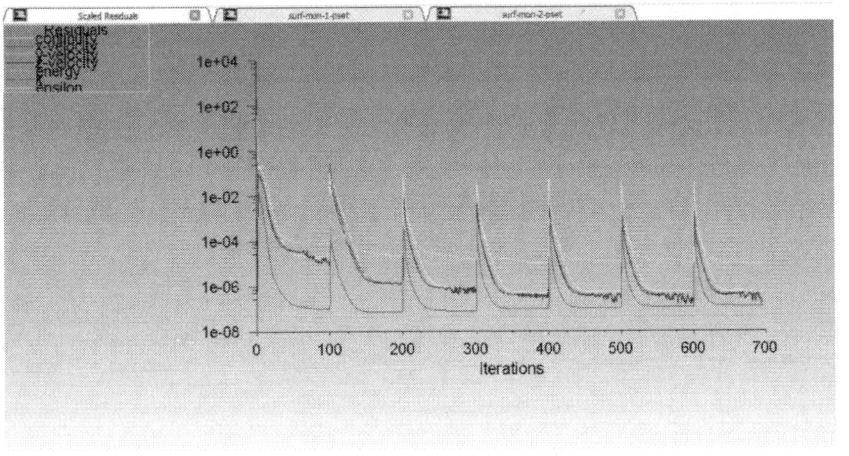

Figure 9. Convergence Data Over Iterations.

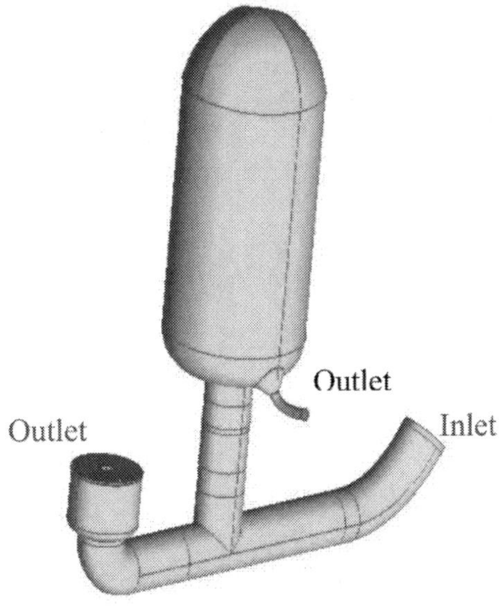

Figure 10. Fixed Inlet and Outlet Condition.

3. Results and discussion

The single most striking observation to emerge from the data comparison was the velocity vector movement as shown in Figure 8 that illustrate behavior pattern across the pipeline that also justify the real condition for every designated case. For case 1 or known as acceleration as state by [7], the waste valve in open condition state and delivery valve in closed condition. Case 1 begins when water with a flowrate of $10m^3/h$ from the source flows through the drive pipe into the ram pump body, fills pipeline gradually at the speed ranging from -9.73m/s to 10.02m/s (negative due opposite direction of water) and begins to exit through the waste valve with a range of speed between 1.224m/s and 7.345m/s. The delivery valve remains in normal closed position by both the attached spring and no pressure and water being delivered to the outlet destination [12]. The simulation continues with pressure analysis (static, dynamic and total pressure) as shown in Figure 9, Fig 10 and Fig 11 for the remaining case.

IOP Conf. Series: Materials Science and Engineering **342** (2018) 012090 doi:10.1088/1757-899X/342/1/012090

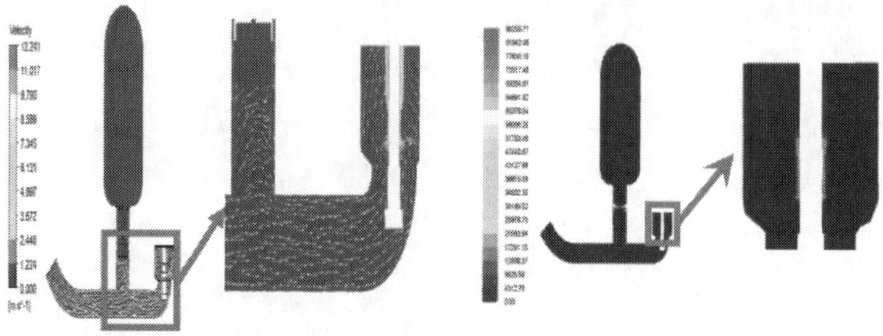

Figure 11. Velocity and dynamics pressure result for case 1.

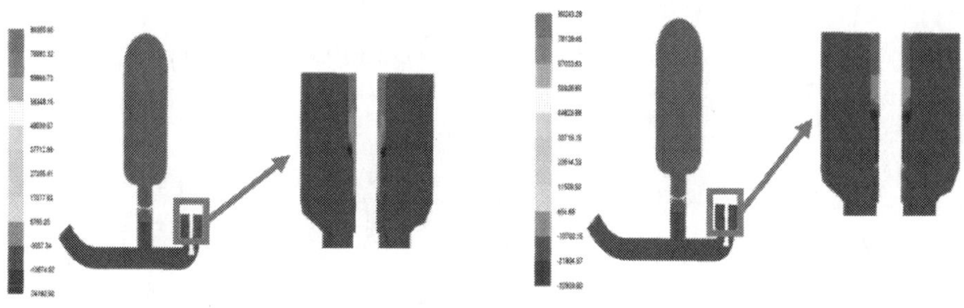

Figure 12. Total and static pressure result for case 1.

Figure 13. Velocity and dynamics pressure result for case 2.

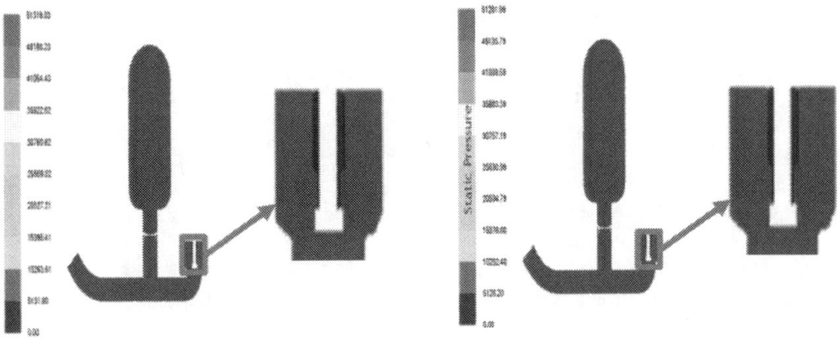

Figure 14. Total and static pressure result for case 2.

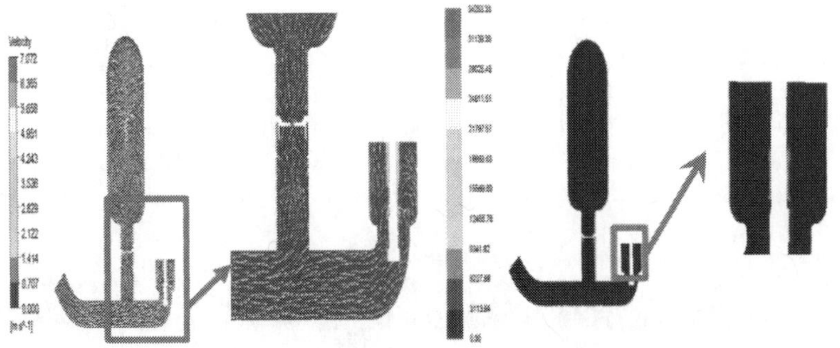

Figure 15. Velocity and dynamics pressure result for case 3.

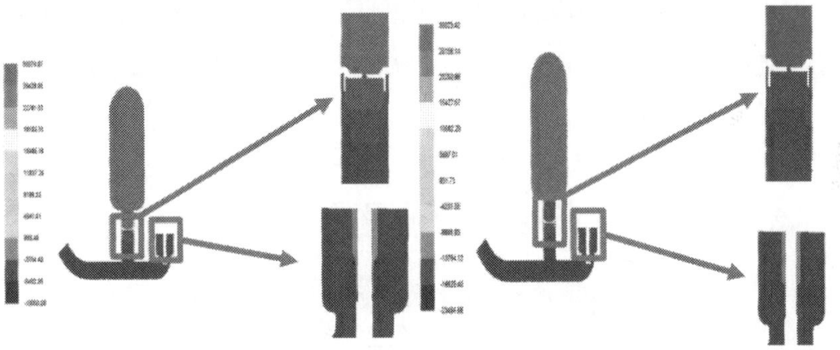

Figure 16. Total and static pressure result for case 3.

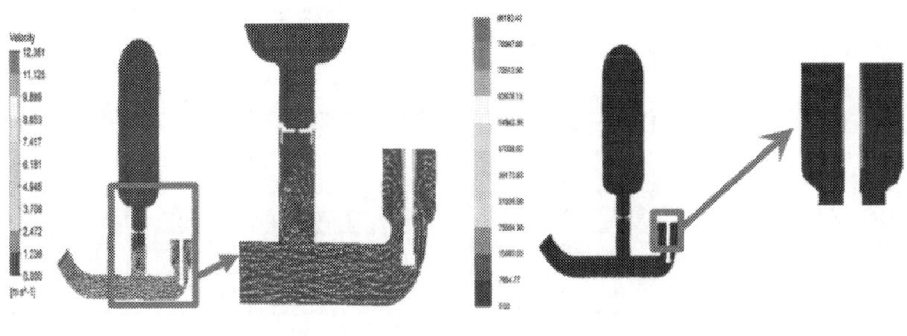

Figure 17. Velocity and dynamics pressure result for case 4.

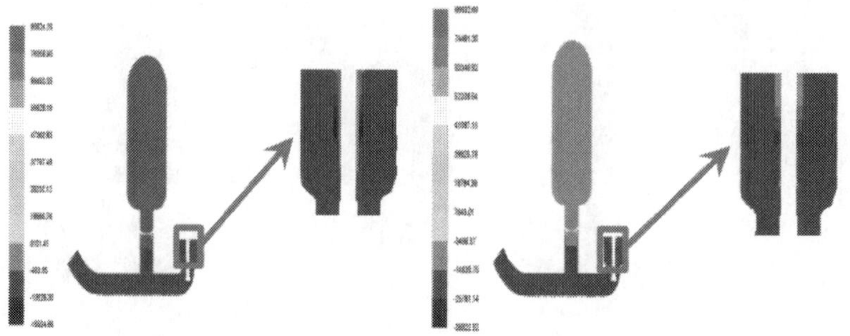

Figure 18. Total and static pressure result for case 4.

Based on the internal velocity projection and colour contour from Figure 11 till Figure 18, it clearly show that each case contibute to difference velocity and pressure contour along the pipeline. The colour appear generated much improve up to 60% enhancement perfomance in term of velocity and pressure and only 40% water losses in hydraulic ram pump as shown in Table 5 compare to existing design in the market that recorded over 60-70% water losses [13]. This analysis successfully minimizes losses generated at waste valve and at significantly improve the pumping capacity of newly design pump.

Table 5. Overall simulation result.

Case	Parameters	Min	Max	Mass flow Rate (kg/s)
Case_1	X-velocity (m/s)	-9.73	10.02	2.8 (waste valve)
	Y-velocity (m/s)	-3.00	14.36	
	Z-velocity (m/s)	-9.36	9.16	
	Static pressure (Pa)	-90019.72	89269.26	
	Dynamic Pressure (Pa)	0	103134.7	
	Total pressure (Pa)	-33084.33	101617.5	
Case_2	X-velocity (m/s)	-4.72	4.61	2.8 (Delivery valve)
	Y-velocity (m/s)	-9.82	2.611	
	Z-velocity (m/s)	-9.03	1.97	
	Static pressure (Pa)	-33755.69	59968.68	
	Dynamic Pressure (Pa)	0	60627.01	
	Total pressure (Pa)	0	60247.15	
Case_3	X-velocity (m/s)	-5.71	5.64	1.67 (waste valve) 1.12 (Delivery valve)
	Y-velocity (m/s)	-3.57	8.58	
	Z-velocity (m/s)	-5.24	5.79	
	Static pressure (Pa)	-40669.32	30126.38	
	Dynamic Pressure (Pa)	1.7649e-08	36889.26	
	Total pressure (Pa)	-19625.42	33314.94	
Case_4	X-velocity (m/s)	-9.619028	9.93	2.8 (waste valve)
	Y-velocity (m/s)	-3.070991	14.32	
	Z-velocity (m/s)	-9.281295	9.09	
	Static pressure (Pa)	-88607.82	85683.18	
	Dynamic Pressure (Pa)	0	102462.8	
	Total pressure (Pa)	-32589.2	101365.1	

4. Conclusion

This project was undertaken to design a new model of hydraulic ram pump and evaluate the internal flow analysis to justify improvement on the design that has been implemented. This study set out to assess the effects of new design especially design of the waste and delivery valve towards internal flow within 4 projected cases and specified boundary condition. The results of the investigation show that the projection of velocity vector for every details case and pressure contour. This projection lead to creation of result data in terms of velocity along 3 axis, static pressure, dynamic pressure, total pressure and mass flow rate. The main findings can be summarized as follow:

- Projection of internal flow within new design hydraulic ram pump. Justification for every lowest and highest analysis data for every case.
- Clear statement for every improvement and designated simulation. Reason and comparison from previous researches.

These findings enhance of understanding regarding the adjustment of design towards internal flow changes and overall performance [14]. It is recommended to develop this propose design to actual modelling and may explore the results for future research.

Acknowledgements

The authors gratefully acknowledge the help of the Ministry of Youth, Pahang state government in providing the fund and Universiti Malaysia Pahang for making available its lab facilities and technical expert.

References

[1] Kumar H, Beyene T, Ofgaa G, Kaso M, Rao P N and Negassa T 2016 Pollution free design and manufacturing of hydraulic ram pump for villages in hill areas. *Int. J. Mech. Eng. Res. and Tech* **2**(2) 15
[2] Denson M, Galad M, and Malamug J 2016 Design and fabrication of a hydraulic diaphragm pump *CLSU International Journal* of **1**(1) 17
[3] Piyush B, Shende S K and Choudhary A P N 2015 Analysis and enhancement of hydraulic ram pump using computational fluid dynamics (CFD) *International Journal for Innovative Research in Science and Technology* **2**(3) 24
[4] Balgude R D, Rupanavar S P, Bagul P S, and Ramteke A R 2015 Designing of hydraulic ram pump *International Journal Of Engineering And Computer Science* **4**(4) 2319–7242
[5] Li J, Yang K, Guo X and Guo Y 2015 Adjustable and high-head model experiment on the hydraulic ram pump *China Institute of Water Resources and Hydropower Research* **46**(1) 6
[6] Dhrupad S, Monotosh D, Bipul B, Deepak P and Sermirlong M 2016 Investigation and parameter optimization of a hydraulic ram pump using taguchi method. *J. Inst. Eng. India Ser* **1**(1) 9
[7] Verspuy C and Tijsseling A S 1993 Hydraulic ram analysis. *Journal of Hydraulic Research* **31**(2) 267–278
[8] Reza Fatahi A, B L A and A K 2015. Determine the efficiency of hydraulic ram-pumps. E-Proceedings of the 36th IAHR World Congress
[9] Bhaskaran R, and Collins L 2005 Introduction to CFD Basics. Cornell University - Sibley School of Mechanical and Aerospace Engineering, Ithaca, USA, 1–17.
[10] XiongY L and C H Bruneau Y D 2017. Numerical study on viscoelastic fluid flow past a rigid body *Applied Mathematical Modelling* **42**(1) 20
[11] Bhaskar K U, Murthy Y R, Raju M R, Tiwari S, Srivastava J K and Ramakrishnan N 2007 CFD

simulation and experimental validation studies on hydrocyclone *Minerals Engineering* **20** 60–71

[12] Ndache Mohammed S 2007 Design and construction of a hydraulic ram pump from leonardo electronic journal of practices and technologies. *Leonardo Electronic Journal of Practices and Technologies* **1**(11) 59–70

[13] Diwan P, Patel A and Sahu L 2016. Design and fabrication of hydraulic ram with methods of improving efficiency **34** 2393–8374

[14] Nambiar P, Shetty A, Thatte A, Lonkar S, and Jokhi V 2015. Hydraulic ram pump: Maximizing efficiency. In 2015 International Conference on Technologies for Sustainable Development (ICTSD) pp 1-4

IOP Conf. Series: Materials Science and Engineering **342** (2018) 012091 doi:10.1088/1757-899X/342/1/012091

Factors influencing the choice of performance measures for the oil and gas supply chain – exploratory study

Masha Menhat[1,2] and Y Yusuf[2]

[1] School of Maritime and Business Management, Universiti Malaysia Terengganu, 21030, Kuala Terengganu, Malaysia
[2] Lancashire Business School, University of Central Lancashire, PR1 2HE, Preston, United Kingdom

E-mail: mashanur@gmail.com

Abstract. The current value of the oil price per barrel has severely impacted the oil and gas industry around the world. This has worsened the situation due to the fact that it has long been the backbone of the country through the energy supply, employability and also its role as the major economic contributor. Due to multiple external factors that affect this industry such as crude oil availability, oil price fluctuation, high transportation cost, as well as exposure to high uncertainties, it is sensible for the supply chain practitioner to shift their focus in managing their resources and capabilities. To maximise the potential of supply chain activities in improving overall company performance, it is important to pay extra attention on their performance management. This includes the design of meaningful performance measurement framework to assess organisational performance. This study will explore the influencing factors in choosing performance measures for the oil and gas supply chain. Five in-depth interviews were conducted with supply chain experts within the industry. Eight influencing factors have been identified through the interviews.

1. Introduction

The drop of oil price per barrel to below USD 35 [1] has severely impacted the oil and gas industry around the world. To make things worse, this industry has been the backbone of the oil and gas operators country through its role as the major economic contributor [2,3]. Even though it receives positive responses from the end user [oil and gas consumer], but the affect that it has on the oil and gas producer and the economic condition of the country are not to be undermined [4].

There are multiple external factors that affect this industry such as crude oil availability, oil price fluctuation, high transportation cost, as well as exposure to high uncertainties [5–9]. Therefore, it is sensible for the supply chain practitioners of the industry to shift their focus into managing their resources and capabilities. This can be conducted through supply chain management. As with other industries, the reliance on outsourcing activities has mushroomed in the oil and gas industry. Research suggested that 30 - 40% of oil and gas activities will be outsourced to accomplish the project [6,10]. Oil and gas organisations normally outsource part of their function to subsectors in order to minimise their operation cost, to overcome their limited manpower and expertise as well as to pass some of the supply chain risk to another party [11]. To maximise the potential of supply chain activities in improving overall company performance [12], it is important to pay extra attention on supply chain performance management. Anatan [13] suggested that supply chain management practices have a significant influence on the achievement of competitive advantage. This study will explore the factors that influencing the choice of performance measures for the oil and gas industry. In-depth interview is conducted with five supply chain experts to explore these factors.

2. Supply chain performance management

Performance management is a highly relevant topic for business organisation and has gained scholars attention since the 1970s [14]. However, the earliest works on performance management of supply chain are primarily focused on the context of manufacturing industry [15,16]. Nowadays, with the increase on reliance of outsourcing activities, companies are more concern on improving their supply chain performance to compete in todays' market [17–19]. In line with this trend, research on supply chain performance management that focuses on various contexts of supply chain management have gained so much attention by both researchers and supply chain practitioners.

The industrial characteristics influence supply chain strategy and structure, where different industry requires different strategy [20]. For instance, due to the competitive market in the garment industry [21], the performance framework should also focus on the agility measures, which were driven by innovative product and unstable demand [22]. On the contrary, the automobile industry's performance rely upon the forecasting accuracy [as a make-to-stock system] and this requires measuring the accuracy of forecasting [23]. The given examples suggest that different supply chain context will provide different views in managing their supply chain, which were important on the knowledge expansion on this subject. This in line with Gunasekaran et al. [24] that emphasised the importance of concentrating on the most critical success factors in developing performance measures. They added, most of the companies are still lacking in designing meaningful measures to assess organisational performance.

Chima and Hill [25] stated that the main challenges face by the oil and gas industry is not the oil and gas resources availability but instead on the production and delivery of refined product to the end consumers at the lowest cost. Similar to this, Varma et al. [26] highlight the industry imposed by high transportations cost in comparison to the other industries. The authors proposed other distinguishing characteristics of the industry including process industry, in-flammability, contamination, bulk volumes, long supply chain and fluctuation of raw materials. Despite presenting performance measures for the downstream oil and gas industry, Varma et al. [26] did not investigate some other important elements of the industry, which are safety and environmental factors.

Among other research conducted in the oil and gas context are the local content issues [27–29] project management of upstream oil and gas in the United Kingdom [30] and the link between agility supply chain and business performance [31]. Asrilhant et al. [30] proposed that the key driver for performance success is through internal business elements but it receives inadequate attention by the management. In particular, some of the key elements such as flexibility, interdependency, and learning and innovative are often neglected. Similarly, Yusuf et al. [31] finds that there is strong positive correlation between supply chain and competitive factors like quality, proactive, delivery, speed and innovation. These findings support similar elements, which are known as key driver to success in Asrilhant et al. [30] work. Despite considerable evidence presented performance management research in the supply chain context, there are still many areas remain unexplored [23,32]. Amongst the areas that are worth further investigation are the understanding of characteristics of various supply chain measures in different approaches and environment, the critical success factor in supply chain integration and partnership, benchmarking in performance measures and also the role of socio–environmental measures in supply chain. In particular, a study on supply chain performance within oil and gas environment is still under researched. As such, this research will explore the factors influence the choice of performance measures in the oil and gas industry.

3. Methodology

An exploratory approach was adopted for this study consisting of in-depth interview with five supply chain practitioners from the oil and gas industry. The same method has been widely used for exploring an area when there is scarce publication on the topic [30,33,34]. Five phone interviews with supply chain managers based in the United Kingdom, United States, Indonesia, and Malaysia were undertaken. The reasons these five participants were chosen are based on accessibility and contact that the researcher had with them when she was working in the oil and gas industry. Voss et al. [35] proposed that the ideal contact to gain access for case study is by approaching the person that could open the door. The participants involved have 15 – 30 years experiences in the supply chain and oil and gas industry. These consisted of those involved in the operator side and also contractor side. Table 1 depicted a brief description about the participants for this study. All the participants except participant A have experiences working directly with oil and gas operators. Nevertheless, as a consultant himself Participant A have various experiences dealing with major oil and gas operators around UK, which positioned him as suitable participant for this study. The phone interviews were recorded, transcribed, and analysed. The analysis processes involved coding and constructing themes by categorizing the similarities in the interviews data [36,37]. This process, which is known as thematic analysis [38] was conducted using Microsoft Excel.

Table 1. Participants demographic.

Participant	Year of experience	Role in company	Country	Educational background	Previous companies	Current company
A	20	President	UK	Engineering	Services advisory – oil and gas operator	Services advisory
B	25	Project procurement manager	USA	Business administration	Oil and gas operator and contractor	Services advisory
C	15	Senior supply chain manager	Indonesia	Engineering	Oil and gas operator and contractor	Oil and gas operator
D	30	Head of supply chain	Malaysia	Quantity surveying	Oil and gas operator and contractor	Oil and gas contractor
E	15	Leader, supply chain	Malaysia	Engineering	Oil and gas operator and contractor	Oil and gas contractor

4. Result and analysis

4.1 Factors that influence the choice of performance measures

There are various factors suggested by participants which are local content, the need for balance priority across measures, the role in supply chain which either operator or joint investor, company's reputations, desire for safety, profitability potential, oil price per barrels, and finally company's objective.

4.1.1 Local content

The local contents, which aim to leverage business wealth into country wealth are seen as part of the factors that influence the choice of performance measures. All the rules and regulation set by the country in operation such as localize supply chain should be incorporated into the supply chain performance management. Increasingly, the local agency whose provide license to the operator has inquired them to engage certain amount of services or equipment from the local companies instead of international companies.

"The percentage of your total bought in of your goods and services is being set for local country provision. And sometimes it might be in absolute amount of spend in pound. But, usually it is in percentage." [Participant A]

Even though other participants did not state the local content as the influencing factor in choosing the performance measures, most of them have suggested that it has affect the way they manage their supply chain performance as qualitative extracts;

"So, if you are working in a remote area, your company might have a desire to hire a certain amount of local people or local companies to work on those jobs. I think that becomes very important." [Participant B]

"The local content required us to deal with the local supplier "[Participant C]

"And another thing on procurement is, if we go to the market, we need to look at whether the market being regulated by the government. Here in Malaysia, the government objective is to maximise local participation." [Participant D]

Qualitative extract from Participant C is one of the classic examples on how they have to consider local content requirement in managing their performance;

"...where most of them are the middle person and do not have sufficient knowledge on product or services they provided." [Participant C]

In order to comply with local content requirement while at the same time ensuring services delivery, company have to go one step further in developing local companies and thoroughly measure their performance in the caveat of the level of competencies of the local providers. In addition to that, as pointed by participant E, there is a need to trade time and cost in order to meet the stringent process set by the country in question. For this reason, other measures like process compliance have to be introduced and implemented to tackle this issue.

"They might need more transparent and more stringent process involves in our supply chain activities which in the same time come with cost and time." [Participant E]

4.1.2 Role of company in supply chain

Participants A and E apprehend that it is important to consider the role that a company have in supply chain which are either operator or joint investor. A company role in supply chain will influence the information required and evaluate in managing supply chain performance.

"So, for example in procurement, if you are a joint venture operator, you would want to make sure, while you are not responsible of the delivery of the procurement, you want to make sure that the operator who is responsible is following a good and fair process and is compliance to the standards and procedures that has been set in place." [Participant A]

"If we look into client perspectives which involve national oil and gas companies, there are other indicators that out weight cost and timeliness which is integrity and transparency because they are representing public entities or Governments Corporation." [Participant E]

The previous interview responses suggest that the role a company have in supply chain will determine the performance measures that are meaningful in assessing their performance.

4.1.3 Reputations of company

One of the participants suggested that the reputation of the company particularly on environmental concerns and corporate social responsibilities are also influence the choice of performance measures. This factor will determine how company managing their operations and what are the measures that important to the company to ensure that company's reputations are maintained.

"The primary effect that I've seen driving measurement one is the reputation of your company. I think companies in the oil and gas business are really concern about the reputations. They want to make sure that they are doing things responsibly, protecting people, and protecting environment." [Participant B]

4.1.4 Desire for safety

Desire for safety in the oil and gas industries is also one of the determinant factors in choosing performance measures. Pertaining to this, the priority of safety measures over other measures is believed to drive the performance measurement of oil and gas companies. Moreover, participants B, C, and E emphasised that safety aspects have higher priority over any other measures including financial measures.

"In a safety critical industry, safety has to be a very important measure. And health and safety supply chain is critically important to the operators." [Participant A]

"I think the next factor is sort of the desire for safety, going to operate safely. And basically not to perform work that is going to harm anyone. So, I think that companies in the oil and gas business place safety and facilities integrity as more important than lowering costs." [Participant B]

"Oil & Gas Industry is high risk industry where safety and quality are considered have priority over price and cost." [Participant C].

"Safety always considered as the first priority for the oil and gas companies. That is why we have to ensure our supply chain always kept safety in mind." [Participant E]

4.1.5 Profitability potentials

Apart from that, most of the participants have commented that profitability potential is considered as one of the important drivers in determining choices of performance measures to be used in managing their supply chain.

"And another determinant factor is profitability potential." [Participant B]

"Whatever performance measures that we choose must be able to contribute to the profit and loss. For example, cost saving would be directly influence the bottom line, and then timely delivery that wouldn't cause any standby would also directly improve the bottom line. That is why I said, all the measures selection should reflect on how to improve the bottom line." [Participant D]

"Since most of the oil and gas companies are running by the profit driven organisation, most of the indicators chosen are influence by cost and delivery reliability aspects." [Participant E]

4.1.6 Risk mitigations

The next factor highlighted by participants A, B, and C is to consider risk mitigation aspect in choosing performance measures. Participant A mentioned about locality issues like crime rate as part of consideration in determining performance measures. While, Participant B elaborates the need for joint venture activities for cost and risk sharing in the oil and gas exploration activities. Thus, the latter suggest the requirement in managing performance of the partnership activities.

"If you are in the midstream distribution in the Nigerian delta where theft is a problem, people tapping into the pipeline and selling out black market refineries then, I imagined crude loss would be a very important measure." [Participant A]

"And the next one I have is risk mitigation. So, you know the oil and gas business requires really large sum of money for projects and at the same time have really large potential profitability. But, there is really large factor of risk of loss and then they have to reserve that in their project. So, you see in the oil and gas business upstream where companies will take on partners and joint ventures caused the potential risk in a one company might be too great if they go in solely by themselves." [Participant B]

"If we were to pass all the risk to our vendor, then obviously they will factor in somewhere in their cost. A lot of people love to have risk pre-contract, but a lot of time we fail to realise whether the premium that we have to pay is worth it or not. Maybe in some cases, we could get better price by managing the risk instead of pass all the risk to the subcontractor."
[Participant C]

4.1.7 Oil prices per barrel

Other than that, the world's oil price per barrel is considered as one of the factors in devising performance measures for supply chain management in the industry. For instance, companies have to relook into their strategy in order to sustain in this industry in a low-price oil environment and revisit their priorities to reflect the situation. Qualitative extracts from the participants explain the situation;

"Again, we get back to the 100$per barrel prices of oil, if we are in supply chain and all we want to measure is cost, the business might be saying that the most important now is the speed of completion. Because if we can get stuff on-board faster, the higher profitability from more quickly obtaining oil & gas sales is greater than focusing on reducing project costs." [Participant B]

"Since the world's oil price is very low, we are trying very hard. There are a lot of areas that we have to squeeze. One thing for example a lot of time we never care less about the brand that we purchase, if we normally use certain brand, we just go for that. Nowadays, the oil and gas price have dropped a lot, we started looking for other options."
[Participant D]

"Take for example, in the current dropped of oil price per barrel, we got to really manage our cost and to get more jobs for the company to sustain. In this case we have to change our sourcing strategy, choose different vendors or equipment in order to get better price." [Participant E]

4.1.8 Company's objectives

Finally, company's objectives are also considered as the influencing factor. Qualitative extracts of Participant D explain how the measures chose have to be aligned with company objective, which in this case he spell out about company's profit target. Other participants mentioned about company's objectives with regards to desire for safety, profitability potential, company's reputation as the drivers in choosing performance measures. These have been presented in the previous sections.

"For example, first of all we want to have 300mil which is the enabler. So, from our revenue stream we found that we can only gain 280mil, we short of 20mil, where can we get this 20 mil? One of the options is through staff reduction, and another 10 mil maybe through procurement saving, so that we go to bid. That is how we normally do it. Performance measures must always starts from objective, whatever PMs you designed, must be able to support you to achieve your objective. It must be aligned with the company objective." [Participant D].

Table 2 summarises the factors influencing the choice of performance measures according to participants.

Table 2. Factors that influence the choice of performance measures.

Influencing factors	Participant A	Participant B	Participant C	Participant D	Participant E
Profitability potential		√		√	√
Company's reputation		√			
Desire for safety	√	√	√		√
Risk mitigation	√	√		√	
Oil prices per barrels		√		√	√
The role in supply chain	√				√
Local content	√	√	√	√	√
Company's objectives	√	√		√	

5. Discussion

Several significant findings can be drawn from this study. First, we have seen that local content has been presented by all participants as an influencing factor in managing supply chain performance. It is interesting to come across similar opinion raised by participants operating in developing countries and also developed countries on their concern with regard to local content. This lead to the next question whether the local content requirements are applied both for the work performed in these two continents. Our follow up question clarified this matter where Participant B explains that they have to comply with local content requirement in USA which concern more on safety, environmental issues and social responsibilities and not particularly concentrated on engaging local services provider. There are a lot of literatures discussing on Local content requirement which has started since 1990s where they focus on the booming industry on that time, manufacturing and automobile industry [16,27,29,39]. Recent article from [28] presents how the international oil and gas companies have to respond to local content requirement to international oil and gas companies, which includes to renegotiate local content requirement with local government proved that it is part of vital components in supply chain performance management.

Other than that, the three influencing factors in choosing performance measures namely role of company in supply chain, profitability potential, and company's objective are recognised amongst the factors that received the same amount of attention in the supply chain of other industry. In the literature review of performance measures and metrics in logistic and supply chain management from 1995 – 2004 by Gunasekaran & Kobu [23], it is highlighted that most of the widely used performance measures are financial measures which consists of 38% of total measures. Not only that, Taticchi et al. [40] have also mentioned about the need to meet conflicting objectives of maximising profitability and at the same time reducing operating cost and a lot other literatures have addressed the same measures [14,32,41]. In this sense, this particular factor (Company's objectives) makes no different between the oil and gas industry and other industries. In the same vein, existing literatures claimed that company's objective is one of the fundamental factors in designing performance measurement framework [14,32,42]. From this analysis it is appealing to investigate further the most common company's priority in this industry. While for role in supply chain on either the operator side or joint investor, it relates back to the information required by the customer. If there is the need for traceability and detail information for the bidding exercise, it has to be part of the performance measures.

As specified by the participants, the reputation areas that they focus on are social responsibilities and also safety issues. Nowadays, the company's reputation and their concern on safety have gained immense attention from the public, non-government organisation, environmentalist, and human right organisation. Not only that, any environmental incident issues resulted from oil and gas activities will attract so much publicities and eventually tarnished their reputation in the industry. This is due to the impact that it has to the people and environments [34,42,43]. Hence, it is essential to consider company's reputation in deciding on performance measures.

In addition, the participants also proposed risk mitigation as part of the influencing factors. In the case of the oil and gas industry, it is important to note that it is impossible to dictate the oil and gas reservoirs for exploration activities. On top of that, there is the risk of low oil and gas content within it after they have incurred so much cost and effort on it. Therefore, risk mitigation is something that requires company consideration in structuring their strategy and performance framework. Apart from risk mitigation, oil price per barrel has the huge impact on their business strategy as it will determine company's profitability potential. Based on the response from one of the participant, company will shift their focus on different performance measures in different oil price environment. For example, he mentioned that at the highest price, they will focus on delivery speed. Whilst, on the lower price they have to focus more on the cost saving to ensure they could manage their margin. In this case, a company has to be flexible to response to the external environment by managing their internal resources and capabilities. The same applies in supply network of other industry. "When business was booming, executives concentrated on maximising speed, and when the economy head south, firms desperately tried to minimise supply costs" [44]. This might infer, for the oil and gas industry, the world oil prices per barrel are one of the driving factors in priorities their performance measures.

6. Conclusion and research limitation

This exploration study has present significant findings on the influencing factors in the choice of performance measures. There are eight influencing factors have been identified for the oil and gas industry. More specifically, with the aid of existing literature, we managed to determine the factors, which are also relevant to other industry and the distinguishing characteristics in those factors. Local contents requirement has found to be the only factor proposed by all participants. Hence, it is very important to manage local content issue in achieving supply chain performance improvements. This finding is in line with previous literature and works report provided by the oil and gas advisory companies. One of the limitation of this study is the exploratory interview was conducted on a small number of respondents. The study on larger number of experts will bring more added values to this exploratory study. Moreover, the study was conducted among respondents that largely involve on the upstream sector and the result could not represent an accurate perspective on the downstream parts of the industry. However, with the aid of existing literature on the downstream part, further research can be performed to validate this result with large number of respondents in order to construct generalisation on the influencing factors in choosing performance measures for the industry.

References

[1] Oil price falls to fresh 11-year low. BBC News (Internet). 2016 Jan 6 (cited 2017 Dec 22); Available from: http://www.bbc.co.uk/news/business-35243442
[2] International Energy Agency. Energy Policies of IEA Countries -The United Kingdom 2012 Review. 2012.
[3] Organisation Petroleum Exporting Countries. OPEC-IEA cooperation and the international oil market outlook (Internet). 2015 (cited 2015 Apr 10). Available from: http://www.opec.org/opec_web/en/886.htm
[4] Hayes A. Companies affected most by low oil prices. Investopedia. 2015.
[5] Urciuoli L, Mohanty S, Hintsa J, Boekesteijn EG. The resilience of energy supply chains: a multiple case study approach on oil and gas supply chains to Europe. Supply Chain Manag Int J. 2014;19(1):46–63.
[6] Yusuf YY, Gunasekaran A, Musa A, Dauda M, El-Berishy NM, Cang S. A relational study of supply chain agility, competitiveness and business performance in the oil and gas industry. Int J Prod Econ. 2014 Jan;147:531–43.

IOP Conf. Series: Materials Science and Engineering **342** (2018) 012091 doi:10.1088/1757-899X/342/1/012091

[7] Price Water House Cooper UK. The impact of lower oil prices on the UK economy. 2015.
[8] Varma S, Wadhwa S, Deshmukh SG. Evaluating petroleum supply chain performance: Application of analytical hierarchy process to balanced scorecard. Vol. 20, Asia Pacific Journal of Marketing and Logistics. 2008. p. 343–56.
[9] Al-Othman WBE, Lababidi HMS, Alatiqi IM, Al-Shayji K. Supply chain optimization of petroleum organization under uncertainty in market demands and prices. Eur J Oper Res. 2008;189:822–40.
[10] Pillai M, Sandelands E, Ashokan G. Developing EPC Value Chain. Oil and Gas business. 2010;
[11] UK Upstream Oil and Gas Supply Chain – Economic Contribution | Oil & Gas UK (Internet). (cited 2017 Dec 22). Available from: https://oilandgasuk.co.uk/uk-upstream-oil-and-gas-supply-chain-economic-contribution.cfm
[12] Shi M, Yu W. Supply chain management and financial performance: literature review and future directions. Int J Oper Prod Manag. 2013 Sep 16;33(10):1283–317.
[13] Anatan L. Factors Influencing Supply Chain Competitive Advantage and Performance. Int J Bus Inf. 2014 Sep 1;9(3):311.
[14] Neely A, Gregory M, Platts K. Performance measurement system design: A literature review and research agenda. Int J Oper Prod Manag. 1995 Apr 1;15(4):80–116.
[15] Beamon BM. Measuring supply chain performance. Int J Oper Prod Manag. 1999 Mar 1;19(3):275–92.
[16] Munson CL, Rosenblatt MJ. The Impact of Local Content Rules on Global Sourcing Decisions. Prod Oper Manag. 1997 Sep 1;6(3):277–90.
[17] Fernandes LJ, Barbosa-póvoa AP, Relvas S. Risk Management Framework for the Petroleum Supply Chain. Risk Manage. 2010;28:157–62.
[18] Lambert DM, Cooper MC. Issues in Supply Chain Management. Ind Mark Manag. 2000;83(29):65–83.
[19] Wildgoose N, Brennan P, Thompson S. Understanding your supply chain to reduce the risk of supply chain disruption. J Bus Contin Emerg Plan. 2012 Jan;6(1):55–67.
[20] Halman Johannes I. M., Voordijk Johannes T. Balanced Framework for Measuring Performance of Supply Chains in House Building. J Constr Eng Manag. 2012 Dec 1;138(12):1444–50.
[21] Sukwadi R, Wee H-M, Yang C-C. Supply Chain Performance Based on the Lean–Agile Operations and Supplier–Firm Partnership: An Empirical Study on the Garment Industry in Indonesia. J Small Bus Manag. 2013 Apr 1;51(2):297–311.
[22] Stratton R, Warburton RDH. The strategic integration of agile and lean supply. Int J Prod Econ. 2003 Aug 11;85(2):183–98.
[23] Gunasekaran A, Kobu B. Performance measures and metrics in logistics and supply chain management: a review of recent literature (1995–2004) for research and applications. Int J Prod Res. 2007 Jun 15;45(12):2819–40.
[24] Gunasekaran A, Patel C, McGaughey RE. A framework for supply chain performance measurement. Int J Prod Econ. 2004 Feb 18;87(3):333–47.
[25] Chima CM, Hills D. Supply-Chain Management Issues. 2007;5(6):27–36.
[26] Varma S, Wadhwa S, Deshmukh SG. Evaluating petroleum supply chain performance: Application of analytical hierarchy process to balanced scorecard. Asia Pac J Mark Logist. 2008 Jul 4;20(3):343–56.
[27] Kazzazi A, Nouri B. A conceptual model for local content development in petroleum industry. Manag Sci Lett. 2012;2(6):2165–74.
[28] Ngoasong MZ. How international oil and gas companies respond to local content policies in petroleum-producing developing countries: A narrative enquiry. Energy Policy. 2014 Oct 1;73(Supplement C):471–9.
[29] Ovadia JS. Local content and natural resource governance: The cases of Angola and Nigeria. Extr Ind Soc. 2014 Nov 1;1(2):137–46.
[30] Asrilhant B, Dyson RG, Meadows M. On the strategic project management process in the UK upstream oil and gas sector. Omega. 2007 Feb;35(1):89–103.

[31] Yusuf YY, Musa A, Dauda M, El-Berishy N, Kovvuri D, Abubakar T. A study of the diffusion of agility and cluster competitiveness in the oil and gas supply chains. Int J Prod Econ. 2014;147(PB):498–513.

[32] Gopal PRC, Thakkar J. A review on supply chain performance measures and metrics: 2000-2011. Int J Product Perform Manag. 2012 Jun 15;61(5):518–47.

[33] Urciuoli L, Mohanty S, Hintsa J, Boekesteijn EG. The resilience of energy supply chains: a multiple case study approach on oil and gas supply chains to Europe. Supply Chain Manag Int J. 2014 Jan 7;19(1):46–63.

[34] Bunse K, Vodicka M, Schönsleben P, Brülhart M, Ernst FO. Integrating energy efficiency performance in production management – gap analysis between industrial needs and scientific literature. J Clean Prod. 2011 Apr 1;19(6):667–79.

[35] Voss C, Tsikriktsis N, Frohlich M. Case research in operations management. Int J Oper Prod Manag. 2002;22(2):195–219.

[36] Barton A. Research Methods: A Practical Guide for the Social Sciences. By B. Matthews and L. Ross (Harlow: Longman, 2010, 490pp. £30.00 pb)Crime: Local and Global. By J. Muncie, D. Talbot and R. Walters (Cullompton: Willan, 2010, 263pp. £75.00 hb, £23.00 pb). Br J Criminol. 2012 Sep 1;52(5):1017–21.

[37] Yin RK. Case Study Research: Design and Methods. SAGE; 2009. 241 p.

[38] Braun V, Clarke V. Using thematic analysis in psychology. Qual Res Psychol. 2006;3(2):77–101.

[39] Qiu LD, Tao Z. Export, foreign direct investment, and local content requirement. J Dev Econ. 2001 Oct 1;66(1):101–25.

[40] Taticchi P, Tonelli F, Pasqualino R. Performance measurement of sustainable supply chains: A literature review and a research agenda. Int J Product Perform Manag. 2013 Oct 28;62(8):782–804.

[41] Otto A, Kotzab H. Does supply chain management really pay? Six perspectives to measure the performance of managing a supply chain. Eur J Oper Res. 2003 Jan 16;144(2):306–20.

[42] Forum OIIE and P. Environmental Management in Oil and Gas Exploration and Production: An Overview of Issues and Management Approaches. UNEP/Earthprint; 1997. 76 p.

[43] Publication: Energy Policies of IEA Countries - The United Kingdom 2012 Review (Internet). (cited 2017 Dec 22). Available from: http://www.iea.org/publications/freepublications/publication/energy-policies-of-iea-countries---the-united-kingdom-2012-review.html

[44] Lee HL. The Triple-A Supply Chain (Internet). Harvard Business Review. 2004 (cited 2017 Dec 22). Available from: https://hbr.org/2004/10/the-triple-a-supply-chain

The study of Phosphorus distribution at Putrajaya Wetland

[1]Nazirul Mubin Zahari, [1]Nur Farzana Fasiha Abdul Malek, [1]Chow Ming Fai, [1]Siti Humaira Haron, [1]Mohd Hafiz Zawawi, [1]Iszmir Nazmi Ismail, [1]Daud Mohamad, [1]Agusril Syamsir, [1]Lariyah Mohd Sidek, [1]Mohd Zakwan Ramli, [2]Norfariza Ismail, [3]Ahmad Zubir Sapian, [3] Normaliza Noordin, [3]Nurliyana Abdul Rahaman, [3]Yahzam Muhamad and [3]Jarina Mat Saman

[1]Institutes Energy Infrastructure, Department of Civil Engineering, Universiti Tenaga Nasional (UNITEN), Jalan IKRAM-UNITEN, 43000 Kajang, Selangor, MALAYSIA
[2] Civil and Structural Engineering Branch, Research and Development Division, Malaysian Public Works Department, Jalan Sultan Salahuddin, Kuala Lumpur, MALAYSIA
[3] Environmental Department, Lake & Wetland, Putrajaya Corporation (Perbadanan Putrajaya), 24 Persiaran Perdana, Presint 3, Putrajaya, MALAYSIA

E-mail : mubinzahari@gmail.com

Abstract. This study is concerning phosphorus distribution in Putrajaya Wetland. Phosphorus is one of the important component in nutrients for living things be it aquatic or non – aquatic organisms. Total phosphorus (TP) results will give some information on the trophic status of surface water in water bodies. The focus of this study is to determine the total phosphorus concentration in Putrajaya Wetland which is in the inlet of the wetland then outlet of the wetland (Central Wetland Lake). The water sample is taken from Putrajaya Wetland and the test was conducted in the laboratory. The result from this study shows the results for total phosphorus according to month, sampling station and cells. Lowest total phosphate at the Central Wetland compare with all the wetland arms cells.

1. Introduction

Phosphorus is one of the important component in nutrients for living things whether aquatic or non – aquatic organisms. Phosphorus is a significant element because it is crucial for nucleotides and ATP (adenosine triphosphate) which are fundamental energy sources for living beings. However, excessive phosphorus content in ecosystem could lead to negative impact. Putrajaya Wetland is one of the tourist attractions in Klang Valley area. High concentration of phosphorus in the wetland could cause negative impacts to the aquatic and non-aquatic organisms in the wetland. This can be a contributing factor to ecosystem problem.

Phosphorus is a fundamental component in all living beings. Genetic materials such as DNA as well as energy producing ATP (adenosine triphosphate) and general compounds such as phospholipids are essential in living cells as a constituent [1]. Phosphorus plays an important role in the ecosystem. It has a predominant function in cellular energetics as ATP and is an important part of many structural and biochemically functional components for the cell growth in plants [2]. Various pollutants including phosphorus (P) May 2017 be absorbed in the sediments that have accumulated at the bottoms of wetlands as well as lakes and rivers. These sediments May 2017 accumulate over long periods of time and the pollutants can be released into the water and being utilized by the plants. Phosphorus concentrations generally are low in seawater and can restrict organic efficiency in some marine biological systems [3,4].

Kalff et. al [5]. stated that P is often the limiting nutrients in freshwater and brackish system. Variation in soil's P content influences the biogeochemical cycle of elements including nitrogen and carbon. Pote et. al [6] found that P is the most often element that limits eutrophication because of too many blue-green algae that are able to utilize atmospheric gases. Sims et. al [7] stated that loss of P from agricultural production systems are known to contribute to acceleration of eutrophication in natural waters. Stewart et. al [8] found that plants primarily acquire the P needed for their development from soil P reserves. Soil P content is mainly affected by factors such as organisms and biogeochemical processes in soil, climate and parent material whereby its distribution has a large spatial heterogeneity.

1.1. Phosphorus distribution

Phosphorus is the main and basic nutrient for marine biological system. Its biogeochemical cycle plays an important role in limiting the primary production process in marine environment. Marine sediment is regarded as a crucial carrier for phosphorus cycling. When the external loading of P becomes higher, the sediments can absorb it as a pool. When the water loading is reduced, the sediments would release the absorbed phosphorus to the aquatic surroundings as a pollution source. Total phosphorus (TP) evaluation May 2017 give some information on the trophic status of sediments in marine biological system. However, not all of the phosphorus fractions can be released from sediment into the overlying water, and the released phosphorus fractions can barely have a similar mobility and bioavailability.

To study phosphorus behavior in sediments in assessing the trophic status, it is more efficient to evaluate based on the phosphorus fractions. Amid sorption process, considering phosphorus portions is additionally beneficial to heighten the phosphorus sorption on dregs, taking the holding method of phosphorus and residue into account. Thus, it May 2017 be useful to assess the portability and bioavailability of the absorbed phosphorus. Phosphorus sorption/desorption in the sediment– water interface is an essential procedure influencing the phosphorus cycling in any marine eco-frameworks. Other than temperature, saltiness and pH of the media, properties of residue particles, for example, molecule measure, particular surface territory, surface charge thickness, can likewise be of noteworthy significance for phosphorus sorption process. Considering the phosphorus sorption kinetic characteristics and phosphorus form changes during sorption process, the sorption could be considered as an important physical process. Increase in temperature benefits the sorption of phosphorus. Sorption was a spontaneous, endothermic and entropy increasing process. It has a dominating function in cell energetic as ATP and is a vital part of numerous auxiliary and biochemical utilitarian parts for cell development in plants. Different contaminations including P might be caught up in the silt accumulated at the bottom of waterways and lakes or being used up by plants. Internal loading in the sediments plays an important role in wetlands. Hence, it is important to evaluate the potential of P release and its utilization. However, not all types of P are available for plants. Therefore, it is necessary to investigate the fractions and bioavailability of P in sediments. Knowledge in P speciation is important to observe environmental changes and P bio availability. Some studied shows P bioavailability in the sediments from the middle and lower reaches of the Yangtze River region using different chemical extraction methods. The results showed that the bioavailable P concentrations can be evaluated by measuring the concentrations of TP, IP and NAIP in the heavily polluted sediments and that bioavailable P can only be evaluated using different chemical extracting methods in slightly polluted sediments [9]. The main objective of this study is to investigate the total phosphorus (TP) distribution in surface water at Putrajaya Wetland.

IOP Conf. Series: Materials Science and Engineering **342** (2018) 012092 doi:10.1088/1757-899X/342/1/012092

2. Methodology

2.1. Putrajaya Wetland

The Putrajaya's constructed wetland system comprises of five arms (refer to one main stream with total 23 cell) will discharge to the Central Wetland which make the 24 cells in all before flows into the Putrajaya Lake. They straddle the water courses of Sungai Chuau, Sungai Bisa and three tributaries. A rock filled weirs was constructed along the five arms of the wetland to divide all cells in wetland. Although all the five arms are connected, they differ in size, depths, plant communities and pollutant loads that each is designed to handle. Putrajaya Wetland is designed with features a multi-cell of multi-stage system with flood mitigation capability to make best use of the space available for colonization by water plants. All plants inside wetland function to intercept any pollutants. These plants act as to facilitate filtering and removing water pollutants where bacteria and microorganisms can flourish [10].

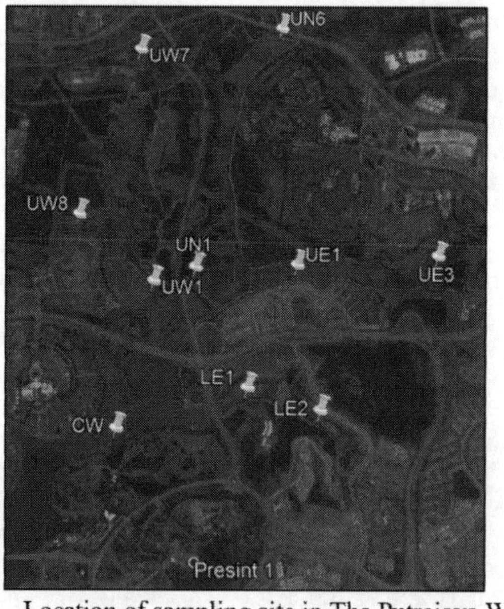

Figure 1. Location of sampling site in The Putrajaya Wetland.

Table 1. Sampling coordinate at Putrajaya Wetland.

POINT	NAME	LATITUDE	LONGITUDE
UW7	Upper West 7	2.972827°	101.692815°
UW8	Upper West 8	2.961730°	101.690493°
UW1	Upper West 1	2.958223°	101.695829°
UN6	Upper North 6	2.975433°	101.701737°
UN1	Upper North 1	2.959439°	101.698333°
UE3	Upper East 3	2.962103°	101.713427°
UE1	Upper East 1	2.960403°	101.704643°
LE2	Lower East 2	2.951822°	101.707165°
LE1	Lower East 1	2.952581°	101.702536°
CW	Central Wetland	2.949129°	101.694874°

2.2. *Sample Collection and Laboratory Test*

All water samples were taken from Putrajaya Wetland. Samples were collected from 10 points within 5 months for this study starting from January 2017 to May 2017. Table 1 shows the coordinate of each sampling location at Putrajaya Wetland. Each sample is taken from individual point for every end of the month. The water samples were collected at three points which are at the inlet, middle and outlet. The raw sample were then being tested immediately within 48 hours. Then, all data is tabulated and interpreted in graphical format. The data is tabulated according to month, station and arm. These total phosphorus values were recorded in mg/l unit. HACH method 8190 is used to determine the total phosphorus. The water samples was collected and kept in a bottle before transport to the lab.

3. Results

3.1. *Distribution of phosphorus at each cell by month*

Figure 2 shows the total phosphorus results by cells from January 2017 – May 2017. The total phosphorus for cell CW shows the decreasing trend compare with other cells from January 2017 to May 2017. The total phosphorus concentration reading for January 2017 and February 2017 is 0.10 mg/l while in March 2017, April 2017 and May 2017 is 0.11 mg/l. There is a slightly increasing amount of total phosphorus concentration from February 2017 to March 2017 which is 0.01 mg/l. Central Wetland has the lowest concentration of total phosphorus among all 12 cells because possible pollutant already treated at wetland arm.

Both LE1 and LE2 are located at the Lower East sector. LE1 is at the middle while LE2 is the inlet and the water will eventually flow to CW. Generally, LE1 concentration of total phosphorus for January 2017 is 0.12 mg/l, and increased rapidly in February 2017 2017 (0.18 mg/l), decreased from 0.04 mg/l to 0.14 mg/l in March 2017 and slightly increase to 0.15 mg/l in April 2017 and May 2017. LE2 in January 2017 is 0.15 mg/l and increased to 0.23 mg/l in February 2017, then decreased to 0.15 mg/l in March 2017 but increased to 0.21 mg/l in April 2017 and then decreased back to 0.18 mg/l in May 2017. The phosphorus concentration pattern in LE2 and LE1 seems fluttered probably due to the inlet water sources at the Lower East sector which is in precinct 12, whereby from observation, currently there is construction project going on. Contaminants in forms of sediment and floatable types are those commonly found in storm water from construction activities.

At Upper East sector there are UE1 and UE3 cells. UE1 is the middle and UE3 is the inlet. From Figure 2, UE3 total phosphorus concentration bar is slightly higher than UE1. As for UE1, total phosphorus concentration in January 2017 and February 2017 is consistent at 0.18mg/l with no changes, then a slight decrease to 0.16 mg/l in March 2017 and gradually increased to 0.17 mg/l and 0.21 mg/l respectively in April 2017 and May 2017. On the other hand, UE3 concentration in January 2017 is 0.21 mg/l and increased by 0.04 mg/l to 0.25 mg/l in February 2017. In March 2017 the concentration further decreased to 0.18 mg/l and 0.14 mg/l in April 2017. However, it further increased by 0.0 mg/l to 0.16 mg/l in May 2017. The main inlet source of Upper East sector is from shopping complex which is the commercial area and other sources are for instance, the ongoing construction sites there. Possible contaminants such as pesticides, herbicides and organic materials can be found from commercial area.

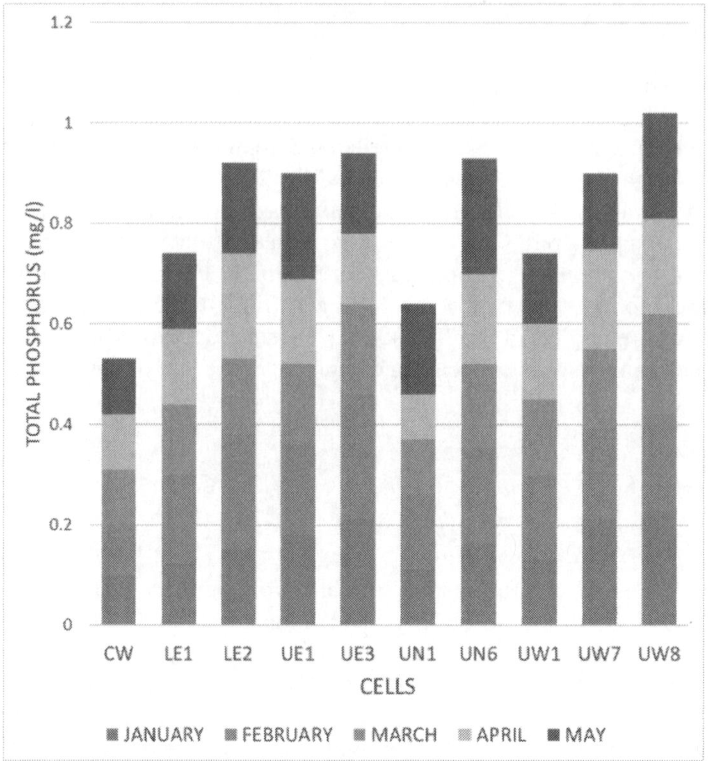

Figure 2. The Total Phosphorus Results by Cells from January 2017 – May 2017

As for the Upper North sector, there are middle cells UN1 and UN6 serves as inlet of the sector. According to Figure 2, UN1 has the lowest bar of concentration while UN6 have equal level of concentration. UN1 concentration in January 2017 is 0.11 mg/l, 0.15 mg/l in February 2017 and decreased to 0.11 mg/l and 0.09 mg/l in March 2017 and April 2017 respectively, then rapidly increased to 0.18 mg/l in May 2017. For UN6 cell, the concentration of total phosphorus in January 2017 is 0.16 mg/l, increased by 0.03 mg/l to 0.19 mg/l in February 2017 and decrease to 0.17 mg/l in March 2017. It increased gradually to 0.18 mg/l and 0.23 mg/l in April 2017 and May 2017 respectively.

At Upper West sector, there are UW1 which serves as the middle, as well as UW7 and UW8 are the inlets. The bar for UW1 is the lowest among the cells in Upper West structure followed by cells UW7 and UW8. In January 2017, total phosphorus concentration for UW1 is 0.13 mg/l and increased by 0.04 mg/l to 0.17 mg/l in February 2017 which then decreased gradually to 0.15 mg/l in March 2017 and April 2017, and 0.14 mg/l in May 2017. For inlet UW7, the concentration in January 2017 is 0.21 mg/l, gradually decreased by 0.02 mg/l in February 2017 to March 2017 with 0.18 mg/l and 0.16 mg/l respectively. In April 2017, the concentration increased by 0.04 mg/l to 0.20 mg/l and decrease to 0.15 mg/l in May 2017. For inlet UW8, the phosphorus in January 2017 is 0.23 mg/l and decreased to 0.19 mg/l in February 2017. In March 2017 it increased by 0.01 mg/l to 0.20 mg/l and decreased back by 0.01 mg/l to 0.19 mg/l in April 2017, but in May 2017 it increased to 0.21 mg/l. The sources of storm water entering from UN6 is from Serdang which is the farm area. It serves as a centre for agro-based events and activities. For UW6, the source of inlet is from Presint 11 which residential area.

IOP Conf. Series: Materials Science and Engineering **342** (2018) 012092 doi:10.1088/1757-899X/342/1/012092

4. Conclusion

The first objective is to investigate the phosphorus (P) distribution in Putrajaya Wetland. The P distribution of all data were determined and fulfill an objective. This research determines the distribution of phosphorus in Putrajaya Wetland according to month, sampling station and flow arm. The objective is to determine the phosphorus (P) content in inlet, middle and outlet of Putrajaya Wetland was successfully using method HACH 8190.

Acknowledgements

This research project was supported by the Universiti Tenaga Nasional (UNITEN) under internal grant (UNIIG 2017) and Perbadanan Putrajaya.

References

[1] Takeshi Yoshimura, J. N. (2007). Distributions of particulate and dissolved organic and inorganic. *Marine Chemistry 103*, 112-121.
[2] Dorich, R. N. (1984.). Availability of phosphorus to algae from eroded soil fractions. *Agric. Ecosyst. Environ. 11*, ,253–264.
[3] Takeshi Yoshimura, J. N. (2007). Distributions of particulate and dissolved organic and inorganic. *Marine Chemistry 103*, 112-121.
[4] Thingstad, T. Z. (1998). P limitation of heterotrophic bacteria and phytoplankton in the northwest. *Limnology and Oceanography 43,*, 88-94.
[5] Kalff, J. (2002). Limnology—Inland Water Ecosystems. *Prentice Hall, Upper Saddle.*
[6] Pote, D. D. (1996). Relating extractable soil phosphorus to phosphorus losses in runoff. *Soil Sci. Soc. Am. J. 60*, 855-859.
[7] Sims, J. S. (1993). Phosphorus loss in agricultural drainage—historical perspective and current research. *J. Environ. Qual. 27,*, 277–293.
[8] Stewart, J. T. (1987). Dynamics of soil organic phosphorus. *Biogeochemistry 41*, 41–60.
[9] Ding Lua, P. G. (2015). Evaluation of phosphorus distribution and bioavailability in sediments of a subtropical wetland reserve in southeast China. *Ecological Indicators 16*, 556-563.
[10] Bahagian Alam Sekitar, Tasik Dan Wetland, Perbadanan Putrajaya. (N.D.). The Putrajaya Lake And Wetland. Retrieved from http://plwmos.ppj.gov.my/v_intro_lake_wetland.asp

Effect of Phosphate levels on vegetables irrigated with wastewater

S O Oladeji[1] and M D Saeed[2]

[1]Polymer Technology Department, Hussaini Adamu Fed. Polytechnic, Kazaure, Jigawa State, Nigeria
[2]Chemistry Department, Bayero University Kano, Kano State, Nigeria

E-mail: saheedilori75@gmail.com

Abstract: This study examined accumulation of phosphate ions in wastewater and vegetables through man-made activities. Phosphate level was determined in wastewater and vegetables collected on seasonal basis along Kubanni stream in Zaria using UV/Visible and Smart Spectro Spectrophotometers for their analyses. Results obtained show that phosphate concentrations ranged from 3.85 – 42.33 mg/L in the first year and 15.60 – 72.80 mg/L in the second year for wastewater whereas the vegetable had levels of 3.80 – 23.65 mg/kg in the year I and 7.48 – 27.15 mg/kg in the year II. Further statistical tests indicated no significant difference in phosphate levels across the locations and seasons for wastewater and vegetables evaluated. Correlation results for these two years indicated negative ($r = -0.062$) relationship for wastewater while low ($r = 0.339$) relationship noticed for vegetables planted in year I to that of year II. Phosphate concentrations obtained in this study was higher than Maximum Contaminant Levels set by Standard Organization such as WHO and FAO for wastewater whereas vegetables of the sampling sites were not contaminated with phosphate ions. Irrigating farmland with untreated wastewater has negative consequence on the crops grown with it.

1. Introduction

Phosphorus is a naturally occurring element in the environment that can be found in all living organisms as well as in water and soils. It is an essential component for many physiological processes related to proper energy utilization in both plants and animals [1]. It normally occurs in nature as part of a phosphate ion $(PO_4)^{3-}$, consisting of a phosphorus atom and 4 oxygen atoms; the most abundant form is orthophosphate. On land, most phosphate is found in rocks and minerals. It can be added to the environment by man's activities as point source discharges or as non-point source runoff. Typical sources include industrial and municipal wastewater discharges or runoff from agricultural lands or urban areas. Livestock derive part of their phosphate needs from plant materials. However, much of the naturally occurring phosphate in grains is in an indigestible form [2]. Therefore, inorganic phosphates sources are added to poultry and swine feeds to ensure adequate nutrition and to prevent rickets. As a result, much of the dietary phosphate passes through livestock and poultry and is excreted in animal manure. Utilizing animal manure as a fertilizer on crop and grazing land can recycle nutrients. Plants uptake phosphate from soil mostly in the orthophosphate forms. Natural soil phosphate level is often low enough to limit crop production. Both inorganic phosphate fertilizers (treated rock phosphate) and organic phosphate sources (animal manures) are equally adept at supplying the orthophosphate ion and correcting phosphate deficiencies in soil [3]. Most of the phosphates in animal manure are in an organic form and must be converted to plant-available forms via soil biological activity, a process known as mineralization. The net effect of this characteristic is that phosphate derived from animal manure act more like a slow-release fertilizer than commercial inorganic fertilizers, which are more soluble and readily available to plants [4].

Phosphorus is an essential nutrient for plants and animals. It is a limiting nutrient for aquatic organisms. Phosphorus forms parts of important life-sustaining molecules that are very common in the biosphere. It does not enter the atmosphere, mostly remaining on land, in rock and soil minerals [5]. Eighty percent of the mined phosphorus is used to make fertilizers. Phosphates from fertilizers, sewage and detergents can cause pollution in lakes and streams [6]. Over enrichment of phosphate in both fresh and onshore marine waters can lead to massive algae bloom which, when they die and decay leading to eutrophication of fresh water. Phosphates rich deposits have generally formed in the ocean or from guano, and over time, geologic processes bring ocean sediments to land. This study examined phosphate accumulation in wastewater and vegetables of sampling area through man-made activities.

2. Materials and Methods

2.1. Sampling

Sampling was done in three seasons (harmattan, dry and rainy) for two years making six (6) samples per site and this was done using composite sampling technique at point of inlet into the river at five different points along Kubanni stream channels in Zaria, Kaduna State, Nigeria. Sample bottles used were firstly rinsed with sampled water three times before filling to the brim at a depth of one meter below the wastewater from each of the five designated sampling points. These sample bottles were labelled accordingly, stored in ice-blocked coolers and transported to the laboratory while in the laboratory; they were stored in the refrigerator maintained at 4 °C prior to the analysis [7]. Vegetable samples of spinach, lettuce, cabbage, carrot, okra, onion and tomato were randomly handpicked from different garden plots along Kubanni stream channels using hand-gloves, bulked together to form a composite sample, wrapped in envelopes and labelled accordingly. Each vegetable samples were washed with tap water, followed by deionized water, air dried in the laboratory, grounded to powder and sieved using 250 µm sieve [8].

2.2. Phosphate determination in wastewater

100 cm^3 of wastewater sample was dispensed inside conical flask; one drop of phenolphthalein indicator was added and thoroughly shaked. The sample turned pink and a few drops of strong acids (HNO_3) were added until the pink colour disappeared. The volume of the solution was adjusted to 1000 cm^3 with deionized water. To this solution, 4 cm^3 of molybdate reagent and ten drops of $SnCl_2.2H_2O$ solution were added and thoroughly mixed for colour development at the temperature of 30 °C. The solution changed to blue colour, it was allowed to stand for twelve minute and the intensity of colour was measured at 690 nm using JENA Model UV/Visible Spectrophotometer. The blank was prepared by taking 40 cm^3 deionized water through the same procedure as in the wastewater sample [7]. Orthophosphate ion combines with ammonium molybdate under acidic condition to form a yellow complex compound known as ammonium phosphor-molybdate. It was then reduced by tin(II)chloride solution in glycerol, the yellow complex is reduced to a blue-coloured compound due to molybdenum blue.

$$PO_4^{3-} + 12(NH_4)_2MoO_4 + 24H^+ \longrightarrow \underset{\text{(yellow complex compound)}}{4(NH_4)_3PO_4.3MoO_3} + 12NH_4^+ + 12H_2O \quad (1)$$

$$4(NH_4)_3PO_4.3MoO_3 + Sn^{2+} \longrightarrow \text{Molybdenum blue} + Sn^{4+} \quad (2)$$

2.3. *Phosphate determination in vegetable samples*
Each of the vegetable samples was chopped into small pieces. The chopped samples were then air-dried. The air-dried samples were ground and sieved with a sieve of 1 mm mesh. Then, 1 g of each of the sieved samples was weighed into acid-washed porcelain crucibles and analyzed as described by [9] using Smart Spectro Spectrophotometer 2000. The results of phosphate in wastewater and vegetables were expressed as bar-charts using Microsoft Excel, these results were subjected to one way Analysis of Variances (ANOVA) and Pearson Product Moment Correlations (PPMC) using Statistical Package for the Social Sciences (SPSS) 20.0 version software.

3. Results and Discussion

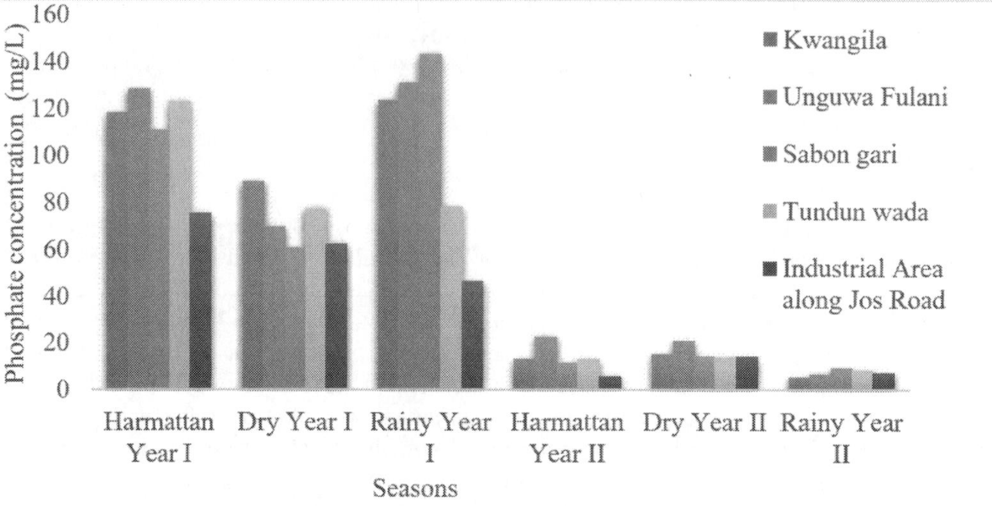

Figure 1. Concentrations of phosphate in wastewater across seasons.

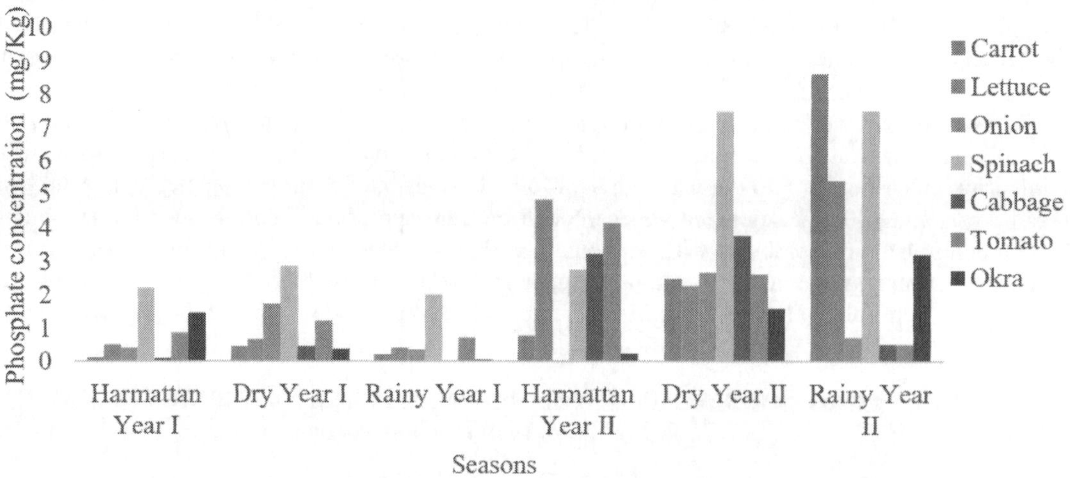

Figure 2. Concentrations of phosphate in vegetables analyzed.

Figure 1 show phosphate concentrations in wastewater from Kubanni stream channels. The concentrations determined were in the range of 46.07 – 143.20 mg/L for the first year. Highest level of 143.20 mg/L was obtained at Sabon-gari during the rainy season followed by 130.70 mg/L at the same season but in Unguwa-fulani and this was closely followed by 128.20 mg/L at the same Unguwa-fulani but in the harmattan season while the least concentration of 46.07 mg/L was recorded at Industrial area along Jos road in the rainy season. Elevated levels of phosphate during harmattan and rainy seasons could be as a result of runoff from agricultural land, sewage and detergents as suggested by [3]. Phosphate levels were also high at these sampling sites; Kwangila (123.54 mg/L) in the rainy season, Tundun-wada (123.20 mg/L) during the harmattan season and Sabon-gari (110.74 mg/L) in the harmattan season. Phosphate concentrations were generally high in wastewater for year I which might be due to heavy discharged of effluents from municipal and industrial areas as suggested by [10]. The concentrations determined were in the range of 5.14 – 22.50 mg/L in second year. Highest concentration (22.50 mg/L) was recorded at Unguwa-fulani during the harmattan season followed by 20.74 mg/L from the same site but in the dry season while the lowest level of 5.14 mg/L was noted at Kwangila sampling site during the rainy season. The chart showed low levels of phosphate in year II which could be attributed to flooding of first year that led to heavy erosion and washed away most of the phosphates in wastewater as suggested by [2]. Other sites with low levels of phosphate were; Kwangila (15.04 mg/L), uniform concentrations of 14.08 mg/L were recorded at both Industrial area along Jos road and Sabon-gari while 13.71 mg/L was observed at Tundun-wada, all these results were obtained in the dry season. Comparing the results obtained in year I with that of year II, it was revealed that harmattan season of first year (74.98 – 128.20 mg/L) had high concentration of phosphate in wastewater than harmattan season of second year (5.74 – 22.50 mg/L), likewise, dry season of year I (60.24 – 88.75 mg/L) showed high level of phosphate than dry season of year II (13.71 – 20.74 mg/L). In the same way, rainy season of year I (46.07 – 143.20 mg/L) had high level of phosphate in wastewater than corresponding rainy season of year II (5.14 – 9.24 mg/L). WHO recommends 5 mg/L as maximum contaminant level for phosphate in wastewater before it could be discharged on river and this indicates wastewaters analyzed are polluted with phosphate ions. [10] reported 103.23 – 164.22 mg/L as phosphate levels in wastewater which was above concentrations obtained in second year while similar to levels obtained in the year one of this study.

Figure 2 presents phosphate levels in vegetables collected along Kubanni stream channels. Determined concentrations ranged from 0.11 – 2.85 mg/Kg for the year one. Highest levels of 2.85 mg/Kg was obtained in spinach planted during the dry season followed by the same spinach (2.20 mg/Kg) in the harmattan season while the least concentration of 0.11 mg/Kg was observed in carrot cultivated during the harmattan season. Spinach showed high level of phosphate than other vegetables analyzed which could be due to its large surface area as suggested by [11]. Other vegetables with low levels of phosphate were; onion (1.72 mg/Kg) planted in the dry season, okra (1.45 mg/Kg) cultivated during the harmattan season and tomato (1.20 mg/Kg) planted in the dry season. Vegetables analyzed had low levels of phosphate in year I, this could be attributed to less solubility of phosphate in water and inability of it to be converted to orthophosphate form that are needed by vegetables despite heavy application of fertilizers as suggested by [12] and [13]. In second year, vegetables studied had concentrations of 0.23 – 8.60 mg/Kg. Highest level of 8.60 mg/Kg was noted in carrot planted in the rainy season followed by spinach (7.50 mg/Kg and 7.48 mg/Kg) both in the dry and rainy season while the least concentration was recorded in okra (0.23 mg/Kg) cultivated in the harmattan season. Dry season of year II showed peak level of phosphate which might be related to heavy application of phosphate enrich fertilizers coupled with the nature of wastewater use for irrigation as stated by [14]. Other vegetables with low levels of phosphate were; lettuce (5.40 mg/Kg) cultivated in the rainy season, tomato (4.14 mg/Kg) planted during the harmattan season and cabbage with level of 3.76 mg/Kg. The chart indicates high level of phosphate in year II than year I which might be related to flooding experienced in year I, this necessitated heavy application of phosphate enriched fertilizer and other chemicals than usual in the following year as observed by [10]. Comparing these years results, it was revealed that harmattan season of year I (0.11 – 2.20 mg/Kg) showed low level of phosphate ions than harmattan season of year II (0.23 – 4.83 mg/Kg). Likewise, dry season of first year (0.37 – 2.85

mg/Kg) had less concentration of phosphate ions than dry season of second year (1.57 – 7.48 mg/Kg). In the same way, rainy season of year I (0.20 – 2.00 mg/Kg) had low level of phosphate ions than corresponding rainy season of year II (0.50 – 8.60 mg/Kg). The phosphate levels analyzed were below the concentrations obtained by other workers including [15] (967.69 – 3390.50 mg/Kg), [10] (32 – 73 mg/Kg) and [6] (33.27 – 187.95 mg/Kg).

Table I presents ANOVA for phosphate levels in wastewater between year one and two. Statistical analysis shows that $p = 0.876 > 0.050$ this means that there is no significant difference in phosphate levels across the sampling locations as observed from their mean. Phosphate concentrations were also examined across the seasons to establish their differences which indicates $p = 0.975 > 0.050$ meaning no significant difference across the seasons. This might be as a result of sampling sites are falling within the same vicinity thereby their source of contaminations with phosphate are similar irrespective of change in seasons as observed by [14]. The same table shows $p = 0.197 > 0.050$ this means that there is no significant difference in phosphate levels among various vegetables analyzed and their mean and standard deviation illustrate this finding. The table also shows $p = 0.952 > 0.050$ meaning no significant difference in phosphate concentrations across the seasons for vegetables evaluated. PPMC conducted to establish the relationship between phosphate levels between years one to two for wastewater as presented in Table 2. Statistical data showed the mean with standard deviation level for phosphate ion to be 95.661±30.993 in year I while 11.955±5.135 was obtained in year II with the Pearson correlation (r) = -0.062, degree of freedom (df) = 13 and $p = 0.827 > 0.050$ meaning negative relationship. The same table shows low relationship for phosphate levels in vegetables as their data indicates 0.810±0.794 for year I and 3.103±2.501 for year II.

Table 1. ANOVA Results for phosphate in wastewater and vegetables.

Analysis of Variances		Sum of Squares	df	Mean Square	F	Sig.
Phosphate in Wastewater (Locations)	Between Groups	3021.049	4	755.262	0.298	0.876
	Within Groups	63345.869	25	2533.835		
	Total	66366.918	29			
Phosphate in Wastewater (Seasons)	Between Groups	2148.175	5	429.635	0.161	0.975
	Within Groups	64218.743	24	2675.781		
	Total	66366.918	29			
Phosphate in Vegetable (Among vegetables)	Between Groups	40.6000	6	6.767	1.532	0.197
	Within Groups	150.146	34	4.416		
	Total	190.746	40			
Phosphate in Vegetable (Seasons) Groups	Between Groups	5.771	5	1.154	0.218	0.952
	Within	184.975	35	5.285		
		190.746	40			
	Total					

Table 2. PPMC Summary for phosphate in wastewater and vegetables.

Variables	*N*	$\bar{x}$	SD	r	df	Sig.
Phosphate Year I	15	95.661	30.993	-0.062	13	0.827
Phosphate Year II	15	11.955	5.135			
Phosphate Year I	21	0.810	0.794	0.339	19	0.133
Phosphate Year II	21	3.103	2.501			

4. Conclusion

Contamination of the stream with phosphate affects the vegetables that are irrigated with the water. This may interfere with food chain and affect the human consumers of these vegetables. The study showed that wastewater and vegetables across the locations and seasons indicated no significant difference in the samples analyzed which majorly due to anthropogenic activities. Periodical monitoring of anion levels along the stream should put in place as to evaluate their environmental impacts in order to assess their possible potential risks.

References

[1] Hu H 2002 *Human Health and Heavy Metals Exposure, In: Life Support: The Environment and Human Health* ed M McCally (London: MIT Press) p 134 – 140
[2] Ademoroti CMA 1996 *Standard Method for Water and Effluents Analysis* (Ibadan: Foludex Press) p 22 – 112
[3] Edward GP, Molof AH and Schneeman P 1990 *JAWWA* Determination of orthophosphate in fresh and saline water **57** 917 – 21
[4] Williams CH and David DJ 1973 *Australian J. Resource* **11** 43 – 56
[5] Alloway BJ and Ayres DC 1993 *Chemical Principles of Environmental Pollution* ed Blackie (London: Chapman and Hall) p 190 – 242
[6] Akan JC, Abdulrahman FI, Sodipo OA and Lange AG 2010 *J Amer. Sci.* **6** 78 – 87
[7] Arnold E and Mary AH 1998 *Standard Methods for the Examination of Water and Wastewater* (Washington DC: APHA Press) p 45 – 60
[8] Munson RD and Nelson WL 1990 *Principle and Practices in Plants Analysis in "Soil Testing and Plant Analysis* (ed) RL Westernan (London: Mandison) p 359 – 87
[9] Radojevic M and Bashkin VN 1999 *Practical Environmental Analysis* (Cambridge: Royal Society of Chemistry Press) p 466 – 75
[10] Akan JC, Abdulrahman FI, Dimari GA and Ogugbuaja VO 2008 *Euro. J. Sci. Res.* **23** 122 – 33
[11] Cantliffe DJ 1973 *Agro J* **65** 563 – 65
[12] Razzaque MS 2009 *Amer. J. Physio. & Renal Physio.* **296** 470 – 76
[13] Singh BR 1994 *Environmental Rev* **12** 133 – 46
[14] Oladeji SO 2017 *Archives of Agri. and Env. Sci.* **2(3):** 141- 47
[15] Uwah EI, Akan JC, Moses EA, Abah J and Ogugbuaja VO 2007 *Agri. J.* **2**: 392 – 9
[16] Farooq M, Anwar F and Rashid U 2008 *Pak. J. Botany* **40**: 2099 – 106

Thermal properties of polyethylene reinforced with recycled–poly (ethylene terephthalate) flakes.

Nik Ruqiyah Nik Hassan[1], Noor Mazni Ismail[1], Suriati Ghazali[2] and Dewan Muhammad Nuruzzaman[1]

[1]Faculty of Manufacturing Engineering, Universiti Malaysia Pahang, 26600 Pekan, Malaysia
[2]Faculty of Chemical and Nature Resources Engineering, University Malaysia Pahang, Lebuhraya Tun Razak, 26300 Gambang Kuantan, Pahang, Malaysia.

E-mail: nikruqiyah77@gmail.com

Abstract In this study, recycled plastic bottles (RPET) were used as a filler in high density polyethylene (HDPE) thermoplastic. The plastic sheet of RPET/HDPE was prepared by using hot and cold press machine. The effects of RPET addition and hot press process to the thermal properties of the composite RPET/HDPE were investigated using differential scanning calorimetry (DSC) and thermogravimetric (TGA). Results from DSC analysis show that the melting point of HDPE slightly shifted to a higher temperature for about 2°C to 4°C with the addition of RPET as a filler. The starting degradation temperature of RPET/HDPE composite examined from TGA analysis also seen to be slightly increased. It was observed that the incorporation of recycled PET flakes into HDPE is achievable using hot press process with slight improvement seen in both melting point and thermal stability of the composite compared to the neat HDPE.

1. Introduction

Nowadays, plastics are getting more popular among the industry of fabrication container compared to other type of materials because of their properties that are light weight, high wear resistance and easy to fabricate. Most important characteristic of the plastic is a low production cost and this make plastic become more valuable. Because of this, the waste from the plastic also increase day by day. The waste of the plastic bottle in Malaysia and plastic industry increase progressively every year, as thus remanufacture of plastic bottle waste can help to reduce the waste and save the environment.

Among many popular thermoplastic polymers, Polyethylene terephthalate (PET) is one of the thermoplastic polymers that has been widely used in manufacturing of fiber, bottles, film and others engineering plastic part. PET has good mechanical properties, chemical resistance, clarity, thermal stability and low cost of production, as well as its process ability and recyclability [1] .

PET bottles are normally made from co-polymer PET because of its lower crystallinity, high ductility, better process ability and clarity [2]. The large production of PET plastic bottles nowadays has exponentially increase waste of PET. PET is a non-degradable plastic in normal conditions as there is no known organism that can consume its relatively large molecules.

Thus, recycled is a choice to deal with a problem to manage the plastic waste and in order to reduce the environmental problems, and to reduce oil usage, carbon dioxide emission and PET waste [3].

Normally, there are two types of recycling process, mechanically and chemically. Mechanical recycling is a physical method also known as material method, the process usually consists of separation of waste, wash for contaminants removal, grinding into flakes size, re-extrusion, and reprocessing for production of new PET goods. The plastic will be formed into flakes, granulated, or pallet, and it will be melt through the process of extrusion, injection moulding or blow moulding [4]. Chemical recycling is a process of the polymers chemically convert into monomers through chemical reaction. The monomers were then can be used for new polymerization to reproduce the original or any related product [5]. Comparing both methods, mechanical recycled and fiber reinforced polymer composites are more attractive because of their ease of fabrication and economic (low manufacturing cost with lots of quality of product produced) [7].In the present study, sheet samples of RPET/HDPE composite were prepared mechanically by hot compressing method using hot and cold press machine. DSC and TGA analysis were then carried out to investigate the thermal formability behavior of the RPET/HDPE composite as in plastic sheet form.

2. Materials and Method

2.1 Materials

The materials used in this experiment were recycled poly (ethylene terephthalate) RPET and HDPE polyethylene. The properties of raw HDPE and RPET are listed in Table 1. The recycled PET was obtained from mineral water bottles.

Table 1. Properties of HDPE and RPET.

Polymer	HDPE	RPET
Melting Temperature	135°C	250°C
Density	1.00g/m^3	0.96 g/m^3
Shape Form	Pallet	Flakes

2.2 Sample preparation

The sample preparation process start by collecting the empty mineral water bottles from the recycle bins around UMP campus. The bottles were then crushed into flake size using crusher machine [8]. After that, the flakes were washed with water to get rid of dust and dirt, followed by drying in oven for 5 minutes at 80°C. Three different percentage of RPET (0% RPET/100% HDPE, 30% RPET/70% HDPE, 50% RPET/50% HDPE) samples were prepared which were then hot-pressed into sheet shape with 1 mm thickness by using hot and cold press machine, Lotus Scientific 25 Ton Hot Cold Molding Press (Model LS22025). The temperature for both the bottom and upper plate compress machine were set at 135°C.

IOP Conf. Series: Materials Science and Engineering **342** (2018) 012094 doi:10.1088/1757-899X/342/1/012094

2.3 Thermogravimetry (TGA) and Differential Scanning Calorimetry (DSC) analysis

In TGA testing, samples with approximately 3 mg were scanned from room temperature to 600 °C under nitrogen atmosphere with 50 ml/min flow rate and heating rate of 10 °C/min. In DSC testing, samples weighing about less than 3 mg in an aluminum pan were heated up to 200 ^{0}c with heating rate of 10 ^{0}c/min under nitrogen gas, flow rate of 50ml/min. Nitrogen gas was used to prevent the oxidation of the samples.

3. Results and Discussion

3.1 DSC test results and discussion

The result of DSC test are shown in table 2, it was observed that the melting point of HDPE with increasing the amount of recycled PET in HDPE. If compared the heat HDPE, the melting point of 30% RPET/HDPE and 50% RPET/HDPE increased for about 2°c and 3°c, respectively. A higher melting point observed in the RPET/HDPE samples can be suggested due to the dislocation locking by RPET flakes [9].

DSC analysis for the polyethylene plastic pellets graph change before and after present of RPET shown in Figure 1a, 1b, 1c below. The PE pellet sample shows a peak of melting at 129^{0}c with heat of melting 11.66 J/g. The increased value of recycled PET flakes percentage into the HDPE, the melting temperature increased because due to RPET destruct the repeating structure.

Table 2. The thermal properties of PE in DSC analysis.

Polymer	Melting point (Tm)
Polyethylene (PE)	129^{0}c
(PE)+ 30%RPET	132^{0}c
(PE)+ 50% RPET	133^{0}c

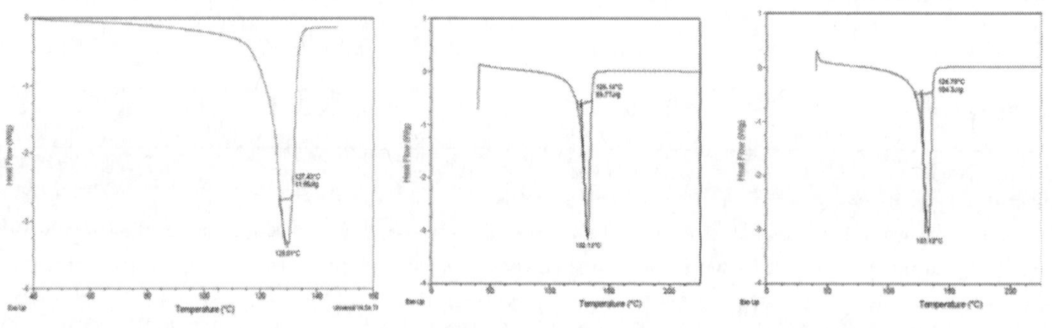

(1a) Polyethylene (PE) (1b) PE+ 30% RPET (1d) PE+50%RPET

Figure 1. The DSC graph

3.2 TGA test results and discussion

Thermogravimetric analysis (TGA) is a thermal analysis technique that measure the weight change of a sample under a certain temperature time [10].

Figure 2 shows the decomposition patterns of HDPE and RPET/HDPE samples. As can be seen in figure 2, thermal decomposition of all 3 samples occurred in are stage pattern, with a maximum decomposition rate occurred between 370°c to 470°c. At 500°c, the HDPE sample undergoes complete degradation. Meanwhile, RPET/HDPE samples show few percentage of residue. The residue seen in RPET/HDPE samples can be assigned to the RPET flake which increased with the increment of RPET content.

As label in Table 3, there are 3 different points were recorded. The points refer to the onset temperature of degradation (T_o), the degradation temperature at 5% of mass loss (Ts) and the end temperature of degradation (T_e). The heat PMMA initially loss about 3% of moisture at 320°c and then it pyrolyzed in a single step between 370°c to 470°c. The heat degradation temperature of HDPE+ 30% RPET and HDPE+50%RPET shown in the table 3, at 5% of mass loss (Ts) and end temperature of degradation (T_e) are at the same temperature which is 370°c to 460 °c.

Addition of RPET into the thermoplastic HDPE degrade thermally at low temperature. Table 3 below show when 30% and 50% of RPET were added the temperature degrade at 460°c. Figure 2 below shows the graph of weight change (%) vs temperature (°c) of heating the sample of polyethylene after added the RPET 30% and 50%. The weight lost probably associate to the join lignin degradation during process of hot press. The orange color of line in graph define as polyethylene (HDPE).The residue weight of sample at different temperature was obtain accordingly. The thermal of HDPE can be divide into 2 stage by refer to the figure 2 below. The first stage, HDPE, HDPE+ 30%RPET and HDPE +50% RPET start to degrade at the same temperature up to 320°c about 3% weight loss is refer to the moisture departure. Thermal degradation of HDPE start at 360°c and continued up to 470°c with weight loss of about 96.7%. The thermal degradation of HDPE with the addition of RPET become more sensitive which are the weight lost become less than HDPE only, for HDPE with 30% RPET about 92.9% weight loss and HDPE with 50%RPET is about 83.6%. The temperature start degrade for this HDPE +30% RPET and HDPE+50%RPET sample are same, as shown in table below.

Table 3. The TGA analysis.

Polymer	**Onset temperature degradation (T_o)**	**Start of Degradation Temperature (Ts)**	**End of degradation temperature (T_e)**	**Weight Loss (%)**	**Residue after 500°C (%)**
		Point 1	Point 2		
HDPE	320°c	370°c	470°c	96.7%	0
HDPE+30%RPET	320°c	370°c	460°c	92.9%	1.5
HDPE+50%RPET	320°c	370°c	460°c	83.6%	10.5

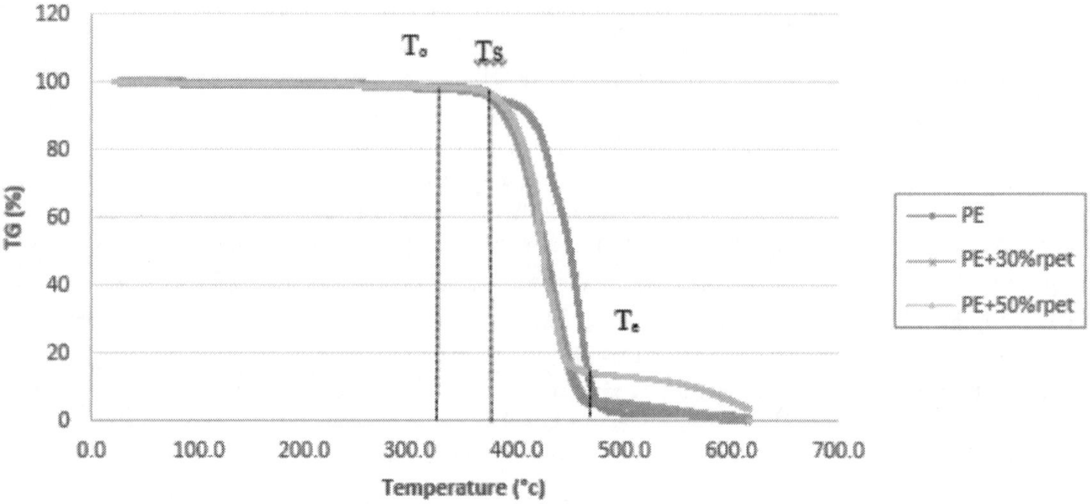

Figure 2. The graph of TGA analysis.

It can be observed that addition of RPET to the HDPE prohibits complete degradation of the composite at 500°C and an increase in ash content was noticed with the increase in RPET content.

Conclusion

The result shows that the melting temperature were increased in the present of RPET in PE matrix. The increased percentage of RPET into PE increased the value melting temperature. In the future, maleic anhydride will be used as compactibalizer in PE/RPET composite in order to study the effect of compactibilizer in improving the bonding of matrix-reinforcement bonding.

Acknowledgements

The authors would like to express gratitude to Universiti Malaysia Pahang for the financial support of this research through RDU130386 grant.

References

[1] Ghanbari, A., Heuzey, M. C., Carreau, P. J. and Ton-That, M. T. 2013. A novel approach to control thermal degradation of PET/organoclay nanocomposites and improve clay exfoliation. *Polymer (United Kingdom)*, *54*(4), 1361–1369.

[2] Gupta, V. B. and Bashir, Z. 2002 PET Fibres, Films, and Bottles. In: *Handbook of ThermoplasticPolyesters.* Ed. By Fakirov S. Vol. 1, Weinheim, Germany: Wiley-VCH,pp. 317 – 388.

[3] Takoungsakdakun, T. and Pongstabodee, S. 2007. Separation of mixed post-consumer PET–POM–PVC plastic waste using selective flotation. *Separation and Purification Technology*, *54*(2), 248-252.

[4] Awaja, F. and Pavel, D. 2005. Injection stretch blow moulding process of reactive extruded recycled PET and virgin PET blends. *European Polymer Journal*, *41*(11), 2614-2634.

[5] Braun, D. 2002. Recycling of PVC. *Progress in polymer science*, *27*(10), 2171-2195.

[6] Wienaah, M.M. 2007, Sustainable Plastic Waste Management – A Case of Accra, Ghana, in Land and Water Resource Engineering. , *Royal Institute of Technology*. p38.

[7] Fu, S. Y., Lauke, B., Mäder, E., Yue, C. Y. and Hu, X. 2000. Tensile properties of short-glass-fiber- and short-carbon-fiber-reinforced polypropylene composites. *Composites Part A: Applied Science and Manufacturing*, *31*(10), 1117–1125.

[8] Inada, K., Matsuda, R., Fujiwara, C., Nomura, M., Tamon, T., Nishihara, I. and Fujita, T. 2001 Identification of plastics by infrared absorption using InGaAsP laser diode. *Resources, Conservation and Recycling,* 33(2), 131–146.

[9] Yu. F. Zabashta. 1974.The dislocation mechanism of polymer deformation: defects of spiral *molecules t. G. Shevchenko Kiev State University. merov*, No. 4, pp. 579-583, July-August,.June 14, 1973.

[10] Joseph D. Menczel, R. Bruce Prime. 2014 Thermal analysis of polymers: Fundamentals and *Applications. John Wiley & Sons*; pp 1-50.

The influence of Span-20 surfactant and micro-/nano- Chromium (Cr) Powder Mixed Electrical Discharge Machining (PMEDM) on the surface characteristics of AISI D2 hardened steel

N A J Hosni and M A Lajis

Sustainable Manufacturing and Recycling Technology, Advanced Manufacturing and Materials Centre (SMART-AMMC), University Tun Hussein Onn Malaysia, 86400 Parit Raja, Batu Pahat, Johor Darul Takzim, Malaysia

E-mail: amri@uthm.edu.my

Abstract. The application of powder mixed dielectric to improve the efficiency of electrical discharge machining (EDM) has been extensively studied. Therefore, PMEDM have attracted the attention of many researchers since last few decades. Improvement in EDM process has resulted in the use of span-20 surfactant and Cr powder mixed in dielectric fluid, which results in increasing machiniability, better surface quality and faster machining time. However, the study of powder suspension size of surface charateristics in EDM field is still limited. This paper presents the improvement of micro-/nano- Cr powder size on the surface characteristics of the AISI D2 hardened steels in PMEDM. It has found that the reacst layer in PMEDM improved by as high as 41-53 % compared to conventional EDM. Also notably, the combination of added Cr powder and span-20 surfactant reduced the recast layer thickness significantly especially in nano-Cr size. This improvement was great potential adding nano-size Cr powder to dielectric for machining performance.

1. Introduction

Electrical discharge machining (EDM) has been used extensively in machining difficult-to-cut materials particularly in mould and die industry. As one of the advanced machining techniques, EDM is able to machine difficult-to-cut or hard materials such as hardened steels, Inconel, titanium, etc. However, due to low cutting speed and hence productivity will be lower, machining practitioners must consider the machining conditions and parameters such as voltage, peak current, pulse duration, flushing and dielectric medium before executing any new machining process. Most research has been studied on the effect of changes of the machining conditions and parameters but there were very few studies on the improvement of the machining speed by implementing the higher current and voltage and at the same time try to maintain the good quality of surface integrity and reduced tool wear in EDM operation. High speed or high performance EDM is really desired by any metal fabrication industries which is very profitable by reducing lead-time of production.

However, the material removal mechanism is achieved by the melting and evaporation of workpiece material at each electrical discharge spot, which are then ejected and flushed away by the dielectric fluid. Due to the rapid high temperature melting and cooling process, subsurface defects such as cracks, spalling, porosity, residual stress, metallurgical transformations, and heat affected zones (HAZ) are easily found on the machined surface of the workpiece. To restore the machined surface properties and remove the surface detects, the technique of fine powder mixed into the dielectric fluid of EDM, called the powder mixed EDM (PMEDM), is thus proposed.

Despite the various improvements offered by PMEDM, the important issue of effects of using powder suspended in dielectric fluid in PMEDM processes on machining time, surface quality, and accuracy still needs to be improved and investigated. The addition of powder suspended in dielectric fluid in PMEDM together with high concentration of powder in the dielectric fluid was found as a viable method to improve surface quality and reduce machining time. The electrical conductivity of powder was linked to increase the overall conductivity of dielectric fluid thus decreasing the dielectric strength of the dielectric fluid and widening the sparking gap distance between the tool electrode and the workpiece. The increase in sparking gap distance together with the high concentration of powder in the dielectric fluid facilitates the discharging process to be more uniform and reduces the need to back off the tool electrode because of short circuiting and arcing, which result in lower machining time. Also, the suspension of powder produced a high sparking gap size, which led to reduction of electric discharge power density, hence low in explosion force due to the powder size.

Powder mixed of EDM (PMEDM) machining is a revolutionary technique originally developed to achieve higher machining rates and better surface finish compared to conventional EDM. Most of the studies were conducted to evaluate the process performance in terms of machinability and surface integrity in relation to the machining process parameters. Sharma (2010) [1] investigated the effect of aluminium powder addition in dielectric during EDM of hastelloy on machining performance using reverse polarity. It was observed that the addition of aluminium powder in the dielectric fluid obtained material removal rate increase sharply. It is because of the fact that with addition of powder particles in dielectric, the spark gap is filled up with additive particles. Kumar and Davim (2011) [2] studied the role of Silicon powder in EDM machining of Al-10%sicp metal matrix composites. It was observed that when Si powder is suspended into the dielectric fluid of EDM, the machining rate improves. The reason for enhancement of machining rate is mainly that the conductive powder particles when added into the dielectric fluid of EDM lower the breakdown strength of the dielectric fluid. Kansal et al (2005) [3] was studied the effect of silicon powder addition in dielectric during electric EDM machining of EN-31 steel. It was observed that the value of material removal rate (MRR) increases with increase in peak current. It was further reported that with increase in concentration of the powder, the MRR tends to increase. This is because the added additive causes bridging effect between both the electrodes, facilitates the dispersion of discharge into several increments and hence increases the MRR. The maximum MRR is obtained at highest level of concentration of added silicon powder (2 g/l) and peak current (12 A). Kung et al (2009) [4] studied the effect of aluminium powder mixed in EDM machining of cobalt-bonded tungsten carbide. It was reported that the MRR generally increases with an increase of aluminum powder concentration. This trend is valid up to a maximum value, after a certain limit, the increase of aluminum powder concentration leads to the decrease of MRR. Both MRR and electrode wear rate (EWR) increase with an increase of the grain size.

Wu et al. (2005) [5] obtained the improvement of surface finish on SKD steel in EDM machining with aluminium and surfactant added dielectric. It was observed that the surface roughness status of the workpiece has been improved up to 60% as compared to that EDMed under pure dielectric with high surface roughness Ra of 0.434 mm. Peças and Henriques (2008) [6] investigated the effect of silicon powder concentration and dielectric flow in the surface morphology in EDM machining on AISI H13 with powder mixed dielectric. It was observed that the crater diameter, crater depth and the white-layer thickness are reduced by the use of silicon powder particles suspended in the dielectric. Chow et. al (2000) [7]studied the performance of EDM in micro-slit machining of titanium alloy using SiC and aluminium as additive in the kerosene dielectric medium. Addition of additives increased the gap distance, which resulted in better MRR and produced better surface finish. Further, it was observed that using SiC, better material removal was obtained in comparison to aluminium powder as additive. Al (5 g/l) and SiC (25 g/l) produced the highest material removal. Klocke et al. (2004) [8] study the effects of aluminum and silicon powder suspended dielectrics on the thermal influenced zone by electrodischarge machining with small discharge energies. They found that powder additives change the thermal material removal mechanism. They used High speed framing camera (HSFC) in this research capture the single discharge EDM. By comparing the intensities of Al and Si powders additives the result is that silicon additive intensity is higher than the aluminum additive.

From the literature, it can be inferred that powder mixed of EDM or PMEDM has the potential of becoming a useful and cheap alternative process for surface modification particularly in machining difficult-tocut materials. In addition, with incorporating magnetic and membrane filtering system, it is highly potential to improve machining speed which leads to higher productivity especially in advanced material processing technology. However, assesing the powder concentration uniformity and powder agglomeration and deposition in the machining tank were the problem that was investigated. Furthermore, powder agglomeration and deposition in the machining tank is costly and varies the powder concentration during machining. In order to avoid powder aggregation and depostion, and to ensure the homogenous suspension of particles in the dielectric fluid, the addition of surfactant to the powder mixed dielectric was proposed [5]. Kolli (2017) [9] found out that the addition of surfactant to dielectric fluid (electrical discharge machining oil and graphite powder) improved the material removal rate and surface roughness. It was noticed to have reduced the recast layer thickness and agglomeration of graphite and sediment particles. Material migrations between the electrode and the workpiece surface were identified, and migration behavior was powerfully inhibited by the mixing of surfactant. Surfactant added into dielectric fluid played an important role in the discharge gap, which increased the conductivity, and suspended debris particles in dielectric fluid reduced the abnormal discharge conditions of the machine and improved the overall machining efficiency.

In this study, the experimental works were explored the effects of Span-20 surfactant and Chromium (Cr) powder mixed electrical discharge machining (PMEDM) of AISI D2 hardened steel. In addition, the effects of powder size influencing the recast layer thickness were also discussed.

2. Experimental methods

The study adopted Sodick EDM AQ55L, a sinking EDM machine developed to perform the experiment as shown in Figure 1 and schematic diagram of high performance electrical discharge machining (HPEDM) are shown in Figure 2 . A self-made cycling system was mounted on the EDM machine, and span-20 surfactant was added to dielectric. The dielectric was filtered and could be used repeatedly by the cycling system. Experiments were conducted in this study by using a 99.9% purified electrolytic copper of 10 mm diameter as an electrode material with 100 mm in length and 8 mm in diameter. AISI D2 steel in the dimensions 50 mm x 50 mm x 10 mm was used as the workpiece. Table 1 showed the chemical compositions of AISI D2 hardened steel material and Table 2 showed the specifications of Cr powder used. The powder used in this experiment was Chromium (Cr) with size 45-55μm and 70-80nm as shown in Table 3.

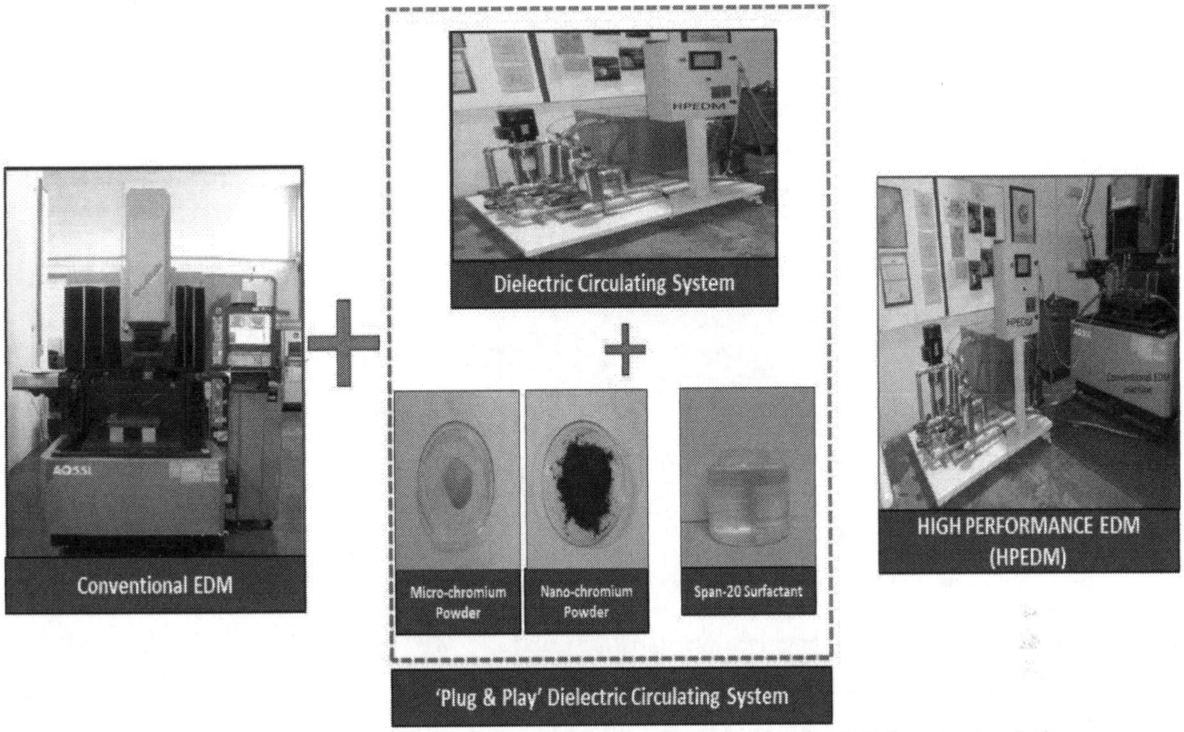

Figure 1. Sodick High Speed EDM Die Sink (AQ 55L) with Powder mixed dielectric circulating system.

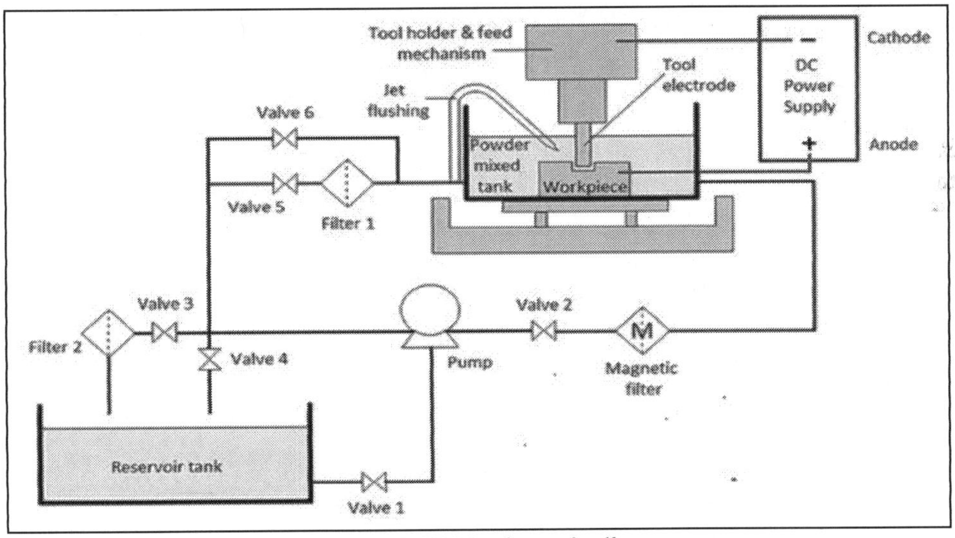

Figure 2. HPEDM schematic diagram.

Table1. The chemical composition of AISI D2 Hardened steel.

Element	C	Cr	Si	Mn	Mo	P<	S<	V<
Content (%)	1.40-1.60	11.00-13.00	0.40	0.60	0.80-1.20	0.030	0.030	0.20-0.50

IOP Conf. Series: Materials Science and Engineering **342** (2018) 012095 doi:10.1088/1757-899X/342/1/012095

Table 2. Specifications of Cr powder used.

Cr%	C%	S%	P%	Si%	Al%	Fe%	Sieve Analysis
99%	0.01%	0.015%	0.090%	0.08%	0.01%	97%	-325 Mesh

Table 3. Shape and profile of different Cr powder particles in PMEDM tests.

Range for particle sizes	Medium-size	Nano-size
Powder sizes	45-55 μm	70-80 nm
Appearance (SEM image)	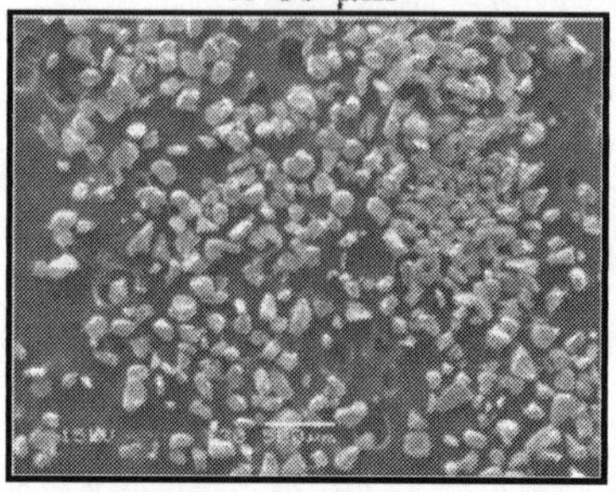	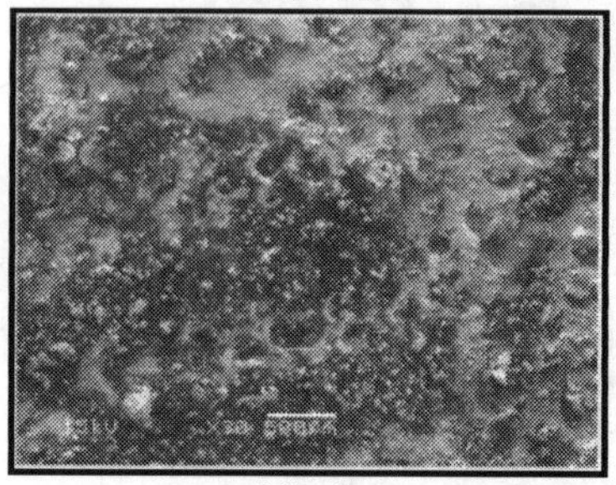
Powder colour	Grey	black
Shapes	angular, spheroidal irregular	agglomerate of particles, clusters

In this study, die-sinking EDM at a depth of 3mm were conducted using 10mm electrode on the surface of AISI D2 hardened steel under different machining condition shown in Table 4.

Table 4. EDM machining parameter.

Working Parameters	**Description**
Workpiece	AISI D2 (50x50x10) mm
Electrode	Copper ,Cu (Ø 10mm)
Dielectric	Kerosene + Cr powder
Grain size of Cr powder	45-55μm (Micro-Cr powder)
	70-80nm (Nano-Cr powder)
Cr powder	0-4 g/L (45-55μm)
concentrations, C_p	0-0.6 g/L (70-80nm)
Span-20 surfactant	0-10 g/L (45-55μm)
concentrations, C_p	0-1.0 g/L (70-80nm)
Polarity, P	Reverse Polarity
Peak Current, I_p	40A
Pulse-on, P_{on}	50μs
Duty Cycle	80%
Voltage	120V
Depth of Cut	3 mm
Flushing rate	3.5lmm^3/hour

3. Results and Discussions

3.1 Comparison of recast layer thickness provided by sapn-20 surfactant and micro-/ nano- Cr powder.

Figure 3 and Figure 4 shows the percentage of improvement of recast layer thickness comparing with conventional EDM of different Cr powder particle sizes. It was found that proper addition of Span-20 surfactant and Cr powder particles helps to enhance the machining efficiency by further stabilising the electric discharge. The improvement in process stability resulted from a reasonably large gap size that reduces the arcing frequency through lower debris concentration and a more even debris distribution in the interspace.

Additionally, when comparing the minimum-maximum recast layer thickness (Cp=0 g/L/Cs=10 g/L - Cp=2 g/L/Cs=10 g/L), the value of recast layer thickness is observed to decreased dramatically from 30.80 μm to 18.00 μm and increased to 22.47 μm ,respectively as shown in Figure 3 for micro-Cr PMEDM. As for comparison conventional EDM, the minimum recast layer thickness peaked at 18.00 μm (Cp=2 g/L/Cs=10 g/L) which shows a reduction of 41.56 % from coventional EDM. Recast layer is an unwanted layer created by an EDM process that is composed by deposition for other elements on the workpiece surface. This layer formed by the deposition of the molten metal solidifying in the crater of machined surface. The molten material is rapidly quenched by the dielectric. The recast layer is hard but brittle contain of micro-crack. The improvement of recast layer thickness value clarify that the potential of addition of span-20 surfactant and Cr powder into dielectric fluid ability to gives a notable potential to be utilized as one of the advanced and innovative technique to eliminate the some of the disadvantages of conventional EDM method and enhanced EDM capability [10, 11].

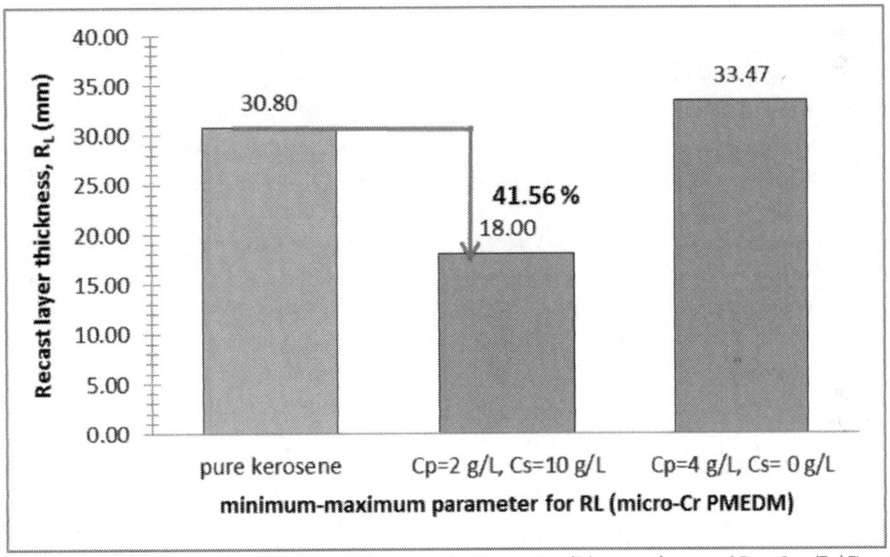

Figure 3. Material removal rate on minimum (Cp=0 g/L/Cs=10 g/L)–maximum (Cp=2 g/L/Cs=10 g/L) recast layer thickness for micro-Cr PMEDM.

As Figure 4 stated above, for the nano-Cr PMEDM lead to 51.67% reduction in recast layer thickness comparing with conventional EDM case when comparing the minimum-maximum recast layer thickness (Cp=0 g/L/Cs=0.5 g/L - Cp=0.6 g/L/Cs=0.5 g/L) from 30.80 μm to 14.27 μm and 27.07 μm , respectively.This results revealed that nano-Cr PMEDM exhibits better reduction in recast layer thickness compared to micro-Cr PMEDM. The highest percentage of reduction in recast layer thickness was at the smaller particle sizes. This can be attributed to the amount of molten metal solidified on the machined surface is much higher in micro-Cr PMEDM. Thus, thicker recast layer thickness formed. This is occurred due to discharge distribution in micro-Cr PMEDM is lower compared to nano-Cr PMEDM, which caused the current density in micro-size particle more than nano-size particle.

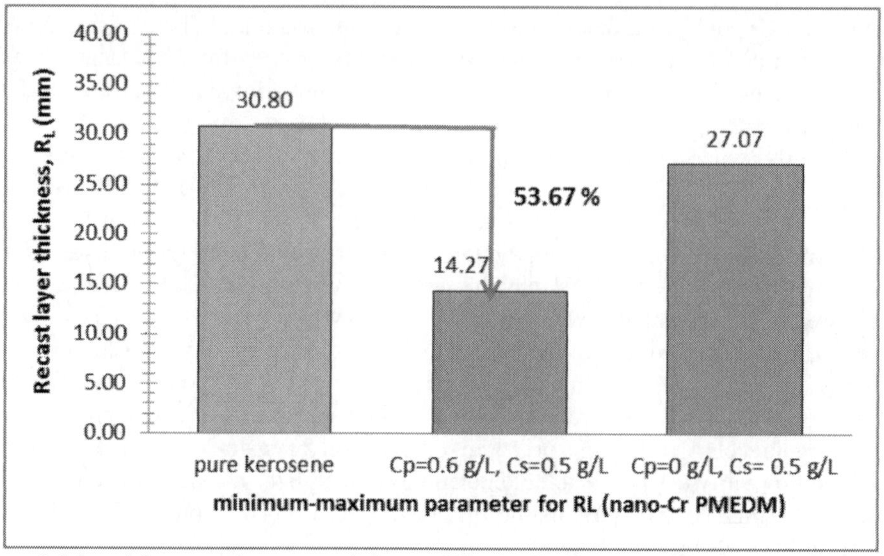

Figure 4. Material removal rate on minimum (Cp=0 g/L /Cs=0.5 g/L) –maximum (Cp=0.6 g/L /Cs=0.5 g/L), recast layer thickness for nano-Cr PMEDM.

Subsequently, Figure 5 and Figure 6 shows surface morphology of machined of AISI D2 surface and sub-surface when minimum-maximum recast layer thickness, it was observed that the EDM produces much damages on the surface and sub-surface like micro-cracks, micro-holes, metallurgical transformations and heat affected zones are happened on the workpiece samples due to rapid melting of work surface during electrical discharge process and then rapid cooling of work surface [5] [4]. As it is observed in Figure 5 and Figure 6, the machined sample using Cr powder, micro-holes are decreased and also the surface density of cracks is relatively less when added of Cr powder. However, the surface density of cracks increased at Cr powder concentration of 4 g/L due to exceeds of Cr powder in dielectric, which increased the rapid melting and cooling of work surface, thus also increased the hardness of EDMed surface [12]. In order to maintain or even improveing these properties and decreasing the surface and sub-surface defects, the technique of addition of Span-20 surfactant and Cr powder has been applied in this investigation. It is observed that the thickness of the recast layer having performed EDM by the dielectric with Cr powder and span-20 surfactant is smaller than that of pure kerosene. This is due to the fact that surfactant can improve the viscosity of tar and the surface tension of kerosene. When dielectric flows easily between electrode and workpiece, the dreg removal rate is improved, and the discharged energy is also evenly distributed. The phenomena of unstable concentrated discharge are also decreased.

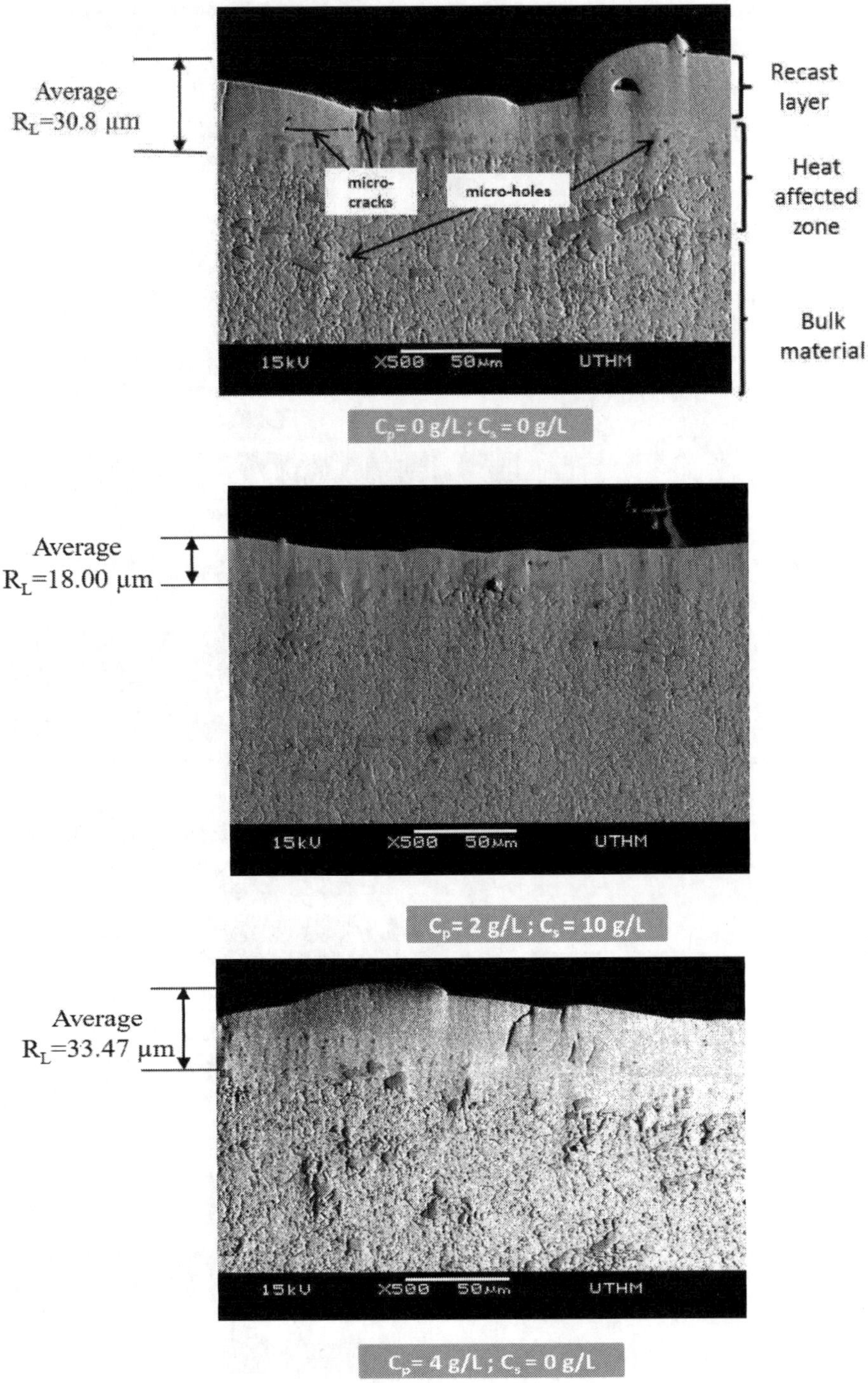

Figure 5. Recast layer of minimum-maximum recast layer thickness for micro-Cr PMEDM; (a) pure kerosene, (b) minimum recast layer (Cp=2 g/L/Cs=10 g/L) , (c) maximum recast layer (Cp=4 g/L/Cs=0 g/L).

IOP Conf. Series: Materials Science and Engineering **342** (2018) 012095 doi:10.1088/1757-899X/342/1/012095

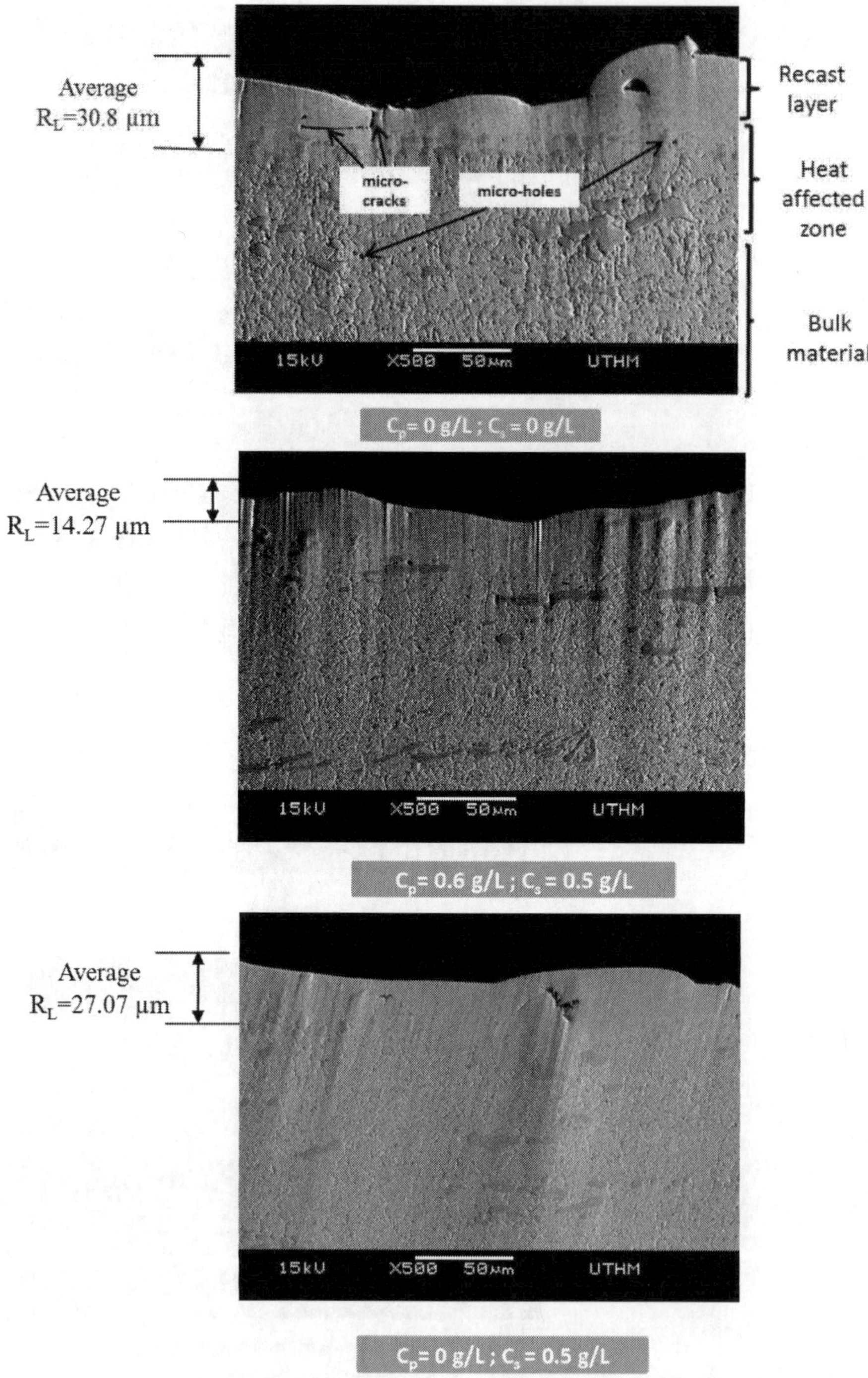

Figure 6. Recast layer of minimum-maximum recast layer thickness for nano-Cr PMEDM; (a) pure kerosene, (b) minimum recast layer (Cp=0.6 g/L/Cs=0.5 g/L) , (c) maximum recast layer (Cp=0 g/L/Cs=0.5 g/L).

4. Concluding remarks

In this study, the influences of Cr powder and span-20 surfactant are significant on the recast layer thickness. It has found that the recast layer thickness in combination Cr powder and span-20 surfactant improved by as high as 41-53 % compared to conventional EDM. In addition, the damages on the surface and sub-surface like micro-cracks, micro-holes, metallurgical transformations and heat affected zones are happened on the workpiece also decreased when Cr powder and span-20 surfactant added into the dielectric.

Acknowledgements

The authors would like to express the deepest appreciation to the Ministry of Higher Education (MOHE), Malaysia, for funding this project through the Exploratory Research Grant Schemes (ERGS-Vote numbers 0873). Additional supports in terms of facilities were also provided by Sustainable manufacturing and Recycling Technology, Advanced Manufacturing and Materials Centre (SMART-AMMC), Universiti Tun Hussein Onn Malaysia (UTHM).

References

[1] Sharma, S., et al., 2010 *Effect of aluminium powder addition in dielectric during electric discharge machining of hastelloy on machining performance using reverse polarity.* International Journal of Advanced Engineering Technology,. **1**(3): p. 13-24.

[2] Kumar, A., et al., 2011. *Analysis of machining characteristics in additive mixed electric discharge machining of nickel-based super alloy Inconel 718.* Materials and Manufacturing Processes, **26**(8): p. 1011-1018.

[3] Kansal, H., S. Singh and P. Kumar, 2005. *Parametric optimization of powder mixed electrical discharge machining by response surface methodology.* Journal of materials processing technology, **169**(3): p. 427-436.

[4] Kung, K.-Y., J.-T. Horng and K.-T. Chiang, 2009. *Material removal rate and electrode wear ratio study on the powder mixed electrical discharge machining of cobalt-bonded tungsten carbide.* The International Journal of Advanced Manufacturing Technology, **40**(1-2): p. 95-104.

[5] Wu, K.L., et al., 2005. *Improvement of surface finish on SKD steel using electro-discharge machining with aluminum and surfactant added dielectric.* International Journal of Machine Tools and Manufacture, **45**(10): p. 1195-1201.

[6] Peças, P. and E. Henriques, 2008. *Effect of the powder concentration and dielectric flow in the surface morphology in electrical discharge machining with powder-mixed dielectric (PMD-EDM).* The International Journal of Advanced Manufacturing Technology, **37**(11-12): p. 1120-1132.

[7] Chow, H.-M., et al., 2000. *Study of added powder in kerosene for the micro-slit machining of titanium alloy using electro-discharge machining.* Journal of Materials Processing Technology, **101**(1): p. 95-103.

[8] Klocke, F., et al., 2004 *The effects of powder suspended dielectrics on the thermal influenced zone by electrodischarge machining with small discharge energies.* Journal of materials processing technology,. **149**(1): p. 191-197.

[9] Kolli, M. and A. Kumar, 2017. *Surfactant and graphite powder–assisted electrical discharge machining of titanium alloy.* Proceedings of the Institution of Mechanical Engineers, Part B: Journal of Engineering Manufacture, **231**(4): p. 641-657.

[10] Ho, K.H. and S.T. Newman, 2003. *State of the art electrical discharge machining (EDM).* International Journal of Machine Tools and Manufacture, **43**(13): p. 1287-1300.

[11] Zhao, W.S., Q.G. Meng, and Z.L. Wang, 2002. *The application of research on powder mixed EDM in rough machining.* Journal of materials processing technology, **129**(1): p. 30-33.

[12] Toshimitsu, R., et al.,2016 *Improvement in surface characteristics by EDM with chromium powder mixed fluid.* Procedia CIRP, **42**: p. 231-235.

Internally Heated Screw Pyrolysis Reactor (IHSPR) heat transfer performance study

S H Teo, H L Gan, A Alias and L M Gan

Faculty of Mechanical Engineering, Universiti Malaysia Pahang, 26600 Pekan, Pahang, Malaysia.

E-mail: raymondtsh90@gmail.com

Abstract. 1.5 billion end-of-life tyres (ELT) were discarded globally each year and pyrolysis is considered the best solution to convert the ELT into valuable high energy-density products. Among all pyrolysis technologies, screw reactor is favourable. However, conventional screw reactor risks plugging issue due to its lacklustre heat transfer performance. An internally heated screw pyrolysis reactor (IHSPR) was developed by local renewable energy industry, which serves as the research subject for heat transfer performance study of this particular paper. Zero-load heating test (ZLHT) was first carried out to obtain the operational parameters of the reactor, followed by the one dimensional steady-state heat transfer analysis carried out using SolidWorks Flow Simulation 2016. Experiments with feed rate manipulations and pyrolysis products analyses were conducted last to conclude the study.

1. Introduction

The surging population size and the necessity of automobiles have directly increased new tyres productions. Nonetheless, all new or used tyres will soon reach their shelf life, termed as end-of-life tyres (ELT) which explain their shocking annual disposal rate of 1.5 billion worldwide. Conventional remedies applied have proven to be less effective given their failure in harnessing the material and energy completely from the ELT. Pyrolysis or thermolysis on the other hand is capable of converting the ELT into useful high value products which are competent as green fuels alternatives to the depleting natural fossil fuels. That said, machineries and technologies capable of pyrolysing the ELT to energy are highly sought after today.

Over the past decades, ELT pyrolysis technologies came in all shapes and sizes but among all, screw or auger reactor is favorable for its simplicity and convincing technology in pyrolysis fuel (oil and solid char) production. The traditional screw reactor comprises an externally heated cylindrical "tube" together with a low-temperature screw shaft responsible in mixing and transporting the feedstocks. The high temperature difference between the heated reactor wall and the cold internal screw shaft has attributed to its lackluster heat transfer performance and risked providing consistent throughput. To overcome the matter, it was suggested that heat should be introduced into the hollow internal screw shaft by means of electric, heated air or fluids or other heating mediums.

Nevertheless, there was limited research on the heat transfer performance across an IHSPR. Meanwhile, a pilot-scaled IHSPR was developed by local renewable energy industry which serves as the research subject in this particular paper.

2. Literature review

2.1. End-of-Life Tyres: Threat and Opportunity

Constructed to withstand severe conditions and having lifespan between 80 and 100 years [1][2], the annual disposal of 1.5 billion [3][4] non-biodegradable, durable and chemically-resilient end-of-life tyres (ELT) globally could bring about devastating effect to the environment. Its mismanagement such as illegal dumpsites pose threats to potential health and environmental hazard, besides the possibility of fire outbreak with high emissions of toxic gases [5][6] and as a breeding ground for disease carrying vectors such as mosquitoes [5][7].

2.2. ELT Pyrolysis

Being an irreversible process simultaneously involving the change of physical phase and chemical composition of the ELT [5], pyrolysis process involved applying significant heat (at temperature 200°C-1200°C) [8][9] on the ELT in an unreactive environment, thereby thermal degrading the ELT to produce high energy density oil, gas and solid char. A typical ELT pyrolysis process generates hydrocarbon gases (5-20 wt%), solid char (30-40 wt%)and oil (40-60 wt%) [9] [10] with commendable calorific values of 30-40 MJ/m^3, 30 MJ/kg and 41–44 MJ/kg, respectively [5][11][12][13]. The products, however, are low-grade and require further refinement prior to reuse [14].

2.3. Screw/ Auger Reactor

Given its relatively simple construction and familiar operation, the technologies of screw conveyors have long been used intensively and extensively in industries to transport and treat solids materials regularly [15][16]. Screw reactor was first applied in biomass pyrolysis field in year 1969 and was found to be applicable in both fast and slow pyrolysis field [17] with its operation needing little or no carrier gas [16]. Generally, the screw reactor comprised of a tubular externally heated wall with a screw within the reactor conveys and stirs the feedstocks homogenously and repeatedly to improve heat transfer between the feedstocks and the heated wall. The used of motorized screw prevents coagulation of the feedstocks and at the same time allows drive speed alteration which regulates the solid residence time of the feedstocks within the reactor [18] thereby mechanically expelling the resulting char to its reservoir for subsequent feedstocks feeding and thermal degradation process to take place [19]. Lastly, the screw reactor is favorable for its convincing technology in pyrolysis oil and solid char productions [20] and has attracted much interest of many industries.

2.4. Governing Parameters

The governing parameters affecting the tyre pyrolysis system would cover the reactor temperature, particle size of feedstock, and feed rate.

2.4.1. Reactor Temperature. The reactor temperature has been known as one of the most influential parameters affecting the performance of any thermochemical process reactor in its products productions. Countless studies have been carried out and it was found that reactor temperature was manipulated mainly to yield desired outputs, with low temperature favouring solid char production and high temperature favouring pyrolysis gas production.

2.4.2. Particle Size of Feedstocks. Islam [21] has discovered that the increase of feedstocks size resulted in an increase in yield of pyrolysis oil up to a certain extent before the yield started to decrease. It was deduced that larger feedstocks have smaller reaction surface and therefore experienced significantly lower heating rate compared to smaller particles which could be completely thermal-degraded under similar pyrolysis environment.

2.4.3. Feed Rate. Increment of feedstock feed rate lead to proportional increment in volatile matter released during the pyrolysis process. This subsequently reduces the volume available in the reactor which reduces the gas residence time and prevented the cracking of gas fractions into permanent gas and consequently yields higher production of liquid oil [11].

2.5. Heat Transfer Theory and Analysis Approach
The newly developed IHSPR was a concentric cylinder with three layers. The interior layer was mainly filled with heated air supplied at a fixed rate from the BFBC. Meanwhile, the exterior layer comprised of recycled heated air aimed to reuse the heated air supplied to maintain the overall temperature of the reactor before it was released into the atmosphere via the exhaust duct. Isolated from the aforementioned layers, the middle layer was where the ELT chips pyrolysis has taken place. It was however only filled with stagnant air at ambient temperature and pressure during ZLHT.

The reactor operated by applying the fundamentals of thermodynamics and heat transfer. Second law of thermodynamics has stated that heat always transfers from high temperature to low temperature. 875 °C of heated air was constantly supplied into the hollow screw shaft (interior layer) while other layers were kept at considerably lower temperature. Meanwhile, the insignificant thickness of the walls separating the vast temperature difference of fluids too has contributed to the relatively large temperature gradient in the radial direction, making it a one-dimensional heat transfer towards the exterior [22]. Concurrently, the considerably constant fluid temperature at their respective layer allows heat transfer through the walls steadily.

On the other hand, the presence of the rotating screw and the concentric cylindrical design of the IHSPR have exacerbated the measurement of pyrolysis temperature. Therefore instead of the pyrolysis region, two k-type thermocouples were installed at the exterior layer of the screw reactor ($T_{exterior,\ inlet}$ and $T_{exterior,\ outlet}$) which were only capable of measuring the temperature of the recycled heated air surrounding the pyrolysis region. The thermocouples were aimed to determine the temperature profile across the pyrolysis reactor and also to indicate the time required to achieve pyrolysis temperature ($T_{pyrolysis}$), which marked the beginning of ELT chips pyrolysis process.

3. Objectives
- To study the heat transfer performance of the pyrolysis reactor via steady-state heat transfer analysis (SSHTA).
- To evaluate the heat transfer performance of the pyrolysis reactor through products characteristics analysis.

4. Research methodology
Zero-load heating tests (ZLHT) were first conducted on the empty screw reactor to determine the operational parameters and temperature distribution across the reactor. It shall meanwhile indicate the time required for the reactor to achieve pyrolysis temperature, $T_{pyrolysis}$, which marked the beginning of ELT chips pyrolysis process. With the help of k-type thermocouples and Graphtec midi LOGGER GL840 data logger, the temperature throughout the pyrolysis reactor was recorded. The test was carried out thrice. At the same time, the test shall provide the necessary information on the boundary conditions of the heated air supplied into the pyrolysis reactor in steady-state heat transfer analysis (SSHTA).The facilities involved were as shown in Figure 1.

The one-dimensional SSHTA was carried out on the SolidWorks model of the INSPR using SolidWorks Flow Simulation 2016 to study and illustrate the temperature profile across the newly developed reactor. Figure 2 and 3 showed the presence of two separated fluid regions present in the IHSPR design. Figure 4 showed the boundary conditions prepared for SSHTA. The SSHTA would provide adequate information on the temperature gradients especially that of the pyrolysis region. The boundary conditions of the heated air namely, temperature, pressure and flow rate to name a few were determined from the ZLHT carried out earlier.

Lastly, experiments involving ELT chips feed rate manipulations were carried out. Non-specific ELT chips mixture of 100 ± 15 square millimeters without both the steel thread and the textile netting were fed into the pyrolysis reactor with the feed rate ranging from 0.3 kg/min to 0.5 kg/min for duration of two hours. The reactor was not purged with any inert gases as research showed it is rather redundant and hence less economically viable [11][23] The pyrolysis char and pyrolysis oil produced were collected from their respective reservoir, weighed and sent for characteristics analysis.

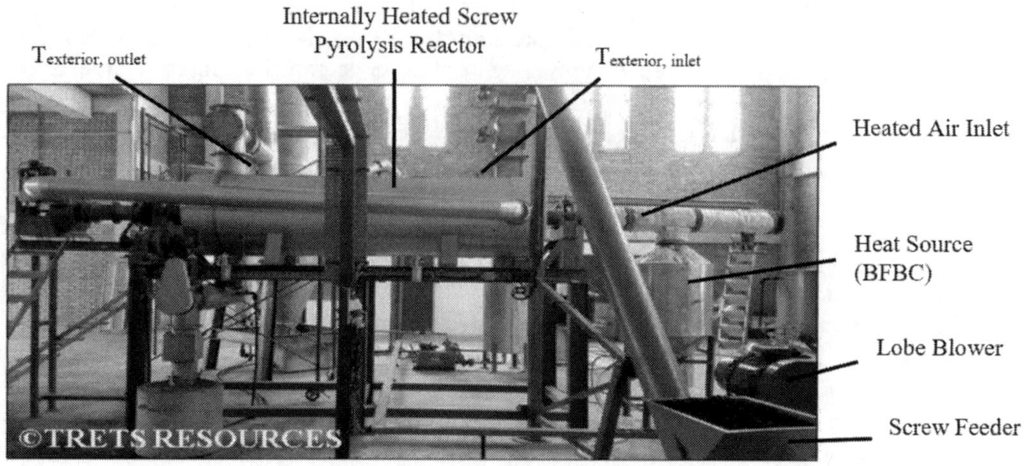

Figure 1. IHSPR system.

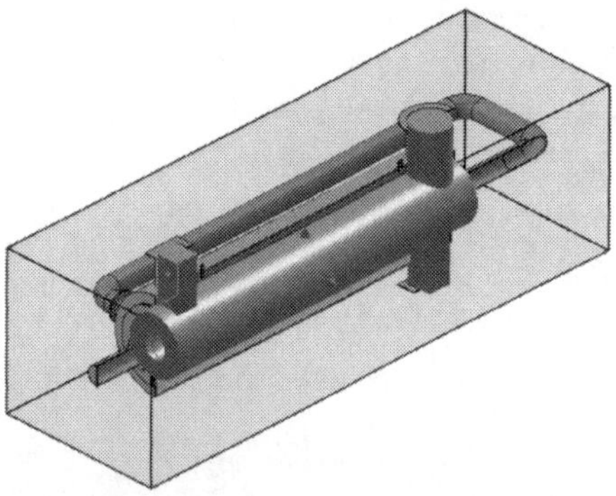

Figure 2. ELT Pyrolysis region in the IHSPR SolidWorks model.

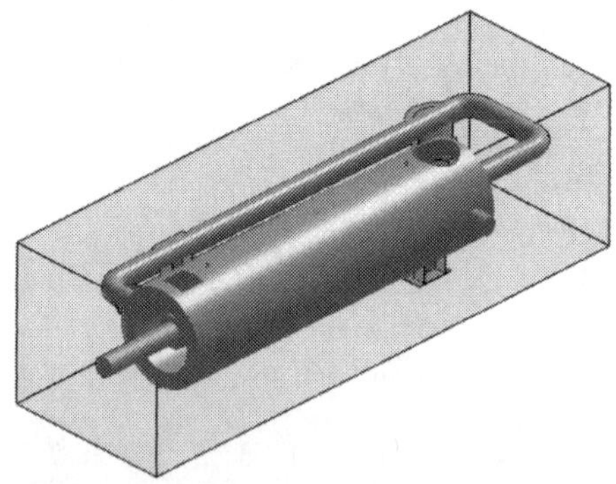

Figure 3. Heated air region in the IHSPR SolidWorks model.

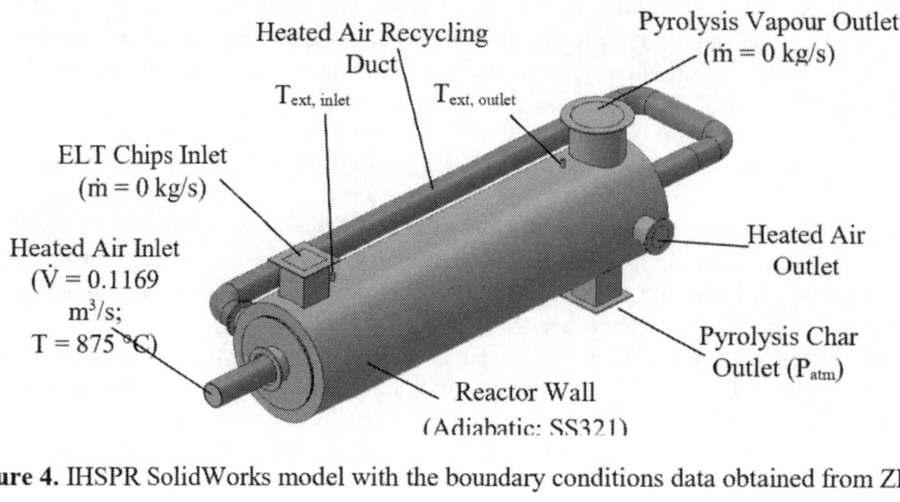

Figure 4. IHSPR SolidWorks model with the boundary conditions data obtained from ZLHT.

5. Results and discussions

5.1. Zero-Load Heating Tests (ZLHT)

Aside from preventing leakages of pyrolysis fuel and hence ensuring the safety of its operations through the isolation of the pyrolysis vapour from the heated air, the presence of the rotating screw and the concentric cylindrical design concept of the internally heated screw reactor have exacerbated the measurement of pyrolysis temperature. Therefore, instead of the pyrolysis region, there were two k-type thermocouples installed at the exterior layer of the screw reactor ($T_{ext,\ inlet}$ and $T_{ext,\ outlet}$) only capable of measuring the temperature of the heated air surrounding the pyrolysis region. Nevertheless, their contributions were proven vital in evaluating the temperature of the inlet and outlet region of the pyrolysis reactor. After three repetitive ZLHT, it was found that the heated air was supplied at a constant temperature of 875°C at a flow rate of 0.1169m^3/s into the hollow internal screw shaft of the pyrolysis reactor. After two hours of continuous supply, the heated air at the exterior layer of the pyrolysis reactor has achieved 200°C which indicated the interior pyrolysis region of the reactor has definitely achieved pyrolysis temperature [8][9] as shown in Figure 5 below.

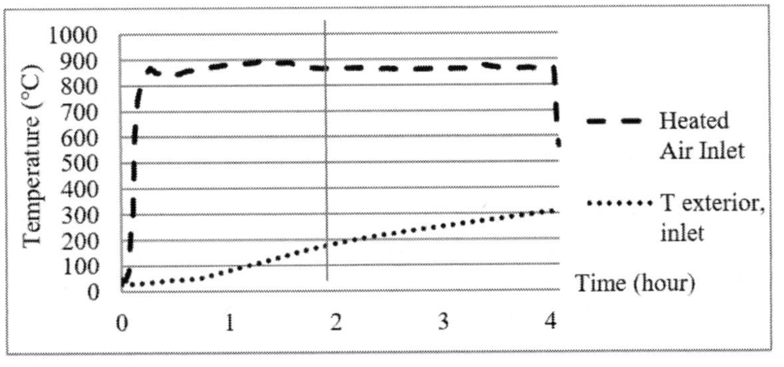

Figure 5. Temperature gradients at both inlets of the developed IHSPR.

5.2. Steady-State Heat Transfer Analysis (SSHTA)

5.2.1. Mesh Setup. The global mesh comprised of level four initial mesh with first degree of refinement level while the local mesh of cuboid shape were of level three refining cells for fluid and solid cells and fluid-solid boundary, with first degree of refinement level. Figure 6 showed the result of mesh setup on the SolidWorks model of the IHSPR before SSHTA was carried out. There were a total of 1121737 cells. The mesh has generally focused on the body of the IHSPR to gain valuable information on the heat transfer and temperature gradient throughout the pyrolysis and heated air regions.

IOP Conf. Series: Materials Science and Engineering **342** (2018) 012096 doi:10.1088/1757-899X/342/1/012096

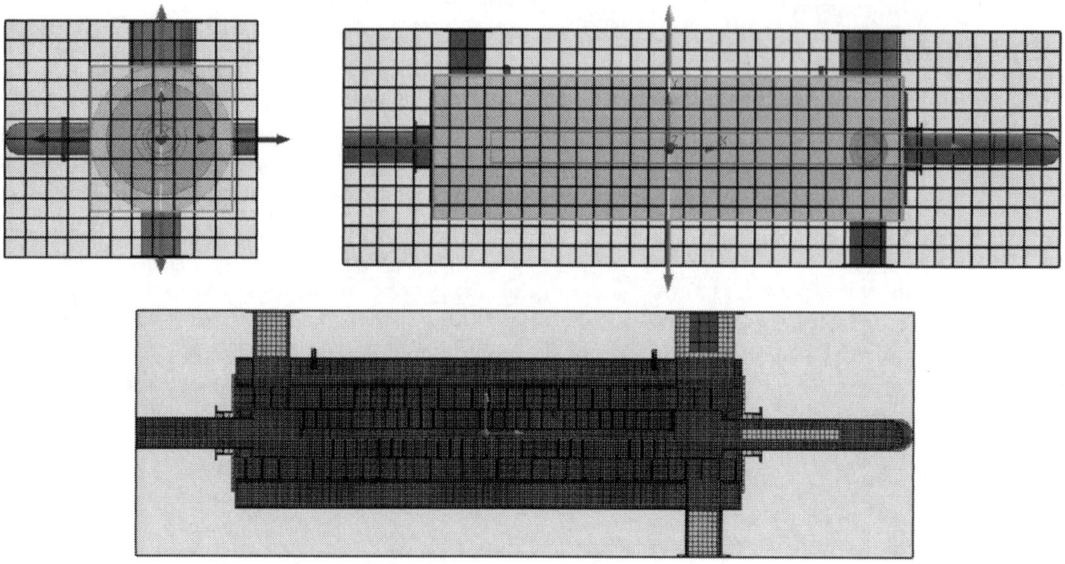

Figure 6. Mesh setup on the IHSPR SolidWorks model. Red, green and blue colours indicated refinery levels 3, 2 and 1 respectively.

5.2.2. Flow Simulations. Figure 7 showed the temperature gradients across the developed pilot-scaled IHSPR after it was heated for two hours with a constant heated air input of 875°C and flow rate of 0.1169m^3/s. The inlet of the exterior layer of the IHSPR, $T_{exterior, inlet}$ as indicated has achieved 474.65K which is equivalent to 201.45°C and meanwhile the screw blade or "flighting" has achieved 912.91K which is equivalent to 639.76°C. The convincing temperature results marked the beginning of ELT chips feed rate manipulations experiments which aimed to evaluate the heat transfer performance of the developed pilot-scaled IHSPR through products characteristics.

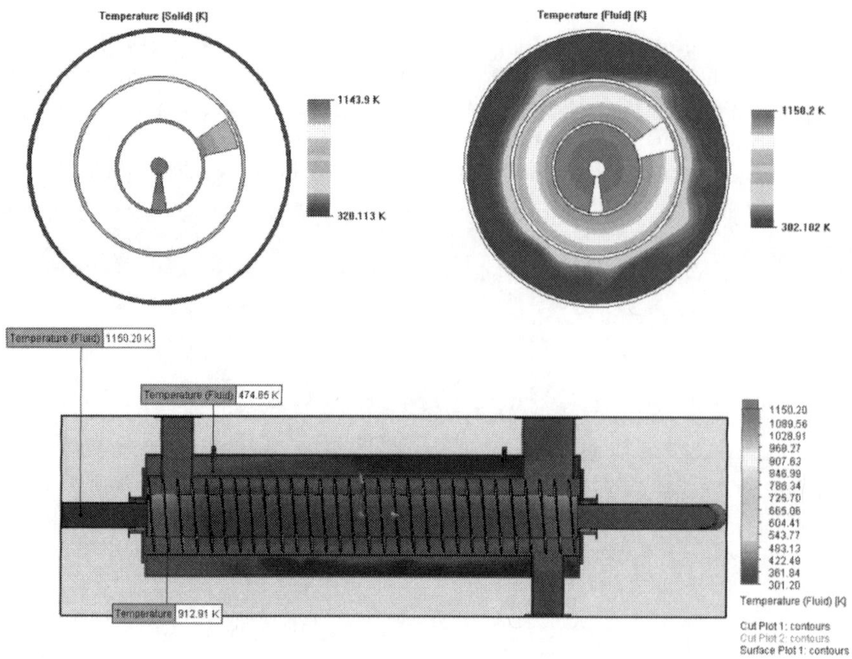

Figure 7. Temperature gradient across the IHSPR SolidWorks model.

5.3. ELT Chips Pyrolysis with Feed Rate Manipulations

For feed rate 0.3 kg/min, the average weight of the products produced were 51.99 % wt char, 25.09 % wt oil and 22.91 % wt gas. In the 0.4 kg/min feed rate experiments, the amount of oil produced was found to increase by 56.2% to 39.19 %wt, with char and gas dropped by 13.46% and 31%, to 15.82%wt and 44.99 %wt, respectively. The trend remain the same in the 0.5 kg/min feed rate experiments, with products char and gas plummeted to 36.18 %wt and 12.06 %wt respectively with the product oil raised to 51.76 %wt. The results clearly indicated increment of feedstock feed rate lead to proportional increment in volatile matter released during the pyrolysis process. Subsequently the volume available in the reactor reduces which reduces the gas residence time and prevented the cracking of gas fractions into permanent gas and consequently yields higher production of liquid oil.

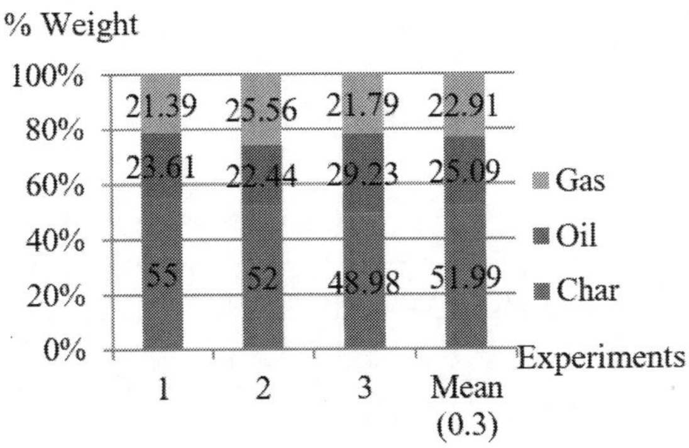

Figure 8. Products distribution at 0.3 kg/min ELT chips feed rate.

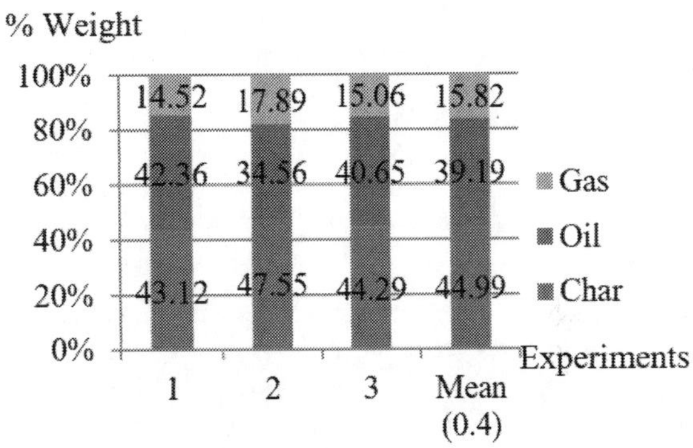

Figure 9. Products distribution at 0.4 kg/min ELT chips feed rate.

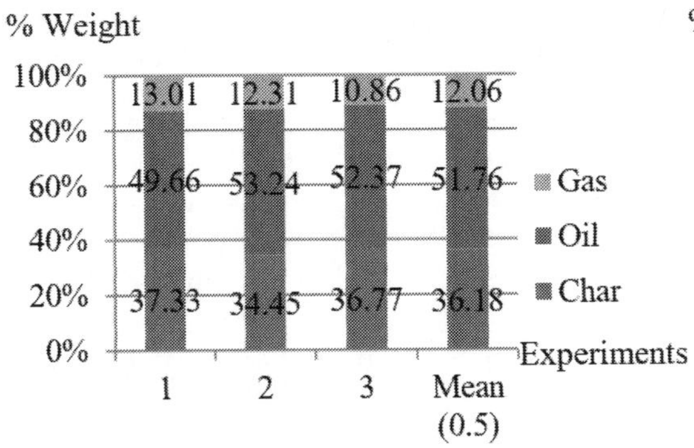

Figure 10. Products distribution at 0.5 kg/min ELT chips feed rate.

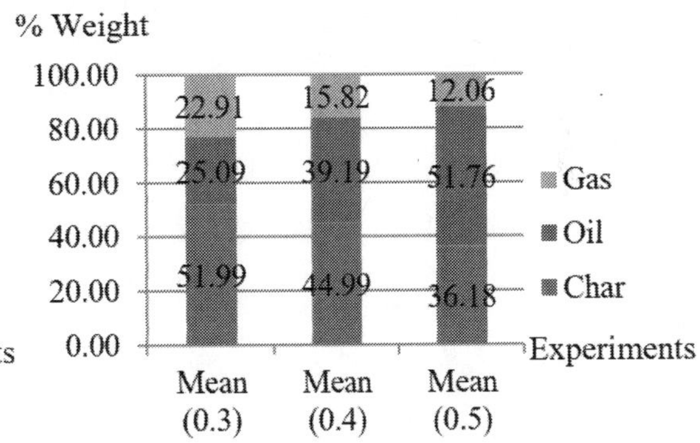

Figure 11. Mean products distribution for manipulated ELT chips feed rate.

5.4. Proximate Analysis Using Thermogravimetric Analyser

To determine the compositions namely moisture, volatiles, fixed carbon and ash present in the products, proximate analysis via Thermogravimetric Analyser (TGA) was conducted in Center of Excellence for Advanced Research in Fluid Flow (CARIFF) Universiti Malaysia Pahang. The temperature range was set to be from 30°C to 900°C with the heating rate set to 50°C/ min and air flow rate of 20ml/min was set. The vast temperature range was set to determine the overall decomposition gradient of the products. High heating rate was set to speed up the thermal degradation process since the detail decompositions of impurities and mineral matter were not required. Knowing that the carbon black would most probably remain unchanged because of its inertness under nitrogen atmosphere [24] and meanwhile to determine the percentages of ash within the products, air was used instead of nitrogen gas [25]. Nonetheless, the results obtained from the TGA analysis were rather astonishing given the decomposition curves not only did not illustrate the compositions present within the products at all, they were no significant difference among products of different feed rate. It was hypothesized that the ELT chips were pyrolysed in an oxygen-rich environment, thereby altering the thermochemical decomposition of the ELT chips from that of the conventional pyrolysis process. The decomposition curves for the products were as shown from Figure 12 to 17.

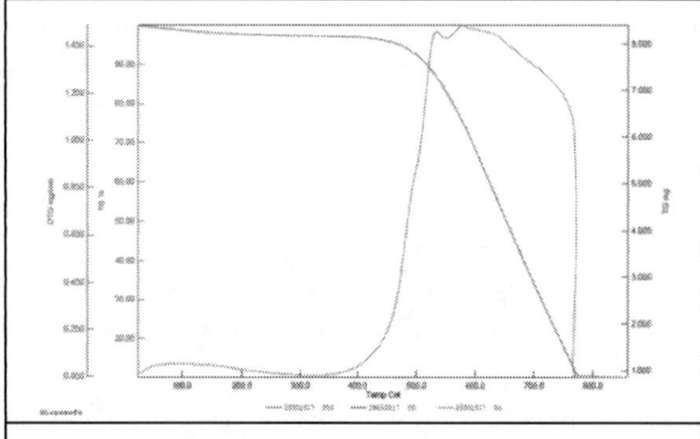

Figure 12. TGA analysis of 0.3 kg/min pyrolysis char.

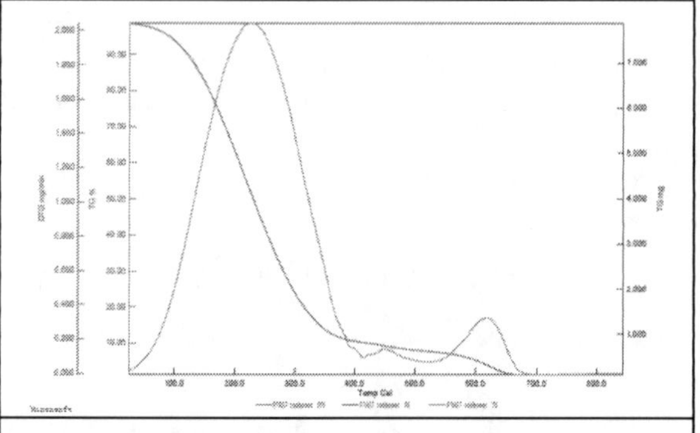

Figure 15. TGA analysis of 0.3 kg/min pyrolysis oil.

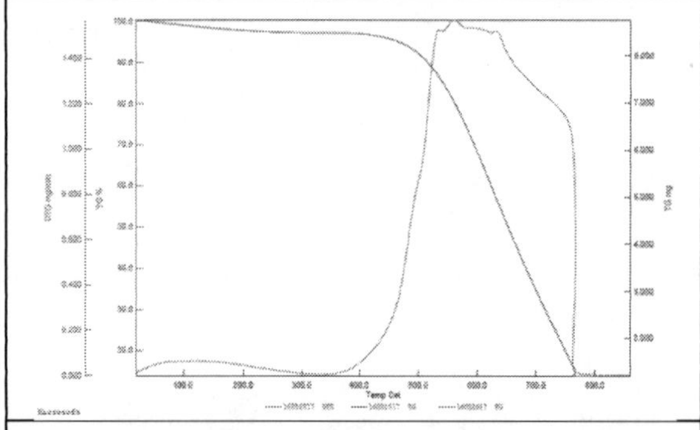

Figure 13. TGA analysis of 0.4 kg/min pyrolysis char.

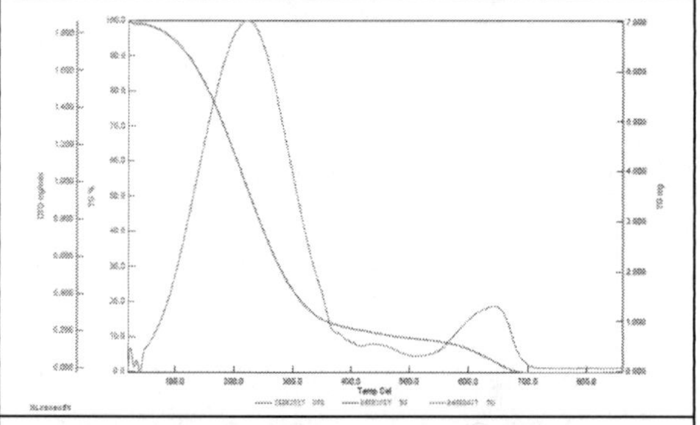

Figure 16. TGA analysis of 0.4 kg/min pyrolysis oil.

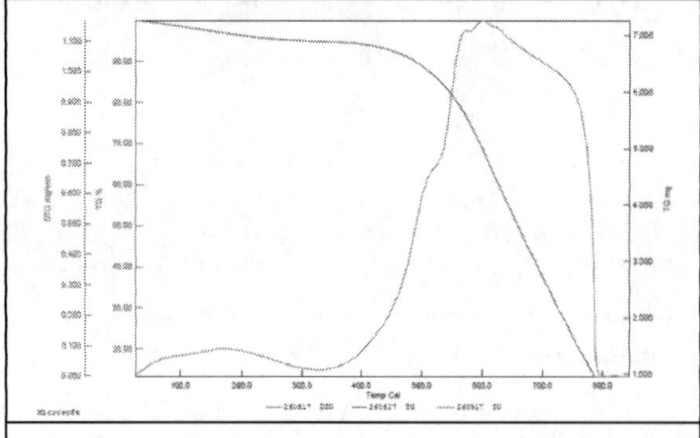

Figure 14. TGA analysis of 0.5 kg/min pyrolysis char.

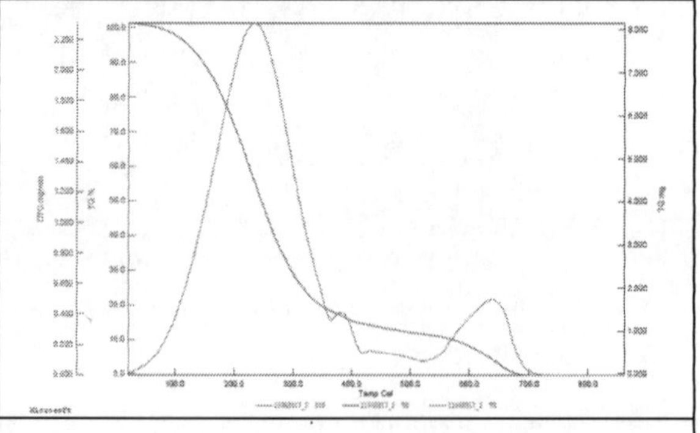

Figure 17. TGA analysis of 0.5 kg/min pyrolysis oil.

6. Conclusion

Heat transfer performance study was carried out on a pilot-scaled IHSPR developed by local renewable energy industry. With the operational parameters obtained from ZLHT, SSHTA results showed convincing temperature gradient across the reactor after it was heated for two hours, which marked the beginning of ELT chips feed rate manipulations experiments. There were noticeable reductions in the productions of pyrolysis gas and char but otherwise for pyrolysis oil as the feed rate increased from 0.3 kg/min to 0.5 kg/min. However, the compositions namely moisture, volatiles, fixed carbon and ash within the pyrolysis oil and char remained unknown as they were undetectable through the set TGA environment. Future experiments shall be conducted in a nitrogen-purged environment to challenge and verify the findings discovered by Agirre, Griessacher, Rösler and Antrekowitsch [23] and Alkhatib [11].

Acknowledgments
Many thanks to Trets Resources Sdn Bhd for lending the developed IHSPR to be the research subject of this particular paper.

References
[1] Alsaleh, A. and Sattler, M. L. (2014). Waste Tire Pyrolysis: Influential Parameters and Product Properties. *Current Sustainable/Renewable Energy Reports*, *1*(4), 129–135. https://doi.org/10.1007/s40518-014-0019-0
[2] Martínez, J. D., Puy, N., Murillo, R., García, T., Navarro, M. V. and Mastral, A. M. (2013). Waste tyre pyrolysis – A review. *Renewable and Sustainable Energy Reviews*, *23*, 179–213. https://doi.org/http://dx.doi.org/10.1016/j.rser.2013.02.038
[3] Rowhani, A. and Rainey T.J. (2016). Scrap Tyre Management Pathways and Their Use as a Fuel—A Review. *Energies*. **9** (888): 1-26. doi:10.3390/en9110888.
[4] Wang, W.C., Bai C., Lin C.T. and Prakash S. 2015. Alternative fuel produced from thermal pyrolysis of waste tires and its use in a DI diesel engine. *Applied Thermal Engineering*. **93**: 330–338. http://dx.doi.org/10.1016/j.applthermaleng.2015.09.056
[5] Osayi J. I. and Iyuke S., O. S. E. (2014). Biocrude Production through Pyrolysis of Used Tyres, *2014*. https://doi.org/Biocrude Production through Pyrolysis of Used Tyres. (n.d.). Retrieved August 16, 2015, from file:///C:/Users/VICENTE/Downloads/386371 (2).pdf
[6] Sienkiewicz, M., Kucinska-Lipka J., Janik H. and Balas A. 2012. Progress in used tyres management in the European Union. *Waste Management* **32**: 1742-1751. http://dx.doi.org/10.1016/j.wasman.2012.05.010
[7] Tiwari, A., Israr, M. and Gangele A. 2014. Utilization of Waste Material in Pyrolysis Plant. *International Journal of Emerging Technology and Advanced Engineering*. **4** (11): 388-392.
[8] Jaminska, N., Brestovic T. and Carnogurska M. (2013) The effect of temperature pyrolysis process of used tires on the quality of output products. *Acta Mechanica Et Automatica* **7** (1): pp 20-25.
[9] Gulzad A. (2011). Recycling and pyrolysis of scrap tire Slovak University of Technology, Bratislava.
[10] Juma, M., Koreňová Z., Markoš J., Annus J. and Jelemenský L. 2006. Pyrolysis and combustion of scrap tire. *Petroleum & Coal*. **48** (1): 15-26.
[11] Alkhatib, R. (2015). Development of an alternative fuel from waste of used tires by pyrolysis. Environmental Engineering. Ecole des Mines de Nantes.
[12] Altayeb, R. K. (2015). Liquid fuel production from pyrolysis of waste tires: process simulation, exergetic analysis, and life cycle assessment. Master's thesis. American University of Sharjah, United Arab Emirates.
[13] Nkosi, N. and Muzenda, E. (2014). A Review and Discussion of Waste Tyre Pyrolysis and Derived Products. *Proceedings of the World Congress En Engineering 2014, II*.
[14] Connor, K. et al. (2013). Developing a sustainable waste tire management strategy for Thailand.http://www.wpi.edu/Pubs/E-project/Available/E-project-030113-050719/unrestricted/IQP_FINAL.pdf. Accessed on 4 December 2016.
[15] Resende, F. L. P. 2014. Reactor configurations and design parameters for thermochemical conversion of biomass into fuels, energy, and chemicals. *Reactor and Process Design in Sustainable Energy Technology*. Elsevier: 1-25. http://dx.doi.org/ 10.1016/B978-0-444-59566-9.00001-6
[16] Aramideh, S. (2014). Numerical simulation of biomass fast pyrolysis in fluidized bed and auger reactors. Master's thesis. Iowa State University, Ames, Iowa.
[17] Roegiers, J. (2016). *Heat and mass transfer modelling of auger reactors* (Master's thesis). Retrieved from http://lib.ugent.be/fulltxt/RUG01/002/275/059/RUG01-002275059_2016_0001_AC.pdf
[18] Sadaka, S., Sharara, M. and Ubhi, G. (2014) Performance assessment of an allothermal auger gasification system for on-farm grain drying. *Journal of Sustainable Bioenergy Systems*, **4**: 19-32. http://dx.doi.org/10.4236/jsbs.2014.41003

[19] Aramideh, S., Xiong Q., Kong S.C. and Brown R.C. (2015). Numerical simulation of biomass fast pyrolysis in an auger reactor. *Fuel* **156**: 234-242. http://dx.doi.org/10.1016/j.fuel.2015.04.038

[20] Brassard, P., Godbout S. and Raghavan V. 2017. Pyrolysis in auger reactors for biochar and bio-oil production: A review. *Biosystems Engineering*, **161**: 80-92. http://dx.doi.org/10.1016/j.biosystemseng.2017.06.020

[21] Islam, M.R., Parveen M., Haniu H.and Islam Sarker. M. R. (2010). Innovation in Pyrolysis Technology for Management of Scrap Tires: A solution of Energy and Environment. *International Journal of Environmental Science and Development* pp:89-96. ISSN:2010-0264.

[22] Çengel, Y. A., and Ghajar, A. J. (2015). *Heat and mass transfer: Fundamentals & applications.* Fifth Edition. New York: McGraw-Hill. ISBN 978-0-07-339818-1

[23] Agirre, I., Griessacher T., Rösler G. and Antrekowitsch J. 2013. Production of charcoal as an alternative reducing agent from agricultural residues using a semi-continuous semi-pilot scale pyrolysis screw reactor. *Fuel Processing Technology* **106**: 114–121. doi:10.1016/j.fuproc.2012.07.010

[24] Chakraborty, S. et al. 2011. Composition analysis of carbon black-filled polychloroprene rubber compound by thermo-oxidative degradation of the compound. *Journal of Elastomers and Plastics* 43: 499-508. DOI: 10.1177/0095244311413442.

[25] Speight, J. G. (2015). *Handbook of Coal Analysis*, Second Edition. Published by John Wiley & Sons, Inc. ISBN 978-1-118-36924-1.

Biomechanical analysis using Kinovea for sports application

Nor Muaza Nor Adnan[1], Mohd Nor Azmi Ab Patar[1,2], Hokyoo Lee[2], Shin-Ichiroh Yamamoto[2], Lee Jong-Young[3] and Jamaluddin Mahmud[1]

[1] Faculty of Mechanical Engineering,Universiti Teknologi MARA, 40450 Shah Alam, Selangor, Malaysia.
[2] College of Systems Engineering and Science, Shibaura Institute of Technology, Fukasaku, Minuma-ku, Saitama, 337-8570 Japan
[3] College of Physical Education, Suwon University, Gyunggi-Do, South Korea

E-mail: azmipatar@gmail.com

Abstract. This paper assesses the reliability of HD VideoCam–Kinovea as an alternative tool in conducting motion analysis and measuring knee relative angle of drop jump movement. The motion capture and analysis procedure were conducted in the Biomechanics Lab, Shibaura Institute of Technology, Omiya Campus, Japan. A healthy subject without any gait disorder (BMI of 28.60 ± 1.40) was recruited. The volunteered subject was asked to per the drop jump movement on preset platform and the motion was simultaneously recorded using an established infrared motion capture system (Hawk–Cortex) and a HD VideoCam in the sagittal plane only. The capture was repeated for 5 times. The outputs (video recordings) from the HD VideoCam were input into Kinovea (an open-source software) and the drop jump pattern was tracked and analysed. These data are compared with the drop jump pattern tracked and analysed earlier using the Hawk–Cortex system. In general, the results obtained (drop jump pattern) using the HD VideoCam–Kinovea are close to the results obtained using the established motion capture system. Basic statistical analyses show that most average variances are less than 10%, thus proving the repeatability of the protocol and the reliability of the results. It can be concluded that the integration of HD VideoCam–Kinovea has the potential to become a reliable motion capture–analysis system. Moreover, it is low cost, portable and easy to use. As a conclusion, the current study and its findings are found useful and has contributed to enhance significant knowledge pertaining to motion capture-analysis, drop jump movement and HD VideoCam–Kinovea integration.

1. Introduction

In this era, gradually plyometric exercises are also being used in rehabilitation, but some coaches are used as a platform to elevate the potential sport performances of their athlete. Plyometric movements arise during sport such as in running, jumping and throwing. The focus on developing the efficiency of the stretch-shortening cycle is what discriminates plyometric exercise from normal strength training [1]. As a part of plyometric exercises, drop jump involves a person dropping to the ground from a raised platform then immediately jumping up [2]. Subsequently, there are research results of jumping up height appear to continue to increase with lighter knee flexion [3].

The usage of motion analysis is expanding due to high demand of accuracy in biomechanics field. It is essential to get relatively similar result based on exact measures. Therefore, countless free motion analysis software developed to accomplish the level where established motion analysis systems such as Cortex are considered as the precise methods for analyses of human movements. However, the methodology of the system is complex and intricate for simple movement analysis [4]–[6]. Nonetheless, every one of these techniques, experiences the ill effects of the 3 potential disadvantages, which they are somewhat costly, which makes them hard to obtain excluding supported college research centres and first class wear, clubs and they are hard to transport, particularly on account of overwhelming stages, which, thusly, limits their applicability [7].

The study presented in this paper examines whether Kinovea software is reliable to analyse the drop jump perform by athletes. The analysis will be done by using motion capture software, Cortex 6.0.0.1645 and Kinovea 0.8.25. The cortex is motion analysis software for handling all phases of motion capture within a single program for initial setup, calibration, tracking and post processing [8] whereas Kinovea is a video analysis, which is a free software application for the analysis, comparison and evaluation of sports and training, especially suitable for physical education teachers and coaches. Some advantages of this software are: observation, measurement, comparison of videos, etc. This software also can perform the analysis without use physical sensors or by means reflective markers and it is uncomplicated to use. The overview function is a summary image of the video. It creates a composite picture where you can see the motion at a glance by sample images from the video at regular interval. Accordingly, this study is directed to verify Kinovea as one of that solid software that can be utilized in this research field. The aim of this project is to explore the reliability of Kinovea for detection and analysis of the movement focused on drop jump movement, its repeatability test and the relation of knee flexion angle due to the drop jump exercise with the maximum jumping up height achieved.

2. Materials And Method

2.1. *System Validation*

The main focus of this study is to produce a good quality of recording that is motion capture of drop jump movement, simultaneously using an established infrared cameras motion capture system and HD VideoCam. Then, the same motion data are tracked and analysed using Cortex and Kinovea. The relative angle of knee during drop jump phases are measured and analysed. These angles are compared to the angles measured using the established motion capture system. Basic statistical analyses are carried out to compute mean, standard deviation and variance. These data are also used to compute the percentage difference and variance. To produce a reliable system, both values should be less than 10% [9]–[11]. Finally, by comparing both results and the outputs from the basic statistical analysis, the reliability and accuracy of HD VideoCam–Kinovea are measured and assessed. In addition, the advantages and limitations of both HD VideoCam–Kinovea and Hawk–Cortex systems are discussed and summarised.

2.2. Parameters and Experimental Setup

As for HD Video Camera and Kinovea Software parameters and experimental setup,a Sony VideoCam was used to record the drop jump movement by the subject. The camera was placed on a level tripod at the height of 0.060m to allow a good capture of a jump from a 0.040m stage height perpendicular to the centre of the pathway at a distance of 3.568m. This setting ensured that the calibration area covered the lower limb of the subject (field of view). Focus and aperture were adjusted until the system produced clear and sharp 2D images [4], [12]. While for HAWK Cameras and CORTEX Software parameters and experimental setup, the 3-dimensional motion data was recorded using 6 Hawk digital infrared motion capture cameras; with a recording speed of 100 Hz, for a duration of 10 sec. Marker #1 of the calibration L-frame was set at the desired origin of the capture volume. A relationship was established between real-world positions (object-coordinates) and the corresponding image-coordinates from the camera view. The 6 Hawk cameras positioned as illustrated in Figure 2. F-Stop, a number of ratio of the focal length of the lens to the diameter of the aperture [13]. F-stop on a Motion Analysis camera was locked at its optimum value. The camera was fully operated in order to see most of the markers.

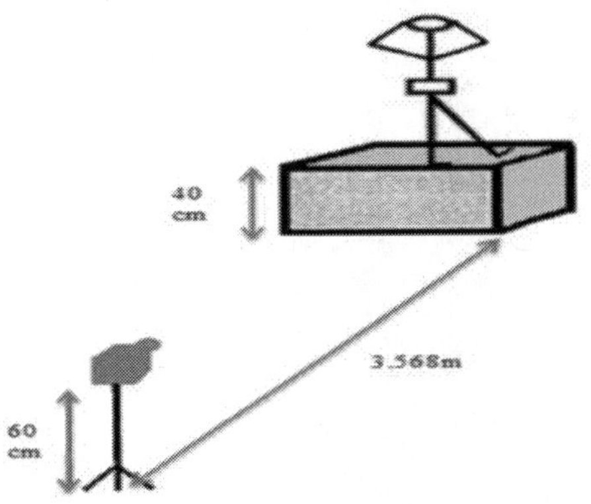

Figure 1. Illustration of the calibrated plane, camera position and field of view.

2.3. Experimental Protocols

Only one subject was used in this case. A healthy subject without any gait disorders, age 26 years (BMI mean 28.60 ± 1.40) was recruited. Markers were placed on lower limb parts of each subject as shown in Figure 3. The subject was asked to drop from a level platform and jump up at his own natural pace and repeating the movement 5 times. The motion was recorded simultaneously using both HD VideoCam–Kinovea and Hawk–Cortex systems. Results and data collected from the two different motion capture systems were compared.

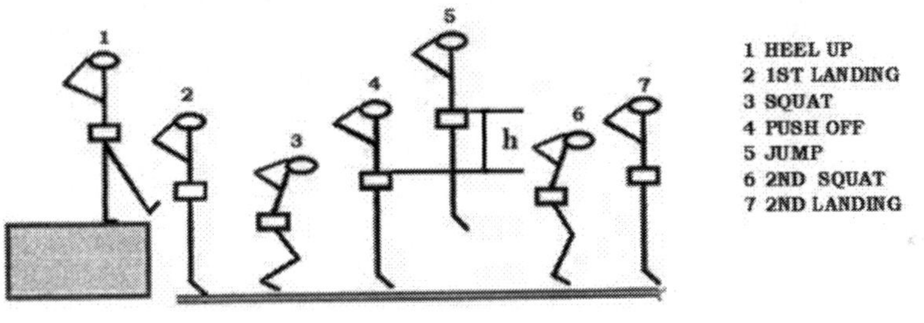

Figure 2. The drop jump movement phases.

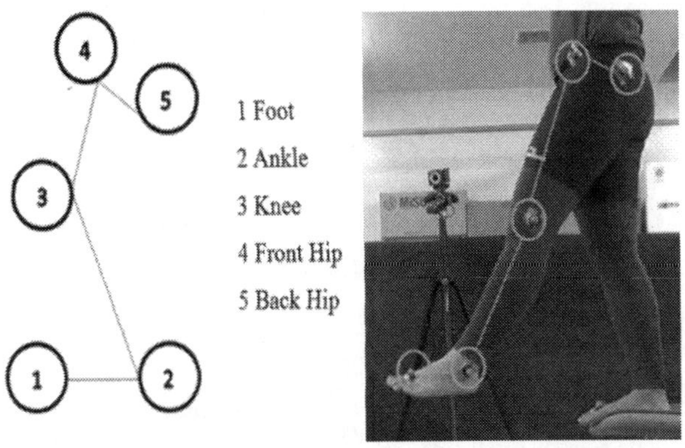

Figure 3. The markers' position on a subject.

Figure 4. Heel up position captured by the HD digital camera and displayed in Kinovea.

2.4. Motion Tracking Using Kinovea

Kinovea was used to retrieve all of the outputs (images and videos) from the Sony VideoCam camera. All of the recorded video (video files) was first playbacked to check for any obvious error. Low quality captures were also discarded. For the accepted videos, Kinovea was used to locate and specify all the five main markers (Figure 2), which the process was very similar for tracking markers using the infrared motion capture system. Kinovea kept track of the marker movement for the whole capture (drop jump movement). The example procedure for tracking and the example output at jump is shown in Figure 4.

2.5. Angle Measurement

Figure 4 shows a sample instance (image) from a complete cycle of a drop jump phase. Using Kinovea, three points (markers) were selected in order to measure the relative angle of ankle [14] for any instance. The main marker under consideration was the knee. The other two markers include the two markers adjacent to the knee. From these 3 points (markers), the relative angles of knee were determined. During drop jumping, the relative angles of knee change with respect to the motion. Instead of time, the important parameters determined in this study are angle changes with respect movement phase [15]. The data is stored according to subject and trials. Since the subjects performed 5 trials, in general 5 videos have been analysed and the data is stored for Repeatability Analysis.

2.6. Repeatability Analysis

To measure the repeatability of the system, the protocol is repeated for ten times, where each for time, the walking gait (motion of walking) was recorded. The percentage difference, mean values, standard deviation (S.D) and variance were computed using Eq. (1) to Eq. (4).

$$\text{Percent Difference (\%)} = \left| \frac{\text{Value}_1 \text{-} \text{Value}_2}{\frac{1}{2}\text{Value}_1 + \text{Value}_2} \right| \times 100\% \quad (1)$$

$$\text{Mean } (\bar{x}) = \frac{\sum x}{n} \quad (2)$$

$$\text{Standard deviation }(\text{SD})=\sqrt{\frac{\sum(x-\bar{x})^2}{n}} \quad (3)$$

$$\text{Variances }(\text{SD}^2)=\frac{\sum(x-\bar{x})^2}{n} \quad (4)$$

3. Results And Discussion

Motion data captured using both systems are tracked and analysed using Cortex and Kinovea. Figure 5 shows the sample of out the capture as could be viewed in the Cortex's interface. In general it displays the skeleton and walking motion.

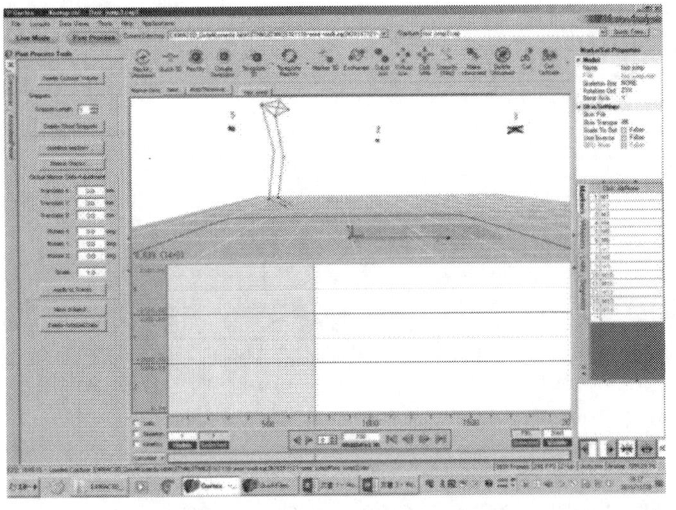

Figure 5. Jump position captured by the 6 Hawk cameras and displayed in Cortex interface.

Figure 4 shows the sample of out the capture as could be viewed in the Kinovea's interface for the same instance as shown in Figure 5. The relative knee angles comprised of 7 gait phases for the total of 5 trials were measured from the analysed videos; and the data are tabulated in Table 1 and Table 2. The associated graphs (Measured Relative Knee Angle,^{0}) of walking gait cycle of Cortex and Kinovea are as shown in Figure 6 and Figure 7.

Table 1. Basic Statistic - Cortex (Relative Knee Angle).

CORTEX RELATIVE KNEE ANGLE

PHASES	HEEL UP	1ST LANDING	SQUAT	PUSH OFF	JUMP	2ND SQUAT	2ND LANDING
TRIAL 1	156.99	137.08	34.60	119.81	158.93	128.73	158.16
TRIAL 2	155.83	132.93	39.01	114.43	170.88	113.97	155.39
TRIAL 3	156.91	138.93	33.12	119.82	155.55	117.16	156.12
TRIAL 4	159.12	137.50	44.11	102.08	157.54	125.65	158.79
TRIAL 5	158.96	133.70	40.68	111.51	160.27	123.20	160.06
AVERAGE	157.56	136.03	38.30	113.53	160.63	121.74	157.71
S.D	1.28	2.31	4.01	6.56	5.35	5.44	1.72
VARIANCE	1.63	5.34	16.08	43.00	28.67	29.54	2.95

Table 2. Basic Statistic - Kinovea (Relative Knee Angle).

KINOVEA RELATIVE KNEE ANGLE

PHASES	HEEL UP	1ST LANDING	SQUAT	PUSH OFF	JUMP	2ND SQUAT	2ND LANDING
TRIAL 1	153.12	175.41	55.74	115.82	169.88	149.19	151.72
TRIAL 2	157.83	171.83	55.97	113.92	169.88	131.48	147.80
TRIAL 3	166.46	175.06	55.24	121.29	163.46	140.13	148.36
TRIAL 4	154.51	176.37	58.79	109.08	162.98	149.14	149.48
TRIAL 5	154.26	176.58	55.43	118.55	169.59	146.61	157.49
AVERAGE	157.24	175.05	56.23	115.73	167.16	143.31	150.97
S.D	4.87	1.71	1.30	4.16	3.22	6.78	3.53
VARIANCE	23.72	2.91	1.69	17.29	10.37	45.92	12.44

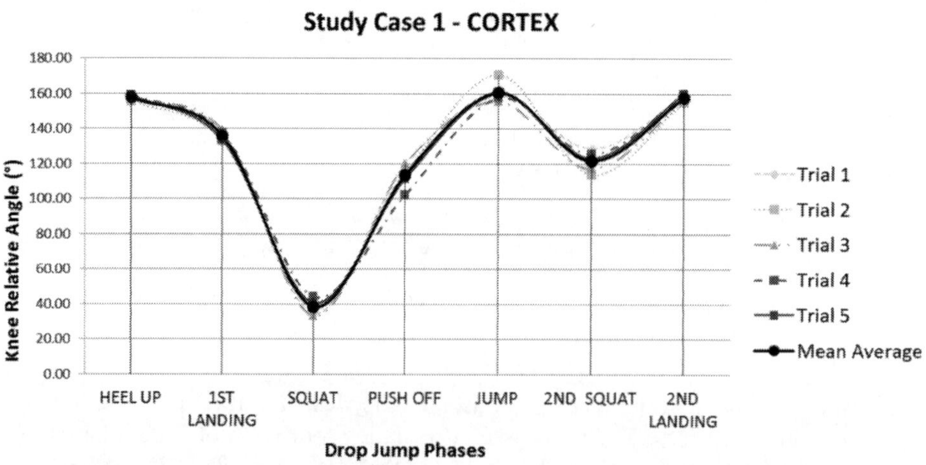

Figure 6. Measured Relative Knee Angle (^{0}) of walking gait cycle (Cortex).

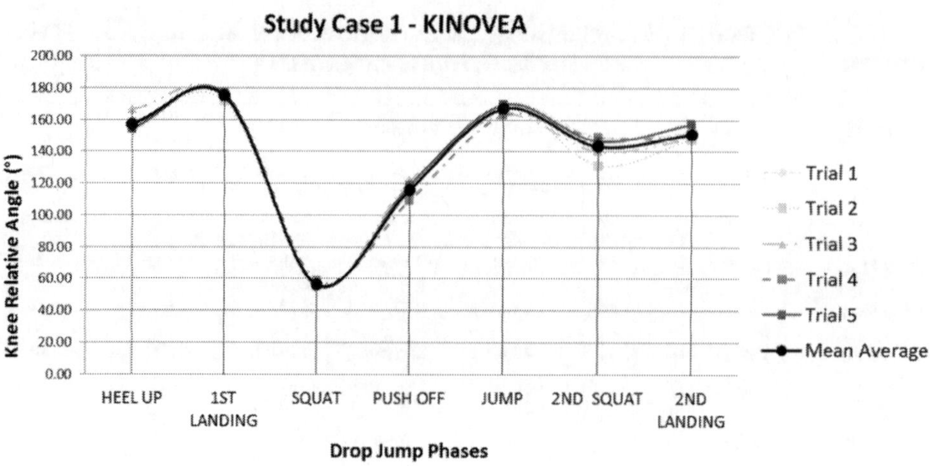

Figure 7. Measured Relative Knee Angle (^{0}) of walking gait cycle (Kinovea).

A graph comparing the measured relative knee angles of drop jump movement using both systems are shown in Figure 8. From the graph, it can be observed that both Kinovea and Cortex display similar pattern of walking gait cycle but show a slight differences in their peak values at each walking gait phases.

On the other hand, the toe off phase peak values shows the most prominent height difference between Kinovea and Cortex that are 56.23^0 and 38.30^0. As the objective of this study for case study 1 is to compare the results of relative ankle angle between Kinovea and Cortex, a tabulated percentage difference, % concerning all the 7 gait phases is shown in Table 3.

From Table 3, four out of seven phases of walking gait proved to have percentage difference less than 10%. The percentage difference should be less than 10% of error. The criterion for repeatability is the variance should be less than 10% [10]. The lowest percentage values is 0.21% occurred at heel up phase, the highest value is 37.93% at squat phase followed by 25.09% at 1st landing phase. These indicate that most of the errors are small and acceptable. Reviewing from the result obtained, the high error values at the 1st landing phases, the squat phase and the 2nd squat phase are might be due to the time taken to pull off the phases are quite shorter than others hence affecting the video recorded. The slow movement of the limb during stance reduces blurriness, which should make it easier to digitize the picture as the speed of capture by HD Camera is lower than Hawk Camera [16], [17]. Based on the graph, the Cortex plotted line of relative knee angle is slightly lower than Kinovea at most of the phases. It is may be contributed by the difference number data obtained where Cortex produces data from 200 frames per second where Kinovea are producing data from 50 frames per second thus the coordinate produced by Cortex are more accurate. For that reason, it is denoted that Cortex produced higher accuracy and precision in analysing motion.

Table 3. Average of Measured Relative Knee Angle (^{0}) of drop jump movement.

PHASES	HEEL UP	1ST LANDING	SQUAT	PUSH OFF	JUMP	2ND SQUAT	2ND LANDING
AVERAGE KINOVEA	157.24	175.05	56.23	115.73	167.16	143.31	150.97
AVERAGE CORTEX	157.56	136.03	38.30	113.53	160.63	121.74	157.71
PERCENTAGE DIFFERENCE (%)	0.21	25.09	37.93	1.92	3.98	16.27	4.37

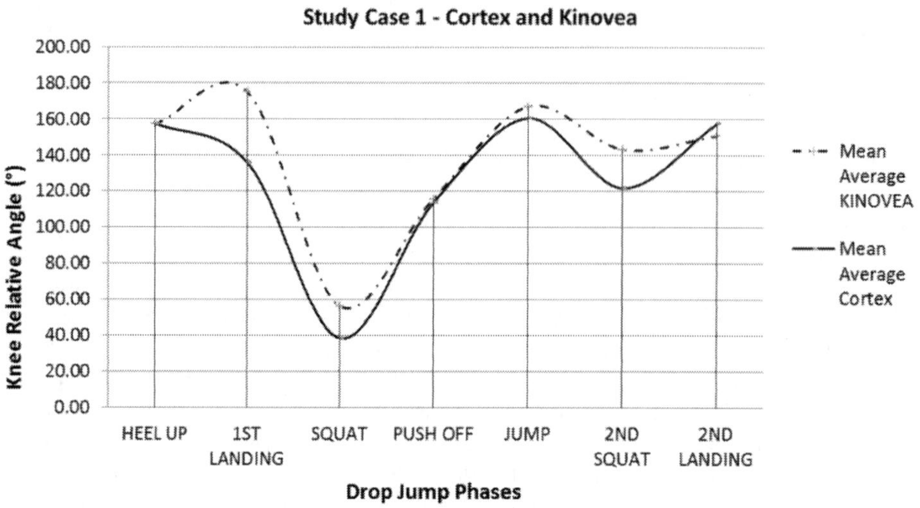

Figure 8. Measured Relative Knee Angle (^{0}) of walking gait cycle (Cortex).

4. Conclusion

In short, this paper focussed on drop jump movement analysis using Kinovea. The acquired results have shown good implications of using Kinovea as a tool to analyse the drop jump movement. The motion capture-analysis system combining HD VideoCam-Kinovea has been validated [18]. The percentage difference between Kinovea and Cortex, an established motion capture system are relatively small thus concluded Kinovea reliability.

Results also show that in general, the protocol is repeatable and when compared to an established motion capture system, Kinovea is reliable. Therefore, it is found that Kinovea has the potential to be used as a motion capture-analysis tool. The capability of Kinovea is currently being further investigated to confirm it accuracy and repeatability for other applications. It could be concluded that this study has contributed significant and new knowledge about drop jump movement analysis using HD VideoCam-Kinovea system.

For future amends, the subject might undergo pre-scheduled training or time range studies are included to generate more consistent results. In other hand, high speed camera is also suggested to be used to obtain a highly digitized camera. It is recommended further exploration on this topic to be made in order to strengthen these findings.

Acknowledgments

This study has been realised via the research collaboration between Universiti Teknologi MARA, Malaysia and Shibaura Institute of Technology, Japan under the "SIT-UiTM Visiting Researchers Programme 2016".

The main research has been funded by the Ministry of Higher Education (MOHE) Malaysia and Universiti Teknologi MARA under the Fundamental Research Grant Scheme, grant no. 600–RMI/FRGS 5/3 (0098/2016). Informed consent was obtained from all the volunteered subjects with ethical approval from Universiti Teknologi MARA Research Ethics Committee, under the part of study related to the development of motion capture-analysis system using Kinect and Matlab (REC/187/16) and 600-RMI/FRGS 5/3 (76/2013)).

The system validation related tasks were conducted at the Biomechanics Lab, Shibaura Institute of Technology, Omiya Campus, Japan where the students were funded by Shibaura Institute of Technology and Universiti Teknologi MARA under the "SIT-UiTM Visiting Researchers Programme 2016".

References

[1] J. Schiffer, "Plyometric Training and the High Jump," *New Stud. Athl.*, no. 3, pp. 9–21, 2012.
[2] S. Ramachandran and B. Pradhan, "Effects of short-term two weeks low intensity plyometrics combined with dynamic stretching training in improving vertical jump height and agility on trained basketball players," *Indian J. Physiol. Pharmacol.*, vol. 58, no. 2, 2014.
[3] J. Schoenecker, J. Cicchese, J. Engholm, and K. Bauer, "Optimal Angle of Knee Flexion for Maximal Vertical Jump in Division III Collegiate Athletes," no. 2007, p. 2007, 2012.
[4] K. Bujang, A. F. A. Nazri, A. F. Ahmad Azam, and J. Mahmud, "Development of a motion capture system using kinect," *J. Teknol.*, vol. 76, no. 11, pp. 75–86, 2015.
[5] D. Winter, *The Biomechanics and Motor Control of Human Gait: Normal, Elderly and Pathological*, vol. 2nd. 1991.
[6] J. Richards, "Biomechanics in clinic and research," no. May, p. 207, 2008.
[7] B. Requena, F. Requena, I. García, E. S. S. de Villarreal, and M. Pääsuke, "Reliability and validity of a wireless microelectromechanicals based system for measuring vertical jumping performance," *J. Sport. Sci. Med.*, vol. 11, no. 1, pp. 115–122, 2012.
[8] Mohd Nor Azmi Ab Patar, T. Komeda, C. Y. Low, and J. Mahmud, "System Integration and Control of Finger Orthosis for Post Stroke Rehabilitation," *Procedia Technol.*, vol. 15, pp. 756–765, 2014.
[9] J. Mahmud, C. A. Holt, and S. L. Evans, "An innovative application of a small-scale motion analysis technique to quantify human skin deformation in vivo," *J. Biomech.*, vol. 43, no. 5, pp. 1002–1006, 2010.
[10] G. Mustapha, M. F. Abd Razak, M. S. Mohd Hamzah, N. H. Mohd Yahya, and J. Mahmud, "The Development Of A Low Cost Motion Analysis System: Cekak Visual 3D v1.0," *Int. J. GEOMATE*, vol. 11, no. 24, pp. 2248–2252, 2016.
[11] M. Centre, "Variance and standard deviation (grouped data)," *Creat. Commans*, p. 2, 2003.
[12] K. M. S. T. Yusuf, A. F. Ahmad Nazri, G. Mustapha, and J. Mahmud, "Analysis of static and dynamic motion accuracy for kinect-virtual Sensei system," *ARPN J. Eng. Appl. Sci.*, vol. 10, no. 17, pp. 7328–7335, 2015.
[13] Motion Analysis, "Cortex Reference Manual," 2010.
[14] M. N. A. A. Patar, T. Komeda, L. C. Yee, and J. Mahmud, "Model-based systems engineering of a hand rehabilitation device," *J. Teknol.*, vol. 76, no. 4, pp. 101–106, 2015.
[15] M. N. A. Ab Patar, A. F. Said, J. Mahmud, A. Majeed, and M. A. Razman, "System Integration and Control of Dynamic Ankle Foot Orthosis for Lower Limb Rehabilitation," *2014 1st Int. Symp. Technol. Manag. Emerg. Technol. (Istmet 2014)*, pp. 82–85, 2014.
[16] D. B. Nielsen, "Comparison of angular measurements by 2D and 3D gait analysis," *Dep. Rehabil. ; Sch. Heal. Sci. Jönköping Univ.*, p. 26, 2008.
[17] C. Damsted, R. O. Nielsen, and L. H. Larsen, "Reliability of video-based quantification of the knee- and hip angle at foot strike during running.," *Int. J. Sports Phys. Ther.*, vol. 10, no. 2, pp. 147–54, 2015.
[18] N. Amirah, H. Hisham, A. Faiz, A. Nazri, J. Madete, L. Herawati, and J. Mahmud, "Measuring Ankle Angle and Analysis of Walking Gait using Kinovea," *Int. Med. Device Technol. Conf.*, pp. 247–250, 2017.

Effect of palm oil fuel ash on compressive strength of palm oil boiler stone lightweight aggregate concrete

K Muthusamy[1], N A Zamri[2], A Kusbiantoro[3], N H A S Lim[4] and M A Mohd Ariffin[5]

[1,2]Faculty of Civil Engineering and Earth Resources, Universiti Malaysia Pahang, 26300 Gambang, Pahang
[3]Faculty of Engineering Technology, Universiti Malaysia Pahang, 26300 Gambang, Pahang
[4,5] Faculty of Civil Engineering, Universiti Teknologi Malaysia, 81310 UTM Skudai, Johor

E-mail: khairunisa@ump.edu.my

Abstract. Both palm oil fuel ash (POFA) and palm oil boiler stone (POBS) are by-products which has been continuously generated by local palm oil mill in large amount. Both by products is usually disposed as profitless waste and considered as nuisance to environment. The present research investigates the workability and compressive strength performance of lightweight aggregate concrete (LWAC) made of palm oil boiler stone (POBS) known as palm oil boiler stone lightweight aggregate concrete (POBS LWAC) containing various content of palm oil fuel ash. The control specimen that is POBS LWAC of grade 60 were produced using 100% OPC. Then, another 4 mixes were prepared by varying the POFA percentage from 10%, 20%, 30% and 40% by weight of cement. Fresh mixes were subjected to slump test to determine its workability before casted in form of cubes. Then, all specimens were subjected to water curing up to 28 days and then tested for its compressive strength. It was found out that utilizing of optimum amount of POFA in POBS LWAC would improve the workability and compressive strength of the concrete. However, inclusion of POFA more than optimum amount is not recommended as it will increase the water demand leading to lower workability and strength reduction.

1. Introduction

Globally, urbanization was driven by increasing numbers of population growth. Rapid urbanization was seen to put high pressure on building and infrastructure expansion. The continuously growing construction activities leads to higher demand for non-renewable resources such as aggregates from the environment to be used in the production of construction building materials such as cement and concrete. Aggregate being one of the materials which continuously extracted from the hills for concrete production is depleting from the face of the earth. One of the researcher, Hainin et al [1] has stated that the extensive use of aggregate would cause depletion of local aggregate supply. At the same time, the cement production which consumes a huge amount of limestone and releases greenhouse gas causes environmental degradation. The negative impact of cement production to the environment has been highlighted by Meyer [2]. Realizing the importance of preserving the natural non-renewable resources for a cleaner and healthier environment, various attempts have been made by researchers in order to identify waste materials that have potential to replace the existing precious natural raw materials. Utilizing locally available industrial by-products as partial aggregate or cement replacement would reduce the construction material cost and decrease amount of waste dumped at the landfill.

In relation to that, Malaysia being one of the world largest palm oil producers also generates increasing amount of by-products which disposed as waste. Palm oil boiler stone (POBS) and palm oil boiler ash are waste produced after the incineration of oil palm shell, fruit bunch and oil palm fibres to produce electricity for the palm oil mill. Palm oil boiler stones are big solid chunks that formed in the incinerator as a result of incineration process. These rock like material are usually dumped as waste. As for palm oil fuel ash, it appears in form of greyish ash. Normally it is dumped at open environment within the palm oil mill vicinity and easy to be carried by the wind thus causing air pollution. The pollution caused by the ashes has been pointed by past researchers [3, 4]. It is seen that integrating these unwanted waste materials as mixing ingredient in concrete would reduce high dependency on natural aggregate and cement as well as reduce amount of wastes disposed to the environment. Therefore, the present research looks into the effect of palm oil fuel ash (POFA) as partial cement replacement towards workability and compressive strength of palm oil boiler stone lightweight aggregate concrete.

2. Methodology

2.1. Materials

Ordinary Portland cement from a single source was used as binder throughout the experimental work. River sand was utilized as fine aggregates. For mixing and curing purpose, clean tap water was used. No granite aggregate is used. Palm oil fuel ash (POFA) which is a by-product obtained by burning of fibers, shells and empty fruit bunches was obtained from nearby palm oil mill in the state of Pahang, West Malaysia. At the laboratory, the ash was dried in an oven at 105 ± 5 °C for 24 h to remove moisture. The ash was sieved through using 300 μm sieves to remove impurities. After the sieving process, palm oil boiler ash was ground to sufficient fineness to be used as partial cement replacement. The palm oil boiler stone (POBS) was obtained from palm oil mill at Pahang. The stone was also dried in an oven at 105 ± 5 °C for 24 h to remove moisture. It was then crushed using jaw crusher to reduce its size before used as coarse aggregate replacement. .

2.2. Mix Proportion and Testing

The concrete in this experiment is designed to achieve a compressive strength of 60 N/mm^2 at 28 days of curing age. Five concrete mixes consists of 0%, 10%, 20%, 30% and 40% POFA by weight of cement were prepared. Concrete mix with 0% POFA was act as control specimen. The slump test is essential in order to produce fresh concrete with good consistency. This test was conducted according to BS EN 12350-2 [5]. To determine the concrete strength, compressive strength test was conducted according to BS EN 12390-3 [6]. The mixes were prepared in form of cubes with size of 100mm x 100mm x 100mm and subjected to water curing for 28 days.

3. Results and discussion

3.1. Workability

Figure 1 demonstrates the influence of various percentage of POFA as partial cement replacement towards the workability of POBS LWAC. From the graph, it was observed that the workability of POBS LWAC with POFA was reasonably satisfactory within the expected range. The slump values of POBS LWAC with different percentage of POFA mixtures were measured between 23 and 97 cm. Slump with POBA replacement between 10 to 30% produces true slump with a medium degree of workability which normally applied for normal reinforced concrete. The best slump is best denoted by POBS LWAC with 10% POFA replacement. The slump flow increase by 4% from control mixes with 97 mm. Similarly Zeyad et al [7] reported that the normal consistency of OPC concrete is lower than that of the concrete with POFA. Inclusion of more than 10% POFA reduces concrete workability. Workability becomes lower as more POFA is used as partial cement replacement. The lower slump result is due to porous particle of POFA that absorbs more water compared to OPC when it is utilized more than the optimum amount. In addition, the high carbon content of the POFA also contributes to lower workability of mix. On overall, addition of POFA at suitable amount would produce concrete mix with the targeted workability.

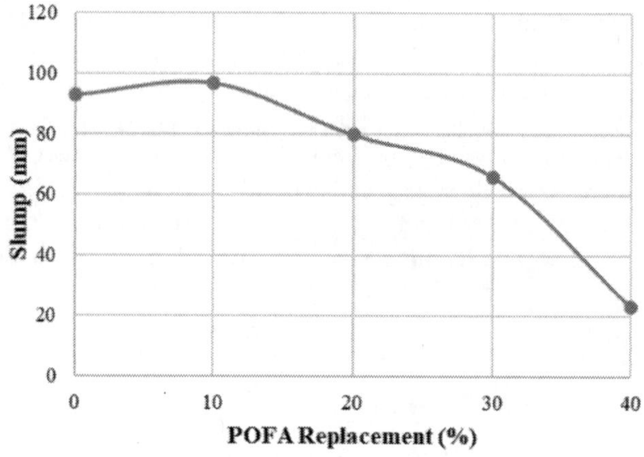

Figure 1. Effect of POFA content on workability of lightweight aggregate concrete mix.

3.2. Dry Density

Figure 2 shows the influence of various percentage of POFA as partial cement replacement towards the density of POBS LWAC the age of 28 days. Generally, the dry density of POBS LWAC with POFA for all percentages falls in the range for LWAC. In this research, 10% POFA replacement in POBS LWAC shows the highest dry density. This is due to the pozzolanic reaction from POFA that produces larger amount of C-S-H gel causing the concrete to be more compact finally producing concete with higher dry density. Studies by [8, 9] have proved that the pozzolanic additives reduce porosity and increase the dry density of concrete. However, when a higher amount of POFA is utilized than optimum amount it produces concrete with lower dry density. This is due to porous POFA particle that absorbs high water when it is utilized of more than optimum amount in the concrete mix. As a result, the mix becomes dry and hard to be compact thus causing segregation. Improper compaction resulted in honeycomb that trapped inside concrete paste, leading to low dry density.

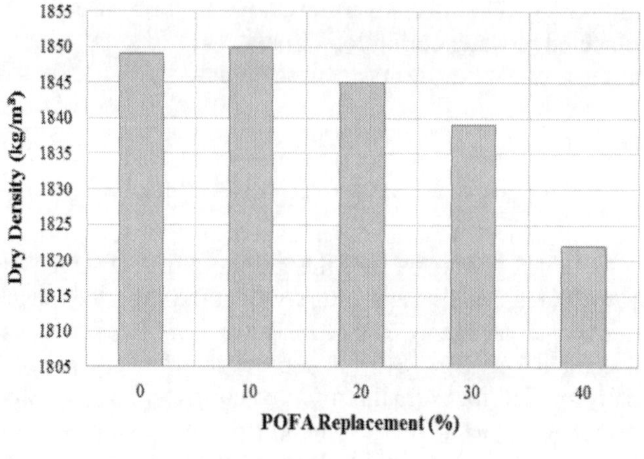

Figure 2. Dry density of lightweight aggregate concrete containing various percentage of POFA.

3.3 Compressive Strength

Figure 3 shows the variation of POC LWAC compressive strength with POFA percentage at the age of 28 days. It can be observed that all the samples excluding POBS LWAC with 40% POBA replacement attained 28 days strength above 60 MPa which in the range for high strength concrete (HSC). HSC is often used in the construction of high load bearing columns and also to produce precast concrete products. The strength gain of the POBS LWAC with POBA is superior to that of the control POBS LWAC when cement is replaced with 10% POFA. The pozzolanic reaction due to presence of POFA generates secondary C-S-H gel which makes the total C-S-H gel in the blended cement lightweight aggregate concrete is higher than control specimen. Unreacted fine POFA also act as filler that fill in the pores inside the concrete structure makes the concrete denser and stronger.

It is also observed that the compressive strength of concrete containing more than 10% POFA replacement were lower than control specimen. The compressive strength of concrete continues to drop as larger amount of POFA is used as partial cement replacement. This is happening due to the porous characteristic of POFA particle as compared to cement. When POFA is used more than the optimum amount, drier mix and difficult to be compacted is produced resulting in a formation of concrete with larger amount of voids as compared to specimen with lesser amount of ash content. Similarly, previous researcher [10, 11] noted the decreasing strength of concrete when POFA used exceeded the optimum amount. Conclusively, utilization of POFA at 10% as partial cement replacement contributes towards strength enhancement of high strength lightweight aggregate concrete.

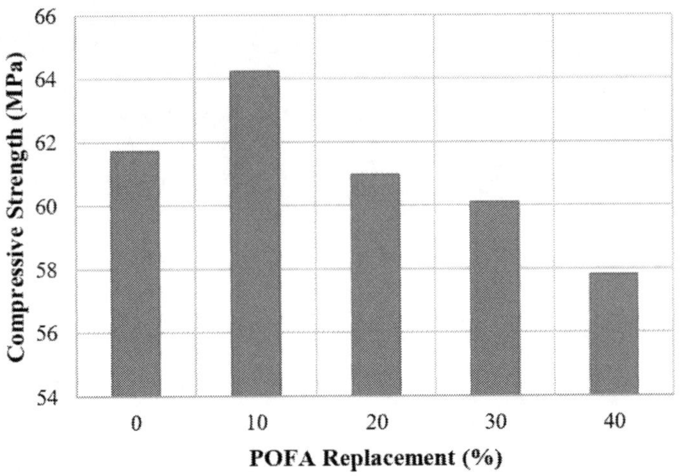

Figure 3. Compressive strength of lightweight aggregate concrete at 28 days.

4. Conclusion

The present findings show that palm oil fuel ash can be used as partial cement replacement to produce palm oil boiler stone lightweight aggregate concrete exhibiting higher compressive strength than the plain specimen. Success in utilizing palm oil fuel ash in lightweight concrete production would reduce the amount of cement used and contribute towards cleaner environment.

Acknowledgments

The authors would like to acknowledge the help and co-operation received from the lecturers and technical staff of Universiti Malaysia Pahang in conducting the experimental work. The financial support received from Universiti Malaysia Pahang through grant RDU1703109 is gratefully acknowledged.

IOP Conf. Series: Materials Science and Engineering **342** (2018) 012098 doi:10.1088/1757-899X/342/1/012098

References

[1] Hainin M R Yusoff N I M Sabri M F M Aziz M A A Hameed M A S and Reshi W F 2012 Steel slag as an aggregate replacement in Malaysian hot mix asphalt. *International Scholarly Research Network*, 1-5.

[2] Meyer C 2009 The greening of concrete industry. *Cement & Concrete Composites*, 31: 601-605.

[3] Hussin M W Muthusamy K and Zakaria F 2010 Effect of mixing constituent towards engineering properties of POFA cement -based aerated concrete. *Journal of Materials in Civil Engineering*, 287-295.

[4] Aprianti E Shafigh P Bahri S and Farahani J N 2015 Supplementary cementitious materials origin from agricultural wastes – A review. *Journal of Construction and Building Materials*, 74 : 176-187.

[5] British Standard. 2000. Testing Fresh Concrete. Slump test. London. BS EN 12350-2.

[6] British Standard. 2002. Testing Hardened Concrete. Compressive strength test of specimens. London. BS EN 12390-3.

[7] Zeyad A M Megat Johari M A Muhammad Bunnori N Ariffin K S and Altwair N M 2013 Characteristics of treated palm oil fuel ash and its effects on properties of high strength concrete. *Advanced Material Research,* 626 : 152-156

[8] Iqbal K M and Alhozaimy O M 2011 Properties of natural pozzolan and its potential utilization in environmental friendly concrete. *Canadian Journal of Civil Engineering*, 38(1) : 71-78.

[9] Gong J Chou K Zheng Y H and Zhao M 2014 A quantitative study on packing density and pozzolanic activity of cementitious materials based on the compaction packing model. *Material Science and Engineering*, 62 : 1-11.

[10] Megat Johari M A Zeyad A M Muhammad Bunnori N and Ariffin K S 2012 Engineering and transport properties of high-strength green concrete containing high volume of ultrafine palm oil fuel ash. *Journal of Construction and Building Material*, 30 : 281–288.

[11] Muthusamy K and Zamri, N A 2016 Mechanical properties of oil palm shell lightweight aggregate concrete containing palm oil fuel ash as partial cement replacement. *KSCE Journal of Civil Engineering*, 20(4), 1473-148.

IOP Conf. Series: Materials Science and Engineering **342** (2018) 012099 doi:10.1088/1757-899X/342/1/012099

Compressive strength performance of OPS lightweight aggregate concrete containing coal bottom ash as partial fine aggregate replacement

K Muthusamy, R Mohamad Hafizuddin, F Mat Yahaya, M A Sulaiman, S M Syed Mohsin, N N Tukimat, R Omar and S C Chin

Faculty of Civil Engineering and Earth Resources, Lebuhraya Tun Razak, Universiti Malaysia Pahang, 26300 Gambang, Kuantan, Pahang

E-mail: khairunisa@ump.edu.my

Abstract. Concerns regarding the negative impact towards environment due to the increasing use of natural sand in construction industry and dumping of industrial solid wastes namely coal bottom ash (CBA) and oil palm shell (OPS) has resulted in the development of environmental friendly lightweight concrete. The present study investigates the effect of coal bottom ash as partial fine aggregate replacement towards workability and compressive strength of oil palm shell lightweight aggregate concrete (OPS LWAC). The fresh and mechanical properties of this concrete containing various percentage of coal bottom ash as partial fine aggregate replacement were investigated. The result was compared to OPS LWAC with 100 % sand as a control specimen. The concrete workability investigated by conducting slump test. All specimens were cast in form of cubes and water cured until the testing age. The compressive strength test was carried out at 7 and 28 days. The finding shows that integration of coal bottom ash at suitable proportion enhances the strength of oil palm shell lightweight aggregate concrete.

1. Introduction

Malaysia is one of the largest palm oil producing country in the world. At the same time, the palm oil processing mills are generating many types of wastes. As the palm oil industry continues to grow, more wastes are generated. These wastes when disposed cause environmental pollution and destroy the habitats of flora and fauna. The importance of preserving the cleanliness of environment for present and future generation, has motivated many researches to be carried out to turn these wastes such as oil palm shell (OPS), palm oil clinker, empty fruit bunches, palm oil fibres and palm oil fuel ash into profit generating element. Oil palm shell is one of the wastes which continuously generated in increasing amount by palm oil mills. Approximately 6.89 million tonnes of OPS were generated annually by Malaysian palm oil mills [1]. This waste which obtained after the oil extraction process usually dumped as waste in the factory's compound. The continuous dumping of these wastes causes larger land consumption in the factory area which can be utilized for more beneficial purposes. From the view of environmental sustainability, the dumping of waste is causing environmental pollution [2, 3]. As one of the solutions to this problem, palm oil mills need to spend extra expenditure to manage these wastes. However, integrating this waste material in production of useful materials would reduce quantity of waste ending at landfill. Besides that, coal bottom ash which is a solid waste generated from local coal industry also causes negative impact to the environment. The dumping of this waste also causes environmental problems such as contamination of ground and surface water. According to

Rafieizonooz et al. [4] the projected forecast for electricity usage in peninsular Malaysia will be produced from coal and gas (58% and 25%) by the year 2024. Thus, it is expected that production of coal bottom ash would increase and poses more environmental problem unless it is converted into beneficial product.

At the same time, the development of construction industry has resulted in increasing use of natural sand for construction. Based on the report by Department of Minerals and Geosciences of Malaysia, the sand mining activities in Malaysia have been trending in increasing mood which made a difference from 37.3 million tonne in 2011 to 46.7 million tonne in 2016. Increasing sand mining activity exposes the environment quality degradation issues such as destruction of natural habitats, soil erosion and reduced river water quality due to excessive mining activity. These environmental problems occur when the rate of extraction of sand, gravel and other materials exceeds the rate at which natural processes generate these materials [5]. Discovery of other alternative material to reduce the use of natural sand would be able to preserve nature resources and contribute to sustainable construction. The environmental problems caused by dumping of solid wastes and increasing sand mining activity have led to the effort of integrating waste material as partial sand replacement in oil palm shell lightweight aggregate concrete. Past researchers [2, 6, 7, 8, 9] has investigated the properties and potential of oil palm shell lightweight aggregate concrete. However, no experimental findings are available on the use of coal bottom ash as partial sand replacement in this lightweight concrete. Thus, the present research aims to investigate the effect of integrating coal bottom ash as partial sand replacement in OPS lightweight concrete.

2. Methodology

2.1. Materials

The materials used in this experiment are ordinary Portland cement, local river sand, water, water reducing admixture, oil palm shell (OPS) and coal bottom ash (CBA). Tap water at the laboratory was used for concrete mixing and curing work. Coal bottom ash (CBA) shown in Figure 2 was supplied by one of the coal power plant located in West Coast Malaysia. Oil palm shell (OPS) in Figure 1 was supplied by nearby palm oil mill located in the state of Pahang, East Coast Malaysia. The properties of OPS used in this research are tabulated in Table 1.

Figure 1. Oil Palm Shell (OPS).

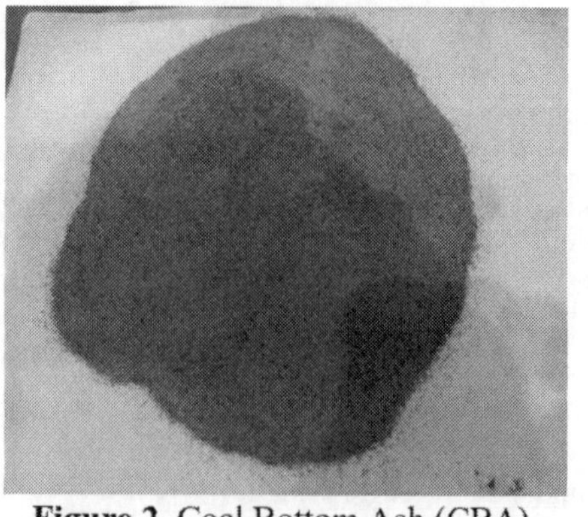

Figure 2. Coal Bottom Ash (CBA).

Table 1. Properties of oil palm shell aggregates.

Properties	**Oil palm shell**
Maximum size (mm)	10
Specific gravity	1.19
Density (kg/m^3)	1481
Water absorption for 24 h (%)	16.57
Aggregate abrasion value, (%)	6.9
Fineness modulus	6.33
Flakiness index (%)	65.17
Elongation index (%)	12.36
Aggregate impact value (%)	6.47
Aggregate crushing value (%)	5.71

2.2 Mix Proportion and Testing Procedure

A total of 5 mixes were used in experimental work. Mix containing 100% OPC was used as control specimen. The rest of the mixes were prepared by varying the percentage of CBA at 5%, 10%, 15% and 20% by weight of sand. Prior to mixing process, all the materials were weighed accurately. Then, all the mixing ingredients were mixed uniformly before cast in form of cubes of 100mm. The cubes were demoulded the next day and subjected to water curing. In order to determine the concrete workability, freshly prepared mixed were immediately subjected to slump test. The slump test was conducted following the procedure in ASTM C143 [10]. The effect of CBA content on the concrete strength is determined through compressive strength test adhering to the procedure in BS EN 12390 [11]. The test was conducted at 7 and 28 days.

3. Results and discussion

3.1 Workability

The result of slump test conducted on mixes containing various percentage of coal bottom ash (CBA) is illustrated in Table 2. It can be observed that the inclusion of CBA affects the workability of concrete. As the percentage of CBA is added in the mix becomes higher, the concrete mix become stiffer resulting in lower slump value. One of the contributing factors is higher water absorption by the porous particles of coal bottom ash. The porous particles of coal bottom ash absorbed water internally causing reduction in free water for lubrication of particles resulting in lower slump. The influence of aggregate characteristic towards concrete workability has been highlighted by Neville [12]. Similar trend has been observed by past researchers [13, 14] who investigated the effect CBA as partial sand replacement on concrete workability.

Table 2. Slump test result

Mix with CBA %	Slump Value	Slump Classification
0%	100	High
5%	85	Medium
10%	75	Medium
15%	60	Medium
20%	5	Very Low

3.2 Compressive Strength

Figure 3 presents the compressive strength test result of specimens containing various content of CBA as partial fine aggregate replacement. Looking at the effect of CBA as partial sand substitute, the strength gain of the concrete with CBA was superior to that of the control mix as the percentage of CBA replacement was increased to up to 20%. However, the highest compressive strength was obtained when the sand is replacement with 15% of CBA by weight. The strength gain of that particular mix is 14% higher than the control specimen. This encouraging result is attributed to the function of CBA as filler which makes the concrete internal structure denser and more compact resulting in higher compressive strength. Similar positive trend has been reported by other researchers [15, 16] who added certain percentage of CBA as partial fine aggregate replacement in different types of concrete.

A decrease in compressive strength was observed as the percentage of sand replacement was more than 15%. This probably attributed to the characteristic of fine CBA possessing higher specific area than sand thus requires more water to coat the CBA particles. As a result, the concrete mix becomes drier, more difficult to mix and to compact when quantity of CBA used is increased. Finally, the concrete exhibits lower strength due to lack of bonding between particles and existence of more voids in the hardened concrete. Conclusively, integration of suitable percentage of CBA has enhanced the strength of OPS lightweight aggregate concrete.

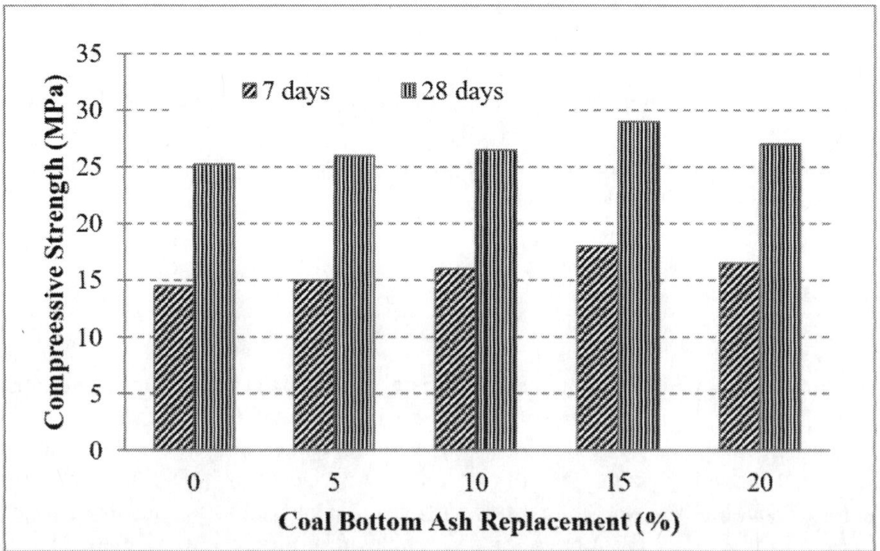

Figure 3. Compressive strength test result.

4. Conclusion

The finding shows that use of suitable amount of coal bottom ash able to produce oil palm shell lightweight aggregate concrete suitable for structural application. Integration of 15% of CBA produces concrete exhibiting the highest compressive strength value as compared to other mixes. This finding is expected to encourage the use of this waste material in lightweight concrete production and at the same time reduce the use of natural river sand.

Acknowledgments
The authors gratefully acknowledge the financial support provided by the university through internal grant RDU170339.

Reference
[1] Chong H L H Chia P S and Ahmad M N 2013 "The adsorption of heavy metal by Bornean oil palm shell and its potential application as constructed wetland media." *Bioresource Technology* 130 pp. 181-186,
[2] Mohammad M U I Kim H M Alengaram U J and Jumaat M Z. 2016. Mechanical and fresh properties of sustainable oil palm shell lightweight concrete incorporating palm oil fuel ash. *Journal of Cleaner Production*. 115 : 307 – 514.
[3] Kim H M Alengaram U J and Jumaat M Z 2016 Assessing some durability properties of sustainable lightweight oilpalm shell concrete incorporating slag and manufactured sand. *Journal of Cleaner Production* 112 763 – 770.
[4] Rafieizonooz M Mirza J Salim M R Hussin M W and Khankhaje E 2016. Investigation of coal bottom ash and fly ash in concrete as replacement for sand and cement. *Construction and Building Materials* 116 15–24.
[5] Ashraf M A Maah M J Yusof I Wajid A and Mahmood K 2011 Sand mining effects, causes and concerns: A case study from Bestari Jaya, Selangor, Peninsular Malaysia. *Scientific Research and Essays* 6(6) : 1216-1231
[6] Mannan, M A and Ganapathy C 2004. Concrete from an agricultural waste-oil palm shell. *Building and Environment* 39(4) 441–448.
[7] Teo D C L Mannan M A and Kurian V J 2010. Durability of lightweight OPS concrete under different curing conditions. *Material Structure* 43: 1 – 3.
[8] Jumaat M Z Alengaram U J Ahmmad R Bahri S and Saiful Islam A B M 2015 Characteristics of palm oil clinker as replacement for oil palm shell in lightweight concrete subjected to elevated temperature. Construction and Building Materials, 101 (1) : 949 – 951.
[9] Kim H M Alengaram U J and Jumaat M Z 2016 Assessing some durability properties of sustainable lightweight oil palm shell concrete incorporating slag and manufactured sand. *Journal of Cleaner Production*. 112 763 – 770
[10] ASTM C143 2015 Standard test method for slump of Hydraulic-cement concrete, ASTM International, West Conshohocken, PA
[11] British Standard. 2002. Testing Hardened Concrete. Compressive strength test of specimens. London. BS EN 12390-3.
[12] Neville A M 2011 Properties of Concrete (5th Edition).Prentice Hall
[13] Sandhya B and Reshma E K 2013 A study of mechanical properties of cement concrete by partial replacement of fine aggregates with bottom ash, *International Journal of Students Research in Technology and Management* 1 (4) 416–430.
[14] Hamzah A F Norwati J Ramadhansyah P J Mohd Fadzil A and Norul Ernida Z A 2015 Fresh characteristic and mechanical compressive strength development of self-compacting concrete integrating coal bottom ash as partial fine aggregates replacement. *International Journal of Engineering & Technology IJET-IJENS* 15(4): 61-67.
[15] Aswathy P U and Paul M M 2015 Behaviour of self compacting concrete by partial replacement of fine aggregate with coal bottom ash. *International Journal of Innovative Research in Advanced Engineering (IJIRAE)* 2 (10). 45-52.
[16] Balasubramaniam T and Thirugnanam G S 2015 An Experimental investigation on the mechanical properties of bottom ash concrete *Indian Journal of Science and Technology*. 8(10): 992-997.

IOP Conf. Series: Materials Science and Engineering **342** (2018) 012100 doi:10.1088/1757-899X/342/1/012100

Critical analysis of procurement techniques in construction management sectors

Suman Tiwari Suresh Tiwari, Shiau Wei Chan and Muhammad Faraz Mubarak

Department of Production and Operation Management, Universiti Tun Hussein Onn Malaysia, Batu Pahat, Johor, Malaysia

E-mail: suman_tiwari24@yahoo.com.my; swchan@uthm.edu.my

Abstract. Over the last three decades, numerous procurement techniques have been one of the highlights of the Construction Management (CM) for ventures, administration contracting, venture management as well as design and construct. Due to the development and utilization of those techniques, various researchers have explored the criteria for their choice and their execution in terms of time, cost and quality. Nevertheless, there is a lack of giving an account on the relationship between the procurement techniques and the progressed related issues, for example, supply chain, sustainability, innovation and technology development, lean construction, constructability, value management, Building Information Modelling (BIM) as well as e-procurement. Through chosen papers from the reputable CM-related academic journals, the specified scopes of these issues are methodically assessed with the objective to explore the status and trend in procurement related research. The result of this paper contributes theoretically as well as practically to the researchers and industrialist in order to be aware and appreciate the development of procurement techniques.

1. Introduction

It is aphoristic that an undertaking of CM project might be viewed as effective if the completion of building is done in the expected timeline, not beyond the allocated budget and the quality which is being standardized in order to attain the maximum level of customer fulfillment. Progressively, the satisfaction of these criteria has been related with the issue of procurement technique for construction. In other words, the determination of the suitable procurement technique can shape the achievement of the project. Comprehensively, the issues that are confronting the building procedure are: (1) Separation of outline plan from construction; (2) absence of team integration; (3) absence of effective communication; (4) vulnerability; (5) evolving environment, (6) changing customers' needs and desires, and (7) increment of project complications [1]. These, together with monetary changes such as recession and inflation, have driven construction experts and the business to offer elective techniques for building procurement, for example, design and build, administration contracting and construction of project management. There are various methods in procuring a venture which are Private Finance Initiative (PFI), Public-Private Partnership (PPP), and so forth which may likewise be contemplated under the procurement technique. Despite that, they are categorised more precisely as a pathway towards procurement and depicted it as a theory and an agreement of trust [2]. For instance, majority partnering arrangements are a general understanding, which incorporates one or a few contracts below one of the three key techniques. The explored research has offered ascend to an extensive number of definitions of procurement technique. The best two procurement techniques definitions are:

i) "a procurement system is an organizational system that assigns specific responsibilities and authorities to people and organizations, and defines the relationships of the various elements in the construction of a project" [3].

ii) "a mechanism for linking and coordinating members of the building team throughout the building process in a unique systematic structure, both functionally and contractually. Functionally via roles, authority and power, contractually via responsibilities and risks. The main aim is to deliver a project that meets its objectives and fulfill the client criteria and expectations"[4].

With the expansion in usage of alternate procurement techniques, various researchers have design charts of decision-making process to examine the criteria for their choice and their rate of achievement regarding time, cost and quality. Throughout the years the selection procedure of procurement has progressed toward becoming progressively complex, primarily because of the prolonged expansion of various procurement techniques for procuring building tasks, the tasks' consistently expanding its complexity and the customer's requirement for a more incentive for money projects. Thus, it is necessary to classify that the factor of time, cost and quality only are currently insufficient in regards to the present complex condition of construction's project. The objective of this paper is to reveal an inclusive review of literature which arbitrates to address the current factors that are related with the determination of current procurement technique. In order to accomplish this objective, a systematic literature review has been conducted for the year of 1980 to 2016, highlighting on the current issues with the purpose of building up an up- to-date decision making chart to choose the suitable procurement technique for the construction's project.

2. Research methodology

In order to accomplish the aim of this paper, a two stage of methodology procedure has been carried out. They are:

2.1 Stage 1 : Determining the relevant materials

The identification process of materials for literature review began by investigating primary and secondary sources. The primary sources included refereed conferences, refereed journals, theses, intermittent papers and government reports. The secondary sources included trade journals, textbooks, magazines, and newspapers. The sources that will be selected should be related to CM. The top five best journals which were reviewed during the exploration are: i) International Journal of Procurement; ii) International Journal of Project Management; iii) Engineering, Construction and Architectural Administration; iv) American Society of Civil Engineers (ASCE); v) Construction Management and Economics. The justifications of choosing these respective five journals are due to the fact that it has been highly ranked in the field of CM by [5]. Subsequently, these five journals are also being identified to have a high frequency of procurement techniques published paper.

Other than that, the selection of another five peer reviewed journals which indicates a high frequency of CM published papers is being done for the literature review. These five peer reviewed journals are from i) Industrial Marketing Management; ii) Construction Innovation; iii) An International Journal; iv) International Journal of Quality and Reliability Management; v) International Journal of Managing Projects in Business. Overall, these top 10 journals are being chosen for the journals source. Besides, four of the top international conferences proceedings related to CM are being reviewed as well. They are i) Organisation and Management of Construction Symposium (known as CIB W65); ii) Procurement Systems Symposium (CIB W92); iii) Building Economics and Construction Management (CIB W90) as well as iv) Association of Research in Construction Management (ARCOM). In addition, reports and occasional paper were also selected as one of the material due to its complete and current information related to CM and procurement. These papers are i) Chartered Institute of Building (CIOB); ii) Royal Institute of Chartered Surveyors (RICS); iii) Royal Institute of British Architects (RIBA); iv) Building Research Establishment (BRE); v) Construction Industry Research and Information Association (CIRIA). The search has been carried out using ten journals and nine conference proceedings and report. Thus, the search indicates 120 CM-procurement related articles and for the aim of this paper, the reference stated below were suitable and being chosen to be discussed extensively.

2.2 Standardized note-taking and assessment

Subsequently, at this particular stage the summary of the literature review were filtered and a literature document was created. The primary scopes that were looked in this research are supply chain, sustainability, innovation and technology development, lean construction, constructability, value management, Building Information Modelling (BIM) and e-procurement. These scopes were selected based on the previous literature in the procurement related fields as well as their connection with CM. In particular, the specific writings of previous authors [6,7,8,9,10,11,12,13]. After assembling the materials for literature review, the selected articles are reviewed critically and systematically in order to identify the resemblance in the literatures of previous researchers, the prevalent issues highlighted, and disparity or disagreement of the facts as well as elucidations made by previous researchers.

3. Critical review of current issues related to procurement methods

3.1 Supply Chain Management (SCM)

Supply chain management (SCM) is defined as the network management of interconnecting businesses ramified in the overall arrangement of goods and service packages demanded by end customers [14]. Therefore it is obvious that SCM encloses all the significant movement and depository of raw materials, work-in-process inventory, as well as finished goods from the origin point to the consumption point. According to [15] "(SCM) is directed toward the minimisation of transaction cost and the "enhancement and transfer of expertise between all parties". Modern procurement method does not make a hierarchical structure to manage the social issue and discontinuity of the building procedure. Researchers recommended that clustering by a technology group approach as a method for handling reconciliation issue. It was contended that SCM is a management theory and that its philosophy is applicable to any procurement techniques; in spite of the fact that management types of procurement contributes a better framework [16].

3.2 Sustainability

Sustainable Procurement (SP) is a procedure whereby associations address their issues and demand for goods, services, administrations, works and utilities in a way that accomplishes the value of money on an entire life premise in order to create genuine long haul benefits, to the association, as well as to society and the economy, while reducing the damage to the environment [17]. Currently, there is an obligatory introduced in Europe which is all new built and refurbished building structures are ought to show consistence with 'Target Carbon Emissions Rates' and in addition with the Building Energy Model (Part L) of the Building Regulations 2006. Researchers agree that the generation of such models demands a few cycles at the design and plan stage and invites close integration between the various experts [6,18,19]. Breakthrough in term of innovation has been thwarted by numerous hindrances, for example, the industry's divided nature, absence of long haul perspective, customers' unwillingness to share load, lack of clear definition and advantages of sustainable construction, constraint related to regulations and conflict in government policies and insufficient fiscal incentives [20]. The research done [21] indicates that there is a huge conflict between SP and cost reduction mainly in central government. Thus, in this manner, this would provoke the inquiry regarding the balance of the design and build contractor will consider among sustainability for presenting a more focused and competitive offer for the works. In addition, it is supported that sustainable construction demands change to the method of construction, the utilization of resources yet more and most importantly in the building procedure [22]. Therefore, to accomplish this, critical change to the organisation, structure and communication channels of the business should be made. It was also highlighted that the usage of traditional procurement techniques creates a professional obstruction to innovative change that is required by sustainability. Besides, [6] explained that the usage of traditional procurement involving design and build is proven to be an extremely difficult task to outline environmentally sensitive buildings as the emphases required are inconsistent with the contractor motivating force to stay away from postponements and additional cost. By following the guideline of sustainable construction, from the government's point of view, it will encourage a genuine social change in the CM towards the selection of collaborating as a procurement process [23].

3.3 Lean Construction

The previous researchers looked for radical changes to enhance production in the construction industry and they identified that the usage of lean techniques are the best practice. This idea has been explained in the articles specifically, [24] that in spite of the fact that the impact of integration on the design and delivery of construction projects have been discussed at length, the effect of this on the teams' organisation deserves detailed investigation. They set forward the case that researchers in lean construction contended that traditional design practices are outdated and have performed ineffectively in dealing with the stream or meeting customers' prerequisites. Therefore, it can be contended that, for the standards of lean construction to be adequately connected, there should be a collaboration associated with all parties which is eventually cultivated through integration. The previous researcher recorded six main components of lean construction which are: i) waste minimisation; ii) procedure which highlights production, planning and control; iii) end of customer's focus; iv) continuous developments; v) strong collaborative relationships; and vi) systems point of view [10]. Lean construction in terms of procurement can be contended that a high emphasise can be done upon the significance of cultivating a powerful collaborative relationships from the initial stage such as partnering and also recognizing continuous improvement. Methodical experimentation, continuous improvement, and as well as constant learning over all technical and organisational levels are significant factor of the lean theory, especially as a method of endeavor to improve the value of customer while minimising waste [25,26]. Nevertheless, there was some evidence available which demonstrates that executing lean practices in the collaborating consortium is not really straight forward as been frequently suggested [27]. A basic change ought to be in the practices and behaviors of the included participants or parties. They noted three sorts of boundaries towards cooperation in particular, industrial, organizational and cultural barriers.

3.4 Innovation and Technology Development

The term innovation directly shows that it is about practicing latest sciences and solutions for construction and it has turned out to be fundamental for development of construction organisations due to the high pressure from customers to enhance quality, minimise cost and accelerate construction processes. These scopes were discussed in the early year of 1960s' [28] who arranged the four different sort of innovation with the logic that they are executed and created. They are i) time or cost: An advancement or innovation which is cost-efficient and time-saving in comparison of the current strategies or techniques; ii) performance development: an innovation which has the best performance in execution; iii) aesthetic: a development of innovation that has different and latest appearance; iv) ersatz: an elective method for doing things that is constrained by particular circumstances (i.e. lack of particular materials or manpower). Subsequently, innovation is considered as one of the element of competitive advantage and can be seen as pre-imperative for an organisation achievement and survival. Egbu [29] contended that the capacity to improve or innovate highly depends on how an organisation manage and exploit their own available resources. In spite of a few hindrances towards innovation in construction, there has been a change currently to settle the divided structure of the CM in industries with critical endeavors to collaborate the design and construction. This is especially obvious in the increments in design and builds ventures, administration contracting and project management [6,13,27,30,31,32].

3.5 Constructability

First and foremost, a design plan which can be built would prompt to time and cost-saving. Nonetheless, regardless of its significance, little advancement has been made to resolve the constructability issue and this is because of the confrontational behavior between customer, contractors and consultants under the conventional procurement settings. The term "constructability" was introduced in the 1960s and 1970s but confronted negative feedbacks for its restriction in scope as it limits to the design procedure [33]. After that, various researches have been carried out with a specific end goal to a better task execution and improvement of constructability. For instance, research in procurement methods embraced the idea that "the continual use of traditional lump sum may stifle technological innovation, particularly the design and constructability of public sector buildings" [8]. Since the year of 2000, the connection among constructability and procurement technique has been broadly discussed, with most researchers in assentation that fully coordinated procurement techniques, for example, Design and Build and venture administration are most suitable for customers setting a high need on the constructability of their project.

3.6 Value management

Value management is a standardized way to deal in delivering a project with the needed functions at optimum whole life cost without affecting negatively to quality, execution and reliability. Thus, it will be the group of team which is systematically used to distinguish and wipe out superfluous expenses of the goods. Superfluous cost is the one which contributes neither the qualities, utilization, life cycle, appearance nor client required demands. The requirement of the clients are normally: Aesthetics (magnificence, colour, design), ergonomics (shape, measurements, ease), Economics (cost of operation, maintenance cost), and Technical (performance). Through the help of workshops of the ventures, value management can guarantee a dynamic support from all stakeholders of project and hence energize integration, innovation enhanced constructability, improved communications through the network of supply chain and the incorporation of sustainability practices [34]. This focuses towards a more incorporated management related approach as the design can be adjusted ahead under the advice of the main contractor by utilising their knowledge and input with a perspective of minimising waste, enhancing buildability and contributing clearer understanding of the brief through the working group, for example, management contracting and partnership [12,24,35,36,37]. By breaking the divided approach of the conventional course and empowering collaboration through an incorporated strategy for procurement, communication and concepts can be shared that can enhance the value for the particular project. The earlier this approach is being carried out, the more the advantages will be available for the customer.

3.7 Building Information Modeling (BIM)

Building Information Modeling (BIM) is a revolutionary innovation and process that has immediately transformed the perspectives and way of buildings are considered, designed, constructed and operated [38]. BIM in its optimal state influences the whole project life-cycle from procurement, feasibility studies through to design, engineering, construction, operation, and demolition. The significance that procurement has as influence of this cycle has for some time been underrated. Much further, it is not just the impacts of procurement on BIM one ought to consider, but overall also depends on the opportunities BIM offers in revolutionising the way projects are procured in the first place. Without a doubt, while looking at world construction industries, one of the most serious issues is the trouble in implementing widespread team working and cooperation, mainly because of social barriers – something that BIM brags as one of its primary qualities. As indicated by previous researchers, this is because of the working BIM demonstrates to be generally available by all team member of projects, whereby configuration changes can be consequently updated [11,39,40]. It is known that clause 2.1 of JCT's Construction Excellence advances coordinated effort as an abrogating guideline. BIM contributes from all individuals from the building group with a specific end goal to plan, model and monitor any genuine circumstances or site conflicts before executing as well as reduce any re-work [41]. BIM tools were created to extensively enhance profitability in the construction industries and make it possible to oversee and manage the information produced through the lifecycle of building structures more efficiently. Therefore, early collaboration is indispensable to guarantee that expertise and information is shared in the beginning. Because of this, it can be contended that the management contract, plan, design, build and partnering approaches are best equipped to manage this as accentuation is on developing a more constructible design which incorporates the association of the supply chain.

3.8 E-Procurement

It has been discussed by researchers that that the need for information frameworks, for example, e-procurement to encourage the coordination of the supply chain is getting more attention and consideration from the construction industries [7,42,43]. The utilization of e-procurement innovated technology decidedly influences administrators' point of view on both procurement practices and procurement performance. There are arrangements at present occurring over the proposed modification of the EU acquisition orders (OJEU) that, once concurred, should be embraced by the law. This amendment is aimed to rearrange and accelerating current procurement practices. One of the key changes to this requirement is the need for mandatory e-procurement with complete electronic procurement, including the online submission of tenders, being required by certain date and year. However, there is a drawback in this new initiative which is, the utilization of e-procurement stages requires noteworthy interest and investments in both specialist software and staff training. It can be contended that the time and speculation required by providers or organizations may produce a "two-tier" framework, with huge numbers of powerful suppliers investing into e-procurement in accordance with the mandates and leaving aside the individuals who are less capable of meeting the required speculation. The idea of e-procurement can be utilized with most procurement techniques with the exception of the partnering arrangement. This is so because of the way that with banding together, works are either let to contractual workers that are on a long haul structure such as strategic partnering and in this manner assessment is for the most part through small scale rivalries or it depends on pre-tender chosen interviews. In addition, banding together connections are typically shaped between two parties based on trust and transparency and along these lines the e-procurement stage would not be the best way to deal and accelerate effective partnering.

4. Conclusion

In a nutshell, during the past three decades procurement techniques have earned extensive consideration and discussion within the construction management in the industries. This research has contributed a critical review for the status of the best in terms of procurement techniques in the academic field with the objective of setting a fundamental platform for researchers and analysts to acquire more valuable experiences into concerns of procurement techniques. It has determined the research trends in procurement techniques which may enable the modern industrial specialists to appreciate the key research area in their improvement of current standards and procedures, for example, supply chain, sustainability, innovation and technology development, lean construction, constructability, value management, Building Information Modelling (BIM) as well as e-procurement. The supported usage of these standards can go far towards battling short-termism and industry discontinuity after some time. This will directly drive change as customers and their project teams encounter the advantages accomplished through these methods. Eventually, this will compare to a shift towards a more coordinated and integrated industry where accomplishing best value and consistent improvement through team incorporation is of fundamental significance [44]. It is tough to get away from the preface that increased cooperation within the industries. This factor will be essential to accomplish future profits, and for the industry to deliver upgrades on the customers triangulated elements of cost, time and quality. The industry which conveys the best incentive for customers in this evolving world, has the best utilisation and standardisation of data innovation. However, in numerous areas, there appear to be obstructions as far as far reaching appropriation of current method while considering the procurement route for a venture. This is due to the related risk and behaviour towards change. In order to meet the managerial, specialized technical and social barriers in the construction industry, both the industry and its participant need to welcome change and enable innovative procurement techniques to develop. As stated [45], this change should be a customer driven process upheld by strong building team.

References

[1] Chan, S.W., Abdullah, M. E., Zaman, I. and Fong, P. P. (2016). Identification of sustainable procurement in manufacturing industry. *Journal of Engineering and Applied Science*, 11(10), 2141 - 2145.

[2] Naoum, S., (2003). An overview into the concept of partnering. *International Journal of Project Management*, 21(1), 71-76.

[3] Love, P., Skitmore, M. and Earl, G., (1998). Selecting a suitable procurement method for a building project, *Construction Management and Economics*,p.221,16(2), 221-233.

[4] Naoum, S., (2011a). *Procurement and management of construction.* MSc Lecture notes (unpublished material). School of Built Environment and Architecture. London South Bank University, pp. 1-15.

[5] Chau, K. W. (1997). The ranking of construction management journals. *Construction Management and Economics*, 15(4), 387–398.

[6] Hamza, N. and Greenwood, D., (2007). The impact of procurement methods on delivering environmentally sensitive buildings. *23rd Annual ARCOM Conference*, 3-5 September 2007, Belfast, UK, Association of Researchers in Construction Management, 723-732.

[7] Kameshwaran, S., Narahari, Y., Rosa, C., Kulkarni, D. and Tew, J., (2007). Multiattribute electronic procurement using goal programming. *European Journal of Operational Research*, 179(2), pp.518-536.

[8] Love, P., Davis, P., Edwards, D., and Baccarini, D., (2008). Uncertainty avoidance: public sector clients and procurement selection. *International Journal of Public Sector Management*, 21(7), pp. 753-776.

[9] Al – Bizri, S. and Gray, C., (2010). Management framework for technology clusters implementation. *Construction Management and Economics*, 28:7, 771-782.

[10] Eriksson, P., (2009). A Case Study of Partnering in Lean Construction. *The 5th Nordic Conference on Construction Economics and Organisation.* Reykjavík University, Iceland.

[11] Lam, P. T. I. and Wong, W. H. (2011). A comparative study of buildability perspectives between clients, consultants and contractors. *Construction innovation.* 11(3), pp. 305-320.

[12] Henjewele C., Sun, M. and Fewings, P., (2012). Analysis of Factors affecting Value for Money in UK PFI Projects, *Journal of Financial Management of Property and Construction*, 17(1), pp. 9-28.

[13] Weisheng, L.; Liu, A., and Hongdi, W., (2013). Procurement innovation for public construction projects: A study of agent-construction system and public-private partnership in China Engineering, *Construction and Architectural Management*, 20(6).

[14] Harland, C. M. (1996). Supply chain management: Relationships, chains and networks. *British Journal of Management*, 7(1).

[15] Khalfan, M, McDermott, P and Cooper, R., (2004). Integrating the supply chain within construction industry. *20thAnnual ARCOM Conference*, 1-3 September 2004, Heriot Watt University. Association of Researchers in Construction Management, p.901, 2(1), 897-904.

[16] Chan, S. W., S. T. Suman Tiwari, M.F. Ahmad, Zaman, I., and Sia, W. L. (in press). Green Procurement Practices and Barriers in Furniture Manufacturing Companies. *Advanced Science Letters.*

[17] Berry, C. and McCarthy, S.,(2011). *Guide to Sustainable Procurement in Construction*, CIRIA.

[18] Bower, D. (2003). *Management of Procurement*, 1st ed., Thomas Telford, London.

[19] Masterman, J., (2005). *Introduction to Building Procurement Systems*, 2nd edition., Spon Press, London

[20] Adetunji, I., Price, A., Fleming, A. and Kemp, P. (2003). Sustainability and the UK Construction Industry - A Review. Proceedings of the Institution of Civil Engineers, *Engineering Sustainability*.

[21] National Audit Office (NAO) (2005). *Sustainable Procurement in Central Government, National Audit Office*, London.

[22] Bullen, P., and Davis, P., (2003). Explorations in sustainable construction and procurement in Western Australia. *19th Annual ARCOM Conference*, 3-5, University of Brighton, 745-54.

[23] Ball, D and Fortune, C., (2000). Building project procurement processes and the development of environmentally friendly housing schemes. *16th Annual ARCOM Conference*, 6-8 Glasgow Caledonian University, 1(27), 1-9.

[24] Forgues, D. and Koskela, L., (2009). The influence of a collaborative procurement approach using integrated design in construction on project team performance. *International Journal of Managing Projects in Business,* 2(3), pp.370-385

[25] Jorgensen, S. and Emmitt, S. (2008). Lost in transition: the transfer of lean manufacturing to construction. Engineering, *Construction and Architectural Management*, 15(4) pp. 383-398

[26] Jorgensen, S. and Emmitt, S., (2007). Investigation the Integration of Design and Construction from a Lean Perspective. *Construction Innovation*, 9(1).
[27] Eriksson, P., Dickinson, M. and Khalfan, M., (2007). The influence of partnering and procurement on subcontractor involvement and innovation, *Facilities,* 25(5) pp. 203 – 214.
[28] Bowley, M., (1966). *The British Building Industry: Four studies in response and resistance to change*. Cambridge University Press, Cambridge.
[29] Egbu, C., (2004). Managing knowledge and intellectual capital for improved organizational innovations in the construction industry: an examination of critical success factors. *Engineering, Construction and Architectural Management,* 11(5), 301–15.
[30] Kumaraswamy, M and Dulaimi, M, (2001). Empowering innovative improvements through creative construction procurement. Engineering, *Construction and Architectural Management*, 8(6), pp. 325-334
[31] Ng, T., Rose, T., Mak, M. and Chen, S., (2002). Problematic issues associated with project partnering – the contractor perspective, *International Journal of Project Management*, 20(6), pp. 437-49.
[32] Shafik, M. and Martin, P., (2006). The impact of procurement methods on the Scottish house building industry. *22nd Annual ARCOM Conference*, 4-6 September 2006, Birmingham, UK, Association of Researchers in Construction Management, 81-90.
[33] Wong, F., Lam, P., Chan, E. and Shen, L., (2007). A study of measures to improve constructability. *International Journal of Quality and Reliability Management,* 24 (6), pp. 586-601.
[34] Egan, J., (1998). *Report on Rethinking Construction. Department of the Environment Transport and Region,* London: HMSO.
[35] Naoum, S., (1994). Critical analysis of time and cost of management and traditional contracts, *Journal of Construction Engineering and Management*, ASCE, 120(4), pp. 687-705.
[36] Cha, H. and O'Connor, J. (2005). Optimizing Implementation of Value Management Processes for Capital Projects. *Journal of Construction and Engineering Management,* 131(2), 239–251.
[37] Doloi, H., (2013). Empirical Analysis of Traditional Contracting and Relationship Agreements for Procuring Partners in Construction Projects. *Journal of Management in Engineering*, 29(3), 224–235.
[38] Hardin, B. (2009) *BIM and Construction Management*, Indianapolis: Wiley Publishing, IN
[39] Dainty, A., Millett, S. and Briscoe, G, (2001). New perspectives on construction supply chain integration, *Supply Chain Management: An International Journal*, 6 (4), pp. 163 – 173.
[40] Du, J., Liu, R., and Issa, R., (2014). BIM Cloud Score: Benchmarking BIM Performance. *Journal of Construction and Engineering Management,* 140(11), 04014054.
[41] Khosrowashahi, F. and Arayici, S., (2014). Roadmap for implementing of BIM in the UK construction Industry.*Engineering Construction and Architectural Management,* 19 (12) pp. 610-635.
[42] Cheng, E., and Li, H., (2001). Development of a conceptual model of construction partnering. *Engineering, Construction and Architectural Management*, 8(4), 292–303.
[43] Presutti Jr, (2003). Supply management and E-procurement: creating value added in the supply chain. *Industrial Marketing Management*, 32, 219-226.
[44] Naoum, S., (2011b). *People and Organisational Management in Construction.* Thomas Telford Publishers, London, ISBN 978-0-7277-4151-6
[45] Ruparathna, R. and Hewage, K., (2013). Review of Contemporary Construction Procurement Practices. *Journal of Management and Engineering*.

IOP Conf. Series: Materials Science and Engineering **342** (2018) 012101 doi:10.1088/1757-899X/342/1/012101

Evaluation of the wind pumped hydropower storage integrated flood mitigation system

Aishah Safi and Firdaus Basrawi

Energy Sustainability Focus Group (ESFG), Faculty of Mechanical Engineering, Universiti Malaysia Pahang, 26600 Pekan, Pahang, Malaysia

E-mail: aishahsafi203@gmail.com

Abstract. As Wind Pumped Hydropower Storage (WPHS) need high cost to construct, it is important to study their impacts on economic and environmental aspects. Thus, this research aims to evaluate their economic and environmental performances. First, Hybrid Optimization Model for Electric Renewable (HOMER) was used to simulate power generation system with and without the flood reservoir. Next, the total amount of emitted air pollutant was used to evaluate the environmental impacts. It was found the wind-diesel with reservoir storage system (A-III) will have much lower NPC than other systems that do not include reservoir for flood mitigation when the cost of flood losses are included in the total Net Present Cost (NPC). The NPC for system A-III was RM 1.52 million and for diesel standalone system (A-I) is RM 10.8 million when the cost of flood losses are included in the total NPC. Between both energy systems, the amount of pollutants emitted by the A-III system was only 408 kg-CO_2/year which is much less than the A-I system which is 99, 754 kg of carbon dioxide per year. To conclude, the WPHS integrated with flood mitigation system seems promising in the aspects of economic and environment.

1. Introduction

Pumped hydropower storage (PHS) is a powerful form of energy storage on the electric grid today. At the moment, it is the most affordable form of bulk energy storage available. Integrating wind energy with pumped hydro energy storage (WPHS) can overcome the intermittency of wind by consuming energy during low-demand periods and supplying energy during high demand. PHS is a proven, low-risk technology with a great track record in their efficiency, has longer facility lifetimes, adjustable and can quickly respond to the fluctuations of electricity [1]. Although WPHS is clean, renewable and can store excess energy during non-peak times but the problem with this storage facility is that they can be expensive to construct [2]. Therefore, other aspects should be included as the added value of WPHS, to make it worth to construct this storage facility. The reservoirs of WPHS can store huge amounts of water. If the reservoir is designed in the similar mechanism of flood reservoir it may capable to mitigate flood during extreme hydrological events. When the effect of a flood can be avoided or at least reduced, there will be a lot of advantages can be gain, especially in the term of economic aspects. Due to the feasibility of WPHS reservoir in flood mitigation and the high cost of those projects, a study should be conducted to evaluate the feasibility of the proposed WPHS integrated with flood mitigation system. Through the evaluation, the economic and environmental aspects of the proposed power system can also be assessed.

IOP Conf. Series: Materials Science and Engineering **342** (2018) 012101 doi:10.1088/1757-899X/342/1/012101

2. Methodology

2.1. System description and specification

In this research, the residential at Kuala Pahang was selected as a targeted study area. It was estimated that this sub division of Pekan district has the total population about 7,936 peoples with 6 villages and consists of 1,133 households. Each village was estimated to have 180 households. Approximately the total electricity usage is 25,290 kWh per month. The daily load profile in Kuala Pahang is shown in Figure 1 and it shows that highest demand of electricity occurs at night.

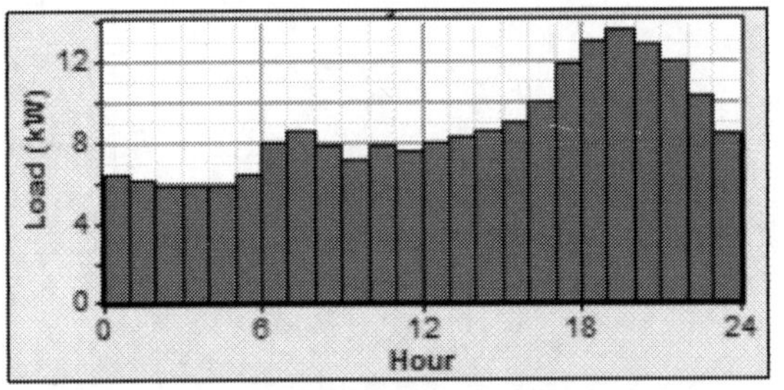

Figure 1. Daily load profile in Kuala Pahang.

In this study, there was three power generation system studied. First is the diesel standalone system as A-I in Figure 2, the second system is wind-diesel without reservoir storage as A-II in Figure 3, the third system is the wind-diesel with pumped hydropower storage as A-III in Figure 4. The proposed PHS system not only will exploit the excess energy from intermittent renewable energy but also can be effectively used for flood mitigation by holding the excess runoff in the reservoirs during annual flood seasons. The storage space of the lower reservoir is kept low for a huge discharge of excess runoff during extreme hydrological events or monsoon seasons.

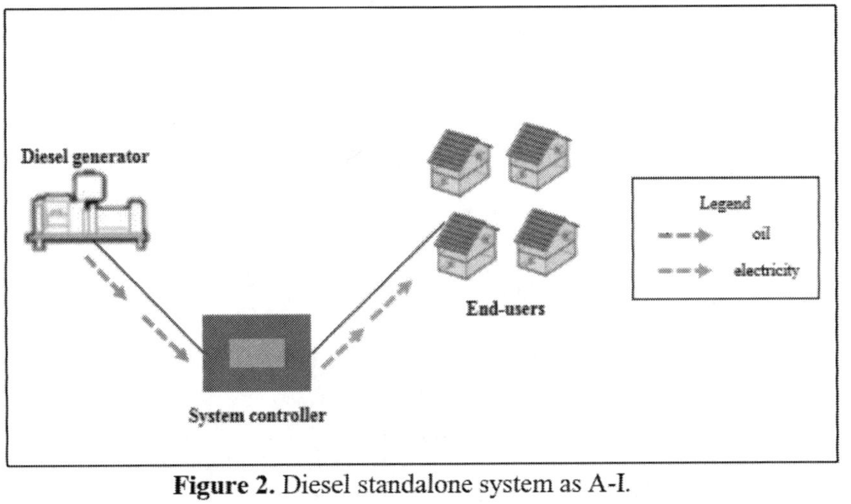

Figure 2. Diesel standalone system as A-I.

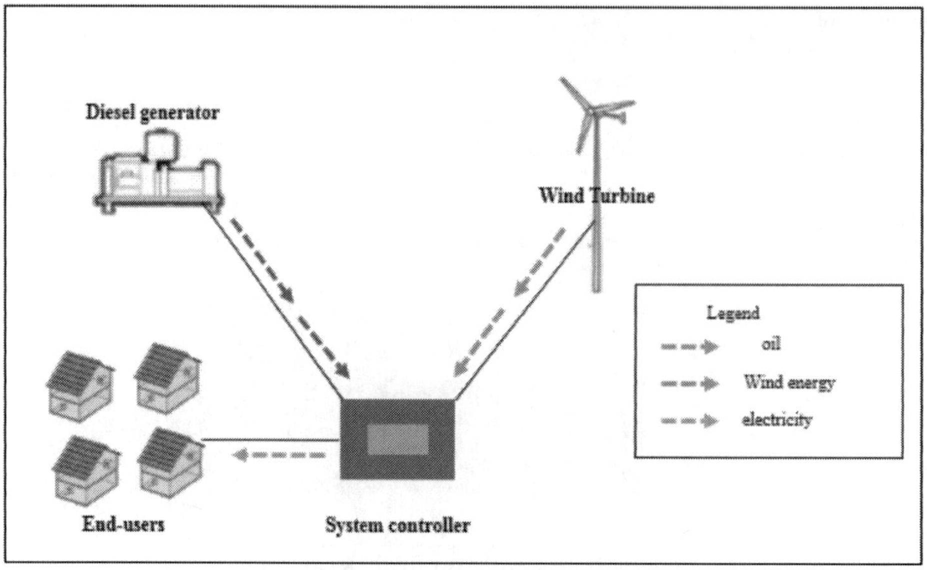

Figure 3. Wind-diesel without reservoir storage as A-II.

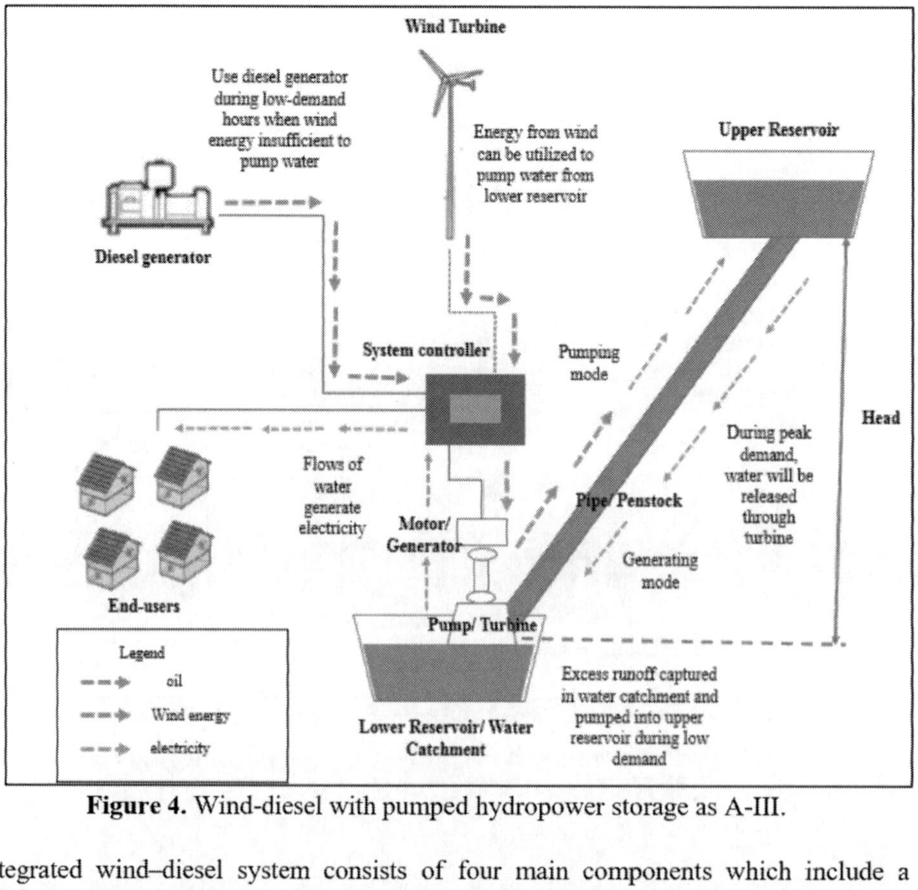

Figure 4. Wind-diesel with pumped hydropower storage as A-III.

The integrated wind–diesel system consists of four main components which include a diesel generator, wind turbines, batteries as water reservoir and converters units. The technical specification of the energy system for the generator, wind turbine, inverter and battery bank, as simulated in Hybrid Optimization Model for Electric Renewable (HOMER) was described in Table 1.

Table 1. Data of selected system component.

Description	Data
Reservoir	
Nominal voltage	240 V
Nominal capacity	2,161,254 Ah
Lifetime throughput	1,471,210,496 kWh
Wind turbine	
Model	Fuhrländer 30
Cut in wind speed	2.5 m/s
Rated wind speed	8 m/s
Cut out wind speed	25 m/s
Lifetime	20 years
Hub height	30 m
Inverter	
Model	Solectria PVI-80kW-480V
Size (step size 50 kW)	0 - 500 kW
Lifetime	15 years
Generator	
Model	Cummins
Size (step size 50 kW)	0 – 500 kW
Rated power	80 kW
Minimum running hours	15,000hr
Minimum load ratio	25%

2.2. Modelling pumped storage hydropower in Hybrid Optimization Model for Electric Renewable (HOMER).

Canales and Beluco had discussed the steps to model pumped hydropower storage (PHS) with a few modification in HOMER [3]. In HOMER, a battery was created as equivalent to electrical storage mechanism of PHS with particular capacity and roundup efficiency. The connection between the battery created and PHS was explained further in next sub-section.

2.2.1. Battery representing the reservoir. The battery is capable of storing excess energy and supply when there is a demand which has the similar capability of PHS reservoirs. The properties of battery modelled in HOMER was assumed to remain the same during its lifespan. Based on explanation by Canales and Beluco [3], the total stored energy in the volume of reservoir can be described as in equation (1).

$$E_S = \frac{9.81\ x\ \eta_{hyd}\ x\ H\ x\ Vol}{3600} \quad (1)$$

Where, E_S is the total stored energy [kWh], H is the gross head [m], η_{hyd} is the efficiency in turbine mode [%] and Vol is the volume of reservoir [m^3]. The discharge current of the battery with a fixed voltage and capacity is considered independent, and the stored energy can be expressed as in equation (2).

$$E_S = \frac{V\,x\,C_B}{1000} \tag{2}$$

Where, E_S is the stored energy [kWh], V is the voltage [V] and C_B is the capacity [A.h]. The power produced by the battery is corresponding to the value of current when the voltage is assumed to be constant and can be calculated as in equation (3).

$$P_{bat} = \frac{V\,x\,I}{100} \tag{3}$$

Where, P_{bat} is the power [kW], V is the voltage [V] and I is the current [A]. According to Canales and Beluco, there are three steps to model PHS in HOMER. The first step is to set a reference voltage of the equivalent of the battery and estimate the capacity of the equivalent battery which proportional to the reservoir volume. The battery also must be set in DC bu. This important to make sure that inputs and outputs of the battery will represent the flow rate of water being stored or leaving the PHS reservoir. The second step is to create an equivalent battery with a selected reference voltage and capacity decided earlier. The round trip efficiency should be set at 100% and 0 % for the minimum state of charge. The third step is to include the converter component, which will represent the various options for hydropower plant capacity. This component can be used to control the conversion efficiency in either pump or turbine modes.

2.3. *Environmental analysis*

To estimate the amount of emission produced, the emission factors [kg/kWh] used for estimating emission in power plants is identified first as in Table 2 [4]. Then, the emission factor of air pollutants was multiplied by the total amount of electricity generated as in order to get the amount of emission release to the environment.

Table 2. Emission factors [kg/kWh] used for estimating amount of emission produced.

Fuel type	CO_2	NO_X	SO_X	CO
Generator	0.85	0.0025	0.0164	0.0002

3. Result and discussion

Figure 5 shows the configuration of proposed system as designed in HOMER. The components of the system consist of a wind turbine, batteries or reservoir, a power converter and generator. The simulated peak demand of primary load was 16 MW and the total energy consumption was 206 kWh/day.

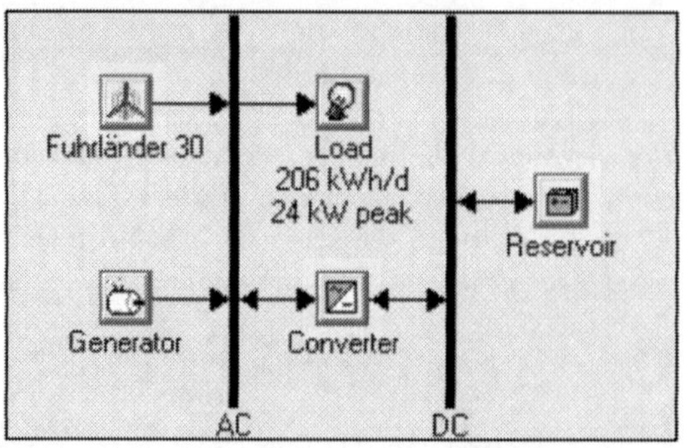

Figure 5. Schematic diagram of the wind-diesel pumped hydropower with generator system.

All possible system configurations are listed in ascending order in the HOMER output. Figure 6 shows the simulation results of the annual production of electrical energy by each system. The production value is the annual energy output of each electrical energy-producing component in the system added to the total electrical production. As in Figure 6, the combination of generator/wind turbine systems (A-II) has the highest production of electricity which was 88,165 kWh/year. Out of this, 77,032 kWh/year comes from the generator component and 11,133 kWh/year produced by the wind turbine (WT). The second highest electricity production is from the standalone generator system (A-I) which was 81,453 kWh/year. Meanwhile, the combination of generator/wind turbines/reservoir system (A-III) produces about 78,233 kWh/year, 77,933 kWh/year comes from the WT and 300 kWh/year produced by generator component. The analysis showed that equipping the energy system with reservoir improves the penetration of renewable sources. Without reservoir storage for the generator/wind turbine system, the penetration of wind was only 12.6 %. But, with the addition of reservoir in the system, the penetration of wind energy was almost 100 %.

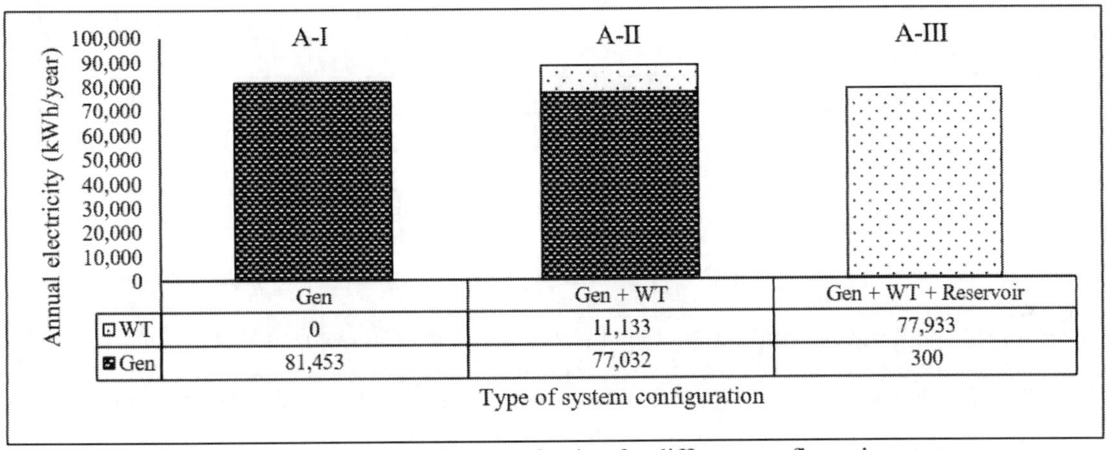

Figure 6. Comparison electricity production for different configuration system.

As expected, there were significant amounts of reduction in emission where the generator system is mixed with renewable energy sources as compared to diesel standalone system. Basically, when less energy produced from combustion of diesel, the less emission of the pollutant is. The reduction of emission for generator/wind turbines/reservoir system (A-III) was reduced by 99.6 % compared to the diesel standalone system (A-I). The emissions from A-III system was low compared to the other systems due to more participant of wind energy in their electricity production as shown in Figure 7. When the proposed PHS is integrated with renewable resources such as wind energy for energy input, tonnes of harmful emission such as CO_2, SO_2, NOx, CO, and PM10 can be avoided from releasing to the atmosphere [5]. If the PHS is awarded Certified Emission Reductions by the United Nation Executive Board of Clean Development Mechanism, tonnes of avoided carbon dioxide can be translated into carbon credits and can be further translated into monetary value. One carbon credit is equivalent to one metric tonne of carbon dioxide and the Malaysian Energy Centre assumes the price range to be US$ 3-10 per tonne of CO_2 [6].

IOP Conf. Series: Materials Science and Engineering **342** (2018) 012101 doi:10.1088/1757-899X/342/1/012101

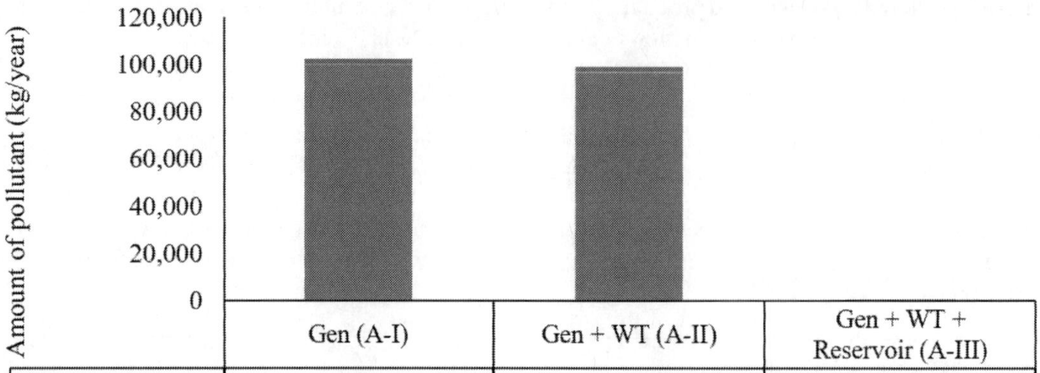

	Gen (A-I)	Gen + WT (A-II)	Gen + WT + Reservoir (A-III)
Nitrogen oxides	2,197	2,125	8.99
Sulfur dioxide	200	194	0.82
Carbon monoxide	246	238	1.01
Carbon dioxide	99,754	96,501	408

Type of configuration system

Figure 7. Amount of pollutants produced by different power system (kg/year).

Figure 8 shows the data on the amount of pollutants produced and Levelized Cost of Energy (LCOE) in a graph. The type of pollutants emitted as listed in Figure 7 which included CO_2, CO, SO_X, and NO_X. The emission of the pollutants into the environment not only causing harmful effects on human health but also the one of the main contributor to the global warming phenomenon [7]. Meanwhile, the LCOE is interpreted by HOMER as the mean cost per kWh of electrical energy unit produced by the system. The LCOE can be a useful tool for comparison of various energy systems. If a system has a reasonably low LCOE, it implies that the electricity is being produced at a low cost. Thus, for the investors, they will have a higher chance of returns with low cost of electricity production [8]. As can be seen in Figure 8, the system such as the generator standalone system (A-I) has the lowest LCOE among all systems. Meanwhile, generator/wind turbine systems (A-II) and generator/wind turbines/reservoir system (A-III) had higher LCOE than average rate of the electricity tariff in Malaysia which is RM 0.38/kWh [9]. But, in term of environmental aspects, generator/wind turbines/reservoir system (A-III) had the least emission of pollutants to the environment. However, A-III system which included reservoir as their storage device in their system had significantly increased the LCOE and the Net Present Cost (NPC) as compared to other systems. The inclusion of reservoir in the proposed system is important for flood mitigation during annual monsoon seasons and also can act as an energy storage device for intermittent renewable energy sources.

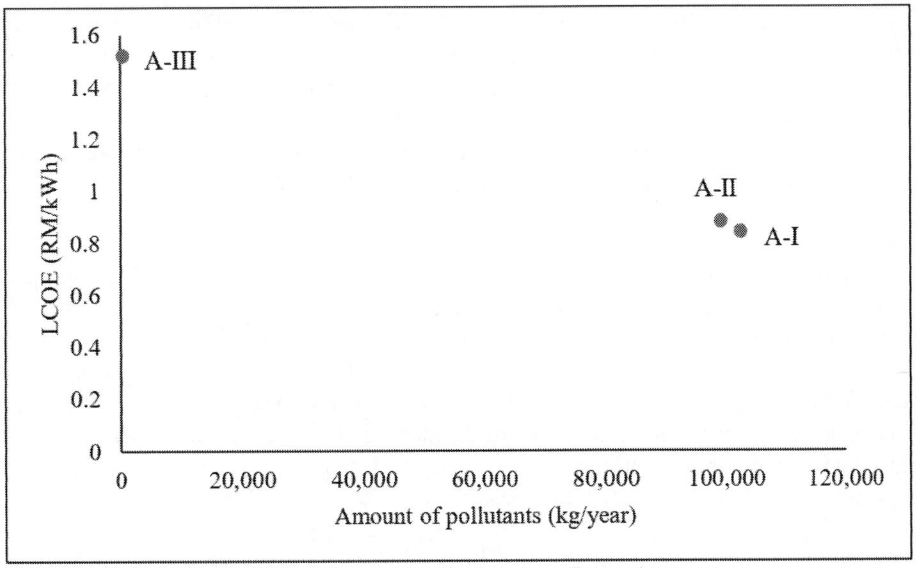

Figure 8. Summary of difference configuration system.

Figure 9 shows the comparison when the cost for flood losses was added into the total NPC of configuration systems without reservoir for flood mitigation. The Figure 9 (a) shows the total NPC without flood losses included in each system and meanwhile Figure 9 (b) shows the total NPC with flood losses included in the overall cost. Only system A-III had included the reservoir in their configuration which may help prevent floods, so the total NPC does not change. The total loss due to floods recorded in 2014 in the Pekan district alone is almost RM 10 million and this cost has been included to the total NPC of the power system that does not have a reservoir in their system.

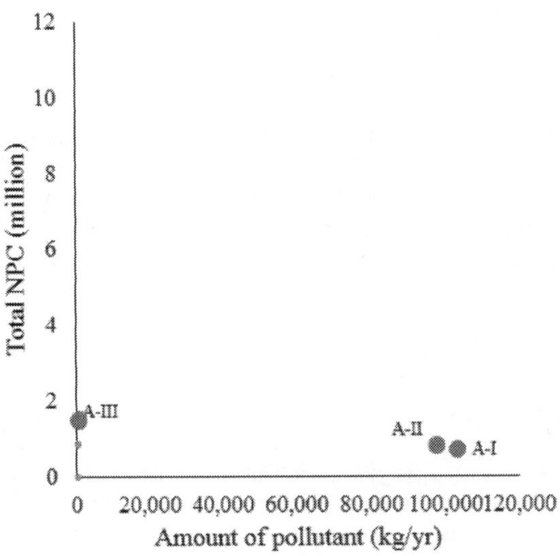

Figure 9 (a)

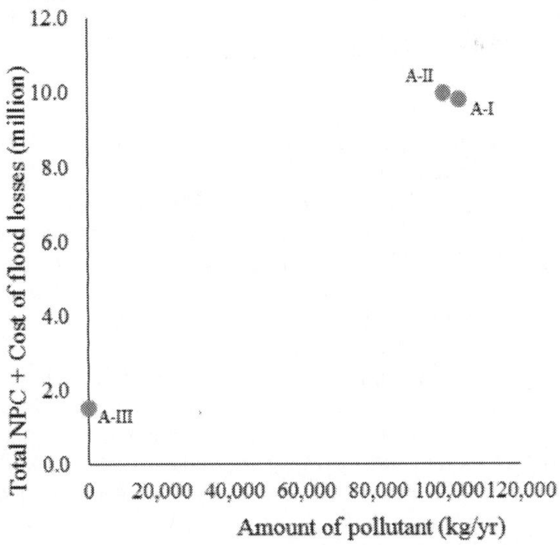

Figure 9 (b)

A total of RM610 million in losses was recorded in the year 2014 flood event that hit Pahang which leads to the destruction of property, agricultural produce, and infrastructure. It is the worst flood disaster after 1971, RM348.5 million losses had been recorded from infrastructure damage involving roads, bridges and public utilities in the state. For the agriculture sector, Pahang had lost over RM65.1 million and 219 houses in seven districts including Jerantut, Kuantan, Lipis, Pekan, and Temerloh were destroyed in the disaster [10]. The cost of flood damage is high as compared to the costs required to construct projects such as A-III system that include reservoirs to overcome floods. The NPC for system A-III is RM 1.52 million which much lower than another system when the cost of flood losses are included in the total NPC. It is better to allocate a small budget to implement projects which can mitigate floods rather than to bear high costs due to flood losses as in the year 2014.

4. Conclusion

The analysis has shown that when the cost of flood losses are included in the total Net Present Cost (NPC), the wind-diesel with reservoir storage system (A-III) will have much lower NPC than other systems that do not include the reservoir for flood mitigation. The addition of reservoir as a storage device in the power generation system has significantly raised the value of NPC and also the levelized COE of the system. But, the inclusion of the reservoir in the proposed system is important for flood mitigation during annual monsoon seasons and at the same time, it can act as energy storage device for intermittent renewable energy sources. Thus, the pumped hydropower storage integrated flood mitigation system offers benefits in both economic and environmental aspects.

Acknowledgement

The authors would like to acknowledge Faculty of Mechanical Engineering (Universiti Malaysia Pahang), Energy Sustainability Focus Group and RDU160310 research grant which contributed to the development of the research presented in this paper.

References

[1] A. Coburn, E. Walsh, P. J. Solan, and K. P. McDonnell, "Combining wind and pumped hydro energy storage for renewable energy generation in Ireland," *Journal of Wind Energy,* vol. 2014, p. 6, 2014.

[2] H. Chen, T. N. Cong, W. Yang, C. Tan, Y. Li, and Y. Ding, "Progress in electrical energy storage system: A critical review," *Progress in Natural Science,* vol. 19, pp. 291-312, 3/10/ 2009.

[3] F. A. Canales and A. Beluco, "Modeling pumped hydro storage with the micropower optimization model (HOMER)," *Journal of Renewable and Sustainable Energy,* vol. 6, p. 043131, 2014.

[4] M. Shekarchian, M. Moghavvemi, T. M. I. Mahlia, and A. Mazandarani, "A review on the pattern of electricity generation and emission in Malaysia from 1976 to 2008," *Renewable and sustainable energy reviews,* vol. 15, pp. 2629-2642, 8// 2011.

[5] N. Strachan and A. Farrell, "Emissions from distributed vs. centralized generation: The importance of system performance," *Energy Policy,* vol. 34, pp. 2677-2689, 11// 2006.

[6] X. Lim, W. H. Lam, and A. H. Shamsuddin, "Carbon credit of renewable energy projects in Malaysia " presented at the 4th International Conference on Energy and Environment 2013 (ICEE 2013), 2013.

[7] M. M. Mokhtar, M. H. Hassim, and R. M. Taib, "Health risk assessment of emissions from a coal-fired power plant using AERMOD modelling," *Process Safety and Environmental Protection,* vol. 92, pp. 476-485, 2014/09/01/ 2014.

[8] R. Millsapp. (2017, 11/07/2017). *Levelized cost of energy*. Available: http://www.renewable-energy-advisors.com/learn-more-2/levelized-cost-of-electricity/

[9] Y. T. Ging, "Energy ministry: No change to 2016 power tariffs, but peninsular users to get smaller rebate," in *The Malay Mail Online*, ed. Petaling Jaya: The Malay Mail, 2015.

[10] BERNAMA. (2015, 27/08/2017). *Banjir: Pahang alami kerugian lebih RM610 juta*. Available: http://www.astroawani.com/berita-banjir/banjir-pahang-alami-kerugian-lebih-rm610-juta-adnan-yaakob-53046

IOP Conf. Series: Materials Science and Engineering **342** (2018) 012102 doi:10.1088/1757-899X/342/1/012102

Experimental investigation on flexure and impact properties of injection molded polypropylene-nylon 6-glass fiber polymer composites

D M Nuruzzaman[1], N M Kusaseh[2], M A Chowdhury[3], N A N A Rahman[2], A N Oumer[4], N Fatchurrohman[2], A K M A Iqbal[2] and N M Ismail[2]

[1,2] Faculty of Manufacturing Engineering, University Malaysia Pahang, 26600 Pekan, Pahang Darul Makmur, Malaysia.
[3] Department of Mechanical Engineering, Dhaka University of Engineering and Technology, Gazipur, Gazipur 1700, Bangladesh.
[4] Faculty of Mechanical Engineering, University Malaysia Pahang, 26600 Pekan, Pahang Darul Makmur, Malaysia.

E-mail: dewan052005@yahoo.com

Abstract. In this research study, glass fiber (GF) reinforced polypropylene (PP)-nylon 6 (PA6) polymer blend composites were prepared using injection molding process. Specimens of four different compositions such as 80%PP+20%PA6, 80%PP+18%PA6+2%GF, 80%PP+16%PA6+4%GF and 80%PP+14%PA6+6%GF were prepared. In the injection molding process, suitable process parameters were selected depending on the type of composite specimen in producing defects free dog bone shaped specimens. Flexure and impact tests were carried out according to ASTM standard. The important flexure properties such as flexural modulus, flexural yield strength, flexural strength and flexural strain were investigated. The obtained results revealed that flexural modulus of 80%PP+20%PA6 polymer blend is the lowest and the polymer blend composite shows steadily improved modulus as the glass fiber content is increased. Results also showed that flexural strength of pure polymer blend is the lowest but it improves gradually when the glass fiber content is increased. Impact test results revealed that impact strength of 80%PP+20%PA6 polymer blend is the highest whereas all the composites show reduced impact strength or toughness. It is noticed that 80%PP+14%PA6+6%GF composite exhibits the lowest impact strength.

1. Introduction

The applications of polymer blend composites for different engineering components are gaining popularity according to industrial requirements. These polymer blend composites are suitable replacement for conventional materials due to high strength and stiffness. In order to enhance the mechanical properties, different thermoplastic polymer blends can be reinforced by different types fibrous materials. In processing of fiber reinforced polymer composites, different manufacturing techniques are suitable depending on the type of polymer composite [1]. In the past decade, research works were carried out on different types of fiber reinforced polymer composites to investigate different mechanical and tribological properties [2-8]. From these investigations, the obtained results revealed that properties of the composites were influenced by a number of operating parameters, types of matrix and reinforced materials.

Flexural properties of injection molded glass fiber reinforced nylon 6-6 composites were investigated under different operating conditions [9]. Results showed that flexural properties were remarkably improved due to the increase in fiber loading whereas the properties were not much improved due to the increase in fiber length. Polypropylene and epoxy matrix composites reinforced by the E-glass fiber were fabricated and the impact properties of these composites were analyzed [10]. It was found that type of resin is an influential parameter affecting the impact performance of the composites. Fiber reinforced polymer composites were fabricated and the mechanical properties of the composites were investigated under different strain rate conditions [11]. The experimental data revealed that flexural strength of the composites was remarkably influenced by the application of different strain rates. Moreover, it was noticed that impact strength of the composite was varied depending on the type of fiber reinforcement. Using a 3-point bending test, flexural strengths of hybrid polymer composites were investigated for different thicknesses of composites [12]. It was found that the strength of the composite was markedly influenced by the thickness of composite. Research works were carried out to investigate the mechanical properties of PA66-PP blend composites reinforced by glass fiber [13]. Results showed that flexure and impact properties of PA66-PP blend composites were much different from that of pure PA66-PP blend.

Research works were carried out to investigate the tensile properties of glass fiber reinforced nylon [14]. The experimental data showed that tensile modulus, yield strength, tensile strength and tensile strain were remarkably influenced by the glass fiber content. The impact strength of the nylon composite was also examined in this study. It was apparent that the percentage of glass fiber content has notable effect on the impact strength of the composite. The injection parameters such as pressure distribution, temperature at flow front, fill time etc. were theoretically analyzed for nylon polymer [15]. Carbon and glass fiber reinforced polyester resin composites were fabricated and mechanical properties were analysed [16]. Based on the experimental data, it was concluded that impact and flexure strengths of these composites were remarkably influenced by the percentage reinforcement and the type of composite.

In the present research study, four different types glass fiber (GF) reinforced polypropylene (PP)-nylon 6 (PA6) polymer blend composites were successfully prepared using injection molding process. Flexure and impact properties of these composites were investigated and the effects of glass fiber content on these properties were analysed.

2. Experimental

Using an injection molding machine (NISSEI PNX 60), dog bone shaped specimens of four different compositions such as 80%PP+20%PA6, 80%PP+18%PA6+2%GF, 80%PP+16%PA6+4%GF and 80%PP+14%PA6+6%GF were prepared. The specimens were prepared following ASTM D638 standard. During the injection molding process, suitable process parameters such as melting temperature, injection pressure etc. were selected depending on the type of composite specimen in producing defect-free dog bone shaped specimens. Three point flexural tests were carried out to investigate the mechanical properties of the composite specimens. The flexure tests were carried out using universal test machine Instron (model:3369, maximum load capacity: 50KN) under a crosshead speed of 4 mm/min. All the tests were carried out following ASTM D790 standard. The test results such as flexural yield strength, flexural strength, flexural modulus and flexural strain were generated using Bluehill 2 software. In order to determine the impact strength for different types of composite specimens, impact tests were carried out using an impact pendulum Instron CEAST 9050. Izod impact tests (unnotched) were carried out according to ASTM D4812 standard for different types of composite specimens. During each impact test, a torque of 5 N.m was applied for holding the specimen with the impact fixture. For all the composite specimens, the hammer energy of 11J was applied in order to quantify the impact strength of the composite.

3. Results and Discussion

The flexural test results for different compositions of glass fiber reinforced PP-PA6 polymer blend composites are shown in Figure 1. These results were generated by the test machine using Bluehill 2 software. The tests were carried out under a crosshead speed of 4 mm/min for all the composite specimens. From the flexural stress-flexural strain diagrams as shown below for different types of composites, the obtained important parameters such as flexural modulus, flexural yield strength, flexural strength and flexural strain are discussed in the following sections.

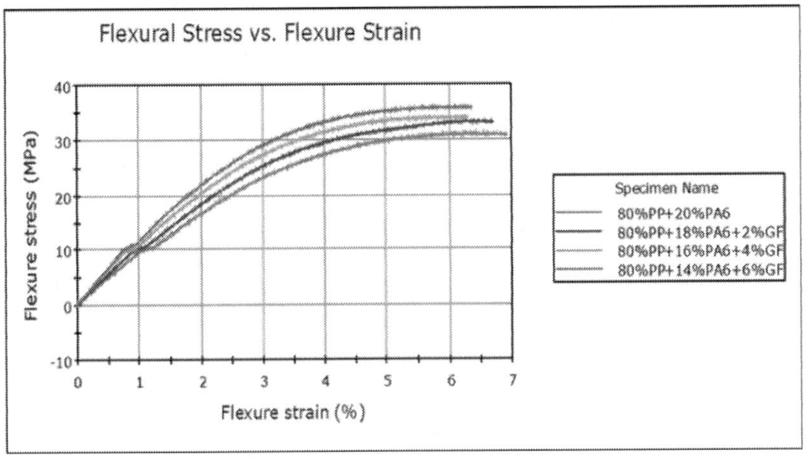

Figure 1. Flexural stress - flexural strain diagrams of glass fiber reinforced PP-PA6 polymer blend composites.

Figure 2 shows the flexural modulus of glass fiber reinforced PP-PA6 polymer blend composites of different compositions. Flexural modulus or bending modulus is the measure of stiffness of the composite to bending deformation. From the figure below, it can be seen that 80%PP+20%PA6 polymer blend exhibits a low flexural modulus of 0.93 GPa. The obtained results show that due to the addition of glass fiber, 80%PP+18%PA6+2%GF composite shows slightly improved modulus of 1.01 GPa which is 8.60% higher in comparison with the modulus of pure polymer blend. The 80%PP+16%PA6+4%GF composite reveals further improved modulus of 1.10 GPa which is 18.28% higher than that of neat polymer blend. Finally, the composite 80%PP+14%PA6+6%GF shows a further increased flexure modulus of 1.24 GPa which is 33.33% higher in comparison with neat PP-PA6. From these results, it is apparent that due to the increased glass fiber content, the stiffness of the composite increases gradually under flexure loading, which means that more resistance to bending deformation as the content of glass fiber increases.

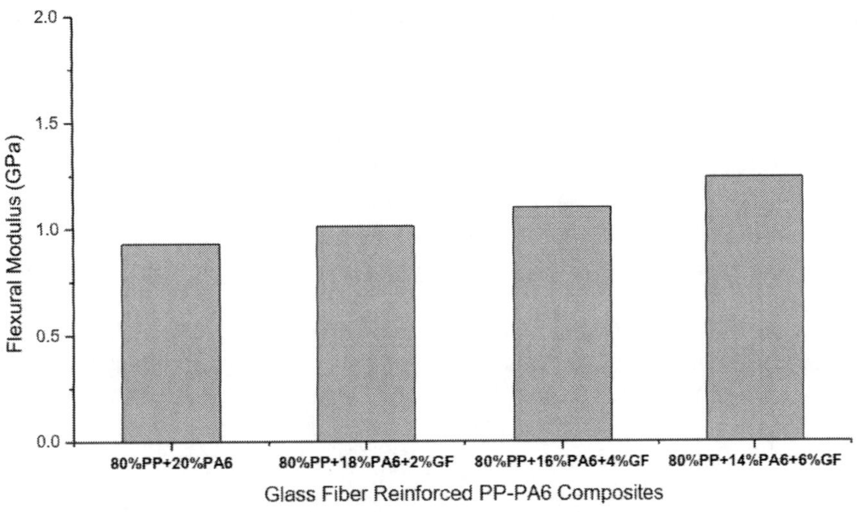

Figure 2. The flexural modulus of glass fiber reinforced polypropylene-nylon 6 composites.

Figure 3 shows the flexural yield strength (offset 0.2%) for different compositions of glass fiber reinforced polypropylene-nylon 6 composites. From the obtained results it was found that beyond the elastic limit, permanent deformation of 80%PP+20%PA6 neat polymer blend took place at low stress level of 30.83 MPa. The obtained data also revealed that plastic deformation of 80%PP+18%PA6+2%GF composite started at 33.20 MPa which is about 7.69% higher than the yield strength of PP-PA6 polymer blend. For 80%PP+16%PA6+4%GF composite, yielding started at slightly higher stress level of 33.92 MPa which is about 10% higher than that of neat polymer blend. Finally, for the 80%PP+14%PA6+6%GF composite, the result showed an improved value of yield strength 35.46 MPa which is almost 15% higher than the yield strength of pure polymer blend. From these results, it is very clear that due to the increased glass fiber content, plastic or nonlinear deformation of the composite started at little higher stress level.

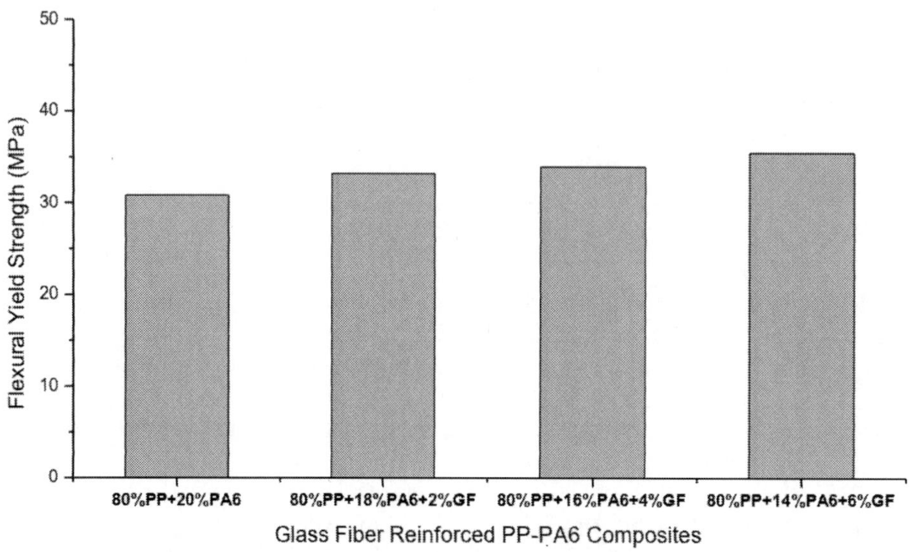

Figure 3. The flexural yield strength (MPa) of glass fiber reinforced polypropylene-nylon 6 composites.

Figure 4 shows the three point bending strength or flexural strength of PP-PA6-GF composites. This strength is the highest stress experienced by the composite specimen at extreme fibers on the tension side of the composite. The obtained result shows that maximum flexural strength of 80%PP+20%PA6 polymer blend is 31.40 MPa. The 80%PP+18%PA6+2%GF composite shows the strength of 33.57 MPa which is 6.9% higher than the pure polymer blend. Flexural strength of 80%PP+16%PA6+4%GF composite is 34.33 MPa which is 9.3% higher as compared with that of pure polymer blend. Lastly, 80%PP+14%PA6+6%GF composite shows a further improved flexural strength of 36.13 MPa which is about 15% higher as compared with that of neat polymer blend. From these experimental results, it is evident that there is a steady improvement in the flexure strength due to the increase in glass fiber content.

Figure 5 exhibits the flexural strain of glass fiber reinforced polypropylene-nylon 6 blend composites. In the experiments, this was the unit deformation (%) at the furthest layer of the composite specimen under a crosshead speed of 4 mm/min. The obtained results reveal that PP-PA6 pure polymer blend undergoes a plastic deformation after reaching the yield point and absorbs much energy. In this case, the flexural strain at break is 6.92%. The obtained results also reveal that for the glass fiber content of 2%, 4% and 6%, the composite experiences the flexural strain 6.70%, 6.29% and 6.34% respectively. These results generally indicate that flexural deformations are nearly equal for different types of composites and not much influenced by the content of glass fiber up to 6%.

IOP Conf. Series: Materials Science and Engineering **342** (2018) 012102 doi:10.1088/1757-899X/342/1/012102

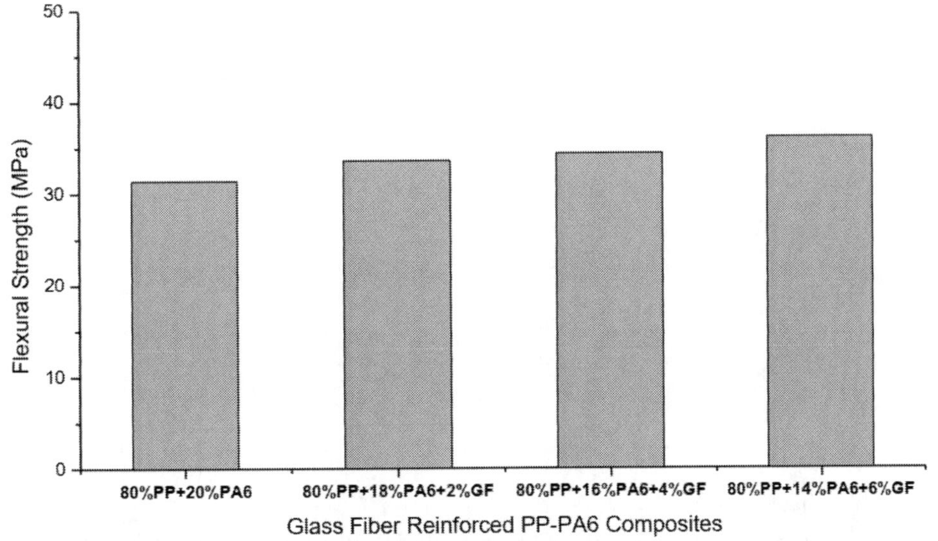

Figure 4. The flexural strength of glass fiber reinforced polypropylene-nylon 6 composites.

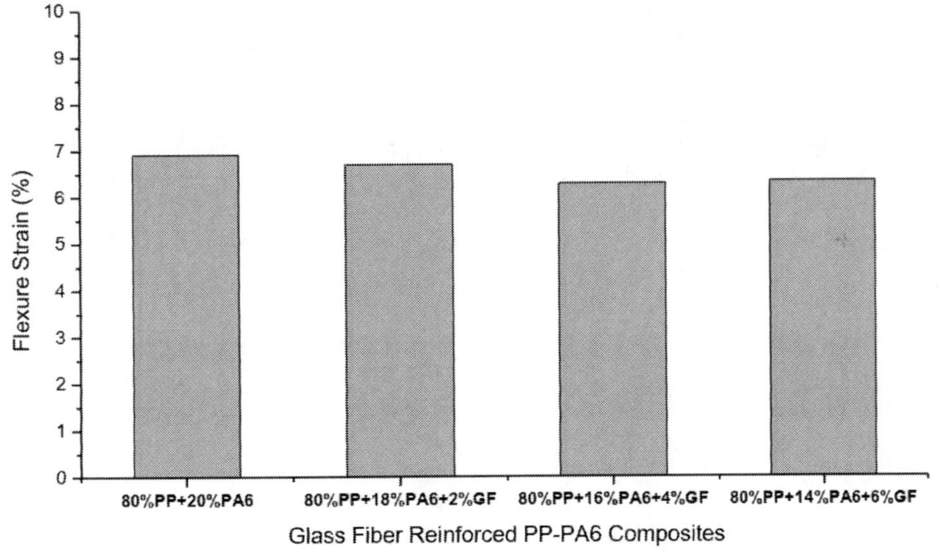

Figure 5. The flexural strain (%) of glass fiber reinforced polypropylene-nylon 6 composites.

Figure 6 shows the impact strength of glass fiber reinforced PP-PA6 blend composites. In the experiments, impact tests were carried out using an applied torque 5 N.m to the specimens. Moreover, the hammer energy of 11 J was applied in order to break the composite specimens. From the obtained results, neat polymer blend 80%PP+20%PA6 shows the impact strength of 25.10 kJ/m^2. The 80%PP+18%PA6+2%GF composite shows a reduced impact strength of 22.80 kJ/m^2 than that of pure polymer blend. For the 80%PP+16%PA6+4%GF composite, the impact strength is again reduced to 22.33 kJ/m^2 when the glass fiber content is increased to 4%. Finally, it can be noticed that the impact strength of the composite is further decreased to 19.81 kJ/m^2 when the fiber content is 6%. From these results, it is obvious that the impact toughness of the composite gradually decreases due to the increase in the glass fiber content.

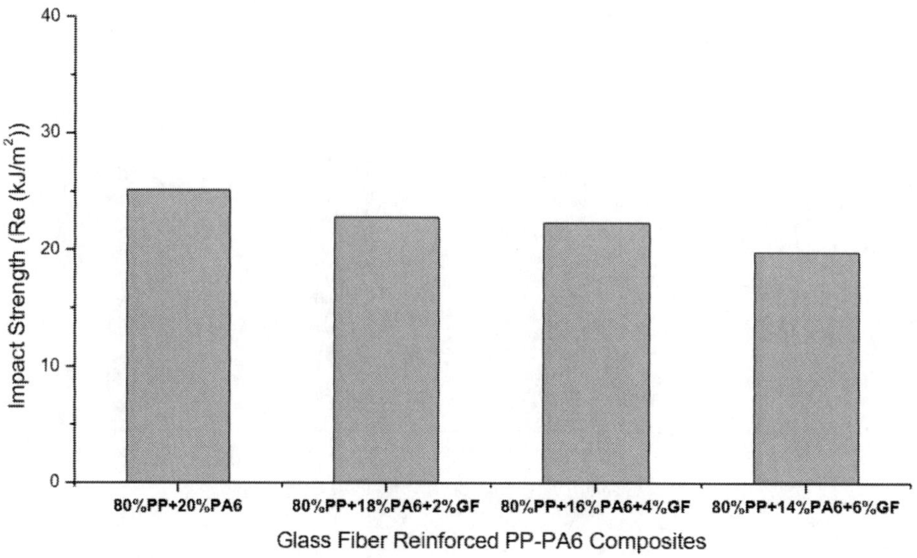

Figure 6. The impact strength, Re (kJ/m^2) of glass fiber reinforced polypropylene-nylon 6 composites (Torque: 5 N.m).

4. Conclusions

In this research, glass fiber reinforced polypropylene-nylon 6 polymer blend composites of different compositions were successfully prepared using injection molding process. Flexural and impact properties of these composites were investigated. The following conclusions are drawn:

1. The flexural modulus of 80%PP+20%PA6 polymer blend is the lowest and the polymer blend composite shows steadily improved modulus as the glass fiber content increases up to 6%.
2. Flexural yield strength of 80%PP+20%PA6 polymer blend is the lowest and it increases steadily for the blend composite as the glass fiber content is increased. The flexural strength of pure polymer blend is the lowest but it improves gradually when the glass fiber content is increased up to 6%.
3. The impact strength of 80%PP+20%PA6 polymer blend is the highest whereas all the composites show a little reduced impact strength. It is noticed that 80%PP+14%PA6+6%GF composite exhibits the lowest impact strength or toughness.

Acknowledgements

The authors would like to thank postgraduate research grant scheme PGRS160374, University Malaysia Pahang (UMP) for the financial support. All the technical staffs of faculty of manufacturing engineering are gratefully acknowledged for their assistance.

References

[1] Divya H V, Naik L L and Yogesha B 2016 Processing techniques of polymer matrix composites – A review *Int. J. Eng. Res. General Sci.* **4** 357-62
[2] Zhou S, Zhang Q, Wu C and Huang J 2013 Effect of carbon fiber reinforcement on the mechanical and tribological properties of polyamide6/polyphenylene sulfide composites *Mater. Des.* **44** 493–9
[3] Nuruzzaman D M, Chowdhury M A and Rahaman M L 2011 Effect of duration of rubbing and normal load on friction coefficient for polymer and composite materials *Ind. Lubr. Tribol.* **63** 320-26

[4] Yuan L, Ma X, Liang G and Yan H 2007 Fibre reinforced organic rectorite/unsaturated polyester composites *Compos. Sci. Technol.* **67** 2311-22
[5] Nuruzzaman D M, Rahaman M L and Chowdhury M A 2012 Friction coefficient and wear rate of polymer and composite materials at different sliding speeds *Int. J. Surf. Sci. Eng.* **6** 231-45
[6] Unterweger C, Bruggemann O and Furst C 2014 Effects of different fibers on the properties of short-fiber-reinforced polypropylene composites *Compos. Sci. Technol.* **103** 49-55
[7] Chowdhury M A, Nuruzzaman D M, Roy B K, Samad S, Sarker R and Rezwan A H M 2013 Experimental investigation of friction coefficient and wear rate of composite materials sliding against smooth and rough mild steel counterfaces *Tribol. Ind.* **35** 286-96
[8] Braga R A and Magalhaes P A A 2015 Analysis of the mechanical and thermal properties of jute and glass fiber as reinforcement epoxy hybrid composites *Mater. Sci. Eng.,C* **56** 269-73
[9] Lafranche E, Krawczak P, Ciolczyk J P and Maugey J 2007 Injection moulding of long glass fibre reinforced polyamide 6-6: guidelines to improve flexural properties *Express. Polym. Lett.* **1** 456–66
[10] Arikan V and Sayman O 2015 Comparative study on repeated impact response of E-glass fiber reinforced polypropylene & epoxy matrix composites *Composites Part B* **83** 1-6
[11] Elanchezhian C, Ramnath B V and Hemalatha J 2014 Mechanical behaviour of glass and carbon fibre reinforced composites at varying strain rates and temperatures *Procedia. Mater. Sci.* **6** 1405-18
[12] Ahmed M N, Kumar P V, Shivanand H K and Muzammil S B 2013 A study on flexural strength of hybrid polymer composite materials (e glass fibre-carbon fibre-graphite) on different matrix material by varying its thickness *Int. J. Mech. Eng. Technol.* **4** 274-86
[13] Lingesh B V, Rudresh B M and Ravikumar B N 2014 Effect of short glass fibers on mechanical properties of polyamide 66 and polypropylene (PA66/PP) thermoplastic blend composites *Procedia. Mater. Sci.* **5** 1231-40
[14] Nuruzzaman D M, Iqbal A K M A, Oumer A N, Ismail N M and Basri S 2016 Experimental investigation on the mechanical properties of glass fiber reinforced nylon *IOP Conf. Ser.: Mater. Sci. Eng.* **114** 1-7
[15] Nuruzzaman D M, Kusaseh N, Basri S, Oumer A N and Hamedon Z 2016 Modeling and flow analysis of pure nylon polymer for injection molding process *IOP Conf. Ser.: Mater. Sci. Eng.* **114** 1-7
[16] Durairaj R B, Mageshwaran G and Sriram V 2016 Investigation on mechanical properties of glass and carbon fiber reinforced with polyester resin composite *Int. J. Chemtech. Res.* **9** 417-23

A prototype for communitising technology: Development of a smart salt water desalination device

F M Fakharuddin, N Fatchurrohman, S Puteh and H M A R Puteri

Faculty of Manufacturing Engineering, Universiti Malaysia Pahang, Malaysia

E-mail: fatchurrohman@ump.edu.my

Abstract. Desalination is defined as the process that removes minerals from saline water or commonly known as salt water. Seawater desalination is becoming an attractive source of drinking water in coastal states as the costs for desalination declines. The purpose of this study is to develop a small scale desalination device and able to do an analysis of the process flow by using suitable sensors. Thermal technology was used to aid the desalination process. A graphical user interface (GUI) for the interface was made to enable the real time data analysis of the desalination device. Arduino™ microcontroller was used in this device in order to develop an automatic device.

1. Introduction

Salt water desalination device is needed especially in remote and rural area. The development of this device could be a life changer as it can provide access to safe and clean water. Desalination is a process that extracts salts and minerals in saline water or a target substance. By doing so, it can help produce water resource for human needs, irrigation or others. There are many types of desalination technique, though however, there is no real time monitoring process. Hence, it is difficult to detect the actual salinity of the sea water feed and the effectiveness of the desalination process.

The objective of this research, is to develop a small scale of salt water desalination device with suitable sensor and Smart System Interface. By which the process aim of this study is to design the desalination device hardware with a built-in Graphical User Interface (GUI), that create a real time data analysis interface and can analyze the effectiveness of the desalination process. A simplest but yet can demonstrate, smart interface device is planned to be used as a monitor panel for the desalination process. This interface device is function to measure the volume of water used, as well as the effectiveness of the desalination process in real time.

This salt water desalination device is developed as a prototype for communitising technology. Communitisation technology is used to adapt the wants and needs of a community [1]. The technology had to suit a large number of people, which can be a challenge in diversity. It was stated to be an initiative for empowering and involving the community in a development [2]. There also link for sustainable development between innovation and community action [3], where it offers a possible approach in the research of the community level action for sustainability.

In general, the thermal type desalination plants had been built in large capacity facility together with a conventional power plant and in such case, the low pressure heating steam can be easily supplied from a power plant. Moreover, Middle Eastern countries traditionally preferred the thermal process to the Reverse Osmosis (RO) process from the past because the raw seawater quality obtained from Red Sea is quite bad, which causes the lifetime of the RO membrane to be short [4].

Desalination has two types which are based on thermal technologies and membrane technologies. For thermal technology it would be Multi Stage Flash distillation (MSF), Multi-Effect Distillation (MED) and Vapour Compression Flash (VCF). While for the membrane technology it would be Electrodialysis (ED), Electrodialysis Reversal (EDR), Capacitive Deionisation (CDI) and Reverse Osmosis (RO).

MSF desalination plant was to evaporate the saline water. Steam that been evaporated from saline water will be condensed on the surfaces of the cooling tube bank, in which it is place at upper section of the evaporators. Furthermore, in the MSF plant when the top brine temperature above 100°C, most of the stages will be maintained in vacuum state [5]. MED was most widely used because of its advantages of simple high thermal efficiency, pre-treatment and flexible operation mode. This system could be operated integrated or independently with other auxiliary technologies such as thermal vapour compression and mechanical vapour compression [6]. Unlike MSF, MED consists of complicated mechanism of heat transfer in which will difficult to deal with. Moreover, there has been no comprehensive study on the dynamic simulation and transient behaviour of MED alongside considering its tube- bundle. The majority of studies have instead been limited to steady-state and thermodynamic performance of the process. Therefore, from this journal a dynamic simulation of MED is been done [7]. Vapour Compression Flash (VCF) was developed to avoid the shortcoming of several existing desalination methods with technical superiority. Moreover, additional thermal energy was not necessary for VCF. During seawater desalination process, the compression system has recovery of the latent condensation heat and higher power-heat conversion efficiency. The seawater pre-treatment requirements were low, and the product water purity was high [5]. The VCF system had a greater recovery of the latent condensation heat and power heat conversion efficiency during seawater desalination. Furthermore, pre-treatment of sea water requirements of VCF are low, and the final product shows high purity of the water [5][8].

2. Methodology

Type of desalination device will be based on thermal technology, which was vapour distillation technique. This vapour distillation technique was the most suitable for making a prototype of smart salt water desalination device, because it is cost effective by comparing to others in which need more money and more complicated parts that need to build. The main objective of this study was to make a smart device in which capable of running a fully automatic desalination process and be able to demonstrate the real time analysis of the desalination process.

An illustration of the device and its flow was take place in order to create a systematic system in which can run automatically and defined in a smart device. The illustration of the device was design by using Catia™ version 5 software. The illustration and its process flow is shown in the figure 1.

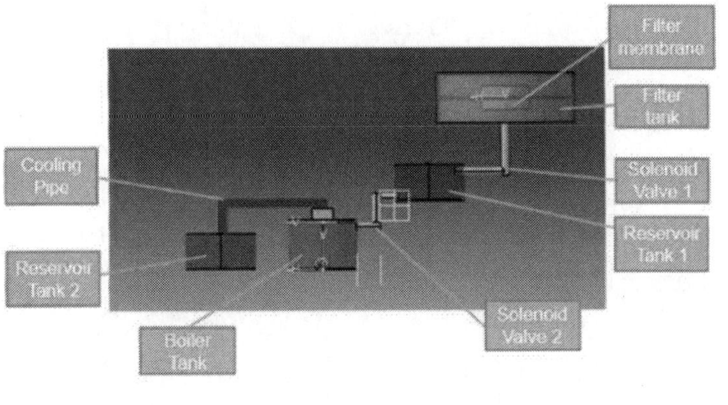

Figure 1. Illustration of the desalination device.

IOP Conf. Series: Materials Science and Engineering **342** (2018) 012103 doi:10.1088/1757-899X/342/1/012103

Designing the desalination device, a simple technique was used but yet it was an effective desalination device in order to full fill all the objectives. When dealing with fluid preventing from leakage of fluid is a crucial part or your system would not be accurate enough and lot of waste product will be produce. Never the less, the main platform was build up from steel frame and plate that would be used to place all the necessary hardware in order to develop the device.

For electrical wiring of this system it divided into two parts. Which were one for controlling the desalination process and another one to control the data transfer getting from the sensor. Thus, it had two Arduino™ microcontrollers that run independently as the programming been set specifically for each one of the Arduino™ microcontroller. Details of the circuit diagram were shown in figure 2 and figure 3.

In this device, the main electrical and electronics component were the two types of sensor and four channels relay module that being implement in this device. The specific sensor used were:

- LM35 is the type of temperature sensor that been used.
- HC-SR04 is the ultrasonic sensor used as water level sensor.

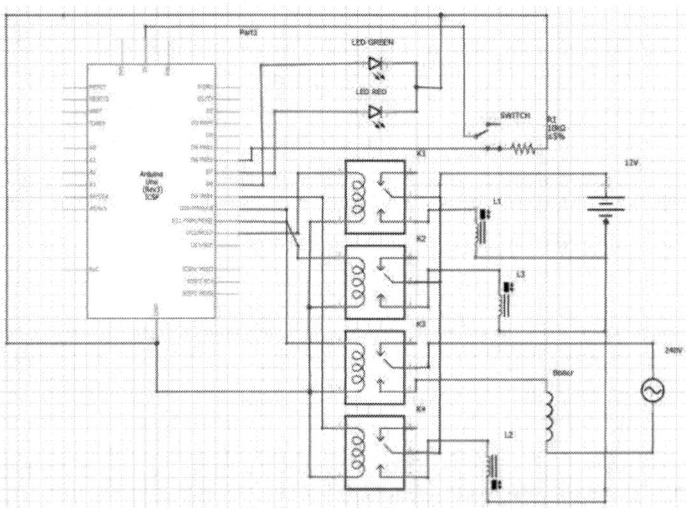

Figure 2. Circuit Diagram for controlling the smart Salt Water Desalination Process Device.

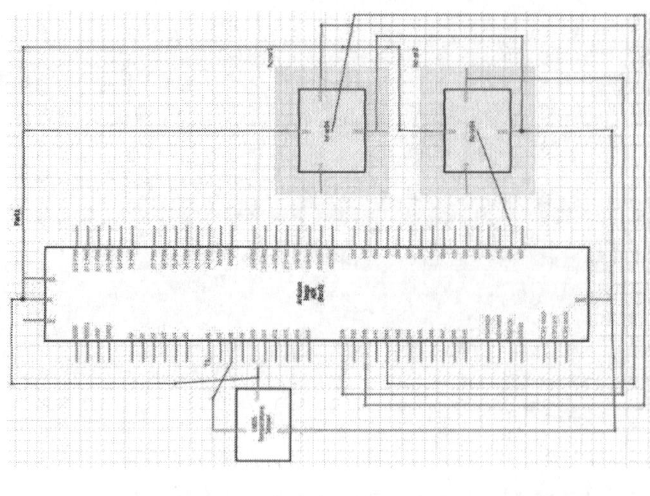

Figure 3. Circuit Diagram of the sensors for smart Salt Water Desalination Process Interface.

From text console, toolbar with buttons for common functions, test editor for writing code and series of menu were used to develop the system. The hardware is connected in order to communicate and upload programs. A GUI was made by using the Visual Studio™ C programming. This GUI provide the platform for displaying the real time data analysis of the desalination process.

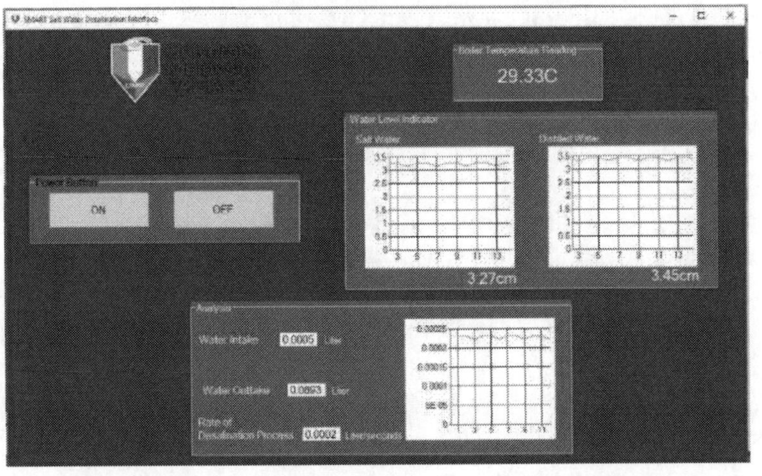

Figure 4. GUI of the real time data analysis interface.

3. Results and discussions

The value of each data was taken and then analyse by using simple mathematical equation in order to interpret the data. Each data was taken according to the specific time of heating given. Therefore, six set of data have been made and analysed in order to know the initial and final state of the sample. The only manipulated variable would be the water input with starting value of 0.5 litre (L) to 1 L with interval of 0.1. While for the fixed would be the time taken to take place for the whole process.

From the figure 5, the water input for each sample were increasing with interval of 0.1 L starting from 0.5 L until 1 L. From the result after the sea water undergo desalination process, the water output from sample number one to six shows increasing of water output. Water sample for number one shows the lowest water output with 0.1078 L while water sample number 6 shows the highest water output with 0.1847 L.

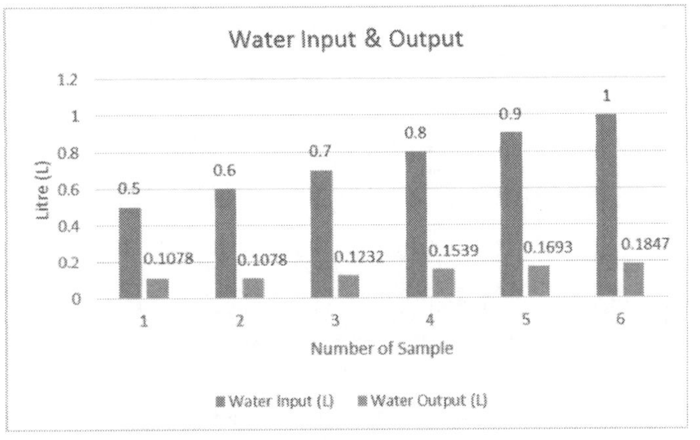

Figure 5. Chart of water input and output.

For a pure distilled it should be pH 7. Nevertheless, due to pure distilled water highly react to CO_2 from surrounding, the actual reading of distilled water would be slightly less pH 7. The targeted pH value for a successful desalination process would be range at pH 6.7 until pH 7. It shows the results that been measured and calculated which are the pH value water output and rate of the desalination process. Moreover, the results showed that the device is able to undergo desalination process although two of the result shows constant and increasing of the pH value at sample number 5 and 6 respectively as shown in figure 6. All of the sample shows the initial reading of pH value almost the same. With range of 8.4 to 8.6 pH, the reading pH of 8.4 for the lowest and 8.6 for the highest. For the water output, number of sample from number 1 to 3 shows the value below the pH 7 while for sample number 4 to 6 shows pH above pH 7.

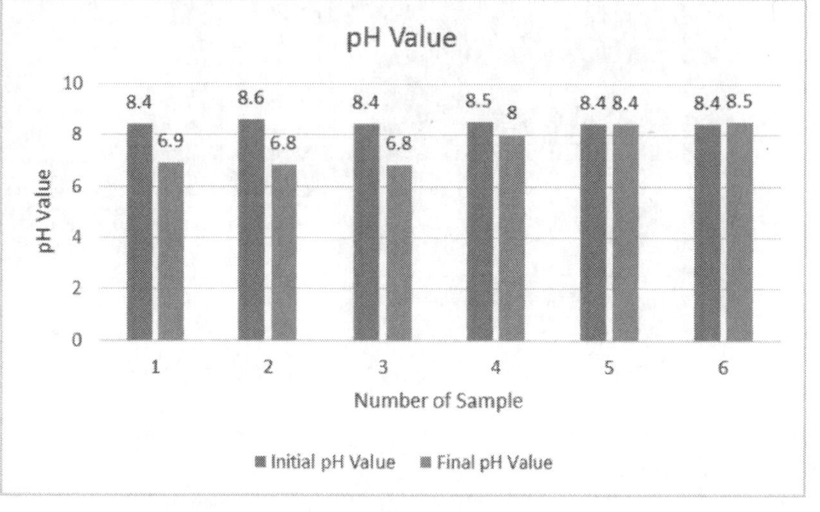

Figure 6. pH value of samples.

Figure 7 shows the rate of desalination process. For sample number 1 shows the lowest value with 2.57E-04 L/s and sample number 6 shows the highest value with 4.40E-04 L/s. Moreover, for sample number 1 and 3 shows the same reading with both 2.57E-04 L/s.

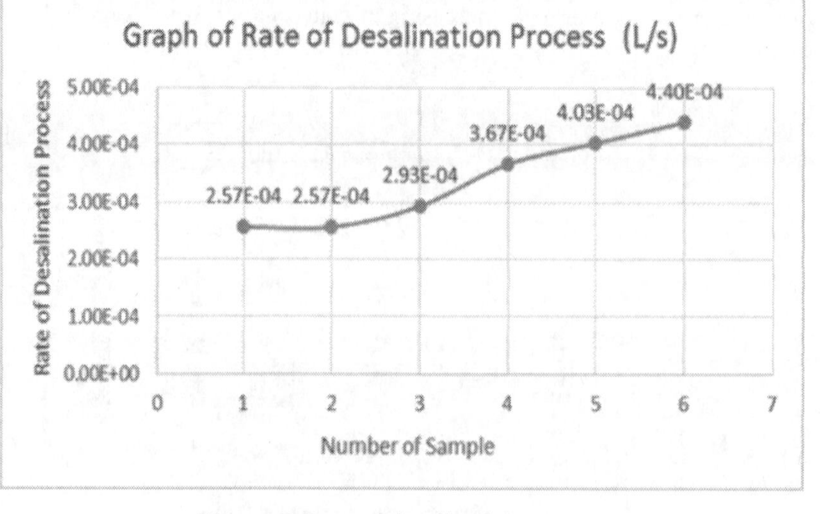

Figure 7. Rate of desalination.

Figure 8 shows the effectiveness of the desalination process. The highest percentage achieved was 21.55% for sample number 1 following with 19.24%, 18.81%, 18.47%, 17.96% and 17.59% which were for the sample number 4, 5, 6, 2 and 3 respectively. From the percentage of effectiveness, it can be concluded that the best water intake should take in this experiment would be 0.5 L which is the sample number 1.

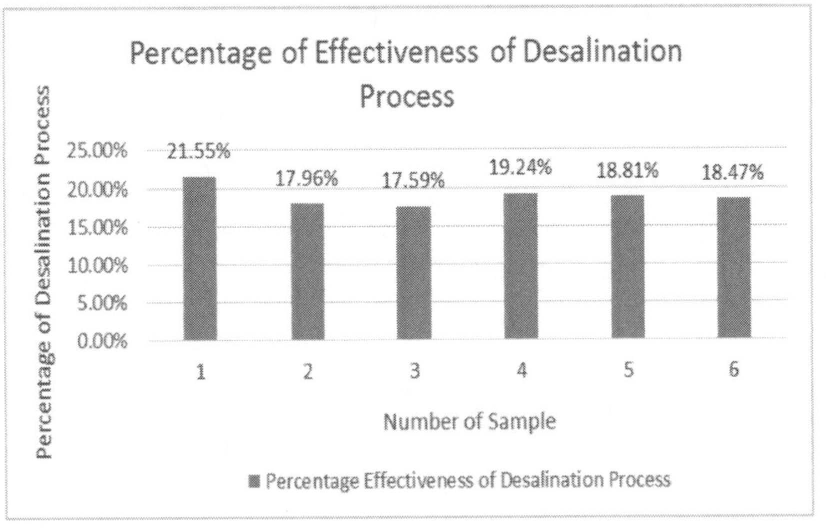

Figure 8. Percentage effectiveness of desalination process.

4. Conclusion

This research was accomplished with the development of the desalination device using suitable sensors and smart device interface which can be implemented on analysing real time data. Suitable sensors were incorporated which were used to get the data and to analyse the state effectiveness of the device. A smart device interface with Graphical User Interface (GUI) was developed using the Visual Studio^{TM} as the platform.. Based on the results obtained, highest water output with 0.1847 L form 1 L input, ideal pH value of 6.8, the highest rate of desalination process of 4.40E-04 L/s and the effectiveness of this device was 21.55% effective by using the water intake of 0.5 L of salt water. Finally, it can be concluded that all the objectives have been achieved and a smart salt water desalination device can be implemented as a prototype for communitising technology.

Acknowledgement

The authors gratefully appreciated Universiti Malaysia Pahang for the financial support through Internal Research Grant RDU150371.

References

[1] Donnelly V and Merrick R 2003 Community portals through communitization *ACM SIGCAPH Computers and the Physically Handicapped* no 73–74 pp 9–14

[2] Kumar S and Singh A K 2016 Communitization Of Institutions And Services In Nagaland *Global Journal of Multidisciplinary Studies* vol 6 no 1

[3] Seyfang G. and Smith A 2007 Grassroots innovations for sustainable development: Towards a new research and policy agenda. *Environmental Politics* vol 16(4) pp 584-603.

[4] Jones A T and Campbell R L 2005 Seawater Desalination: A Generalized Model for Feedwater Intakes *Proc. MTS/IEEE Ocean* pp 1–5

[5] Choi S H 2016 On The Brine Re-Utilization of a Multi-Stage Flashing (MSF) Desalination Plant *Desalination* vol 398 pp 64–76

[6] Han B Liu Z Wu H and Li Y 2014 Experimental Study on a New Method for Improving the Performance of Thermal Vapor Compressors for Multi-Effect Distillation Desalination Systems *Desalination* vol 344 pp 391–395
[7] Azimibavil S and Dehkordi A J 2016 Dynamic Simulation of a Multi-Effect Distillation (MED) process *Desalination* vol 392 pp 91–101
[8] Jin C Z Chou Q L Jiao D S and Shu P C 2014 Vapour Compression Flash Seawater Desalination System and its Energy Analysis *Desalination* vol 353 pp 75–83

IOP Conf. Series: Materials Science and Engineering **342** (2018) 012104 doi:10.1088/1757-899X/342/1/012104

Effect of kenaf fiber in reinforced concrete slab

S M Syed Mohsin, A O Baarimah and G A Jokhio

Faculty of Civil Engineering and Earth Resources, Universiti Malaysia Pahang, Lebuhraya Tun Razak, 26300 Gambang, Kuantan, Pahang, Malaysia

E-mail: maszura@ump.edu.my

Abstract. The effect of kenaf fibers in reinforced concrete slab with different thickness is discusses and presented in this paper. Kenaf fiber is a type of natural fiber and is added in the reinforced concrete slab to improve the structure strength and ductility. For this study, three types of mixtures were prepared with fiber volume fraction of 0%, 1% and 2%, respectively. The design compressive strength considered was 20 MPa. Six cubes were prepared to be tested at 7th and 28th day. A total of six reinforced concrete slab with two variances of thickness were also prepared and tested under four-point bending test. The differences in the thickness is to study the potential of kenaf fiber to serve as part of shear reinforcement in reinforced concrete slab that was design to fail in shear. It was observed that, addition of kenaf fiber in reinforced concrete slab improves the flexural strength and ductility of the reinforced concrete slab. In the slab with reduction in thickness, the mode of failure change from brittle to ductile with the inclusion of kenaf fiber.

1. Introduction

Reinforced concrete is widely used in the construction industry due to its advantages such as strong, robust, economical and durable. Even so, ongoing research was carried out to improve the characteristic of the concrete as concrete is known to be weak in tension. Amongst of the investigation is to add short discreet fiber into the concrete as fiber is known to have benefits of improving the shear and flexural strength, adding ductility, delaying cracking propagation, increasing energy absorption and altering the mode of failure of the concrete [1-5]. Fiber reinforced concrete have obtained recognition and is used in many types of structures and infrastructures such seismic building, pre-cast product, tunnel lining, shotcrete and pavement slab [6-7].

Kenaf fiber is seen as a promising green material as it reused natural resources in the concrete. Furthermore, owning to the benefits of fiber's tensile properties, inclusion of kenaf fiber in concrete resulted in better flexural and shear strength and ductility of the reinforced concrete structure [8-12]. However, to ensure good performance, kenaf fiber, similar like other type of natural fibers need to undergo some treatment to reduce high water absorption characteristic of the fiber. One of the treatments recommended is to use a chemical such as sodium hydroxide (NaOH) to reduce the hydrophobic characteristic of the fiber, thus enhancing the adhesion between the fiber surface and the matrix. This was done by removing the hydroxyl group in cellulose and increasing the surface roughness which resulted in the improvement of the tensile properties of kenaf fibers as compared to untreated kenaf fibers [13-15].

Previous work focused on the effect and behaviour of kenaf fiber in reinforced concrete beams. The potential of using kenaf fiber to serve as part of shear reinforcement is yet to be studied. Promising results was observed in kenaf fiber reinforced concrete and its benefit in concrete. Thus, present work aims to study the behaviour of reinforced concrete slab added with kenaf fiber and its potential in compensating the loss in concrete shear capacity.

2. Materials and Methods

For this study, the concrete mixture was designed in accordance to British Standard (BS EN 206-1, 2000) to achieved at least 20 MPa on the 28th day and is considered as the control mix. Three concrete mix designs were prepared as shown in Table 1. Three amounts of fibers in volume fraction of 0%, 1% and 2% were considered to study the effect of kenaf fiber content in the concrete. Mix 1 (C0) was taken as control mixture as the amount of fiber content is 0%.

Table 1. Concrete mix design.

Properties	Mix 1 (C0)	Mix 2 (K1)	Mix 3 (K2)
V_f (%)	0	1	2
Cement (kg/m^3)	280	280	280
Fine Aggregate (kg/m^3)	556	556	556
Coarse Aggregate (kg/m^3)	1292	1292	1292
Water (Liter/m^3)	162	162	162
W/C ratio	0.58	0.58	0.58
SP (Liter/m^3)	5.6	5.6	5.6
Kenaf fiber (kg/m^3)	0	6.5	13

Table 2 and Figure 1 shows the properties and picture of kenaf fiber, respectively, considered in this study. As kenaf fiber is obtained from the bast fiber in kenaf plant, it is considered as natural fiber and have high water absorption characteristic. This will cause the concrete mixture to have low workability. Therefore, to maintain water cement ratio of 0.58 and achieve the required slump, super plasticizer (SP) was added into the concrete mixtures. The super plasticizer used in this study was Sika ViscoCrete®-2199. Furthermore, the kenaf fiber was treated with 1% of NaOH in order to clean and remove impurities from the fiber surface as well as enhancing the tensile properties of kenaf fiber.

Table 2. Properties of kenaf fiber.

Properties	Kenaf Fiber
Length, L (mm)	30
Diameter, D (mm)	0.1-0.5
Aspect Ratio, L/D	-
Tensile Strength (MPa)	400-550
Unit Weight (kg/m^3)	650

Figure 1. Kenaf fiber.

Six cubes of 150 x 150 x 150 mm were casted for each mixture to study the effect of kenaf fiber on the compressive strength of the concrete. The test was conducted on 7th day and 28th day as recommended by British Standards BS EN 12390-3 (2009). Whereas, two series of slab with different thickness were prepared for this study. The slab has same length and width of 1000 mm and 500 mm, respectively. However, for the first series of slab, the thickness considered was 120 mm, and for the second series of the slab, was 100 mm. The slab thickness in the second series was reduced by 17% and was designed to fail in shear. A total of two slabs were casted for each mixture in both series and tested under four-point bending test. The loading arrangement on the slab is given in Figure 2.

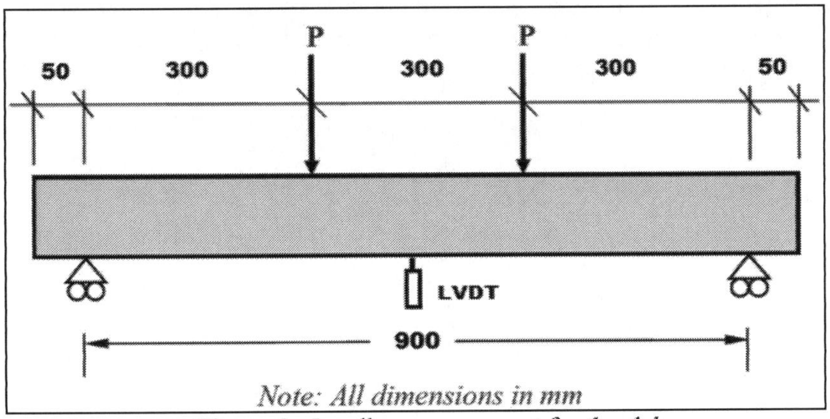

Figure 2. Loading arrangement for the slab.

3. Results and Discussion

This section presents the compressive strength results, load and deflection curve of the slab, cracking pattern and mode of failure.

3.1 Compressive Strength Test

Table 3 tabulates the average compressive strength results of concrete cubes added with kenaf fibers. As can be seen, addition of kenaf fiber decreased the compressive strength of the concrete at 7th and 28th days. At 28th day, the results observed a decrease of 17.5% and 24.0% in K1 and K2, respectively, in comparison to the concrete design strength of 20 MPa. This is due to the reduction in density of the concrete added with kenaf fibers, and potentially due to the cubes was not fully dry and hardened as it was cured inside the water for 7 days. Increasing the amount of kenaf fiber will reduce further the density of the concrete, therefore, the compressive strength results decreases with the increase of fiber content.

Table 3. Compressive strength of kenaf fiber concrete mixtures.

Mixture ID	Vf (%)	Compressive Strength (MPa) 7 Days	28 Days
C0	0	17.40	22.30
K1	1	12.60	16.50
K2	2	11.30	15.20

3.2 Load versus Deflection Curves

Figure 3 shows the load versus deflection curve of the first series of slab tested under four-point bending test. RC1-0, KF1-1, and KF1-2 is the slab identification with 0%, 1% and 2% fiber content, respectively. RC1-0 was taken as the control slab for comparison. The key parameters extracted from the figure is presented in Table 4. Py and δy is the load at yield and its respective deflection, Pu and δu is ultimate load at failure and its respective deflection, whereas Pmax and δmax is maximum load carrying capacity and its respective deflection. The ductility ratio (μ) is calculated by dividing ultimate deflection and deflection at yield. The highest ductility was observed in slab KF1-1 with ductility ratio of 3.10. Whereas, in slab KF1-1 with lowest ductility, the structure became stiffer and failed in a sudden manner. From the results, addition of the kenaf fiber increases the load carrying capacity as higher loading is required to bridge the crack opening. The increase was up to 13% and 21% for KF-1 and KF-2, respectively. In the first series of the slab, addition of 2% of kenaf fiber caused the slab to become stiffer and failed suddenly after resisting higher load to produce larger crack opening width. The pattern of failure exhibits that the slab was over reinforced for KF1-2. For KF1-1, the structure had higher ductility and strength. Therefore, it can be justified that the optimum amount of fiber for this series can be taken at 1% as highest ductility ratio was observed at 3.10.

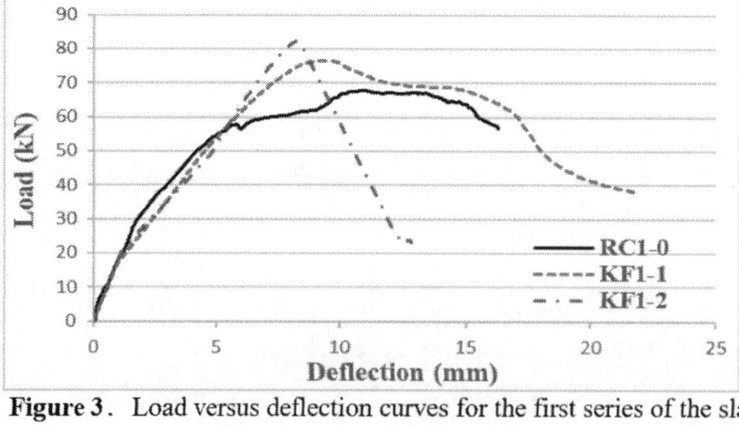

Figure 3. Load versus deflection curves for the first series of the slab.

Table 4. Critical data for first series of kenaf fiber reinforced concrete slab.

Specimen	Py (kN)	δy (mm)	Pu (kN)	δu (mm)	Pmax (kN)	δmax (mm)	μ= δu / δy
RC1-0	57.75	5.83	57.53	16.15	67.68	10.87	2.77
KF1-1	54.97	5.14	65.07	15.94	76.56	9.45	3.10
KF1-2	49.05	4.76	69.83	9.21	82.15	8.21	1.94

The load versus deflection curves for second series of slab and its respective key parameters are given in Figure 4 and Table 5, respectively. For this series, RC2-0 is considered as the control slab with 0% of fiber content and 20 mm reduction in slab thickness. KF2-1 and KF2-2 is the kenaf fiber reinforced concrete slab with 20 mm reduction in slab thickness added with 1% and 2% fiber content, respectively. From the figure, addition of fiber caused the slab to become stiffer as higher loading was required to induce first cracking and propagates the opening of the crack. As the amount of fiber increased, the load at yield and maximum load carrying capacity increases up to 26% and 9%, respectively. In term of ductility, an upward trend was observed with the incrementation of fiber content. The control slab failed in shear as designed, and inclusion of fiber improves the ductility of the slab, changing the failure mode to a more ductile manner.

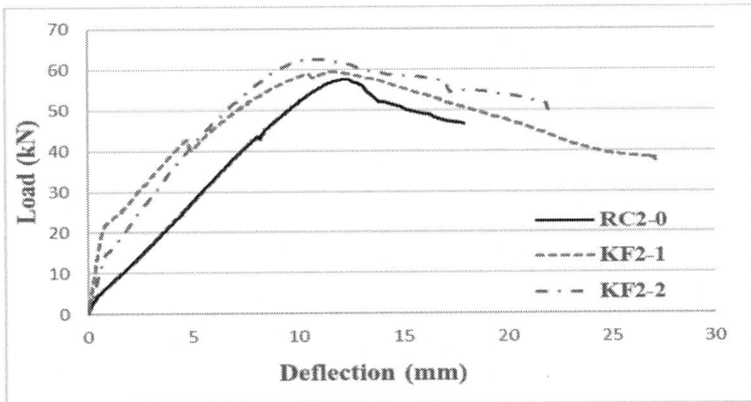

Figure 4. Load versus deflection curves for the second series of the slab.

Table 5. Critical data for second series of kenaf fiber reinforced concrete slab.

Specimen	Py (kN)	δy (mm)	Pu (kN)	δu (mm)	Pmax (kN)	δmax (mm)	μ= δu / δy
K2-0	43.53	8.12	49.21	15.93	57.50	12.46	1.96
K2-1	51.30	8.77	50.41	18.04	59.30	11.88	2.06
K2-2	54.80	8.68	52.70	19.98	62.53	10.68	2.30

Comparison of second series of slab with the control slab from the first series depicted that kenaf fiber was not able to fully compensate the loss in concrete shear capacity due to the thickness reduction. As can be seen in Table 5, the highest maximum load carrying capacity was 62.53 kN for KF2-2, which were about 8% lower than the control slab, RC1-0. Interestingly, the ultimate deflection of KF2-2 slab was 24% higher than the control slab (RC1-0), showing better ductility in the concrete structure.

3.3 Cracking pattern and mode of failure

Figure 7 shows the cracking pattern for the slabs at failure. In the first series of slab (K1), the cracking propagates at the mid-span, and for K1-2, due to higher dosage of fiber, the slab became stiffer and failed in shear-bending. For the second series of slab (K2), the slab failed in shear as intended design. This shows that, kenaf fiber have potential to improve the ductility of the structure and delayed the cracking propagation.

Specimen	Cracking pattern	Failure mode
K1-0		Bending
K1-1		Bending-Shear
K1-2		Shear-Bending
K2-0	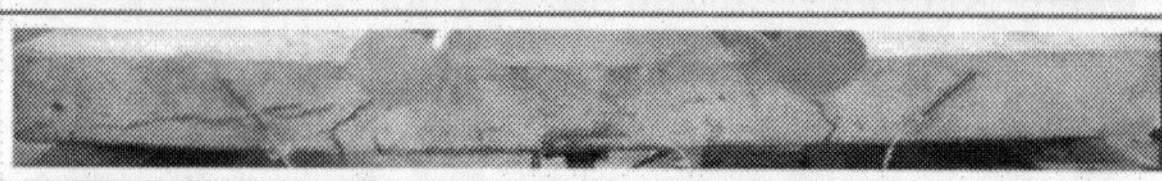	Shear
K2-1		Bending
K2-2		Bending

Figure 5. Cracking pattern and failure mode of slab.

4. Conclusion
Based on the finding, it was observed that kenaf fiber have the capability to enhances the flexural strength of the slab, reduces the cracking propagation and improves the ductility of the reinforced concrete slab. However, the loss of the concrete shear capacity due to the reduction in the thickness of the slab was not compensate by the addition of kenaf fiber. This is probably due to insufficient treatment of the kenaf fiber. Further investigation can be carry out to improves the properties of the kenaf fiber so that it will performs better once added into concrete.

Acknowledgement
The authors would like to acknowledge Universiti Malaysia Pahang (UMP) in financially supporting this research through internal grant RDU 170340.

References
[1] Hannant D J 2003 Fibre-reinforced concrete *Advanced Concrete Technology–processes* ed. J Newman and B S Choo (Oxford: Butterworth-Heinemann, Elsevier) chapter 6 pp 1 - 17
[2] Khaloo A R, Afshari M 2005 Flexural behaviour of small steel fibre reinforced concrete slabs, *Cem. and Concr. Compos.* **27** 141-149
[3] Ding Y, You Z, Jalali S 2011 The composite effect of steel fibres and stirrups on the shear behaviour of beams using self-consolidating concrete, *Eng. Struct.* **33** 107-117
[4] Minelli F, Conforti A, Cuenca E, Plizzari G 2014 Are steel fibres able to mitigate or eliminate size effect in shear?, *Mater. Struct.* **47** 459-473
[5] Baarimah A O, Syed Mohsin S M 2017 Behaviour of reinforced concrete slabs with steel fibers *IOP Conf. Ser.: Mater. Sci. Eng.* **271** 012099
[6] Swamy R N, Lankard D R 1974 Some practical applications of steel fibre reinforced Concrete, *Proc. of the Instit. of Civil Eng.* **56** 235-256
[7] Brandt A M 2008 Fibre reinforced cement-based (FRC) composites after over 40 years of development in building and civil engineering, *Compos. Struct.* **86** 3-9
[8] Mansur M A, Aziz M A 1982 A study of jute fibre reinforced cement composites, *Int. J. of Cem. Compos. and Lightweight Concrete* **4** 75-82
[9] Gouveia N D, Fernandes N A, Faria D M, Ramos A M, Lúcio V J 2014 SFRC flat slabs punching behaviour–Experimental research, *Compos. Part B: Eng.* **63** 161-171

[10] Elsaid A, Dawood M, Seracino R, Bobko C 2011 Mechanical properties of kenaf fiber reinforced concrete *Constr. and Build. Mater.* **25** 1991-2001

[11] Syed Mohsin S M, Azimi S J, Namdar A 2014 Behaviour of Oil Palm Shell Reinforced Concrete Beams Added with Kenaf Fibres. *In Appl. Mech. and Mater.* **567** 351-355

[12] Syed Mohsin S M, Manaf M F, Sarbini N N, Muthusamy K 2016 Behaviour of reinforced concrete beams with kenaf and steel hybrid fibre, *ARPN J. of Eng. and Appl. Sci.* **11** 5385-5390

[13] Li X, Tabil L G, Panigrahi S 2007 Chemical treatments of natural fiber for use in natural fiber-reinforced composites: a review *J. of Polym. and the Environ.* **15** 25-33

[14] Yousif B F, Shalwan A, Chin C W, Ming K C 2012 Flexural properties of treated and untreated kenaf/epoxy composites *Mater.and Des.* **40** 378-385

[15] Mahjoub R, Yatim J M, Sam A R M, Hashemi S H 2014 Tensile properties of kenaf fiber due to various conditions of chemical fiber surface modifications *Constr. and Build. Mater.* **55** 103-113

Effect of fly ash content towards Sulphate resistance of oil palm shell lightweight aggregate concrete

K Muthusamy, M Y Fadzil, A Z Muhammad Nazrin Akmal, S Wan Ahmad, Z Nur Azzimah, H Mohd Hanafi and R Mohamad Hafizuddin

Faculty of Civil Engineering and Earth Resources, Universiti Malaysia Pahang, Lebuhraya Tun Razak, 26300 Gambang Pahang

E-mail: khairunisa@ump.edu.my

Abstract. Both oil palm shell (OPS) and fly ash are by-product generated from the industries. Disposal of these by-product as wastes cause negative impact to the environment. The use of both oil palm shell and fly ash in concrete is seen as an economical solution for making green and denser concrete. The primary aim of this research is to determine the effects of FA utilization as sand replacement in oil palm shell lightweight aggregate concrete (OPS LWAC) towards sulphate resistance. Five concrete mixes containing fly ash as sand replacement namely 0%, 10%, 20%, 30% and 40% were prepared in these experimental work. All mixes were cast in form of cubes before subjected to sulphate solution for the period of 5 months. It was found that addition of 10% fly ash as sand replacement content resulted in better sulphate resistance of OPS LWAC. The occurrence of pozzolanic reaction due to the presence of FA in concrete has consumed the vulnerable Calcium hydroxide to be secondary C-S-H gel making the concrete denser and more durable.

1. Introduction

The global demand of material construction is significantly increasing due to population growth worldwide [1]. The robust growth has increased people request towards housing and infrastructure. As a result, concrete which is the most widely consumed material in the world due to its versatility is in high demand. The high dependence has significantly natural aggregates such as natural sand and granites. Therefore, using alternative sources as replacement for natural aggregates appears to be extremely important. Industrial waste materials such as fly ash (FA) can be used as alternative sources in concrete as they can assist in solving some environmental concerns. Used of recycled industrial waste diminish the problem of waste disposal and reduce the intensive use of energy and natural resources [2]. Concurrently, oil palm shell (OPS) by-product that generated abundantly from palm oil industry are creating an eyesore. Fortunately, these two by-product has huge potential to be utilized as fine aggregate and coarse aggregate replacement in lightweight aggregate concrete. FA is identified as pozzolanic material and can be utilize as partial sand replacement while OPS can replace coarse aggregate in lightweight aggregate concrete. These sustainable building material attempt is seen to benefits both coal and palm oil industry in terms of waste management cost and at the same time assist towards greener construction building material.

2. Method of Research

2.1. Materials

For binding purpose, Portland Cement brand that was chosen to be used throughout this experimental research is Orang Kuat cement. This type of cement was conforming to [3] for Portland cement specification and suitable for structural concreting precast, brickmaking and all general purpose applications. Oil palm shell (OPS) obtained from FELCRA palm oil mill factory was used to fully replace conventional coarse aggregate to produce lightweight aggregate concrete. Upon taking from palm oil mill, the oil palm shells were washed thoroughly using tap water to remove impurities, oil and unwanted materials that may affect the concrete strength. After that, the OPS was oven-dried at 110 ± 5°C for 24 hours to remove moisture and prevent from damage. The physical properties of processed OPS is tabulated in Table 1.

Table 1. Physical properties of processed OPS.

Physical Properties	OPS
Specific gravity (kg/m^3)	1.37
Water absorption (24 hour) (%)	12.47
Moisture content (%)	12.45
Aggregate abrasion value, Los Angeles (%)	7.6
Bulk density (kg/m^3)	568
Fineness modulus (FM)	6.53
Flakiness index (%)	38.63
Elongation index (%)	98.74
Aggregate impact value (%)	18.18
Aggregate crushing value (%)	14.84

In this experiment, tap water was used for concrete mixing purpose. Distilled water was used to prepare chemical solution for specimen immersion. The superplasticizer used throughout this research is classified as Type A water-reducing admixtures in accordance to ASTMC494 [4]. Local river sand that was used as fine aggregate. Fly ash (FA) was used as partial lightweight fine aggregate to produce lightweight aggregate concrete in this research. The FA was collected from a coal fire plant located in west Malaysia. FA taken was stored in a clean and dry room to ensure its purity. Table 2 shows the chemical composition of FA. Following the standard in [5], this FA was classified as pozzolanic material.

Table 2. Chemical composition of fly ash.

Chemical Composition	Percentage (%)
Silicon dioxide (SiO_2)	39.0
Aluminium oxide (Al_2O_3)	28.7
Iron Oxide (CaO)	20.3
Potassium oxide (K_2O)	2.1
Iron (Fe_2O_3)	8.9
Copper (CuO)	0.1
Manganese (MnO)	0.1
Loss of Ignition	2.0

2.2. *Preparation of Testing Samples*
Five concrete mixes consists of 0%, 10%, 20%, 30% and 40% FA by weight of cement were casted in 100 x 100 x 100 mm standard cubes. Plain OPS LWAC with 0% FA was act as control specimen. The concrete mixes were prepared in concrete mixer. Then the mixes were compacted using a vibrating table to achieve better compaction. Specimens were covered by wet gunny sack and left overnight. After 24 h, the concrete was demoulded and water cured for 28 days before ready for durability test.

2.3 *Testing Method*
Sulphate resistance test was conducted in order to determine the durability performance of OPS LWAC containing FA as partial sand replacement performance towards sulphate attack. The sulphate solution was prepared according to ASTM C1012 [6]. The degree of sulphate attack was evaluated by measuring mass change of samples of 100 x 100 x 100 mm concrete cubes. Then, six specimens from each mix were immersed in 5% sodium sulphate (Na_2SO_4) solution as shown in Figure 1. Mass measurement and visual assessment of the concrete were conducted every week until the age of 5 months.

Figure 1. Specimens immersed in sulphate solution.

3. Analysis and Discussion

3.1. *Sulphate Attack*
The sulphate resistance of OPS LWAC containing various percentage of fly ash were evaluated by measuring the mass change. The mass change of OPS LWAC with FA upon subjected to sulphate solution is illustrated in Figure 2 and Figure 3. As can be seen in Figure 2, the concrete mass for all specimens keep on increasing until the age of 20 weeks. The least mass change was denoted by 10% sand replacement by FA with 1.89%, The value is lower than control OPS LWAC with 23.7%. Compared to other specimens, OPS LWAC with 10% FA mass starts affected at 10 weeks. Utilization of FA as optimum amount enables pozzolanic reaction that consumes vulnerable calcium hydroxide that can easily be attacked by sulphate ion. The generation of secondary C-S-H gel fills the voids further strengthen the concrete making it stronger, denser, and durable during its service life. Concrete containing pozzolan have better performance in sulphate solutions since the pozzolanic reactions reduce the quantity of calcium hydroxide and increase calcium silicate hydrate gel [7].

It was observed, that mixes containing FA beyond 10% replacement exhibit higher mass change value as amount of FA added increases. The highest mass change was denoted by 40% FA with 2.92. This significant increment in the weight is caused by the ease of sulphate ion invasion into the concrete causing more ettringite and gypsum is produced within the concrete internal structure. This makes concrete containing higher amount of FA exhibit highest value of total mass changes as shown in Figure 3. Conclusively, optimum amount of FA as partial sand replacement is needed to produce OPS LWAC with enhanced durability than control specimen. The result is in line by research done by [9].

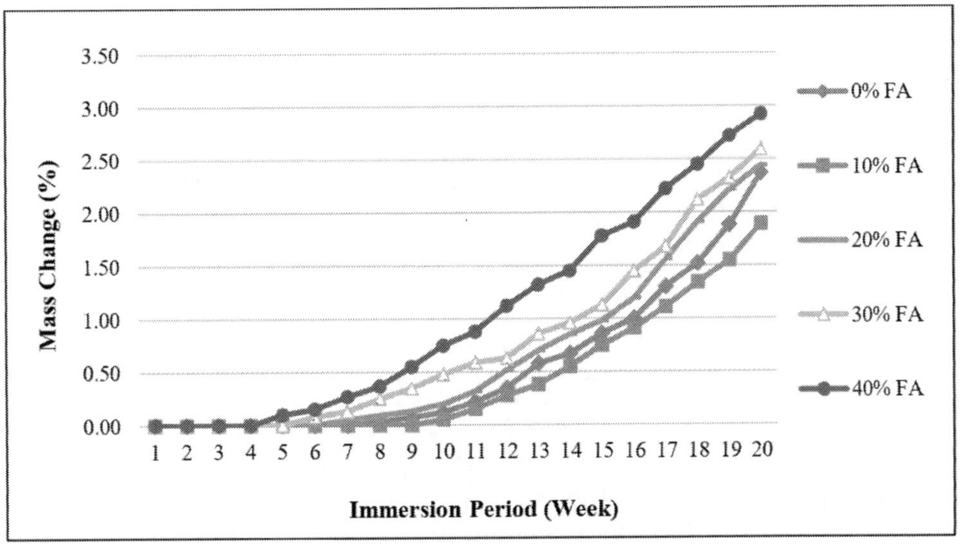

Figure 2. Mass change vs immersion period of OPS LWAC with FA.

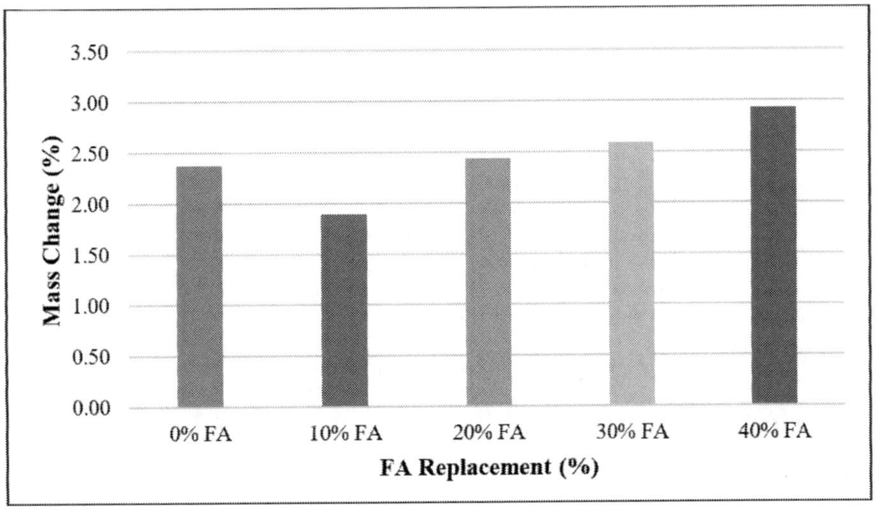

Figure 3. Mass change of OPS LWAC with FA.

4. Conclusion

The results obtained demonstrate that the partial sand replacement by FA in OPS LWAC is advantageous when optimum amount of FA is used. Replacement of 10% FA would assist towards increasing OPS LWAC durability towards sulphate attack owing to the pozzolanic reaction that takes place between the silica of FA with calcium hydroxide produced from the hydration process. However, incorporation of high amount of FA reduces the concrete durability against sulphate attack.

Acknowledgements
The authors would like to acknowledge the help and co-operation received from the lecturers and technical staff of Universiti Malaysia Pahang in conducting the experimental work. The financial support received from Universiti Malaysia Pahang through grant RDU1703173 are also gratefully acknowledged.

References
[1] Gonçalves M C and Margarido F 2015 *Materials for Construction and Civil Engineering: Science, Processing, and Design*. Springer, London.
[2] Dash M K, Patro S K and Rath A K 2016 Sustainable use of industrial-waste as partial replacement of fine aggregate for preparation of concrete – A review. *International Journal of Sustainable Built Environment*, **5**(2) 484-516
[3] Malaysian Standard 2003 Portland cement (ordinary & rapid-hardening): Part 1. Specification. Malaysia
[4] American Society of Testing and Materials. 2005. Standard Specification for Chemical Admixtures for Concrete. Philadelphia, ASTM C494.
[5] American Society of Testing and Materials. 2005. Standard Specification for Fly Ash and Raw Material or Calcined Natural Pozzolan for Use as a Mineral Admixture in Portland Cement Concrete. Pennsylvania, ASTM C618.
[6] American Society of Testing and Materials. 2004. Standard Method for Length Change of Hydraulic Cement Mortars Exposed to A Sulphate Solution. Philadelphia, ASTM C1012.
[7] Chang Z T, Song X J, Munn RM and Marosszeky M 2005 Using limestone aggregate and different cements for enhancing resistance of concrete to sulfuric acid attack. *Cement and Concrete Research,* **35**(8) 1486-1494
[8] Merida A and Kharchi F Pozzolan 2015 Concrete Durability on Sulphate Attack. *Procedia Engineering,* **114** 832-837

IOP Conf. Series: Materials Science and Engineering **342** (2018) 012106 doi:10.1088/1757-899X/342/1/012106

The development of Zirconia and Copper toughened Alumina ceramic insert

Nur Amalina Sabuan[1], Nurfatini Zolkafli[1] , Mebrahitom A[1], Azmir Azhari[1] and Othman Mamat[2]

[1] Faculty of Manufacturing, Universiti Malaysia Pahang (UMP), Pekan, Pahang, Malasyia
[2]Mechanical Engineering Department, Universiti Teknology PETRONAS (UTP), Seri Iskandar, Perak, Malaysia

E-mail: nuramalinasabuan@gmail.com; mebrahitoma@ump.edu.my

Abstract. Ceramic cutting tools have been utilized in industry for over a century for its productivity and efficiency in machine tools and cutting tool material. However, due to the brittleness property the application has been limited. In order to manufacture high strength ceramic cutting tools, there is a need for suitable reinforcement to improve its toughness. In this case, copper (Cu) and zirconia (ZrO_2) powders were added to investigate the hardness and physical properties of the developed composite insert. A uniaxial pre-forming process of the mix powder was done prior to densification by sintering at 1000 and 1300°C. The effect of the composition of the reinforcement on the hardness, density, shrinkage and microstructure of the inserts was investigated. It was found that an optimum density of 3.26 % and hardness 1385HV was obtained for composite of 10wt % zirconia and 10wt% copper at temperature 1000 °C.

1. Introduction

The first attempt in using ceramic as cutting tools was made in early 1930's for turning of gray cast iron. High wear resistance, compressive strength, and its chemical inertness of ceramics promised success. Initially, only Al_2O_3 was used as cutting tools but in early 1970's, ceramic composite cutting tools containing alumina/titanium carbide was discovered [1, 2]. Aluminum oxide/alumina (Al_2O_3) composite cutting tools are entirely used as indexable insert. Lower bending strength and toughness of this material compared to carbide made it rarely used. There are doubts rising in using ceramic as insert cutting tools since ceramic exhibit properties of high brittleness and lower strength that cause low tool life performance. In fact, ceramics can resist a temperature exceed 2000°C vs 870°C compared to insert made from carbide powder. Apart from that, this high temperature helps in softening of the work piece material which allows deeper and cleaner cuts. These advantages cannot be achieved by using carbide material since cobalt binder has lower melting point [3]. Through last few years, remarkable improvements had been done in strength and toughness of Al_2O_3 ceramic composites [4-6].

A.K.Dutta [7] studied on Al_2O_3 incorporated with 5vol% of silver as insert cutting tool. The Al_2O_3/silver (ASC) composite was successfully produced as cutting tool insert with higher toughness compared to monolithic alumina. The hardness of ASC compared to zirconia toughened alumina (ZTA) were found to be 6.5 GPa higher, while the fracture toughness measured was found to be 7.05MPa m1/2. Addition of yttrium also showed improvement in alumina based composite [8]. Effects of yttrium on the mechanical property and the cutting performance of Al_2O_3/Ti(C,N) composite ceramic tool material have been studied and the results showed that by adding certain amount of yttrium the mechanical properties of the composite material can be improved. As the result, flexural strength and fracture toughness value was 1010MPa and 6.1MPa m1/2, respectively. The developed ceramic tools also showed higher fracture resistance when machining hardened C45 steel which improved by 20% compared to ceramic composite without yttrium. Jef Vleugels [9] reported on fabrication, wear and performance of ceramic cutting tools listed on requirements for cutting tools materials that includes cutting speed, cooling conditions and thermal properties of two contact materials that determine the temperature at cutting edge. The observation showed that zirconia (ZrO_2) composites were suitable for turning of DIN 42 $CrMo_4$ steel at 300 and 500 m/min with additional of 40vol% of TiCN. Dong Wang [10] reported on fabrication and cutting performance of Al_2O_3/TiC/TiCN ceramic cutting tool of an ultra-high-strength steel and discovered that, under condition with sintering temperature of 1650 °C, holding time of 15mins and sintering pressure of 35MPa, tool material of AT10N20 exhibit the best comprehensive mechanical properties with hardness, flexural strength and fracture toughness measured was 20.8GPa, 881.4MPa, and 7.8MPa respectively . In this work copper (Cu) and zirconia (ZrO_2) powders were added to investigate the preliminary values of hardness and physical properties of the developed composite insert.

2. Experimental

The raw materials for this experiment are readily prepared from supplier: Alumina powder (Al_2O_3), Zirconia powder (Zr_2O), copper (Cu) powder, polyvinyl alcohol (PVA) as a binder and polyethyl glycol (PEG) as a plasticizer. During mixing, few drops of binder and plasticizer were added and mixed thoroughly before undergo drying process. Binders surrounds the ceramics particles and provide lubricants during compacting and act as a temporary bonds after pressing since it will be decomposed at high temperature which was during sintering process which required temperature over 1000°C.Plasticizer was added into the binders to improve the flexibility and functionality of the green compact since the plasticizer had brittle properties at room temperature and was not conducive for further processing steps. Conventional mixing is practiced in developing the composite. Al_2O_3 powder is added with Zr_2O to help toughened the Al_2O_3. Different composition of Cu 10 (wt%) and 20 (wt%) is added in the mixture before PVA and PEG is added. The mixture is mixed thoroughly before drying at 200°C for 3 hours and left in vacuum condition for overnight to stabilize the mixture.

A diamond shaped green compact was produced by uniaxial compaction of 500MPa with holding time of 10 minutes per sample. The mould and the green compact is shown in Figure 1. The green compact was then sintered under two different sintering temperature: 1000°C and 1300°C. The hardness value of the sintered inserts was measured using Vickers hardness for a three samples of each and the average values were recorded. The density was also measured for each respective sample. The microstructure of the sintered samples was observed using optical microscope.

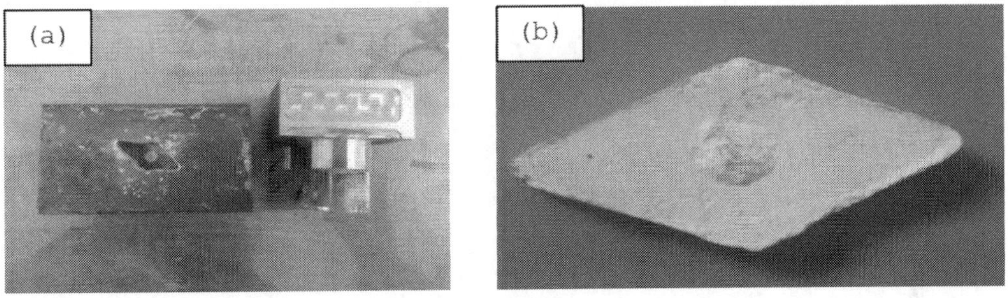

Figure 1. (a) The mould for the green compact of the ceramic insert (b) green compact .

Table 1. Parameters for sample preparation.

Parameters	Condition
Powder content (mixing)	90wt%Al_2O_3,10wt%ZrO_2 (90AlZr)
	80wt%Al_2O_3,10wt%ZrO_2,10wt%Cu (80AlZr10Cu)
	70wt%Al_2O_3,10%wtZrO_2,20wt%Cu (70AlZr20Cu)
Drying	200°C for 3 hours
Compaction	500MPa
Sintering	1000°C and 1300°C

3. Result and Discussion

Three different ceramic inserts as shown on the Table 1 whas subjected to 1000°C and 1300 °C sintering temperature and each sintered composite inserts were investigated for hardness, density and microstructure. The hardness value of the three compositions is shown in Figure 2. It was observed that the average hardness at temperature 1000 °C for 90AlZr was 1089 HV0.5N and a slightly increase to about 1385 HV0.5N for 80AlZr10Cu and decrease to a value 1207 HV0.5N for 70AlZr20Cu was observed. However, the hardness value at sintering temperature 1300°C was observed decreasing as the composition of copper powder was increasing. At temperature 1300 °C the hardness was higher for 90AlZr compared to 1000 °C. Generally higher sintering temperature gave high hardness value for ceramic reinforced alumina composites where the sintering may not be effective at 1000 °C. However, it was generally observed that the hardness values were seen decreasing as the copper composition was increasing at each sintering temperatures. This may be due to the low melting temperature of cupper which may be a case for the liquid phase sintering that lowers the hardness of the composites. The addition of cupper and zirconia played a great role for increasing the toughness of the insert which consequently decrease the hardness value [11].

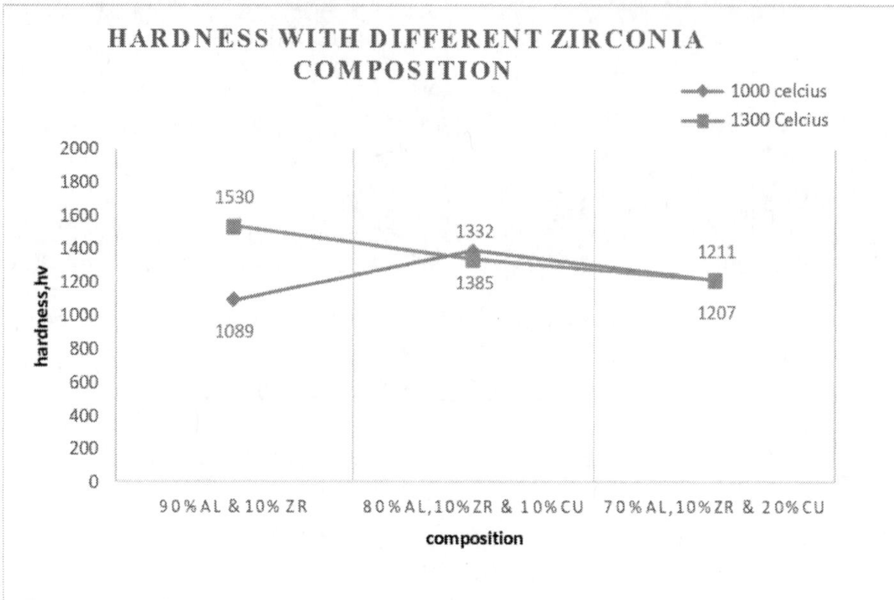

Figure 2. Hardness of samples with different composition and sintering temperature.

The results of the density after sintering were given in Figure 3. It was observed that density at 1000 °C for 90AlZr showed the lowest value which is 2.95 g/cm^3 and increased dramatically for 80AlZr10Cu which was 3.26 g/cm3 but then as the cupper percentage by weight increased to 20%, the density value was measured to 3.14 g/cm^3. At temperature 1300 °C, all compositions of the inserts show lower density than the at 1000 °C sintering temperature. This may be due to the high sintering temperature which the cupper particle may thermodynamically have changed to non-solid state which resulted into some porosity inside the samples.

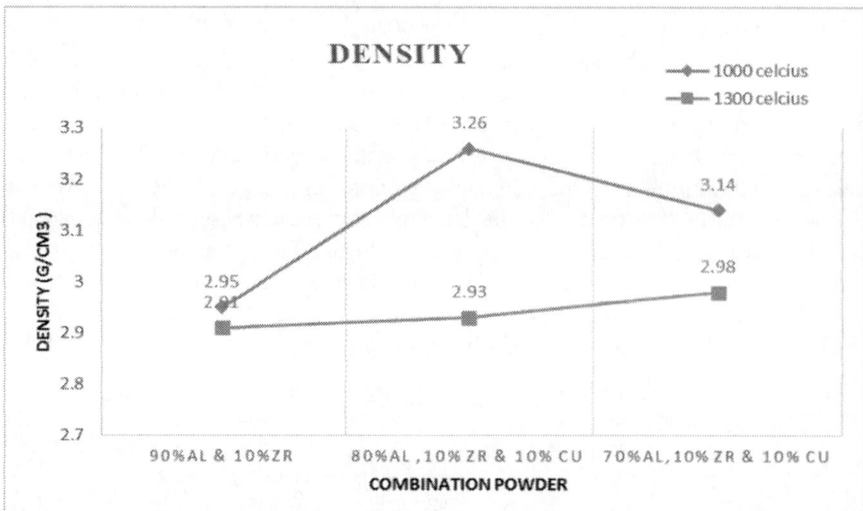

Figure 3. Density of sample at different sintering temperature and Cu composition.

Shrinkage was significantly observed when Cu was incorporated in Al_2O_3 matrix during the sintering cycle. Figure 4 shows the shrinkage value increases as the weight percentage of the composition of Cu goes from 10% by weight to 20 % by weight. This may be due to the presence of the Cu liquid which helps the grain boundaries slide and create more adhesion as a result volumetric size decreased.

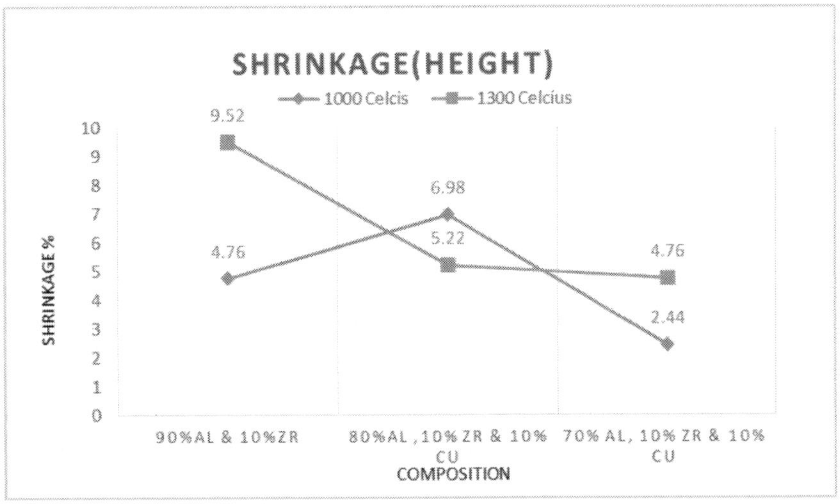

Figure 4. Shrinkage of sample at different sintering temperature and Cu composition.

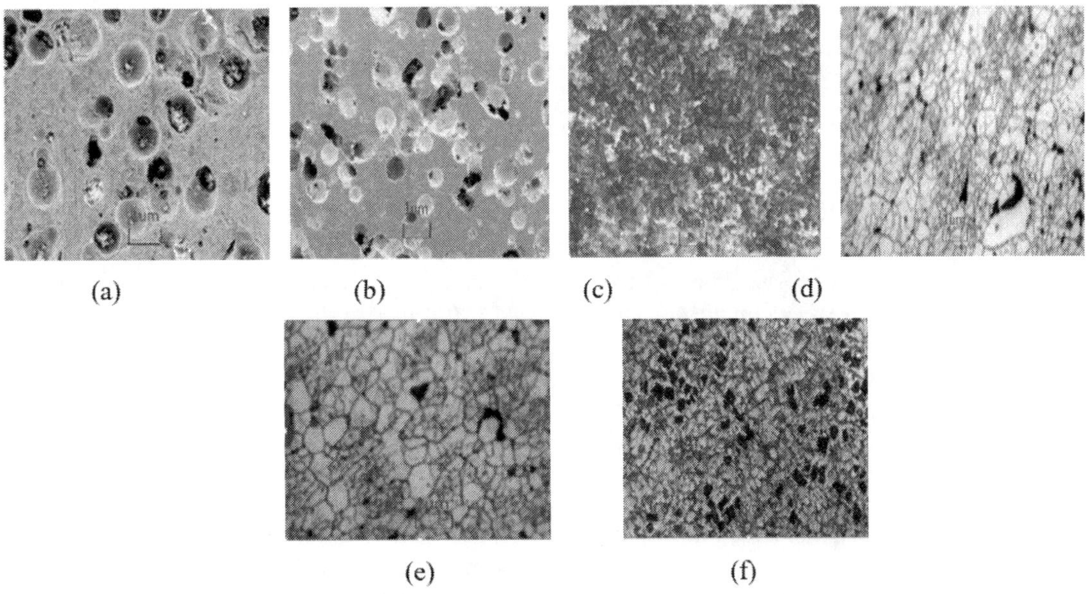

Figure 5. Microstructural changes for **90AlZr** at a) 1000°C b) 1300°C, **80AlZr10Cu** at c)1000°C d)1300°C and **70AlZr20Cu** at e)1000°C and f)1300°C.

Figure 5 a) and b) shows the presence of Al_2O_3 with ZrO_2 particle using optical microscope. The microstructure shows that the zirconia phase was dissolved in the alumina matrix uniformly in all cases, while the Cu were within the interface of the grain boundary to form an interface solid solution. This showed that Cu assisted in the liquid phase sintering of the sintered composite. Alumina matrix is expected to be toughened by the zirconia particle. From the observation high number of porosity is seen at 1300 °C compared at 1000 °C due to the melting of the cupper particles. Observations of the surface using the optical microscope show relatively good bonding between Al_2O_3 and Cu of the insert 70AlZr20Cu at 1000 °C than at **1300 °C.** High number of porous was seen at the higher sintering temperature due to phase change of the cupper particle as can show with dark regions between alumina This phenomenon is especially seen in the regions of clusters of alumina and zirconia phases when liquid Cu has no possibility to infiltrate the porous structure. However these observations should be more studied with controlled experiment and high resolution microstructural study.

4. Conclusion

A preliminary experimental study was performed to develop alumina matrix insert reinforced with zirconia and cupper. From this study, hardness, density and microstructural properties were investigated. It was shown that increasing Cu content generally improves the density at sintering temperature of 1000 °C. However, the density of samples sintered at temperature of 1300 °C was decreasing when Cu content increases. This was due to the melting temperature of the Cu is much lower than the Al_2O_3 which led to phase changes which is not favourable for the good property the sintered inserts produced. It was also found that an optimum density of 3.26% and hardness 1385HV was obtained for composite of 10wt % zirconia and 10wt% copper at temperature 1000°C.

Acknowledgement

This research was supported by research grant **RDU150121** from Ministry of Higher Education of Malaysia and **RDU1703193** of Universiti Malaysia Pahang.

References

[1] R. L. Landingham, "Novel compositions for oxide ceramics," Google Patents, 1988.
[2] A. Hosseini and H. A. Kishawy, "Cutting tool materials and tool wear," in *Machining of titanium alloys*: Springer, 2014, pp. 31-56.
[3] E. E. Timm and D. B. Schwarz, "Cobalt-bound tungsten carbide metal matrix composites and cutting tools formed therefrom," Google Patents, 1990.
[4] G. M. Asmelash, O. Mamat, and F. Ahmad, "Toughening mechanisms of Al2O3-SiO2-ZrO2 composite materials," *Ceramics SilikÃ¡ty*, vol. 56, p. 4, 2012.
[5] G. Mebrahitom Asmelash and O. Mamat, "Pressureless Sintering and Characterization of Al2O3-SiO2-ZrO2 Composite," in *Defect and Diffusion Forum*, 2012, pp. 113-128.
[6] G. Mebrahitom Asmelash, O. Mamat, F. Ahmad, and A. K. Prasada Rao, "Thermal shock and fatigue behavior of pressureless sintered Al2O3â€"SiO2â€"ZrO2 composites," *Journal of Advanced Ceramics*, vol. 4, pp. 190-198, September 01 2015.
[7] A. K. Dutta, A. B. Chattopadhyaya, and K. K. Ray, "A study on alumina - 5 vol.% silver composite as cutting tool insert," *Journal of Materials Science Letters*, vol. 19, pp. 1501-1503, September 01 2000.
[8] C. Xu, C. Huang, and X. Ai, "Mechanical property and cutting performance of yttrium-reinforced Al 2 O 3/Ti (C, N) composite ceramic tool material," *Journal of materials engineering and performance*, vol. 10, pp. 102-107, 2001.
[9] J. Vleugels, "Fabrication, wear and performance of ceramic cutting tools," in *Advances in Science and Technology*, 2006, pp. 1776-1785.
[10] D. Wang, C. Xue, Y. Cao, and J. Zhao, "Fabrication and cutting performance of an Al2O3/TiC/TiN ceramic cutting tool in turning of an ultra-high-strength steel," *The International Journal of Advanced Manufacturing Technology*, vol. 91, pp. 1967-1976, July 01 2017.

[11] W. A. Logsdon and P. K. Liaw, "Tensile, fracture toughness, and fatigue crack growth rate properties of zirconium copper," *Journal of Materials Engineering,* vol. 9, pp. 63-69, March 01 1987.
[12] M. Munz, M. Risto, R. Haas, R. Landfried, F. Kern, and R. Gadow, "Machinability of ZTA-TiC Ceramics by Electrical Discharge Drilling," *Procedia CIRP,* vol. 6, pp. 77-82, 2013.

Analysis of acoustic emission during abrasive waterjet machining of sheet metals

Nazrin Mokhtar[1], MA Gebremariam[1], H Zohari[2] and Azmir Azhari[1]

[1]Faculty of Manufacturing Engineering, Universiti Malaysia Pahang, 26600, Pekan, Malaysia
[2]Faculty of Mechanical Engineering, Universiti Malaysia Pahang, 26600, Pekan, Malaysia

E-mail: nazrinmokhtar91@gmail.com; azmir@ump.edu.my

Abstract. The present paper reports on the analysis of acoustic emission (AE) produced during abrasive waterjet (AWJ) machining process. This paper focuses on the relationship of AE and surface quality of sheet metals. The changes in acoustic emission signals recorded by the mean of power spectral density (PSD) via covariance method in relation to the surface quality of the cut are discussed. The test was made using two materials for comparison namely aluminium 6061 and stainless steel 304 with five different feed rates. The acoustic emission data were captured by Labview and later processed using MATLAB software. The results show that the AE spectrums correlated with different feed rates and surface qualities. It can be concluded that the AE is capable of monitoring the changes of feed rate and surface quality.

1. Introduction

Abrasive waterjet (AWJ) machining today have become more common in wide variety of precision machining process. AWJ itself can compete and complement other types of non-traditional machining process such as laser cutting, electro-discharge machining and plasma cutting due to its unique advantage [1]. AWJ works by focusing a highly pressurised waterjet and abrasive are subsequently forced to accelerate at a high speed. The focused stream would impinge on the work surface and caused a focused erosion in which it cut the intended surface. Apart from cutting application, AWJ has been used in other processes such as drilling, turning, milling and shot-peening [2]. Due to the nature of the process, it's a unique technology which greatly benefits form negligible cutting force, zero thermal distortion and high machining flexibility [3,4].

Typically, in AWJ machine environment due to nature of using water and abrasive sensor and monitoring devices suffers from the harsh corrosive environment. Sensors such as force sensor, vertical force sensor are expensive and impractical to be used in the continuous monitoring of AWJ process in which the sensor suffers from the erosive nature of AWJ [5–7]. In addition, the need for an automated control system in which it enables the monitoring of the health or the process is highly required in the development of fully automated machining system. This is where a more practical sensor which can withstand the harsh corrosive environment is needed in order to enable the development of a fully automated AWJ system. Such system would be capable of detecting on-line, the disturbance in the process, register the changes and in case of failure occurs, stop the machine[8].

Acoustic emission (AE) can be defined as 'transient elastic waves generated by the rapid release of energy from within a material of by a process' by the standard BS EN 1330-9:2000 for non-destructive testing. AE in particular have been applied in various field of manufacturing process due to its sensitivity to changes in process parameter, and can be applied in processes involving friction, impact, plastic deformation and fracture [9]. In addition, typically AE sensor captures frequency content in excess of 100kHz, it well beyond the frequency generally associated with the dynamic behaviour of machine tools and cutting tools, the AE frequency response taken can be associated with the cutting process itself [10]. In AWJ particularly the sources of AE can be found from several sources such as the elastic impact of abrasive particles, ploughing without material removed, crack formation/propagation, material removal and single grain fracture [3].

2. Experimental Setup

The experiment for the study was conducted at Faculty of Manufacturing Engineering, Universiti Malaysia Pahang. The experimental set-up consists of three major components.

2.1. Waterjet machine

A self-developed waterjet machine is used to perform the experiment. The developed CNC waterjet machine is shown in Figure 1. The system consists of an air-driven water pump that is capable of producing up to 30,000 psi (~ 206 MPa) of pressurised water stream. The waterjet cutting head consisted of the typical system of a sapphire orifice (0.15 mm in diameter), mixing chamber (0.76 mm in diameter), abrasive flow-controller and nozzle. The position and movement of the cutting head are controlled by a computer numerical control (CNC) positioning table.

Two different sheet metals were used for the experiment namely aluminium 6061 and stainless steel 304 with a thickness of 1.5 mm and 1.25 mm respectively. The traverse rate was varied to 5 different speeds at 5 m/s, 10 m/s, 15 m/s 20 m/s and 25 m/s respectively. The standoff distance was fixed at 5mm, while pressure and abrasive flow rate remained constant at 137 MPa and 2.0 g/s respectively. The abrasive used in the experiment was 80 mesh garnet. The signal was collected when the cutting reaches the midsection of the cutting line as illustrated in Figure 2. The cut quality was measured in terms of surface roughness and kerf taper widths (i.e. top and bottom widths). The surface roughness was measured by using Surfcom130A which has a resolution of 0.1 to 10 nanometres. The kerf widths were measured using an optical video measurement system.

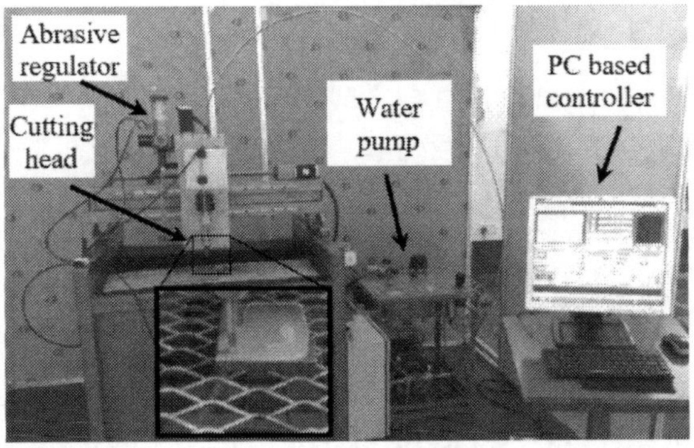

Figure 1. The self-developed CNC waterjet machine (The inset shows the cutting process).

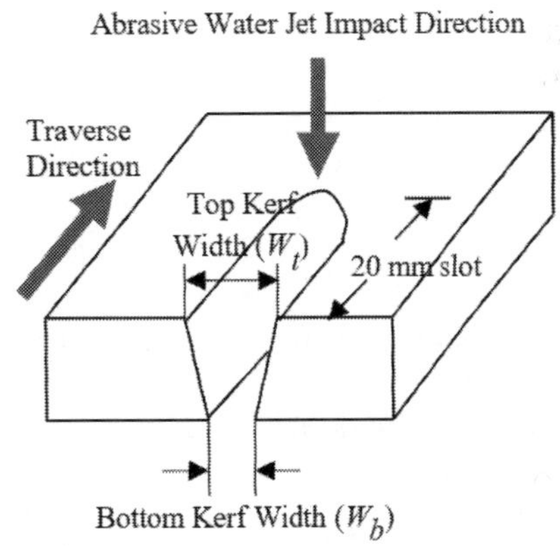

Figure 2. Kerf Geometry of an AWJ cut [11].

2.2. Acquisition system

National Instrument Compact USB DAC 9171 system coupled with National Instrument BNC 9223 module is used to acquire AE raw signal from the sensors. Labview software is used to collect the AE signals generated during the cutting process. The captured AE signals were further analysed in MATLAB software.

2.3. *AE sensor*
Physical Acoustic piezo-electric transducer PK15I AE sensor mounted on the work specimen are used to measure the acoustic signal. The sensor pre-equipped with an integrated pre-amplifier and an operating frequency range of 100 to 450 kHz

3. Results and discussion
Figure 3 shows the effect of traverse rate on surface roughness and kerf taper width for both aluminium 6061 and stainless steel 304. It can be seen that an increase in traverse speed increases the surface roughness of the cut as shown in Figure 3 (a). This result can be compared with the previous research done by Derzija Begic-Hajdarevica [12] which found that the increase of transverse speed has a great effect on the surface roughness especially the bottom of the cut. This is due to less overlap machining actions and fewer abrasive particles to impinge the surface at a higher traverses rate thus increasing the roughness of the surface. As a result, the abrasives could not erode the target material properly thus resulting in a rough surface. At a lower traverse rate, the abrasives particles will have enough time to impinge on the target material thus producing a well penetrated cut with a smoother surface. A slower traverse rate also gives a cleaner cut with a smoother surface [13]. For better quality of cut, the traverse rate has to be as low as possible as to give enough time for the jet to impact the surface thus causing the sufficient penetration.

The increase of traverse speed is found to be inversely proportional to the top kerf and bottom kerf width as shown in Figure 3 (b). This result can be compared with the findings of Wang and Jun [14,15] which reported that the top kerf width increases with the nozzle transverse speed because a slower pass allows more abrasive particle to impact on the work specimen and open a wider slot. This reason was further supported by Hascalik and Viscal Gupta [11,16]. In other words, the increase in the exposure time that was caused by decreasing traverse rate resulted in the increment in both of the kerf top and bottom widths.

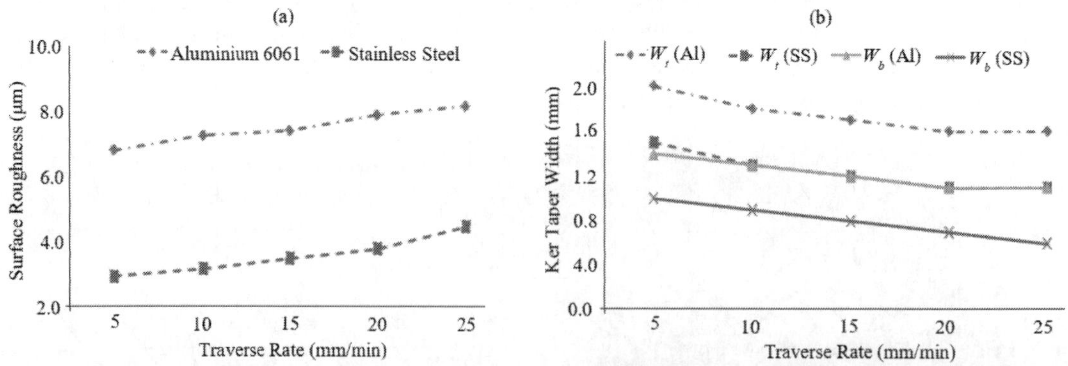

Figure 3. Effect of traverse rate on, (a) surface roughness and (b) kerf taper width.

The power spectral density (PSD) is a statistical estimation method that shows the strength variation of a signal in frequency domain. It is the statistical results of the structure response which excited by random dynamic loads and is a relationship curve with the values of PSD and frequency [17]. Figure 4 and Figure 5 show the PSD computed by Welch method for different traverse rates. It is difficult to interpret the data as the PSD signals generate spikes without any inclination to the frequency. This is also similarly found by Maia [18] which states that the signals generated by the PSD technique behaved as for random signals would generate lines without inclination. While for periodic signals, it would generate spikes with recurring frequency characteristic of the signal. Therefore, another approach has been used to differentiate the signal by using PSD via covariance method. This method causes the elimination of noise and the magnification of the cutting mechanism through the signal sum. Figure 6 and Figure 7 shows the PSD via covariance method. Based on both figures, the average PSD increases in amplitude as the transverse speed increases. This demonstrates that the PSD via covariance method has shown a more consistent trend with regard to varying the traverse rate. This is believed that the acoustic emission analysis using PSD via covariance method can be used to monitor the surface quality cut in abrasive waterjet machining process.

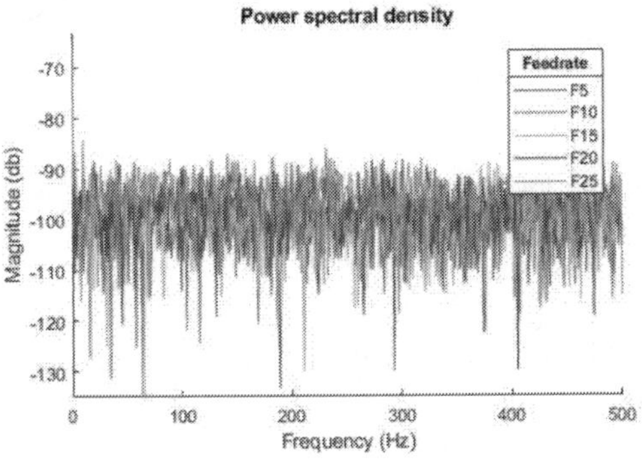

Figure 4. PSD of aluminium 6061.

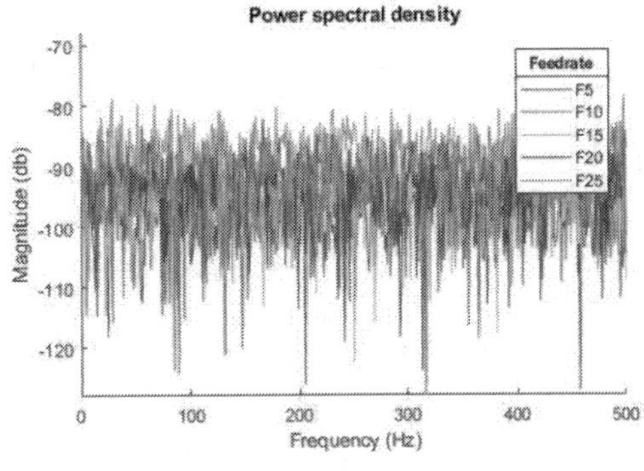

Figure 5. PSD of stainless steel 304.

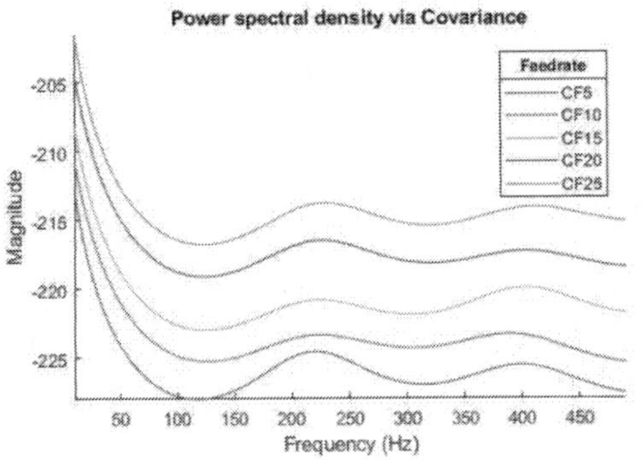

Figure 6. PSD via covariance of aluminium 6061.

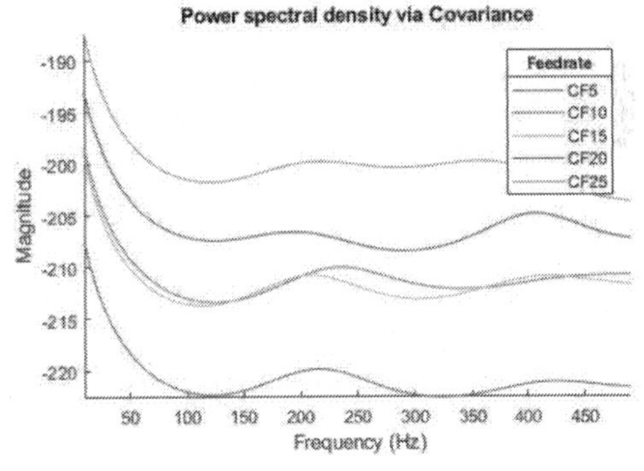

Figure 7. PSD via covariance of stainless steel 304.

4. Conclusion

This work reports the analysis of acoustic emission on the surface quality of aluminium 6061 and stainless steel 304 during abrasive waterjet machining process. Surface quality and traverse rate were assessed and the AE signal was collected during the experiment and further processed using PSD and PSD via covariance method. Thus, it can be concluded that monitoring the quality of the machined surface of sheet metals by the mean of PSD via covariance method is highly possible and the average PSD increases as the traverse speed increase.

Acknowledgments

Authors would like to gratefully acknowledge the financial support from the Universiti Malaysia Pahang through RDU1703313 and PGRS160307.

References

[1] Nanduri M, Taggart D G and Kim T J 2002 The effects of system and geometric parameters on abrasive water jet nozzle wear *Int. J. Mach. Tools Manuf.* **42** 615–23

[2] Arola D, McCain M L, Kunaporn S and Ramulu M 2001 Waterjet and abrasive waterjet surface treatment of titanium: A comparison of surface texture and residual stress *Wear* **249** 943–50

[3] Rabani A, Marinescu I and Axinte D 2012 Acoustic emission energy transfer rate: A method for monitoring abrasive waterjet milling *Int. J. Mach. Tools Manuf.* **61** 80–9

[4] Chen L 1998 Optimising abrasive waterjet cutting of ceramic materials *J. Mater. Process. Technol.* **74** 251–4

[5] Momber A W 2001 Energy transfer during the mixing of air and solid particles into a high-speed waterjet: an impact-force study *Exp. Therm. Fluid Sci.* **25** 31–41

[6] Kovacevic R, Mohan R and Zhang Y M 1995 Cutting Force Dynamics as a Tool for Surface Profile Monitoring in AWJ *J. Eng. Ind.* **117** 340

[7] Kovacevic R 1992 Monitoring the depth of abrasive waterjet penetration *Int. J. Mach. Tools Manuf.* **32** 725–36

[8] Dolinšek S and Kopač J 1999 Acoustic emission signals for tool wear identification *Wear* **225–229** 295–303

[9] Hassan A I, Chen C and Kovacevic R 2004 On-line monitoring of depth of cut in AWJ cutting *Int. J. Mach. Tools Manuf.* **44** 595–605

[10] Dornfeld D 1992 Application of acoustic emission techniques in manufacturing *NDT E Int.* **25** 259–69

[11] Gupta V, Pandey P M, Garg M P, Khanna R and Batra N K 2014 Minimization of Kerf Taper Angle and Kerf Width Using Taguchi's Method in Abrasive Water Jet Machining of Marble *Procedia Mater. Sci.* **6** 140–9

[12] Begic-Hajdarevic D, Cekic A, Mehmedovic M and Djelmic A 2015 Experimental Study on Surface Roughness in Abrasive Water Jet Cutting *Procedia Eng.* **100** 394–9

[13] Azmir M A and Ahsan A K 2008 Investigation on glass/epoxy composite surfaces machined by abrasive water jet machining *J. Mater. Process. Technol.* **198** 122–8

[14] Wang J 1999 A machinability study of polymer matrix composites using abrasive waterjet cutting technology *J. Mater. Process. Technol.* **94** 30–5

[15] Wang J and Wong W C K 1999 A study of abrasive waterjet cutting of metallic coated sheet steels *Int. J. Mach. Tools Manuf.* **39** 855–70

[16] Hascalik A, Çaydaş U and Gürün H 2007 Effect of traverse speed on abrasive waterjet machining of Ti–6Al–4V alloy *Mater. Des.* **28** 1953–7

[17] Chen D, Fan J and Zhang F 2013 Extraction the unbalance features of spindle system using wavelet transform and power spectral density *Meas. J. Int. Meas. Confed.* **46** 1279–90

[18] Maia L H A, Abrao A M, Vasconcelos W L, Sales W F and Machado A R 2015 A new approach for detection of wear mechanisms and determination of tool life in turning using acoustic emission *Tribol. Int.* **92** 519–32

IOP Conf. Series: Materials Science and Engineering **342** (2018) 012108 doi:10.1088/1757-899X/342/1/012108

Effect of magnetic polarity on surface roughness during magnetic field assisted EDM of tool steel

A.M. Efendee[1,2], M. Saifuldin[2], MA Gebremariam[1] and A Azhari[1]

[1]Faculty of Manufacturing Engineering, Universiti Malaysia Pahang, 26600 Pekan, Malaysia

[2]Faculty of Manufacturing Engineering Technology, TATI University College, 24000 Kemaman, Malaysia

E-mail: efendee@tatiuc.edu.my; azmir@ump.edu.my

Abstract. Electrical discharge machining (EDM) is one of the non-traditional machining techniques where the process offers wide range of parameters manipulation and machining applications. However, surface roughness, material removal rate, electrode wear and operation costs were among the topmost issue within this technique. Alteration of magnetic device around machining area offers exciting output to be investigated and the effects of magnetic polarity on EDM remain unacquainted. The aim of this research is to investigate the effect of magnetic polarity on surface roughness during magnetic field assisted electrical discharge machining (MFAEDM) on tool steel material (AISI 420 mod.) using graphite electrode. A Magnet with a force of 18 Tesla was applied to the EDM process at selected parameters. The sparks under magnetic field assisted EDM produced better surface finish than the normal conventional EDM process. At the presence of high magnetic field, the spark produced was squeezed and discharge craters generated on the machined surface was tiny and shallow.Correct magnetic polarity combination of MFAEDM process is highly useful to attain a high efficiency machining and improved quality of surface finish to meet the demand of modern industrial applications.

1. Introduction

The electrical discharge machining (EDM) is a process of metal removal from a conducting material by a repetitive spark discharge between tool and work piece separated by a small spark gap immersed in dielectric fluid. The tool moves toward workpiece at a desired gap distance to initiate ionization process when electric current is established between both. A continuous and scattered outburst spark from the tool create small craters on material surface[1]. Machinability of EDM on certain materials depends on suitability of the selected tools and parameters because these factors are going to determine the output characteristics such as surface roughness, electrode wear and material removal rate. Peak current is among the most important parameter in EDM machining. This parameter is measured in amperage. In EDM roughing process, some higher currents normally are selected because of greater material removal rate but it will give poor surface finish and high tool wear. The current will flow into the system and voltage will increases until it creates an ionization path within the dielectric. As the current is flowing, the voltage will drop and stabilize the working gap level[2]. Another essential set up in EDM is sparking distance. In EDM processes the value of voltage prefixed determines the width of the spark gap between the leading edge of the electrode and the workpiece.

Low small gap width between workpiece and the electrode makes the machining debris from EDM process is difficult to remove out. The excessive debris from EDM machining suspended within the dielectric fluid would accumulate in the machining gap tend to induce abnormal electrical discharge, so the stability of EDM progress would be disturbed.

Several authors have introduced magnetic fields in EDM to transport ferro-magnetic material debris through the gap and to improve debris circulation. The first study was done in 1978 by De Bruijnet al.[3]. Other researchers also agreed that the resultant magnetic force was efficient in transporting debris out from the machining area during material removal process[4-7]. Magnetic field assisted EDM would facilitate the stability of machining progress that occurs in between the spark gap. The machining performance would be improved by expelling debris from the machining gap fast and easily. The experimental results suggested that the magnetic field assisted EDM process was one of the most promising processes that can be regarded as an attractive and excellent alternative for surface finishing process[8].

The general direction for the magnetic flux flow is from the north (N) to the south (S) pole of the magnet. In addition, these magnetic lines form closed loops that leave at the north pole of the magnet and enter at the south pole. Magnetic poles are always in pairs. However, magnetic flux does not actually flow from the north to the south pole or flow anywhere for that matter as magnetic flux is a static region around a magnet in which the magnetic force exists. In other words, magnetic flux does not flow or move it is just there and is not influenced by gravity[9]. Researchers also had introduced and invented new EDM machining processes such as dry EDM and near dry EDM, powder mixed EDM, and ultrasonic vibrations assisted EDM to solve EDM machining limitations. In this study, magnetic field assisted EDM (MFAEDM)process with selected magnetic polarity was developed to explore the effects of magnetic polarity on EDM machining characteristics. Excellent combination of EDM machining techniques and parameters would satisfy the surface finish of machined components for better industrial application.

2. Experimental method

The Charmilles Roboform22 EDM machine was used in this investigation to conduct a series experiments in this work. A set of permanent magnetic (magnetic flux density: 18T) device was attached to the EDM machine. The experimental setup was used to explore the performance of the MFAEDM on Stavax ESR tool steel (AISI 420 mod.) in kerosene dielectric by using graphite electrode. The schematic diagram of the experimental setup is shown in Figure 1. The two magnets were arranged to face each other to achieve the desired magnetic polarity. The workpiece material was tool steel (AISI 420 mod.) which is widely used in die and mould manufacturing industry for the core component. The workpiece specimen was milled into cube shape of 30mm X 30mm X 30 mm. The specimen has mass elemental composition of Fe – 84.32%, C [10]– 0.38%, Si – 0.9%, V – 0.3%, Cr – 13.6% and Mn – 0.5% [11]. Cylindrical graphite was selected as the electrode material. The electrode size used to conduct electrical discharge was 25mm in diameter and 40mm length to accommodate machining depth of cut and clamping purpose. The essential properties of graphite electrode are summarized in table 2. In addition, kerosene was adopted as a dielectric fluid in this study. The kerosene in the EDM container serves several functions such as to initiate sparks, as an insulator between the tool and the workpiece, prevent shorting and acts as a coolant.

The experiments were carried out in normal condition as the datum result and MFAEDM were conducted for different magnet polarity. The permanent magnets were clamped side by side to the stavax ESR samples so that the same pole facing each other (N-N) to achieve the repulsive magnetic force. In another set up difference pole combination (N-S) was arranged to create an attractive magnetic force. The machining output such as material removal rate, MRR (g/min), electrode wear rate, EWR(g/min), and surface roughness (Ra/mm) were adopted to evaluate the effects of machining parameters on the MFAEDM and conventional EDM processes. For every experiment, the electrode and the samples material were weighed before and after removal process to calculate MRR and EWR. The samples of EDM will undergo for surface roughness measurementto evaluate surface quality of the machined surface. The value of surface roughness was obtained by averaging four measurements on different positions of each machining condition. Then, those samples were part-off and the cross-section surfaces were well-polished as a preparation for cross section observation. The surface then etched in a combination of HCL and picric for eight minutes before undergoing for SEM observation.

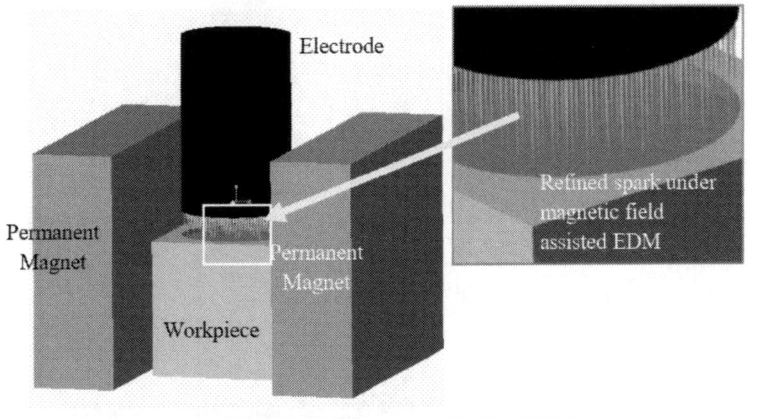

Figure 1. Schematic diagram of MFAEDM.

3. Result and discussion

In general, EDM machining with or without magnetic field assistance produces a higher surface roughness at a higher value of peak current as shown in Figure 2. This is expected since higher discharge current results in higher thermal loading on the both electrode and workpiece thus removing a higher amount of materials. As a result, surface finish becomes rougher at a higher peak current.

In case of magnetic field assistance, the magnetic field drives the magnetic lines according the arrangement and combination of magnets. Machining debris is attracted to the attached magnet near the cutting area. As a result, the machining area will be cleared from debris thus reducing disturbance for subsequent cutting cycles. Therefore, a smoother surface can be expected during MFAEDM process. Figure 2 shows the effect of magnetic polarity and peak current of surface roughness of Stavax ESR tool steel. It is clear that the surface roughness is reduced with the use magnetic field assistance by about 16% and 20% respectively for peak current of 8A and 24 A. Magnetic field assistance restricts the spread of plasma thus increasing the energy at the machining area[4]. Consequently, the material removal process is improved with the combination of enhanced spark energy and magnetic force.

It can be seen from Figure 2 also, the polarity combination of N-S produces better surface finish as compared to N-N magnet arrangement. The surface roughness was reduced as much as 10% for peak current of 8 A and 8% for peak current of 24 A. This circumstance has a connection to the magnetic force direction and magnetic field lines where the sparks are affected by the magnetic field [12]. In N-S magnet arrangement, the magnetic field acting perpendicular to the electrode [13]. The resultant magnetic field which is in tangential action improves those disengage spinning electrons and ions from their core atom thus increasing ionization process [6]. The process increases the plasma pressure and plasma density. The plasma pressure acts as a piston on the melt pool that exerts downward force [4].

Furthermore, it is evident that N-S MFAEDM produces better surface quality than N-N MFAEDM as shown in Figure 3. It can be seen that crater and recast layers were formed on the machined surface as a result of continuous melting and evaporation processes during the machining. The valleys produced from the spark in normal EDM (Figure 3 (a)) are deeper and wider than N-N MFAEDM (Figure 3 (b)). N-S MFAEDM produces craters which are shallow and small as shown in Figure 3 (c) thus resulting in better surface quality.

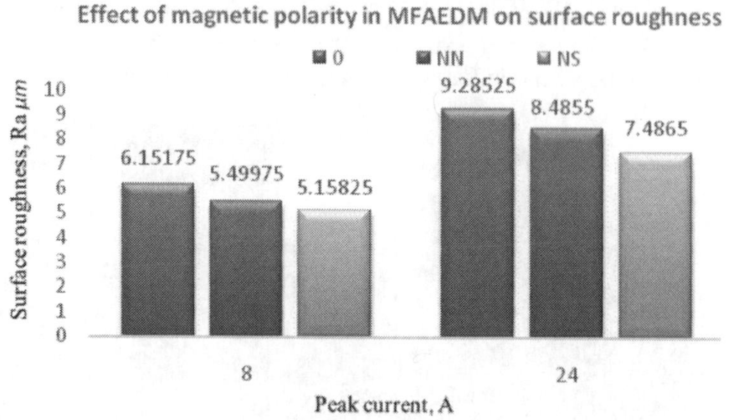

Figure 2. Effect of magnetic polarity and peak current on surface roughness.

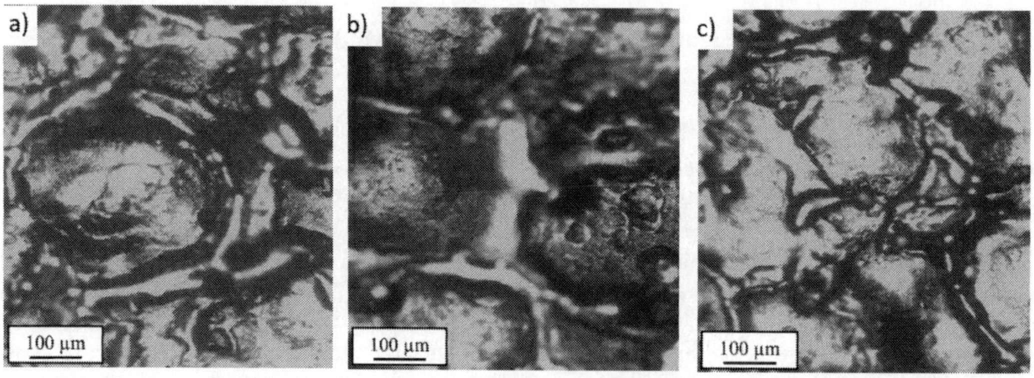

Figure 3. Surface microstructure of (a) normal EDM, (b) N-N MFAEDM and (c) N-S MFAEDM at 200X magnification of optical microscope (24A,50µs).

In N-N magnetic pole combination, there is a neutral spot in the middle of the magnetic flux line which is contrary to N-S magnetic pole combination. At this neutral spot, the resultant magnetic force is zero[14]. So, the plasma pressure for ionization process in N-N polarity is lower than N-S polarity. Whilst in N-S MFAEDM, the magnetic force acts tangential to the plasma channel. It is known that N-S MFAEDM process increases the ionization process thus the formation of plasma channel is accelerated. In the meantime, the spark produced was squeezed resulting in tiny and shallow craters on the machined surface thus reducing the its roughness[15]. Therefore, N-S magnetic polarity combination minimizes the ignition delay time and maximizes the spark intensity thus improving the surface quality of EDM samples.

4. Conclusions

As a conclusion, it is obvious that magnetic field assisted EDM provides better quality of machined surfaces. In term of magnetic polarity application, it was found that it is best to use north to south combination of magnetic poles. Therefore, a correct magnetic polarity combination of MFAEDM process is highly useful to attain a high efficiency machining and improved quality of surface finish to meet the demand of modern industrial applications.

Acknowledgments
Authors would like to gratefully acknowledge the financial support from the TATI University College and Universiti Malaysia Pahang through STG9001-1508 and PGRS160342 respectively.

IOP Conf. Series: Materials Science and Engineering **342** (2018) 012108 doi:10.1088/1757-899X/342/1/012108

References

[1] Bojorquez B, Marloth R and Es-Said O 2002 Formation of a crater in the workpiece on an electrical discharge machine *Engineering Failure Analysis* **9** 93-7

[2] Mahendran S, Devarajan R, Nagarajan T and Majdi A 2010 A review of micro-EDM. *In: Proceedings of the international multi conference of engineers and computer scientists* **2**

[3] De Bruijn H, Delft T and Pekelharing A 1978 Effect of a magnetic field on the gap cleaning in EDM *CIRP Ann* **27** 93-5

[4] Singh Bains P, Sidhu S S and Payal H 2016 Study of Magnetic Field-Assisted ED Machining of Metal Matrix Composites *Materials and Manufacturing Processes* **31** 1889-94

[5] Tomura S and Kunieda M 2009 Analysis of electromagnetic force in wire-EDM *Precision engineering* **33** 255-62

[6] Gholipoor A, Baseri H, Shakeri M and Shabgard M 2016 Investigation of the effects of magnetic field on near-dry electrical discharge machining performance *Proceedings of the Institution of Mechanical Engineers, Part B: Journal of Engineering Manufacture* **230** 744-51

[7] Khan A A, Ndaliman M B, Ali M Y, Muath D, Haazin A, Ibrahim A and Mustapa M 2013 The effect of EDM with external magnetic field on surface roughness of stainless steel. *In: Int. Conf. Mech. Aut. Aer. Engg*

[8] Kim J-D 2003 Polishing of ultra-clean inner surfaces using magnetic force *The International Journal of Advanced Manufacturing Technology* **21** 91-7

[9] Giancoli D C 2008 *Physics for scientists and engineers*(Pearson Education International)

[10] Ge M-Z, Xiang J-Y, Yang L and Wang J T Effect of laser shock peening on the stress corrosion cracking of AZ31B magnesium alloy in a simulated body fluid *Surface and Coatings Technology* **310** 157-165

[11] Han L-L, Lin C-M and Shih Y-S 2013 Corrosion Resistance of ASSAB Stavax ESR Stainless Steel by Heat and Cold Treatment *Materials Transactions* **54** 833-8

[12] Schaeffer H E 1908 *A study of the electric spark in a magnetic field*(Bryn Mawr College.)

[13] Beeteson J S 2001 *Visualising magnetic fields: numerical equation solvers in action* vol 1 (Academic Press)

[14] Serway R A and Jewett J W 2018 *Physics for scientists and engineers with modern physics* (Cengage learning)

[15] Lin Y-C and Lee H-S 2008 Machining characteristics of magnetic force-assisted EDM *International journal of machine tools and manufacture* **48** 1179-86

AUTHOR INDEX

AUTHOR INDEX

AUTHOR INDEX

AUTHOR INDEX